Chemical Process Retrofitting and Revamping

Dedication

In memory of my late parents:
Gade Sakuntala Devi and Gade Gopala Swami,
who nurtured and educated me

and

To my dearest wife:
Puvvada Krishna Kumari,
for her selfless support and devotion to our family

Chemical Process Retrofitting and Revamping

Techniques and Applications

Edited by

GADE PANDU RANGAIAH

Department of Chemical & Biomolecular Engineering
National University of Singapore
Singapore

WILEY

Registered office
John Wiley & Sons Ltd, The Atrium, Southern Gate, Chichester, West Sussex, PO19 8SQ, United Kingdom

For details of our global editorial offices, for customer services and for information about how to apply for permission to reuse the copyright material in this book please see our website at www.wiley.com.

Library of Congress Cataloging-in-Publication Data

Chemical process retrofitting and revamping : techniques and applications / edited by G.P. Rangaiah.
pages cm
Includes bibliographical references and index.
ISBN 978-1-119-01633-5 (cloth)
1. Chemical plants–Remodeling. 2. Chemical plants–Maintenance and repair. 3. Chemical plants–Equipment and supplies. I. Rangaiah, Gade Pandu, editor.
TP155.5.C5128 2016
660–dc23

2015033565

Cover images: Main image: Courtesy of the editor and authors. Background image: from GettyImages, at http://www.gettyimages.com/photos/187591800?license=rf&family=creative&phrase=187591800&sort=best&excludenudity=true

A catalogue record for this book is available from the British Library.

Set in 10/12pt Times by Aptara Inc., New Delhi, India.

1 2016

Contents

List of Contributors

Catherine Azzaro-Pantel, Professor of Chemical Engineering, Université de Toulouse, Laboratoire de Génie Chimique, LGC UMR CNRS 5503, France (email: catherine .azzaropantel@ensiacet.fr)

Raman Balajee, Process Safety Lead – Asia, Air Products Singapore Private Limited, Singapore (email: balajeeraman@yahoo.com)

Russell F. Dunn, Department of Chemical and Biomolecular Engineering, Vanderbilt University, Nashville, USA
and Polymer and Chemical Technologies, LLC, Nashville, USA (email: russell.dunn@ vanderbilt.edu)

Ajit K. Ghosh, AKG Process Consulting, 33 McFarlane Court, Highett, Australia (email: akgprocessconsulting@gmail.com)

Moonyong Lee, School of Chemical Engineering, Yeungnam University, Republic of Korea (email: mynlee@ynu.ac.kr)

Nguyen Van Duc Long, School of Chemical Engineering, Yeungnam University, Republic of Korea (email: allenthelong@yahoo.com)

Le Quang Minh, School of Chemical Engineering, Yeungnam University, Republic of Korea (email: minh2510@ynu.ac.kr)

W. Niu, Department of Chemical and Biomolecular Engineering, National University of Singapore, Singapore (email: judger1984@gmail.com)

S. (Shruti) Pandey, Department of Chemical and Biomolecular Engineering, National University of Singapore, Singapore (email: shrutip89@gmail.com)

G.P. (Gade Pandu) Rangaiah, Department of Chemical & Biomolecular Engineering, National University of Singapore, Singapore (email: gprangaiah@gmail.com)

D.P. Rao, Formerly Professor of Chemical Engineering, Indian Institute of Technology, Kanpur, India
and Managing Partner, Process Intensification Consultants, India (email: dprao@iitk.ac.in)

C.C.S. Reddy, Lead Process Design Engineer, Singapore Refining Company Private Limited, Singapore (email: cchandrasreddy@gmail.com)

Jarrid Scott Ristau, Department of Chemical and Biomolecular Engineering, Vanderbilt University, Nashville, USA
and Polymer and Chemical Technologies, LLC, Nashville, USA (email: jarrid.s.ristau@vanderbilt.edu)

Shivom Sharma, Department of Chemical & Biomolecular Engineering, National University of Singapore, Singapore (email: shivomsharma.iitr@gmail.com)

Joonho Shin, Basic Materials & Chemicals R&D, LG Chem, Moonji-dong, Yuseong-gu, Republic of Korea (email: deepest@lgchem.com)

B.K. (Bhargava Krishna) Sreepathi, Department of Chemical & Biomolecular Engineering, National University of Singapore, Singapore (email: bhargava@u.nus.edu)

X.Z. Tan, Department of Chemical and Biomolecular Engineering, National University of Singapore, Singapore (email: xuezhetan@gmail.com)

Preface

New chemical plants are continually being built around the world to produce useful and valuable products for society's well-being and development. They add to the numerous plants already in operation. Performance of the existing processes in these plants must be monitored regularly, and process improvements have to be made for their continued economic and environmental sustainability. Hence, process retrofit and revamp, which refer to changes in existing process equipment and/or structure, are essential to improve the operation of existing chemical plants. Process retrofitting and revamping are likely to have been occurring since the first chemical plant was built. However, there are only a few books dedicated to process retrofit and revamp.

The main stages in the industrial retrofit/revamp projects are: (1) establishing objectives and basis for retrofit/revamp, (2) conceptual process design, (3) front-end engineering design (FEED), (4) engineering, procurement and construction (EPC), and (5) commissioning/start-up and completion. Conceptual process design for a retrofit/revamp develops an improved process using the existing equipment as much as possible. This is different from the conceptual process design for a new plant, where the process flowsheet is developed first followed by equipment design in two sequential steps. The requirement of using existing process equipment and structure substantially increases constraints and potential combinations, thus making the conceptual process design of retrofit/revamp projects much more challenging. Most of the research on process retrofit/revamp as well as many chapters in this book are focused on this stage of conceptual process design, which requires techniques and tools for process simulation, optimization and control. Stages (3) to (5) in industrial retrofit/revamp projects are similar to those in the engineering, construction and commissioning of new plants. However, activities in these stages should consider existing plant layout, safety, utilities and other facilities as well as the impact of the retrofit/revamp on them. The retrofit/revamp should have no or minimal impact on the existing plant operation.

This book on process retrofitting and revamping is divided into three parts. The first part has three chapters, starting with the Introduction chapter to provide an overview of process retrofitting and revamping. The next two chapters in the first part are on project engineering, management and safety aspects of process retrofitting and revamping, which are important in industrial implementation. The focus of the second part of the book, Chapters 4 to 8, is on computational techniques for solving process retrofit/revamp problems. One of them is on process intensification technology for retrofitting/revamping of chemical processes. The techniques described in Chapters 4 to 8 can be employed for retrofit/revamp in a variety of process industries. Chapters 9 to 13 form the third part of the book and cover retrofit/revamp applications in different chemical process industries. They describe the options considered and/or analysis employed for the retrofit/revamp of the process studied, which are useful for related applications.

The current book will be useful for researchers, practitioners and postgraduate students interested and working in process retrofit and revamp. Researchers and practitioners can adapt and apply available techniques in the chapters to their processes or specific problems. Chapters of this book can be used for projects in advanced courses on separation processes, modeling and optimization for senior undergraduate and postgraduate students. In general, readers can choose chapters of interest and read them independently.

Industrial projects on process retrofit and revamp continue to occur and are likely to increase. As a result, research on process retrofit and revamp is expected to grow and be an active area in chemical engineering. I hope that this book will contribute to further developments in techniques and tools for process retrofit and revamp including their successful implementation in industrial projects.

I am grateful to all the contributors and reviewers for their cooperation in meeting the requirements and schedule to finalize this book on process retrofit and revamp. In particular, I would like to thank Prof. S.K. Gupta, Prof. A.F.A. Hoadley and Prof. K. Mitra for their timely help in organizing the review of some chapters co-authored by me. My deep appreciation goes to Shivom Sharma, Sreepathi Bhargava Krishna and C.C.S. Reddy for all their help and suggestions in the preparation of this book. Finally, I am grateful to the Department of Chemical & Biomolecular Engineering at the National University of Singapore for all the facilities, support and freedom to publish the present book and my previous four books on process optimization and control.

I have had the idea to prepare this book on process retrofit and revamp for many years. I am very happy that it has materialized now, partly helped by the enthusiastic and productive research of my recent graduate students – C.C.S. Reddy, Sreepathi Bhargava Krishna, Shivom Sharma, Shruti Pandey and Wendou Niu.

In the early stage of this book's preparation, I was affected by sciatica, which made it difficult to walk or stand for more than five minutes. This cast dark clouds on the preparation and completion of this book. The three ladies in my adult life (my dearest wife, Krishna Kumari and loving daughters – Jyotsna and Madhavi) took care of me and advised me on the best treatment and management. Thanks to their efforts, the effects of sciatica have gradually diminished and are now almost absent. I am grateful to all my family members for their deep affection and unwavering support to me.

Gade Pandu Rangaiah
National University of Singapore
Singapore

September 2015

Additional Online Resources

Additional resources for the book can be found at http://booksupport.wiley.com

Please enter the book title, editor name or ISBN to access files.

Part I
Overview

1
Introduction

G.P. Rangaiah
Department of Chemical & Biomolecular Engineering,
National University of Singapore, Singapore

1.1 Chemical Process Plants

There are a wide range of chemical process industries such as agrochemicals, ceramics, cement, cosmetics, fragrances and flavors, food and drinks, glass, industrial gases, industrial/inorganic chemicals, leather, mineral processing, nuclear, oil and gas, paper and pulp, paints and pigments, petrochemicals, pharmaceuticals, polymers, rubber, soap and detergent, specialty chemicals, synthetic fibers, sugar, vegetable oils and water. Many of these involve continuous processes whereas some are batch processes. New process plants continue to be designed and built, relatively more in developing countries, to produce useful and valuable products required by the society. These are usually designed and their economic viability assessed assuming a plant life of 10 to 20 years. However, chemical plants, once built, continue to operate for very much longer than this assumed plant life.

Thus, chemical process plants in operation have been increasing steadily in the world. They were designed in the past few years or even decades, perhaps optimally for the economic, technological and societal conditions at that time. Obviously, technological knowledge has been advancing since the existing plants were designed. In addition, economic and societal conditions are dynamic and change over time for one reason or other. For example, energy prices and global warming concerns have increased substantially; new and better technologies (such as catalysts, process equipment and their internals), separation processes and intensified processes as well as simulation and optimization techniques are being continually developed and improved through research and industrial implementation.

Chemical Process Retrofitting and Revamping: Techniques and Applications, First Edition.
Edited by Gade Pandu Rangaiah.

Hence, it is imperative to review regularly the performance of the existing plants and assess the possibilities for their improvement. This can be for one or more of the following objectives (Grossmann *et al.*, 1987; Rong *et al.*, 2000).

- To reduce energy required and/or operating cost
- To improve conversion and/or selectivity of reactions
- To increase production/throughput of the process
- To use feed of different quality and/or alternative feed
- To meet new specifications of product(s)
- To produce new products
- To enhance the control of the process
- To improve the safety, reliability and flexibility of the process
- To reduce the adverse impact of the process on the environment.

The first and relatively simple step is to optimize and set the operating conditions such as temperature, pressure and flow rate in the existing process for the chosen objective (for example, energy required and operating cost). This is often referred to as **operation optimization**, and involves analysis of the process and use of optimization techniques. In operation optimization, there is no change in process configuration or equipment. It can be performed off-line or on-line because of frequent changes in the operating environment of the process (such as product requirements and prices). On-line optimization is also known as real time optimization. Process improvement by operation optimization is limited because of constraints imposed by current process configuration, equipment and/or technology employed. It is thus necessary to consider modifications in all these for improving the current process substantially.

1.2 Process Retrofitting and Revamping

Process retrofitting and revamping refers to making suitable changes and/or additions to existing process configuration and equipment. It may involve new technology such as membrane separation or reactive distillation to supplement or replace distillation. One example of equipment and configuration changes in a process is heat exchanger network retrofitting, which involves area additions in existing exchangers and/or installation of new exchangers for increased energy recovery and re-use. Obviously, process retrofitting and revamping should maximize the use of the existing equipment in the plant as much as possible.

Analysis, simulation and optimization techniques could be used to achieve the chosen objective of process retrofitting/revamping. However, process retrofitting/revamping is more than operation optimization because the former considers changes to the process configuration and equipment in addition to operating conditions. Hence, solving a process retrofitting/revamping problem and implementing the solution found are more complex and challenging than those in the case of operation optimization.

Is there some difference between process retrofit and revamp? According to oxforddictionaries.com, revamp is to give a new and improved form, structure or appearance (to something), and retrofit is to provide (something) with a component or accessory not fitted during manufacture. Although these two meanings seem to be similar, process retrofit perhaps refers to adding new equipment to the existing process and so its scale, complexity and capital cost are relatively lower. On the other hand, process revamp involves changes in configuration and so its scale, complexity and capital cost are more. Currently, retrofit and revamp are often used synonymously in the chemical engineering literature although some practitioners use retrofit for smaller projects (that is, investment) and revamp for bigger projects.

In the technical literature, Rong *et al.* (2000) state that the main objectives of process retrofits include increasing the production capacity, efficiently processing new feedstock, utilizing new process technologies, reducing environmental impact, and reducing operating costs. According to Smith (2005), the motivation to retrofit (or revamp) an existing plant could be to increase capacity, allow for different feed or product specifications, reduce operating costs, improve safety and reduce environmental emissions. Kemp (2007), in the glossary of terms, defines retrofit or revamp as any change to an existing chemical process. On the other hand, Towler and Sinnott (2012) state that revamps fall into two categories: debottlenecking (discussed below) and retrofitting, which implies the former is not within the scope of retrofitting.

Should we distinguish retrofit/retrofitting and revamp/revamping? It is perhaps desirable for clarity and consistency. As suggested by Rao in Chapter 5, retrofitting can be used to mean adding to or replacing the whole or part of one type of equipment with a better alternative, and revamping for reorganizing the process involving several process steps (thus different equipment types). This indicates that retrofitting is smaller in scale, complexity and capital cost compared with revamping. Thus, the suggested distinction between retrofit and revamp is consistent with the use of these terms by some practitioners.

One of the purposes of process changes is to increase the plant throughput (that is, increasing production rate without any change in feed, process performance or product quality). This specific activity is referred to as ***debottlenecking***. This common term in process industries is related to a bottle's neck, which is generally narrow and limits the flow rate through it. Obviously, the neck has to be widened for increasing the flow rate. Similarly, in debottlenecking of an operating process, equipment/operation limiting the throughput is identified and then it is suitably modified for increasing the production rate. In this way, process capacity can be increased by 5 to 20%, with much smaller capital investment compared with building new facilities. This is possible because of the spare size/capacity available in many items of the existing equipment because of design margins used at the time of their design and fabrication. It is possible to increase the throughput by modifying one (type of) equipment or several types of equipment. Thus, debottlenecking can be achieved by retrofit or revamp. In other words, one objective of retrofit or revamp can be debottlenecking. Recall that retrofit and revamp can be for any of the reasons listed in the previous section. For specificity and clarity, debottlenecking (and not retrofit or revamp) should be used if the sole objective of process changes is an increase in the plant throughput. Note it is different from plant expansion, wherein there are no existing items of equipment to be considered and generally no space restrictions.

How is process retrofit/revamp different from operation optimization? Unlike the operation optimization, retrofit/revamp design will have more degrees of freedom (that is, variables related to existing equipment changes and to new equipment) and more combinations to be considered for optimization. Hence, it is much more challenging than operation optimization. However, retrofit/revamp design can improve the process significantly compared with operation optimization.

Analysis and solution of a process retrofit/revamp problem will require simulation and optimization. Hence, computational techniques for process simulation and optimization find applications in process retrofit and revamp. However, appropriate models have to be developed for process retrofit/revamp, and the resulting optimization problems have more constraints and are more challenging than are operation optimization problems.

1.3 Stages in Process Retrofitting/Revamping Projects

There are five main stages in the industrial retrofit/revamp projects. These are: (1) establishing objectives and basis for retrofit/revamp, (2) conceptual process design, (3) front-end engineering design (FEED), (4) engineering, procurement and construction (EPC), and (5) commissioning/start-up and completion. The main activities in these five stages are summarized in Table 1.1. See Chapter 2 in this book for more details on these, and Chapter 3 for safety aspects of retrofit/revamp projects.

The first stage is to establish objectives for the retrofit/revamp project, which can be one or more of the purposes listed in Section 1.1. This should be carried out comprehensively by considering plant performance and operating experience; current product demand, quality and price as well as their future projections; raw material availability, quality and price; and promising technologies and improved equipment for the process. The feasibility of achieving the set objectives for retrofit/revamp should be assessed for obtaining management approval to proceed to the next stage. Chemical engineers are the main players in the first stage, and they need to obtain required data on raw materials and product(s) from the supply and marketing departments.

In the second stage, conceptual process design is performed by chemical engineers having experience in process development and design. It can be divided into two main steps. In the first step, existing plant design and operating data are collected, a simulation model of the process is developed and validated against the design/operating data, and rating studies are performed to identify the limits of existing equipment. Thus, this step involves extensive data collection and analysis, process simulation and equipment modeling, all within the domain of chemical engineering. In the second step, several options for retrofit/revamp should be developed and assessed for their techno-economic feasibility. These options are developed by considering heat, mass and process integration (for example, heat integration by pinch analysis or optimization; see Chapter 7 in this book), as well as promising new technologies and equipment (for example, membrane separations by themselves or along with the existing distillation column; see Chapter 10 in this book), and process intensification (for example, reactive distillation) described in Chapter 5. Further, retrofit/revamp options should use existing equipment as much as possible (to reduce capital investment) and also meet space constraints in the plant. The chosen option for retrofit/revamp should be assessed for controllability and also discussed with the operators of the existing process. Most of the research on process retrofit/revamp, as well as many chapters in this book, are focused on this step of conceptual process design, and this is discussed further in the next section. The two steps in the conceptual process design require techniques and tools for process simulation, optimization and control.

Stages 3 to 5 in industrial retrofit/revamp projects are similar to those in the engineering, construction and commissioning of new plants. However, activities in these stages (outlined

Table 1.1 *Stages in retrofit/revamp projects and main activities in each of them.*

Stage	Main Activities
1. Objectives and Basis for Retrofit/Revamp	• Study product demand, quality requirements and price • Review changes in availability, quality and prices of raw materials • Review plant performance and operating experience • Study promising technologies and improved equipment for the process • Set objective(s) and basis for retrofit/revamp, and assess their feasibility
2. Conceptual Process Design	• Collect plant design and operating data • Develop and validate simulation model of the process • Perform rating studies to identify equipment limits • Develop retrofit/revamp options considering heat/process integration, promising technologies, process intensification etc. • Assess techno-economic feasibility of retrofit/revamp options to choose the best one • Evaluate controllability of the chosen option and also discuss with the operators of the existing process
3. Front-End Engineering Design (FEED)	• Prepare process flow diagrams, piping and instrumentation diagrams, equipment data/specification sheets etc. • Prepare plot, equipment and pipe layout drawings • Request quotations for equipment, construction, instrumentation, electrical work etc. • Prepare drawings of control room, shelter, plant lighting etc. • Design fire-fighting system
4. Engineering, Procurement and Construction (EPC)	• Completion of all engineering work • Procurement of all equipment • Review of constructability and related documents, and approvals • Project monitoring and construction coordination • Management of change
5. Commissioning and Completion	• As built documents review and approval • Operators training • Pre-startup safety review • Commissioning procedures and checks • Plant performance test • Documentation on project highlights, lessons learnt and retrofit/revamp opportunities in the future

in Table 1.1) should consider existing plant layout, safety, utilities and other facilities as well as the impact of the retrofit/revamp on them. The retrofit/revamp should have no or minimal impact on the existing plant operation. This means the retrofit/revamp should preferably be implemented during the routine maintenance period. In any case, the time available for construction, installation and start-up for retrofit/revamp will be limited, and so prior planning and scheduling are essential to the timely ordering and receipt of new equipment. Careful consideration should be given to available space and access for construction and installation of new equipment as well as safety of the existing plant and all personnel during the retrofit/revamp implementation. Activities in stages 3 to 5 involve engineers from many disciplines, namely, chemical, civil, electrical, environmental, instrumentation, mechanical, safety and reliability engineering. Hence, chemical engineers will have to work with engineers from other disciplines within the company and also in other companies (for example, EPC and equipment vendors).

1.4 Conceptual Process Design for Process Retrofit/Revamp Projects

As mentioned in the previous section, the conceptual process design stage of process retrofit/revamp projects is in the chemical engineering domain, and chemical engineers are responsible for this stage. Conceptual process design is also encountered in the design and development of new chemical processes and plants, referred to as grassroots or greenfield design. There is one significant difference between these two conceptual process designs. In the conceptual process design for grassroots design, a process flowsheet is developed and simulated followed by design of the individual equipment. In effect, these two major steps are carried out sequentially, which reduces complexity and computational effort. Further, there are more degrees of freedom in each of these steps.

On the other hand, conceptual process design for retrofit/revamp will have to develop a process flowsheet using (most of) the existing equipment and latest technologies. This requirement substantially increases the constraints and also potential combinations, thus making the conceptual process design of retrofit/revamp projects much more challenging. Further, the need to reuse the existing equipment in the plant in order to reduce investment makes the conceptual process design for the retrofit/revamp project plant-specific. This is because the type/size of existing equipment are unlikely to be identical in different plants.

In general, evolution and optimization are two major approaches for determining design modifications to an existing plant for achieving the retrofit/revamp objectives (Grossmann *et al.*, 1987). In the former, process engineers propose design and structural changes using targets, bounds, heuristics, insights from plant operation, knowledge on the latest technologies and experience. The proposed changes are then analyzed and evaluated through rigorous simulation. Thus, the evolution approach starts from the existing system, and the process engineer is actively engaged in developing solutions, which are likely to be more practical and acceptable to the plant personnel. It seems to be a natural approach and probably the commonly used method in the industry. However, the evolution approach involves coming up with many alternatives and their analysis. Its success depends on the experience and ingenuity of the process engineers involved.

The optimization approach requires formulation and solution of an optimization problem for determining optimal design of the process while using the existing equipment and the latest technology as much as possible. The solution involves development and evaluation of many alternatives, not by the engineer but automatically by the optimization algorithm. Although this approach is simple in principle and provides a general framework, both the formulation and solution of the optimization problem are challenging, particularly when structural changes are involved. Hence, one likes to combine the evolution and optimization approaches judiciously, for tackling conceptual process design for retrofit/revamp projects. For example, process engineer can propose a suitable process flowsheet/structure incorporating the existing equipment, which can then be simulated and optimized. Application of the evolution and optimization approaches together can be seen in Chapters 9 to 11 of this book, whereas the optimization approach alone is used and described in Chapters 7 and 12.

1.5 Research and Development in Process Retrofit/Revamp

The earliest systematic procedure for tackling retrofit/revamp problems was by Douglas and co-workers (Fisher *et al.*, 1987; Douglas, 1988). It consisted of a hierarchy of several steps: estimating bounds on the benefits from retrofitting, estimating the benefits of replacing the existing plant by the same or better process alternative, and estimating the investment and operating cost savings by making suitable changes in the process. The last step considers modifications to the main process without heat exchangers, followed by heat exchanger network design and then refining the complete process with exchangers.

In the past few decades, there have been research efforts to develop systematic methods for solving retrofit/revamp problems. For example, Uerdingen *et al.* (2003) presented an indicator-based methodology for screening the cost-saving potential of continuous processes. It consisted of decomposition analysis, retrofit solution generation and approximate economic evaluation. Jensen *et al.* (2003) extended it to include environmental and safety indicators for sustainable retrofit targets. Carvalho *et al.* (2008) developed an Excel-based software (Sustain-Pro) and added an indicator-based sensitivity analysis algorithm into the original methodology in order to accelerate the generation of retrofit solutions.

Retrofit/revamp of a particular section of processes has received significant attention. For example, many methods have been developed for heat exchanger network (HEN) retrofitting. These methods utilize pinch analysis and/or optimization techniques; see Chapter 7 in this book for more details on HEN retrofitting, and Sreepathi and Rangaiah (2014) for a review of studies on HEN retrofitting. Optimization techniques have also been applied to retrofitting batch plants; Chapter 8 in this book provides an overview of batch plant retrofitting by optimization. Although there are many studies on process retrofit/revamp techniques and applications, they are substantially fewer than those on grassroots design of chemical processes. Possible reasons for this are more complexities and challenges in tackling retrofit/revamp problems, and the possibility of applying the knowledge and expertise gained from solving grassroots design problems to retrofit/revamp problems. This can be seen from the development and application of pinch analysis and optimization techniques for HEN synthesis (that is, design) followed by their extension to HEN retrofitting.

To quantify studies on retrofit/revamp methods and applications, the Scopus database was searched for journal papers, reviews and book chapters containing retrofit, retrofitting, revamp, revamping and/or debottlenecking in the title, abstract and/or keywords. This search was carried out in three groups of journals of interest to chemical engineering. The three groups are chemical engineering main, chemical engineering-related and chemical engineering practice; and the period covered is from 1976 to 2014. The journals included in each of these groups and the total number of journal articles containing retrofit, retrofitting, revamp, revamping and/or debottlenecking in the title, abstract and/or keywords are presented in Figure 1.1.

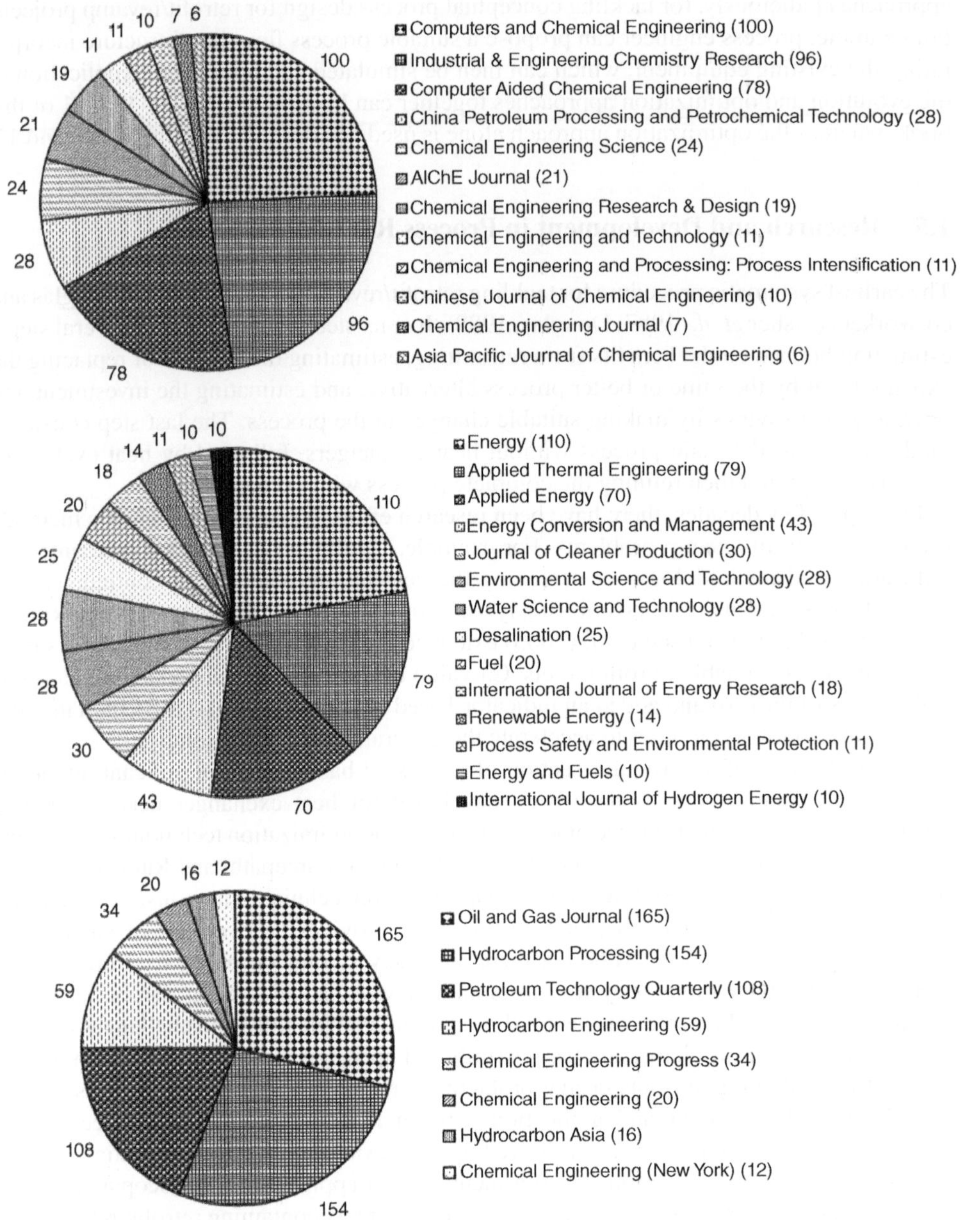

Figure 1.1 *The number of articles published from 1976 to 2014 mentioning retrofit/retrofitting, revamp/revamping and/or debottlenecking in chemical engineering main (top), -related (middle) and practice (bottom) journals.*

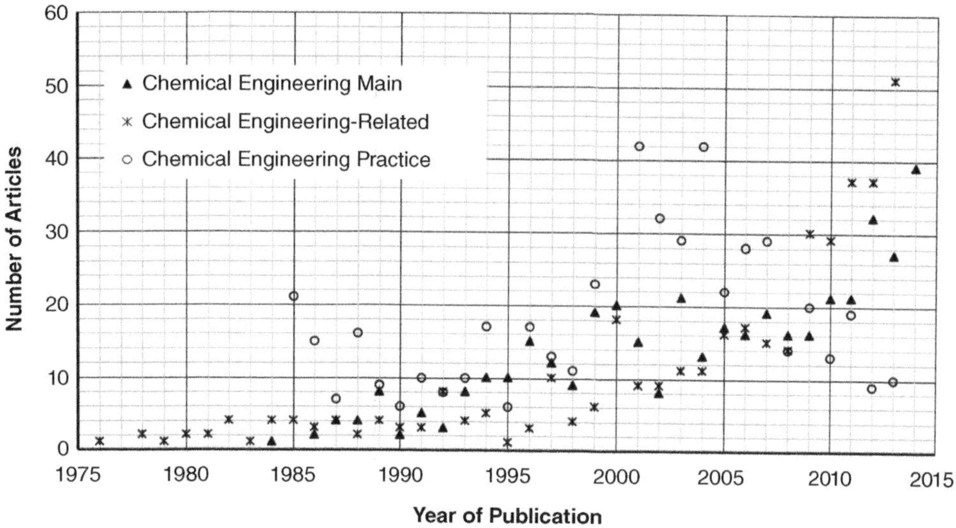

Figure 1.2 *The trend of number of articles mentioning retrofit/retrofitting, revamp/revamping and debottlenecking in the three groups of chemical engineering journals, in each year from 1976 to 2014. Note that the number of articles in chemical engineering practice in 2000 is 72, and that in chemical engineering-related in 2014 is 93; neither of these large numbers is shown in the Figure for clarity.*

The trend of the number of articles published in the three groups of journals in each year over the period covered is shown in Figure 1.2. It is clear that the number of articles related to process retrofit/revamp is increasing in chemical engineering main and -related journals. However, the number of articles mentioning process retrofit/revamp in chemical engineering practice journals went through a peak around the year 2000. Perhaps, this was due to an anticipated oil price increase after many years, and advancements in process technology, simulators and optimization software, besides optimism at the beginning of a new century.

How many of the journal articles in Figure 1.2 use retrofit/retrofitting, revamp/revamping and debottlenecking? This can be seen in Figure 1.3. Chemical engineering practitioners employ revamp more than retrofit whereas researchers mostly use retrofit. Perhaps, researchers are following the previous papers such as Grossmann *et al.* (1987), Douglas (1988) and Rong *et al.* (2000). Going by the words employed in the articles, objectives of retrofit/revamp are not just throughput increase but others listed in Section 1.1. There seems to be more articles on debottlenecking in chemical engineering main and practice journals than those in chemical engineering-related journals.

Figures 1.1 and 1.2 indicate increasing interest in process retrofit/revamp from chemical engineering researchers and substantial retrofit/revamp activity in chemical process industries. However, there seems to be only a few books and book chapters devoted to process retrofit and revamp. The book by Lieberman (2010) has many practical strategies to reuse and also improve the performance of existing equipment (such as pumps, compressors, heat exchangers, heaters and distillation columns in chemical process industries) and of several

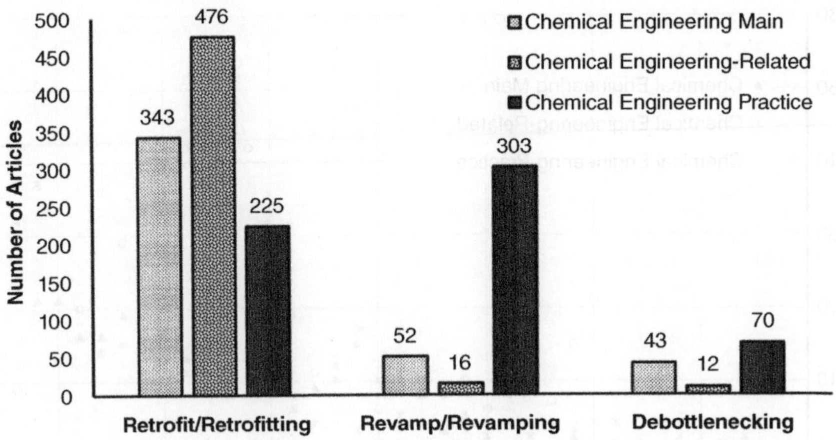

Figure 1.3 *The number of articles using retrofit/retrofitting, revamp/revamping and debottlenecking in the three groups of chemical engineering journals from 1976 to 2014.*

petroleum refining processes. It also has one chapter on summary checklist for re-using and improving common process equipment. The process design book by Douglas (1988) has one chapter on process retrofit procedure with an application to hydrodealkylation (HDA) process. The chemical engineering design book by Towler and Sinnott (2012) covers process revamps in one section of a chapter. Smith (2005) discusses retrofit of heat exchangers and their networks and distillation sequences in different sections of his book on process design and integration. Retrofit of heat exchanger networks is described in one chapter in Kemp (2007).

1.6 Scope and Organization of this Book

This book on process retrofitting and revamping is divided into three parts, each with a certain focus, for the convenience of the reader. Whereas this chapter provides an overview to process retrofitting and revamping, the next two chapters in Part I are on project engineering/management and safety aspects of process retrofitting and revamping, which are crucial for success in industrial implementation. The focus of Chapters 4 to 8 in Part II is on computational techniques for solving process retrofit/revamp problems. One of them is on process intensification technology for retrofit/revamp of chemical processes. The techniques described in Chapters 4 to 8 can be employed in a variety of process industries. Finally, Chapters 9 to 13 in Part III are on retrofit/revamp applications in chemical process industries. They describe the options considered and analysis employed for the retrofit/revamp of the process studied, which are useful for related applications.

Chapter 2 by Reddy is on project engineering and management of retrofit/revamp projects. Retrofitting and revamping projects are very common in the process industries, especially in old and aging facilities. These are increasing even in relatively new plants. However, there is very limited literature covering project engineering and management of such projects of practical importance. In Chapter 2, Reddy presents a comprehensive methodology for project engineering and management of retrofit/revamp projects, and describes important elements of retrofit/revamp project management. The significance of plant test runs, detailed pre-FEED (Front-End Engineering Design), FEED studies and project management are emphasized. Finally, retrofit/revamp options for common process equipment and systems are outlined in numerous tables, which are useful for convenient and quick reference in retrofit/revamp studies.

In **Chapter 3**, Balajee and Reddy present a comprehensive overview of all process safety aspects required for sustainable revamps. Many process safety incidents with significant impact are mainly due to unrecognized and unmitigated process safety risks. Process safety aspects are even more important for revamp projects because of complex modifications involved and/or effects of new equipment on the existing plant. Use of comprehensive process safety techniques is essential for successful implementation and operation of a revamped plant. The overview of safety aspects of revamp projects in Chapter 3 includes the application of hazard analysis methods and process hazard analysis (PHA) techniques, during Pre-FEED and FEED stages for risk assessment and risk mitigation. The importance of updating process safety information (PSI) and implementation of management of change (MOC) procedure after PHA study are emphasized. Chapter 3 is useful as a quick reference on process safety aspects for revamp projects, for both practicing engineers and students.

Part II of the book begins with **Chapter 4**, where Sharma and Rangaiah describe the modeling, simulation and optimization steps as well as software and tools for design, retrofitting and revamping of chemical processes. A general modeling procedure for process units/operations is illustrated using membrane separation for a gas mixture. This separation process is not readily available in process simulators (for example, Aspen Plus and Aspen HYSYS), and so its model is implemented in Aspen Custom Modeler (ACM) and subsequently included in process simulators. Chapter 4 discusses both deterministic and stochastic optimization techniques employed to obtain the optimal design and operating conditions of chemical processes, for single or multiple objectives. Global optimization methods have been implemented in Excel, MATLAB and C++. Therefore, interfacing a process simulator with a global optimization program is required to improve the process performance. Chapter 4 also describes interfacing of ACM, Aspen Plus and Aspen HYSYS with Excel using simple examples. Finally, the optimization of a membrane separation process for CO_2 removal from natural gas for simultaneous maximization of both methane purity and recovery, using an Excel based MOO program and Aspen Plus is presented.

In **Chapter 5**, Rao describes many opportunities of process intensification (PI) in process retrofit/revamp projects. PI, an emerging area of chemical engineering, represents a paradigm shift in process design and equipment and provides a break away from the concepts of unit operations and unit processes. It aims at orders of magnitude reduction in the sizes of equipment by enhancing the rates of transport and reaction, and in the plant footprint by combining the elementary steps such as separation and reaction. These reductions lead to inherently safer design, minimization of solvents, diluents and energy requirements. Chapter 5 also presents equipment that makes use of centrifugal fields as alternatives to distillation, absorption and extraction columns, and recently proposed modified channel geometry equipment as alternatives to stirred reactors in retrofit and revamp. Further, methods for PI like reactive and hybrid separations, heat and mass integrations (pinch technologies), conversion of processes from batch to continuous and process-specific integration are highlighted for use in the revamp. Process synthesis with PI has also been highlighted.

Dunn and Ristau, in **Chapter 6**, present graphical and mathematical design methodologies for heat and mass integration, which have been used over the past two decades to identify energy conservation and water recovery process designs in the chemical process industry. In large existing chemical plants, one is interested in the application of these design tools to identify retrofit designs where the challenge is to identify process retrofit options associated with larger systems (that is, identifying design options from the simultaneous analysis of a large number of streams). Graphical design tools allow a visualization of the design problem, not always possible when using mathematical optimization approaches. Chapter 6 focuses on two visual design strategies for solving large process integration problems. The first strategy uses visual tools for identifying heat integration designs that allow heat recovery in the industry. It uses three graphical tools, namely, temperature-interval diagram, heat pinch composite curves and the enthalpy-mapping diagram. The second strategy uses visual tools for identifying water recycle network designs that allow water use conservation and wastewater reduction in industry. This strategy uses two graphical tools: the material recycle pinch diagram and the source-sink mapping diagram. Examples of both graphical design strategies are provided.

Heat exchanger networks (HEN) retrofitting is the subject of **Chapter 7** by Sreepathi and Rangaiah. HENs are employed in process industries to decrease the external utilities required. Retrofitting existing HENs can achieve significant energy savings for a relatively small investment. Mathematical programming methods are increasingly used to solve HEN retrofitting problems. Of these, stochastic global optimization (SGO) methods such as genetic algorithm and differential evolution are gaining popularity. HEN retrofitting can be *via* modifying the HEN structure that includes adding new heat exchangers, additional area in existing exchangers, re-piping/re-sequencing existing heat exchangers and/or using heat transfer enhancements. HEN retrofitting at simple, moderate and complex levels, which are relatively easy to more difficult for implementation, and application of multi-objective optimization (MOO) are described and illustrated by applying them to retrofitting the HEN of an industrial problem having 10 hot and 5 cold streams. The objectives used are cost of retrofit and utility cost after retrofitting. MOO using the elitist non-dominated sorting genetic algorithm (NSGA-II) provides many Pareto-optimal solutions for each retrofitting level. An optimal solution can be chosen from one of these solutions based on the situation (for example, budget, manpower and time available for retrofitting).

Chapter 8 by Azzaro-Pantel deals with a number of retrofit issues dedicated to batch plants. Retrofitting is particularly important for batch processes, which have to face fluctuations in production demand, new product manufacturing and shortened time-to-market periods. Despite its significance and importance, review of the dedicated literature shows that retrofitting has received much less attention than have design problems, which can be considered as a paradoxical situation since process retrofit occurs more frequently than does process design in industrial practice. The motivation for the retrofit problem has been first provided by the industrial environment. Yet, the development of retrofit design strategies is challenging for academic research, with specific features compared with the design problem. Chapter 8 is focused on the methods and tools used to solve the batch plant retrofit problems; these methods and tools have been developed by the process systems engineering community in the last few decades. Similarities and differences with the design problem are emphasized. Conclusions and directions for future research are proposed highlighting the need for more robust retrofit strategies embedding a multi-criteria decision approach.

Part III of the book begins with **Chapter 9** by Lee *et al.* on retrofitting a side stream distillation column (SSC) for separating a ternary mixture. SSC is a common energy-intensive process and has received much scientific attention due to growing industrial demand. Recent studies and industrial applications have shown that process intensification using a dividing-wall column (DWC) can significantly reduce the energy required for the separation of mixtures by distillation. Lee *et al.* review the design and application of SSCs and DWCs in both academia and industry. A systematic procedure for the retrofit of an SSC to a DWC is presented to overcome the limited use of SSC due to the high purity requirements and low energy efficiency. Several industrial examples are presented to show the effectiveness of the procedure for retrofitting SSCs to improve the energy efficiency, thus mitigating CO_2 emissions. Various issues such as techno-economic analysis, column modification, construction and operability, are discussed briefly to provide more insight into the industrial application of DWC in the retrofit of SSCs.

Chapter 10 by Tan *et al.* is on retrofitting distillation columns for separation of olefins from paraffins in a petrochemical industry. Ethylene and propylene are the most important olefins. Owing to their similiarites in physicochemical properties with their paraffinic counterparts, their separation by cryogenic distillation requires high capital and operating costs. Hence, various alternative technologies including membrane separations have been studied by researchers. Recent studies on Carbon Molecular Sieve (CMS) membranes have shown feasibility for handling extreme operating conditions in ethylene plants. Techno-economic viability of retrofitting selected distillation columns for olefin/paraffin separation in an ethylene plant, with CMS-based membrane separation is investigated in Chapter 10. The retrofit design is optimized for two simultaneous objectives: minimize capital cost of the additional membrane unit and maximize utility cost savings of the hybrid system. Results show that the hybrid membrane-distillation system can significantly reduce the utility cost of the propylene fractionator.

In **Chapter 11**, Reddy and Rangaiah cover retrofit of vacuum systems for improving their energy efficiency. Vacuum operation is required in many process industries for a number of reasons. For this, steam jet ejectors (SJEs) are widely used due to their high reliability and medium pressure steam availability; however, they are highly inefficient. Owing to high energy prices, there is increased retrofitting of SJEs to reduce the energy cost. In Chapter 11, vacuum generation methods and options for developing energy efficient retrofit design of vacuum generation systems are analyzed. The requirements, benefits and constraints for implementation of these options are described, and a flow chart for the selection of a suitable retrofit option is presented. Practical considerations in retrofitting (for example, re-use of equipment and cooling water, plot space requirement and reliability improvement) are presented. Feasible retrofit options for both hazardous and non-hazardous applications are illustrated with one case study each. In general, numerous opportunities exist in process industries for retrofitting vacuum systems as they are widely used and need to be upgraded for revamping the process unit.

In the chemical industry, water is extensively used as a reactant, separation solvent and heating/cooling medium, and therefore the water integration is required due to both economic and environmental reasons. In **Chapter 12**, Sharma and Rangaiah present a mathematical model that can be used for grassroots and retrofit design of water networks.

Then, a large water network example in a typical petroleum refinery is analyzed for design, retrofit and revamp, using the MOO approach. Initially, MOO is employed for the design problem to explore the trade-off between fresh water consumption and total water flow rate through all regeneration units, which gives better insight and also provides many alternative optimal solutions for water network design. Two optimal water network designs are selected, and then their retrofit/revamp is studied to meet the revised environmental regulations and increased hydrocarbon load from the refinery processes. In the water network retrofit, it is assumed that there is no change in the network topology, and only capacities of the regeneration units can change. On the other hand, in the water network revamp, changes in the existing network topology along with capacities of different regeneration units are allowed. This revamp is relatively more complex and costly compared with retrofit. In this chapter, the ε-constraint method is used to obtain Pareto-optimal fronts for MOO problems.

In **Chapter 13**, Ghosh describes industrial application of debottlenecking and retrofitting of a chemical pulp refining process for paper manufacturing. The complexity of paper manufacturing depends upon the types of paper manufactured, end user applications and starting raw materials. It is a capital-intensive industry and comprises multiple process steps. Use of new technology in the form of plant and equipment to *debottleneck* many rate-determining steps and to increase throughput is often financially prohibitive. Identifying critical steps and *retrofitting* with appropriate components in such steps can be very cost-effective in the optimization of existing plants. The chemical pulp refining process is one of the most important stages for manufacturing papers and boards. In many paper mills, the installed refining system may be sub-optimal in the form of capacity, the type of refiners used, the mode of operation and the type of refiner tackles fitted. Chapter 13 presents realization of tangible benefits for two paper mills from retrofitting the refining process of paper machines. These include substantial savings in energy required for refining, increase in throughput and improvement in product quality for one mill. For the second mill, retrofitting of the existing double disc refiner by a high consistency refiner followed by a low consistency refining system enabled the mill to manufacture premium grade 'semi-extensible' sack kraft papers with significant improvement in the cross-directional profiles and absolute values of the critical properties of such papers.

1.7 Conclusions

Existing chemical process plants are continually increasing in number. Their performance has to be reviewed regularly, and their sustainability improved by making suitable changes in process equipment and/or structural changes. Hence, interest and activity in process retrofitting, revamping and debottlenecking are expected to grow. This book, one of the few on process retrofitting and revamping, will be useful to chemical engineering students, practitioners and researchers. Each chapter in this book can be read independently of other chapters. Hence, readers of this book can choose to read a few or all chapters, depending on their background and interest. However, they are encouraged to read the first two chapters before reading others. Many chapters in this book have exercises to solve in order to gain expertise. Most of these exercises can be given as projects in chemical engineering courses at advanced level.

This chapter introducing process retrofit/revamp and the next two chapters on project engineering and safety aspects of retrofit/revamp projects are valuable, particularly for chemical engineering students and young engineers in chemical process industries. For example, modification options for common process equipment/systems in numerous tables at the end of Chapter 2 are useful as a convenient and quick reference in retrofit/revamp studies. Computational techniques and process intensification technology for retrofit/revamp of chemical processes, described in Chapters 4 to 8, are applicable to a broad range of chemical process industries. The last five chapters (Chapters 9 to 13) describe the improvements considered and analysis employed for the retrofit/revamp of the process studied in each of them, which are useful for related applications.

The number of articles mentioning retrofit/retrofitting, revamp/revamping and debottlenecking in the chemical engineering main and -related journals, has been increasing, especially rapidly in the past five years. This is an indication of increasing research and development in process retrofit/revamp. Reviews such as Chapter 5 on process intensification technologies, Chapter 8 on batch plant retrofitting and the recent article on heat exchanger network retrofitting by Sreepathi and Rangaiah (2014) are relevant and helpful in further advancing the computational techniques and technologies for process retrofit/revamp and in increasing their applications. There will be more studies on process retrofit/revamp and successful applications in chemical process industries.

References

Carvalho, A., Gani, R., and Matos, H. (2008) Design of sustainable chemical processes: systematic retrofit analysis generation and evaluation of alternatives. *Process Safety and Environment Protection*, **86**, 328–346.

Douglas, J.M. (1988) *Conceptual Design of Chemical Processes*, McGraw-Hill, New York.

Fisher, W.R., Doherty, M.F. and Douglas, J.M. (1987) Screening of Process Retrofit Alternatives, *Industrial & Engineering Chemistry Research*, **26**, 2195–2204.

Grossmann, I.E., Westerberg, A.W., Biegler, L.T., and Carnegie Mellon University Engineering Design Research Center (1987) Retrofit Design of Processes, Department of Chemical Engineering. Paper 110. http://repository.cmu.edu/cheme/110.

Jensen, N., Coll, N., and Gani, R. (2003) An integrated computer-aided system for generation and evaluation of sustainable process alternatives. *Clean Technology Environment Policy*, **5**, 209–225.

Kemp, I.C. (2007) *Pinch Analysis and Process Integration*, 2nd edn, Elsevier, Oxford.

Lieberman, N.P. (2010) *Process Engineering for a Small Planet: How to Reuse, Re-Purpose and Retrofit Existing Process Equipment*, John Wiley, Hoboken, New Jersey.

Rong, B.G., Han, F.Y., Kraslawski, A., and Nystrom, L. (2000) Study on the methodology for retrofitting chemical processes. *Chemical Engineering and Technology*, **23**, 479–484.

Smith, R. (2005) *Chemical Process Design and Integration*, John Wiley & Sons Ltd, Chichester, England.

Sreepathi, B.K. and Rangaiah, G.P. (2014) Review of heat exchanger network retrofitting methodologies and their applications. *Industrial & Engineering Chemistry Research*, **53**, 11205–11220.

Towler, G. and Sinnott, R. (2012) *Chemical Engineering Design: Principles, Practice and Economics of Plant and Process Design*, 2nd edn, Elsevier, Oxford.

Uerdingen, E., Fischer, U., and Hungerbühler, K. (2003) Screening for profitable retrofit options of chemical processes: a new method. *AIChE Journal*, **49**, 2400–2418.

2

Project Engineering and Management for Process Retrofitting and Revamping

C.C.S. Reddy
Lead Process Design Engineer, Singapore Refining Company Private Limited, Singapore

2.1 Introduction

Process engineers often need to perform and manage retrofitting and revamping, especially in old and aging facilities. Owing to increased competition, energy prices, environmental concerns and technological advances, this activity is increasingly occurring even in relatively new plants. Generally, there will be some design margins available for the installed equipment in the process plants. For example, these are: 10% for pumps and heat exchangers, 20% for control valves (Mukherjee, 2012) and separators, and 15% for motors. Distillation columns are usually designed to limit vapour velocity to 80% of flooding velocity. Pumps are generally selected with impeller size lower than the maximum possible, which provides a cheap option of increasing impeller size to increase their exit pressure/capacity. Equipment design pressure and temperature are usually specified to meet the worst case conditions. This provides room for increasing operating pressure and temperatures during the revamp. Also, after equipment specification by design engineers, equipment fabricators supply equipment with additional design margins. Thus, installed capacity of the equipment is often more than the specified capacity. These extra margins provide many revamp opportunities for improving process efficiency.

Chemical Process Retrofitting and Revamping: Techniques and Applications, First Edition.
Edited by Gade Pandu Rangaiah.

Revamp/retrofit projects may have one or more objectives such as to reduce energy consumption, meet new separation requirements, increase throughput, process new raw materials, produce products with different specifications, improve reliability, improve safety and reduce environmental emissions. As described in the Introduction chapter, retrofit and revamp are often used synonymously. Hence, revamp and revamping used in this chapter refer to process retrofit and debottlenecking as well. There are numerous revamp projects executed in the process industries. However, there is very limited literature covering process design and project management of such projects of practical importance. The scope of revamp projects is generally complex, and takes more design and evaluation time as many options are available to achieve the project objective(s). For successful design and implementation of revamp projects, interactions with the existing/operating plant have to be evaluated systematically. For multi-objective revamp projects, it may be challenging to satisfy all the objectives and so a suitable trade-off solution may have to be chosen.

This chapter presents a comprehensive methodology for project engineering and management of revamp projects, describes important elements of revamp project management and outlines revamp options for process equipment and systems. Section 2.2 presents key differences between revamp and grassroots projects. A comprehensive revamp methodology is presented in Section 2.3. The subsequent section outlines project/process engineering scope for the revamp projects. Key elements of revamp project management are presented in Section 2.5. Then, Section 2.6 summarizes revamp options for many process equipment, process systems and OSBL (outside battery limit) systems. This chapter ends with conclusions in Section 2.7. Contents of this chapter are useful for practising process engineers, project engineers and in educational institutions.

2.2 Key Differences between Revamp and Grassroots Designs

One important role of process/chemical engineers in operating plants is to perform revamp studies and execute them to improve business competitiveness. Depending on the project scope, revamp projects can range from very simple to very complex. They generally require more effort compared with that for grassroots projects. Although there are some similarities between revamp and grassroots projects, there are many key differences, as summarized in Table 2.1. Revamp projects require more front end project activities, especially in conceptual design and the subsequent FEED (front end engineering design) works. These activities have significant impact on revamp project cost and success.

2.3 Revamp Design Methodology

Process revamp studies are generally initiated by operation/process engineers. Every plant is different, and revamp opportunities vary widely from one plant to another. However, a systematic revamp methodology can be used in any plant. Such a detailed revamp methodology is shown in Figure 2.1. It starts with setting revamp project objectives, project framing (basis, assumptions and boundary conditions), preliminary project cost (with $\pm 50\%$ accuracy) and economics in the pre-feasibility phase. Next, conceptual design or pre-FEED stage begins. Plant operating and design data are collected, and plant performance tests are conducted.

Table 2.1 *Comparison of revamp and grassroots projects.*

Item	Revamp projects	Grassroots projects
Project scope	Project scope may not be well defined at the beginning as there may be many options and issues. It requires design and analysis of all alternatives for choosing the most attractive option and defining the project scope	Generally, project scope will be well defined by the technology provider and detailed engineering contractor
Design constraints	There are generally more constraints due to re-use and modifications of the existing plant equipment within limited plot space and project execution time	Fewer design options and constraints
Quality of process engineers	Requires very experienced process engineers with good understanding of process operations, technology and economic trade-offs	It can be performed by process engineers with some process knowledge and expertise
Conceptual design	Conceptual design needs to be calibrated with plant operating data and also performance test data. It has to be performed within existing equipment constraints	Office-based conceptual design is performed as there is no requirement to check any plant data
Data collection and verification	Up-to-date equipment, piping, civil, structural and electrical design drawings and process simulations may not be available. So, data collection and verification of existing plant equipment/facilities will require substantial time. Test runs may also be required to generate required data	Data collection is unlikely to be required
Interactions with existing plant equipment and system	It is essential to study interactions and effects on the existing process equipment and utility systems. These interactions may extend to other plants in the same facility	This is not applicable
Review of existing plant's OSBL (outside battery limit) systems	This is generally required since the revamp may require more utilities, pressure relief, storage facilities etc.	No such review is required as all the required OSBL facilities will be provided in the new plant's OSBL
Design margins	Design margins of original equipment/ plant may be reduced	Good design margins can be provided

(continued)

Table 2.1 *(Continued)*

Item	Revamp projects	Grassroots projects
Generation of quickly implementable non-CAPEX project options	Revamp study can identify non-CAPEX (capital expenditure) improvement ideas, which can be implemented immediately	Not applicable
Process safety risks	Detailed study/verification is required to avoid overlooking safety risks to existing equipment/facilities. There may be more fire/explosion/toxic chemicals exposure risks during project execution as the revamp work may be executed in close proximity to existing facilities, handling flammable and toxic substances	Process safety risk is relatively lower as all the equipment/ facilities are designed and installed at about the same time
Project team	Relatively smaller team as most of the activities are performed by in-house project team	Generally, very big project and engineering contractor teams will be involved
Equipment delivery	Equipment delivery and installation need to meet turnaround/plant stoppage schedule	Equipment delivery and installation are flexible within the project schedule
Equipment installation	It is mostly during plant shutdown period, and hence installation may be constrained by the shutdown period	It can be installed any time
Plot space	Plot space is limited and constrained by maintenance access, fire, safety and emergency handling requirements	Sufficient plot space is generally available
Maintenance and emergency access	Access may be limited as more equipment is added to the existing plot	Sufficient access will be provided
Engineering cost	Owing to plant and equipment constraints, generally higher at 15-20% of project cost (Elshout and Gracia, 2009)	Generally lower at about 10% of the project cost (Elshout and Gracia, 2009).
Project cost	Generally less compared with the grassroots project	Generally more compared with the revamp project

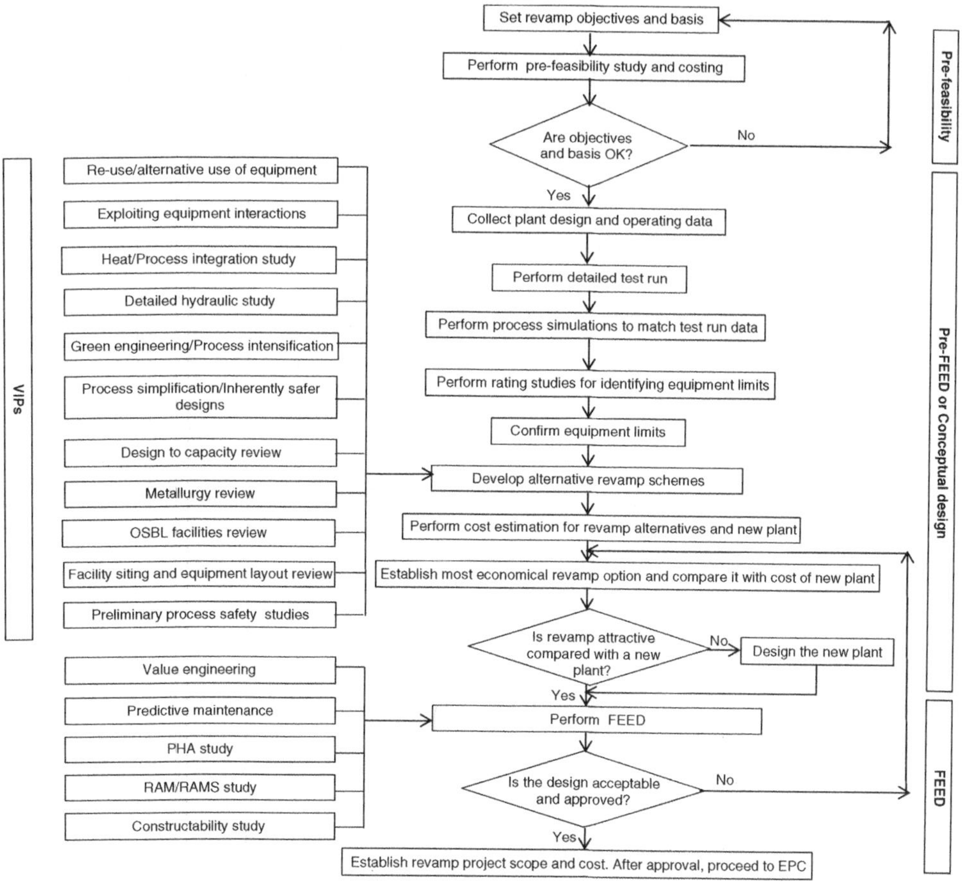

Figure 2.1 *A detailed and systematic methodology for revamp projects.*

Process simulations are calibrated to match performance test data and also plant operating data. Once the calibrated simulation is available, re-simulations and equipment rating studies are conducted to identify various bottlenecks and develop many feasible revamp options for achieving project objectives. As shown on the left side of Figure 2.1, several studies (referred as Value Improvement Practices, VIPs) are required to explore all revamp options. Block flow diagrams or PFDs (process flow diagrams) are prepared for each of the revamp alternatives and also for building a new plant. A cost estimate is performed to establish project cost (with ±30% accuracy) for all options. The process engineer will then consolidate all revamp options and their relative advantages, constraints and cost data. After thorough screening, the process engineer and/or engineering contractor will determine the most promising revamp option. For the revamp to be deemed beneficial, capital expenditure (CAPEX) of the best revamp option should be marginally lower, say, 60–70% of the cost of building a new facility. Detailed option analysis and summary of the recommended revamp option will be submitted to key stake holders (usually the board of directors) of the company for approval. Many chapters in this book describe methods/techniques and their applications for developing and evaluating process alternatives in the conceptual process design stage of retrofit/revamp projects for process industries.

After approval of conceptual design or Pre-FEED, the revamp project moves into the FEED stage, where the selected option is fully developed; this includes preparing process and instrumentation diagrams (P&IDs), equipment layout, piping layout, equipment datasheets and so on. A detailed process safety study such as PHA (process hazard analysis), reliability, availability, maintenance and constructability reviews are conducted. The design is modified to incorporate any action items that are a result of these safety, reliability and constructability studies. A detailed project cost (with ±10% accuracy) for the final design is generated by getting quotations from equipment vendors and engineering contractors. The finalized FEED study will be submitted to the key stake holders again for their approval. Then, finalized FEED is used for detailed engineering, procurement and construction (EPC).

2.4 Project/Process Engineering and Management of Revamp Projects

2.4.1 Revamp Objectives and Pre-Feasibility Study

Revamp projects generally have more than one objective; this may include increasing unit capacity, reducing energy consumption, reducing environmental emissions, improving process safety and so on. However, the initiator of the project needs to identify the main objective of the project. For example, in a revamp project involving a fired heater, energy reduction is likely to be the main objective, and there could be minor objectives such as reducing emissions and heat radiation from a flue gas stack. For better justification of the revamp project, it is important to highlight such minor objectives also, as these contribute to the viability of the project. The process engineer needs to prepare the list of project objectives, key business drivers, basis of design (for example, raw material and product and energy prices) and assumptions (for example, business plan, project boundary conditions, payback period and necessary government approvals). S/he needs to circulate them with rough project cost (with ±50% accuracy) and economic analysis to the key stake holders for their support. This activity is carried out in the pre-feasibility stage (Figure 2.1).

2.4.2 Conceptual Design (Pre-FEED)

At the initial stage of a revamp project, process engineers may not realize the benefits and constraints of various revamp options. After approval of the pre-feasibility stage, they will perform conceptual design in the Pre-FEED stage for generating and analysing feasible revamp design options, meeting project objectives and constraints. Revamp project objectives can be met by many revamp design options. For example, energy consumption of two distillation columns separating a three-component mixture can be reduced by implementing some of the following revamp options:

- Increasing feed pre-heat
- Reducing column pressure
- Reducing reflux ratio
- Changing column trays/packing (for high efficiency/capacity)

- Installing a pre-flash tower
- Implementing heat pumps (from pinch analysis)
- Implementing a dividing-wall column

Comprehensive conceptual design is required to select the best revamp option. It is more detailed compared with the grassroots projects as many existing equipment and facility interactions need to be considered in the design. Conceptual design and FEED define nearly 80% of the capital cost and 98% of the operating cost (Mendez, 2007), and eventually the project success. Conceptual design will cost less than 5% of the project cost even if it is performed by an engineering contractor (Golden *et al.*, 2003). There will be many design options available as most of the original equipment can be re-used with or without modifications and also new equipment can be added. For revamp projects, installed cost of the equipment is 4–8 times equipment purchase cost (Barletta *et al.*, 2002). So, failure to identify the correct number of pieces of new equipment in conceptual design will increase the project cost in detailed engineering stage, leading to overrun of the budget. If conceptual design is not performed properly, revamp project scope can expand by 30–70% when the project proceeds to the next stages (Barletta *et al.*, 2002). So, comprehensive conceptual design is required although it takes more time and effort. It reduces the project risk significantly.

Conceptual design should be performed by a small team led by an experienced process engineer as its success is largely dependent on his/her ability. So, s/he should have the following attributes:

- Sound technical knowledge combined with operating experience to understand the viability of the options
- Critical and analytical thinking
- Creative and innovative thinking
- Concepts and techniques of value improvement practices
- Good cross-functional and inter-discipline domain knowledge
- Familiarity with proven revamp ideas
- Knowledge on the latest technological developments

These qualities are important so that s/he will not miss any feasible option or decide the option based on limited alternatives and simulations.

As a first step, all the relevant plant simulations, operating and design information need to be collated. The plant data collection and review are essential for successful design and implementation of revamp projects. Review of existing plant data and process safety information (PSI) is very important to an understanding of capacity limitations and constraints of the plant systems. Similarly, thorough review of an existing plant's OSBL facilities such as utility system, flare and fuel gas system is essential for defining project scope accurately. Typical data required for a revamp study are summarized in Table 2.2.

Plant operating data are essential to validate and adjust process simulations to mimic the plant operation. If available DCS (distributed control system) trending data for 2–3 years will be very useful. Original equipment might have been installed many years ago, several modifications might have been made to them over the years, and they may not be performing according to the details available in the data sheets. Verification of plant's operating data will quickly enable a designer to identify data mismatch. For example, a

Table 2.2 *Typical data required for the revamp study.*

Process flow diagrams	Process and instrument diagrams
Cause and effect charts	Process interlocks and alarms list
Operation procedures	Operating envelope
Process plant simulations	Process and mechanical data sheets of process equipment
Equipment drawings	Equipment installation details
Equipment inspection and maintenance reports	Simulation of flare network
Pressure relief valves and rupture disk design data	Hazardous area classification drawings
Fire circles; over-pressure contours from explosion modelling	Design details of OSBL facilities
Plot plans	Piping isometric drawings
Details of pipe racks	Substation, DCS and SIS system capabilities
Details of underground cables	Motor and other drivers data sheets
Control valve data sheets	Control system narrative
Fire fighting system	Facility's emergency handling systems
Plant turnaround or stoppage plans	Facility's work permit system

heat exchanger's fouling cannot be predicted accurately by using correlations or process simulators.

Long-term trending of maintenance data is very useful to enable an understanding of equipment reliability issues. Plant simulations will not identify some of the critical concerns such as corrosion, erosion, liquid and gas maldistribution, and potential slug formation in piping. A detailed review of historical equipment inspection and problem/repair history is required to understand various maintenance, metallurgy and reliability issues.

A detailed review of existing equipment's design and construction drawings will also provide insights into various equipment and plot limitations. For example, increased feed rate, improper internals and level controls can adversely affect a separator's efficiency. A review of these details in the existing separators will be very useful for revamp projects. Similarly, careful review of column internal details is required to check possible modification to increase capacity/efficiency and reduce flooding/weeping.

Also, the process designer needs to review and understand the piping and equipment layout drawings; for larger plants, 3D design technology integrated with laser scanning of piping can be very useful (Farrow, 2013). Plant walk down is very useful. These piping and layout reviews will help in the following aspects:

- Reducing piping and equipment installation costs
- Estimating accurate piping hydraulics
- Eliminating potential piping design issues such as asymmetric piping manifolds (for equipment installed in parallel arrangement)
- Eliminating liquid slugs
- Eliminating improper equipment installations (for example, flow meters, de-superheaters, gas/liquid injection nozzles and sample points)

To reduce the cost of manufacture, process engineers must continuously monitor and optimize the operations. This involves implementation of changes in operating variables and/or equipment. In the older plants, by the time a new revamp study is initiated, many changes might have been implemented and hence the present operation can be very much different from the original design. By using original plant simulation only, the process engineer may fail to identify the most profitable revamp design. Therefore, validation and adjustment of process simulation is very important for successful and profitable revamp design. Comprehensive plant performance test runs (near revamp conditions) should be conducted to gather vital data that are useful for calibrating the existing plant simulations.

For example, accurate analysis of a heat exchanger network performance requires values of flow rate, inlet and outlet temperatures of hot and cold streams for each heat exchanger from test runs as well as exchanger areas. Simulated duties of hot and cold streams of the heat exchanger are compared with test run data. If there is mismatch, simulation input data should be compared with the test run data, and suitable adjustments of stream flow rates and/or temperatures are required to minimize the mismatch.

Once the calibrated simulation model is available, simulations can be re-run and equipment rating studies can be performed to identify equipment limitations and to generate many options to meet the revamp objectives. Spreadsheet and vendor models are also required to perform conceptual design and options. Heat and material balances, and preliminary equipment sizes, are estimated. Detailed process simulations, block flow diagrams, PFD, pipe and process equipment sizes are generated for each revamp design option; these are useful for detailed analysis of benefits/constraints and CAPEX estimation for each option.

The following VIPs and techniques are useful to develop holistic revamp options. These VIPs are also proven to significantly improve profitability of the projects (Green and Perry, 2008).

I. **Re-use of Existing Equipment:** For revamps, it is required to critically review, debottleneck, retrofit and re-use the existing plant equipment. If the revised capacity is less than 20–30% above the name-plate capacity, there will be a pretty good chance of re-using the existing equipment in the revamp (Litzen and Bravo, 1999). Martin and Cheatham (1999) have demonstrated this for the revamp of a crude distillation unit (of 58,000 barrels/day or 384 m^3/h capacity), by increasing crude unit capacity by 20%. Maximizing existing equipment re-use will reduce capital cost of the revamp project.

II. **Exploiting Equipment Interactions:** The designer needs to consider total system interactions before finalizing scope for any equipment. Consider a revamp project with the main objective of increasing the duty of a column's feed pre-heater; it will require assessment of the condenser's maximum capacity and top rectifying tray section. Simulation of the whole process is needed to understand system interactions. Lee *et al.* (2007) and Zhu (2014) described ethylene plant retrofit cases to emphasize the importance of system interactions in revamps.

III. **Heat/Process Integration Study:** Pinch analysis is well established for finding optimal utilities, heat transfer area, optimal fresh water consumption, minimum cooling water demand, reduced emissions targets and so on (Smith, 2005; Kemp, 2007). One application of pinch analysis to retrofitting the heat exchanger network of a crude distillation column is described by Milosevic *et al.* (2013). They have shown fuel savings of 17.3 MW from modifying exchangers in the crude pre-heat train; investment cost and payback period for this case study are US$10.2 million and 1.7 years respectively. Optimization techniques provide more flexibility to simultaneously target and design the process; there are commercial optimization software and papers on their applications (Jianwei *et al.* 2011; Serna-Gonzalez and Ponce-Ortega, 2011; Reddy *et al.* 2013b). Reddy *et al.* (2013a) illustrated the use of mixed integer linear programming for maximizing power generation from a steam network, involving many steam turbine drivers and a large steam turbine power generator. Sreepathi and Rangaiah (2015) described the application of multi-objective optimization to retrofitting heat exchanger networks involving process streams with variable heat capacity.

IV. **Detailed Hydraulic Study:** Often, detailed hydraulic study is neglected in conceptual design. It can lead to unnecessary requirement of new pumps, compressors, high pressure design of process vessels and/or excessive process relief flows. Also, installing new piping is challenging and expensive in revamp projects. A good hydraulic study based on plot and equipment layout is essential to correctly define revamp project scope. If system pressure drop is not carefully estimated from piping and valves, flow and pressure drop across heat exchangers may drop and can severely decrease their performance.

V. **Green Engineering/Process Intensification:** Green engineering deals with techniques to reduce energy, raw materials consumption, waste generated and operation at high temperatures, and process intensification is one approach to them. Green engineering practices are effective for revamp projects to improve process sustainability. There are many green engineering ideas in the literature (Allen, 2007; Mendez, 2007; Contreras and Bravo, 2011). They include the following elements for revamp projects.
   - Choose operating conditions to minimize energy costs, increase productivity and reduce waste products
   - Consider internally/externally heat integrated distillation columns and/or low energy separation processes such as membrane distillation
   - Use pinch and/or optimization approach for reducing energy and water consumption
   - Use high selectivity/yield catalyst to boost production and reduce recycle flows
   - Reduce waste generation instead of waste treatment. For example, waste/foul water recycle with or without treatment reduces fresh water consumption and also avoids the need to expand waste water treatment plant
   - Wherever possible, re-use existing equipment and/or waste products (for example, use waste hydrocarbon gases as fuel, and use treated fouled water for washing and desalting in petroleum refineries)
   - For new equipment required, choose high efficiency equipment
   - Explore process intensification. Stankiewicz and Moulijn (2000, 2004) discussed process intensifying equipment/methods and their applications
   - Use advanced control systems and real time optimizers for operating and sustaining plant operations

VI. **Process Simplification/Inherently Safer Design:** Process simplification refers to a systematic value-based methodology to reduce investment and operating costs by eliminating or combining the process and utility systems (Green and Perry, 2008).

It simplifies the process by eliminating non-essential equipment/process steps. See Koolen (2002) for a comprehensive coverage of process simplification techniques and examples. An inherently safer plant relies on the quantity, properties and usage conditions of process materials to prevent injuries, environmental damage and property damage rather than relying on control systems, interlocks, alarms and procedures to stop incipient incidents (CCPS, 1993). It is mainly based on principles of process intensification, substitution, attenuation, limitation of effects (that is, designing facilities that minimize the impact of a release of hazardous material or energy) and simplification/error tolerance. Several guidelines, examples and inherently safer checklists are presented by CCPS (1993). Applying these concepts will be very useful to simplify the process and also make it safer.

VII. **Design-to-Capacity Review:** Process engineers tend to over-design equipment and piping due to a number of reasons, which include:
- Flexibility to handle more than the intended capacity of the revamp project
- Concerns on mass/heat transfer efficiencies and catalyst activity
- Capability to perform even under heavy fouling conditions
- Uncertainty on final design hydraulics as pipe layouts, equipment and control systems are usually selected in the next phase of the project

Although reasonable design margins are useful, too large margins can severely affect performance of some equipment. For example, too much design margin can foul a heat exchanger very rapidly (due to reduced fluid velocity) and hence reduce its duty. Oversized pumps require bigger impellors and motors, and consume more electrical power. Thus, a project's capital and operating costs increase with large margins, affecting its attractiveness. Hence, it is important to conduct a systematic review to remove unnecessary excessive design margins, to control project scope and cost; also, it will help to ensure sustainable performance from the revamp project. Woods (2007) has presented comprehensive guidelines for application, selection, troubleshooting and cost data for 350 pieces of equipment, which can be used for proper equipment selection. For example, he has given a guideline to maintain tube velocity of 3 to 4.5 m/s of a liquid stream for minimizing heat exchanger fouling. One can use this guideline to check tube-side velocity of an exchanger during the revamp study. It will ensure minimum fouling and hence helps to sustain the revamp benefits.

VIII. **Metallurgy Review:** Selection of the correct material of construction is important to minimize hazards due to corrosion, erosion, high temperature sulfidation, hydrogen or low temperature embrittlement and chloride stress cracking. It is vital for ensuring the safe and reliable operation of the equipment after the revamp. During the project, equipment operating variables such as flow rate, composition, temperature and pressure may change significantly. A critical review is required to carefully evaluate metallurgy upgrades, coatings, weld overlays and so on. Alternatively, a revamp study can look into opportunities to change process conditions to avoid expensive metallurgy upgrades. Also, it is very important to understand any requirements for refractory, insulation, coating and protection wrapping, post-weld heat treatment and so on, as they may inadvertently increase the revamp project cost in later project stages.

Table 2.3 List of OSBL facilities.

Steam generation/distribution systems	Cooling/refrigerated water systems
Plant and instrument air system	Flare system
Substation facilities	Fire water system
Hot oil systems	Waste water treatment plants
Sour water treatment	Amine treatment plants
Incinerators	Sulfur-recovery units
Air separation plants or import of N_2/O_2	Power generation units or import power
DCS (distributed control system)	PLC (programmable logic controller)

IX. **OSBL Facilities Review:** Capacities of OSBL facilities summarized in Table 2.3 should be reviewed thoroughly if the revamped project requires more utilities and other OSBL services. Use of more utilities for revamp projects may require additional piping and also may need a review to check the adequacy of plant and utility area interconnecting pipe racks. If additional piping and pipe racks are required, revamp project cost may escalate. Also, economics of energy-saving projects are strongly influenced by costs of fuel, electricity, steam and power generation. Unless the whole steam and energy system is reviewed, it is not possible to accurately verify savings from such projects. If a revamp project has several energy-saving ideas, the sequence of their implementation can result in totally different savings. Zhu (2014) illustrated this with a case study, for total site-wide energy optimization via five implementation sequences. Key outcomes of the study are: (a) implementing energy-saving options could have big impacts on steam balances, which must be taken into account in making decisions for project selection; the benefit estimates obtained from steam balances reflect the true benefits; and (b) energy improvement ideas must be assessed in OSBL in terms of potential effects on steam, fuel, and water balances to obtain the true benefits and capital costs. Reddy *et al.* (2013a) presented a spreadsheet-based steam system optimization model, which can be used for assessing steam-saving projects; there are also several steam system optimization software packages available on the market.

A facility's flare relief system analysis is required to ensure that relief capacity from the revamp project does not exceed the existing relief system design capacity. Proper checks need to be performed at substation facilities for installing switch gears, variable speed drives (VSDs) and other OSBL systems. If not properly evaluated, OSBL system limitations can escalate revamp project cost to such an extent that the whole project may become unattractive. A 20–30% escalation in the project cost is not uncommon if existing OSBLs are not properly evaluated in conceptual design.

X. **Facility Siting and Equipment Layout Review:** These are crucial for revamp projects as they have a strong influence on equipment selection, risk mitigation, piping and pressure relief system costs. If the revamp site is selected far away from the existing plant, new equipment will require lots of additional piping and new pipe racks, which in turn require even more plot space. Toxic or flammable compounds present at high pressures inside new piping will introduce additional safety risks and hence require more risk mitigation systems, as required by QRA (quantitative risk assessment) studies. If a liquid storage vessel is installed on the ground floor within the fire circle[1] of the plant, contributing most relief load during a plant fire scenario, additional fire-case relief flow from the new vessel adds to the relief header's peak flow. If the relief header's sizing is hydraulically very tight, it may require relief header and flare system modifications, which can be very expensive. Fire safety practices dictate the distance between various process equipment and systems. Proper equipment layout is necessary to minimize fire/explosion risks. Further, facility siting will decide the maintenance and emergency response access. Hence, proper site selection is very important during conceptual design.

XI. **Preliminary Process Safety Studies:** A simple what-if study is generally performed to identify major safety concerns, if any, with the project scope. FMEA (failure modes and effects analysis) or HAZOP (Hazard and operability) study can also be performed if sufficient design documents are available. Otherwise, these detailed studies will be conducted in FEED. Necessary changes are made to the project scope for mitigating any significant process safety concerns.

Finally, cost estimation for each design option is performed, mostly using in-house cost data or preliminary vendor quotations. Accuracy of such cost estimate can be ±30%. Process/project engineers will highlight the advantages and disadvantages of various design options, and recommend the most promising revamp design for implementation.

2.4.3 FEED (Front End Engineering Design)

FEED is carried out after approval of the best revamp option in the conceptual design (pre-FEED) stage. For large revamp projects, an engineering company will be engaged to perform the FEED work. For smaller and moderate projects, FEED work will be carried out by the company's in-house staff. In addition to process engineers, the FEED team will generally have engineers from other disciplines/departments.

Process engineers will prepare PFDs, P&IDs (which include all the new pipe sizes, manual and automatic control valves, control loops, process interlocks, tie-in points[2], new equipment, old equipment (re-used with modification), highlight piping and equipment to be demolished), plot plan, piping layout (with the assistance of piping engineers), process and mechanical specifications of new piping and equipment, process control narrative[3], process interlock list/matrix, cause-and-effect charts, functional design specifications for SIS (safety instrumented system), piping and equipment lists. They also perform all the calculations for pressure relief valves and relief system, and prepare utility system modifications. After developing these details, further studies are performed to make the revamp project safer, more sustainable, more reliable, more operable and easier to maintain.

RAM (Reliability, Availability and Maintainability) Study: This systematic methodology predicts the performance of the plant, with the given configuration, in terms of reliability, availability and maintainability. Reliability is the probability that the plant/equipment will perform the required functions, under specified conditions, for a stated period of time.

---

[1] Fire circle (or fire impact area) generally refers to a circular area, surrounding a potential flammable release source (such as process vessels, tanks and pipes). It is typically a ground area of 230–460 m^2 (API Standard 521, 2014).

[2] A tie-in point is the place in the existing piping or a flange on equipment where new piping needs to be connected.

[3] Control narrative refers to a document which includes complete details of process control systems such as process, advanced control logics and set points, alarm limits.

Availability, in general, is defined as the ability of the plant/equipment to perform its required function over a stated period of time. Maintainability is the probability that a failed item can be restored to operation effectiveness within a given period of time when repair action is performed as per the specified procedure (Smith, 2011). Software is available for performing RAM studies. For smaller projects, spreadsheets can be used. Reliability and process safety are interlinked, and so combined RAM and safety (RAMS) studies can be performed with the RAMS software (Sikos and Klemes, 2010). It considers many factors affecting the plant performance such as equipment performance, redundancy, demand requirements and logistics. RAM analysis is based on statistical failure data such as mean time between failures (MTBF), mean time to repair (MTTR), mean time to failure (MTTF) and mean down time (MDT). Wherever possible, failure data available within the company should be used for RAM/RAMS study. If not, typical failure data available in the literature/software can be used.

Constructability Study: This is important as it allows engineers to know construction issues of the revamp project, especially for removal or installation of large equipment. It reviews safety aspects of construction, especially the space required for construction, equipment maintenance activities and operations access, before detailed engineering and construction. Plot and equipment layout drawings or 3D models can be very useful for quick review and correcting design for proper equipment lifting and installation. 3D models can also be used to improve ergonomic aspects in the design by reviewing and installing manual valves and instruments at easily accessible heights and by providing sufficient pathways for operator movements during plant operations. All the action items from RAM/RAMS and constructability study need to be quickly addressed.

Predictive Maintenance Study: This study involves consideration of proactive sensors and other controls for monitoring and controlling equipment conditions during their operation. It will help to extend equipment availability by performing some on-stream operating parameters adjustment. It is also useful in effective planning of equipment shutdown. Some examples are use of corrosion probes, vibration sensors (Green and Perry, 2008), pH measurement, instruments to monitor heat exchanger fouling and packed beds used for reaction and/or separation.

Value Engineering: Value engineering VIP is a flexible, organized and multi-disciplinary team effort (Green and Perry, 2008) to critically examine the project scope by careful consideration of project objectives and removal of unnecessary job scope. Its main purpose is to minimize project cost without any compromises on the project objectives. Some examples are:

- Eliminate use of more expensive metals than the minimum required metal specification
- Elimination of unnecessary spare equipment, valves and so on
- Reduce over specifications
- Use of alternative equipment and/or vendor
- Avoid using customized designs
- Maximize use of vendor's standard design packages

PHA (Process Hazard Analysis) Study: PHA study during FEED stage is crucial as it may require changing some or a major part of the project due to process safety risks. It has potential to escalate the project cost significantly. PHA study is conducted by the PHA leader, who is a highly qualified process/safety engineer with many years of operations and design experience. The leader will decide the PHA study members such as design engineers, process safety, operations, maintenance, reliability, occupational safety, environmental and laboratory staff. Depending on the revamp project size, PHA study duration can be one week to a few months. PHA study may contain some or all of the following studies:

- Review of the original plant PHA and all plant modifications from the plant start-up
- HAZOP study
- What-if study
- FMEA
- Pressure relief system study to check and verify pressure relief system model
- SOA (safety objective analysis)
- Consequence analysis (for example, spill, leaks, vapour/liquid release modelling, fire radiation and pollutants dispersion studies, and explosion analysis)
- Risk analysis and mitigation (that is, fault tree and event tree analysis)
- Layer of protection analysis[4]
- SIS (safety instrumented system) or SIL (safety integrity level as per IEC 61508/IEC61511)

At the completion of the PHA study, the leader will publish all the action items and also list all the risk mitigation modifications and changes for the revamp project. Process engineers and other engineers will implement the required modifications within the agreed timeline. They also need to update all the required PSI documents such as PFDs and P&IDs.

Thorough conceptual design and FEED are essential to finalize the firm project scope. Every dollar spent in project scoping returns three dollars in reduced project cost, schedule and risk (Gebert *et al.*, 2004). Lagace Jr. (2006a) has presented a check list for good scope definition. After finalizing the revamp design, vendor quotations are obtained for all equipment, piping, instrumentation and control system, civil work, electrical system and an EPC contract. Purchases and services for smaller revamp projects may be handled by the procurement department of the company. After receiving the quotations, process engineers will lead a careful review of vendor proposals, and will rank them according to their suitability for meeting project objective, cost, delivery, reliability and other evaluation criteria. After the project team's agreement on the final vendors and corresponding quotation costs, process/design engineers will prepare the project cost estimate. Appropriate contingency must be included in this cost estimate, whose accuracy will be ±10%. Finally, FEED study with project cost will be presented for the company's approval to proceed to EPC phase and award the contracts.

2.4.4 Detailed Engineering, Procurement and Construction

Generally, process engineers have relatively lower involvement in detailed engineering as the majority of the process engineering work will be completed in conceptual design and FEED design stages. The detailed engineering stage completes the balance of the engineering work required for the plant construction such as site preparation, equipment foundation, equipment installation drawings, civil and structural drawings, electrical and

[4] LOPA analysis is explained in Chapter 3 of this book.

instrument cable layout drawings, piping isometrics, and applying for necessary government approvals. Process engineers need to verify the design calculations (such as hydraulic and relief calculations) if equipment layout is changed from the FEED stage. They will participate in field equipment and piping installation checks and coordination to complete all the necessary tie-in works. Another constructability study is performed in this stage, if there are significant design changes from the FEED stage, to ensure all construction-related concerns are addressed properly. Issued for construction (IFC) documents are generated at the end of the detailed engineering phase. Procurement of equipment will be performed during the detailed engineering once sufficient details are available.

There will be several plant walks (including for pre-start-up safety review, PSSR conducted after completion of construction and before pre-commissioning) by multidisciplinary teams. Process/project engineers will prepare punch lists for follow-up of any design and/or construction deficiencies identified during these plant walks, and also ensure that all of them are properly addressed before the plant start-up. IFC documents are updated for any design changes performed during the construction stage, for converting to As Built documents. Further, process/project engineers are required to review, amend and approve the IFC (issued for construction) and As Built versions of the following documents.

- Drawings and data sheets of all equipment including motors and instruments
- Piping isometrics
- PFDs, P&IDs, Cause-and-effect matrices and SIS design
- All the process safety information
- Electrical wiring diagrams
- Relief system design
- Fire safety system design
- Training materials
- Operating and control manuals

Process engineers (or EPC vendor) will prepare pre-commissioning and commissioning procedures, and also check lists for these activities. They will also train operators, and coordinate between the EPC contractor, operations and project departments. Pre-commissioning is performed in a systematic way to complete the following activities.

- New plant piping and equipment are free from fabrication defects, any loose parts (for example, improperly tightened nuts and bolts in pressure vessels), construction debris, corrosion scales and so on
- Flushing of all the piping
- Installing all the instrumentation and performing the loop checks
- Filling necessary catalysts
- Rotation check of pumps and compressors (but no load tests)
- Final gas leak checks

After completion of pre-commissioning, the revamped plant will be commissioned by introducing raw materials and adjusting the plant operating parameters to the desired targets (Lawson, 1999).

2.4.5 Project Completion

After project completion, performance test and post-project process evaluation are conducted.

Performance Test: It is generally conducted for at least 24 hours after smooth plant start-up and stable operation. Before the test, process/operations engineers will brief operation staff on the key activities involved. Depending on the project scope, the technology provider, equipment vendor and/or engineering contractor will also participate in performance testing. Process engineers will adjust the operating conditions to meet the revamp design/guaranteed conditions. They will coordinate all the data and sample collection activities. If any problem is identified during the test, the performance test will be aborted until such problem is fully rectified. Thereafter, another performance test will be conducted. Finally, process engineers will prepare the performance test report for circulation to various company stakeholders. This report will highlight any shortcomings and bottlenecks for future revamping of the plant beyond the current revamp objectives.

Post-Project Process Evaluation: After the plant performance test is satisfactory, process engineers will monitor the plant for assessing revamp benefits. If required, they will adjust operating variables to ensure stable operation and minimize maintenance activities. They will also try to improve profitability through continuous innovation approaches. After the mechanical guarantee period (generally 1 year), process/project engineers will prepare the post-project evaluation report, which will contain the following information.

- Measure of achievement of the project objectives and summary of other project achievements
- Comparison of actual spending vs. approved budget for the project, and reasons for any deviations from the approved budget
- List of project change variations and reasons for them
- Summary and analysis of revamped plant performance data
- Summary and analysis of any problems/bottlenecks identified during performance test and plant operation, and their mitigation plans
- Comparison of allocated project duration against the actual duration
- Analysis of adequacy of key project staffing and other supporting staff
- Summary of best practices used in the project
- Highlights of what worked well and what needs improvement
- Key safety performance of the project
- List of key lessons learned
- Summary of scope/methods for further performance improvement/revamp
- Photos of key equipment, their internals and key construction activities

2.5 Key Elements of Project Management

This section presents key elements for execution of a successful revamp project. A typical revamp project has five stages: pre-feasibility, conceptual design, FEED, EPC and start-up. A complete revamp project execution sequence is shown in Figure 2.2. Key project resources required, cost estimation methods, their accuracy and deliverables at every project stage

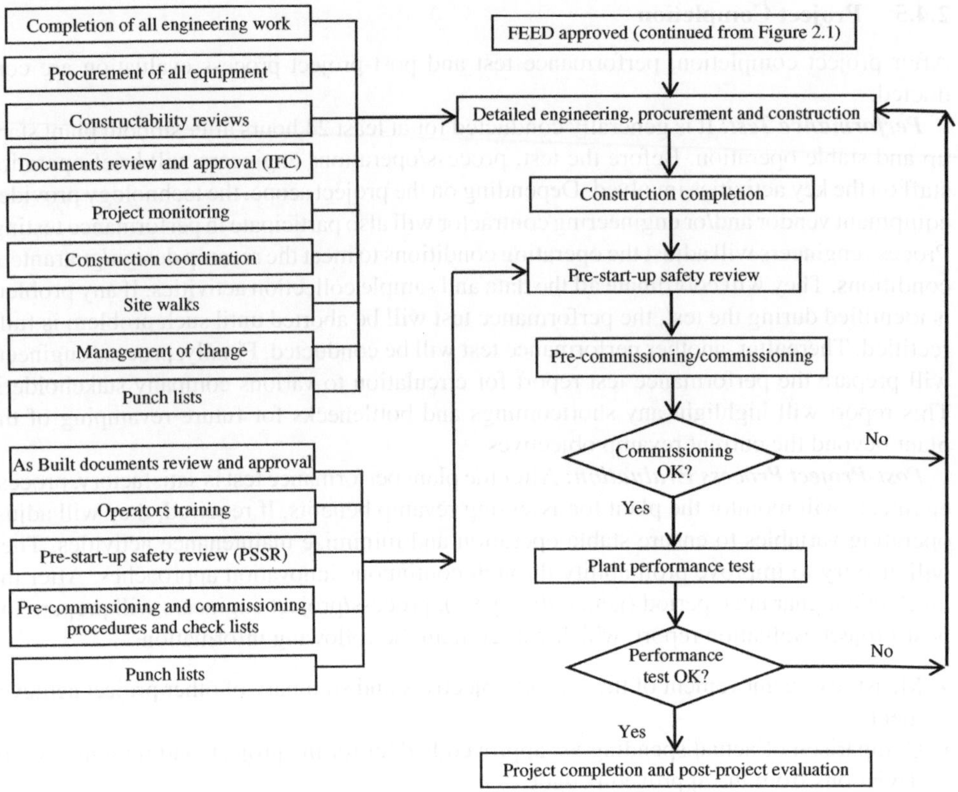

Figure 2.2 *Revamp project execution strategy.*

are summarized in Table 2.4. Some of these stages are combined in very small projects. For very large projects, there could be substages in some of the project stages such as conceptual design and FEED. So, companies may have a lower or higher number of project stages, depending on the project size, nature and the company's project management policies.

Experienced engineers/managers from all key disciplines are important for a successful revamp project. Without a complete team, it is not possible to create a comprehensive list of project options. Depending on the project scope, process, operations, project, electrical, instrumentation, environmental, process safety, civil, piping, costing, inspection and reliability engineers/managers are required. For larger projects, an engineering contractor may be required for project development and implementation. During pre-FEED, FEED and EPC stages, a lot of coordination is required from various discipline engineers as many activities such as equipment specifications preparation, selection and installation requires inputs from them in a sequential manner. Good team-work culture within the project team (including owner's and engineering contractor's project teams) is a prerequisite for quick decision making and also successful project execution. Based on the nature of the project, it is essential to select study team and other technical resources, which can support the study on an ad-hoc basis. Furthermore it is important to identify them at the beginning of

Table 2.4 *Summary of revamp project stages, resources required, cost estimation and key deliverables.*

	Pre-feasibility	Conceptual design or Pre-FEED	FEED	EPC	Project start-up
Purpose	To set the project objectives, basis, assumptions and preliminary economics	To generate alternative revamp options and establish the best revamp option	To fully develop the revamp option	To complete engineering, procurement and construction	To measure success of the revamp project and document key highlights and lessons learned
Resources required	Process and operations engineers	Process and operation engineers; engineering contractor (if required); ad-hoc basis: Project engineer, mechanical engineer, electrical engineer, civil engineer, environmental engineer, laboratory analyst and document controller	Process and operations engineers, project engineer, mechanical engineer, civil engineer, electrical engineer, safety engineer, reliability engineer, environmental engineer, engineering contractor (if required) and document controller	Project manager/engineer, EPC contractor (with his own project team), process and operations engineers, mechanical engineer, inspection engineer, reliability engineer, safety engineer, civil engineer, environmental engineer, electrical engineer, construction and operations supervisors	Process and operations engineers, project engineer, operations supervisors/operators, engineering contractor, equipment vendors, and laboratory staff
Cost estimation method and accuracy[a]	Conceptual cost estimate using in-house (past) projects cost data and published cost data; ± 50%	Cost estimate based on conceptual design. Some major equipment quotations are used; ±30% to 25%	Cost estimation based on finalized quotations; ±10%	NA	NA
Deliverables	• Project note highlighting objectives • Project framing (basis, assumptions and boundary conditions)	• Review report of plant design, operating, inspection and maintenance data • Plant test report	• PFDs and P&IDs • Cause-and-effect charts • Equipment process and mechanical data sheets • Equipment and piping lists	• EPC weekly and monthly progress reports. • Complete project as built documents (see Table 2.5) • Operating procedures	• Summary of project highlights and lessons learned. • Summary of revamp opportunities in the future

(continued)

Table 2.4 *(Continued)*

	Pre-feasibility	Conceptual design or Pre-FEED	FEED	EPC	Project start-up
	• Preliminary project economics • Block flow diagram • Resources and activities planning for next phase • Preliminary project schedule	• Summary of equipment bottlenecks, retrofit and debottleneck opportunities • Equipment rating study reports • Revamp options and calibrated plant simulations for all options. • Summary of OSBL modifications • PFDs for all options • Summary of VIPs used for establishing revamp options • Summary of revamp options, benefits and drawbacks • Cost estimation for all revamp options • Major equipment quotations • Recommended revamp option • Summary of project risks and mitigation plan • Resources and activities planning for next phase • Updated project schedule	• Material selection diagrams • Instruments list • DCS I/O point list • List of tie-in points • Request for quotations for equipment • Vendor quotations for equipment, civil, instrumentation, electrical works, EPC contact etc. • Pipe layout drawings • Plot layout and equipment layout drawings • Interlock lists • PHA and RAM reports • Control philosophy • Constructability report • Critical long-lead equipment list • Detailed instrument and cable sizes and routing drawings • Civil, structural, control room and shelter drawings • Soil survey report • Plant lighting drawings • Quotation for EPC work • Fire fighting system design • Revised project risks and mitigation plan • Revised cost estimation • Resources and activities planning for next phase • Updated project schedule	• Operations and maintenance training manuals • All the construction, PSSR, pre-commissioning and commissioning punch lists with action items closed • Summary of project scope variations and cost impact • Management of Change documents • Plant performance test report • Mitigation plans if a performance gap is identified	

Note: [a] A comprehensive discussion on cost estimation methods is available in Green and Perry (2008)

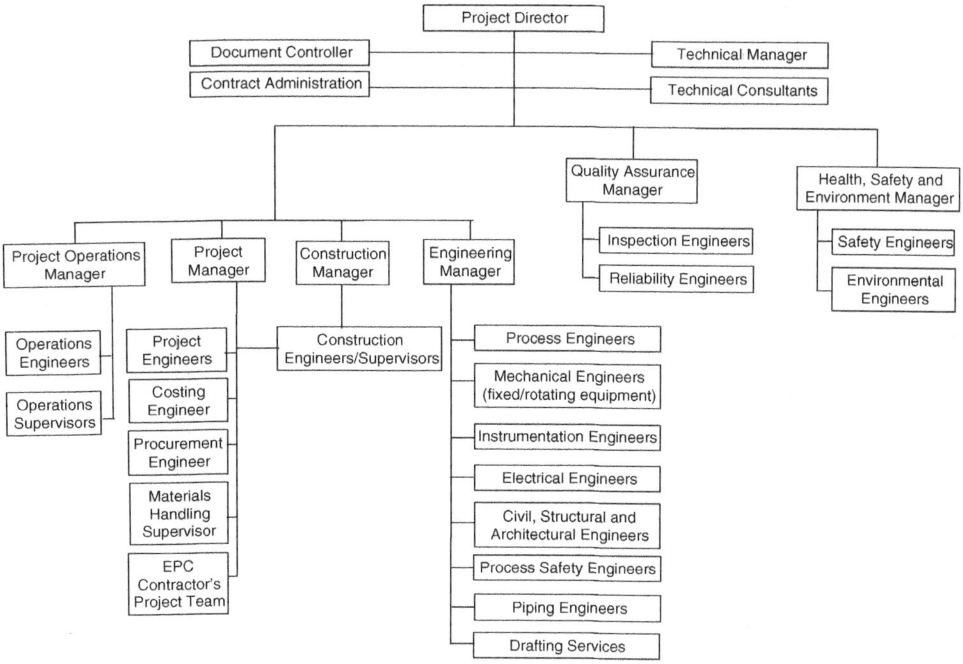

Figure 2.3 *Typical organization chart for a revamp project.*

the project. As the project transitions from one stage to another, this resource plan needs to be updated appropriately. A typical organization chart of a revamp project is shown in Figure 2.3.

2.5.1 Project Schedule

Depending on its size, the project schedule will have many major and minor activities. A high level schedule for the whole revamp project is shown in Figure 2.4, which is generally prepared at the beginning of the project. During project execution of each stage, a detailed project schedule is prepared to update key developments from that stage. As the project enters the EPC phase, a very detailed project schedule is prepared to track the engineering, procurement and construction activities. Prompt action needs to be taken if any activity misses its completion target. This will ensure execution of the revamp project in the planned turnaround. So, the project schedule is a very powerful tool to monitor and control the revamp project.

2.5.2 Project Execution and Progress Monitoring

Revamp project execution should be well planned to avoid excessive downtime requirement during the plant turnaround that would quickly erode the project economics. Refineries and petrochemical plants, for instance, typically have 4–5 years between major turnarounds with an actual turnaround duration of 20–40 days. Often, time will not be sufficient to complete all the revamp equipment installations, tie-in connections, pre-commissioning

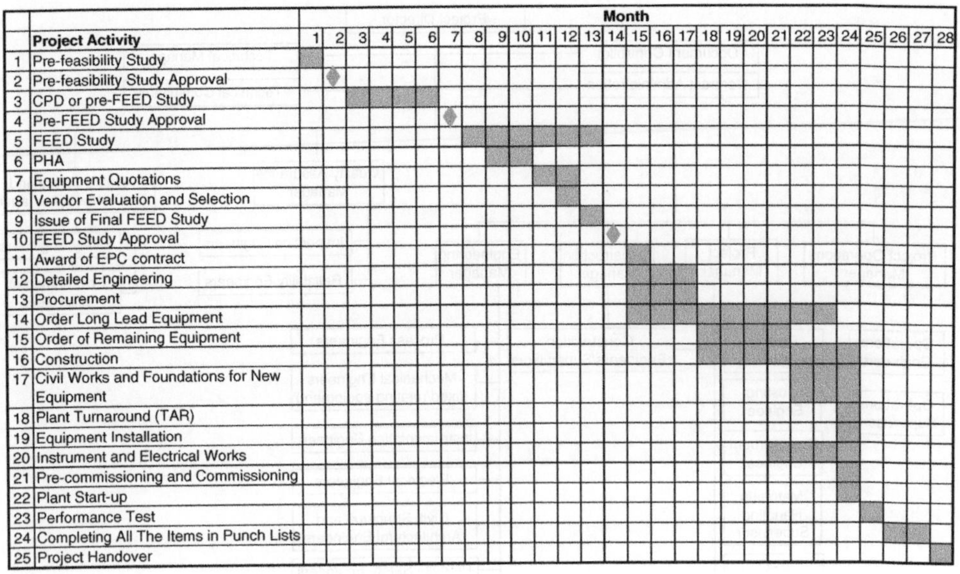

Figure 2.4 *Typical schedule of a revamp project.*

and commissioning activities. Hence, it will be advantageous to install some equipment and systems (whichever can be installed) before turnaround, so that remaining construction activities and tie-in jobs can be completed during the turnaround. Since revamp project execution generally takes more than a year, advanced planning is critical (Elshout and Gracia, 2009). Long lead equipment should be ordered as early as possible. Good coordination and team-work between the owners and contractor staff is essential for successful project completion within the turnaround duration.

2.5.3 Project Cost Control

Project complexity is the most influential factor affecting the project cost. Lagace Jr. (2006b) presented many useful factors influencing the project cost. By studying a comprehensive list of options during early phases and throughout the project, project teams can reduce capital cost by 5–10% and keep scope changes to no more than 1% of capital cost (Brown and Singh, 2014). The consequences of the decisions made in early phases of the project dictate the economic viability of the project, and they also have long term effects on operations and maintenance costs (Dipaolo, 2007). Too much over-design, low project risk tolerance, slow project decision making, selecting improperly developed options or unproven technology, and stringent local statutory regulations will increase the project cost. Use of VIPs, choosing experienced design teams and good engineering contractors will help to reduce the project cost. Properly developed scope in pre-FEED and FEED will reduce the project scope changes and hence control the project cost. Once project scope is fixed, the project team should minimize further changes. Poor project scope change control is one important factor for escalating costs and hence its failure (Harding, 2012).

2.5.4 Risk Management

Every project has some risk, and so it is very important to maintain a risk register for the project. At the beginning of every revamp project, all possible project risks should be identified and recorded in the project framing document. The team should also identify and document risk mitigation measures. As the project moves from one stage to another, risks and mitigation measures need to be revisited, and efforts should be made to minimize risks by implementing mitigation measures. 20% of the project activities accounts for roughly 80% of the potential risks. Proper decision making to address the key risks is necessary. It is important to understand the plant's work permit and safety work procedures. If not properly addressed, project risks or work permit and safety procedures can extend the construction duration and seriously affect the project. Plan and apply early for necessary statutory permits during the EPC phase, as a delay in getting them can stop construction activities and also plant start-up.

2.5.5 Final Project Deliverables

At the end of the construction phase, the EPC contractor will prepare the final project documents (Table 2.5) and submit them to the company. Generally, these will be split into engineering and vendor-related document dossiers. The project team will check the completeness and accuracy of the dossiers, and send comments so that EPC contractor can update these documents quickly. This activity is very important as these documents will be used for the entire life of the revamped plant and will also provide a basis for future revamp projects.

2.6 Revamp Options for Process Equipment

Apart from utilizing the excess of capacity in the plant equipment, the following revamp options that do not involve capital expenditure are very useful. For quick benefits, some of these options can be implemented before implementing the complete revamp project.

- Optimize reactor operating conditions to increase conversion and reduce recycle
- Use of better catalyst to increase conversion and reduce by-products and recycle
- Operation of parallel pumps and compressors
- Opening bypass valves of control valves
- Increase (reduce) operating pressure of distillation columns if top trays or overhead condenser (bottom trays or reboiler) is limiting
- Optimize reflux ratio of distillation columns
- Optimize feed preheating for distillation columns; if column is flooding above (below) the feed section, decreasing (increasing) feed preheat will be beneficial
- Increase the amount of impurities in the products up to the specification limits
- Use of fouling-reduction chemicals and/or mechanical cleaning of heat exchangers
- Use better advanced control systems and on-line optimization tools
- Maintain compressor inter-stage and process heat exchangers, condensers, reboilers and fired heaters with minimum fouling
- Maintain high (low) suction pressure (temperature) for the compressors

Table 2.5 *Final documents from a revamp project.*

General:
- List of project specifications
- Major equipment specifications and data sheets
- Material take-offs
- Training material

Mechanical Engineering:
- Equipment List
- Material selection diagrams
- Mechanical data sheets for the equipment
- Equipment mechanical design calculations
- Equipment inspection reports
- Insulation, fireproofing and painting specifications
- Spare parts and consumables list
- Equipment fabrication and testing reports
- Equipment drawings
- Equipment installation drawings
- RAM model study report
- Vendor manuals
- Equipment factory acceptance test (FAT) reports
- Hydrostatic test and other tests report

Process Engineering:
- Design basis for the revamp design
- Process block diagrams
- Process Flow Diagrams (PFDs)
- Process & Instrumentation diagrams (P&IDs)
- Tie-in and line lists
- Utility summary for the project
- Cause-and-effect charts
- Relief devices summary and calculations
- Relief system design calculations
- Utilities flow diagram
- Process design calculations including process simulations
- Hydraulic calculations
- Equipment sizing calculations
- Equipment, instrument and motor data sheets
- List of process alarms and interlock set points.
- Operating manual
- PHA report and compilation of PSI data
- Operations assurance report

Piping Engineering:
- Line list
- Piping and supports design calculations
- Stress analysis report for piping
- Piping layout and isometric drawings
- Pipe rack drawings
- Miscellaneous piping accessories
- Hydrostatic test and other tests report
- Tie-in list

Fire Protection:
- Firewater system design basis
- Fire water piping drawing
- Gas detectors layout diagram
- Fire sprinklers, fire monitors and deluge system design details
- Details of other fire fighting system such as fire extinguishers

Electrical Engineering:
- Single line diagrams
- Electrical load summary
- Electrical equipment (such as motor, heaters, VSD, transformers etc.) specifications and data sheets.
- Electrical cable schedule
- Voltage drop calculations
- Electrical area classification
- Electrical instrument interface drawings
- Underground electrical cable and cable trench arrangement
- Electrical equipment layout drawings
- Motor control center and power layout
- Load and short-circuit analysis
- VSDs configuration
- Plant lighting plan
- Grounding and lightning arrestor plan

Instrumentation:
- Instrument index
- Specification and datasheets for both new and modified existing control valves and other instruments
- Instrumentation, controls and metering philosophy
- Instrument cable layout plan and tray drawings
- Instrument air piping and tubing tie-in
- PLC architecture and control philosophy narrative
- Logic diagram for key shut downs and interlocks
- Junction box termination drawings
- Cabinet layout and termination drawings
- Loop drawings
- Conduit schedule
- DCS input/output list

(continued)

Table 2.5 (Continued)

Civil and Structural Engineering:
• Soil investigation report
• Equipment foundation design
• Survey and preliminary grading plan
• Piling and structural design report
• Key plot plan
• Plan of underground facilities including drainage
• Sewerage connection drawing
• Road, pavement plan and access review
• Design of buildings, equipment rain shelters
Project performance guarantees

- Open hand valves in steam turbines
- Minimize feed and product storage requirements by a more flexible pumping system and optimized supply chain activities
- Optimize use of steam turbines for maximizing power generation and also minimizing steam venting
- Minimize steam venting and slow rolling of steam turbines
- Reduce recycle of gas from compressor's discharge to suction

Potential revamp options requiring capital cost are listed in Tables 2.6 to 2.16 for different process equipment and systems.

Table 2.6 *Revamp options for hydraulic systems, separators and control valves.*

Equipment/ System – Objective	Options
Hydraulic system – increase capacity	Replace column and reactor internals (such as trays/packing, liquid distributers, gas spargers, catalyst retaining and holding screens) with low pressure drop designs Reduce operating pressure of distillation columns and reactors Replace exiting piping with bigger piping, use of economic pipe diameter or installing parallel piping header Change heat exchangers from series to parallel Use non-contact type flow meters such as clamp-on ultrasonic flow meters Operate feed tanks at high level and product tanks at low level Use flash drums in the feed pipes to distillation columns
Gas–liquid separators – increase capacity	Use the available ~20% design margin. If further capacity is required, increase nozzles size, install demister/vane separator and better feed inlet device such as schoepentator. If these improvements are not sufficient, replace existing separator with a new separator with all the above improvements or install another separator or gas–liquid coalescer parallel to the existing separator.

(continued)

Table 2.6 *(Continued)*

Equipment/ System – Objective	Options
Liquid–liquid and gas–liquid–liquid separators – increase capacity	Use the available ~20% design margin. If further capacity is required, increase nozzles size, install better feed inlet device, calming baffle or turbulence isolation plate, coalescer pad and/or vane pack. If these are not sufficient to meet the required capacity, replace existing separator with a new separator with all the above improvements or install another separator or liquid–liquid/ gas–liquid–liquid coalescer parallel to the existing separator.
Pneumatic system – increase capacity	Control conveying air velocity, optimize air-to-solid ratio, install new parallel blowers, replace existing blower with bigger capacity, replace rotary valves, increase speed of blowers, screw feeders and rotary valves Replace conveying header with bigger diameter if conveying pressure drop is more than 0.8 bar for the positive pressure system and 0.4 bar for the vacuum system (Mills, 2004)
Control valves – increase capacity	Avoid control valves by controlling pump/compressor by adjusting driver speed (for example, use of VSD for a motor and use of steam turbine governor control valve) Install additional control valve parallel to the existing control valve Replace high differential (DP) control valve with low DP control valve.

Table 2.7 *Revamp options for centrifugal compressors.*

Objective	Options
Increase capacity/head	Modify compressor internals such as impellers, stationary gas passage parts, shafts, seals and bearings (Bloch, 2006) Increase compressor speed. Electrical motors are seldom oversized. So, capacity and/or head increase (< 10%) may require a new, bigger motor. Compared with electric motors, steam/gas turbines have more design margin and may not require replacement Compressor cleaning Increase/decrease molecular weight of suction gas if compressor is limited by motor speed/power (Lieberman, 2010) Increase suction pressure Modify gears to increase the power If suction load contains condensable vapour, installation of a trim condenser and separator using cooling/chilled water will be beneficial Install additional compressor in parallel or series; latter will only increase pressure head Avoid gas recycle from compressor discharge to suction
Increase energy efficiency	Use guide vanes for suction flow rate control Use variable speed drivess (VSDs) Use high efficiency motors or back pressure steam turbine driver Avoiding condensing steam turbine drivers Cool suction gas Avoid gas recycle from compressor discharge to suction

Table 2.8 *Revamp options for reciprocating compressors, centrifugal fans, blowers, centrifugal pumps and electric motors.*

Equipment – Objective	Options
Reciprocating compressors – increase capacity/head	Increase speed Change cylinders and liners Re-bore unlined cylinders and change pistons rods Replace pulsation dampeners Reduce cylinder clearance Avoid gas recycle from compressor discharge to suction Install additional compressor in parallel or in series; latter will only increase pressure head Reduce pressure drop and duty of pre- and inter-stage coolers
Reciprocating compressors – increase energy efficiency	Reduce gas temperature to various suction stages, using cooling/chilled water Use suction valve's automatic capacity control methods such as Hoerbiger's Hydrocom system (http://www.hoerbiger.com/upload/file/hydrocom_en.pdf) Use of high energy efficiency valves at compressor's suction and discharge ports
Fans – increase capacity	Install bigger impeller and/or additional fans in parallel. Reduce pipe/duct pressure losses
Fans – reduce energy consumption	Impeller trimming, use VSD or steam turbine driver
Air blowers –increase capacity	Clean and modify inlet filters, reduce suction air temperature, avoid or fully open discharge pipe control valve, reduce discharge pipe pressure losses by installing a parallel header, replace internals such as roots and lobes, install bigger blowers, increase blower speed by gear/driver replacement
Centrifugal pumps – increase capacity/head	Increase impeller size and/or pump speed; replace impeller wear rings; increase (decrease) temperature of more (less) viscous fluids; replace motor with bigger capacity; increase pressure and/or elevation of suction vessel Install new pumps in parallel or in series
Centrifugal pumps – minimize energy consumption	Reduce impeller size and/or pump speed. Minimize pump recirculation (from discharge to suction) flow. Increase (decrease) temperature of more (less) viscous fluids. Reduce system pressured drop
Electric motors – increase capacity/ efficiency	Use spare motors and/or install new high efficiency motors

Table 2.9 *Revamp options for heat exchangers and fin fan (or air) coolers.*

Equipment – Objective	Options
Heat exchangers – increase duty	Install additional heat exchangers in parallel/series/series–parallel combinations Use efficient exchangers such as plate and spiral type Tube-side improvement: use tube inserts, twisted tubes and tubes with internal fins Shell-side improvement: use helical baffles, EM baffles, externally finned tubes, twisted tubes and shell-side seal strips (Lieberman, 2010) Replace tube bundle with low fin tubes if both shell and tube sides have clean fluids (Lieberman, 2010) Increase number of tubes and/or change square to triangular pitch, for shell and tube exchangers; increase number plates for plate heat exchangers
Heat exchangers – reduce fouling	Increase number of passes Use shell and tube exchangers with twisted tubes, plate and/or spiral heat exchangers Use automatic/manual back flush system Use acoustic techniques for fouling reduction Use fouling reduction chemicals and solvents Increase number of baffles and/or baffle cut Install filters on process hot and cold streams
Fin fan coolers – increase duty	Increase fan pitch angle and/or fan speed, reduce tip clearances, install inlet bells, fan seal disk and/or high efficiency/capacity fans, replace fan driver with bigger capacity, reduce fouling, re-tubing with higher number of small diameter and higher efficiency finned tubes. Humidify air cooler inlet air using water mist. Install additional fin fan coolers in parallel to the existing cooler
Fin fan coolers – reduce energy consumption	Use VSD on fan motors or steam turbine drivers. Install high efficiency fans

Table 2.10 *Revamp options for fired heaters.*

Objective	Options
Increase capacity	Install additional fired heater in parallel Increase convection heat recovery Install bigger impellers in forced draft fans and induced draft fans (FDF and IDF), bigger FDF/IDF fans and/or motors Install high pressure feed pumps to increase coil velocity to 10 ft/s (3 m/s) (Whatley *et al.*, 2007) Furnace re-tubing and/or coil metallurgy upgrades Radiant box modifications
Increase efficiency and reduce environmental emissions such as SO_x, NO_x, CO_2 and particulate	Convert natural draft to forced draft Reduce dampeners' throttling at fan suction and discharge by using VSDs Installation/modifying air pre-heaters Reduce air leakages Use O_2, CO and combustible analyzers for reducing excessive combustion Change oil firing to gas firing Install better refractory and/or ceramic coatings on furnace tubes Install larger capacity burners, low and ultra-low NO_x burners Modifications/replacement of flue gas stack Install selective catalytic reduction (SCR) system Modify convection section for better heat recovery. This may include: Addition of more tubes in the available free space in the convection section Installation of better finned tubes Installation of steam generation/superheating tubes

Table 2.11 *Revamp options for reactors, mixers, distillation columns and vacuum systems.*

Equipment – Objective	Options
Reactors – increase capacity	Increase gas/liquid hourly space velocity to allowable limits (Towler and Sinnott, 2012) Reduce inerts Improve heat removal/addition using better cooling/heating utilities, better coolers/heaters or installing parallel coolers/heaters in the recycle loops Use low pressure drop internals and catalyst beds Install additional reactors in parallel Change series arrangement to parallel Use a pre-reactor in series Use higher capacity/head feed pumps and recycle compressors
Mixers – increase capacity	Use static mixers and/or retrofit/replace existing mixers
Distillation columns – reduce energy requirements	Use pre-flash, feed conditioning, feed-product heat exchange, heat pumps, dividing-wall columns (Premkumar and Rangaiah, 2009), intermediate heat exchangers, optimize pump around flows, Organic Rankine Cycle (ORC) and Kalina Cycle (KC) to recover power (Chew *et al.*, 2014) Use alternate separation techniques and/or more efficient heat usage techniques such as heat cascading (Branan, 2012)
Distillation columns – increase capacity	Use high capacity (for example, multi-pass) trays or packing Use pre-flash column and/or better liquid distributors Modify existing reboiler/condenser; install a new reboiler/condenser in parallel. Additional condensers can also be installed in series at the expense of more reboiler duty
Vacuum systems – increase capacity or reduce energy requirements	Install pre-condensers cooled with cooling/chilled water. Use multi-stage steam ejectors arranged in series/series–parallel arrangement. Use liquid ring or dry vacuum pumps

Table 2.12 *Revamp options for pressure relief and flare systems.*

Equipment – Objective	Options
Pressure relief systems – minimize relief load	Design new equipment to withstand worst case conditions of temperature and pressure
	Minimize use of pressure relief valves by using inherently safer design concept
	Insulate or fire proof the equipment
	Install new process drums/vessels at least 25 feet (7.6 m) above the ground or any platform that can accumulate flammable liquid
	Provide sufficient operator response time (~30 minutes) for the case of overfill in the process drums/vessels
	Install new process drums/vessels so that they are away from the fire circle that is generating maximum relief during fire case scenario
	Use dynamic simulation for the relief rate estimation (Haribabu *et al.*, 2010)
	Replace flare gas header with one of bigger diameter and/or install parallel flare gas headers
	Minimize/avoid safety valves by careful review and combining process equipment/vessels for protection with common safety valve(s)
	Avoid oversized control valves for supplying hot utility to reboilers. Limit maximum heat input to the reboiler
	Pressure relief valves on cooling water lines can be avoided by locking the cooling water inlet and outlet valves for the heat exchanger
	Consider use of reliable and independent power supply sources and utility systems such as instrument air, cooling water and steam
	Install critical pumps/compressors in pairs (with one in operation and another on standby). Install electric motor for the first pump/compressor, and consider an alternative driver such as steam/gas turbines and diesel engines
	Lock close the bypass valves of control valves.
Flare system – increase capacity	Replace flare stack with a bigger diameter or taller stack, increase size of relief header; install a parallel relief header
Flare system – debottleneck	Implement SIS system in the units. Perform dynamic simulation to reduce the relief load from the units (Ha *et al.*, 2014), if the original relief load is obtained from steady-state simulation. 20–30% load reduction is possible.
Flare system – increase energy efficiency	Install flare gas recovery compressors
	Use low heating value buffer gas for flare header purging

Table 2.13 *Revamp options for plant air system, power generation, fuel system, waste heat recovery and waste water treatment.*

Process system – Objective	Options
Plant air system – reduce energy consumption	Optimize compressor operation, reduce friction and air leakages in air distribution system, low pressure drop air coolers and instruments, reduction of air leaks, use of heat of compression dryers, cool compressor suction air, waste heat recovery at compressor inter-stage coolers, better control
Plant air system – increase capacity	Install additional compressors in parallel. Use of heat of compression dryer
Power generation – increased power generation	Use cogeneration, ORC, KC or (hydraulic) power recovery turbines
	Cool inlet air to gas turbines (Green and Perry, 2008)
	Use steam turbines as drivers for pumps/compressors and also for steam pressure reduction
	Use gas turbine exhaust gases to supply waste heat to the fired heaters (Branan, 2012)
Fuel gas/oil system – debottlenecking/revamping	Minimize fuel consumption by better heat integration at the process units.
	Install purge/flare gas recovery systems
	Import/blend natural gas into fuel gas
	Install more fuel gas/oil storage
Waste heat recovery (WHR) – reduce energy consumption	Use flash steam, hot water or low temperature process streams for generating chilled water (using absorption chillers) or power (using ORC or KC)
	Steam condensate recovery; blowdown heat recovery at boilers and process steam generators.
	Flue gas heat recovery for increased process heating and/or steam generation. Reddy *et al.* (2013c) summarized possible WHR techniques. Ankur *et al.* (2012) presented a cogeneration targeting method for site-wide low-grade heat recovery
Waste water treatment facilities – increase capacity/debottlenecking	Install/replace bioreactors with membrane bioreactors
	Increase re-use of treated fouled water
	Reduce water usage to minimum required
	Segregate different types of waste water streams and use localized treatment such as coagulation/flocculation followed by ultra/micro filtration/reverse osmosis membrane units. This can recycle ~50% of the waste water as fresh water

Table 2.14 *Revamp options for cooling towers and cooling systems.*

Process system – Objective	Options
Cooling towers – increase capacity/heat load	Increase cooling water return temperature (within cooling tower fill maximum operating temperature) from process heat exchangers. This can be achieved by changing cooling water use in heat exchangers from parallel to series or series–parallel arrangements (Reddy *et al.*, 2013b)
	Increase fan pitch angle, install larger capacity and high efficiency fans and new motors
	Clean cooling tower fill
	Install additional cooling tower cell in parallel or additional cooling water pumps
	Reduce air-side static pressure loss by replacing tower fill with better fill, arranging fill parallel to the air flow and installing low pressure drop drift eliminators
	Increase active heat transfer area for air and water by increasing the amount of fill or installing better fill
	Install better and finer water spraying system
Cooling towers – reduce energy consumption	Install variable speed drives on cooling water pumps and fans
	Install velocity stacks
	Use of high efficiency fans and motors
Cooling system – increase capacity	Reduce cooling duties by better heat integration in the plant; use cooling water in series arrangement of coolers; debottleneck cooling towers, cooling/sea water pumping system, air coolers and refrigeration systems
	Install additional cooling towers, sea water cooling system, air coolers and refrigeration systems; install additional cooling headers and pumps
	Use waste heat at absorption chillers, for producing chilled water

Table 2.15 *Revamp options for deaerators, boilers, steam turbines and steam system.*

Equipment/system – Objective	Options
Deaerators – increase capacity	Increase temperature of feed water and/or steam condensate flow to the deaerator
	Increase number of spray nozzles and/or replace trays/packing to handle more flow. If this is not sufficient, replace the deaerator with a bigger capacity
	Increase size of piping for feed water, steam condensate and steam. Replace control valves with bigger size and/or install new control valves parallel to the existing control valves
Boilers – reduce energy consumption	Reduce excessive air or O_2, use economizer, blowdown heat recovery, reduce air leaks, improve firing controls, VSDs for boiler water pumps and air blowers, use high efficiency boilers, improveg insulation
Boilers – increase capacity	Install additional boilers
Steam turbines – reduce energy consumption	Use high efficiency steam turbines, avoid using condensing turbines, and replace low efficiency steam turbines with electric motors
Steam turbines – increase capacity	Increase governor control valve opening to 100% and/or size of steam nozzles (Lieberman, 2010).
	Increase steam supply pressure, temperature and/or reduce letdown pressure
	For condensing turbines, increase surface condenser duty to increase vacuum
Steam system – revamping	Reduce steam requirement by better heat integration in the plant, avoid slow rolling of steam turbines, replace steam turbine gland packing and ejector system with mechanical seals, replace steam jet system with liquid ring or dry vacuum pumps
	Plant-wide steam network optimization; repair passing steam traps; improved steam condensate recovery, use better insulation
	Use duct firing of heat recovery steam generators. Increase steam generation through WHR from fired heater's convection section, heat recovery steam generators and process heat exchangers/steam generators
	Install absorption heat transformers
	Install additional boilers, WHR steam generators
Steam condensate recovery	Reduce steam condensate draining by installing steam traps; reduce flashing of steam condensate in condensate headers by condensate cooling; separating flash steam; install additional steam condensate recovery systems

Table 2.16 *Revamp options for debottlenecking and revamping storage facilities.*

Install pumping system to enable product supply (feed stock receiving) from (to) several storage tanks to (from) many loading locations such as ship/truck loading arms, pallets and drum loading. This will minimize the need for more storage tanks
Increase loading pump capacity by increasing impeller size, install parallel pumps, reduce piping pressure losses by reducing control valve DP, and install additional piping headers in parallel
Reduce viscous fluid's viscosity by heating and also by installing steam/electrical heat tracing and insulation
Modify tank vents to facilitate higher liquid filling and withdrawal rates
Build new storage tanks/silos
Install additional loading terminals and/or packaging facilities

2.7 Conclusions

This chapter presented many key differences between the revamp projects and grassroots projects followed by a systematic methodology for a revamp project. The key steps and recommendations for successful revamp projects are summarized below.

- Perform plant test runs and calibrate simulation models for establishing realistic revamp design and scope
- In the pre-FEED stage, use value improvement practices for the development of feasible and sustainable revamp project options
- In the FEED stage, conduct process hazard analysis, RAM, predictive maintenance, value engineering and constructability studies for accurate scoping of the most profitable revamp option
- Carefully plan and execute key steps in the revamp project
- Implement comprehensive project management for minimizing project delays and risks

Finally, in Section 2.6, many options for revamping common process equipment and systems are provided in Tables for ready reference to practicing engineers, students and academicians.

Acronyms

CAPEX	Capital expenditure
DCS	Distributed control system
DP	Differential pressure
EPC	Engineering, procurement and construction
FAT	Factory acceptance test
FDF	Forced draft fan
FEED	Front end engineering design
FMEA	Failure mode and effects analysis
HAZOP	Hazard and operability
IDF	Induced draft fan

IFC	Issued for construction
KC	Kalina cycle
MDT	Mean down time
MTBF	Mean time between failures
MTTF	Mean time to failure
MTTR	Mean time to repair
ORC	Organic Rankine cycle
OSBL	Outside battery limit
PFD	Process flow diagram
PHA	Process hazard analysis
P&ID	Process and instrumentation diagram
PLC	Programmable logic controller
PSI	Process safety information
PSSR	Pre-start-up safety review
QRA	Quantitative risk assessment
RAM	Reliability, availability and maintainability
RAMS	Reliability, availability, maintainability and safety
RFQ	Request for quotation
SCR	Selective catalytic reduction
SIS	Safety instrumented system
SOA	Safety objective analysis
TAR	Turnaround
VSD	Variable speed drive
VIP	Value improvement practice
WHR	Waste Heat Recovery

References

Allen, D.T. (2007) Green engineering and the design of chemical processes and products, *Chemical Engineering*, 36–40.

Ankur, K., Bulatova, I., Smith, R., and Kim, J.K. (2012) Site-wide low-grade heat recovery with a new cogeneration targeting method, *Chemical Engineering Research and Design*, **90**, 677–689.

API Standard 521 (2014) *Pressure-Relieving and Depressuring Systems*, 5th edn, American Petroleum Institute.

Barletta, T., Martin, G.R., and Visser, A. (2002) Consider comprehensive CPD efforts to cut costs, *Hydrocarbon Processing*, 53–62.

Bloch, H.P. (2006) *A Practical Guide to Compressor Technology*, 2nd edn, John Wiley, Hoboken, New Jersey.

Branan, C. (2012) *Rules of Thumb for Chemical Engineers*, 5th edn, Gulf Professional Publishing, Burlington.

Brown, T. and Singh, S. (2014) Project optimization through engineering, *Chemical Engineering*, 51–58.

CCPS (1993) *Guidelines for Engineering Design for Process Safety*, American Institute of Chemical Engineers, New York.

Chew, J.C., Reddy, C.C.S., and Rangaiah, G.P. (2014) Improving energy efficiency of dividing-wall columns using heat pumps, Organic Rankine cycle and Kalina cycle, *Chemical Engineering and Processing: Process Intensification*, **76**, 45–59.

Contreras, C.D. and Bravo, F. (2011) Practice green chemical engineering, *Chemical Engineering*, 41–44.
Dipaolo, C.R. (2007) Target costing, *Chemical Engineering*, 66–69.
Elshout, R. and Gracia, D. (2009) Revamps: strategies for a smooth turnaround, *Chemical Engineering*, 34–39.
Farrow, G. (2013) Laser surveying a revamp, *PTQ Revamps*, 29–31.
Gebert, T., Villarreal, R., and Whitsura, F.R. (2004) Revamping automation systems, *PTQ Revamps & Operations*, 9–12.
Golden, S.W., Moore, T., and Nigg, J.M. (2003) Optimize revamp projects with a logic based approach, *Hydrocarbon Processing*, 75–83.
Green, D.W. and Perry, R.H. (2008) *Perry's Chemical Engineers' Handbook*, 8th edn, McGraw- Hill, New York.
Ha, H.Z., Harji, A., and Webber, J. (2014) Accurate prediction of tower relief, PTQ, pp. 51–57, 2nd Quarter.
Harding, J. (2012) Avoiding project failures, *Chemical Engineering*, 51–54.
Haribabu, C., Valli, A., Khanna, V., and Bhattacharya, D. (2010) Calculating column relief loads, PTQ, pp. 55–65, 2nd Quarter.
Hoerbiger, http://www.hoerbiger.com/upload/file/hydrocom_en.pdf
Jianwei, Z., Yuhang, L., Nan, Z., and Guy, K. (2011) Optimizing and revamping a refinery hydrogen network, PTQ revamps, www.digitalrefining.com/article/1000417, accessed in July 2014.
Kemp, I.C. (2007) *Pinch Analysis and Process Integration*, 2nd edn, Elsevier Butterworth-Heinemann, Burlington.
Koolen, J.L.A. (2002) *Design of Simple and Robust Process Plants*, Wiley-VCH Verlag GMBH, Weinheim.
Lagace, J.C. Jr. (2006a) Project success builds on a well-defined scope, *Chemical Engineering*, 36–38.
Lagace, J.C. Jr. (2006b) Making sense of your project cost estimate, *Chemical Engineering*, 54–58.
Lawson, G. (1999) *Project Management for the Process Industries*, IChemE, Warwickshire.
Lee, J.J., Jeong, H.Y., Alanis, F.J., Sinclair, I., and Park, N.S. (2007) Reduce revamp cost by optimizing design and operations, *Hydrocarbon Processing*, 77–81.
Lieberman, N.P. (2010) *Process Engineering for a Small Planet: How to Re-Use, Re-Purpose and Retrofit Existing Process Equipment*, John Wiley, New Jersey.
Litzen, D.B. and Bravo, J.L. (1999) Uncover low-cost debottlenecking opportunities, *Chemical Engineering Progress*, 25–32.
Martin, G.R. and Cheatham, B.E. (1999) Keeping down the cost of revamp investment, PTQ, pp. 99–107, Summer 1999.
Mendez, M. (2007) The role of chemical engineers in green engineering, what we can do to support its goals, *Chemical Engineering*, 41–44.
Mills, D. (2004) *Pneumatic Conveying Design Guide*, 2nd edn, Elsevier Butterworth-Heinemann, Burlington.
Milosevic, M., Rudman, A., and Brown, R. (2013) Are you using pinch technology effectively in your daily operations: parts 1 and 2, *Hydrocarbon Processing*, 99–102 (June 2013), 77–81(July 2013].
Mukherjee, S. (2012) Revamping your process plant, *Chemical Engineering*, 36–41.
Premkumar, R. and Rangaiah, G.P. (2009) Retrofitting conventional column systems to dividing-wall columns, *Chemical Engineering Research and Design*, **87**, 47–60.
Reddy, C.C.S., Naidu, S.V., and Rangaiah, G.P. (2013a) Optimization of a steam network, *Chemical Engineering*, 54–59.
Reddy, C.C.S., Rangaiah, G.P., Lim, W.L. and Naidu, S.V. (2013b) Holistic approach for retrofit design of cooling water networks, *Industrial & Engineering Chemistry Research*, **52**, 13059–13078.
Reddy, C.C.S., Rangaiah, G.P. and Naidu, S.V. (2013c) Waste heat recovery methods and technologies, *Chemical Engineering*, 28–38.
Serna-Gonzalez, M. and Ponce-Ortega, J.M. (2011) Total cost target for heat exchanger networks considering simultaneously pumping power and area effects, *Applied Thermal Engineering*, **31**, 1964–1975.

Sikos, L. and Klemes, J. (2010) Reliability, availability and maintenance optimization of heat exchanger networks, *Applied Thermal Engineering*, **30**, 63–69.

Smith, D.J. (2011) *Reliability, Maintainability and Risk, Practical Methods for Engineers*, 8th edn, Butterworth-Heinemann, Oxford.

Smith, R. (2005) *Chemical Process Design and Integration*, John Wiley, Chichester.

Sreepathi, B.K. and Rangaiah, G.P. (2015) Retrofitting of heat exchanger networks involving streams with variable heat capacity: Application of single and multi-objective optimization, *Applied Thermal Engineering*, **75**, 677–684.

Stankiewicz, A.I. and Moulijn, J.A. (2000) Process intensification: transforming chemical engineering, *Chemical Engineering Progress*, 22–34.

Stankiewicz, A.I. and Moulijn, J.A. (2004) *Re-engineering the Chemical Processing Plant: Process Intensification*, Marcel Dekker, New York.

Towler, G. and Sinnott, R. (2012) *Chemical Engineering Design: Principles, Practice and Economics of Plant and Process Design*, 2nd edn, Elsevier, Oxford.

Whatley, M., Golden, S., and Nigg, J. (2007) Revamping atmospheric crude heaters, *PTQ Revamps*, 27–29.

Woods, D.R. (2007) *Rules of Thumb in Engineering Practice*, Wiley-VCH Verlag GMBH, Weinheim.

Zhu, F. (2014) *Energy and Process Optimization for the Process Industries*, John Wiley, New Jersey.

3

Process Safety in Revamp Projects

Raman Balajee[1] *and C.C.S. Reddy*[2]

[1]*Process Safety Lead – Asia, Air Products Singapore Private Limited, Singapore*

[2]*Lead Process Design Engineer, Singapore Refining Company Private Limited, Singapore*

3.1 Introduction

The main objective of this chapter is to describe all relevant process safety techniques and procedures applicable and important for developing revamp projects. Many process safety incidents with significant impacts were due to unrecognized and unidentified process safety management elements right from the design stage and all the way up to various stages of plant operating cycle. Process system design should be closely coupled with process safety design requirements as the hazardous consequences could seriously harm the society around, which may impair the corporate reputation and business continuity. The safety review process is centered to minimize the risk of accidental releases of toxic/flammable materials, fire, explosion and adverse environmental impacts associated with the operation of process equipment and systems by identifying, assessing and controlling process hazards.

The following important process safety review processes should be performed thoroughly during the process revamp project[1] execution stages for a successful and sustainable project.

[1] Process safety aspects for revamp projects discussed in this chapter are also applicable for retrofit and debottlenecking projects. See Chapter 1 for description of process retrofit, revamp and debottlenecking.

Chemical Process Retrofitting and Revamping: Techniques and Applications, First Edition.
Edited by Gade Pandu Rangaiah.

A. Preliminary Studies and Reviews (during the conceptual design or Pre-FEED)
   (i) Preliminary Hazard Review
   (ii) Plot Plan and Layout Review
   (iii) Electrical Area Classification Reviews
   (iv) Pressure Relief Design Considerations
   (v) Fire Safety Design Considerations
B. Process Hazard Analysis (PHA) during Front-End Engineering Design (FEED)
C. Post PHA Studies and Reviews during Engineering Procurement and Construction (EPC) stage
   (i) Instrumented Protective System Integrity Reviews
   (ii) Revision of PSI and Operator Induction.
   (iii) Pre-Start-up Safety Review (PSSR)
   (iv) Management of Change (MOC)

A complete set of process safety information (PSI) should be available prior to the process safety review. The hazard review leader makes the plan on documentation requirements and team selection for a successful and effective review. The safeguards identified during the PHA are highly important and valid throughout the plant's lifetime. The PHA also identifies any new hazards introduced due to plant modifications including facility changes. It also recommends appropriate solutions to mitigate risks to acceptable levels due to these hazards. Necessary additional quantitative risk assessments (as dictated either from PHA or from local regulatory requirements) have to be performed for any major risk identified. Where the risk is not tolerable even after economical design of mitigation measures such as safer design practices, pressure relief systems, basic process control systems (BPCSs) and administrative layers, the process system design needs to incorporate a safety instrumented system (SIS). SIS design needs to satisfy the required Safety Integrity Level (SIL) or to the maximum allowable average probability of failure on demand (PFD_{avg}) so that the overall risk is within the target set by the corporate/local authority or relevant codes, whichever is more stringent.

It is highly recommended to adopt inherent safety principles in the process (revamp) design right from the conceptual design phase to sustain the revamp project benefits and safety KPIs (key performance indicators) of the organization on a long–term basis. A comprehensive safety review should be performed prior to the commissioning and start-up of the (revamped) plant to ensure all design and safety-related actions are completed, documented and communicated. One fundamental but often ignored activity in revamp projects is the PSI updates, particularly revision of standard operating procedures and training the operators on the revised procedures. MOC is a crucial work process to be followed diligently in the entire project execution phases, as every overlooked change has some safety implications necessitating revision of the hazard review documentation basis.

First, the importance of learning lessons from past process safety incidents is highlighted in Section 3.2. The subsequent section presents preliminary hazard review procedure, risk matrix, what-if method, plot plan and layout review, pressure relief system review and fire safety design aspects. Section 3.4 presents PHA techniques and procedures: hazards and operability analysis (HAZOP), failure modes and effects analysis (FMEA), instrumented protective system (IPS) design, fault trees, event trees, layer of protection analysis (LOPA) and finally SIS life cycle. The importance of revision of PSI is highlighted in Section 3.5.

PSSR and MOC procedures are described in Sections 3.6 and 3.7 respectively. Finally, this chapter concludes by highlighting the main points on process safety for successful revamp projects.

3.2 Lessons from Past Process Safety Incidents

Learning lessons from past process safety incidents are very helpful for recognizing the importance of process safety aspects and hence developing safe revamp design. For example, the explosion incident at the Nypro site, Flixborough, UK on 1st June 1974, revealed several gaps in design safety aspects during development and execution of temporary or permanent modifications. During this incident, which occurred during a minor retrofit, 28 workers were killed and about 90 on-site employees and public suffered significant injuries. The number of casualties would have been more if the incident had occurred on a weekday instead of a Saturday, as employees at the main office block were not working. Out of 28 deaths, 18 fatalities happened inside the control room due to windows shattering and the roof collapsing; no-one in the control room escaped this disaster.

In the Flixborough incident, there was initially a leak of cyclohexane vapour from the stirrer gland of the 5th reactor. Without detailed assessment, industrial grade water was sprayed on the reactor to condense the leaking vapour. This industrial grade water contained nitrates and caused stress corrosion cracking at the 5th reactor (Kletz, 1999). The plant was subsequently shut down for investigation and repair. A serious problem was identified with the reactor and the decision was taken to remove the 5th reactor from service and to connect the 4th and 6th reactors through a temporary bypass pipeline (as shown in Figure 3.1), so that the plant could continue to operate. The decision of such a retrofit and MOC were done without full assessment of the potential consequences. No engineering drawings or calculations were performed for this temporary pipe section with bellows, to ensure that it could safely hold and transfer the contents of the reactor system at varying process conditions. No considerations were sought as to the original layout of the plant. The resulting accident at Flixborough highlights the importance of process safety

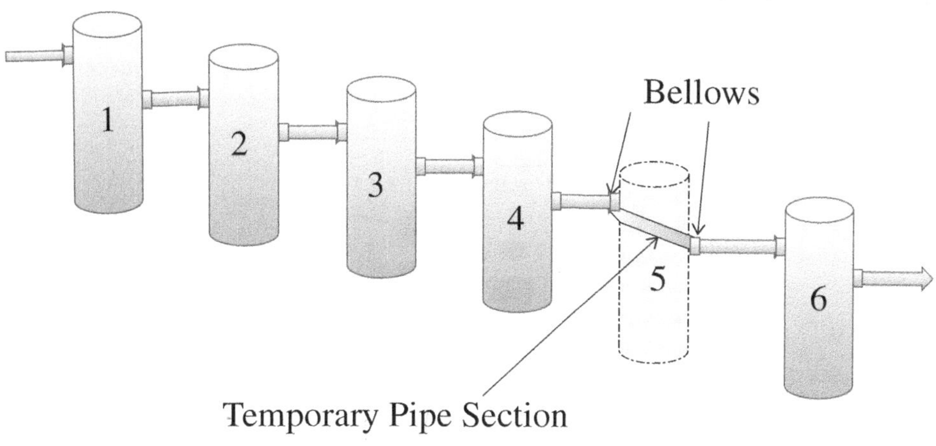

Figure 3.1 Temporary pipeline to bypass 5th reactor at Flixborough in 1974.

during design and execution of revamp projects. Sanders (2005) has presented many cases of serious industrial accidents that were due to improper review of process modifications.

3.3 Preliminary Hazard Review during Conceptual Design

Preliminary hazard review is intended to evaluate major hazards at the early stages of technology development and/or proposal development and project planning. Inherently safer design, high risk process elimination, appropriate safeguards and the need for further studies are identified at this stage. This would help the project team to understand major on- and off-site risks and the kind of domino effects from events, which may impact the environment and corporate reputation.

Hazards are contained by multiple protective barriers, which may have weaknesses or 'holes'. When these holes align in a straight line (as illustrated by a Swiss cheese model in Figure 3.2), hazard energy is released. Barriers may be physical, engineered, containment or administrative controls, and holes can be latent or incipient. Hazards can be mitigated most cost-effectively if they are identified in early phases of revamp projects (for example, during conceptual design). Preliminary hazard review is a multi-step process where the expertise of the team members is critical for successful hazard and risk identification. Identification of hazards for each section shall be carried out by a team consisting of a minimum of people with knowledge of process design, plant operations, local legislations and process safety. The following minimum documentation is required to be reviewed prior to the start of each hazard identification session, and the relevant details have to be briefed to the team members.

- Any previous hazard review study material relevant to the process
- Most credible initiating events and their frequencies can be gathered either from the company's incidents library or from agencies such as CSB (Chemical Safety Board in US), CIMAH (Control of Industrial Major Accident Hazards in US), COMAH (Control of Major Accident Hazards in UK), European Seveso II Directive and HSE (Health and Safety Executive in UK).

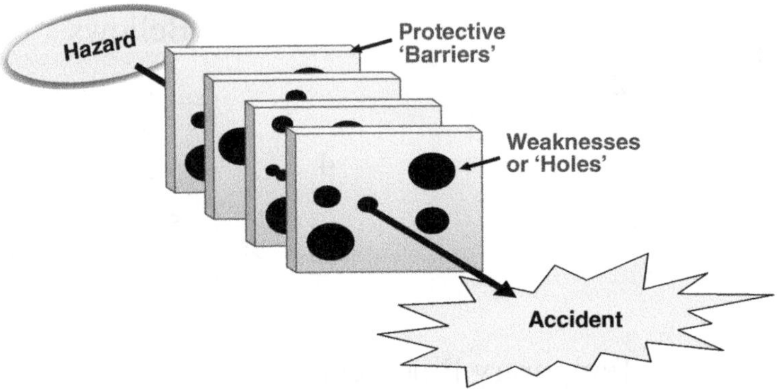

Figure 3.2 *Swiss cheese model on how the hazard penetrates through holes of barriers.*

Table 3.1 *Hazard categories and their characteristics.*

Major hazard categories – subcategories	Cause of incidents
Fire – Pool Fire	Occurs when a liquid from a spill accumulates on the ground in sufficient quantity to sustain a fire if ignition occurs
Fire – Flash Fire	Occurs when volatile components of a flammable release form a vapour cloud in air. When this cloud is ignited, often some distance from the release location, a flame front travels through the flammable mixture and then stabilizes at the release point as a pool fire
Fire – Jet Fire	Continuous release of highly pressurized flammable fluid that ignites on or near the point of release, resulting in a jet fire and continuous thermal radiation
Explosion – BLEVE (Boiling liquid expanding vapour explosion)	Occurs when a significant quantity of flammable material is released suddenly from a pressurized system and is immediately ignited, for example, due to vessel bursting under fire attack
Explosion – Vapour Cloud Explosion	Occurs when, following the formation of a vapour–air mixture, the flame travels through the mixture at high speed, generally due to confinement and/or a high degree of turbulence within the cloud, and generates an overpressure wave
Explosion – Pressure Vessel Burst	Occurs when a vessel fails suddenly and catastrophically because of an uncontrolled pressure build-up inside the vessel and/or mechanical failure
Toxic – Toxic Gas Cloud	Release of toxic liquid or gas may generate a toxic gas/vapour cloud which disperses downwind, thus resulting in a hazard

- Plot plan layout indicating location of vessels, occupied buildings and equipment
- Process flow diagram with basic process information and key safety features
- Process description
- Anticipated physical and chemical changes from normal processing and deviations.
- Vents and emissions from processes
- List of dangerous substances used and their safety data sheets

It is necessary to perform the review with an open mind set to identify the risks and consequences. A major accident would generally occur from one of these three categories – a fire or an explosion or a toxic release. These three major hazard categories are further subdivided in Table 3.1 that could be helpful in developing relevant scenarios.

The outcome of the preliminary hazard review is also useful to document into formal permits or notification to the authorities.

3.3.1 Risk Matrix for Qualitative Judgments

Risk Matrix is the most widespread tool used in process industries for risk evaluation. It is mainly used to determine the magnitude of a risk and hence determine whether or not the risk is sufficiently controlled. The severity and likelihood of an unwanted event will

Table 3.2 Example of a risk matrix for qualitative risk judgment.

Severity / Likelihood	Societal fatal risk/Community evacuation/ Shelter in place (1)	Risk of fatality onsite/Fire and explosion resulting > 25 K US$ damages [API RP 754], Release of materials higher than CCPS Tier 1 threshold quantities (2)	Serious injury to employees/ Contractors, business reputation issues/ Serious environmental impact (3)
HIGH (H) "Possible" (more frequent than 1 incident every 100 years)	H1[a]	H1[b]	H2[c]
MEDIUM (M) "Occasional" (less frequent than 1 incident every 100 years to 10,000)	H1[b]	M[d]	L[e]
LOW (L) "Seldom" (1 incident every 10,000 to 1,000,000 years)	M[d]	L[e]	
NIL (N) "Unlikely"	Perform QRA		

[a]Minimum two SIL2 SIFs (or equivalent safeguards to achieve this Risk Reduction Factor, RRF) are required.
[b]Two SILs required of which one should be SIL2 SIF.
[c]One SIL2 SIF is required.
[d]One SIL1 SIF is required.
[e]SIL is not required.

determine the risk associated with the revamp design. As an example, a 4 × 3 risk matrix is depicted in Table 3.2. The four probabilities (likelihood) and three levels of severity are categorized into four risk levels (RL). H1 (High 1), being the most serious risk category, should be addressed as quickly as possible once identified. The H2 (High 2) category comes to the next level of urgency, followed by M (medium) and L (Low). Events having a potential severity (Category 1), even if their probability is very low, should be analysed by applying appropriate quantitative risk assessment methods. Determination of the required SIL for a SIF (safety instrumented function) from this risk matrix requires the categorization of consequence and likelihood. For example, using Table 3.2, a typical chemical company may decide SIL[2] selection.

3.3.2 What-If and Process Safety Check Lists

For process plants involving physical or chemical processes, a formal hazard identification technique such as what-if or a safety checklist is deployed to identify the hazard scenarios

[2] SIL and SIF are explained in detail in Section 3.4.3.

arising from process deviations outside the safe design and operating limits. A sample what-if check list is presented in Table 3.3 for the preheater section of the process shown in Figure 3.3. A what-if question on exceeding the design limit of furnace tube skin temperature, initiated by higher burner heat load is illustrated. Adequacy of safeguards is analysed in accordance with the risk criteria (as shown in Table 3.2) and necessary recommendations are made.

3.3.3 Plot Plan and Layout Review

Project plot siting is very important for revamp projects. A careful review of the surrounding facilities is also necessary. Some of the important considerations are:

- Review of surrounding plants/facilities
- Existing pipe racks
- Maintenance and operations access
- Sewers and storm water drainage
- Existing fire and radiation circles
- Hazardous chemical releases from upwind locations
- Existing electrical area classification
- Close access to substations
- Close access to raw materials, utilities, flare network and product storage area
- Future population encroachment
- Noise generated at the surrounding facilities

Consider the topography in site selection and layout as it can have an impact on the potential consequences of an event (CCPS, 2003a). Soil investigation is required to estimate the need for soil improvement or piling. Project sites with low load-bearing capacity may have higher settling of soil, which may lead to uneven settlement of equipment, cracks in equipment, increase in piping stress and failure/leaks in piping. Soil settlement is very important for tank form. Many incidents of tank bottom failures are reported due to soil settlement. Project economics favour sites with minimum piling.

Analyse the project site for potential flooding due to undersized sewers. This is very important if the revamped plant requires below grade/ground level piping and/or vessels to be maintained hot. Rainwater floods can increase heat loss from the piping, which may reduce the flow rate, particularly if the piping contains viscous liquid, and seriously affect the operation. Wind direction is another important consideration. Potential flammable, toxic chemical and/or particulate matter release points shall be sited downwind of potential ignition sources/incipients. New site selection shall be reviewed carefully for any adverse effect of these releases. For example, fin-fan coolers installed next to fine-catalyst handling areas may get fouled very quickly and impair their cooling performance; toxic chemical releases will harm operators; and flammable releases may cause fire/explosion.

Access to the plant is very important for maintenance, operations and emergency handling considerations. Easy access to each area and to the process equipment shall be provided for fire-fighting personnel and their equipment. A new project site shall not restrict maintenance and/or emergency access for the existing plant. The new plant layout shall include adequate surface slope to facilitate drainage and a means of preventing the spread of flammable liquids from one operating area to another. Fire hydrants, fire-fighting equipment and fire monitors

Table 3.3 Example of what-if worksheet for preliminary hazard analysis.

What-If Safety Review Log Sheet						Page of		
Study Name:	Capacity expansion at Plant 1	Team Leader:				X		
Node:	3	Team Members:	A,B,C					
Date:	15/11/2014	Ref. #:						
What-If Question	**Cause**	**Consequence**	**Severity**	**Likelihood**	**RL**	**Safeguards**	**Rec#**	**Recommendations**
Tube skin temperature of furnace exceeds its design limit	Excessive heat load by over-firing at the furnace burners	Meltdown and failure of heater tubes and resulting fire due to loss of containment	2	M	M	High tube skin temperature interlock to cut-off fuel gas on–off valve	1.1	1. To consider auto cut-off of the hydrocarbon feed to furnace upon high tube skin temperature activation 2. To consider installing restriction orifices in burner's gas pipeline to restrict maximum fuel to the burner

[a]Risk Level (RL) is based on the Risk Matrix classification as per Table 3.2.

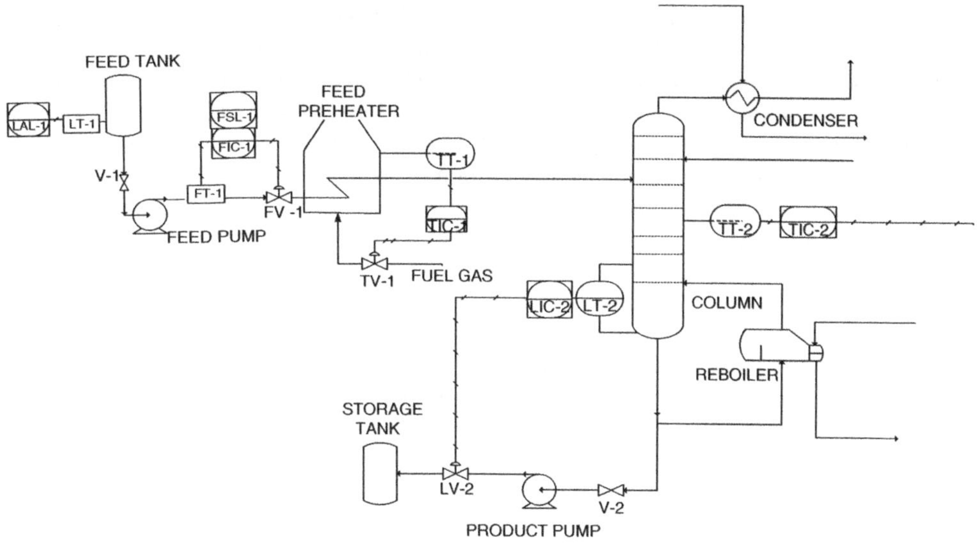

Figure 3.3 *Process flow diagram of a typical fractionator with a feed preheater.*

shall be placed in readily accessible locations. If the new facility requires trucks for raw material and/or product transport, sufficient access roads shall be considered. Close siting to utilities and tank farms will reduce the piping length and also safety risks. Reliable access to utilities is required for safe shutdown and operation of the plant. It will also improve project economics. The new site shall be close to the existing control room. This will minimize instrument and communication cable lengths. It is also preferable to install new facilities close to the existing substation. However, the facility siting guidelines need to be followed.

The project team shall make a survey of current day and night noise level data in the surrounding facilities. This will help to assess whether permitted noise levels will be achievable, even with the new equipment, at reasonable cost. Otherwise, alternative site selection is required. Heat radiation is another consideration. New facilities shall be installed at places sufficiently far away from heat radiation sources such as flare and flue gas stacks. Heat radiation calculations shall be performed to ensure that the new site is located well within the permitted heat radiation intensity limits. Such limits are given in API 521 (2014). Luminosity is another important consideration for new plant siting; this may require taller flare stacks.

3.3.4 Area Classification Reviews

Area classification is employed to identify locations where electrical equipment could provide a source of ignition to flammable liquids and gases/vapours that may be present. It is also used to prevent dust explosions. Area classification is necessary for selecting and installing safer electrical and instrumentation equipment, and its review is very important for revamp projects as it may influence site selection and also project cost. In some cases it may require a change in electrical area classification for existing facilities and hence require replacement of some electrical and instrumentation equipment.

There are several area classification standards such as API RP 500 (2012), API RP 505 (1997), NFPA (National Fire Protection Association) publication 70 (2013), NFPA publication 497, NFPA 70, and local standards in practice. They define classified areas surrounding the individual equipment, based on characteristics and probability of leaked/released liquid/gas/solids. This approach reduces fire/explosion risk due to the simultaneous occurrence of a flammable atmosphere and electrical ignition to an acceptable level. The project team can use an appropriate area classification standard acceptable to the company and local government authorities.

During the revamp project, area classification drawings for the existing plant need to be reviewed first. Later, thorough analysis shall be conducted to decide the area classification for new and existing equipment. After this assessment, the project team can develop/modify area classification diagrams for the new/existing project sites. If required, the layout plan needs to be modified and re-optimized to control project cost. Once project location has been classified, requirements for electrical equipment and associated wiring shall be determined from applicable publications such as API Recommended Practice 14F (API RP 14F) (2008), NFPA 70 (2013) and local regulations.

3.3.5 Pressure Relief System Considerations

Pressure relief systems are very important to protect process equipment, prevent fire, and reduce toxic releases and explosion hazards. Guidelines for their design are available in the Centre for Chemical Process Safety publication CCPS (1993), Parry (1994), Mannan (2005), API 520 Parts 1 and 2 (2014), and API 521 (2014). During the revamp design, it is very common to modify or add a new relief system section to the existing relief system. Before modifying or designing new relief systems, explore the following inherently safer design practices (CCPS, 1993).

- Intensification: Using small quantities of flammable, toxic substances
- Substitution: Replacing a material with a less hazardous substance
- Attenuation: Using less hazardous conditions or a less hazardous form of a material
- Limitation of Effects: Designing facilities that minimize the impact of release of hazardous material or energy
- Simplification/Error Tolerance: Designing facilities that make operating errors less likely, and that are forgiving of errors that are made

A process plant has a number of layers of protection as depicted in Figure 3.4. In this diagram, the three inner layers from the centre: process design; controls, alarms, operator supervision; and automatic action, SIS and ESD (emergency shutdown) systems prevent the pressure or vacuum relief. The outer four layers: relief devices, physical protection, plant and community emergency responses will mitigate the effects of relief. For a revamp project, the three inner layers should be designed carefully so that revamp of the relief system can be minimized or avoided. All new equipment in the revamp project shall be reviewed for relief using relevant overpressure causes. API 521 (2014) has listed 16 overpressure causes and relief estimation methods. Some of the useful guidelines for revamp projects are described in the following paragraphs.

In the revamp project, the maximum allowable working pressure (MAWP) of existing equipment may have to be either increased or decreased. Similarly, some new pressure

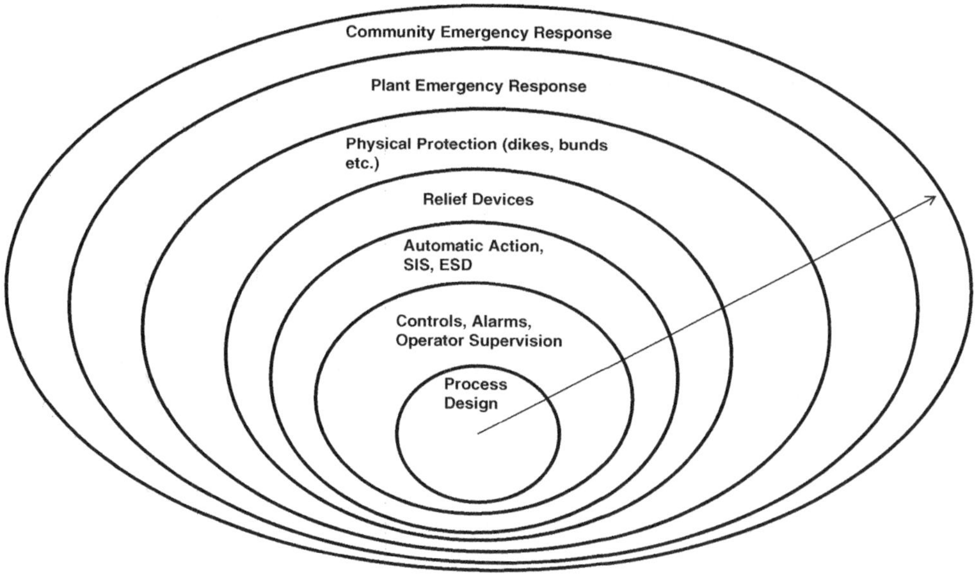

Figure 3.4 *Layers of protection for a process plant (CCPS, 1993).*

vessels may need to be added to the existing system, protected by a pressure relief system. Care shall be taken not to bottleneck the existing pressure relief system. If MAWP of a pressure vessel/heat exchanger is increased, it's PRD (pressure relief device) set point can be increased to the new MAWP; here, PRD refers to a pressure relief valve (PRV), vacuum relief valve or rupture disk (RD). If the PRD is also protecting other vessels connected to the main vessel, then it may cause exceeding of the MAWP of some of the vessels. This possibility should be considered and suitable changes should be made.

Increasing PRV set pressure will increase the flow through the existing safety valve. It will cause the PRV's inlet pressure drop to increase, and may violate allowable limits (PRV inlet pipe pressure drop <3% of PRV set pressure) as per API RP 520, Part 2 (2008). It may also increase the PRV's discharge header pressure drop, and hence exceed the allowed back pressure for PRV (conventional relief valve: 10% of PRV set pressure; balance relief valve ~50% of PRV set pressure). Pilot operated safety valves may be required if PRV inlet and outlet piping pressure drops are very high or if the PRV set point is close to the operating pressure; generally, pilot operated safety valves are used if operating pressure is >90% of PRV set pressure. Back flow prevention may be required in pilot operated safety valves, to prevent potential backflow from flare/relief header.

If MAWP of the vessel/heat exchanger and consequently PRD set point are reduced, it will reduce the PRD capacity. Care shall be taken to check the new required relief capacity against the PRD's capacity. Increasing the number of pressure vessels in the existing plant may affect the existing fire circles and increase relief load during fire scenarios. If the new pressure vessels are designed for full vacuum pressure, a vacuum relief valve is not required. Otherwise, suitable vacuum relief valves shall be installed. If the PRV's inlet or outlet piping needs to be changed, their pressure drop shall be checked at the rated capacity of the PRV. If they do not meet API 521/520 (2014) requirements, PRV's may chatter and become damaged.

During retrofit, it is often required to increase the existing centrifugal pump capacity and/or head. Pump capacity can be increased by impeller replacement (if the pump casing has allowance for a bigger impeller), pump replacement or installation of more pumps in parallel. Pump head can be increased by impeller, pump replacement or installation of more pumps in series. If the pump head and/or capacity are increased, its downstream piping and equipment shall be reviewed for over-pressure protection.

Occasionally, pump drivers may need to be changed (for example, from motor to steam turbine). Changes in driver speed need to be carefully evaluated so that a pump's pressure/flow will not pose an overpressure threat to the downstream equipment, including steam piping. In some cases, pump capacity/head increase may increase relief requirements due to rupture of tubes in a heat exchanger, if present in the downstream of the pump. Tube rupture relief can be neglected when the test pressure of the low-pressure side is greater than or equal to MAWP of the high-pressure side. Rotating equipment driver changes shall be considered carefully. Failure of drivers may increase relief loads due to loss of process cooling, steam, feedstock, quench, dilution streams and so on. Reliability of utilities is very important as loss of electricity, cooling water, instrument air and so on may impair functioning of safety systems.

During revamp, demister or coalescer pads may be installed in the separators. However, they may become chocked, and hence potential overpressure scenarios shall be evaluated. For the process equipment, proper positive isolation (double block and bleed arrangement or blinding) is required to avoid unintended flows. Any chance of reverse flow needs to be critically evaluated. Removal of insulation or fire proofing material from the process vessels may increase relief requirements, if there is the possibility of fires in that area. Many chemical plants have sections without any significant fire hazards. Addition of steam tracing, heating coils or steam jackets may also require increased relief rates. If any of the existing restriction orifices (RO) installed in piping is deleted or the RO size is increased, its impact on the relief system needs to be checked. Addition of isolation valves on some equipment may require additional PRDs due to blocked outlet or fire case scenarios. Increasing heat recovery or increasing burner capacity in fired heaters may require more relief in the downstream equipment.

New on–off valves' closing speeds shall be reviewed and adjusted to prevent hydraulic shock waves or water hammer. Also, piping velocity shall be checked so that it is well within erosion velocity. The possibility of multi-phase mixtures inside pipes shall be evaluated carefully to apply proper hydraulic equations and estimate adequate pipe sizing.

New process equipment installed within 7.6 meters from the grade/ground level will require pressure relief devices due to fire scenario. For revamp cases, it may be advantageous to place vessels on platforms at >7.6 meters if fire case relief load is most credible and bottlenecks the existing flare system. Fire circle or zone is defined as the maximum affected area during any equipment fire in the facility. API 521 (2014) defines its area as 230 to 460 m^2. Addition of process equipment inside an existing fire circle may increase the fire circle size. Hence, care shall be taken to review the fire circle size with each equipment addition. It will impact the peak relief load during the fire scenario.

In some process units, loss of process control can result in a significant change in temperature and/or pressure, which can result in exceeding the design limits of the materials used in the process equipment. Adequate over-pressure and temperature protection needs to be established and implemented. Addition of PRDs in the revamp project may increase back pressure of some existing PRDs and may exceed velocity and temperature limits in some headers in the flare network. Thorough flare network calculations using software tools such as Aspen Flare Analyser, Flaretot, and Visual Flare are recommended to assess and modify the flare system, for eliminating the design limit deviations.

Change of control valves (to reduce pressure drop and/or increase capacity) and opening the bypass valve of the control valve may overpressure their downstream equipment. Hence, such modifications should be reviewed with detailed hydraulic calculations. Credit can be taken (in relief system design) for operator response if response time is between 10 and 30 minutes, depending on the consequences of overpressure. Often, it requires additional PRDs on the existing PRD inlet piping manifold. Check to ensure that inlet and outlet header pressure drops do not exceed the limits specified in API 520/API 521 (2014). New PRD set pressures shall be set higher compared with the existing PRD set pressures, as per limits specified in API standard 520 (2014). This will prevent chattering and damage of PRDs.

Thermal expansion and fire cases are not required to be checked, if the existing equipment is re-used, with the same service and also the same level control setting. Overpressure relief requirements due to each utility failure, fire cases and any other combination scenarios need to be estimated. API 521 (2014) has a comprehensive list of effects for utilities failure. All the PRV manifolds shall be checked to estimate back pressures at the PRVs. PRD overpressure calculations for equipment shall be documented as shown in Table 3.4. Vacuum relief (if the vessel/s is/are not designed to withstand full vacuum) shall also be documented. All the flare scenarios and flare network shall be properly documented. An example of PRV sizing calculations for the system shown in Figure 3.5 is presented in Table 3.4.

Figure 3.5 shows a typical demineralized (DM) water injection system in a petrochemical plant. DM water from the utility section is pumped to the feed vessel (V-1) using the centrifugal pump P-1. This DM water is pumped to the reactor at high pressure using pump P-2. There is one flow control valve at the outlet of P-2. Pressure of V-1 is controlled by nitrogen/vent split range control system. MAWP and set pressure for safety valve (PRV-1) of V-1 is 500 kPa. V-1 is designed for full vacuum and so vacuum relief is not required. A summary of PRV-1 relief calculations for the present case is shown in Table 3.4. During a revamp project, it is intended to increase the DM water flow rate through P-1 by increasing impeller size by 15%. The characteristic curves for P-1 in the original and revamped cases are shown in Figure 3.6. With suction pressure of 30 kPa, 48 m of head corresponds to a discharge pressure of 500 kPa at P-1. From Figure 3.6, one can see that pressure relief rate for the scenario of a blocked outlet of V-1 (diameter 4 meters and height 6 meters) increases significantly from 55.5 m^3/h (present case) to 164.7 m^3/h (due to P-1 impeller size increase by 15% during the revamp). So, PRV requires to be changed and the new required capacity for PRV-1 is shown in Table 3.4. For detailed calculation procedures, see API standard 520 Part 1 and 521 (2014).

Branan (2005) has given a simplified method for quick estimation of manifold pressure drops. Flare knockout drum sizing needs to be checked for proper vapour–liquid separation and hence prevention of carryover of liquid to the flare stack. Steam injection to the flare stack also needs to be reviewed to prevent potential black smoke and noise. Relief headers'

Table 3.4 *Summary of overpressure relief calculations.*

Process Unit	ABC		Date: 01-11-2014				
PRD Number	PRV-1		Equipment Protected: V-1				
Prepared By	XYZ		P&ID Number: P&ID-1				
Overpressure Scenarios	Applicable		Present requirements		Revamped case requirements		Comments
	Yes	No	Capacity (kg/h)	Orifice Area (mm^2)	Capacity (kg/h)	Orifice Area (mm^2)	
Accumulation of non-condensables		No	NA	NA	NA	NA	
Check valve failure		No	NA	NA	NA	NA	
Chemical reaction		No	NA	NA	NA	NA	
Cleaning of equipment		No	NA	NA	NA	NA	
Closed outlets on vessels	Yes		55500[a]	830	164700[a]	2323	P-2 outlet closed
Excessive inlet flow		No	NA	NA	NA	NA	
Excessive process heat input		No	NA	NA	NA	NA	
Failure of automatic controls (BPCS failure)	Yes		250	71	250	71	Nitrogen control valve fail open
Failure of heat exchanger tubes		No	NA	NA	NA	NA	
Fire (external)	Yes		1967	506	1967	506	
Flooding of condensing tubes		No	NA	NA	NA	NA	
Inadvertent Mixing		No	NA	NA	NA	NA	
Loss of cooling medium		No	NA	NA	NA	NA	
Loss of electrical power		No	NA	NA	NA	NA	
Operator error	Yes		55500[a]	830	164700[a]	2323	P-2 outlet closed
Overfilling of vessels		No	NA	NA	NA	NA	
Pressure regulator failure		No	NA	NA	NA	NA	
Reflux failure on towers		No	NA	NA	NA	NA	
Special start-up conditions		No	NA	NA	NA	NA	
Thermal expansion of fluid contents		No	NA	NA	NA	NA	
Upset conditions during operation		No	NA	NA	NA	NA	
Others		No	NA	NA	NA	NA	
Governing case for PRV-1 sizing			Closed outlet case (API size J)		Closed outlet case (API size M)		

[a]DM water density is assumed to be 1000 kg/m^3.

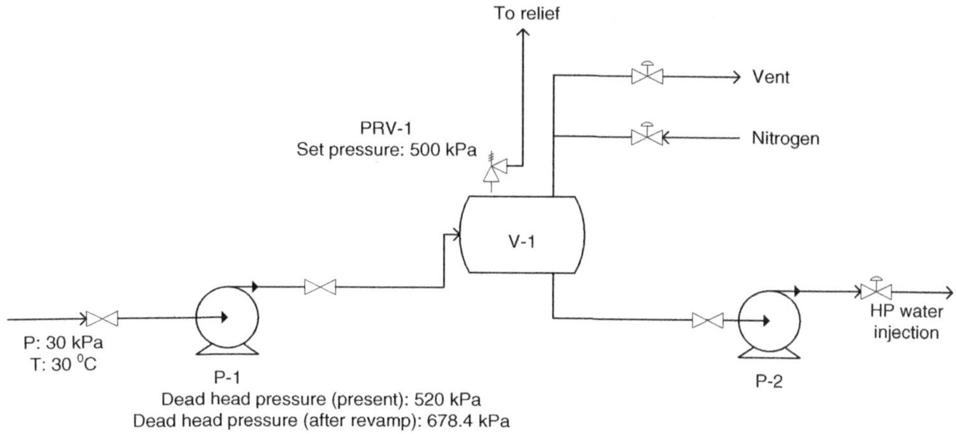

Figure 3.5 *Demineralized (DM) water injection system.*

sizing should be checked to ensure that back pressure does not exceed the lowest allowable back pressure for the PRDs. It is recommended that the velocity at the flare stack tip be limited to 0.5 Mach for peak, short-term and infrequent flow, and 0.2 Mach for normal and frequent conditions. If the flare quantity increases substantially in the revamped case, flare stack height needs to be reviewed to minimize heat radiation effects. Madlani (2012) presented one case study on revamping a flare network.

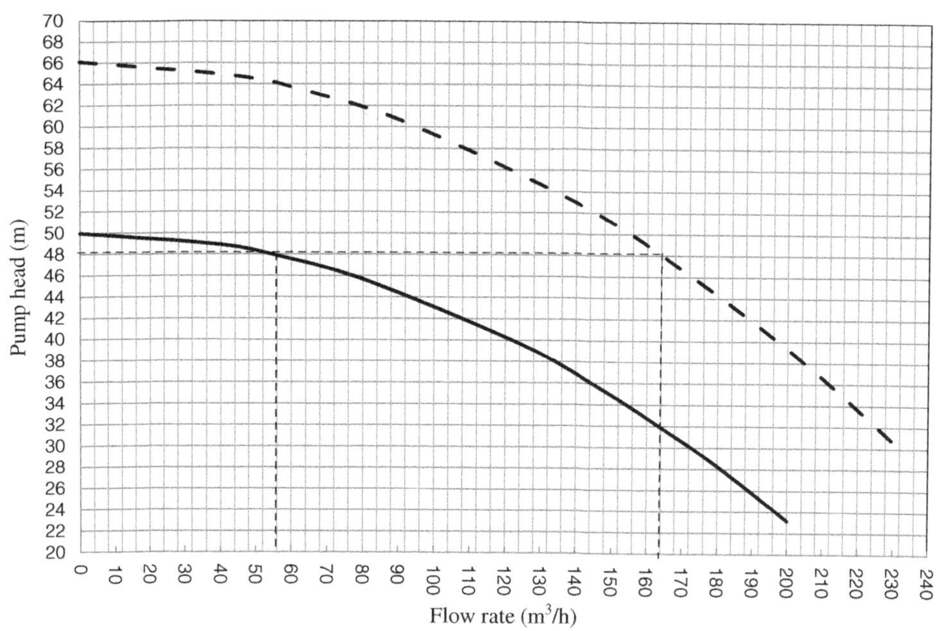

Figure 3.6 *Characteristic curves of P-1 with present (lower curve) and larger (upper curve) impeller.*

3.3.6 Fire Safety for Revamp Projects

Fire safety should be best addressed during the Pre-FEED or FEED stage. Although it is possible to limit the extent of fire damage, it is good to minimize fires by using inherently safer design practices. Before finalizing the revamp project scope, designers/project team should study the effect of the proposed modifications on the fire causes and evaluate the adequacy of the existing fire prevention features of the plant. The following fire safety aspects shall be reviewed during the development of revamp projects.

- **Siting, Spacing and Access**: New equipment siting should minimize the fire safety risks. There should be sufficient space to prevent spread of fire, and also suitable access for fire-fighting. Company-approved spacing guidelines should be followed. If fire risk is valid, sufficient fire-fighting systems need to be provided
- **Drainage**: A properly designed drainage system reduces fire risks by draining spilled flammable material away from potential fire/ignition sources, removes fire water quickly during any fire-fighting scenario, reduces pressure relief rate during fire scenario, and also prevents spread of fire
- **Isolation**: New equipment handling flammable material should have remote isolation to safely isolate the feed during fire scenarios. Any instrumented system used for remote isolation may require fire proofing
- **Fire-resistant Materials**: Where economically justified, new facilities should use fire-resistant materials. If there is fire risk, apply fire proofing materials for equipment protection
- **Release of Flammable Vapour**: Design should incorporate sufficient measures to minimize the escape of flammable vapour. Where vapour release is necessary, sufficient mitigations such as dilution, release at very high locations using flame arrestor and/or guard against ignition sources are required
- **Accumulation of Electrostatic Charges**: Avoid accumulation of static charges by adequate design practices. Applicable design methods are well presented by Pratt (2000)
- **Ignition Sources**: Sources of ignition in potentially hazardous locations should be identified, and reasonable safeguards should be provided to minimize the fire/explosion risks
- **Flammable Vapour Detectors**: Flammable vapour detectors should be considered in locations where there is potential for leakage of hydrocarbon vapour
- **Fire and Smoke Detection**: Fire detection systems and alarms should be considered in hazardous locations. These include gas, thermal, smoke detectors, thermal detectors, as well as IR/UV flame scanners
- **Fire-fighting System**: Fire water systems with foam, fire monitors, hose, and sprinkler systems are widely used in large facilities. Care should be taken to ensure uninterrupted supply of sufficient fire water all the time. For extinguishing small localized fires, fire extinguishers (dry chemical, CO_2, halogen etc.) can be used. Deluge systems which operate by automatic fire water opening on detection of smoke or heat are useful for automatic fire fighting. Large nitrogen-pressurized dry chemical power systems can be effectively used to automatically or manually (by remote operation) control fires at chemical storage facilities involving pyrophoric materials
- **Equipment Fire Safety Considerations**: For pumps handling flammable liquids, use steel or alloy materials for casing, and install a remotely operated emergency isolation

valve in the suction piping. Further, a double mechanical or tandem seal with leak detection instrumentation shall be used to minimize seal leakages. Critical pumps should be installed with minimum flow protection and also low suction pressure trips. An adequate Net Positive Suction Head (NPSH) margin needs to be provided for pumps. Deluge systems are effective for pump fires. A pump/compressor's base should be designed to prevent oil accumulation. Minimize small diameter piping in compressors, and avoid any chance of liquid compression in reciprocating compressors. Distance piece vents of reciprocating compressors shall be routed to a safe location. Prevent surging of centrifugal compressors and deadheading of reciprocating compressors. Install properly sized spillback controls and safety relief systems, and remote isolation valves for critical compressors

Avoid overheating of furnace tubes. Consider spare pumps for furnace heaters. Install remote isolation valves for stopping feed and fuel during fire scenarios. Design automatic shutdown of furnace on loss of feed and low fuel gas/oil pressure. API RP 556 has published guidelines for safe instrumentation of fired heaters. Storage tanks should be constructed with secondary containment bunds, such that material released through failure events can be contained. Common bunding is acceptable if the bunded region can contain contents of all tanks. Incompatible chemicals should not be stored in the same bunding or dike. There are some serious incidents due to incompatible chemicals leaking and reacting within the same dike (Sanders, 2010).

Tanks should have safety systems to prevent formation of explosion mixtures during tank emptying or draining. Tanks should be installed with pressure/vacuum protection devices and flame arrestors in vents. A proper grounding system is necessary at storage tanks to prevent lightning strikes. Locate pumps and ignition sources out of tank bund walls. Use double seals for the roof of floating-roof tanks. Consider nitrogen blanketing of low-pressure storage tanks. Avoid flexible couplings on piping. Install frangible tank shell-to-roof welding for API 650 tanks. Install overflow protection for all storage tanks.

PRD discharges should be routed to the flare system. Release to atmosphere, if necessary, should be routed to a safe location with a safe disposal system such as sufficiently sized knockout drums (to prevent liquid release), steam or nitrogen dilution and so on. Always use proper material of construction to prevent corrosion, erosion, hydrogen embrittlement, chloride stress cracking, high-temperature sulfidation and so on. Design piping to avoid erosion and minimize use of hoses and temporary piping. Consider adequate piping supports to prevent vibrations. Install adequately sized stress relief loops in piping. Minimize the number of pipe flanges and joints. Avoid threaded joints in piping. Avoid pressure surge in pipes by reducing pipe velocity and avoiding quick-close on–off valves.

Gas turbines should be installed with automatic fuel shutoff valves. Locate cooling towers far away from process units. Whenever possible, maintain cooling water pressure more than process stream pressure, at the coolers. Install gas detectors at the top of cooling towers, and an automatic shutdown system for cooling tower fans, in case of gas detection. Boiler and other utilities should be located far away from process facilities so that they will not be exposed to fires and explosions originating in the process plants. Install deflagration or detonation arrestors in low-pressure fuel gas systems connected to atmosphere or fired systems.

Install critical power, control, instrument, and alarm wiring and pneumatic tubing outside the fire-exposed envelope. Cable trays and conduit banks should be routed underground or on the upper levels of elevated pipe racks at least 30 feet above the ground and outside the drainage path of hydrocarbon spills. Install fire proofing of critical systems and equipment if they are inside the fire-exposed envelope.

Comprehensive coverage of many fire safety design and implementation guidelines is available in Nolan (1996) and CCPS (2003b).

3.4 Process Hazard Analysis (PHA)

PHA is an integral part of project management. It is a systematic effort to identify and analyse the significance of hazardous situations associated with a process or activity. It is an experience-based evaluation conducted by a team tasked to identify hazards associated with a process, evaluate the risks associated with those hazards, and make recommendations to reduce or eliminate the risks. The focused hazards in PHA are mainly the acute events that can result in unplanned and catastrophic chemical releases, fires or explosions. It ensures that an appropriate safe design basis is implemented and documented.

The objective of PHA is to minimize (a) fatalities and injuries of company/contractor employees and the surrounding community, (b) events resulting in major environmental issues, and (c) major losses to the company. The success of hazard review is dictated by (i) availability of process safety information, (ii) multi-disciplinary team with the right level of knowledge and techniques, and (iii) commitment of the team and the management. PHA may contain some or all of the following studies: What-if, HAZOP, FMEA, LOPA, QRAs (quantitative risk assessments) like fault/event tree analysis, SIS analysis, dispersion and heat radiation modelling, review of previous plant modification PHAs, PSI verification, human factors, facility siting reviews, and past incidents review. Based on the revamp project scope, the hazard review leader will decide the PHA study scope, which may increase or decrease during the study phase depending on the identified risks associated with the process. PHA study timing will be decided by the hazard review leader after checking the availability of PSI information and necessary experts.

3.4.1 Process Plant Hazard Review using HAZOP

The most common technique applied in PHA of process plants is HAZOP. It provides a method to systematically review the design and operational philosophy, and maintenance procedures, of a system so as to identify the potential occurrence of hazardous events (including their impact on people, property and the environment). It is also helpful to address some of the operability problems which impact process efficiency/productivity. A HAZOP study should be performed for all/major lines of revamp process and utility areas, all the way up to the facility boundaries, which are critical to plant safety and operations. The gaps identified as part of HAZOP are documented in an action register, for further investigation. Dedicated person(s)/group(s) will be assigned to find solutions within an agreed time frame. All the solutions/action items need to be completed in accordance with the risk matrix set by either the corporate or local regulatory bodies.

Accurate, up-to-date P&IDs (piping and instrumentation diagrams) shall be available to conduct a PHA. The hazard review leader shall determine if the P&ID or PFD is accurate enough to conduct a meaningful hazard review. The hazard review should be delayed until the P&ID or PFD quality is adequate. HAZOP is conducted by a team of individuals led by a person knowledgeable in the HAZOP technique. One member (typically, process safety engineer) of the team should be trained and knowledgeable in the HAZOP technique. The other members are usually selected for their knowledge of the process and/or technical contribution to the team while keeping the team size to a manageable level, say 3 to 7.

A sample hazard review team consists of team leader, scribe, process, operations, process safety, maintenance, instrumentation, electrical and environmental engineers. The actual composition of a specific HAZOP team will depend upon the revamp project scope, objectives and other considerations. The team leader is responsible for the study preparation, guiding/managing the team and supervising the study documentation and report generation. Team members provide required information about the system and participate in the study to identify deviations, causes, consequences and likelihood. A dedicated technical secretary or scribe is assigned to record results and prepare reports. It is important to plan and communicate the agenda of reviews in advance, and consolidate alike nodes of the process in a sequence so that the relevant subject matter expert is engaged effectively during his/her availability within the HAZOP study timeframe. This approach keeps the core team to an efficient size. Dave (2004) has presented a comprehensive coverage of the HAZOP technique and its application. As a minimum, the HAZOP leader should obtain the following documents prior to the PHA of retrofit project: hazard review documents of the original plant, approved and frozen P&ID diagrams, shutdown tables, PFDs (for all possible modes), alarm and trip summary with set points, relief device specification and sizing documents, SIS tables, summary of MOCs from the previous hazard reviews up to this retrofit proposal, summary of incidents and safety near misses (relevant to process) from the incident tracking database of the site. If the hazard review leader identifies the quality of PSI as compromising the team's ability to identify hazards, he/she should postpone the review until the quality of information reaches a satisfactory level.

Nodal Classification: The HAZOP leader identifies nodes prior to the initial meeting. With some assistance from the process engineer, the leader will break the revamp process into smaller and manageable nodes. Selection and sequence of nodes during the study are generally decided by the team leader. The node boundaries usually occur where process parameters change such as change in design temperature or pressure or phase. The study starts with primary focus around the process lines and vessels that contain hazardous substances and energy. The auxiliary/utility systems are also reviewed subsequently as they can significantly affect the process safety. Nodal classification for a process fractionator with a feed preheater is illustrated in Figure 3.7. The feed tank up to the valve V-1 is classified as node 1, the feed pump up to the flow control valve is classified as node 2. Feed from the feed control valve into the preheater and up to the process column inlet is classified as node 3, the burner (fuel supply and control) and flue gas system (including draft fans) are classified as node 4, fractionator is classified as node 5, and bottom pump and product storage tanks are classified as nodes 6 and 7 respectively.

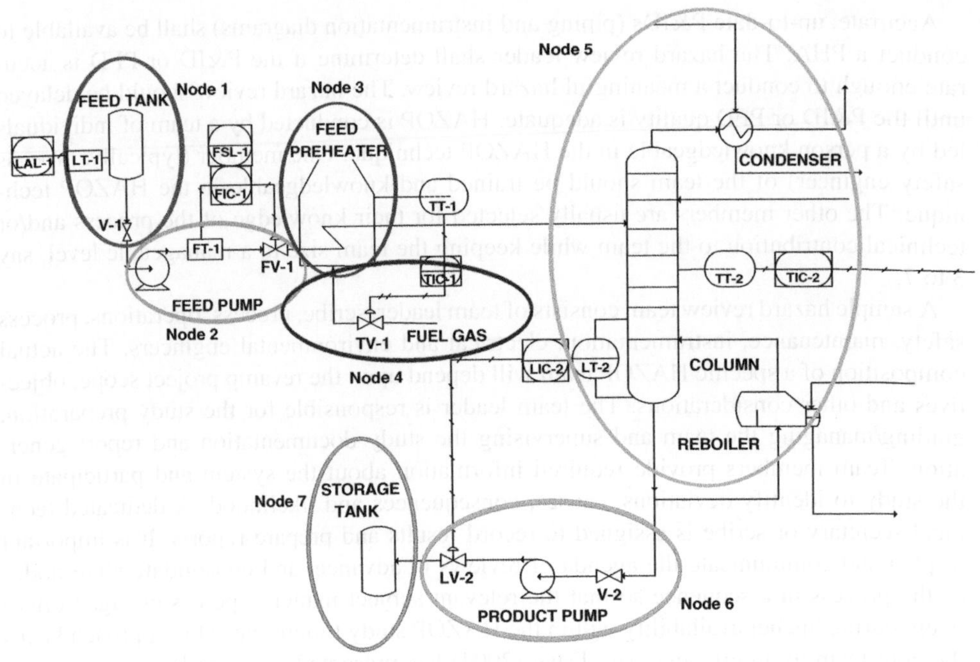

Figure 3.7 *P&ID with nodal classification for a process fractionator.*

HAZOP Checklist Consolidation: HAZOP reports serve the entire life of the revamp project and subsequently the operating plant. The hazard review leader should make a clear judgment on the appropriate hazard review methodology for the revamp project and appropriate checklists (assisted by either via software and/or templates) that fit into the hazard analysis practices of the existing plant.

Hazard Review Process: The subject expert (usually the process engineer who designed the proposal) will explain the system and operational philosophy at the beginning of each nodal review. Usually, the first node of a session takes longer than the rest as the focus and attention of the team are not centred at the beginning. Once the momentum picks up, the flow of the review will be smooth. The hazard review leader initiates the brainstorming process using deviations relevant to the process node and also ensures that all other columns of the checklist are defined and completed adequately. Guided deviations for specific parameters are listed in Table 3.5.

The hazard review leader fosters open communication across all functional members during the discussions. The team should ensure that minimum safeguard requirements exist in the design, failing which a recommendation should be made. The responsible person/team is assigned for each of those recommendations and a time frame is stipulated in accordance with hazard review team's agreement with the management.

A sample HAZOP worksheet for the feed preheater process (Figure 3.7) is presented in Table 3.6; this is for a specific deviation in node 3 of Figure 3.7. The hazard review leader should ensure that basic information such as session date, skid name, node number,

Table 3.5 List of deviations from intended operations for specific parameters.

Parameter	Deviations from intended operation	Parameter	Deviation from intended operation
Flow	No flow, High flow, Low flow, Reverse flow, Misdirected flow	Reaction	No reaction, High reaction rate, Low reaction rate, Additional/Incomplete/ Reverse/Wrong reactions
Temperature	High temperature, Low temperature	Containment	Total Loss of containment, Partial Loss of Containment
Pressure	High pressure, Low pressure	Vibration	High vibration
Procedure	Additional step, Omitted step	Services/Utilities	Loss of services/utilities
Level	High level, Low level	Corrosion/Erosion	High corrosion rate
Composition	High concentration, Low concentration, Contaminated composition, Wrong composition	Start-up	Improper start-up
Phase	Additional phase, Loss of phase	Shutdown	Improper shutdown
Mixing	No mixing, Excessive mixing, Poor mixing	Maintenance	Improper maintenance

node description, associated drawing numbers and intent of the node are captured correctly. Items shown in the columns in Table 3.6 are the minimum HAZOP requirements to be captured and analysed. The significance of these items/columns is as follows.

1. Deviation: Parameter which departs from the design intention
2. Cause: Initiating events or reasons why such a deviation might occur
3. Consequence: Potential realistic effects of the deviation. They include major process upsets, damage to equipment, hazardous material release, fires and explosions
4. Category: Categories of consequences such as safety, reliability, operability and equipment protection
5. Safeguards: Existing features designed to prevent causes or to mitigate consequences
6. Safeguard Status: Available currently or unavailable currently
7. Severity: Severity of the result of an unwanted event based on the risk matrix in Table 3.2
8. Likelihood: Likelihood of an unwanted event based on the ranges in the risk matrix (Table 3.2)
9. Risk Level (RL): The combination of likelihood and severity will give any event a place on a risk matrix as depicted in Table 3.2

Table 3.6 *Sample HAZOP worksheet for the feed preheater.*

HAZOP Worksheets

Date and Session:	20/10/2014
Skid:	Feed Preheater
Node Number:	3
Node Description:	Liquid from feed tank via the feed pump into preheater and up to the inlet line of column
Drawings:	ABC. 000.xx Sht 1 Rev. 3
Intention:	Liquid from feed pump at 3000 kPa and 25 °C enters into this preheater and heated up to 400 °C. This mixed-phase feed enters into column C105

Deviation	Cause	Consequence	Category	Safeguards	Severity	Likelihood	RL	SIF	Recommendation	By	Remarks
No Flow	Loss of Liquid from feed tank	Preheater coils overheated, thus resulting in pipe and furnace integrity failure and fire spread	S	Coils are designed with Incolloy 800H material (can withstand temperatures up to 800°F).	2	H	H1[a]	SIL2	To configure this low low flow switch signal from feed tank outlet to trip the fuel supply to the heater	Process Engr/ Controls Engr	
			S	FSL- 1 (Flow Switch Low Alarm)							
			S	LAL-1 (Low Level Alarm) at the feed tank with operator's response							

[a]Risk Level (RL) and SIF are classified per Risk Matrix in Table 3.2.

10. Classification of SIF for the triggered safeguards out of risk matrix (Table 3.2) should be designed to comply with the risk target set by the company/local codes. The next section covers the methodology of Safety Instrumented System design
11. Recommendation: Suggested corrective actions to reduce the unwanted event's frequency (to meet the acceptable frequency target), in the form of design changes, procedural changes, or further studies and associated retrofits
12. Recommendation Responsibility: 'By' column allows the team to designate the party responsible for addressing the recommendations

Severity and likelihood helps to make a semi-quantitative or qualitative judgment during the hazard review, to trigger a requirement of QRA or SIL assessment and verification as per ANSI/ISA-84.00.01 (2004). SIL assessment and verification is detailed in Section 3.4.3. QRA is not covered in this chapter; for a good coverage of this topic, interested readers can refer to CCPS (2003a). Consolidated actions from the HAZOP study should be presented to the business and management representatives for agreement and commitment to implement the actions within the agreed time frame.

3.4.2 Failure Modes and Effects Analysis (FMEA) Tool

FMEA is a systematic, proactive alternative method to HAZOP for evaluating a process to identify and mitigate process safety, environmental and asset integrity issues. It employs different failure modes and effects in order to identify key risks and necessary mitigation measures. It is very useful for proper assessment of highly instrumented or electrical system. FMEA can be deployed to evaluate processes for possible failures and to prevent them by correcting the processes proactively rather than reacting to adverse events after failures have occurred. It includes a review of the following steps in the process:

- Failure modes to identify what could go wrong
- Failure causes on why would the failure happen
- Failure effects about the consequences of each failure

FMEA Procedure:

- Describe the product/process and its function. An understanding of the product/process under consideration is important for performing FMEA (Crow, 2002)
- Block Diagrams, PFDs or P&IDs should show all process systems, equipments and utility streams with adequate detail of design
- Identify failure modes. A failure mode is defined as the manner in which a component, system or process could potentially deviate from the design intent. A failure mode in one component can be the cause of a failure mode in another component. See the example in Table 3.7
- Identify the causes for each failure mode. A failure cause is defined as a design weakness that may result in a failure. Potential causes for each failure mode should be identified and recorded
- Describe the effects/consequences of failure modes. For each failure mode identified, determine what the ultimate effects will be. A failure effect is defined as the result of a failure mode on the function of the product/process as perceived by the customer

Table 3.7 Example of FMEA worksheet.

FMEA Safety Review Log Sheet									Page of
Study Name:	Capacity expansion at Plant 1	Team Leader:							
Node:	2	Team Members:							
Date:	15/11/2014								
Ref #:									
Tag No.	Failure Mode	Consequence	Safeguards	Severity	Likelihood	RL	Rec#	SIF	Recommendation
Feed pump	Pump discharge/ downstream over-pressurized by block in fluid	Fluid pressure at pump or the pressure at furnace tubes exceeds the design, resulting in failure of pump seal, impeller and furnace tubes	Pressure controller at pump recycle line Relief device at downstream	2	H	H1*	1.1	SIL 2	To install a High-Pressure trip at the pump discharge, which shuts down the feed pump, isolates the feed and fuel supply into furnace in the event of high pressure detection

- Establish numerical rankings for the consequences and also likelihood of the effect. For example, use risk matrix classification as per Table 3.2
- Enter the risk level based on the classification in Table 3.2
- Identify current controls preventing the cause of the failure mode from occurring or detecting the failure before it results in the consequences. Each of these controls should be assessed to determine how well it is expected to identify or detect failure modes
- Determine the likelihood of detection. This is an assessment of the likelihood that the current controls will detect the cause of the failure mode or the failure mode itself
- Review risk level. If risk is significant, QRA and/or SIS studies may be required
- Determine recommendations to address potential failures that would have high RL. Assign responsibility and target completion date for each of these actions. This makes responsibility clear-cut and facilitates tracking
- Update the FMEA as the design/process/assessment changes or new information becomes available

FMEA in Table 3.7 illustrates an overpressure scenario of the pump by the spurious closing of on–off valve at its discharge. This can generate unacceptable risk for the pump, furnace and people around. A SIF of SIL2 is at least necessary to mitigate this risk when high pressure is detected at the pump discharge.

3.4.3 Instrumented Protective System Design

Process industries rely upon many types of instrumented protective systems (IPSs) to respond to abnormal operations/situations that might otherwise have serious process safety consequences. Such systems are commonly identified as safeguards during PHAs (particularly during What-if, HAZOP and FMEA studies). They are classified based on the IPS type and the expected risk reduction. Ultimately, risk criteria are required to support the specification of the risk reduction requirements for each identified IPS. CCPS (2007) discusses the lifecycle management of IPS implemented to achieve a specified risk reduction strategy. IPS should be designed and managed to meet appropriate industry practices. For example, SIS is a type of IPS, which acts to achieve or maintain a process safe state when abnormal operation occurs. Consensus standards such as ANSI/ISA 84.00.01 (2004) provide technical requirements for the design and management of a SIS to achieve a specified reliability based on four safety integrity levels: SIL1, SIL2, SIL3 and SIL4. Hazards can lead to loss of life, production and profitability, damage to assets, environment and company's reputation. A careful application of international standards on risk assessment and risk reduction will provide necessary protection.

SIF is defined as a collection of sensors, logic solver and final elements that protect against a particular hazard. There may be multiple sensors and final elements for each safety function as shown in Figure 3.8. The fundamental work process for SIL evaluation includes the following.

- Identify potential hazards
- Identify available safeguards
- Determine risk tolerances
- Determine if SIF is required; this may require event tree, fault tree or LOPA analysis
- Evaluate if other safeguards can be upgraded

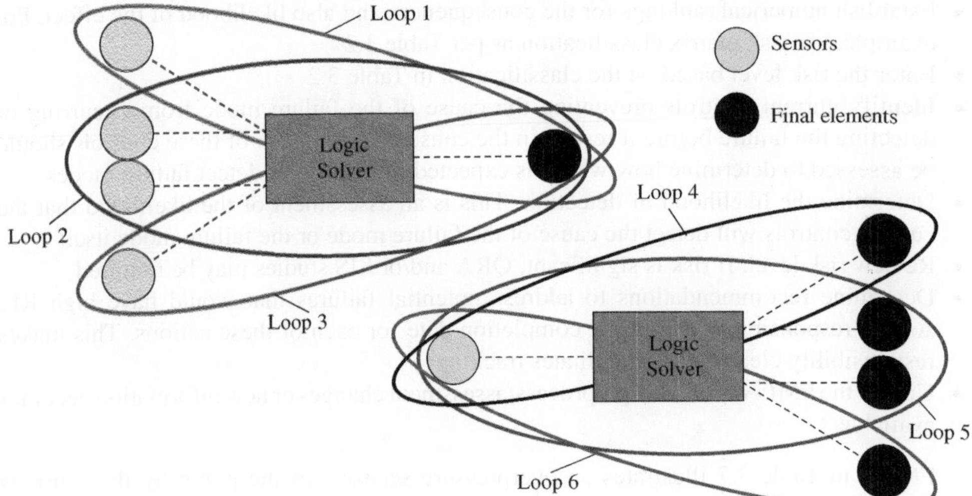

Figure 3.8 *Configuration of safety instrumented function (SIF).*

- If SIF is still required, determine its required risk reduction as a SIL or maximum allowable PFD_{avg}

The required safety integrity level of the instrumented function shall be derived by taking into account the required risk reduction that is to be provided by that function. For those SILs, the target PFD_{avg} on demand and the target frequencies of dangerous failure are listed in Table 3.8 [ANSI/ISA-84.00.01(2004) Part 3] for each SIF. Several risk analysis methods ranging from qualitative to fully quantitative can be deployed based on the severity and complexity of the scope, as listed in Table 3.9.

3.4.4 Fault Tree Analysis

Fault tree analysis is a rigorous method that can be used to determine the PFD_{avg} or to supplement better estimates of individual initiating causes or independent protection layers (IPLs) in LOPA. Fault tree is a deductive method for identifying ways in which hazards can

Table 3.8 *Safety integrity levels: probability of failure on demand, target risk reduction factor and target frequency of dangerous failure to perform the SIF.*

Safety integrity level (SIL)	Target average probability of failure on demand (PFD_{avg})	Target risk reduction Factor (RRF)	Target frequency of dangerous failures to perform the SIF (per hour)
4	$\geq 10^{-5}$ to $<10^{-4}$	>10,000 to ≤100,000	$\geq 10^{-9}$ to $<10^{-8}$
3	$\geq 10^{-4}$ to $<10^{-3}$	>1000 to ≤10,000	$\geq 10^{-8}$ to $<10^{-7}$
2	$\geq 10^{-3}$ to $<10^{-2}$	>100 to ≤1000	$\geq 10^{-7}$ to $<10^{-6}$
1	$\geq 10^{-2}$ to $<10^{-1}$	>10 to ≤100	$\geq 10^{-6}$ to $<10^{-5}$

Table 3.9 Risk analysis methodologies.

Risk analysis method	Type
Risk matrix	Qualitative
Layer of protection analysis (LOPA)	Semi-quantitative
Fault tree/Event tree analysis	Quantitative
Consequence analysis	Quantitative

lead to accidents. The approach starts with a well-defined accident (top event) and works backward to various scenarios that can cause the accident. The top event in a fault tree is the event of concern as shown in Figure 3.9. Fault tree components are classified as active (continuous) or passive (on demand). Logic gates describe the combined outcome of two or more inputs. They describe the logical relationships between components, sub-events and the top event. The 'AND' gate is used where all components input to the gate must be true in order for the output from that gate to be true. The 'OR' gate is used where at least one of the components input to the gate must be true in order for the output from that gate to be true (Crowl and Louvar, 2002). Values for PFD_{avg} are calculated for the protective layers in a fault tree. Fault tree analysis for a pipeline rupture event by the failed pressure regulator and relief device is depicted in Figure 3.9.

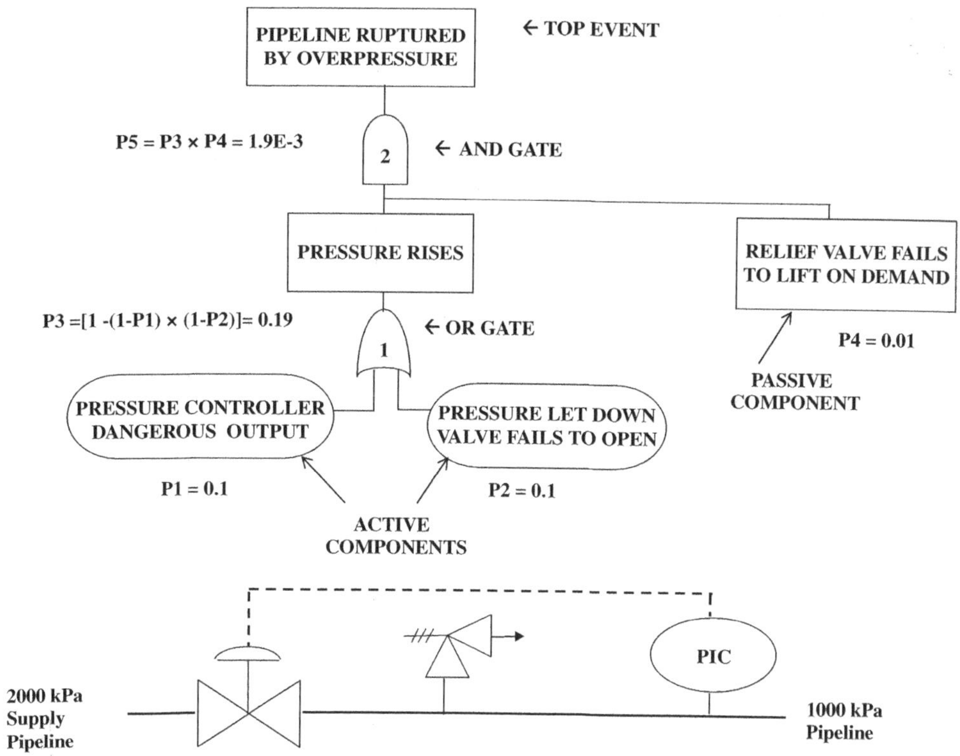

Figure 3.9 Fault tree analysis for pipeline rupture by failure of the pressure regulator shown in the rectangular box at the bottom.

3.4.5 Event Tree Analysis

Event tree analysis achieves the same result as a fault tree analysis but via an inverse approach. These analyses are performed to evaluate the potential for an accident that is the result of every initiating event. We start with an initiating event and work inductively towards the result, such as from an equipment failure or a process upset. The possible sequence of events that leads to potential accidents is captured, by taking into account both the success and failure of any associated safety functions as the events progress. Event sequences are systematically recorded and the relationships between the initiating event and subsequent events are appropriately defined. Every branch of the tree illustrates a possible sequence of events. Protective layers designed to mitigate the initiating events are identified (through their PFD_{avg} values) as an outcome of this event tree analysis. A sample event tree analysis with the calculation steps for mitigated failures per demand is depicted in Figure 3.10. The initiating event considered for this event tree analysis is 'failed pressure controller', and an additional layer of pipeline rupture probability by over-pressurization is also considered.

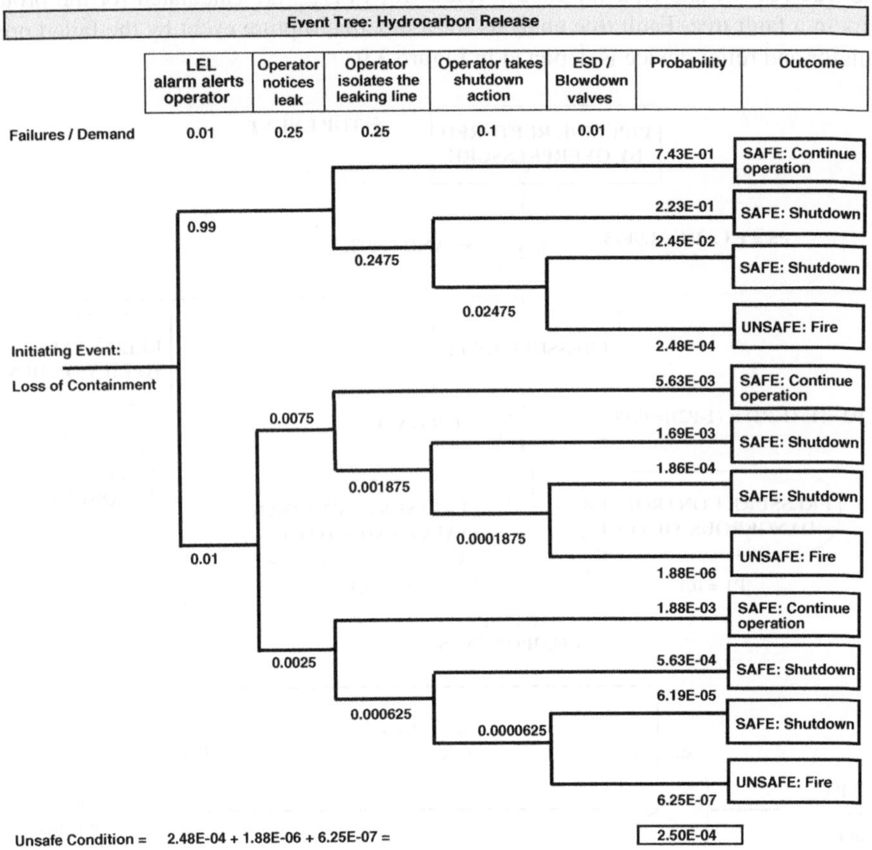

Figure 3.10 *Event tree analysis for pipeline rupture by failure of the pressure controller shown in the rectangular box in Figure 3.9.*

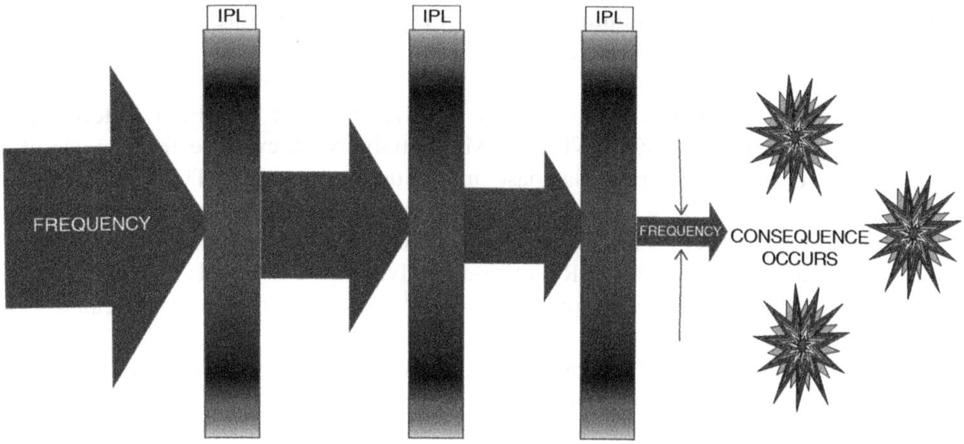

Figure 3.11 Graphical representation of reduced frequency by independent protection layers (IPLs).

3.4.6 Layer of Protection Analysis (LOPA)

LOPA, a commonly applied method, analyses the effect of adding independent protective layers to reduce the frequency of the consequence of an event. This layer of protection concept can be thought of as setting up a system to address as many events and outcomes as possible. The more obstacles in the path to a hazard, the more effective they are in preventing a series of events from leading to a harmful outcome. It is all about designing a process so that the consequences of an event are kept as small as possible, through the application of several protective systems, beginning with inherent safety and continuing with additional layers of protection. The reduced frequency of the consequences is shown in Figure 3.11 by the reduced size of the graphic. Note that, although the frequency of the consequence is reduced, the hazard can still occur with the same consequential severity. When performing LOPA, every independent protection layer (IPL) must be capable of preventing or mitigating the hazardous outcome, and must be completely independent of the initiating event and other IPLs (Dowell and Williams, 2005). In addition, it needs to be auditable. As a result, all IPLs are safeguards but not all safeguards are IPLs.

Effectiveness and independence of IPLs: Every IPL must be relevant to the initiating cause, with the ability to prevent the consequence or mitigate the consequence regardless of the cause (Dowell and Hendershoot, 2002). The operation of the protection layer must be completely independent from all other protection layers and the initiating event. Equipment in one protection layer cannot be shared with other protection layers. A common mistake is to declare the control system to be an IPL when the same field equipment is used for both the instrumented protection and the basic process control system. It is important to understand when a safeguard can and cannot be claimed as an IPL in LOPA. The other common issue is giving more than one credit to the same operator on response to multiple alarms for the same initiating event. If the operator is failing to respond to one alarm as he is away from the control room for some reason, he is not going to respond to other alarms within that response time itself. A very good coverage of selecting valid IPLs, and LOPA technique and applications, is presented by CCPS (2001).

Steps to determine the maximum allowable PFD_{avg} for a SIF using LOPA methodology are presented below (Frederickson, 2006).

Step 1 – Complete the LOPA without taking any credit for the SIF. First, determine the initiating events from HAZOP/What-if/FMEA study. Next, evaluate frequencies of all initiating events from company database and industry experience. Then, determine the probability that each IPL will function successfully from an industrial database. PFD_{avg} of some typical protection layers are (CCPS, 2000): BPCS control loop = 0.10; Operator's response to alarm = 0.10; Relief safety valve = 0.01 to 0.001; and vessel failure probability at maximum design pressure = 10^{-4}. Finally, compare the calculated risk with the tolerable risk target

Step 2 – SIL Requirements: If the calculated risk is less than the tolerable risk target, the instrumented protection need not be classified with any SIL rating. Otherwise, back calculate the PFD_{avg} necessary to achieve the tolerable risk target. This represents the maximum allowable PFD_{avg} for SIL rating

Step 3 – Selection of SIF: SIF of the proposed layer should be chosen for the necessary PFD_{avg} shortfall. The output from the SIL architecture is verified for correctness and consistency with respect to the products and standards provided as input

Step 4 – Repeat the Analysis: Repeat steps 1 to 3 using the revised PFD_{avg} value of the proposed instrumented modification, to recalculate the overall incident frequency meeting the risk target

Step 5 – Documentation: Summarize the findings of the LOPA study and, if appropriate, document that the proposed change meets the corporate risk tolerance criteria/local code targets. Safety Requirements Specification (SRS) is defined to undertake appropriate cost-effective designs for SISs in order to achieve optimal test and maintenance strategies for the lifespan of the plant

LOPA Exercise: A LOPA exercise for the process in Figure 3.12 is illustrated in Figure 3.13. It is performed to determine whether the risk meets the target at a hydrocarbon separator flash drum for an overpressure hazard scenario initiated by pressure sensor blockage (PC-1). Assume the target mitigated likelihood as 10^{-5} events/year. The pressure controller uses only one sensor; when it fails, the pressure is not controlled. The same pressure sensor is used for control and alarming purposes. Therefore, the alarm provides no additional protection for this initiating cause. No safety relief valve is provided in the design, and no SIS is provided for the system, as shown in LOPA in Figure 3.12 and Figure 3.13.

Apart from the failure frequency of the pressure sensor and the credit by virtue of process design, we can't take any other instrumented credits in the initial design as discussed earlier; the controller and alarm are taking a response from the same source, which has failed. Hence columns 5 to 8 in Figure 3.13 are given a factor of 1 (No credit). Since the target mitigated likelihood = 10^{-5} events/year and actual mitigated likelihood = 0.01 (by multiplying values in columns 3 to 8 in Figure 3.13), an additional safety performance of magnitude 0.001 (=10^{-5}/0.001) is necessary to meet the target. As the mitigated risk is far away from the target, we should proceed to improve the design by adding the following IPL options.

- Independent pressure tapping on the flash drum with a transmitter and alarm (PFD_{avg} = 0.1)

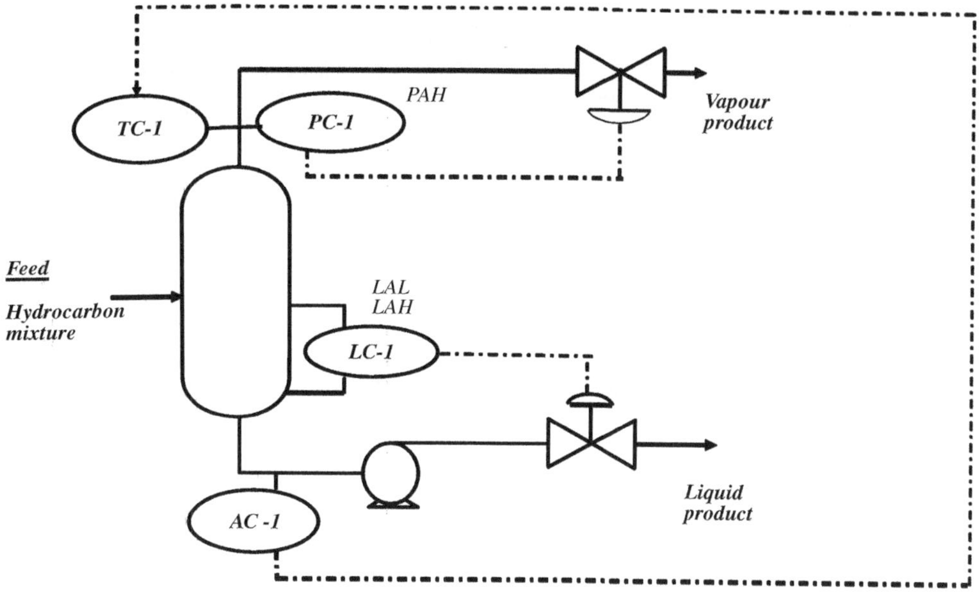

Figure 3.12 P&ID of hydrocarbon separation system.

- Pressure relief valve sized for the credible demand installed on the overhead pipe outlet of the flash drum ($PFD_{avg} = 0.01$)

LOPA is performed on the enhanced design and the solution is shown in Figure 3.14. Now, actual mitigated likelihood = 0.00001 (by multiplying columns 3 to 8 in Figure 3.14), which meets the target mitigated likelihood (10^{-5}). The hazard review document should be revisited for these changes to check any other impacts due to the changes, and those should be addressed without compromising the event likelihood target.

	1	2	3	4	5	6	7	8	9	10
				Protective layers						
#	Initial event description	Initiating cause	Cause likelihood	Process design	BPCS	Alarm	SIS	Additional mitigation (safety valves, dikes, restricted access, etc.,)	Mitigated likelihood	Notes
1	High pressure	Connection (tap) for pressure sensor P1 becomes plugged	0.1	0.1	1.0	1.0	1.0	1.0	0.01	Process sensor does not measure the drum pressure
Much higher than target! We need to improve the design.										

Figure 3.13 LOPA Exercise: Solution for the initial design.

	1	2	3	4	5	6	7	8	9	10
				Protective layers						
#	Initial event description	Initiating cause	Cause likelihood	Process design	BPCS	Alarm	SIS	Additional mitigation (safety valves, dikes, restricted access, etc.,)	Mitigated likelihood	Notes
1	High pressure	Connection (tap) for pressure sensor P1 becomes plugged	0.1	0.1	1.0	0.1	1.0	0.01	0.00001	Process sensor does not measure the drum pressure

Enhanced design includes an independent pressure sensor with alarm and also a pressure relief valve.

The enhanced design achieves the target mitigated likelihood.

Figure 3.14 LOPA Exercise: Solution for the enhanced design.

3.4.7 Safety Instrumented System (SIS) Life Cycle

The phases involved in SIS Life Cycle with their objectives, typical input and output are summarized in Table 3.10 (ANSI/ISA-84.00.01, 2004). Qualified and experienced SIL experts are essential to validate the results of SIS.

3.5 Revision of PSI and Operator Induction

Not requiring an operating procedure update was one of the vital causal factors at various industrial incidents associated with revamp projects. A case study (CCPS, 2008) illustrates the lack of operating procedure administration as a vital cause for the event of solvent mischarging in a batch plant. PSI documents need to be updated. If the changes made to the process during revamp project are significant and impact the training programme, then the operating personnel as well as the employees engaged in routine and non-routine work in the process area should be given a refresher or additional training in light of the changes. Any incident investigation recommendations, compliance audits and PHA recommendations need to be reviewed to see what impacts they may have on the process before beginning the start-up.

Ideally, we need robust SIS, inherently safer processes, equipments and people with the right level of training and safety culture to have incident-free plant operations. Operating procedures play a major role to enhance the understanding of operator-dependent safeguards in the event of process upsets. Operating procedures should be updated following a revamp [Section F - OSHA PSM 1910.119 (2010)] by addressing the following aspects as a minimum.

- Emergency shutdown procedures
- Emergency operations
- Operating limits (consequences of deviations and the steps required to avoid deviations)

Table 3.10 *Life cycle phases of safety instrumented systems (SISs).*

SIS life cycle phase	Objectives	Input	Output
Hazard and risk assessment	To determine hazards and hazardous events of the process and associated equipment, consequences, requirements for risk reduction and safety functions required to achieve the necessary risk reduction	Process design, layout, manning arrangements and safety targets	List of hazards and additional performance requirements
Allocation of safety functions to protection layers	Allocation of safety functions to protection layers and for each safety instrumented function, the associated safety integrity level	Description of the required SIFs and SIL requirements	Description of allocation of safety requirements
SIS safety requirements specification	To specify the requirements for each SIS in terms of the required SIFs and their associated safety integrity, in order to achieve the required functional safety	Description of allocation of safety requirements	SIS safety requirements and software safety requirements
SIS design and engineering	To design the SIS to meet the requirements for SIFs and safety integrity	SIS safety requirements and software safety requirements	Design of the SIS, and plan for integration test
SIS installation commissioning and validation	To integrate and test the SIS, and to validate that SIS meets in all respects the requirements for safety in terms of required SIFs and the required safety integrity	SIS design, SIS integration, test, SIS safety requirements and plan for the safety validation of SIS	Results of installation, commissioning and validation activities (including integration test results)
SIS operation and maintenance	To ensure that the functional safety of SIS is maintained during operation and maintenance	SIS requirements, SIS design and plan for SIS operation and maintenance	Results of operation and maintenance activities
SIS modification	Corrections or enhancements to SIS so that the required SIL is achieved and maintained	Revised SIS safety requirements	Results of SIS modification

(continued)

Table 3.10 (Continued)

SIS life cycle phase	Objectives	Input	Output
Decommissioning	To ensure proper review, sector organization, and ensure SIF remains appropriate	As built safety requirements and process information	SIF placed out of service
SIS verification	To test and evaluate outputs of a given phase to ensure correctness and consistency	Plan for verification of SIS for each phase	Results of verification for each phase
SIS functional safety assessment	To investigate and arrive at a judgment on the functional safety achieved by SIS	Planning for SIS functional safety assessment	Results of SIS functional safety assessment

- Safety and health considerations (precautions necessary to prevent exposure including engineering controls and administrative controls)
- Safety systems and their functions

3.6 Pre-Start-up Safety Review (PSSR)

PSSR is essential to ensure safe start-up and operation of a revamped plant. It is an organized process of inspection, review and verification to ensure that a modification to an existing plant complies with design, construction and testing standards. It is the most important element in OSHA's (Occupational Safety and Health Administration, USA) Process Safety Management (PSM) regulations. It states that PSSR is needed whenever PSI is changed. In effect, this requirement is such that virtually all changes will have to be reviewed in PSSR. These regulations require that the employer shall perform PSSR for new facilities and for modified facilities when the modification is significant enough to require a change in PSI. The revamp project manager organizes the pre-PSSR activities to determine the exact work process for PSSR, its timing and composition of the review team. The safety engineer assigned to the project reviews the status of project progress which has been called up for PSSR, based on construction progress/document accomplishments, original safety plan status, PHA open item closures and so on., and then supplies feedback to the project manager to postpone if they are not foreseen to be ready on time. As part of PSSR, safety systems are checked as to whether they are installed in accordance with engineering and design philosophies. This reconfirms the performance of hazard review, installation of safeguards listed during hazard review at site, and documentation for the changes including all vendor-supplied packages. It also confirms that a system exists to ensure that a hazard review is done if any changes that occur during commissioning and start-up require modifications to the PSI. PSSR confirms that all recommendations identified during hazard reviews, QRA and construction punch lists are completed, and that the associated safeguards are included in the hazard review documentation on site.

In order to carry out PSSR, the team will require a suitable checklist. Relevant process safety items must be addressed in the inspections by having key topics defined fully within the PSSR checklist (CCPS, 2004). A typical checklist to be completed by the PSSR team includes the following but not limited to.

- The construction and equipment have been checked for conformance with design specifications and applicable codes and standards
- Applicable process safety information has been developed or updated as necessary to reflect the new/modified process
- Plant-wide safe work procedures are adequate considering the new/modified process conditions
- Standard operating procedures and emergency procedures have been developed or updated as needed to ensure safe operation of the new/modified process
- Maintenance procedures and a preventative maintenance schedule have been developed or updated as needed to ensure safe operation of the new/modified process
- For new processes, the modification has been subjected to management of change review and all recommendations have been resolved or implemented before start-up
- Training of each employee involved in operating and/or maintaining the process has been completed and documented

PSSR is essential to ensure that the new/modified facility will perform as envisioned. The larger and more complex the project, the greater is the need to perform PSSR with care and in detail. Large revamp projects tend to fall behind schedule and, near the end, there is usually pressure to cut corners to speed up the schedule. This is also the point in time where the potential for mistakes may be greatly increased. Under these circumstances, it is even more important to step back and conduct a thorough PSSR.

3.7 Management of Change (MOC)

The MOC procedure ensures that any modification made to the process does not introduce new hazards or increase magnitude of risks associated with the existing hazards (CCPS, 2007). It is an independent systematic review and authorization by a multi-disciplined management team (usually consisting of technical, operations, maintenance, inspection, reliability, occupational safety, process safety and plant general managers) before any change is implemented in the process. The quality of the MOC procedure mainly depends on the competency of the reviewers and management's commitment. The team evaluates any modification or operating envelope changes using an MOC checklist on safety, hygiene, asset integrity, fire/explosion aspects, environmental emissions and ergonomic issues. For revamping modifications, the following are the important causes requiring MOC procedure/process, if any changes are made during the EPC stage after completion of the PHA stage (in FEED):

- Any modifications in the plant design either in process, equipment, piping, instrumentation or electrical components
- Change of plot layout
- Any changes in safety relief valve type, setting, capacity or installation

- Any changes in critical process control systems
- Any design and implementation changes in safety alarm settings, interlocks and process trips
- Any repair or replacement (except replacement in kind). It covers materials, testing, fabrication techniques and/or change of applicable codes
- Any changes in operations outside the limits specified in standard operating procedures

They are many types of MOC checklists currently in use in the industry. The following key aspects are used in developing robust MOC procedure (CCPS, 2007):

- Maintain a dependable practice
- Identify potential change situations
- Evaluate possible impacts
- Decide whether to allow the change
- Complete follow-up activities

A process change initiator fills the manual or electronic MOC forms by describing the modification/change intent and also attaching all the supporting technical documents. This form is circulated in a progressive way to the predetermined review and approval authorities. Each reviewer assesses the modification/change using his own area of expertise, and then approves the modification/change if s/he is convinced that the suggested change is safe. Otherwise, s/he will return the MOC form to the initiator, for making proper amendments to the modification scope, to make it safer or acceptable. The reviewer will also assign a specific person to address each of his recommendations. The MOC initiator will send the MOC recommendations to all assigned persons and will ensure that they are completely addressed. After completion of the required actions, the initiator will send the updated MOC form to the reviewer again. If the reviewer approves the form it will be sent to the next reviewer. The initiator needs to ensure that all recommendations by all reviewers are completely addressed. All the revised PSI needs to be attached to the MOC form. This procedure is repeated in sequential fashion if the MOC process is using paper forms, so that all the designated reviewers complete the review and approve the modification. If an online MOC process is used, reviews can be done concurrently. MOC software generates necessary communication mails and alerts assigned persons for their action, and also reviewers for their review and approval.

PSSR (conducted before the modifications are commissioned) is also a part of the MOC procedure. Any new items identified during the PSSR may require further rounds of review and approvals by the reviewers. Completed MOC is an important part of process safety management; it needs to be filed in the facility's process safety and project files, and stored for the lifetime of plant operations. These forms are required for future MOC audit and also are useful during future PHA studies. CCPS (1995) and CCPS (2007) presented MOC process and useful check lists. Note that a proper MOC procedure could have avoided the Flixborough incident outlined in Section 3.2 (CCPS, 2007).

3.8 Conclusions

This chapter presented a comprehensive overview of all process safety aspects applicable for revamp projects. It provides a quick reference for practicing engineers and also students. Note the following key points for achieving sustainable revamps.

- Proper application of hazard analysis methods and PHA (What-if, HAZOP, FMEA) in pre-FEED and FEED stages is very important to develop safer and cost-effective designs
- Plot plan, layout, area classification, pressure relief and fire safety design reviews are crucial to make revamp projects safer
- Based on any unacceptable and unmitigated risk identified during hazard analysis, further risk assessment and risk mitigation techniques need to be applied. LOPA and conceptual SIS designs based on Risk Matrix can be employed if a qualitative to semi-quantitative method is preferred. Fault tree and event tree analyses with a robust LOPA can be applied if a quantitative method is essential
- Implementation of MOC procedure after PHA study is essential to capture safe implementation of any design and procedure changes performed during the EPC stage
- Complete update of PSI documents is required for safe plant operation and also future plant revamps

Acronyms

AC	Analytical Controller
ANSI	American National Standards Institute
API	American Petroleum Institute
BLEVE	Boiling Liquid Expanding Vapour Explosion
BPCS	Basic Process Control System
CCPS	Centre for Chemical Process Safety
CIMAH	Control of Industrial Major Accident Hazards
COMAH	Control of Major Accident Hazards
CSB	Chemical Safety Board
DM	Demineralized
EPC	Engineering, Procurement and Construction
ESD	Emergency Shutdown
FEED	Front-End Engineering Design
FIC	Flow Indicator and Controller
FMEA	Failure Modes and Effects Analysis
FSL	Flow Switch Low
FT	Flow Transmitter
FV	Flow Control Valve
HAZOP	Hazards and Operability Analysis
HSE	Health and Safety Executive
IEC	International Electrotechnical Commission
IPL	Independent Protection Layer
IPS	Instrumented Protective System
KPI	Key Performance Indicator
LAH	Level Alarm High
LAL	Level Alarm Low
LC	Level Controller
LOPA	Layer of Protection Analysis
LT	Level Transmitter

MAWP	Maximum Allowable Working Pressure
MOC	Management of Change
NFPA	National Fire Protection Association
NPSH	Net Positive Suction Head
OSHA	Occupational Safety and Health Administration
PAH	Pressure Alarm High
PC	Pressure Controller
PIC	Pressure Indication and Controller
P&ID	Piping and Instrumentation Diagram
PFD	Process Flow Diagram
PFDavg	Average probability of failure on demand
PHA	Process Hazard Analysis
PRD	Pressure Relief Device
Pre-FEED	Pre-Front End Engineering Design
PRV	Pressure Relief Valve
PSI	Process Safety Information
PSM	Process Safety Management
PSSR	Pre-Start-up Safety Review
QRA	Quantitative Risk Assessment
RD	Rupture Disk
RL	Risk Level
RO	Restriction Orifices
RRF	Risk Reduction Factor
SIF	Safety Instrumented Function
SIL	Safety Integrity Level
SIS	Safety Instrumented System
SRS	Safety Requirements Specifications
TC	Temperature Controller
TIC	Temperature Indicator and Controller
TT	Temperature Transmitter
TV	Temperature Control Valve

Exercises

3.1 Perform LOPA for the scenario of 'liquid carry over to vapour product line resulting in fire and explosion downstream', for the system shown in Figure 3.12. Assume the probability of occurrence is of medium category (likelihood ~ 0.001/yr) and severity is category 2 as per Risk Matrix Table 3.2. Propose the minimum IPLs required in the design so as to meet the target mitigated likelihood below 10^{-5} event/year. PFD_{avg} of SIF arrived from Risk Matrix can be referenced to Table 3.8.

3.2 Perform event tree analysis for the scenario in Exercise 1. Capture the sequence of events that lead to the consequence by taking into account both the success and failure of every associated safety function considered.

3.3 Compare the results obtained from Exercises 1 and 2 to understand the differences between a conservative, quicker solution compared with the less conservative, quantitative approach.

References

ANSI/ISA-84.00.01 (2004) Functional Safety: Safety Instrumented Systems for the Process Industry Sector – Part 3: Guidance for the Determination of the Required Safety Integrity Levels, ISA.

API RP 14F (2008) Design, Installation, and Maintenance of Electrical Systems for Fixed and Floating Offshore Petroleum Facilities for Unclassified and Class 1, Division 1 and Division 2 Locations, 5th edn, American Petroleum Institute.

API RP 500 (2012) Recommended Practices for Classification of Locations for Electrical Installations at Petroleum Facilities Classified as Class I, Division I and Division 2, 3rd edn, American Petroleum Institute.

API RP 505 (1997) Recommended Practice for Classification of Locations for Electrical Installations at Petroleum Facilities, 1st edn, American Petroleum Institute.

API standard 520 (2014) Sizing, Selection and Installation of Pressure-Relieving Devices in Refineries, Part 1&2, 9th edn, American Petroleum Institute, July 2014.

API standard 521 (2014) Pressure-Relieving and Depressuring Systems, 6th edn, American Petroleum Institute, January 2014.

API RP 526 (2011), Instrumentation, Control, and Protective Systems for Gas Fired Heaters, 2nd edn, American Petroleum Institute, April 2011.

API RP 754 (2010) Process Safety Performance Indicators for Refining and Petrochemical Industries, 1st edn, American Petroleum Institute, April 2010.

Branan, C.R. (2005) *Rules of Thumb for Chemical Engineers, A Manual of Quick, Accurate Solutions to Everyday Process Engineering Problems*, 4th edn, Elsevier Inc.

CCPS (1993) *Guidelines for Engineering Design for Process Safety*, AIChE, New York.

CCPS (1995) *Plant Guidelines for Technical Management of Chemical Process Safety*, AIChE, New York.

CCPS (2001) *Layer of Protection Analysis: Simplified Risk Assessment*, AIChE, New York.

CCPS (2000) *Guidelines for Chemical Process Quantitative Risk Analysis*, 2nd edn, AIChE, New York.

CCPS (2003a) *Guidelines for Facilities Siting and Layout*, AIChE, New York.

CCPS (2003b) *Guidelines for Fire Protection in Chemical, Petrochemical, and Hydrocarbon Processing Facilities*, AIChE, New York.

CCPS (2004) *Guidelines in Preventing Human Error in Process Safety*, AIChE, New York.

CCPS (2007) *Guidelines for Risk Based Process Safety*, AIChE, New York.

CCPS (2008) *Guidelines for Hazard Evaluation Procedures*, 3rd edn, Wiley.

Crow, K. (2002) Failure Modes and Effects Analysis (FMEA), DRM Associates, www.npd-solutions.com.

Crowl, D.A. and Louvar, J.F. (2002) *Chemical Process Safety Fundamentals with Applications*, 2nd edn, Prentice Hall International Series.

Dave, M. (2004) *Practical HAZOPs, Trips and Alarms*, 1st edn, Elsevier, Burlington.

Dowell, A. and Hendershoot, D. (2002) Simplified Risk Analysis – Layer of Protection Analysis, AIChE National Meeting, Indianapolis, Paper 281a, Nov. 3–8, 2002.

Dowell, A. and Williams, T. (2005) Layer of protection analysis: generating scenarios automatically from HAZOP data. *Process Safety Progress*, **24** (1), 38–44.

Frederickson, A. (2006) Layer of Protection Analysis, www.safetyusersgroup.com, May 2006.

Kletz, T. (1999) *What Went Wrong? Case Histories of Process Plant Disasters*, 4th edn, Elsevier, Burlington.

Madlani, D. (2012) Flare for Design, *Hydrocarbon Engineering*, May 2012.

Mannan, S. (2005) *Lee's Loss Prevention in the Process Industries*, 3rd edn, Elsevier Butterworth–Heinemann.
NFPA 70 (2013) National Electric Code, National Fire Protection Association.
NFPA 497 (2012) Recommended Practice for the Classification of Flammable Liquids, Gases, or Vapors and of Hazardous (Classified) Locations for Electrical Installations in Chemical Process Areas, National Fire Protection Association.
Nolan, D.P. (1996) *Handbook of Fire and Explosion Protection Engineering Principles for Oil, Gas, Chemical, and Related Facilities*, Noyes Publications, New Jersey.
OSHA Process Safety Management of Highly Hazardous Chemicals 1910.119, Occupational Safety & Health Administration, Washington D.C., Jan 2010.
Parry, C. (1994) *Relief Systems Handbook, Institution of Chemical Engineers*, VCH Publishers, New York.
Pratt, T.H. (2000) *Electrostatic Ignitions of Fires and Explosions*, American Institute of Chemical Engineers, New York.
Sanders, R.E. (2005) *Chemical Process Safety – Learning from Case Studies*, 3rd edn, Elsevier Butterworth–Heinemann, Burlington, USA.
Sanders, R.E. (October 2010) Treat tanks with care. *Chemical Processing*, 29–32.

Part II

Techniques for Retrofitting and Revamping

4

Mathematical Modeling, Simulation and Optimization for Process Design

Shivom Sharma and G.P. Rangaiah
Department of Chemical & Biomolecular Engineering, National University of Singapore, Singapore

4.1 Introduction

A mathematical model (or simply model) of a chemical process is a simplified representation of physicochemical behavior of the real process. It is used to predict values of output variables for given input variables and process design (including operating) variables (such as heat exchanger area, compressor outlet pressure, reactor size and column pressure). Hence, a model can be used for what-if studies and process troubleshooting. Moreover, models find many applications for process optimization, process control and operator training. Interestingly, optimization is sometimes required and employed in model development (for example, for estimating values of unknown parameters in a model based on first principles and for developing an empirical model based on data). Models are often difficult to solve analytically, and so they are mostly solved numerically. The words: modeling and simulation are related to a model but they are not synonymous. Modeling refers to all the steps in developing and validating a model for the process, whereas simulation refers to the use of the developed model for studying the process behavior/response for one or more sets of input and design variables.

In general, modeling and simulation are used to optimize the process operation and design. Operation optimization improves the performance of a process by changing the operating conditions such as temperature, pressure and flow rate of process streams but without changing the size of any equipment or process flow sheet. Recall from Chapter 1

Chemical Process Retrofitting and Revamping: Techniques and Applications, First Edition.
Edited by Gade Pandu Rangaiah.

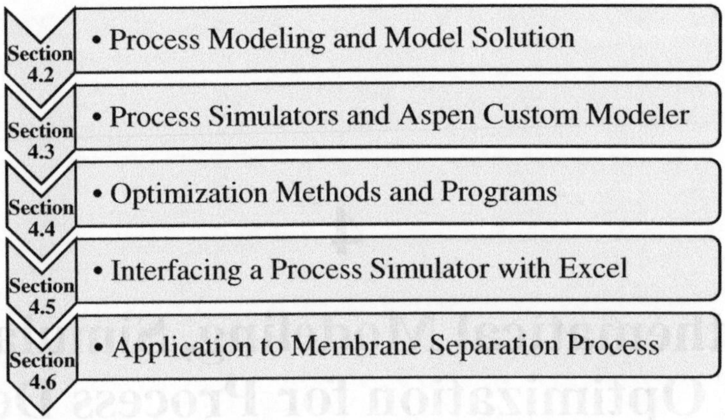

Figure 4.1 *Organization of this chapter showing modeling, simulation and optimization steps for the design of new chemical processes and for retrofitting/revamping existing processes.*

that process design can be for new processes or for retrofitting and revamping of existing processes. Process retrofitting and revamping refer to the re-design of a plant for the specified objective(s) such as increased throughput, decreased energy consumption and revised product quality. This is achieved by changes in existing equipment and/or addition of new equipment (leading to a new process configuration) besides changes in operating conditions. See Chapter 1 for further discussion on process retrofitting, revamping and debottlenecking, and Chapters 7, 8, 10 and 12 for applications of process modeling, simulation and optimization in process retrofitting and revamping.

This chapter has two main objectives. One objective is to provide an overview to modeling, simulation and optimization as well as software and tools for them. Another objective is to describe model implementation for use with a commercial simulator and interfacing of the simulator with Excel for comprehensive studies including optimization. These are illustrated using a membrane separation process for a gas mixture. As shown in Figure 4.1, this chapter covers modeling, simulation and optimization steps for design of new chemical processes and for retrofitting/revamping existing processes. Section 4.2 discusses the modeling procedure for process units/operations, and then presents modeling of membrane separation of a gas mixture as an example. The developed model for a process unit can be solved analytically or numerically.

Nowadays, several process simulators such as Aspen Plus and Aspen HYSYS are commercially available for simulating complete chemical processes. Common process units and a property database for numerous chemicals are available in such simulators. However, models for less common and/or new process units (for example, membrane separation) are not readily available in the simulators, but they may be available in the literature or can be developed from first principles. Mathematical model for a new process unit can be implemented in Aspen Custom Modeler (ACM), and then it can be exported to (included in) Aspen Plus or Aspen HYSYS for simulating processes having a new process unit besides common process units such as heat exchangers, compressors, reactors and columns. Process simulators for simulation and ACM for implementing models of new process units are described in Section 4.3. Here, implementation of a membrane separation model in ACM is discussed.

Optimization has been successfully used to obtain optimal design and operating conditions of chemical processes. Hence, optimization methods and software are briefly reviewed in Section 4.4. Single objective optimization (SOO) methods can be classified into two broad types; namely, deterministic (for example, quasi-Newton and sequential quadratic programming, SQP) and stochastic (for example, genetic algorithm, GA and differential evolution, DE). It is often desirable to consider several criteria for evaluating the performance of a process. Multi-objective optimization (MOO) methods (for example, ε-constraint, non-dominated sorting genetic algorithm-II, NSGA-II, and improved multi-objective differential evolution, I-MODE) can be used to explore the trade-off among conflicting objectives. Process simulators have some in-built optimization methods; for example, SQP is available in Aspen Plus. Generally, SOO methods available in process simulators are not global optimization methods. Global optimization methods for SOO and MOO have been implemented in Excel, MATLAB (Matrix Laborarory), C++ and so on. Hence, interfacing a process simulator with a global optimization program is required to improve the plant design and operation. In Section 4.5, interfacing of ACM/Aspen Plus/Aspen HYSYS with Excel is described using simple examples.

Acid gases (CO_2 and H_2S) have to be removed from natural gas before its transportation. Although an amine absorption process is commonly used to remove acid gases, a membrane separation process can be used as an alternative process or for retrofitting an amine absorption process (Niu and Rangaiah, 2014). Section 4.6 presents simulation of a gas separation process (consisting of a mixer, membrane units, compressors and heat exchangers) in Aspen Plus, for removing CO_2 from a mixture of methane and CO_2. The simulated process is interfaced with I-MODE program in Excel. The gas separation process is then optimized for simultaneous maximization of both CH_4 purity and CH_4 recovery. Finally, important points in process modeling, simulation and optimization for process design are summarized in Section 4.7.

4.2 Process Modeling and Model Solution

As stated earlier, process modeling refers to the development and validation of a model for the process of interest. Successful modeling provides a process model (usually, a set of equations) that can be used in a number of applications such as what-if study, troubleshooting, optimization, control and operator training. Complexity of the developed model depends on the intended use and desired accuracy of the model as well as on available knowledge, understanding and measured data of the process.

4.2.1 Process Modeling

Consider a simplified chemical process shown in Figure 4.2, which converts raw material in the feed stream into a desired product through reaction, heat exchange and separation operations. A process model for each of these operations will be in the form of $\mathbf{y} = \mathbf{g}(\mathbf{u})$ where the output variables ($\mathbf{y}$) are related to the input variables ($\mathbf{u}$) through a set of algebraic/differential equations ($\mathbf{g}$). Input variables, here, include inlet streams conditions and

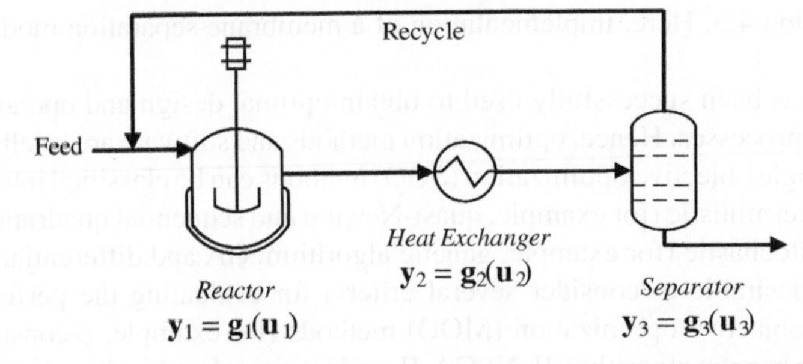

Figure 4.2 *A simple chemical process consisting of a reactor, heat exchanger and separator.*

also equipment-related data such as type and size. In the Chemical Engineering discipline, there is extensive knowledge on fluid mechanics, heat transfer, thermodynamics and separation (for example, by distillation). Hence, models for the related basic operations can be based on fundamental principles such as mass and energy balances. Such models are known as first principles (or mechanistic or phenomenological) models, which may have parameters to be determined. On the other hand, knowledge on reaction kinetics may be limited. Reactor operating data, if available, can be used to develop an empirical (or black box) model for the reactor. If some knowledge about reactions is available, then it can be used together with operating data to develop a semi-empirical model. An empirical model may not be applicable for conditions outside the range of operating data employed in developing it, whereas a first-principles model is likely to be valid for a wider range of operating conditions.

A systematic modeling approach has emerged from numerous models developed for studying chemical processes. Figure 4.3 shows the main steps involved in process modeling. A process may have several models depending on the purpose (for example, steady-state and dynamic models for optimization and control, respectively). The modeling goal significantly affects the types of mathematical equations in the process model and the required

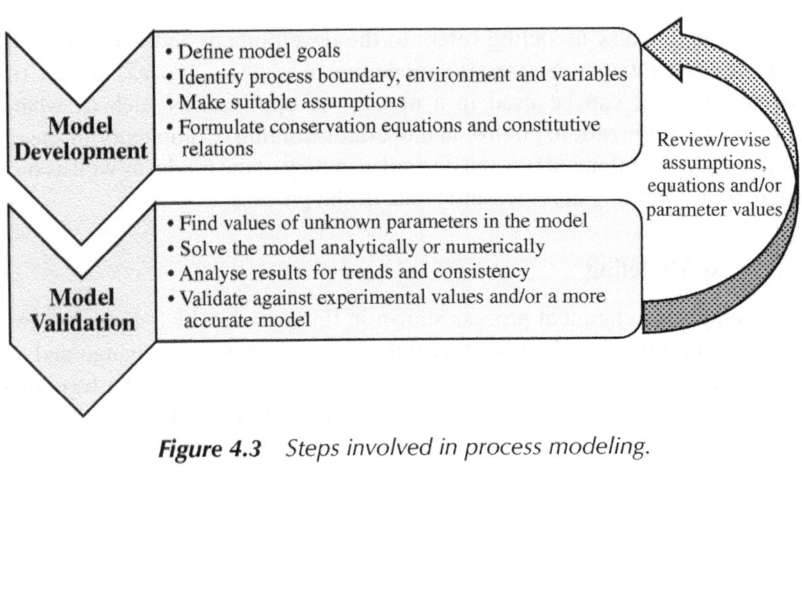

Figure 4.3 *Steps involved in process modeling.*

information to develop it. A clear statement of the physical system, its environment and boundaries is necessary for modeling. Input, state, output and disturbance variables for the process have to be identified carefully. Suitable assumptions are made to decrease the model complexity without losing its reliability and desired accuracy. Mathematical equations are formulated based on the conservation of intrinsic (for example, mass, energy, electrical charge, etc.) quantities. In addition to these balances, there may be equations for heat/mass transfer rate, reaction rate expressions, equations for equilibrium, relations between thermodynamic properties and state variables (such as temperature and pressure); all these are known as constitutive relations or equations.

After the model development, it is necessary to find values of output variables for several sets of values of input variables. If the model has unknown parameters, suitable values for them are found from the literature and/or via parameter estimation. An appropriate analytical or numerical method is selected and used for solving the model equations. The obtained results are analyzed for their trends and consistency with our understanding and knowledge about the process. Effects of the solution method should also be studied to choose the right method and/or tolerance. Finally, model predictions are validated against experimental results and/or a more accurate model. If model results, their trends and/or prediction accuracy are not acceptable, the assumptions made and parameter values used are reviewed and revised, and subsequent steps in Figure 4.3 are repeated.

Process modeling is described in detail in the books by Denn (1986) and by Hangos and Cameron (2001). In particular, Denn (1986) shows that volume balance, used in some industries, is satisfactory for many liquid systems. He highlights a common mistake in unsteady-state energy balance when heat capacity is not constant, and lists the published articles with the incorrect energy balance.

4.2.2 Model Solution

A process model usually consists of three types of equations: algebraic equations (AE), ordinary differential equations (ODE) and/or partial differential equations (PDE). AEs mainly arise in steady-state models, ODEs occur in steady-state models with spatial variation and also in dynamic models, and PDEs appear in models with variations in two or more dimensions including time. These equations have to be solved for model validation (Figure 4.3) and then in the subsequent applications of the model.

Models with linear AEs and linear ODEs can be solved analytically, whereas numerical methods are required to solve all other models. Numerical methods are iterative in nature, and convergence to the correct solution is not guaranteed. Hence, one should try different methods, tolerances and/or scenarios for the model/problem under study, and review the results obtained for consistency. Figure 4.4 summarizes types of equations in mathematical models and their possible solutions. Models for the real world problems mainly include non-linear AE, ODE and/or PDE. A single non-linear AE can be solved using bisection (interval halving), secant or Newton–Raphson method. Multiple AEs are often solved by the generalized Newton method. Non-linear ODEs can be solved using Euler, Runge–Kutta or multi-step methods. ODEs with boundary conditions (for example, temperatures at both ends of a fixed bed reactor), known as boundary value problems, can be solved by shooting, collocation or the Galerkin method. Generally, PDEs are converted into ODEs for solving using analytical or numerical methods. Hence, they are not shown separately in Figure 4.4.

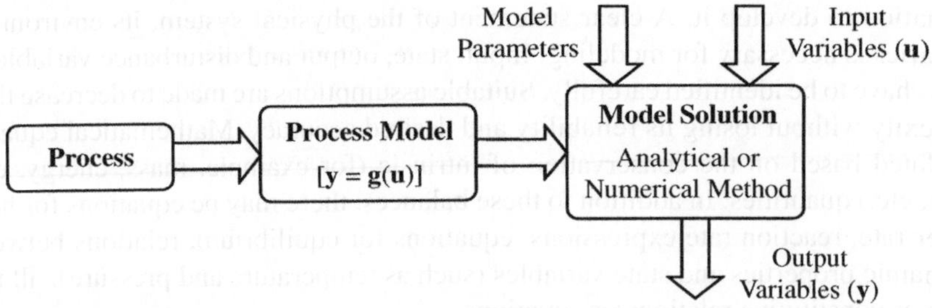

Type of Equation	Remarks on Solution
Linear Equation(s)	Analytical Solution
Quadratic Equation	Analytical Solution, Two Solutions
Non-linear Equation(s)	Numerical Solution, One or more Solutions
Linear Differential Equation(s)	Analytical Solution
Non-linear Differential Equation(s)	Numerical Solution

Figure 4.4 *Process model and its solution using analytical or numerical method.*

See Gerald and Wheatley (2003), and Rice and Do (2012) for details and applications of analytical and numerical methods for solving model equations.

A number of commercial software packages are available for solving the mathematical models. In Excel, the Goal Seek tool is useful for solving a single AE, whereas a system of linear AEs can be solved using the MINVERSE and MMULT functions. MATLAB has linsolve and fsolve commands/subprograms for solving a system of linear and non-linear AEs, respectively. fsolve includes three different numerical methods: trust region dogleg, trust region reflective and Levenberg–Marquardt methods. Further, in MATLAB, several techniques/commands are available to solve non-stiff and stiff ODEs (for example, ode45 and ode15s, respectively). The International Mathematics and Statistics Library (IMSL) in FORTRAN and C provides numerous mathematical and statistical functions for solving and analyzing models. Numerical Recipes is a series of books on programs for numerical methods in Pascal, FORTRAN, C and C++ languages (http://www.nr.com/). COMSOL Multiphysics is another software package or model development and its solution; it is mainly used for studying coupled phenomena. CFD (computational fluid dynamics) software such as ANSYS Fluent has physical modeling capabilities, especially for applications in heat transfer, turbulent flow, combustion and aerodynamics.

4.2.3 Model for Membrane Separation of a Gas Mixture

Membrane separation is finding increasing applications in process industries, and it is used for illustrating the model development step of process modeling (Figure 4.3). A membrane allows selective permeation of one or more of the components present in a mixture (for example, CH_4 and CO_2). Feed enters on one side of the membrane, some of it permeates

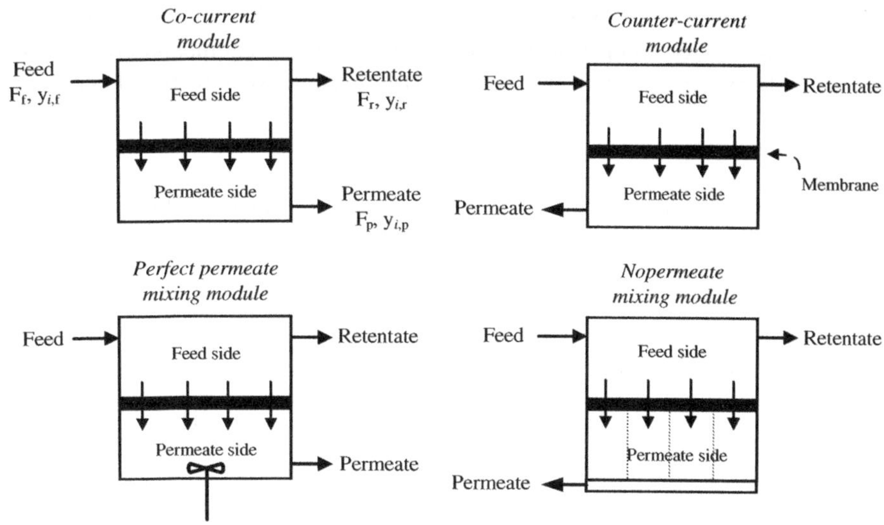

Figure 4.5 Different flow arrangements in membrane modules.

through the membrane and so the stream leaving from the other side of the membrane is called permeate (Figure 4.5); the remaining unseparated mixture on the feed side is called retentate. The extent of separation and required membrane area depend on membrane characteristics and components in the feed as well as on the flow pattern on both sides of the membrane. Several flow arrangements, such as co-current, counter-current, perfect permeate mixing and no permeate mixing are possible in the design of a membrane module (Figure 4.5). In co- and counter-current operations, feed and permeate flow in the same and opposite directions, respectively. In perfect and no permeate mixing operations, permeate has, respectively, perfect and no mixing along the membrane.

Membrane modules are based on flat, tubular, spiral and hollow fiber configurations. Flat (or plate and frame) membrane modules are used for waste water treatment and electrodialysis, whereas tubular membrane modules are used for waste water treatment, recovery processes and membrane bioreactors. Spiral wound membrane modules are used for desalination and recovery processes. Finally, hollow fiber membrane modules are generally used for water treatment and recovery processes (www.onlinembr.info/index.htm; access date: March 2015). Multi-stage membrane processes with or without recycle can be designed for improving the separation performance. See Mulder (1996) for basic principles of membrane technology, and Baker (2004) or Li *et al.* (2011) for advanced membrane technology and applications.

The goal of the model for membrane unit for gas separation is to predict the flow rate and composition of retentate and permeate streams, for a given feed stream containing n components, membrane type and area, and permeate pressure. Here, the process boundary and variables are limited to one of the membrane modules shown in Figure 4.5. In this section, the solution-diffusion mechanism is used to predict the separation behavior of the membrane. In the development of a membrane model, it is assumed that the process is at steady state, pressure is constant on feed side, and permeability of a component through the membrane is independent of pressure, which are all reasonable. Further, the temperature difference between feed and permeate sides due to the Joule–Thompson effect is assumed to be negligible, which is valid if the pressure difference between feed and permeate sides is not large.

For the assumptions made above, the governing equations for the membrane model are as follows. It is a good practice to sketch the process being modeled and write down variables of interest (for example, Figures 4.5 and 4.6) before formulating the model equations (4.1)–(4.4).

Overall Mass Balance:

$$F_f = F_p + F_r \tag{4.1}$$

Component Mass Balances:

$$F_f y_{i,f} = F_p y_{i,p} + F_r y_{i,r} \qquad [i = 1,\ 2, \ldots,\ n - 1] \tag{4.2}$$

Mole Fraction Summation for Feed, Retentate and Permeate:

$$\sum_{i=1}^{n} y_{i,f} = 1; \quad \sum_{i=1}^{n} y_{i,r} = 1; \quad \sum_{i=1}^{n} y_{i,p} = 1 \tag{4.3}$$

Rate Equations based on Membrane Permeance (Constitutive Relations):

$$F_p y_{i,p} = A_m Q_i \left(P_f y_{i,f} - P_p y_{i,p}\right) \qquad [i = 1, 2, \ldots,\ n] \tag{4.4}$$

In the above equations, F_f, F_p and F_r are molar flow rates of feed, permeate and retentate streams, respectively; $y_{i,f}$, $y_{i,p}$ and $y_{i,r}$ are mole fractions of the ith component in the feed, permeate and retentate streams, respectively; P_f and P_p are feed-side and permeate-side pressures, respectively; A_m is the membrane area and Q_i is the permeance (= permeability/effective membrane thickness) of the ith component through the membrane. Note that balance for component n is not included in Equation 4.2 to avoid redundant equations with Equation 4.1.

If the gas mixture has two components (A and B), then there are 7 equations and 14 variables in the membrane separation model (see Figure 4.6). Of these variables, F_f, P_f, $y_{A,f}$, $y_{B,f}$, Q_A and Q_B are known variables. Further, equation: $y_{A,f} + y_{B,f} = 1$ is redundant as $y_{A,f}$ and $y_{B,f}$ are known variables for the feed stream. Hence, there are 8 unknown variables (F_p, $y_{A,p}$, $y_{B,p}$, F_r, $y_{A,r}$, $y_{B,r}$, P_p and A_m) and 6 equations in this model, and there are 2 degrees of freedom in the membrane separation model for a binary mixture. For a gas mixture of n components, there are $2n + 3$ equations and $4n + 6$ (F_f, F_p, F_r, $y_{1,f}$, ... $y_{n,f}$, $y_{1,p}$, ... $y_{n,p}$,

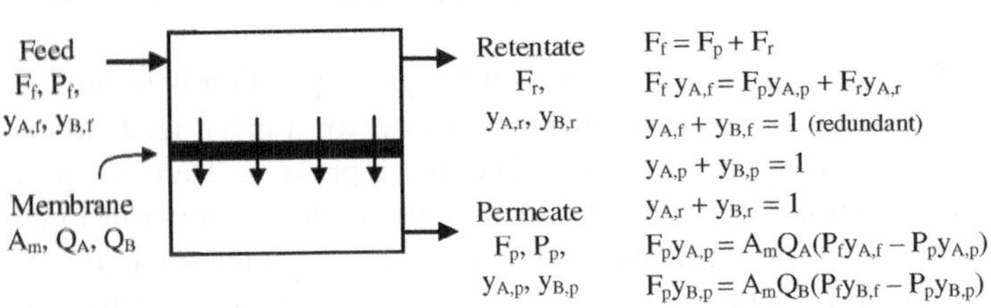

Figure 4.6 *Membrane separation model for a mixture of two components.*

$y_{1,r}, \dots y_{n,r}$, $Q_1, \dots, Q_n$, A_m, P_f and P_p) variables in the above model. Of these variables, F_f, $y_{1,f}, \dots y_{n,f}$, P_f are for the feed stream and are assumed to be given; $Q_1, \dots, Q_n$ are related to membrane-specific variables to be found from the literature. Then, the remaining number of variables is $2n + 4$ (namely, F_p, F_r, $y_{1,p}, \dots y_{n,p}$, $y_{1,r}, \dots y_{n,r}$, P_p and A_m). As $y_{1,f}, \dots y_{n,f}$ are known, mole fraction summation for feed in Equation 4.3 becomes redundant. Now, there are $2n + 2$ equations in the model, and so the model has 2 degrees of freedom; for this, the user provides values of A_m and P_p.

AE in the above model can be solved using the generalized Newton method. The model output (that is, predictions) should be verified against the experimental results to validate the model and values of $Q_1, \dots, Q_n$ used. The above membrane model is already validated in the literature, and hence model validation is not discussed here. Interested readers are referred to Chapter 10 in this book, where a gas separation model is validated, and Ahmad *et al.* (2015) for experimental validation of a hollow fiber membrane model for a gas separation system.

Model equations become numerous for complex processes such as distillation columns. One should carefully check the number of equations, number of variables, known/given variables and degrees of freedom. Equations such as mole fraction summation may be overlooked with the thinking that mole fractions should add up to unity. However, mole fraction variables can take any values during the calculations and it is necessary to ensure that sum of mole fractions of a stream is 1.0. If they are given (for example, for the feed), care should be taken to ensure mole fractions of all components in the stream sum up to unity.

4.3 Process Simulators and Aspen Custom Modeler

Process simulators such as Aspen Plus, Aspen HYSYS, PRO/II, UniSim, ProSim, Belsim Vali and CHEMCAD are available and used in both academia and industry to simulate chemical processes. They contain: (1) mathematical models for many common process units (for example, reactor, distillation column, heat exchanger, pump and compressor), (2) different numerical methods for solving process models, (3) property database for numerous chemicals, and (4) thermodynamic models for predicting properties of chemical mixtures. In order to analyze or design chemical processes, relevant components and thermodynamic models from the database are chosen, and process units in the simulator are put together according to the process flow diagram. The user needs to provide some design/operating conditions of each process unit as per its degrees of freedom. See Hamid (2013) for an introduction to process simulation using Aspen HYSYS, and Luyben (2013) on using Aspen Plus and Aspen Dynamics for design and control of distillation systems.

As mentioned earlier, some process models are not available in the simulators such as Aspen Plus. In such situations, ACM can be used to implement models available in the literature or newly developed models. Subsequently, these ACM models can be included in Aspen Plus and/or Aspen HYSYS for use like built-in models in any process. The above model for gas separation using membrane (Section 4.2.3) can be implemented and solved in ACM; see Appendix 4A for more details on the ACM model for no permeate mixing membrane module. In order to implement the membrane model in ACM, all chemicals are defined from the component list in the Aspen Properties User Interface program, and then a property simulation is run to generate the Aspen property definition file, which will be accessed by ACM. In ACM code (see Figure 4.A.3), fixed variables or inputs (for example, feed temperature, pressure, composition and membrane area) are defined, and the process parameters (for example, permeance) and variables are declared. Further, inlet and outlet streams are linked with the model inlet and outlet, respectively, using ports.

In ACM, all the model equations are solved simultaneously, and so degrees of freedom are important. The ACM code is compiled using function key F8. As described in Appendix 4A, an icon can be drawn to represent the membrane model.

This membrane model in ACM can now be used with other process models in ACM itself. In order to use it in Aspen Plus, the membrane model in ACM has to be saved as a *.msi file using the export wizard in ACM and Microsoft C++ compiler; ACM V7.3 requires use of the Professional/Premium/Ultimate edition of Microsoft Visual Studio 2008, whereas ACM V8.0-8.6 require the Professional/Premium/Ultimate edition of Microsoft Visual Studio 2010. Subsequently, the ∗.msi file can be installed in Aspen Plus and/or Aspen HYSYS library for use like built-in models in simulating chemical processes.

4.4 Optimization Methods and Programs

Optimization is used to find the values of decision variables, which minimize or maximize the value of a given objective function and also satisfy the specified constraints. Mathematically, a SOO problem can be defined as shown in Equations (4.5)–(4.8).

$$\text{Objective:} \quad \text{Min. or max. } f(\mathbf{x}) \tag{4.5}$$

$$\text{Subject to:} \quad \mathbf{x}^L \leq \mathbf{x} \leq \mathbf{x}^U \tag{4.6}$$

$$\mathbf{h}(\mathbf{x}) = 0 \tag{4.7}$$

$$\mathbf{g}(\mathbf{x}) \leq 0 \tag{4.8}$$

Here, $\mathbf{x}$ is the vector of decision variables, $\mathbf{x}^L$ and $\mathbf{x}^U$ are, respectively, vectors of lower and upper bounds on decision variables, and $\mathbf{g}$ and $\mathbf{h}$ are sets of inequality and equality constraints, respectively. An optimization problem will have different characteristics depending on:

- search space (continuous or discrete)
- objective function (convex or concave, unimodal or multimodal, differentiable or non-differentiable), and
- presence of constraints (equality or inequality, linear or non-linear)

Some problems may have multiple optima. Although a local optimum may be acceptable in some situations and industrial practice, only the global optimum is the correct solution in certain applications (for example, equilibrium calculations by free energy minimization). Hence, an optimization method should be selected carefully for solving the optimization problem based on its characteristics and need to find the global optimum.

Methods for solving optimization problems can be classified into two main types; namely, deterministic and stochastic. Deterministic methods often require derivatives of objective functions and constraints, and so they can only be applied to solve optimization problems with continuous objective functions and constraints. They generate a deterministic sequence of estimates (of the optimum) in the search space based on values of objective, constraints and their derivatives. Deterministic methods are computationally efficient and can locate the optimum exactly, but they may not be able to solve optimization problems having discontinuous and non-smooth objective function and constraints. Some examples of deterministic methods for solving non-linear optimization problems, which are often encountered in process optimization, are: quasi-Newton methods for unconstrained optimization, SQP and generalized reduced gradient (GRG) method for constrained optimization, and branch and bound (B&B) method for global optimization. Deterministic methods generally give only one optimum, which can be a local or global optimum. For more details on deterministic methods, see the books on optimization and their applications in Chemical Engineering by Floudas (2000), Edgar *et al.* (2001) and Ravindran *et al.* (2006).

The Solver tool in Excel, developed by Frontline Systems, has three optimization methods; namely, simplex linear programming (LP) for linear problems, GRG nonlinear for smooth non-linear problems and evolutionary for non-smooth problems. MATLAB has fminsearch and fmincon commands/subprograms to solve the unconstrained and constrained optimization problems, respectively. fminsearch is based on the Nelder–Mead simplex direct search method, whereas fmincon has four different methods; namely, interior-point, trust-region-reflective, SQP and active-set. The General Algebraic Modeling System (GAMS) has a large number of optimization solvers for different types of optimization problems, which can be found at http://www.gams.com/dd/docs/solvers/allsolvers.html. Finally, Numerical Recipes has several programs for minimization/maximization of functions such as linear programing, downhill simplex, conjugate gradient and direct set methods.

Stochastic methods, also known as meta-heuristic algorithms, require little or no assumptions about the optimization problem and its characteristics, and so they can be employed for solving any type of optimization problem, including those having black box models (that is, without explicit model equations and their characteristics). They can locate the global optimum with high reliability, generally require more computational effort and the solution found may not be as accurate as that found by deterministic methods. Generally, stochastic algorithms perform exploration (that is, explore the entire search space) in the initial stage of search followed by exploitation (that is, search the promising region thoroughly) in the later stage of search.

Some well-known stochastic methods for solving SOO problems are simulated annealing (SA), GAs,DE and particle swarm optimization (PSO). These were initially proposed and developed for optimization problems with bounds only [that is, unconstrained problems without Equations (4.7) and (4.8)]. Subsequently, they were extended to constrained problems by incorporating a strategy for handling constraints. One relatively simple and popular strategy is the penalty function, which involves modifying the objective function (Equation 4.5) by the addition (in the case of minimization) of a term which depends on constraint violation. For example, see Equation (4.9),

$$\text{Min. } \mathrm{F}(\mathbf{x}) = \mathrm{f}(\mathbf{x}) + \mathrm{R}\left[\sum_{i=1}^{n_{ec}} \left|\mathrm{h}_i(\mathbf{x})\right| + \sum_{j=1}^{n_{ic}} \max\left\{0, \mathrm{g}_j(\mathbf{x})\right\}\right] \tag{4.9}$$

where F is the modified objective function, R is the user-defined penalty parameter, and n_{ec} and n_{ic} are, respectively, the number of equality and inequality constraints. See Rangaiah (2010) for stochastic global optimization techniques and their applications in Chemical Engineering. More recent stochastic algorithms can be found in the book by Yang (2010).

Process optimization problems may have more than one objective [for example, min. $f_1(\mathbf{x})$ and min. $f_2(\mathbf{x})$] related to performance, cost, energy, environmental impact, safety and so on. For example, conversion and selectivity of a reactor are important and conflicting; there is trade-off between capital and operating costs; and process cost and safety can be conflicting. In such situations, it is better to perform MOO for objectives of interest. Unlike the single optimal solution from SOO, MOO gives many optimal solutions if the objectives are conflicting. These optimal solutions are known as Pareto-optimal or non-dominated or non-inferior solutions. They are equally good with respect to the objectives chosen for MOO, which means each Pareto-optimal solution is better than another in at least one objective and simultaneously worse in at least one other objective. In other words, there is some trade-off in objective function values among the Pareto-optimal solutions. Figure 4.9(a) in Section 4.6 shows the non-dominated solutions obtained for simultaneous maximization of CH_4 recovery and CH_4 purity. MOO has found numerous applications in Chemical Engineering and related areas (Bhaskar *et al.*, 2000; Rangaiah, 2009; Rangaiah and Bonilla-Petriciolet, 2013).

A number of approaches are available for solving MOO problems (Rangaiah, 2009). One common approach (known as scalarization) is to transform the MOO problem into a SOO problem, which can then be solved using established SOO methods. Solution of the resulting SOO problem gives one Pareto-optimal solution, and hence the SOO problem has to be solved many times for obtaining other Pareto-optimal solutions. Two techniques for the scalarization approach are weighted sum and ε-constraint methods.

In the weighted sum method, the original objectives [for example, min. $f_1(\mathbf{x})$ and min. $f_2(\mathbf{x})$] are combined together to form a new objective [for example, $f(\mathbf{x}) = w_1 f_1(\mathbf{x}) + w_2 f_2(\mathbf{x})$]. In this, w_1 and w_2 are suitable weights, and the sum of all weights should be equal to 1. There will be k weights in the weighted sum method for k objectives. On the other hand, the ε-constraint method optimizes the MOO problem for the most important objective, while other objectives are considered as additional constraints in the optimization problem, as shown in Equations (4.10)-(4.12).

$$\text{Objective:} \quad \text{Min. } \mathbf{f}_1(\mathbf{x}) \tag{4.10}$$

$$\text{Subject to:} \quad \mathbf{x}^L \leq \mathbf{x} \leq \mathbf{x}^U, \ \mathbf{h}(\mathbf{x}) = 0, \ \mathbf{g}(\mathbf{x}) \leq 0 \tag{4.11}$$

$$f_2(\mathbf{x}) < \varepsilon \tag{4.12}$$

For k objectives, there will be $k - 1$ additional constraints, each with a different ε variable. For obtaining one Pareto-optimal solution, the SOO problem has to be solved with suitable values for weights (in the case of weighted sum method) or ε variables (in the case of ε-constraint method). Both these methods are classical methods for solving MOO problems.

Instead of transforming a MOO problem into a SOO problem and solving it repeatedly, researchers have modified stochastic SOO methods for solving MOO problems. MOO methods such as NSGA-II, I-MODE and Multi-objective Particle Swarm optimization (MOPSO) can generate many Pareto-optimal solutions in a single run even for problems with many

Table 4.1 *Stochastic SOO and MOO programs with their source.*

Software/program (first 5 are for SOO, and others are for MOO)	Source/Institution
GA in C	IIT Kanpur
GA, SA and PSO	MATLAB
DE with taboo list in FORTRAN	NUS Singapore
DE program in several languages	www1.icsi.berkeley.edu/ ~storn/code.html#matl
Bare-Bones PSO in Excel	NUS Singapore
NSGA-II in C	IIT Kanpur
EMOO in Excel	NUS Singapore
gamultiobj	MATLAB
NGPM	MATLAB
GANetXL	University of Exeter
MULTIGEN	Université de Toulouse
SPEA2	ETH Zurich
I-MODE in Excel	NUS Singapore
jMetal in Java	University of Malaga
MOEA framework in Java	www.moeaframework.org
NIMBUS via internet	University of Jyvaskyla
ACADO Toolkit	http://acado.github.io/

objectives. They are generally population-based techniques; in each iteration/generation, they choose better solutions considering values of all objective functions (for example, by non-dominated sorting and crowding distance calculations). See Deb (2001) for details on GA and its adaptation for MOO. A number of stochastic global optimization programs are readily available for interested engineers and researchers; these are mentioned in Table 4.1.

Chemical process optimization problems often involve extensive calculations for simulating the process (for example, using a process simulator); once the process is simulated, computation of objectives and/or constraints involves a few calculations. Further, algorithms for a stochastic optimizer for SOO (for example, GA) and its modification for MOO (for example, NSGA-II) are similar in most of the steps. Hence, GA for SOO problems and NSGA-II for MOO problems are likely to take comparable computational time for solving process optimization problems with one or more objectives. If a MOO problem is converted into a SOO problem using either the weighted sum or ε-constraint method, then computational time will depend on the required number of Pareto-optimal solutions, number of objectives and the method used for solving the converted SOO problem. As stated earlier, deterministic methods are computationally efficient compared with stochastic methods for solving SOO problem, but they may give local optimum and so several trials may be required to find the global optimum. On the other hand, stochastic global optimization methods are likely to find the global optimum. Hence it is not clear which approach, that is, converting into SOO problem or modified stochastic methods, is better for solving MOO problems. If an efficient and reliable deterministic method is available, a weighted sum or ε-constraint method may be faster than stochastic global optimization methods for finding some Pareto-optimal solutions of bi-objective problems involving explicit model equations and many equality constraints. On the other hand, stochastic global optimization methods modified for MOO since around 1995 have found numerous applications in Chemical Engineering (Masuduzzaman and Rangaiah, 2009; Sharma and Rangaiah, 2013a; Rangaiah et al., 2015). They are particularly suitable for MOO problems involving process simulation using a process simulator.

4.5 Interfacing a Process Simulator with Excel

Global SOO and MOO optimization methods, particularly recent ones, have been implemented and are readily available in Excel, MATLAB, C++, FORTRAN, and so on (Table 4.1). But, they may not be available in process simulators. So, interfacing between a process simulator (for example, Aspen HYSYS, Aspen Plus or ACM) and a global optimization program (for example, I-MODE in Excel) is necessary for improving the design of chemical processes. In particular, students and practitioners are familiar with Excel, and can easily use worksheets in Excel for computations, data analysis, plotting, and so on.

Interfacing a process simulator with Excel (using Visual Basic for Applications, VBA coding) facilitates: (1) transfer of values of decision variables from an Excel worksheet to a process simulator, (2) activation or deactivation of the process simulation from Excel, and (3) transfer of selected stream and/or unit data from the process simulator to Excel worksheet (see Figure 4.7). Further, interfacing a simulator with Excel is useful in performing a simulation multiple times for design and operational changes in the process, and saving the process simulation results in Excel worksheets for subsequent analysis. The user can

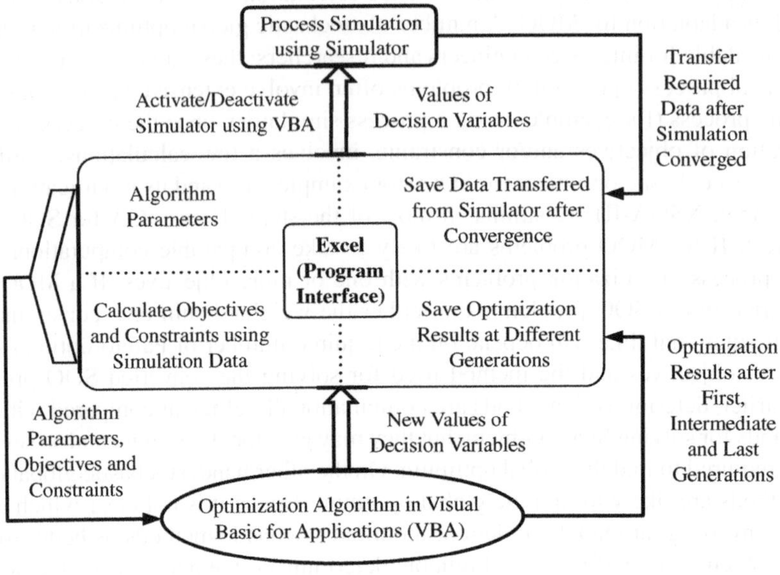

Figure 4.7 Interfacing between a process simulator and Excel (containing an optimization program).

provide the decision variables, bounds on decision variables and code objective function and constraints in Excel worksheets. The optimization algorithm is usually coded in VBA as a macro. In the current simulators, design variables such as number of stages, feed stage and side draw in distillation column are not available as decision variables for optimization by the built-in optimizer, whereas an external stochastic optimizer interfaced with a process simulator can use both discrete and continuous variables as decision variables. Interfacing of Aspen Plus and Aspen HYSYS with MATLAB can be found in Fontalvo (2014) and Khan *et al.* (2011), respectively.

Aspen Plus and Aspen HYSYS are widely used for simulating industrial chemical processes. For illustrating interfacing of Aspen Plus with Excel, a process for separating an acetone, methanol and toluene mixture using three distillation columns is selected from Aspen Plus examples, and then it is interfaced with the I-MODE program; see Appendix 4B for the process flow sheet and detailed steps for interfacing. For illustration, values of five decision variables (feed flow rate, feed heater outlet temperature, reflux ratio for shortcut column 1 and feed stage for column 2 and number of stages for column 3) are transferred from Excel to Aspen Plus. Also, values of several variables (energy input for feed heater, columns 2 and 3 reboiler duty and flow rates of all three product streams) from the converged process simulation are transferred back to Excel. In Aspen Plus, Variable Explorer can be used to identify the VBA syntax for all variables in process streams or units. Interfacing of Excel with ACM is similar to that with Aspen Plus.

Interfacing between Aspen HYSYS and Excel is described, in Appendix 4C, using the ammonia synthesis process (from Aspen HYSYS examples), which converts a mixture of nitrogen and hydrogen into ammonia. First of all, the Aspen HYSYS library is added to VBA, and then Object Browser in VBA is used to access the Aspen HYSYS library, which gives VBA syntax for all variables in process streams or units; see Appendix 4C for the process flow sheet and detailed steps for interfacing. For illustration, values of several decision variables (feed flow rate, length and diameter of each reactor, and temperatures of inlet streams to both separators) are transferred from Excel to Aspen HYSYS. After the convergence of the process simulation, values of selected variables (product flow rate and duties of both coolers) are transferred from Aspen HYSYS to Excel for analysis and further computations.

4.6 Application to Membrane Separation Process

In natural gas, methane is the main component, and other hydrocarbons such as ethane, propane and butane may be present in smaller quantities. Water, H_2S and CO_2 are also present as impurities in natural gas, and they have to be removed before pumping the natural gas through pipelines. An amine absorption process is commonly used for removing these impurities. Nowadays, membranes are being explored for natural gas processing (Baker and Lokhandwala, 2008; Ahmad *et al.*, 2012; Niu and Rangaiah, 2014). This section presents simulation and optimization of such a membrane separation process for two objectives, using Aspen Plus, ACM, I-MODE and interface between Aspen Plus and Excel. As mentioned earlier, chemical processes can be simulated in Aspen Plus. In Section 4.3, a membrane model has been implemented in ACM, which can be added to Aspen Plus.

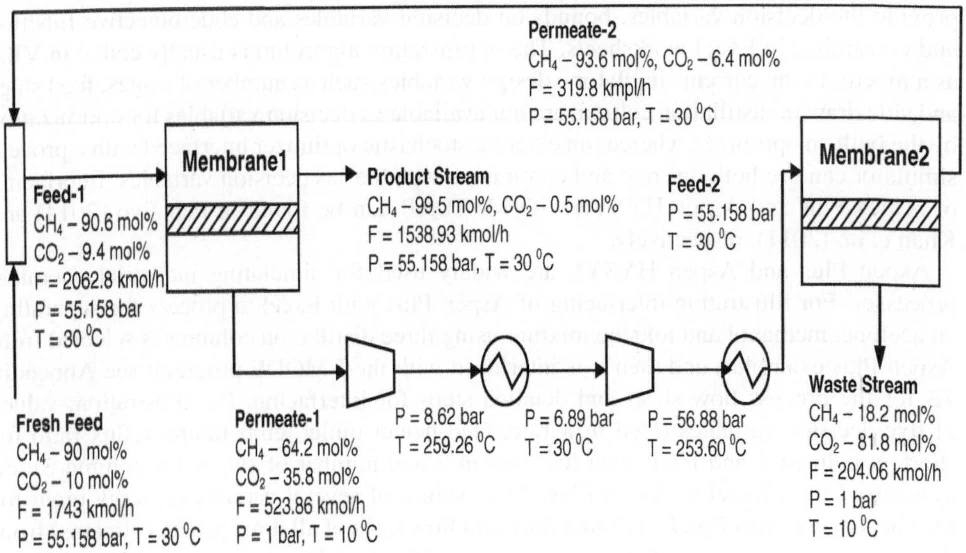

Figure 4.8 *Process flow diagram of two-stage membrane separation for CO_2 removal from natural gas; stream data correspond to the optimal solution (♦) in Figure 4.9.*

A cellulose acetate (CA) membrane selectively removes CO_2 from natural gas; it has a CO_2 permeance of 0.3 m^3(STP)/(m^2.h.bar) and a CO_2/CH_4 selectivity of 21 (Peters *et al.*, 2011). For this study, it is assumed that permeance and selectivity for the CA membrane do not change within the temperature and pressure ranges used. A process flow diagram of two-stage membrane separation for CO_2 removal from natural gas is presented in Figure 4.8. Natural gas feed has 90% (molar basis) CH_4 and the remaining component is CO_2 (Baker and Lokhandwala, 2008). Note that this is a simplified process for illustration; it becomes complex if other components/impurities are present in the natural gas. See Niu and Rangaiah (2014) for a realistic study on retrofitting the amine absorption process for the natural gas sweetening via hybridization with membrane separation.

Feed and permeate side pressures in each stage of the membrane unit are 55.158 bar and 1 bar, respectively; temperature difference between feed and permeate due to the Joule–Thomson effect is assumed to be 20 °C (Niu and Rangaiah, 2014). The non-random two-liquid model (NRTL) thermodynamic model is used for predicting properties of the gas mixture. In order to avoid very high temperatures, two-stage compression (with cooling after each stage) is used to increase the pressure of the permeate-1 stream from 1 bar to 55.158 bar. The separation process in Figure 4.8 is simulated in Aspen Plus for several design and operating conditions to check convergence of the simulation and to understand the effect of possible decision variables on objectives, before proceeding to the optimization step.

In this membrane separation process, CH_4 recovery and CH_4 purity are two important and conflicting objectives [see Equations (4.13) and (4.14)], and also they are free from fluctuating costs. The area of membrane in each stage is selected as the decision variable, and the constraint is that CH_4 mole fraction in the natural gas retentate has to be more than 0.99. The formulated MOO problem for the membrane separation process is as follows [Equations (4.13)–(4.17)].

Objective functions:

$$\text{Max. } CH_4 \text{ recovery} = \frac{\text{Molar flow rate of } CH_4 \text{ in product stream}}{\text{Molar flow rate of } CH_4 \text{ in fresh feed}} \text{ [fraction]} \quad (4.13)$$

$$\text{Max. } CH_4 \text{ purity} = \frac{\text{Molar flow rate of } CH_4 \text{ in product stream}}{\text{Molar flow rate of product stream}} \text{ [mole fraction]} \quad (4.14)$$

Subject to

$$500 \text{ m}^2 \leq \text{Area of Membrane 1} \leq 10,000 \text{ m}^2 \quad (4.15)$$

$$500 \text{ m}^2 \leq \text{Area of Membrane 2} \leq 10,000 \text{ m}^2 \quad (4.16)$$

$$CH_4 \text{ Purity} \geq 0.99 \quad (4.17)$$

The I-MODE program of Sharma and Rangaiah (2013b) has been used to optimize the gas separation process for two objectives. It has taboo list, termination criterion and self-adaptation of algorithm parameters. In this study, a taboo radius of 0.01 and population size of 100 are used, and search is terminated based on the maximum number of generations (MNG = 100). The I-MODE program in Excel is interfaced with Aspen Plus via VBA (Section 4.5).

Figure 4.9(a) shows the Pareto-optimal front obtained for maximizing CH4 recovery and CH4 purity. As expected, these objectives are conflicting in nature, and the trade-off between them is mainly due to area of membrane-2 (Figure 4.9b). Area of membrane-1 is constant and near to its upper bound, for the entire range of Pareto-optimal front (Figure 4.9b). Although CH4 recovery and CH4 purity are expected to be conflicting, the quantitative trade-off in Figure 4.9(a) gives greater understanding for selecting one of the optimal solutions. The most suitable optimal solution for the membrane separation process is the

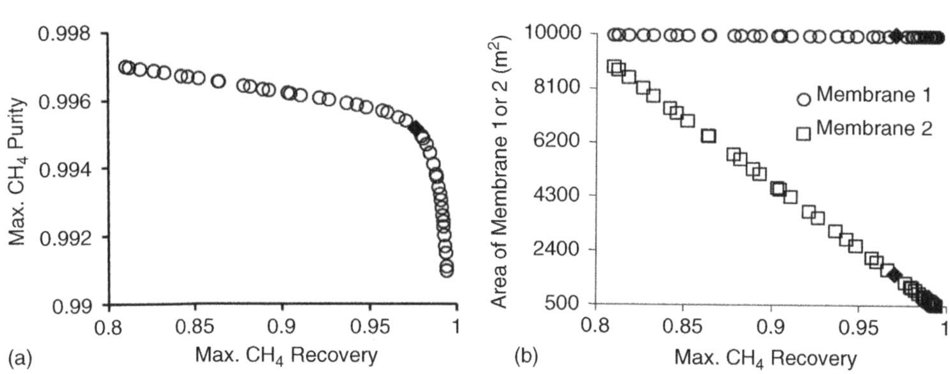

Figure 4.9 MOO results for simultaneous maximization of CH_4 recovery and CH_4 purity: Pareto-optimal front (left plot), and optimal values of decision variables (right plot).

corner point (CH4 Recovery = 0.976, CH4 purity = 0.995, Area of Membrane 1 = 9999.55 m2 and Area of Membrane 2 = 1307.58 m2) in the Pareto-optimal point. This selection is because of the significant drop in CH4 purity after the corner point, for a small increase in CH4 recovery. Alternatively, net flow or rough set method (Thibault, 2009) can be used to rank the Pareto-optimal solutions obtained and then select one of them for implementation. These ranking methods require inputs/preferences from the decision maker.

4.7 Conclusions

Process modeling is necessary to improve the understanding of process behavior. The developed model can then be used for simulating the process at various conditions and/or optimization, which play a vital role in improving the design and operation of new as well as existing chemical processes. In reality, a process is likely to have more than one objective; in such situations, MOO is valuable for generating many optimal solutions showing the quantitative trade-off among the objectives. This chapter provides an overview to process modeling, optimization and methods/programs for them. It also describes model implementation in ACM and interfacing a process simulator with Excel, which is familiar to engineers and students. Model development and its implementation in ACM are illustrated for a membrane unit separating a gas mixture. Simulation and optimization of a membrane-based separation process for two objectives demonstrate the use of Aspen Plus, I-MODE optimizer in Excel and simulator–Excel interface. The contents of this chapter are useful to both engineers and students on their exciting journey for simulating and optimizing chemical processes.

Acronyms

ACM	Aspen Custom Modeler
AE	Algebraic Equation
B&B	Branch and Bound
CA	Cellulose Acetate
CFD	Computational Fluid Dynamics
DE	Differential Evolution
EMOO	Excel-based Multi-objective Optimization
FORTRAN	Formula Translating System
GA	Genetic Algorithm
GAMS	General Algebraic Modeling System
GRG	Generalized Reduced Gradient
I-MODE	Improved Multi-objective Differential Evolution
IMSL	International Mathematics and Statistics Library
LP	Linear Programming
MATLAB	Matrix Laboratory
MOEA	Multi-objective Evolutionary Algorithm
MOO	Multi-objective Optimization

MOPSO	Multi-objective Particle Swarm Optimization
MNG	Maximum Number of Generations
NRTL	Non-Random Two-Liquid Model
NSGA-II	Non-dominated Sorting Genetic Algorithm – II
ODE	Ordinary Differential Equation
PDE	Partial Differential Equation
PSO	Particle Swarm Optimization
SA	Simulated Annealing
SOO	Single Objective Optimization
SPEA2	Strength Pareto Evolutionary Algorithm 2
SQP	Sequential Quadratic Programming
VBA	Visual Basic for Applications

Appendix 4A: Implementation of Membrane Model in ACM

The following steps describe the implementation of the membrane model developed in Section 4.2.3, in ACM. Some familiarity with Aspen Plus and ACM is required for understanding these steps.

- Open Aspen Properties User Interface program, and include the required chemical components (New – Chemical Processes – Chemicals with Metric Units; All Items – Components – Specifications – Selection). Run the simulation (using F5 key) and save it as Aspen Properties Backup file (name: NGSep.aprbkp)
- Open a new ACM file and save it with a suitable name (for example, NGSep.acmf). Click Configure Properties to open Physical Properties Configuration window, and import the NGSEP.aprbkp file. Click Default to select the components (that is, CH_4 and CO_2) from the available components; see Figure 4.A.1
- Add the new model with a suitable name (for example, NGSep). Use Model Assistant (Ctrl + M) to get the syntax of different variables for ACM code (see Figure 4.A.2).

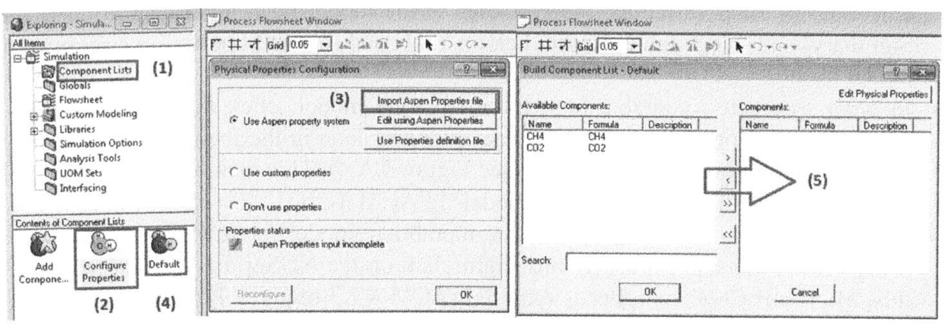

Figure 4.A.1 *Physical properties configuration and addition of components; sequence of steps to be executed are indicated by numbers in parentheses.*

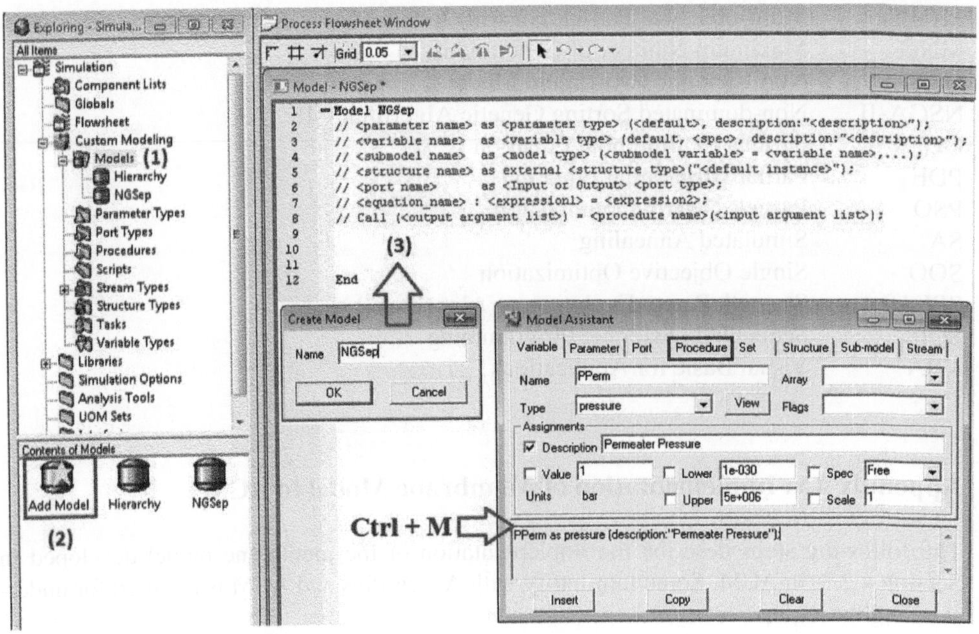

Figure 4.A.2 *Model addition and use of Model Assistant; sequence of steps to be executed are indicated by numbers in parentheses.*

In ACM, fixed variables are held constant, whereas free variables are calculated by the model. Procedure tab of Model Assistant (Figure 4.A.2) is used to compute thermodynamic properties of chemical components and their mixtures, as necessary. Copy the ACM code (containing model equations) shown in Figure 4.A.3 to NGSep Model in Figure 4.A.2. Use function key F8 to compile the ACM code

- Drag the NGSep Model to flow sheet window; this gives universal icon B1 (Figure 4.A.4). The model form provides all variables in the membrane model, which can be seen using Ctrl + D. In order to reduce the variables in the model form, open new form (name: Summary) and select Table. Right click on Table and select Properties to customize the model form
- In order to create a suitable icon for membrane model, click the Add Icon (name: MemMod); this opens Icon Editor. Create the new icon in the Icon Editor, and add the inlet and outlet ports to the new icon (see Figure 4.A.4). This membrane model in ACM can now be used with other process models in ACM itself
- For using the model in Aspen Plus, the membrane model in ACM has to be saved as a *.msi file using the export wizard (right click on the NGSep in All Items pane). For this, Microsoft C++ compiler is required (ACM V7.3 requires Professional/Premium/Ultimate edition of Microsoft Visual Studio 2008, whereas ACM V8.0-8.6 requires Professional/Premium/Ultimate edition of Microsoft Visual Studio 2010). After this, the *.msi file can be installed in Aspen Plus and/or Aspen Hysys library for use like built-in models in simulating chemical processes

```
-Model NGSep

  // Parameters and variables
    NCells                              as IntegerParameter (Description:"Number of cross flow cells", 100);
    ACell                               as Area             (Description:"Area per cross flow cell (m2)");
    FRet([0:NCells])                    as Flow_Mol         (Description:"Retentate mole flow from cell (kmol/h)");
    FPerm([1:NCells])                   as Flow_Mol         (Description:"Permeate mole flow from cell (kmol/h)");
    ZRet(ComponentList,[0:NCells])      as Molefraction     (Description:"Retentate mole fraction in cell");
    ZPerm(ComponentList,[1:NCells])     as Molefraction     (Description:"Permeate mole fraction in cell");
    PPerm                               as Pressure         (Description:"Permeate pressure (bar)", Fixed);
    A                                   as Area             (Description:"Total membrane area (m2)", Fixed);
    L(ComponentList)                    as Notype           (Description:"Permeability in m3(STP)/(m2 h bar)", Fixed);
    Lmol(ComponentList)                 as Notype           (Description:"Molar permeability (kmol/(m2 h bar))");

    Lmol  = L*1.01325 / (273.15*0.0831433);

    Inlet     as Input  MoleFractionPort;
    Retentate as Output MoleFractionPort;
    Permeate  as Output MoleFractionPort;

  // Feed Input
    FRet(0) = Inlet.F;
-   For comp in ComponentList Do
      ZRet(comp,0) = Inlet.z(comp);
    EndFor

  // Model Equations
    ACell = A/NCells;
-   For k in [1:NCells] Do
      FRet(k-1) = FRet(k) + FPerm(k);
-     For comp in ComponentList Do
        FPerm(k) *ZPerm(comp,k)  = ACell * Lmol(comp) * (Retentate.P*ZRet(comp,k) - Permeate.P*ZPerm(comp,k));
        FRet(k-1)*ZRet(comp,k-1) = FRet(k)*ZRet(comp,k) + FPerm(k)*ZPerm(comp,k);
      EndFor
      sigma (foreach (comp in componentlist) ZPerm(comp,k)) = 1;
    EndFor

  // Retentate Output
    Retentate.T = Inlet.T;
    Retentate.P = Inlet.P;
    Retentate.F = FRet(NCells);
-   For comp in ComponentList Do
      Retentate.z(comp) = ZRet(comp,NCells);
    EndFor

  // Permeate Output
    Permeate.T  = Inlet.T;
    Permeate.P  = PPerm;
    Permeate.F = Sigma(FPerm);
-   For comp in ComponentList Do
      Permeate.F * Permeate.z(comp) = Sigma(Foreach (cell in [1:NCells]) FPerm(cell)*ZPerm(comp,cell));
    EndFor

End
```

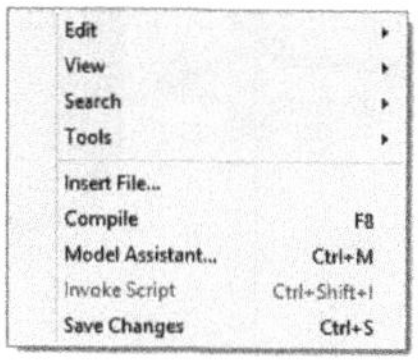

Figure 4.A.3 *ACM code for no permeate mixing membrane module; Model Assistant in the pop-up window (obtained by right click on the mouse) can be used to obtain the syntax of ACM code.*

Appendix 4B: Interfacing of Aspen Plus v8.4 with Excel 2013

The interfacing of a process simulator (for example, Aspen Plus and Aspen HYSYS) with Excel is used to simulate the process for design and operational changes. The process output can be saved in Excel for further analysis, such as preparing plots and surrogate modeling. Further, interfacing is required to optimize the performance of the process using a global optimization program developed in Excel.

To illustrate the interfacing of Aspen Plus v8.4 with Excel 2013, a process for separating an acetone, methanol and toluene mixture using three distillation columns (Figure 4.B.1) is taken from simulation examples available in Aspen Plus. In this Aspen Plus simulation, rigorous distillation columns are used. The main steps involved in this interfacing are described below. Some familiarity with Aspen Plus, Excel and VBA is required for adopting/using them. Note that sample Aspen Plus and Excel files for interfacing can be downloaded from this book's website.

- Open a new Excel file, and then create a new macro (by going through Excel – VIEW – Macros – View Macros). Give a suitable name to the new macro (for example,

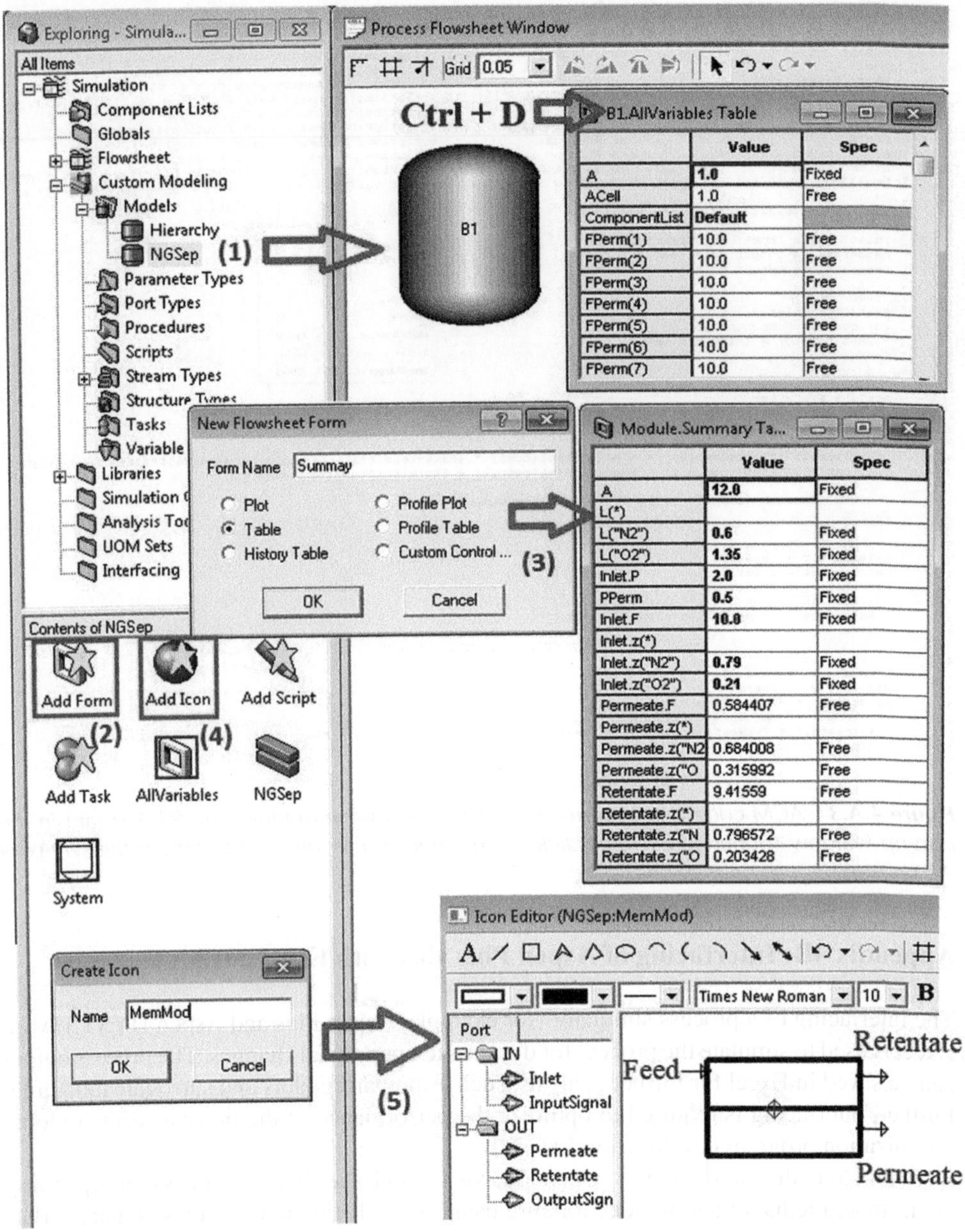

Figure 4.A.4 *Add form and new icon to the membrane model; sequence of steps to be executed are indicated by numbers in parentheses.*

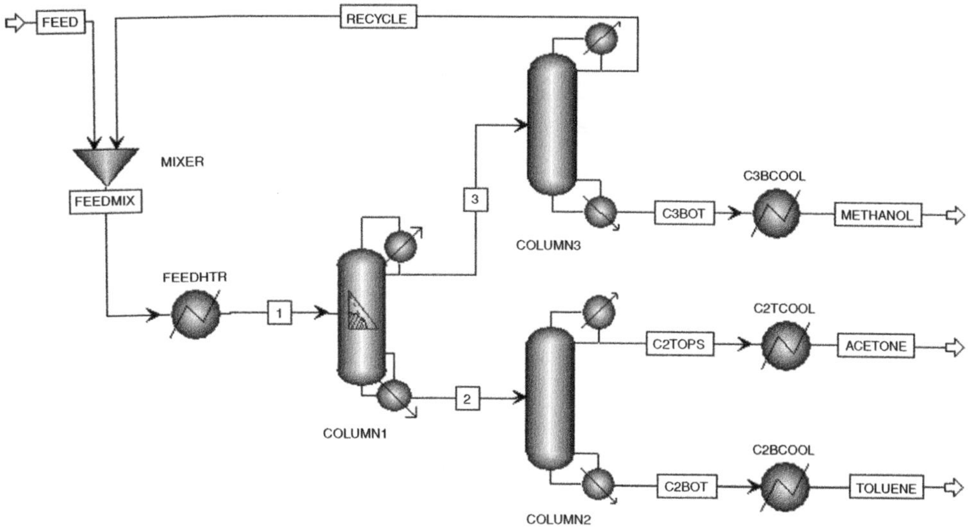

Figure 4.B.1 *Separation of an acetone, methanol and toluene mixture using three distillation columns: example in Aspen Plus V8.4.*

Distillation_Synthesis). Add Aspen Plus library (that is, Aspen Simulation Workbook – V8.4) to VBA (Excel – VBA – Tools – References). Note that the user needs to install Aspen Plus on his/her computer before adding Aspen Plus library. In Aspen Plus, Variable Explorer is used to identify the VBA syntax for all variables in process streams or units (Aspen Plus – Tools – Variable Explorer)

- As shown in Figure 4.B.2, the Aspen Plus simulation file is opened first. In VBA, Dim statement is used for declaration of variables, whereas Set statement is used for creating new objects. Then, values of feed flow rate, feed heater outlet temperature, reflux ratio for column 1 and feed stage for column 2 and number of stages for column 3 are transferred from cells C3 to C7 in Excel worksheet named 'DV to Aspen Plus', to Aspen Plus
- Aspen Plus simulation is now activated from VBA. After the convergence of process simulation, values of energy input for feed heater, columns 2 and 3 reboiler duty, and flow rates of all three product streams are transferred from the simulator to cells C3 to C8 in Excel worksheet named 'Data from Aspen Plus'. VBA code for these variables and/or data transfer is given in Figure 4.B.2
- Place Excel file with the created macro and Aspen Plus simulation file in the same directory/folder. Update the directory address and simulation file, as required, in the VBA code (see Figure 4.B.2). In Excel, the control button can be used to run the simulation; for this, create a control button by going through Excel – Developer – Insert – Form Control, and then assign the new macro (that is, Distillation_Synthesis) to the control button by a right click on it

See the last paragraph in the next appendix for some pointers for reducing the computational time and for handling non-convergence of process simulation during the optimization.

```
Sub DISTILLATION_SYNTHESIS()

'OPEN SIMULATION FILE IN ASPEN PLUS

Dim Distillation_Process As Ihapp
Set Distillation_Process = GetObject("C:\Users\cheshiv\Desktop\Interfacing\Distillation Synthesis.bkp")
Distillation_Process.Visible = True
Distillation_Process.SuppressDialogs = True
Application.DisplayAlerts = False

'DECISION VARIABLES TRANSFER FROM EXCEL TO ASPEN PLUS

Dim Feed_Flow_Rate As IHNode, Feed_HTR_Temp As IHNode
Dim COL1_RR As IHNode, COL2_NOS As IHNode, COL3_FS As IHNode

Set Feed_Flow_Rate = Distillation_Process.Tree.FindNode("\Data\Streams\FEED\Input\TOTFLOW\MIXED")
Set Feed_HTR_Temp = Distillation_Process.Tree.FindNode("\Data\Blocks\FEEDHTR\Input\TEMP")
Set COL1_RR = Distillation_Process.Tree.FindNode("\Data\Blocks\COL1\Input\MOLE_RR")
Set COL2_NOS = Distillation_Process.Tree.FindNode("\Data\Blocks\COL2\Input\FEED_STAGE\2")
Set COL3_FS = Distillation_Process.Tree.FindNode("\Data\Blocks\COL3\Input\NSTAGE")

Feed_Flow_Rate.Value = Sheets("DV to Aspen Plus").Range("C3")
Feed_HTR_Temp.Value = Sheets("DV to Aspen Plus").Range("C4")
COL1_RR.Value = Sheets("DV to Aspen Plus").Range("C5")
COL2_NOS.Value = Sheets("DV to Aspen Plus").Range("C6")
COL3_FS.Value = Sheets("DV to Aspen Plus").Range("C7")

Distillation_Process.Reinit      'REINITIALIZE ASPEN PLUS
Distillation_Process.Run         'ACTIVATE ASPEN PLUS

'DATA TRANSFER FROM ASPEN PLUS TO EXCEL

Dim Feed_Heater_Q As IHNode, Col2_RebQ As IHNode, Col3_RebQ As IHNode
Dim Methanol_Flow_Rate As IHNode, Acetone_Flow_Rate As IHNode, Toluene_Flow_Rate As IHNode

Set Feed_Heater_Q = Distillation_Process.Tree.FindNode("\Data\Blocks\FEEDHTR\Output\QNET")
Set Col2_RebQ = Distillation_Process.Tree.FindNode("\Data\Blocks\COL2\Output\REB_DUTY")
Set Col3_RebQ = Distillation_Process.Tree.FindNode("\Data\Blocks\COL3\Output\REB_DUTY")

Set Methanol_Flow_Rate = Distillation_Process.Tree.FindNode("\Data\Streams\METHANOL\Output\TOT_FLOW")
Set Acetone_Flow_Rate = Distillation_Process.Tree.FindNode("\Data\Streams\ACETONE\Output\TOT_FLOW")
Set Toluene_Flow_Rate = Distillation_Process.Tree.FindNode("\Data\Streams\TOLUENE\Output\TOT_FLOW")

Sheets("Data from Aspen Plus").Range("C3") = Feed_Heater_Q.Value
Sheets("Data from Aspen Plus").Range("C4") = Col2_RebQ.Value
Sheets("Data from Aspen Plus").Range("C5") = Col3_RebQ.Value
Sheets("Data from Aspen Plus").Range("C6") = Methanol_Flow_Rate.Value
Sheets("Data from Aspen Plus").Range("C7") = Acetone_Flow_Rate.Value
Sheets("Data from Aspen Plus").Range("C8") = Toluene_Flow_Rate.Value

End Sub
```

Figure 4.B.2 *VBA code for interfacing Aspen Plus file (for simulating the separation process in Figure 4.B.1) with Excel.*

Appendix 4C: Interfacing of Aspen HYSYS v8.4 with Excel 2013

To illustrate the interfacing between Aspen HYSYS and Excel, an ammonia synthesis process (Figure 4.C.1) is taken from simulation examples available in Aspen HYSYS. This process consists of several reactors, separators and heat exchangers/coolers. The main steps of this interfacing are shown below. Some familiarity with Aspen HYSYS, Excel and VBA is required for adopting/using them. Note that sample Aspen HYSYS and Excel files for interfacing can be downloaded from this book's website.

- Open a new Excel file, and create a new macro (name: Ammonia_Synthesis). Add Aspen HYSYS library (that is, HYSYS 8.4 Type Library for Aspen HYSYS v8.4) to VBA (by going through Excel – VBA – Tools – References). Note that the user needs to install Aspen HYSYS on his/her computer before this addition. Object Browser in VBA is used to access the Aspen HYSYS library, which gives VBA syntax for all variables in process streams or units

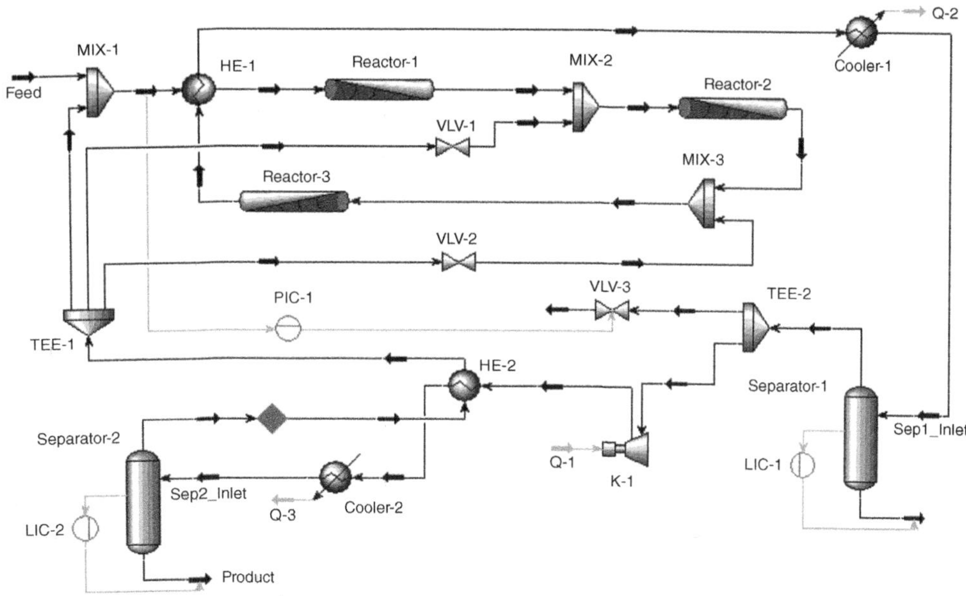

Figure 4.C.1 *Ammonia synthesis process example in Aspen HYSYS V8.4.*

- As shown in Figure 4.C.2, Aspen HYSYS simulation file is opened first. In VBA, Dim statement is used for declaration of variables, whereas Set statement is used for creating new objects. Then, values of feed flow rate, length and diameter of each reactor, and temperatures of inlet streams of both separators are transferred from cells C3 to C11 in Excel worksheet named 'DV to Aspen HYSYS', to Aspen HYSYS. To reduce computational time, remember to deactivate HYSYS solver before this transfer, and then activate it after transferring values of all decision variables to proceed with the calculations
- After the convergence of the simulation, values of product flow rate and duties of both coolers are transferred from the simulator to cells C2 to C4 in Excel worksheet named 'Data from Aspen HYSYS'. VBA code for these variables and/or data transfer is given in Figure 4.C.2; include this code in the new macro
- The Excel file with the created macro and HYSYS simulation file should be placed in the same directory/folder. In Excel, the control button can be used to run the simulation; for this, create a control button by going through Excel – Developer– Insert– Form Control, and then assign the new macro (that is, Ammonia_Synthesis) to the control button by a right click on it

In order to reduce the computation time, the Aspen Plus/Aspen HYSYS simulation has to be put on hold and then pass values of all decision variables to the simulation file. After passing values of all decision variables, process simulation should be activated. Sometimes, process simulation may require a considerable amount of time for convergence, and Excel shows a waiting message, which can be avoided by including Application.DisplayAlerts = False in the macro (as shown in Figures 4.B.2 and 4.C.2). The optimization program in

```
Sub Ammonia_Synthesis()

  ' OPEN SIMULATION FILE IN ASPEN HYSYS
  Set hyApp = CreateObject("Hysys.Application.latest")
  Currentpath = ThisWorkbook.Path
  Set hycase = hyApp.SimulationCases.Open(Currentpath & "\Ammonia Synthesis.HSC")
  hycase.Activate
  hycase.Visible = True
  Application.DisplayAlerts = False

  ' DECISION VARIABLES TRANSFER FROM EXCEL TO ASPEN HSYYS

  Dim Feed_Flow_Rate As ProcessStream
  Dim Reactor1 As PFReactor, Reactor2 As PFReactor, Reactor3 As PFReactor
  Dim Separator1_Inlet As ProcessStream, Separator2_Inlet As ProcessStream

  Set Feed_Flow_Rate = hycase.Flowsheet.MaterialStreams("Feed")
  Set Reactor1 = hycase.Flowsheet.Operations.Item("Reactor-1")
  Set Reactor2 = hycase.Flowsheet.Operations.Item("Reactor-2")
  Set Reactor3 = hycase.Flowsheet.Operations.Item("Reactor-3")
  Set Separator1_Inlet = hycase.Flowsheet.MaterialStreams("Sep1_Inlet")
  Set Separator2_Inlet = hycase.Flowsheet.MaterialStreams("Sep2_Inlet")

  hycase.Solver.CanSolve = False    'DEACTIVATE THE ASPEN HYSYS

  Feed_Flow_Rate.MolarFlow = Sheets("DV to Aspen Hysys").Range("C3")
  Reactor1.TubeLength = Sheets("DV to Aspen Hysys").Range("C4")
  Reactor1.TubeDiameter = Sheets("DV to Aspen Hysys").Range("C5")
  Reactor2.TubeLength = Sheets("DV to Aspen Hysys").Range("C6")
  Reactor2.TubeDiameter = Sheets("DV to Aspen Hysys").Range("C7")
  Reactor3.TubeLength = Sheets("DV to Aspen Hysys").Range("C8")
  Reactor3.TubeDiameter = Sheets("DV to Aspen Hysys").Range("C9")
  Separator1_Inlet.Temperature = Sheets("DV to Aspen Hysys").Range("C10")
  Separator2_Inlet.Temperature = Sheets("DV to Aspen Hysys").Range("C11")

  hycase.Solver.CanSolve = True    'ACTIVATE THE ASPEN HYSYS

  ' DATA TRANSFER FROM ASPEN HSYYS TO EXCEL

  Dim Product As ProcessStream
  Dim Cooler_1 As CoolerOp, Cooler_2 As CoolerOp

  Set Product_Stream = hycase.Flowsheet.MaterialStreams("Product")
  Set Cooler_1 = hycase.Flowsheet.Operations.Item("Cooler-1")
  Set Cooler_2 = hycase.Flowsheet.Operations.Item("Cooler-2")

  Sheets("Data from Aspen Hysys").Range("C2") = Product_Stream.MassFlow.GetValue("KG/H")
  Sheets("Data from Aspen Hysys").Range("C3") = Cooler_1.EnergyStream.HeatFlow.GetValue("KJ/H")
  Sheets("Data from Aspen Hysys").Range("C4") = Cooler_2.EnergyStream.HeatFlow.GetValue("KJ/H")

End Sub
```

Figure 4.C.2 *VBA code for interfacing Aspen HYSYS file (for simulating ammonia synthesis process in Figure 4.C.1) with Excel.*

Excel calls the simulation file for each set of decision variables, and converged simulation file from the previous call is activated automatically, instead of opening a new file for each new call.

It is possible that the process simulation fails to converge for a particular set of decision variables. In such cases, worst values should be assigned to objectives and constraints so that that particular solution will be eliminated from the population for the next generation of stochastic search. Additionally, the unconverged simulation file should be closed (using hyCase.Close), and the saved file should be opened before calling the simulation file for a new set of decision variables. See Figure 4.C.3 for VBA code for checking convergence of a recycle loop and a distillation column present in an Aspen HYSYS simulation file. A similar VBA code can be developed for checking the convergence in an Aspen Plus simulation file. If optimization stopped due to some unknown reasons (for example, Aspen HYSYS file

```
Sub Chemical_Process()
    ' OPEN SIMULATION FILE IN ASPEN HYSYS
    ' DECISION VARIABLES TRANSFER FROM EXCEL TO ASPEN HSYYS & DEACTIVATE THE ASPEN HYSYS
    ' ACTIVATE THE ASPEN HYSYS

    ' CHECK CONVERGENCE OF RECYCLE RCY-1 PRESENT in CHEMICAL PROCESS FLOWSHEET
      Dim RCY As Recycle
      Set RCY = hycase.Flowsheet.Operations.Item("RCY-1")

      If RCY1.RecycleConvergence = Converged Then
       ' DATA TRANSFER FROM ASPEN HSYYS TO EXCEL
      Else
       hycase.Close
      End If

End Sub
```

Figure 4.C.3 *Convergence check for recycle loop (RCY-1) and distillation column (COL-1) present in an Aspen HYSYS simulation file.*

closed), then an error message appears and the remaining VBA code cannot be executed. In this case, a GoTo Line'N' statement can be included to skip the remaining VBA code and continue the optimization by calling the simulation file for a new set of decision variables. This helps in continuing the optimization run for subsequent generations.

Exercises

4.1 Write all model equations for a mixture of methane, ethane, carbon dioxide and H2S, and then confirm degrees of freedom given in Section 4.2.3.

4.2 Based on the example in Section 4.6, simulate one- and three-stage natural gas separation processes in Aspen Plus. Interface the Aspen Plus simulation file with the Excel-based optimization program, and then optimize the natural gas separation for single and/or multiple objectives.

4.3 Simulate the separation of acetone, methanol and toluene process (Figure 4.B.1 in Appendix 4B) in Aspen Plus. Identify important objectives, decision variables and constraints for this separation process. Interface the simulation file with the optimization program (in Excel or MATLAB), and then optimize the separation process.

4.4 Simulate the ammonia synthesis process (Figure 4.C.1 in Appendix 4C) in Aspen HYSYS. Identify important objectives, decision variables and constraints for this ammonia process. Interface the simulation file with the optimization program (in Excel or MATLAB), and then optimize the ammonia synthesis process.

References

Ahmad, F., Lau, K.K., Lock, S.S.M., *et al.* (2015) Hollow fiber membrane model for gas separation systems: experimental validation, process simulation and module characteristics study. *Journal of Industrial and Engineering Chemistry*, **21**, 1246–1257.

Ahmad, F., Lau, K.K., Shariff, A.M., and Murshid, G. (2012) Process simulation and optimal design of membrane separation system for CO_2 capture from natural gas. *Computers & Chemical Engineering*, **36**, 119–128.
Baker, R.W. (2004) *Membrane Technology and Application*, 2nd edn, John Wiley & Sons, Chichester.
Baker, R.W. and Lokhandwala, K. (2008) Natural gas processing with membranes: an overview. *Industrial & Engineering Chemistry Research*, **47**, 2109–2121.
Bhaskar, V., Gupta, S.K., and Ray, A.K. (2000) Applications of multi-objective optimization in chemical engineering. *Reviews in Chemical Engineering*, **16**, 1–54.
Deb, K. (2001) *Multi-objective Optimization using Evolutionary Algorithms*, Wiley, Chichester.
Denn, M.M. (1986) *Process Modeling*, Longman Scientific & Technical, New York.
Edgar, T.F., Himmelblau, D.M., and Lasdon, L.S. (2001) *Optimization of Chemical Processes*, 2nd edn, McGraw Hill, New York.
Floudas, C.A. (2000) *Deterministic Global Optimization – Theory, Methods and Applications*, Kluwer Academic Publishers, Dordrecht.Fontalvo, J. (2014) Using user models in MATLAB within the aspen plus interface with an excel link. *Ingenieria E Investigacion*, **34** (2), 39–43.
Gerald, C.F. and Wheatley, P.O. (2003) *Applied Numerical Analysis*, 7th edn, Addison Wesley.
Hamid, M.K.A. (2013) *Aspen HYSYS: An Introduction to Chemical Engineering Simulation*, Lambert Academic Publishing.
Hangos, K.M. and Cameron, I. (2001) *Process Modelling and Model Analysis*, Academic Press, London.
Khan, M.S., Husnil, Y.A., Kwon, Y.S., and Lee, M. (2011) Automated Optimization of Process Plant using Particle Swarm Optimization. Proceedings of the 2011 4th *International Symposium on Advanced Control of Industrial Processes*, Thousand Islands Lake, Hangzhou.
Li, N.N., Fane, A.G., Ho, W.S.W., and Matsuura, T. (2011) *Advanced Membrane Technology and Applications*, Wiley, Hoboken, New Jersey.
Luyben, W.L. (2013) *Distillation Design and Control Using Aspen Simulation*, 2nd edn, Wiley, Hoboken, New Jersey.
Masuduzzaman and Rangaiah, G.P. (2009) Multi-objective optimization applications in chemical engineering, in *Multi-objective Optimization: Techniques and Applications in Chemical Engineering* (ed. G.P. Rangaiah), World Scientific, Singapore.
Mulder, M. (1996) *Basic Principles of Membrane Technology*, Kluwer Academic Publishers, Dordrecht, The Netherlands.
Niu, M.W. and Rangaiah, G.P. (2014) Retrofitting amine absorption process for natural gas sweetening via hybridization with membrane separation. *International Journal of Greenhouse Gas Control*, **29**, 221–230.
Peters, L., Hussain, A., Follmann, M., *et al.* (2011) CO_2 removal from natural gas by employing amine absorption and membrane technology – a technical and economical analysis. *Chemical Engineering Journal*, **172**, 952–960.
Rangaiah, G.P. (ed.) (2009) *Multi-objective Optimization: Techniques and Applications in Chemical Engineering*, World Scientific, Singapore.
Rangaiah, G.P. (ed.) (2010) *Stochastic Global Optimization: Techniques and Applications in Chemical Engineering*, World Scientific, Singapore.
Rangaiah, G.P. and Bonilla-Petriciolet (eds) (2013) *Multi-Objective Optimization in Chemical Engineering: Developments and Applications*, John Wiley & Sons, Chichester.
Rangaiah, G.P., Sharma, S., and Sreepathi, B.K. (2015) Multi-objective optimization for the design and operation of energy efficient chemical processes and power generation. *Current Opinion in Chemical Engineering*, **10**, 49–62.
Ravindran, A., Ragsdell, K.M., and Reklaitis, G.V. (2006) *Engineering Optimization Methods and Applications*, 2nd edn, Wiley, Hoboken, New Jersey.
Rice, R.G. and Do, D.D. (2012) *Applied Mathematics and Modeling for Chemical Engineers*, 2nd edn, Wiley, Hoboken, New Jersey.
Sharma, S. and Rangaiah, G.P. (2013a) Multi-objective optimization applications in chemical engineering, in *Multi-Objective Optimization in Chemical Engineering: Developments and Applications* (eds G.P. Rangaiah and A. Bonilla-Petriciolet), John Wiley & Sons, Chichester.

Sharma, S. and Rangaiah, G.P. (2013b) An improved multi-objective differential evolution with a termination criterion for optimizing chemical processes. *Computers & Chemical Engineering*, **56**, 155–173.

Thibault, J. (2009) Net flow and rough set: two methods for ranking the pareto domain, in *Multi-Objective Optimization: Techniques and Applications in Chemical Engineering* (ed. G.P. Rangaiah), World Scientific, Singapore.

Yang, X.S. (2010) *Engineering Optimization: An Introduction with Metaheuristic Applications*, Wiley, Hoboken, New Jersey.

5

Process Intensification in Process Retrofitting and Revamping

D.P. Rao

Formerly Professor of Chemical Engineering, Indian Institute of Technology, Kanpur, India
Managing Partner, Process Intensification Consultants, India

5.1 Introduction

The revamping and retrofitting in the chemical, pharmaceutical and fine-chemical industries involve the innovative use of mostly conventional equipment and processes. An emerging area, namely, process intensification, offers additional methods. This chapter deals with this emerging area and its role in retrofitting and revamping.

5.1.1 Retrofitting and Revamping

The technological developments within the life span of a plant may render some of its original equipment obsolete and may require replacement with their improved versions, or need the reorganization of process steps to stay competitive. Conforming to the new environmental regulations, the change in feedstock and energy mix, increasing energy costs and/or tighter specification of products require changes in the plant. Incremental changes may be needed in the plant from time to time, as capital expenditure would be too high for total replacement of the plant.

The required changes in a plant could be refurbishing, retrofitting, revamping or de-bottlenecking. Refurbishing involves replacing the old parts with the new to enhance

Chemical Process Retrofitting and Revamping: Techniques and Applications, First Edition.
Edited by Gade Pandu Rangaiah.

reliability in performing its original function. The words retrofitting and revamping are sometimes used synonymously. We use here 'retrofitting' to mean adding or replacing of the whole or a part of an equipment with a better alternative, and 'revamping' for reorganizing the process involving several process steps. In some cases, the production can be increased significantly but for the performance of one or two pieces of equipment that form a bottleneck for increasing the production. Replacing or modifying that equipment to remove the bottleneck is referred to as debottlenecking. The debottlenecking may involve either or both retrofitting and revamping.

5.1.2 Evolution of Chemical Industries and Process Intensification

The evolution of chemical processes and process equipment is closely related to the methods and apparatus used in the chemistry laboratory. At the early stage of evolution of chemical industries, process steps in the manufacture of a chemical mimicked the steps used in the chemistry lab in its preparation. Most of these processes were batch processes. Some of these evolved into continuous processes as the production volumes increased. Batch processes occupy the preeminent position, even today, in the pharmaceutical and fine-chemical industries. Some of the process equipment – stirred vessels, packed towers, filters, and so on – are the up-scaled versions of the apparatus used in the chemistry laboratory of yesteryear. Process intensification (PI), which represents a paradigm shift in equipment as well as in process design, takes advantage of advances in reaction engineering and transport phenomena in the design of equipment and processes (as opposed to the mere scale-up of the apparatus of the chemistry lab and mimicking the step in the laboratory preparation).

PI aims at the design of equipment and devising of processes that could bring about a dramatic reduction in equipment size or footprint. The reduction could be in the range from 10 to 100 times. Such a reduction would lead to inherently safer design and a reduction in capital and operating costs.

5.1.3 Flow Chemistry

The advent of microchannel heat exchangers, reactors and separators is paving the way for continuous flow processes even in the laboratory preparation of chemicals and for screening of new catalysts for their effectiveness. These flow processes are referred to as Flow Chemistry, which is closely related to the PI. The flow chemistry is likely to stimulate interest in continuous processes and thus may lead to retrofitting and revamping in the pharmaceutical and fine-chemical industries. Further, some of these devices are being scaled up for use in industrial applications and would be of use in retrofitting and revamping. Several review chapters and papers have appeared from time to time on various aspects of PI. The appendix at the end of this chapter lists both monographs and the latest reviews on aspects relevant to PI. These will facilitate the readers, new to PI, with a quick entry to this fast growing area. This chapter presents a comprehensive review of all aspects of PI with a focus on revamp and retrofit.

5.2 Methods of Process Intensification

Figure 5.1 shows the two methods of achieving PI. The intensification of rates leads to new equipment whereas process integration leads to new processes.

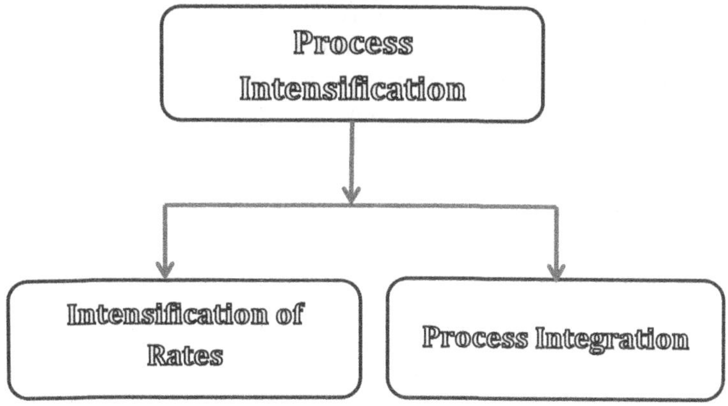

Figure 5.1 Two facets of process intensification.

5.2.1 Intensification of Rates

The sizes of heat exchangers, separators and reactors are dictated by the allowable throughputs and the rates of heat and mass transfer and reaction. These are rates may be expressed as follows:

$$\text{Heat and mass transfer rate } \alpha \text{ Specific/interfacial area}$$
$$\alpha \text{ Transfer coefficients}$$
$$\alpha \text{ Driving force}$$
$$\text{Intrinsic reaction rate} = f\text{ (Concentration of reactants, } T \text{ and } P)$$

The enhancement of one or more of the above aspects leads to intensification of rates and hence the size reduction. These aspects are presented below.

Interfacial area. The method of enhancing the specific area (surface area per unit volume) is one of the ways of achieving process intensification. The finned-tube, plate or spiral heat exchangers, structured packing for distillation and absorption columns, wash-coated monoliths and structured supports for catalytic reactors, and capillary-tube membranes, which provide high specific areas (about 500 to 5000 m^2/m^3), are in use. A recent addition to this list is a variety of meso- and micro-channel devices and metal or ceramic foams. These came to the forefront in the last decade. They offer a very high surface area for unit volume of 10^3 to 10^5 m^2/m^3, rivaling the area for oxygen transfer in the lungs (about 10^5 m^2/m^3).

Transfer coefficients. In countercurrent flow of phases, as in the case of distillation, absorption and extraction columns, the heavy phase flows down under the influence of gravity. Therefore, the allowable flow rates (limited by flooding) and the interfacial transport in these devices are controlled by Earth's gravity. To enhance them, the gravitational force could be replaced with centrifugal force as in the case of the centrifuge. We will present later the separation devices that take advantage of centrifugal fields to enhance flooding limits and interfacial heat and mass transfer.

Channel geometry. Invariably the geometry used for plug flow reactors is the circular cross section. In carrying out many reactions, intense radial mixing with plug flow in the axial direction is desired. Recently, many channel geometries have been proposed that promote mixing in the direction perpendicular to flow; these are presented later. Such a mixing enhances the heat transfer through the channel wall and mass transfer between the phases. The devices with channels of noncircular geometries are the alternatives to stirred tank reactors (STR). They facilitate conversion of a batch process into a continuous process.

High shear flows. The single and multiphase flow in narrow meso- and micro-channels formed by the rotating tube in a stationary tube or rotating disc between stationary discs leads to high shear-flow fields, which, in turn, generate a fine gas–liquid or liquid–liquid dispersion with high interfacial area between the phases. In these devices, the specific interfacial areas are as high as 2.2×10^5 m^2/m^3 against 300 to 500 m^2/m^3 in packed beds. Moreover, they facilitate tight control of the temperature of the dispersion, as the volumetric heat transfer rates are high.

Driving force. The driving force dictates the rate of transport across the phases and hence the size of the equipment. Therefore, the countercurrent flow of the phases is the preferred option over the concurrent and crosscurrent flow. When a solid phase is involved, the countercurrent flow of phases may pose practical problems like attrition and circulation of solids as in adsorption. Therefore fixed beds are used. However, PI is achieved by simulating countercurrent flow and devising suitable cycles for the thermal and pressure swing adsorption. The plug flow reactors are preferred in many cases over stirred tank reactors as the driving force for the reaction in the former is high.

Making use of the above methods, the items of equipment have been devised in the recent past to achieve PI. These are described in Section 3. They serve as alternatives to conventional separators, reactors and mixers, and are eminently suitable for retrofit and revamp. In fact, their efficacies are evaluated by retrofitting in the existing plants.

5.2.2 Process Integration

Process integration, the other facet of PI, is akin to revamp. The principles of heat integration (for example, heat exchanger network design using pinch technology) and reactive separation (that is, combining reaction and separation into a single step) had been developed even before process intensification came into vogue. They are the subset of process integration. In addition, it is also possible to combine several process steps. This kind of process integration, referred to hereafter as process-specific integration, could lead to a huge reduction in size/footprint of the plant. The principles of this type of process integration have not fully developed. Case studies have been presented later to illustrate the potential of the process-specific integration.

5.3 Alternatives to Conventional Separators

In this section, we describe the equipment devised using the PI methods to replace the conventional separators in retrofitting and revamping.

Podbielniak [1] way back in 1935 proposed a centrifugal contactor to shorten the distillation column, recognizing that the gravitational acceleration dictates its height. Several patents were granted for a number of variants of centrifugal contactors consisting of rotors with concentric cylindrical or spiral perforated shells, enclosed in a cylindrical casing. One of these patents led to the commercial exploitation for extraction. The extractor is the well known Podbielniak centrifugal extractor [2]. In 1960, Pilo and Dahlbeck [3] obtained a US patent for a rotating packed bed (RPB) for carrying out distillation, absorption and extraction. It was at about the same time that the pressure swing adsorption and divided-wall column were proposed. It took about four decades for their commercial exploitation. Pilo and Dahlbeck's patent failed to evoke much interest. Ramshaw and Mallinson [4] came up with more or less the same device and obtained a US patent. The commercial scale demonstration and the attention-catching name 'HIGEE' for rotating packed beds not only caught the imagination of the chemical engineering community but also propelled PI to the forefront.

5.3.1 Rotating Packed Beds (HIGEE)

Figure 5.2 shows a sketch of a HIGEE unit for carrying out countercontact of gas and liquid phases. The rotor is an annular packed bed. It is fixed onto a vertical shaft connected to a drive. The rotor is enclosed in a stationary casing. The rotor speed is set between 500 and 2000 RPM (rotations per minute). The shaft can be oriented in the horizontal or vertical position or at any angle without affecting its performance as the centrifugal acceleration is much higher than 100 times gravitational acceleration (100*g*).

The gas enters through the casing. It passes through the rotor, from its outer periphery to the inner periphery, and leaves from the eye of the rotor through the stationary pipe. The liquid enters through the stationary pipe fitted concentric to the gas pipe. The liquid is distributed at the inner periphery of the packing as jets or spray of droplets. A high-porosity metal foam or wire-mesh packing is used for distillation and absorption. Low-porosity packing, like those of beads, is used for reactor or adsorption applications. A HIGEE

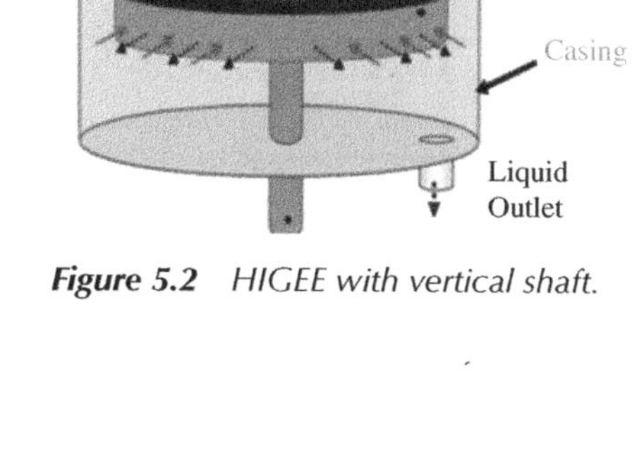

Figure 5.2 HIGEE with vertical shaft.

can replace a simple distillation or absorption column. A multi-rotor HIGEE can replace complex columns like interlinked or divided-wall columns as will be shown later. Though it has potential as an extractor, this aspect has not yet been adequately explored. A size of reduction of 20 to 100 times has been claimed in the case of distillation and absorption. What is responsible for the PI? The centrifugal force permits high throughputs of gas and liquid even with high specific area of the packing in excess of 2000 m^2/m^3, which is about 5–8 times higher than that of the packing used in conventional columns. The liquid clings to the rotating packing and flows over it as thin films and fine flying droplets under the influence of the centrifugal force. This leads to 2–6 times increase in the liquid-side coefficient.

However, the gas undergoes a solid-body-like rotation and the slip velocity between liquid and gas is in the same range as in a conventional packed bed. Hence, for the cases in which the controlling resistance is on the gas side, the expected size reduction is only about 5–8 times, which is due to the high specific area of the packing. In most distillation columns, the dominant resistance is on the vapor side. To enhance the slip velocity and hence the gas-side mass-transfer coefficient, a rotor with split packing was proposed by Chandra *et al.* [5]. Its brief description is given below.

5.3.2 HIGEE with Split Packing

Figure 5.3 shows a HIGEE with split packing. One set of alternating annular rings of packing is mounted on one side disc of the rotor and the other set of rings is mounted on the

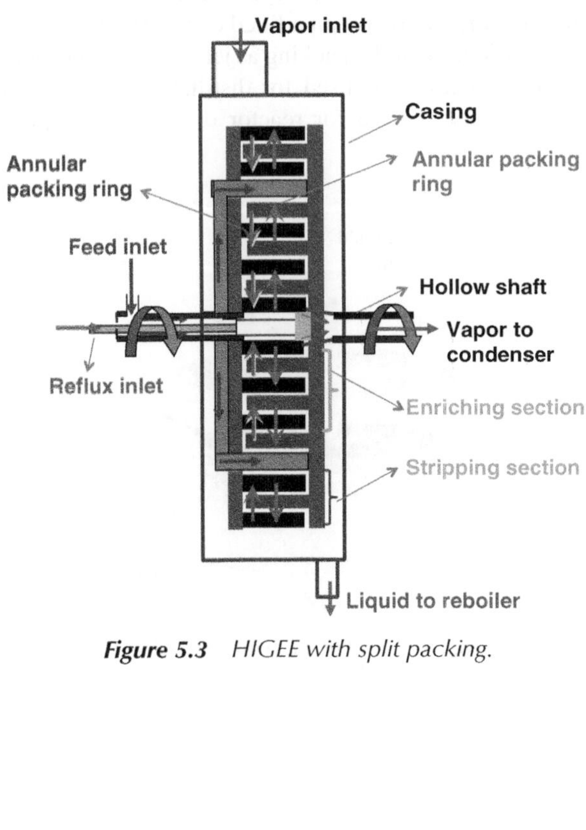

Figure 5.3 HIGEE with split packing.

other disc. The discs are attached to separate shafts.The discs can be set in co- or counter-rotation. When they rotate in counter-rotation, the tangential slip velocity between the gas and liquid is expected to be nearly double that of the tangential velocity of the outer annular ring. This leads to the enhancement of vapor-side mass-transfer coefficient. The reflux and vapor entry and their withdrawal from the bed are also shown in Figure 5.3. The feed enters somewhere along the radius as shown in the Figure. The liquid-side mass transfer is enhanced as the liquid undergoes vigorous break-up and coalescence at the entrance region of each packing ring. Further, the counter-rotation of alternate rings ensures a uniform liquid distribution even if there is a maldistribution at the inner eye. Though very high transfer rates are possible with split packing, the fabrication is complex and needs considerable ingenuity. A device, namely, zigzag HIGEE which is simple to fabricate, is given below.

5.3.3 Zigzag HIGEE

Figure 5.4 shows a zigzag HIGEE [6]. The rotor has no packing. A stationary top disc (5-stationary disc) with concentric circular fins (4-baffle) forms a part of the casing. The lower disc (1-rotational disc) too is provided with circular fins. The upper half of these fins is provided with holes or slots unlike the fins on the top disc. The lower disc is attached to a shaft driven by a motor.

Liquid enters the innermost ring on the lower disc. It acquires the tangential velocity of the lower disc and hence climbs up the fin due to the centrifugal force. A part of the liquid stream flows through the holes the jets and spray of droplets and the rest flows over the fin. These liquid jets and spray collect on the next stationary fin and flow down onto the rotating disc due to gravity, only to flow similarly on the next set of rings and so on up to the outer periphery. The vapor enters the casing and from there it goes through the zigzag path between the stationary and rotating fins. The liquid and vapor are in some sort of crosscurrent flow between the adjacent fins. The major portion of the mass transfer takes place during this crosscurrent flow of vapor and the droplets and jets of the liquid.

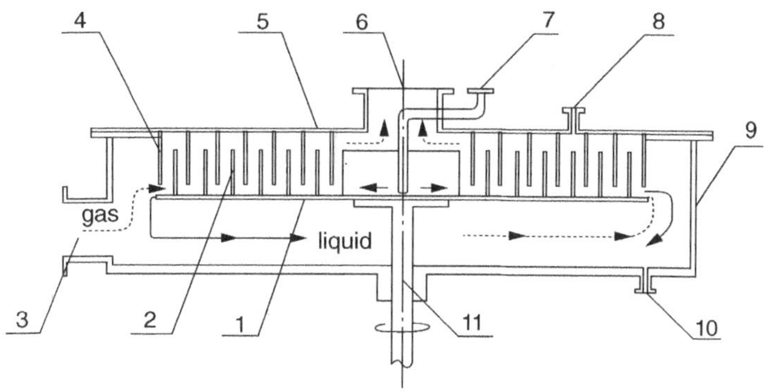

1-Rotating disc 2-Baffle 3-Gas inlet 4-Baffle 5-Stationary disc
6-Gas outlet 7-Liquid inlet 8-Intermediate feed 9-Rotor casing
10-Liquid outlet 11-Rotating shaft

Figure 5.4 *Zigzag HIGEE. Source: Reprinted from Ref [6] © 2007, with permission from Elsevier B. V.*

Since the upper disc is stationary, the vapor can be drawn off as side streams, or multiple feeds can be introduced. This facility for side streams and multiple feeds represents a major development in HIGEE technology. However, the energy requirement is expected more than that required with rotors having monoblock or split packing since the liquid is brought to a halt and accelerated again to the speed of the fin at each set of rotating fin and its adjacent stationary fin. The gas also suffers a higher pressure drop due to its zigzag path. The number of theoretical stages may be less than 15 for a 50-cm radius rotor, which is lower than the required stages for most commercial applications. Wang *et al.* [6] proposed a multi-rotor zigzag HIGEE to overcome this limitation, which is described in the following section.

5.3.4 Multi-rotor Zigzag HIGEE

Figure 5.5 shows a zigzag HIGEE with three rotors in a single casing. The liquid enters through the topmost rotor's stationary disc and flows through the liquid guide pipe onto the lower rotating disc. It flows from there as in the single–rotor form of zigzag HIGEE. The vapor flows from the lower rotor to the upper one through the space formed between the liquid guide pipe and the shaft. A close look at Figure 5.5 reveals that several multiple feeds and multiple side streams are possible. The flow of liquid and vapor streams can be configured in a multiple rotor to achieve the pump-arounds and even more complex configurations as such as in divided-wall and crude columns.

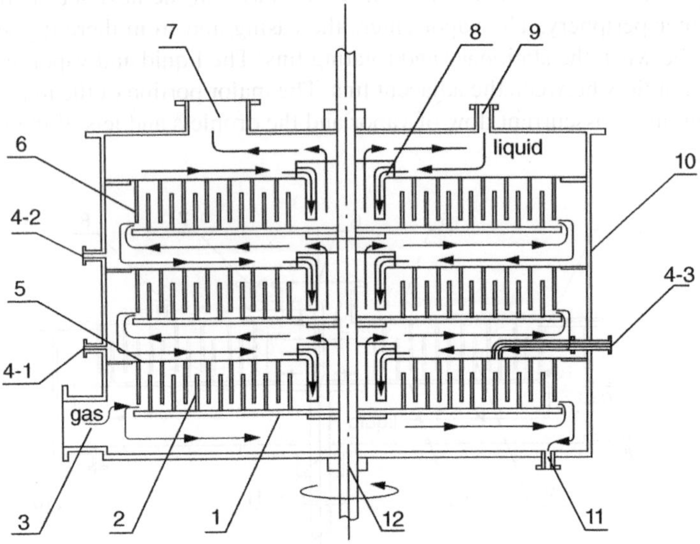

1-Rotating disc 2-Baffle 3-Gas inlet 4-1,2,3-Liquid inlets 5-Stationary disc 6-Baffle7-Gas outlet 8 -Guide pipe 9 -Reflux pipe 10-Rotor casing11-Liquid outlet 12-Rotating shaft

Figure 5.5 *Zigzag HIGEE with multiple rotors. Source: Reprinted from Ref [6] © 2007, with permission from Elsevier B. V.*

Table 5.1 *Comparison of RPB and packed column for selective absorption of H_2S (abridged Table 2 of Qian et al. [11].*

	Simple HIGEE	Packed column
Packing volume (m^3)	0.031	14
Solvent circulate rate (ton/h)	21	30
Power consumption (kW)	217	250
Steam for regeneration (ton/h)	8.5	9.5
Absorption of H_2S (as % in Feed)	~99.9	~99.9
Absorption of CO_2 (as % in Feed)	8.9	79.9

To overcome the disadvantages of crosscurrent contact and lower overall transfer coefficients, Luo *et al.* [7] replaced the rotating circular fines of zigzag HIGEE with circular fins of packing (calling it TSCC-RPB). They claim that a theoretical stage per 2 cm radial width can be obtained. Future work would focus on multi-rotor HIGEE (specially the liquid distribution in the rotors) as it can meet most of the requirements of distillation and absorption encountered in industry. Besides zigzag HIGEE, a variety of rotor packing have been proposed by Lin and Jian [8] and Sung and Chen [9]. If the trend continues, more variants of the rotor are likely to appear in the near future.

5.3.5 Applications of HIGEE for Retrofitting

In principle, any distillation column, absorber or stripper can be replaced with a HIGEE. A HIGEE with monoblock packing is here referred to as a simple HIGEE. Its earliest application was in deoxygenating seawater for use in flooding oil fields [10]. A simple HIGEE with a capacity of 100 ton/h has been in use in China. This unit has replaced bulky desorption columns on an offshore platform. Dow chemicals has built and put a simple HIGEE to commercial use for the reactive stripping in the production of HOCl [10]. Qian *et al.* [11] reported its commercial application for the selective absorption of H_2S over CO_2 from a gas stream in a petroleum refinery. Table 5.1 presents a comparison of some of the features of a simple HIGEE and the packed column it replaced.

It is a common perception that electrical power consumption is higher with HIGEE compared with the packed column that it replaces. Table 5.1 shows that electrical energy consumption and the steam requirement are less as the solvent rate is lower and the selective absorption of H_2S is much higher with the HIGEE than with the packed column.

A HIGEE with split packing is in operation for desulfurization of a gas stream in an oil refinery in Visakhapatnam, India. Another similar unit with split packing has been in use for the demonstration of its efficacy for carrying out distillation in pharmaceutical industries (www.solfirsttechnologies.com). The rotor is accommodated inside its reboiler. A unit of the same design of a larger capacity is being installed in a pharmaceutical industry for fractionation of the methanol–water system.

Wang *et al.* [12] reported that more than 40 Zigzag HIGEE units are in commercial operation in the Chinese pharmaceutical industries. The potential of HIGEE can be gauged from Table 5.2, which is an abridged version of Table 2 of Luo *et al.* [7]. The potential of HIGEE as an alternative to conventional columns appears to be impressive and deserves

Table 5.2 Comparison of performances of different HIGEEs and packed columns (abridged Table 2 of Luo et al. [7].

Equipment	ro (m)	ri (m)	Z (m)	RPM	System	NPM[a]
Packed column (height = 19.5 m)	–	–	–	–	Acetone–water	4
Simple HIGEE	0.40	–		1500–3000	Ethanol–isopropanol	50–57
Simple HIGEE	0.055	0.03	0.06	0–1830	Alcohol–water	20–136
Zigzag HIGEE	0.32	0.10	0.08	600–1400	Ethanol–water	20–38
TSCC-RPB[b]	0.18	0.073	0.05	600–1400	Acetone–water	24–58

[a] NPM: Number of theoretical plates per metre of packing.
[b] TSCC-RPB HIGEE reported by Luo et al. [7].

serious consideration in retrofit and revamp. The estimated number of theoretical plates per metre of packing for the HIGEE with split packing ranged from 77 to 95 [13]. However, the methods of eliminating shaft seals using magnetic drives or locating the motors inside the casing make HIGEE an attractive option where fire hazards are a major concern.

HIGEE has been looked upon as a unit for distillation and absorption. However, Chen's group at Beijing University of Chemical Technology has explored its efficacy as a reactor in improving yield and selectivity, and for the production of nanoparticles (by reactive precipitation) and in making polymers with tighter molecular mass distribution [14]. These involve fast reactions in which micromixing and short residence time play a pivotal role. The HIGEE offers excellent micromixing and affords short residence time in processing gas–liquid, liquid–liquid and gas–liquid–solid phases [14]. They claimed that the production of nanoparticles of $CaCO_3$ by HIGEE in five enterprises has reached a total capacity of 36 000 T/y. The use of HIGEE for the production of polymers and active pharmaceutical ingredients (API) has also been reported [14]. The HIGEE as an alternative to trickle bed reactor has been evaluated for solid–liquid–gas reactions where mass transfer is controlling the reaction rate [15]. A size reduction of about 25- to 50-fold appears to be feasible.

5.3.6 Podbielniak Centrifugal Extractor

Podbielniak [2] devised a centrifugal liquid–liquid extractor. It is known as a Podbielniak extractor (POD, for short) and is perhaps the first commercial exploitation of a centrifugal field for separation. Figure 5.6 shows a cutaway view of the extractor. The rotor consists of perforated concentric circular shells as the packing. The heavy liquid enters at the center of the packing (HLI in Figure 5.6) and flows to the outer periphery under the influence of centrifugal force. The light liquid fed into the casing of the rotor (LLI) flows into the center under the influence of a pressure gradient. These two immiscible liquids are in countercurrent flow to each other. The body force – the density difference (of the phases) times, g, the gravitational acceleration – dictates the flooding velocity and hence allowable throughputs in an extractor. The centrifugal acceleration takes the place of the gravitational acceleration in a centrifugal field. A good solvent with a density close to that of the feed cannot be used with a conventional extractor as the throughputs would be too low (due to flooding) to be of any practical value. Such a solvent could be used with a centrifugal extractor since the centrifugal acceleration can be set in excess of 100g. Four to five

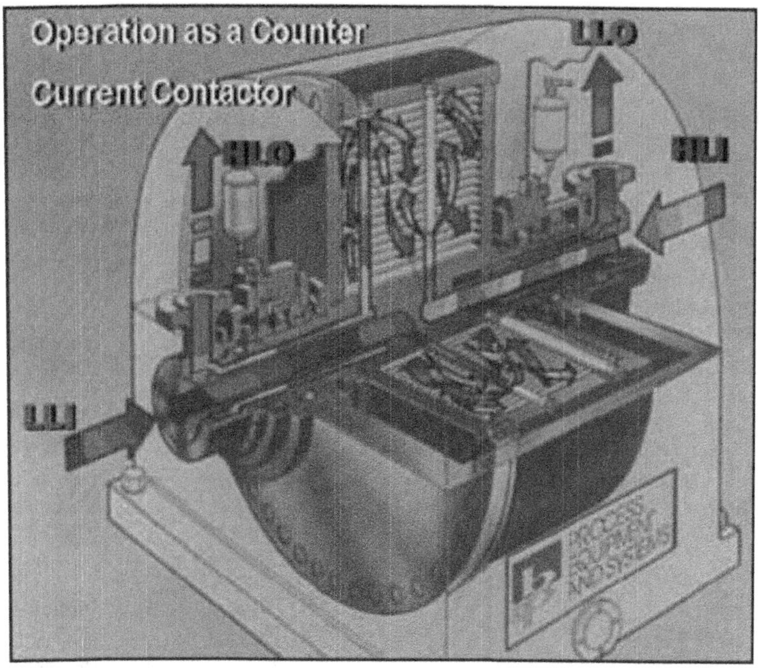

Figure 5.6 Podbielniak centrifugal extractor. Source: With permission from B&P Process Equipment and Systems LLC.

theoretical stages can be realized in a single unit. A large number of units are in service in Chemical, pharmaceutical and fine-chemical manufacture [16].

Comparison of HIGEE and POD. Studies on the extraction in a HIGEE have not been reported. If the heavy liquid is made to flow on the packing of HIGEE as films, the interfacial area would be higher than in a POD. Further, the radial dispersion in a HIGEE is likely to be negligible whereas it is substantial in a POD. Therefore, the performance of the HIGEE could be higher than that of the POD. There is a need to study the application of HIGEE for liquid–liquid extraction.

5.3.7 Annular Centrifugal Extractor

Figure 5.7 shows an annular centrifugal extractor. The extractor consists of a cylinder rotating in a stationary cylindrical casing. The feed and solvent are fed into the annular spacing. Owing to the high shear rate in the annular region, a dispersion of two immiscible liquids is formed. It flows into the rotating cylinder through its bottom opening. Inside the rotating cylinder, the dispersion separates dues to the centrifugal force as it rises through the rotating cylinder. The heavy liquid collects at the outer wall as a thick layer and is drawn from the top of the cylinder. The lighter liquid flows over the layer of heavy liquid as another thick layer. The two layers are drawn separately as shown in the Figure. Vedantam and Joshi [17] presented a literature review of this extractor.

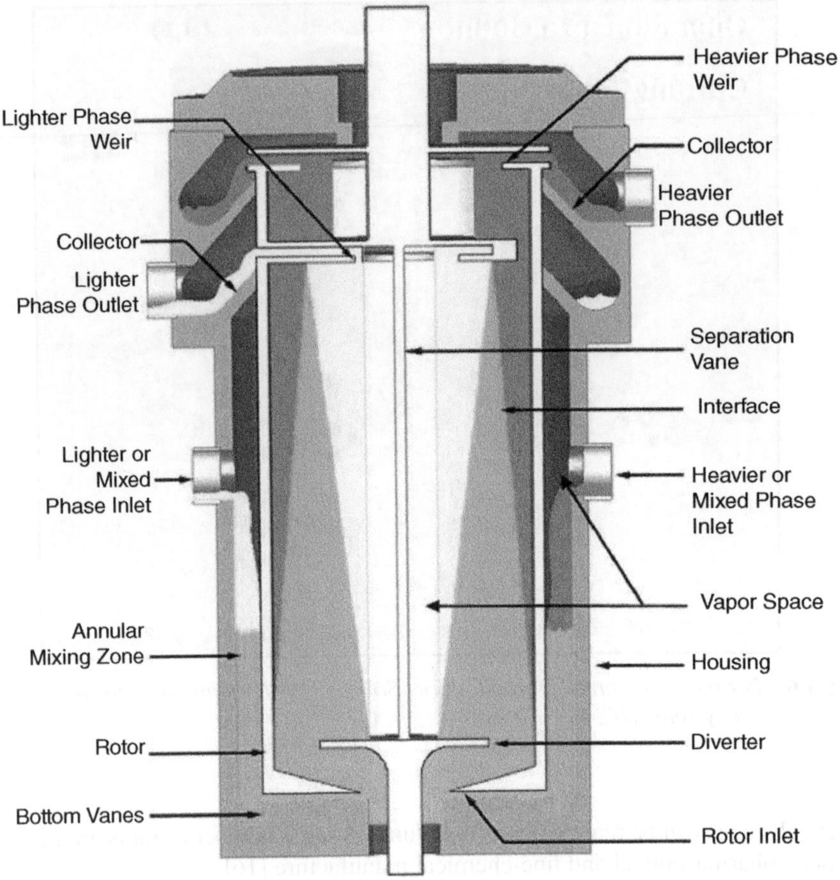

Figure 5.7 *Annular centrifugal extractor. Source: Courtesy of CINC Deutschland GmbH & Co. KG.*

The holdups of the two phases are very small, which is advantageous in certain applications. As the phases are in the co-current flow, they can attain equilibrium. Therefore, the extractor is at best one equilibrium stage. However, multiple units can be configured to form a multi-stage countercurrent extraction unit. There is a need to evaluate the relative merits of the multiple units' configuration visàvis the Podbielniak extractor, which offers 4 to 6 stages in a single unit.

5.3.8 Adsorbers

The 'bulk separation' of mixtures (in contrast to the removal of trace components) into high-purity products by adsorption requires a countercurrent flow of phases as in other equilibrium-based separations. The moving-bed adsorber, which offers the countercurrent flow of the phases, was used for the separation of a gaseous mixture of hydrocarbons using activated carbon as adsorbent.It is known as a hypersorber. Though it performed well, it was abandoned after preliminary efforts as the circulation of adsorbent led to its loss due to attrition, and thus became economically unviable [18].

The separation of liquid mixtures of isomers like those of xylenes by distillation and extraction is difficult as the separation factor is close to unity. UOP Inc. (now a part of Honeywell) devised a simulated moving-bed (SMB) adsorber for the bulk separation of liquid mixtures using chromatographic processes. These are known as SORBEX™ processes. In an SMB, the adsorbent is held in a fixed bed, which is provided with inlet or outlet streams at equal intervals along the bed. Rotary valves switch the inlet and output streams to the bed to simulate the discrete movement of the bed in the opposite direction to the liquid flow. SMBs have been in operation for over 4 decades for the separation of various liquid mixtures [19]. The SORBEX process ranks high among the outstanding chemical engineering achievements. These innovative processes represent the early developments in field of process intensification.

So far SMBs have been used for the separation of liquid mixtures, in which regeneration of adsorbent is done by elution. Efforts are being made to extend the use SMB for thermal- and pressure-swing processes [20]. A simulated moving-bed heat exchanger has been proposed for regenerative heat exchange using a fixed bed of solids as an alternative to bulky blast furnace stoves. A significant size reduction appears to be possible [21].Pressure-swing adsorption (PSA) is used for the drying of air, recovery of hydrogen from refinery off-gases, production of oxygen or nitrogen from air, and upgrading lean natural and landfill gases. These are cyclic processes and two or more fixed beds are used. In most of the PSA cycles the adsorption step is the limiting step. At the beginning of the step, a mass-transfer zone (MTZ) forms and moves through the bed. Figure 5.8 shows the mass-transfer zone in a bed. The separation takes place within the MTZ, which is a small fraction of the bed. One either side of the MTZ the two phases are in equilibrium and hence the driving force on either side of the MTZ is zero. This leads to a very low 'average' driving force in the MTZ, and therefore the adsorbent requirement is large. Furthermore, most of the PSA processes yield only one product of high purity and the other a 'waste product'. Thermal-swing regeneration is preferable over pressure swing because the energy cost would be less. The former is particularly preferable where large volumes are involved, as in CO_2 capture from flue gases and the upgrading of lean natural gas. This scenario offers vast scope for process intensification; that is, to reduce the size of the beds, to obtain both products of high purity, and to reduce energy cost by switching from pressure swing to thermal swing. An outline of a few methods of process intensification is given below.

Duplex PSA [22], also called Dual reflux PSA [23], and its modified version [20] have been proposed to obtain both products of high purity. It involves only two beds and requires a moderate vacuum. PSA units consisting of 3 beds and 4 beds have been proposed which mimic the moving-bed adsorber to obtain high-purity products and to achieve a high average driving force. The existing units can be retrofitted to achieve process intensification [23]. Developments in the synthesis of adsorbents, like tailoring the openings of molecular

Figure 5.8 *Mass transfer zone in a fixed bed.*

sieves, grafting of solvents onto adsorbents, and structured and monolith adsorbents [24] provide opportunities for retrofitting and revamping. Furthermore, the PSA units too can be retrofitted or revamped to modify their cycles to improve their performance by inserting and removing the valves used in switching of the streams.

5.4 Alternatives to Stirred Tank Reactor (STR)

The stirred tank with a jacket for heating or cooling (also called autoclave or digester) is the workhorse of the pharmaceutical, fine-chemicals, mineral and paper industries. It is used as a reactor, mixer, decanter, heater and cooler. The bubble and slurry-bubble columns too are akin to the STR. It is used as a batch reactor or semi- or fed-batch reactor for those reactions requiring a large residence time, as the tubular reactor would be too long and unwieldy. The STR is bulky, and the yield and selectivity could be low.

If a tubular reactor is used in place of an STR for those reactions with slow reaction rates, the flow could be laminar. Therefore, the desired perfect radial mixing and yet axial plug flow is not possible with the tubular reactor. Owing to this limitation, the STR is used for the lack of alternatives. If two phases are involved, the interfacial mass-transfer rate could be low. For highly exothermic and fast reactions, the stirred tank is first charged with one of the reactants and the other is added at a slow rate, as dictated by the heat-transfer rate through the jacket of the stirred tank to control the temperature within the required limits. Therefore, this semi- or fed-batch mode of operation is time consuming. Furthermore, scale-up of STR is very difficult. The volumetric heat- and mass-transfer coefficients are very small for large tanks and hence they require a long time to heat or cool the reaction mixture. Several types of reactors are now on offer to replace the STR. Some of these are:

1. HEX reactor (plate heat exchanger as reactor)
2. Advanced Flow^{TM} reactor (AFR)
3. Agitated cell reactor or Agitated tube reactor (ACR or ATR)
4. Oscillatory-flow baffled reactor (OBR)
5. Spinning disc reactor (SDR)
6. Spinning tube-in-tube reactor (STTR)
7. Stator-rotor spinning disc reactor (Stator-rotor SDR)

5.4.1 HEX Reactor

The plate heat exchangers were the early process intensification devices used as alternatives to the shell-tube heat exchangers. They have been in commercial use for several decades. They are also used as reactors requiring heat exchange. These are called HEX reactors and sometimes known as multifunctional reactors. Figure 5.9 shows a typical plate heat exchanger used as a HEX reactor. The reactants flow through one set of channels (shown in green) whereas the cooling/heating fluid flows through the other set of channels. The channels are corrugated or provided with inserts to generate vortices to promote the mixing and heat transfer. The reactive channels packed with the catalyst particles can be used for the catalytic reactions. Thonon and Tochon [25], Reay *et al.* [26] and, recently, Anxionnaz *et al.* [27] have presented comprehensive reviews of the literature on HEX reactors. Heat exchangers with complex geometries are being made by the '3-D printing' method, since

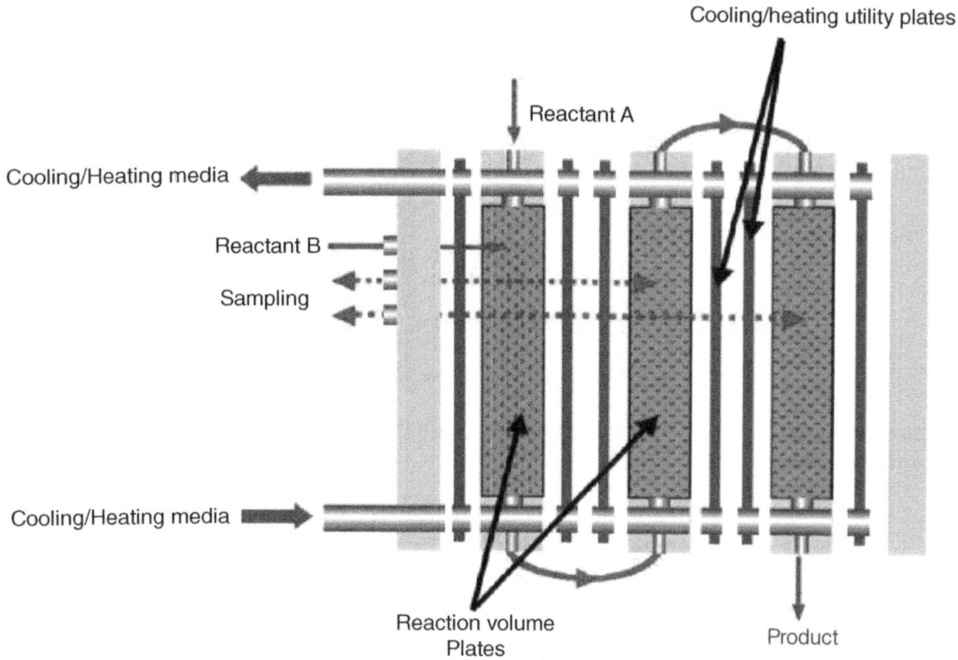

Figure 5.9 *Plate heat exchange reactor. Source: Reproduced from Ref [64] © 2014 with permission from Elsevier B. V.*

they could not be fabricated otherwise by conventional techniques. These may lead to wider applications.

5.4.2 Advanced-flow^{TM} Reactor (AFR)

Corning Inc., the well-known glass and ceramic manufacturer, is offering glass and ceramic reactors (Figure 5.10) with heart-shaped channels to enhance mixing of reactants, rate of heating or cooling of the reacting mixture. These are named as Advanced-Flow Reactors [28]. Figure 5.11 shows the reactor channel configuration. An AFR module consists of four layers of glass sheets stuck together so as to form a reaction channel and cooling or heating channels on either side of the reaction channel. The heart-shaped chambers do the splitting and combining of the streams to promote mixing. Modules with variants of channel geometries are also on offer from Corning Inc. These are similar to the bonded plate heat exchangers. Similar ones in metal should be available soon.

Table 5.3 presents capacities of different modules. The AFR can be used over temperature and pressure ranges of −60 to 200 °C and 0–18 bar, respectively. The glass and ceramic construction eliminates contamination and corrosion in most cases. These are in a way similar to plate heat exchangers. The volumetric heat transfer coefficient is about 1.5 MW/m^3-K, which is about 1000 times higher than that in the jacketed vessels. Likewise the estimated overall mass-transfer coefficient between the phases for liquid–liquid systems is about 2–40 s^{-1}, which is more than 100 times those in a jacked vessel [29]. The

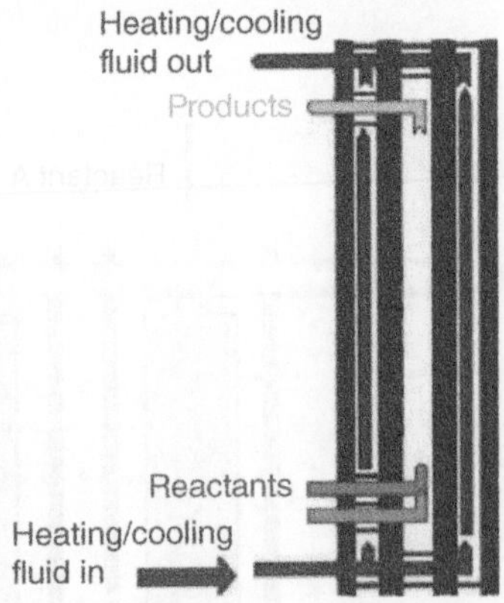

Figure 5.10 Advanced-flow reactor module. Source: Reprinted from Ref [29] © 2012 with permission from the American Chemical Society.

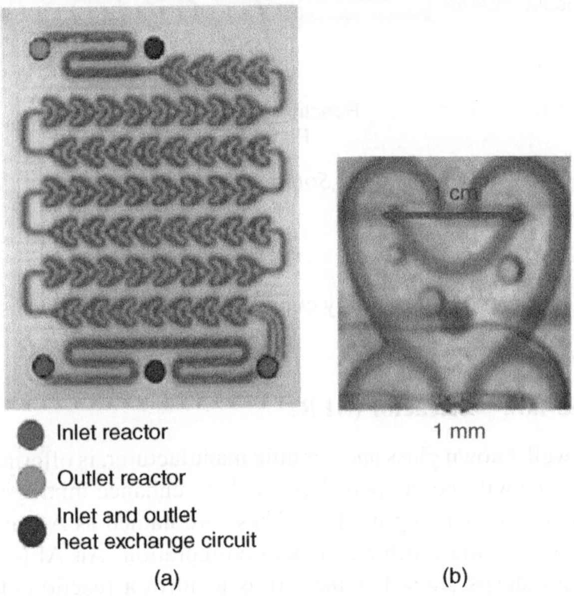

Figure 5.11 Heart-shaped cells of reaction channel. Source: Reprinted from Ref [29] © 2012 with permission from the American Chemical Society.

Table 5.3 Some characteristics of Corning Advanced Flow Reactor modules.

Generation	G1	G3	G4
Material	Glass	Glass	Ceramic
Holdup of reactants (ml)	8–11	55–65	200–600
Flow rate (ml/min)	30–200	400–2000	1000–5000
Production (ton/year)	15	70	1400

modules can be connected in parallel for scaling up the production. AFR is an alternative to STR. The potential commercial applications can be gauged from the characteristics of the modules given in Table 5.3. The use of AFR has been proposed for a variety of reactor applications involving liquid–liquid, liquid–solid, and gas–liquid–solid phases. Recently, Bogdan *et al.* [30], Braune *et al.* [31] and Buisson *et al.* [32] studied the use of AFR in synthesis of pharmaceuticals.

As the volumetric mass-transfer coefficient of liquid–liquid systems is very large, the module can be used as a single-stage extractor but requires a separator. The modules can be configured to form a countercurrent extractor using suitable phase separators.

5.4.3 Agitated Cell Reactor (ACR)

AM™ Technology [33] is offering three types of equipment, namely agitated cell reactor, agitated tubular reactor (ATR) and variable-channel reactor (VCR). These may be viewed as a number of stirred tank reactors connected in series.

Figure 5.12 shows a cutaway view of an ACR with its parts labelled. It consists of three plates: one faceplate with sight glasses, a thick plate with big enough holes and engraved channels to form interconnected cells when closed with a back plate. The sight glasses are provided to enable the operator to view the cell contents. Cylindrical plugs are placed to allow stirring of the contents of the cells. Hollow cylinders or springs are used as the

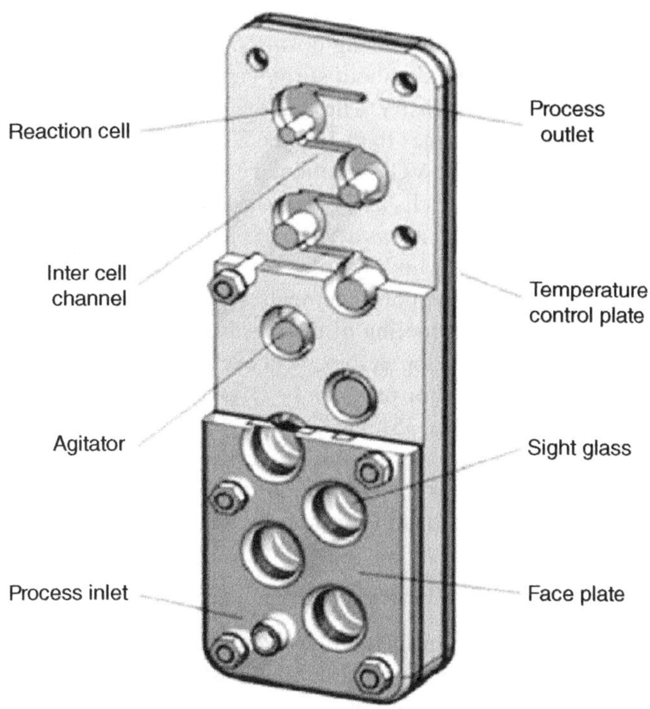

Figure 5.12 Agitated cell reactor. Source: Reprinted from Ref [35] © 2012 with permission from The Institution of Chemical Engineers.

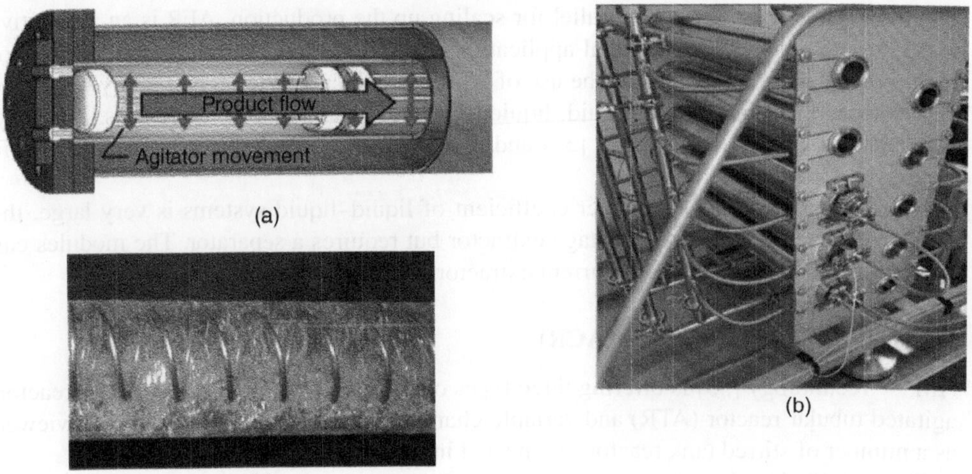

(c)

Figure 5.13 *Agitated tubular reactor: (a) Tube with agitator, (b) Assembly and (c) Spring as agitator. Source: Reprinted from Ref [34] © 2012 with permission from the American Chemical Society.*

stirrers. The back plate has a channel for circulating heat-transfer medium for cooling or heating the cell contents. The whole module is mounted on a shaker. Flexible tubes are used to transfer the reactants and products using dosing pumps.The module forms well-mixed stirred reactors connected in series without stirrers and seals. These are meant for micro-flow reactors for use in flow chemistry. The volume of the plugs used for stirring can be varied to obtain variable cell volumes to realize the STRs of varying volumes connected in series. These are referred to as the VCR mentioned earlier.

For industrial applications, the cells are replaced with tubes and hence the name agitated tubular reactor is used. Figure 5.13 shows a photograph of an ATR. The agitator or stirrer is a solid cylindrical plug or a hollow cylindrical plug made of perforated shell or simply a coiled spring. The ATR relies on loose agitator elements and mechanical shaking of the reactor body for mixing. The heating or cooling of the reaction mixture in the tubes can be done using double-walled tubes as in a double-pipe heat exchanger. To carry out fast exothermic reactions, the volume of each cell or tube can be tailored to regulate the temperature rise. Gasparini *et al.* [34], Jones *et al.* [35] and Gómez-Quero *et al.* [36] reported studies on the use of these reactors in converting batch processes into continuous processes. These studies involved slurries in synthesizing *N*-iodomorpholinium hydroiodide salt, gas–liquid biosystems and 3-phase gas–liquid–solid systems.

5.4.4 Oscillatory-flow Baffled Reactors (OBR)

Figure 5.14 shows a sketch of an OBR, flow patterns, and a picture of a reactor. The reactor consists of tubes fitted with circular baffles. The tubes are connected in series in a serpentine geometry for easy mounting on a rack. The flow is set to an oscillatory motion superimposed over a net flow, using a piston-cylinder arrangement in the range of 0.5 to 10 Hz. This oscillatory flow ensures good mixing between the baffles. The flow would

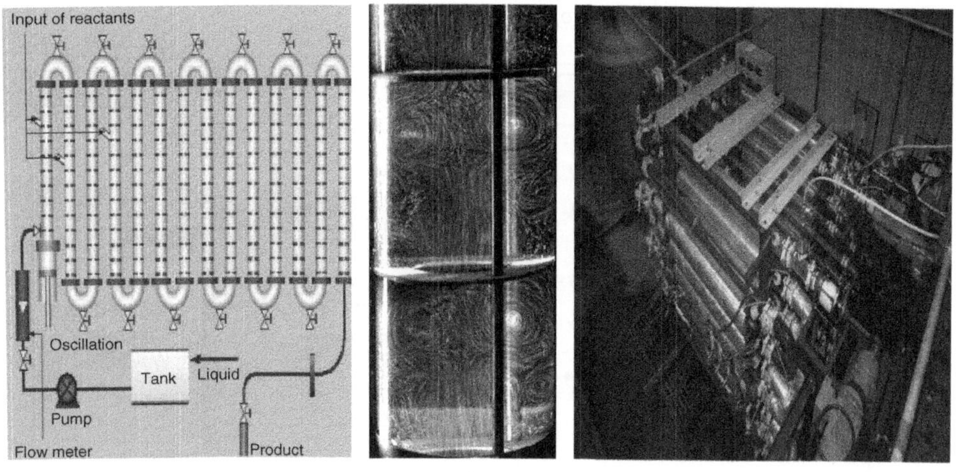

(a) (b) (c)

Figure 5.14 *Sketch Oscillatory flow baffle reactor, flow visualization and industrial unit Source: Courtesy of http://www.nitechsolutions.co.uk*

be laminar but for the oscillatory motion. The residence time of the fluid can be varied from almost infinite to a low value as dictated by the permissible maximum velocity. Flows involving liquid–liquid, gas–liquid, and gas–liquid–solid systems have been employed. A variety of inserts in place of circular baffles have been used depending on the fluid properties.

The OBR can be viewed as a series stirred cells or a tubular flow reactor. The tube can be jacketed to provide cooling or heating. Its potential has been demonstrated for processing in the pharmaceutical, fine-chemicals and biochemical industries. An account of these can be found elsewhere [26, 37].

5.4.5 Spinning Disc Reactor (SDR)

Figure 5.15 shows an SDR with a horizontal disc in a casing [38]. The disc rotates at an rpm ranging from a few hundred to several thousand to obtain the desired centrifugal force. The liquid is introduced at the center. It flows out radially as a thin film, collects on the vertical wall of the casing, and drains to the bottom of the casing from where it is withdrawn. The surface of the disc could be smooth, corrugated or perforated to promote turbulence. The performance of the perforated disc is superior to the other two. Conical discs can also be used.

The SDR may be viewed as a HIGEE with a simple geometry, in which the packed bed is replaced with a horizontal disc. The liquid flowing over the disc can be cooled or heated by passing fluid through the channel provided in the disc. Thus, the SDR affords tight control of the temperature of the liquid film, unlike HIGEE. It can be used as an alternative to a stirred tank reactor for the desktop manufacture of pharmaceuticals and fine chemicals, or for polymers for a low volume of production. Multiple discs can be accommodated for high

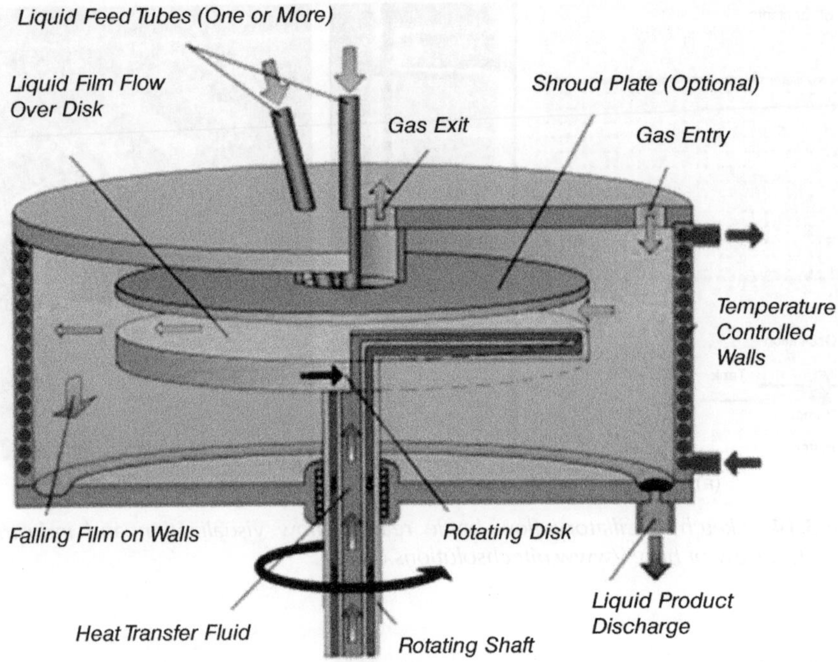

Figure 5.15 *Spinning disc reactor. Source: Reproduced from Ref [42] © 2012 with permission from The Royal Society of Chemistry.*

production volumes. The theory of SDR and its applications have been presented by Reay *et al.* [26], Ramshaw [39], Boodhoo [40], Feng *et al.* [41], and Pask *et al.* [42].

5.4.6 Spinning Tube-in-tube Reactor (STTR)

The mixing of phases and heat removal might be of paramount importance in the case of very fast exothermic reactions, to ensure safety, selectivity and yield. The good dispersion of the phases ensures high interphase mass transport. In such cases, the Couette flow through narrow channels offers an advantage. Reactions can be carried out in a stream flowing through the narrow space formed between a spinning tube and a stationary tube [43]. Figure 5.16 shows a sketch of the SSTR. A variable-speed motor drives the spinning tube. The reactants flow through the annular gap (indicated in red). The outer side of the stationary tube is jacketed to control the temperatures of different sections of the reactor. The inner tube can also be used for heating or cooling. The SSTR is useful in carrying out highly exothermic and explosive reactions without the use of a diluent, which requires separation after the reaction. It has been used to carry out liquid–liquid, liquid–gas, and gas–liquid–liquid reactions. A similar proprietary device named Raptor^{TM} has been reported by Visscher *et al.* [44].

Chen et al. [45] reported studies on mass transfer between the gas and liquid dispersions in a channel formed between two concentric stationary tubes. More importantly, the gas was introduced through a short microporous tube at the liquid entry end into flowing liquid

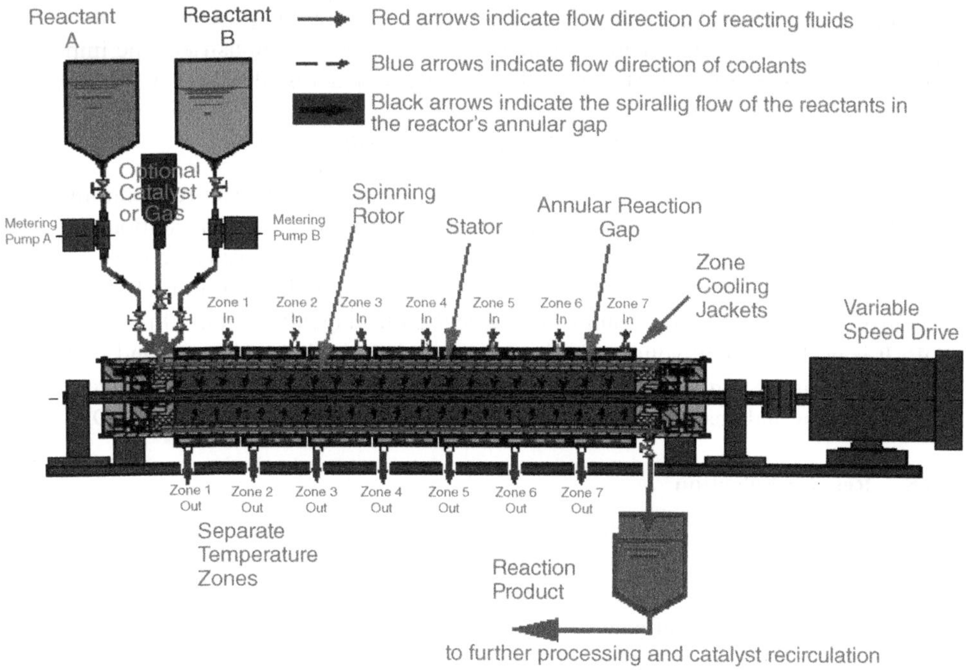

Figure 5.16 *Spinning tube in stationary tube reactor. Source: Reproduced with permission from Strategy License LLC.*

through annular channels of width 250, 500, 750 and 1000 μm. They compared their data of volumetric mass-transfer coefficient ($k_L a$) and interfacial area (a) with those of other gas–liquid contactors in a tabular form, which is reproduced below (Table 5.4). Refer to the original table in reference [45] for the sources of the data. The volumetric mass-transfer coefficient is two orders of magnitude higher than that in conventional gas–liquid contactors like the STR, one order of magnitude higher compared with that in the static mixer and

Table 5.4 *Mass transfer coefficients ($k_L a$) and interfacial area (a) of different contactors [45].*

Type of contactor	$k_L a$ ($s^{-1} \times 10^2$)	a (m^2/m^3)
Gas–liquid microchannel contactor (1000 μm deep, 500 μm wide)	30–2100	3400–9000
Microporous tube-in-tube microchannel reactor	100–1435	62 421–220 400
Bubble column	0.5–24	50–600
Impinging jet absorbers	2.5–122	90–2050
Stirred tank	3–40	100–2000
Spray column	1.5–2.2	75–170
Couette–Taylor flow reactor	3–21	700–1200
Static mixer	0–250	100–1000

comparable to that in the microchannel contactor. The throughputs are about 60 times higher than those of micro-channels. Chen *et al.*'s [45] work demonstrates the importance of gas injection in the SSTR. However, the pressure drop across the reactor is a concern.

5.4.7 Stator-rotor Spinning Disc Reactor (Stator-rotor SDR)

Figure 5.17 shows a sketch of a multi-disc stator and rotor SDR. The reaction mixture is passed through the channel formed between the spinning discs and the stationary casing. The tangential velocity of the fluid at the surface of the spinning disc is the same as that of the disc, and it is zero at the surface of the stationary casing, leading to high shear across the narrow channel. This is in contrast to the negligible tangential liquid film velocity in an SDR. It is useful for low-volume production involving highly exothermic and multiphase reactions.Meeuwse *et al.* [46, 47], Visscher et al. [48] and de Beer *et al.* [49] have reported studies on mass and heat transfer and residence characteristics of stator-rotor SDR.

5.4.8 Reactor Selection

The selection of a reactor, from those described above as an alternative to the STR, depends on the heat reaction, the nature of the phases involved and the rate of production. The safety, reliability, energy consumption and cost have to be considered. The volumetric heat- and mass-transfer coefficients can be viewed as measures of process intensification. The attainable coefficients in the reactors dictate the yield, selectivity and size reduction.

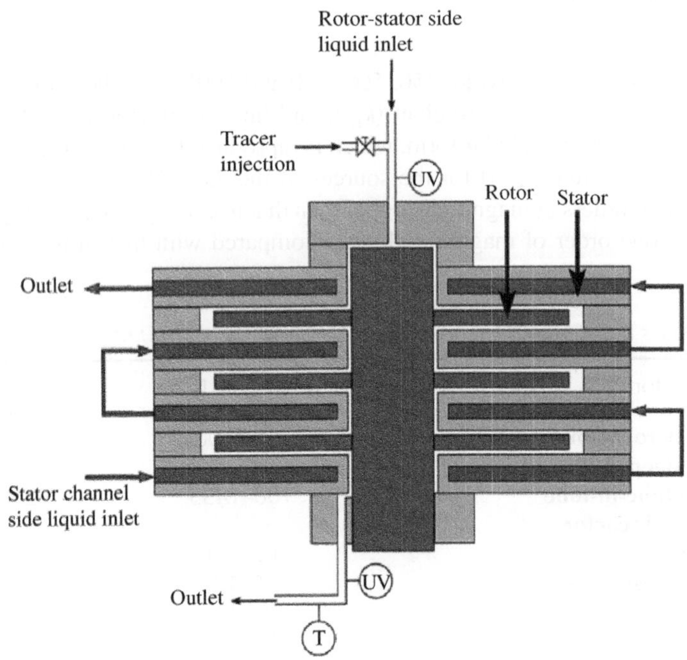

Figure 5.17 *Disc stator-rotor disc reactor. Source: Reproduced from Ref [49] © 2014 with permission from Elsevier B. V.*

Unfortunately, the available data are sparse at present. From the heat transfer point of view, the suitability of the reactors may be put in deceasing order as SSTR, stator-rotor SDR, HEX reactors, AFR, SDR, ACR/ATR, OBR and HIGEE. If the interphase transfer dictates the reaction rate, the volumetric mass-transfer coefficient plays a dominant role. The micro-mixing and residence time also are of importance in the case of fast reactions and thermally unstable products. The above ordering (but for the SDR) could be used for gas–liquid or liquid–liquid phases. If solid–liquid or solid–liquid–gas are involved, the decreasing order could be OBR, ACR/ATR, AFR, HIGEE, stator-rotor SDR, and SSTR. Studies are underway on the characteristics of the reactors. These would help to ascertain their relative merits quantitatively in the future.

5.4.9 Microchannel Devices

Microchannel devices are tools for the study of flow chemistry and chemical sensors.The size of the channels ranges from 10 to 500 microns. These are made using the same techniques for making electronic devices such as microprocessor chips. Ehrfeld [50] presents some details of the fabrication of these devices. Microchannels lead to high surface area per unit volume, as in lungs for oxygen/CO_2 transfer, and thus, high heat- and mass-transfer rates though the flow is laminar. These devices can be used as heat exchangers, liquid–liquid, gas–liquid, solid–liquid–gas reactors, separators (micro-distillation, micro-extractors, and so on) and mixers [51]. These are alternatives to round-bottom flasks, mantle heaters, condensers and mixers. The microchannel devices can be assembled to form a tabletop or rack-mounted chemical plant for continuous operation. These can be used to undertake chemical synthesis and for screening catalysts. Advanced flow reactors, the millimeter channel devices like stator-rotor devices are the scaled-up versions of microchannel devices. The former are for the production in the pharmaceutical and fine-chemical industries, whereas the latter are for the synthesis of chemicals and screening of catalysts.

5.5 Process Integration

The structure of most of chemical plants is based on the 'separation–reaction–separation' sequence in processing the raw materials into products. The process steps are carried out sequentially, mimicking the steps carried out in the chemistry laboratory. However, the integration of two or more process steps or a revamp of the whole process using conventional or PI equipment (presented in the earlier sections) could lead to a significant reduction in the footprint and in the material and energy requirements. Such integration is referred to here as process integration. This phrase has been used in connection with heat and mass integration for the entire plant even before PI gained prominence [52]. Reay *et al.* [26] discussed the methods adopted for process integration and presented a number of case studies. Bakker [53] presented an industry perspective of process integration with some examples of PI implementation. Borukhova and Hessel [51] gave an account of process integration techniques along with the methods of 'transport, chemical and process design intensification' in microchannel devices. Van Gerven and Stankiewicz [54] presented a broader view of process integration.

There is a vast scope of process integration in the pharmaceutical and fine-chemical industries.

Here, we refer to process integration to encompass the following:

- Heat and mass integration (pinch technology)
- Reactive separations (integration of separation and reaction)
- Hybrid separations (combining two separations)
- Conversion of crosscurrent process into countercurrent process
- Process-specific methods (reengineering the process)
- In-line processing (processing during transport).

Brief outlines of the above techniques are given below.

5.5.1 Heat and Mass Integration

Linnhoff proposed the heat integration and synthesis of heat exchanger networks (HENs) [52]. It is well known as pinch technology. The graphical and numerical techniques have been well developed [55]. This technique has been extended for mass-exchange networks like water networks and hydrogen networks [52]. These are not considered here.

5.5.2 Reactive Separations

There are two distinct categories of reactive separations. In one category, the reaction aids separation and in the other separation aids the reaction. In the former, decreasing or eliminating the mass-transfer resistance on the liquid side by allowing the solute to react with a non-volatile 'non-diffusing' component as in the absorption of CO_2 into an amine solution enhances the mass-transfer rate. Absorption with reaction has been practised for the removal of CO_2 and H_2S over the past several decades, and an extensive literature is available. Similarly, in distillation, reactive entrainers are used to separate close boiling or azeotropic mixtures [19]).

In the latter category of separation aiding the reaction, a reversible reaction is driven to completion, circumventing the thermodynamic equilibrium constraint, by the removal of one of the products by separation from the reaction mixture. Consider the reversible reaction:

$$A + B \leftrightarrow C + D$$

The removal of either C or D as it forms can drive the reaction to completion. Several methods like distillation, adsorption, stripping and another chemical reaction are available for the removal one of the components. Several esterification reactions can be driven to completion by stripping of the water as it forms. If the relative volatility of water is high, as the in production of dibutyl phthalate from phthalic anhydride and n-butanol, the water can be removed by simple distillation.The water could be preferentially adsorbed (taking advantage of the smaller size of water using a molecular sieve like 3A or 5A zeolite) or by extraction using a suitable solvent.

Knaebel and Cussler [56] proposed a novel pressure-swing adsorption for ammonia synthesis. The gases are recycled (for a part of the cycle) through an ammonia reactor and an adsorption bed at the reactor temperature and pressure. The ammonia formed in the reactor is held back in the adsorption bed and the remaining nitrogen and hydrogen are recycled. The adsorbed ammonia is recovered by blow down. Then the bed is rinsed and pressurized for use in the next cycle. The authors showed that ammonia at high purity can be produced at a relatively low pressure of 10 to 100 bar. Comprehensive reviews of various conventional separations with reaction and membrane-based reactive separation are available [57, 58]. Gallucci *et al.* [59] presented an introductory review of membrane reactive separations.

5.5.3 Hybrid Separation

A suitable combination of two separation techniques like distillation and adsorption leads to a desired separation of a mixture which is either infeasible or expensive with a single technique. This combination of the different separation techniques is called hybrid separation. The techniques of hybridization are viewed as the techniques of process integration. Stankiewicz [60] presented the examples of extractive distillation, adsorptive distillation, membrane distillation, membrane absorption/stripping, and adsorptive membranes. A detailed discussion on hybrid separation processes involving distillation and one of the separation processes, namely, absorption, desorption, extraction, adsorption and membrane processes is available [61].

5.5.4 Conversion of Crosscurrent into Countercurrent Process

In pharmaceutical industries, it is common to use a stirred tank reactor, after completing the reaction step, for 'washing' with a solvent to remove impurities. Single washing is equivalent as a batch single-stage extraction and repeated washing in the same tank with batches of solvent is equivalent to batch crosscurrent extraction. The solvent is generally water, ethyl acetate, dichloromethane, or a similar compound. These solvents are recovered, or discharged if it is water, as an effluent after treatment. In many processes, 2 to 8 washings are carried out to reduce the impurities to the desired level. This entails loss of valuable material due to its solubility in the solvent and in decanting, and thus lowers the yield. The driving force can be enhanced using countercurrent processes and the required solvent can be reduced, leading to higher yield and lowering of the separation costs of solvents.

Example 1: In an API manufacture, a stirred tank of 5 m^3 capacity is used as a reactor. It is charged with a solvent and cooled to 5 °C from the ambient temperature. The reactants are added and the reaction is carried about for 8 h under constant stirring. At the end of the reaction, the reaction mixture is 'washed' with the same volume of water at 10 °C and then the wash water is decanted. The washing is repeated 6 times. Then, the vessel is cleaned for the next operation to meet regulatory requirements. The total batch time is about 40 h. Thus the vessel is occupied for 40 h. These washings amount to batch crosscurrent multistage extraction. There is a loss of the API as well as of the solvent (in which the API is held) in each of these washing. The yield is only about 60%.

Alternative: The washing can be done in a Podbielniak centrifugal extractor, which is continuous countercurrent extraction. Alternately, the same can be done in a cascade of annular centrifugal extractors configured in the countercurrent mode. It is known that the amount of cold water required and the loss of the API would be far less compared with values for the crosscurrent process [18].

Washing and adsorption: The belt filters are used for washing and dewatering of fine solids in the manufacture of catalysts, zeolites, alumina and other crystalline substances, and so on. The belt filter requires the associated equipment — several vacuum pumps and separation drums. Through-circulation tray driers and through-circulation rotary driers are used to dry solids. All these are continuous crosscurrent processes. These could be replaced with a moving-bed, Hildebrand (screw) conveyer or Kennedy extractor [18], each of which is a countercurrent process. Their potential for PI has not been explored.

5.5.5 Process-specific Integration

Example 2: Revamp of the methyl acetate process by the Eastman Chemical Company is a classic case of revamps and process intensification [62]. The earlier process had one reactor and 9 distillation and extraction columns. These were replaced with a single reactive distillation column. Though it falls under reactive distillation, the process integration is somewhat specific to the components involved.

The above reactive distillation column for the production 23 ton/h of methyl acetate has been in operation for more than two decades. This example illustrates the potential of process integration. Though the plant footprint reduction achieved by process integration is astounding, the column is about 80 m tall and 8 m in diameter. Further size reduction is desirable. In the reaction section of the column, the bubble cap trays are used to provide large residence time, as the reaction rate is the controlling step. Such a high residence is not possible in HIGEE. On the other hand, the top and bottom sections involve only simple distillation and these could be replaced by two HIGEE units to reduce the size of the column.

Example 3: Agreda and Cwirko, in their presentation accessed from the internet, mention their recent PI efforts at the Eastman Chemical Company in the process for the manufacture of poly(ethylene terephthalate). They did not reveal the details but said that they removed three process steps and intensified the remaining process steps to double the capacity and to improve quality.

In many pharmaceutical and fine-chemical industries, a section or whole plant can be revamped to convert a batch process into a continuous process by making use of the PI equipment described earlier. This could lead to an improved yield, quality, and a reduced footprint and utilities. A few examples are presented to illustrate the possibilities.

Example 4 (A fine-chemical manufacture): To produce tributyl phosphate (TBP) at a rate of 2 ton/day, a 5-step batch process is followed for the manufacture. TBP is made by reaction of butyl alcohol with $POCl_3$. The by-product is HCl. It is a highly exothermic and very fast reaction. The presence of HCl could lead to side products and it is desirable to remove it as early as possible. Figure 5.18 presents the current 5-step batch process. The details are as follows:

Step-I: Butyl alcohol is charged into a glass-lined stirred vessel, which is cooled to about 4 °C by circulating brine through its cooling jacket. Then, $POCl_3$ is charged at such a rate so as not to allow the temperature to rise above 7–8 °C, which is dictated by the heat-removal rate through the jacket. The other side products are mono- and di-butyl phosphates. A vacuum of 400 mm Hg is applied to draw the vapor from the reactor. However, there was no presence of HCl from the discharge of the vacuum pump. After all the $POCl_3$ has been added, a vacuum of 8–10 mmHg is applied to strip off the HCl. The temperature of

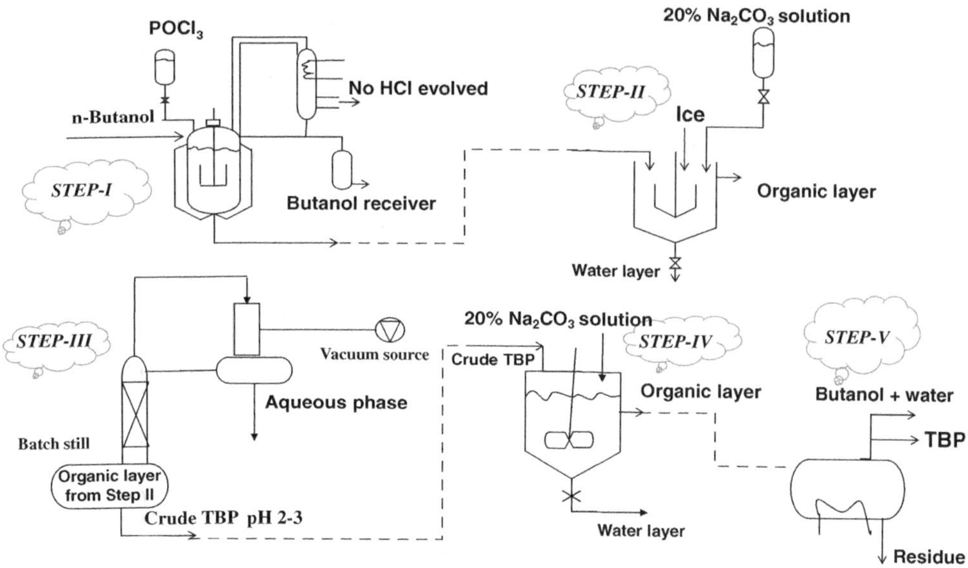

Figure 5.18 *Original 5-step TBP batch process.*

the reaction mixture is raised to 40 °C by circulating hot water through the jacket of the glass-lined reactor. The vapor is drawn through a scrubber to remove the HCl with aqueous sodium carbonate solution. After four hours of this operation the HCl content is reduced to 85% of the value that had originally formed.

Step-II: The contents of the glass-lined vessel are transferred into an open tank, which is half filled with crushed ice. Then, 18 mass-% aqueous sodium carbonate solution at 10 °C is added to neutralize the remaining HCl. The contents are stirred during the addition and then the two-phase dispersion is allowed to settle to form organic and aqueous layers. The aqueous layer is discharged to an effluent-treatment plant.

Step-III: The organic layer is transferred to a batch distillation still equipped with a packed tower. The mixture is distilled. As the water–n-butanol system forms a heterogeneous azeotrope, the condensate separates into two layers in the receiver. The organic layer is refluxed, and the water layer is drawn as distillate and discharged as effluent.

Step-IV: The residue (TBP, butanol and HCl) in the still is charged into a tank. The *p*H is about 2–3. Therefore, it is again neutralized with 18 mass-% aqueous sodium carbonate solution to attain a *p*H of 6–7. The organic layer is then decanted.

Step-V: The decanted organic layer is subjected to simple vacuum distillation to remove first the water and later the butanol. The remaining residue is distilled to obtain the TBP product.

As is the case with most fine-chemical and pharmaceuticals industries, the reaction kinetics is not known. However, we may treat the reaction to be instantaneous. $POCl_3$ is known to react violently with water and NaOH, but not with Na_2CO_3. HCl is highly soluble in the TBP–butanol mixture. With these data at hand, it is required to revamp the process to reduce steps, convert the batch process into a continuous process and increase the yield to be in excess of 90% using process intensification techniques.

Proposed revamp: Obviously, the heat transfer through the jacket of the glass-lined stirred tank is one of the constraints. Stripping of HCl from the reaction mixture from a stirred vessel is another constraint as the interfacial area for stripping is small. Neutralizing the organic layer with aqueous sodium carbonate solution leads to contamination of TBP with water. These limitations need to be overcome in converting the batch process into a continuous process.

The Corning Advanced flow reactor appears to be suitable as the overall heat-transfer coefficient is about 1.7 MW/m^3K, compared with 1×10^{-3} MW/m^3K in a jacketed tank, which is about 3 orders of magnitude higher. Consider using the crude TBP obtained in the process as diluent, in place of butanol, to prevent a temperature rise above 6–7 °C, and an excess of butanol to consume $POCl_3$ completely in a single pass through the reactor. The required production rate of 2 ton/day amounts to 1.4 kg/min. Let us set the crude TBP rate at 3-times the feed rate of butanol and $POCl_3$. Therefore, the total flow rate would be less than 6 kg/min.The heat of reaction to be removed is about 1.5 kW for producing 1.5 kg/min of TBP. One ceramic module (G4) or 3 glass modules (G3) connected in parallel can be used. Preliminary runs can be used to explore the possibility of reducing the amount of diluent and the number of modules. Alternatively, a suitable ATR, SDR, in-line mixture or HEX reactor can be used.

The reactor exit stream would consist of TBP, butanol, HCl and by-products. Several alternatives exist to remove the HCl and recover butanol and refine TBP.

Alternative 1: Make use of the existing batch still (of Step-III) and the vacuum distillation still (used in Step-V). Operate the batch still under vacuum. Scrub the vapor from the still with water to obtain HCl solution or with NH_3 solution to obtain NH_4Cl (which could be used elsewhere). If the liquid contains a significant amount of butanol, it would separate into an organic layer and an aqueous layer. The aqueous layer containing HCl could be used elsewhere or discharged after neutralization. The organic layer could be recycled to the reactor. However, the butanol (organic) layer may contain water. If it is recycled as the diluent, the water in the recycle will consume the $POCl_3$. Therefore, it is desirable to dehydrate the organic layer before recycling to the reactor. Depending on the amount of the organic layer, it may be discarded or recycled to the reactor after dehydration. It can be dehydrated using an adsorption bed of 3A zeolite.

The residue in the batch still is split into two parts. One part is recycled to the reactor as the diluent.The rest is subjected to vacuum distillation first to recover butanol and later the TBP as product as in Step-V of the original process. Note that the residue is not contaminated with water, unlike in the original process.

Alternative 2: It is desirable to avoid the water contamination even in scrubbing the vapor from the still (as was done in alternative 1). Instead, pass the reaction mixture through a bed of granular Na_2CO_3 or $CaCO_3$ to remove HCl. Use two beds; while one is in use for neutralizing the solution, the other bed is used in the making of a fresh bed. The neutralized TBP and butanol mixture is subjected to simple distillation under vacuum to recover and recycle the butanol and to refine the TBP as done earlier. Many other schemes are possible.

Alternative 3. Figure 5.19 presents a process flow sheet. The reactor feed is a mixture of crude TBP as the diluent, $POCl_3$ and an excess of butanol to consume $POCl_3$ completely in a single pass. The reactor exit stream is then subjected to two flash distillations, under vacuum if required. The first is to recover HCl and the other is to separate butanol from the TBP. An SDR with a provision for heating could be used in place of flash distillation. The HCl could

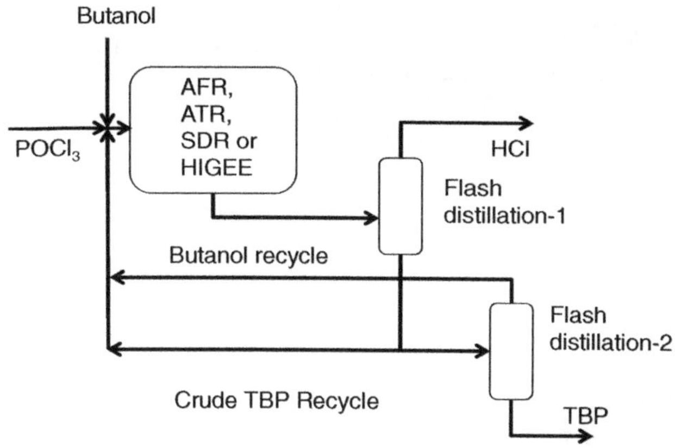

Figure 5.19 *Revamped TBP process with PI equipment.*

be used elsewhere, or it could be scrubbed with water to obtain HCl solution (hydrochloric acid) or with aqueous sodium carbonate solution for sending to an effluent treatment. Thus using PI equipment (AFR and SDR) the process can be made continuous.Note that the contamination of TBP with water and $NaCO_3$ has been avoided.

This example presents several possibilities for the use of PI in revamping. It appears that a vast majority of pharmaceutical and fine-chemical industries present such opportunities.

5.5.6 In-line Processing

The tube inserts for the in-line mixing have been in use for a long time. They have been used in revamping heat exchangers to enhance the heat-transfer rate and to enhance mixing of streams being fed to a reactor. The mixing of reactants, before they enter the reactor, is desired in some cases. The in-line mixing of reactants is achieved as they pass through the pipe connecting the reactor and the storage tanks of the raw materials, and which is provided with static mixing inserts. The in-line inserts represent ideal PI equipment as there are no moving parts and they do not require ancillary equipment like motors.

Figure 5.20 shows some of the static-mixer elements. On the left is a double-pipe heat exchanger with inserts (inside the inner pipe) to create turbulence and thus enhance the heat-transfer coefficient. On the right, the top picture shows the white and green liquid streams mixing to yield the uniform liquid stream of light green. The middle picture shows the dispersion of gas stream (tiny bubbles) in the liquid stream. The bottom one shows dispersion of a red stream in a white stream that is immiscible with the other liquid. Very comprehensive reviews of static mixers, including the design parameters of the various mixers, have been presented by Thakur *et al.* [63] and Ghanem *et al.* [64].

The structured packing for column and ceramic monoliths has been in use for several decades. The latter is used in catalytic converters for automobile exhausts for the in-line conversion of CO and oxides of nitrogen into harmless oxides. It is also used for catalyst and adsorbent supports in process industries. They offer higher specific surface area and low pressure drop. A comprehensive review of monolith reactors as alternatives to trickle-bed, slurry, and slurry bubble column reactors is available in Roy *et al.* [65]. The monolith reactor

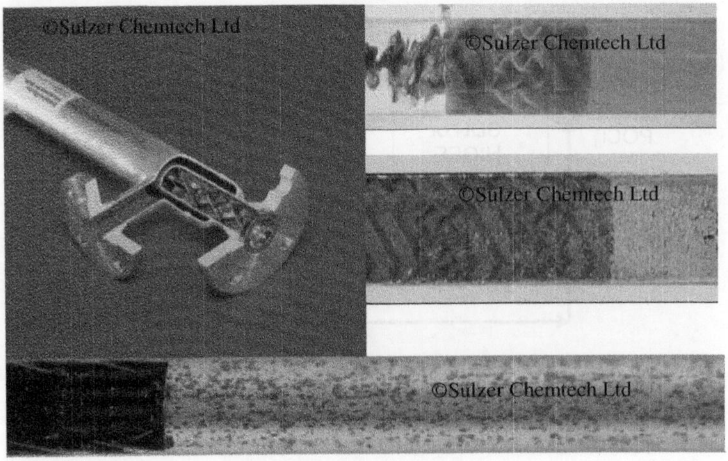

Figure 5.20 In-line static mixers. Source: Reproduced with permission from Sulzer Chemtech Ltd.

deserves serious consideration in revamp involving the above reactors. The monoliths are used in fast-pressure-swing adsorption for fractionation of gas mixtures. The combination of the Xebec^{TM} proprietary rotary valve and structured adsorbents has been claimed to reduce the unit size by 10-fold compared with the conventional PSA unit [66]. The upcoming 3-D printing technology offers a vast potential to create static-mixer elements and monoliths of complex geometries that are otherwise infeasible to fabricate. They may make a difference in the equipment available to the process industries.

5.5.7 Twister® - A Supersonic Separator

Natural gas, after separation of the associated liquid (gas oil), contains moisture. If this gas is put into a pipeline for transportation over a long distance, the life of the pipeline reduces due to corrosion by the moisture. Therefore, the gas is dehydrated before it is put into a pipeline. Generally, triethylene glycol is used to dehydrate the gas in conventional absorption and stripping processes. Twister BV [67] developed a supersonic separator

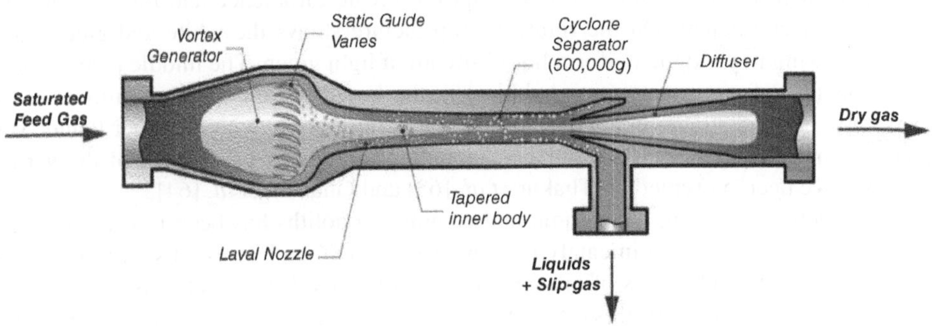

Figure 5.21 Twister and typical operating conditions. Source: With permission from Twister BV.

technology for the in-line dehydration and 'dewpointing' of natural gas streams at high pressure without the use of glycols. Figure 5.21 shows a sketch of the Twister. The gas can enter the unit under a broad pressure range of 60–180 bar and leaves at a pressure which is typically 25–35% lower than the inlet pressure. The isentropic expansion leads to cooling from typically 20–40 °C at the inlet to −25 to−40 °C inside the Laval Nozzle. In milliseconds the water vapor and natural-gas liquids condense into droplets. Owing to centrifugal forces of the swirling supersonic flow, the droplets move to the periphery. A scoop at the periphery draws the gas rich in the droplets of water and oil. A unit of 2 m length could process about 1 million standard (at 1 bar and 0 °C) m^3/day. The relative merits and cost analysis of the natural-gas dehydration by the Joule–Thomson cooling and low-temperature separation and the Twister technologies are available (Machado *et al.* [68]). Two Twister dehydration trains of 300 million standard ft^3/day (8.5 million m^3/day) capacity are in service on platforms offshore Malaysia and are operating at 98% availability.

5.6 Fundamental Issues of PI

PI has been a collection of innovatively designed equipment and processes. Attempts are being made to identify the unifying principles of PI to set it as a discipline and to provide rational bases for equipment and process design. Notable among these attempts are those by Van Gerven and Stankiewicz [54]. They have put forward four basic principles and four approaches to achieve PI. These aid in the development of PI equipment; however, creativity seems to be the need of the hour in their design. A few guidelines, mentioned earlier, are available for process integration. However, a methodology needs to evolve for 'process-specific' process integration. A related aspect that is receiving attention is the process synthesis incorporating PI (for instance see References [69] and [70]). More on these aspects is likely to appear in the future. The modeling, optimization and techno-economic evaluation of PI applications is warranted. A beginning has been made in this direction recently. Machado *et al.* [68], Krishna *et al.* [71], and Sudhoff *et al.* [72] reported such studies involving rotating packed beds.

5.7 Future of PI

The chemical industries are capital intensive. Furthermore, any unscheduled interruption in the production of a continuous process leads to a huge financial loss. Therefore, reliability assumes prime importance in the choice of technology. Hence, it is not surprising that 'from the concept to commercialization' takes about five decades as in the case of pressure-swing adsorption and dividing-wall distillation columns. Most of the PI technologies are of recent origin and hence are not fully developed or understood. Therefore, the risk element is high with high-volume production and is moderate for low-volume production as in the pharmaceutical and fine-chemical industries. There is a need for someone, within the industry, to champion the cause of PI and against the status quo as was done in the manufacture of HOCl by Dow Chemicals several years ago, using then new HIGEE technology [10]. While these are the hurdles for the absorption of PI technologies, the inherently safer design could be a single point for its adoption when safety is a major concern.

However, most of the pharmaceutical industries are batch processes. These processes are developed in shrouds of secrecy without much regard to economy in the use of solvents or energy, as the profit margins are very high. It is likely that the majority of processes can be made efficient in terms of the use of solvents and energy by retrofitting and revamping by adopting PI technologies, specially by the process-specific integration and converting batch into continuous processes using alternatives to stirred-tank reactors and separation columns. The PI technologies will find their way into commercial application through retrofit and revamp; however, it may take a few more decades for their widespread use if one goes by the experience with the past technologies. Offering courses in the chemical engineering curriculum could accelerate the acceptance of PI by industries.

5.8 Conclusions

The methods of achieving process intensification, which subsumes retrofitting and revamping, are outlined. Separation equipment like the HIGEE, Podbielniak extractor and annular centrifugal extractor is presented as an alternative to conventional distillation, absorption and extraction columns. As alternatives to stirred tanks, HEX reactors, advanced flow reactors, agitated tubular reactors, oscillatory-flow baffled reactors, spinning disc reactors and stator-rotor reactors consisting of tubes or discs are described. The selection of reactors has to be based on the severity of heating and cooling and on the weight of the interphase transport on the reaction rate. These PI equipments offer alternatives to conventional equipment for retrofitting. The scope for retrofit in PSA is highlighted.

Various methods of process integration, namely heat and mass integration, reactive and hybrid separations, conversion of crosscurrent into countercurrent processes, in-line processing and process-specific integration are considered. A few examples of process integration are presented. These techniques could be exploited in revamping of the processes. The PI equipment together with process integration would usher in a shift in practices in the pharmaceutical and fine-chemical industries.There is a certain reluctance in the selection of new equipment, but fortune favors the brave.

Acknowledgement

The assistance of Srikanth, Mrunalini and Prelatha of IIT Hyderabad is gratefully acknowledged.

Appendix 5A: Monographs, Reviews and Some Recent Papers

Monographs, and review and recent papers on various aspects of Process Intensification are listed below for the benefit of readers new to this area.

Monographs

Boodhoo, K. and Harvey, A. (eds) (2013) Process Intensification for Green Chemistry: Engineering Solutions for Sustainable Chemical Processing, John Wiley & Sons, Ltd, Chichester.

Stankiewicz, A.and Moulijn, J.A. (eds) (2004) Re-engineering the Chemical Processing Plant: Process Intensification, Marcel Dekker, New York.

Reay, D., Ranshaw, C. and Harvey, A. (2013) Process Intensification: Engineering for Efficiency, Sustainability and Flexibility, 2nd edn, Elsevier Ltd, New York.

Wang, Y. and Holliday, J.D. (eds) (2005) Microreactor technology and process intensification, American Chemical Society Symposium (226 meeting), New York.

Keil, F.J. (ed.) (2007) Modeling of Process Intensification, Wiley-VCH, Weinheim, Germany.

Koch, M.V., VandenBussche, K.M. and Chrisman, R.W. (eds) (2007) Micro Instrumentation for High Throughput Experimentation and Process Intensification – a Tool for PAT, Wiley-VCH Verlag GmbH & Co. KGaA, Weinheim, Germany.

Hessel, V., Renken, A., Schouten, J.C. and Yoshida, J.-I. (eds) (2009) Micro Process Engineering: A Comprehensive Handbook, Wiley-VCH Verlag GmbH & Co. KGaA, Weinheim, Germany.

El-Halwagi, M.M. (2006) Process Integration, Elsevier Inc., New York.

European Roadmap for Process Intensification, http://www.efce.info/efce_media/European_Roadmap_PI-p-531.pdf (accessed on 2015 Feb.7)

Compact Heat Exchangers

Southall, D., Le Pierres, R. and Dewson, S.J. (2008) Proceedings of ICAPP '08, Anaheim, CA, USA, June 8–12, Paper 8009.

Microreactors and Separators

Kashid, M.N., Renken, A. and Kiwi-Minsker, L. (2011) Gas–liquid and liquid–liquid mass transfer in microstructured reactors. *Chem. Eng. Sci.,* **66**, 3876–3897.

Yue, J., Rebrov E.V. and Schouten, J.C. (2014) Gas–liquid–liquid three-phase flow pattern and pressure drop in a microfluidic chip: Similarities with gas–liquid/liquid–liquid flows. *Lab Chip*, **14**, 1632–1649.

Kockmann, N. and Roberge, D.M. (2009) Harsh Reaction Conditions in Continuous-Flow Microreactors for Pharmaceutical Production. *Chem. Eng. Technol.*, **32**, (11), 1682–1694.

Kenig, E.K., Su, Y., Lautenschleger, A. *et al.* (2013) Micro-separation of fluid systems: A state-of-the-art review. *Sep. Purif. Technol.*, **120,** 245–264.

Lam, K.F., Sorensen, E. and Gavriilidis, A. (2013) Review on gas–liquid separations in microchannel devices. *Chem. Eng. Res. Des.,* **91**, 1941–1953.

Sonoreactors

Gogate, P.R. (2008) Cavitational reactors for process intensification of chemical processing applications: A critical review. *Chem. Eng. Process.*, **47**, 515–527.

Rotating Packed Beds

Rao, D.P., Bhowal, A. and Goswami, P.S. (2004) Process intensification in rotating packed beds (HIGEE): An appraisal. *Ind. Eng. Chem. Res.*, **43**, 1150–1162.

Zhao, H., Shao, L. and Chen, J.-F. (2010) High-gravity process intensification technology and application.*Chem. Eng. J.,* **156** (3), 588–593.

Wang, G.Q., Xu, Z.C. and Ji, J.B. (2011) Progress on Higee distillation — Introduction to a new device and its industrial applications. *Chem. Eng. Res. Des.*, **89** (8), 1434–1442.

Vedantam, S., Wardle, K.E., Tamhane, T.V. *et al.* (2012) CFD Simulation of Annular Centrifugal Extractors, *Int. J. Chem. Eng.,* Article ID 759397, 31 pages doi:10.1155/2012/759397.

HEX Reactors

Anxionnaz, Z., Cabassud, M., Gourdon, C. and Tochon, P. (2008) Heat exchanger/reactors (HEX reactors): Concepts, technologies: State-of-the-art. *Chem. Eng. Process.*, **47** (12), 2029–2050.

Static Mixers

Ghanem, A., Lemenand, T., Dell Valle, D. and Peerhossaini, H. (2014) Static mixers: Mechanisms, applications, and characterization methods – A review. *Chem. Eng. Res. Des.*, **92** (2), 205–228.

Rotating Reactors

Visscher, F., van der Schaaf, J. and Nijhuis, T.A. (2013) Rotating Reactors – A review. *Chem. Eng. Res. Des.*, **91** (10), 1923–1940.

Pask, S.D., Nuyken, O. and Cai, Z. (2012) The spinning disk reactor: an example of a process intensification technology for polymers and particles. *Polym. Chem.*, **3**, 2698–2707.

Process Integration

Dunn, R.F., and El-Halwagi, M.M. (2003) Process integration technology review: background and applications in the chemical process industry. *J. Chem. Technol. Biotechnol.*, **78**, 1011–1021.

Membrane Separations

Bernardo, P., Drioli, E. and Golemme, G. (2009) Membrane Gas Separation: A Review/State of the Art.*Ind. Eng. Chem. Res.*, **48**, 4638–4663.

Gallucci, F., Basile, A. and Hai, F.I. (2011) Introduction - A review of membrane reactors. In A. Basile and F. Gallucci (eds), *Membranes for membrane reactors: preparation, optimization and selection,* John Wiley & Sons U. K. pp. 1–61.

Foundations and Process Synthesis

Freund, H. and Sundmacher, K. (2008) Towards a methodology for the systematic analysis and design of efficient chemical processes Part 1. From unit operations to elementary process functions. *Chem. Eng. Process.*, **47**, 2051–2060.

Van Gerven, T. and Stankiewicz, A. (2009) Structure, energy, synergy, time – The fundamentals of process intensification. *Ind. Eng. Chem. Res.*, **48** (5), 2465–2474.

Lutze, P., Babi, D.K., Woodley, J.M. and Gani, R. (2013) Phenomena Based Methodology for Process Synthesis Incorporating Process Intensification. *Ind. Eng. Chem. Res.*, **52**, 7127–7144.

Techno-economic Evaluation

Krishna G., Rangaiah, G.P. and Lakshminarayanan, S. (2013) Modeling and Optimization of Reactive HiGee Stripper-Membrane Process for Methyl Lactate Hydrolysis. *Ind. Eng. Chem. Res.*, **52**, 7795–7802.

Sudhoff, D., Leimbrink, M., Schleinitz, M. *et al.* (2015) Modelling, design and flexibility analysis of rotating packed beds for distillation. *Chem. Eng. Res. Des.*, **94**, 72–89.

References

(1) Podbielniak, W.J. (1935) Centrifugal countercurrent contact apparatus. US patent 2,004,011, filed Aug. 18, 1930 and issued Jun. 4, 1935.

(2) Podbielniak, W.J. (1954) Centrifugal countercurrent contact apparatus. US patent 2,670,132, filed Aug 19, 1949.,and issued Feb. 23, 1954.

(3) Pilo, C.W. and Dahlbeck, S.W. (1960) Apparatus for intimate contacting of two fluid media having different specific weight. US patent 2,941,872, filed June 9, 1959 and issued June 21, 1960.

(4) Ramshaw, C. and Mallinson, R.H. (1981) Mass Transfer Process. US patent 4,383,255, filed Nov.27, 1979 and issued Aug. 11, 1981.

(5) Chandra, A., Goswami, P.S. and Rao, D.P. (2005) Characteristics of flow in a rotating packed bed (HIGEE) with split packing. *Ind. Eng. Chem. Res.*, **44** (11), 4051–4060.

(6) Wang, G.Q., Xu, Z.C., Yu, Y.L. *et al.* (2008) Performance of a rotating zigzag bed - A new HIGEE. *Chem. Eng. Process.*, **47** (12), 2131–2139.

(7) Luo, Y., Chu, G.-W., Zou, H.-K. *et al.* (2012) Characteristics of a two-stage counter-current rotating packed bed for continuous distillation. *Chem. Eng. Process.*, **52** (17), 55–62.

(8) Lin, C.C. and Jian, G.S. (2007) Characteristics of a rotating packed bed equipped with blade packing. *Sep. Purif. Technol.*, **54** (1), 51–60.

(9) Sung, W.D. and Chen, Y.S. (2012) Characteristics of a rotating packed bed equipped with blade packings and baffles. *Sep. Purif. Technol.*, **93**, 52–58.

(10) Trent, D.L. (2004) Chemical processing in high-gravity fields, in *Re-Engineering the Chemical Processing Plant: Process Intensification* (eds A. Stankiewicz, and J.A. Moulijn), Marcel Dekker, New York, USA pp. 46–80.

(11) Qian, Z., Li, Z.-H. and Guo, K. (2012) Industrial Applied and Modeling Research on Selective H_2S Removal Using a Rotating Packed Bed. *Ind. Eng. Chem. Res.*, **51** (23), 8108–8116.

(12) Wang, G.Q., Xu, Z.C. and Ji, J.B. (2011) Progress on Higee distillation - Introduction to a new device and its industrial applications. *Chem. Eng. Res. Des.*, **89** (8), 1434–1442.

(13) Agarwal L., Pavani, V., Rao, D.P. *et al.* (2010) Process Intensification in HiGee Absorption and Distillation: Design Procedure and Applications. *Ind. Eng. Chem. Res.*, **49** (20), 10046–10058.

(14) Zhao, H., Shao, L. and Chen, J.-F. (2010) High-gravity Process Intensification Technology and Application. *Chem. Eng. J.*, **156** (3), 588–593.

(15) Dhiman, S., Verma, V., Rao, D.P. *et al.* (2005) Process intensification in a trickle-bed reactor: Experimental studies. *AIChE J.*, **51** (12), 3186-3192.

(16) Dubal, G., Podbielniak Contactor: A Unique Liquid-Liquid Extractor http://www.pharmaceuticalonline.com/doc/podbielniak-contactor-a-unique-liquid-liquid-0001(accessed on November 2014).

(17) Vedantam, S. and Joshi, J.B. (2006) Annular centrifugal contactors — a review. *Chem. Eng. Res. Des.*, **84**(A7), 522–542.

(18) Treybal, R.E. (1981) *Mass-transfer operations*, 3rd edn, McGraw Hill Inc., Singapore.

(19) Seader, J.D. and Henley, E.J. (2006) *Separation Process Principles*, 2nd edn, Wiley-New Delhi India.

(20) Sivakumar, S.V. and Rao, D.P. (2012) Adsorptive separation of gas mixtures: Mechanistic view, sharp separation and process intensification. *Chem. Eng. Process.*, **53**, 31–52.

(21) Murthy, D.S., Sivakumar, V., Keshav K. et al. (2008) Process Intensification in a Simulated Moving-bed Heat Regenerator. *J. Heat Transfer*, **130** (9), 091801-8.

(22) Leavitt, F.W. Duplex adsorption process. U.S. Patent 5,085,674, filed Oct 25, 1990 and issued Feb. 4, 1992.

(23) Hirose, T. (1991) A simple design method of a new PSA process consisting of both rectifying and stripping sections. Proceedings of the 2nd China-Japan-USA Symposium on Adsorption.

(24) Xebec PSA technology http://www.xebecinc.com/technology-xebec-psa.php (accessed on November 2014).

(25) Thonon, B. and Tochon, P. (2004) Compact Multifunctional Heat Exchangers: A Pathway to Process Intensification in *Re-engineering the Chemical Processing Plant: Process Intensification*; (eds A. Stankiewicz and J.-A. Moulijn), Marcel Dekker Inc., New York, USA pp. 132–176.
(26) Reay, D., Ramshaw, C. and Harvey, A. (2013) *Process Intensification Engineering for Efficiency Sustainability and Flexibility*, 2nd edn, Elsevier, Ltd., New York USA
(27) Anxionnaz, Z., Cabassud, M., Gourdon, C. and Tochon, P. (2008) Heat exchanger/reactors (HEX reactors): Concepts, technologies: State-of-the-art. *Chem. Eng. Process.*, **47** (12), 2029–2050.
(28) Corning® Advanced-Flow ™ Reactors, www.corning.com/reactors (accessed on November 2014).
(29) Nieves-Remacha, M.J., Kulkarni, A.A. and Jensen, K.F. (2012) Hydrodynamics of Liquid–Liquid Dispersion in an Advanced-Flow Reactor. *Ind. Eng. Chem. Res.*, **51** (50), 16251–16262.
(30) Bogdan, A.R., Poe, S.L., Kubis,D.C.*et al.* (2009) The Continuous-Flow Synthesis of Ibuprofen. *Angew. Chem. Int. Ed.*, **48** (45), 8547 –8550.
(31) Braune, S., Pöchlauer, P., Reintjens, R. *et al.* (2009) Selective nitration in a microreactor for pharmaceutical production under cGMP conditions. *Chim Oggi.*, **27** (1), 26–29.
(32) Buisson, B., Donegan, S., Wary, D. *et al.* (2009) Slurry hydrogenation in a continuous flow reactor for pharmaceutical application. *Chim Oggi.*, **27** (6), 12–16.
(33) Scaling up Flow Chemistry http://www.amtechuk.com/ (accessed on 20-11-2014).
(34) Gasparini, G., Archer, I., Jones, E. *et al.* (2012) Scaling Up Biocatalysis Reactions in Flow Reactors. *Org. Process Res. Dev.*, **16** (5), 1013–1016.
(35) Jones, E., McClean, K., Housden, S., *et al.* (2012) Biocatalytic oxidase: Batch to continuous. *Chem. Eng. Res. Des.*, **90** (12), 726–731.
(36) Gómez-Quero, S., Cárdenas-Lizana, F. and Keane, M. A. (2012) Liquid phase catalytic hydrodechlorination of 2,4-dichlorophenol over Pd/Al_2O_3: Batch vs. continuous operation. *Chem. Eng. J.*, **166** (3), 1044–1051.
(37) Harvey, A. (2013) Green Chemistry in Oscillatory Baffled Reactors in *Process Intensification for Green Chemistry, Engineering Solutions for Sustainable Chemical Engineering,* (eds K. Boodhoo, and A. Harvey), John Wiley & Sons. Chichester pp. 157–172.
(38) http://www.rsc.org/ej/PY/2012/c2py20237a/c2py20237a-f2.gif (accessed on 20 Nov. 2014).
(39) Ramshaw, C. (2004) The Spinning Disc Reactor, in *Re-Engineering the Chemical Processing Plant: Process Intensification* (eds A. Stankiewicz and J.A. Moulijn), Marcel Dekker, New York pp. 81–131.
(40) Boodhoo, K. (2013) Spinning disc reactor for green processing and synthesis, in *Process Intensification for Green Chemistry Engineering Solutions for Sustainable ChemicalProcessing* (eds K. Boodhoo and A. Harvey), John Wiley & Sons Ltd., Chichester U. K. pp. 59–87.
(41) Feng, X., Patterson, D.A., Balan, M. *et al.*, (2013) The spinning cloth disc reactor for immobilized enzymes: A new process intensification technology for enzymatic reactions. *Chem. Eng. J.*, **221**, 407-417.
(42) Pask, S.D., Nuyken, O. and Cai, Z. (2012) The spinning disk reactor: an example of a process intensification technology for polymers and particles. *Polym. Chem.*, **3**, 2698–2707.
(43) R.C. Costello & Assoc., Inc. Process Intensification Rethinking Chemistry http://rccostello.com/spinning_tube.html (accessed on 25 Nov. 2014).
(44) Visscher, F., van der Schaaf, J. and Nijhuis, T.A. (2013) Rotating reactors—A review. *Chem. Eng. Res. Des.*, **91** (10), 1923–1940.
(45) Chen, L.-F., Chen, G.-Z. and Wang, J.-X. (2011) High-Throughput Microporous Tube-in-Tube Microreactor as Novel Gas–Liquid Contactor: Mass Transfer Study. *AIChE J.*, **57** (1), 239–249.
(46) Meeuwse, M., van der Schaaf, J., Kuster, B.F.M. *et al.* (2010) Gas–liquid mass transfer in a rotor–stator spinning disc reactor. *Chem. Eng. Sci.*, **65** (1), 466–471.
(47) Meeuwse, M., Lempers, S., van der Schaaf, J. et al. (2010) Liquid–solid mass transfer and reaction in a rotor–stator spinning disc reactor. *Ind. Eng. Chem. Res.*, **49** (1), 10751–10757.
(48) Visscher, F., van der Schaaf, J. and de Croon, M. (2012) Liquid–liquid mass transfer in a rotor–stator spinning disc reactor. *Chem. Eng. J.*, **185–186**, 267–273.
(49) de Beer, M.M., Keurentjes, J.T.F., Schouten, J.C. and van der Schaaf, J. (2014) Engineering model for single-phase flow in a multi-stage rotor–stator spinning disc reactor. *Chem. Eng. J.*, **242**, 53–61.

(50) Ehrfeld, W. (2004) Process Intensification Through Microreaction Technology in *Re-Engineering the Chemical Plant Process Intensification* (eds A. Stankiewicz and J.A. Moulijn), Marcel Dekker Inc., New York pp.177–200.
(51) Borukhova, S. and Hessel, V. (2013) Micro process technology and novel process Windows — Three intensification fields, in *Process Intensification for Green Chemistry: Engineering Solutions for Sustainable Chemical Processing* (eds K. Boodhoo and A. Harvey), John Wiley & Sons Ltd., New York pp. 91–141.
(52) El-Halwagi, M.M. (2006) *Process Integration*, Elsevier Inc., New York.
(53) Bakker R.A. (2004) Process Intensification in Industrial Practice: Methodology and Application in *Re-Engineering the Chemical Plant Process Intensification* (eds A. Stankiewicz and J.A. Moulijn) Marcel Dekker Inc., New York pp. 456–501.
(54) Van Gerven, T. and Stankiewicz, A. (2009) Structure, energy, synergy, time–: The fundamentals of process intensification. *Ind. Eng. Chem. Res.*, **48** (5), 2465–2474.
(55) Seider, W.D., Seader, J.D. and Lewin, D.R. (2004) *Product & Process Design Principles, Synthesis, Analysis and Evaluation*, 2nd edn, John Wiley & Sons Inc. New York
(56) Knaebel, K.S. and Cussler, E.L. (1996) A novel pressure swing adsorption system for ammonia synthesis, in *Fundamentals of Adsorption* (ed. M.D. Le Van), Kluwer Academic Publishers, New York USA.
(57) Kulaprathipanja, S. (ed.) (2002) *Reactive Separation Processes*, Taylor and Francis, New York.
(58) Boodhoo, K. and Harvey, A. (eds) (2013) *Process Intensification for Green Chemistry, Engineering Solutions for Sustainable Chemical Engineering,* John Wiley & Sons, Chichester U.K.
(59) Gallucci, F., Basile, A. and Hai, F.I. (2011). Introduction - A review of membrane reactors. In A. Basile and F. Gallucci (eds), *Membranes for membrane reactors: preparation, optimization and selection*, John Wiley & Sons, U.K. pp. 1–61.
(60) Stankiewicz, A. (2004) Reactive and Hybrid Separations: Incentives, Applications, Barriers (eds A. Stankiewicz, and Jacob A. Moulijn), in *Re-Engineering the Chemical Processing Plant: Process Intensification*, Marcel Dekker, New York pp. 271–318.
(61) Stichlmair, J.G. and Fair, J.R. (1998) *Distillation Principles and Practice*, Wiley-VCH, New York.
(62) Agreda, V.H., Partin, L.R. and Heise, W.H. (1990) High-Purity Methyl Acetate via Reactive Distillation. *Chem. Eng. Prog.*, Feb., 40–46.
(63) Thakur, R.K., Vial, Ch. and Nigam, K.D.P. (2003) Static mixers in the process industries - A review. *Chem. Eng. Res. Des.*, **81** (6), 787–826.
(64) Ghanem, A., Lemenand T., Dell Valle, D. and Peerhossaini, H. (2014) Static mixers: Mechanisms, applications, and characterization methods – A review. *Chem. Eng. Res. Des.*, **92** (2), 205–228.
(65) Roy, S., Bauer, T. and Al-Dahhan, M. (2004) Monoliths as Multiphase Reactors: A Review. *AIChE J.*, **50** (11), 2918–2938.
(66) XEBEC ADSORPTION INC. A world powered by clean energy http://www.xebecinc.com (accessed on 15 Nov. 2014).
(67) Breaking Barriers in Gas Processing http://twisterbv.com (accessed on 10 Nov. 2014).
(68) Machado, P.B., Monteiro, J.G.M. and Medeiros, J.L. (2012) Supersonic separation in onshore natural gas dew point plant. *J. Nat. Gas Sci. Eng.*, **6**, 43–49.
(69) Freund, H. and Sundmacher, K. (2008) Towards a methodology for the systematic analysis and design of efficient chemical processes Part 1. From unit operations to elementary process functions. *Chem. Eng. Process.*, **47**, 2051–2060.
(70) Lutze, P., Babi, D.K., Woodley, J.M. and Gani, R. (2013) Phenomena Based Methodology for Process Synthesis Incorporating Process Intensification. *Ind. Eng. Chem. Res.*, **52**, 7127–7144.
(71) Krishna, G., Rangaiah, G.P. and Lakshminarayanan, S. (2013) Modeling and Optimization of Reactive HiGee Stripper-Membrane Process for Methyl Lactate Hydrolysis. *Ind. Eng. Chem. Res.*, **52**, 7795–7802.
(72) Sudhoff, D., Leimbrink, M., Schleinitz, A. *et al.*, (2015) Modelling, design and flexibility analysis of rotating packed beds for distillation. *Chem. Eng. Res. Des.*, **94**, 72–89.

6

Using Process Integration Technology to Retrofit Chemical Plants for Energy Conservation and Wastewater Minimization

Russell F. Dunn[1,2] *and Jarrid Scott Ristau*[1,2]
[1]*Department of Chemical and Biomolecular Engineering, Vanderbilt University, Nashville, USA*
[2]*Polymer and Chemical Technologies, LLC, Nashville, USA*

6.1 Introduction

In response to the staggering environmental and energy problems associated with manufacturing facilities, the process industry has recently dedicated much attention to mitigating detrimental impacts on the environment, conserving resources, and reducing the intensity of energy usage. These efforts have gradually shifted from a unit-based approach to a systems-level approach. As such, the past two decades have seen significant industrial and academic efforts devoted to the development of process systems design methodologies that target energy conservation and waste reduction.

Process integration is a holistic approach to process design and operation that emphasizes the unity of the process [1, 2]. Heat integration involves the use of the heating and cooling potential of process streams within a chemical plant for heat recovery. Mass integration involves the optimal allocation, separation, and generation of mass streams and species within a chemical plant [3]. In the environmental context, heat integration for energy conservation has been driven by demand for expensive utilities within chemical industries, and

Chemical Process Retrofitting and Revamping: Techniques and Applications, First Edition.
Edited by Gade Pandu Rangaiah.

mass integration for waste reduction has been driven by the promulgation of more stringent environmental regulations coupled with the desire to improve industrial competitiveness.

In the past, the industrial goal for waste reduction tasks was to identify a recovery system that would effectively allow the recycle and reuse of certain wastes. This goal was generally accomplished by postulating a variety of process system configurations and operating conditions and then individually screening these alternatives to evaluate their overall economic impact to the company (operating cost, capital investment, etc.). Recently, significant progress has been made toward developing systematic design methodologies that not only identify a system that accomplishes the waste reduction task, but also a system that represents the most cost-effective approach. Specific focus has been devoted toward the development of systematic design methodologies for identifying cost-effective wastewater minimization systems. The other important category of process integration is energy integration, which deals with the global allocation, generation, and exchange of energy throughout the process [4,5]. An overview of some of the process integration design tools for addressing heat integration for energy recovery and water recycle networks for wastewater reduction is provided in the following sections.

6.1.1 Heat Integration Networks

During the past few decades, rising energy costs have required operating companies to look for ways to improve energy conservation. Industrial heat exchange networks, 'HENs', are of particular importance because of their role in recovering process heat. A HEN is a network consisting of one or more heat exchangers that collectively satisfy the energy conservation task. Therefore, in most chemical process industries it is necessary to synthesize cost-effective HENs that can transfer heat among the hot and cold streams (where temperature conditions refer to the initial stream state). During the design stage, temperature specifications for the hot and cold streams must be met and a decision must be made regarding the use of a process stream or an external utility (for example, steam, cooling water, etc.) to accomplish the required heat duty. Even in relatively simple situations, the problem of pairing and sequencing of exchanger streams becomes a large one and the use of systematic techniques is necessary.

For a given system, the synthesis of HENs entails answering several questions including:

- Which heating or cooling utilities should be employed?
- What is the optimal heat duty to be removed or added by each utility?
- How should the hot and cold streams be matched (that is, stream pairings)?
- What is the optimal system configuration (that is, how should the heat exchangers be arranged?, is there any stream splitting and mixing?, etc.)?

Numerous methods have been developed for the synthesis of HENs. These methods have been reviewed by Linnhoff [4], Shenoy [5], Douglas [6] and Gundersen and Naess [7]. Synthesis of HEN's can be directed toward grassroots or retrofit designs. Retrofit design of HENs has frequently occurred as many existing plants are periodically modified to meet increased capacity demands. HEN retrofits allow the designer many unique opportunities for a variety of design scenarios such as:

1. Designing a retrofit HEN with no changes to the existing plant heat exchangers that are already integrated. This reduces the number of streams included in the HEN analysis. The primary limit of this approach is that the existing heat integration in a chemical

process plant is often not optimal and additional energy savings may be realized if modifications to the existing integrated heat exchangers were considered. Of course, this approach is often taken since the existing integrated heat exchangers have already been purchased and are operational

2. Designing a retrofit HEN using existing integrated heat exchanger heat transfer surface area but considering a rearrangement of these heat exchangers to a more optimal location within the plant. This can allow larger surface area heat exchangers (increased capacity) to replace existing integrated heat exchangers and may also allow the smaller surface area heat exchanger to be relocated and used in another part of the plant
3. Designing a retrofit HEN with forbidden matches of particular streams within a heat exchanger. Stream matches are often forbidden based on safety considerations from incompatible streams (leaking of the stream on one side of the exchanger into the stream on the other side of the heat exchanger), location issues (distance between the streams to be integrated), environmental issues (contamination of a stream from heat exchanger leakage) and many other similar design criteria
4. Designing a retrofit HEN with a limited capital budget. Often a chemical plant has a limited capital budget for heat exchanger purchase, construction, and installation. Economy of scale can be used, as later discussed in this chapter, to identify the individual heat exchangers in a retrofit design that provides the highest profitability to the company

Many other scenarios can also be envisioned for retrofit HEN designs. These scenarios can be incorporated in and guide the use of the graphical design tools provided in this chapter.

6.1.2 Water Recycle Networks

Wastewater reduction and water conservation are becoming increasingly more important issues in process industries. More stringent environmental regulations, concerns over long-term health effects on humans and Nature, and the future availability of 'clean' water resources are just a few of the factors that are driving efforts toward improvements in water conservation and wastewater reduction in manufacturing processes. As these issues continue to receive intense government scrutiny and heightened concern from community-organized environmental groups, the ability of industry to address these issues may soon impact their right to operate within these communities and the sustainability of their future operations.

These critical concerns have driven efforts for the development of design methodologies that allow the identification of cost-effective wastewater reduction and water conservation process designs, involving direct recycle and reuse of water, that can be implemented within a variety of process industries. These design methodologies and approaches cover a variety of techniques ranging from graphical-based approaches, including 'water pinch' analysis [8, 9] and the source–sink graphical methodology [10–12], to mathematical optimization-based approaches [13, 14].

For a given system, the synthesis of wastewater reduction and water conservation networks entails answering several questions including:

- Which wastewater streams should be recycled or reused?
- What is the optimal load of each wastewater stream to be recycled or reused?
- What is the optimal allocation of wastewater streams that are to be routed to process water users?

- What is the optimal system configuration (for example, how should the water allocation system be arranged?, is there any stream splitting and mixing?, etc.)?

In addition, several review articles on process integration design tools for energy conservation and wastewater reduction design have been published [15–17]. Finally, several books solely devoted to these topics in detail have recently been published [18, 19].

Graphical design methodologies are provided in this chapter for identifying retrofit heat integration and water recycle opportunities in industry. While other mathematical optimization and complex mathematical approaches also exist, the graphical methodologies presented are particularly useful for solving large industrial retrofit problems. The methodologies presented use existing graphical tools and new graphical techniques that have been identified and applied by the lead author over the past two decades to identify retrofit heat integration and water recycle opportunities in more than 20 large chemical and polymer-manufacturing companies in the United States.

6.2 Graphical Design Tools for Retrofitting Process for Energy Conservation by Designing Heat Exchange Networks

Energy usage for heating or cooling often makes up a sizable fraction of a chemical plant's annual costs and most processes require heating or cooling for separation, reactions, or other energy-intensive chemical processes. The key concept of process integration and pinch analysis is to transfer energy in *hot streams* that need to be cooled to *cold streams* that need to be heated. Figure 6.1 provides an illustration of the concept of using heat integration

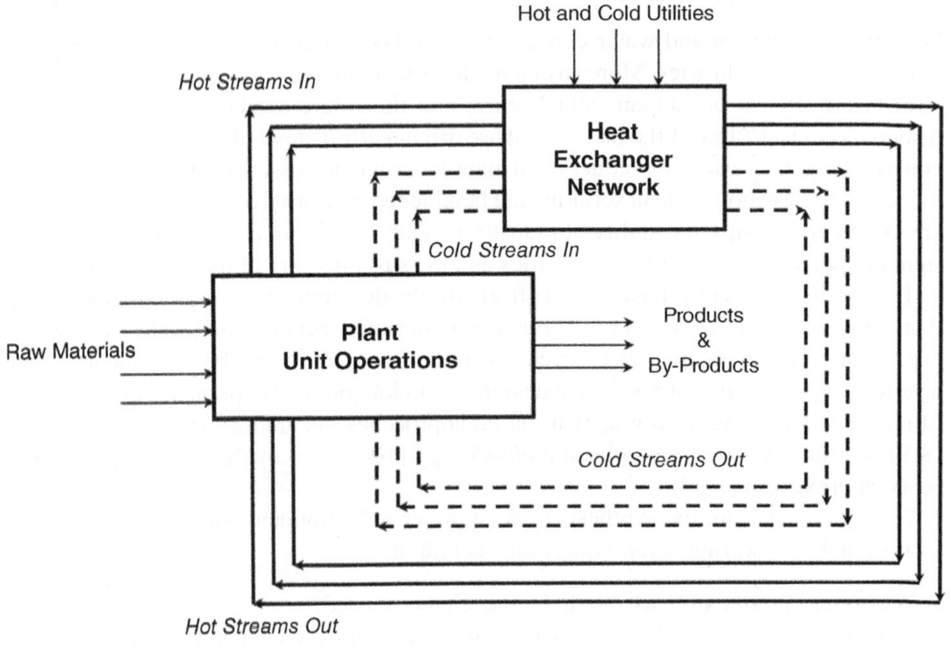

Figure 6.1 Illustration of the design of heat exchange networks for chemical process plants.

Table 6.1 *Example hot and cold stream data.*

Stream	T_{in} (°C)	T_{out} (°C)	mC_p (kW/°C)	Q (kW)
Hot Stream 1 (H1)	127	71	15.9	890.4
Hot Stream 2 (H2)	121	54	8.0	536.0
Cold Stream 1 (C1)	49	113	10.6	678.4
Cold Stream 2 (C2)	82	116	21.2	720.8

to identify heat integration networks. By creating a network of heat exchangers within a chemical plant that facilitates this transfer of heat, the traditional utility requirements (for example, steam, cooling water, etc.) of a process can often be significantly reduced. Numerous books and publications, as previously mentioned, have been devoted to HEN design methodologies that can be used to identify optimal and near-optimal heat exchange networks [4–7].

A shortcut graphical technique for energy conservation design for retrofit process is provided. Three important graphs are proposed to create initial heat exchange networks for consideration in retrofit processes. Of particular importance is the usefulness of this technique for solving large industrial problems. The three key graphs used for this shortcut graphical technique are:

1. The temperature–interval diagram (TID) [4]
2. The heat pinch composite curves [4]
3. The enthalpy-mapping diagram (EMD)

These three graphs will be illustrated using the hot and cold stream data provided in Table 6.1.

6.2.1 The Temperature–Interval Diagram (TID)

First, the TID is a temperature plot, not to scale, of all hot and cold streams to be analyzed. Two temperature scales are shown on the TID, the hot temperature scale and the cold temperature scale. A minimum temperature driving force, ΔT_{min}, pre-selected by the designer, separates the hot and cold temperature scales numerically and this minimum temperature driving force is the minimum temperature difference between a hot and cold stream to be allowed on either end of a counter-current heat exchanger. Each hot and cold steam is represented on the TID as an individual arrow whose tail is the supply temperature and whose head is the target temperature. Figure 6.2 shows an example temperature–interval diagram with two hot streams and two cold streams plotted.

Initial observations are that the TID graphically shows the possible matches between the hot and the cold streams that will not violate the minimum approach temperature. This is noted by ensuring the thermodynamic constraint that heat is only passed from a hot stream at a temperature interval equivalent to or higher than a cold stream temperature interval. Thus, the TID provides an overall quick, qualitative view of potential stream matches without an excess of information.

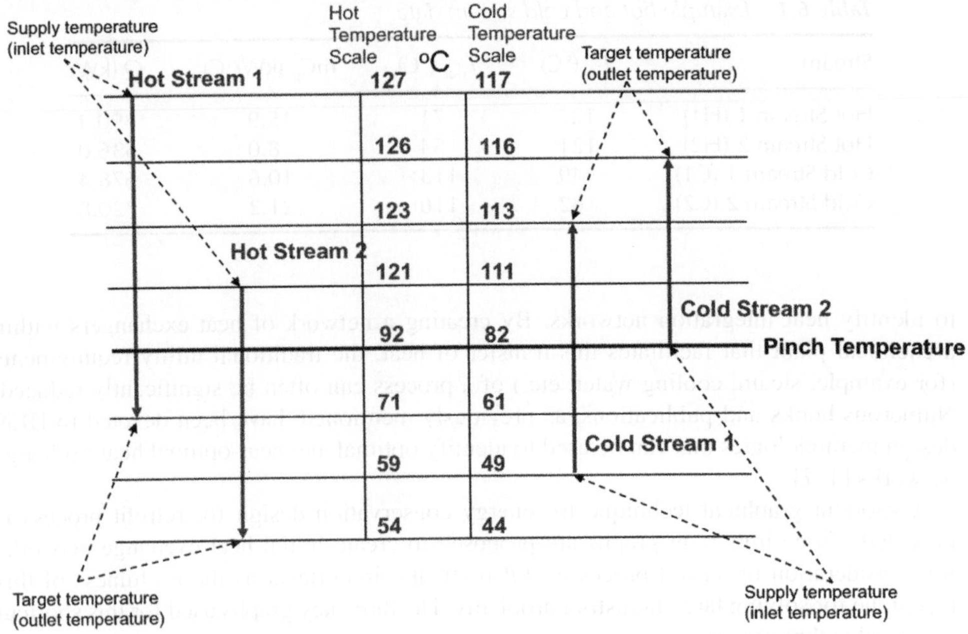

Figure 6.2 *Temperature–interval diagram (TID) with shifted temperature scales.*

6.2.2 The Heat Pinch Composite Curves (Temperature–Enthalpy Diagrams)

Hot and cold streams can be individually plotted on a Temperature–Enthalpy diagram. There may be no phase changes associated with the streams (sensible heat effects only) or there may be a phase change such as evaporation or condensation of these respective streams. Figures 6.3 and 6.4 are provided to illustrate these concepts of the Temperature–Enthalpy diagram.

All of the hot streams are combined by superposition to generate a hot composite curve and all the cold streams are combined to generate a cold composite curve. Simultaneous plotting of the hot composite curve and the cold composite curve provides a heat pinch diagram. An example heat pinch curve provided as Figure 6.5 was generated using the two hot and two cold streams provided earlier in the TID. The composite curve provides a comprehensive picture of the heat supply and heat demand for the process over the entire temperature range. The point where both curves are closest to each other is called the pinch point, and the corresponding temperatures are the pinch temperatures. The pinch temperatures are indicated by the middle dotted line on Figure 6.3 and the temperature difference between the hot composite curve and the cold composite curve shown at the pinch point is the minimum temperature driving force that was pre-selected by the designer (10 °C for this example). The pinch point separates the process into two separate sub-networks of potential heat exchangers. Above the pinch, no cooling utilities should be used. Conversely, below the pinch, no heating utilities should be used. The overlap region between the cold and the hot curves is the total heat integration potential in the process.

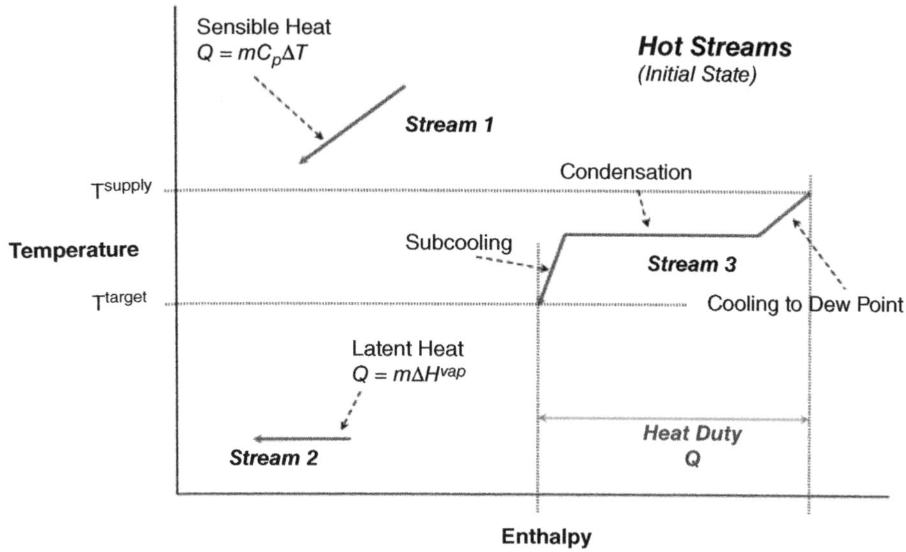

Figure 6.3 *Some basic definitions and values for hot streams.*

The portion of the hot composite curve and the cold composite curve shows the minimum cooling and heating utilities, respectively, that must be supplied. Increasing the minimum approach temperature will shift both curves farther apart, reducing the total heat integration potential.

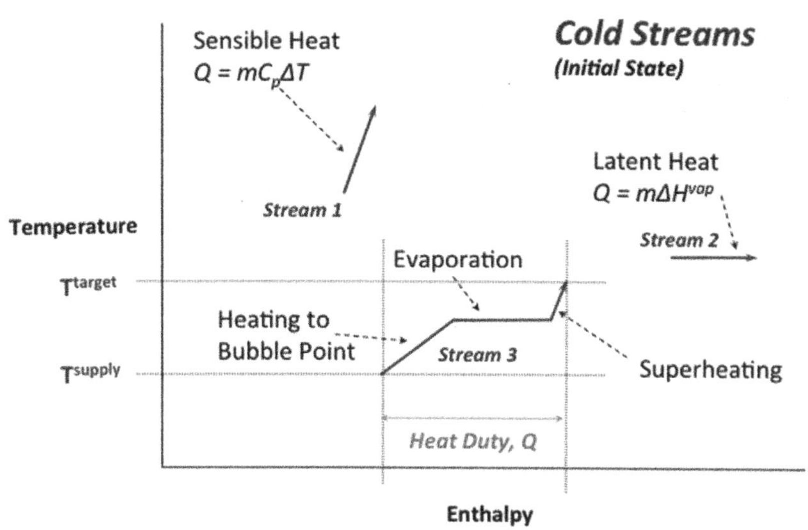

Figure 6.4 *Some basic definitions and values for cold streams.*

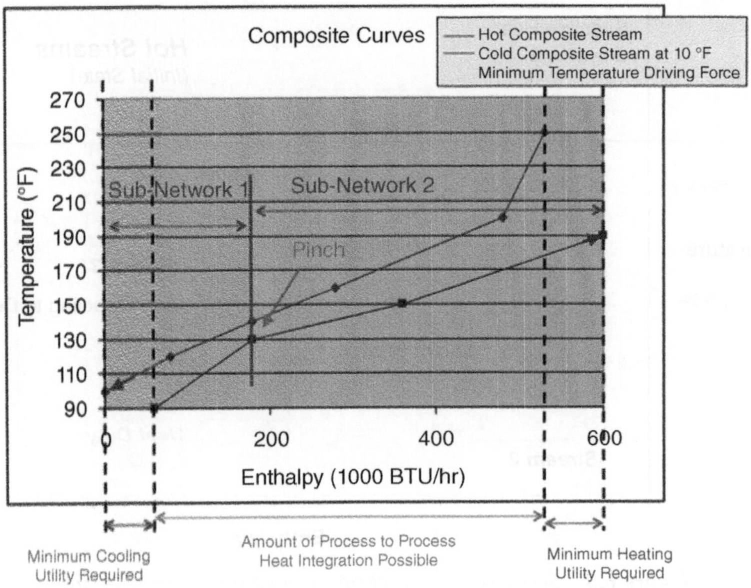

Figure 6.5 Hot and cold composite curves and definitions.

6.2.3 The Enthalpy-Mapping Diagram (EMD)

EMDs should be generated above and below the pinch since there are separate sub-networks above and below the pinch. The EMD is a bar graph type representation of the individual heat duties of each hot and cold stream. The length of a line segment represents the total heat duty for that stream. These diagrams are used simultaneously with the TID above and TID below the pinch point temperatures. Again, for the example provided, Figure 6.6 simultaneously shows the EMD and the TID for the portions of the hot and cold streams above the pinch. An EMD and TID can similarly be generated for the hot and cold streams below the pinch.

6.2.4 Identifying Heat Integration Matches Using the TID and EMD

The composite heat pinch diagram allows the designer to identify the maximum heat integration potential, the minimum cold utility usage and the minimum hot utility usage targets. The TID/EMD provide visual tools to aid the designer in identifying which hot stream(s) should be matched with which cold streams in order to meet or approach the targets identified in the composite heat pinch diagram. The designer can identify possible matches based on the heat duty magnitudes shown on the EMD at the top of the window and then check for a positive temperature driving force using the TID at the bottom of the window. For example, it is clear from the EMD that streams H1 and C2 are the largest hot and cold stream duties above the pinch. Matching these two streams together in an integrated heat exchanger represents the largest single opportunity for heat exchange and takes advantage of the economy of scale. Also, it is clear by then evaluating the supply and target temperatures for these streams using the TID that the entire heat duty of stream H1 (the maximum heat

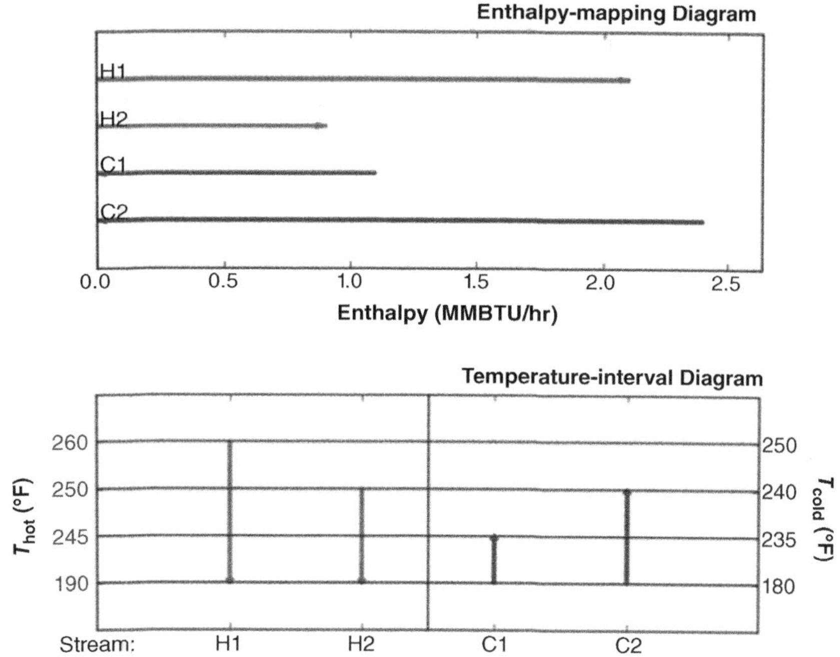

Figure 6.6 *Sub-network above the pinch: enthalpy-mapping diagram and temperature–interval diagram.*

that could possibly transfer between streams H1 and C2, for example, the minimum of the two heat duty values) can be transferred without violating the minimum temperature driving force. This is seen by the fact that the target temperature of cold stream C2 does not exceed the supply temperature of hot stream H1. This graphical strategy facilitates the identification of matches that take advantage of the economy of scale associated with integrating streams with the largest heating and cooling duties and is particularly helpful for analyzing more realistic large industrial problems for heat integration. There is no limit to the number of hot and cold streams that can be represented simultaneously on the EMD and TID.

6.2.4.1 Tradeoffs between the Operating and Capital Cost of a Heat Exchange Network

Adjusting the minimum temperature driving force for a heat exchange network design allows for tradeoffs between operating and capital cost. For example, increasing the minimum temperature driving force (shifting the composite curves farther apart) results in a lower heat integration potential, as observed by a reduced overlap region between the hot and cold composite curves. Thus, fewer utilities are saved with a HEN that uses a larger ΔT_{min} when compared with a HEN that uses a smaller ΔT_{min}. However, the HEN with a larger ΔT_{min} also uses less overall heat transfer surface area in the combined heat exchangers and, thus, results in lower capital costs for the heat exchangers. Thus, there is an optimal ΔT_{min} that results in the lowest total annualized cost for a heat exchange network. While this is true in theory, the implementation of retrofit industrial heat exchange networks is often driven by the desire to maximize energy savings (lowest ΔT_{min} that is practical) or by capital expense limitations in the form of a constrained budget for the total revenue that will be spent on heat exchangers. Either scenario can be easily addressed using the shortcut graphical technique.

6.2.4.2 *Water Savings Achieved by Implementing Heat Exchange Networks*

Numerous examples of water saving can be seen through the implementation of heat exchange networks for energy conservation. For instance, the installation of any integrated heat exchanger will reduce the demand on the existing utility systems, primarily boilers for steam supply and cooling towers for cooling water supply. Reduced demand on these systems results in reduced water discharges from blowdown and evaporation as well as a reduced flow rate of fresh water makeup to accommodate for these losses.

6.2.4.3 *Suggested Guidelines for Using the EMD/TID Graphical Approach for Identifying Retrofit Heat Exchange Networks*

Some guidelines for identifying heat exchange networks for matching hot and cold streams in retrofit processes using the EMD/TID graphical shortcut method are provided below.

1. Identify all hot streams that need to be cooled and all cold streams that need to be heated along with their respective supply and target temperatures.
2. Obtain the physical properties of the streams involved. The properties include: heat capacity, latent heat (for phase changes), and boiling points of the streams at their respective pressures.
3. Obtain the inlet and outlet temperatures, flow rates and, if necessary, the operating pressures of the streams.
4. Select a ΔT_{min}. Start with the lowest temperature (5 °C). This allows the designer to evaluate the maximum utility savings that could possibly be achieved during the first pass design.
5. Generate a TID. The hot and cold pinch temperatures decompose the problem into a network above the pinch and a network below the pinch.
6. Using specific process knowledge, establish any forbidden matches. Forbidden matches are established by the designer and represent matches between hot and cold streams that would be undesirable and are thus forbidden. Forbidden matches are often justified based on stream locations, stream properties, safety considerations and any other key engineering criteria.
7. Eliminate hot and cold streams with heat duty below a minimum threshold. The potential savings from the reduction in use of a hot and cold utility that would be realized by integrating a hot or cold stream drives the selection of this minimum threshold. Consider a single cold stream that is currently heated by a steam utility in a heat exchanger in a chemical process and where the heat duty of this exchanger is known. Annual steam costs associated with this exchanger can easily be calculated based on the current value of steam. A threshold for minimum annual savings is then multiplied by ten years to determine a starting capital cost that would be available (10-year depreciation) for retrofit to purchase a new heat exchanger in order to realize those

steam savings. Thus, a cold stream that requires $2,000/yr of steam in an existing heat exchanger provides a capital budget of only $20,000 installed cost for a retrofit heat exchanger and the designer may decide *a priori* that this is too low a capital budget (too small a heat exchanger) to be considered for the overall retrofit project. Thus, this can establish a minimum threshold value for heat duty.

8. Use the EMD to compare the amount of duty in each stream. Match streams initially with the largest heat duty, Q, to take advantage of the economy of scale, a concept commonly referred to in chemical equipment design economics as the six-tenths factor rule.
9. Make sure that the matched streams have a positive temperature driving force. The inlet temperature of the hot stream must be at a higher or at the same temperature interval than the outlet temperature of the cold stream to be matched.
10. Maximize the temperature driving force of the matched streams to minimize the heat exchanger area.
11. For the match selected, transfer the maximum amount of heat available (the smaller of the two Q values).
12. Only split a hot or cold stream if you see a second possible match that requires the respective hot or cold stream supply temperature to ensure a thermodynamically feasible match. Using the entire flow rate of the hot and cold stream will maximize the temperature driving force of the heat exchanger.
13. Calculate mass flow rates that represent the amount of splitting and the target temperature for each split segment of a stream.
14. Calculate any unknown inlet and outlet temperature if the stream is not split.
15. Evaluate the trade-off between the size of the heat exchangers needed for stream splitting compared with non-stream splitting (series matches) for process simplification or operability.
16. Calculate the fraction of total heat integration possible and the return on investment. From this information, determine whether the selected heat integration is desirable.
17. Consider a possible match for the remaining streams.
18. Evaluate profitability of each match based on the cost of a heat exchanger and the amount of utilities saved. Stop the process when the next heat exchanger identified has lower profitability than desired.
19. Consider adjusting ΔT_{min}.
20. Consider exiting and entering streams for each unit operation to evaluate if pre-heating or pre-cooling the streams not currently heated or cooled would offer any benefit.
21. Consider changing the pressure drop.

6.2.5 Graphical Tools Facilitate HEN Design for Large-scale Industrial Problems

In addition to offering tools that simplify HEN design, the graphical approach has the added benefit of allowing the engineer to focus on the problem from a larger perspective. This enables reduction of the system to those streams with the most potential for energy recovery. Frequently, a significant fraction of the energy recovery possible in a plant having many hot and cold streams can be achieved with only two or three heat exchangers. Once this fraction of the maximum recoverable energy satisfies the designer's needs, additional

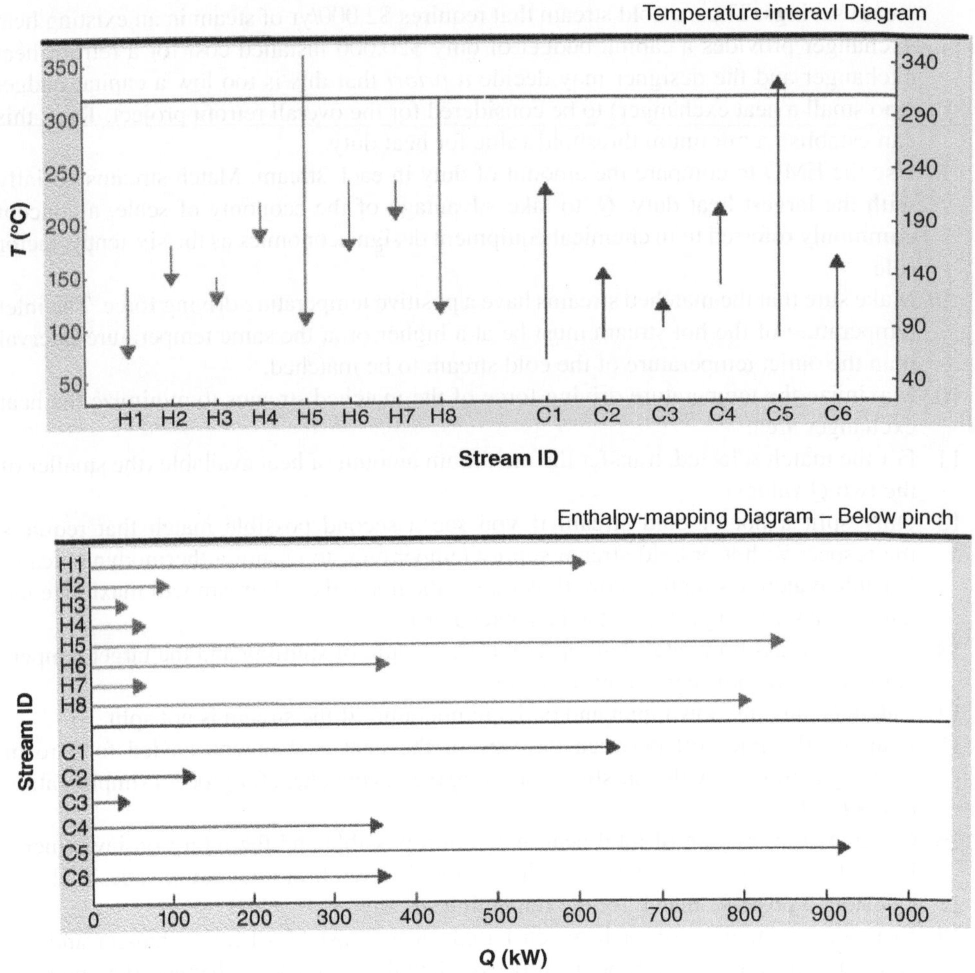

Figure 6.7 *Temperature–interval diagram and enthalpy-mapping diagram for a moderately sized industrial problem.*

stream matches need not be made, and the HEN to be implemented may be significantly streamlined when compared with the optimal network as determined by an optimization approach. Figure 6.7 shows the TID and EMD for a moderately sized HEN and illustrates this point.

For the case above, it is possible to achieve greater than 70% of the maximum energy recovery with the addition of only three heat exchangers below the pinch. A total of 2590 kW are available for recovery for this system. Matching H5 and C5 accounts for 840 kW, and matching H8 with C1 recovers an additional 640 kW. With only two matches, already 57% of the energy target has been met. One more heat exchanger – pairing H6 and C6 – contributes 360 kW more, and the sum of these three duties constitutes 71% of the maximum recovery.

This is a key strength of the graphical approaches when compared with mathematical optimization alternatives. Whereas optimization guarantees identification of a network that meets the energy targets, it often overcomplicates matters when applied to larger problems. While the optimal network may achieve 100% recovery, it may also be much more difficult to implement on a practical level. If a large fraction of the recovery may be achieved with only a very small number of heat exchangers, this is frequently a more attractive option for a plant considering retrofit than a plan that recovers all of the energy but requires many more exchangers. Optimization approaches do not generally allow for such considerations and thereby restrict the engineer to an appealing design on paper, but one that is wholly impractical.

6.3 Graphical Design Tools for Retrofitting Processes for Wastewater Reduction by Designing Water Recycle Networks

The concern for the depletion of clean water resources and the effect of disposing of chemical waste to the environment is also a major challenge for chemical industries. A shortcut graphical technique for water conservation design for retrofit processes is provided. Two important graphs are proposed to create initial water recycle networks for consideration in retrofit processes. Of particular importance is the usefulness of this technique for solving large industrial problems. The two key graphs used for this shortcut graphical technique are:

1. Material recycle pinch diagram [2]
2. The source-sink mapping diagram [2]

These two graphs will be illustrated using the water stream data provided in Table 6.2 below. A limitation of the graphical approach is that it can only be used to identify water recycle networks for streams containing only a single contaminant. As indicated previously, mathematical optimization approaches can be used to identify water recycle networks for streams involving multiple contaminants [14].

6.3.1 The Material Recycle Pinch Diagram

The material recycle pinch diagram is a graphical procedure for targeting the minimum amount of fresh feed usage and waste discharge as well as the maximum possible recycle

Table 6.2 *Example water recycle network stream data [8].*

Process number	Mass load of contaminant (kg/h)	Inlet concentration (ppm)	Outlet concentration (ppm)	Water flow (10^3 kg/h)
1	2	0	100	20
2	5	50	100	100
3	30	50	800	40
4	4	400	800	10

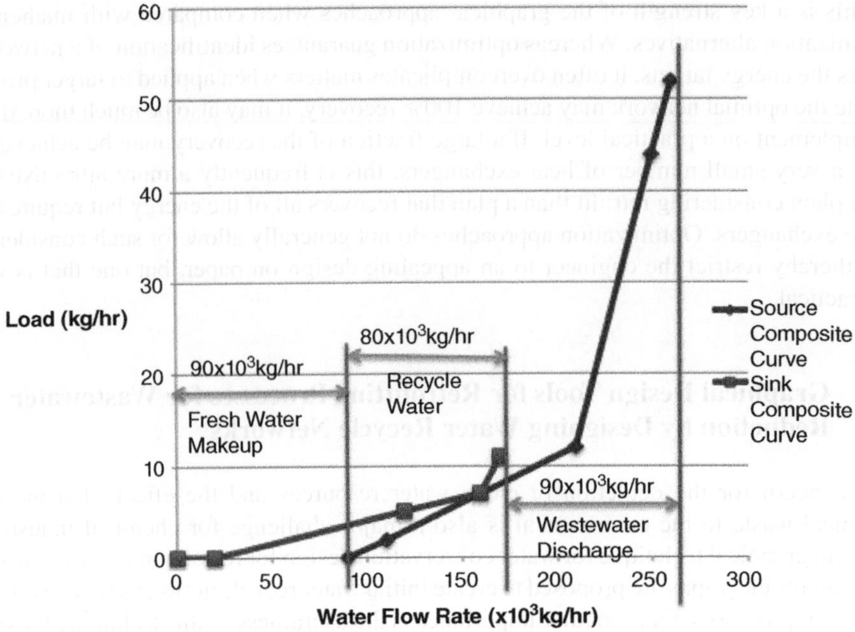

Figure 6.8 *The shifted material recycle pinch diagram for Table 6.2.*

amount. Source streams are defined as exit wastewater streams from a chemical process and some examples include, but are not limited to, blowdown from cooling towers and boilers, wash water, and purge water streams. They represent 'sources' of water that have the potential to be used for water recycle. Sink streams are defined as entering water streams to unit operations within chemical processes and some examples include, but are not limited to, water makeup to cooling towers and boilers, fresh wash water, and dilution water streams. They represent 'users' of water that can potentially use recycled water instead of freshwater makeup.

The source composite curve in Figure 6.A.1 has not been shifted. To determine the minimum amount of fresh feed usage and waste discharge as well as the maximum amount of recycle, the user has to move the source composite curve horizontally until the sink composite curve touches the source at the pinch point and the remainder of the source composite curve exists entirely to the right of the sink composite curve (that is, curves cannot overlap). Pinch point is the point where the load of recycled sources matches with that of the sinks [18]. Figure 6.8 shows the shifted material recycle pinch diagram.

The source composite curve in Figure 6.8 has been shifted by 90×10^3 kg/h to the sink. Now, the user can analyze the diagram using pinch analysis. The points below the pinch at which there are no sources (the region denoted 'fresh water makeup' by the horizontal arrows) show the minimum amount of fresh water. The points above the pinch where there are no sinks (the region denoted 'wastewater discharge' by the horizontal arrows) represent the minimum amount of waste discharge. The points where the sinks and sources overlap (the region denoted 'recycle water' shown by the horizontal arrows correspond to the maximum amount of recycle.

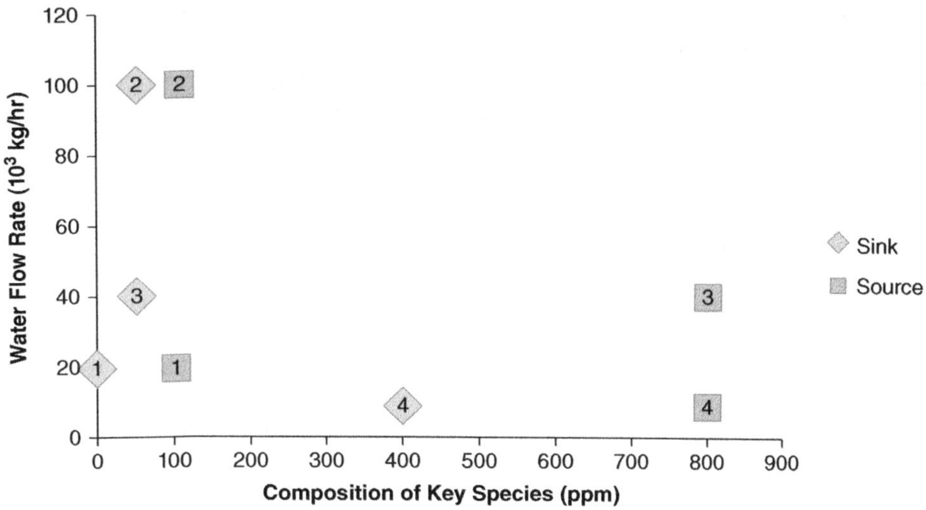

Figure 6.9 *The source–sink mass mapping diagram.*

6.3.2 The Source–Sink Mapping Diagram

All source and sink water streams can be plotted on a single diagram at their respective water flow rates and their stream compositions. Of particular use is the mass mapping diagram, which shows the water flow rate versus mass composition. Another form of source–sink mapping diagram is the thermal mapping diagram that shows the water flow rate versus stream temperature. For the thermal mapping diagram, 'temperature' is considered to be the stream contaminant or constraint. An example mass mapping diagram using the data provided previously in Table 6.2 is shown in Figure 6.9.

6.3.3 Suggested Guidelines for Identifying Water Recycle Matches Using the Material Recycle Pinch Diagram and Source–Sink Mapping Diagrams

The material recycle pinch diagram provides the maximum water recycle, minimum freshwater usage and minimum wastewater discharge targets. The mass mapping diagram provides a visual tool to aid the designer in identifying which source stream(s) should be recycled and to which sink in order to meet or approach the targets identified in the material recycle pinch diagram. Some guidelines for identifying water recycle networks for matching source and sink streams in retrofit processes using the material recycle pinch diagram/source–sink mapping diagram graphical shortcut method are provided below.

1. Identify all water discharge and water usage streams.
2. Separate streams into sources, wastewater that can be potentially recycled, and sinks, water usage streams that can potentially accept a portion of recycled wastewater.
3. Obtain flow rates and compositions of sources and sinks.
4. Generate the Material Recycle Pinch Diagram containing source composite curve and sink composite curve.

5. Generate the shifted Material Recycle Pinch Diagram by moving the source composite curve horizontally to the right to create a pinch point (the composite curves must touch) between the source composite curve and sink composite curve. This allows the identification of the minimum fresh water usage, the minimum wastewater discharge target, and the maximum water recycle potential.
6. The pinch point(s) divides the network into water recycle sub-networks. No waste should be discharged from sources below the pinch. No freshwater should be used in any sink above the pinch.
7. Determine the pinch point composition. This is the slope of the line segment that goes through the pinch. This slope will correspond to a source with the same composition.
8. Sources with compositions above the pinch should not be used with sinks below the pinch unless the source has the pinch composition. This source can be used in both networks, that above and below the pinch.
9. Design the network below the pinch before designing the network above the pinch. Source streams exiting the network below the pinch may be recycled to the network above the pinch.
10. Generate the Mass Mapping Diagram.
11. Supply fresh water to any sinks with 'no contamination' composition requirement.
12. Match the lowest concentration source with the lowest concentration sink. If contaminant balance is not complete, match the next lowest concentration source with this sink.
13. Add flow to complete the overall sink flow rate requirement if all potential sources streams have been used.
14. Analyze the Residual Mass Mapping Diagram to choose the next source and sink matching.
15. Complete the network to achieve the minimum fresh water usage target and the minimum wastewater discharge target.

An illustration of these concepts is provided in Appendix 6A.

6.4 Conclusions

A shortcut graphical technique for energy conservation design for retrofit process has been provided. Three important graphs are provided to create initial heat exchange networks for consideration in retrofit processes. Of particular importance is the usefulness of this technique for solving large industrial problems. The three key graphs used for this shortcut graphical technique are:

1. The TID
2. The heat pinch composite curves
3. The EMD

A shortcut graphical technique for water conservation design for retrofit processes has also been provided. Two important graphs are provided to create initial water recycle networks for consideration in retrofit processes. Of particular importance is the usefulness of this technique for solving large industrial problems. The two key graphs used for this shortcut graphical technique are:

1. Material recycle pinch diagram
2. The source–sink mapping diagram

Examples of these visual approaches have been illustrated in these chapters at an introductory level. Furthermore, the incorporation of these approaches in education through undergraduate chemical process design courses taught between 2011–2015 to chemical engineering students and the use of these approaches in solving industrial retrofit consulting projects have reinforced the usefulness of these graphical methodologies.

Appendix 6A: Illustrating the Water Recycle Network Design Guidelines

1. Identify all water discharge and water usage streams. See Table 6.A.1 Separate streams into **sources**, wastewater that can be potentially recycled, and **sinks**, water usage streams that can potentially accept a portion of recycled wastewater
2. Obtain flow rates and compositions of **sources** and **sinks**. See Table 6.A.1
3. Generate the **Material Recycle Pinch Diagram** containing **source composite curve** and **sink composite curve**. See Figure 6.A.1
4. Generate the shifted **Material Recycle Pinch Diagram** by moving the **source composite curve** horizontally to the right to create a pinch point between the **source composite curve** and **sink composite curve**. This allows the identification of the minimum fresh water usage, the minimum wastewater discharge target, and the maximum water recycle potential. See Figure 6.A.2
5. The pinch point(s) divides the network into water recycle sub-networks

Table 6.A.1 Data streams.

		Sink		Source
Process Number	Massload of contaminant (kg/hr)	Inlet contaminant concentration (ppm)	Outlet contaminant concentration (ppm)	Water flow rate ($\times 10^3$ kg/hr)
1	2	0	100	20
2	5	50	100	100
3	30	50	800	40
4	4	400	800	10
Total	41			170

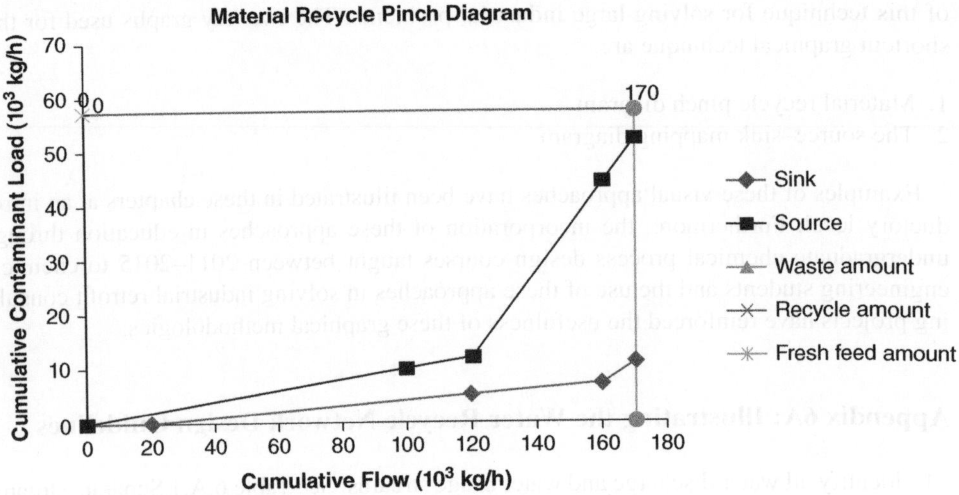

Figure 6.A.1 *The non-shifted material recycle pinch diagram.*

6. Determine the pinch point composition. This is the slope of the line segment that goes through the pinch. This slope will correspond to a source with the same composition. See Figure 6.A.3
7. **Sources** from above the pinch should not be used with **sinks** below the pinch unless the **source** has the pinch composition. This **source** can be used in both networks, that above and below that pinch. See Figure 6.A.4

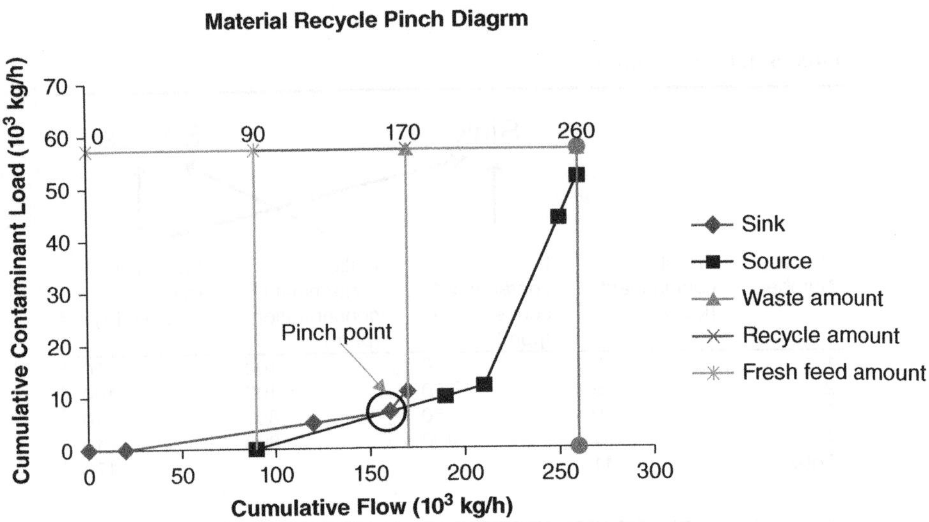

Figure 6.A.2 *The shifted material recycle pinch diagram.*

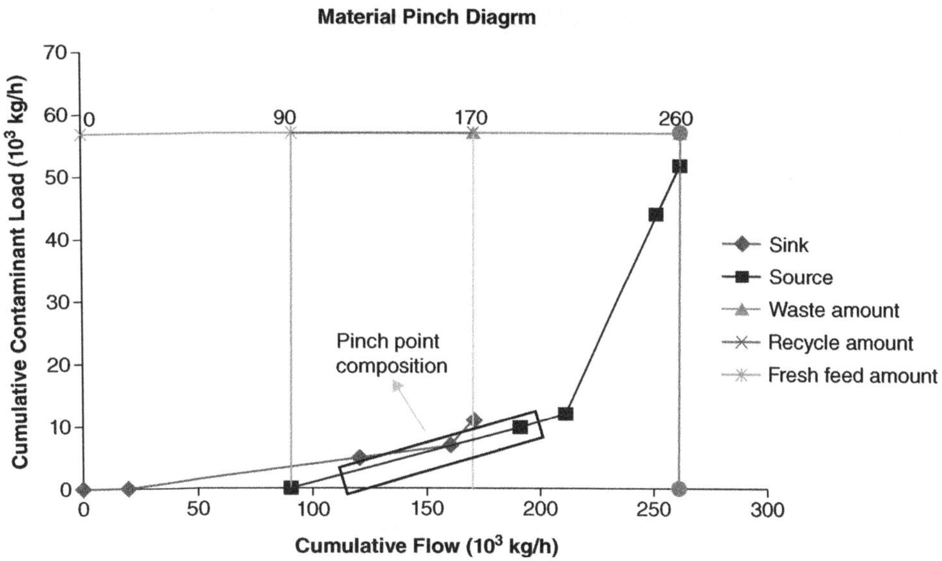

Figure 6.A.3 *The pinch point composition corresponds to the slope of the line segment that goes through the pinch point.*

8. Design the network below the pinch before designing the network above the pinch. **Source** streams exiting the network below the pinch may be recycled to the network above the pinch
9. Generate the **Mass Mapping Diagram**. See Figure 6.A.5
10. Supply fresh water to any **sinks** with 'no contamination' composition requirement. See Figure 6.A.6

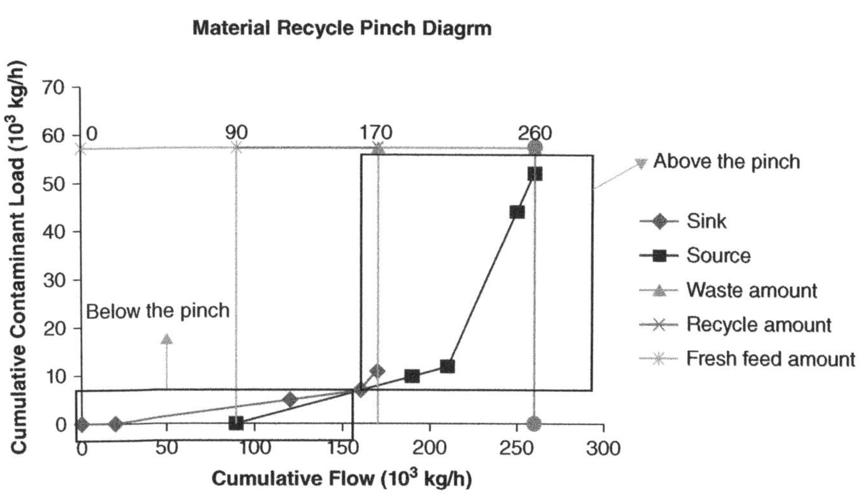

Figure 6.A.4 *The regions above and below the pinch point on the material recycle pinch diagram.*

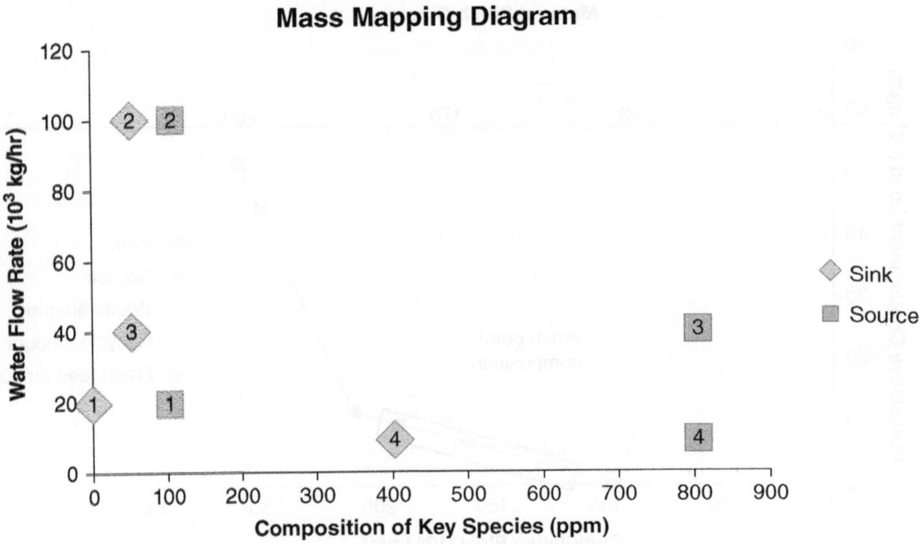

Figure 6.A.5 The mass mapping diagram.

11. Match the lowest concentration **source** with the lowest concentration **sink**. If contaminant balance is not complete, match the next lowest concentration **source** with this **sink.** See Figure 6.A.7
12. Add fresh water flow to complete the overall **sink** flow rate requirement if all potential **sources** streams have been used. Calculate the overall material balance and component material balance for this match

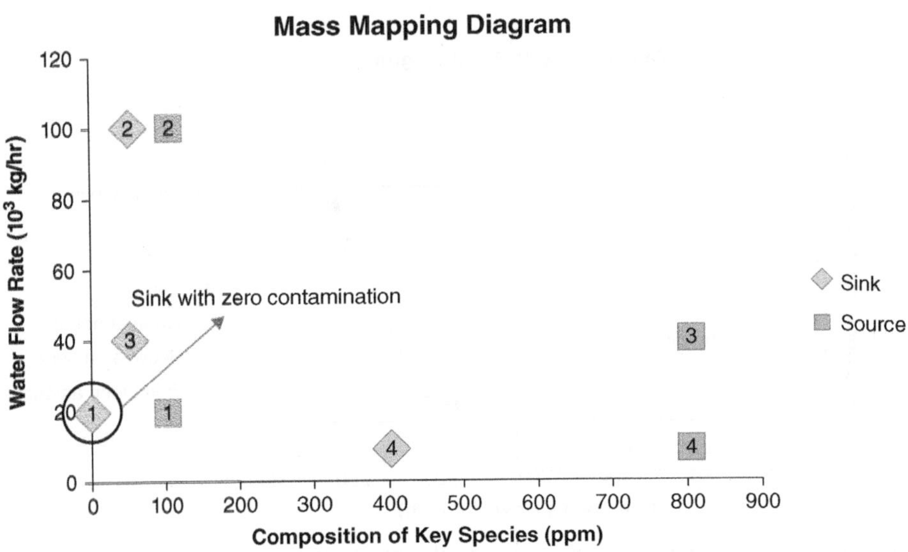

Figure 6.A.6 An example of the sink with zero contamination.

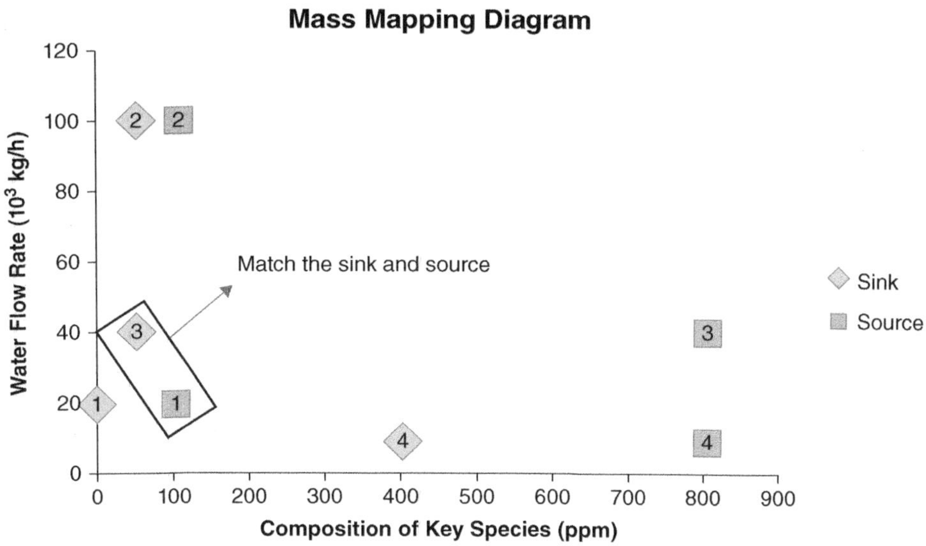

Figure 6.A.7 *The source and sink matching.*

13. Draw the **Residual Mass Mapping Diagram** as shown in Figure 6.A.8 with the selected source–sink match identified in step 12 eliminated. Analyze this diagram using step 11 to choose the next **source** and **sink** matching
14. Complete the network to achieve the minimum fresh water usage target and the minimum wastewater discharge target. See Figure 6.A.9

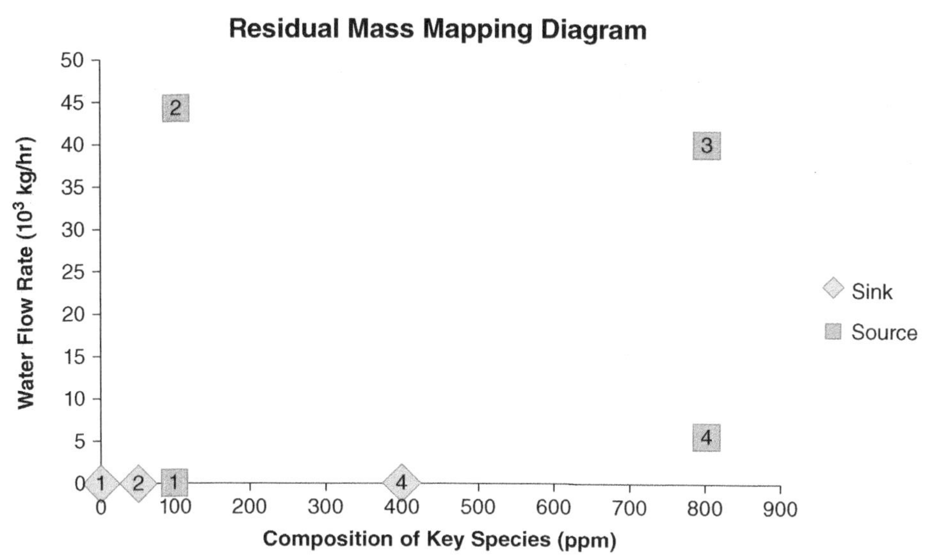

Figure 6.A.8 *The residual mass mapping diagram.*

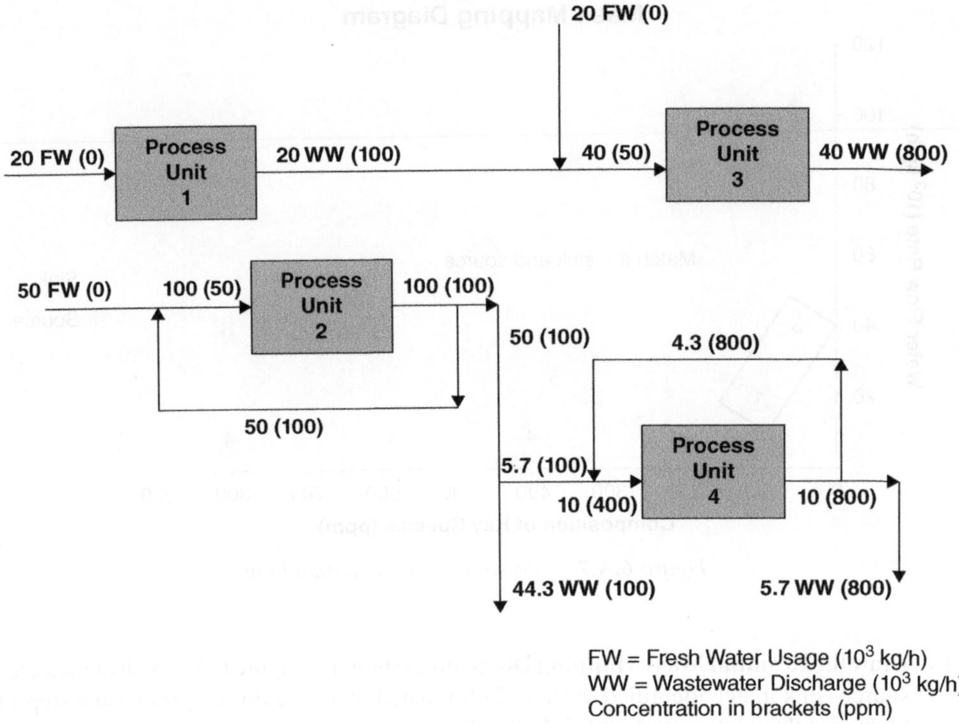

Figure 6.A.9 *A sample network solution. Ten solutions that meet the maximum water recycle target have been reported in the literature [11].*

Exercises

6.1 Given the stream data provided in Table 6.3, identify the energy targets and pinch temperatures. Design a HEN that achieves the maximum energy recovery.

6.2 A polystyrene manufacturing facility has the following streams that are currently heated (cold streams) using high-pressure steam (450 psi) and cooled (hot streams) using cooling water (20 °C). These utilities are available at \$14.50/1000 kg and \$0.02/m^3, respectively.

Table 6.3 *Process stream data for Exercise 1.*

	T_{in} (°C)	(°C)	Q (kW)
H1	240	120	240
H2	160	60	300
Cl	60	150	180
C2	90	180	360

Table 6.4 *Process stream data (hot streams) for Exercise 2.*

	T_{in} (°C)	T_{out} (°C)	Q (kW)
H1	360	180	540
H2	220	120	200
H3	240	120	240
H4	220	210	20
H5	180	60	480
H6	300	60	480
H7	150	80	210

(a) Which streams, if any, can potentially be omitted from consideration when designing a HEN for the facility? Why?
(b) Identify the pinch temperatures and energy targets for all streams in Tables 6.4 and 6.5. How are these affected if certain streams are omitted as suggested in part a?
(c) What are the potential savings on hot and cold utilities if a HEN meeting the maximum energy recovery is implemented?
(d) Design a HEN for achieving the maximum energy recovery.

6.3 A wastewater recycle network is desired for the following process units in a chemical plant (shown in Figure 6.E.1).
(a) Develop the sink and source composition curves. Determine the target for the maximum water recycle possible as well as the minimum fresh water required target and wastewater produced target. Identify the pinch composition.
(b) Design a wastewater recycle scheme that meets these targets.

6.4 Using the stream data provided in Tables 6.6 and 6.7, identify the recycle targets and design a water recycle network to minimize wastewater production.

Table 6.5 *Process stream data (cold streams) for Exercise 2.*

	T_{in} (°C)	T_{out} (°C)	Q (kW)
C1	60	120	180
C2	45	60	30
C3	120	240	240
C4	80	160	320
C5	80	120	120
C6	100	180	160
C7	120	200	320
C8	90	150	120
C9	60	80	40

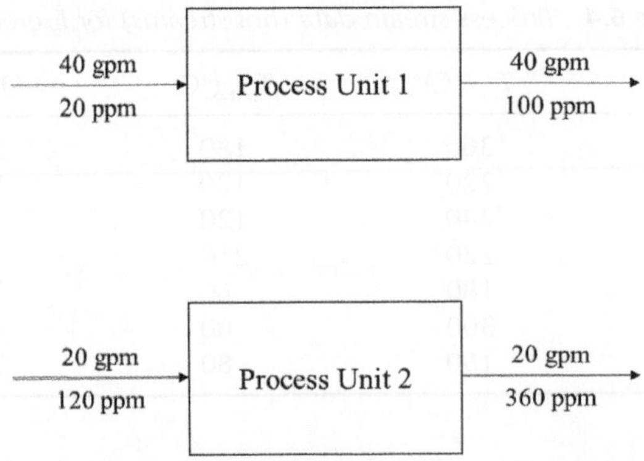

Figure 6.E.1 *Water flow and composition data for exercise 6.3.*

Table 6.6 *Process stream data (source streams) for Exercise 4.*

Source name	Flow rate (kg/h)	Composition (ppm)
R1	120	220
R2	100	150
R3	150	60

Table 6.7 *Process stream data (sink streams) for exercise 4.*

Sink name	Flow rate (kg/h)	Composition (ppm)
S1	200	30
S2	90	50
S3	110	40
S4	80	0

References

(1) El-Halwagi, M.M. (1997) *Pollution Prevention Through Process Integration: Systematic Design Tools*, Academic Press, San Diego.
(2) El-Halwagi, M.M. (2006) *Process Integration*, Academic Press, San Diego.
(3) El-Halwagi, M.M. and Spriggs, H.D. (1998) Solve design puzzles with mass integration. *Chemical Engineering Progress*, **94**, 25–44.
(4) Linnhoff, B. (1993) Pinch analysis — a state of the art overview. *Transactions of the Institution of Chemical Engineers, Chemical Engineering Research and Design.*, **71** (Part A5), 503–522.
(5) Shenoy, U.V. (1995) *Heat Exchange Network Synthesis: Process Optimization by Energy and Resource Analysis*, Gulf Publishing Co, Houston.
(6) Douglas, J.M. (1988) *Conceptual Design of Chemical Processes*, McGraw-Hill, New York.
(7) Gundersen, T. and Naess, L. (1988) The synthesis of cost optimal heat exchanger networks: an industrial review of the state of the art. *Computers & Chemical Engineering*, **12**, 503–530.
(8) Wang, Y.P. and Smith, R. (1994) Wastewater minimization. *Chemical Engineering Science*, **49**, 981–1006.

(9) Bagajewicz, M. (2000) A review of recent design procedures for water networks in refineries and process plants. *Computers & Chemical Engineering*, **24**, 2093–2113.

(10) Dunn, R.F. and Dobson, A.M. (1999) Water conservation and reuse at a major CPI facility, National Association of Corrosion Engineering Annual Conference Proceedings, paper 317, 1–18.

(11) Dunn, R.F. and Wenzel, H. (2001) Process integration design methods for water conservation and wastewater reduction in industry, part 1: design for single contaminants. *Clean Products and Processes*, **3**, 307–318.

(12) Parthasarathy, G., Dunn, R.F., and El-Halwagi, M.M. (2001) Development of heat-integrated evaporation and crystallization networks for ternary wastewater systems. 2. Interception task identification for the separation and allocation network. *Industrial & Engineering Chemistry Research*, **40**, 2842–2856.

(13) Dhole, V.R., Ramchandi, N., Tainsh, R.A., and Wasilewski, M. (1996) Make your process water pay for itself. *Chemical Engineering*, **103** (1), 100–103.

(14) Dunn, R.F., Wenzel, H., and Overcash, M. (2001) Process integration design methods for water conservation and wastewater reduction in industry, part 2: design for multiple contaminants. *Clean Products and Processes*, **3**, 319–329.

(15) Parthasarathy, G. and Dunn, R.F. (2005) *Minimization of Environmental Discharge Through Process Integration, Towards Zero Discharge: Innovative Methodology and Technologies for Process Pollution Prevention*, John Wiley & Sons, Inc., New Jersey.

(16) Dunn, R.F. (2001) *Challenges for Applying Process Integration Design Methodologies in the Chemical Process Industries, Process Design Tools for the Environment*, Taylor & Francis Publishers, New York.

(17) Dunn, R.F. and El-Halwagi, M.M. (2003) Process integration technology review: background and applications in the chemical process industry. *Journal of Chemical Technology and Biotechnology*, **78**, 1011–1021.

(18) Kemp, I.A. (2006) *Pinch Analysis and Process Integration (Second Edition) A User Guide on Process Integration for the Efficient Use of Energy*, Elsevier.

(19) El-Halwagi, M.M. (2012) *Sustainable Design through Process Integration: Fundamentals and Applications to Industrial Pollution Prevention. Resource Conservation, and Profitability Enhancement*, Elsevier.

7

Heat Exchanger Network Retrofitting: Alternative Solutions via Multi-objective Optimization for Industrial Implementation

B.K. Sreepathi and G.P. Rangaiah
Department of Chemical & Biomolecular Engineering,
National University of Singapore, Singapore

7.1 Introduction

Chemical processes involve reactions and/or separations for transforming raw materials into valuable products required by society. These processes often occur at elevated temperatures, whereas raw materials are mostly at ambient temperature and products are often stored at about room temperature. Hence, energy is required for heating the material streams before reaction and/or for separation, and then outlet streams have to be cooled using cold utilities such as water and air. There is potential to recover thermal energy from hot process streams and to reuse it for heating cold process streams. For this, heat exchanger networks (HENs) are employed in process plants; each network consists of more than one heat exchanger. Thus, energy integration and HEN help to improve the energy efficiency of chemical processes. This led to numerous studies on the synthesis and design of HENs in the last four decades after the oil crisis of the 1970s.

The existing plants were designed and established many years ago according to the then prevailing economic and technological conditions. Energy resources are depleting, and environmental concerns and energy prices are increasing. Hence, improving energy

Chemical Process Retrofitting and Revamping: Techniques and Applications, First Edition.
Edited by Gade Pandu Rangaiah.

integration in existing process plants is essential for enhancing energy efficiency of chemical processes since most of their energy requirement is for heating and cooling of process streams. Energy integration can be increased by retrofitting of HENs in existing process plants. HEN retrofitting involves area addition, use of heat-transfer enhancements, installing new exchangers and/or relocating existing heat exchangers to reduce the hot and cold utilities required. The ideal scenario would be 'zero-utilities', which is likely to be infeasible because of both the temperature-driving force required for heat transfer (that is, heat exchanger approach temperature) and the energy available in hot process streams compared with that required by cold process streams.

The reader is referred to books by Shenoy (1995), Smith (2005) and Kemp (2007) for basics and methods for heat integration and their applications. Klemeš *et al.* (2013) described the recent developments in process integration in general. The phrase 'HEN revamping' is sometimes used for HEN retrofitting, but the latter is common in the literature on heat integration, and hence HEN retrofitting is used throughout this chapter. Sreepathi and Rangaiah (2014a) reviewed over 80 journal papers published from 1993 to 2013 on HEN retrofitting methodologies and their applications. There has been an increasing interest in research and development on HEN retrofitting in recent years, as can be seen from the number of journal papers published since 1993 in Figure 7.1.

HEN retrofitting methods can be broadly classified into three groups: (a) pinch analysis-based methods, (b) mathematical programming-based methods, and (c) hybrid methods. Pinch analysis methods, as the name indicates, employ pinch analysis and heuristics to solve retrofit problems. These include composite curves, grand composite curves, energy/area targeting, grid diagrams and/or visualization techniques (Smith, 2005; Kemp, 2007). In pinch analysis methods, retrofitting is performed based on the recognition of a pinch point

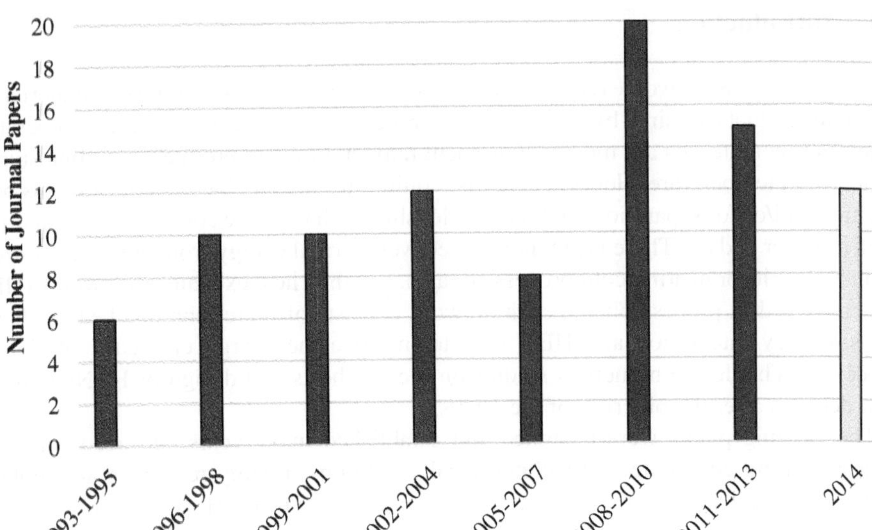

Figure 7.1 *Trend of the number of journal papers on HEN retrofitting, published from 1993 to 2014. Note that the last bar for 2014 is for one year only.*

and following the three basic principles; namely, no cold utility above the pinch point, no hot utility below the pinch point, and no process heat exchange across the pinch. The match that limits the heat recovery in an existing HEN is known as the 'pinching match' and the point at which this occurs is known as 'network pinch' (Asante and Zhu, 1996). In some retrofitting methods, this information is used to relax the network pinch via process changes like flow rate and temperature changes. For recent studies on pinch-based retrofitting, see Nordman and Berntsson (2009), Li and Chang (2010) and Abbood *et al.* (2012).

Mathematical programming-based methods formulate a model of the retrofitting problem and utilize an optimization method to generate solutions; hence, they can also be called optimization-based methods. These can be divided into two types based on the nature of the optimization method used: deterministic and stochastic. Deterministic methods use nonlinear programming (NLP) or mixed integer nonlinear programming (MINLP) techniques along with superstructure representation of HEN, to solve the optimization problem for retrofitting. Stochastic methods use techniques such as simulated annealing (SA), genetic algorithm (GA) or differential evolution (DE) to solve the retrofitting optimization problem. Mathematical programming-based methods can be step-wise/sequential or simultaneous. The step-wise approach is relatively faster, but may give a sub-optimal solution. The simultaneous approach, in theory, can find the optimal solution but requires more computational time. So, these are chosen based on necessity and the scenario at hand. For recent research on mathematical programming-based HEN retrofitting, see Ponce-Ortega *et al.* (2008), Pan *et al.* (2014) and Sreepathi and Rangaiah (2014b, 2014c).

Hybrid methods make use of both pinch analysis and mathematical programming methods, to exploit the strengths of both. These provide some user interaction and can also solve large problems. Studies on hybrid methods include Asante and Zhu (1996), Briones and Kokossis (1999) and Smith *et al.* (2010). See Sreepathi and Rangaiah (2014a) for a review of journal papers from 1993 to 2014 on HEN retrofitting methods and their applications.

Multi-objective optimization (MOO) has found numerous applications in chemical engineering (Rangaiah, 2009; Rangaiah and Bonilla-Petriciolet, 2013). It provides many optimal solutions (known as Pareto-optimal solutions/front or non-dominated solutions) with different trade-offs among the specified objectives. These solutions are equally good with respect to the objectives used in the MOO problem; that is, when two of them are compared, one solution is better in at least one objective and simultaneously worse in at least one other objective, than another solution. Thus, they show a quantitative trade-off among the objectives used in the MOO problem. The Pareto-optimal solutions generated by MOO can be reviewed, and one of them can be chosen based on user preference. For example, where investment cost is constrained, a suitable retrofit solution under this constraint can be chosen. Recently, MOO was applied to three HEN retrofit problems by Sreepathi and Rangaiah (2014b). It is desirable to know the quantitative trade-off among the objectives for a deeper understanding and for selection of an optimal solution for implementation. Further, discontinuities in the Pareto-optimal front may correspond to installation of a new exchanger or a change in the HEN structure. Knowing all these, the optimal solution, meeting the practical considerations not included in the MOO problem, can be chosen from among a few nearby optimal solutions.

In this chapter, we will study the potential of MOO for HEN retrofitting in three levels. The first level deals with retrofitting by area addition in the existing heat exchangers to improve energy recovery and reuse. Area addition is relatively easy to implement, and so this level is referred to as simple retrofitting in this chapter. The second level of retrofitting considers both area addition in the existing exchangers and installation of new heat exchangers; this level is referred to as moderate retrofitting. The third level involves structural modifications (i.e., relocation of existing heat exchangers) along with area addition in existing exchangers and installation of new heat exchangers. Since structural modifications are more difficult, the third level is referred to as complex retrofitting in this chapter. MOO provides many optimal solutions for each level of retrofitting, and it is interesting to compare them and to analyse possible additional benefit from increased extent of retrofitting.

The rest of this chapter is structured as follows. The next section describes HENs and the structural representation used in this study for HEN retrofitting. Section 7.3 discusses the improvement techniques that can be used to retrofit HENs. Section 7.4 describes the methodology to retrofit HENs, which includes MOO, HEN model and exchanger reassignment strategy (ERS). The case study solved by the methodology is described in Section 7.5. Section 7.6 discusses the results obtained for simple, moderate and complex retrofitting. Finally, conclusions of this study are presented in Section 7.7.

7.2 Heat Exchanger Networks

HEN and its representations are described in this section. For this, consider a typical process flow sheet, shown in Figure 7.2 (top plot), having a reactor, flash separator, distillation column and a number of heat exchangers including heaters and coolers. In this process, HEN consists of all heat exchangers as well as hot and cold streams going through them. Design and operating conditions of this HEN (including number of exchangers) can be changed for better energy integration; very importantly, this should be performed without changing the inlet and outlet conditions of other process units; namely, reactor, flash separator and distillation column. For example, the stream 'Effluent' has to be cooled from 300 °C at the reactor exit to 40 °C at the 'Flash Separator' inlet; this can be achieved by heat exchange with process streams and/or cold utility in one or more exchangers, as part of HEN design or retrofitting.

In Figure 7.2, hot streams are shown by solid arrows and cold streams by dashed arrows. The 'Effluent' from the reactor coming out at 300 °C is split and passed through two heat exchangers to increase the temperature of the 'Feed' and 'Recycle' streams. Then, to meet the requirement for the inlet temperature of the 'Flash Separator', the 'Effluent' stream is finally cooled to 40 °C in a cooler. Similarly, the 'Product' stream at 280 °C is used to pre-heat the 'Flash Liquid' before it enters the distillation column. It is then cooled in a cooler to the specified outlet temperature (40 °C) for storage.

The process flow diagram (such as the top plot in Figure 7.2) shows the complete process including heat exchangers. To focus on HEN analysis and design, a 'grid diagram' to represent the network was introduced by Linnhoff and Flower (1978). In the grid diagram (bottom plot in Figure 7.2), all hot and cold streams are drawn as horizontal lines, and each exchanger between them is represented by two circles joined with a vertical line. Hot streams are placed in the top of the diagram whereas cold streams are at the bottom of the diagram. Since heat exchangers are mostly counter-current type, streams are presented accordingly, that is, hot streams flowing from left to right, and cold streams flowing from right to left, thus making it easier to check approach temperatures of each exchanger. Each

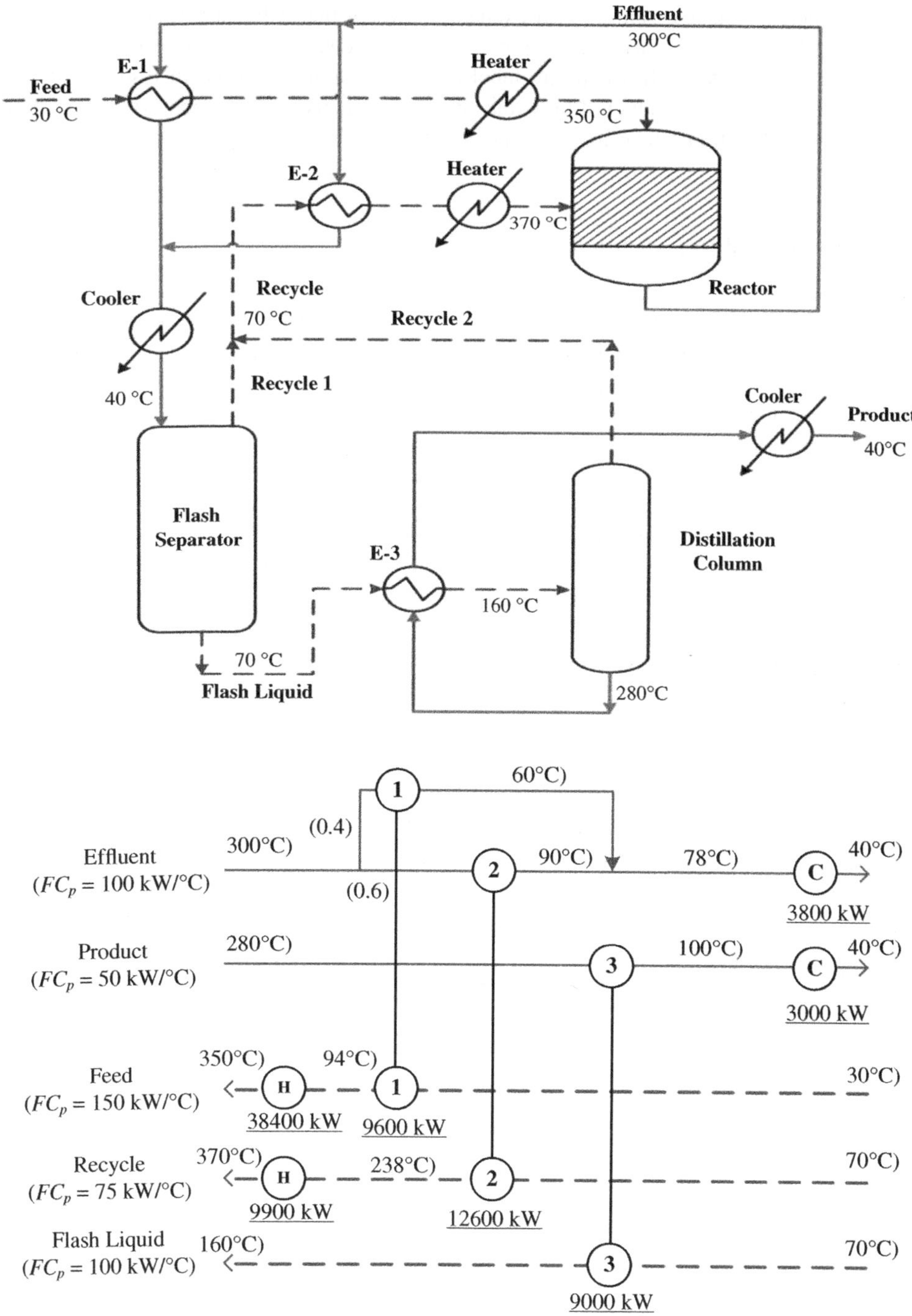

Figure 7.2 *Simple process flow sheet with hot streams shown by continuous arrows and cold streams shown by dashed arrows (top plot), and corresponding grid diagram of HEN (lower plot).*

heater/cooler, which is essentially an exchanger except that it has a utility stream on one side instead of a process stream, is shown in a single circle; utility streams are often not shown for brevity. Apart from stream label and temperatures (shown in Figure 7.2), heat capacity flow rate (i.e., product of flow rate and heat capacity, FC_P) of each stream is shown near/below its label, and heat load of an exchanger is given below its bottom circle in the grid diagram.

7.2.1 Structural Representation

Grid representation such as Figure 7.2 is useful to design HEN and also to review heat recovery in an existing plant. HEN in the grid diagram needs to be coded/represented using a suitable scheme and numbers for manipulation by optimization (mathematical programming) methods on a digital computer. This sub-section describes this representation of HEN structure, used in this chapter. This representation is used to model the existing HEN and then to optimize to generate retrofit solutions.

The structural representation used in this chapter is adapted from Bochenek and Jezowski (2010). It is called 'nodal representation' as it uses nodes for identifying the potential sites for heat exchangers, stream splitting and mixing. A Structural matrix (**SM**) is used to encapsulate the topological features of HEN; that is, it has hot and cold stream numbers as well as node numbers on them for each heat exchanger. For this, each hot stream is numbered starting from 1 onwards; split streams, if present, are numbered towards the end. The number of nodes on each hot stream is assumed. Similarly, cold streams and their split streams are numbered, and the number of nodes on each cold stream is assumed. All these can be seen in Figure 7.3. For example, one of the hot streams in this figure has a split stream, and so main hot streams are numbered H1 and H2, followed by H3 for the split stream. H3 is the split stream on H1. Note that streams are numbered without H or C in SM.

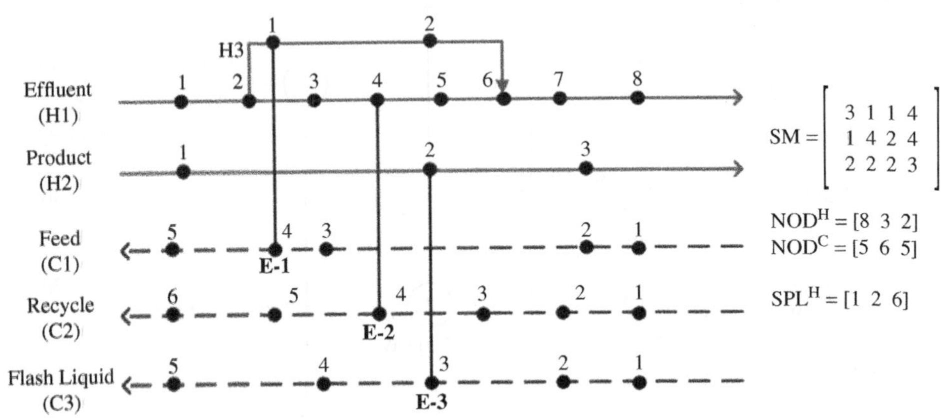

Figure 7.3 Nodal representation of HEN corresponding to the process shown in Figure 7.2. In this HEN, E-1, E-2 and E-3 are heat exchangers, and heaters/coolers are not shown. HEN structure is represented by matrices **SM**, $\mathbf{NOD^H}$, $\mathbf{NOD^C}$ and $\mathbf{SPL^H}$ on the right side. See Section 7.2.1 for more details.

For example, consider the following SM [Equation (7.1)]:

$$\mathbf{SM} = \begin{bmatrix} 3 & 1 & 1 & 4 \\ 1 & 4 & 2 & 4 \\ 2 & 2 & 3 & 3 \end{bmatrix} \tag{7.1}$$

Each row of this matrix represents one heat exchanger in the HEN. The first column of this matrix has the number of the hot stream flowing through this exchanger, and the second column has the node number on that hot stream; similarly, the third and fourth columns represent the cold stream number and the node number on that stream, respectively. Thus, the first heat exchanger (E-1) corresponds to the first row $[3 \quad 1 \quad 1 \quad 4]$ of the **SM** in Equation (7.1); it means there is a heat exchanger between the 1st node of hot stream 3 and the 4th node of cold stream 1. Similarly, the second row of SM in Equation (7.1) is for E-2, which is between the 4th node of hot stream 1 and the 4th node of cold stream 2.

The number of nodes on all hot streams is in a row vector, $\mathbf{NOD^H}$ and the number of nodes on all cold streams is in another row vector, $\mathbf{NOD^C}$. For example, $\mathbf{NOD^H} = [8 \quad 3 \quad 2]$ means there are 8, 3 and 2 nodes, respectively, on hot streams 1, 2 and 3, as shown in Figure 7.3. Similarly, $\mathbf{NOD^C} = [5 \quad 6 \quad 5]$ means there are 5, 6 and 5 on cold streams 1, 2 and 3, respectively. The number of nodes on a stream can be the same or different; for a retrofit problem, these can be chosen based on the existing heat exchangers; if a stream has more heat exchangers in the original network, then more nodes are chosen for it. Also, more nodes should be assumed for streams with large enthalpy changes since they may require more exchangers.

Two matrices, $\mathbf{SPL^H}$ and $\mathbf{SPL^C}$, represent all possible splits on hot and cold streams, respectively. In each of these matrices, the first column contains the number of the stream (having the split), the second and third columns have, respectively, the starting and ending nodes of the stream, where split occurs. For example, $\mathbf{SPL^H} = [1 \quad 2 \quad 6]$ signifies a split stream on a hot stream starting from its 2nd node and ending at its 6th node. Splitting is specified on streams having large values of FC_P and, in retrofitting problems, on streams already having splits. Recall that the numbering of streams is such that all the main streams are numbered first, followed by split streams.

In solving retrofit problems using optimization and nodal representation, decision variables are structural and parameter changes, which can be optimized. Structural changes include, but are not limited to, heat exchanger re-piping (where either one or both of the streams involved in the heat exchanger are changed), removal of an existing exchanger or addition of a new heat exchanger, and shifting of a heat exchanger on the same stream (referred as re-sequencing). Re-sequencing and re-piping together are referred to as re-location. Parameter changes include the split ratios and heat exchanger areas. Structural changes are discrete variables (in **SM** matrix) whereas parameter changes are continuous variables, thus resulting in a complex mixed integer non-linear programming (MINLP) problem.

7.3 HEN Improvements

Consider the heat transfer rate equation and possible improvements that can be done for improving energy recovery in HEN. Heat transfer rate (Q) is the product of overall heat transfer coefficient (U), heat transfer area (A) and log-mean temperature driving force (ΔT_{lmtd}), that is, $Q = UA\Delta T_{\text{lmtd}}$. Hence, the amount of heat transferred in the HEN can be improved by increasing U, A and/or ΔT_{lmtd}. Since U depends mainly on the film heat transfer coefficient and fouling factor of hot and cold streams in the heat exchanger, U can be improved by enhancing heat transfer coefficients and/or decreasing fouling. The second parameter, A, is one of the common parameters varied in HEN retrofit. An existing exchanger achieves greater heat exchange between the same streams for an increment in the heat transfer area. With increased area, tube velocity will reduce (for the same volumetric flow rate), which leads to a lower U value and increased fouling rate; if required, fluid velocity can be maintained by increasing the tube passes or using tube inserts. ΔT_{lmtd} can be varied by changing the inlet temperatures of the streams via re-sequencing or re-location of the heat exchanger. All these potential improvements for increasing heat exchange in a heat exchanger and consequently in HEN are discussed in the rest of this section.

Area Addition: HENs can be retrofitted by adding area in the existing exchangers, thus increasing the amount of heat transfer between the two streams already matched. The amount of area that can be added to a heat exchanger depends on its type. For example, in a plate-and-frame heat exchanger, heat transfer area can be increased even by 100% based on the space available to add more plates. On the other hand, the area of a shell-and–tube (ST) heat exchanger can be increased by only 10–20% without adding an extra shell; this is by inserting additional tubes within the space available in the existing tube bundle. The cost of area addition is generally of the form: $a(\Delta A)^b$ where ΔA is the additional area. For example, in Shenoy (1995), capital cost for area addition is $750 \times (\Delta A)^{0.81}$. The change in pressure drop is negligible for area addition of 10–20%. In this chapter, area addition of up to 15% is considered to avoid the effects of pressure drop increase.

Heat Transfer Enhancements: Heat transfer enhancements increase the film heat transfer coefficient, thus improving U and consequently heat transfer in the exchanger. In the case of the ubiquitous ST heat exchanger, heat transfer enhancement can be achieved on the shell and/or tube sides as required. Tube-side enhancements help in improving the film heat transfer coefficient on the tube side, and are useful if the exchanger involved has lower film heat transfer coefficient on the tube side. Tube-side enhancements include, but are not limited to, twisted-tape inserts, coiled-wire inserts and internal fins. Similarly, shell-side enhancements are used to improve the heat transfer coefficient on the shell side. They include helical baffles, external fins and Expanded Metal (EM) baffles. More details on heat transfer enhancements are available in Pan *et al.* (2013).

Heat transfer enhancements can be implemented without much effect on the pressure drop in the exchanger, though this may involve slight modifications in the exchanger. In such cases, enhancements behave in a similar manner to area addition to an existing exchanger. Cost of enhancement has a fixed cost term and an area-dependent term, that is, $a + b \times$ Area (Pan *et al.*, 2013). The improvement achieved by enhancements in an exchanger may not be limited to 10–20%, as is the case with area addition to an ST exchanger. It can be up to 100% in the film heat transfer coefficient using coiled-wire inserts and internal fins together (Pan *et al.*, 2013), though pressure drop will increase significantly, requiring retrofit of the associated pump/compressor or a lower pressure on the downstream. In this chapter, only area addition is considered in the application studied; if necessary, the corresponding increase in UA can be achieved by using heat transfer enhancements to improve U.

Addition of New Exchangers: A new heat exchanger can be added to the existing HEN to increase the heat recovery and reuse. It can be in series or parallel to an existing exchanger. Parallel arrangement is recommended only if the additional heat exchanger duty is associated with increased cold and hot stream flow rates compared with the original case. If cold/hot stream flow rates do not change, parallel arrangement leads to a decrease in fluid velocity and consequently increased fouling. In such a situation, parallel arrangement of new exchangers may not be sustainable, and care should be taken to ensure that fluid velocity does not drop below the recommended velocity for minimizing fouling. In series combination, a higher cold stream outlet temperature is expected for the same flow rate of hot/cold streams. The outlet temperature will not change much in parallel combination but it can handle more flow rate. Thus, in the case of throughput increase, parallel combination is preferred to maintain the same outlet temperature and to avoid lower fluid velocity. Here, the case study considered has the same throughput, and so only series addition of a new exchanger is considered. In Figure 7.4(b), exchanger N is the new one added to HEN in Figure 7.4(a). Note that the addition of a new exchanger in series causes additional pressure drop, which may require retrofit of the associated pump/compressor. Other options for reducing pressure drop are to replace existing piping (both upstream and downstream of the heat exchanger) with bigger piping, installing a pipe header in parallel and/or replacing control valves with bigger valves.

Re-sequencing of Exchangers: Re-sequencing of a heat exchanger refers to moving the heat exchanger on the same two streams, and so its inlet/outlet temperatures of hot/cold streams change (i.e., there is no change in the streams involved in the heat transfer). In Figure 7.4(a), heat exchanger 3 is present between streams H1 and C2 before the split occurs on H1, whereas it is after the split merges in Figure 7.4(b). This shifting of exchanger 3 is called re-sequencing of a heat exchanger. During re-sequencing of heat exchangers, care should be taken not to violate the design temperature and pressure of heat exchangers. Also, there may be changes in pressure relief system requirements due to tube rupture. This needs to be reviewed before selecting the final re-sequencing scheme.

Re-piping of Exchangers: Re-piping of a heat exchanger refers to shifting the heat exchanger to different streams, that is, change of cold and/or hot streams in that exchanger. For example, in Figure 7.4(a), exchanger 2 is present on the main stream of H1 during split and C2, whereas it is between streams H2 and C2 in Figure 7.4(b). One should ensure that the re-piped heat exchanger meets the material of construction, temperature and pressure requirements of new stream(s) expected to flow through it. If re-piping necessitates pipe rack modifications, piping cost can be very significant. Hence, piping cost should be considered for selecting the most profitable HEN configuration. Both re-piping and re-sequencing together are referred to as re-location of heat exchangers.

Pressure Drop Considerations: HEN modifications discussed above may increase the pressure drop in exchangers. Existing pumps and compressors may have been designed with some margin to handle increased pressure drop and/or flow rate. If the pressure drop caused by the use of additional area/heat transfer enhancement in an existing exchanger or addition of a new exchanger is more than the design capacity of the pump/compressor, then a new pump/compressor may have to be installed.

New centrifugal pumps are generally specified with a 10% safety margin on flow rate. Further, a selected impeller will be of a medium size between the maximum and minimum impeller sizes possible for the pump casing. Thus, there will be opportunity to replace the

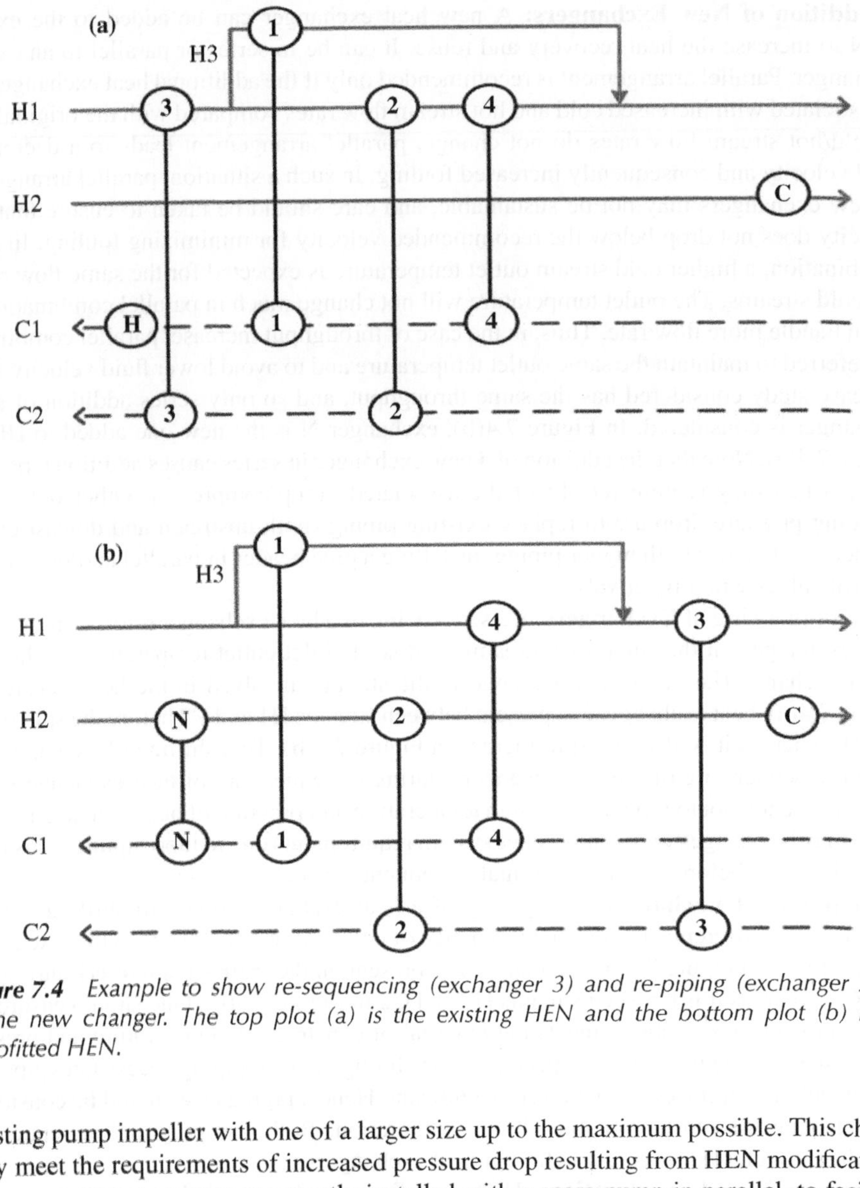

Figure 7.4 Example to show re-sequencing (exchanger 3) and re-piping (exchanger 2); N is the new changer. The top plot (a) is the existing HEN and the bottom plot (b) is the retrofitted HEN.

existing pump impeller with one of a larger size up to the maximum possible. This change may meet the requirements of increased pressure drop resulting from HEN modifications. Moreover, process pumps are mostly installed with a spare pump in parallel, to facilitate the pump maintenance and improve reliability. Parallel operation of both the pumps can increase the pump head for the same flow rate. This can easily overcome the additional pressure drop arising from HEN modifications. One drawback of this approach is that, during maintenance of any of these pumps, heat exchanger capacity will decrease. This may be acceptable for some plants with budget constraints. For increasing capacity/head of existing compressors, see Tables 2.7 and 2.8 in Chapter 2 of this book for several options.

In summary, increased pressure drop in an exchanger can be offset by making simple modifications like control valve type and pipe size to reduce their pressure drop. Area addition within 10–20% causes a relatively small increase in pressure drop, which can be handled by the design margin of pumps present. For heat transfer enhancements in an ST exchanger, where the pressure drop can be large, some modifications such as decreasing the number of tube passes and changing baffles on the shell side can be considered and implemented to manage with existing pumps. Hence, pressure drop constraints are not considered in the HEN model used for retrofitting in this study, which is applicable to both liquid and gas streams in the process.

7.4 MOO Method, HEN Model and Exchanger Reassignment Strategy

It is evident from the previous section that there are many options and combinations for retrofitting HENs to improve the energy recovery, thus requiring an optimization method to find the best combination. Further, investment is often limited, and it is better to have many optimal solutions with different investment requirements. For this, MOO methodology, HEN model calculations and ERS are described in this section.

7.4.1 Multi-objective Optimization

In process design and retrofitting, optimization is used for better performance of the process for given conditions. This optimization is with respect to continuous and/or discrete (decision) variables, and involves complex, non-linear equations along with equality and inequality constraints arising from governing equations and equipment limitations. Optimization of chemical processes has been focused primarily on single objective optimization, that is, dealing with only one objective function. In reality, these problems involve two or more objectives that are generally conflicting such as investment cost, utility cost, and safety. Such problems require MOO. See Chapter 4 in this book for more details on optimization methods and programs.

In this work, elitist non-dominated sorting genetic algorithm (NSGA-II) is used for MOO of HEN retrofitting. In this algorithm, the initial population is first generated randomly between the specified bounds on decision variables, and the corresponding values of objective function and constraints are calculated (Figure 7.5). This step is followed by many generations until convergence. In each generation, mutation and crossover operations are performed on the current/parent population to generate new/offspring population followed by calculations of objective functions and constraints. The parent and offspring populations are then combined, and non-dominated sorting is performed on the total population to choose the better individuals; if there are constraints, constraint dominance criteria are used instead of non-dominated sorting. The crowding distance is employed to guide the selection towards a well-distributed Pareto-optimal front. For more details about these operations and NSGA-II, see the book by Deb (2001).

The NSGA-II program, coded in MATLAB, is available with the name 'NGPM' on the Mathworks website (http://www.mathworks.com/matlabcentral/fileexchange/31166-ngpm-a-nsga-ii-program-in-matlab-v1-4, access date: November 2014). The feasibility approach, which employs constraint dominance criteria, is used to handle constraints in HEN retrofit problem. In this approach, for selection between two solutions A and B, if both of them are feasible (i.e., given constraints are satisfied), then the solution with the

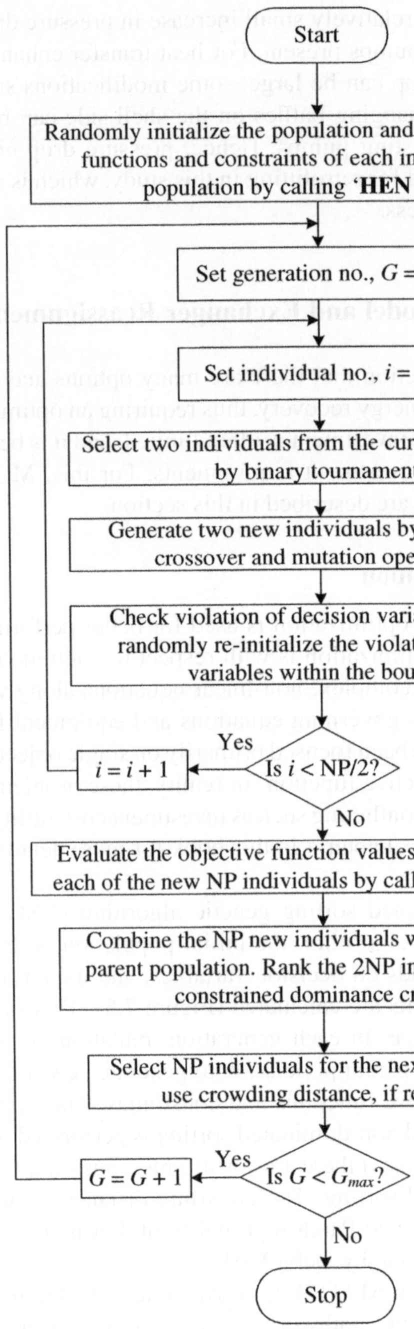

Figure 7.5 Flowchart of NSGA-II for MOO; NP is the population size, G is the generation number and G_{max} is the maximum number of generations allowed. See Figure 7.6 for the flowchart of HEN model.

better objective function is chosen. If A is feasible and B is infeasible, then A is chosen; and if both A and B are infeasible, then the solution with the least total constraint violation is chosen. Thus, the feasibility approach places higher priority on constraint satisfaction in the comparison of two solutions. The NGPM program for HEN retrofit optimization is run on an Intel Core i7-4770 desktop computer (3.4 GHz and 16 GB RAM) using 'parallel-computing' toolbox.

7.4.2 HEN Model

In HEN retrofitting by MOO via NSGA-II, decision variables are HEN structure (i.e., elements in **SM**, which are discrete in nature), corresponding exchanger areas and split ratios of streams (which are all continuous in nature). The HEN structure, areas and split ratios are sent to the HEN model (Figure 7.6) for performing model calculations including objectives and constraints. The generated structure (**SM**) is first tested for feasibility. It is infeasible if it has (a) two heat exchangers on the same node, (b) a heat exchanger where the split occurs or mixes, or (c) a heat exchanger is present on the node exceeding the

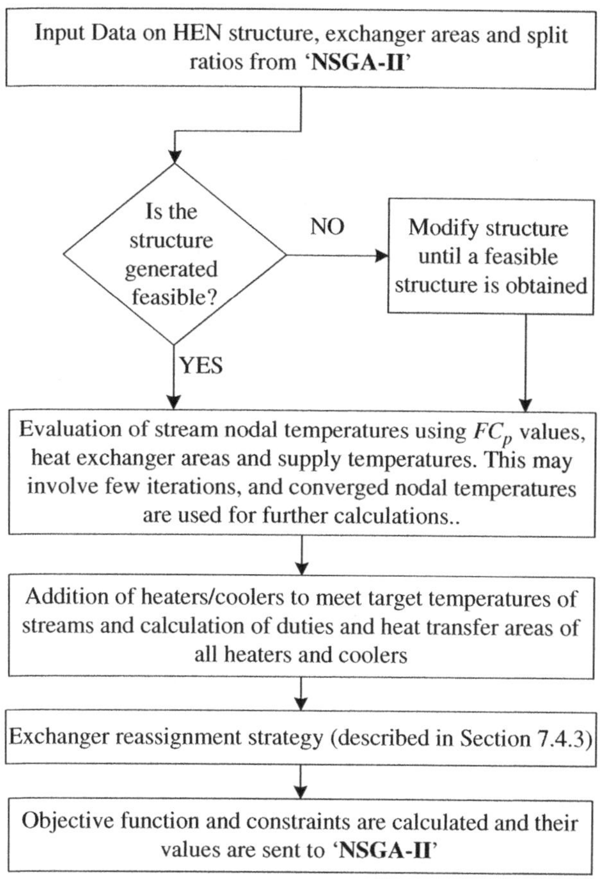

Figure 7.6 *HEN model calculations used for solving retrofit problems.*

number of nodes specified for that stream. An infeasible **SM** is modified by re-initializing and regenerating the particular row having the infeasibility within the specified bounds, and then testing for feasibility; this procedure is repeated until a feasible **SM** is generated (Figure 7.6). Then, for the heat exchanger areas and split ratios given by the optimizer, nodal temperatures of all streams are calculated using the model Equations (7.2) to (7.7) and the procedure in Appendix 7A. After this, the hot or cold utility required for each stream and the corresponding area for heater or cooler are calculated. Once areas of all heat exchangers, heaters and coolers in the generated network are calculated, an ERS, described in the next section, is employed to assign the existing heat exchangers to the calculated areas. This is followed by calculation of objectives and constraints.

The following equations are used to calculate outlet temperatures of each heat exchanger present in the HEN. Counter-current flow is assumed in each exchanger; and h_{hot} and h_{cold} represent the film heat transfer coefficient on the hot and cold side, respectively.

$$Q = (FC_P\Delta T)_{hot} = (FC_P\Delta T)_{cold} = UA\Delta T_{lmtd} \tag{7.2a}$$

$$\frac{1}{U} = \frac{1}{h_{hot}} + \frac{1}{h_{cold}} \tag{7.2b}$$

$$\Delta T_{lmtd} = \frac{\left(T_{in}^{hot} - T_{out}^{cold}\right) - \left(T_{out}^{hot} - T_{in}^{cold}\right)}{\ln\left\{\left(T_{in}^{hot} - T_{out}^{cold}\right) / \left(T_{out}^{hot} - T_{in}^{cold}\right)\right\}} \tag{7.3}$$

By combining the above equations, the HE model can be represented as follows [Equations (7.4a) and (7.4b)]:

$$T_{out}^{hot} = \alpha^{hot} T_{in}^{hot} + (1 - \alpha^{hot}) T_{in}^{cold} \tag{7.4a}$$

$$T_{out}^{cold} = \alpha^{cold} T_{in}^{cold} + (1 - \alpha^{cold}) T_{in}^{hot} \tag{7.4b}$$

where Equations (7.5)–(7.7) hold.

$$\alpha^{hot} = \frac{(\gamma^{cold} - \gamma^{hot}) e^{\gamma^{cold}}}{(\gamma^{cold} e^{\gamma^{cold}} - \gamma^{hot} e^{\gamma^{hot}})} \tag{7.5}$$

$$\alpha^{cold} = \frac{(\gamma^{cold} - \gamma^{hot}) e^{\gamma^{hot}}}{(\gamma^{cold} e^{\gamma^{cold}} - \gamma^{hot} e^{\gamma^{hot}})} \tag{7.6}$$

$$\gamma^{hot} = \frac{UA}{(FC_P)^{hot}} \text{ and } \gamma^{cold} = \frac{UA}{(FC_P)^{cold}} \tag{7.7}$$

7.4.3 Exchanger Reassignment Strategy (ERS)

ERS is used to assign the existing heat exchangers to the areas (i.e., heat exchangers) generated by the optimizer. It is required for complex retrofitting, and not for simple and moderate retrofitting, not involving re-location of existing exchangers. The ERS strategy should focus on the utilization of the existing exchangers in the best possible way to decrease the investment needed for retrofitting. It should also be practical to provide solutions acceptable for implementation in the industry. The ERS used in this work is adapted from Sreepathi and Rangaiah (2014b); it includes a practical limit on the amount of area that can be added to a heat exchanger. Assuming all heat exchangers are of the ST type, area addition to an existing exchanger is limited to 15%. Let $\mathbf{H_e}$ is a vector containing areas of existing heat exchangers in the network (provided by the user). $\mathbf{H_g}$ is another vector, which has areas of exchangers generated by the optimizer. Sizes of $\mathbf{H_e}$ and $\mathbf{H_g}$ need not be the same. Note that ERS does not consider the constraints on minimum approach temperature, which are handled separately in the optimizer (namely, feasibility approach in NSGA-II).

The steps involved in the ERS are as follows (Sreepathi and Rangaiah, 2014b).

1. Arrange elements of $\mathbf{H_e}$ in descending order. This is to re-use larger existing exchangers first
2. Select an element of $\mathbf{H_g}$ and compare with the elements of $\mathbf{H_e}$
3. If the difference in area of the two elements is within ±1% of the element of $\mathbf{H_e}$, then assign the selected element of $\mathbf{H_g}$ to the element of $\mathbf{H_e}$ compared. Make these elements of $\mathbf{H_e}$ and $\mathbf{H_g}$ unavailable for subsequent assignment (by making them zero). If there is more than one element from $\mathbf{H_e}$ that can be assigned, then the element which is closer is assigned; if there are two elements of $\mathbf{H_e}$, which are equally close, then the element which comes first in $\mathbf{H_e}$ is assigned to the selected element of $\mathbf{H_g}$
4. Repeat steps 2 and 3 for all elements of $\mathbf{H_g}$. This completes assignment of generated exchangers, whose areas are very close to those of some existing exchangers. There will remain some generated exchangers yet to be assigned, and some existing exchangers yet to be used
5. Select a non-zero element of $\mathbf{H_g}$ and compare with non-zero elements of $\mathbf{H_e}$
6. If the element of $\mathbf{H_g}$ is within −5% of the element from $\mathbf{H_e}$, then assign the former to the latter. Make these elements of $\mathbf{H_e}$ and $\mathbf{H_g}$ unavailable for subsequent assignment. If there is more than one element from $\mathbf{H_e}$ that can be assigned, then the element which is closer is assigned; if two elements of $\mathbf{H_e}$ are equally close, then the element which comes first in $\mathbf{H_e}$ is assigned
7. If the element of $\mathbf{H_g}$ is within +5% of the element from $\mathbf{H_e}$, then assign the former to the latter. This involves area addition to the existing exchanger and consequently additional investment cost
8. Repeat steps 6 and 7 for all elements of $\mathbf{H_g}$
9. Repeat steps 5 to 8 by increasing the percentage to ±10% and then ±15%. This completes assignment of generated exchangers, whose areas are within a specified percentage of some existing exchangers; some assignments may require area addition. There may remain some generated exchangers yet to be assigned, and some existing exchangers yet to be used
10. Update elements of $\mathbf{H_e}$ and $\mathbf{H_g}$ to exclude the elements that have already been assigned
11. Repeat steps 5 and 6 for all elements of $\mathbf{H_g}$ by varying the percentage to −25%, −50%, −75%, −90% and −99%. This allows assignment of generated exchangers, whose areas are (much) less than those of some existing exchangers
12. Update elements of $\mathbf{H_e}$ and $\mathbf{H_g}$ to exclude the elements that have already been assigned
13. The remaining elements of $\mathbf{H_g}$ are satisfied by installing new heat exchangers and are considered in the investment cost

ERS is employed for every feasible HEN structure generated. Then, the cost of additional area and re-location, if any, are calculated. The cost of area reduction in an existing exchanger is relatively small, and so it is neglected in this study.

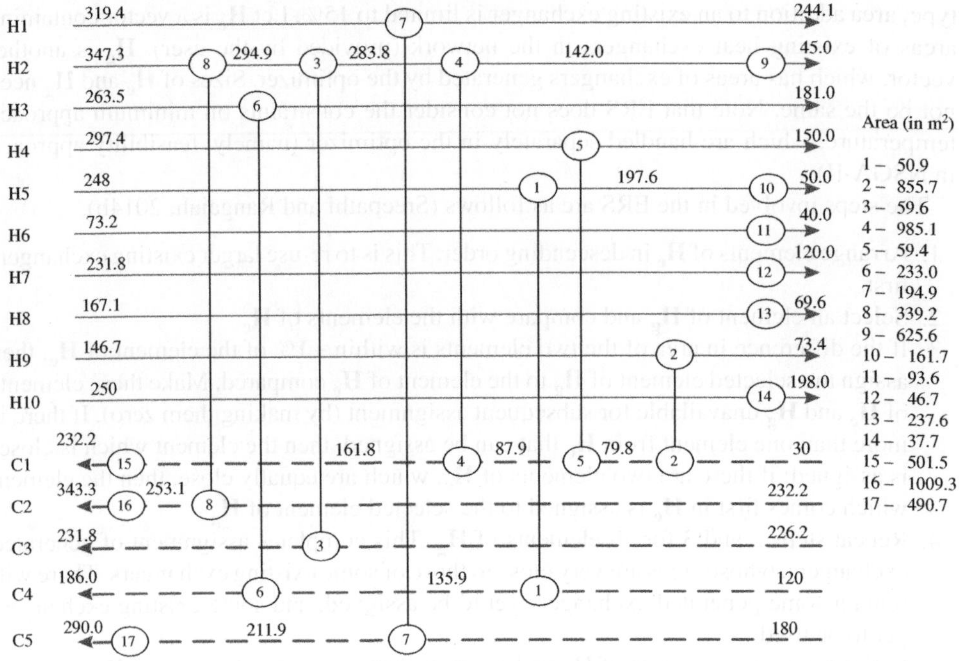

Figure 7.7 Existing heat exchanger network in the case study.

7.5 Case Study

The case study employed in this chapter is a modified version of the pre-heat train in the crude distillation unit, discussed in Nguyen *et al.* (2010). The HEN has 11 hot streams and 5 cold streams, and 17 exchangers including heaters and coolers. The existing network and exchanger areas are shown in Figure 7.7, and the stream and cost data of the problem are provided in Table 7.1. Film heat transfer coefficients in this table include an appropriate fouling factor. The minimum approach temperature in the network is 20 °C in the cooler numbered '11' on H6. The annual utility/operating cost of the existing network is US$13,930,596 for a hot utility of 95,366 kW and a cold utility of 57,923 kW. Capital cost correlations given in Table 7.1 are adapted from Turton *et al.* (2013) for the range of heat exchanger areas used in this work and using the Chemical Engineering Plant Cost Index (CEPCI) of 574 for April 2014. The energy cost of flue gas (US$140/kW.year) is taken from Nguyen *et al.* (2010), and the cooling water cost (US$10/kW.year) is from Turton *et al.* (2013).

7.6 Results and Discussion

In the following sub-sections, MOO results for simple, moderate and complex retrofitting of HEN in the case study are presented and discussed. The objectives used in MOO are cost of retrofit and utility cost, and the approach temperature on each heat exchanger in HEN

Table 7.1 *Stream and cost data for the HEN of a crude distillation unit.*

Stream	Supply temperature, T^S (°C)	Target temperature, T^T (°C)	FC_p (kW/°C)	h (kW/m²K)
H1	319.4	244.1	136.2	1.3
H2	347.3	45	194.5	0.76
H3	263.5	181.0	123	1.4
H4	297.4	150	20.7	1.2
H5	248	50	63.2	1.2
H6	73.2	40	57.7	1.3
H7	231.8	120	48.5	1.4
H8	167.1	69.6	165.3	1.4
H9	146.7	73.4	253.6	1.2
H10	250	198	120.5	1.5
C1	30	232.2	373.3	0.6
C2	232.2	343.3	488.1	0.79
C3	226.2	231.8	392.6	3.2
C4	120	186	200.4	1.5
C5	180	290	320.6	1.2
Flue gas (HU)	1000	500		0.12
Water (CU)	20	40		2.0

Note: capital cost (US$) = 54672 + 622.6 A for each new exchanger of A m²; capital cost (US$) = 622.6 ΔA for additional area of ΔA m² in an existing exchanger; and energy cost of flue gas and cooling water is US$140/kW.year and US$10/kW.year, respectively.

is constrained to be greater than 15 °C for all retrofitting levels. Note that the minimum approach temperature is decreased from 20 °C in the existing HEN to 15 °C in order to improve energy recovery in the network. For moderate and complex retrofitting, the user has to choose the number of nodes on each stream. For the case study, the number of nodes assumed is 10 on each stream; this is to provide enough locations for the optimizer to insert new exchangers and/or relocate exchangers within reasonable computational time.

7.6.1 Simple Retrofitting

In simple retrofitting, the existing HEN structure is fixed and only area addition of up to 15% in the existing exchangers is allowed. The 8 decision variables in MOO of this retrofitting in the case study are the heat exchanger areas, which are continuous in nature. Areas of coolers and heaters are calculated based on the target temperatures during the HEN model calculations, and hence they are not included in the decision variables. This may sometimes require the installation of a new heater/cooler in the network but this is unlikely since the aim of retrofitting is to reduce utilities. Simple retrofitting does not involve ERS as the existing HEN structure is not altered.

The MOO was carried out with a population size of 100, and the non-dominated solutions found at generations 100, 250, 500 and 1000 are presented in Figure 7.8. As can be seen, there is significant improvement in the front from generation 100 to 250, and acceptable Pareto-optimal front is obtained at generation 250 in about 60 s of computational time. As expected, MOO results in Figure 7.8 confirm that the two objectives are conflicting

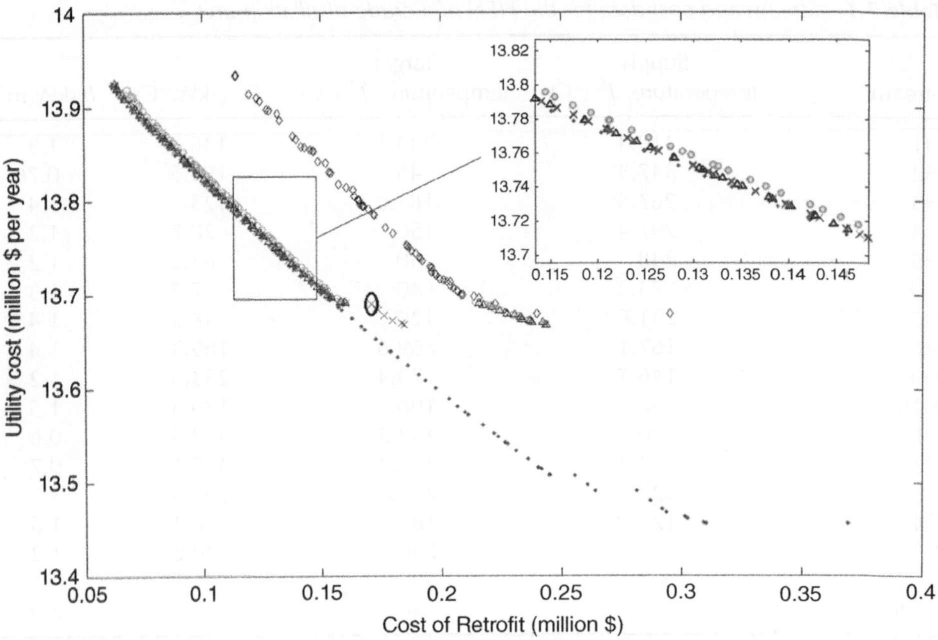

Figure 7.8 *Comparison of Pareto fronts for simple retrofitting via 15% area addition at various generations with population size of 100; diamonds, circles, triangles and crosses are the solutions at 100, 250, 500 and 1000 generations, respectively. The solution selected for discussion is in the oval. Dots are the front for simple retrofitting with up to 30% area addition.*

in nature: a decrease in utility cost is accompanied by an increase in the investment cost. The optimal front has nearly constant slope and a few discontinuities. Investment cost of the front does not go beyond US$200,000, probably because of the 15% limit on area addition. To confirm, this limit is increased to 30% and the non-dominated solutions found after 1000 generations are shown as dots in Figure 7.8. The front for 30% area addition is similar the 15% front for retrofit cost below US$150,000, and extends to higher retrofit cost to US$400,000 as more area is available for heat recovery. However, utility cost does not decrease much for retrofit cost of more than US$300,000

One solution (circled in Figure 7.8) from the front is chosen, and the network is presented in Figure 7.9. Cost of retrofitting is US$158,048, which results in an annual utility cost of US$13,693,518 (i.e., 1.7% decrease) and payback period of 0.67 years; hot and cold utilities used are, respectively, 93,778 kW and 56,454 kW. Minimum approach temperature in the retrofitted network is 20 °C in the cooler numbered 11 on H6, as in the original HEN. In Figure 7.9, additional area required or area reduction in each of the existing exchangers is provided in brackets. Many changes in areas are small, and may not be required in practice; only exchangers 4 and 8 require additional area, and exchangers 9 and 15 have more than the required area. The network requires a new exchanger (cooler) N with a small area of 1.6 m^2 to meet the target temperature of hot stream H3. This cooler can be avoided by increasing the area in exchanger 6 by 10 m^2, subsequently reducing area in exchanger 1 by 4.4 m^2 and increasing the area in cooler 10 by 1.4 m^2; note that units 6, 1, 10 and N constitute a

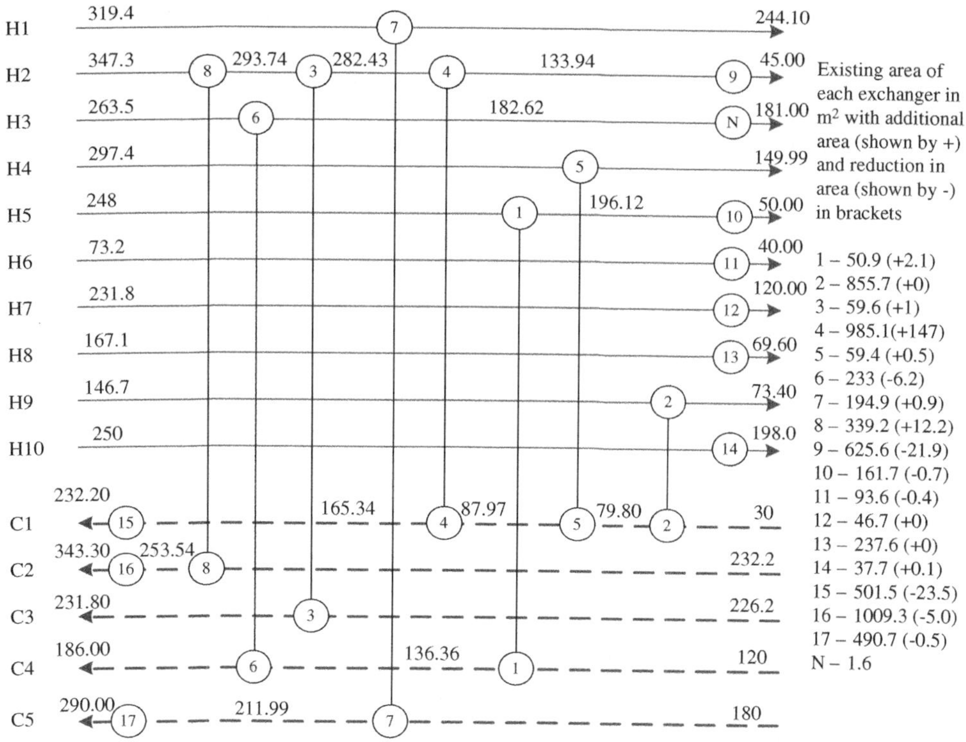

Figure 7.9 *HEN after simple retrofitting according to the selected optimal solution in Figure 7.8.*

loop employed in pinch analysis (Smith, 2005). Hence, only five exchangers 4, 6, 8, 9 and 15 require retrofitting. The updated network involves a lower retrofit cost by US$56,508 but incurs a utility cost increase of $2,000 per year. It still achieves the target temperatures and satisfies the specified minimum approach temperature. The updated solution is better than the solutions on the Pareto-optimal front as it is beneath the front in Figure 7.8; this indicates that the optimizer may not always find the best solution. Hence, the selected solution from the Pareto-optimal front should be reviewed, and possible minor changes (like removing small heat exchangers) need to be assessed to develop an improved solution for implementation.

7.6.2 Moderate Retrofitting

In moderate retrofitting, new heat exchangers are added to the existing HEN structure along with area addition of up to 15% in the existing exchangers. The decision variables for MOO of moderate retrofitting are the discrete variables of the HEN structure (new exchangers) and heat exchanger areas, which are continuous in nature. For the case study assuming up to 4 new exchangers, there are 16 discrete variables corresponding to HEN structure and 12 continuous variables for exchanger areas (i.e., 8 for existing exchangers and 4 for new heat exchangers). Areas of heaters/coolers are calculated based on the target temperatures during model calculations, and hence they are not included in the decision variables. This may sometimes require installation of a new heater/cooler in the network, but it is unlikely since retrofitting is to reduce utilities. Moderate retrofitting also does not involve ERS. MOO for moderating retrofitting of HEN in the case study was carried out with a population size of 500 (compared with 100 for simple retrofitting because of the increased number of decision variables). As can be seen from the non-dominated solutions found at different generations (Figure 7.10), there is significant improvement in the front from generations 100 to 500, and an acceptable front is obtained at generation 500 in about 850 s of computational time. The optimal front is smooth, and has a gradual change in the slope.

A solution from the circled region in the front in Figure 7.10 is chosen for analysis. The retrofitted network for this solution is presented in Figure 7.11 along with existing exchanger areas and required area addition/reduction. Retrofitting cost is US$1,124,584, which reduces annual utility cost by 17% to US$11,530,232 for a payback period of 0.47 years. Hot and cold utilities used are 79,405 kW and 41,752 kW, respectively, and a minimum approach temperature of the network is 20 °C in the cooler 11 on H6, as in the original HEN. The retrofitted network has four new heat exchangers (A, B, C and D) and also two new coolers E and F, which are very small and can be eliminated by making slight changes in some connected exchangers. Among the existing exchangers/heaters/coolers, exchanger 7 requires additional area, and units 9, 12–15 and 17 require area reduction.

One notable exchanger in the HEN in Figure 7.11 is cooler 14 with an area requirement of only 0.2 m^2 after retrofitting. This cooler can be removed using the pinch analysis. There is a 'path' from cooler 14 to heater 17 via exchanger B, and so cooler 14 can be eliminated

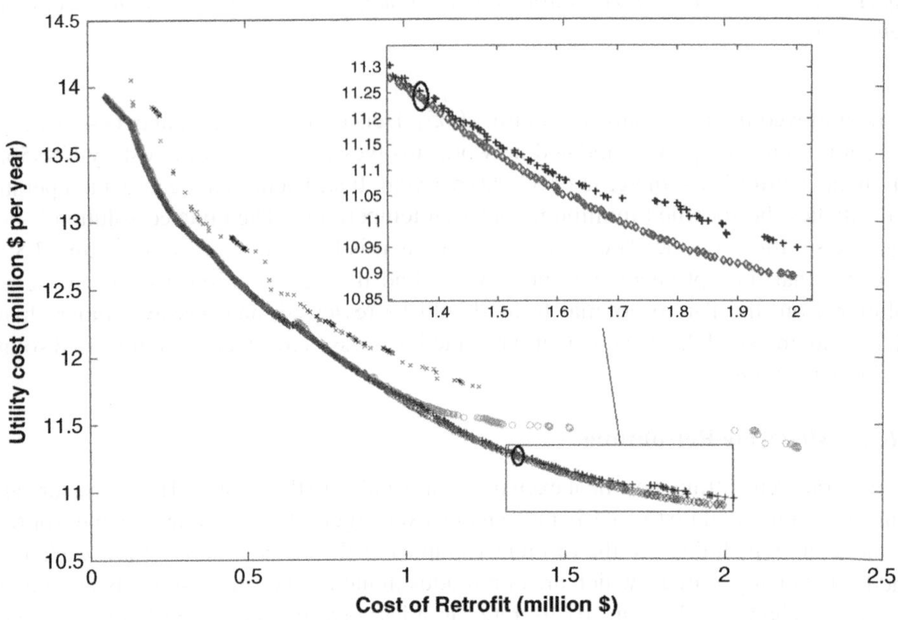

Figure 7.10 Comparison of Pareto fronts for moderating retrofitting at various generations with population size of 500; crosses, circles, pluses and diamonds are the solutions at 100, 250, 500 and 1000 generations, respectively. The solution selected for discussion is in the oval.

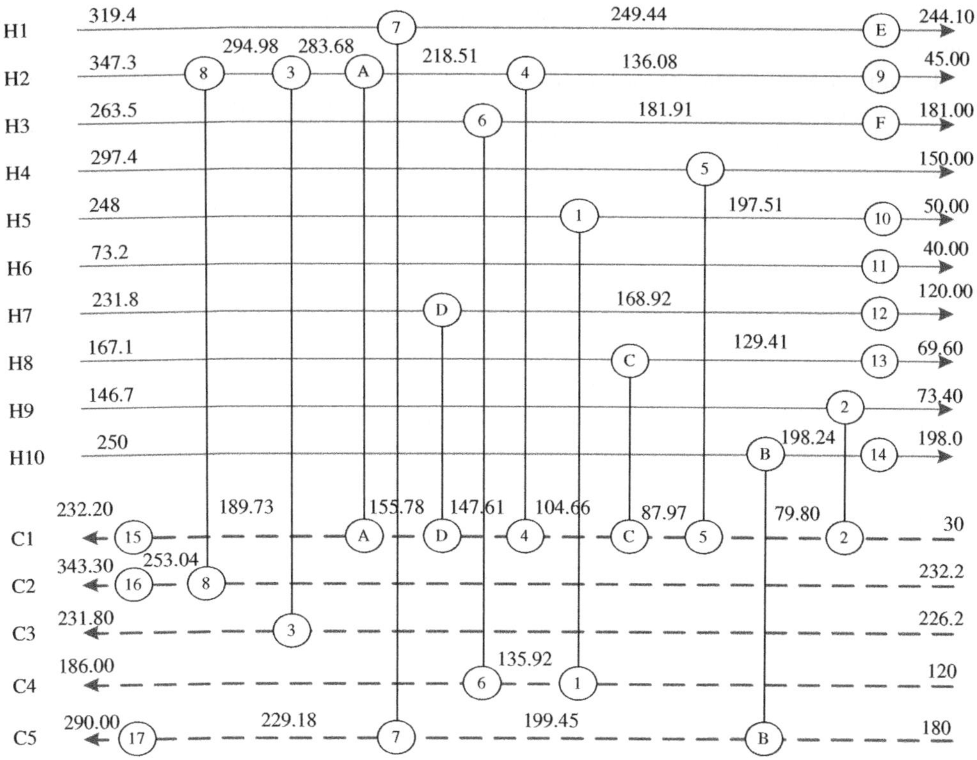

Existing area of each exchanger in m² with additional area (shown by +) and reduction in area (shown by –) in brackets

1 – 50.9 (+0.2)	6 – 233 (-3.5)	11 – 93.6 (-0.3)	16 – 1009.3 (+0)	D – 168.8
2 – 855.7 (+0)	7 – 194.9 (+29.2)	12 – 46.7 (-21.4)	17 – 490.6 (-98.4)	E – 4.3
3 – 59.6 (-0.3)	8 – 339.2 (+0)	13 – 237.6 (-59.9)	A – 489.1	F – 0.9
4 – 985.1 (+0.5)	9 – 625.6 (-16.4)	14 – 37.7 (-37.5)	B – 295.2	
5 – 59.4 (+0.5)	10 – 161.7 (+0)	15 – 501.5 (-187.6)	C – 289.6	

Figure 7.11 *HEN after moderate retrofitting according to the selected solution in Figure 7.10.*

by increasing the duty (and area) of exchanger B and decreasing the duty (and area) of heater 17. The decrease in the utility cost due to these changes is US$4,338/year, coupled with a 0.2% increase in the cost of retrofit (i.e., by US$2,240). Further, new exchanger F in the HEN (Figure 7.11) is very small, and can be eliminated using the loop involving cooler F, exchanger 6, exchanger 1 and cooler 10. A 'loop' in the HEN means there are more units (i.e., exchangers, heaters and coolers) in the network. The loop can be broken by reducing the duty of cooler F to zero (thus eliminating cooler F) and modifying duties of exchanger 6 (increases), exchanger 1 (decreases) and cooler 10 (increases). Removing this loop in the selected solution saves US$53,177 in the cost of retrofit (i.e., by 4.7%) for the same utility cost as the utility load of cooler F is transferred to that of cooler 10. Thus, the benefits of eliminating coolers 14 and F are a US$50,937 reduction (i.e., 4.5%) in the cost of retrofit and utility cost savings of US$4,338/year (i.e., 0.04%). Note that cooler E is relatively small in size, and it can be eliminated using the path from it to heater 17 via exchanger 7. However, this results in an area increase of exchanger 7 by more than 15%, and hence elimination of cooler E is not considered.

In summary, retrofitting according to the chosen optimal solution along with elimination of new exchanger F requires five new exchangers (A, B, C, D and E) and area changes in existing exchangers 7, 9, 12, 13, 15 and 17. Although these changes are more than those for the simple retrofitting solution discussed in the previous section, the utility cost is significantly lower. Further, the existing cooler 14 becomes available for some other use in the plant.

7.6.3 Complex Retrofitting

In complex retrofitting, the relocation of existing exchangers, area addition and the use of new heat exchangers are all allowed. The decision variables in MOO of complex retrofitting are the discrete variables of the HEN structure and heat exchanger areas, which are continuous in nature. For the case study, there are 60 discrete variables corresponding to HEN structure and 15 continuous variables for heat exchanger areas. Note that the (maximum) number of total exchangers (i.e., 15 excluding heater and coolers) in the network is chosen by the user, based on the existing network. The areas of heaters/coolers are calculated based on the target temperatures during model calculations, and hence they are not included in the decision variables. The MOO was run with a population size of 2000, and the non-dominated solutions found at generations 500, 1000, 1500 and 2000 are shown in Figure 7.12. These results show that the non-dominated solutions at 500 generations are satisfactory. However,

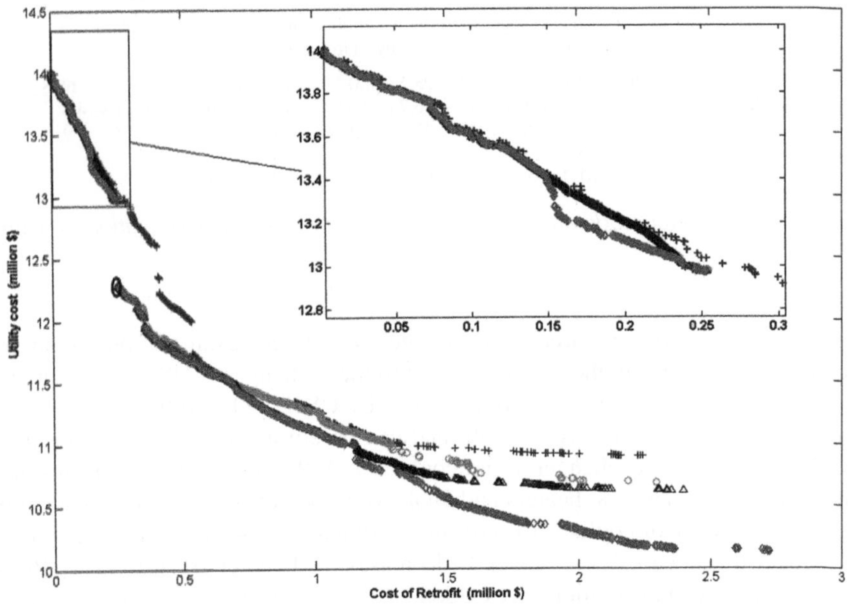

Figure 7.12 Comparison of Pareto fronts for complex retrofitting at various generations with a population size of 2000; pluses, circles, triangles and diamonds are the solutions at 500, 1000, 1500 and 2000 generations. The solution selected for discussion is in the oval.

owing to the large number of decision variables, around 2000 generations are required for convergence; computational time for 500 and 2000 generations is 6700 s and 24 800 s, respectively.

The entire front obtained after 2000 generations in Figure 7.12 can be divided into three regions based on their slopes. In region 1, utility cost reduces by 8% from 14 million to 12.9 million for a small investment US$0.25 million. Region 2 is with utility cost ranging from 12.3 million to 11 million. The final region 3 is with utility cost from 11 million to 10.2 million, requiring a very high investment. Closer analysis of optimal solutions at the discontinuity between regions 1 and 2 showed that the discontinuity was because all the existing heat exchanger areas have been fully used (i.e., very small changes in their area) and a new heat exchanger was installed. Note that there are solutions in region 1, which have new heat exchangers, but the existing exchangers are not completely exhausted (i.e., high amount of area reduction is involved).

The HEN corresponding to the non-dominated solution at the discontinuity in the front (circled in Figure 7.12) is presented in Figure 7.13. Cost of retrofitting is US$253,328, and utility cost of this network is US$12,238,781/year; payback period for this retrofitting is just 0.15 years. The hot and cold utilities used are 84,055 kW and 47,085 kW, respectively. Minimum approach temperature of the network is 18.9 °C (compared with the specified constraint of 15 °C) in exchanger 8 between streams H5 and C1; these are different

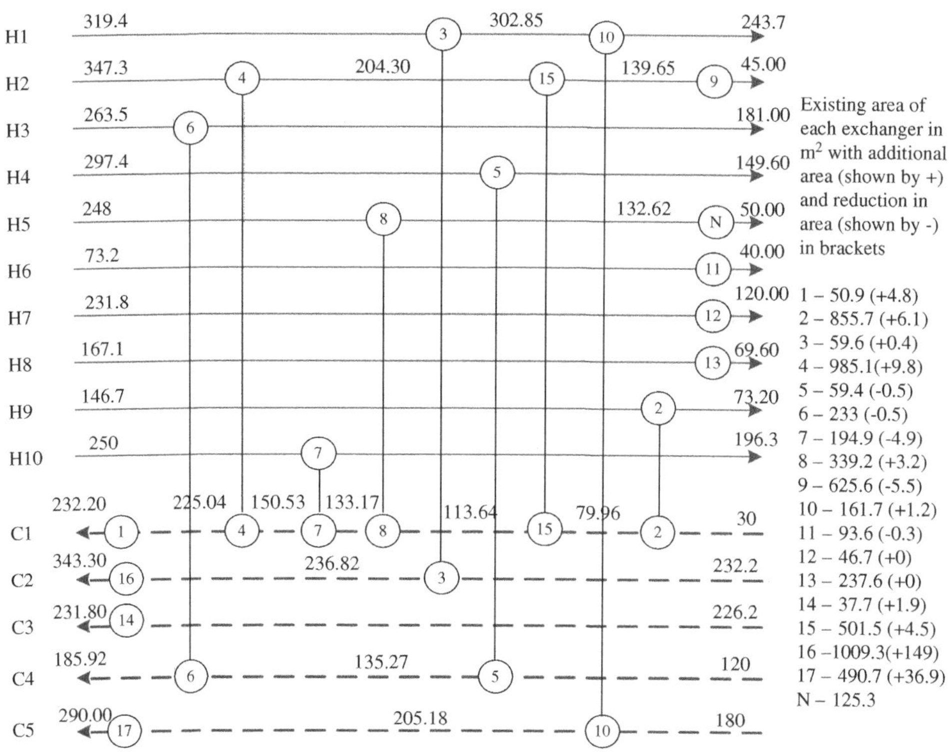

Figure 7.13 *HEN after complex retrofitting according to the selected solution in Figure 7.12.*

from the original HEN. The retrofitted network has one new cooler 'N', and uses all existing exchangers (including heaters and coolers). Areas of existing heat exchangers used, areas added and reduced are provided in Figure 7.13. In this figure, numbering of heat exchangers/cooler/heaters is based on the original network. Of all existing heat exchangers used, 8 units numbered 1, 3, 5, 7, 8, 10, 14 and 15 require re-piping (i.e., at least one stream in the heat exchanger is changed); and exchanger 4 between streams H2 and C1 is re-sequenced (i.e., it is present between the same streams but moved to a higher temperature range). There are 7 existing heat exchangers numbered 2, 6, 9, 11, 12, 13, 16 and 17, which are not re-located. As re-location is allowed in complex retrofitting, exchangers are used to their full capacity except for exchangers 7 and 9, where the difference is less than 3% of the existing area. Additional area is required in existing exchangers 1, 2, 4, 8, 14, 15, 16 and 17, and slight reduction is needed in existing exchangers 7 and 9.

7.6.4 Comparison and Discussion

The non-dominated solutions obtained for simple retrofitting with 15% area addition, simple retrofitting with 30% area addition, moderate retrofitting and complex retrofitting are plotted together in Figure 7.14. For a fair comparison, all these are obtained using a population of 2000 and 2000 generations. Simple retrofitting (with both 15% and 30% limit)

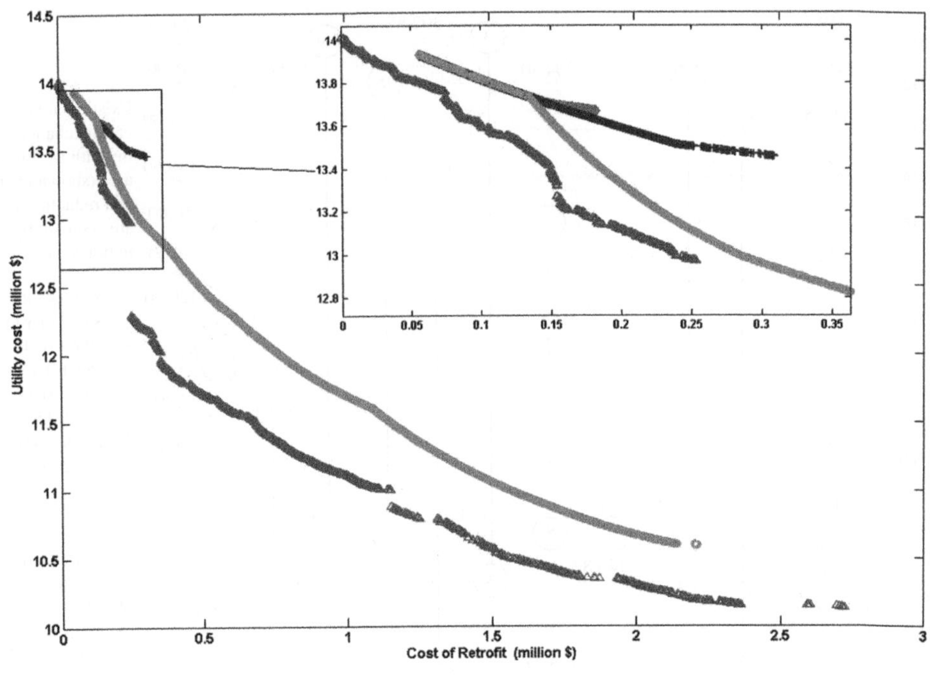

Figure 7.14 Comparison of Pareto fronts for simple, moderating and complex retrofitting of HEN in the case study; diamonds and pluses are the solutions for simple retrofitting with 15% and 30% area addition, respectively; and circles and triangles are the solutions for moderate and complex retrofitting, respectively.

and moderate retrofitting are similar in the retrofit cost range of US$50,000 to US$140,000, as they have fixed structure and no new heat exchangers were installed for low retrofit cost. From retrofit cost of US$140,000 to US$300,000, moderate retrofitting performs better than simple retrofitting, showing the beneficial effect of new heat exchangers. As expected, complex retrofitting outperforms simple and moderate retrofitting over the entire retrofit cost range, that is, existing exchanger relocation helps to achieve lower utility costs for the same investment compared with retrofitting with fixed structure. Though relocating exchangers may not be easy to implement, it should be considered for some exchangers where possible.

To have a fair comparison, economics of selected optimal solutions for various retrofitting levels are analysed for the same retrofit cost. The results are compiled in Table 7.2. For a retrofit cost of around US$100,000, simple retrofitting and moderate retrofitting have similar payback periods, whereas complex retrofitting has nearly one-third of this payback period. For simple retrofitting, 30% proves to be better in terms of both utility savings and payback period, mainly due to additional available area. Overall, complex retrofitting has better utility cost savings and payback period compared with other levels, mainly due to re-location of heat exchangers. Also, for higher cost of retrofit, moderate and complex retrofitting have similar payback periods (Table 7.2), showing the increased influence of new heat exchangers over re-location of existing heat exchangers.

NSGA-II algorithm in NGPM is a stochastic optimization technique using random numbers for exploration and exploitation of the search space. Hence, its results may be affected by the random number initiator. To assess this, MOO problem for each of simple, moderate and complex retrofitting of the HEN in the case study was run 10 times (each time with a different random number initiator). The best front obtained for each case of retrofitting from all these runs is similar to that at the highest number of generations shown in Figures 7.8, 7.10 and 7.12. It is possible to improve slightly the final

Table 7.2 *Economic analysis of selected optimal solutions for various retrofitting levels.*

Cost of retrofit (US$)	Retrofitting level	Utility cost (US$/year)	Utility cost savings (US$/year)	Payback period (year)
100,000	Simple (15%)	13,820,320	110,276	0.91
	Simple (30%)	13,820,320	110,276	0.91
	Moderate	13,820,320	110,276	0.91
	Complex	13,602,450	328,146	0.31
178,000	Simple (15%)	13,685,420	245,176	0.73
	Simple (30%)	13,654,290	276,306	0.65
	Moderate	13,503,769	426,827	0.42
	Complex	13,230,456	700,140	0.25
253,328	Moderate	13,093,625	836,971	0.30
	Complex	12,238,781	1,691,815	0.15
1,124,584	Moderate	11,530,232	2,400,364	0.46
	Complex	11,092,478	2,838,118	0.40

non-dominated solutions in these figures by local optimization with respect to continuous variables (i.e., with the HEN structure as given by MOO). This can be done for selected non-dominated solutions, and it may eliminate exchangers with small areas and/or small area addition/reduction.

7.7 Conclusions

In this work, three HEN retrofitting levels, namely, simple, moderate and complex, and MOO for them are described and tested on a case study involving 10 hot streams, 5 cold streams and 17 heat exchangers including heaters and coolers. Potential benefits of area addition, relocation of existing exchangers and addition of new heat exchangers are explored and discussed. Simple retrofitting with 15% and 30% limit on area addition in existing exchangers provides a relatively easy option to decrease utility consumption. Results in the case study show that simple retrofitting can decrease utility cost by ~1.7% for a retrofit cost of ~US$190,000. Although this is attractive with a payback period of 8 months, utility consumption may not decrease much. Addition of new heat exchangers without any existing exchanger relocation, as in moderate retrofitting, provides better and more options in terms of both wider spread of non-dominated solutions and lower retrofit cost for similar utility cost, compared with fixed-structure solutions from simple retrofitting. In the case study, moderate retrofitting can decrease utility cost by ~23% for a retrofit cost of ~US$2,250,000. Complex retrofitting involving relocation of existing exchangers besides new heat exchangers and area changes in existing exchangers gives even better solutions with a lower payback period. In the case study, complex retrofitting can decrease utility cost by ~28% for a retrofit cost of ~US$2,750,000.

As the level of retrofitting increases, computational time for MOO increases. However, this is not a serious concern since MOO of HEN retrofitting can be performed on the ubiquitous desk/laptop computers. Different levels of HEN retrofitting and MOO together can provide numerous optimal solutions for quantitative understanding of trade-offs in the chosen objectives, analysis and selection of a suitable solution for implementation. The optimal solutions, as seen in the case study, will require low to large investment for retrofitting and result in utility cost savings accordingly. Payback period in the case study is very attractive at less than one year. Obviously, retrofit results from optimization are case-specific, and hence each application needs to be investigated as described in this chapter.

Appendix 7A: Calculation of Nodal Temperatures

In this appendix, calculation of nodal temperatures in HEN is described using an example. Consider a HEN with two hot streams and one cold stream, whose data are in Table 7.A.1. Assume the number of nodes is 4 on each of the three streams, and nodal temperatures supplied by the user for all streams are given by the following vector [Equation (7.A.1)]:

$$\tau_o = \begin{bmatrix} 348 & 300 & 250 & 180 & 120 \\ 146 & 140 & 100 & 100 & 100 \\ 30 & 80 & 150 & 160 & 200 \end{bmatrix} \tag{7.A.1}$$

Table 7.A.1 *Stream data for the example problem.*

Stream	Supply T (°C)	Target T (°C)	FC_P (kW/°C)
H1	348	45	194.5
H2	146.7	50	253.6
C1	30	230	373.3

In the above, the first, second and third rows correspond to streams H1, H2 and C1, respectively. The HEN structure generated by the optimizer can take a maximum value of 4 for nodes, but the temperature of the stream leaving the 4th node needs to be calculated. This is the reason for having 5 nodal temperatures for all streams in $\boldsymbol{\tau}_0$ when the maximum number of nodes is only 4. Initial nodal temperatures can be any values within the stream supply and target temperatures, and they need not include target/supply temperatures (as seen in the third row of $\boldsymbol{\tau}_o$).

Note: $U = 0.3$, 0.4 and 0.4 kW/m^{2}K for all heat exchangers, heaters and coolers, respectively.

Assume the HEN structure generated by optimizer $= \begin{pmatrix} 2 & 2 & 1 & 2 \\ 1 & 3 & 1 & 4 \end{pmatrix}$; here, the first row indicates that there is a heat exchanger between the 2nd node of H2 and the 2nd node of C1, and the second row is for a heat exchanger between the 3rd node of H1 and the 4th node of C1. This means that the nodal temperatures used for calculation of heat exchanger 1 are of the 2nd node of H2 and the 2nd node of C1, and for calculation of exchanger 2 are of the 3rd node of H1 and the 4th node of C1. These nodal temperatures are the inlet temperatures to the respective exchanger. The generated HEN structure is feasible. Areas generated by the optimizer are 860 m^2 and 780 m^2 for the first and second exchanger, respectively. All the temperatures until each heat exchanger is reached are updated to either the supply temperature or the outlet temperature of the preceding exchanger on that stream, as there is no heat exchange taking place in between to affect the nodal temperatures.

Now, following the flowchart (Figure 7.A.1), τ_i is initiated as τ_0 given above. Calculations for outlet temperatures of heat exchanger 1 using Equations (7.4) to (7.7) are:

$$\gamma^{hot} = \frac{UA}{(FC_P)^{hot}} = \frac{0.3 \times 860}{253.6} = 1.017; \ \gamma^{cold} = \frac{UA}{(FC_P)^{cold}} = \frac{0.3 \times 860}{373.3} = 0.692$$

$$\alpha^{hot} = \frac{(\gamma^{cold} - \gamma^{hot})e^{\gamma^{cold}}}{(\gamma^{cold}e^{\gamma^{cold}} - \gamma^{hot}e^{\gamma^{hot}})} = \frac{(0.692 - 1.017)e^{0.692}}{(0.692e^{0.692} - 1.017e^{1.017})} = 0.4542$$

$$\alpha^{cold} = \frac{(\gamma^{cold} - \gamma^{hot})e^{\gamma^{hot}}}{(\gamma^{cold}e^{\gamma^{cold}} - \gamma^{hot}e^{\gamma^{hot}})} = \frac{(0.692 - 1.017)e^{1.017}}{(0.692e^{0.692} - 1.017e^{1.017})} = 0.6286$$

$$T_{out}^{hot} = 0.4542T_{in}^{hot} + (1 - 0.4542)\,T_{in}^{cold} = 82.68$$

$$T_{out}^{cold} = 0.6286T_{in}^{cold} + (1 - 0.6286)\,T_{in}^{hot} = 73.08$$

After heat exchanger 1 calculations, τ_{i+1} is given by Equation (7.A.2):

$$\boldsymbol{\tau}_{i+1} = \begin{bmatrix} 348 & 300 & 250 & 180 & 120 \\ 146 & 146 & 82.68 & 82.68 & 82.68 \\ 30 & 30 & 73.08 & 73.08 & 73.08 \end{bmatrix} \tag{7.A.2}$$

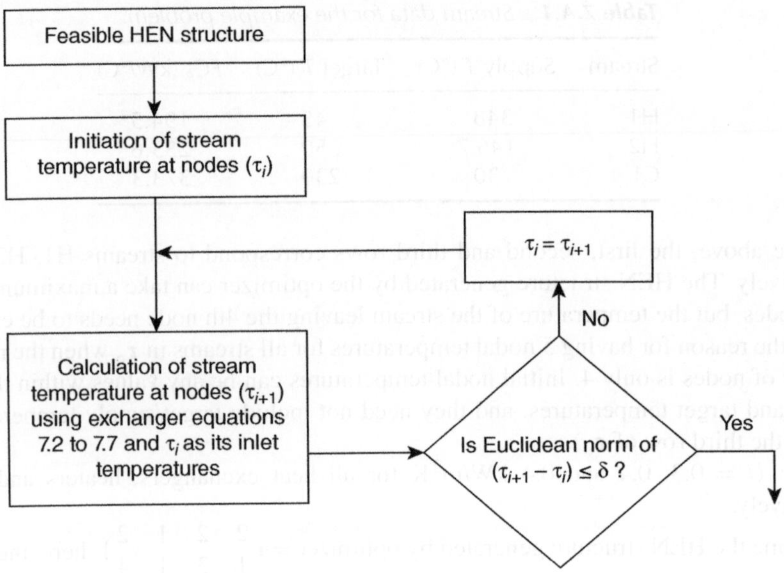

Figure 7.A.1 *Flowchart employed in this work for nodal temperature calculations.*

Similarly, calculations for outlet temperatures of heat exchanger 2 are:

$$\gamma^{hot} = \frac{UA}{(FC_P)^{hot}} = \frac{0.3 \times 780}{194.5} = 1.203; \quad \gamma^{cold} = \frac{UA}{(FC_P)^{cold}} = \frac{0.3 \times 780}{373.3} = 0.627$$

$$\alpha^{hot} = \frac{(\gamma^{cold} - \gamma^{hot})e^{\gamma^{cold}}}{(\gamma^{cold}e^{\gamma^{cold}} - \gamma^{hot}e^{\gamma^{hot}})} = \frac{(0.627 - 1.203)e^{0.627}}{(0.627e^{0.627} - 1.203e^{1.203})} = 0.3807$$

$$\alpha^{cold} = \frac{(\gamma^{cold} - \gamma^{hot})e^{\gamma^{hot}}}{(\gamma^{cold}e^{\gamma^{cold}} - \gamma^{hot}e^{\gamma^{hot}})} = \frac{(0.627 - 1.203)e^{1.203}}{(0.627e^{0.627} - 1.203e^{1.203})} = 0.6772$$

$$T_{out}^{hot} = 0.3807T_{in}^{hot} + (1 - 0.3807)\,T_{in}^{cold} = 177.74$$

$$T_{out}^{cold} = 0.6772T_{in}^{cold} + (1 - 0.6772)\,T_{in}^{hot} = 161.82$$

After heat exchanger 2 calculations, τ_{i+1} is as shown in Equation (7.A.3):

$$\tau_{i+1} = \begin{bmatrix} 348 & 348 & 348 & 177.74 & 177.74 \\ 146 & 146 & 82.68 & 82.68 & 82.68 \\ 30 & 30 & 73.08 & 73.08 & 161.82 \end{bmatrix} \tag{7.A.3}$$

This completes one iteration. Figure 7.A.2 shows the HEN structure (under consideration) along with the nodal temperatures for this iteration.

Now, the Euclidean norm of $(\tau_i - \tau_{i+1})$ is calculated to be 169.98. This does not satisfy the convergence condition in the flowchart (Figure 7.A.1); so, τ_i is updated with τ_{i+1}, and the above calculations for exchangers 1 and 2 are repeated. The iterations are repeated until the Euclidean norm of $(\tau_i - \tau_{i+1})$ is less than a small value (δ).

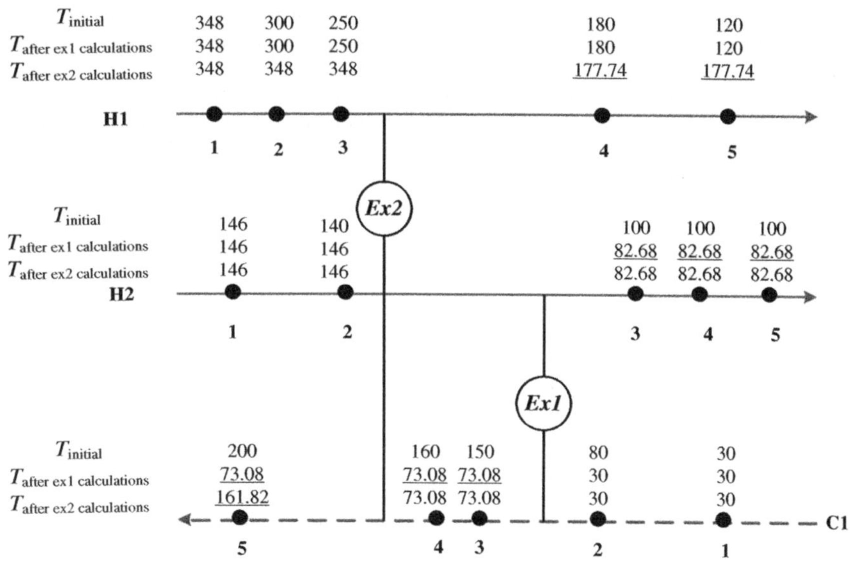

Figure 7.A.2 *HEN structure of the discussed example with nodal temperatures initially and in subsequent calculations in one iteration; heaters and coolers (not shown) are to be added to meet the target temperatures. The updated temperatures (after each calculation) are underlined.*

Exercises

7.1 The retrofitted HEN by MOO may have small heaters/coolers, mainly because stochastic optimization methods do not give a precise optimum. Study the effect of removing these heaters/coolers in Figure 7.9 on the utility and retrofit costs.

7.2 Using the stream data in Table 7.1 and given heat exchanger areas, calculate nodal temperatures in the existing HEN and compare them with those shown in Figure 7.7. For more practice, calculate and confirm nodal temperatures in retrofitted HENs in Figures 7.9, 7.11 and 7.13.

7.3 A minimum approach temperature of 15 °C is used for optimizing simple, moderate and complex HEN retrofitting in the case study in this chapter. Study the effect of changing the minimum approach temperature from 15 °C to 10 °C, 20 °C and 30 °C for simple, moderate and/or complex retrofitting. Note that this requires methods/programs similar to those described in this chapter.

References

Abbood, N.K., Manan, Z.A., and Wan Alwi, S.R. (2012) A combined numerical and visualization tool for utility targeting and heat exchanger network retrofitting. *Journal of Cleaner Production*, **23** (1), 1–7.

Asante, N.D.K. and Zhu, X.X. (1996) An automated approach for heat exchanger network retrofit featuring minimal topology modifications. *Computers and Chemical Engineering*, **20** (Suppl. 1), S7–S12.

Bochenek, R. and Jezowski, J.M. (2010). Genetic algorithms formulation for retrofitting heat exchanger network, in (ed. G.P. Rangaiah), *Stochastic Global Optimization: Techniques and Applications in Chemical Engineering*, World Scientific, Singapore, pp. 545–589.

Briones, V. and Kokossis, A.C. (1999) Hypertargets: a conceptual programming approach for the optimisation of industrial heat exchanger networks—II. Retrofit design. *Chemical Engineering Science*, **54** (4), 541–561.

Deb, K. (2001). *Multi-Objective Optimization using Evolutionary Algorithms*, John Wiley, Chichester

Kemp, I.C. (2007) *Pinch Analysis and Process Integration. A User Guide on Process Integration for Efficient Use of Energy*, Elsevier, Amsterdam, The Netherlands.

Klemeš, J.J., Varbanov, P.S., and Kravanja, Z. (2013) Recent developments in Process Integration. *Chemical Engineering Research and Design*, **91** (10), 2037–2053.

Li, B.H. and Chang, C.T. (2010) Retrofitting heat exchanger networks based on simple pinch analysis. *Industrial & Engineering Chemistry Research*, **49** (8), 3967–3971.

Linnhoff, B. and Flower, J.R. (1978) Synthesis of heat exchanger networks—1. Systematic generation of energy optimal networks. *AIChE Journal*, **24** (4), 633–642.

Nguyen, D.Q., Barbaro, A., Vipanurat, N., and Bagajewicz, M.J. (2010) All-at-once and step-wise detailed retrofit of heat exchanger networks using an MILP Model. *Industrial & Engineering Chemistry Research*, **49** (13), 6080–6103.

Nordman, R. and Berntsson, T. (2009) Use of advanced composite curves for assessing cost-effective HEN retrofit I: Theory and concepts. *Applied Thermal Engineering*, **29** (2–3), 275–281.

Pan, M., Bulatov, I., and Smith, R. (2014) Efficient retrofitting approach for improving heat recovery in heat exchanger networks with heat transfer intensification. *Industrial & Engineering Chemistry Research*, **53** (27), 11107–11120.

Pan, M., Bulatov, I., Smith, R., and Kim, J.K. (2013) Optimisation for the retrofit of large scale heat exchanger networks with different intensified heat transfer techniques. *Applied Thermal Engineering*, **53** (2), 373–386.

Ponce-Ortega, J.M., Jimenez-Gutierrez, A., and Grossmann, I.E. (2008) Simultaneous retrofit and heat integration of chemical processes. *Industrial & Engineering Chemistry Research*, **47** (15), 5512–5528.

Rangaiah, G.P. (ed.) (2009) *Multi-Objective Optimization: Techniques and Applications in Chemical Engineering*, World Scientific, Singapore.

Rangaiah, G.P. and Bonilla-Petriciolet, A. (eds) (2013) *Multi-Objective Optimization in Chemical Engineering: Developments and Applications*, John Wiley, Chichester.

Shenoy, U.V. (1995) *Heat Exchanger Network Synthesis: Process Optimization by Energy and Resource Analysis*, Gulf Publishing Company, Houston, TX.

Smith, R. (2005) *Chemical Process Design and Integration*, John Wiley and Sons, Manchester, UK.

Smith, R., Jobson, M., and Chen, L. (2010) Recent development in the retrofit of heat exchanger networks. *Applied Thermal Engineering*, **30** (16), 2281–2289.

Sreepathi, B.K. and Rangaiah, G.P. (2014a) Review of heat exchanger network retrofitting methodologies and their applications. *Industrial & Engineering Chemistry Research*, **53** (28), 11205–11220.

Sreepathi, B.K. and Rangaiah, G.P. (2014b) Improved heat exchanger network retrofitting using exchanger reassignment strategies and multi-objective optimization. *Energy*, **67**, 584–594.

Sreepathi, B.K. and Rangaiah, G.P. (2014c) Retrofitting of heat exchanger networks involving streams with variable heat capacity: application of single and multi-objective optimization. *Applied Thermal Engineering*, **75**, 677–684.

Turton, R., Bailie, R.C., Whiting, W.B., and Shaeiwitz, J.A. (2013) *Analysis, Synthesis and Design of Chemical Processes*. Pearson Education International, Ann Arbor, MI.

8

Review of Optimization Techniques for Retrofitting Batch Plants

Catherine Azzaro-Pantel
Professor of Chemical Engineering, Université de Toulouse, Laboratoire de Génie Chimique, LGC UMR CNRS 5503, France

8.1 Introduction

Retrofit design refers to the redesign of an existing facility so as to improve its performance taking into account its agility to consider new process constraints due to a competitive market and environmental legislation. Retrofitting is particularly important for batch processes, which have to face fluctuations in production demand, new product manufacturing and shortened time-to-market periods. In that context, methodologies to address systematically the retrofit of batch processes are required in order to generate and assess promising retrofit alternatives. Analysis of the dedicated literature shows that, although the retrofit problem is of great importance in batch processes, it has received less attention than design problems (the so-called "grass-roots" design), which can be considered as a paradoxical situation since process retrofit occurs more frequently than does process design in industrial practice. The motivation for this problem was initially provided by the industrial environment. Yet, analysis of the dedicated literature shows that the development of retrofit design strategies is also identified as a difficult problem for academic research, with specific features compared with the design problem.

This chapter is organized to deal with a number of retrofit issues dedicated to batch plants. First, the typical features of batch plants are presented in Section 8.2. This is essential for the reader to understand the formulation of the batch design/retrofit problem in which they are intrinsically captured, as highlighted in Section 8.3. Section 8.4 is then devoted to

Chemical Process Retrofitting and Revamping: Techniques and Applications, First Edition.
Edited by Gade Pandu Rangaiah.

methods and tools used to solve the batch plant retrofit problem that have been developed by the process systems engineering community in the last few decades. The similarities and differences with the design problem are emphasized. Conclusions and directions for future research are proposed in the final section.

8.2 Batch Plant Typical Features

The production of high-value-added products (pharmaceutical active principles, agrochemicals, etc.) has stimulated investigations on batch processes for several decades. Generally, a batch plant can manufacture several products sharing standard equipment in a series of production campaigns, with the ability to adapt itself to variations of raw materials and to rapid market fluctuations. A batch process involves discrete tasks embedded in an interconnected predefined sequence from raw materials to final products. The predefined sequence is commonly known as a recipe. The recipe consists of the amount of raw materials that are necessary to be involved in the individual tasks as well as their durations. It also includes the operating conditions of the tasks as well as additional constraints of the process (safety features for instance).

Two types of process can be distinguished. In *flowshop or multiproduct* plants, all the products follow the same sequence through all the process steps (Figure 8.1). In *jobshop* or *multipurpose* batch plants, not all the products require all steps and/or might follow a different sequence of steps. The operational flexibility is particularly adapted to situations where product demands and formulations change rapidly, or where small production runs do not involve dedicated lines or plants. These features lead to a high level of complexity in the management of such plants, mainly due to the changes in the demand pattern that may occur (Figure 8.2).

Furthermore, batch plants can operate under different storage policies. Five main storage policies can be identified (Shah, 1998; Voudouris and Grossmann, 1996); some of these are illustrated in Figure 8.3.

Zero-Wait (ZW): the material is unstable and so it must be processed immediately after being produced.

Unlimited Intermediate Storage (UIS): the intermediate product is stable and can be stored without any problem. There are one or more dedicated vessels with unlimited storage capacity.

Finite Intermediate Storage (FIS): the product is stable, and there are one or more dedicated vessels with finite capacity for storage. This corresponds to a practical situation in which the number and capacity of storage tanks is usually limited.

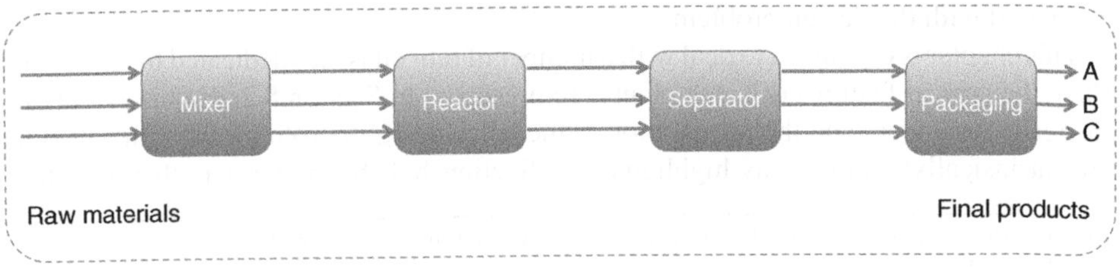

Figure 8.1 *Multiproduct batch plant structure.*

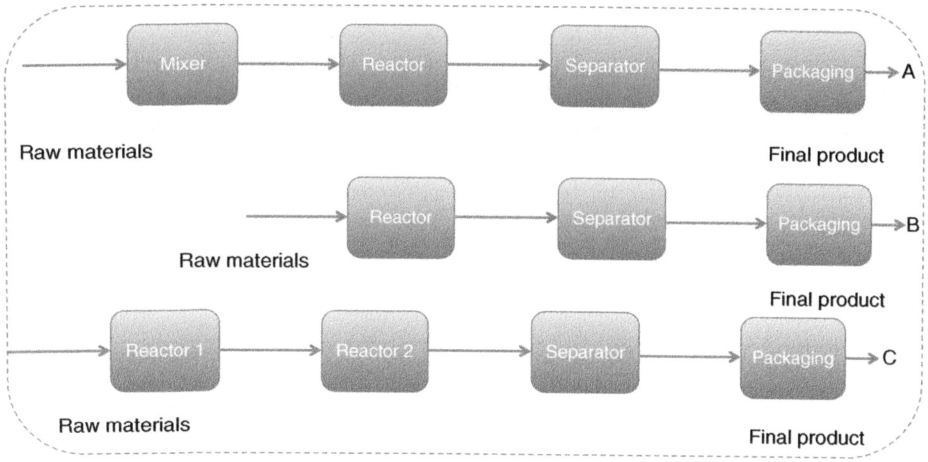

Figure 8.2 Multipurpose batch plant structure.

No Intermediate Storage (NIS): the material is stable but no storage vessel is available. However, it may reside temporarily in the equipment unit, which has been used for the active section of the process. In that case, the process unit is used for storage which can be viewed as detrimental to process profitability.

Common Intermediate Storage (CIS): the material is stable and may be stored in one or more vessels that may be shared by other materials although not simultaneously.

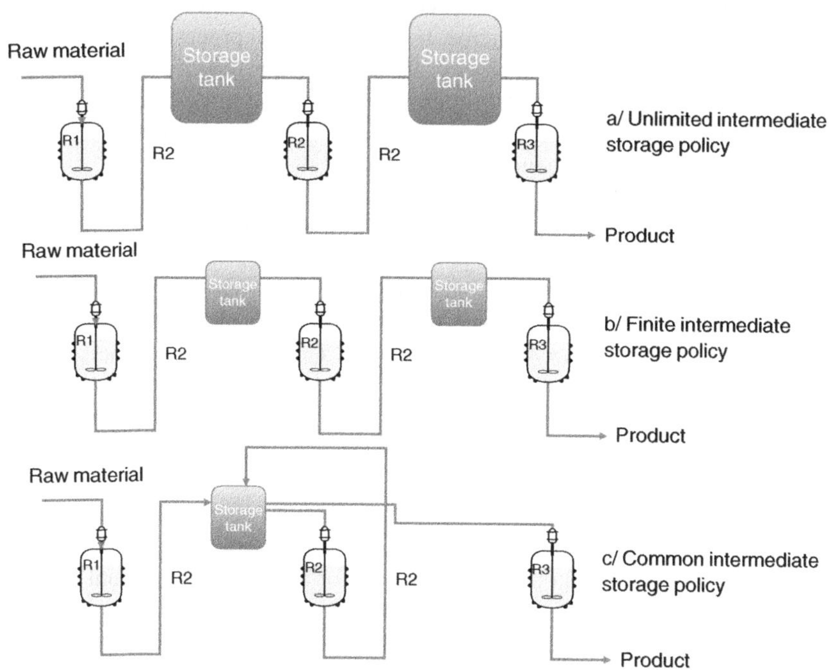

Figure 8.3 Different storage policies for batch plants.

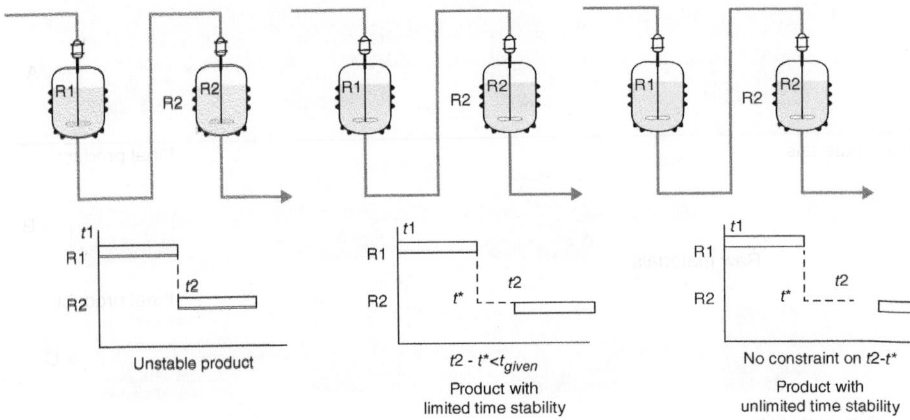

Figure 8.4 *Scheduling vs. product stability.*

These operational policies are not generally taken unilaterally; they are generally mixed, which means that the aforementioned operating philosophies coexist in one process, leading to the mixed intermediate storage (MIS) strategy. These situations may correspond to the stable, unstable or stable in limited time nature of the intermediate product as illustrated in Figure 8.4.

In general, a product is produced over a period of time by using repeatedly a sequence of operations, that is, a campaign of production. There are several ways to repeat the sequence of tasks needed, in order to produce the required quantity of product. Of course, the so-called *non-overlapping* campaign scheme, in which the total processing time is the number of batches multiplied by the time to process a single batch, is not efficient and is not applied practically. The strategy in which the total time to process all the batches can be reduced by starting a batch before the preceding batch has finished is more attractive. This leads to the concept of *overlapping* sequencing scheme.

The organization of the campaign is also a key component of the batch plant management. Let us consider 3 products: A, B and C, each requiring to be run n_A, n_B and n_C times in a given period. Two strategies can be performed to produce multiple products using the same equipment. Either campaigns of the same product followed by campaigns of the next product and so on.... that is, run A – n_A times followed by B – n_B times followed by C – n_C times – AAAAAAAABBBBBBBBCCCCCCC, or the products can be mixed leading to ABCABCABC... or ACBACBACB..., or AABCAABCAABC for instance. The advantage of using the mono-product campaign is that it is somewhat simpler to run, and cleaning is simplified since the risk of cross contamination is reduced except at the end of a campaign. Yet, storage is needed to meet a constant demand. A mixed campaign strategy discards the product storage issue but requires more rigorous cleaning procedures.

In a batch plant, all the stages have their own processing time and have a maximal capacity leading to a maximal possible batch size for a product. The actual size of a batch being processed in a plant then depends on the capacity of the size-limiting stage, and the time elapsed between the processing of consecutive batches cannot be shorter than that required by the so-called time-limiting stage. A very important concept is the limiting cycle time for a product, corresponding to the longest processing time among all the stages,

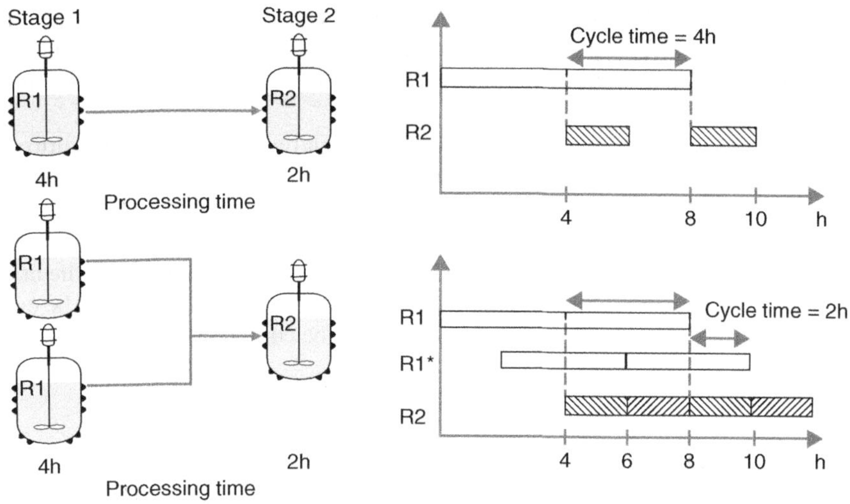

Figure 8.5 *Original plant (top) and two out-of-phase parallel units in stage 1 (bottom).*

and this determines the time between two successive batches for an identical sequence release strategy from a product to the following one. Two types of bottlenecks can thus occur, determined by size and time. Units can be added at a stage in order to debottleneck the plant; this situation is particularly important for retrofit purposes. These units can be operated either in-phase or out-of-phase as illustrated in Figure 8.5.

On the one hand, parallel units operating out-of-phase are useful to eliminate a time bottleneck in a series of batch stages. With two out-of-phase units, the cycle time is reduced by a factor of two compared with that for one unit. For a limiting stage, the addition of out-of-phase units reduces the limiting cycle time, so that the required unit sizes are reduced as the equipment utilization of the remaining units is increased. On the other hand, the addition of in-phase parallel units promotes capacity debottlenecking. The batch from the previous stage is split and assigned to all the in-phase units of that stage. The units process all the material at the same time and the completed batches are then merged together and transferred to the next stage. In that case, the limiting cycle time is not affected but the largest batch size that can be processed in the plant is increased.

All the above features may be encountered in batch plants so that scheduling strategies need to be considered at the design or retrofit stage. An analysis of the literature shows that, if they are not incorporated or integrated very well, this may lead to over- or under-design. Therefore, it is important to consider design or retrofit and scheduling simultaneously, especially if multipurpose batch plants are involved.

As aforementioned, the market demand for products manufactured through batch processes is usually variable. At the stage of conceptual design of a batch plant, it is almost impossible to obtain precise information on the future product demand over the lifetime of the plant. However, decisions must be made on the plant capacity. For this reason, design is generally performed with product portfolio and production scheme that are representative of the future product list, but it must be highlighted that the operated plant will probably have to be managed with other product candidates. The capacity of the plant should be able to balance the product demand satisfaction and the extra plant capacity in order to reduce the loss on the excessive investment cost or that on market share due to varying demands on products. This typical feature also justifies why retrofit actions are so important in batch plants.

The size factor is particularly important in the design/retrofit problem. Traditionally, S_{ip} relates the size of the equipment unit at stage i required to produce a unit of product p at the end of the process. The size factor determined from the recipe of each product at each production stage is a proportionality constant which gives the equipment size at stage i needed to produce a unit amount of final product. The stage batch size B_{ip} is the amount of final product p that stage i is able to produce. The limiting batch size can thus be computed from a set of different products so that the volume of the equipment unit at a given stage of the process can be determined.

8.3 Formulation of the Batch Plant Retrofit Problem

8.3.1 Design versus Retrofitting Problem

For a batch plant design problem, the aim is to determine the number and size of the necessary equipment items for every unit operation involved in the manufacturing sequence of the different product sets to obtain a given production level, while optimizing one or several performance objectives. Similarly, the decisions involved in retrofitting projects may be of the discrete type, the continuous type or a combination of both. More precisely, the retrofitting of an existing batch plant consists of finding design modifications due to new market conditions such as an increasing demand for some products or a revised product slate. It can be briefly stated as: (i) determine which equipment items are to be modified; and (ii) find the number and size of the new processing units and storage vessels. Like in the design case, the retrofitting is carried out according to one or several objectives. The two retrofitting actions have been clearly highlighted in Carvalho and Soletti (2000).

- Retrofit for changes in the demand of products
  This retrofit/expansion is applied when plants are required to satisfy changes in the demand of one or more products. The operational strategy should not be drastically changed, and the bottlenecks of the process must be identified and the feasibility to add new equipment in or out of phase must be considered
- Retrofit for a revised product slate
  In this case, new products can be added or removed, leading to changes in the production configuration. This condition increases the complexity of the problem, because the addition of new products leads to new production routes

Several mathematical formulations of the problems can be found in the dedicated literature (Papageorgaki and Reklaitis, 1993; Papageorgaki *et al.*, 1992).

An important aspect of batch plant modelling relates to the representation of the recipe. The most common representation of recipe is the so-called state task network (STN) that was proposed by Kondili *et al.* (1993), which involves two types of nodes; that is, state nodes and task nodes. The former represent all the materials that are processed in the plant, that is, feeds (raw materials, solvents etc.), intermediate and final products. The latter nodes are relative to unit operations that are performed throughout the process: tasks are carried out by units and transform one material from one state to another. The fraction of a state consumed or produced by a task, if not equal to one, is indicated in the graph. Several examples are provided in Floudas and Lin (2004). Other networks were also proposed such as the state sequence network (SSN), but the distinction between both formulations is rather subtle and is far beyond this chapter. More information concerning the distinction between the STN and SSN networks can be found in Majozi and Zhu (2001).

All formulations for design/retrofit and scheduling of batch processes can be classified into two groups based on the time representations: discrete-time vs. continuous-time models. In the former case, two cases can be encountered. First, the time horizon is divided into a number of time intervals of uniform duration and events such as the beginning and ending of a task are associated with the boundaries of these time intervals. Generally, a time interval based on the greatest common factor (GCF) of the processing times is used. This strategy usually leads to very large combinatorial problems. This explains why an alternative based on a non-uniform discretization of time characterized by the times at which an event occurs, evolving by event jump is more attractive.

The continuous-time representation has the same ambition: events are allowed to take place at any point in the continuous domain of time. This is carried out by introducing the concept of variable event time that can be defined either globally or for each unit. For this purpose, additional variables are necessary to determine the timings of events. Since a major fraction of the inactive event-time interval assignments is eliminated with the continuous-time approach, the resulting mathematical models require less computational effort for their solution of the associated scheduling task involved in design/retrofit problems. However, the mathematical models involved in the continuous-time approach may embed more complicated structures compared with their discrete-time counterparts.

8.3.2 Design/Retrofit Problems: A Four-Level Framework

Since the retrofit problem can be viewed as a particular case of the design problem with specific constraints that capture the existing situation of the plant, the conceptual batch process retrofit problem proposed by Reklaitis (1989) can be decomposed into four decision levels.

1. Determination of the **modified** processing network that involves the definition of the product recipe and task partitions for each **new** product
2. Selection of the best operating strategy: this concerns the decision on the operating mode to be used for manufacturing the different products involved. The choice of the operating strategy may be modified at this level for the initial set of products, while respecting the nature of the products (ZW, FIS, UIS, CIS and MIS)
3. Allocation of equipment items to tasks and storage of material, **taking into account the initial situation of the plant**
4. Sizing of the **new** various pieces of equipment (processing units, intermediate vessels and connection items)

This particular situation means that a configuration obtained from a retrofit situation can be viewed as non-optimal from a design point of view, meaning that if the expansion/modification could be anticipated, a better solution could be obtained.

The generic optimal retrofit problem that is directly derived from a design case can be expressed as follows.

Given:

- A set of products, the current production requirement for each product and its selling price, and an initial configuration of equipment item used for the manufacture of these products
- A set of changes in the demands and prices of the existing products over a pre-specified time period (horizon) and/or a set of modifications in the product slate in the form of addition of new products, elimination of old products, or modification of the product recipe of existing products
- The product recipes describing the production of the new and old products over a single or multiple campaign structures
- All possible equipment units to be installed in the plant and the associated suitability to perform the different operational tasks (for example, processing, storage and transport)
- The time horizon of planning
- The resource utilization over the time horizon (for example, utilities consumption)
- Inventory availability for each material involved
- Demands and deliveries along the time horizon
- Storage policies
- Operating and capital cost data involved in the plant/process installation and operation

Determine:

- The new optimal plant configuration (that is, additional number and type of interconnected equipment items including the auxiliary equipment); that is, the sizes of new processing units and intermediate storage vessels
- A process schedule making use of the selected resources to achieve the new required level of production (that is, timing of all tasks, storage policies, batch sizes, amounts transferred and allocation of tasks to equipment)

So as to optimise plant performance objective(s).

8.4 Methods and Tools for Retrofit Strategies

8.4.1 General Comments

Whatever the case of batch plants, either of multiproduct or multipurpose nature, few works have addressed the retrofit problem compared with the batch plant design problem. Most of the mathematical formulations used to model the retrofit problem usually lead to optimization-oriented problems. It must be also emphasized that scheduling is intrinsically associated with batch plant design/retrofit, and is part and parcel of the formulation of both issues. It can be considered as a sub-problem of the design/retrofit problem. As highlighted in the compared formulation of these companion problems, the retrofit problem appears to be a more constrained problem. Yet, the basic solution strategy involves the same mathematical approaches and tools.

The main contributions dedicated to the retrofit problem for multiproduct and multipurpose plants are summarized in Tables 8.1 and 8.2, respectively. The main problem

Table 8.1 *Main works on multiproduct batch plant retrofit.*

Reference	Main characteristics of the problem	Solution strategy	Optimization objectives	
Vaselenak *et al.* (1987)	Recipes for all the products are given with fixed processing times Addition of in-phase and out-of phase parallel units Overlapping production schedule Processing vessels can act as their own storage tanks Continuous range of equipment sizes, and number of batches is permitted to be non-integer No semi-continuous equipment	MINLP	Maximization of net income from selling the products minus the annualized investment cost	Mono-objective strategy
Espuña and Puigjaner (1989)	Single product campaigns No restriction on utilities No Intermediate storage Each unit is used no more than once per batch Parallel equipment units are always assigned to the same task In-phase and out-of-phase modes Out-of-phase mode not allowed in semi-continuous tasks In-phase units operating out-of-phase have the same overall capacity Overlapping mode of operation is assumed	Heuristic rules	Extra plant investment cost and overall net profit from the additional production	

(*continued*)

Table 8.1 *(Continued)*

Reference	Main characteristics of the problem	Solution strategy	Optimization objectives	
Fletcher *et al.* (1991)	Same framework as in Vaselenak *et al.* (1987) The restriction that any new unit must be operated in the same manner for all products is removed	MINLP	Maximization of net income from selling the products minus the annualized investment cost	Mono-objective strategy
Yoo *et al.* (1999)	An operating mode different from an existing mode can be considered Introduction of the "group" concept: the out-of-phase mode and the in-phase mode operations are merged into a generalized superstructure	MINLP	Same as above	
Van den Heever and Grossmann (1999)	Multiperiod optimization model	Multiperiod MINLP model Disjunctive formulation Logic-based OA	Minimization of total cost including operating and expansion costs	
Montagna (2003)	Extension of the explicit model by Yoo *et al.* (1999) Intermediate storage tanks for the retrofit problem	MINLP (DICOPT++)	Maximization of the annual profit of the plant	
García-Ayala *et al.* (2012)	Disjunctive programming model and a rolling horizon algorithm for optimal multiperiod capacity expansion in a multiproduct batch plant	MINLP (DICOPT)	Minimization of investment cost	

Goel *et al.* (2004)	Retrofit strategy of Yoo *et al.* (1999) with consideration of inherent reliability and maintainability characteristics of existing and new equipment	MINLP	Maximization of the overall expected net profit: tradeoff between the cost of unavailability and the extra capital investment needed to increase the size and/or inherent availability of new equipment
Dietz *et al.* (2005)	Extension of a design model including equipment modelling (operating conditions) Validation with an industrial plant	Discrete event simulation and Multi-objective Genetic Algorithm (MOGA)	Minimization of the investment cost and of the number of different sizes for the units of the plant
Simon *et al.* (2008)	Identification of improvement opportunities in a batch plant by considering first the product market situation Analysis of the batch plant according to three levels (the plant, the process, and the unit operation level) based on indicators which are of various types and specific for each analysis level. These indicators are linked to heuristics, which are used to identify the retrofit actions	Systematic indicators, heuristics, and model-based retrofit framework	Overall process performance indicators, and general unit operation performance indicators

Multi-objective strategy

(continued)

Table 8.1 *(Continued)*

Reference	Main characteristics of the problem	Solution strategy	Optimization objectives	
Halim and Srinivasan (2008)	Integrated simulation–optimization framework for process and cleaning waste minimization (BATCH-ENVOPExpert)	Coupling of Grafcet and unit-based simulation (process flowsheeting) embedded in a multi-objective optimization based on simulated annealing	Minimization of environmental impact from process waste and amount of cleaning liquid, maximization of production and minimization of batch production time	Multi-objective strategy
Bumann *et al.* (2010, 2011)	A three-stage framework Stage 1: Process analysis of boundary conditions—Process simulation and flowsheeting (mass and energy balances) Stage 2: Generation of process retrofit alternatives Stage 3: Evaluation of process retrofit results Case study of a mono-product batch plant	A systematic path flow decomposition-based methodology State Task Network	Computation of path flow indicators (economic and environmental criteria)	

Table 8.2 *Main works on multipurpose batch plant retrofit.*

Reference	Main characteristics of the problem	Solution strategy	Optimization objectives	
Papageorgaki *et al.* (1992)	H1- Product changes H2- Addition of new units (in-phase or out-of-phase mode) H3- Multiproduct campaign H4- No intermediate storage H5- No reuse of equipment except for the case of consecutive tasks H6- Processing time is a function of the batch size H7- Production demands are treated as variables with predetermined upper and lower bounds H8- Resource constraints: renewable resources	MINLP Decomposition strategy	Maximization of net profit which involves capital costs, operating costs and sales revenue.	Mono-objective strategy
Papageorgaki and Reklaitis (1993)	H1 to H7 + H9- Constraints on the availability of utilities (for example, electricity, water and steam)	Non-convex MINLP Decomposition strategy	Same as above	
Barbosa-Póvoa and Macchietto (1994)	Maximal State Task Network equipment developed to account for the process/plants interaction in a single representation. The process recipes and the plant structure (units and connections) are combined into a single framework Cleaning-in-place; renewable and non-renewable resources Plant topology	MILP	Maximization of annualized profit	

(continued)

Table 8.2 (Continued)

Reference	Main characteristics of the problem	Solution strategy	Optimization objectives
Carvalho and Soletti (2000)	Changes in the product and demand revisions in the product scale Addition and/or elimination of equipment units Batch size dependent on processing times	MILP	Minimization of cost of acquisition of new equipment
Pinto *et al.* (2005)	Resource Task Network unified framework Binary variables are used to represent the operational and structural choices that have to be made General storages policies General assignment case	MILP	Maximization of annualized profit
Dedieu *et al.* (2003)	Discrete event simulation model as a slave problem: raw materials, utilities, final and intermediate products, renewable resources (equipment and storage vessels), are modelled using the formalism of finite state automata MOGA is considered as a master problem	Discrete event simulation and MOGA	Minimization of investment cost, of the number of different sizes for the units of the plant and of the number of the production campaigns to reach a steady-state or oscillatory regime (Tri-criteria case)

Multi-objective strategy

characteristics and typical features of the solution strategy as well as the type of objective functions are indicated, following the guidelines proposed in Barbosa- Póvoa (2007). Several comments can be highlighted.

- Independent of the nature of the plant, either multiproduct or multipurpose, the most-used mathematical approaches are MILP (Mixed-Integer Linear Programming) or MINLP (Mixed-Integer Non-Linear Programming). The predominant objective function is based on an economic criterion.
- In all of the reported works concerning multiproduct plants, the production capacity of the multiproduct batch plant is specified by only two parameters: the limiting batch size and the limiting cycle time. The retrofitting strategy considered in these approaches usually focuses on adding new equipment to either increase the limiting batch size or decrease the limiting cycle time for each product or for both. The available characteristics of the existing plant and of the new retrofitted plant are not considered in these approaches except in the work of Goel *et al.* (2004).
- Concerning multiproduct batch plants, a few works (Van den Heever and Grossmann, 1999; García-Ayala *et al.*, 2012) take into account multiperiod capacity expansion in the multiperiod retrofit problem. This is more adequate to approach design and investment decisions compared with the typical retrofit problem, where just one period is taken into account. But it increases the size and complexity of the problem. To mitigate this effect, a disjunctive model is developed by transforming the original problem using the convex hull relaxation over a disjunctive set.
- The design and retrofit of batch plants was previously analysed and critically reviewed in Barbosa-Póvoa (2007). It can be said that the design of multiproduct batch plants has been an active area of research over the past decade with a majority of works involving deterministic approaches, wherein the problem parameters are assumed to be known with certainty. However, in reality there can be uncertainty in a number of factors such as processing times, costs, demands, and not all the requirements placed by the technology of the process and the properties of the substances are defined. To cope with this, there has been increased interest in the development of different types of probabilistic models that explicitly take into account the various uncertainties (Sahinidis, 2003). For example, Wellons and Reklaitis (1989) proposed an MINLP model for the design of batch plants under uncertainty with staged capacity expansions. Based on the structure of multiproduct batch plants, Straub and Grossmann (1992) developed an efficient procedure to evaluate the expected stochastic flexibility, embedded within an optimization framework for selecting the design (size and number of items of parallel equipment). Two-stage stochastic programming approaches have also been applied for design under uncertainty (Ierapetritou and Pistikopoulos, 1996; Cao and Yuan, 2002). The use of probabilistic models that describe the uncertain parameters in terms of probability distributions in an optimization framework can be viewed as greedy in computational time because of the large number of scenarios involved in the discrete representation of the uncertainty or the need to use complex integration techniques when uncertainty is modelled by continuous distributions. Besides, the use of probabilistic models is realistic only when a historic data set is available for uncertain parameters, which is rarely the case at the preliminary design stages in new product development. Fuzzy concepts and arithmetic constitute an alternative to describe the imprecise nature of product demands

as presented in Aguilar-Lasserre *et al.* (2009). Although modelling imprecision and uncertainty have been recognized as key issues in design problems, they are rarely considered in the retrofit problems, perhaps because retrofitting actions can be viewed as a solution to compensate for the discrepancy that has been observed between the preliminary design problem and its evolution that has been occurring so far.

- It can be reported from a literature analysis that surprisingly very few works mention the use of batch plant flowsheeting tools as a response to retrofit actions. It must be highlighted that such tools can be useful to cycle time reduction and debottlenecking, not only from a scheduling point of view but also from a retrofit one. Some flowsheeting software tools such as SuperPro Designer (http://www.intelligen.com/debottlenecking.html) have been specifically developed for batch plants and can be used to identify and analyse the capacity and time utilization of each piece of equipment, and to identify opportunities for increasing throughput with the minimum possible capital investment.

8.4.2 Key Approaches in Batch Plant Retrofitting: Deterministic vs Stochastic Methods

The formulation of the batch plant retrofit problem generally involves either deterministic methods based on mathematical programming such as LP (Linear Programming), NLP (Non-Linear Programming), MILP or MINLP or stochastic ones such as evolutionary algorithms. The non-linearity aspect may result from the nature of the constraints and/or the objective function. The mixed feature means that continuous and integer variables are involved in the formulation.

Since the design/retrofit problem embeds batch plant scheduling, it systematically includes the determination of the production sequence and the equipment sizes based on a performance criterion. Equipment sizes are considered either as continuous variables or as discrete ones; and so the problem can involve either discrete variables or a set of mixed ones. Most of the existing literature has focused on single objectives involving a cost criterion typically based on capital investment. This criterion is generally expressed as a non-linear function of the size of the equipment, following the six-tenths rule.

Let us recall that the quickest way to make an order-of-magnitude estimate of either a plant cost or an equipment item is to scale it from the known cost of an earlier plant that used the same technology or from published data. This requires no design information other than the production rate of a plant or a typical size/feature of an equipment. The capital cost of either a plant or an equipment item is related to capacity by Equation (8.1):

$$C_2 = C_1 \left(\frac{S_2}{S_1} \right)^n \tag{8.1}$$

Here, C_i is the capital cost of the plant or of an equipment item with capacity S_i. The exponent n is about 0.6, which corresponds to an average value across the whole chemical industry.

For design/retrofit problems, profitability analysis is scarcely performed on net present value assessment, since most often a given production level has to be reached leading to fixed revenue. NLP formulations are developed that are later on reformulated as MILP problems such as those proposed by Voudouris and Grossmann (1992). This was possible by assuming discrete sizes for the potential equipment to be installed in the plant.

The main interest of using rigorous deterministic methods is that an optimum can be reached: some mathematical properties relative to the objective function and constraints, such as continuity or derivability, have to be checked. This verification can turn out to be difficult for complex problems, for instance when the objective function is computed from a flowsheeting simulation tool. The most popular deterministic methods are: Branch & Bound methods (Gupta and Ravindran, 1985; Ryoo and Sahinidis, 1995), the Generalized Benders Decomposition (Geoffrion, 1972) and the Outer Approximation (Duran and Grossmann, 1986) algorithms, the Extended Cutting Plane method (Westerlünd and Petterson, 1995) and Disjunctive Programming methods (Raman and Grossmann, 1994). Some computational programs are *CPLEX, SBB, BARON, DICOPT++* and *LOGMIP* solvers within the *GAMS* modelling environment (Brooke *et al.*, 1998), *MINLP_BB* (Leyffer, 1999) and a*ECP* (Westerlünd and Lundqvist, 2003).

The discrete decisions involved in both design and retrofit situations (for example, equipment number and allocation, task allocation) lead to an inherent combinatorial aspect, which is challenging from the computational complexity point of view (Pekny and Reklaitis, 1998). These problems are identified as NP-hard (**N**on-deterministic **P**olynomial-time hard) problems (Garey and Johnson, 1979), so that there is no known algorithm that can solve the problem in a polynomial time. This feature makes stochastic methods particularly attractive to tackle such problems.

The stochastic or metaheuristics methods involve the evaluation of the objective function at different points of the search space, chosen through the use of a set of heuristics, combined with generations of random numbers. A major drawback of them is that they cannot guarantee to obtain an optimum. Yet, engineering problems often involve a complex formulation for which a good alternative to an empirical solution as measured by a significant performance improvement is generally sufficient. Stochastic methods can be divided into neighbourhood techniques; for example, Simulated Annealing (Kirkpatrick *et al.*, 1983) and Tabu Search (Teh and Rangaiah, 2003) and evolutionary algorithms such as genetic algorithms (GAs) (Holland, 1975), evolutionary strategies (Beyer and Schwefel, 2002) and evolutionary programming (Yang *et al.*, 2006). See Rangaiah (2010) for details on these and other methods as well as their applications in Chemical Engineering.

The interested reader is referred to the abundant literature dedicated to the presentation of the methods implemented in the GAMS solvers. Only the basic principles of some algorithms that are different in nature and used recurrently for the solution of design/retrofit problems are outlined below. More information can be found in Ponsich and Azzaro-Pantel (2007).

DICOPT++: This solver is based on the outer-approximation (OA) algorithm (Duran and Grossmann, 1986). This problem is decoupled into one continuous NLP sub-problem and one MILP master problem. In the continuous sub-problem, the discrete variables are set to fixed values, corresponding to the result of the previous master problem. If feasible, the solution of the NLP (by use of *CONOPT3*) provides an upper bound Z_U of the initial problem. If the termination criterion is not reached, the integer master problem is built by linearizing the constraints at the NLP solution and adding them to all the linearizations computed in the previous iterations. The resulting MILP problem is solved with the *CPLEX* solver (that is, the continuous variables are set to the last NLP solution values) in order to get a new lower bound Z_L of the global problem. The accumulation of constraint linearizations allows a more precise description of the feasible region. Consequently, the

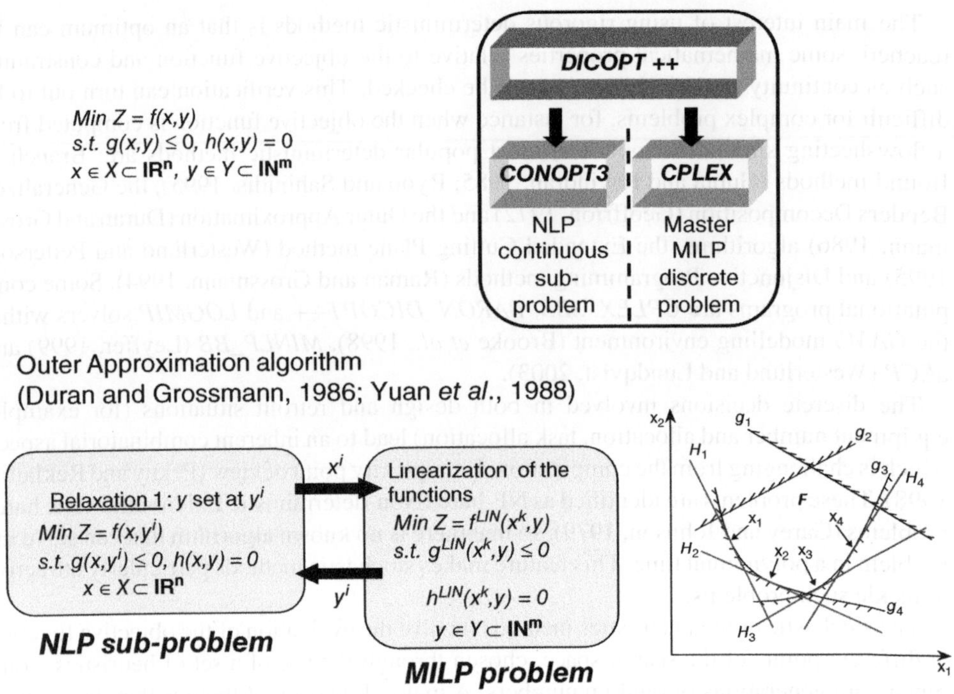

Figure 8.6 *Typical strategy involved in the DICOPT++.*

upper and lower bounds, respectively, decrease and increase monotonically at each major iteration of the algorithm. The termination criterion is reached when the objective function deteriorates. Since non-convexities could disrupt the process, either by cutting regions of the feasible space with an invalid constraint linearization or by generating invalid lower bounds, local and global convexity tests (Kocis and Grossmann, 1988) and augmented penalty (Viswanathan and Grossmann, 1990) have been proposed. The strategy of DICOPT++ is illustrated in Figure 8.6.

SBB: This solver is similar to a classical Branch & Bound algorithm following the principle of separation of the initial problem by successively branching on the discrete variables and tightening their respective bounds (Figure 8.7). The resulting sub-problems are represented by nodes and relaxed to get continuous problems. Solving those relaxations provides either a lower bound to the global optimum or an upper bound when the problem is mixed integer and continuous, respectively. The bound intersection is the termination criterion. The slight difference between a classical Branch & Bound method and *SBB* lies in the MINLP nature of the initial problem, which leads to non-linear relaxation for each node. These are solved with the NLP solver *CONOPT3*. *SBB* does not need any MILP solver since the discrete variables are handled by the Branch & Bound procedure.

Stochastic methods: Stochastic algorithms have proven their efficiency to tackle high combinatorial problems such as those encountered in design/retrofit problems. Stochastic methods (such as simulated annealing and tabu search) or techniques based on social intelligence (ant colonies and particle swarms) or cultural concepts have been largely

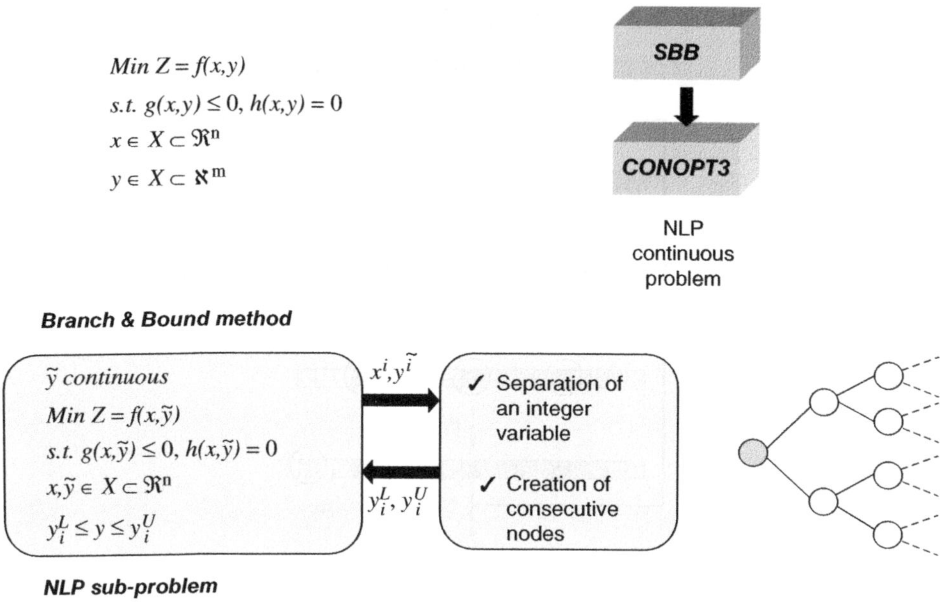

Figure 8.7 *Typical strategy involved in the SBB method.*

used for the batch plant design problem. Since the underlying principle is quite similar, the objective of this chapter is not to present an exhaustive description of each of these methods but to highlight the specific interest of using such a method compared with a deterministic one. The principles of GAs (Holland, 1975) that have been used in numerous applications in process systems engineering, and in particular for batch plant retrofit, are described for the sake of illustration.

GAs (Figure 8.8) lie on the analogy between a population of individuals and a set of solutions of some optimization problem. The algorithm makes the solution set evolve towards a good quality, or adaptation, and mimics the rules of natural selection stated by Darwin: the weakest individuals will disappear while the best ones will survive and be able to reproduce themselves. By way of genetic inheritance, the characteristics that make these individuals "stronger" will be preserved generation after generation. The mechanisms implemented in the GAs try to reproduce this natural behaviour by creating numerical selection rules, crossover and mutation operators so that a population constituted by a set of good solutions to the optimization problem can be obtained at the end of the optimization process. A major difficulty in GAs use lies in parameter tuning, which greatly depends on the user's experience and problem knowledge. A sensitivity analysis is necessary to choose suitable values for GA parameters such as population size, maximum number of generations, crossover and mutation probabilities.

An alternative proposed in previous works (for example, Bérard *et al.*, 2000) consists in coupling a Discrete Event Simulator (DES) (in order to evaluate the feasibility of the production at medium term scheduling) with a master optimization procedure generally based on stochastic techniques such as GA (to take into account the combinatorial feature due to the large number of discrete variables in the optimization problem). These ideas

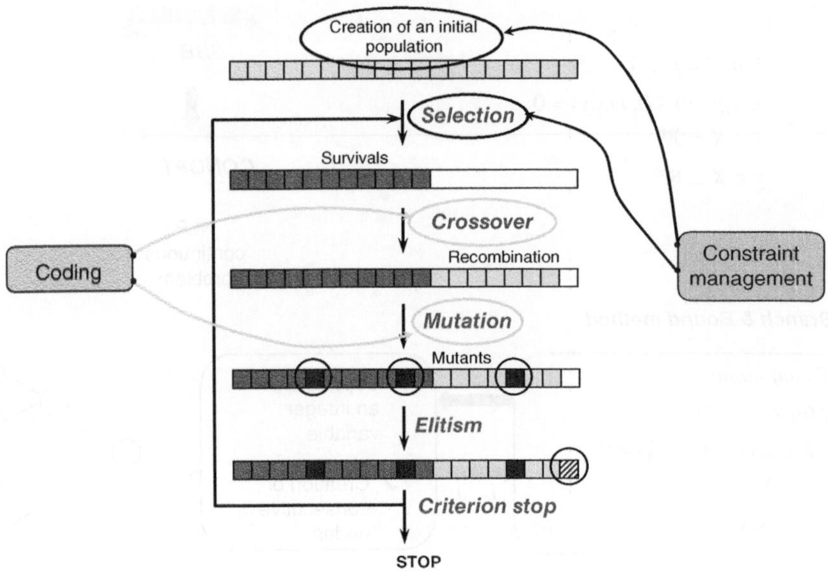

Figure 8.8 *Typical strategy involved in an evolutionary GA method.*

were generalized in Dietz *et al.* (2005) within a framework for multicriteria design and retrofitting taking into account both the batch plant structure and the operating conditions as optimization variables.

8.4.3 New Trends in Batch Plant Retrofitting: Steps for More Sustainable Processes

The review paper by Barbosa-Póvoa (2007) covered various approaches for both design and retrofit design problems. It is noteworthy that the approaches reported to the date of this publication took into account only retrofitting analysis regarding the scheduling of batch operations, and Barbosa-Póvoa (2007) did not consider the benefits that can be achieved in terms of sustainable design.

A strong incentive for retrofitting is to take into account the more and more stringent environmental regulations, with special focus given on global warming potential and thus on CO_2 emissions. Sustainable process design has been identified as a key topic for many years (Azapagic, 1999). As far as retrofitting is concerned, it can be viewed as the consideration of environmental aspects as constraints or objectives in a multi-objective problem formulation. The concept of sustainability is larger than the consideration of economic and environmental issues, since the societal component has also to be included in the three pillars of sustainability. Yet, the inclusion of social indicators is not straightforward to implement (Azapagic *et al.*, 2006). For this purpose, the focus is most generally given to the operational contribution of sustainable development, that is, economic and environmental issues. Some examples of proposed environmental metrics for chemical processes have been based on material and energy intensity, water consumption, toxic emissions, pollutant and greenhouse gas emissions. Several studies focused on minimization of waste generation and waste treatment costs in order to highlight sustainable alternatives (Cabezas *et al.*, 1999; Halim and Srinivasan, 2008). Dietz *et al.* (2005) investigated the pollution global index as an environmental incentive and presented Pareto-optimal solutions for a protein production case study by minimization of investment cost and pollution global index.

Systematic methodologies to generate new design alternatives have also been proposed (Halim and Srinivasan, 2008, 2009). A different approach where indicators, heuristics, and process model are combined to provide decision support in retrofitting of chemical batch processes has also been developed (Simon *et al.*, 2008). The developed framework considers the identification of improvement opportunities in a batch plant by considering first the product market situation. The batch plant analysis is structured on three levels: the plant, the process, and the unit operation level, respectively. The analysis of these levels is based on indicators which are of various types and specific for each analysis level. These indicators are linked to heuristics which are used to identify the retrofit actions.

According to Carvalho *et al.* (2009), retrofit analysis for the generation of more sustainable options can be performed in two kinds of software tools. The first group involves tools that evaluate process performance in terms of sustainability, life cycle assessment and environmental impact (for example, SimaPro and GaBi software). The second group are those that determine new design alternatives in order to reduce the environmental impact (for example, AquoMin of Relvas and Matos, 2008, and ENVOPExpert of Halim and Srinivasan, 2008). Following these guidelines, a generic and systematic methodology to generate new sustainable design alternatives for a given process, which might operate in batch or in continuous mode, is presented in Carvalho *et al.* (2009). The methodology presented in Carvalho *et al.* (2008, 2009) involves the following steps: (1) data from simulations or real plant data are collected; (2) flowsheet decomposition is performed so that the process is decomposed into paths; (3) a set of mass and energy indicators is calculated for each path, in order to detect the process bottlenecks; (4) through a sensitivity-based analysis, the target indicators for improvements are identified; (5) a sensitivity analysis is then carried out in order to determine the target variables for improving the process; and (6) a set of synthesis methods are used to generate new sustainable process alternatives. These alternatives are then evaluated through the performance criteria (for example, Sustainability Metrics and Safety Indices). The analysis and the automatic generation of process alternatives that are more sustainable are performed through the software tool called SustainPro.

8.5 Conclusions

The retrofit problem is a companion topic to the design issue, since batch manufacturing requires the frequent adaptation of production facilities to take into account internal and external developments since the plant was designed. The concept of flexibility in the design of plants is intrinsically linked to the nature of batch plants and can be viewed as a cornerstone for the design and retrofit of batch processes along the lifecycle of the plant. The retrofit potential of a multiproduct/purpose batch plant would ideally be taken into account at the design stage. This is a complex problem that needs to be considered in which uncertainty plays a major role. An interesting contribution was recently proposed in Moreno-Benito and Espuña (2014) by integrating process synthesis and plant allocation decisions, such as feedforward control trajectories of processing conditions or the selection of operating modes and equipment items. For this purpose, a modelling strategy based on Mixed-Logic Dynamic Optimization and two-stage stochastic formulation are used to maximize the expected profit taking into account several uncertain scenarios.

The retrofit analysis regarding the scheduling of batch operations must be thoroughly revisited, integrating the benefits that can be obtained in terms of sustainable design, particularly considering economic and environmental pillars. There is also a need to develop more robust retrofit strategies embedding a multiple criteria decision problem. In addition, there is a trend to take into account more integrated and connected facilities within the supply chain, particularly for multiproduct batch plant design for which flexibility is the essential quality. This supports the need for supply chain design frameworks, thus encompassing a holistic viewpoint, so that significant benefits can be obtained by addressing the interactions among different decision levels. This offers an important and open issue as far as batch plants are concerned as highlighted in Fumero *et al.* (2013).

References

Aguilar-Lasserre, A.A., Pibouleau, L., Azzaro-Pantel, C., and Domenech, S. (2009) Enhanced genetic algorithm-based fuzzy multiobjective strategy to multiproduct batch plant design. *Applied Soft Computing*, **9** (4), 1321–1330.

Azapagic, A. (1999) Life cycle assessment and its application to process selection, design and optimisation. *Chemical Engineering Journal*, **73**, 1–21.

Azapagic, A., Perdan, S., and Clift, R. (2006) *Sustainable Development in Practice – Case Studies for Engineers and Scientists*, John Wiley and Sons.

Barbosa-Póvoa, A.P. (2007) A critical review on the design and retrofit of batch plants. *Computers & Chemical Engineering*, **31** (7), 833–855.

Barbosa-Póvoa, A.P. and Macchietto, S. (1994) Redesign of a multipurpose batch pilot-plant with cleaning in-place (CIP) integration. *Computers & Chemical Engineering*, **18**, S277–S281.

Bérard, F., Azzaro-Pantel, C., Pibouleau, L., *et al.* (2000) Towards an incremental development of discrete-event simulators for batch plants: Use of object-oriented concepts. *Computers & Chemical Engineering*, **23**, S565–S568.

Beyer, H.G. and Schwefel, H.P. (2002) Evolution Strategies, a comprehensive introduction. *International Journal of Natural Computing*, **1**, 3–52.

Brooke, A., Kendrick, D., Meeraus, A., and Raman, R. (1998) GAMS User's Guide; GAMS Development Corporation: Washington, D.C.

Bumann, A.A., Papadokonstantakis, S., Fischer, U., and Hungerbuhler, K. (2010) Optimization of chemical batch processes within a systematic retrofit framework including evaluation of historical process data. *Chemical Engineering Transactions*, **21**, 919–924.

Bumann, A.A., Papadokonstantakis, S., Fischer, U., and Hungerbuhler, K. (2011) Investigating the use of path flow indicators as optimization drivers in batch process retrofitting. *Computers & Chemical Engineering*, **35** (12), 2767–2785.

Cabezas, H., Bare, J.C., and Mallick, S.K. (1999) Pollution prevention with chemical process simulators: the generalized waste reduction (WAR) algorithm. *Computers & Chemical Engineering*, **23** (4–5), 623–634.

Cao, D.M. and Yuan, X.G. (2002) Optimal design of batch plants with uncertain demands considering switch over of operating modes of parallel units. *Industrial & Engineering Chemistry Research*, **41** (18), 4616–4625.

Carvalho, S.H.V. and Soletti, J.I. (2000) Retrofit design of multipurpose batch plants with multiple production routes. *Brazilian Journal of Chemical Engineering*, **17** (4–7), 1015–1022.

Carvalho, A., Gani, R., and Matos, H. (2008) Design of sustainable chemical processes: Systematic retrofit analysis generation and evaluation of alternatives. *Process Safety and Environmental Protection*, **86**, 328–346.

Carvalho, A., Matos, H.A., and Gani, R. (2009) Design of batch operations: Systematic methodology for generation and analysis of sustainable alternatives. *Computers & Chemical Engineering*, **33** (12), 2075–2090.

Dedieu, S., Pibouleau, L., Azzaro-Pantel, C., and Domenech, S. (2003) Design and retrofit of multiobjective batch plants via multicriteria genetic algorithm. *Computers & Chemical Engineering*, **27**, 1723–1740.

Dietz, A., Schirlin, M., Azzaro-Pantel, C., Pibouleu, L., Domenech, S. (2005) Multicriteria optimization of batch plant design and retrofit: application to an industrial fine chemistry case study. In: The seventh Italian Conference on Chemical and Process Engineering, 15–18 May 2005, Giardini Naxos, Italy.

Duran, M.A. and Grossmann, I.E. (1986) An outer-approximation algorithm for a class of mixed-integer nonlinear programs. *Mathematical Programming*, **36**, 307–339.

Espuña, A. and Puigjaner, L. (1989) On the solution of the retrofitting problem for multiproduct batch semicontinuous chemical plants. *Computers and Chemical Engineering*, **13** (4–5), 483–490.

Fletcher, R., Hall, J.A.J., and Johns, W.R. (1991) Flexible retrofit design of multiproduct batch plants. *Computers and Chemical Engineering*, **15** (12), 843–852.

Floudas, C. and Lin, X. (2004) Continuous-time versus discrete-time approaches for scheduling of chemical processes: a review. *Computers & Chemical Engineering*, **28**, 2109–2129.

Fumero, Y., Corsano, G., and Montagna, J.M. (2013) Integrated modeling framework for supply chain design considering multiproduct production facilities. *Industrial & Engineering Chemistry Research*, **52** (46), 16247–16266.

García-Ayala, G., Ríos-Mercado, R.Z., and Chacón-Mondragón, O.L. (2012) A disjunctive programming model and a rolling horizon algorithm for optimal multiperiod capacity expansion in a multiproduct batch plant. *Computers & Chemical Engineering*, **46**,29–38.

Garey, M.R. and Johnson, D.R. (1979) *Computers and Intractability: A Guide to the Theory of NP-Completeness*, W.H. Freeman, New York.

Geoffrion, A.M. (1972) Generalized benders decomposition. *Journal of Optimization Theory and Applications*, **10**, 237–260.

Goel, H.D., Weijnen, M.P.C., and Grievink, J. (2004) Optimal reliable retrofit design of multiproduct batch plants. *Industrial and Engineering Chemistry Research*, **43** (14), 3799–3811.

Gupta, O.K. and Ravindran, V. (1985) Branch and bound experiments in convex nonlinear integer programming. *Management Science*, **31**, 1533–1546.

Halim, I. and Srinivasan, R. (2008) Designing sustainable alternatives for batch operations using an intelligent simulation-optimization framework. *Chemical Engineering Research and Design*, **86**, 809–822.

Halim, I. and Srinivasan, R. (2009) Sequential methodology for scheduling of heat-integrated batch plants. *Industrial & Engineering Chemistry Research*, **48** (18), 8551–8565.

Holland, J.H. (1975) *Adaptation in Natural and Artificial Systems*, University of Michigan Press, Ann Arbor, MI.

Ierapetritou, M.G. and Pistikopoulos, E.N. (1996) Batch plant design and operations under uncertainty. *Industrial & Engineering Chemistry Research*, **35** (3), 772–787.

Kirkpatrick, S., Gelatt, S.D. Jr., and Vecchi, M.P. (1983) Optimization by simulated annealing. *Science*, **220**, 671–680.

Kocis, G.R. and Grossmann, I.E. (1988) Global optimisation of nonconvex mixed-integer non linear programming (MINLP) problems in process synthesis. *Industrial & Engineering Chemistry Research*, **27** (8), 1407–1421.

Kondili, E., Pantelides, C.C., and Sargent, R.W.H. (1993) A general algorithm for short-term scheduling of batch operations—1. Mixed integer linear programming formulation. *Computers & Chemical Engineering*, **17**, 211–227.

Leyffer, S. (1999) User manual for MINLP_BB; University of Dundee Numerical Analysis Report NA/XXX: Dundee, UK.

Majozi, T. and Zhu, X.X. (2001) A novel continuous-time MILP formulation for multipurpose batch plants. 1. Short-term scheduling. *Industrial & Engineering Chemistry Research*, **40** (25), 5935–5949.

Montagna, J.M. (2003) The optimal retrofit of multi-product batch plants. *Computers and Chemical Engineering*, **27** (8–9), 1277–1290.

Moreno-Benito, M. and Espuña, A. (2014) Flexible batch process and plant design using mixed-logic dynamic optimization: single-product plants. *Industrial & Engineering Chemistry Research*, **53** (44), 17182–17199.

Papageorgaki, S. and Reklaitis, G.V. (1993) Retrofitting a general multipurpose batch chemical plant. *Industrial & Engineering Chemistry Research*, **32** (2), 345–362.

Papageorgaki, S., Tsirukis, A.G., and Reklaitis, G.V. (1992) The influence of resource constraints on the retrofit design of multipurpose batch chemical plants. NATO Conference on Batch Processes.

Pekny, J.F. and Reklaitis, G.V. (1998) Towards the convergence of theory and practice: A technology guide for scheduling/planning methodology, in (eds J.F. Pekny and G.E. Blau) *Proceedings of the Third International Conference on Foundations of Computer-Aided Process Operations*, Snowbird, Utah, pp. 91–111.

Pinto, T., Barbosa-Póvoa, A.P., and Novais, A.Q. (2005) Optimal design and retrofit of batch plants with periodic mode of operation. *Computers and Chemical Engineering*, **29**, 1293–1303.

Ponsich, A. and Azzaro-Pantel, C. (2007) Mixed-integer nonlinear programming optimization strategies for batch plant design problems. *Industrial & Engineering Chemistry Research*, **46** (3), 854–863.

Raman, R. and Grossmann, I.E. (1994) Modelling and computational techniques for logic based integer programming. *Computers & Chemical Engineering*, **18**, 563–578.

Rangaiah, G.P. (ed.) (2010) *Stochastic Global Optimization: Techniques and Applications in Chemical Engineering, Vol. 2 in the Advances in Process Systems Engineering*, World Scientific, Singapore.

Reklaitis, G.V. (1989) Progress and issues in computer aided batch process design. In Third International Conference on Foundations of Computer-Aided Process Design (FOCAPD), pp. 241–276.

Relvas, S. and Matos, H.A. (2008) AquoMin: a software tool for mass-exchange networks targeting and design. *Computers and Chemical Engineering*, **32**, 1085–1105.

Ryoo, H.S. and Sahinidis, N.V. (1995) Global optimization of nonconvex NLPs and MINLPs with applications in process design. *Computers & Chemical Engineering*, **19**, 551–566.

Sahinidis, N.V. (2003) Optimization under uncertainty: state-of-the-art and opportunities, In Proceedings of the Conference on Foundations of Computer Aided Process Operations, FOCAPO 2003, Coral Springs, USA, pp. 153.

Shah, N. (1998) Single- and multisite planning and scheduling: Current status and future challenges. 3rd International Conference on Foundations on Computer Aided Process Operations, American Institute of Chemical Engineers, 75–90.

Simon, L.L., Osterwalder, N., Fischer U., and Hungerbühler, K. (2008) Systematic retrofit method for chemical batch processes using indicators, heuristics, and process models. *Industrial and Engineering Chemistry Research*, **47** (1), 66–80.

Straub, D.A.S. and Grossmann, I.E. (1992) Design optimization of stochastic flexibility. *Computers & Chemical Engineering*, **17**, 339–354.

Teh, Y.S. and Rangaiah, G.P. (2003) Tabu Search for global optimization of continuous functions with application to phase equilibrium calculations. *Computers & Chemical Engineering*, **27**, 1665–1679.

Van den Heever, S.A. and Grossmann, I.E. (1999) Disjunctive multiperiod optimization methods for design and planning of chemical process systems. *Computers and Chemical Engineering*, **23** (8), 1075–1095.

Vaselenak, J.A., Grossmann, I.E., and Westerberg, A.W. (1987) Optimal retrofit design of multiproduct batch plants. *Industrial & Engineering Chemistry Research*, **26** (4), 718–726.

Viswanathan, J. and Grossmann, I.E. (1990) A combined penalty function and outer-approximation method for MINLP optimisation. *Computers & Chemical Engineering*, **14**, 769–782.

Voudouris, V.T. and Grossmann, I.E. (1992) Mixed-integer linear-programming reformulations for batch process design with discrete equipment sizes. *Industrial & Engineering Chemistry Research*, **31** (5), 1315–1325.

Voudouris, V.T. and Grossmann, I.E. (1996) MILP model for scheduling and design of a special class of multipurpose batch plants. *Computers & Chemical Engineering*, **20** (11), 1335–1360.

Wellons, H.S. and Reklaitis, G.V. (1989) The design of multiproduct batch plants under uncertainty with staged expansion. *Computers and Chemical Engineering*, **13**, 115–126.

Westerlünd, T. and Lundqvist, K. (2003) Alpha-ECP, Version 5.04. An interactive MINLP-solver based on the extended cutting plane method; Report 01-178-A, Process Design Laboratory, Abo Akademi University:Abo, Finlande.

Westerlünd, T. and Petterson, F. (1995) A cutting plane method for solving convex MINLP problems. *Computers & Chemical Engineering*, **19**, S131–136.

Yang, Y.W., Xu, J.F., and Soh, C.K. (2006) An evolutionary programming algorithm for continuous global optimization. *European Journal of Operational Research*, **168** (2), 354–369.

Yoo, D.J., Lee, H.K., Ryu, J.H., and Lee, I.B. (1999) Generalized retrofit design of multi-product batch plants. *Computers and Chemical Engineering*, **23** (6), 683–695.

Yuan, X., Zhang, S., and Pibouleau, L. (1988) A Mixed-Integer Nonlinear-Programming Method for Process Design. *RAIRO - Recherche Opérationnelle-Operations Research*, **22**, 331–346.

Part III

Retrofitting and Revamping Applications

9

Retrofit of Side Stream Columns to Dividing Wall Columns, with Case Studies of Industrial Applications

Moonyong Lee[1], *Le Quang Minh*[1], *Nguyen Van Duc Long*[1], *and Joonho Shin*[2]
[1]*School of Chemical Engineering, Yeungnam University, Republic of Korea*
[2]*Basic Materials & Chemicals R&D, LG Chem, Moonji-dong, Yuseong-gu, Republic of Korea*

9.1 Introduction

A distillation column typically separates a feed mixture into two product streams based on the differences in boiling point or relative volatility of the constituent components (Kiss, 2014). Distillation developed rapidly in the 19th and 20th centuries, prompted by the oil and petrochemical industry and by the chemical and pharmaceutical industries (Fair, 1984). Currently, distillation is the most widely used separation method on an industrial scale with more than 40,000 columns in operation worldwide (Humphrey, 1995). Remarkably, almost every product on the market contains chemicals that have been passed through a distillation column (Kiss, 2014). If an ideal binary mixture is introduced into a distillation column, the two components in the mixture can be separated into arbitrarily high purities in the two product streams. On the other hand, real-world separation problems often involve multicomponent feed streams. To separate multicomponent mixtures into more than two products, a sequence of distillation columns is mostly used. At least two columns are employed when separating three components, such as the direct and indirect sequences (Figures 9.1a and b) (Yildirim *et al.*, 2011). Such multicolumn distillation arrangements are often referred to as distillation sequences, distillation configurations, distillation trains, or distillation schemes (Gorak and Sorensen, 2014).

Chemical Process Retrofitting and Revamping: Techniques and Applications, First Edition.
Edited by Gade Pandu Rangaiah.

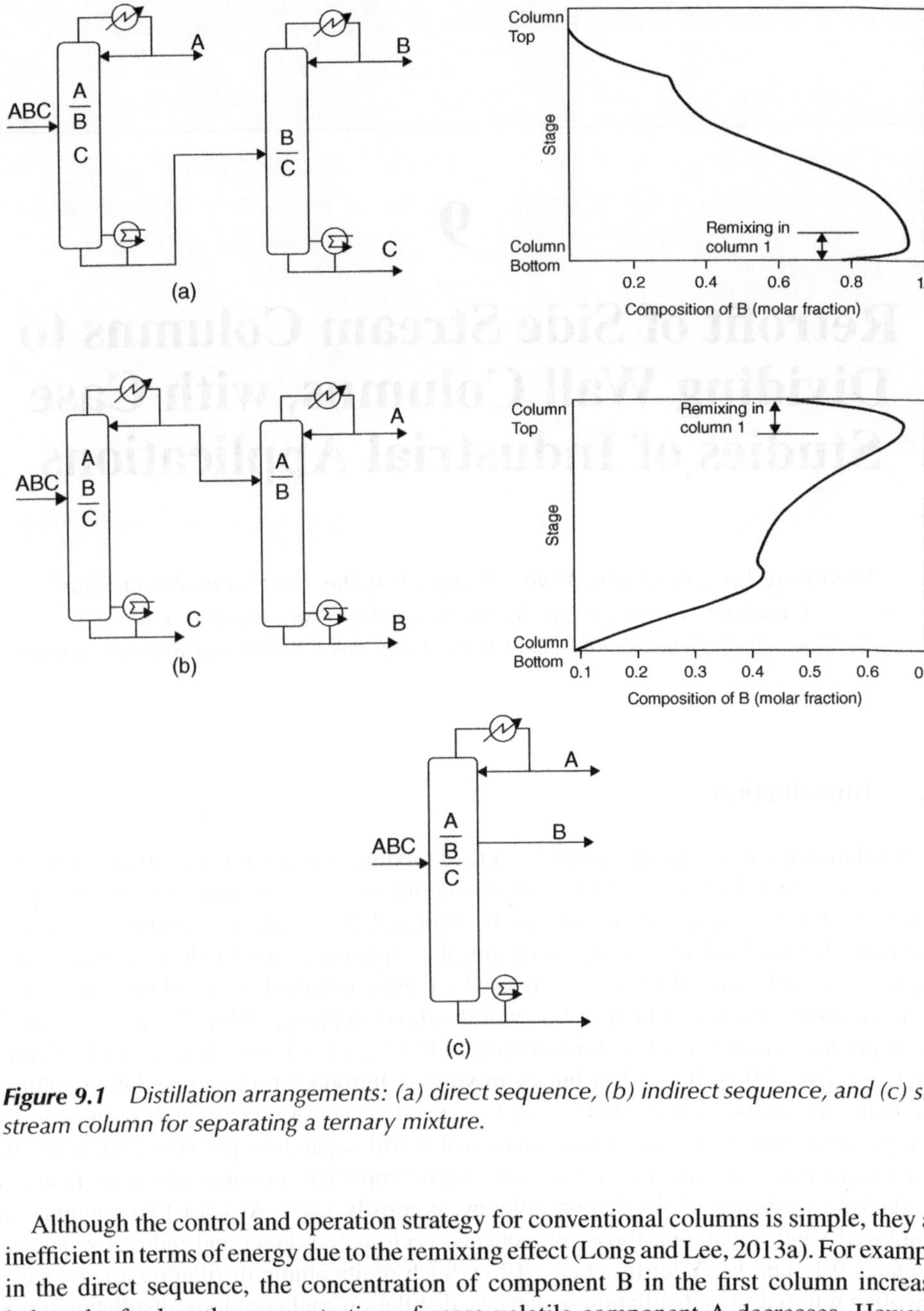

Figure 9.1 *Distillation arrangements: (a) direct sequence, (b) indirect sequence, and (c) side stream column for separating a ternary mixture.*

Although the control and operation strategy for conventional columns is simple, they are inefficient in terms of energy due to the remixing effect (Long and Lee, 2013a). For example, in the direct sequence, the concentration of component B in the first column increases below the feed as the concentration of more-volatile component A decreases. However, moving further down the column, the concentration of B decreases as the concentration of less-volatile component C increases. Thus, the concentration of B reaches a peak only to be remixed (Smith, 2005). Similarly, with the first column in the indirect sequence, the concentration of B first increases above the feed as that of component C decreases. It reaches a maximum only to decrease as the concentration of component A increases. Again, the concentration of B reaches a peak only to be remixed. This remixing that occurs in both sequences is the main source of inefficiency in the separation. Thus, side stream columns (SSCs), in which the intermediate component is withdrawn by a side draw (Figure 9.1c), are considered as more energetically favorable alternatives to avoid this phenomenon possibly leading to substantial energy consumption in the two-column sequences. In addition, the SSCs are used widely in a cost-effective way, where a highly pure product is not required.

At the end of the 1980s, retrofit projects comprised 70–80% of capital investment projects in the processing industry (Liu and Jobson, 2004). Because distillation is an energy-intensive process and requires considerable capital investment, the retrofit of distillation columns is carried out more often than is the installation of new equipment (Gadalla *et al.*, 2003). Retrofit projects of distillation processes focus on the re-use of existing equipment more efficiently to increase profit (Long and Lee, 2013b). The most straightforward designs are grassroots designs because they have the greatest freedom to choose the design options and size of equipment, while the retrofit design must attempt to work within the constraints of the existing equipment (Smith, 2005).

As an interesting option for retrofit, process integration has been successfully applied to reduce the energy requirements compared with processes, where all the units are configured with little or no integration (Long *et al.*, 2010). With energy consumption in focus, there has been significant interest in thermally coupled distillation column (TCDC) sequences and dividing wall columns (DWCs) for liquid mixtures of three or more components (Fidkowski and Krolikowski, 1986; Agrawal and Fidkowski, 1999; Segovia-Hernandez *et al.*, 2007; Tamayo-Galván *et al.*, 2008; Bravo-Bravo *et al.*, 2010; Long and Lee, 2013c; Minh *et al.*, 2014). The TCDC sequences were obtained by the implementation of interconnecting streams (one in the vapor phase and the other in the liquid phase) between two columns (Long and Lee, 2012a). Each interconnection also replaces one condenser or one reboiler from one of the columns, providing potential cost savings (Long and Lee, 2012a). Figure 9.2 shows three known TCDC configurations for ternary separation (Agrawal and Fidkowski, 1999), which are characterized as side rectifier, side stripper, and fully coupled (so-called Petlyuk column) configurations. Among them, the Petlyuk column provides the maximum energy reduction, which can avoid the repeated separation path defined as no part of separation being carried out twice (Poth *et al.*, 2004; Long and Lee, 2012b).

A DWC is a more intensified form of TCDC. Many studies have examined its advantages including the great potential for reduced utilities consumption (Errico *et al.*, 2009; van Diggelen *et al.*, 2010; Kiss and Bildea, 2011; Long and Lee, 2012b; Minh *et al.*, 2012). In a DWC, a vertical wall is installed in the central section of the column, dividing it into a prefractionator and main section (Amminudin and Smith, 2001), which is conceptually the same as the Petlyuk column, given their thermodynamically equivalent arrangements. Up to 30% of the energy requirement can be saved by using a DWC, while also lowering the engineering and hardware costs compared with those of conventional direct and indirect distillation sequences (Triantafyllou and Smith, 1992; Sotudeh and Shahraki, 2008; Kiss and Rewagad, 2011; Long *et al.*, 2013). Another attractive aspect of the DWC is that it requires only a small plot area for its implementation, which often becomes a critical constraint, particularly in a retrofit project.

The use of SSCs to separate ternary mixtures has attracted considerable scientific interest with the growing industrial demand. According to Glinos and Malone (1985), these columns

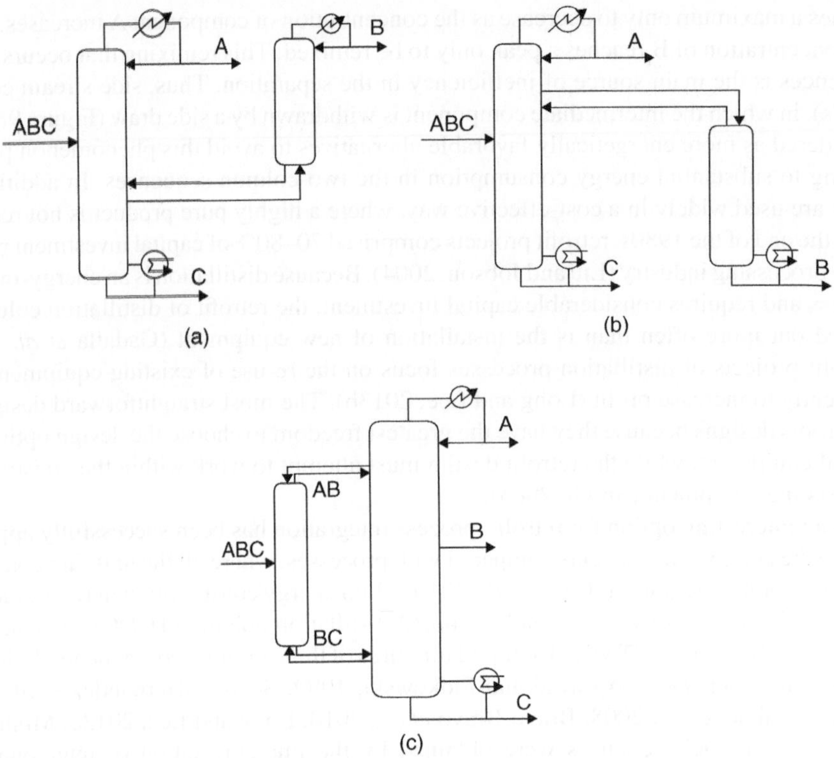

Figure 9.2 *TCDC configurations for ternary separation: (a) side rectifier, (b) side stripper, and (c) fully thermally coupled (Petlyuk column).*

are normally appropriate either as prefractionators, where the side stream is fed to another column for further separation, or to generate recycle streams when there are no strict requirements for the recycle composition. This is often the case in the petroleum industry. The limitation on the purity of the intermediate component product stream causes the main problem of SSCs. A high-purity side stream might require a large reflux ratio and a large number of stages that require a large overall plant footprint, as well as the associated energy requirements (Rooks *et al.*, 1996). Consequently, better process alternatives will be needed to retrofit SSCs to reduce the operating cost and/or enlarge the column capacity. This chapter first gives an overview of the SSCs and DWCs in both academia and industry, which covers a literature overview from the basic design to applications. The methodologies and procedures for the retrofit of SSCs to DWCs are then introduced, followed by case studies of industrial applications.

9.2 Side Stream Column

9.2.1 Side Stream Configuration

An increasing number of SSCs are used in the chemical and petroleum industries. They offer a cost-effective way of producing three products from a single column in some separations

(Papastathopoulou and Luyben, 1991). Basically, SSCs are convenient alternative schemes to sequences of simple columns when some limits on product purity are acceptable or when those limits correspond to a very low impurity content of light- and heavy-boiling components (Niesbach *et al.*, 2013; Brambilla, 2014). The main idea comes from the composition profiles of the first column of the two sequences. In the direct sequence, the concentration of intermediate component B in the first column increases below the feed as that of the more volatile component A decreases (Brambilla, 2014). On the other hand, moving further down the column, the less volatile component increases in concentration. Therefore, the concentration of component B reaches a peak only to be remixed. At this highest point, one side stream is withdrawn from the vapor phase (Figure 9.3a) (Brambilla, 2014). The second column of the sequence is replaced by only one stripping section with limited separation capability between component B and less volatile component C.

Similarly, the concentration of B first increases above the feed as the less volatile component C decreases in the first column of the indirect sequence. The concentration of

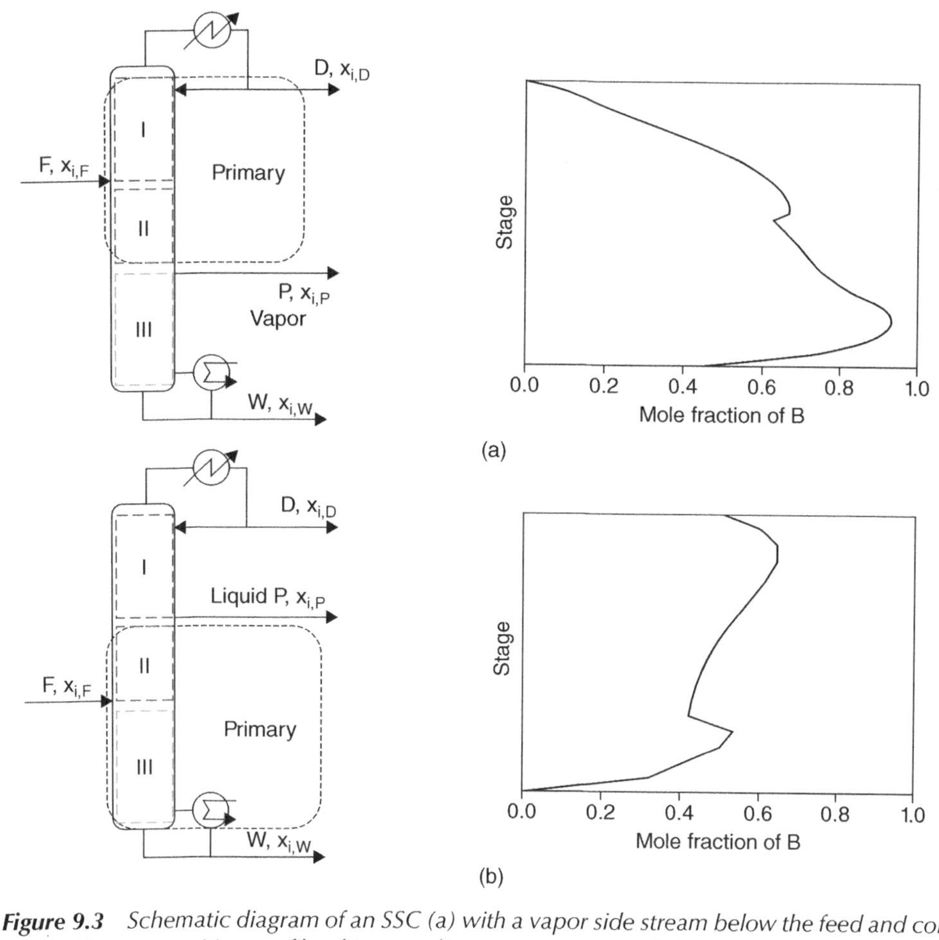

Figure 9.3 *Schematic diagram of an SSC (a) with a vapor side stream below the feed and corresponding composition profile of intermediate component; and (b) with a liquid side stream and corresponding composition profile of intermediate component.*

component B reaches a maximum, then only to decrease with a concomitant increasing concentration of the more volatile component A. A liquid stream is suggested in the rectifying section to replace the second column in the indirect sequence with limited separation capability between component B and the more volatile component A (Figure 9.3b).

9.2.2 Heuristic Rules for the Use of SSCs

The target of most commercial SSCs has focused mainly on the separation of zeotropic mixtures. The criteria for the use of side streams for the separation of ideal mixtures with the expected feed composition regions of SSC arrangements are (Tedder and Rudd, 1978):

- If the middle product dominates in the feed (that is, >50%) and the heaviest product is in small quantity (that is, <5%), then favor the side stream below the feed
- If the middle product dominates in the feed (that is, >50%) and the lightest product is in small quantity (that is, <5%), then favor the side stream above the feed

The SSCs can be used in some certain ranges of feed compositions and/or volatilities, and the following list of heuristics provides useful guidelines for SSC application (Glinos and Malone, 1988).

- SSC can be considered if the mole fraction of A and/or C < 0.2
- Consider using a side stream when the column is used to recycle the reactants back to the reactor from a liquid-separation system in a reaction–separation–recycle process
- Consider a side stream above the feed when the intermediate component is more difficult to separate from the heavier than from the lighter components. Otherwise, consider a side stream below the feed. In either case, heuristic 1 must be satisfied

The purity of the side stream product is restricted by thermodynamics and the nature of the distillation process, which cause these columns to be appropriate either as prefractionators (that is, the side stream is fed to another column for further separation) or to generate recycle streams when there is no strict requirement on the recycle composition (Glinos and Malone, 1985). In addition, SSCs might be useful when the middle-boiling components are trace components or even the main product, as in a pasteurization column (Rooks *et al.*, 1996). Note that the pasteurization column is a side stream column used to remove the light-ends in the vapor phase by using a full-reflux condenser.

In summary, in the ideal case, SSCs can be attractive either when the middle product dominates in the feed and one of the other components is in only minor quantities, or when the middle-boiling components are trace components, or when they are used as prefractionators. Thus, SSCs can only be applied under special circumstances of feed composition and product specification (Smith, 2005).

Although their application to non-ideal cases is relatively rare, SSCs can also be economical under certain process conditions and design requirements, with some of the criteria that favor the use of SSCs as follows (Rooks *et al.*, 1996):

- There is an imbalanced feed, such as a very small amount of a middle-boiling component
- The side stream can be taken with flexible purity requirements, such as side streams going to waste treatment, recycle and as azeotropic feeds to other columns

9.2.3 Pros and Cons of SSC

Compared with the single two-product columns, an SSC can be used in similar designs to reduce the costs when achieving the desired side purity without a large increase in the vapor rate above that required for primary separation (Nikolaides and Malone, 1987). Furthermore, an SSC can be used to enhance the energy performance of the conventional configurations for ternary mixture separations (Glinos *et al.*, 1986; Emtir and Etoumi, 2009). Particularly, a side stream is introduced at the peak point composition of the middle components as a transfer stream to the second column to reduce the remixing effects in the first column (Emtir and Etoumi, 2009). For a non-ideal four-component mixture (acetaldehyde, methanol, ethanol, and water), Rooks *et al.* (1996) reported that a distillation sequence with three columns including two SSCs can reduce the total vapor rate by approximately 20% compared with the distillation process with five columns (base case) when the total number of stages used was the same. SSCs can also be used in binary separation if different purity levels are desired (Papastathopoulou and Luyben, 1991). For example, a single column can be used to produce two grades of propylene products, where the bottom stream is propane, while the medium-purity propylene and high-purity polymer-grade propylene are collected from the side and distillate streams, respectively.

On the other hand, the limitation of the purity of the intermediate component product stream is the common disadvantage of SSCs (Nikolaides and Malone, 1987). This is because achieving a high-purity side stream may require high reflux ratio and a large number of stages, which also lead to higher purity in the distillate and bottom products (Glinos and Malone, 1985; Wachter *et al.*, 1988). This high reflux ratio and large number of stages leading to a very high operating and investment costs is another limitation to the implementation of an SSC for the recovery of high-purity product in ternary mixture separation.

9.2.4 Design of SSC

An SSC with a liquid draw above the feed (Figure 9.3b), where three relatively pure products are obtained from an ideal ternary mixture (A is the lightest component and C is the heaviest component), is considered. The main advantage of an SSC is that while performing the primary separation, that is, B/C, the vapor rate required for this split to affect additional separation at small incremental cost, say A/B, is taken advantage of (Glinos and Malone, 1985). On the other hand, A/B separation is the primary separation if the feed is above the side stream (Figure 9.3a). Several preliminary design methods for an SSC are available for ideal mixture separation characterized by a constant-volatility model (Glinos and Malone, 1985), whose purpose is mainly on shortcut evaluation of the preliminary design, quick screening of alternatives, and rapid estimation of the design sensitivity. Basically, these design methods were developed for "sharp splits", where the major products were essentially free of at least one of the components in the feed (Rooks *et al.*, 1996). The initial design of an SSC can be solved based on the shortcut algorithms reported elsewhere (Glinos and Malone, 1985), and can be referred to for details.

For non-ideal mixtures, particularly in azeotropic mixtures, a simple method cannot be used to design an SSC in a shortcut manner. A common approach is to use repeated simulations, for example, by estimating the side stream location and flow rate, and then attempting to converge on a solution (Rooks *et al.*, 1996). This method, however, is time-consuming and provides few insights. To overcome these drawbacks, efforts have been made to develop more suitable methods for the design of non-ideal mixture separation (Kohler *et al.*, 1994; Rooks *et al.*, 1996; Marquardt *et al.*, 1999; Antonio and Gutierrez, 2007; Adiche and Vogelpohl, 2011).

9.3 Dividing Wall Column

9.3.1 Introduction

In contrast to the heat-integrated distillation column (HIDiC), which is still a great extent in the academic field, the DWCs have been implemented widely in industry (Olujic *et al.*, 2009). The main advantages of a DWC are the considerable savings in energy consumption and investment cost. The cost reduction in investment and operation is approximately 30% over conventional sequences, depending on the feed composition (Dejanović *et al.*, 2011; Rong, 2011). Therefore, the number of installed units has increased continuously since the 1980s, and this technique is now considered to be well established in the chemical industry (Staak *et al.*, 2014).

Being thermodynamically equivalent to a Petlyuk column, a DWC can be considered a single shell realization of the Petlyuk column shown in Figure 9.2c (Amminudin *et al.*, 2001). The energy requirement of these configurations is less than that of their conventional counterparts, because they provide the operating conditions that avoid or minimize the entropy of mixing, which, in the case of conventional arrangements, occurs more or less pronounced at the feed stage, the ends of the columns, and due to the remixing of internal flows (Dejanović *et al.*, 2010). Additional benefits arise from the fact that only one reboiler and condenser are involved instead of using two reboilers and two condensers in a conventional distillation sequence (Premkumar and Rangaiah, 2009).

On the other hand, a large temperature difference between the reboiler and condenser in a DWC, and the heat transfer across the dividing wall, might cause some limitations in DWC application. Furthermore, conventional or other configurations would be better suited to separation of a mixture with non-ideal behavior (Long and Lee, 2012c).

For ternary mixture separation, three different DWC configurations can be applied (Figure 9.4) (Long and Lee, 2014). The first type (Figure 9.4a), which is the most popular

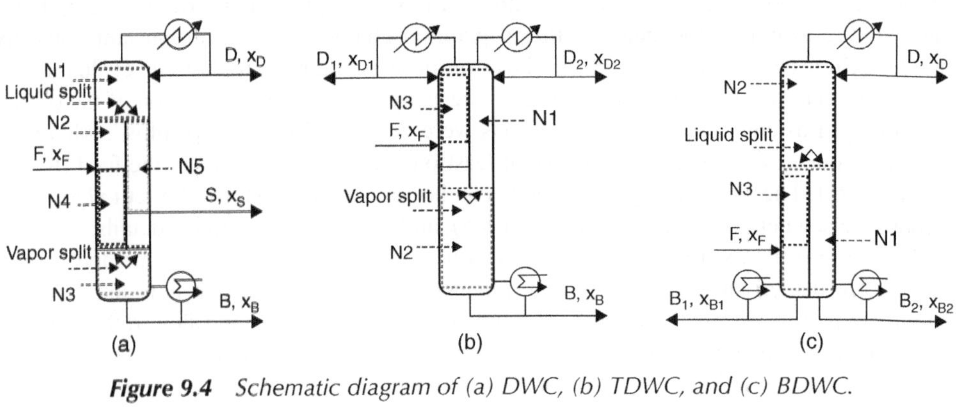

Figure 9.4 *Schematic diagram of (a) DWC, (b) TDWC, and (c) BDWC.*

(Wright, 1949), has the dividing wall, feed and side draws normally placed close to the middle of the column (Aspiron and Kaibel, 2010). Figures 9.4b and 9.4c present the second and third configurations. The wall is located either at the upper or lower part of the column, respectively. These configurations were proposed by Monro (1938), and first employed in 2004 (Kaibel *et al.*, 2006). The column in Figure 9.4b is referred to as a top dividing wall column (TDWC), whereas the column shown in Figure 9.4c is called a bottom dividing wall column (BDWC) (Long and Lee, 2014).

Designing a DWC is much more complicated than designing conventional arrangements because it has more design variables (Amminudin *et al.*, 2001), such as the number of plates in each column section, liquid and vapor splits into the dividing wall section, the feed stage location, and the side-draw location, which interact with each other and need to be optimized simultaneously for the best column design (Long and Lee, 2012c). In DWC design, estimating the number of stages in each section is commonly difficult. Because any variable associated with the column structure is an integer variable, column optimization is a typical mixed integer non-linear programming problem (MINLP), which cannot be solved using commercially available process simulators (Dejanović *et al.*, 2010). Coupling external optimization routines with process simulators is needed.

9.3.2 Design and Optimization of DWC

9.3.2.1 Shortcut Methods

The design and optimization of a fully thermally coupled distillation column (FTCDC) are solved by a shortcut method, which uses a three-column model (Triantafyllou and Smith, 1992). The number of stages, minimum reflux, stage requirement, and feed stage are established by using the conventional Fenske–Underwood–Gilliland–Kirkbride (FUGK) method. The best reflux ratio is then determined for a given cost scenario. The methods are restricted in their applications by simplifying assumptions, such as constant molar flow and constant relative volatility. A new design procedure for an FTCDC was proposed to overcome the limitations of the previous approach (Amminudin *et al.*, 2001). The new procedure adopted the equilibrium stage composition concept to define a liquid composition profile by solving the mass balance rigorously on each equilibrium stage, while restricting the procedure to the assumption of constant molar flow. An assumption of constant relative volatility is not needed because of its semi-rigorous nature.

9.3.2.2 V_{min} Diagram Method

Halvorsen and Skogestad (2003) proposed the use of a minimum vapor-flow, or so called V_{min} diagram, method, which is a graphical visualization of the minimum energy. In their study, a constant pressure, constant relative volatility and constant molar flow under a limiting case with an infinite number of stages were assumed (Halvorsen and Skogestad, 2003). Fidkowski and Krolikowski (1986) and Carlberg and Westerberg (1989a, 1989b) presented the minimum-energy expressions based on Underwood equations for the Petlyuk arrangements with three components, and Halvorsen and Skogestad (2003) added the minimum-energy requirement for the general multicomponent case. An important finding was that the minimum-energy requirement ($V_{min}^{Petlyuk}$) and the detailed vapor-flow requirements may be found based on the V_{min} diagram.

9.3.2.3 Response Surface Method

A practical method based on the shortcut and response surface methodology (RSM) was developed to design and optimize the structure of DWCs (Long and Lee, 2012c). A shortcut design based a conventional column configuration was employed to initialize the DWC structure (Lee *et al.*, 2011a), which was then optimized efficiently using RSM. Because of the large number of mixed type of variables in the design of a DWC, a statistical experimental design was found to be a powerful tool for determining the effects of the factors and their interactions (Long and Lee, 2012c).

9.3.2.4 Other Rigorous Methods

Several rigorous methods for the design and optimization of DWCs have been proposed, such as a rigorous design method based on models of the detailed column superstructures (Dunnebier and Pantelides, 1999), a rigorous optimization method, where Aspen Plus™ software is linked with an external optimization routine based on an evolutionary algorithm (Wenzel and Rohm, 2003), a sequential quadratic programming-based method implemented in Aspen Plus™ (Kiss and Suszwalak, 2012), and a MILP approach based on the use of two superstructure levels (Caballero and Grossmann, 2014).

9.4 Retrofit of an SSC to a DWC

9.4.1 Introduction

In general, a retrofit project in a distillation process covers a wide range of modifications and utilizations of existing equipment. It may include only a simple modification or replacement of column internals up to large modifications of the column configuration, partial enlargement of the shell diameter and height by replacing a part of a column or increasing the number of trays, and the modification of auxiliary equipment (Long *et al.*, 2010). Amminudin and Smith (2001) claimed that the key to a successful retrofit lies in exploiting existing hardware by maximizing the use of existing equipment while, at the same time, minimizing the need for new hardware to minimize the capital cost.

The aim of most retrofit projects for distillation columns is to reduce the energy requirements while maintaining the purity and recovery of the main product. As described above, SSCs consume a considerable amount of energy because of the inherent thermodynamic restrictions and the nature of the distillation process (Glinos and Malone, 1985). The retrofit of an SSC to a DWC (Figure 9.5) can save energy significantly. The successful exploitation of DWC in retrofitting was also mentioned by ExxonMobil (Ennenbach *et al.*, 2000; Kolbe and Wenzel, 2004; Slade *et al.*, 2006).

9.4.2 Design and Optimization of Retrofitted DWC

9.4.2.1 Initial Structure Design

A shortcut design procedure based on a conventional column configuration is used for the initial design of a DWC structure (Figure 9.6) (Triantafyllou and Smith, 1992; Long *et al.*, 2010; Lee *et al.*, 2011a). The first column in this conventional configuration corresponds to

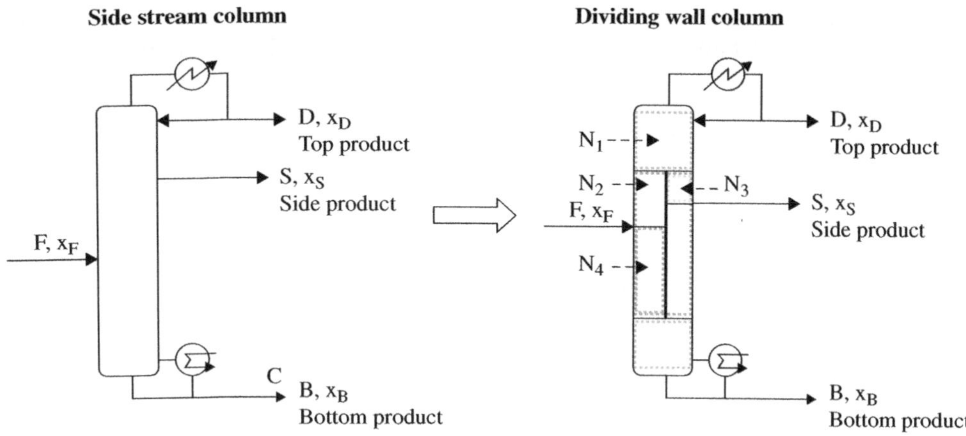

Figure 9.5 Schematic diagram of SSC and retrofitted DWC.

the prefractionator section in the DWC, while the rectifying section of the second column and the stripping section of the third represent the top and bottom sections of the DWC, respectively. The stripping section of the second column and the rectifying section of the third column are equivalent to the dividing wall section of the DWC (Lee *et al.*, 2011a). The combination of the bottom stream from the second column and the top stream from the third column represents the side stream of the DWC. Consequently, four sections including a prefractionator section for the feed mixture, top and bottom sections above and below the dividing wall section, and a dividing wall section form a DWC. The initial estimates of the DWC structure can be implemented easily by applying a conventional shortcut design method based on this structural similarity.

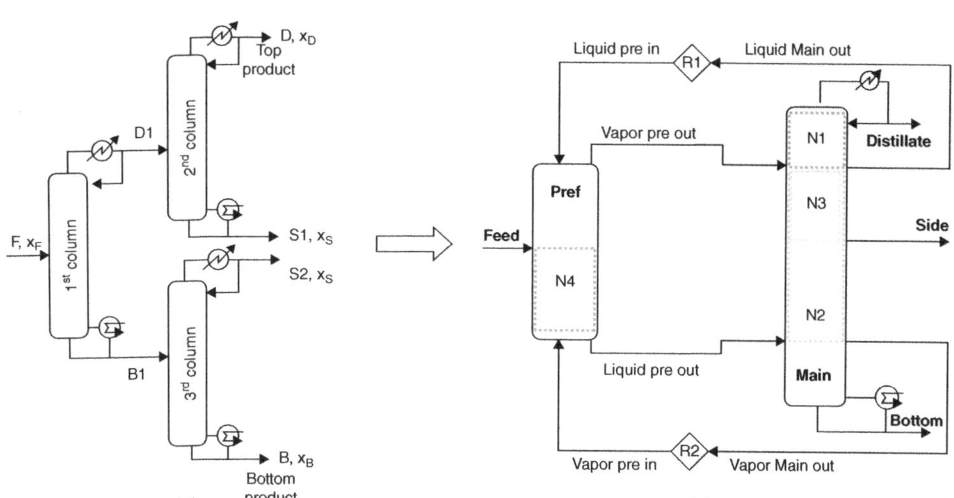

Figure 9.6 Schematic diagram of (a) three-column distillation system, and (b) Petlyuk column.

For example, the structure of each column can be determined using the well-known Fenske–Underwood–Gilliland (FUG) method (Long and Lee, 2013d) as follows [Equation (9.1)]:

$$\frac{N-N_m}{N+1} = 0.75[1-\left(\frac{R-R_m}{R+1}\right)^{0.5688}] \tag{9.1}$$

where the minimum number of theoretical stages at the total reflux (N_m) and the minimum reflux for an infinite number of theoretical equilibrium stages (R_m) are estimated using the Fenske equation (Fenske, 1932), and Underwood equation (Underwood, 1948), respectively. The actual reflux ratio (R) is normally chosen to be $1.2R_m$ (Rangaiah *et al.*, 2009).

The feed tray location is determined from Equation (9.2) by assuming that the relative feed location is constant as the reflux ratio is changed from the total reflux to a finite value (Wankat, 2012):

$$\frac{N_F}{N} = \frac{N_{F,m}}{N_m} \tag{9.2}$$

where $N_{F,m}$ is the feed stage at the total reflux estimated using the Fenske equation as follows [Equation (9.3)]:

$$N_{F,m} = \frac{\ln\left[\left(\frac{x_{D,LK}}{x_{D,HK}}\right)/\left(\frac{x_{F,LK}}{x_{F,HK}}\right)\right]}{\ln \alpha_{LK-HK}} \tag{9.3}$$

9.4.2.2 *Rigorous Simulation*

A simulation of a DWC based on the equivalent Petlyuk column has two recycle loops, which requires two recycle blocks (R1 and R2), as shown in Figure 9.6b. The prefractionator column (Pref) has neither reboiler nor condenser, which can be simulated using an absorber column. This column has two recycle streams, which includes a recycled vapor stream to a prefractionator (Vapor pre in) at the bottom, and the recycled liquid stream to the prefractionator (Liquid pre in) at the top. These two streams need the initial estimates to solve the Pref column. A distillation unit for the main column, the material streams, Feed, Vapor pre out, Liquid pre out, Liquid Main out, Vapor Main out, Distillate, Side, and Bottom for Main column are added to the flow sheet. Two recycle blocks need to be added to the flow sheet: one recycle block (R1) for the Liquid Main out and another (R2) for Vapor Main out to initialize the recycle stream. Once the Pref is solved, the main column is simulated with an estimated number of stages and stage locations from the shortcut calculation.

The main column has five degrees of freedom, and so five specifications should be provided. The purity and recovery of the main product are normally used as specifications. The flow rates of Liquid Main out and Vapor Main out are also considered as specifications. The streams Liquid Main out and Vapor Main out are calculated and available for further calculations. Consequently, the complete flow sheet in Figure 9.6b can converge due to two recycle blocks. Whenever the initial estimates obtained from the shortcut simulation do not lead to the convergence of a rigorous simulation, small changes in these estimates need to be attempted to achieve convergence.

9.4.2.3 RSM-based Optimization

Because of the large number of mixed variables associated with the DWC design, a statistical experimental design was found to be a powerful tool for determining the effects of the factors and their interactions (Long and Lee, 2012c). To identify the most cost-effective solution within the constraints of a particular system is the aim of any process design (Kolbe and Wenzel, 2004). After initially setting the DWC structure, the main design variables including the internal vapor (F_V) and liquid (F_L) flows to the prefractionator, and the number of trays in the top (N_1), middle (N_2), side product (N_3), and feed stripping (N_4) sections need to be optimized (Long and Lee, 2013d). Figure 9.5 presents the main design variables of the DWC. A simplification of the assumption of the same number of trays in each side of the dividing wall section is often employed but is not mandatory. A Box–Behnken design (Box and Wilson, 1951) under the RSM is used to determine how the variables interact and to optimize the system in terms of the reboiler duty after determining the preliminary ranges of the variables through single-factor testing (Long and Lee, 2013d). A second-order polynomial model is used to fit the simulation data, which can be expressed as Equation (9.4),

$$Y = \beta_0 + \sum_{i=1}^{k} \beta_i X_i + \sum_{i=1}^{k} \beta_{ii} X_i^2 + \sum_{i=1}^{k} \sum_{j>i}^{k} \beta_{ij} X_i X_j + \varepsilon \tag{9.4}$$

where Y is the predicted response (reboiler duty), X_i are the uncoded or coded values of the variables, β_0 is a constant, β_i, β_{ii} and β_{ij} are the coefficients of the linear, quadratic and interactive terms, respectively, and ε is an error term (Long and Lee, 2012c). MINITAB software is used to obtain the second-order polynomial coefficients and optimal reboiler duty.

9.4.3 Column Modification and Hardware

The retrofit of an SSC to a DWC includes removing the existing trays and adding new trays (if necessary) as well as a new dividing wall into the middle section (Long and Lee, 2013d). This might take time but the existing trays can be replaced with new high-performance trays or packing to increase the capacity. Another approach is to cut the existing column shell into three pieces, which are then welded in the prefabricated new column section (Kolbe and Wenzel, 2004). In their report, the internals for the new section are installed and leveled on-site. In addition, to match the new debottlenecking capacity, some of the pump sets are replaced. Interestingly, they claim that the modification is planned in such a way as to allow the complete reconstruction from the "ready for opening" to the "ready for feed-in" stage in a plant shutdown period of only 10 days.

Another approach is to cut the existing column shell into two pieces (top and bottom sections), and then weld the prefabricated new middle section in, which increases the total number of trays. By carrying out this approach, energy can be reduced not only from the effect of the DWC but also the effect of an increase in the number of trays. Note that the heat losses from the shell and heat transfer across the dividing wall can be ignored in most cases.

In the case that the middle section is still used and the optimized feed and side streams are different from the existing SSC's, consideration should be given to keeping their nozzles constant instead of moving to an optimized location, which would require some modifications. Although the optimal point provides better operating cost savings, keeping the same location of the feed and side stream might be preferred because of the fewer modifications required (Long and Lee, 2013d). A trade-off between the operating cost and investment cost needs to be considered in a retrofit design.

In the DWC, a dividing wall can be implemented normally by welding a metal plate along the inner walls of the column (Dejanović *et al.*, 2010). On the other hand, a welding job might be very difficult for a small column and it also adds strength to the column, which leads to additional mechanical and thermal stress. To remedy this problem, a partially welded or non-welded dividing wall system can be considered, which depends on the effects of internal flow leaks during operation (Dejanović *et al.*, 2010).

An important device in the DWC is an internal reflux splitter to split the liquid coming from the column overhead into sections divided by a wall. The energy efficiency of a DWC can be affected by small deviations in the internal flows. Only a liquid split can be adjusted during operation while a vapor split cannot be manipulated arbitrarily once the column is constructed. Several types of reflux splitters are available in commercial DWC implementations. The moving-bucket-type reflux splitter reported by Montz as a first-generation reflux splitting system provides a simple and more precise solution (Asprion and Kaibel, 2010). This splitter consists of a casing that is subdivided into three chambers (Figure 9.7a). The split ratio is controlled by adjusting the time period of the hollow inner bucket onto the prefractionation or the main column chamber. The main potential problem is that the moving elements can be damaged during operation by the constant movement of the mechanical parts. For more reliable and robust liquidsplitting operation, a manual-type reflux splitter with no moving parts (Figure 9.7b) was devised (Lee *et al.*, 2010) and implemented successfully on several commercial DWCs. The latest development in reflux-splitter systems was made by a simple modification to the conventional distributor and collector tray (Lee and Kim, 2012). The in-the-column-type reflux splitter (Figure 9.7c) consists of a modified collector tray, a collector box, and two transfer lines. This type of splitter is simpler and cheaper than the two types considered above, and can reduce the column height.

These different types of internal reflux splitters have to be chosen depending on the specific retrofit case, based on their features listed in Table 9.1.

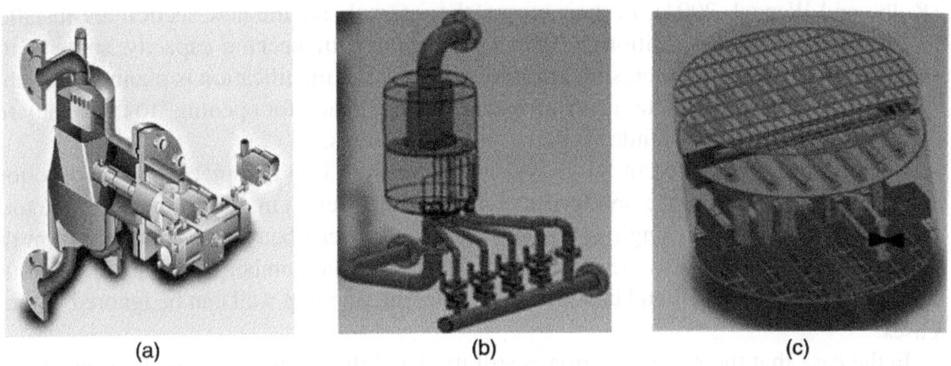

Figure 9.7 Various internal reflux splitters: (a) moving-bucket type, (b) manual type and (c) in-the-column type.

Table 9.1 *Comparisons of several reflux-splitter systems (Shin* et al.*, 2013).*

Splitter type	Operating window/Control	Reliability/Maintenance	Cost/Structure
Moving-bucket type	Wide/Continuous	Fair/Overhaul needed	High/Complex
Manual type	Narrow/Discrete	Excellent/Regular inspection	Moderate/Simple
In-the-column type	Narrow/Discrete	Excellent/Regular inspection	Low/Simple

One of the main practical barriers in DWC implementation might be the concerns regarding the potentially risky new technologies. A DWC column may also be required to switch from normal DWC operation mode to conventional operation mode due to unforeseen circumstances or maintenance. A dual-mode DWC configuration (Shin *et al.*, 2011), which can switch its mode of operation by simply varying its valve positions, can be an attractive option to mitigate plant workers' concerns regarding the potential risks in DWC implementation (Figure 9.8).

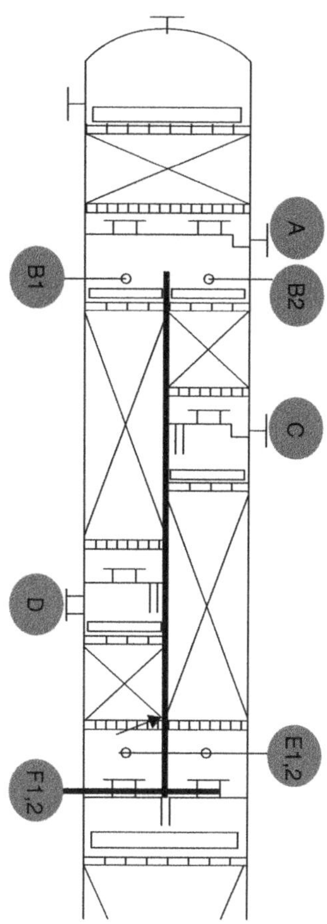

Figure 9.8 *DWC capable of two operation modes (Shin* et al.*, 2011). Source: Reproduced from Shin* et al.*, 2011, with permission from Gulf Publishing Company.*

All these modifications of an existing SSC to a DWC incur significant plant downtime (Long *et al.*, 2010). In many retrofit projects, downtime is the largest economic factor, and a significant amount can lead to a loss of production, an interruption in product supply, as well as a disruption of the other downstream processes, where plants depend on each other (Amminudin and Smith, 2001). On the other hand, chemical/petrochemical plants normally have an annual maintenance period of 10 to 14 days for major repair and replacement, where the retrofitting discussed above can be accomplished within this short maintenance period (Premkumar and Rangaiah, 2009). Therefore, production loss can be negligible during the retrofitting time. In some retrofit projects, owing to the tight turnaround schedule and high labor cost for column modification, replacement by a new column is more cost-effective than is modifying an existing column body.

9.5 Case Studies of Industrial Applications

9.5.1 Acetic Acid Purification Column

9.5.1.1 Process Description

Acetic acid (AA) is an important industrial chemical with many industrial uses and an annual worldwide demand of almost six million tonnes (Jane, 2001). The industrial method based on the carbonylation of methanol (MeOH) is widely preferred for its manufacture, accounting for approximately 60% of the total world manufacturing capacity (PEP Report, 1994). Figure 9.9 outlines the AA production process by MeOH carbonylation.

The process consists of three main units: a catalyst preparation/regeneration unit, a MeOH carbonylation unit and an AA purification unit. In the purification unit, a mixture containing the crude AA and contaminants from the carbonylation unit is separated in a series of conventional distillation columns (Figure 9.10).

The purification unit consists of three valve-tray columns. The production medium for AA in the purification stage at 130–200 °C contains up to 16% water, 26% methyl iodide, and other components, such as methyl acetate (MA), methanol (MeOH), hydrogen iodide (HI), formic acid (FA), and propionic acid (PA) (PEP Report, 1994). The fractionation column removes the light components and portions of water in the mixture, and the dehydration column treats both water and FA. The last column, which is an SSC, produces the final AA product from the side draw by cutting off the remaining light and heavy components from

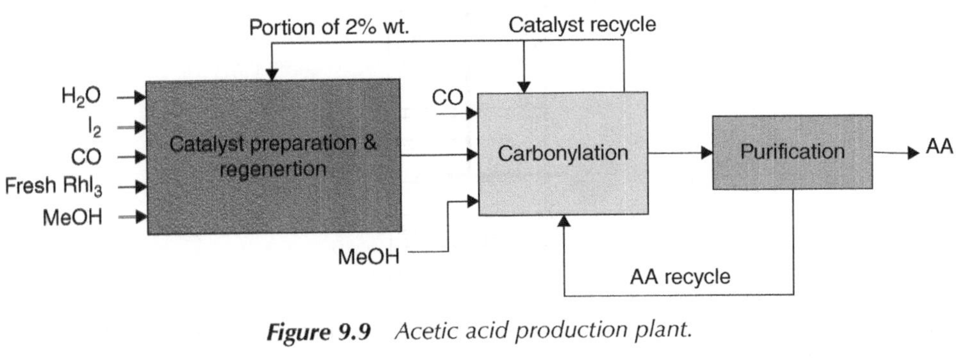

Figure 9.9 *Acetic acid production plant.*

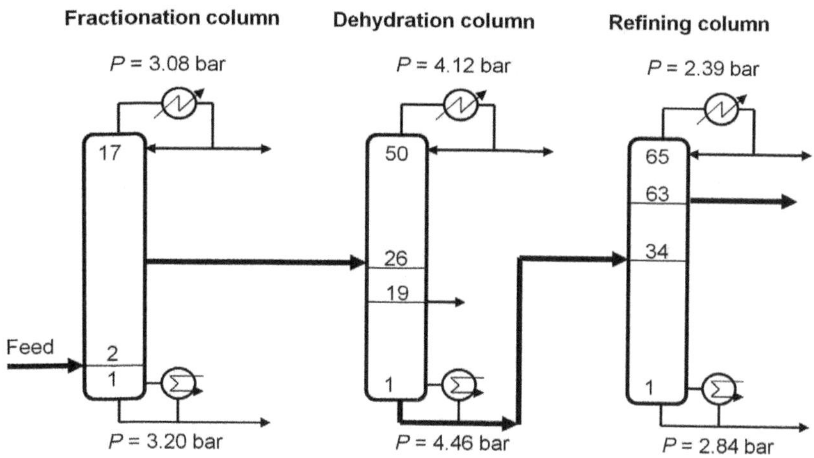

Figure 9.10 *Acetic acid purification unit of three conventional columns (Long and Lee, 2011). Source: Reproduced from Long and Lee, 2011, with permission from John Wiley & Sons.*

the top and bottom. The existing SSC, which is 3.6 m in diameter with 65 trays, requires 20.25 MW in the reboiler.

A retrofit to a DWC was studied for the SSC in the AA purification unit to reduce the energy requirements (Long and Lee, 2011, 2013d). This case study reports the retrofit studies focusing mainly on the design and optimization methodology aspect for the retrofit to a DWC with no change in the diameter or the total number of stages of each column, in order to highlight the optimal use of existing columns.

9.5.1.2 Initial Structure Design

A shortcut design method was used to estimate the initial DWC structure. A conventional three-column configuration in Figure 9.6a was used to find the initial number of trays of each section of the DWC as well as the feed and side stream locations by assuming the recovery of components and specifying the actual reflux ratio. For example, the 1st column was simulated by assuming 0.02% light component (FA) at the bottom (B1) and 0.02% heavy component (PA) at top (D1). The 2nd column was then simulated by specifying the middle (AA) and top (FA) component according to the required purity (0.99 and 0.0008 molar fractions, respectively). Similarly, the 3rd column was simulated with the heavy component (PA) at the bottom (0.6 molar fraction of PA) and light component (AA) at the top specifications (0.99 molar fraction of AA). The specifications of FA in B1 and PA in D1 of the 1st column were varied by trial and error to achieve an almost equal composition of S1 and S2. The actual reflux ratio (R) for the three columns was chosen as $1.2R_m$. Note that, in the three-column model, the middle product is represented by the addition of two material streams, namely “S1” and “S2” in Figure 9.6a (Triantafyllou and Smith, 1992; Amminudin *et al.*, 2001; Lee *et al.*, 2011a).

To predict the vapor–liquid equilibrium (VLE) of these simulations, the NRTL-HOC property model, which uses the Hayden–O’Connell equation of state as the vapor-phase model and the NRTL for the liquid phase, was employed (Aspen Technology, 2009).

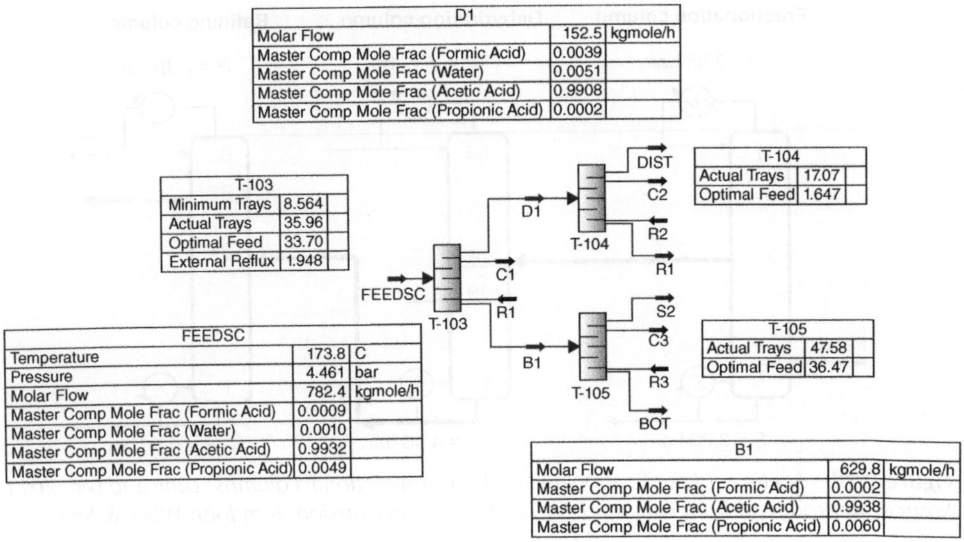

Figure 9.11 *Three-column shortcut simulation for initial design of DWC.*

Figure 9.11 presents the resulting estimated values of the main design parameters. After the number of stages for each section was designed initially, they were adjusted relatively to satisfy the total number of stages of the column being the same as for the existing SSC. From this shortcut model, the number of stages for the prefractionator and top and bottom sections of the main column were calculated to be 36, 2, and 27, respectively.

9.5.1.3 Rigorous Simulation

Once the shortcut estimates are completed, the DWC needs to be simulated rigorously to obtain accurate results. The approaches for a rigorous simulation involve fewer assumptions, and are more realistic than the FUG equations (Premkumar and Rangaiah, 2009). The rigorous simulation used in this case study was based on the equilibrium-stage model. The simulation model of a DWC based on the equivalent Petlyuk column has two recycle blocks (R1 and R2), as shown in Figure 9.12.

In this case, the flow rate and AA purity of the side stream and the distillate rate are used as three specifications. The flow rate of "Liquid Main out" and "Vapor Main out" should be specified as the remaining two specifications. The mass- and energy-balance of the base case were obtained by activating this configuration (Figure 9.12).

9.5.1.4 Structure Optimization Using RSM

After the design using a shortcut and rigorous simulation, the main design variables associated with the column structure were optimized by the RSM using a simplified assumption of the same number of trays in each side of the dividing wall section. It is necessary to determine the preliminary ranges of the variables by single-factor testing. The variables' interaction and optimization of the system in terms of the reboiler duty were found by RSM

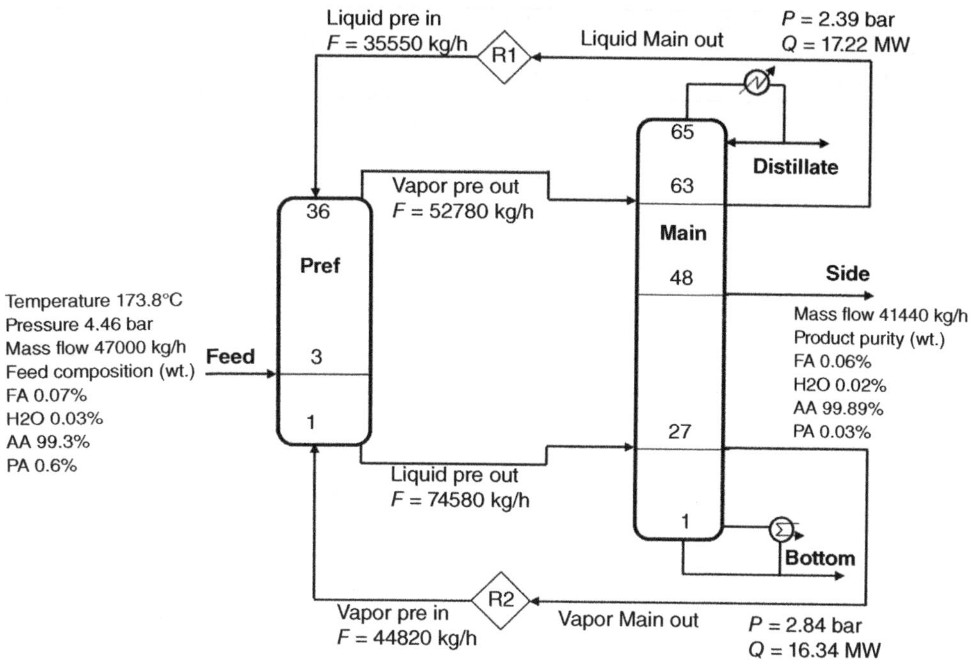

Figure 9.12 *DWC configuration for rigorous simulation and converged results.*

with a Box–Behnken design. The factors and levels used in this case study are listed in Table 9.2. Factors are defined as design variables which affect the objective. In the RSM, each factor is defined on a coded scale from −1 to +1 (the low and high levels of factors).

To optimize these four parameters associated with the DWC structure, twenty-seven simulations were run. For each simulation, to minimize the reboiler duty, the internal vapor and liquid flows to the prefractionator were changed. The resulting second-order polynomial model is given in Equation (9.5):

$$Y = 36.85 - 0.73N_1 - 1.09N_2 + 0.49N_3 + 0.50N_4 + 0.008N_1^2 + 0.01N_2^2 + 0.008N_3^2 + 0.003N_4^2 + 0.019N_1N_2 + 0.004N_1N_3 - 0.008N_1N_4 - 0.01N_2N_3 - 0.01N_2N_4 + 0.005N_3N_4 \quad (9.5)$$

Table 9.2 *Coded levels of factors.*

	Levels		
Factors	−1	0	1
Top section (N1)	1	5	9
Middle section (N2)	36	43	50
Side product location (N3)	1	3	5
Feed stripping location (N4)	1	5	9

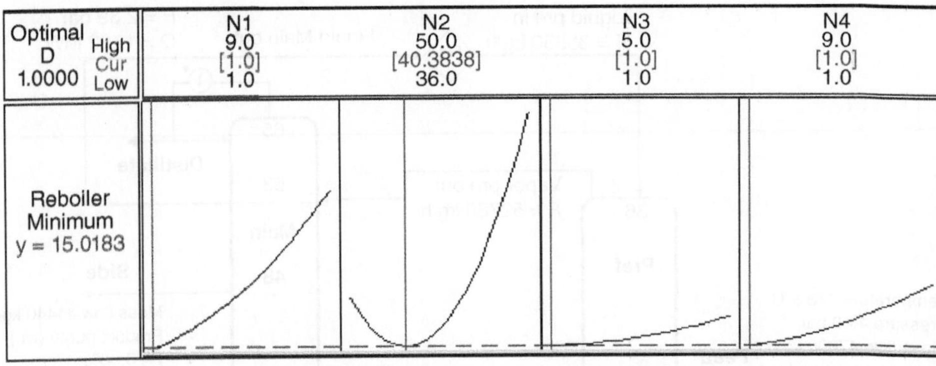

Figure 9.13 *Optimization plot by RSM.*

As can be seen from Equation (9.5), in this case, the middle section had a significantly greater influence on the reboiler duty than did the top section, followed by the feed stripping location and the side product location. The quality of the polynomial model was expressed by the coefficient of determination, that is, R^2. Based on the analysis results from MINITAB, the model showed a high R^2 value of 0.93. The value of R^2 is close to 1.0, which indicates a strong correlation between the simulated and predicted values. The relationship between the independent variables and the response is well explained by the regression model.

Using this RSM model, the minimum reboiler duty was observed at the number of 1, 40, 1 and 1 tray in the top, middle, side product, and feed stripping sections, respectively, and was predicted to be 15.02 MW, which is equivalent to 25.83% in energy savings (Figure 9.13). F_V and F_L were then optimized to minimize the reboiler duty, whilst maintaining sufficient product purity and recovery.

Figure 9.14 compares the schematics of the existing SSC and the resulting DWC. The energy requirements from a rigorous simulation were also determined to be 15.16 MW at the optimal design variables (equivalent to 25.14% in energy savings), which is well matched to the prediction by the RSM.

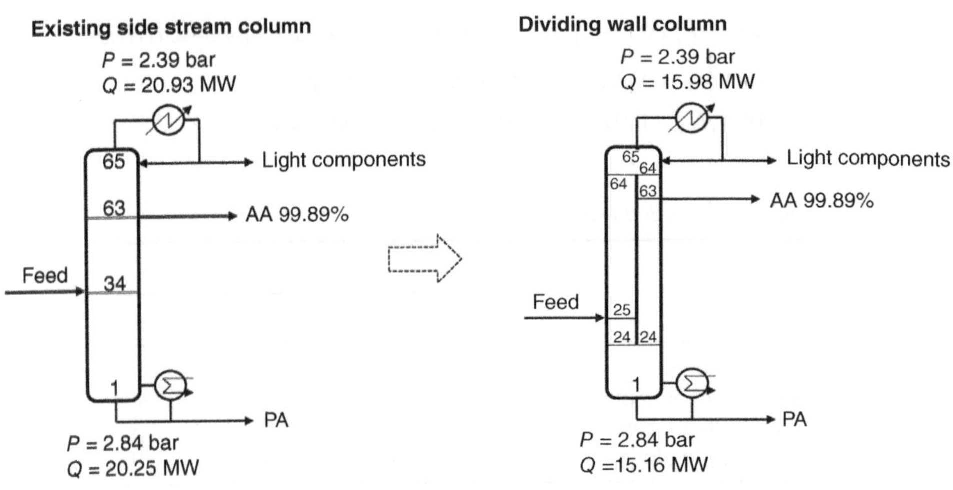

Figure 9.14 *Existing SSC and retrofitted DWC.*

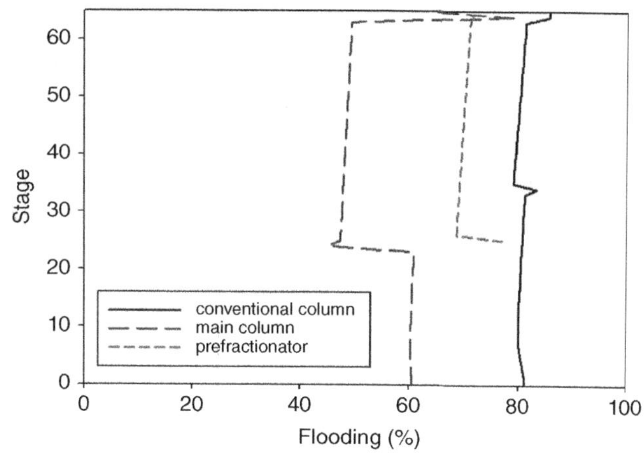

Figure 9.15 *Flooding profile.*

9.5.1.5 Column Hydraulics

To modify an SSC, a wall was installed in the middle section with no change in the total number of trays. The column hydraulics and flooding need to be checked carefully during the retrofit design. To determine the maximum flooding of the columns considered, the rating mode was simulated with the column internal specifications, such as the type of trays, column diameter, tray spacing, and number of passes (Long and Lee, 2011). As shown in Figure 9.15, the maximum flooding decreased from 86% to 79% through a retrofit to the DWC, which suggests that the plant's production capacity can be increased after retrofit. After 8 years of operation, the cost savings by the retrofitted DWC was estimated to be 17.1 million USD. Note that the utility-cost data used in this study are listed in Table 9.3 (Turton *et al.*, 2012). The simple payback period was found to be at only 4.6 months because of high energy-costs savings and the re-use of most of the existing equipment.

9.5.2 n-BuOH Refining Column

9.5.2.1 Process Description

The first commercial DWC in Korea was developed through the retrofit of an SSC and was installed successfully in a plant of LG Chem. Ltd. (Lee *et al.*, 2011b). Their case study reports this industrial implementation of the DWC with particular attention paid to the practical aspects of the retrofit project. Figure 9.16 shows an alcohol-purification unit in a 2-ethylhexanol (2-EH) production plant. This unit consists of three main distillation

Table 9.3 *Utilities cost data [Turton* et al.*, 2012]*

Utility	Price ($/GJ)
Cooling Water	0.35
Steam (Low Pressure)	13.28
Steam (Medium Pressure)	14.19

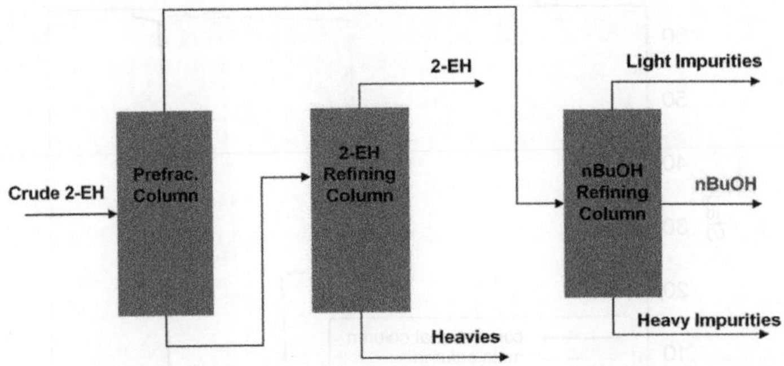

Figure 9.16 Alcohol-purification unit in a 2-EH production plant.

columns: a prefractionation column, a 2-EH refining column and a normal butyl alcohol (nBuOH) refining column. The nBuOH refining column is a side stream column, where nBuOH is recovered from the side draw, while the light and heavy impurities are cut off from the top and bottom of the column.

9.5.2.2 *Preliminary Feasibility Study*

The retrofit project was targeted to the nBuOH refining SSC to improve the energy efficiency of the column. A preliminary feasibility study revealed considerable energy savings through the retrofit to a DWC with a reasonable cost of installation. The feasibility study's results were reviewed with field engineers and plant operators to improve their understanding of the basic concept and benefits of the DWC.

9.5.2.3 *Basic Design and Optimization*

Motivated by the promising results from the preliminary feasibility study, more rigorous studies were performed for the design and optimization of a retrofitted DWC. The DWC structure, such as total number of theoretical trays, and feed and product locations, was optimized rigorously while maintaining the same column height as the existing column. The liquid and vapor split ratios into the dividing wall section were also optimized. The construction costs were estimated by considering the site-specific concerns. An internal change from an existing random packing to an advanced structured packing was considered to keep the same column height as well as improve the separation performance. Figure 9.17 shows the SSC and the retrofitted DWC configurations.

Process simulation software such as Aspen Plus^{TM}, Aspen Hysys^{TM} and PRO II are used extensively for the retrofit studies. In this study, the Aspen Hysys^{TM} was used to simulate the process. The plant operation data were also used to ensure the reliability of the simulation results. Installing a dividing wall does not always guarantee enhanced energy efficiency. The benefits are system-specific and dependent on the properties of the separation system. The potential benefits and inherent problems were thoroughly estimated by carrying out an extensive simulation study. In particular, the vapor-splitting ratio into the dividing wall section cannot be adjusted once the column is installed, while the liquid-splitting ratio can be manipulated easily during operation.

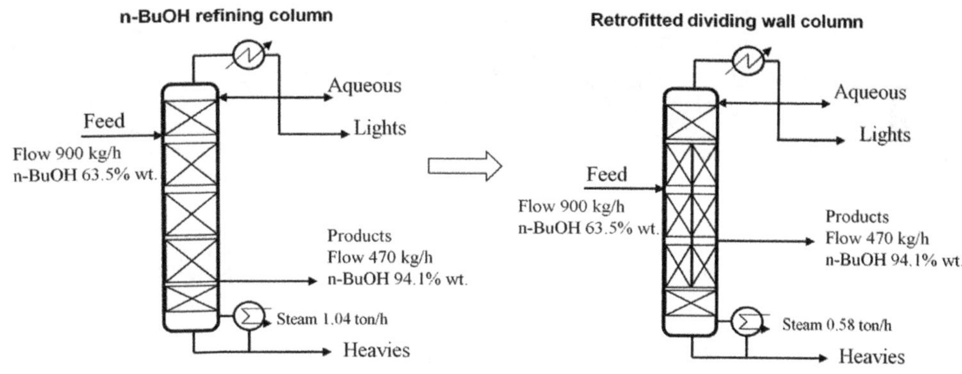

Figure 9.17 *n-BuOH refining column: (a) before and (b) after retrofit to DWC.*

At the top of the dividing wall section, the liquid coming down from the rectifying section can be split in a controlled manner between the two sides of the wall by using a total liquid trap-out tray and by sending a part of the total liquid to the prefractionator side and the remainder to the side stream side (Kiss and Bildea, 2011). An internal liquid splitter is available for adjusting the liquid flow split ratio. Meanwhile, at the bottom of the divided wall section, the vapor is split in proportion to the cross-sectional area of each side and the hydrodynamic conditions (above all, pressure drop and flow resistance). Overall friction of each side is fixed by the internal geometry, which is already set at the design stage and cannot be changed later during operation. The vapor split is thus not adjustable during operation for control purposes. However, one can still design and implement the internals to give a desirable vapor split ratio by adjusting the friction of the bottom end section of the dividing wall. Therefore, to assess the effects of unexpected deviations from its normal design conditions, extensive sensitivity studies were carried out to examine the possible variations of the actual vapor-splitting ratio. The effects of changes in the feed composition, liquid split ratio and other operating variables were modeled to ensure that performance and product quality could be maintained within acceptable ranges.

9.5.2.4 *Detailed Design and Fabrication*

According to the basic design study results, a detailed design was carried out, including a fabrication plan for the column, internal, and several subsidiary devices. In the column internal design, special care was taken on the internals in the dividing wall section to ensure that the vapor split ratio was as designed. The pressure drops in the dividing wall section were predicted precisely and analyzed by the hydraulic correlations from the column internal suppliers and by rigorous computational fluid dynamics (CFD) analysis. A simple correlation was satisfactory for predicting the pressure drops with reasonable accuracy.

One important device in a DWC is an internal reflux splitter to manipulate the split of the liquid coming from the column overhead into the dividing wall section. A moving-bucket-type splitter was designed and fabricated to provide a precise split adjustment. Because of the potential risks from restless movement of its mechanical parts, it was replaced with a manual-type reflux splitter in the next turnaround period.

Figure 9.18 *Foundation reinforcement for DWC installation. Source: Reproduced with permission from Moonyong Lee, Le Quang Minh, Nguyen Van Duc Long, and Joonho Shin.*

The key to a successful retrofit normally lies in exploiting the existing equipment (Amminudin and Smith, 2001). To evaluate the possibility of reusing the original column, its flooding factor was checked to analyze its hydraulic performance. The diameter of the existing column was large enough to use it for the new DWC. On the other hand, a decision was made to fabricate a new one in the shop in advance to replace the existing column because of the tight turnaround schedule and dominant labor cost. As the arrangements of nozzles and pipes are complicated and potentially confusing, a 3-D visual model was created to help reduce human error in construction.

9.5.2.5 *Foundation and Instrumentation*

For a new column installation, sufficient structural strength is needed for the increased height and weight of the column after retrofitting. The retrofitted column consists of 5 packing sections, which are the same as in the previous column. But the different type of packing and additional internals installed in the DWC increased the total weight of the column. Therefore, the necessary civil work was performed to reinforce the foundation in the turnaround period (Figure 9.18). The PIS (Plant Information System) and the screens of the distributed control system (DCS) were modified along with the additional instruments and control loops. The increased number of temperature sensors could help the engineers monitor the column operation and detect any abnormal behavior. The installation work was completed successfully within the scheduled maintenance period.

9.5.2.6 *Start-up and Commissioning*

A column should be operated with a considerably lower feed rate during start-up or shut down. DWCs normally use less energy, indicating lower vapor and liquid loadings, than do conventional columns, which might cause a turndown problem. The DWC was operated successfully at less than half of the normal load during start-up by increasing the boil-up and reflux rates to a necessary level. The column temperature profile during normal

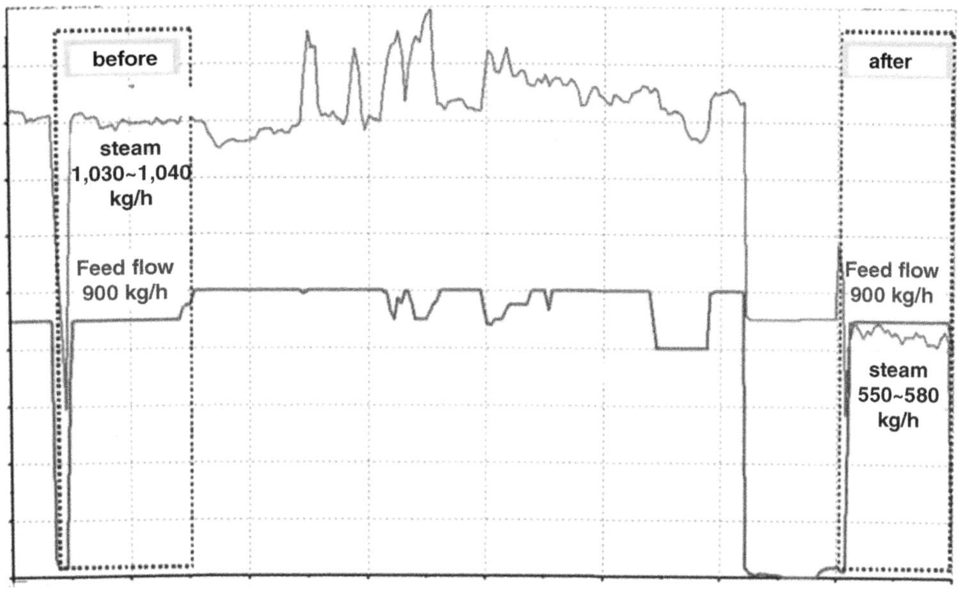

Figure 9.19 *Utility consumption before and after the retrofit.*

operation matched the simulation results quite well. The performance of the DWC was quite successful. Steam consumption in the reboiler was reduced significantly (by 45%) after the retrofit to the DWC (Figure 9.19). Purity improvement of nBuOH was an additional benefit.

From the preliminary study to actual implementation, the assistance of plant engineers and operators is essential for the success of the project. In many cases, process engineers who design DWC columns are unaware of the site-specific concerns. Thorough and clear communication with them reduces unnecessary time loss and risks during implementation. The concepts and operation modes of the DWC and new devices as well as the start-up and shutdown procedures were explained to the plant engineers and operators. Their questions were clarified with 3-D models and simulation results.

9.6 Other Case Studies

9.6.1 Ethylene Dichloride (EDC) Purification Column

Poly(vinyl chloride) (PVC) is an important product in the petrochemical industry. Vinyl chloride monomer (VCM) for the PVC raw material is normally produced by a thermal reaction of ethylene dichloride (EDC), which is synthesized from ethylene, oxygen and chlorine by oxychlorination and direct chlorination (Porscha, 2001). Figure 9.20 outlines the VCM production plant. The EDC routed to the pyrolysis furnace requires a high purity sufficient to achieve a high-quality VCM and to prevent side reactions that lead to fouling in the pyrolysis tubes. The EDC-purification unit uses an SSC to achieve this separation task. The existing SSC operates at 1.5 bar with the main EDC product (99.5 wt.%) being collected in the side stream withdrawing from the column. A retrofit project was carried out to improve the energy efficiency of the energy-intensive SSC.

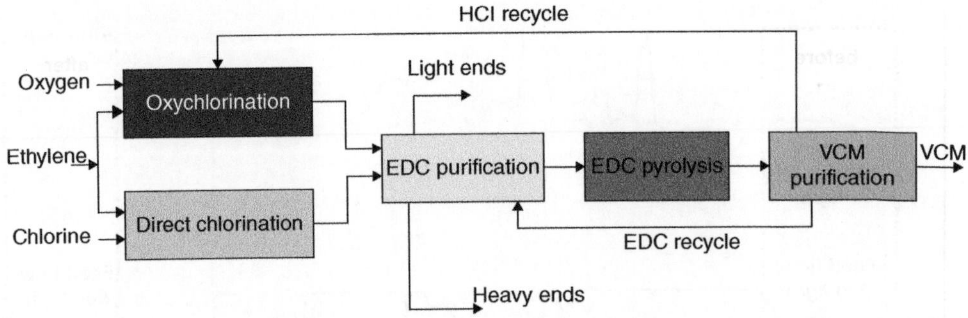

Figure 9.20 VCM production plant.

A feasibility study for installing the DWC was conducted. The design of the DWC obtained by the shortcut calculations and the rigorous simulation was then optimized. The retrofit study results showed that the retrofitted DWC can reduce the reboiler duty significantly (by 45.3%). The re-use of most existing equipment including the column body allowed the retrofit to be achieved with reasonable installation cost. The total modification cost was estimated to be 460,000 USD for installing a new dividing wall, which brings a small payback period of 2 months.

9.6.2 Diphenyl Carbonate (DPC) Purification Column

Polycarbonate (PC) is an important engineering thermoplastic with excellent mechanical, optical, electrical, and heat-resistance properties (Cheng *et al.*, 2013). Diphenyl carbonate (DPC), a precursor for the production of polycarbonate, is typically synthesized by the so-called "phosgene process", which uses phosgene prepared from carbon monoxide (CO) and chlorine (Cl_2) as a monomer (Cheng *et al.*, 2013). To overcome the number of environmental and economic shortcomings of the phosgene process, a new non-phosgene carbonate-production process was developed recently (Fukuoka *et al.*, 2003) and commercialized by the Chimei-Asahi Corporation. Figure 9.21 outlines the Asahi Kasei's non-phosgene process for producing polycarbonate.

In the DPC-production step, at least two reactive distillation columns combined with each other must be used (Fukuoka *et al.*, 2003). High production rate and selectivity are obtained using this distillation sequence. A vacuum SSC was proposed to remove the light and heavy components to obtain high-purity DPC with 99.95% wt. yield. A further retrofit

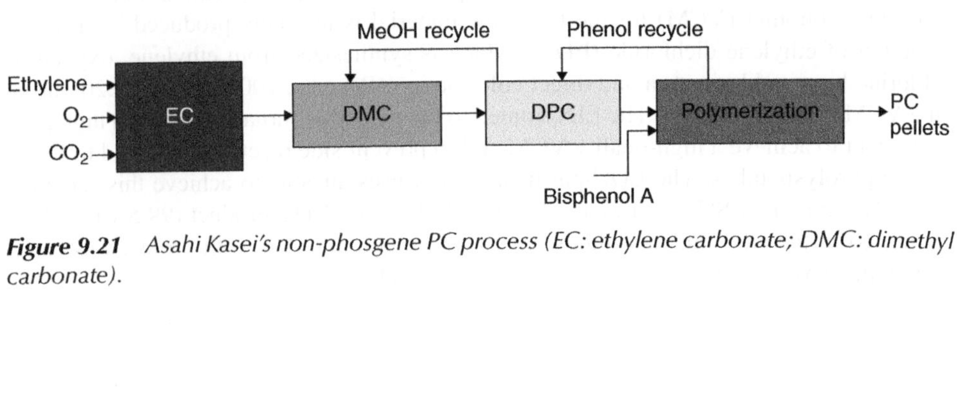

Figure 9.21 Asahi Kasei's non-phosgene PC process (EC: ethylene carbonate; DMC: dimethyl carbonate).

Table 9.4 Energy-saving effect of the retrofitted DWCs.

System		EDC	DPC	HC cut	Paraffin	BTX	LPG
Feed component	A	Lights	Lights	C1–C3	n-Pentane	Benzene	C3
	B	EDC	DPC	C4–C5	n-Hexane	Toluene	C4
	C	Heavies	Heavies	C6–C8	n-Heptane	*p*-Xylene	C5
Feed composition (%wt.)	A	0.7	47.5	9.1	28.5	27.9	14.5
	B	98.5	52.2	54.8	33.1	33.0	38.2
	C	0.8	0.3	36.2	38.4	39.1	47.3
Product purity (%wt.)	D	46.5	98.6	87.5	99.4	99.4	98.7
	S	99.5	99.95	91.0	89.3	92.1	81.8
	B	2.94	11.0	96.4	89.6	93.1	91.8
Reboiler duty (MW)	SSC	14.8	4.92	0.44	1.95	4.29	0.33
	DWC	8.1	3.74	0.39	1.17	1.80	0.29
Energy saving (%)		45.3	24.1	9.6	39.9	58.0	8.6

The product purity in D, S, and B streams is for component A, B, and C, respectively.

by DWC was studied to improve the energy efficiency in the high-purity DPC separation by the SSC. Similarly, the study assumed no change in either the diameter or the total number of stages of the column. After the design and optimization procedure, the final result showed the retrofitted DWC can offer a 24.1% saving in energy.

9.6.3 Other SSCs

Rangaiah *et al.* (2006) studied the application of SSCs to the separation of multicomponent mixtures in the oil and gas, petrochemical, and other chemical process industries. In their study of SSCs, the number of stages and the locations of the feed and side stream were varied to minimize the total annual cost of the SSC units. The preliminary studies for the retrofit to a DWC were accomplished for several SSCs introduced in their study. This opens an opportunity to reduce the utilities further by using a DWC configuration without changing the height and diameter of the existing column. Most of the DWCs gave enhanced performance with 8.6 to 58% energy savings compared with the existing SSCs. The energy savings by the retrofitted DWC depend on many complicated factors such as feed composition, relative volatilities, ease of separation, and product purity. Depending on each case, the energy saving of retrofitting an SSC to a DWC can be different over a wide range. It is still an open and interesting issue as to when and how much the retrofitted DWC can achieve this type of benefit.

Table 9.4 summarizes the feed and product conditions as well as the energy comparisons for all the case studies described in this section.

9.7 Conclusions

Retrofit of SSCs to DWCs, which is an important breakthrough in both grassroots and retrofit designs, is becoming a promising trend. A retrofit design procedure, which utilizes shortcut, rigorous and RSM-based optimization methodologies, was described. The retrofitted DWC structure was designed and optimized efficiently with little computational effort. The results predicted by the RSM and by a rigorous simulation showed good agreement. The resulting DWC can significantly reduce energy costs over an SSC. This book chapter presents several industrial examples to implement the DWC technology from the design stage to the construction and operation stages. The existing SSC can be retrofitted easily to a DWC column by adding a dividing wall, or by adding or exchanging the middle section of an existing SSC with a new dividing wall section in a short period. These results show the benefit of retrofitting an SSC to a DWC. The design and optimization of DWC structure, and engineering in retrofit of the SSC, as well as their operation were found to be practicable without many difficulties. Thus, retrofit of SSCs to DWCs has huge potential to achieve substantial improvement in energy efficiency in a very cost-effective manner in energy-intensive distillation.

Acknowledgements

This study was supported by a grant from the Gas Plant R & D center funded by the Ministry of Land, Transportation and Maritime Affairs (MLTM) of the Korean government.

Nomenclature

2-EH	2-Ethylhexanol
3-D	Three-dimensional
A	Lightest component
AA	Acetic acid
B	Intermediate component
BDWC	Bottom dividing wall column
BTX	Benzene, Toluene, *p*-Xylene
C	Heaviest component
C1	Methane
C3	Propane
C4	Butane
C5	Pentane
C6	Hexane
C8	Octane
CFD	Computational Fluid Dynamics
Cl_2	Chlorine
CO	Carbon monoxide
CO_2	Carbon dioxide
DCS	Distributed control system
DPC	Diphenyl carbonate
DWC	Dividing wall column
EDC	Ethylene dichloride
FA	Formic acid

F_L	Internal liquid flow to the prefractionator
FTCDC	Fully thermally coupled distillation column
FUGK	Fenske–Underwood–Gilliland–Kirkbride
F_V	Internal vapor flow to the prefractionator
HC	Hydrocarbon
HI	Hydrogen iodide
HIDiC	Heat-integrated distillation column
LPG	Liquefied Petroleum Gas
MA	Methyl Acetate
MeOH	Methanol
MILP	Mixed integer linear programming
MINLP	Mixed integer non-linear programming
n-BuOH	n-Butanol
N_m	Minimum number of theoretical stages
PA	Propionic acid
PC	Polycarbonate
PIS	Plant information system
Pref	Prefractionator section
PVC	Poly(vinyl chloride)
R1	Recycle block 1
R2	Recycle block 2
R_m	Minimum reflux ratio
RSM	Response surface methodology
SSC	Side stream column
TCDC	Thermally coupled distillation column
TDWC	Top dividing wall column
VCM	Vinyl chloride monomer
V_{min}	Minimum vapor flow rate
$x_{D,HK}$	Mole fraction of heavy key component in distillate
$x_{D,LK}$	Mole fraction of light key component in distillate
$x_{F,HK}$	Mole fraction of heavy key component in feed
$x_{F,LK}$	Mole fraction of light key component in feed
$\alpha_{LK\text{-}HK}$	Relative volatility between light key and heavy key components

References

Adiche, C. and Vogelpohl, A. (2011) Short-cut methods for the optimal design of simple and complex distillation columns. *Chemical Engineering Research and Design*, **89**, 1321–1332.

Agrawal, R. and Fidkowski, Z.T. (1999) New thermally coupled schemes for ternary distillation. *AIChE Journal*, **45**, 485–496.

Amminudin, K.A. and Smith, R. (2001) Design and optimization of fully thermally coupled distillation columns. Part 2: application of dividing wall columns in retrofit. *Chemical Engineering Research and Design*, **79**, 716–724.

Amminudin, K.A., Smith, R., Thong, D.Y.C. and Towler, G.P. (2001) Design and optimization of fully thermally coupled distillation columns. Part 1: preliminary design and optimization methodology. *Chemical Engineering Research and Design*, **79**, 701–715.

Antonio, C.G. and Gutierrez, A.J. (2007) Design of side-stream azeotropic distillation columns. *Chemical Engineering Research and Design*, **85**, 1384–1389.

Aspen Technology (2009) Aspen Physical Property System, version number V7.1.

Asprion, N. and Kaibel, G. (2010) Dividing wall columns: fundamentals and recent advances. *Chemical Engineering and Processing*, **49**, 139–146.

Box, G.E.P. and Wilson, K.B. (1951) On the experimental attainment of optimum conditions. *Journal of the Royal Statistical Society. Series B*, **13**, 1–45.

Brambilla, A. (2014) *Distillation Control and Optimization: Operation Fundamentals Through Software Control*, McGraw-Hill Professional, New York, US.

Bravo-Bravo, C., Segovia-Hernandez, J., Gutierrez-Antonio, C. *et al.* (2010) Extractive dividing wall column: design and optimization. *Industrial & Engineering Chemistry Research*, **49**, 3672–3688.

Caballero, J.A. and Grossmann, I.E. (2014) Optimal synthesis of thermally coupled distillation sequences using a novel MILP approach. *Computers & Chemical Engineering*, **61**, 118–135.

Carlberg, N.A. and Westerberg, A.W. (1989a) Temperature-heat diagrams for complex columns. 3. Underwood's method for the Petlyuk configuration. *Industrial & Engineering Chemistry Research*, **28**, 1386–1397.

Carlberg, N.A. and Westerberg, A.W. (1989b) Temperature-heat diagrams for complex columns. 2. Underwood's method for sidestrippers and enrichers. *Industrial & Engineering Chemistry Research*, **28**, 1379–1386.

Cheng, K., Wang, S.J. and Wong, D.S.H. (2013) Steady-state design of thermally coupled reactive distillation process for the synthesis of diphenyl carbonate. *Computers & Chemical Engineering*, **52**, 262–271.

Dejanović, I., Matijasevićc, L.J., Halvorsen, I.J., *et al.* (2011) Designing four-product dividing wall columns for separation of a multicomponent aromatics mixture. *Chemical Engineering Research and Design*, **89**, 1155–1167.

Dejanović, I., Matijašević, LJ. and Olujić, Ž. (2010) Dividing wall column—A breakthrough towards sustainable distilling. *Chemical Engineering and Processing*, **49**, 559–580.

Dunnebier, G. and Pantelides, C.C. (1999) Optimal design of thermally coupled distillation columns. *Industrial & Engineering Chemistry Research*, **38**, 162–176.

Emtir, M. and Etoumi, A. (2009) Enhancement of conventional distillation configurations for ternary mixtures separation. *Clean Technologies and Environmental Policy*, **11**, 123–131.

Ennenbach, F., Kolbe, B. and Ranke, U. (2000) Divided-wall columns—a novel distillation concept. *Petroleum Technology Quarterly*, **5**, 97–103.

Errico, M., Tola, G., Rong, B.G. *et al.* (2009) Energy saving and capital cost evaluation in distillation column sequences with a divided wall column. *Chemical Engineering Research and Design*, **87**, 1649–1657.

Fair, J.R. (1984) Historical development of distillation equipment. *AIChE Symposium Series*, **79**, 1–14.

Fenske, M.R. (1932) Fractionation of straight-run Pennsylvania gasoline. *Industrial and Engineering Chemistry*, **24**, 482–485.

Fidkowski, Z. and Krolikowski, L. (1986) Thermally coupled system of distillation columns: Optimization Procedure. *AIChE Journal*, **32**, 537–546.

Fukuoka, S., Kawamura, M., Komiya, K. *et al.* (2003) A novel non-phosgene polycarbonate production process using by-product CO_2 as starting material. *Green Chemistry*, **5**, 497–507.

Gadalla, M., Jobson, M. and Smith, R. (2003) Shortcut models for retrofit design of distillation columns. *Chemical Engineering Research and Design*, **81**, 971–986.

Glinos, K.N. and Malone, M.F. (1985) Design of sidestream distillation columns. *Industrial & Engineering Chemistry Process Design and Development*, **24**, 822–828.

Glinos, K. and Malone, M.F. (1988) Optimality regions for complex column alternatives in distillation systems. *Chemical Engineering Research and Design*, **66**, 229–240.

Glinos, K.N., Nikolaides, I.P. and Malone, M.F. (1986) New complex column arrangements for ideal distillation. *Industrial & Engineering Chemistry Process Design and Development*, **25**, 694–699.

Gorak, A. and Sorensen, E. (2014) *Distillation: Fundamentals and Principles*, Elsevier, Oxford, UK.

Halvorsen, I.J. and Skogestad, S. (2003) Minimum energy consumption in multicomponent distillation. 2: three-product Petlyuk arrangements. *Industrial & Engineering Chemistry Research*, **42**, 605–615.

Humphrey, J. (1995) Separation processes: playing a critical role. *Chemical Engineering Progress*, **91**, 31–41.

Jane, J.H. (2001) The Cativa™ process for the manufacture of acetic acid. *Platinum Metals Review*, **44**, 95–105.

Kaibel, B., Jansen, H., Zich, E. and Olujic, Z. (2006) Unfixed dividing wall technology for packed and tray distillation columns. Distillation & Absorption Symposium 2006, IChemE, Rugby, pp. 252–266.

Kiss, A.A. (2014) Distillation technology—still young and full of breakthrough opportunities. *Chemical Technology and Biotechnology*, **4**, 479–498.

Kiss, A.A. and Bildea, C.S. (2011) A control perspective on process intensification in dividing wall columns. *Chemical Engineering and Processing*, **50**, 281–292.

Kiss, A.A. and Rewagad, R.R. (2011) Energy efficient control of a BTX dividing-wall column. *Computers & Chemical Engineering*, **35**, 2896–2904.

Kiss, A.A. and Suszwalak, D.J.-P.C. (2012) Innovative dimethyl ether synthesis in a reactive dividing-wall column. *Computers & Chemical Engineering*, **38**, 74–81.

Kohler, J., Kuen, T. and Blass, E. (1994) Minimum energy demand for distillations with distributed components and sideproduct withdrawals. *Chemical Engineering Science*, **49**, 3325–3330.

Kolbe, B. and Wenzel, S. (2004) Novel distillation concepts using one-shell columns. *Chemical Engineering and Processing*, **43**, 339–346.

Lee, B. and Kim, G. (2012) Liquid splitter. Korean Patent Pending, 10-2012-0030223.

Lee, B., Kim, G., Lee, M. *et al.* (2010) Liquid splitter. Korean Patent Pending, 10-2010-0120467.

Lee, S.H., Shamsuzzoha, M., Han, M. *et al.* (2011a) Study of structural characteristics of a divided wall column using the sloppy distillation arrangement. *Korean Journal of Chemical Engineering*, **28**, 348–356.

Lee, S.K., Shin, J.H., and Lee, J.K. (2011b) Divided wall distillation column for producing high purity normal butanol, and normal butanol distillation method. US Pat. US20110303526.

Liu, Z.Y. and Jobson, M. (2004) Retrofit design for increasing the processing capacity of distillation columns: 1. A Hydraulic performance indicator. *Chemical Engineering Research and Design*, **82**, 3–9.

Long, N.V.D. and Lee, M. (2011) Improved energy efficiency in debottlenecking using a fully thermally coupled distillation column. *Asia-Pacific Journal of Chemical Engineering*, **6**, 338–348.

Long, N.V.D. and Lee, M. (2012a) Improvement of the deethanizing and depropanizing fractionation steps in NGL recovery process using dividing wall column. *Journal of Chemical Engineering of Japan*, **4**, 285–294.

Long, N.V.D. and Lee, M. (2012b) Improvement of natural gas liquid recovery energy efficiency through thermally coupled distillation arrangement. *Asia-Pacific Journal of Chemical Engineering*, **7**, 71–77.

Long, N.V.D. and Lee, M. (2012c) Dividing wall column structure design using response surface methodology. *Computers & Chemical Engineering*, **37**, 119–124.

Long, N.V.D. and Lee, M. (2013a) Design and optimization of heat integrated dividing wall columns for improved debutanizing and deisobutanizing fractionation of NGL. *Korean Journal of Chemical Engineering*, **30**, 286–294.

Long, N.V.D. and Lee, M. (2013b) Optimal retrofit design of extractive distillation to energy efficient thermally coupled distillation scheme. *AIChE Journal*, **59**, 1175–1182.

Long, N.V.D. and Lee, M. (2013c) Debottlenecking the retrofitted thermally coupled distillation sequence. *Industrial & Engineering Chemistry Research*, **52**, 12635–12645.

Long, N.V.D. and Lee, M. (2013d) Optimal retrofit of a side stream column to a dividing wall column for energy efficiency maximization. *Chemical Engineering Research and Design*, **91**, 2291–2298.

Long, N.V.D. and Lee, M. (2014) Review of retrofitting distillation columns using thermally coupled distillation sequences and dividing wall columns to improve energy efficiency. *Journal of Chemical Engineering of Japan*, **47**, 87–108.

Long, N.V.D., Kwon, Y.S. and Lee, M. (2013) Design and optimization of thermally coupled distillation schemes for the trichlorosilane purification process. *Applied Thermal Engineering*, **59**, 200–210.

Long, N.V.D., Lee, S.H. and Lee, M. (2010) Design and optimization of a dividing wall column for debottlenecking of the acetic acid purification. *Chemical Engineering and Processing*, **49**, 825–835.

Marquardt, W.R., Watzdorf, R.V. and Bausa, J. (1999) Shorcut methods for nonideal multicomponent distillation: 2. Complex columns. *AIChE Journal*, **45**, 1615–1628.

Minh, L.Q., Long, N.V.D. and Lee, M. (2012) Energy efficiency improvement of dimethyl ether purification process by utilizing dividing wall columns. *Korean Journal of Chemical Engineering*, **29**, 1500–1507.

Minh, L.Q., Husnil, Y.A., Long, N.V.D. and Lee, M. (2014) Retrofit and debottlenecking of naphtha splitter process to thermally coupled distillation sequence with a side reboiler. *Journal of Chemical Engineering of Japan*, **47**, 687–692.

Monro, D.A. (1938) Fractionating apparatus and method of fractionation. US Pat. 2134882.

Niesbach, A., Adams, T.A. II and Lutze, P. (2013) Semicontinuous distillation of impurities for the production of butyl acrylate from bio-butanol and bio-acrylic acid. *Chemical Engineering and Processing*, **74**, 165–177.

Nikolaides, I.P. and Malone, M.F. (1987) Approximate design of multiple-feed/side-stream distillation systems. *Industrial & Engineering Chemistry Research*, **26**, 1839–1845.

Olujic, Z., Jodecke, M., Shilken, A., Schuch, G. and Kaibel (2009) Equipment improvement trends in distillation. *Chemical Engineering and Processing*, **48**, 1089–1104.

Papastathopoulou, H.S. and Luyben, W.L. (1991) Control of a binary sidestream distillation column. *Industrial & Engineering Chemistry Research*, **30**, 713–721.

PEP Report (1994) Acetic acid by low pressure carbonylation of methanol, November, 1994.

Porscha, P. (2001) Process and unit for production of 1,2-dichloroethane. US Pat. 6252125

Poth, N., Brusis, D. and Stichlmair, J. (2004) Minimaler Energiebedarf von Trennwandkolonnen. *Chemie Ingenieur Technik*, **76**, 1811–1814.

Premkumar, R. and Rangaiah, G.P. (2009) Retrofitting conventional column systems to dividing wall columns. *Chemical Engineering Research and Design*, **87**, 47–60.

Rangaiah, G.P., Chan, J.T.S. and Hidajat, K. (2006) Optimization of complex distillation columns. *Chemical Engineering (New York)*, 60-64 (September).

Rangaiah, G.P., Ooi, E.L. and Premkumar, R. (2009) A simplified procedure for quick design of dividing-wall columns for industrial applications. *Chemical Product and Process Modeling*, **4**, 1–42.

Rong, B.G. (2011) Synthesis of dividing-wall columns (DWC) for multicomponent distillations – a systematic approach. *Chemical Engineering Research and Design*, **89**, 1281–1294.

Rooks, R.E., Malone, M.F. and Doherty, M.F. (1996) A geometric design method for side-stream distillation columns. *Industrial & Engineering Chemistry Research*, **35**, 3653–3664.

Segovia-Hernandez, J.G., Hernandez-Vargas, E.A. and Marquez-Munoz, J.A. (2007) Control properties of thermally coupled distillation sequences for different operating conditions. *Computers & Chemical Engineering*, **31**, 867–874.

Shin, J., Lee, S., Lee, J. and Lee, M. (2011) Manage risks with dividing-wall column installations. *Hydrocarbon Processing*, **90**, 59–62.

Shin, J., Lee, J., Lee, S. *et al.* (2013) Enhance operation and reliability of dividing-wall columns. *Hydrocarbon Processing*, **92**, 85–88.

Slade, B., Stober, B. and Simpson, D. (2006) Dividing wall column revamp optimizes mixed xylenes production. *Distillation & Absorption IChemE., Symposium Series*, **152**, 1-10.

Smith, R. (2005) *Chemical Process Design and Integration*, John Wiley & Sons, Chichester, West Sussex, England, pp. 17–33.

Sotudeh, N. and Shahraki, B.H. (2008) Extension of a method for the design of divided wall columns. *Chemical Engineering & Technology*, **31**, 83–86.

Staak, D., Grutzner, T., Schwegler, B. and Roederer, D. (2014) Dividing wall column for industrial multipurpose use. *Chemical Engineering and Processing*, **75**, 48–57.

Tamayo-Galván, V.E., Segovia-Hernández, J.G., Hernández, S. *et al.* (2008) Controllability analysis of alternate schemes to complex column arrangements with thermal coupling for the separation of ternary mixtures. *Computers & Chemical Engineering*, **32**, 3057–3066.

Tedder, D.W. and Rudd, D.F. (1978) Parametric studies in industrial distillation: Part 1. Design comparisons. *AIChE Journal*, **24**, 303–315.

Triantafyllou, C. and Smith, R. (1992) The design and optimization of fully thermally coupled distillation. *Transactions of the Institution of Chemical Engineers*, **70**, 118–132.

Turton, R., Bailie, R.C., Whiting, W.B. and Shaeiwitz, J.A. (2012) *Analysis, Synthesis and Design of Chemical Processes*, 4th edn, Prentice Hall, Upper Saddle River, New Jersey.

Underwood, A.J. (1948) Fractional distillation of multicomponent mixtures. *Chemical Engineering and Processing*, **44**, 603–614.

van Diggelen, R.C., Kiss, A.A. and Heemink, A.W. (2010) Comparison of control strategies for dividing-wall columns. *Industrial & Engineering Chemistry Research*, **49**, 288–307.

Wachter, J.A., Ko, T.K.T. and Andres, R.P. (1988) Minimum reflux behavior of complex distillation columns. *AIChE Journal*, **34**, 1164–1184.

Wankat, P.C. (2012) *Separation Process Engineering. Includes Mass Transfer Analysis*, 3rd edn, Prentice-Hall, Upper Saddle River, New Jersey.

Wenzel, S. and Rohm, H. (2003) Design of complex distillation columns by overall-cost optimization. *Chemical Engineering & Technology*, **27**, 484–490.

Wright, R.O. (1949) Fractionation apparatus. US Pat. 2471134.

Yildirim, O., Kiss, A.A. and Kenig, E.Y. (2011) Dividing wall columns in chemical process industry: A review on current activities. *Separation and Purification Technology*, **80**, 403–417.

10

Techno-economic Evaluation of Membrane Separation for Retrofitting Olefin/Paraffin Fractionators in an Ethylene Plant

X.Z. Tan, S. Pandey, G.P. Rangaiah, and W. Niu
Department of Chemical and Biomolecular Engineering, National University of Singapore, Singapore

10.1 Introduction

Olefin/paraffin separation is often categorized as one of the difficult separations in the petrochemical industry due to small differences in their boiling points. Various technologies have been developed to replace the traditional low-temperature distillation with methods such as extractive distillation, physical adsorption using molecular sieves, chemisorption using complexing metals like copper and silver ions, absorption (physical and chemical), and membrane separation. In the last few decades, research has gathered momentum in the area of membrane applications for olefin/paraffin separation. This was after significant developments in membrane technologies for dehydration of organic solvents and aromatic/paraffin separation. Membrane separators may not be able to replace distillation columns but the two can be coupled in the form a hybrid membrane-distillation (HMD) system for improved separation.

Different kinds of membranes were tested to identify their permeability and selectivity for olefins and paraffins. Azhin *et al.* (2008) reviewed the application of facilitated

Chemical Process Retrofitting and Revamping: Techniques and Applications, First Edition.
Edited by Gade Pandu Rangaiah.

transport membrane (FTM) technology to olefin/paraffin separation, and mentioned its general mechanisms under varying conditions. However, FTM has not yet been commercialized due to its instability during long-term exposure caused by carrier poisoning and short membrane life. Polymeric membranes without a carrier for olefin/paraffin separation are also under extensive study for large-scale applications. Within this class of polymers, polyimides are the most attractive for membranes because of their relatively better separation characteristics, decent thermal and chemical resistance, and easy processing. Fluorinated polyimides are found to have relatively higher thermal and chemical stability (Schmeling *et al.*, 2010). However, dense polymeric membranes produce less permeate due to their relatively high membrane thickness. The potential of carbon molecular sieve membranes for olefin/paraffin separation has recently been reviewed (Faiz and Li, 2012). Despite their high cost of production and brittleness, efforts are being made to improve their separation performance for industrial applications.

A number of studies have shown that significant savings can be achieved by integrating a membrane module with a distillation column. Such arrangements are able to provide equivalent or better purities at lower utility costs. Gottschlich and Roberts (1990) carried out a study to identify general principles for choosing hybrid separation systems over distillation columns; they concluded that high product purities resulted in lower efficiencies and higher processing costs for all systems, and that the thermodynamic extent of separation is a key parameter for a hybrid system. Davis *et al.* (1993) developed a hybrid FTM–distillation system to perform pilot plant experiments on propylene/propane separation and ethylene recovery from purge gas. Estimated benefits from the hybrid system are a capacity increase of the existing splitter by 80% with no increase in utilities for the production of chemical-grade propylene, and energy savings of 500 billion BTU/year for a 10,000 bbl/day grassroots facility producing polymer-grade propylene. Pettersen *et al.* (1996) investigated a parallel configuration for propylene/propane separation, and the optimum side draw location in the column for feeding into the membrane unit was found to be near the column feed stage, where distillation is the least efficient. Pressly and Ng (1998) investigated the effect of possible HMD configurations through screening calculations, and concluded that series or parallel configurations are preferred over top or bottom configurations for propylene/propane separation.

Caballero *et al.* (2009) proposed a mathematical programming approach to optimize and retrofit an HMD system for ethylene/ethane separation with parallel configuration; their study showed a potential energy saving of up to 30%. Bernardo and Drioli (2010) focused on the application of membrane separation technology in oil-refining and petrochemical sectors, and concluded that membranes should be able to perform adequately under conditions of exposure to organic vapors, especially C3+ hydrocarbons, which are common in gas fields, refineries and petrochemical plants. Benali and Aydin (2010) carried out optimization and economic analysis of various HMD configurations to scrutinize their feasibility for application to C2 and C3 splitters. Motelica *et al.* (2012) presented techno-economic evaluation for determining the increased energy efficiency and debottlenecking of an existing ethylene/ethane separation, in relation to the required membrane performance. It was found that high membrane selectivity (> 60) and/or ethylene permeance of at least 1×10^{-4} mol/(m^2 s kPa) are required for considerable savings. Ploegmakers *et al.* (2013) studied retrofitting of an existing distillation column with a membrane unit for ethylene/ethane separation. Membrane parameters like surface area, feed and permeate pressures were optimized to understand the effect of ethylene permeance and ethylene/ethane selectivity on the energy requirements of HMD configurations.

The prime reason for analyzing retrofitting distillation columns[1] to HMD systems is the reduction of energy consumption for separation, thus improving the economic and environmental sustainability of existing plants. A techno-economic evaluation can reveal the viability of the HMD system for retrofitting different columns in existing ethylene-separation processes. Ploegmakers *et al.* (2013) established the limits of ethylene permeance and ethylene/ethane selectivity for economical retrofitting of an ethylene/ethane splitter with a membrane unit. However, these limits are far removed from the observed parameter values of membranes used for olefin/paraffin separation as reported by Faiz and Li (2012). It is, henceforth, important to carry out optimization considering bounds on membrane selectivity and permeance, which are reported or expected from near-future developments in membrane technologies. Moreoever, Ploegmakers *et al.* (2013) considered the reboiler duty of the ethylene fractionator as a cost; however, this duty should be considered as a credit since cold energy can be produced during the vaporization of the bottom stream (at −8 °C) in the reboiler.

Multi-objective optimization (MOO) of a process gives a set of optimal solutions for process design and operation, in the form of a Pareto-optimal front. It not only shows the trade-off between the chosen objectives, but also identifies the effect of decision variables considered on the objective functions. See Section 4.4 in this book for more details on MOO. The reviews by Masuduzzaman and Rangaiah (2009), Sharma and Rangaiah (2013) and Rangaiah *et al.* (2015) show that MOO has not been applied for retrofitting columns into HMD systems in petrochemical plants. Hence, in this study, retrofitting four selected columns for the olefin/paraffin separation of an ethylene plant with a membrane unit is analzyed for maximizing annual utility cost savings and minimizing capital cost simultaneously.

The rest of this Chapter is organized as follows. Section 10.2 outlines a typical olefin/paraffin separation scheme in a conventional ethylene plant, which includes the selected distillation columns for retrofitting with a membrane unit. Section 10.3 describes the mathematical modeling of the membrane unit and the various assumptions that were made. Section 10.4 presents the procedure of simulating a retrofitted HMD system and the preliminary techno-economic analysis that was carried out. Section 10.5 covers the formulation of MOO problem, which includes selection of objectives, decision variables and constraints in the optimization problems studied. In Section 10.6, results from the optimization of two objectives for various cases are presented and discussed. Finally, conclusions of this study are given in Section 10.7.

10.2 Olefin/Paraffin Separation in an Ethylene Plant

Hydrocarbon feedstocks such as ethane, propane, butane and naphtha are sent together with superheated steam to the cracking furnace, where endothermic reactions occur at high

[1] Retrofitting a column often refers to changing its internals like trays and packing. Here, it is used for adding a membrane unit to an existing column to form an HMD system, as in Motelica *et al.* (2012) and Ploegmakers *et al.* (2013).

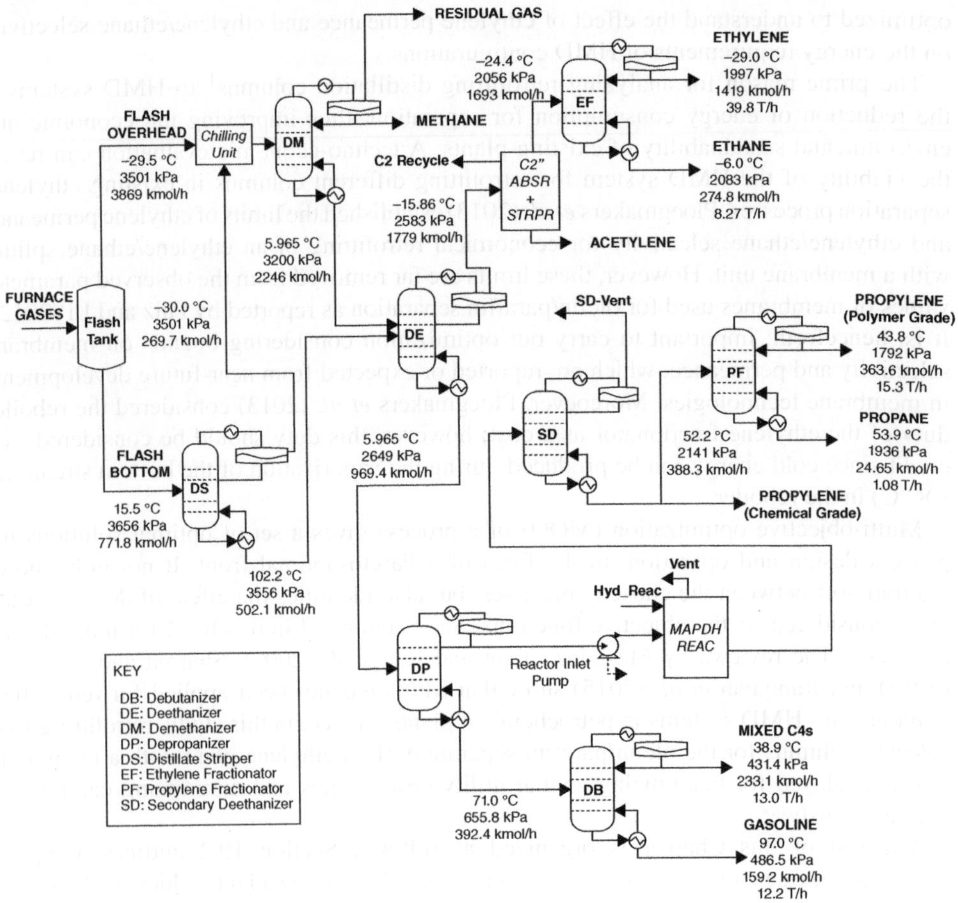

Figure 10.1 *Olefin/paraffin separation in a conventional ethylene plant.*

temperature, producing olefins. The furnace effluent is quenched to separate fuel, C6 and heavier components, and water. The gases are compressed up to 3.5 MPa and washed with alkali to remove sulfur compounds. The stream is then cooled to about 15 °C, sent to a flash drum and subsequently dried to remove moisture before further processing as shown in Figure 10.1. The liquid stream from the flash is sent to the distillate stripper (DS). The vapor streams from the flash and the DS overheads are sent to a chilling train (cold box), which contains a series of heat-exchangers and flash drums. The liquid streams from these flash drums are fed into the demethanizer (DM). The bottoms from the DS and DM are sent to the deethanizer (DE), operating at 2.4 to 2.8 MPa. The DE overhead stream is mainly C2s, namely, ethylene, ethane and acetylene, and its bottoms stream contains C3s and higher.

The DE overhead stream is passed through the acetylene-recovery section or acetylene converter, and the resulting acetylene-free gas is sent to the ethylene fractionator (EF), which produces 99.9 mol% ethylene in the overheads and more than 99 mol% ethane in the bottoms. The latter is recycled for steam cracking into valuable products. The DE bottoms stream is fed to the depropanizer (DP). The overhead stream from the DP is C3 hydrocarbons including propylene, propane, methylacetylene and propadiene. This stream is sent to the hydro-converter, MAPDH REAC (Figure 10.1) with catalysts that selectively convert methylacetylene and propadiene into propene and propane, respectively. The DP's bottoms with C4 and heavier hydrocarbons is fed into the debutanizer (DB), operating at 0.4 to 0.5 MPa, where the overheads comprises mainly of C4s and the bottoms is C5s and higher. The hydrogenated stream from MAPDH REAC is typically sent to the secondary deethanizer (SD), whose vent is recycled and the bottoms is supplied to the propylene fractionator (PF) operating at 1.8 to 2 MPa with more than 200 trays. PF produces polymer-grade propylene (> 99.5%) in the overheads and propane with purity of more than 95% in the bottoms. The latter can either be recycled to the steam cracker or used as fuel (Calamur and Carrera, 2005).

10.3 Membrane Model Development

The membrane module used for the simulation of the HMD process in Aspen HYSYS is developed using Aspen Custom Modeler (ACM) v8.4, and then compiled using Microsoft Visual Studio 2010.

10.3.1 Membrane Modeling

In the simulation of the HMD process, the gas-permeation membrane used is assumed to be an asymmetric hollow-fiber membrane. For this type of membrane, gas permeation does not depend on the flow pattern on the permeate side as the porous supporting layer prevents mixing of the permeate fluxes (Pan, 1986). A schematic of the flow pattern in an asymmetric hollow-fiber membrane is shown in Figure 10.2. Hence, a simple cross-flow model is sufficient to describe the membrane behavior.

The model assumes that the driving force for diffusion depends on the difference between the partial pressures of each component across the membrane as described in the following equation for steady-state operation [Equation (10.1)].

$$F_{perm}Z_{perm,i} = Q_i^G A_m(p_{ret,i} - p_{perm,i}) \tag{10.1}$$

Here, F_{perm} is the total molar flow rate of the permeate stream, $Z_{perm,i}$ is the mole fraction of component i in the permeate stream, Q_i^G is the permeance of component i through the

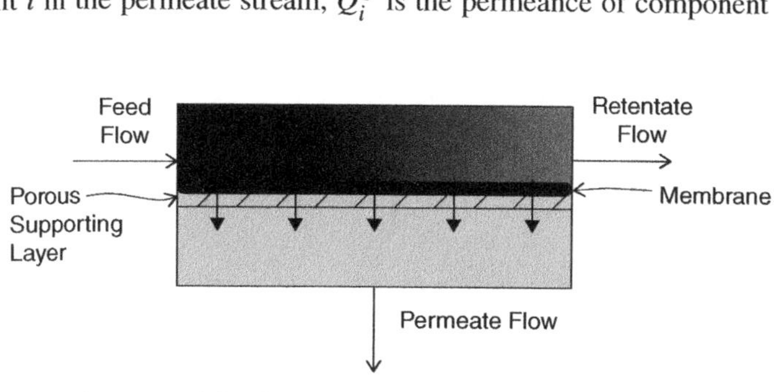

Figure 10.2 *Schematic of the flow pattern in an asymmetric hollow-fiber membrane.*

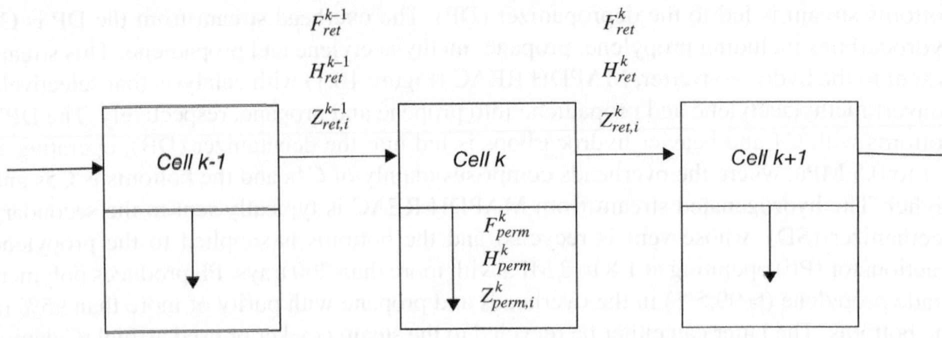

Figure 10.3 *Schematic of three discretized membrane cells with quantities used for mass and energy balances for membrane cell k.*

membrane, A_m is the membrane area, $p_{ret,i}$ is the partial pressure of component i in the retentate stream, and $p_{perm,i}$ is the partial pressure of component i in the permeate stream.

Many studies on HMD systems in the current literature did not consider the Joule–Thompson effect caused by the expansion of permeate gas due to the pressure difference between the high retentate pressure and the low permeate pressure, also known as trans-membrane pressure. This expansion leads to a decrease in the permeate temperature, which in turn decreases the membrane permeance. So, ignoring the Joule–Thomson effect may result in a wrong estimation of membrane separation performance and consequently of the reboiler/condenser duties and utility savings obtained from an HMD system. The membrane model employed in the present study takes into account the Joule–Thompson effect by including the following energy balance [Equation (10.2)]:

$$F_{feed}H_{feed} = F_{ret}H_{ret} + F_{perm}H_{perm} \tag{10.2}$$

Here, F and H are, respectively, the total molar flow rate and total enthalpy of feed, retentate or permeate stream indicated by the subscript.

The membrane module is solved by discretizing the membrane area into 100 cells of equal area, and each cell model is given by Equations (10.1) and (10.2). The final set of mass and energy balances for cell k (Figure 10.3) of the membrane unit are as shown in Equations (10.3)–(10.7):

$$F_{ret}^{k-1} = F_{ret}^{k} + F_{perm}^{k} \tag{10.3}$$

$$F_{ret}^{k-1}Z_{ret,i}^{k-1} = F_{ret}^{k}Z_{ret,i}^{k} + F_{perm}^{k}Z_{perm,i}^{k} \tag{10.4}$$

$$F_{perm}^{k}Z_{perm,i}^{k} = Q_i^G A_m\left(\frac{p_{ret,i}^{k-1} + p_{ret,i}^{k}}{2} - p_{perm,i}\right) \tag{10.5}$$

$$F_{ret}^{k-1}H_{ret}^{k-1} = F_{ret}^{k}H_{ret}^{k} + F_{perm}^{k}H_{perm}^{k} \tag{10.6}$$

$$\sum_{i=1}^{Components} Z_{ret,i}^{k} = 1 \tag{10.7}$$

Here, F^k, Z_i^k and H^k are, respectively, the molar flow rate, mole fraction of component of i and total enthalpy, all in cell k; their subscripts, *ret* and *perm*, refer to retentate and permeate streams. Equations (10.4) and (10.5) are mass balances and mass-transfer equations for each of the components present in the membrane feed. The cross-flow model [Equations (10.3)–(10.7)] was implemented in ACM v8.4 and validated against the experimental data in Pan (1986) and the predicted values of Davis (2002). The Joule–Thompson effect was validated by simulating adiabatic throttling of permeate gas through a valve in Aspen Hysys. Both these validations are described in detail in Appendix 10A.

10.3.2 Assumptions for Membrane Separation Simulation

In order to simulate an HMD system, the following assumptions on membrane separation are made.

1. Membrane is assumed to be of carbon molecular sieve (CMS) type due to its ability to handle high pressures and large feed flowrates without any plasticization, making it viable for use in an ethylene plant (Vu *et al.*, 2002). Membrane performance is assumed to not deteriorate within its expected life of a conservative 3 years, compared with 5 years in the study of Ploegmakers *et al.* (2013). Compressor and heat exchangers are assumed to have a life of 10 years
2. Membrane permeance and selectivity data are based on those of polymeric membranes obtained from Faiz and Li (2012). This is due to a lack of studies on CMS membranes for the separation of olefins/paraffins. The assumption of the use of polymeric membrane data is conservative since CMS membranes have been shown to give better permeance and selectivity than polymeric membranes for olefin/paraffin separation (Xu et al., 2012)
3. The permeance ratio of propylene to ethylene is kept at 0.472 owing to the higher permeability of C2 components compared with C3 components, as deduced from averaging the propylene/ethylene selectivity ratio obtained from Faiz and Li (2012). The permeance ratio of 1,3-butadiene to propylene is deduced from averaging the 1,3-butadiene/propylene selectivity ratio data obtained from Faiz and Li (2012), and was found to be 3.48, despite the larger molecular size of C4 components compared with C3 components. This is due to the rigid and straight molecular shape of 1,3-butadiene, which results in higher permeability across the membrane (Okamoto *et al.*, 1997). Permeance of ethylacetylene and propadiene are assumed to be the same as that of propylene, whereas *cis*/*trans*-2-butene (*Z*/*E*-2-butene) and 1,3/1,2-butadiene are assumed to permeate at the same rate as isobutene
4. In general, when selectivity is plotted against the permeance for a given olefin/paraffin set on a log–log plot, a linear upper bound with negative slope can be observed, as mentioned in Burns and Koros (2003) and Rungta *et al.* (2013). This leads to the following correlation between selectivity and permeance [Equation (10.8)]:

$$\alpha_{AB} = \frac{\beta_{AB}}{\left(Q_A^G\right)^{\lambda_{AB}}} \tag{10.8}$$

Here, Q_A^G is the permeance of fast-permeating component A, α_{AB} is the selectivity of A shown by the membrane as compared with slow-permeating component B, and β_{AB} and λ_{AB} are, respectively, front factor and slope of the linear upper bound. The parameter

Table 10.1 *Values of front factor and slope of the upper bound for olefin/paraffin membrane separation.*

Olefin/Paraffin	Fast-permeating component	β_{AB}	λ_{AB}
Ethylene/Ethane	Ethylene	7.2364	0.212
Propylene/Propane	Propylene	25.294	0.244
Isobutylene/Butane	Isobutylene	61.977	0.242

values used (Table 10.1) for each set of olefin and paraffin are based on the literature data for different olefin/paraffin sets compiled by Faiz and Li (2012). Thus, Equation (10.8) was used to relate selectivity with permeability using the parameter values in Table 10.1, for each set of hydrocarbons.

5. Lie *et al.* (2007) assumed $15/m^2 for an in-house tailored CMS membrane with a bare module factor (for installation) of 3.5. In the current study, a total module cost of $100/m^2 is assumed for CMS membranes, which is similar to that used in studies on column retrofitting (Caballero *et al.*, 2009; Ploegmakers *et al.*, 2013). This cost of CMS membranes includes purchase cost and all installation costs
6. The experimental permeance values obtained from Faiz and Li (2012) are only valid for the temperature at which the experiment was carried out and differs for each experiment. Hence, to ensure permeance values obtained from the literature are justified for use in the simulation, it is assumed that the permeance values are valid for the membrane temperature range of 20 to 50 °C, where most experimental values were obtained
7. In this study, permeance values are considered as decision variables and will be varied over a range. So, considering permeance as a single average value across the membrane is easier. In other words, the temperature effect on permeance is neglected. However, the temperature drop due to the Joule–Thompson effect is limited to 10 °C to 20 °C, and the effect of this temperature drop is considered in heating the permeate stream, if required.

10.4 Retrofitting a Distillation Column with a Membrane Unit

Four distillation columns, namely, DE, DP, EF and PF in Figure 10.1 have been chosen for analyzing economics of retrofitting the existing column with a membrane unit.

10.4.1 HMD Modeling and Simulation

For each HMD system, stream and column specifications are given in Table 10.2; they are from the design data of an ethylene plant. All simulations of processes are based on the parallel HMD configuration, and were carried out in Aspen HYSYS v8.4. Although the feed streams for DE, DP and PF are in liquid and can be processed by a membrane through pervaporation, there is, however, not enough literature on pervaporation experiments using CMS membranes. Therefore, series configuration is not studied. Only gas-separation membranes are considered in the following case studies, with parallel arrangement where side

Table 10.2 *Feed and product specifications for DE, DP, EF and PF columns; number of ideal trays in each column is excluding condenser and reboiler; stage numbers are counted from top with condenser as zeroth stage.*

Column	Feed conditions	Feed and its composition (Mole fraction)	Product purity or other specs	Other conditions
DE	*DM Bottom Feed* 2246 kgmol/h (73 036 kg/h) at 3200 kPa and 6 °C; $N_F = 15$	Methane (0.0025), Acetylene (0.0110), Ethylene (0.6346), Ethane (0.1145), Methy[acetylene (0.0029), Propadiene (0.0031), Propene (0.1677), Propane (0.0075), 1,2-Butadiene 0.0017), 1,3-Butadiene (0.020), Isobutane (0.0077), n-Butane (0.0077), Isobutene (0.0073), *cis*-2-Butene (0.0015), *trans*-2-Butene (0.0016), n-Pentane (0.0061), n-Hexane (0.0025)	Ethylene mole fraction in overheads = 0.8259; Ethane in bottoms = 155 kg/h	Column top pressure = 2583 kPa; Pressure drop/tray = 0.7 kPa; No. of ideal trays = 51; Full reflux condenser; Peng–Robinson model
	DS Bottom Feed 502 kgmol/h (26720 kg/h) at 3556 kPa and 102.2 °C; $N_F = 33$	Acetylene (0.0028), Ethylene (0.0877), Ethane (0.0499), Methylacetylene (0.0084), Propadiene (0.0076), Propene (0.2989), Propane (0.0156), 1,2-Butadiene (0.0157), 1,3-Butadiene (0.1201), Isobutane (0.0392), n-Butane (0.0466), Isobutene (0.0410), *cis*-2-Butene (0.0116), *trans*-2-Butene (0.0113), n-Pentane (0.1165), n-Hexane (0.1271)		

(continued)

Table 10.2 *(Continued)*

Column	Feed conditions	Feed and its composition (Mole fraction)	Product purity or other specs	Other conditions
DP	969 kgmol/h (49 398 kg/h) At 2649 kPa And 89 °C; $N_F = 13$	Ethane (0.00053), Methylacetylene (0.0111), Propadiene (0.0111), Propene (0.5426), Propane (0.0255), 1,2-Butadiene (0.0121), 1,3-Butadiene (0.1086), Isobutane (0.0382), n-Butane (0.0421), Isobutene (0.0382), *cis*-2-Butene (0.0095), *trans*-2-Butene (0.0095), n-Pentane (0.0744), n-Hexane (0.0716)	Propylene mole fraction in overheads = 0.9116; Methylacetylene mole fraction in bottoms = 0.0013	Column top pressure = 576 kPa; Pressure drop/tray = 0.7 kPa; No. of ideal trays: 42; Total condenser; Peng–Robinson model
EF	1693 kgmol/h (48 056 kg/h) at 2056 kPa and −24.4 °C; $N_F = 86$	Methane (0.003), Ethylene (0.8388), Ethane (0.1604), Propene (0.0005)	Ethylene mole fraction in Overheads = 0.999; Ethane mole fraction in bottoms = 0.995	Column top pressure = 1997 kPa;Pressure drop/tray= 0.7 kPa; No. of ideal trays = 125; Total condenser; Peng–Robinson model
PF	388 kgmol/h (16 392 kg/h) at 2141 kPa and 52.2 °C; $N_F = 102$	Propene (0.9390), Propane (0.06), Isobutane (0.001)	Propylene mole fraction in overheads =0.99; Propane mole fraction in bottoms = 0.95	Column top pressure = 1792 kPa; Pressure drop/tray = 0.7 kPa; No. of ideal trays = 200; Total condenser; Peng–Robinson model

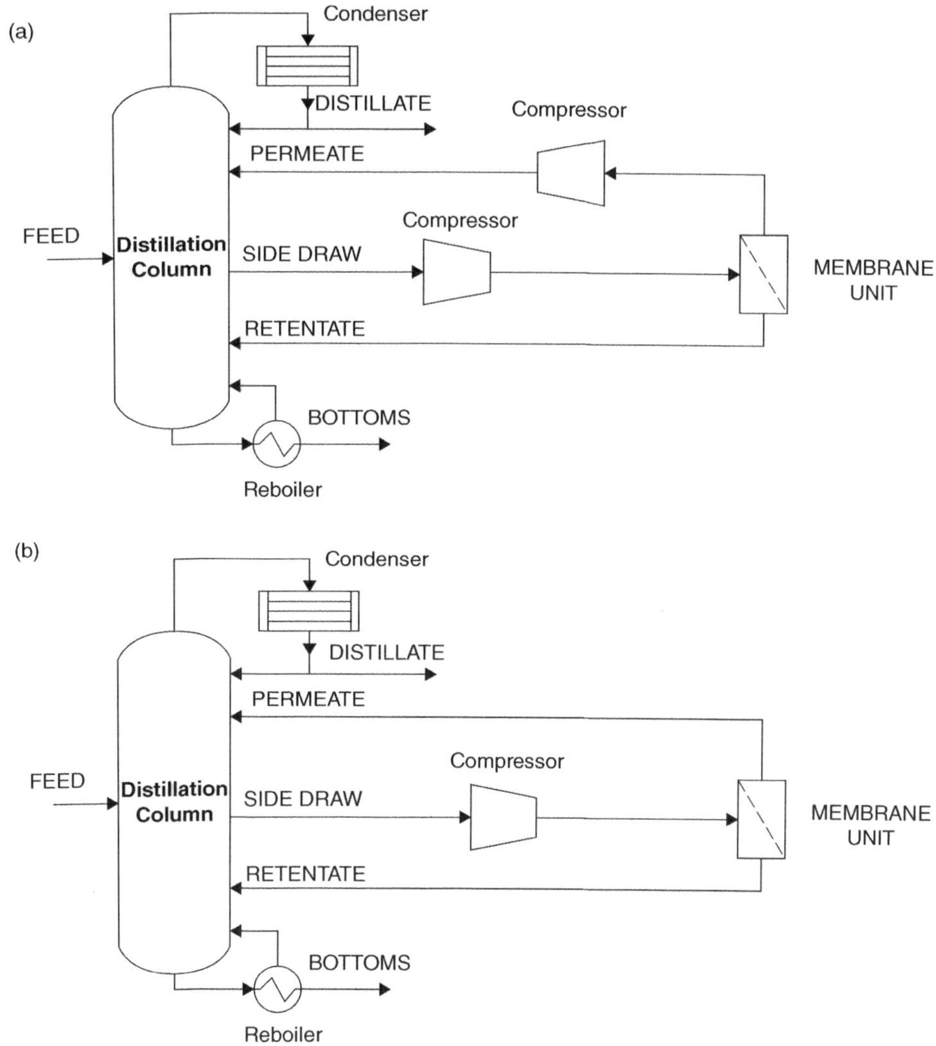

Figure 10.4 *Process flow diagrams of the two parallel HMD systems used in this study.*

draw of a vapor stream is taken from a suitable stage in the column. The stage for side draw is kept near the feed stage for DP, EF and PF, whereas the side draw stage for DE is kept in between its two feed stages. The feed stages for returning the respective permeate and retentate streams to the column are selected based on preliminary analysis.

The two parallel HMD configurations in Figure 10.4, however, cannot be applied to the distillation columns as they are. This is due to the side draw temperature of different columns, which results in the membrane feed temperature outside the temperature range of the experimental permeance values. As seen from Table 10.3, the side draw temperature of DE and EF is well below the range of 20 to 50 °C assumed for realistic membrane separation. Hence, additional process modifications, specific to the distillation column, are

Table 10.3 *Condenser, reboiler and side draw temperatures of the four distillation columns.*

Distillation column	Condenser temperature (°C)	Reboiler temperature (°C)	Side draw temperature (°C)
Deethanizer (DE)	−15.7	88.4	3.3
Depropanizer (DP)	−0.69	67.8	24.6
Ethylene Fractionator (EF)	−29.1	−5.7	−24.5
Propylene Fractionator (PF)	43.7	55.7	45.8

made to each parallel HMD configuration to ensure that the membrane feed stream enters at a suitable temperature and the temperature drop across the membrane caused by the Joule–Thompson effect is not too large.

Additionally, heat integration can be implemented, wherever possible, to cool or heat a stream to a temperature that matches the tray temperature it re-enters at. This will in turn reduce the additional energy required by the condenser or reboiler to accommodate for the temperature of the re-entering stream. These process modifications are discussed in detail in the next section.

10.4.2 Techno-economic Feasibility of Retrofit Operation

Before performing MOO on the HMD systems, a preliminary techno-economic evaluation is helpful in understanding the viability of an HMD system and in deciding the choice of parallel HMD configuration for the MOO. Reasonable values of design/decision variables such as membrane permeance, membrane area, and pressure difference across membrane were chosen to conduct the preliminary techno-economic evaluation with the exception of membrane feed flow rate, which was varied over a range. The capital cost of retrofitting with a membrane unit and the operating cost savings of the resulting HMD case are estimated using the equations and data given in Appendix 10B.

10.4.2.1 DE Retrofitting

Two different HMD configurations, similar to those in Figure 10.4, were assessed for retrofitting the DE column. Configuration A in Figure 10.5 is the first parallel HMD configuration with heat integration after the side draw. The heat exchanger raises the temperature of the side draw to ensure that the membrane feed stream enters at a temperature within the range of the experimental data reported for the membrane unit. Configuration B in Figure 10.5 is based on the second parallel HMD configuration in Figure 10.4; here, heating/cooling of streams is not required. Results of preliminary techno-economic evaluation of these two configurations are summarized in Table 10.4. Configuration B results in mostly negative utility savings; and configuration A, though consistently producing positive utility savings, has a much higher annualized capital cost which makes a DE column retrofitting into an HMD uneconomical. This is due to the use of olefin-selective membrane in the HMD system, resulting in the separation of olefinic and paraffinic compounds instead of separating C2 hydrocarbons from heavier hydrocarbons (Xu et al., 2012). Hence, a DE column is not suitable for retrofitting with membrane separation.

Note that the side draw temperatures in different configurations in Figure 10.5 and other similar figures might have slight deviations from each other, and this can contribute to

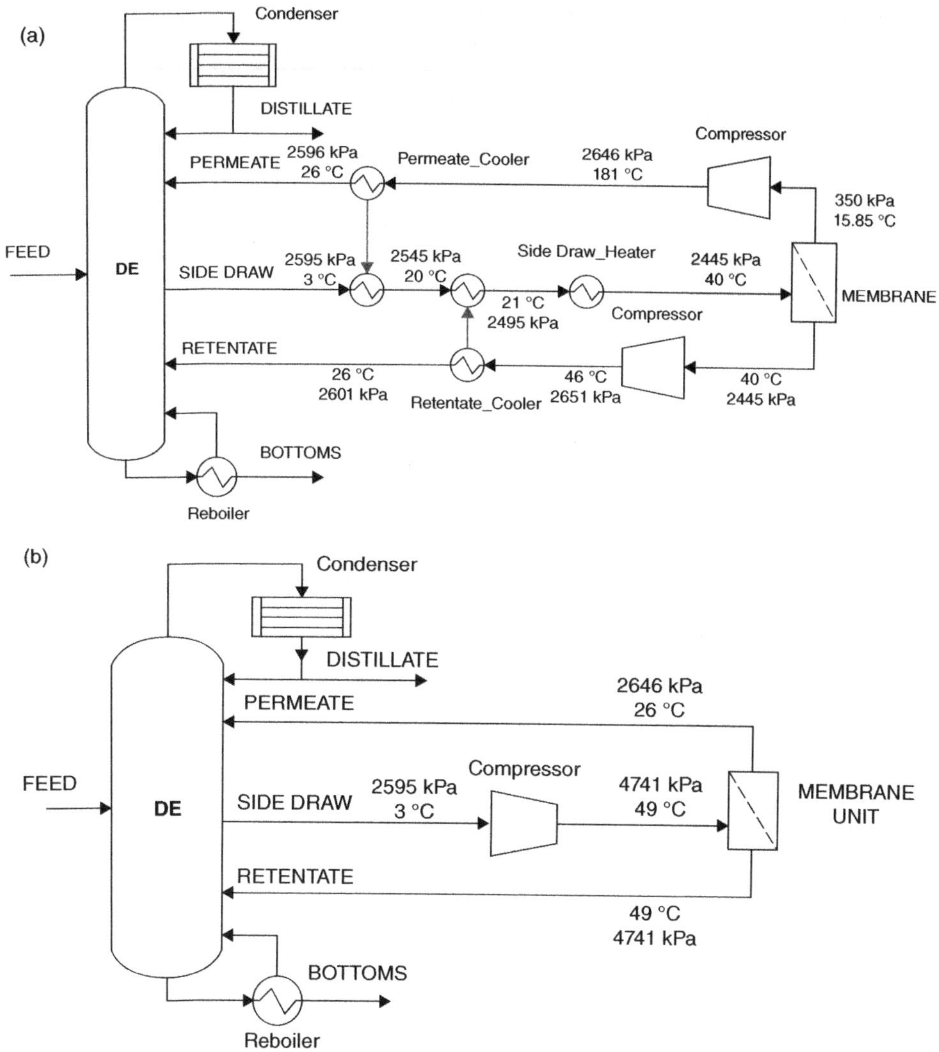

Figure 10.5 *Process flow diagrams of the two parallel HMD configurations used for preliminary techno-economic evaluation of DE retrofitting.*

the difference in compositions and temperatures of the retentate and permeate streams re-entering the column.

10.4.2.2 DP Retrofitting

Two parallel HMD configurations studied for DP are shown in Figure 10.6. No process modification is required for these configurations as the temperature of the side draw is in the range of the experimental data for the membrane unit. Similar to DE, the HMD system for DP produced negative or insufficient utility savings (Table 10.5). This is due to the reason explained in the DE's case. Hence, retrofitting a DP with membrane separation is not economical.

Table 10.4 *Preliminary techno-economic evaluation of parallel HMD systems for DE retrofitting at membrane feed flowrates of 900, 1600 and 2000 kmol/h.*

	Design/Decision variables					
Case	Membrane feed flow (kmol/h)	Ethylene permeance (kmol/h-m^2-bar)	Membrane area (m^2)	Pressure difference across membrane (bar)	Utility cost savings ($/yr)	Annualized capital cost of retrofitting ($/yr)
A.1	900	0.0002	2000	20.95	(2,628)	96,862
B.1					(5,099)	349,432
A.2	1600	0.0002	2000	20.95	14,440	115,024
B.2					(10,689)	1,226,397
A.3	2000	0.0002	2000	20.95	9,038	128,537
B.3					(2,339)	2,229,593

Note: Ethylene permeance in GPU is 1.64 (calculated by the method of Faiz and Li, 2012); and negative utility savings in this and other Tables are shown in parentheses.

Table 10.5 *Preliminary techno-economic evaluation of parallel HMD systems for DP retrofitting at membrane feed flowrates of 100, 350 and 600 kmol/h.*

	Design/Decision variables					
Case	Membrane feed flow (kmol/h)	Ethylene permeance (kmol/h-m^2-bar)[a]	Membrane rea (m^2)	Pressure difference across membrane (bar)	Utility cost savings ($/yr)	Annualized capital cost of retrofitting ($/yr)
A.1	100	0.015	1500	4.88	(23,534)	66,884
B.1					(30,774)	16,876
A.2	350	0.015	1500	4.88	(9,496)	140,114
B.2					(52,467)	134,854
A.3	600	0.015	1500	4.88	22,130	145,227
B.3					(140,588)	287,618

[a] Ethylene permeance in GPU is 12.3 (calculated by te method of Faiz and Li, 2012).

10.4.2.3 EF Retrofitting

In the case of EF, reboiler duty is considered as utility credit because the bottoms stream temperature entering the reboiler is −6 °C, which can cool the propylene utility or produce cold energy, and so it shall be deducted from the total utility cost of both the base and retrofitted cases. For the two parallel HMD configurations considered (Figure 10.7), heat integration is included to heat the membrane feed stream to a high enough temperature. Preliminary techno-economic analysis results in Table 10.6 indicate that Configuration B yields better utility savings compared with Configuration A. The capital cost of retrofitting is, however, more than the utility savings achieved. To confirm economic viability of retrofitting, MOO is carried out to maximize the utility savings and minimize the capital

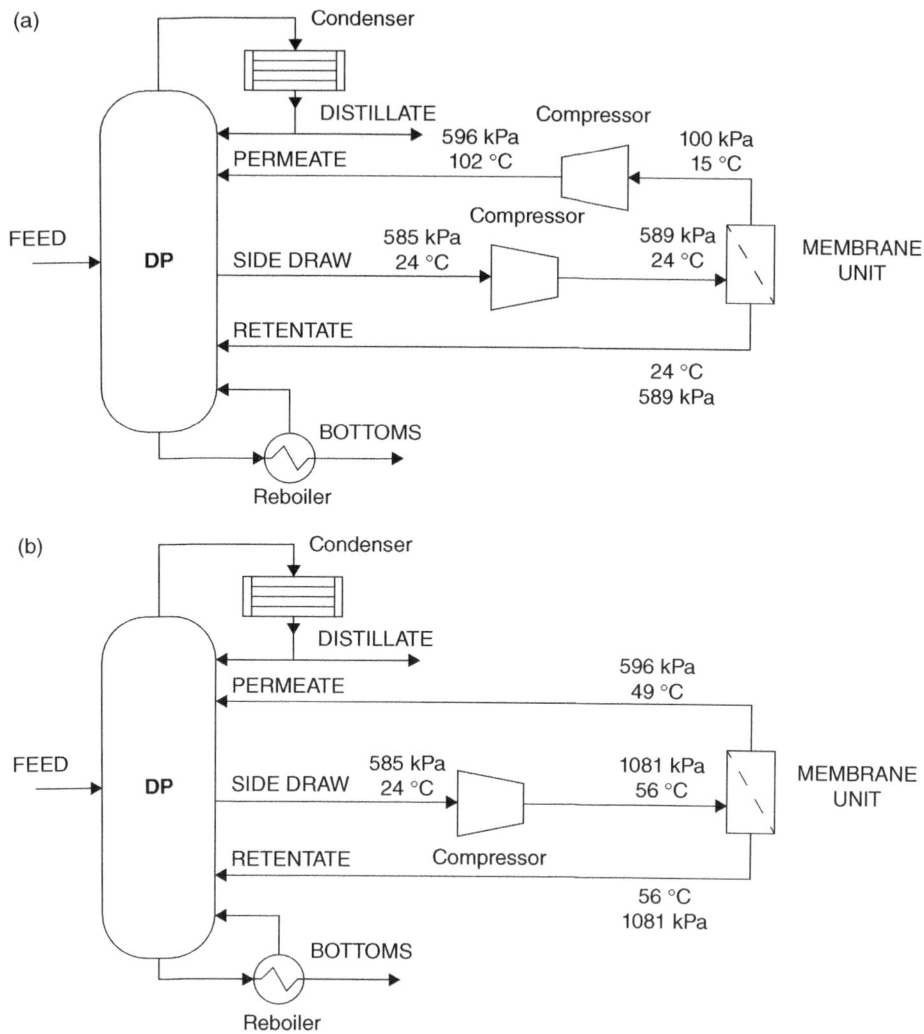

Figure 10.6 *Process flow diagrams of the two parallel HMD configurations used for preliminary techno-economic evaluation of DP retrofitting.*

cost of retrofitting. Another set of MOO is also performed taking reboiler duty as a utility cost as in Ploegmakers et al. (2013).

10.4.2.4 PF Retrofitting

In the case of PF, process modification was made only to the second parallel HMD configuration, as shown in Figure 10.8. A cooler was installed after the compressor to cool the membrane feed stream to a suitable temperature. However, it was found that the cooling of the membrane feed stream results in condensation of the stream. This renders the second parallel HMD configuration infeasible for the gas permeation membrane considered in

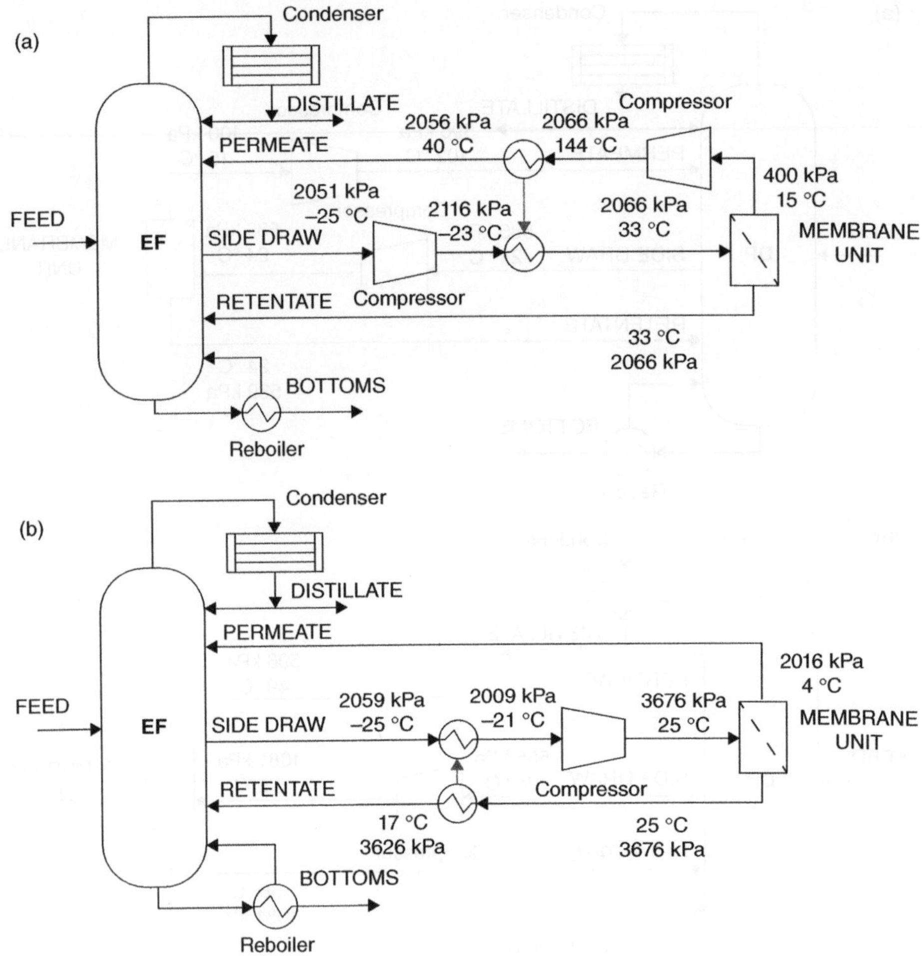

Figure 10.7 *Process flow diagrams of the two parallel HMD configurations used for preliminary techno-economic evaluation of EF retrofitting.*

this study. Results of preliminary techno-economic analysis of the first configuration are summarized in Table 10.7. It can be seen that PF retrofitting to HMD yields good utility cost savings with a much lower capital cost of retrofitting. Hence, MOO is carried out to determine the Pareto-optimal solutions for this retrofitting.

10.5 Formulation of Multi-objective–Optimization Problem

There are two major costs that play an important role in the feasibility of retrofitting a distillation column to HMD. One of them is the capital cost of the membrane module along with compressor, exchangers and associated installation, piping and labor costs (that is, cost of retrofitting). The other is the utility cost, which includes condenser and reboiler

Table 10.6 *Preliminary techno-economic evaluation of parallel HMD systems for EF retrofitting at membrane feed flowrates of 100, 350 and 600 kmol/h.*

	Design/Decision variables					
Case	Membrane feed flow (kmol/h)	Ethylene permeance (kmol/h-m^2-bar)[a]	Membrane area (m^2)	Pressure difference across membrane (bar)	Utility cost savings ($/yr)	Annualized capital cost of retrofitting ($/yr)
A.1	100	0.0015	1500	16.66	(65,936)	211,012
B.1					29,670	39,294
A.2	350	0.0015	1500	16.66	(34,437)	203,189
B.2					32,817	61,619
A.3	600	0.0015	1500	16.66	(4,286)	161,231
B.3					(100,731)	314,799

[a] Ethylene permeance in GPU is 12.3 (calculated by the method of Faiz and Li, 2012).

duty costs, fuel cost for compressor and utility cost for heaters/coolers. Clearly, both the capital and utility costs need to be minimized in order to obtain more benefit from the retrofitted HMD system. At the same time, production rates and quality of products from the retrofitted system should remain the same as before retrofitting so that they do not affect downstream columns/units in the plant.

In order to minimize utility costs, design variables of the membrane unit can be chosen to give better separation. For example, a larger membrane area will lead to more separation for a given feed flowrate and permeate pressure. This may lead to a reduction in condenser and reboiler duties of the associated column. While there is a decrease in utility cost of the column, there is a corresponding increase in the capital cost of the membrane. This results

Table 10.7 *Preliminary techno-economic evaluation of parallel HMD systems for PF retrofitting at membrane feed flowrates of 50, 200 and 300 kmol/h.*

	Design/Decision variables					
Case	Membrane feed flow (kmol/h)	Ethylene permeance (kmol/h-m^2-bar)[a]	Membrane area (m^2)	Pressure difference across membrane (bar)	Utility cost savings ($/yr)	Annualized capital cost of retrofitting ($/yr)
A.1	50	0.01	700	17.7	497,895	76,981
A.2	200	0.01	700	17.7	1,957,050	177,271
A.3	300	0.01	700	17.7	1,893,440	180,190

[a] Propylene permeance in GPU is 81.8 (calculated by the method of Faiz and Li, 2012).

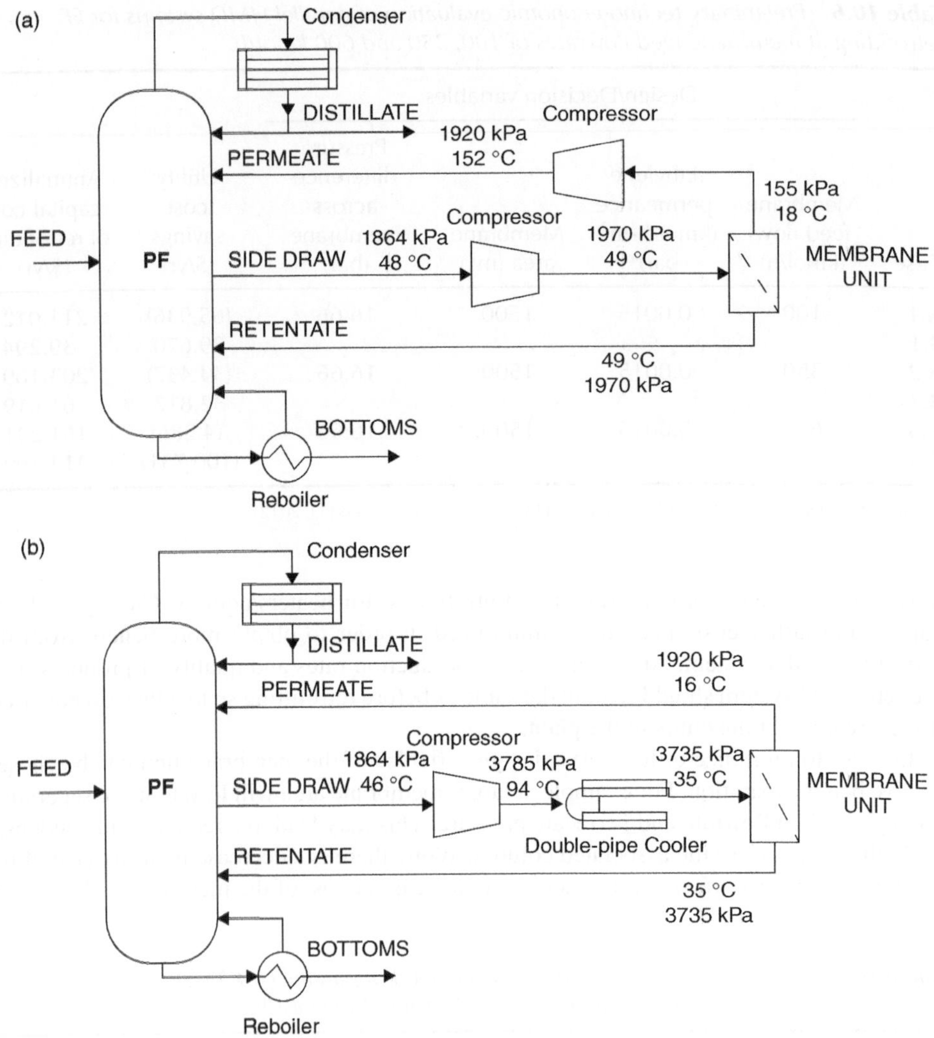

Figure 10.8 *Process flow diagrams of the two parallel HMD configurations used for preliminary techno-economic evaluation of PF retrofitting.*

in a conflict between the two cost objectives, which can be studied using MOO. The unit prices of utilities used in various equipment are shown in Table 10.B.2 (Appendix 10B). The compressor used in the HMD system is driven by an internal combustion engine which requires Fuel Oil No. 2 given in Table 10.B.2. Note that the utility used in the EF reboiler is propylene refrigerant leaving at −2 °C, which can be considered as utility credit (with unit price of \$5.4/GJ) and is multiplied by −1 in calculating the utility cost objective for the EF.

The MOO is performed for maximizing utility cost savings and minimizing capital cost (that is, cost of retrofitting) simultaneously for retrofitting EF (Case 1 assuming reboiler duty as credit), EF (Case 2 assuming reboiler duty as cost), and PF (Case 3), with a membrane unit. The equations for these objective functions are as follows [Equations (10.9) and (10.10)],

$$\text{Maximize Utility Cost Savings} = \text{OPEX}_{\text{hyb}} - \text{OPEX}_{\text{base}} \tag{10.9}$$

where OPEX is the sum of utility cost of reboiler, condenser, permeate compressor and heater/cooler present. In the base case, permeate compressor and heater/cooler are absent, and so utility cost for them is zero,

$$\begin{aligned}\text{Minimize Annualized Capital Cost, CAPEX} &= (\text{Cost of Membrane Unit/Membrane Life Expectancy}) \\ &+ (\text{Compressor Cost} + \text{Exchangers Cost})/\text{Equipment Life Expectancy}\end{aligned} \tag{10.10}$$

where exchangers include heater and cooler, if present.

Decision Variables: Important variables affecting the performance of the HMD system are considered as design/decision variables. As mentioned before, membrane area is an important factor determining the capital and utility costs, and is one decision variable. The permeances of slow- and fast-permeating components in every case can be varied [as per Equation (10.1)], and they affect HMD performance. Hence, permeance is chosen as the second decision variable. The change in permeate flowrate brought by the change in permeance affects the load on the following compressor, to recompress the permeate stream to the column pressure. More compression results in higher operating cost of the compressor. Therefore, there is trade-off between better separation and utility cost in employing a membrane with high permeance.

The third decision variable is the side draw rate from the column. The feed flowrate to the membrane determines the extent of separation affected by the membrane and its impact on the column duties. It also depends upon the amount of feed entering the column, and one has to be careful while choosing its value so that the hydrodynamic conditions in the column are not disturbed significantly. The fourth and last decision variable is the pressure difference across the membrane. With a higher pressure difference, permeate flowrate will increase, as given by Equation (10.1), which may lead to better separation. However, this will also increase the compressor duty, similar to the case of increasing membrane permeance, resulting in a trade-off in utility cost. Bounds on the permeances of components are determined by the values available in the literature (Faiz and Li, 2012). Bounds for side draw depend upon the vapor flowrate inside the column. Membrane areas are reasonably bounded for attaining positive net savings. Table 10.8 summarizes decision variables with their respective bounds.

Constraints: A rotary compressor has been considered for the permeate stream compression. It requires a minimum operating power requirement of 18 kW, which is given as a constraint in the given problem.

Optimizer: For MOO of the olefin/paraffin separation process, the elitist non-dominated sorting algorithm (NSGA-II) implemented in MS Excel using binary coding was employed (Sharma *et al.*, 2012). This optimization program is referred to as EMOO for Excel-based MOO. The optimizer in Excel generates trial solutions, each of which is sent to Aspen

Table 10.8 *Decision variables for MOO of retrofitting EF and PF with membrane separation.*

Column	Ethylene fractionator		Propylene fractionator	
Lower/Upper bound (LB/UB)	LB	UB	LB	UB
Membrane feed flowrate (kgmol/h)	200	900	50	300
Membrane area (m^2)	100	1700	100	1500
Ethylene (for EF) and Propylene (for PF) permeance (kgmol/h-m^2-bar)[a]	0.00032 (2.62)	0.032 (262)	0.006 (49.1)	0.015 (123)
Pressure difference across membrane (kPa)	1566	1966	1570	1870

[a] Ethylene permeance values are calculated using propylene/ethylene selectivity given in the assumptions in Section 3.2. Permenace value in GPU is given in parentheses.

HYSYS through the Excel-HYSYS interface for rigorous simulation of the olefin/paraffin separation process including membrane unit. HYSYS simulation provides results, through the same interface, for computing objectives to the optimizer within Excel. These steps of generating a trial solution (by the optimizer) and process simulation (in HYSYS) are repeated many times for the specified maximum number of generations. More details on NSGA-II and its implementation in Excel can be found in Sharma *et al.* (2012). It has been successfully used in conjunction with process simulation in HYSYS, for example, by Lee and Rangaiah (2009) and Al-Mayyahi *et al.* (2013). The optimization run was carried out up to 200 generations to find the Pareto-optimal front accurately. Other algorithm parameters used in the optimizer are: two-point crossover with probability = 0.8, bit-wise mutation with probability = 0.05, tournament selection, random seed = 0.5 and population size = 100.

10.6 Results and Discussion

10.6.1 Case 1: HMD System for EF (Assuming Credit for Reboiler Duty)

The EF produces 99.9 mol% ethylene as the top product and 99.5 mol% ethane as the bottom product. Decision variables for the MOO of HMD system for EF retrofitting are given in Table 10.8. The side draw is taken from stage 91 and retentate and permeate streams are returned to stages 101 and 46, respectively. The condenser uses propylene at −35 °C as utility, and the reboiler uses propylene at −2 °C; reboiler duty is considered as utility credit. The non-dominated solutions obtained by EMOO after 60, 80 and 100 generations for maximizing utility cost savings and minimizing associated capital cost of the membrane unit (Figure 10.9a) show that convergence of solutions is already achieved at 60 generations. Utility cost savings range from \$ −0.021 Million/yr to \$0.022 Million/yr with the corresponding annualized capital cost ranging from \$0.026 Million/yr to \$0.085 Million/yr. It is observed that utility savings are minimal and the annualized capital cost is more than utility savings. One main reason for this is the consideration of utility credit for reboiler duty; this is further discussed later.

Optimal values of decision variables corresponding to the Pareto-optimal front after 100 generations are shown in Figure10.9b–e. Optimal values of membrane feed flowrate

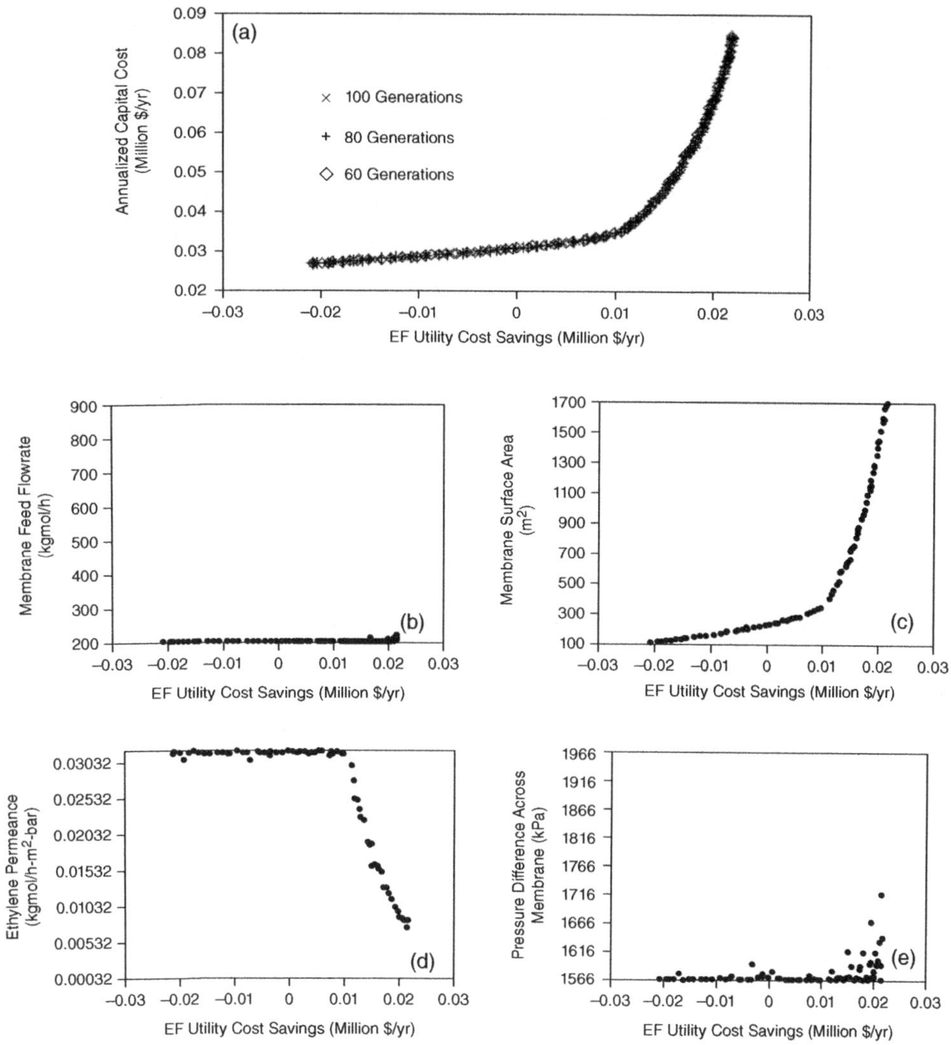

Figure 10.9 *Non-dominated solutions for maximization of utility cost savings and minimization of capital cost for retrofitting EF to an HMD system, considering reboiler duty as utility credit (plot a); optimal values of decision variables corresponding to the Pareto-optimal front are in plots b–e.*

(Figure 10.9b) are found near the lower bound of 200 kmol/h. Optimal membrane area increases with a gentle gradient initially until utility savings of $0.01 Million/yr, after which optimal membrane area increases with a steeper gradient till its upper bound (Figure 10.9c). This trend is also observed in the Pareto-optimal front (Figure 10.9a). Optimal membrane permeance values are scattered near its upper bound, and decrease linearly after utility savings of $0.01 Million/yr (Figure 10.9d). Optimal pressure difference across membrane values are generally scattered near its lower bound (Figure 10.9e). Hence, in this case of retrofitting EF, it can be concluded that both membrane area and membrane permeance contribute to the conflict of objectives in the Pareto-optimal front.

The change in gradient of the optimal values of membrane area (Figure 10.9c), which led to the change in gradient of the Pareto-optimal front in Figure 10.9a, can be explained by the decrease in membrane permeance. The decrease in membrane permeance results in less permeate flow across the membrane. Hence, there is a need for a corresponding increase in membrane area to increase the amount of permeate flow across the membrane to achieve the required utility savings. The decrease in permeance can be explained by the selectivity correlation [Equation (10.8)], whereby low permeance results in higher selectivity which is required for better separation of components.

Ploegmakers *et al.* (1998) claimed 16% savings on total annualized cost (TAC) from retrofitting an EF with a membrane unit in series. They used cooling water in the reboiler as utility cost and a membrane selectivity of 300, which is beyond the upper bound of the selectivity/permeance ratio considered in this study. In the current study, assuming reboiler duty as utility credit, even though there is some positive utility cost saving towards the end of the Pareto-optimal front (Figure 10.9a), the corresponding capital cost is high and so the economics of EF retrofitting to an HMD system are not attractive.

10.6.2 Case 2: HMD System for EF (Assuming Reboiler Duty as Cost)

Another optimization run was carried out by considering the reboiler duty of EF as a utility cost with cooling water as the utility, as in Ploegmakers *et al.* (1998). The non-dominated solutions obtained by EMOO after 60, 80 and 100 generations for maximizing utility cost savings and minimizing associated capital cost of the membrane unit are given in Figure 10.10a; 60 generations are enough for convergence in this case also. Optimal values of decision variables corresponding to the Pareto-optimal front are shown in Figure10.10b–d. It can be seen from Figure 10.10a that utility cost savings increase from 0.04 to 0.32 Million \$/yr while the annualized capital cost increases from 0.03 to 0.44 Million \$/yr. Since the capital cost increases non-linearly with respect to utility cost savings, the most suitable optimal solution for retrofitting is near the center of the Pareto-optimal front; it is with annualized capital cost of 0.093 Million \$/yr and utility cost savings of 0.198 Million \$/yr giving net savings of 0.105 Million \$/year.

For this chosen optimal solution, membrane feed flowrate is 372 kmol/h, membrane area is 703 m^2, membrane permeance is 0.0108 kgmol/m^2-h-bar, and pressure difference across membrane is 1964 kPa. The resulting HMD system provides net savings (Equation 10.B.1) of around 2% of utility cost in the base case, which is relatively low compared with the value of 16% obtained in Ploegmakers *et al.* (2013). The large difference in net savings obtained is due to differences in the selectivity values used. Ploegmakers *et al.* (2013) assumed a membrane selectivity range of 3–300 in their study of EF retrofitting, whereas a realistic range of 18–47, following the permeance/selectivity upper bound, is used in the current study. Payback period (PBP) for recovering capital cost of retrofitting can be calculated by application of Equation (10.11):

$$PBP = \frac{CAPEX_{hyb} \times 10}{OPEX_{bas} - OPEX_{hyb}} \tag{10.11}$$

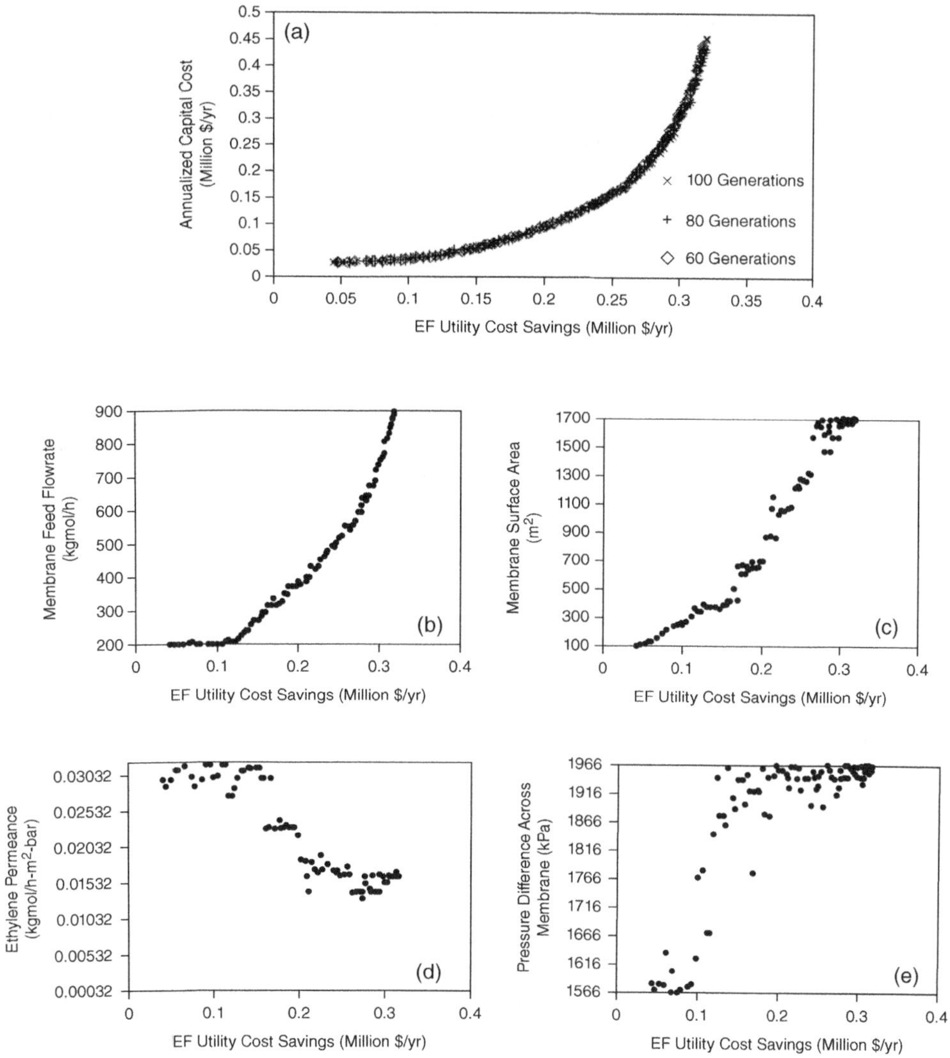

Figure 10.10 *Non-dominated solutions for maximizing utility cost savings and minimizing capital cost for retrofitting EF to an HMD system, considering reboiler duty as cost (plot a); optimal values of decision variables corresponding to the Pareto-optimal front are in plots b–e.*

Here, annualized capital cost given by Equation 10.B.2 is multiplied by 10 years to find the total capital cost for the plant life. Payback period for the chosen optimal solution is 4.7 years (= 0.093 × 10/0.198), which may be acceptable.

The optimal value of membrane feed flowrate (Figure 10.10b) is initially near the lower bound before increasing non-linearly for increasing utility cost savings. The optimal values of membrane area and membrane permeance (Figure 10.10c and 10.10d) follow the trend observed in Case 1. The optimal values of pressure difference across the membrane

(Figure 10.10e) are scattered around the lower bound initially before jumping to the upper bound. Hence, it can be concluded that all the decision variables affect the Pareto-optimal front to a certain extent.

The difference in trend for membrane feed flowrate and pressure difference across the membrane between Cases 1 and 2 can be accounted for by the calculation method for utility cost savings. Case 1 considers the reboiler duty to be a credit. Therefore, better separation achieved by the membrane will result in lesser reboiler duty and, consequently, lower utility cost savings, which is undesirable. In Case 2, reboiler duty is considered as a cost. Thus the system requires better separation by the membrane unit to increase utility cost savings. By increasing membrane feed flowrate and pressure difference across the membrane, the membrane unit would be able to separate larger amounts of components, leading to higher utility cost savings.

10.6.3 Case 3: HMD System for PF

The Propylene fractionator(PF) produces propylene (99 mol% purity) in the overheads and propane (95 mol% purity) in the bottoms. Decision variables for the MOO of PF retrofitting to an HMD system are as per Table 10.8. The side draw for feeding the membrane is taken from stage 102, which is the same as the feed stage for PF, and retentate and permeate streams are fed at stages 146 and 31, respectively. The condenser uses cooling water as utility whereas the reboiler uses low-pressure steam. Non-dominated solutions obtained by EMOO at 60, 80 and 100 generations for maximizing utility cost savings and minimizing associated capital cost of the membrane unit, show that the Pareto-optimal front remains the same with no significant changes after 60 generations (Figure 10.11a). Hence, 60 generations are sufficient to find the Pareto-optimal front in this case as well. The Pareto-optimal front is smooth, continuous and nearly linear except for a spike at the right end. The utility cost savings increase from \$0.157 to \$4.63 Million/yr as the capital cost of retrofitting increases from \$0.017 to \$0.78 Million/yr (Figure 10.11a).

The extreme point of the Pareto-optimal point before the spike (\$4.55 Million/yr utility cost savings at \$0.462 Million/yr capital cost) is the best choice for retrofitting the PF to an HMD system. This optimal solution corresponds to a membrane feed flowrate of 300 kmol/h, membrane area of 1497 m^2, membrane permeance of 0.0149 kgmol/m^2-h-bar, and pressure difference across the membrane of 1560 kPa. It gives a net savings of \$4.16 Million/yr (that is, 67% savings on the current utility cost), which is very attractive with a payback period of just about one year.

Optimal values of decision variables corresponding to the Pareto-optimal front after 100 generations are shown in Figure 10.11b–d. Membrane feed flowrate is scattered near the upper bound with an increasing trend (Figure 10.11b). This is expected as a higher membrane feed rate would mean separation of larger amount of components, leading to greater separation and higher utility savings. Optimal value of the membrane area (Figure 10.11c) increases linearly with utility cost savings. The increase in membrane area results in associated capital cost increase but also generates comparatively higher utility cost savings. The optimal value of propylene permeance (Figure 10.11d), which is the controlling factor of permeance of all components and selectivity of the membrane, is near its upper bound of 0.015 kgmol/m^2-h-bar.

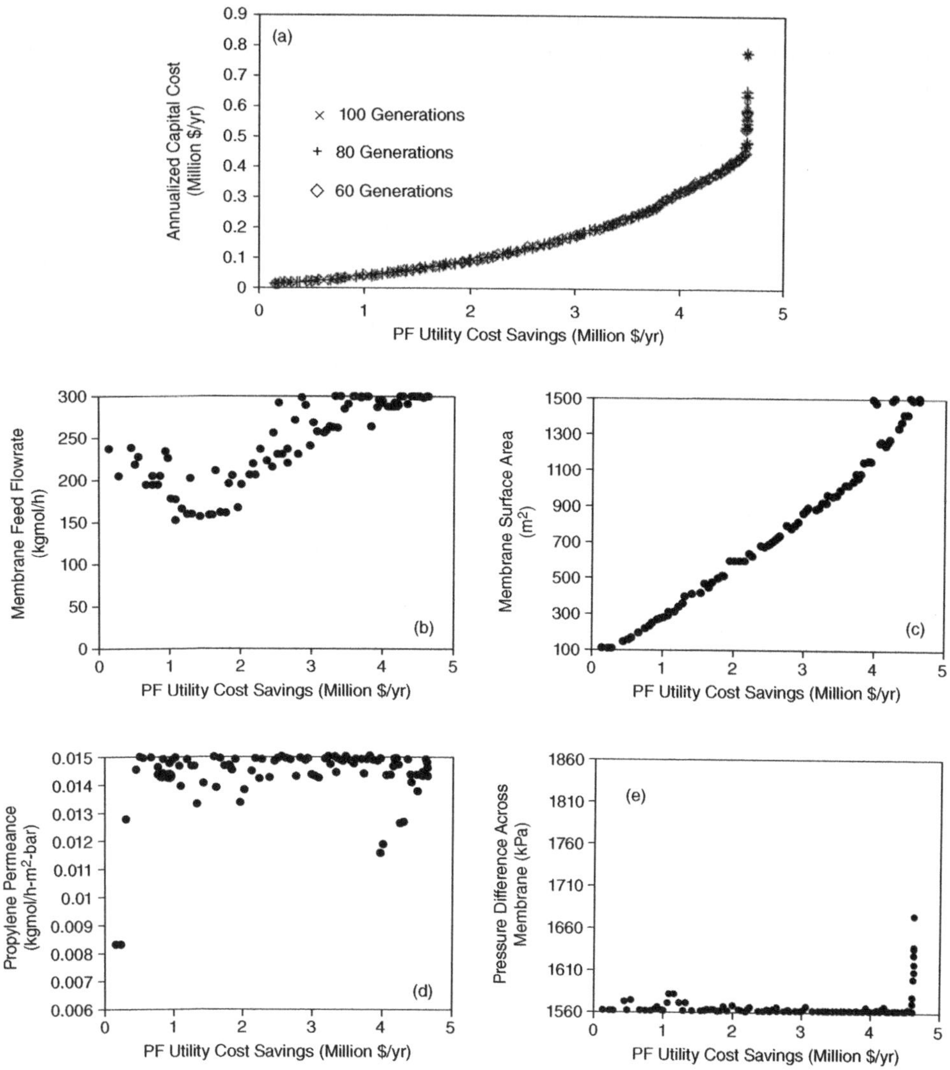

Figure 10.11 *Non-dominated solutions for maximizing utility cost savings and minimizing capital cost for retrofitting PF to an HMD system (plot a); optimal values of decision variables corresponding to the Pareto-optimal front are in plots b–e.*

The optimal values of pressure difference across the membrane are scattered near the lower bound with a spike at the end (Figure 10.11e). The spike is due to the rest of the decision variables reaching their respective upper limit. Hence, to further increase the utility cost savings, the pressure difference across the membrane has to be increased to allow more components to be separated by the membrane. However, it can be observed that varying the pressure difference across the membrane has a weak effect on utility cost savings compared with other decision variables such as membrane area. Increasing the pressure difference across the membrane contributes to a more than proportionate increase in capital cost as seen in Figure 10.11a because of the need for a larger compressor and more energy required for compression.

10.7 Conclusions

The olefins/paraffins separation of an industrial ethylene plant was simulated using Aspen HYSYS, and four distillation columns in this plant were selected for analysis of their retrofitting to HMD systems in parallel configuration. The elitist non-dominated sorting genetic algorithm implemented in Excel and Excel-HYSYS interface were employed for simultaneously maximizing the utility cost savings and minimizing the associated capital cost of retrofitting. Preliminary techno-economic analysis shows that retrofitting DE and DP with a membrane unit are not economical due to the use of olefin-selective membranes. On the other hand, retrofitting EF and PF are promising, and so they were investigated through MOO. PF retrofitting with a membrane unit gives 67% utility savings, which is very attractive. EF retrofitting showed negative net savings; this can be attributed to the utility credit generated by its reboiler, which is deducted from the utility cost in both base and HMD cases. When EF reboiler duty is considered to be a cost, the HMD system yielded net savings of 2%, which is relatively small; however, EF retrofitting may still be attractive with a payback period of 4.7 years. Comparing the HMD systems for EF and PF, PF retrofitting is able to achieve higher net savings compared with EF retrofitting, partly due to the nature of the membrane having a higher selectivity for propylene/propane compared with ethylene/ethane.

Appendix 10A: Membrane Model Validation

The membrane model in the implemented ACM module was first validated with the experimental results of Pan (1986). In the experiments, the effect of varying stage cut (that is, the fraction of feed that permeates through the membrane) on mole fraction of H_2 in the permeate was studied using asymmetric hollow fiber membranes in co- and counter-current configurations. Parameters of the membrane module can be found in Table 10.A.1; stage cut varies from 0.34 to 0.55 in the counter-current configuration and from 0.35 to 0.6 in the

Table 10.A.1 *Membrane and feed properties for hydrogen recovery from simulated purge gas of ammonia plant (Pan, 1996).*

Feed pressure (kPa)		6964
Feed temperature (°C)		25
Feed composition (mole fraction)		H_2 = 0.5178; N_2 = 0.2469; CH_4 = 0.1957 and NH_3 = 0.0396
H_2 Permeance (mol/m^2-s-Pa)		2.84×10^{-8}
Selectivity	H_2/CH_4	100
	N_2/CH_4	1.04
	NH_3/CH_4	2.71
Permeate pressure (kPa)		1123

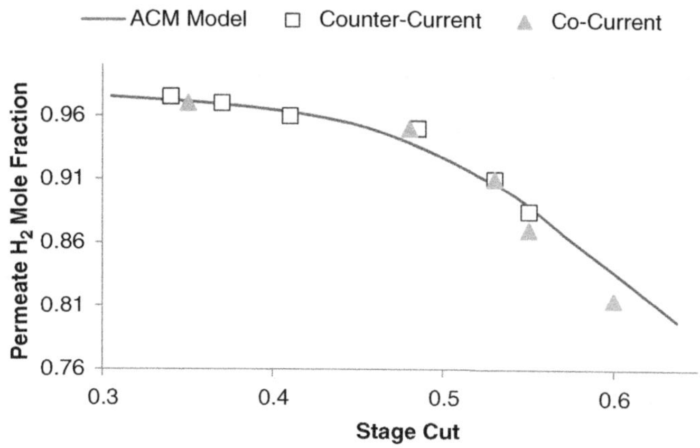

Figure 10.A.1 *Comparison of permeate H_2 mole fraction between ACM module predictions and experimental data for counter- and co-current configuration (Pan, 1986) at various stage cuts.*

co-current configuration. Experimental data shown by points in Figure 10.A.1 indicate that results for the two configurations are similar, as expected from the nature of asymmetric hollow-fiber membranes (Pan, 1986). Simulations were carried out using the developed ACM model and parameters in Table 10.A.1, and the predictions are shown by a line in Figure 10.A.1. It can be observed that the ACM model produces a very good fit of the experimental results over a large range of stage cut. The maximum deviation of ACM model predictions from the experimental values is 3.5%, which occurs when the stage cut is 0.6. This deviation may be due the assumption of negligible back diffusion made in the ACM model; but back diffusion may occur in the porous supporting layer at extreme operating conditions (Pan, 1986).

The developed membrane model/ACM module was also validated against the results of Davis (2002). In his study, a mathematical model of a hollow-fiber membrane was developed in Aspen HYSYS for air separation. Parameters used in the simulation of this separation are given in Table 10.A.2. As can be seen in Table 10.A.3, the ACM model predictions are very close to the simulation results of Davis (2002), with a maximum difference of 0.41% in permeate O_2 concentration.

Table 10.A.2 *Membrane and feed properties for air separation.*

Feed pressure (kPa)	1034
Feed temperature (°C)	21
Feed composition (mole fraction)	$O_2 = 0.21$ and $N_2 = 0.79$
O_2 Permeance (kmol/m^2-h-kPa)	7.14×10^{-7}
N_2 Permeance (kmol/m^2-h-kPa)	1.37×10^{-7}
Membrane area (m^2)	50 000
Permeate pressure (kPa)	103.4

Table 10.A.3 *Comparison of ACM model predictions with the simulation results of Davis (2002).*

	Davis (2002)	ACM Model	Difference (%)
Permeate flowrate (kmol/h)	10.68	10.65	0.32
Retentate flowrate (kmol/h)	89.32	89.35	0.04
Permeate O_2 molar composition	0.4985	0.4965	0.41
Retentate O_2 molar composition	0.1755	0.1759	0.21

Table 10.A.4 *Permeate gas conditions for separation of O_2 from N_2, obtained from Davis (2002).*

Feed pressure (kPa)	1034
Feed temperature (°C)	21
Permeate composition (mole fraction)	$O_2 = 0.496$ and $N_2 = 0.504$
Permeate pressure (kPa)	103.4

Table 10.A.5 *Permeate gas conditions for separation of O_2 from N_2, obtained from the PF HMD system.*

Feed pressure (kPa)	1970
Feed temperature (°C)	49.34
Permeate composition (mole fraction)	Propylene = 0.9877, Propane = 0.0122 and Isobutane = 0.0001
Permeate pressure (kPa)	155.1

As mentioned earlier, the Joule–Thompson effect is the result of expansion of the permeate gas from higher feed pressure to lower permeate pressure (Ahmad *et al.*, 2013). Hence, the Joule–Thompson effect predicted by the ACM module was validated using the simulation of adiabatic throttling of a gas stream through a valve in Aspen HYSYS v.8.4. The gas used in this simulation is the permeate stream from the air separation of Davis (2002); the permeate gas conditions are given in Table 10.A.4. The developed ACM module predicted a permeate gas temperature after expansion of 18.62 °C, which is very close to the Hysys result of 18.88 °C. An additional simulation of adiabatic throttling of a gas stream was conducted using the permeate stream obtained from the PF HMD system; the permeate gas conditions are given in Table 10.A.5. The developed ACM model predicted a permeate gas temperature after expansion of 18.03 °C, which is very close to the Hysys result of 17.96 °C.

Appendix 10B: Costing of HMD System

For the techno-economic evaluation of retrofitting a distillation column to an HMD system, one important indicator is the net savings (NS %). It is the percentage of difference in the

Table 10.B.1 *Values of coefficients in Equations (10.B.4) and (10.B.5) for compressor and drive (Turton et al., 2002).*

Equipment	K_1	K_2	K_3	Unit	Min	Max	F_{bm}
Compressor (Rotary, Carbon steel)	5.0355	−1.8002	0.8253	kW	18	950	2.4
Drive (Internal combustion)	2.7635	0.8574	−0.0098	kW	10	10000	2

Note: Electric or steam-driven drives are used conventionally in industries for their reliability. However, internal combustion drive is chosen in the present study due to the availability of its cost correlation over a wider operating range.

capital and operating costs of the base case and of the HMD case, to the operating cost of the base case. It can be calculated using Equation (10.B.1).

$$NS\,\% = \frac{OPEX_{bas} - (CAPEX + OPEX)_{hyb}}{OPEX_{bas}} 100 \tag{10.B.1}$$

Since retrofitting is considered in this study, $CAPEX_{bas}$ is set to 0 $/yr and not included in the above equation. $CAPEX_{hyb}$ is given by Equation (10.B.2),

$$CAPEX_{hyb} = \frac{(C_{comp} + C_{drive} + C_{cool} + C_{heat} + C_{ex})}{10} + \frac{100A_m}{3} \tag{10.B.2}$$

where A_m is the surface area (m^2) of the membrane and C ($) is the cost of an equipment like compressor, drive, cooler, heater and heat exchangers as indicated by the subscript. The life expectancy of this equipment is assumed to be 10 years[2], and that of the membrane unit to be 3 years[3].

The OPEX is calculated using utility requirements of the equipment and utility prices based on an annual operating time of 8760 hours [Equation (10.B.3)].

$$OPEX_{hyb} = OPEX_{cond} + OPEX_{reb} + OPEX_{comp} + OPEX_{ex} + OPEX_{heat} + OPEX_{cool} \tag{10.B.3}$$

Turton *et al.* (2009) provide the following correlation [Equation (10.B.4)] for estimating the purchase cost of equipment (PCE) for a compressor and its drive.

$$log\left(PCE_{\frac{comp}{drive}}\right) = K_1 + K_2 log\,(S) + K_3\,[log\,(S)]^2 \tag{10.B.4}$$

Here, S (kW) is the power input required by the compressor/drive, and K_1, K_2 and K_3 are coefficients, whose values are summarized in Table 10.B.1. The total module cost is given by Equation (10.B.5),

$$C = (1 + 0.15 + 0.03) \times F_{bm} \times (PCE)\left(\frac{650}{397}\right) \tag{10.B.5}$$

where 15% is for contingency and 3% for contractor's fees. F_{bm} accounts for equipment erection, piping, instrumentation, electrical, buildings, and process design and engineering.

[2] Process equipment is used for more than 10 years. However, a conservative 10 years is used in this study.

[3] Ethylene plants normally operate for 4–5 years without turnaround. In such a case, the membrane unit needs to be isolated and membranes replaced on-stream.

The Chemical Engineering Plant Cost Index (CEPCI) is taken as 650. Its value was 397 in the period May to September 2001 when the PCE data were obtained (Turton *et al.*, 2009).

For calculating the sizes of coolers, heaters, and heat exchangers, logarithmic mean temperature difference (LMTD) is computed with cooling water entering at 30 °C and leaving at 40 °C, and the process stream leaving the cooler at 35 °C. the value of U is assumed to be 350 W/m^2.K for a cooler with hot fluid as light oil and cold fluid as water, 100 W/m^2.K for heat exchangers for heat integration with the exchange medium as high-pressure gases, and 30 W/m^2.K for a heater with hot fluid as steam and cold fluid as atmospheric pressure gases (from http://www.engineeringpage.com/technology/thermal/transfer.html, access date: 01-11-2014).

Then, area of the cooler/heater/heat exchangers (A_c) is obtained from Equation (10.B.6):

$$Q = UA_cLMTD \tag{10.B.6}$$

Assuming a double-pipe heat exchanger (for heat exchange area in the range 2 to 200 ft^2), PCE is calculated using Equation (10.B.7) (Seider *et al.*, 2010):

$$PCE_{cool} = \exp(7.1460 + 0.16 \times \ln(A_c)) \tag{10.B.7}$$

F_p is given by Equation (10.B.8) (Seider *et al.*, 2010):

$$F_p = 0.8510 + 0.1292 \times \left(\frac{P}{600}\right) + 0.0198 \times \left(\frac{P}{600}\right)^2 \tag{10.B.8}$$

Material factor F_m is 2 for an outer pipe of carbon steel and an inner pipe of stainless steel. Since the CEPCI value is 500 for the PCE data in Seider *el al.*, the total module cost of a cooler is given by Equation (10.B.9):

$$C_{ex} = F_mF_p(PCE_{ex})\left(\frac{650}{500}\right) \tag{10.B.9}$$

Utility data and prices used in this study are given in Table 10.B.2.

Table 10.B.2 *Utility data and prices used in the present study.*

S. No.	Utility	Temperature (°C)	Unit price	Reference
1.	Refrigerant (Ethylene)	−101	21 \$/GJ	Values are interpolated using the cost data in Seider *et al.* (2013)
2.	Refrigerant (Propylene)	−35	10.6 \$/GJ	
3.	Refrigerant (Propylene)	−20	8.2 \$/GJ	
4.	Refrigerant (Propylene)	−2	5.4 \$/GJ	
5.	Cold energy	5	4.4 \$/GJ	
6.	Low-pressure steam (1.03 barg)	120	29.3 \$/t	Turton *et al.* (2002)
7.	Cooling water	30–40	0.0148 \$/m^3	
8.	Fuel Oil No. 2		14.2 \$/GJ	

Exercises

10.1 Simulate one or more distillation columns, whose data are given in Table 10.2, using a process simulator such as Aspen HYSYS. Using the cost data in Appendix 10B, calculate the operating cost of each of these columns.

10.2 Choose one of the optimal solutions for Case 1 in Figure 10.9, and simulate the HMD system for these conditions using a process simulator. Note that this requires a membrane module (described in Section 10.3.1); if it is not available, simulate the membrane unit using the simple component splitter, readily available in a process simulator. Estimate objective function values, and compare them with those in Figure 10.9.

10.3 Simulate the chosen optimal solution for PF in Case 3 (Section 10.6.3) using a process simulator; if a membrane module (described in Section 10.3.1) is not available, simulate the membrane unit using the simple component splitter. Estimate objective function values, and compare them with those in Figure 10.9.

10.4 Investigate optimization of solutions in the previous two exercises for one or more objectives; decision variables and their bounds can be changed suitably. This optimization requires a suitable optimizer available in the simulator or an external program (with interface to the simulator).

References

Ahmad, F., Lau, K.K., Shariff, A.M. and Yeong, Y.F. (2013) Temperature and pressure dependence of membrane permeance and its effect on process economics of hollow fiber gas separation system. *Journal of Membrane Science*, **430**, 44–55.

Al-Mayyahi, M.A., Hoadley, A.F.A. and Rangaiah, G.P. (2013) CO_2 Emissions targeting for petroleum refinery optimization, in *Multi-Objective Optimization in Chemical Engineering: Developments and Applications* (eds G.P. Rangaiah and A. Bonilla-Petriciolet), John Wiley, Chichester, West Sussex.

Azhin, M., Kaghazchi, T. and Rahmani, M. (2008) A review on olefin/paraffin separation using reversible chemical complexation technology. *Journal of Industrial and Engineering Chemistry*, **14**, 622–638.

Benali, M. and Aydin, B. (2010) Ethane/ethylene and propane/propylene separation in hybrid membrane distillation systems: Optimization and economic analysis. *Separation and Purification Technology*, **73** (3), 377–390.

Bernardo, P. and Drioli, E. (2010) Membrane gas separation progresses for process intensification strategy in the petrochemical industry. *Petroleum Chemistry*, **50** (4), 271–282.

Burns, R.L. and Koros, W.J. (2003) Defining the challenges for C_3H_6/C_3H_8 separation using polymeric membranes. *Journal of Membrane Science*, **211**, 299–309.

Caballero, J.A., Grossmann, I.E., Keyvani, M. and Lenz, E.S. (2009) Design of hybrid distillation-vapor membrane separation systems. *Industrial & Engineering Chemistry Research*, **48** (20), 9151–9162.

Calamur, N. and Carrera, M. (2005) *Propylene, in Kirk-Othmer Encyclopedia of Chemical Technology* (eds A. Seidel and M. Bickford), John Wiley, Hoboken, NJ.

Davis, R.A. (2002) Simple gas permeation and pervaporation membrane unit operation models for process simulators. *Chemical Engineering & Technology*, **25** (7), 717–722.

Davis, J.C., Valus, R.J., Eshraghi, R. and Velikoff, A.E. (1993) Facilitated transport membrane hybrid systems for olefin purification. *Separation Science and Technology*, **28**, 463–476.

Faiz, R. and Li, K. (2012) Polymeric membranes for light olefin/paraffin separation. *Desalination*, **287**, 82–97.

Gottschlich, D.E. and Roberts, D.L. (1990) SRI Project 6519 for DOE Contract No. DE-AC07-76ID01570, Energy minimization of separation process using conventional/membrane hybrid systems.

Lee, E.S.Q. and Rangaiah, G.P. (2009) Optimization of recovery processes for multiple economic and environmental objectives. *Industrial & Engineering Chemistry Research*, **48** (16), 7662–7681.

Lie, J.A., Vassbotn, T., Hägg, M.B. *et al.* (2007) Optimization of a membrane process for CO_2 capture in the steelmaking industry. *International Journal of Greenhouse Gas Control*, **1** (3), 309–317.

Masuduzzaman and Rangaiah, G.P. (2009) Multi-objective optimization applications in chemical engineering, in *Multi-Objective Optimization: Techniques and Applications in Chemical Engineering*, (ed. G.P. Rangaiah), World Scientific, Singapore.

Motelica, A., Bruinsma, O.S.L., Kreiter, R. *et al.* (2012) Membrane retrofit option for paraffin/olefin separation - a technoeconomic evaluation. *Industrial & Engineering Chemistry Research*, **51** (19), 6977–6986.

Okamoto, K., Noborio, K., Hao, J. *et al.* (1997) Permeation and separation properties of polyimide membranes to 1, 3-butadiene and n-butane. *Journal of Membrane Science*, **134** (2), 171–179.

Pan, C.Y. (1986) Gas separation by high-flux, asymmetric hollow-fiber membrane. *AIChE Journal*, **32** (12), 2020–2027.

Pettersen, T., Argo, A., Noble, R.D. and Kovalb, C.A. (1996) Design of combined membrane and distillation processes. *Separations Technology*, **6** (3), 175–187.

Ploegmakers, J., Jelsma, A.R.T., Ham, A.G.J.v.d. and Nijmeijer, K. (2013) Economic evaluation of membrane potential for ethylene/ethane separation in a retrofitted hybrid membrane-distillation plant using unisim design. *Industrial & Engineering Chemistry Research*, **52** (19), 6524–6539.

Pressly, T.G. and Ng, K.M. (1998) A break-even analysis of distillation-membrane hybrids. *AIChE Journal*, **44**, 93–105.

Rangaiah, G.P., Sharma, S. and Sreepathi, B.K. (2015) Multi-objective optimization for the design and operation of energy efficient chemical processes and power generation. *Current Opinion in Chemical Engineering*, **10**, 49–62.

Rungta, M., Zhang, C., Koros, W.J. and Xu, L. (2013) Membrane-based ethylene/ethane separation: The upper bound and beyond. *AIChE Journal*, **59**, 3475–3489.

Schmeling, N., Konietzny, R., Sieffert, D. *et al.* (2010) Functionalized copolymide membranes for the separation of gaseous and liquid mixtures. *Beilstein Journal of Organic Chemistry*, **6**, 789–800.

Seider, W.D., Seader, J.D., Lewin, D.R. and Widagdo, S. (2010) *Product and Process Design Principles – Synthesis, Analysis and Evaluation*. 3rd edn, John Wiley, Hoboken, NJ.

Sharma, S. and Rangaiah, G.P. (2013) Multi-objective optimization applications in chemical engineering, in *Multi-Objective Optimization in Chemical Engineering: Developments and Applications* (eds G.P. Rangaiah and A. Bonilla-Petriciolet), John Wiley, Chichester.

Sharma, S., Rangaiah, G.P. and Cheah, K.S. (2012) Multi-objective optimization using MS Excel with an application to design of a falling-film evaporator system. *Food and Bioproducts Processing,* **90**, 123–134.

Turton, R., Bailie, R.C., Whiting, W.B. and Shaeiwitz, J.A. (2009) *Analysis, Synthesis and Design of Chemical Processes*, 3rd edn, Pearson Education, Inc., Upper Saddle River, NJ.

Vu, D.Q., Koros, W.J. and Miller, S.J. (2002) High pressure CO_2/CH_4 separation using carbon molecular sieve hollow fiber membranes. *Industrial & Engineering Chemistry Research*, **41** (3), 367–380.

Xu, L., Rungta, M., Brayden, M.K. *et al.* (2012) Olefins-selective asymmetric carbon molecular sieve hollow fiber membranes for hybrid membrane-distillation processes for olefin/paraffin separations. *Journal of Membrane Science*, **423–424**, 314–323.

11

Retrofit of Vacuum Systems in Process Industries

C.C.S. Reddy[1] *and G.P. Rangaiah*[2]
[1]*Lead Process Design Engineer, Singapore Refining Company Private Limited, Singapore*
[2]*Department of Chemical & Biomolecular Engineering, National University of Singapore, Singapore*

11.1 Introduction

Vacuum systems are used in process industries, mainly to facilitate distillation, drying and evaporation at low temperatures to avoid product degradation and/or for better separation. They are also used in power generation. In petroleum refineries, vacuum distillation prevents coking and degradation of products, and also reduces energy requirements for separation. Evaporation at low temperature and under vacuum is employed in food processing to minimize product degradation. Process industries generally use rough vacuum from 1 mmHg to 760 mmHg. Steam jet ejectors (SJEs) alone or in combination with liquid ring vacuum pumps (LRVPs) are the most common equipment for vacuum generation in process industries. Dry vacuum pumps (DVPs) are also used in non-hazardous applications and those requiring very low capacity. There are several papers and books describing the merits and limitations of vacuum-generation methods (for example, Aliasso, 1999; Ryans and Bays, 2001; Bannwarth, 2005; Power, 2005; Ryans and Croll, 1981).

SJEs are widely used in process industries due to their high reliability and medium pressure (MP) steam availability. However, they are highly inefficient. Owing to high energy prices, process revamp projects are increasing in number. Revamping of vacuum distillation columns requires retrofitting of the overhead vacuum systems. Retrofit of a

Chemical Process Retrofitting and Revamping: Techniques and Applications, First Edition.
Edited by Gade Pandu Rangaiah.

surface condenser's vacuum system is essential to increase the power generation of the condensing steam turbines. In general, there is water shortage in many parts of the world. Vacuum-assisted desalination is one of the most economical methods to generate industrial water, and increasing its capacity requires revamp of its vacuum system. Examples of industrial vacuum system retrofit[1] can be found on the internet.[2]

There are a number of challenges in retrofitting industrial vacuum systems. The main challenge is to maximize re-use of the existing equipment, especially condensers, as this avoids the requirement of additional project space. Another challenge is to develop the retrofit design within the available supply of utilities [for example, MP steam, low-low pressure (LLP) steam and cooling/chilled water], electrical infrastructure at substations, and also spare power and instrument cables at the retrofit site. Another important challenge is to meet the operational reliability, equivalent to the original vacuum system. Common constraints for vacuum system retrofits are to limit the retrofit design within size limitations of vacuum-producing equipment and also within the existing plot size. Good understanding of various retrofit options and their trade-offs are essential for developing a profitable vacuum system revamp design.

The main objective of this chapter is to analyze various vacuum-generation methods for developing energy-efficient retrofit design of vacuum-generation systems. For this, benefits and constraints of vacuum-generation practices in the industry are discussed in Section 11.2. The subsequent section presents design principles and utility requirements for SJEs and LRVPs. Section 11.4 outlines methods for generation of chilled water, which may be required for improving efficiency of the vacuum system. Section 11.5 describes the design strategy for retrofitting an existing vacuum system followed by the optimization problem for minimizing the vacuum system operating cost. Two case studies of industrial vacuum-generation systems and retrofit options for reducing energy consumption and/or increasing their capacity are outlined in Sections 11.6 and 11.7. In case study 1, retrofit options (namely, use of chilled water in the pre-condenser and use of LRVP to replace the last stages of a multi-stage SJE system) for a refinery vacuum system are investigated and analyzed. Case study 2 is on the retrofit using DVP to reduce energy consumption of a condensing steam turbine's surface condenser. Selection and implementation of the retrofit for the two case studies are discussed in Sections 11.6 and 11.7. The chapter ends with conclusions in Section 11.8. Contents and case studies of this chapter are useful for a better understanding and selection of an energy-efficient revamp option for vacuum generation as well as in retrofit design of industrial vacuum-generation systems.

11.2 Vacuum-generation Methods

This section reviews various vacuum-generation methods applicable to process industries. There are many types of equipment that can be used for vacuum generation in industries.

---

[1] Note that retrofit/retrofitting and revamp/revamping are used interchangeably in this chapter; see Chapter 1 for discussion on retrofit, revamp and debottlenecking.

[2] Websites for vacuum system retrofit examples:

- http://www.everestblowers.com/technical-articles/vacuum-booster-for-chemical-process.pdf
- https://repository.tamu.edu/bitstream/handle/1969.1/94258/ESL-IE-82-04-143.pdf?sequence=1&isAllowed=y
- http://www.graham-mfg.com/usr/pdf/TechLibVacuum/28.PDF

Among them, SJEs and LRVPs are generally used in process industries. In addition to these, other industries use rotary vacuum pumps, which can be oil-lubricated (vane, piston or plunger type) or without oil lubrication (that is, dry vacuum pumps such as lobe roots, claws or screw type). Oil-lubricated rotary vacuum pumps are generally not suitable for process applications, mainly due to contamination of the product and disposal problems of contaminated oil. See Ryans and Bays (2001) for an excellent review of various vacuum-generation equipment capacities, operating ranges and efficiencies. The main types of vacuum technologies, applicable for the process industries, are discussed below.

Steam Jet Ejectors: They produce vacuum using expansion of motive steam. There can be one, two or more stages of SJEs in series or series–parallel combination depending on the level of vacuum required, utility optimization and operational flexibility for various plant loads. SJEs are usually installed with surface (non-contact type, using cooling water on tube side) or barometric (direct contact of cooling water and ejector outlet gas/vapour) condensers in between the stages (inter-condensers) and after the final stage (after-condenser). They may also use pre-condensers before the first-stage ejector to reduce suction gas by condensing volatile components in the inlet stream to the SJE. Series installation of multi-stage SJEs is used to reduce the operating cost. Series–parallel and parallel installations are mainly to cater for very high suction loads or varying operating loads. Turndown operation is often achieved by installing smaller SJEs in parallel, and by stopping some of them at low flow rates of suction gas.

SJEs are highly reliable, can be fabricated using different materials for very large capacities, and their installed cost is not high. Also, easy availability of steam in petroleum refineries and petrochemical plants makes them the natural choice. They generally operate well at a single design point but become unstable if the load (suction) gas flow rate decreases. SJEs are highly inefficient, at <10% (Ryans and Bays, 2001), mainly due to lack of moving parts to efficiently convert fluid velocity into pressure (Power, 2005). Multi-stage SJE systems with water-cooled surface condensers are generally used to reduce the steam and cooling-water consumption. For economic operation, almost all of the used steam (typically ~90% in the first-stage condenser and up to 98% in other-stage condensers) should be condensed and separated before it reaches the next stage/ejector (James *et al.*, 2008) to reduce suction load on the subsequent stages/ejectors. In general, condensation of used steam results in lot of energy wastage, and steam condensate needs to be treated as foul/waste water and requires treatment. In petroleum refineries, SJEs are commonly employed for crude vacuum distillation, distillation of visbreaker residue and diesel drying. Generally a 3–4-stage SJE system is used for these applications requiring a vacuum of 10–40 torr (mmHg).

Liquid Ring Vacuum Pumps: These use a liquid ring formed by a seal liquid to compress gases in a nearly isothermal process. Water is the most common seal liquid in petroleum refineries since it can be used, separated and re-used safely. LRVPs require electrical power for their operation, and their suction capacity is limited by the temperature of the seal liquid. Hence, they are generally used to replace the last one or two stages of the multi-stage SJE system. LRVPs have higher efficiency compared with SJEs but are not suitable for low vacuum (<33 torr) and also for very high suction flows.

Dry Vacuum Pumps (DVPs): These use the operating principles of rotary-lobe roots blowers, claw compressors or screw compressors, and hence require electrical power. They run dry and hence at very high temperature due to the heat of compression. Usually,

DVPs are installed with cooling water jackets to control pump temperature. They operate at higher efficiencies compared with SJEs and LRVPs (Ryans and Bays, 2001). DVPs are highly energy efficient and widely used in the steel, chemical and bio-fuel industries. However, they are seldom used in petroleum refining and petrochemical industries due to their limited capacity of 50–1400 actual cubic feet per minute, acfm (Ryans and Bays, 2001), safety concerns (such as potential explosion and may require flame arrestors in the suction and discharge pipes) and reliability concerns (Tunna, 2005). However, DVPs are very beneficial for a condensing steam turbine's vacuum system and other applications requiring low capacities and involving non-flammable/explosive gas.

SJE–LRVP Combination: For greater utility savings, LRVPs, which are more efficient than SJEs, are sometimes used in petroleum refineries to replace the last one or two stages of a multi-stage SJE system. Cooling water (at ~30 °C) can be used as the seal liquid for such a combination, and suction gas load (mainly non-condensable components) will be lower at the last stages. Combining larger multi-stage SJE systems with LRVPs will generally result in less capital expenditure and more utility cost savings. For non-hazardous applications, SJE–DVP combinations can be used to reduce operating costs. These combinations are promising for retrofitting existing SJE systems, which are analyzed later in this chapter.

The annual cost of operating a vacuum system is several times the equipment cost. So, a detailed review of the operating cost of a vacuum system is essential. For example, use of chilled water instead of cooling water can reduce steam/power consumption of the vacuum system. It also eliminates cavitation problems associated with LRVPs used at low suction pressures. Although chilled water is costlier than cooling water, it can be attractive if the electricity price is low or waste heat is available for chilled water production. Selection of SJE, LRVP, DVP, SJE–LRVP or SJE–DVP combination depends on many factors listed in Table 11.1.

11.3 Design Principles and Utility Requirements

This section presents useful design principles and utility requirement estimations for SJEs, LRVPs and DVPs.

11.3.1 Suction Load of Vacuum System

The suction flow rate or "load" to a vacuum system consists of non-condensable gases and condensable process vapour. Non-condensable gases may come from:

- Air leakage into the system due to vacuum operation (from flanged joints, valve stems, pump and agitator shaft seals, instrument connections, and so forth)
- Gases dissolved in the feed
- Instrument flush/purge gases
- By-product gases from chemical reactions in the process

Air leakage is almost unavoidable in vacuum systems. Its rate cannot be accurately predicted as it depends on the system design, installation and maintenance. Minimal-leakage systems require high maintenance cost and system downtime for repairs. On the other hand, very high design air leakage requires larger and expensive vacuum pumps.

Table 11.1 *Relative merits of SJE, LRVP and DVP.*

Factor	SJE	LRVP	DVP
Suction and discharge pressures	They are suitable for almost any suction pressure. Capital and operating costs increase with decrease in the suction pressure and also with increase in the discharge pressure	Suction pressure must be greater than vapour pressure of the fluid used for liquid ring. Power consumption increases with decrease in the suction pressure and with increase in the discharge pressure	Power consumption and capital cost increase with decrease in the suction pressure and also with increase in the discharge pressure
Suction flow rate	They can be designed for almost any suction flow rate	Limited to 22 000 acfm (Ryans and Bays, 2001)	Limited to 1400 acfm
Gas molecular weight and temperature	The higher the gas molecular weight is, the lower are the capital and operating costs. Higher gas temperatures require more capital and operating costs		
Main utilities	They require steam at ~10.5 barg pressure	Electricity for LRVP and closed-loop seal liquid pump	Electricity for blower/compressor
Turndown operation	Turndown operation is difficult. If turndown requirements are known at the design stage, smaller SJEs can be installed in parallel. Steam can be saved by stopping some of them for turndown operation	Turndown operation is possible within limits imposed by the manufacturer, by installing variable-frequency drive for motors (Skelton, 1998). Power can be saved at low loads. Parallel arrangement can also be used	

(continued)

Table 11.1 *(Continued)*

Factor	SJE	LRVP	DVP
Capital and operating costs	They incur high operating cost, and also require foul water stripper or wastewater disposal facility. Equipment cost of SJEs is less but require inter/after condensers and elevated structures	If suction pressure is high and suction flow is low, then LRVPs incur lower operating cost than do SJEs. They are more expensive than SJEs but do not require after-condensers. However, LRVP requires a relatively smaller seal water cooler	If suction pressure is high and suction flow is low, then DVP incurs a lower operating cost than do SJE and LRVP. But, DVPs are more expensive than SJEs and LRVPs (Ryans and Bays, 2001)
Foul/waste water treatment facilities	Steam used will be converted into foul water, which needs to be treated in foul/wastewater treatment facility	Foul water generation will be little as there is no steam condensation	They do not produce foul water
Supply and discharge temperatures of cooling water	Higher cooling water supply and return temperatures require larger ejector system, more steam and cooling water, and also generates more foul water	Higher cooling water supply and return temperatures increase the power requirement and also size of the LRVP	They require a small amount of cooling water in jackets, for pump temperature control. So, they are unaffected by cooling water temperatures
Space constraint	Require elevated platforms for surface condensers, (semi-) barometric drain pipes for condensate removal, and separators for separating foul water and oil. Non-barometric installation is possible by combining SJEs and LRVPs	Surface condensers and elevated structures are not required. Instead, they require closed-loop liquid coolers for reducing water consumption and gas–liquid–liquid separators.	They are compact and require less plot space
Reliability concerns	They are very reliable, do not have any moving parts and require very minimal maintenance	They have good reliability but require some maintenance due to the presence of moving parts and seals	They have reliability issues due to close clearances of moving parts and dry operation

Designers usually estimate air-leakage rates using values available in the literature and/or industrial experience. Air leakage is mainly estimated by using the system volume method (Milton, 1986; Coker, 2007; HEI, 2007; Sieder *et al.*, 2009; Couper *et al.*, 2012) or the system component method (Coker, 2007; Couper *et al.*, 2012).The latter method estimates air leakage by assigning air-leakage rates to each and every system component (that is, flanges, connections, valves, sight glasses, seals, stuffing boxes of agitators, pumps, safety valves and vacuum breakers), and does not use the system volume. It uses published average air-leakage values for these components (for example, those available in Coker, 2007; Couper *et al.*, 2012). Air leakage from all fittings is added to obtain an estimated air-leakage rate. This value is often multiplied by 2 or 3, to provide a safety margin (Coker, 2007). On the other hand, the system volume method considers the volume of the main vessels (for example, distillation column with reboiler and condenser, and surface condenser of a condensing turbine) on which a vacuum is applied.

Air-leakage rate based on the system volume method is given by Sieder *et al.* (2009), as shown in Equation (11.1),

$$W_a = 5 + \left\{0.0298 + 0.03088[\ln(P_s)] - 0.0005733[\ln(P_s)]^2\right\} V_s^{0.66} \qquad (11.1)$$

where W_a is weight of air leakage in lb/h, P_s is system pressure in torr, and V_s is system volume in ft^3. Coker (2007) and Couper *et al.* (2012) gave useful tables for estimating air leakage by the system component method. Safety factors are normally applied to the estimated air-leakage rates in order to improve reliability; for this, an over-design factor of between 2 and 3 is recommended (Coker, 2007). Sometimes, a hybrid approach is used by combining these methods. Such a method adds up air-leakage rates estimated by both system volume method and system component method. However, a lower safety factor of between 1.5 and 2 times is used in this case (Coker, 2007).

11.3.2 Steam Jet Ejectors

Usually, 3–4 stage SJEs with inter-stage surface condensers are installed to reduce the operating cost. The number and size of the SJEs and surface condensers depend on suction and discharge pressures, suction temperature, vapour quality (amount of non-condensable gases and their molecular weight), cooling-water temperature, and cost of steam and electricity. The vacuum produced is limited by the temperature of the cooling water; the colder the temperature of the cooling water, the lower is the vacuum produced. A pre-condenser placed between the vacuum vessel and the SJE system can recover valuable process vapour and reduce vapour load to the SJE system, thus lowering the capital and operating costs of the vacuum system. Similarly, an inter-condenser installed between ejector stages will condense the motive steam and process vapour, to reduce vapour load to and consequently capital and operating costs of the downstream ejectors.

Gupta *et al.* (1979) have compared two models: the Hougen, Watson and Ragatz model, and the Dotterweich and Mooney model for computing motive steam consumption in a single-stage, single-fluid SJE as a function of nozzle and diffuser efficiencies, and concluded that the former model closely predicts the empirical curves better than does the other model. As the suction gas to the SJE may contain many components, the Heat Exchange Institute (HEI) has developed a method, which converts suction gas to dry air equivalent (DAE). HEI (2007) published curves to convert suction gas streams into DAE using molecular weight entrainment ratios (MW_{COG} for gases other than water vapour based on average molecular weight, and MW_{CWV} is the molecular weight entrainment ratio of water vapour) and temperature entrainment ratios (TC_a for air or TC_w for water vapour). From these curves, Reddy *et al.* (2013a) have derived Equations (11.2)–(11.5) for estimating TC_a, TC_w and MW_c. These are convenient for use in computer programs.

$$TC_a = -4\times10^{-10}T^3 + 3\times10^{-7}T^2 - 0.0005T + 1.0131 \tag{11.2}$$

Error: 0 to 1.73%

$$TC_w = -1\times10^{-13}T^4 - 7\times10^{-12}T^3 + 8\times10^{-8}T^2 - 0.0006T + 1.015 \tag{11.3}$$

Error: 0 to – 0.4%

For $M = 0$ to 60,

$$MW_c = 1\times10^{-5}M^3 - 0.0013M^2 + 0.0642M + 0.0161 \tag{11.4}$$

Error: – 7.2 to 2.4%

For $M = 60$ to 150,

$$MW_c = -2\times10^{-5}M^2 + 0.0077M + 0.9464 \tag{11.5}$$

Error: + 0.17 to 2.18%

In the above equations, T is the suction gas temperature in °C, MW_c is either MW_{COG} or MW_{CWV}, and M is the molecular weight of suction gas (of water vapour for MW_{CWV} or of suction gas other than water vapour for MW_{COG}). Water vapour and other components in the suction gas can be converted into DAE using the correction factors from Equations (11.2) to (11.5) and the following Equation (11.6),

$$\textit{DAE of suction gas or vapour} = \frac{W_{OG}}{TC_a \times MW_{COG}} + \frac{W_W}{TC_W \times MW_{CWV}} \tag{11.6}$$

where W_{OG} is the flow rate of gases/vapour other than water vapour and W_W is the flow rate of water vapour.

The amount of motive steam required to compress a unit DAE mass of suction gas/vapour in a steam ejector is defined as R_a. Power (2005) has published curves for estimating R_a values, with suction pressure on abscissa and discharge pressure on ordinate. These curves cover suction pressures from 0.1 to 1000 torr, discharge pressures from 1 to 1000 torr, and R_a from 0.5 to 20. R_a Values from this plot are accurate for steam ejectors with a single steam nozzle and low compression ratios associated with values of R_a less than 4. The steam usage of larger stages using multiple nozzles may be as much as 10 percent lower than the R_a value obtained from the plot. The R_a value from the published curves needs to be corrected using multipliers (available as separate plots in Power, 2005) for motive steam pressure different from 10.4 barg, for closeness of discharge pressure to the motive steam pressure, for stability at no load, for design discharge pressure at no load and for steam mass flow rate in a stage. Alternatively, for rough estimation of the ejector's steam consumption, one can simply use the following Equation (11.7) (Trambouze, 1999):

$$R_a = \left[\frac{Pd}{Pa}\left(0.434 - \frac{1.338}{Pa} + 0.475\times10^{-3}Pa\right) - 0.187\right]\left(1.2 - \frac{P_v - 10.2}{20}\right) \tag{11.7}$$

Here, *Pa* and *Pd* are, respectively, the suction and discharge pressures in torr, and P_v is the motive steam pressure in barg. For most cases, R_a estimated by the above equation is ~25% higher compared with that from the plots in Power (2005). Finally, steam requirement for ejectors can be estimated multiplying the DAE with R_a (kg of motive steam/kg of DAE equivalent of load gas).

For condenser calculations involving air-and-water mixtures, the overall heat transfer coefficient, U W/(m^{2}K) can be estimated by using Equations (11.8)–(11.9) developed based on the chart in Power (2005). For a gas–vapour mixture with non-condensable vapour mole percentage, *NC*, from 1 to 50%:

$$U = 5.678\left\{220.0417 + 1.6919\ln(NC) - 2.67975\,[\ln(NC)]^2 - 1.5465\,[\ln(NC)]^3\right\} \quad (11.8)$$

Error: 0 to 4.5%

For a gas/vapour mixture with *NC* from 50 to 95%:

$$U = 5.678\left\{\begin{array}{l} -245896 + 233845.3\ln(NC) - 83300.5\,[\ln(NC)]^2 \\ +13183.62\,[\ln(NC)]^3 - 782.58\,[\ln(NC)]^4 \end{array}\right\} \quad (11.9)$$

Error: − 2 to + 2%

11.3.3 Liquid Ring Vacuum Pumps

Suction pressure of LRVPs is dictated by the temperature of seal liquid/water. As the temperature of the seal water increases, some seal liquid vaporizes and occupies the space between the impeller blades and the liquid ring, which reduces the suction capacity of the LRVP (Bannwarth, 2005). The effect of cooling-water temperature on suction capacity of an LRVP decreases at higher suction pressures. So, with a normal cooling-water temperature of ~30 °C, LRVPs are used to replace steam ejectors operating at suction pressures > 150 torr. For suction pressures <200 torr, two-stage LRVPs are generally required (Birgenheier and Wetzel, 1988). Seal water absorbs the heat of compression, friction, gas cooling and vapour condensation. To remove this heat with minimum foul-water generation, partial or complete seal-water recirculation through a heat exchanger to cool the seal liquid is generally used.

LRVPs are generally more expensive compared with SJEs but they do not require large heat exchangers to condense the vapour at their outlet. Hence, the capital cost of an LRVP is slightly more than the combined cost of SJEs and condensers. Further, the operating cost of LRVPs is generally lower than that of SJEs. Power required by an LRVP can be estimated using Equation (11.10) for isothermal compression (Bannwarth, 2005):

$$P = \frac{Pa \times V_s \times \ln \dfrac{Pd}{Pa}}{27000 \times \eta_e} \quad (11.10)$$

Here, *P* is the isothermal compression power in kW, *Pa* and *Pd* are, respectively, suction and discharge pressures of the LRVP in torr, V_s is suction capacity at suction pressure in m^3/h, 27000 is the conversion factor (for obtaining *P* in kW with *Pa* and *Pd* in torr and V_s in m^3/h), and η_e is the isothermal efficiency, typically 0.25 to 0.4 (Bannwarth, 2005).

11.3.4 Dry Vacuum Pumps

DVPs use screws, roots, lobes or vanes for compressing load gases. They are energy efficient, compact and environmentally friendly. However, safety is a concern in handling flammable vapour/gases because of potential explosion risk, for example, from a spark generated by contact between the rotating parts and the casing. DVPs may require flame arrestors in the suction and discharge pipes (Tunna, 2005). Autoignition is also a concern. DVPs are applicable for suction pressures greater than 0.1 mmHg.

Purchase cost of DVPs is generally higher compared with that of SJEs or LRVPs alone. However, they do not require large heat exchangers to condense the vapour at the inter-stage and outlet of SJEs. They also do not require a closed circulation cooler and seal liquid, as are required for LRVP. Hence, the installed cost of a DVP is only marginally more than the combined cost of an SJE and its condensers or LRVP. Further, the operating cost of DVPs is generally much lower than that of SJEs or LRVPs. Power required by a DVP can be estimated using Equation (11.11) (Ryans and Bays, 2001),

$$P = \frac{W \times \left(\left(\frac{Pd}{Pa} \right)^{0.286} - 1 \right)}{12 \times \eta_d} \tag{11.11}$$

where W is gas/vapour flow rate in lb/h, Pa and Pd are, respectively, suction and discharge pressures in torr, η_d is the adiabatic thermal efficiency of DVP and P is the power required in hp. Ryans and Bays (2001) presented typical values of η_d, which are in the range 0.05 to 0.5.

11.4 Chilled-water Generation

Chilled water can be generated using heat pumps. They can be either absorption or mechanical chillers. Absorption chillers, which use waste heat (as a source of thermal energy) rather than mechanical/shaft energy for operation, can be economical. Heat source for an absorption chiller can be LLP steam (at ~1.5 barg) or a hot process stream. Lithium bromide (LiBr) absorption pumps are frequently used due to their lower cost and application range down to the freezing point of water.

The pressure–temperature profile of a single-stage absorption chiller is shown in Figure 11.1. Water is used as the refrigerant in LiBr chillers. Low-temperature (waste) heat (for example, LLP steam) is accepted in the generator to vaporize water from the dilute LiBr solution. This concentrates the LiBr solution, which is returned to the absorber, after exchanging heat at the economizer. Concentrated LiBr solution loses heat by exchange with cold dilute LiBr solution, in the economizer. Water vapour from the generator is condensed in the condenser by rejecting heat to cooling water. High-pressure water from the condenser is depressurized and sent to the evaporator, where water evaporates by taking heat from the process system requiring cooling. Vapour from the evaporator will be absorbed in the concentrated LiBr solution at the absorber, which makes the LiBr solution dilute. Heat released during absorption is removed by heat exchange with cooling water. The dilute LiBr solution at the absorber is pumped back to the generator to complete the cycle. A common

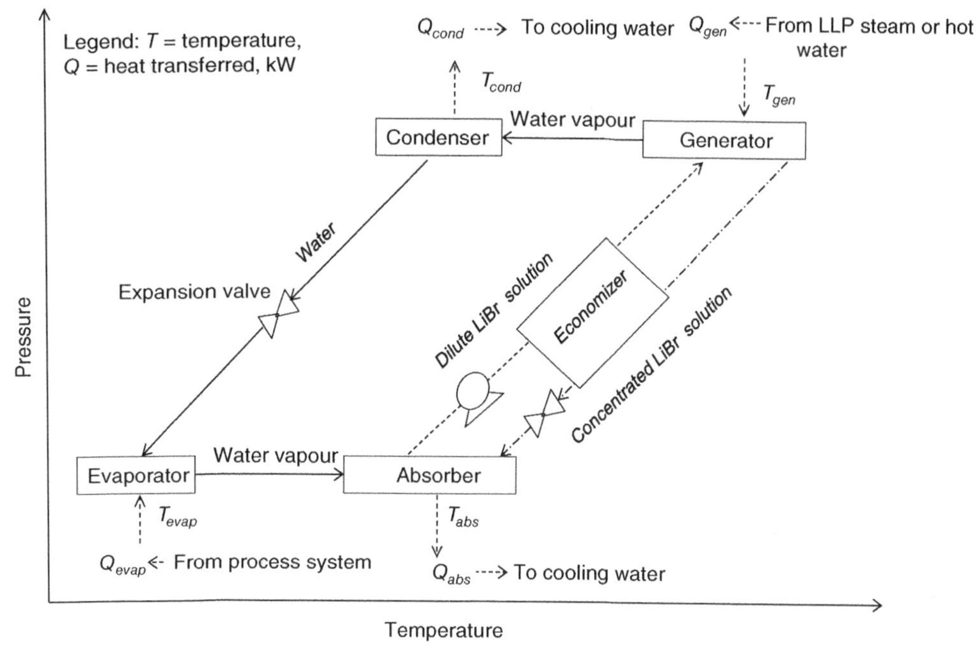

Figure 11.1 *Pressure—temperature profiles of the absorption chiller unit.*

measure for the performance of an absorption chiller is the Coefficient of Performance (COP), defined as shown in Equation (11.12).

$$COP_{AHP} = \frac{Q_{evp}}{Q_{gen}} \tag{11.12}$$

A large COP indicates more cooling by using external heat.

A mechanical chiller, whose operating principle is the same as that of air conditioners and domestic refrigerators, uses a compressor (as shown in Figure 11.2) instead of the generator and absorber system in the absorption chiller. Heat from the process system is used in the evaporator, which operates at a lower pressure. Refrigerant vapour generated at the evaporator is compressed by a compressor. Compressed vapour is condensed at the condenser by exchanging heat with cooling water. Cool high-pressure liquid from the condenser is depressurized, which causes liquid to partially flash and cool. Cooled vapour–liquid mixture flows to the evaporator, where it gains heat (from the process system requiring cooling) and vaporizes completely, to complete the cycle.

The COP of the mechanical chiller is given by Equation (11.13),

$$COP_{AHP} = \frac{Q_{evp}}{W} \tag{11.13}$$

where W is the external mechanical/shaft energy input to the chiller. Compared with mechanical chillers, absorption chillers have low COP values. Nonetheless, their operating cost can be substantially lower because they are energized by low-grade waste heat, while

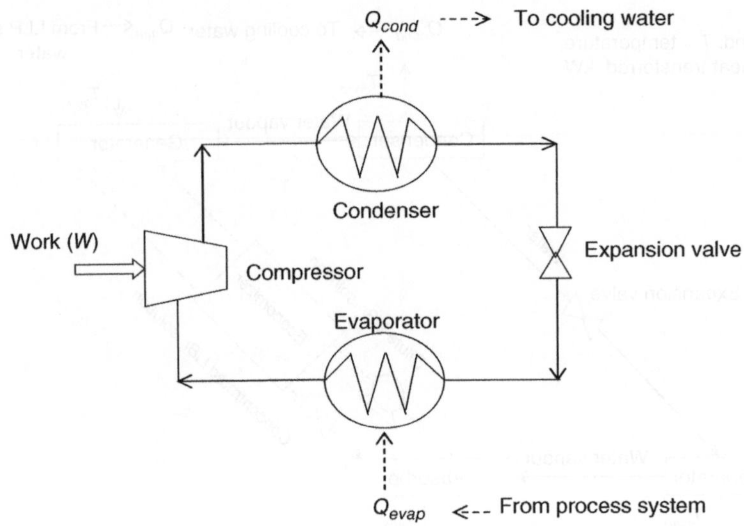

Figure 11.2 *Schematic of a mechanical chiller.*

mechanical chillers must be motor driven. At lower electricity prices, mechanical chillers can be attractive for chilled-water generation. Typical operating requirements of absorption and mechanical chillers were presented by Reddy *et al.* (2013b).

11.5 Optimization of Vacuum System Operating Cost

Retrofit strategy and procedure for minimizing the operating cost of multi-stage SJE, LRVP, DVP, SJE–LRVP and SJE–DVP systems are discussed in this section. They are illustrated with two case studies in the subsequent sections. There are several ways to reduce the energy cost or increase capacity of a vacuum system. Engineers often find it difficult to make correct revamp decisions as most of the information is vendor specific. This chapter illustrates the use of simple techniques, based on available literature, to revamp a vacuum system easily and quickly.

A strategy for optimizing revamp design of an existing vacuum system is presented in Figure 11.3; it can also be used for the design of a new vacuum system. First of all, the nature of the vapour/vapour–gas mixture needs to be examined. If some vapour can be condensed using cooling water, an additional pre-condenser needs to be installed in parallel to the existing pre-condenser, or the existing pre-condenser needs to be replaced by a larger version.

Then, the design choice depends on the explosion and autoignition risk of the vapour. If the existing SJE system cannot handle the additional non-explosive, non-autoignitable vapour/vapour–gas load, then additional DVPs can be considered in parallel arrangement to the existing SJE system (for example, a single train of multi-stage DVPs in parallel if the additional load is <1400 acfm, and several trains of multi-stage DVPs in parallel if the additional load is >1400 acfm). If the existing SJE system cannot handle the additional

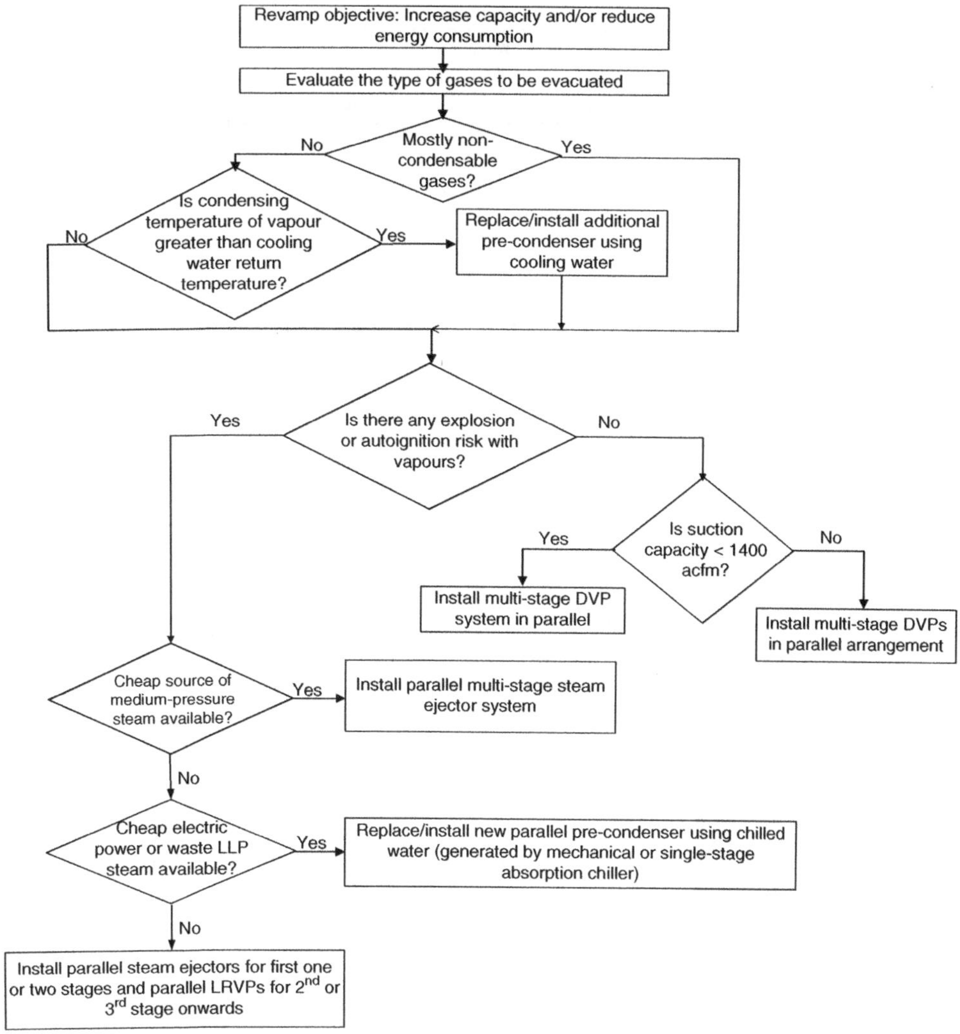

Figure 11.3 *Design strategy for retrofitting an existing vacuum system.*

explosive/autoignitable vapour/vapour–gas load, then the availability of cheaper utilities needs to be evaluated. If cheap MP steam is available, install multi-stage SJEs in parallel. If cheap electrical power or LLP steam is available, install a pre-condenser cooled by chilled water (using mechanical refrigeration or single-stage absorption chiller). Otherwise, use a parallel SJE system for the first one/two stages and LRVPs for the second/third stage onwards. Process simulators such as Aspen Hysys, Aspen Plus, Petrosim and Pro/II can be used to estimate the condensing temperature at which a majority (~90%) of vapour condenses, for various stages of the vacuum system. For revamp design, it is very important to minimize capital and operating costs. Capital cost can be reduced by re-using the existing plant equipment (such as vacuum condensers, water–oil separator and pumps) as much as possible. Cooling water, after use in pre-condensers, can be re-used in inter-condensers.

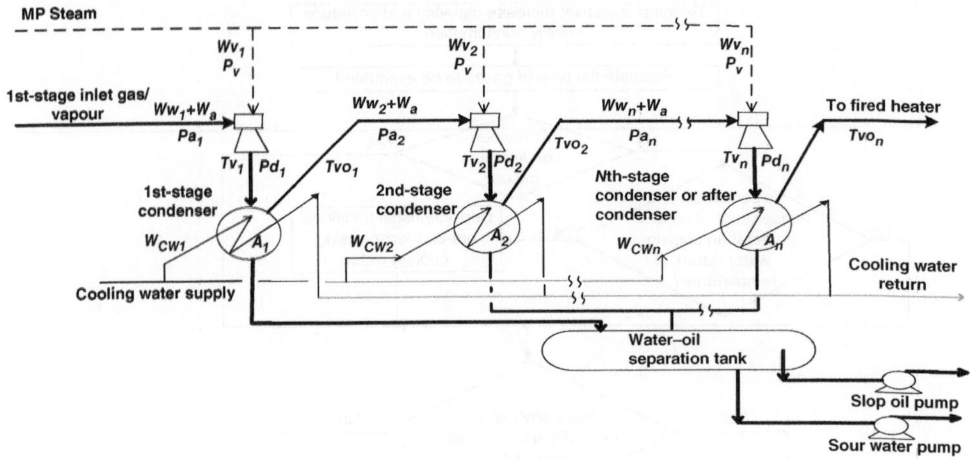

Figure 11.4 *Schematic of a multi-effect steam ejector system.*

Once the vacuum system design is selected, optimum discharge pressures and steam consumption for each stage of the multi-stage SJE system (shown in Figure 11.4) can be obtained by solving the following optimization problem, shown in Equation (11.14); for example, by using the Solver tool in MS Excel. The objective function is to minimize overall steam consumption for the multi-stage SJE system (Reddy *et al.*, 2013a). This objective function is used for retrofitting as operating cost is many times the annualized capital cost for the multi-stage SJE system.

$$\text{Minimize } Wv = \sum_{i=1}^{n} Wv_i = \sum_{i=1}^{n} \left[\frac{Pd_i}{Pa_i} \left(0.434 - \frac{1.338}{Pa_i} + 0.000475 Pa_i \right) - 0.187 \right] \times \left(1.2 - \frac{P_v - 10.2}{20} \right) \left(\frac{W_{OG}}{MW_{COG}} + \frac{Ww_i}{MW_{CWV}} \right) \tag{11.14}$$

The quantities in this equation can be calculated as follows. See Nomenclature for the significance of various symbols in the following Equations (11.15)–(11.25).

Suction pressure for $(i + 1)$th stage considering pressure drop in the condenser:

$$Pa_{i+1} = Pd_i - Dp_i \tag{11.15}$$

Water vapour flow rate to the inlet of $(i + 1)$th stage:

$$Ww_{i+1} = Ww_i - \frac{M_{WV} \times Pv_i \times W_{OG}}{M_{OG} \times (Pa_{i+1} - Pv_i)} \tag{11.16}$$

In Equation (11.16), it is assumed that mole fractions of water vapour and other gases (except water vapour) will be in proportion to their partial pressures. The saturation pressure of water vapour corresponding to the vent temperature of the *i*th-stage condenser is a function of the *i*th-stage condenser vent temperature:

$$Pv_i = f(Tvo_i) \tag{11.17}$$

Steam properties in Equations (11.17) and (11.22) can be calculated using the Excel spreadsheet, freely available at http://www.x-eng.com.

Vent temperature for the *i*th-stage condenser considering the approach temperature for its vent:

$$Tvo_i = T_{wco} + TAP_i \tag{11.18}$$

Mole fraction of water vapour at the *i*th-stage condenser inlet:

$$nw_i = \left[\frac{Ww_i + Wv_i}{M_{WV}}\right] \Big/ \left[\frac{Wa}{M_{OG}} + \frac{(Ww_i + Wv_i)}{M_{WV}}\right] \tag{11.19}$$

Mole percentage of non-condensable gases at the *i*th-stage condenser:

$$NC_i = (1 - nw_i) \times 100 \tag{11.20}$$

Partial pressure of water vapour at the *i*th-stage condenser inlet:

$$Ppw_i = Pd_i \times nw_i \tag{11.21}$$

Saturation temperature of water vapour corresponding to its partial pressure at the *i*th-stage condenser inlet is a function of its partial pressure at the *i*th-stage condenser inlet:

$$Tv_i = f(Ppw_i) \tag{11.22}$$

Log mean temperature difference for the *i*th-stage condenser:

$$LMTD_i = \frac{(Tv_i - T_{wco}) - (Tvo_i - T_{wc})}{\ln \dfrac{(Tv_i - T_{wco})}{(Tvo_i - T_{wc})}} \tag{11.23}$$

Overall heat transfer coefficient for the *i*th-stage condenser, U, is given by Equations (11.8) and (11.9). Area and cooling water required for the *i*th-stage condenser are given by Equations (11.24) and (11.25), respectively:

$$A_i = \frac{L_i \times (Ww_i + Wv_i - Ww_{i+1}) \times 1000}{U_i \times LMTD_i \times 3600} \tag{11.24}$$

$$Wcw_i = \frac{L_i \times (Ww_i + Wv_i - Ww_{i+1}) \times 0.239}{(T_{wc} - T_{wco})} \tag{11.25}$$

Here, 0.239 is the conversion factor from kJ to kcal.

In the above optimization problem, decision variables are: Pd_i for $i = 1, 2, \ldots, n-1$, and their bounds are $Pa_1 < Pd_i < Pd_n$. Other constraints are $Wv_i > 0$; $TAP_i > TAP_{min}$ and $A_i < A_{ie}$ for $i = 1, 2, \ldots, n-1$. The constraint: $A_i < A_{ie}$ is included in order to re-use the existing condensers in the retrofit design. The optimization problem [Equation (11.14)] and the model equations [Equations (11.15)–(11.25)] are used in the case studies discussed later.

The above optimization problem is applicable to systems involving air and water only. It ignores vapour superheat at the ejector inlet, vapour sub-cooling and liquid sub-cooling in condenser area calculations, and also assumes a simple log-mean temperature driving force (LMTD) without any correction factor. Hence, condenser area calculations are approximate but sufficient for optimization. Actual condenser design is quite complex for systems involving multi-component vapours, and requires good heat exchanger design software such as HTRI, Aspen EDR or HTFS. However, the optimization problem in Equation (11.14) is very useful to arrive at preliminary design/retrofit and/or verify vendor's proposals. For large systems involving other vapours, the optimization problem can be solved by including simulation data from process simulators such as Aspen Hysys, Aspen Plus and Pro/II.

Reddy *et al.* (2013a) have tested the accuracy of the above optimization procedure for one example in Power (2005). They found that the condenser's vent pressures tally well with those given in Power (2005), with error in the range +14% to −8%. Steam flow rate estimated by the above procedure is up to 25% more for 3- and 4-stage SJE, and lower by ~20% for 2-stage SJE, compared with the reported values (Power, 2005). This is mainly due to the difference in R_a values from the graphs in Power (2005) and Equation (11.7) used in the optimization procedure. For accurate estimation of steam flow rates, one can use the optimized vent pressures and the graphs given in Power (2005) but the optimization becomes partly manual and more tedious.

11.6 Case Study 1: Retrofit of a Vacuum System in a Petroleum Refinery

Case Study 1 is on retrofitting the vacuum system of a crude distillation column in a typical petroleum refinery. Design details for the present operation (Case 1A) are shown in Figure 11.5. Pressure at the top of the distillation column is 48 torr, vacuum system suction flow rate is 8500 kg/h and discharge pressure is 865 torr. The suction stream contains 8000 kg/h of water vapour and 500 kg/h of non-condensable gases (molecular

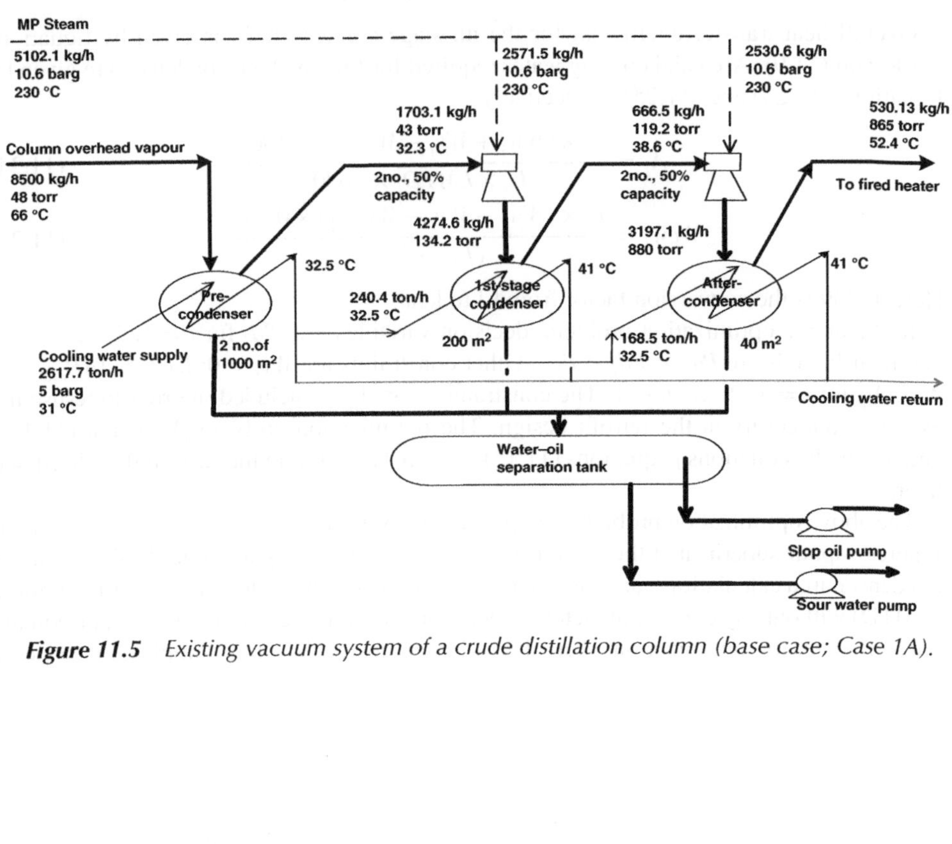

Figure 11.5 Existing vacuum system of a crude distillation column (base case; Case 1A).

weight = 41). Note that there are 2 ejectors in parallel in each of the stages 1 and 2. Case 1A is consuming 5.1 ton/h of MP steam and 2618 ton/h of cooling water. The revamp objectives are to reduce the column overhead pressure to 35 torr and also handle increased water vapour and non-condensable loads of 11 200 kg/h and 650 kg/h, respectively. The following retrofit options are considered to minimize the steam consumption and hence energy consumption of the existing vacuum system in the base case (Case 1A). All the retrofit options are designed to re-use the existing vacuum condensers, oil–water separation tank, and oil and water pumps. New steam ejectors are considered for all the revamp options as the SJE turndown for revamp operation is not generally reliable. Cooling water is re-used as much as possible in all the revamp options. Equations (11.3) and (11.15)–(11.25) given earlier are employed to calculate the required steam, cooling water and condenser areas in this case study.

Case 1B: A 3-stage SJE system without a pre-condenser (Figure 11.6) is considered as it is not possible to condense the suction vapour with cooling water supplied at 31 °C. Cooling water is used at all inter- and after-condensers. After SJE optimization for steam consumption, this option requires 12.91 ton/h of MP steam and 7201 ton/h of cooling water. Compared with Case 1A, cooling water consumption increased substantially due to the increase in steam demand. Pre-condenser in Case 1A can be re-used in Case 1B as 1st condenser. As the fluid velocity on both shell and tube sides is now higher compared with Case 1A, care should be taken to check and ensure that fluid velocity on both shell and tube sides does not exceed the recommended velocities (Serth, 2007). Minor modifications in this condenser are required to install sealing strips (to reduce bundle bypass flow), and to minimize fluid impingement and tube vibrations. 1st-stage and after-condensers in Case 1A

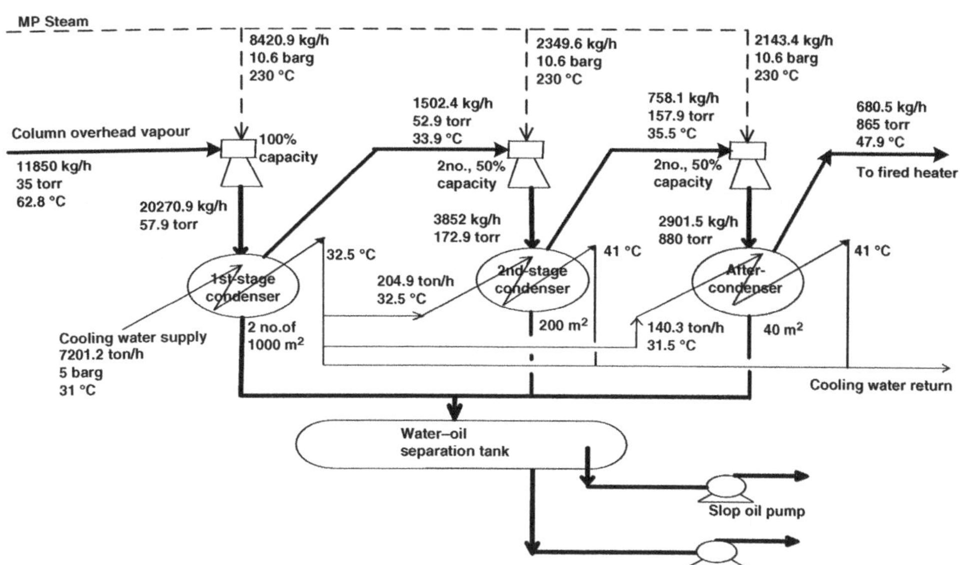

Figure 11.6 Three-stage steam ejector system with water-cooled inter- and after-condensers (Case 1B).

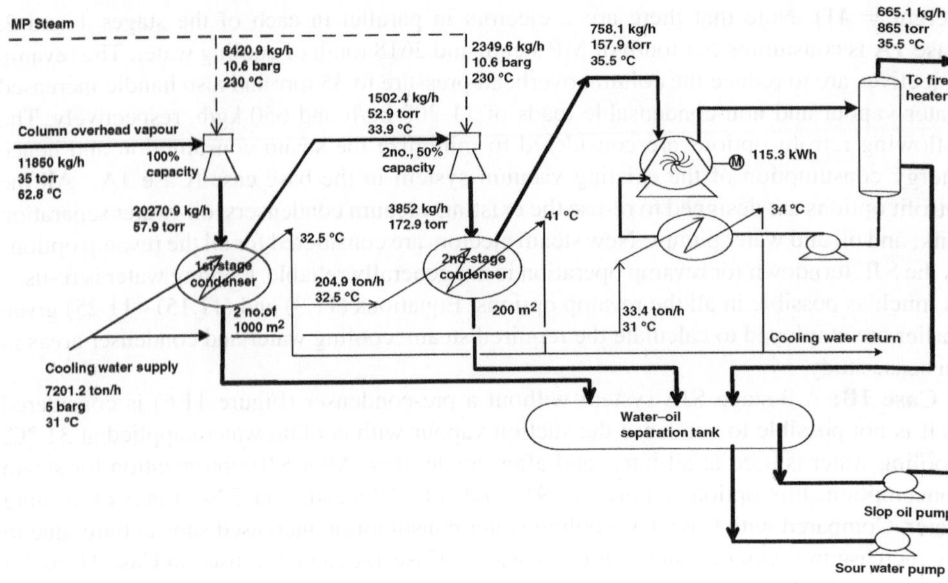

Figure 11.7 Two-stage steam ejector system and LRVP in series (Case 1C).

can be re-used respectively as 2nd-stage and after-condensers in Case 1B.The only new equipment required for retrofit to Case 1B is installation of 5 (= 1 + 2 × 2) ejectors for all the three stages.

Case 1C: In the optimal design of the 3-stage SJE system in Case 1B, the last/third stage is replaced with one LRVP, as shown in Figure 11.7. This SJE-LRVP design requires 10.77 ton/h of MP steam (that is, 2.14 ton/h lower than in Case 1B) and 115.3 kW/h of electricity. This option requires the same amount of cooling water as in Case 1B. The new equipment required for this case is one LRVP and 3 (= 1 + 2) ejectors for use at 1st and 2nd stage. All the existing condensers are re-used from Case 1A.

Cases 1D and 1E: Both these cases have a two-stage SJE system with one pre-condenser (at the inlet of the 1st-stage ejector), cooled by chilled water (Figure 11.8). A single-stage absorption chiller is used in Case 1D for the generation of chilled water using LLP steam; COP of this absorption chiller is 0.685. In Case 1E, a mechanical chiller with R125/R143a as refrigerant is used for the generation of chilled water; COP of this mechanical chiller is 4.18. After optimization of the SJE system (that is, finding the vent temperatures of all condensers including the pre-condenser by optimization using the bounds for condenser area and also for approach temperatures), cooling water consumption for Cases 1D and 1E is, respectively, 2151 ton/h (shown in Figure 11.8) and 1036 ton/h. Chilled water and MP steam consumption for each of these cases is 1074.1 ton/h and 3.92 ton/h, respectively. MP steam consumption for these cases is significantly lower compared with the previous cases, mainly due to the use of chilled water in the pre-condenser. This option requires one LRVP and 4 (= 2 × 2) ejectors. The absorption chiller in Case 1D requires 17.7 ton/h of LLP steam and 2151 ton/h of cooling water. Cooling water, after use in the absorption chiller, is re-used in the 1st- and 2nd-stage condensers. The mechanical chiller in Case 1E

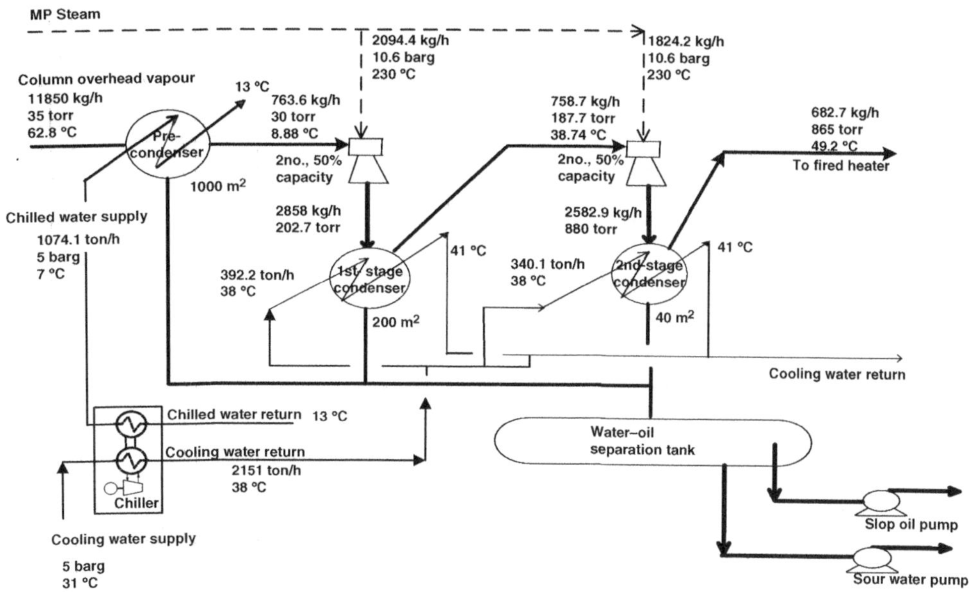

Figure 11.8 Two-stage steam ejector system with one pre-condenser (at the inlet of first-stage steam ejector), cooled by chilled water (Cases 1D and 1E); values shown are for Case 1D.

requires 1987 kWh of electricity and 1036 ton/h of cooling water. Similar to Case 1D, cooling water is re-used in the 1st- and 2nd-stage condensers after its use in the mechanical chiller. New items of equipment required for Cases 1D and 1E are absorption/mechanical chiller and SJEs. For each of these cases, 1st-stage and 2nd-stage condensers from Case 1A are re-used.

Cases 1F and 1G: These have a pre-condenser using chilled water generated by a single-stage absorption chiller (Case 1F) and mechanical chiller (Case 1G); they are obtained by replacing the last/second stage SJE in Cases 1D and 1E by one LRVP. This modified system is shown in Figure 11.9. Chilled water consumption in each of Cases 1D, 1E, 1F and 1G is the same at 1074.1 ton/h. MP steam consumption for each of Cases 1F and 1G is the lowest (2.09 ton/h) compared with all the previous cases, mainly due to the use of chilled water at the pre-condenser and the use of LRVP as the last/second stage. However, compared with Cases 1D and 1E, electricity consumption and cooling water for each of Cases 1F and 1G increase by 101.5 kWh and 14.3 ton/h, both required for LRVP. Total cooling water consumption for Cases 1F and 1G is 2166 ton/h (shown in Figure 11.9) and 1050 ton/h, respectively. All the condensers are re-used from Case1A, and new items of equipment required are absorption/ mechanical chiller, two ejectors and LRVP.

11.6.1 Analysis of the Results

Capital costs of SJEs are estimated from the chart in Power (2005). All capital costs are adjusted for inflation using the Chemical Engineering's Plant Cost Index (CEPCI) of 577 for July 2014. Capital costs for absorption/mechanical chillers, LRVP and DVP are based on

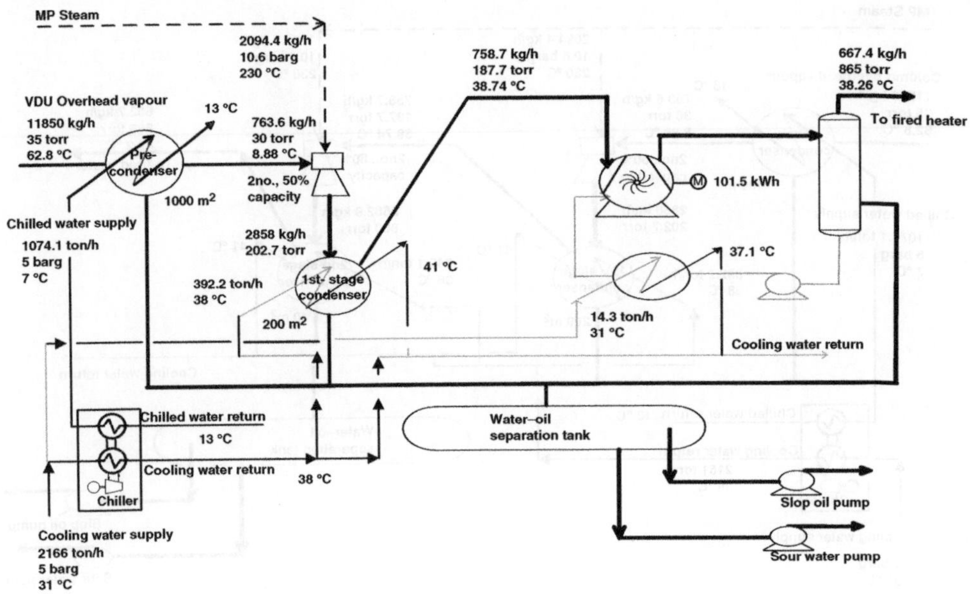

Figure 11.9 *Improved vacuum system with a pre-condenser, first stage with steam ejector and last stage replaced by one LRVP (Cases 1F and 1G); values shown are for Case 1F.*

vendor quotations; installed cost for mechanical chiller, LRVP and DVP are taken as twice the purchase cost whereas the installed cost for absorption (single stage using LiBr solution) chiller is taken as 1.5 times the purchase cost since the absorption chiller is a packaged unit and so involves less expense for installation. Utility requirements for foul-water treatment (0.345 MM BTU of LLP steam, 2.73 kWh of electricity and 135 Imperial Gallons of cooling water per ton of foul water) from Prakash (2003) are used. Reference utility costs assumed are: MP steam = US\$37.7/ton, LLP steam = US\$31.65/ton, cooling water = US\$0.05/ton and electric power = US\$0.15/kW, from an industrial facility. Equation (11.26) is used for calculating payback period:

$$\textit{Payback period} = \frac{\textit{Incremental capital cost for revamp case}}{\begin{array}{c}(\textit{operating cost for base case} - \textit{operating cost for revamp case} + \\ \textit{cost savings due to vaccum improvement})\end{array}} \quad (11.26)$$

Note that cost savings due to vacuum improvement are from the crude distillation column operation; they are in addition to operating cost savings from vacuum system revamp, given by the first two terms in the denominator of Equation (11.26).

Cost results for all Cases 1A to 1G in Case Study 1 are summarized in Table 11.2. Operating cost and payback period are calculated for five scenarios. As can be seen in Table 11.2, annual operating cost of a vacuum system is very significant compared with the installed cost, particularly revamp projects where the existing equipment are re-used as much as possible. The majority of the operating cost arises from the first stage, which

Table 11.2 Analysis of alternatives for retrofitting vacuum system of a distillation column in a petroleum refinery.

Utilities, costs and payback period for retrofit alternatives		Two-stage SJE + pre-condenser	Three-stage SJE	Two-stage SJE + LRVP	Two-stage SJE + pre-condenser using absorption (1D) or mechanical (1E) chiller		SJE + pre-condenser using absorption (1F) or mechanical (1G)	
		Case 1A	Case 1B	Case 1C	Case 1D	Case 1E	Case 1F	Case 1G
SJEs	MP steam (ton/h)	5.10	12.91	10.77	3.92	3.92	2.09	2.09
Condenser area	Total area (m^2)	2240	2240	2200	1240	1240	1200	1200
LRVP	Power consumption (kW)	0	0	115	0	0	101	101
	Cooling water (ton/h)	0	0	33	0	0	14.3	14.3
Single-stage absorption chiller	Cooling water (ton/h)	0	0	0	2151	0	2151	0
	LLP steam (ton/h)	0	0	0	17.69	0	17.69	0
	Power consumption (including power for chilled water pumps) (kW)	0	0	0	213.2	0	213.2	0
Mechanical chiller	Power consumption (kW)	0	0	0	0	1987	0	1987
	Cooling water (ton/h)	0	0	0	0	1036	0	1036
Total utility requirements	MP steam (ton/h)	5.10	12.91	10.77	3.92	3.92	2.09	2.09
	LLP steam (ton/h)	0	0	0	17.7	0	17.7	0
	Chilled water (ton/h)	0	0	0	1074	1074	1074	1074
	Cooling water (ton/h)	2618	7201	7234	2151	1036	2166	1050
	Power requirement (kW)	0	0	115	213	1987	315	2088

(continued)

Table 11.2 *(Continued)*

Utilities, costs and payback period for retrofit alternatives		Two-stage SJE + pre-condenser	Three-stage SJE	Two-stage SJE + LRVP	Two-stage SJE + pre-condenser using absorption (1D) or mechanical (1E) chiller		SJE + pre-condenser using absorption (1F) or mechanical (1G)	
		Case 1A	Case 1B	Case 1C	Case 1D	Case 1E	Case 1F	Case 1G
Savings at foul water stripper	LP steam saved (ton/h)	0	−1.34	−0.97	0.20	0.20	0.52	0.52
	Power saved (kW)	0	−21.33	−15.48	3.23	3.23	8.21	8.21
	Cooling water saved (ton/h)	0	−4.79	−3.48	0.73	0.73	1.85	1.85
Operating cost, Million US$/year	Reference cost case	2.83	7.82	7.17	7.36	4.30	6.80	3.74
	Considering LLP steam for absorption chiller is zero cost	2.83	7.82	7.17	2.46	4.30	1.90	3.74
	Considering electric power cost is 75% of reference cost	2.83	7.81	7.13	7.29	3.65	6.70	3.06
	Considering electric power cost is 125% of reference cost	2.83	7.83	7.21	7.43	4.95	6.90	4.42
	If MP steam cost is 75% of reference cost due to credit from power generation at steam turbines	2.41	6.75	6.28	7.04	3.97	6.63	3.57
	If MP steam cost is 75% of reference cost and LLP steam to absorption chiller is at zero cost	2.41	6.75	6.28	2.13	3.97	1.73	3.57

Operating profit, Million US$/year	Owing to crude distillation column capacity increase	0	5	5	5	5	5	5
Incremental capital cost, Million US$	Steam ejectors	0.00	0.41	0.39	0.06	0.06	0.06	0.06
	LRVP	0.00	0.00	0.40	0.00	0.00	0.39	0.39
	Single-stage absorption chiller	0.00	0.00	0.00	5.01	0.00	5.01	0.00
	Mechanical chiller	0.00	0.00	0.00	0.00	2.51	0.00	2.51
	Total installation cost	0.00	0.41	0.79	5.08	2.57	5.47	2.96
Payback period (years)	Reference cost case	—	38.42	1.19	10.75	0.73	5.31	0.72
	Considering LLP steam for absorption chiller is zero cost	—	38.42	1.19	0.94	0.73	0.92	0.72
	Considering electric power cost is 75% of reference cost	—	23.25	1.11	9.38	0.61	4.84	0.62
	Considering electric power cost is 125% of reference cost	—	110.49	1.27	12.59	0.89	5.89	0.87
	If MP steam cost is 75% of reference cost due to credit from power generation at steam turbines	—	0.63	0.70	13.55	0.75	7.00	0.77
	If MP steam cost is 75% of reference cost and LLP steam to absorption chiller is at zero cost	—	0.63	0.70	0.96	0.75	0.96	0.77

handles the maximum flow rate of suction gas. The only way to reduce this cost is to condense the vapour as much as possible before they reach the 1st-stage ejector. Depending on the suction pressure of the 1st-stage, chilled water may be required for this condensation. Cases 1E and 1G using mechanical chillers have a payback period of ~0.73 years, which will reduce to ~0.62 years if the power cost is lower at 75% compared with the reference case (Table 11.2). If waste LLP steam (~1.5 kg/cm^{2}g) is available at zero cost (which often occurs in petroleum refineries), then Cases 1D and 1F using a single-stage absorption chiller will have a payback period of 0.94 and 0.92 years, respectively. Thus, benefits of pre-condensers using chilled water depend on the availability of waste LLP pressure steam or low electricity prices.

Accurate estimation of the vacuum system's suction load for various plant operating scenarios is difficult. Hence, a considerable safety margin with respect to suction load is often allowed in the system design. It is very difficult to reduce steam consumption of SJEs if the operating suction gas load is lower than the design rate. However, for the case of the SJE system with chilled water pre-condenser, electrical power required for the chiller can be reduced by capacity control or by operating a few chillers arranged in parallel operation. For the case of LRVP and DVP, using variable-frequency drive and/or operating a few parallel units can reduce the energy consumption. So, operating cost benefits can be further increased. Availability of plot space and maintenance costs are the other critical issues for chiller installation. Mechanical chiller, LRVP and DVP installation may require modification costs at power-intake substations.

For the given utility costs, Cases 1C and 1G are the attractive revamp options for the scenarios with low and high capital cost spending, respectively. If the cost of MP steam (used in SJE) is lower at ~75% compared with the base case (due to credit from power generation at steam turbines), then the attractiveness of Cases D, E, F and G reduces. For this scenario, Cases 1B and 1C are attractive if lower capital options are preferred.

For low electricity costs (~75% compared with the reference case), Cases 1E and 1G will be more attractive compared with the other cases. These cases are still attractive even with higher electricity cost (125% compared with the reference case), with a payback period of ~0.88 years. As expected, a higher electricity price reduces the attractiveness of all cases. Waste LLP steam can be recovered economically from the steam condensate system. Alternatively, waste heat can be recovered as hot water from process streams. Petroleum refineries often discard a lot of waste heat through furnace stacks, fin-fan coolers and product coolers, which can be used to generate LLP steam (Reddy *et al.*, 2013b). In such situations, where LLP steam is available at zero cost, a single-stage absorption chiller (Cases 1D and 1F) becomes feasible and attractive with a payback period of ~0.93 years (Table 11.2).

The option involving pre-condenser (cooled by chilled water), first stage with steam ejector and last stage replaced by one LRVP (Case 1F if LLP steam for absorption chiller is available at zero cost and Case 1G) is very attractive with a payback period 0.72 to 0.92 years (with incremental capital cost of US$2.96 million to US$5.47 million). Thus, replacement of the last stage/s of a multi-stage SJE with LRVP is very much beneficial. Based on both this study and also findings from Reddy *et al.* (2013a), it is possible to generalize this for all multi-stage SJEs handling hazardous vapours.

Case 1C is recommended for cases where capital is limited. This case requires three ejectors and one LRVP. For this, it is assumed that MP steam header, steam generation system, foul-water pumps and treatment system have sufficient spare capacity to meet the increased MP steam usage and foul-water generation. As the 1st- and 2nd-stage condensers, oil–water separator and pumps are re-used, there is little requirement for structural and piping modifications. It is also assumed that the substation has sufficient spare capacity to supply 115.3 kWh of electricity. An LRVP is compact and can be installed at ground level or on an elevated platform. Depending on the electrical classification of the installation area, LRVPs may require an explosion-proof motor. As LRVPs use water as the working fluid, there are no additional safety hazards associated with their operation. Installing another LRVP or SJE in parallel (standby) will increase the reliability of the operation. In general, see Chapter 2 for details on project engineering and management of process revamp projects.

Case 1G (or 1F if LLP steam for absorption chiller is available at zero cost) is recommended if more capital investment and plot space are available. These cases require 2 ejectors, one mechanical (or single-stage absorption) chiller and one LRVP. Case 1G requires substantial amounts of electricity, and so may incur additional cost for modification at the substation. If the refrigerant selected is a hydrocarbon or toxic substance, it will introduce new hazards to the site. Careful evaluation is needed in selecting the best refrigerant to maximize COP. Case 1F requires waste-heat source in terms of LLP steam or hot water. If it is not readily available, additional capital investment is required to recover and supply it. As the single-stage absorption chiller does not require a compressor, its electricity consumption is very low and substation modification is minimal for Case 1F. This chiller operation is controlled by a DVP to operate the evaporator and absorber at vacuum level. Reliability of this chiller depends mostly on reliability of the DVP. Preventing LiBr crystallization and minimizing corrosion are other important considerations. Proper care needs to be taken during the design to avoid these issues. Cases 1F and 1G require significant plot space to cater for safe operation and maintenance. They also require significant additional piping in terms of chilled/cooling water, and hence demand sufficient space on pipe racks. These considerations may increase investment cost and so increase payback period. Additionally, these cases also involve the issues related to LRVP stated in Case 1C.

11.7 Case Study 2: Retrofit of a Surface Condenser of a Condensing Steam Turbine

Surface condensers are heat exchangers used to condense flash steam (under vacuum conditions) from condensing steam turbines in petroleum, petrochemical, power plants and other industries. Case study 2 considers the surface condenser for condensing flash steam from a condensing steam turbine, driving a very large centrifugal compressor. Operating conditions of the vacuum system are shown in Figure 11.10. The steam turbine uses 70 000 kg/h of high-pressure steam and exhausts it in partial wet condition. Steam condensate (6991 kg/h) is separated from the turbine's discharge header by gravity and is pumped to the steam condensate drum. The surface condenser is cooled with cooling water, to condense flash steam (Figure 11.10).

Two SJEs are installed in parallel on the surface condenser (Figure 11.10), mainly to remove air leakage and uncondensed gases/vapour (50 kg/h of dry air and 373 kg/h of water vapour). The two ejectors require 1254.7 kg/h of MP steam, which after use is condensed by

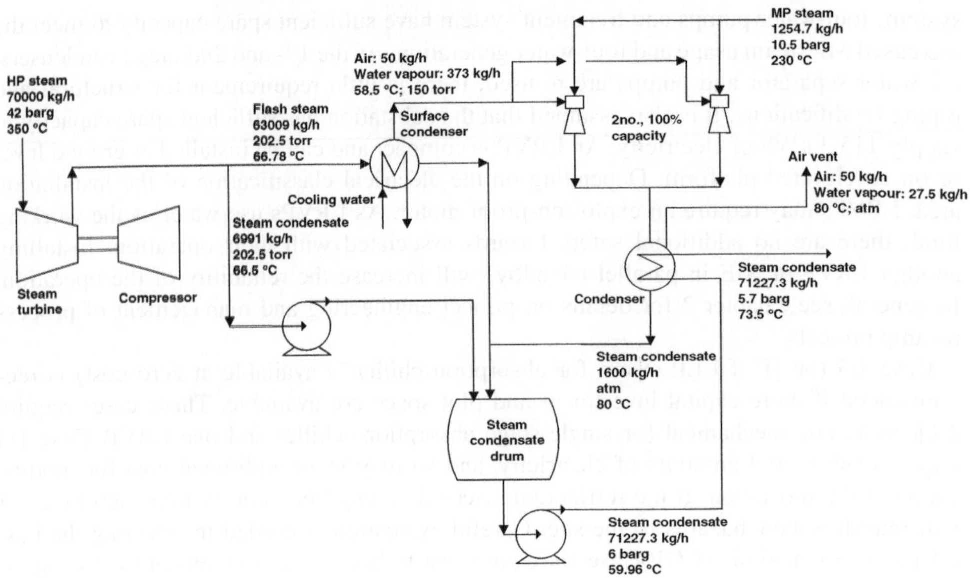

Figure 11.10 Surface condenser installed with steam jet ejectors; values shown are the current operating conditions.

exchanging heat with relatively colder steam condensate (Figure 11.10). As steam ejectors are least efficient, and this service involves non-explosive and non-hazardous gases/vapour, two DVPs (with one in operation and another in standby mode) are considered for retrofit to reduce cost of the vacuum system operation. This modification is shown in Figure 11.11. Note that LRVPs can also be used instead of DVPs. However, DVPs are more efficient, occupy less space and do not produce foul water. So they are selected for this revamp case.

For Case 2, each DVP requires 72.66 kWh of electricity, and operating cost savings due to this modification are US$257 801/year. Installed cost of two DVPs, based on vendor quotation, is US$400 000. So, payback period for this retrofit modification is 1.6 years. Thus, replacing the existing SJEs by DVPs is attractive for low capacities and non-hazardous applications such as a steam turbine's surface condenser vacuum system. DVPs are efficient, they require less electricity and so modification required at the substation is minimal. They are very compact and can be installed at ground level or on elevated platforms. Depending on the area classification of their location, explosion-proof motors may be required.

11.8 Conclusions

This chapter discusses an additional SJE and pre-condenser with chilled water, a combination with/without replacing last stage(s) with LRVP, and use of DVP for increasing capacity, reducing energy consumption and/or increasing reliability of existing vacuum systems in process industries. Requirements, benefits and constraints for implementation of these options are described, and a flow chart for selection of a suitable retrofit option is presented. Practical considerations in retrofitting (for example, re-use of equipment and cooling water, plot space requirement and reliability improvement) are presented. Feasible

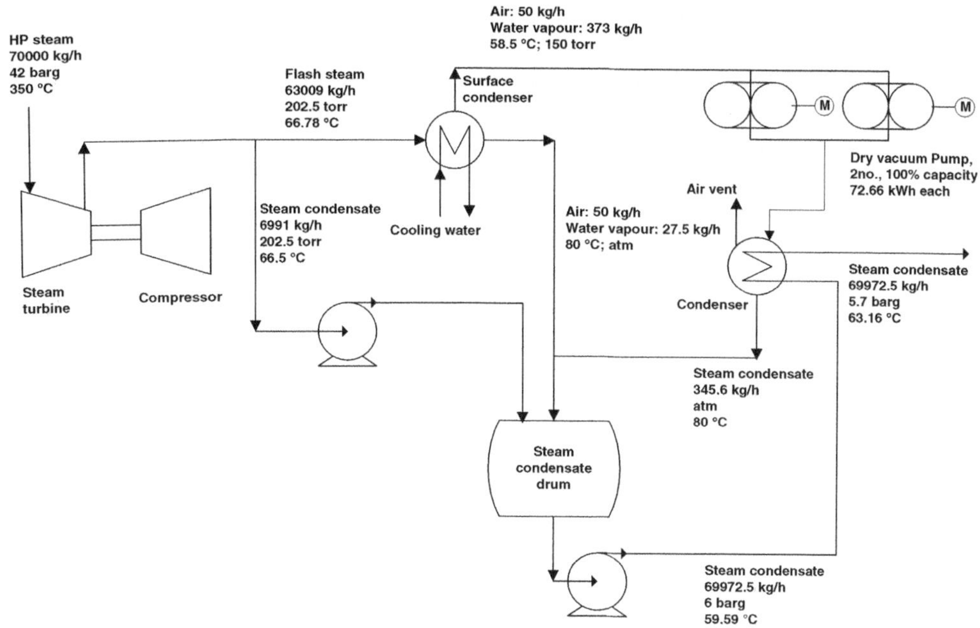

Figure 11.11 *Surface condenser retrofitted with dry vacuum pumps; values shown are for the revamp case.*

revamp options for both hazardous and non-hazardous applications are illustrated with one case study each. Results of Case Study 1 on a hazardous application show that use of a pre-condenser (cooled by chilled water) increases capacity and also reduces operating cost of the vacuum system. The higher electricity cost reduces the economic attractiveness of all options, especially those requiring more electricity such as a mechanical chiller. Lower electricity costs favour options with a mechanical chiller and/or LRVP. If waste heat is available at no cost, options involving an absorption chiller are favourable. Interestingly, there are a few options for retrofitting SJEs in Case Study 1 with payback period of about one year in all utility-cost scenarios considered. Case Study 2 illustrates the case of a non-hazardous vacuum application, where substitution of SJEs with DVPs is attractive, with a payback period of 1.6 years. Practical considerations in the implementation of the selected retrofit options are discussed. In general, numerous opportunities exist in process industries for retrofitting vacuum systems as they are widely used and need to be upgraded for revamping the process unit. Economics of retrofitting vary from one site to another as costs of steam, electricity, plot space and waste heat availability vary greatly. Critical evaluation as presented in this chapter is required to select the most beneficial retrofit option.

Nomenclature

A_i	area for the *i*th-stage condenser, m^2
$A_{i,e}$	area in the existing *i*th-stage condenser, m^2
COP_{AHP}	Coefficient of performance for the absorption heat pump

Dp_i	process gas pressure drop across the condenser at the outlet of *i*th steam jet ejector, torr
L_i	latent heat of vapour in the *i*th stage condenser, kJ/kg
$LMTD_i$	log-mean temperature driving force for the *i*th-stage condenser, °C
M	molecular weight
M_{WV}	molecular weight of water vapour
M_{OG}	molecular weight of gases other than water vapour
MW_{COG}	molecular weight entrainment ratio of gases/vapour other than water vapour
$MWcwv$	molecular weight entrainment ratio of water vapour
N	number of steam ejector stages
NC_i	mole percentage of non-condensable gases in *i*th-stage condenser inlet
nw_i	mole fraction of water vapour in the inlet of *i*th-stage condenser
P	power, kW or hp
P_s	system pressure, torr
Pa_i	suction pressure of *i*th-stage steam ejector, torr
Pd_i	discharge pressure of *i*th-stage steam ejector, torr
Ppw_i	partial pressure of water vapour in the *i*th-stage condenser inlet, torr
P_v	motive steam pressure, barg
Pv_i	saturation pressure of water at vent temperature of *i*th-stage condenser, torr
Q_{evp}	Heat load from process at the evaporator
Q_{gen}	Heat supplied to the generator by the LLP steam
R_a	amount of motive steam required to compress a unit DAE mass of suction gas/vapour
TAP_i	approach temperature for the *i*th-stage condenser vent, °C
TAP_{min}	minimum approach temperature at the condenser, °C
TC_a, TC_w	temperature entrainment ratios of air and water, °C
Tv_i	saturation temperature of water vapour corresponding to Ppw_i, °C
Tvo_i	gas temperature at the *i*th-stage condenser vent, °C
Twc	cooling water supply temperature, °C
$Twco$	cooling water return temperature, °C
U_i	overall heat transfer coefficient of the *i*th-stage condenser, W/m^2K
V_s	system volume, ft^3
W	air leakage, lb/h
W_a	air flow rate in load gas, lb/h
Wcw_i	cooling water flow rate to *i*th-stage condenser, kg/h
W_{OG}	flow rate of gases/vapour other than water vapour, kg/h
Wv_i	steam flow rate to the *i*th-stage steam ejector, kg/h
Ww_i	water vapour flow rate at the inlet of *i*th stage steam ejector, kg/h

Exercises

11.1 Estimate utilities consumption and condenser areas for a 3-stage SJE system using the following data: suction pressure = 30 torr, discharge pressure = 820 torr, water vapour flow rate = 6000 kg/h, flow rate of non-condensable hydrocarbon gases (of molecular weight: 44) = 350 kg/h; air leakage rate = 200 kg/h; motive steam pressure = 10.5 barg, cooling water supply temperature = 30 °C, cooling water return temperature = 40 °C, pressure drop for 1st condenser = 5 torr, pressure drop for remaining condensers = 10 torr, approach temperature for 1st- and 2nd-stage condensers = 2 °C, and approach temperature for remaining condensers = 10 °C.

11.2 Estimate utilities consumption and condenser areas by increasing the approach temperature for 1st and 2nd condensers by increments of 2 °C, between 2 °C and 10 °C; use remaining data given in Problem11.1. Also, estimate capital costs of SJEs and condensers. What conclusion(s) can be made from these results?

11.3 Estimate flow rates of utilities and condenser areas for the following cases and compare the results:

(a) By applying additional 100% safety factor on air leakage (that is, air leakage: 400 kg/h; use all the remaining data from Problem 11.1)

(b) Applying safety factor of 5 torr on vacuum system's suction pressure (that is, new suction pressure is 25 torr; use all the remaining data from Problem 11.1)

Which safety factor will require more operating cost penalty? Which safety factor do you recommend?

11.4 Optimize energy consumption in Problem 11.1 by using an SJE-and-LRVP combination. What is the % reduction in utilities consumption?

11.5 Considering the non-condensable gas in Problem 11.1 to be non-hazardous, optimize the utility consumption by considering the SJE-and-DVP combination. Compare the utility consumptions in Problems 11.1, 11.4 and 11.5.

References

Aliasso, J. (1999) Choose the Right Vacuum Pump. *Chemical Engineering (New York)*, **106** (3), 96–100. Available at http://www.graham-mfg.com/usr/pdf/TechLibVacuum/222.PDF.

Bannwarth, H. (2005) *Liquid Ring Vacuum Pumps, Compressors and Systems*, Wiley-VCH Verlag GmbH & Co. KGaA, Weinheim.

Birgenheier, D.B. and Wetzel, L. (1988) Design of Vacuum Systems for Crude Oil Vacuum Tower Distillation, Hydrocarbon Processing Symposium. Available at http://www.graham-mfg.com/usr/pdf/TechLibVacuum/22.PDF.

Coker, A.K. (2007) *Ludwig's Applied Process Design for Chemical and Petrochemical Plants*, vol. **1**, 4th edn, Elsevier, Burlington, MA, USA.

Couper, J.R., Penney, W.R., Fair, J.R. and Walas, S.M. (2012) *Chemical Process Equipment – Selection and Design*, 3rd edn, Butterworth-Heinemann, Waltham, USA.

Gupta, S.K., Singh, R.P. and Dixit, R.S. (1979) A comparative parametric study of two theoretical models of single stage, single fluid, steam jet ejectors. *Chemical Engineering Journal*, **18**, 81–85.

James D.S.J. and Pujado P.R. (2008) *Handbook of Petroleum Processing*, 2nd edn., Springer, Dordrecht, The Netherlands.

HEI (2007) *Standards for Steam Jet Vacuum Systems*, 6th edn, Cleveland, Ohio.

Milton, P.E. (1986) *Handbook of Evaporation Technology*, Noyes Publications, Park Ridge, New York Jersey

Power, R.B. (2005) *Steam Jet Ejectors for the Process Industries*, 2nd edn, McGraw-Hill, Charleston, USA.

Prakash, S. (2003) *Refining Processes Handbook*, Gulf Professional Publishing, Burlington, USA.

Reddy, C.C.S., Naidu, S.V. and Rangaiah, G.P. (2013a) Energy-efficient Vacuum Systems. *Petroleum Technology Quarterly*, (April) pp. 125–135.

Reddy, C.C.S., Rangaiah, G.P. and Naidu, S.V. (2013b) Waste heat recovery methods and technologies. *Chemical Engineering (New York)*, **January**, 28–38.

Ryans, J.L. and Bays, J. (2001) Run clean with dry vacuum pumps. *Chemical Engineering Progress*, **97** (10), 32–41.

Ryans, J.L. and Croll, S. (December 1981) Selecting Vacuum-Systems. *Chemical Engineering,* 72–90.

Serth, R.W. (2007) *Process Heat Transfer Principles and Applications*, 1st edn, Elsevier Ltd, Burlington, USA.

Sieder, W.D., Seader, J.D. Lewin, D.R. and Widagdo, S. (2009) *Product and Process Design Principles: Synthesis, Analysis and Evaluation*, 3rd edn, John Wiley and Sons, Inc., New York, USA.

Skelton, K. (1998) Variable Frequency Drives for a Vacuum Pump System. *Pumps and Systems Magazine*, January 1998. Available at http://www.graham-mfg.com/usr/pdf/TechLibVacuum/219.PDF.

Trambouze, B. (1999) *Petroleum Refining. Vol. 4. Materials and Equipment*, Editions Technip, Paris.

Tunna, C. (2005) Pumping potentially explosive atmospheres, *The Chemical Engineer (TCE)*, **May**, 30–31.

12

Design, Retrofit and Revamp of Industrial Water Networks using Multi-objective Optimization Approach

Shivom Sharma and G.P. Rangaiah
Department of Chemical & Biomolecular Engineering,
National University of Singapore, Singapore

12.1 Introduction

Water is used as a reactant, separation solvent and/or utility in many chemical processes. Most of the existing chemical plants have been designed for once-through water use, and end-of-pipe treatment of waste water is carried out to meet prevailing environmental regulations. In a chemical plant, some process units (Ps) require nearly pure water, whereas slightly contaminated water can be used in many others. If required, waste water from process units can be partially purified using a treatment unit, and then it can be reused in the plant (El-Halwagi, 2012). Treatment units can reduce the concentration of selected contaminants below certain limits, and treated water is cheaper than fresh water. An optimal water network reduces the consumption of fresh water by efficient reuse and recycling of treated water within the plant. There are many studies on the design and operation of industrial water networks (for example, Wang and Smith, 1994; Mann and Liu, 1999; Dunn and Wenzel, 2001; and Smith, 2005). See also Chapter 6 in this book.

Water requirement for different process units varies in terms of flow rate, temperature and contaminant concentration (CC). The mixing of water from different process units with

Chemical Process Retrofitting and Revamping: Techniques and Applications, First Edition.
Edited by Gade Pandu Rangaiah.

fresh water fulfils the flow rate and contaminant concentration requirements of the process (Bogataj and Bagajewicz, 2008; Faria and Bagajewicz, 2011a). In order to achieve the target temperature, the water stream to a process unit exchanges heat with hot/cold utility or other process streams. In the literature, heat-integrated water networks are designed using pinch analysis (Manan *et al.*, 2009) or a mathematical approach (Bogataj and Bagajewicz, 2008). Therefore, water integration is beneficial from the point of both economic and environmental impacts. Water networks are generally synthesized based on a single performance criterion such as total annual cost and fresh water consumption. However, industrial problems have several performance objectives, which may be conflicting in nature.

In recent years, Faria and Bagajewicz (2009) studied grassroots design and retrofit of water networks, to maximize the net present value or return on investment. Furthermore, Faria and Bagajewicz (2011b) developed a planning model for design and retrofit of industrial water systems; the model can address plant scenarios in the future, such as change in environmental regulations, increase in the cost of fresh water, variability in the quality of fresh water, and bottlenecks due to expansion of plant capacity. Poplewski and Jezowski (2010) used adaptive random search for optimizing the industrial water allocation problem. Jezowski (2010) provides a comprehensive review of water network design methods. The water network model is an MINLP (mixed-integer non-linear programming) and MILP (mixed-integer linear programming) problem for a water network system with multiple and single contaminant, respectively (Bagajewicz and Savelski, 2001).

Aviso *et al.* (2010) designed a water exchange network for several plants, using fuzzy mathematical programming. They found that the optimum solution for several plants together may not coincide with the individual optima for them. Kirthick *et al.* (2010) combined graphical and mathematical approaches for designing water networks; graphical approach was used to provide an initial guess for the mathematical approach. Sotelo-Pichardo *et al.* (2011) retrofitted the water network by minimizing the total annual cost of the retrofit process. They optimized the water network for stricter process and environmental constraints, by reconfiguring the piping, changes in the capacities/performance of existing treatment units and installation of new treatment units.

Ahmetovic and Grossmann (2011) used superstructure optimization for designing a water network, which has multiple fresh water sources, water-using processes, pre-treatment and waste water treatment operations. They optimized the water network for total cost, which includes freshwater cost, piping cost, pumping water cost, investment and operating costs of treatment units. Castro and Teles (2013) compared several global optimization algorithms for designing water networks. The proposed relaxation method (multiparametric disaggregation concept) for solving MINLP has better performance compared with BARON (Branch-And-Reduce Optimization Navigator) and GloMIQO (global mixed-integer quadratic optimizer) commercial solvers. Deng *et al.* (2013) optimized a water network for four objectives: minimum fresh water flow rate, intercepted flow rate, intercepted mass load and number of connections. They have studied one simple water network (having five processes and a single contaminant) and a sour water network from a refinery (involving three contaminants, seven water-using processes and multiple water sources).

Yang *et al.* (2014) optimized a water network for total cost, which consists of annualized investment and operating costs; here, shortcut models for treatment units are developed instead of using a fixed contaminant removal model.

Multi-objective optimization (MOO), in spite of its many applications in chemical engineering (Masuduzzaman and Rangaiah, 2009; Sharma and Rangaiah, 2013), did not receive much attention in the design of water networks. Boix *et al.* (2011) proposed a framework for MOO optimization of water networks; they considered fresh water consumption, water flow rate at the inlet of regeneration units (Rs) and the number of interconnections in the network as objective functions (OFs). Boix *et al.* (2012) simultaneously minimized the water and energy/utility consumptions in a water network, using mathematical programming. Then, the selected design was further improved for energy consumption by indirect heat exchange (that is, heat integration) in the water network. Ramos *et al.* (2014) optimized industrial water networks for three objectives: total fresh water flow rate, number of interconnections and total energy consumption, using goal programming. They considered case studies with 10–12 processes, one contaminant and 1–4 regeneration units. Sharma and Rangaiah (2014) investigated the potential of MOO for the design of heat-integrated water networks. Here, a mathematical approach was used for the water network design, and direct mixing of different water streams was allowed to achieve the required flow rates and contaminant concentrations. The water network involving six processes, four regeneration units and four contaminants was first optimized for fresh water flow rate and total water flow rate at the inlet of all regeneration units, simultaneously. For this, the ε-constraint method along with BARON solver in GAMS (general algebraic modeling system) was used to obtain the Pareto-optimal front. All the non-dominated solutions have different water network designs, and each design was considered for heat integration. In petroleum refineries, a lot of waste heat is available at about 100 °C, and it can fulfil heating requirements of a water network. Study of heat-integrated water networks can be important in other industries too, such as pulp and paper (Bonhivers and Stuart, 2013).

In this chapter, a new and relatively larger example of water network in a typical petroleum refinery is analysed for design, retrofit and revamp using the MOO approach. Initially, MOO is employed for the design problem to explore the trade-off between fresh water consumption and total water flow rate through all regeneration units, which gives better insights and also provides many alternative optimal solutions for water network design. Two water network designs are selected from these optimal solutions. Assuming that each of them has been in operation, their retrofit/revamp is studied to meet the revised environmental regulations and increased hydrocarbon load from the refinery processes. In water network retrofitting, it is assumed that there is no change in the network topology, and only flow rates through existing interconnections and capacities of the regeneration units can change. On the other hand, in water network revamping, changes in the existing network topology along with capacities of different regeneration units are allowed. This revamping is relatively more difficult and costly compared with retrofitting.

This chapter presents a mathematical model that can be used for grassroots, retrofit and revamp design of water networks via optimization. Furthermore, MOO is employed for all these designs. Each MOO problem is solved using the ε-constraint method, along with BARON solver (in GAMS) for single-objective optimization (SOO); see Chapter 4 for brief details on SOO and MOO methods and software. The next section of this chapter presents the mathematical model of a water network. Sections 12.3 and 12.4 describe the water network example in a petroleum refinery, and its MOO problem formulation, respectively. Then, Section 12.5 presents results for design, retrofitting and revamping of the water network in the refinery. Finally, conclusions for this study are presented in Section 12.6.

12.2 Mathematical Model of a Water Network

In the water network shown in Figure 12.1, water-using processes (P_i) and regeneration units (R_j) are interconnected using mixers (MIXs) and splitters (SPTs). Fresh water (FW) is free of contaminant, and can only be used in water-using processes. Waste water (WW) from a process or treated water from a regeneration unit can be used in any other process and/or regeneration unit, but no recycling within the process/regeneration unit is allowed. The interconnections recycle water and/or heat to other processes and regeneration units. All the processes and regeneration units can produce waste water, and the contaminant concentration in the mixed waste water stream leaving the system should be below the discharge limit. The outlet stream from each mixer must satisfy the flow rate and contaminant requirements of the respective water-using processes. Heaters/coolers (HE in Figure 12.1) are used to achieve the target temperature.

In Figure 12.1, $F_{inlet,i}$ ($= F_{outlet,i}$) and $F_{inlet,j}$ ($= F_{outlet,j}$) are water flow rates through ith process and jth regeneration unit, respectively. $Z^k_{inlet,i}$ and $Z^k_{outlet,i}$ are, respectively, the kth contaminant concentration at the inlet and outlet of the ith process, whereas $Z^k_{inlet,j}$ and

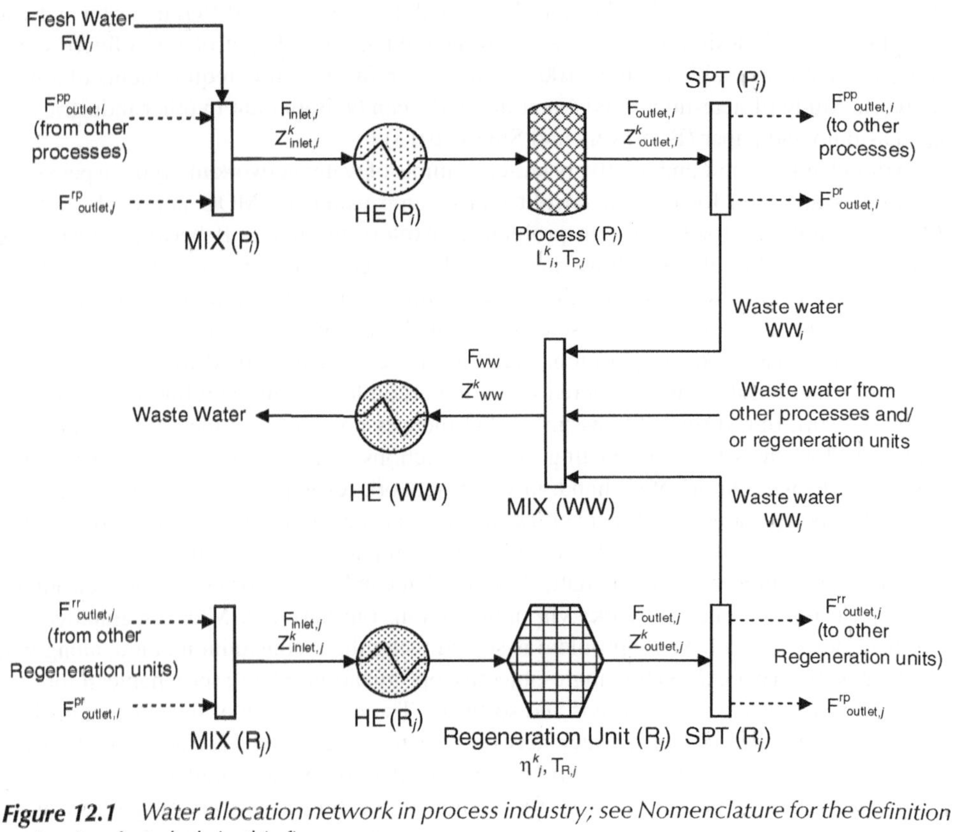

Figure 12.1 Water allocation network in process industry; see Nomenclature for the definition and units of symbols in this figure.

$Z^k_{\text{outlet},j}$ are, respectively, the kth contaminant concentration at the inlet and outlet of the jth regeneration unit. $F^{pp}_{\text{outlet},i}$ and $F^{pr}_{\text{outlet},i}$ are water recycle flow rates from the ith process to other processes and regeneration units, respectively, whereas $F^{rr}_{\text{outlet},j}$ and $F^{rp}_{\text{outlet},j}$ are water flow rates from the jth regeneration unit to other regeneration units and processes, respectively. L^k_i and η^k_j are, respectively, load and removal of the kth contaminant in the ith process and jth regeneration unit. All symbols employed are defined and their units are given in the Nomenclature section towards the end of this chapter. It also includes acronyms used in this chapter.

In modeling the water network, balances around processes, regeneration units, mixers and splitters are considered. Model equations for them are shown below; in these, NP, NR and K are the number of processes, the number of regeneration units, and the number of contaminants, respectively.

- Mixers for processes: water and contaminant balances (for $i = 1, 2, \ldots \text{NP}$ and $k = 1, 2, \ldots \text{K}$)

$$F_{\text{inlet},i} = \sum_{pp=1(pp \neq i)}^{NP} F^{pp}_{\text{outlet},i} + \sum_{rp=1}^{NR} F^{rp}_{\text{outlet},j} + FW_i \tag{12.1}$$

$$F_{\text{inlet},i}.Z^k_{\text{inlet},i} = \sum_{pp=1(pp \neq i)}^{NP} F^{pp}_{\text{outlet},i}.Z^k_{\text{outlet},i} + \sum_{rp=1}^{NR} F^{rp}_{\text{outlet},j}.Z^k_{\text{outlet},j} \tag{12.2}$$

- Processes: water and contaminant balances (for $i = 1, 2, \ldots \text{NP}$ and $k = 1, 2, \ldots \text{K}$)

$$F_{\text{inlet},i} = F_{\text{outlet},i} \tag{12.3}$$

$$F_{\text{inlet},i}.Z^k_{\text{inlet},i} + L^k_i = F_{\text{outlet},i}.Z^k_{\text{outlet},i} \tag{12.4}$$

- Splitters for processes: water balance; no change in concentrations (for $i = 1, 2, \ldots \text{NP}$)

$$F_{\text{outlet},i} = \sum_{pp=1(pp \neq i)}^{NP} F^{pp}_{\text{outlet},i} + \sum_{pr=1}^{NP} F^{pr}_{\text{outlet},i} + WW_i \tag{12.5}$$

- Mixers for regeneration units: water and contaminant balances (for $j = 1, 2, \ldots \text{NR}$ and $k = 1, 2, \ldots \text{K}$)

$$F_{\text{inlet},j} = \sum_{pr=1}^{NP} F^{pr}_{\text{outlet},i} + \sum_{rr=1(rr \neq j)}^{NR} F^{rr}_{\text{outlet},j} \tag{12.6}$$

$$F_{\text{inlet},j}.Z^k_{\text{inlet},j} = \sum_{pr=1}^{NP} F^{pr}_{\text{outlet},i}.Z^k_{\text{outlet},i} + \sum_{rr=1(rr \neq j)}^{NR} F^{rr}_{\text{outlet},j}.Z^k_{\text{outlet},j} \tag{12.7}$$

- Regeneration units: water and contaminant balances (for $j = 1, 2, \ldots \text{NR}$ and $k = 1, 2, \ldots \text{K}$)

$$F_{\text{inlet},j} = F_{\text{outlet},j} \tag{12.8}$$

$$F_{\text{inlet},j}.Z^k_{\text{inlet},j}(1 - \eta^k_j) = F_{\text{outlet},j}.Z^k_{\text{outlet},j} \tag{12.9}$$

- Splitters for regeneration units: water balance; no change in concentrations (for $j = 1, 2, \ldots NR$)

$$F_{\text{outlet},j} = \sum_{rp=1}^{NP} F^{rp}_{\text{outlet},j} + \sum_{rr=1(rr\neq j)}^{NR} F^{rr}_{\text{outlet},j} + WW_j \quad (12.10)$$

- Mixers for waste water: water and contaminant balances

$$F_{WW} = \sum_{i=1}^{NP} WW_i + \sum_{j=1}^{NR} WW_j \quad (12.11)$$

$$F_{WW}.Z^k_{WW} = \sum_{i=1}^{NP} WW_i.Z^k_{\text{outlet},i} + \sum_{j=1}^{NR} WW_j.Z^k_{\text{outlet},j} \quad (12.12)$$

12.3 Water Network in a Petroleum Refinery

In this study, the water network problem from a petroleum refinery is used to demonstrate the benefits of MOO in water network design, retrofitting and revamping. The selected water network has 11 processes and 4 contaminants (namely, salts, organics, H_2S and NH_3). Table 12.1 presents operating and limiting data for different processes in the water network; L^k, $Z^{k,\,max}_{\text{inlet},i}$ and $Z^{k,\,max}_{\text{outlet},i}$ are chosen based on Faria and Bagajewicz (2011a), and these numbers are also confirmed by our contact in a petroleum refinery. Three regenerations units [that is, reverse osmosis, water–oil mixture separator (American Petroleum Institute) (API separator) followed by ammoniacal copper arsenate (ACA), and Chevron waste water treatment] are available for waste water treatment; see Table 12.1 for performance data of regeneration units (Faria and Bagajewicz, 2011a). Initially, limits for all contaminants in waste water are assumed to be 400 ppm, and so end-of-pipe treatment may be required before discharge of waste water.

12.4 Multi-objective Optimization Problem Formulation

The above refinery water network is optimized for minimizing both the quantity of fresh water (OF_1) and total flow rate at the inlet of all regenerator units (OF_2), simultaneously. Minimization of fresh water consumption is necessary to preserve this natural resource, whereas total flow rate through regeneration units decides the water treatment cost. In short, both objectives are important as they affect investment and operating costs of the water network. In the water network design/retrofit/revamp, fresh water consumption conflicts with the total flow rate through regeneration units.

Table 12.1 *Operating and limiting data for water network in a petroleum refinery.*

Process	*k*th Contaminant	Mass load L^k, (kg/h)	$Z^{k,max}_{inlet,i}$ (ppm)	$Z^{k,max}_{outlet,i}$ (ppm)
CDU-1	Salts	22.87	1000	9500
	Organics	91.49	1000	6500
	H_2S	0.29	150	450
	NH_3	0.00	200	400
CDU-2	Salts	27.06	1000	9500
	Organics	108.25	1000	6500
	H_2S	0.34	150	450
	NH_3	0.00	200	400
CDU-3	Salts	12.44	1000	9500
	Organics	49.75	1000	6500
	H_2S	0.16	150	450
	NH_3	0.00	200	400
Amine treatment-1	Salts	0.01	10	1000
	Organics	0.44	1	3500
	H_2S	0.02	0	2000
	NH_3	0.01	0	3500
Amine treatment-2	Salts	0.07	10	1000
	Organics	3.43	1	3500
	H_2S	0.17	0	2000
	NH_3	0.11	0	3500
Hydrocracker	Salts	0.10	10	350
	Organics	1.21	1	1800
	H_2S	0.03	0	1500
	NH_3	0.05	0	1000
Naphtha hydrotreating	Salts	0.05	10	350
	Organics	0.56	1	1800
	H_2S	0.01	0	6500
	NH_3	0.03	0	1000
Diesel hydrofiner-1	Salts	0.16	10	350
	Organics	1.85	1	1800
	H_2S	0.05	0	1500
	NH_3	0.08	0	1000
Diesel hydrofiner-2	Salts	0.30	10	350
	Organics	3.50	1	1800
	H_2S	0.09	0	1500
	NH_3	0.16	0	1000
Kerosene Merox-1	Salts	0.48	100	400
	Organics	14.00	200	6000
	H_2S	0.19	50	2000
	NH_3	0.24	1000	3500
Kerosene Merox-2	Salts	0.51	100	400
	Organics	14.50	200	6000
	H_2S	0.20	50	2000
	NH_3	0.25	1000	3500

(continued)

Table 12.1 (Continued)

Reverse osmosis (RU1)	Reduces salts to 20 ppm
API separator and ACA (RU2)	Reduces organics to 50 ppm
Chevron waste water treatment unit (RU3)	Reduces H_2S and NH_3 to 5 and 30 ppm, respectively

	Investment cost coefficient (IC), \$/{year.(t/h)$^{0.7}$}	Operating cost coefficient (OC), \$/t
RU1	9500	0.04
RU2	4800	0
RU3	12 600	0.0067

The MOO problem for the water network is presented below Equations (12.13)–(12.19).

Objectives:

$$\text{Min. OF}_1 = \sum_{i=1}^{NP} \text{FW}_i \tag{12.13}$$

$$\text{Min. OF}_2 = \sum_{j=1}^{NR} \text{F}_{\text{inlet},j} \tag{12.14}$$

Decision variables:

$$\text{FW}_i,\ \text{F}^{pp}_{\text{outlet},i'},\ \text{F}^{pr}_{\text{outlet},i'},\ \text{F}^{rr}_{\text{outlet},j'},\ \text{F}^{rp}_{\text{outlet},j'},\ \text{Z}^k_{\text{inlet},i'},\ \text{Z}^k_{\text{oulet},i'},\ \text{Z}^k_{\text{inle},j'},\ \text{Z}^k_{\text{outlet},j} \tag{12.15}$$

Constraints:

$$\text{Z}^k_{\text{inlet},i} \le \text{Z}^{k,\max}_{\text{inlet},i} \tag{12.16}$$

$$\text{Z}^k_{\text{outlet},i} \le \text{Z}^{k,\max}_{\text{outlet},i} \tag{12.17}$$

$$\text{Z}^k_{\text{WW}} \le \text{Z}^{k,\max}_{\text{WW}} \tag{12.18}$$

$$\text{F}^{pp}_{\text{outlet},i'},\ \text{F}^{pr}_{\text{outlet},i'},\ \text{F}^{rp}_{\text{outlet},j'},\ \text{F}^{rr}_{\text{outlet},j} \ge \text{F}^{\min}_{\text{IC}} = 1\text{t/h} \tag{12.19}$$

Equations (12.1)–(12.12)

In this optimization problem, decision variables are fresh water flow rate to each process, water flow rates through interconnections between processes and/or regeneration units, and contaminant concentrations in water streams at the inlet and outlet of processes and regeneration units. Equations (12.16) and (12.17), respectively, limit the maximum inlet and outlet concentrations of contaminants in water streams for different processes, whereas Equation (12.18) limits the concentrations of contaminants in the waste water stream. Minimum flow rate in each interconnection is maintained by Equation (12.19). Values of $\text{Z}^{k,\max}_{\text{inlet},i}$ and $\text{Z}^{k,\max}_{\text{outlet},i}$ are given in Table 12.1, whereas the value of $\text{F}^{\min}_{\text{IC}}$ is assumed to be 1.0 t/h. Further, the optimal solution must satisfy the water and contaminant balances shown in Equations (12.1)–(12.12).

The above bi-objective optimization problem can be transformed into a SOO problem by the ε-constraint method via making OF_2 as an additional constraint. Besides this method, goal programming and the weighted-sum method can also be used to convert the MOO problem into a SOO problem. Interested readers are referred to Chapter 4 for more details on the ε-constraint method, and to Rangaiah (2009) for more details on MOO methods. The SOO problem resulting from the application of the ε-constraint method to the above bi-objective problem for water network is described as follows [Equations (12.20) and (12.21)]:

$$\text{Objective:} \qquad \text{Min. OF}_1 = \sum_{i=1}^{NP} FW_i \tag{12.20}$$

$$\text{Additional Constraint:} \qquad OF_2 = \sum_{j=1}^{NR} F_{inlet,j} < \varepsilon \tag{12.21}$$

Decision variables and all other constraints are the same as those in the MOO problem. The SOO problem is a MINLP problem; this problem for water network design (Section 12.5.1) has 396 continuous variables, 182 integer variables and 586 constraints. Its solution for one value of ε gives one Pareto-optimal (non-dominated) solution, and so the SOO problem has to be solved for different values of ε to find many Pareto-optimal solutions, as discussed in Chapter 4.

The converted SOO problem can be solved using any deterministic or stochastic global optimization method. Deterministic techniques such as branch-and-bound, decomposition-based methods and extended cutting-plane method can be used to solve MINLP problems. Although stochastic methods have been used to solve many MOO problems in chemical engineering, they are not effective for solving water network problems, due to the presence of numerous equality constraints arising from water/contaminant balances. Recall that the present study involves a water network model with multiple contaminants, which is an MINLP problem. On the other hand, the water network model for a single contaminant case is an MILP problem (Bagajewicz and Savelski, 2001), which is easier to solve compared with an MINLP problem in the case of multiple contaminants.

The SOO problem for water network is solved using BARON solver in GAMS, which is allowed to run for 4000 seconds to obtain one Pareto-optimal solution at a time. BARON solver can handle different types of non-linear functions such as exponential, logarithmic and power. The water network problem has bilinear terms, which arise from the multiplication of flow rate with concentrations of contaminants. In the water network model, bounds for all variables are provided to achieve faster convergence and also to obtain better solutions. These bounds on variables keep the values of non-linear expressions within the finite ranges. For the water network problem in this study, BARON solver sometimes gave local solutions after 4000 seconds (see Section 12.5.4), and so it was allowed to run for 8000 seconds in such cases. BARON may stop before the maximum allowable CPU time, using default termination criterion (that is, tolerance or precision). Besides BARON, DICOPT solver in GAMS was also tried in order to solve the water network problem, but it did not give a global optimum.

12.5 Results and Discussion

12.5.1 Water Network Design

Figure 12.2 shows the results obtained for two objectives, using the ε-constraint method and different limits [that is, ε values in Equation (12.21)] on the total flow rate through

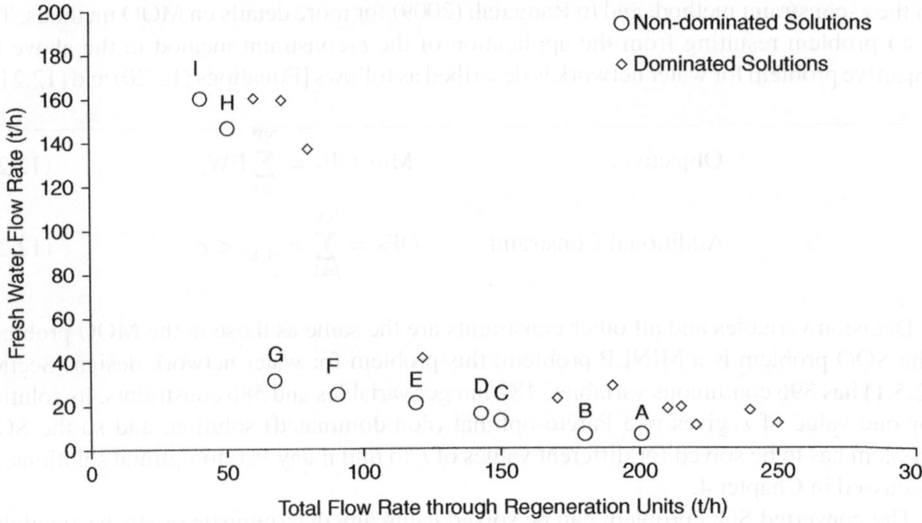

Figure 12.2 *Simultaneous minimization of fresh water flow rate and total flow through regeneration units (concentration of each contaminant in waste water < 400 ppm), for design of the water network in a petroleum refinery.*

regeneration units. In this figure, circles are the non-dominated solutions (that is, Pareto-optimal front). Solvers in GAMS require gradients of objective function and constraints, and their performance may depend on the starting point. For different ε values, BARON may give non-dominated solutions, dominated solutions or no solution yet all within the time specified. To show this, selected dominated solutions obtained from many trials are shown in Figure 12.2. Hereafter, only non-dominated solutions are presented and discussed in the following sections. As expected, non-dominated solutions confirm that fresh water flow rate is conflicting with the total flow rate through the regenerations units (Figure 12.2).

The water network obtained for each non-dominated solution is different, and important design and operating data for them are summarized in Table 12.2. Here, FW, TRU and NIC are, respectively, fresh water flow rate, total flow rate through regeneration units and number of interconnections between processes and/or regeneration units. Furthermore, P-to-P, P-to-R, R-to-R and R-to-P are, respectively, the number of interconnections between processes to processes, processes to regeneration units, regeneration units to regeneration units and regeneration units to processes. In general, the number of interconnections decreases with decrease in the total flow rate through regeneration units, which leads to an increase in fresh water consumption. In the water network design, the concentration of each contaminant in waste water is kept below 400 ppm, which means that the produced waste water may still require end-of-pipe treatment before discharge of waste water. In Figure 12.2, non-dominated solutions H and I require significantly large amounts of fresh water compared with the remaining non-dominated solutions (see Table 12.2), and water treatment for these two solutions is minimum. Overall, these water network designs are likely to incur high cost and may not be sustainable because of the large amount of fresh water required.

Table 12.2 *Important design and operating data for water network designs (that is, non-dominated solutions in Figure 12.2).*

Non-dominated solution	A	B	C	D	E	F	G	H	I
FW (t/h)	7.03	7.59	13.82	16.82	22.01	25.91	31.88	146.80	160.12
TRU (t/h)	200	180	150	142.61	118.51	90	67.23	50	39.91
NIC	28	19	23	16	16	18	13	12	7
P-to-P	6	5	3	2	6	9	4	6	5
P-to-R	12	7	10	7	6	4	4	2	2
R-to-R	2	3	2	2	1	1	2	1	–
R-to-P	8	4	8	5	3	4	3	3	–
Water network cost (1000 $/y)									
C_{FW} = 1 $/t, AF = 0.1	134	122.6	169.3	182.8	212	238.5	278.6	1184.5	1287.3
C_{FW} = 0.38 $/t, AF = 0.1	99.1	85	100.7	99.4	102.9	110	120.4	456.4	493.1
C_{FW} = 1 $/t, AF = 0.2	179.4	162.1	208.8	210.5	234.9	258	294.9	1193.7	1293.7
Concentration (ppm) of contaminants in waste water leaving the system									
Salts	172.51	325.30	400	92.24	60.55	400	400	400	400
Organics	400	400	400	400	400	400	400	400	400
H_2S	53.41	57.32	37.32	92.18	70.42	59.82	43.44	10.56	9.68
NH_3	52.75	57.13	44.41	55.31	42.25	35.89	26.66	6.34	5.81

There are some systematic approaches, such as net flow and rough set methods (Thibault, 2009), and TOPSIS (Boix *et al.*, 2011), which can be used for selecting one final design from the Pareto-optimal front. The selected objectives: fresh water consumption and total flow rate through regeneration units are fundamental measures of water network performance, and they can be related to costs using suitable unit prices. For illustration, fresh water cost (C_{FW}) is assumed to be 1.0 \$/t, and investment and operating cost coefficients (IC_j and OC_j) for different regeneration units in Table 12.1 (Faria and Bagajewicz, 2011a) are used. Assuming the annualization factor (AF) for capital cost is 0.1, and plant operating time is 8000 hours in a year, the total cost of each water network design in Figure 12.2 can be calculated using Equation (12.22):

$$\text{Water Network Cost (\$/year)} = 8000C_{FW}\sum_{i=1}^{NP} FW_i + 0.1\sum_{j=1}^{NR}\left\{F_{\text{inlet},j}^{0.7} IC_j\right\} + 8000\sum_{j=1}^{NR}\left\{F_{\text{inlet},j} OC_j\right\} \quad (12.22)$$

The total cost of each Pareto-optimal water network is also reported in Table 12.2. However, this cannot be the only basis for selecting a particular water network design, as preferences of the decision maker, such as number of interconnections, are important and costs depend on plant location and time. This is illustrated in Table 12.2 using two different fresh water costs (1.0 and 0.38 \$/t) and two different AF for capital cost (0.1 and 0.2).

In Figure 12.2, each solution on the Pareto-optimal front is equally good for the objectives in the optimization problem, and the decision maker can select one solution based on his/her requirements. As each non-dominated solution has a different water network design, water network designs corresponding to solutions C and G are selected for retrofit/revamp studies for a decrease in the concentrations of contaminants in waste water (Sub-Sections 12.5.2 and 12.5.4) and an increase in hydrocarbon load from processes (Sub-Sections 12.5.3 and 12.5.5). Here, solution G is the corner solution (after the initial sharp drop in fresh water flow rate) on the Pareto-optimal front and involves relatively fewer interconnections, and solution C requires a relatively low fresh water flow rate. Solutions A and B are not selected as they have significantly larger total flow rate through regeneration units compared with solution C, and hence their cost will increase with increasing cost of regeneration. Figures 12.3 and 12.4 show water flow rates of all streams in the network design corresponding to solutions C and G.

12.5.2 Retrofitting Selected Water Networks for Change in Environmental Regulations

As mentioned in the previous sub-section, water network designs corresponding to solutions C and G are used to study the effect of change in environmental regulations, which may occur in the future, on the existing water network (that is, solution C or G). In this retrofitting for a decrease in contaminant concentrations in waste water, no new interconnections between processes and/or regeneration units are allowed; only flow rates through existing interconnections and capacities of regeneration units can change. If flow rates through some existing interconnections are below 1 t/h [that is, Equation (12.19) for minimum flow rate], then those interconnections will be removed during optimization. Hence, NIC may decease

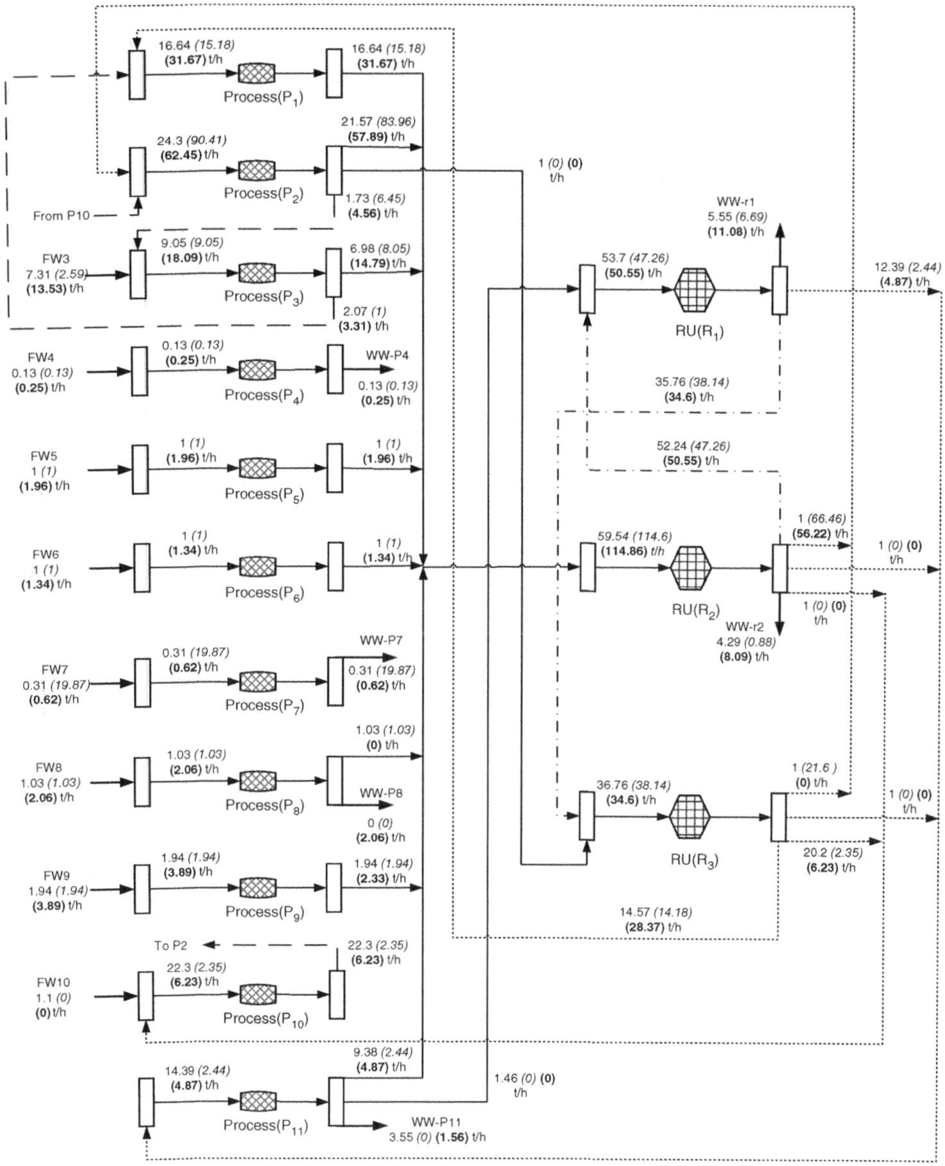

Figure 12.3 Flow rates of different water streams corresponding to solution C in Figure 12.2, solution c2 in Figure 12.5 (shown in first set of parentheses) and solution c3′ in Figure 12.6 (shown in second set of parentheses).

for the retrofit solutions. In other words, changes in the existing water network are fewer. As environmental regulations may change after the original design, waste water generated as per the original design [Figures (12.2)–(12.4)] may require end-of-pipe treatment; note that some contaminant concentrations are closer to 400 ppm (Table 12.2). Here, a lower limit (50 ppm) on concentration of contaminants in waste water is considered.

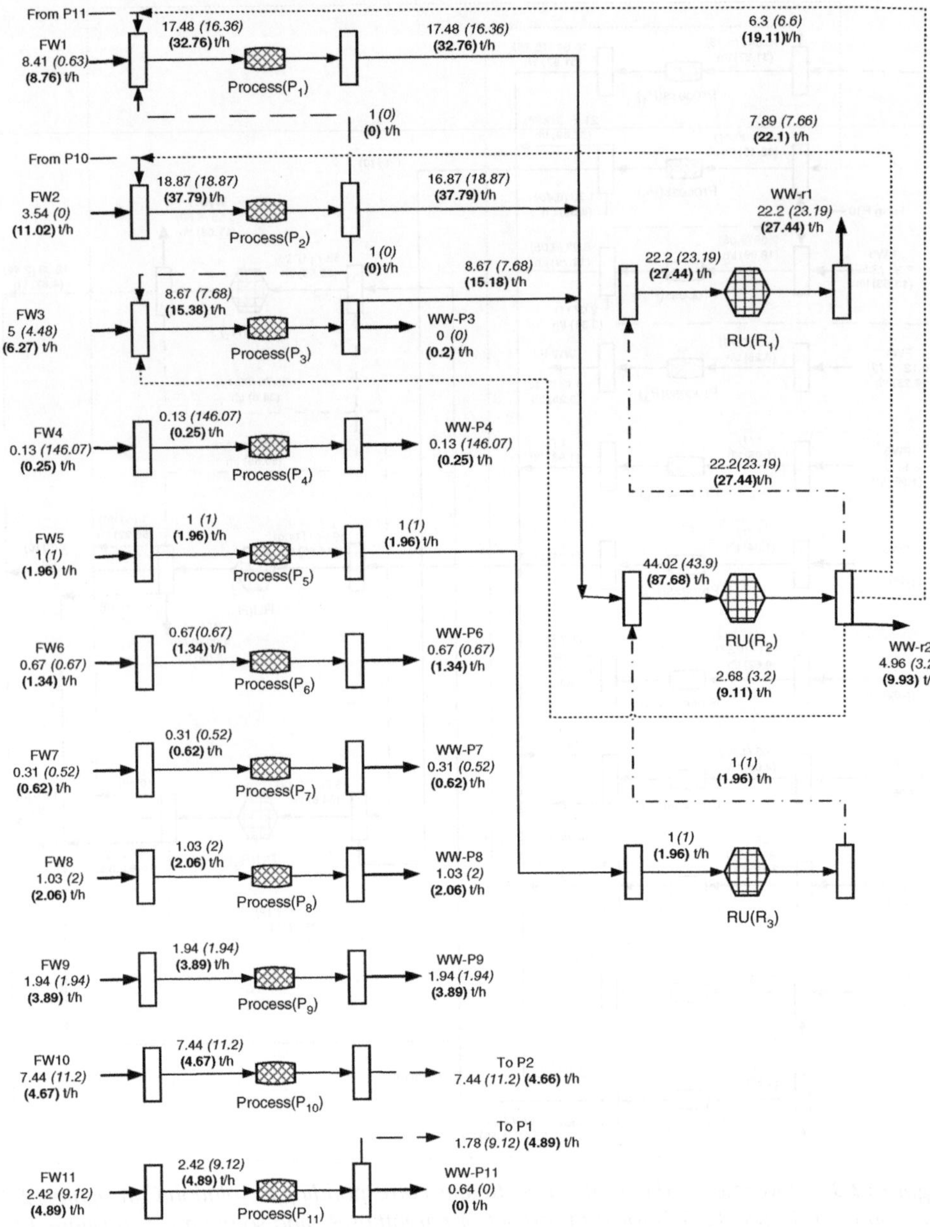

Figure 12.4 *Flow rates of different water streams corresponding to solution G in Figure 12.2, solution g in Figure 12.5 (shown in first set of parentheses) and solution g′ in Figure 12.6 (shown in second set of parentheses).*

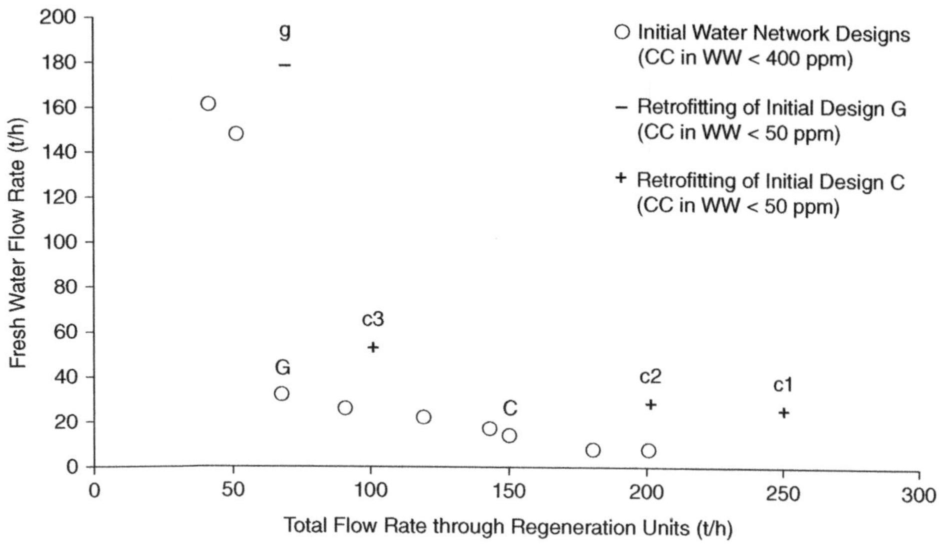

Figure 12.5 *Retrofit solutions of original solutions C and G for decrease in contaminant concentration in waste water; interconnections are the same as those in the original solution.*

The existing water network corresponding to both the solutions C and G is optimized for a lower limit (50 ppm) on contaminant concentrations in waste water, and the obtained non-dominated solutions are shown in Figure 12.5. Here, the MOO approach is used as it provides several alternatives. This optimization problem is smaller than the original refinery water network design problem; this retrofitting problem has 396 continuous variables, 23 (in case of solution C) or 13 (in case of solution G) integer variables and 586 constraints compared with 396 continuous variables, 182 integer variables and 586 constraints in the water network design problem. Note that solutions C and G have 23 and 13 interconnections, respectively, and NIC for the retrofit solutions may remain the same or decrease due to flow rate through an interconnection falling below 1 t/h.

After a number of trials via ε-constraint method, three non-dominated solutions (c1, c2 and c3) are obtained for retrofitting the original solution C, whereas only one solution (g) is obtained for retrofitting the original solution G (Figure 12.5). Important design and operating data (FW, TRU, NIC, P-to-P, P-to-R, R-to-R, R-to-P and contaminant concentrations in waste water) for retrofit solutions c1, c2, c3 and g are presented in Table 12.3. For all retrofit solutions (c1, c2, c3 and g), flow rates through some interconnections have changed to zero compared with the original solutions C and G, and so they have lower NIC values. Flow rates through some interconnections are small for original design solutions C and G in Figures 12.3 and 12.4. Since there is a limit on the minimum flow rate through each interconnection (1 t/h), NIC may decease for the retrofit solutions. Water flow rates for different streams corresponding to solutions c2 (close to solution C) and g are given in Figures 12.3 and 12.4, respectively.

Table 12.3 *Important design and operating data for the non-dominated solutions in Figure 12.5. There is no flow through interconnections in curly brackets, and so NIC has decreased for retrofit solutions (c1, c2, c3 and g) compared with NIC for original solutions C and G in Table 12.2.*

Non-dominated solution	c1	c2	c3	g
FW (t/h)	24.97	27.57	52.89	177.64
TRU (t/h)	250	200	100	68.09
NIC	18	18	14	11
P-to-P	3	3	3	2 {P_2-P_1, P_2-P_3}
P-to-R	8 {P_2-R_3, P_{11}-R_1}	8 {P_2-R_3, P_{11}-R_1}	8 {P_2-R_3, P_{11}-R_1}	4
R-to-R	2	2	1 {R_1-R_3}	2
R-to-P	5 {R_2-P_{11}, R_3-P_2, R_3-P_{11}}	5 {R_2-P_{10}, R_2-P_{11}, R_3-P_{11}}	2 {R_2-P_{10}, R_2-P_{11}, R_3-P_1, R_3-P_2, R_3-P_3, R_3-P_{11}}	3
Concentration (ppm) of contaminants in waste water leaving the system				
Salt	50	50	50	50
Organics	50	50	50	50
H_2S	7.7	11.4	29.3	7.8
NH_3	10.3	13.7	17.6	4.8

12.5.3 Retrofitting Selected Water Networks for Increase in Hydrocarbon Load

Over the years, refineries have processed crudes with different qualities such as hydrocarbon composition, sulfur content and salt content. This may increase the load of some contaminants in water outlet streams from different refinery processes. Hence, the existing water networks corresponding to solutions C and G are studied for increase in hydrocarbon load in all processes. In this retrofitting for removing the additional hydrocarbon load from each process to water stream, only flow rates through existing interconnections and capacities of regeneration units can change. Here, concentration of each contaminant in waste water is limited at 400 ppm (the same as in the original designs/solutions C and G).

The existing water network corresponding to each of the solutions C and G is optimized for 100% increase in the hydrocarbon load in each process. This retrofitting problem has 396 continuous variables, 23 (in case of solution C) or 13 (in case of solution G) integer variables and 586 constraints compared with 396 continuous variables, 182 integer variables and 586 constraints in the water network design problem. Note that solutions C and G have 23 and 13 interconnections, respectively and NIC for the retrofit solutions may decrease or remain the same. After a number of trials via the ε-constraint method, five retrofit solutions (c1′, c2′, c3′, c4′ and c5′) are obtained for the original solution C, and only one retrofit solution (g′) is obtained for the original solution G (Figure 12.6).

Important design and operating data (FW, TRU, NIC, P-to-P, P-to-R, R-to-R, R-to-P and contaminant concentrations in waste water) for solutions c1′, c2′, c3′, c4′, c5′ and g′ are presented in Table 12.4. For retrofit solutions c1′, c2′, c3′, c5′ and g′, flow rates for some interconnections have changed to zero compared with the original solutions C and G, and so they have lower NIC values. But retrofit solution c4′ and original solution C have the

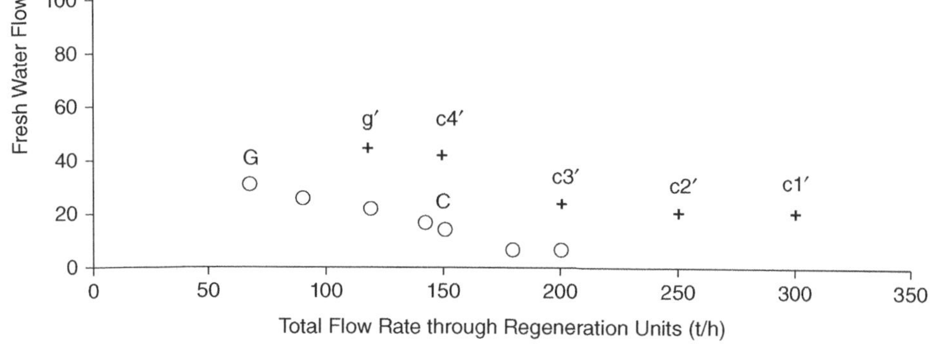

Figure 12.6 *Retrofit solutions of original solutions C and G for 100% increase in hydrocarbon load (HL) in each process; interconnections are the same as those in the original solution.*

Table 12.4 *Important design and operating data for the non-dominated solutions in Figure 12.6. There is no flow through interconnections in curly brackets, and so NIC has decreased for retrofit solutions (c1′, c2′, c3′ and g′) compared with NIC for original solutions C and G in Table 12.2.*

Non-dominated solution	c1′	c2′	c3′	c4′	c5′	g′
FW (t/h)	19.98	20.27	23.65	41.7	109.37	45.73
TRU (t/h)	300	250	200	150	100	117.08
NIC	18	18	16	23	10	11
P-to-P	2 {P_3-P_1}	3	3	3	3	2 {P_2-P_1, P_2-P_3}
P-to-R	9 {P_{11}-R_1}	8 {P_8-R_2, P_1-R_1}	7 {P_2-R_3, P_8--R_2, P_{11}-R_1}	10	4 {P_2-R_3, P_5-R_2, P_6-R_2, P_8-R_2, P_9-R_2, P_{11}-R_1}	4
R-to-R	2	2	2	2	1 {R_1-R_3}	2
R-to-P	5 {R_1-P_{11}, R_2-P_{10}, R_2-P_{11}}	5{R_2-P_{10}, R_2-P_{11}, R_3-P_{11}}	4 {R_2-P_{10}, R_2-P_{11}, R_3-P_2, R_3-P_{11}}	8	2 {R_2-P_{10}, R_2-P_{11}, R_3-P_1, R_3-P_2, R_3-P_{10}, R_3-P_{11}}	3
Concentration (ppm) of contaminants in waste water leaving the system						
Salt	400	400	400	400	400	400
Organics	400	400	400	400	400	400
H_2S	24.04	29.19	29.13	26.1	14.17	30.39
NH_3	35.93	36.02	34.67	25.42	8.5	19.22

same NIC (that is, 23). Water flow rates for different streams corresponding to solutions c3′ and g′ are also given in Figures 12.3 and 12.4, respectively.

It can be seen that values of both objectives for retrofit designs c2 and c3′ do not change much compared with the initial design C, whereas retrofit design g has significantly higher fresh water consumption compared with the initial design G (Tables 12.2, 12.3 and 12.4). Hence, design C is preferable over design G for implementation. In Figures 12.3 and 12.4, some flow rates through interconnections have changed significantly (for example, P_2-R_2 and R_2-P_2), and so additional and/or larger pipes may be required. Waste water produced by retrofit design c3′ has more salt and organics, and so it requires treatment processes (reverse osmosis, API separator with ACA) to reduce the salt and organics concentrations below 50 ppm, before discharge. It is worth mentioning that values of $F_{inlet,j=1}$, $F_{inlet,j=2}$ and $F_{inlet,j=3}$ for solution C are, respectively, 53.7, 59.5 and 36.8 t/h, whereas average values of $F_{inlet,j=1}$, $F_{inlet,j=2}$ and $F_{inlet,j=3}$ for all retrofit solutions (c1, c2, c3, c1′, c2′, c3′, c4′ and c5′) are 46.7, 116.9 and 30.1 t/h, respectively. Flow rate through the second treatment unit ($F_{inlet,j=2}$) increased significantly for the retrofit solutions; this treatment unit (API separator with ACA) can reduce hydrocarbon concentration in waste water, which has increased due to 100% increase in the hydrocarbon load for each process.

12.5.4 Revamping Selected Water Networks for Change in Environmental Regulations

In the case of retrofitting presented in the previous two sub-sections, capacities of regeneration units and flow rates through existing interconnections are allowed to vary but interconnections are assumed to be fixed as in the original design. In this and the next sub-section, interconnections between processes and/or regeneration units are also allowed to change along with capacities of regeneration units for meeting the requirements (namely, change in environmental regulations and increased hydrocarbon load), and this is referred to as revamping of water network. The optimization problem for this revamping is identical to that for a new design.

In the water network revamping for change in environmental regulations, two different limits (50 and 100 ppm) on concentration of contaminants in waste water are considered. As before, the MOO approach is used as it provides several alternatives. The number of decision variables and constraints in the optimization problem for revamping are the same as those in the original design (that is, the revamping problem has 396 continuous variables, 182 integer variables and 586 constraints). A number of trials were made via the ε-constraint method to find the non-dominated solutions. Here, BARON solver gave local solutions after 4000 seconds for the 50 ppm limit on contaminant concentrations in waste water, and so it was allowed to run for 8000 and 4000 seconds, respectively, for the 50 and 100 ppm limits on contaminant concentrations in waste water. It can be seen in Figure 12.7 that BARON solver found fewer non-dominated solutions for the lower limit on concentrations of contaminants in waste water, which is probably due to the decrease in the feasible search space.

Figure 12.7 shows the Pareto-optimal fronts obtained for three different limits on contaminant concentrations in waste water. Table 12.5 presents important design and operating data (FW, TRU, NIC, P-to-P, P-to-R, R-to-R, R-to-P and concentrations of contaminants

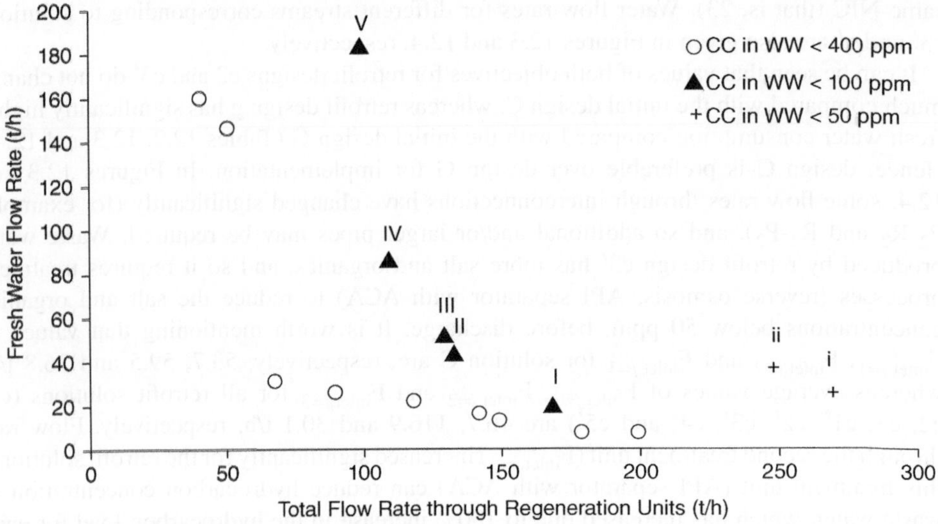

Figure 12.7 Simultaneous minimization of fresh water flow rate and total flow rate through regeneration units, for water network design/revamping for different limits on contaminant concentrations in waste water (CC in WW).

in waste water) for non-dominated solutions obtained with both 50 and 100 ppm limits on contaminant concentrations in the waste water stream. Data on non-dominated solutions obtained with a 400 ppm limit on contaminant concentrations in waste water are already presented in Table 12.2. Three different limits (that is, 50, 100 and 400 ppm) on the concentrations of contaminants in waste water are useful to understand the shift of the Pareto-optimal front with change in environmental regulations.

Table 12.5 Important design and operating data for the non-dominated solutions in Figure 12.7.

Non-dominated solution	I	II	III	IV	V	i	ii
FW (t/h)	19.56	43.59	51.16	85.39	184.32	25.97	36.56
TRU (t/h)	170	133.46	130	110	100	271.37	250
NIC	20	23	16	18	16	23	19
P-to-P	8	14	8	8	7	6	6
P-to-R	6	3	5	3	4	8	8
R-to-R	1	3	1	1	1	2	1
R-to-P	5	3	2	6	4	7	4
Concentration (ppm) of contaminants in waste water leaving the system							
Salt	100	100	100	100	100	25.31	50
Organics	100	100	100	100	100	50	50
H_2S	41.08	34.06	30.30	18.15	8.41	6.92	42.39
NH_3	37.42	22.28	18.18	10.89	5.05	11.46	25.44

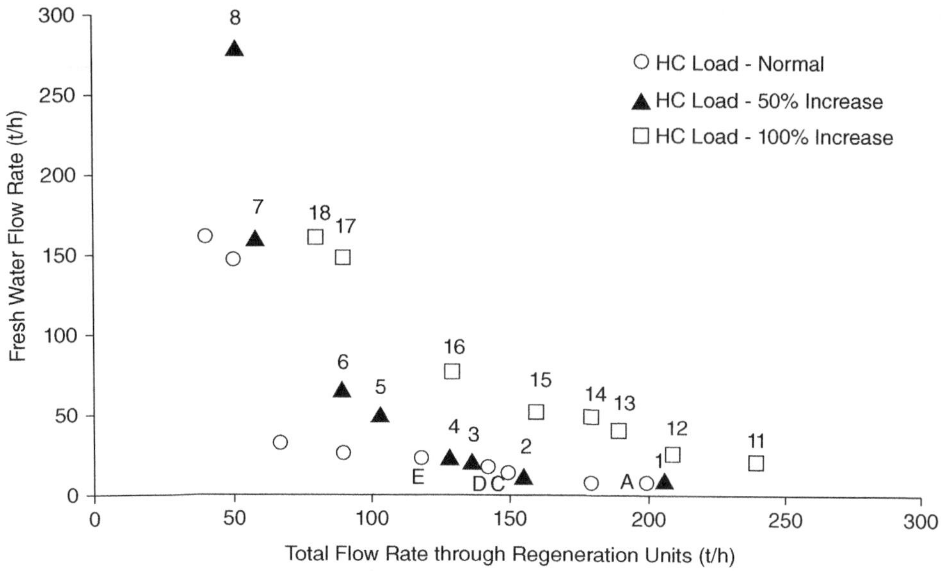

Figure 12.8 Simultaneous minimization of fresh water flow rate and total flow rate through regeneration units, for water network design/revamping for 50% and 100% increase in hydrocarbon load in the refinery processes considered.

In general, FW and TRU for the non-dominated solutions obtained with 100 and 400 ppm limits on contaminant concentrations in waste water stream are comparable (Tables 12.2 and 12.5), and TRU is significantly larger for the non-dominated solutions obtained with 50 ppm limit on contaminant concentrations in waste water stream. Furthermore, H_2S and NH_3 in waste water are below 50 ppm, and salts and organics in waste water are more than 50 ppm, for all the non-dominated solutions obtained with 100 ppm limit on contaminant concentrations in waste water. Hence, only two additional treatment units (reverse osmosis, and API separator with ACA) are required to treat the waste water before discharge. On the other hand, concentrations of all four contaminants are more than 50 ppm for some of the non-dominated solutions obtained with 400 ppm limit on concentrations of contaminants in waste water stream (see Table 12.2), and so all three treatment units will be required as end-of-pipe treatment.

12.5.5 Revamping Selected Water Networks for Increase in Hydrocarbon Load

In the water network revamping for increase in hydrocarbon load, two different values for increase in hydrocarbon load (50 and 100% for each process) are considered. In total, three different values of hydrocarbon load in each process unit (that is, normal, 50% increase, 100% increase) are useful to understand the shift of Pareto-optimal front with change in hydrocarbon load. Here, concentration of each contaminant in waste water is limited at 400 ppm (the same as in the original design).

Figure 12.8 shows the Pareto-optimal fronts obtained for three different hydrocarbon loads. Four non-dominated solutions (1, 2, 3 and 4), obtained with 50% increase in hydrocarbon load, have similar values of both objective functions as the non-dominated solutions

Table 12.6 *Important design and operating data for the non-dominated solutions in Figure 12.8.*

Non-dominated solution for 50% increase in HC load	1	2	3	4	5	6	7	8
FW (t/h)	10.47	12.05	20.89	23.91	49.78	65.54	160.13	277.43
TRU (t/h)	206.35	155.32	136.73	129.21	103.74	90	57.84	50
NIC	34	22	20	22	16	17	6	6
P-to-P	15	8	7	15	7	8	3	4
P-to-R	5	7	7	4	4	4	3	2
R-to-R	5	2	1	1	1	1	–	–
R-to-P	9	5	5	2	4	4	–	–
Concentration (ppm) of contaminants in waste water leaving the system								
Salt	400	293.08	109.57	400	400	400	400	400
Organics	400	400	400	400	400	400	400	400
H_2S	45.27	128.62	74.21	64.83	31.04	23.65	9.68	5.59
NH_3	46.55	77.17	44.53	38.90	19.07	14.19	5.81	3.35
Non-dominated solution for 100% increase in HC load	11	12	13	14	15	16	17	18
FW (t/h)	20.51	25.37	39.73	47.93	51.81	76.42	147.29	160.12
TRU (t/h)	240	210	189.96	180	160	129.39	90	80
NIC	27	21	15	18	11	13	26	6
P-to-P	8	11	6	9	2	7	14	2
P-to-R	8	4	5	4	5	4	6	3
R-to-R	3	1	1	1	1	1	1	–
R-to-P	8	5	3	4	3	1	5	1
Concentration (ppm) of contaminants in waste water leaving the system								
Salt	400	400	304.58	296.49	400	400	400	400
Organics	400	400	400	400	400	400	400	400
H_2S	36.95	57.58	39.02	31.31	29.92	20.28	10.45	9.68
NH_3	29.72	36.04	23.41	19.44	17.95	12.17	6.69	5.81

obtained with normal hydrocarbon load in all processes; solutions 1, 2, 3 and 4 are close to solutions A, C, D and E, respectively (see Figure 12.8). On the other hand, the Pareto-optimal front for 100% increase in hydrocarbon load is clearly away from the other two Pareto-optimal fronts in Figure 12.8. Table 12.6 presents important design and operating data (FW, TRU, NIC, P-to-P, P-to-R, R-to-R, R-to-P and contaminant concentrations in waste water) for non-dominated solutions obtained for 50% and 100% increase in hydrocarbon load. Data for non-dominated solutions obtained with normal hydrocarbon load are available in Table 12.2.

In this study, it is assumed that waste water having a concentration of each contaminant below 50 ppm can directly be discharged. For the case of 50% increase in the hydrocarbon load, non-dominated solutions 2, 3 and 4 require all three treatment units (reverse osmosis to decrease salt, API separator with ACA to remove organics, and Chevron waste water treatment to reduce H_2S and NH_3) as end-of-pipe treatment to meet the water discharge regulations. Waste water corresponding to the remaining non-dominated solutions 1, 5, 6, 7 and 8 has to be treated to reduce salts and organics concentrations using reverse osmosis and API separator with ACA, before its discharge.

All non-dominated solutions (except solution 12) for 100% increase in the hydrocarbon load require only two treatment units (reverse osmosis, and API separator with ACA) as end-of-pipe treatment. In general, FW and TRU are significantly larger for the non-dominated solutions obtained with 100% increase in hydrocarbon load compared with those obtained with 50% increase. Owing to this, H_2S and NH_3 concentrations are well below the discharge limit (that is, 50 ppm). Hence, in this case, only two treatment units are required as end-of-pipe treatment.

12.5.6 Comparison of Retrofitting and Revamping Solutions

Revamping involves more changes than does retrofitting and so it is expected to give better solutions compared with those obtained for retrofitting. In the case of 100% increase in the hydrocarbon load in each process, solutions obtained from retrofitting and revamping are comparable (for example, retrofit solution c2′ in Table 12.4: FW = 20.27 t/h and TRU = 250 t/h, and revamp solution 11 in Table 12.6: FW = 20.51 t/h and TRU = 240 t/h). However, solutions obtained for retrofitting (for example, c1 in Table 12.3: FW = 24.97 t/h and TRU = 250 t/h) are slightly better than those for revamping (for example, i in Table 12.5: FW = 25.97 t/h and TRU = 271.37 t/h), for 50 ppm limit on the concentrations of contaminants in waste water leaving the water network.

The second treatment unit (API separator with ACA) can reduce the hydrocarbon concentration to 50 ppm only, and so it is difficult to solve the retrofitting/revamping problem with a 50 ppm limit on the concentrations of contaminants in waste water. Furthermore, the retrofitting problem has fewer binary variables (solution C = 23, and solution G = 13) compared with the revamping problem (182), and also there is a limit on the minimum flow rate through each interconnection (1 t/h). It can be seen that revamp problems are more difficult to solve compared to retrofit problems, which leads to local solutions for revamp problems. Hence, revamp solutions are inferior than retrofit solutions, for stringent environmental regulations. If the minimum flow rate through interconnections is 0.001 t/h, solutions obtained for revamp (solution 1 with FW = 13.74 t/h and TRU = 236.46 t/h; and solution 2 with FW = 23.15 t/h and TRU = 200 t/h) are better than those for retrofit, for 50 ppm limit on the concentrations of contaminants in waste water. There are 24 NIC for each of revamp solution 1 and 2; these two solutions have, respectively, 7 and 8 interconnections with flow rate less than 1.0 t/h.

12.6 Conclusions

In this chapter, the design and retrofitting of a water network are optimized for both fresh water consumption and total water flow rate at the inlet of regeneration units, simultaneously. The obtained Pareto-optimal front gives better insights and choices by providing a range of alternative designs. Selected initial water network designs are further studied for stringent environmental regulations and increased hydrocarbon/contaminant loads from process units. In this retrofitting, interconnections between processes and/or regeneration units are fixed based on the initial design. The results show that additional and/or larger pipes are required for some of the interconnections, and capacities of regeneration units have to be increased significantly. Results obtained for both grassroots design and retrofit can be considered in the selection of one water network design for implementation. Subsequently, revamping of the refinery water network (which is similar to grassroots design) is also explored for different environmental regulations and hydrocarbon loads from process units. Hence, in the water network design, it is recommended to select a few Pareto-optimal solutions and to explore the changes required in the water network designs, through multi-objective optimization, for future requirements in the plant capacity and/or operation.

Acknowledgement

This research grant is supported by the Singapore National Research Foundation under its Environmental & Water Technologies Strategic Research Programme and administered by the Environmental & Water Industries Programme Office (EWI) of the Public Utilities Board (PUB).

Nomenclature

Acronyms

AF	Annualization factor for capital cost
API separator	Separator for water–oil mixture (American Petroleum Institute)
ACA	Ammoniacal copper arsenate
BARON	Branch-and-reduce optimization navigator
CC	Contaminant concentration
GAMS	General algebraic modeling system
HC	Hydrocarbon
K	Number of contaminants
MILP	Mixed-integer linear programming
MINLP	Mixed-integer non-linear programming
MIX	Water mixer
MOO	Multi-objective optimization
NIC	Number of interconnections between processes and/or regeneration units
NP	Number of processes
NR	Number of regeneration units
OF	Objective function
P	Process unit

R	Regeneration unit
SOO	Single-objective optimization
SPT	Water splitter

Symbols

C_{FW}	Cost of fresh water [$/t]
F_{IC}	Minimum flow rate through interconnections [t/h]
F_{inlet}	Water flow rate at the inlet of process/regeneration unit [t/h]
F_{outlet}	Water flow rate at the outlet of process/regeneration unit [t/h]
F^{pp}	Water flow rate from one process to another process [t/h]
F^{pr}	Water flow rate from process to regeneration unit [t/h]
F^{rp}	Water flow rate from regeneration unit to process [t/h]
F^{rr}	Water flow rate from one regeneration unit to another regeneration unit [t/h]
FW	Fresh water flow rate [t/h]
F_{WW}	Total waste water flow rate from the network [t/h]
IC	Investment cost coefficient [$/{year.(t/h)$^{0.7}$}]
L	Contaminant load in different processes [kg/h]
OC	Operating cost coefficient [$/ton]
TRU	Total flow through regeneration units [t/h]
WW	Waste water flow rate from process or regeneration unit [t/h]
Z_{inlet}	Contaminant concentration at the inlet of process/regeneration unit [ppm]
Z_{outlet}	Contaminant concentration at the outlet of process/regeneration unit [ppm]
Z_{WW}	Contaminant concentration in waste water stream from the network [ppm]
η	Contaminant removal ratio in regeneration units [%]

Subscripts/Superscripts

i	Process unit number
j	Regeneration unit number
k	Contaminant type
max	maximum
min	minimum

Exercises

12.1 Write down the model equations for water network involving only processes, and no regeneration units. Calculate the fresh water consumption for each process, using loads of different contaminants in each process and maximum outlet concentrations of different contaminants in water streams from each process (given Table 12.1).

12.2 Consider a smaller refinery water network problem by ignoring some processes in Table 12.1, and perform design, retrofitting and revamping via multi-objective optimization for two objectives, namely, fresh water consumption and total flow through the regeneration units, as in this chapter.

12.3 For the refinery processes in Table 12.1, design the water network by optimization for three important objectives, namely, fresh water consumption, total flow through the regeneration units and number of interconnections between processes and/or regeneration units. Compare the obtained results with those given in Figure 12.2 for two objectives. This will require many trials using the ε-constraint method or a multi-objective optimization program for several objectives.

References

Ahmetovic, E. and Grossmann, I.E. (2011) Global superstructure optimization for the design of integrated process water networks. *AIChE Journal*, **57** (2), 434–457.

Aviso, K.B., Tan, R.R. and Culaba, A.B. (2010) Designing eco-industrial water exchange networks using fuzzy mathematical programming. *Clean Technologies and Environmental Policy*, **12** (4), 353–363.

Bagajewicz, M.J. and Savelski, M. (2001) On the use of linear models for the design of water utilization systems in process plants with a single contaminant. *Chemical Engineering Research and Design*, **79** (5), 600–610.

Bogataj, M. and Bagajewicz, M.J. (2008) Synthesis of non-isothermal heat integrated water networks in chemical processes. *Computers and Chemical Engineering*, **32** (12), 3130–3142.

Boix, M., Montastruc, L., Pibouleau, L. *et al.* (2011) A multi-objective optimization framework for multicontaminant industrial water network design. *Journal of Environmental Management*, **92**, 1802–1808.

Boix, M., Pibouleau, L., Montastruc, L. *et al.* (2012) Minimizing water and energy consumptions in water and heat exchange networks. *Applied Thermal Engineering*, **36**, 442–455.

Bonhivers, J.C. and Stuart, P.A. (2013) Application of process integration methodologies in the pulp and paper industry, in *Handbook of Process Integration: Minimization of Energy and Water Use, Waste and Emissions*, (ed. J.J. Klemes), Woodhead Publishing, Cambridge.

Castro, P.M. and Teles, J.P. (2013) Comparison of global optimization algorithms for the design of water using networks. *Computers and Chemical Engineering*, **52**, 249–261.

Deng, C., Feng, X. and Wen, Z. (2013) Optimization of water network integrated with process models. *Clean Technologies and Environmental Policy*, **15** (3), 473–487.

Dunn, R.F. and Wenzel, H. (2001) Process integration design methods for water conservation and waste water reduction in industry. *Clean Products and Processes*, **3** (3), 307–318.

El-Halwagi, M.M. (2012) *Sustainable Design through Process Integration*, Butterworth-Heinemann Ltd./Elsevier, Oxford, U.K.

Faria, D.C. and Bagajewicz, M.J. (2009) Profit based grassroots design and retrofit of water networks in process plants. *Computers and Chemical Engineering*, **33** (2), 436–453.

Faria, D.C. and Bagajewicz, M.J. (2011a) Global optimization of water management problems using linear relaxation and bound contraction methods. *Industrial and Engineering Chemistry Research*, **50** (7), 3738–3753.

Faria, D.C. and Bagajewicz, M.J. (2011b) Planning model for the design and/or retrofit of industrial water systems. *Industrial and Engineering Chemistry Research*, **50** (7), 3788–3797.

Jezowski, J. (2010) Review of water network design methods with literature annotations. *Industrial and Engineering Chemistry Research*, **49** (10), 4475–4516.

Kirthick, R., Kumaraprasad, G. and Sruti, B. (2010) Hybrid optimization approach for water allocation and mass exchange network. *Resources, Conservation and Recycling*, **54** (11), 783–792.

Manan, Z.A., Tea, S.Y. and Alwi, S.R.W. (2009) A new technique for simultaneous water and energy minimization in process plant. *Chemical Engineering Research and Design*, **87** (11), 1509–1519.

Mann, J. and Liu, Y.A. (1999) *Industrial Water Reuse and Waste Water Minimization*, McGraw Hill Professional.

Masuduzzaman and Rangaiah, G.P. (2009) Multi-objective optimization applications in chemical engineering, in *Multi-objective Optimization: Techniques and Applications in Chemical Engineering*, (ed. G.P. Rangaiah), World Scientific, Singapore.

Poplewski, G. and Jezowski, J.M. (2010) Application of adaptive random search optimization for solving industrial water allocation problem, in *Stochastic Global Optimization: Techniques and Applications in Chemical Engineering* (ed. G.P. Rangaiah), World Scientific, Singapore.

Ramos, M.A., Boix, M., Montastruc, L. and Domenech, S. (2014) Multi-objective optimization using goal programming for industrial water network design. *Industrial and Engineering Chemistry Research*, **53** (45), 17722–17735.

Rangaiah, G.P. (ed.) (2009) *Multi-Objective Optimization: Techniques and Applications in Chemical Engineering*, World Scientific, Singapore.

Sharma, S. and Rangaiah, G.P. (2013) Multi-objective optimization applications in chemical engineering, in *Multi-Objective Optimization in Chemical Engineering: Developments and Applications* (eds G.P. Rangaiah, and A. Bonilla-Petriociolet), Wiley, Chichester.

Sharma, S. and Rangaiah, G.P. (2014) Multi-objective Optimization of Heat Integrated Water Networks in Petroleum Refineries. Proceedings of the 24th European Symposium on Computer Aided Process Engineering, Budapest, Hungary, pp. 1531–1536.

Smith, R. (2005) *Chemical Process: Design and Integration*, Wiley, Chichester.

Sotelo-Pichardo, C., Ponce-Ortega, J.M., El-Halwagi, M.M. and Frausto-Hernandez, S. (2011) Optimal retrofit of water conservation networks. *Journal of Cleaner Production*, **19**, 1560–1581.

Thibault, J. (2009) Net flow and rough set: two methods for ranking the Pareto domain, in *Multi-Objective Optimization: Techniques and Applications in Chemical Engineering* (ed. G.P. Rangaiah), World Scientific, Singapore.

Wang, Y.P. and Smith, R. (1994) Waste water minimization. *Chemical Engineering Science*, **49** (7), 981–1006.

Yang, L., Salcedo-Diaz, R. and Grossmann, I.E. (2014) Water network optimization with waste water regeneration models. *Industrial and Engineering Chemistry Research*, **53** (45), 17680–17695.

13

Debottlenecking and Retrofitting of Chemical Pulp Refining Process for Paper Manufacturing – Application from Industrial Perspective

Ajit K. Ghosh
AKG Process Consulting, 33 McFarlane Court, Highett, Australia

13.1 Introduction

No manufactured product plays a more significant role in every area of human activity than does paper and paper products. Its importance in everyday life is obvious from its use in recording to the storage and dissemination of information. Virtually all writing and printing is done on paper. It is most widely used in wrapping and packaging material, and is important for structural applications. The uses and applications of paper and paper products are virtually limitless. Apart from the products and services that it provides, the paper and pulp industry is one of the major manufacturing industries in the world, providing employment for a vast number of people and contributing to the national economy.

Paper manufacturing is a complicated process depending upon the types of paper manufactured, end user applications and starting raw materials. However, there are three basic steps in the manufacturing process, namely forming, pressing and drying. In the forming stage, a very dilute slurry of pulp (less than 1% solid) forms a wet fibre mat to a solid level between 15% and 25% through very large drainage operations. Mechanical compression in the press section removes water to leave a solid level of between 33% and 55%, depending

Chemical Process Retrofitting and Revamping: Techniques and Applications, First Edition.
Edited by Gade Pandu Rangaiah.

on the paper grade and press section design. The remaining water is removed by the drying process to give a final water content of between 6% and 8%. Prior to the forming stage, wood fibres present in the pulp slurry are treated by a mechanical process, known as pulp refining. This is particularly important for virgin fibres (that have never been used earlier in papermaking). In this process, fibres are 'modified' by the action of rotating bars opposite to a stationary bedplate as the fibres flow past perpendicularly to the bars. In the pulp refining process, electrical energy applied through a drive motor is transferred to the fibre for fibre treatment or modification. The amount of net energy required for fibre development depends upon the type of fibre source (such as hardwood or softwood) and on the type of paper grades manufactured (such as linerboards for packaging, sack kraft for bags and sacks, fine papers for printing and writing and newsprints). Per ton of paper manufactured, the refining energy requirement could vary between 50 kWh and 250 kWh. For a modern paper machine, producing 400,000 ton/year of finished product, the yearly cost of refining energy could be potentially millions of dollars.

Besides its high energy cost, the refining operation of the paper mill may be a bottleneck in terms of increasing throughput. The finished paper product must meet the specifications of end-user requirements. Quite often, the mill may find it very difficult to achieve target specifications of finished products or/and to increase throughput, as the installed refining capacity may be sub-optimal, even though other unit operations of the paper-making process may have an excess of capacity. The addition of extra refining capacity may be financially prohibitive, as the cost of a new installation could be in excess of millions of dollar per refiner, including associated control hardware. In some cases, the refiner tackles may not be optimally designed, and retrofitting such components may debottleneck production output. In other cases, the mill may decide to manufacture different grade of papers where refining requirements might be different to those for the current product mix. In such cases, the mill would like to use as much as possible of the existing refining system, but still needs to retrofit different kinds of refiners to satisfy the specific requirements. In this chapter, details of two case studies will be discussed where: (i) existing refiner tackles were retrofitted to improve product quality, increased throughput and energy saving and; (ii) the mill switched to manufacture a high-value premium grade of paper by upgrading some refining system and retrofitting one existing refiner.

13.2 Fundamentals of Chemical Pulp Refining

The refining of chemical pulps is a mechanical treatment process that modifies the fibres. The main objective of refining is to improve the bonding ability of fibres so that they can form a strong and smooth paper sheet with good printing properties. The treatment of fibres is carried out in the presence of water and metallic bars inside a refiner where both mechanical and hydraulic forces are used to alter fibre characteristics. The refiner plates or fillings are grooved so that the bars that treat fibres and the grooves between bars allow transportation through the refining machine or refiner. Various stages of the refining process are described by Espenmiller (1969). Initially, fibre flocs are collected at the leading edge of the bars where the consistency is typically 3–5%, the remainder being water. In the next stage, when the leading bar of the rotor approaches that of the stator, the fibre floc is compressed and receives a strong hit, causing water expulsion and likely peeling of shorter fibres that flow into the grooves between the bars. Only the remaining fibres in the floc are compressed between two metallic bars and are 'refined'. After this, both leading edges slide along the fibre floc and press it against the flat bar surface. Most refining is performed during this edge-to-surface stage when bar edges give mechanical treatment and friction between fibres give fibre-to-fibre treatment inside the floc. This stage continues until the leading edges reach the tailing edges of the opposite bars. After this stage, the fibre bundle is still pressed between the flat bar surface until the tailing edge of the rotor bar has passed the tailing edge of the stator bar. All the above stages combined comprise one refining impact on the fibre bundle, and the length of the refining impact depends on the width and intersecting angles of the bars (Lumiainen, 1997).

When rotor bars move across the stator bars, there are quite strong vortex flows in the grooves between the bars. This phenomenon causes the fibre to become stapled on bar edges during the fibre pick-up stage. If the grooves are too narrow, the fibre or fibre flocs cannot rotate in the groove and do not get stapled on the bar edge. Those fibres remain in the grooves and pass the refiner without receiving any refining impact. The leading bar edge catches fibres when the bar edge ploughs through the pulp suspension, and the moving rotor orientates those fibres so that they form fibre bundles on the bar edge. Consequently, the part of the fibres which remains on the side of the bar surface will be sheared and compressed between the surfaces of opposing bars. The stapling of fibres on the bar edges and the behaviour of fibres in the flocs during refining impact greatly in determining the refining results. The type of chemical pulp, for example, long-fibred softwood pulp or short-fibred hardwood pulp, plays a critical role in determining the tackle parameters for optimal refining.

13.2.1 Refining Effects on Various Chemical Pulp Types

The major effects of refining (Page, 1989; Smook, 1989) on individual fibres are summarized in Table 13.1. The first effect is the partial removal of the primary cell wall. Removal of this wall exposes the secondary wall and promotes hydration and fibre flexibility. The further action of fibrillation involves loosening of the fibrils and the raising of finer micro-fibrils

Table 13.1 *Effect of refining on fibre characteristics.*

Type	Effects
Primary	1. Removal of primary wall; formation of fibre debris or fines 2. Penetration of water into cell wall ('bruising' or 'swelling') 3. Breaking of some intra-fibre hydrogen bonds; replacement by water–fibre-hydrogen bonds 4. Increased fibre flexibility 5. External fibrillation and foliation 6. Fibre cutting or shortening
Secondary	1. Fractures (cracks) in the cell wall 2. Fibre stretching and/or compression 3. Partial solubilization of surface hemicelluloses into gels 4. Straightening of fibres (LC refining) 5. Curling of fibres (HC refining)

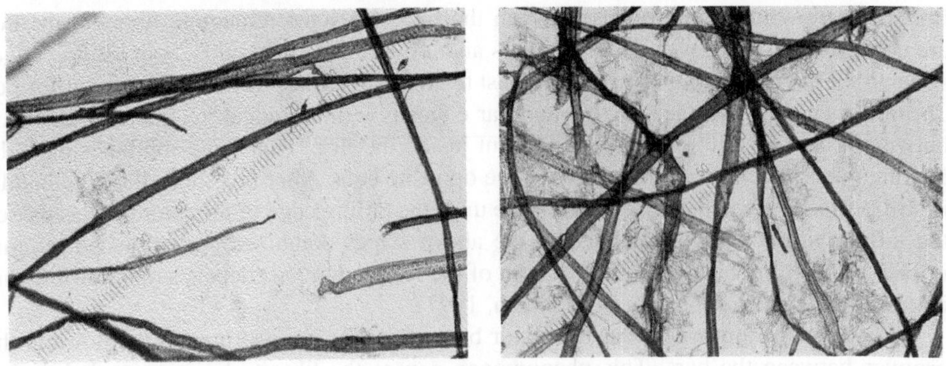

(a) (b)

Figure 13.1 *(a) Before refining (magnification 100); (b) After refining (magnification 100).*

on the surface of the fibres, resulting in a very large increase in surface area of the refined fibres.

The main effects are visible in Figures 13.1a and b which show unrefined and refined pulp after applying a net refining energy of 200 kWh/t for unbleached kraft softwood (*Pinus radiata*) pulp. Micrographs of bleached softwood kraft pulp at higher magnifications of 350 and 900 are shown in Figures 13.2a and b. In these two magnified micrographs, the proof of extensive fibrillation is clearly evident.

13.2.2 Effects of Refining on Pulp and Paper Properties

The speed at which a paper machine runs, and hence its throughput, is largely dictated by the de-watering process in the forming stage of papermaking. This depends upon the type of pulp used and the degree of refining undertaken in the stock preparation. The parameter that is commonly used by papermakers in characterizing the refining level is called *freeness*, the unit of which is millilitre (ml). The higher the freeness value of the pulp, the easier it is to dewater, but it has a lower fibre development potential.

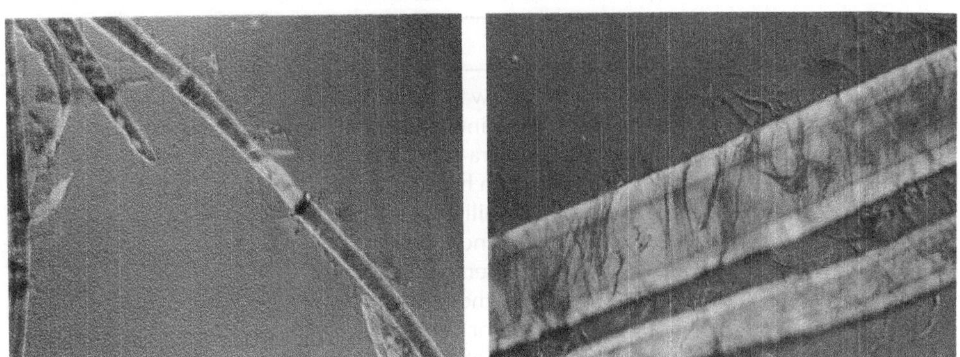

(a) (b)

Figure 13.2 *(a) After refining (magnification 350); (b) After refining (magnification 900).*

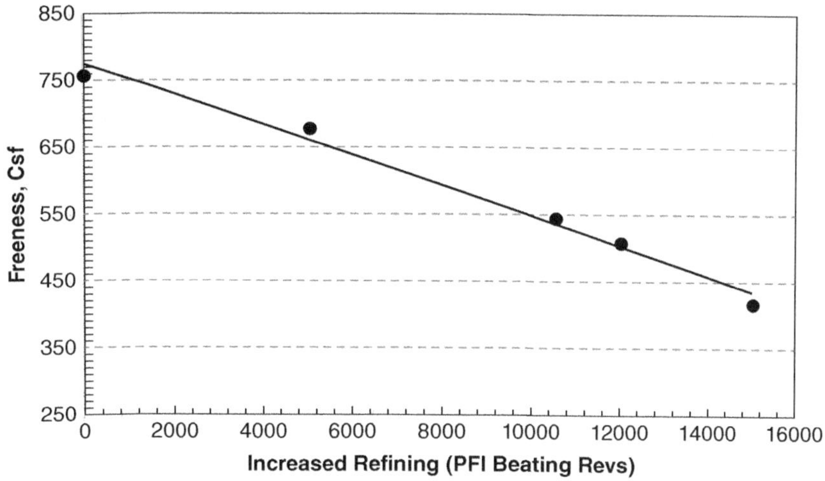

Figure 13.3 *Effect of refining on pulp freeness.*

With increased refining, the freeness value of pulps decreases due to hydration resulting from increased surface area through fibrillation. The decrease in freeness of unbleached softwood (*Pinus radiata*) kraft pulps measured during refining using a laboratory refiner or beater (often called a 'PFI mill') is shown in Figure 13.3. The development of tensile and burst strength of paper made from the same pulp through refining is shown in Figures 13.4 and 13.5, respectively. In both these figures, the strength values are indexed to grammage. The data represented by the larger square symbol correspond to a condition when the same pulp sample was not refined. Both tensile and burst strength values increased by more than 250% by refining to 450 freeness value. Porosity values of paper made from refined pulps are shown in Figure 13.6. Papers made from pulp samples with increased refining would be less porous as fibre 'fines' and fibrils would fill the voids of the fibre network. Increased refining can significantly improve key strength properties of paper, but will decrease machine speed due to lower pulp freeness and slower drainage or filtration.

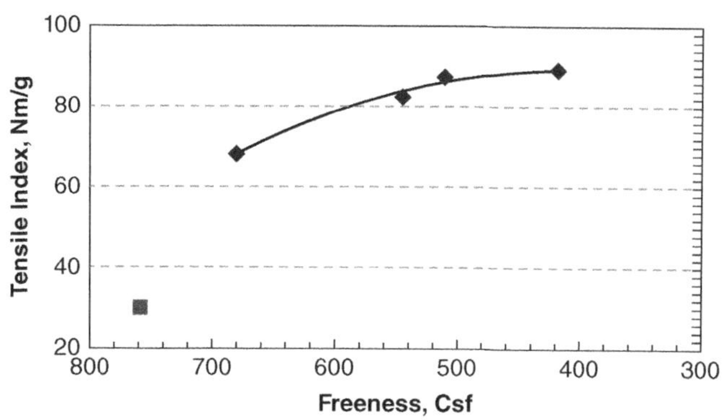

Figure 13.4 *Tensile strength development during refining (square symbol corresponds to unrefined).*

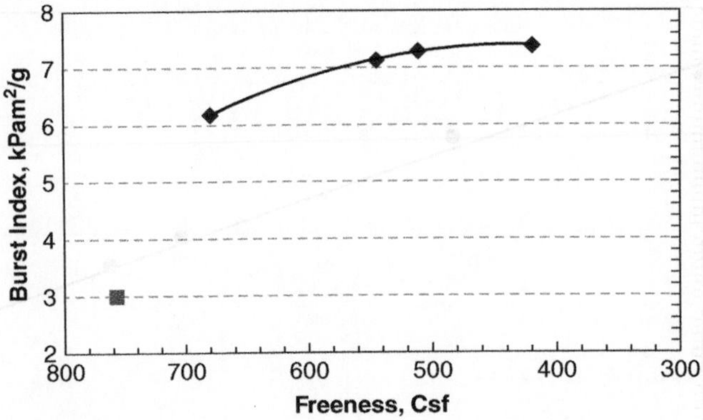

Figure 13.5 *Burst strength development during refining.*

The papermakers always make a trade-off between fibre development and machine speed or throughput. However, the ultimate objective of the paper mills is to produce a paper product that must satisfy the end-users' requirements.

13.3 Theories of Chemical Pulp Refining

The refining results are affected by a number of factors and this has been graphically depicted by Lumiainen (1995). For any paper mill, new or existing, the product mix is well defined. If the mill is fully integrated, for example, starting from wood yard, pulping and papermaking, the end products determine the required types of pulps, wood source and pulping process. Thus all the fibre-based factors would determine the refining results. Although process conditions have some influence on the refining results, designing of the mill's refining system for a given pulp or pulp type is critical. All the other factors are centred around the refiner hardware and the amount and mode of energy applied on the refiner.

Since the start of papermaking, fibre treatment has been almost a qualitative art form dependent on the skill and artistry of the operator. Fibre development has been described

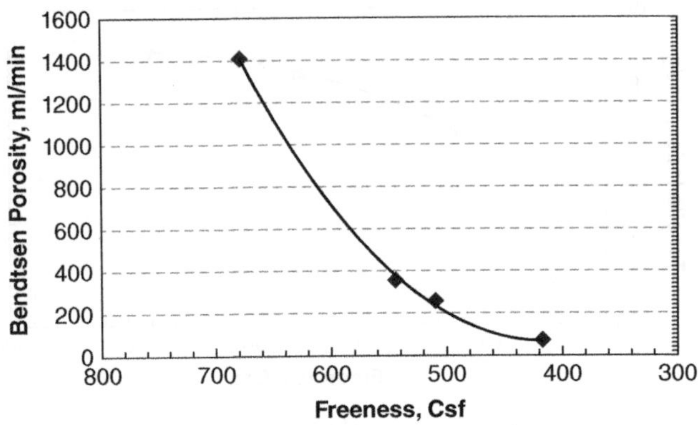

Figure 13.6 *Effect of refining on porosity.*

in terms of wetness development and there was no equivalent method for quantitative description of different types of refining processes. With the advancement of modern papermaking technology, several refining theories have been proposed (Baker, 1995) to develop an understanding of the refining process in order to improve efficiency and optimize operation; to enable quantification of the refining process; and to compare very different types of refiners under different conditions. Ebeling (1980) has presented an excellent review of those theories. A brief discussion of four such theories is given here.

13.3.1 Specific Edge Load Theory

This theory is one of the best known and widely used in the paper industry. The specific edge load (SEL) theory is based on the consideration of fibre development predominantly carried out by the edge of the bar. Wultsch and Flucher (1958) introduced the term 'refining intensity' as the quotient of effective refiner load and edge length per second. Then Brecht and Siewert (1958) defined the term 'refining intensity' as specific edge load. This theory can be described as a two-parameter characterization of chemical pulp refining. Two factors are used to describe the refining process: (i) first, how much the fibres are treated; that is, how much mechanical energy applied; and (ii) secondly, how intensively the fibres are treated; that is, how the total energy is applied. While walking with a high-heeled shoe, the impact on the ground at the point of contact of the heel is stronger compared with that when wearing a runner (sole) without a high heel, although the total pressure (or energy) applied is the same in both cases. Likewise, giving an object a single blow by hitting it with a larger hammer has more impact compared with hitting multiple times with smaller hammers. In both cases, the higher impact or intensity will create more damage compared with the low-intensity situations. In a pulp refining situation, the total energy applied to the fibres can be transferred in different modes of intensities.

The two parameters that describe the refining process by the specific edge load theory are SEL or intensity of refining and specific refining energy (SRE) and these are defined below.

$$\text{SEL} = \text{Net Power/Cutting Edge Length (CEL) per second, Ws/m} \tag{13.1}$$

$$\text{Net Power} = (\text{Applied Power} - \text{Idling Power}), \text{kW} \tag{13.2}$$

Idling load or no-load power is defined as the power drawn by the refiner motor when the refiner plates are widest apart and the refiner motor was running at normal operating rpm with stock flowing through the refiner. It can be measured or calculated. For a double-disc refiner, the idling load can be calculated from Equation (13.3) (Demler and Silveri, 1995):

$$\text{Idling Load} = (2.299 * 10^{-13}) * (D^{4.249}) * (RPM^3) \tag{13.3}$$

where D = Refiner Plate Diameter in inches, and RPM = Refiner Rotor Speed (Motor Speed).

$$\text{Cutting Edge Length (CEL)} = RPS * (n_s.n_r.L),\ \text{km/s} \tag{13.4}$$

where RPS = Revolutions Per Second, n_s = Number of moving bars, n_r = Number of stationary bars and L = Average length of one bar in km.

$$\text{SRE} = \text{(Net Power in kW)/(Fibre Flow in t/h), kWh/t} \tag{13.5}$$

Irrespective of the refiner type (for example, double-disc or conical), cutting edge length per revolution can be calculated from the geometry of the refiner plate. Normally, the vendor of the plate supplier can provide this value. In spite of several limitations of the SEL theory, such as its failure to include factors such as refining consistency, bar width, the condition of fillings, gap clearance and fibres stapling on bar edges, this theory is commonly used, due to simplicity in calculation and the ready availability of all the relevant data.

13.3.2 Specific Surface Load Theory

Specific surface load (SSL) theory is an extension of SEL theory and it assumes that the energy is transferred to the fibre bundles during the short edge-to-edge contact phase as well as during the edge-to-surface phase (Lumiainen, 1990). By combining these two, SSL is defined as shown in Equation (13.6), where Equation (13.7) holds:

$$\text{SSL} = \text{SEL} / \text{Bar width factor} \tag{13.6}$$

$$\text{with Bar width factor} = \{(W_r + W_s)/2\}/\{\cos(\alpha/2)\} \tag{13.7}$$

where, W_r = width of rotor bars in mm, W_s = width of stator bars in mm, and α = average intersecting angle in degrees.

The surface edge load theory was developed in order to account for claimed differences between conical and disc refiners, where the former gives a lower number of long bar impacts and the later gives a much greater number of short bar impacts. The theory also helps to explain the behaviour of fillings with very different widths of bar and gap. The SSL theory works quite well with coarse fillings when bars are wider than the length of the fibre flocs. The specific surface load theory still has many weak points much like the specific edge load theory, but these two theories offer very practical tools for papermakers when selecting fillings and other refining parameters for various applications.

13.3.3 Frequency and Intensity or Severity of Impact

Danforth (1969) developed two independent expressions to quantify the treatment of fibres. According to this theory, refining is expressed as a function of two factors, namely number (*N*) and severity (*S*), of the refining impacts between the bar edges [Equations (13.8) and (13.9)].

$$N = (L_r.L_s.RPM.C.K_1)/(X.R) \tag{13.8}$$

$$S = \{(HP_a - HP_b).A_t.K_2/(L_r.L_s.RPM.C)\} \tag{13.9}$$

where N = relative number of impacts, S = relative severity of impacts, HP_a = total horsepower, HP_b = no-load horsepower, A_t = total area of refining zone, L_r = total length of rotor edges, L_s = total length of stator edges, RPM = rotor speed, X = average bar contact length, R = throughput in ton/day, C = stock consistency and K_1 and K_2 are constants.

This theory includes most of the factors which affect refining. However, the main problem is the use of constants, which precludes quantification unless a considerable amount of trial work is undertaken to assign values to the two constants.

13.3.4 The 'C' Factor

The 'C' factor theory developed by Kerekes (1990) is based on the same starting point as that of other refining theories: the effective refining energy may be directly related to the number of impacts and the intensity or energy content of each impact as shown below in Equation (13.10):

$$E = N * I \tag{13.10}$$

The 'C'-factor represents the capacity of the refiner to impose impacts upon pulp fibres passing through. The C-factor links the power input, *P*, and pulp mass flow rate, *F*, through a refiner to the average number, *N*, and intensity, *I*, of impacts imposed on fibres. *N* and *I* are derived from the relationships shown in Equations (13.11).

$$\begin{aligned} N &= \text{C-factor}/F \\ I &= P/\text{C-factor} \end{aligned} \tag{13.11}$$

The C-factor itself is a function of filling geometry, rotation speed, consistency, fibre length and fibre coarseness. For a disc refiner, the C factor is given by Equation (13.12).

$$\mathrm{C} = \{8\pi^2 GD\rho C_f ln^3\omega(1 + 2\tan\varphi)(R_2^3 - R_1^3)\}/(3w(l + D) \tag{13.12}$$

For a conical refiner, the C-factor is given as follows [Equation (13.13)]:

$$\mathrm{C} = \{8\pi^2 GD\rho C_f ln^3\omega(1 + 2\tan\varphi)\} * \{LR_1^2 + L^2R_1\sin\theta + (L^3/3)\sin^2\theta\}/(w(l + D) \tag{13.13}$$

where

N	=	number of impacts/mass of pulp, kg^{-1}
F	=	pulp mass flow rate, kg/s
I	=	energy/impact, J
P	=	net power applied to refiner, W
G	=	width of grooves, m
D	=	depth of grooves, m
ρ	=	density of water, kg/m^3
C_f	=	pulp consistency, fraction
l	=	fibre length, m
n	=	number of rotor and stator bars on circle circumference in refiner
ω	=	rotational velocity of refiner, degree s^{-1}
φ	=	bar angle from radius, degree
R_1	=	inner radius of refining zone, m
R_2	=	outer radius of refiner zone, m
w	=	coarseness of fibre, kg/m
θ	=	angle of conical refiner, degree
L	=	length of refining zone, m
W	=	width of bar surface, m

The C-factor theory is perhaps the most rigorous and comprehensive of all the refining theories developed to date. However, this theory is only of academic interest and is not of much use for papermakers due to its complexity and since the values of many of the terms present in the equations are very difficult to obtain.

Owing to its simplicity and ease to use, the specific edge load theory is universally used by papermakers. This theory has been used in establishing the incorrect use of refiner tackles in the refining system of a paper mill that produces kraft liner board products. It also subsequently helps in retrofitting the existing refining system that produces tangible benefits for the mill in terms of increased throughput and substantial savings in the cost of refining energy. Details will be discussed in the case study presented in this chapter.

13.4 Types of Commercial Refiners

Two major types of continuous refiners are used for the stock-preparation system of paper machines: disc refiners and conical refiners. The *conical refiners* can be further differentiated into low-angle types (Jordans) and high-angle types (Clafins). The newest member in the conical refiner group is the medium-angle Conflo©-type refiner with a 20° cone angle. In conical refiners, the rotating plug (rotor) and the housing (stator) are fitted with metal bars orientated lengthwise. The fibres flow parallel to the bars. The position of the rotor determines the clearance of the bars and controls the amount of work done on the fibres.

The *disc refiner* group comprises three types: single-disc, double-disc and multiple-disc type refiners. Dingle-disc refiners are almost entirely used for high-consistency (HC) refining, while multi-disc refiners are used for very-low-intensity refining with extremely fine plate pattern. Double-disc refiners are the most common low-consistency (LC) (3.5–5%) refiners used nowadays in the stock-preparation system of paper machines using chemical pulps. The disc refiners are claimed to have a certain advantage over conical refiners. However, within certain limits and when properly applied, there is not a great difference between conical and disc refiners regarding their ability for fibre development. Detailed descriptions of different kinds of refiners with pictures are available in the literature (Smook, 1989; Paulapura, 2000).

13.5 Laboratory and Pilot-scale Refining Investigation

A key step in the stock-preparation phase is refining. For a given fibre, refining plays an important role in stock preparation as well as throughout the complete papermaking process (Ghosh, 2002). Next to raw material (fibre) selection, it has the greatest influence on final product quality. This high ranking in the process makes it all the more important to use correct refiner fillings and other optimum refining parameters. Application of incorrect refiner tackles can lead to the use of excessive power and the non-optimization of fibre quality.

Execution of refiner trials at mills is often difficult without disrupting normal production. Moreover, it is very expensive to undertake such a trial in terms of hardware requirements and machine time. Traditionally, studies to evaluate the impact of refining on pulp quality have been undertaken in the laboratory using the PFI mill (Welch and Kerekes, 1994). This type of study provides little information in terms of defining the optimal refining parameters

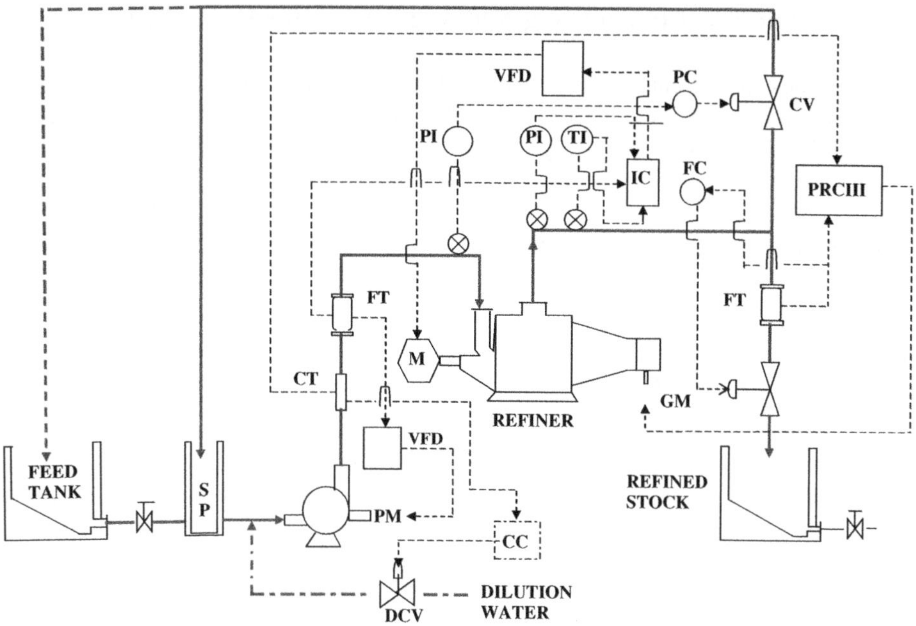

CC : Consistency Controller, CT : Consistency Transmitter, CV : Control Valve
DCV : Dilution Control Valve, FC : Flow Controller, FT : Flow Transmitter
GM : Refiner Gear Motor, IC : Interlock Controller, M : Refiner Drive Motor
PC : Pressure Controller, PI : Pressure Indicator, PM : Pump Motor, PRCIII : Proportional Controller
SP : Stand Pipe, TI : Temperature Indicator, VFD : Variable-frequency Drive

Figure 13.7 *Schematic and control system of a pilot refiner used for lab and mill trials.*

such as specific edge load, net specific energy, refiner fillings, and recirculation, along with other operating variables of a practical nature. The use of a pilot refiner is ideal for this purpose (Ghosh, 2004).

The size of the pilot refiners normally used for pulp refining trials is either 12″ or 16″ (described by the diameter of the refiner plate). Most pilot refiner trials are undertaken using hydra cycle mode, where fibres are passed through the refiner multiple times. In such a mode, the refined pulp is sent back to the feed tank and total refining energy is the accumulation of power used over the refining period. From the pulp flow rate, consistency and the accumulated power, the specific refining energy is calculated. This mode of refining is better than the PFI mill, as the plate pattern of the pilot refiner can be exactly the same as that of the mill refiners. The best mode of pilot refining trial would be taking the pilot refiner to the mill and connecting it to the mill refiner feed chest and running it the same way as that of the mill refiner (single pass with 10% flow recirculation). Figure 13.7 shows the schematics of a 16″ pilot refiner with control system used in the lab (hydra cycle mode) and mill trials. Such a trial was undertaken in a paper mill producing flat sack kraft paper to establish retrofitting requirements of the mill refiners in anticipation of upgrading to a superior product grade of semi-extendable sack kraft papers. Details of the retrofitting are discussed in the second case study presented in this chapter.

13.6 Case Studies of Retrofitting Refining Process for Paper Mills

Two case studies are presented in this section where retrofitting and debottlenecking of the chemical pulp refining process has been described. In the first case study, debottlenecking of the total throughput and retrofitting of a critical stage of papermaking was undertaken. The tangible benefits realized by the mill were improvement in product quality (hence customer satisfaction), increase in mill output and saving in refining energy. In the second case, retrofitting was more related to change in product grade with more added value in the finished products.

13.6.1 Case A: Retrofitting of Existing Refiners to Debottleneck Output of a Modern Paper Machine

In this case, the paper machine of the mill manufactures kraft liner board paper using unbleached virgin kraft pulp, made on the pulp mill on-site from long-fibre softwood. The finished paper products are used for fibre packaging (converted into corrugated boxes). Three key strength properties important for converting the paper during corrugation in the box plants are burst strength, internal ply strength and compression strength. The compression strength measured in the mill test laboratory is short span compression strength (SCT) in the cross direction. Paper manufactured in a commercial paper machine is primarily orientated in both cross direction (across the width of the machine) and machine direction (from wet end towards the dry end or the reeler when the finished product is wound in a massive jumbo roll that could weigh in excess of thirty tons).

The main problems encountered by the mill were struggling to achieve the target paper strength specifications and the machine throughput was below the design output. The mill was finding it difficult to manufacture paper in the specification range between the aim and upper control limit. Most of the time the quality of the paper was barely higher than the minimum specification. The consequence of the first problem was frequent complaints from customers for failure of the papers in the box plant during conversion. Consequential claims for materials and machinery down-time in the box plant were significant. Moreover, the paper mill itself was required to downgrade substantial tonnage of paper that failed to satisfy the minimum product specifications. Without improving the quality to product specification, the mill was unable to increase throughput.

It was established by the mill that the process units that were responsible for not achieving the aimed-for product specifications, namely burst strength, compression strength and internal bond strength of finished paper products, were the paper machine refiners. The paper machine produces 3-ply linerboard papers and the stock approach system of each ply has two refiners (double-disc, operating on duo-flow mode) operating in series as shown in Figure 13.8. Each refining line is equipped with on-line measurement of pulp freeness that is used by the paper-machine operators as a control tool. A common chest of unrefined pulp is used to feed each refining line through individual feed pumps and each refining line has some level of recirculation. As indicated earlier, refiners play the most critical role in fibre development, and the choice of refiner tackle is very important. In any integrated paper mill, such as the one in this case, where pulps are also produced in the pulp mill located on the same site, the paper machine has to use the pulps delivered by the pulp mill. It is the responsibility of the paper-machine personnel to make best use of such pulps to

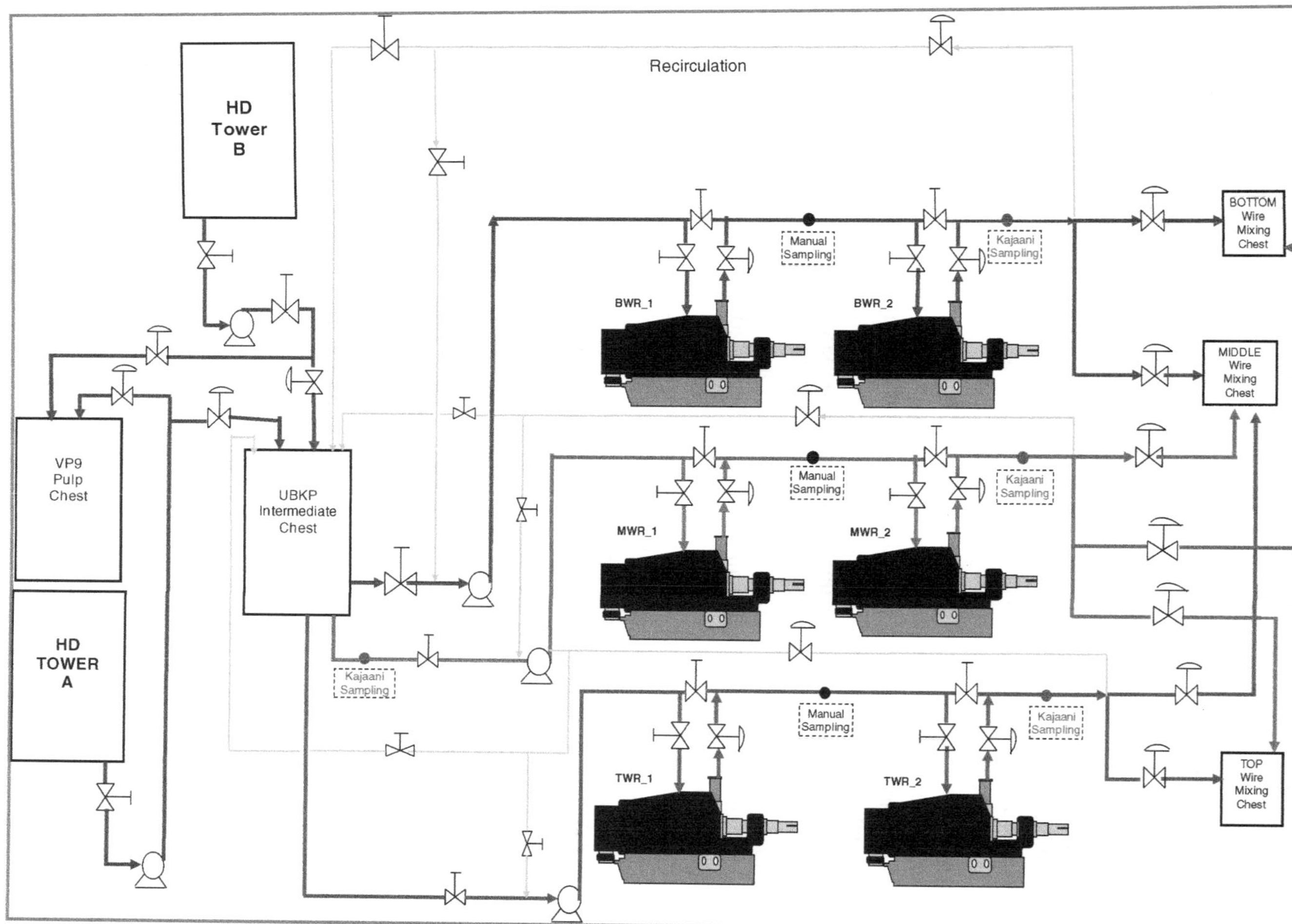

Figure 13.8 Layout of the paper machine's refining system.

manufacture papers that will achieve the target product specification and satisfy customers' end-use requirements.

The choice of the type of refiners, such as double-disc or conical, is generally made at the design stage. There is no significant difference in the type of refiners used for low-consistency refining (between 3.5% and 5% consistency) in terms of fibre development. What is critically important is the choice of correct refiner tackles such as refiner plate (type, pattern, intersecting angles, and so on). Incorrect use of such tackles not only can destroy the fibres, which leads to inadequate fibre development, but also can use excessive specific refining energy. The same plates patterns were in use in each of the six refiners and were selected by the mill on advice from the vendor and based on refining experiences with typical unbleached softwood kraft pulp. However, the type of pulp made in the pulp mill of this case and used by the mill's paper machine is unique in terms of pulp yield (based of wood mass) and residual lignin content of the final pulp.

The pulp made and used in this mill was of high-yield type and of higher Kappa number, between 90 and 95. The Kappa number is a measure of residual lignin after cooking in the digester and washing prior to feeding the refiner and paper machine. The higher the Kappa number, the higher the lignin content in the pulp and the higher the pulp yield. A typical Kappa number range for unbleached softwood kraft pulp used in manufacturing kraft linerboard papers is between 45 and 55. For fully bleached pulp, the Kappa number is less than one or close to zero. The characteristics and refining response of fibres are greatly influenced by the Kappa number of pulps (Ghosh, 2010). The fibre morphology of high-Kappa unbleached softwood kraft pulp, similar to the pulp used in this case study, is completely different to that of pulp with a typical Kappa range of 45 to 55. The primary cell wall of high-Kappa (90 to 95 range) pulp is stiffer and required high–intensity (HI) refining to open up the cell wall prior to fibrillation. Consequently, use of refiner tackles that are suitable for typical unbleached softwood kraft pulp is not expected to give optimum fibre development.

Retrofitting of the refiner tackles was considered. It is imperative that any recommendation of retrofitting a key process component to the mill and subsequent implementation must be supported by concrete evidence as to its efficacy. Extensive investigation on the evaluation of fibre development through in-situ refining trials was undertaken in support of such evidence. These trials comprised (i) changing the refiner plate pattern of only the first refiner of the middle wire (MW) refining system over a wide range of specific edge load, while the plate patterns of all the other refiners remained the same; (ii) plate patterns of both the first and second refiners of one refining line were changed to the trial plates and (iii) a full commercial production trial with the trial plates. Such a trial plate was able to provide maximum refining intensity of 5.9 Ws/m at full motor load.

13.6.1.1 Initial In-situ Trial to Establish Retrofit Criteria

Specific edge load theory was used in selecting the initial trial plate. The plate pattern being used by the mill prior to this investigation ('existing' plate pattern) had the following tackle parameters: 64.59 km/rev cutting edge length and 19.8 degree intersecting angle. At full-load refiner motor power of 1700 kW, the maximum specific edge load for this pattern was 2.8 Ws/m. This value is at the lower end of the recommended specific edge load range of between 2.5 and 4.5 Ws/m for typical unbleached softwood kraft pulp (UBSWK) with Kappa number between 50 and 60. Since the UBSWK pulp used by the mill is of a significantly higher Kappa, the desired refiner plate should be capable of impacting higher intensity. Based on this argument, a new plate pattern was 'selected/designed' with the tackle parameters of 32 km/rev cutting edge length and 28.5 degree intersecting angle.

A request was made to the mill to procure a set of the 'selected' refiner plates so that a sighter in-situ trial could be undertaken by replacing the 'existing' plates of the first of the two refiners in the middle wire refining line. The objectives of this initial trial were (i) to see the effects of refining intensity over a wide range of between 1.6 and 5.9 Ws/m on fibre development; (ii) to generate sufficient 'real world' data; (iii) to convince the mill to undertake a subsequent large-scale trial for HI refining; and (iv) eventually to retrofit the refiner plate patterns to achieve mill target product specification of finished papers and increase throughput without the addition of extra refining capacity that is quite expensive (in excess of one million dollars per refiner).

13.6.1.1.1 Key Pulp Properties

Enhanced fibre development through refining can be quantified from three important pulp properties: (i) freeness change and freeness drop per kWh/t of applied net specific refining energy; (ii) water retention value (WRV); and (iii) fibre length (FL) distribution. Measured values of these properties of pulps from in-situ refined pulp, using both the new trial and 'existing' plates, are used to demonstrate the benefit of HI refining and potential retrofitting of the mill refiners.

Figure 13.9 shows the changes in freeness (Canadian Standard Freeness) (Csf) values as a function of net specific energy (NSE) when refined at various SEL or refining intensities. In the same Figure, the freeness values of refined pulps after both the refiners of Top Wire (TW) and Bottom Wire (BW) that were fitted with the 'existing' plate were also tested, are also included for comparison. It is evident from this Figure that at lower SEL (up to 4.0 Ws/m), changes in freeness with increase in net specific energy was very slow. The freeness changes of refined pulp after these two refiners are linear and almost identical

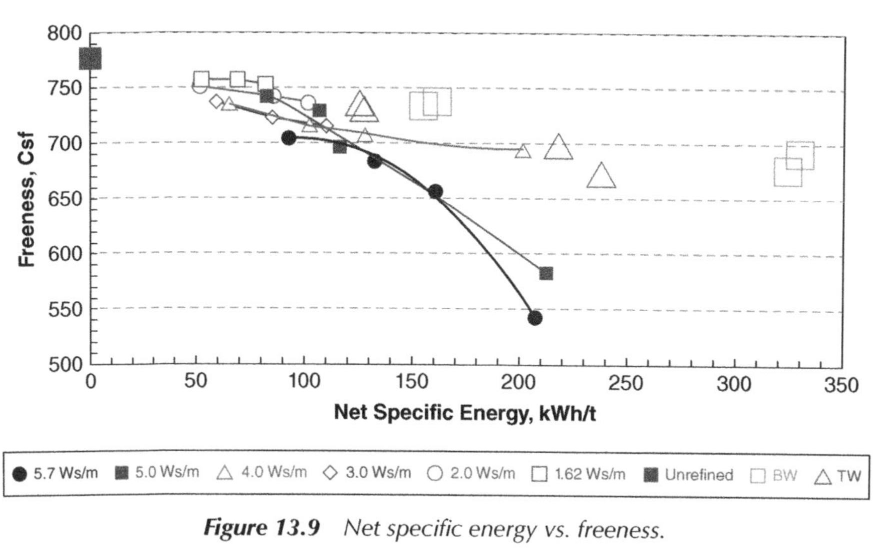

Figure 13.9 Net specific energy vs. freeness.

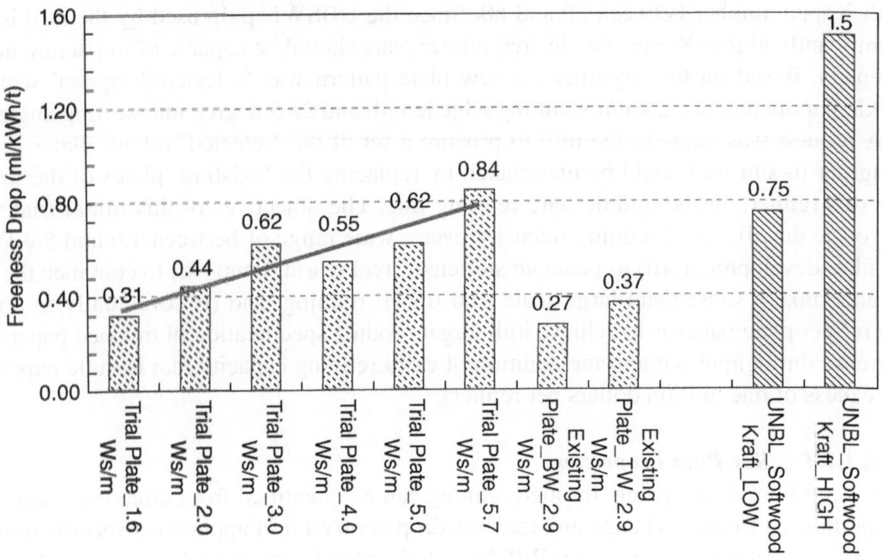

Figure 13.10 *Freeness drop per kWh/t.*

with that of refined pulp from the 1st Middle Wire (MW) operating at 2.0 Ws/m. Freeness decreases of refined pulp using the trial plate at SEL values of 5.0 and 5.7 Ws/m were much higher. At 200 kWh/t net specific energy, the pulp freeness dropped from 780 Csf to 600 and 560 Csf at 5.0 Ws/m and 5.7 Ws/m, respectively. With the current plate, pulp freeness could not be lowered below 690 Csf at net specific energy in excess of 300 kWh/t. This suggests that with the current pattern of refiner plate, the high-Kappa pulp of the mill could not be refined to a freeness level sufficiently low enough to achieve strength targets. However, if the pattern of the trial plates is used, freeness reduction (and strength development, see below) is easily achievable.

The average freeness drop per kWh/t of refining energy at various refining-intensity levels is shown as a bar chart in Figure 13.10. Corresponding values for the existing plates used in TW and BW and also the recommended values for UBSWK (50–60 Kappa range) are also shown in this bar chart. The freeness drop linearly increased with an increase in refining intensity (shown by the solid line) when the trial plate was used. For a high-Kappa pulp, it is expected that the range of freeness drop per kWh/t of NSE will be slightly lower than the recommended values shown in Figure 13.10 (most likely between 0.70 and 1.20, typical for medium-Kappa unbleached softwood pulp). At 5.7 Ws/m refining intensity, the measured freeness drop was 0.84 and this value was within the recommended range.

Water retention values (WRV) is a measure of the degree of hydration (Mohlin, 2002). With increased refining, WRV increases due to fibrillation. The measured values are shown in Figure 13.11. At lower refining-intensity levels of 1.62 and 2.0 Ws/m, the WRV of refined pulp did not change and was close to that of unrefined pulp, suggesting the SEL level must be higher than 2.0 Ws/m. At a given NSE (higher than 100 kWh/t) the WRV value increased with an increase in SEL. For the TW and BW refined pulps (with current refiner plate pattern) the WRV value was the same as that of pulp refined using the trial

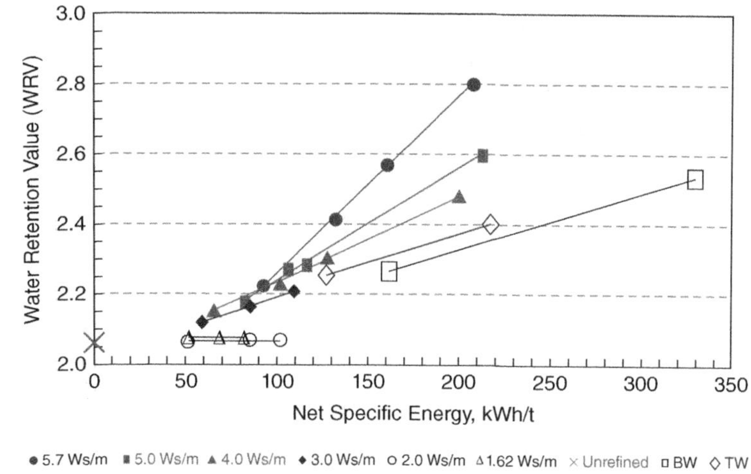

Figure 13.11 Effect of NSE on water retention.

plate but operating at 3.0 Ws/m SEL. At higher refining intensity, the water retention values were higher at the same net specific refining energy, compared with lower SEL. This is evident in Figure 13.11.

Fibre-length distribution is also a good indicator of the level of refining. The effect of refining intensity on fibre-length distribution at constant net specific energy of 210 kWh/t is shown in Figure 13.12. At a constant NSE of 210 kWh/t, refining at 5.0 and 5.7 Ws/m intensity showed significant broadening of the FL distribution curve compared with that of TW pulp refined at 3.0 Ws/m.

Fibre micrographs are qualitative indications of fibre development through refining and are shown in the following Figures at 100x magnification. As expected, for unrefined pulp, the fibre micrograph shown in Figure 13.13 is very clean, without signs of fibrillation

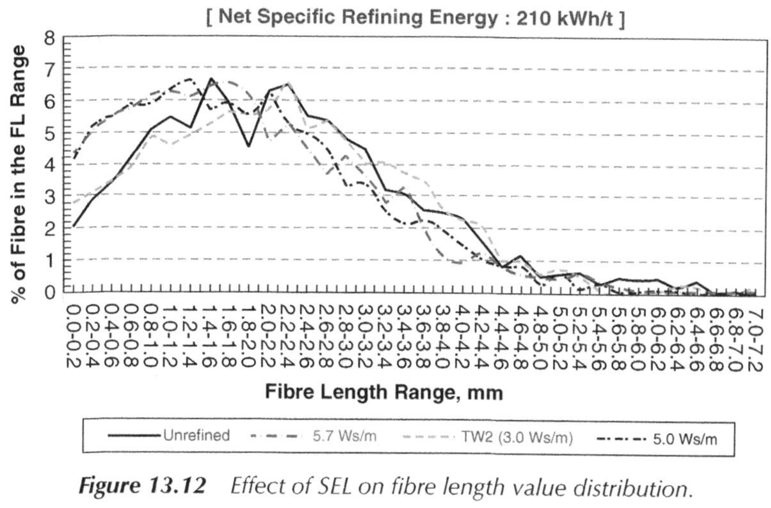

Figure 13.12 Effect of SEL on fibre length value distribution.

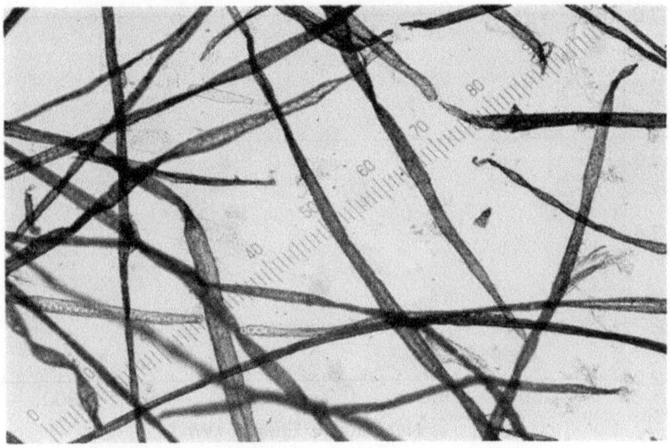

Figure 13.13 Micrograph of unrefined pulp.

or hairy structures. For pulps refined using the 'existing' plate at 217 kWh/t, the fibre micrographs showed poor fibrillation (Figure 13.14) at a maximum obtainable refining intensity of 2.9 Ws/m. At 329 kWh/t net specific energy, good fibrillation occurred, as shown in Figure 13.15, although at significantly higher refining energy (about 59% more) compared with that of the trial plate. Extremely good fibrillation was achieved with the trial plate at 207 kWh/t net specific refining energy operating at a specific edge load or refining intensity of 5.7 Ws/m (Figure 13.16). This clearly supports the notion that 'existing' plate patterns be retrofitted with the trial plates. This is further supported by the improvement of paper-strength properties discussed below.

13.6.1.1.2 Key Paper Strength Properties

Figure 13.17 shows the indexed values of burst strength as a function of NSE at various levels of refining intensity. Similar plots for internal bond-strength values are shown in

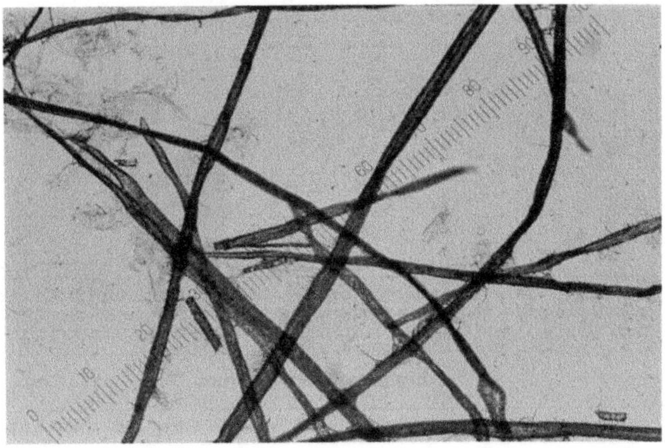

Figure 13.14 Micrograph of refined pulp ('Existing' plate at 217 kWh/t).

Figure 13.15 Micrograph of refined pulp ('Existing' plate at 329 kWh/t).

Figure 13.16 Micrograph of refined pulp trial plate at 207 kWh/t and at 5.7 Ws/m).

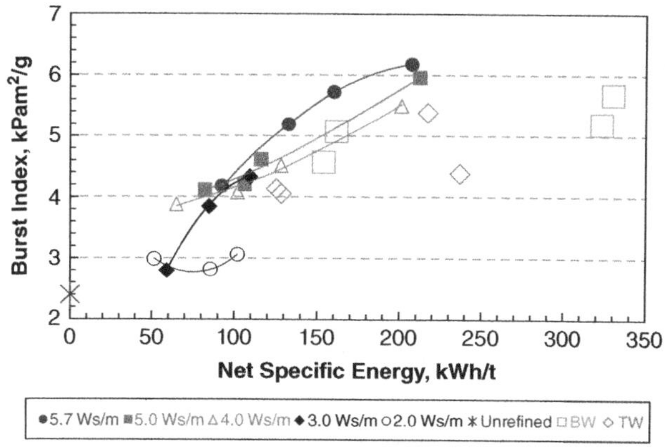

Figure 13.17 Net specific energy vs. burst strength.

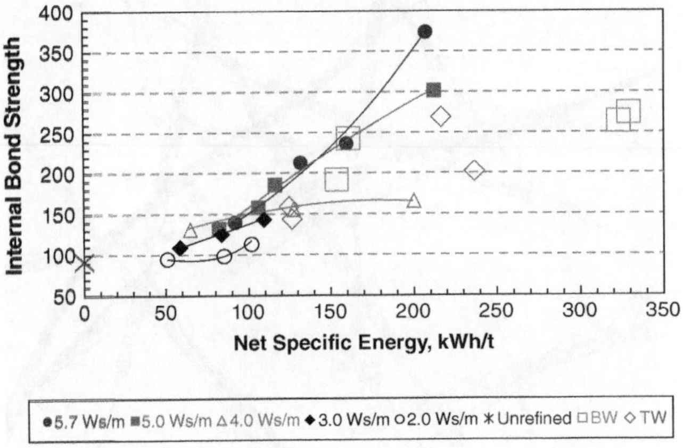

Figure 13.18 *Net specific energy vs. internal bond strength.*

Figure 13.18. Developments of these two important strength properties were achieved with the trial plates at a maximum refining intensity of 5.7 Ws/m and NSE range between 100 and 200 kWh/t. With the 'existing' plate pattern (BW and TW results), a similar or lower level of strength development was achieved at more than twice the net specific refining energy.

The development of tensile strength values of paper sheets made from refined pulps using both the trial and 'existing' plates, operating at different refining intensity and net specific refining energy levels, is shown in Figure 13.19. Similar developments of short span compression strength (SCT) are shown in Figure 13.20. At constant refining intensity (higher than 3.0 Ws/m), tensile and SCT strength increased with increasing net specific energy, although not linearly. When compared with the current plate pattern used on both BW and TW refiners, the same level of tensile and SCT strength could be achieved when an HI plate is used (at 5.0 Ws/m SEL), but at significantly lower net specific energy.

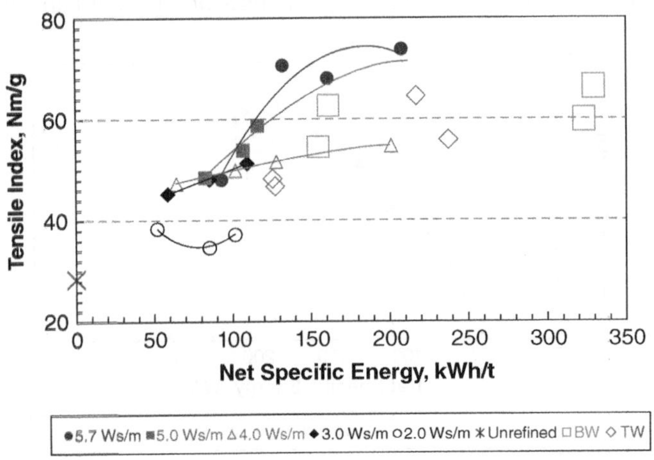

Figure 13.19 *Net specific energy vs. tensile strength.*

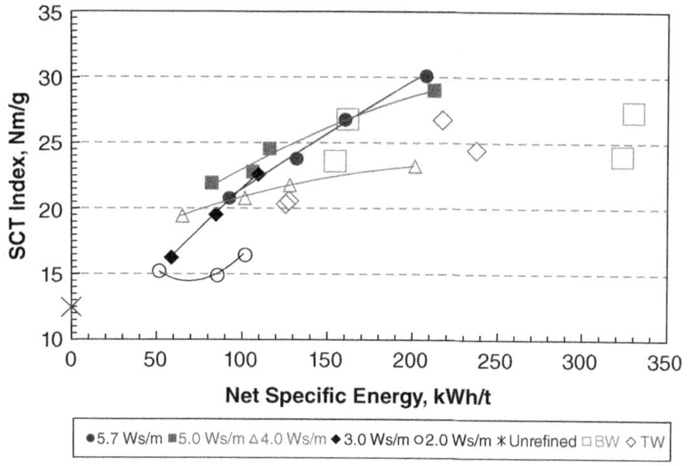

Figure 13.20 *Net specific energy vs. SCT strength.*

It is evident from the above Figures that significant energy saving is likely possible, if HI trial plates are used instead of the 'existing' plate pattern operating at lower intensity, to achieve the target level of key properties. Table 13.2 shows the potential energy saving with the trial plate pattern to achieve target values of key strength properties.

13.6.1.2 Second In-situ Trial after Retrofitting 'Existing' Plates

The initial in-situ refining trial was based on retrofitting only the first of the two refiners in series of the middle wire line, while all the other five refiners of the paper machine refining system had the 'existing' plates. Following the successful first trial that demonstrated that the new trial plate provided much superior fibre development, it was necessary to prove that retrofitting the second refiner in series with an HI plate is also beneficial for the mill. Consequently this second trial was undertaken with the following plate arrangements:

- First refiners of the series (R1) of all three lines had 'trial' HI plates
- Second Refiner (R2) of BW Line changed to HI plate
- R2 Refiners of TW and MW Lines had existing low-/medium-intensity plates

Table 13.2 *Potential net specific energy saving for retrofitting 'existing' plate pattern.*

		Specific energy, kWh/t		
Key properties	Value	Trial plate	'Existing' plate	Saving
Internal bond strength, Scot unit	275	170	329	159
Burst index, $kPam^2/g$	5.5	155	329	174
Tensile index, Nm/g	65	130	329	199
SCT index, Nm/g	26	160	329	169

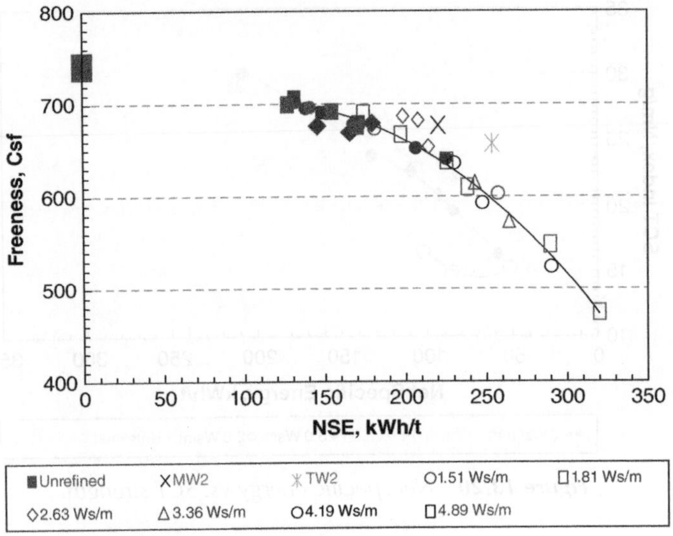

Figure 13.21 Net specific energy vs. freeness.

During the trial, the R1 refiner of the Bottom Wire was loaded at maximum available power, while that of R2 were changed to six different levels. At each motor loading of R2 of the Bottom Wire, throughput changed at three different levels to change the net specific refining energy. The operating conditions of both Top Wire and Middle Wire refiners were not changed and were maintained under pre-trial conditions. Pulp samples were collected after the R1 and R2 refiners of BW at each throughput and refining power setting. Pulp samples were collected after the R2 refiners of TW and MW lines for comparative evaluations.

13.6.1.2.1 Key Pulp Properties (All R1: HI Plate)

The effects of net specific refining energy on pulp freeness after R1 and R2 of BW refiners that were fitted with HI trial plates are shown in Figure 13.21. Data of the TW and the MW after R2 of both the lines that had R1 with HI plate and R2 as 'existing' plate are also shown in the same graph. The 'filled' symbols correspond to values after the R1 refiner, while unfilled symbols represent data after the R2 refiner of BW. Similar plots for WRV are shown in Figure 13.22.

Both freeness and water-retention values of pulps refined using the HI trial plate lie on a single curve. At the same NSE level, freeness reduction of pulps refined on both TW and MW refining lines and using an HI/MI (medium-intensity) combination on R1/R2 refiners on these lines is lower than that of BW pulps refined using HI/HI plates, suggesting that the HI/HI plate combination provides a better refining of this kind of pulp compared with that of HI/MI combination. This is also reflected in WRV values. Lower WRV represents poor hydration or fibrillation as evident in the micrographs of refined pulps shown in Figures 13.23 and 13.24. At similar NSE (254–260 kWh/t), the fibrillation of fibres refined on the TW line with an HI/MI plate combination on R1/R2 refiners is significantly lower compared with that of pulps after BW2 where HI/HI plates used.

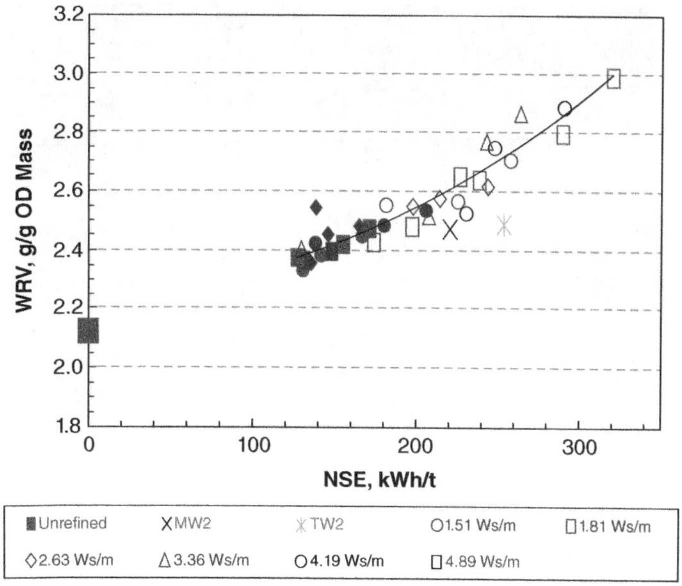

Figure 13.22 *Net specific energy vs. WRV.*

13.6.1.2.2 Key Paper Strength Properties: HI/HI Plate Strategy

The superior fibre development through the HI/HI plate combination, as reflected in key pulp/fibre properties, namely freeness, WRV and fibrillation, and discussed above, is also realized in key paper properties. Figure 13.25 shows the internal bond strength of paper sheets made from refined pulps using both HI/HI (BW line) and HI/MI (TW and MW lines) plate combinations for R1/R2 refiners. Similar plots for burst-strength index values are shown in Figure 13.26. For both these important paper strength properties, when refined using the HI/HI strategy, strength developments were superior compared with those from

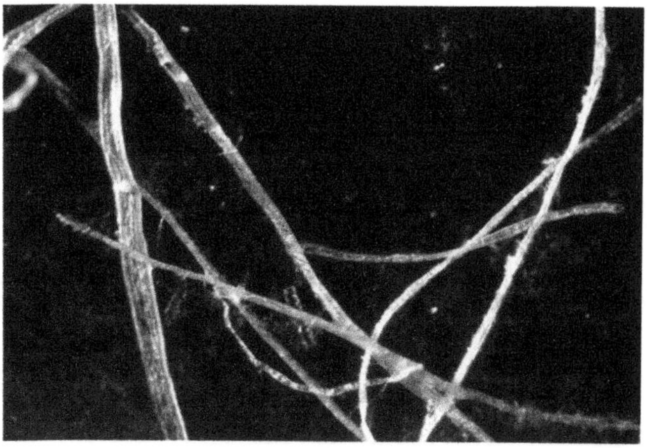

Figure 13.23 *Micrograph at 254 kWh/t (after TW2 HI plate at R1, MI plate at R2).*

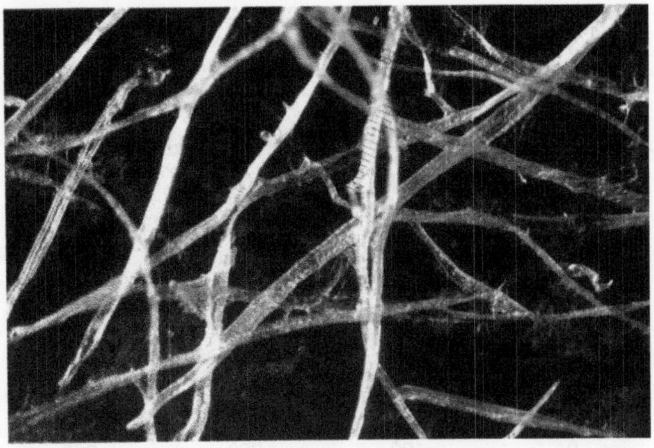

Figure 13.24 Micrograph at 260 kWh/t (after BW2 HI plates at both R1 and R2).

the HI/MI plate combination at the same net specific refining energy. This is particularly true for internal bond strength. NSE vs. tensile strength and SCT compression strength plots are shown in Figures 13.27 and 13.28, respectively. Similar to internal strength and burst-strength values, both tensile and compression strength of paper sheets made from pulps refined with HI/HI plates were higher than those formed with the HI/MI combination, when compared at the same net specific refining energy level. Similar to fibre/pulp properties, the key paper strength properties could be represented by a single polynomial, when the HI/HI plate combination was used during refining.

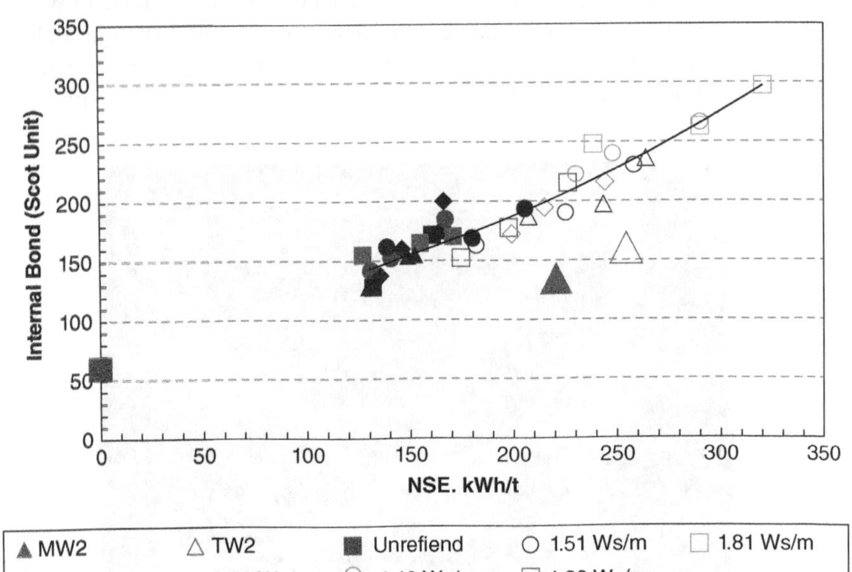

Figure 13.25 Net specific energy vs. internal strength (HI/HI and HI/MI pattern).

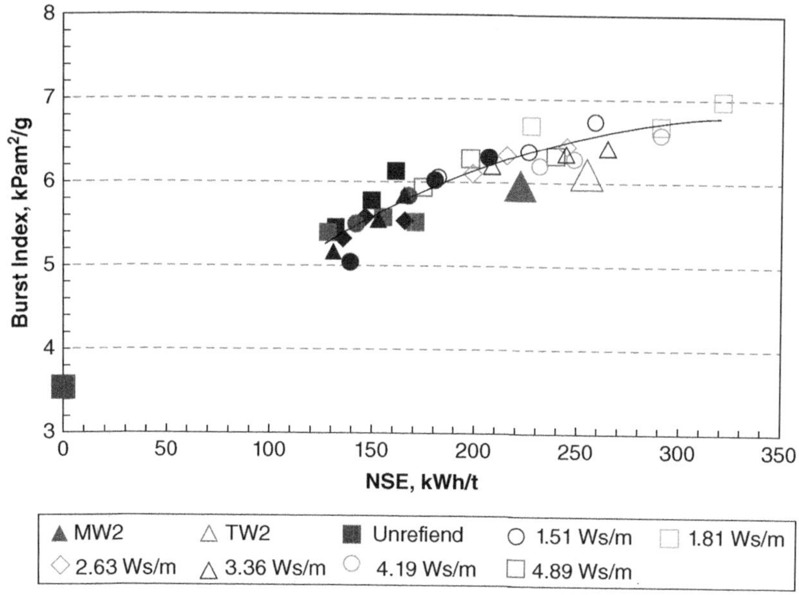

Figure 13.26 Net specific energy vs. burst strength (HI/HI and HI/MI pattern).

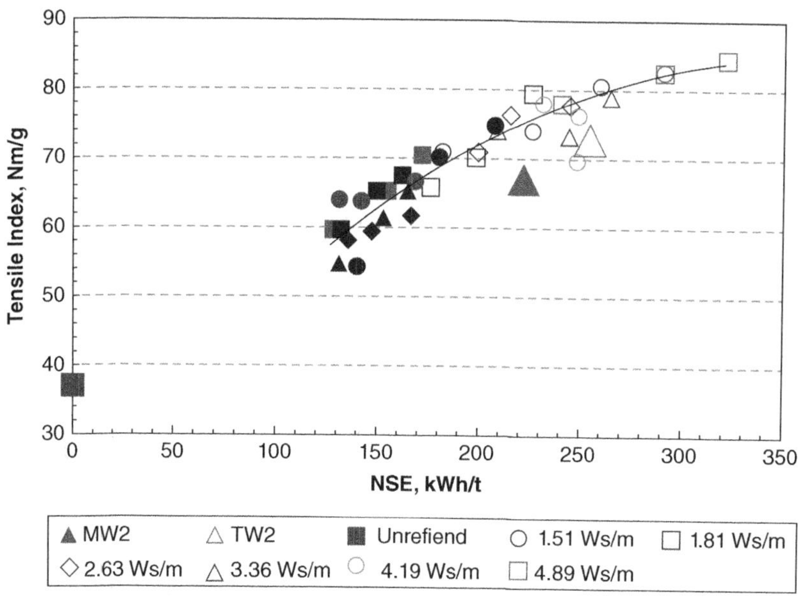

Figure 13.27 Net specific energy vs. tensile strength (HI/HI and HI/MI pattern).

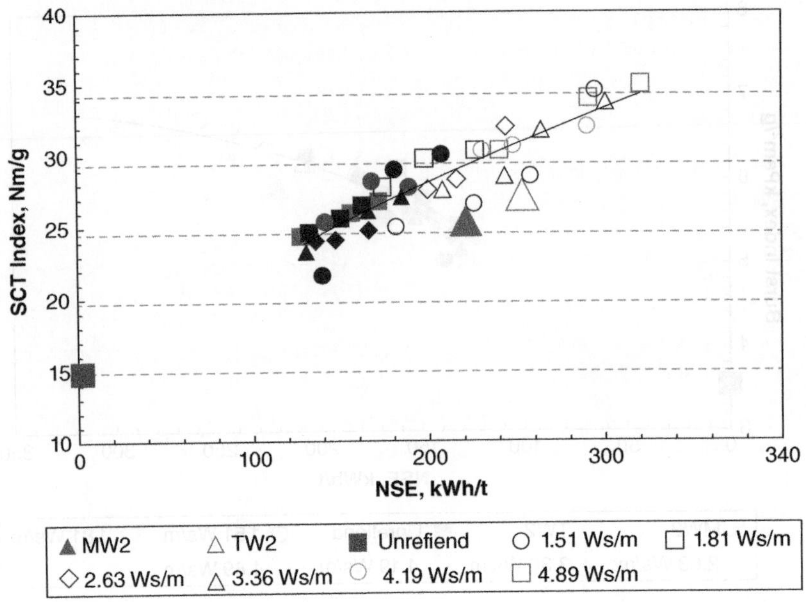

Figure 13.28 *Net specific energy vs. SCT strength (HI/HI and HI/MI pattern).*

It is noted from the above Figures of pulp and paper properties that if the refining strategy of the mill that was using the 'existing' MI plates in both the R1 and R2 refiners are retrofitted by the trial plates in both the refiners in series and in all the three refining lines, superior fibre developments could be achieved. The results presented above are for paper sheets made in the laboratory using pulps that were refined in-situ in the mill. These results helped in further convincing the mill for the need to retrofit the existing gears that were inadequate to achieve target product specification, costing excessive refining energy usage, and a source of the bottleneck for increasing machine throughput. It was still necessary to prove beyond any doubt the tangible benefits to the mill from such a retrofit under production conditions. A full-scale production trial with HI/HI plate combination in all three refining lines was subsequently undertaken.

13.6.1.3 Full-scale Production Trial after Retrofitting 'Existing' Plates

Following the first two series of trials, which established the optimal refining intensity and the mode of operation of the paper machine refiners operating in series, a final refining trial was undertaken, when commercial papers were made. During this trial the refiner parameters were set to the optimum values established from the earlier trials. The objective of this trial was to quantify the realization of potential benefits. Figure 13.29 shows the percentage change in SCT compression strength with respect to the base line value before and after the implementation of the refining strategy. Each data point in this graph corresponds to the test result of one full jumbo reel and the total numbers of data belong to one full years' production of the major grade of linerboard paper manufactured on the paper machine. The vertical line indicates the date when the paper machine refiners were operating at the optimal parameter and when the retrofitting was completed.

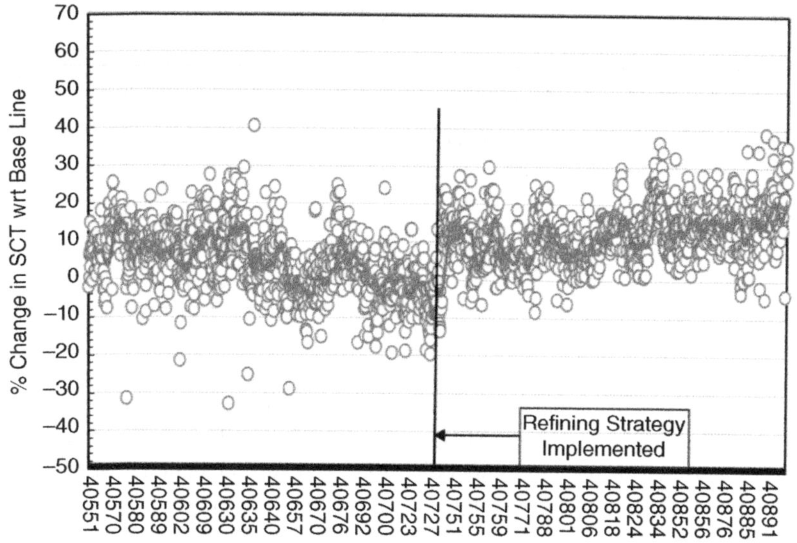

Figure 13.29 Percentage change in SCT on machine-made paper (wrt = with respect to).

Prior to the implementation of the optimal refining strategy, the SCT strength varied between −20% and 20% around the base line value. Moving average (24-points) results shown by the solid line suggest there was a declining trend in SCT value. Following implementation of the refining strategy, a step increase in SCT values is evident, along with an increasing trend in this important property. Figure 13.30 shows a similar plot for internal bond strength, generally expressed by paper mills as Scot ply bond (SPB). Although

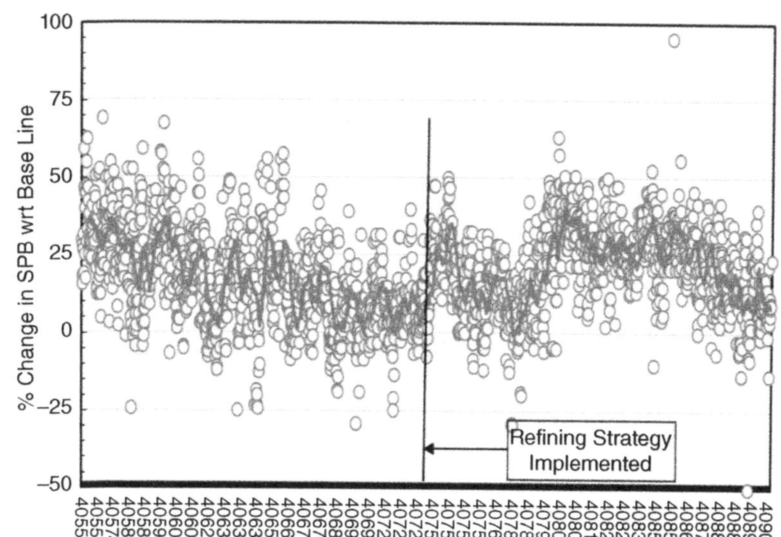

Figure 13.30 Percentage change in SPB on machine-made paper (wrt = with respect to).

Table 13.3 *Tangible benefits realized following retrofitting of refiner tackles (wrt = with respect to).*

	Refining power, kW			Output, OD t/h		
Refining line	Pre-	Post-	Saving	Pre-	Post-	Increase
Bottom Wire (BW)	3334	2571	**763**	8.5	11.4	**2.9**
Middle Wire (MW)	2922	2152	**770**	8.9	9.3	**0.4**
Top Wire (TW)	3341	2689	**652**	11.7	12.3	**0.6**
Total	**9597**	**7412**	**2185**	**29.1**	**33.0**	**3.9**

the positive change in SPB is not as obvious as that of SCT strength, the values for post-implementation period are general higher than the earlier period. The level of improvements of both the SCT and Scot ply bond (SPB) strength of finished machine-made products was deliberately restricted to the target specifications and further benefits of retrofitting were realized in the form of saving of refining energy and increase in throughput. This is shown in Table 13.3. Close to a total 2.2 MW of refining power was saved along with a 3.9 t/h increase in throughout in refined pulps. Considering the fact that paper machines run twenty-four hours and seven days a week with only total ten to fifteen days of planned and maintenance shut-down in a year, the financial benefit is enormous (several million dollars a year). It is important to note that the total cost associated with the retrofitting of the refiner tackles was close to nothing, as the cost of newly designed/selected trial plates was the same as that of the 'existing' plate patterns. The only indirect cost was associated with the mill trials and laboratory evaluations.

The results and methodology use in this case study showed that in the paper-manufacturing process that comprises many individual steps, a single stage could be a real bottleneck. Such a bottleneck could cause serious problems for the mill in terms of achieving target quality specifications, use of excessive energy and being an impediment to increase in output. Identifying the bottlenecking stage and subsequent retrofitting such a step with appropriate equipment/gear cost-effectively is very important in a capital-intensive industry such as paper manufacturing. Quite often a series of investigations and mill trials are required to generate data in order to convince mill management to debottleneck and retrofit. In this case study, a series of trials, including the final production trial after retrofitting and identifying the critical unit stage, were undertaken and the outcome of such trials was so convincingly positive that the mill accepted the recommendation that was flagged immediately after the initial trial.

13.6.2 Case B: Retrofitting of Existing Refiners of a Paper Machine to Switch from 'Flat' to 'Semi-extendable' Sack Kraft Papers

In this case, the paper machine manufactures flat sack kraft paper from a blend of virgin pine (long-fibre) kraft pulp with typical Kappa range of 55–65 and eucalyptus (short-fibre) neutral sulfite semi-chemical (NSSC) pulp. *The* most important properties for sack papers are *stretch* and *tensile energy absorption* (TEA) in both cross direction (CD) and machine direction (MD). The major problems with the quality of sack papers currently made on this machine were poor stretch and TEA profile in the cross direction as shown in

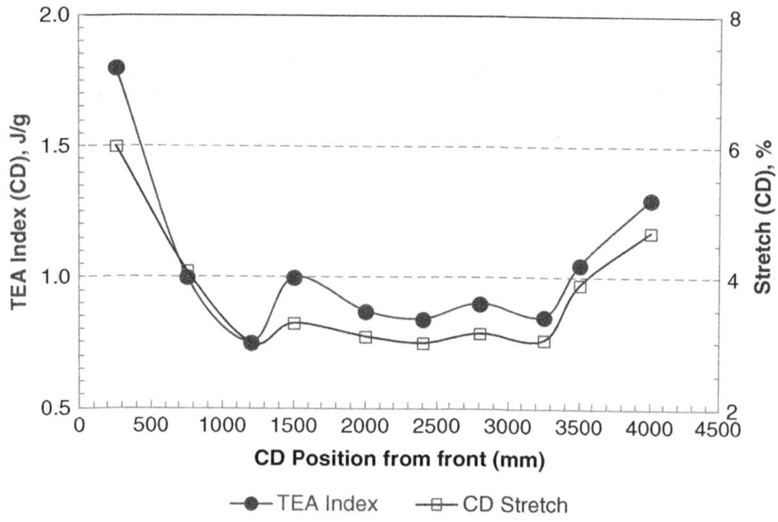

Figure 13.31 *Poor stretch and TEA profiles.*

Figure 13.31 and also lower values of these two key properties. Both CD stretch and CD TEA were significantly higher at the edges compared with the rest of the deckle positions. Consequently, paper bags made from the deckle (width of the jumbo roll made) region of 1000–3500 mm experienced very poor performance at the end-user customers' filling plants.

It is generally expected that CD stretch at the edge would be slightly higher than at the middle of the deckle due to higher shrinkage at the edges ('edge' effect), but not to the level experienced in this machine's products. It was suspected that the poor CD stretch and TEA profiles were related to the poor humidity profile in the cross direction that in turn induced uneven shrinkage in the cross direction, resulting in poor CD stretch and TEA values (Ghosh, 2009). Besides the poor CD profiles and lower values of the key properties, other problems identified with this machine were in the stock preparation in the form of appropriate refiner tackles and the refiners. The refining system of the machine comprised two separate refining lines: one single double-disc refiner for the long-fibre pine pulp and four older single-disc refiners operating in series for the short-fibre NSSC pulp. This is shown schematically in Figure 13.32. The net energy transferred to pulp during refining of NSSC pulps using these older refiners was very low due to excessive idling or no-load power as shown in Table 13.4. Per ton of pulp refined, the idling energy was more than the net specific refining energy. Moreover, the specific edge load of the refiner tackles used was higher than the require range of 0.3 and 0.5 for NSSC pulp (Lehrman, 2001).

The mill had the choice either to shut the machine or upgrade to improve the stretch and TEA values and the profiles in the cross direction. The mill acted to address the CD profile problems by upgrading the dryer section of the paper machine to improve dryer pocket humidity profiles. At the same time, the mill also decided to upgrade the product quality from 'flat' sack to 'semi-extensible' sack papers. Semi-extensible sack paper is an ideal sustainable packaging solution where high pressure and high-speed filling are

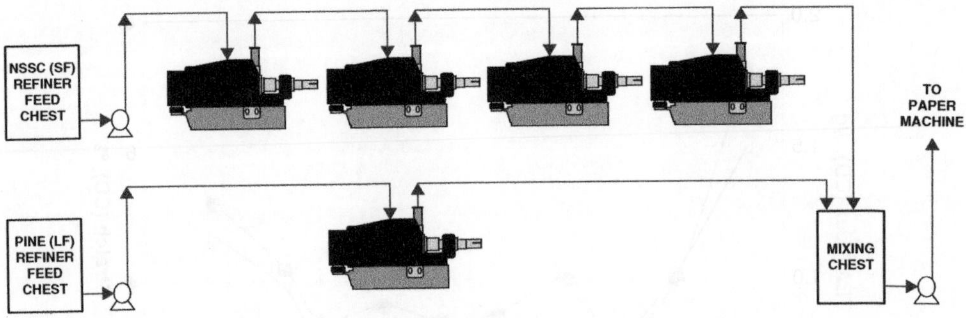

Figure 13.32 *Layout of existing refiners for sack paper.*

required. It is also recommended for industrial products that require high strength and stretch applications. Extensive retrofitting of the refining system of both the long-fibre pine and NSSC pulps was required to achieve the mill's objectives.

Although the existing pine refining system seemed adequate for 'flat' sack papers to manufacture 'semi-extensible' sack papers, high-consistency refining (at 30–35% consistency) is typically used when the end product must have TEA and high porosity (at low air resistance). Typical end products include semi-extensible sack kraft paper because sacks to be filled with powder (for example, cement, milk powder) must be strong enough while simultaneously allowing air removal through the paper layers. As high-consistency refining creates curled/kinky fibres with low bonding ability, the HC stage must be followed by an LC refining stage for straightening of the fibres and increasing their bonding ability (Paulapura, 2000). The existing low-consistency refining system for long-fibre pine pulp was retrofitted with an HC refining stage followed by an LC refining stage. The existing NSSC pulp refining system that comprised four refiners operating in series was retrofitted/replaced with a single double-disc refiner fitted with an appropriate refiner plate that provided refining intensity or specific edge load values in the range of 0.3 and 0.6 Ws/m. The retrofitted refining system for the paper machine that is currently producing 'semi-extensible' sack kraft paper is shown in Figure 13.33.

Pulp and fibre properties following HC and LC refining, along with those of unrefined pulp, are shown in Table 13.5. HC refining has practically no effect on fibre properties, other than fibre length, which seems to be shortened due to curling. After a relatively low-energy

Table 13.4 *Details of existing NSSC refiners prior to retrofitting.*

Existing refiners	Idle load, kW	Net power, kW	SEL, Ws/m	Idling energy, kWh/t	Net energy, kWh/t
1	160.0	129.6	0.90	80.0	64.8
2	356.7	16.8	0.82	178.3	8.4
3	164.6	125.0	0.87	82.3	62.5
4	150.9	123.5	0.86	75.4	61.7
Total	**832.2**	**394.8**		**416.1**	**197.4**

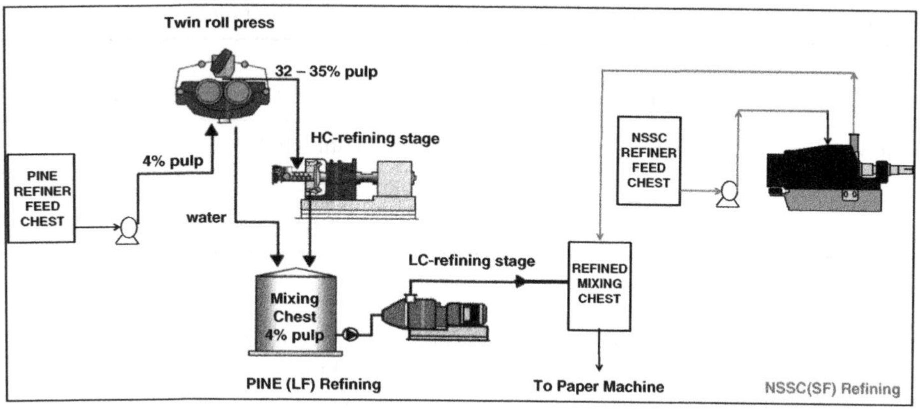

Figure 13.33 Layout of retrofitted refining system for semi-extensible sack kraft paper.

Table 13.5 Effect of HC and LC refining on fibres for sack papers.

Pulp/Paper property	Unrefined	After HC	After LC
Freeness, mL	705	700	550
Bulk, cm^3/g	2.23	1.93	1.68
Fibre length, mm	2.30	2.11	2.25
Tensile index, Nm/g	46.0	44.0	84.0
Tear index, mNm^2/g	19.3	18.8	12.2
TEA, J/g	0.9	1.0	2.0

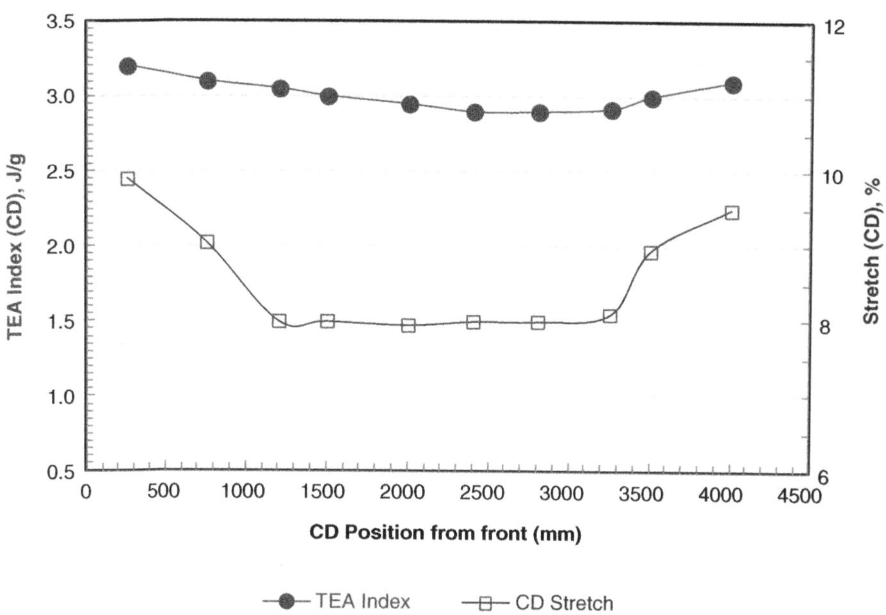

Figure 13.34 Improved profiles following retrofitting.

Table 13.6 Key properties of sack papers before and after retrofitting.

Situation	% Variation at edges with respect to middle		Deckle/CD Average	
	CD Stretch	CD TEA	CD Stretch, %	CD TEA index, J/g
Before	120	134	3.75	1.08
After	**70**	7	**7.95**	**3.00**

HC refining, bulk and strength properties were slightly reduced, because collapsed and curly fibres have less bonding capabilities than do straight unrefined never-dried fibres. LC refining after HC refining straightens the fibres, decreases bulk, and increases fibrillation, tensile strength, TEA and air resistance.

The quality of sack kraft papers currently being made on the paper machine following the retrofits of both the long-fibre pine and short-fibre NSSC refining system, along with the upgrade of the dryer section, has significantly improved. The improved profiles of CD stretch and CD TEA are shown in Figure 13.34. The deckle average values and the percentage variation at edges with respect to the middle of the two key properties are shown in Table 13.6. Both stretch and TEA value post-retrofit was more that doubled, whereas the variation drastically reduced.

13.7 Conclusions

Paper manufacturing is a capital-intensive industry and comprised of multiple process steps. Use of new technology in the form of plant and equipment, to *debottleneck* many rate-determining steps and to increase throughput, is quite often financially prohibitive. Identifying critical steps and *retrofitting* with appropriate gears in such aspects can be very cost-effective in the optimization of existing plant components. The chemical pulp refining process is one of the most important stages for manufacturing papers and boards. In many paper mills, the installed refining system may be sub-optimal in the form of capacity, the type of refiners used, the mode of operation and the type of refiner tackles fitted. In this chapter, tangible benefits were realized for two paper mills when the refining process of the stock-approach system of each of the paper machines was retrofitted. A summary of *retrofits* and their consequential benefits realized is given below:

- The refining system of one paper machine producing kraft liner board papers for packaging applications had refiner tackles that were grossly inadequate in terms of fibre development through refining and usage of excessive net specific refining energy. The mill often experienced difficulties in achieving target specifications and an increase in throughput
- A series of *in-situ* refining trials that involved both mill and laboratory evaluation were necessary to undertake to generate sufficient data to convince the mill, beyond any reasonable doubt, to *retrofit* the refiners with appropriate tackles

- The cost associated with the *retrofit* was almost negligible, apart from the trial and laboratory evaluation costs. However, the benefits were substantial for refining of the high-Kappa pulp that was used to manufacture the kraft liner board papers. These included substantial savings in refining energy, an increase in throughput and improvement in product quality
- The HI refining strategy for high-Kappa pulps through *retrofitting* of existing refiners with proper refiner fillings optimized the mill's paper mill refining operation
- For the mill producing sack kraft paper, four existing refiners operating in series to refine short-fibre neutral sulphite semi-chemical pulps used excessive idling load or energy. *Retrofitting* these refiners with a single double-disc refiner and appropriate refiner tackle saved substantial amounts of refining energy for the mill
- *Retrofitting* of the existing double-disc refiner used to refine unbleached long-fibre *Pinus radiata* kraft pulp to make 'flat' sack kraft paper by an HC refining system followed by LC refining enabled the mill to manufacture premium-grade 'semi-extensible' sack kraft papers
- The combined effects of *retrofitting* of the refining system and the dryer hood of the sack kraft machine delivered significant improvement in the cross directional profiles and absolute values of the two key properties of sack kraft papers, namely CD stretch and CD TEA.

Exercises

Two basic excercises covering refining of short-fibre hardwood and long-fibre softwood pulps which are used in the paper machines producing fine paper and packaging-grade papers are included in this chapter.

13.1 A paper machine producing fine printing and writing-grade paper uses bleached hardwood kraft fibres. The refining system of the machine comprised two 34" double-disc refiners operating in *parallel* at 600 rpm. The thoughput of each refiner is 78 metric ton per day (mtpd). The gross power applied on each refiner is 750 kW. The freeness of unrefiner pulp is 650 Csf and the target freeness of refined pulp is 450 Csf. The refiner plate pattern of both of the refiners is the same and the cutting edge length value is 21.1 km/rev. For bleached hardwood kraft pulp, the recommended range of freeness drop is 2.2–2.5 Csf/kWh/t and specific edge load range is 0.40–0.80 Ws/m. Answer the following questions.

- What is the idling load of each refiner?
- How much NSE is being applied to the hardwood pulp by each refiner?
- What is the freeness drop per kWh/t?
- What is the refining intensity used in each refiner?
- Are the two refiners operating optimally?
- Is there any need to retrofit the refiner plates to a different pattern? If, so what would be the recommended cutting edge length of such plates?

13.2 The paper machine in this exercise manufactures two-ply kraft liner board grade packaging paper using unbleached long-fibre kraft pulps. The refining operation of the stock approach system of the top ply comprised two 42" double-disc refiners operating in *series* at 420 rpm. The volumetric flow rate of pulp slurry at 4% consistency through the refiners is 4150 litre/min. The gross power applied on the first refiner is 1500 kW, while that of the second refiner is 1200 kW, although the maximum available gross power for each refiner is 1700 kW. The freeness of unrefiner pulp is 750 Csf and the target freeness of refined pulp is 600 Csf. The refiner plate pattern of each refiner is different and the cutting edge length values are 64.5 and 59.3 km/rev, respectively, for the first and the second refiner. For the type of unbleached long-fibre kraft pulp refined, the recommended freeness drop is between 0.75 and 0.90 Csf/kWh/t and specific edge load more that 3.5 Ws/m. Answer the following questions.

- Is the overall freeness drop sufficient for the type of pulp refined?
- What is the amount of net specific refining energy applied on the top ply?
- What is the refining intensity or specific edge load through each of the refiners under the current operating conditions?
- What will be the refining intensities if both refiners are run at the maximum available motor power?
- Are the refiners operated at optimal intensity? If not, what would be the recommended plate pattern in terms of cutting edge length?

References

Baker, C.F. (1995) Refining Seminar, SAPPI, Johannesburg, South Africa, PIRA International.

Brecht, W. and Siewert, W.H. (1966) *Das Papier*, **20** (1), 4–14.

Danforth, D.W. (1969) *Southern Pulp and Paper Manufacturer*, **32** (7), 52–53.

Demler, C. and Silveri, L. (1995) Strength development of mechanical and deinked newsprint pulp using low intensity refining. Proceedings of the 49th Appita General Conference, 1995.

Ebeling, K. (1980) A critical review of current theories for the refining of chemical pulp. International Symposium on Fundamental Concept of Refining, The Institute of Paper Chemistry, Appleton, USA, 1980.

Espenmiller, H.P. (1969) The theory and practice of refining. *Southern Pulp and Paper Manufacturer.* **32** (4), 50–57.

Ghosh, A.K. (2002) Use of pilot refiner and screening system to improve stock quality. *Appita Journal*, **55** (2), 112–117.

Ghosh, A.K. (2004) Refining Optimization of Secondary Fibre – use of Bijective Diagram Technique, Peer-reviewed Paper presented in the APPITA/Pan Pacific Conference held at Canberra, Australian on April 19–21, 2004.

Ghosh, A.K. (2009) A Systematic Approach to Optimize Dryer Performance and Energy Savings – Case Studies. Tappi PaperCon09 Conference, May 31–June 03 2009, St. Louis, MO, USA.

Ghosh, A.K. (2010) Optimal Refining Strategy of High Yield Unbleached Pinus Radiata Kraft Pulp: An in-situ investigation. ABTCP-Tappi 2010 Conference, October 2010, San Paulo, Brazil.

Kerekes, R.J. (1990) Characterization of pulp refiners by a C-factor. *Nordic Pulp and Paper Research Journal*, **5** (1), 3–8.

Lehrman, M. (2001) Low Consistency Refining Seminar, J & L Fibre Services, Melbourne, May 01 2001.

Lumiainen, J.L. (1990) New approach to the critical factors effecting refining intensity and intensity results in low consistency refining. *Tappi Papermakers Conference Proceedings*, Tappi Press, Atlanta, USA.

Lumiainen, J.L. (1995). Comparison of the mode of operations between conical and disc refiners, The 1st CALAR Symposium PTS_IZP in Heidenau, Germany, December 1st, 1995.

Lumiainen, J.L. (1997) Comparison of the mode of operations between conical and disc refiners *Paper Technology*, **38** (2), 53–58
Mohlin, U. (2002) Industrial refining of unbleached kraft pulps – the effect of pH and refining intensity. Tappi 2002 Paper Summit.
Page, D.H. (1989) The beating of chemical pulp – the action and effect. 9th Fundamental Research Symposium Notes, Pulp and Paper Fundamental Research Society, Cambridge, UK, 1989.
Paulapura, H. (2000) *Papermaking Science and Technology: Papermaking Part 1 – Refining of Chemical Pulp*, Finnish Pulp and Paper Publication.
Smook, G.A. (1989) *Handbook for Pulp and Paper Technologist*, Tappi Press, Atlanta, USA.
Welch, L.B. and Kerekes, R.J. (1994) Characterization of the PFI mill by the C-Factor. *Appita Journal*, **47** (5), 387–390.
Wultsch, F. and Flucher, W. (1958) *Das Papier*, **12** (13), 334–342.

Index

Chemical Process Retrofitting and Revamping: Techniques and Applications, First Edition.
Edited by Gade Pandu Rangaiah.

Printed and bound by CPI Group (UK) Ltd, Croydon, CR0 4YY
17/09/2026
14964302-0005

Compendium of Drug Degradation Pathways

Compendium of Drug Degradation Pathways

Min Li
Ringoes
NJ 08551, USA

WILEY

This edition first published 2026

Registered Office(s)
John Wiley & Sons, Inc., 111 River Street, Hoboken, NJ 07030, USA
John Wiley & Sons Ltd, New Era House, 8 Oldlands Way, Bognor Regis, West Sussex, PO22 9NQ, UK

For details of our global editorial offices, customer services, and more information about Wiley products visit us at "http://www.wiley.com"www.wiley.com.

The manufacturer's authorized representative according to the EU General Product Safety Regulation is Wiley-VCH GmbH, Boschstr. 12, 69469 Weinheim, Germany, e-mail: Product_Safety@wiley.com.

Wiley also publishes its books in a variety of electronic formats and by print-on-demand. Some content that appears in standard print versions of this book may not be available in other formats.

Library of Congress Cataloging-in-Publication Data applied for:

Hardback ISBN: 9781118832936

Cover Design: Wiley
Cover Image: © klepach/stock.adobe.com, Courtesy of Min Li

Set in 9.5/12.5pt STIXTwoText by Straive, Pondicherry, India
Printed and bound by CPI Group (UK) Ltd, Croydon, CR0 4YY

C9781118832936_031125

To Beihong and Emily

Contents

Preface

My first book, *Organic Chemistry of Drug Degradation*, was published in September 2012, and it was intended to be a systematic summary of drug degradation chemistry based on the types of degradation pathways such as hydrolytic, oxidative, and photochemical degradation. Each type of drug degradation discussed in a particular chapter of that book is substantiated by examples of the specific degradation pathway(s) of a drug or drugs related to the subject of the chapter. Hence, for the drug molecules discussed in those examples, there are no summary or compilation of *all* of their known or reported degradation pathways. In other words, the degradation chemistry of a particular drug was not intended to be presented in its entirety in that book. Although *Organic Chemistry of Drug Degradation* can be used as a textbook for graduate students majoring in pharmaceutical sciences or anyone who is interested in a systematic study of drug degradation chemistry, it has occurred to me that in many occasions the practicing scientists, particularly those working in the pharmaceutical industry, would also like to have a convenient reference book in which the degradation chemistry of many commonly prescribed drugs can be readily found. Hence, for the past 10 years or so, it has been my goal to try to fulfill that need by summarizing the known or reported degradation pathways of representative small molecule drugs in another book, as such a book would be complementary to *Organic Chemistry of Drug Degradation* from a different perspective. Furthermore, such a book might also be supplementary to official pharmacopeias, where the degradation chemistry of drugs is usually not included, despite the fact that the supplementation would be quite limited.

During the course of pondering the contents of such a book, I decided to choose approximately 300+ representative small molecule drugs and then to summarize their degradation pathways in respective monographs that are organized in alphabetical order of the generic names of these drugs. The selection of these representative drugs in the book was based upon the following several factors: the numbers of their prescription, sales status, and inclusion in WHO List of Essential Medicines (the ones that "satisfy the priority health care needs of the population" per World Health Organization's definition). Many drugs in the same therapeutic categories share the same or similar core structures and functional groups and thus would have the same or similar degradation pathways. For this reason, these drugs are typically covered in the same respective monographs, with the best known or most widely used drugs as the main entries of the monographs. Such examples include, but not limited to, the monograph for benzodiazepines, the monograph for omeprazole and related azoles, the monograph for opioids, and the monograph for penicillins.

Preceding the monographs, the main body of the current book *Compendium of Drug Degradation Pathways*, a general chapter entitled "Introduction" is included, in which a brief overview of the regulatory perspective on drug stability and impurities, as well as an overview of drug degradation chemistry is provided. In addition, the role and limitation of stress testing, often referred to as forced degradation study as well, is also discussed. In the ensuing monographs, each one is started with the structure(s) of the drug molecule(s), and compilation of their key information, including chemical names, brand names, formulas, protonated monoisotopic mass-over-charge ratios, average molecular weights, and CAS numbers, as applicable. The above information is mostly derived from the following publicly accessible resources: Pubchem (pubchem.ncbi.nlm.nih.gov), Wikipedia (wikipedia.org), Drugbank online (go.drugbank.com), BMRB (Biological Magnetic Resonance Data Bank, https://bmrb.io/metabolomics/mol_mass.php), and Chemspider (chemspider.com). In the main body of each monograph, the known or reported degradation chemistry of a drug or drugs is summarized in text, which is followed by schematic presentation of the drug degradation pathway(s). Quite often, the reported degradation chemistry of a drug was based upon stress or forced degradation studies of the drug. As it is discussed in Introduction of this book, the degradation chemistry predicted from the forced degradation studies may not necessarily represent the one under real-life scenario, i.e., the degradation chemistry observed under the long-term storage conditions as represented by the stability conditions recommended by the International Council on Harmonisation of Technical Requirements for Registration of Pharmaceuticals for Human Use (ICH) and the World Health Organization (WHO). The quality for the prediction of real-life degradation chemistry relies upon a clear understanding of the principles of drug degradation chemistry, so that the forced degradation study would be conducted under the conditions that are relevant to the drug degradation in real-life scenarios. The degradation pathways obtained by utilizing these conditions, sometimes referred to as purposefully designed conditions or mechanism-based conditions, would most likely to contain the real-life degradation profiles. In other words, the real-life degradation pathways of a drug would more likely be a subset of its degradation pathways obtained under the forced degradation conditions purposefully designed by taking into consideration of the perceived degradation mechanisms and kinetics. To borrow the terms from the quality-by-design paradigm, the degradation pathways of a drug obtained under the purposefully designed forced degradation conditions would be the design space, while the real-life degradation pathways would be the control space, a subset of the design space. Due to the limitation that many forced degradation studies reported in the literature were not necessarily performed under purposefully designed conditions, readers are cautioned in places where artificial degradation may occur in these forced degradation studies.

To facilitate organization of the book contents, the alphabetically arranged monographs are further sequentially grouped into eight sections, following Section 1, Dedication and Preface, and Section 2, Introduction.

It has been quite a long time since my initiation for the writing of this book, which far exceeds the originally proposed timeline for the completion; this is primarily due to the change of my career path and increasing responsibilities, particularly during the last 10 years at Huahai Pharmaceutical Co., Ltd. For this reason, I sincerely apologize for the delay, while being very grateful for the patience of the editors of Wiley. Meantime, I am also very appreciative for the opportunities provided by Huahai that allowed me to create Center of Excellence for Modern Analytical Technologies (CEMAT), the analytical core group within Huahai that is dedicated to investigating and solving the most technically challenging problems occurring in the stages of pharmaceutical research and development, scale up, regulatory filing, and commercial production. CEMAT was modeled after my former organizations at Merck & Co., Inc. (Department of Regulatory and Analytical Sciences–Analytical, or RAS-A), Schering-Plough Corp. (Department of Global Quality Services–Analytical Sciences, or GQS-AS), and Merck again (Analytical Chemistry in Development and Supply–Supply Analytical Sciences, or ACDS-SAS). CEMAT was formally established in early 2016, and it currently has a staff of approximately 150 scientists and is equipped with state-of-art analytical instrument and equipment. Over the years, members of CEMAT have performed thousands of investigations, many of which are related to elucidating drug degradation pathways and mechanisms. The results obtained from these investigations have further expanded my own knowledge in drug degradation chemistry. Among the many young and energetic scientists who have made significant contributions toward the establishment, operation, and continuous improvement of CEMAT, I would like to extend my special gratitude to Mr. Wenquan Zhu, who helped me from the very beginning for the establishment, organization, daily operation, as well as future expansion of CEMAT. My appreciation also goes to the following CEMAT members, some of whom have moved on to other career opportunities: Mr. Wenbin Chen, Dr. Jinsheng Lin, Mr. Jian Ye, Mr. Jianyang Jin, Dr. Lewei Zheng, Ms. Dan Li, Dr. Jichao Wang, Dr. Jian Ma, Mr. Wenping Hu, Ms. Qianqian Lv, Mr. Zijian Kuang, Mr. Shaolan Wang, Mr. Zijian Wang, Mr. Qiang Zhou, Ms. Hong Cai, and Ms. Qiaohong Guo.

Between June 2020 and May 2023, I was also responsible for managing Shanghai Synergy Pharmaceutical Sciences, the Huahai subsidiary dedicated to small molecule innovative drug research. During that three-year tenure, my team advanced two novel drug candidates into clinical stage and developed several preclinical drug candidates. Toward these achievements, I am very much indebted to the following team members at Shanghai Synergy: Dr. Yu Huang, Dr. Jia Chen, Dr. Chengjun Deng, Dr. Xin Xu, Dr. Zhan Cai, and Dr. Zhezhou Yang. Over the past 10 years or so, I also have had the opportunities to work with many capable and hard-working colleagues at different functional areas of Huahai and its subsidiaries. Their expertise, contributions, and friendship have made my journey of the last 10 years more fruitful and enjoyable. Among them, the most notable ones are Mr. Jun Du, Ms. Linda (Lihong) Lin, Dr. Xiaodi Guo, and Mr. Jay (Jie) Wang. Jun was the founder and CEO of Huahai US and also a longtime friend of mine for 20 years, who not only introduced me to Huahai but also gave me firm support and trust throughout the time we had worked together. Linda and her Regulatory Affairs team have worked closely with my CEMAT team on a great number of projects, particularly during their regulatory filing stages. Together, we have overcome many challenges in dealing with pharmaceutical impurities in terms of their assessment, control strategies, analytical methods, and specifications. Xiaodi and his formulation team performed formulation development for HHT-120, a novel orally available anticoagulant developed by my teams initially at Huahai US and later at Shanghai Synergy. HHT-120 demonstrated comparable antithrombotic efficacy but significantly less bleeding risk as compared to an approved drug in animal models. Jay, a congenial colleague often with a sense of subtle humor, was a vice president for Huahai's API Sales Department. In 2016, I worked with him to help resolve analytical issues associated with a key API product for a major client, during which process I introduced the first liquid chromatography – triple quadruple mass spectrometer into Huahai to ensure the method for a trace impurity with sufficient sensitivity, specificity, and robustness.

My special thanks go to the editors of Wiley for their excellent job in the production of this book, in addition to their patience. Last but not the least, my profound gratitude goes to my family, especially my wife Beihong, for her love, support, and unwavering confidence in me for more than 30 years.

Min Li
Ringoes, New Jersey
First draft in October 2022.
Final draft on 1 January 2025.
minli88@yahoo.com

1

Introduction

The stability of a drug product, which includes chemical and physical stability, is a key attribute for its efficacy and safety. This book focuses on the chemical aspect of drug stability. To have a drug product with desirable stability during production, shipment, storage, and throughout its expiration requires a clear understanding of drug degradation chemistry, so that a reasonably stable molecule can be selected as the pharmaceutical development candidate in the first place. This candidate will then be formulated with excipients to impart adequate pharmacokinetic properties onto the drug product. The excipients may stabilize, do nothing with, or destabilize the drug molecule. It would be desirable that the excipients would stabilize the drug molecule, or at least would not destabilize it. To achieve that goal, it is necessary to understand the chemical and physical interactions between the drug molecule and excipients, which also fall into the scope of drug degradation chemistry, or at least the part of the chemical interactions. In my previous book, *Organic Chemistry of Drug Degradation*, drug degradation chemistry was systematically summarized based on the types of degradation pathways such as hydrolytic, oxidative, and photochemical degradation. The current book, *Compendium of Drug Degradation Pathways*, is intended to be complementary to *Organic Chemistry of Drug Degradation* from the perspective of summarizing the known or reported degradation pathways of representative drugs in their entirety. There are more than 300 drugs that are discussed in the ensuing monographs. Before proceeding to the monographs, it would be helpful to have a brief overview of the regulatory perspective on drug stability, impurities, as well as drug degradation chemistry. Readers may need to refer to *Organic Chemistry of Drug Degradation* from time to time for a more comprehensive discussion on a particular degradation mechanism or pathway.

Drug Stability and Regulatory Requirements

The chemical stability of a drug is characterized by the persistence of its active ingredient(s), i.e., not degrading into degradation products, over its intended or registered shelf-life. The study of drug stability probably dates back to when

Compendium of Drug Degradation Pathways, First Edition. Min Li.

modern drug discovery and development started. The first publications on drug stability were released in the 1950s, according to Jamrógiewicz and Merchel [1]. In 1978, the pioneering book, *Chemical Stability of Pharmaceuticals*, was first published, and the second edition was released in 1986; the book provided "a collection of selected drug-stability data from the pharmaceutical literature" [2]. Nevertheless, it was not until 1987 when the US FDA issued the first guidance that required stability data in the submission of investigational new drug applications (INDs) as well as new drug applications (NDAs) [3]. Subsequently in April 1990, representatives of the regulatory agencies and industry associations from Europe, United States, and Japan convened in Brussels to inaugurate the international organization for pharmaceutical regulatory guidance – ICH, which stands for International Council for Harmonisation of Technical Requirements for Pharmaceuticals for Human Use (formerly known as International Conference on Harmonisation of Technical Requirements for Pharmaceuticals for Human Use) [4]. In the early 1990s, ICH started to issue a series of guidance documents for performing the stability study and submission of stability data during the registration of new drug substances and products, starting from Q1, *Stability Testing of New Drug Substances and Products*. And the code Q1 was later revised to Q1A (and the most updated version is Q1A(R2)), as additional guidance documents on stability, Q1B through Q1E, were issued. The other ICH guidance documents that directly relate to the drug stability and drug impurities (including both process impurities and degradation products) are Q3A(R2), *Impurities in New Drug Substances*, and Q3B(R2), *Impurities in New Drug Products*. The above ICH guidance documents are summarized in Table 1.1 [5].

The main environmental factors influencing the drug stability over time are temperature, humidity, and light. Accordingly, Q1A(R2) stipulates the temperature and humidity requirements for typical long-term and accelerated stability studies to be carried out at 25°C, 60% relative humidity (RH) and 40°C, 75% RH, respectively, in order to assess the thermal stability of drug substances and products. For the drug

Table 1.1 ICH Guidance documents on drug stability and impurities.

Code	Document title
Q1A(R2)	Stability Testing of New Drug Substances and Products
Q1B	Stability Testing: Photostability Testing of New Drug Substances and Products
Q1C	Stability Testing for New Dosage Form
Q1D	Bracketing and Matrixing Designs for Stability Testing of New Drug Substances and Products
Q1E	Evaluation of Stability Data
Q3A(R2)	Impurities in New Drug Substances
Q3B(R2)	Impurities in New Drug Products

Table 1.2 Storage conditions and time periods in Q1A(R2) and Q1B(R2) for stability studies of drug substances and products at the time of NDA submission.

Study[a]	Storage condition[b]	Minimum time period required at NDA submission
Long term	25°C, 60%RH	12 months
Intermediate	30°C, 65%RH	6 months
Accelerated	40°C, 75%RH	6 months

[a] Applicants of NDA may choose the intermediate condition as the long-term condition as appropriate, due to climate zone differences.
[b] The temperature variation allowed is ±2°C and the humidity variation allowed is ±5%.

substances or products that undergo significant degradation under the accelerated conditions of 40°C, 75% RH, an intermediate condition of 30°C, 65% RH is recommended for the stability study. Likewise, Q1B outlines the photolysis requirements, i.e., sources of light and duration of exposure, for photostability studies of drug substances and products, in order to assess the photostability of drug substances and products with and without packaging materials. The requirements of Q1A(2R) and Q1B are outlined in Table 1.2 [5].

The World Health Organization (WHO) also issued guidance on drug stability, particularly for Climate Zone III (hot and dry) and Climate Zone IV (hot and humid) regions, which provides the options of selecting 25°C/60%RH, 30°C/65%RH, or 30°C/75%RH as the long-term stability condition dependent upon the climate zones, where the drug products are intended to be marketed [6].

In order to help elucidate the intrinsic stability of drug molecules, stress testing, also referred to as forced degradation, is required in Q1A(R2), among which photostability testing is an integral part of the stress testing. The subject of stress testing will be discussed in detail later in this introduction.

On the other hand, Q3A(R2) and Q3B(R2) set the thresholds for impurities that need to be reported, identified, or qualified, once they exceed the respective thresholds, and these thresholds are summarized in Tables 1.3 and 1.4 [6].

Table 1.3 Thresholds for impurities in drug substances.

Maximum daily dose	Threshold (% of drug substance)		
	Reporting	Identification	Qualification
≤2g	0.05%	0.10% or 1.0 mg/day intake (whichever is lower)	0.15% or 1.0 mg/day intake (whichever is lower)
>2g	0.03%	0.05%	0.05%

Table 1.4 Thresholds for degradation products in new drug products[a].

	Threshold (% of drug substance or total daily intake)		
Maximum daily dose	**Reporting**	**Identification**	**Qualification**
≤1g	0.1%	–	–
>1g	0.05%	–	–
<1 mg	–	*1.0% or 5 μg TDI[b]*	–
1–10 mg	–	*0.5% or 20 μg TDI[b]*	–
>10 mg–2 g	–	*0.2% or 2 mg TDI[b]*	–
>2g	–	*0.10%*	–
<10 mg	–	–	**1.0% or 50 μg TDI[b]**
10–100 mg	–	–	**0.5% or 200 μg TDI[b]**
>100 mg–2 g	–	–	**0.2% or 3 mg TDI[b]**
>2g	–	–	**0.15%**

[a] The reporting thresholds are set based on two ranges of maximum daily doses (i.e., ≤ 1 g/day and >1 g/day), while both the identification threshold and qualification threshold are set based on four ranges of maximum daily doses.
[b] Whichever amount is lower; TDI, total daily intake.

Brief Overview of Drug Degradation Chemistry

The most frequently occurring drug degradation types are hydrolytic and oxidative in nature, followed by various other types, such as isomerization/rearrangement, elimination, decarboxylation, cyclization, dimerization, and photochemical degradation. The categorization of the degradation types is not absolute, as they may be interrelated, e.g., oxidation may trigger decarboxylation, cyclization, and dimerization, while oxidation, isomerization/rearrangement, elimination, decarboxylation, cyclization, and dimerization can often occur under photochemical degradation conditions.

Types of Functional Groups or Structural Moieties Susceptible to Hydrolytic Degradation

The susceptibility to hydrolytic degradation may be estimated based on the hydrolysis activation energy (Ea) or enthalpy of activation ($\Delta H^{\neq}$) of a compound bearing a hydrolyzable functional group [7]. Although Ea and $\Delta H^{\neq}$ are obtained from the different models describing the same reaction process, their numerical values are usually quite close to each other [8]. For this reason, these two types of values may be used interchangeably to evaluate the hydrolytic susceptibility of different hydrolyzable drug molecules. The lower the activation energy or enthalpy of activation, the easier it would be for the drug molecule to undergo hydrolytic degradation, and vice versa. The common functional groups or structural moieties in this category include, but are not limited to, derivatives of carboxylic acid (esters, lactones, amides, imides, lactams, etc.), derivatives of phosphoric acid (esters and amides), carbamates, epoxides, ethers, polysaccharides, and nucleosides. The key factors that have impact on the intrinsic susceptibility of an organic compound to hydrolytic degradation are the steric and electronic factors surrounding the hydrolyzable moieties. While steric hindrance makes hydrolysis substrates of all types more stable, the presence of electron-withdrawing groups adjacent to the carbonyl groups of the hydrolysis substrates renders these carboxylic derivatives more susceptible to hydrolytic degradation [7]. From a different perspective, any factor destabilizing the transition states of hydrolytic degradation renders the hydrolysis substrates more resistant toward hydrolytic degradation [9, 10], while those (derivatives of both carboxylic and phosphoric acid) containing an easy-to-leave group are more susceptible to hydrolytic degradation [11, 12].

Carboxylic Esters and Lactones

Carboxylic esters, especially those without steric hindrance surrounding the ester linkage, are usually quite susceptible to hydrolytic degradation to produce carboxylic acids and alcohols, particularly under alkaline conditions (Scheme 1).

With increasing steric hindrance on either R_1 or R_2 group adjacent to the carbonyl group, carboxylic esters become more resistant toward hydrolysis. For example, while ethyl acetate only has a reported *Ea* value of 10 kcal/mol, hexyl acetate and *t*-butyl acetate have reported *Ea* values of 11.4 and 27 kcal/mol, respectively [13]. Aromatic carboxylic esters appear to be somewhat more stable than their alkyl counterparts, for example, the $\Delta H^{\neq}$ for hydrolysis of ethyl benzoate in a 0.1 M NaOH solution was reported to be approximately 82–84 kJ/mol (~20 kcal/mol) in the temperature range of 20–40°C, while the *Ea* of ethyl acetate was reported to be only 11.56 kcal/mol in a solution of 0.02 M NaOH [14, 15]. Lactones are cyclized esters, and particularly for lactones with a five- and six-membered lactone ring, their hydrolytic degradation under acidic conditions would usually reach an equilibrium point because the reverse reaction

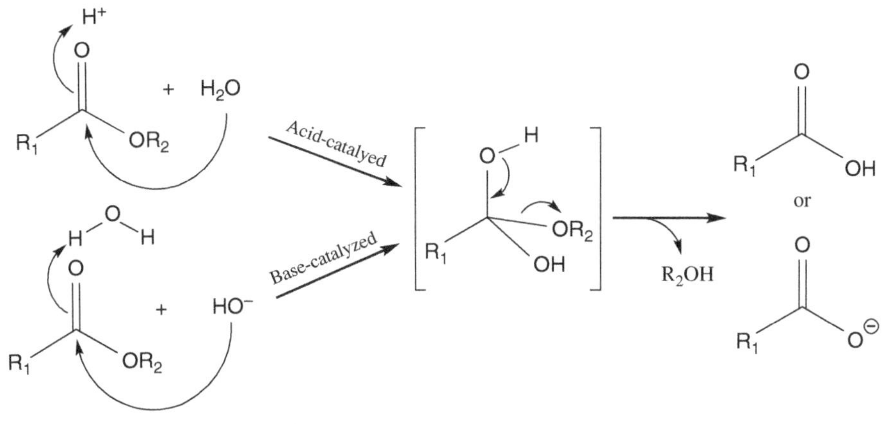

Scheme 1 Hydrolysis of carboxylic esters.

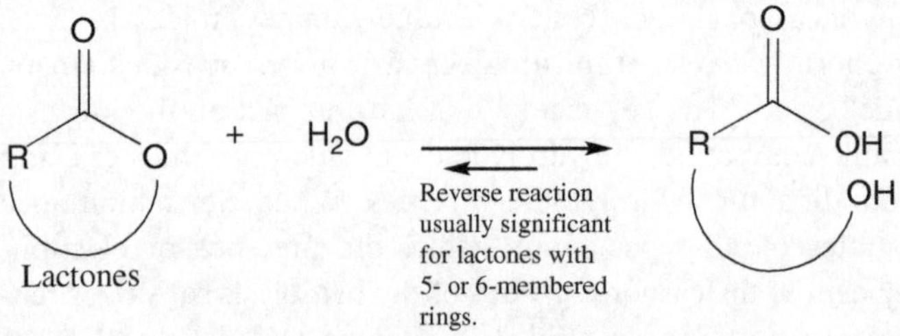

Scheme 2 Hydrolysis of lactones, and the reverse reaction would be significant for lactones with five- or six-membered rings, especially under acidic conditions.

(i.e., re-formation of the lactone ring) becomes increasingly significant as the hydrolysis goes on (Scheme 2) [16].

Carboxylic Amides and Lactams

Carboxylic amides and lactams are usually significantly more stable than the corresponding esters and lactones with regard to hydrolytic degradation, despite the fact that their hydrolysis mechanisms are similar to those of carboxylic esters and lactones, respectively (Scheme 3).

Consistently, typical hydrolysis activation energies of carboxylic amides and lactams are meaningfully higher, usually ~20 kcal/mol or higher. Nevertheless, if an electron-withdrawing group is present at the position alpha to the carbonyl, the amide becomes more susceptible to hydrolysis, because the carbonyl is more electron-deficient and thus more susceptible to attack by water or hydroxide ion. For example, the hydrolysis rate of trifluoroacetamide was reported to be several orders of magnitude higher than that of acetamide [17]. For this reason, the trifluoroacetyl group can be used as a protection group for amines in organic syntheses [18].

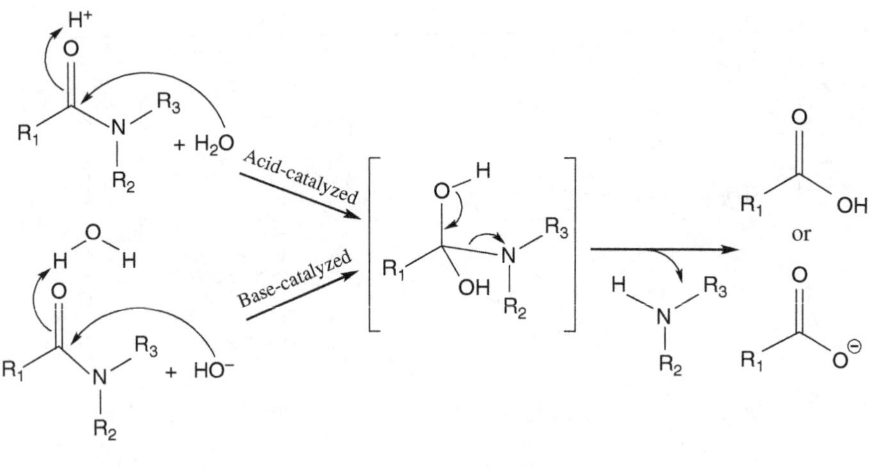

Scheme 3 Hydrolysis of carboxylic amides. R_1 = alkyl or aryl; R_2 = H, alkyl, or aryl; R_3 = H, alkyl, or aryl.

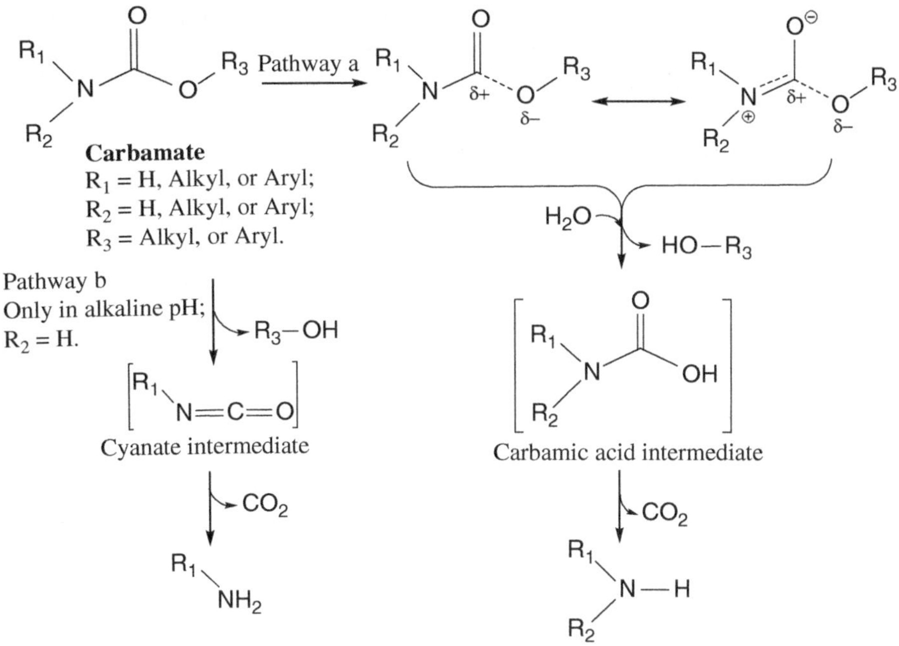

Scheme 4 Hydrolysis of carbamates.

Carbamates

Carbamate is another carbonyl-based functionality often utilized in drug design [19]; structurally, it can be viewed as a hybrid of carbonyl ester and amide. A study by Dittert and Higuchi seemed to be indicated that the hydrolysis of a carbamate in alkaline solution could proceed via two mechanisms: the first one would start with the removal of the ester moiety, resulting in the formation of the unstable intermediate, carbamic acid, which quickly decomposes to an amine degradant and carbon dioxide (Pathway a, Scheme 4). The second one would involve the removal of the amide hydrogen, resulting in the formation of the isocyanate intermediate (Pathway b, Scheme 4) [20]. According to a hydrolytic stability study by Chapman in which *N*-phenyl-carbamates and *N*-cyclohexyl-carbamates were compared to their respective amide analogs, the carbamates were significantly more hydrolytically stable than the amides under acidic conditions, while the carbamates were slightly less stable than the amides under neutral conditions [21]. In a hydrolytic stability study of estramustine, an anticancer drug based on estradiol with a carbamate linkage to a nitrogen mustard, the hydrolytic enthalpy of activation was found to be 89.3 kJ/mol (21.3 kcal/mol) in the pH range of 1–9 [22]. This result is consistent with those obtained by Chapman, suggesting that the hydrolytic stability of carbamates is at least comparable to that of amides in general.

Phosphoryl Esters and Phosphoramidates

The phosphoryl functional group is often utilized to impart aqueous solubility to a drug molecule that contains either a hydroxyl or an amino group. In the former case, the resulting drug molecule is a phosphoryl monoester, while in the latter case, it is an

N-substituted derivative of phosphoramidic acid (amidophosphoric acid). Usually, the phosphoryl monoesters prepared from alkyl alcohols are quite stable hydrolytically. The hydrolysis activation energies of such phosphoryl monoesters are close to ~30 kcal/mol [23, 24]. On the other hand, *N*-substituted derivatives of phosphoramidic acid, or more commonly *N*-substituted phosphoramidates, are generally less hydrolytically stable than the structurally similar phosphates. For example, in a hydrolytic stability study of six 2,4-dichlorophenyl methyl *N*-alkylphosphoramidates, the activation energies for the hydrolysis of the *N*-alkyl moiety (Scheme 5) in 0.19 M HCl solutions were found to be in the range of 10.1–13.2 kcal/mol [25]. Another example is fosaprepitant dimeglumine, which is the phosphoramidic pro-drug of aprepitant (Scheme 6). The activation energy for the hydrolysis of the phosphorylamidic linkage was determined to be 91 kJ/mol (22 kcal/mol) [26]. The above activation energies for the hydrolysis of the phosphorylamidic linkage are significantly lower than the activation energies of typical alkyl phosphoryl esters.

Nevertheless, the phosphoryl esters prepared from phenol or its analogs are not as stable hydrolytically as their alkyl counterparts, because phenol and its analogs are good leaving groups. For this reason, the phenyl phosphoryl functional group is usually used in the design of prodrugs where the free phosphate functionality needs to be released in vivo, such as in the case of anti-viral drug remdesivir (Scheme 7) [27].

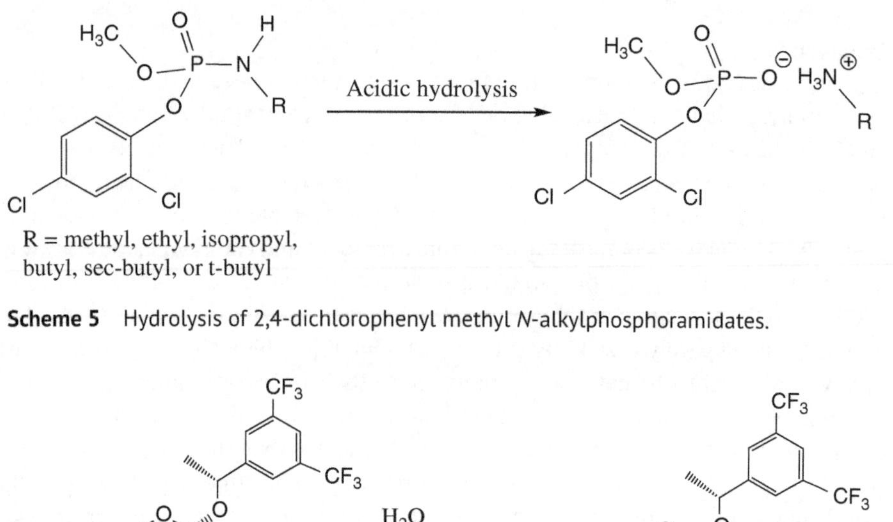

Scheme 5 Hydrolysis of 2,4-dichlorophenyl methyl *N*-alkylphosphoramidates.

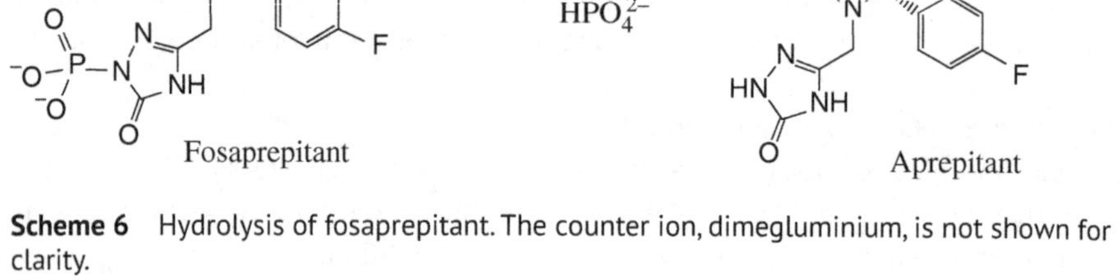

Scheme 6 Hydrolysis of fosaprepitant. The counter ion, dimegluminium, is not shown for clarity.

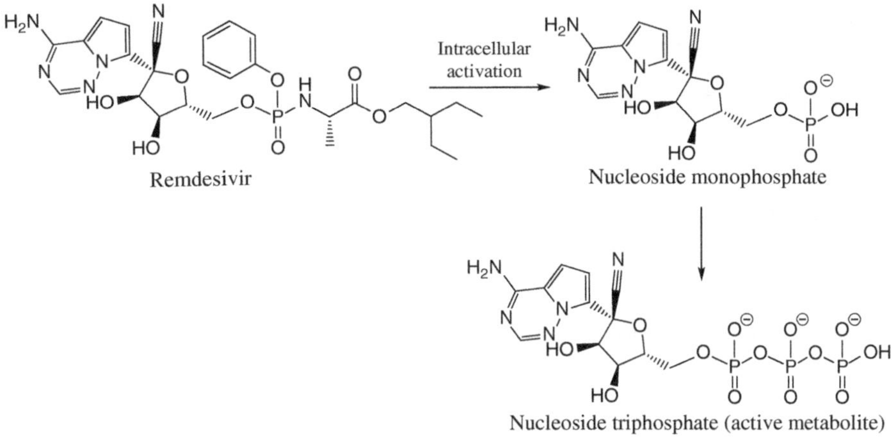

Scheme 7 Bio-activation of remdesivir, a prodrug containing phenyl phosphoramidate moiety.

Sulfonamides and Urea Derivatives
Theoretically, sulfonamide linkage is susceptible to hydrolysis; nevertheless, it is usually extremely stable under both acidic and alkaline hydrolytic degradation conditions, even under some very harsh conditions. The activation energy for a typical sulfonamide was estimated at greater than 30 kcal/mol [28]. Urea and its derivatives are generally resistant toward hydrolytic degradation, as evidenced by the fact that urea, the parent compound, was reported to have hydrolysis activation energies at approximately 30 kcal/mol in water and weakly to modestly acidic solutions [29]. Urea displayed pH-independent hydrolytic behavior under these conditions.

Imides and Sulfonylureas
Imides, particularly cyclic imides, are among the common building blocks utilized in drug design. The activation energies for hydrolysis of cyclic imides, such as phenobarbital and its derivatives (Scheme 8), were reported to be ~20 kcal/mol in

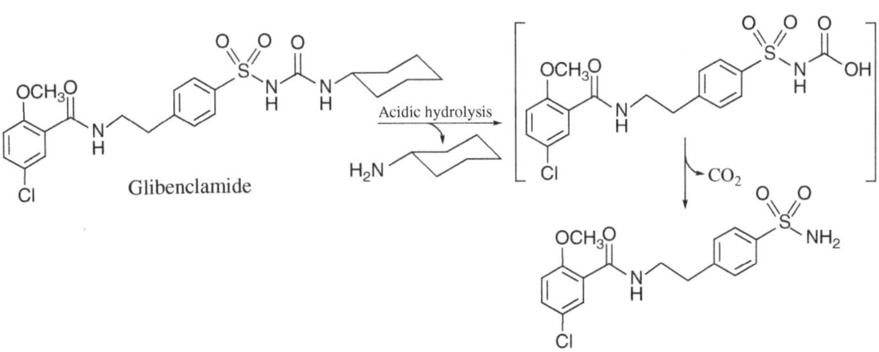

Scheme 8 Hydrolysis of glibenclamide under acidic conditions.

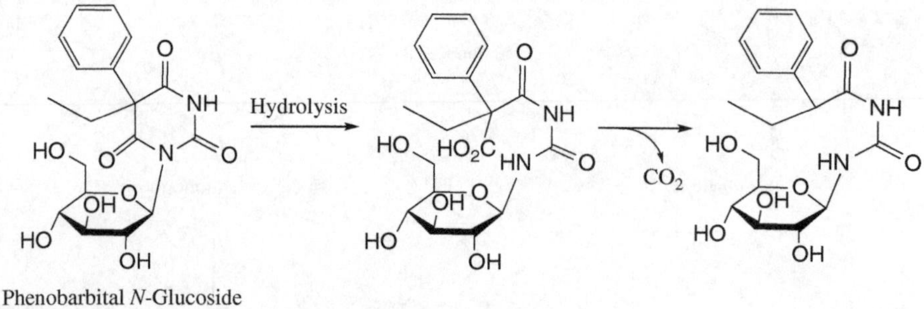

Scheme 9 Hydrolysis of phenobarbital *N*-glucoside.

solutions of neutral pH [30], suggesting that they would be reasonably stable with regard to hydrolytic degradation under neutral conditions. On the other hand, sulfonylureas, such as glibenclamide, may be viewed as a class of hybrid imides. Due to the presence of the sulfonyl group, a strong electron-withdrawing group, the urea functionality becomes less resistant toward hydrolysis, particularly under acidic conditions (Scheme 9) [31]. Structurally similar sulfamylureas are also quite susceptible to acidic hydrolysis [32].

Epoxides and Ethers

Epoxides are electrophiles owing to the constrained tricyclic ether ring, which can be hydrolyzed or hydrated, particularly under acidic or alkaline catalysis, giving rise to *trans*-1,2-diol products (Scheme 10). The hydrolytic stability of epoxides depends upon the substituents surrounding the epoxide group [33]. For propylene oxide, a simple epoxide model compound, the hydrolysis activation energy was reported to be ~19.0–19.5 kcal/mol [34], suggesting it would be reasonably stable under uncatalyzed conditions. Ethers of other types are usually hydrolytically stable, except in cases where a relatively good leaving group coincides with a stabilizable transition state of the hydrolysis in the same molecular entity. The hydrolysis of duloxetine under acidic conditions is such a case in which 1-naphthalenol is a relatively good leaving group, while the resulting carbocation transition state can be stabilized by the neighboring thiophene group through conjugation; in addition to the straight hydrolytic products, the other two main degradants were isomers of duloxetine formed via rearrangement (Scheme 11) [35].

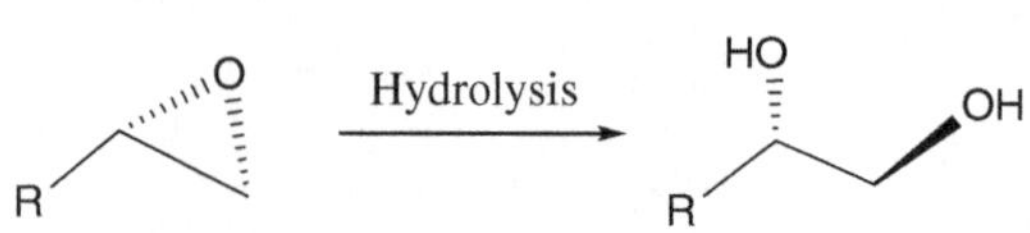

Scheme 10 Hydrolysis of epoxides.

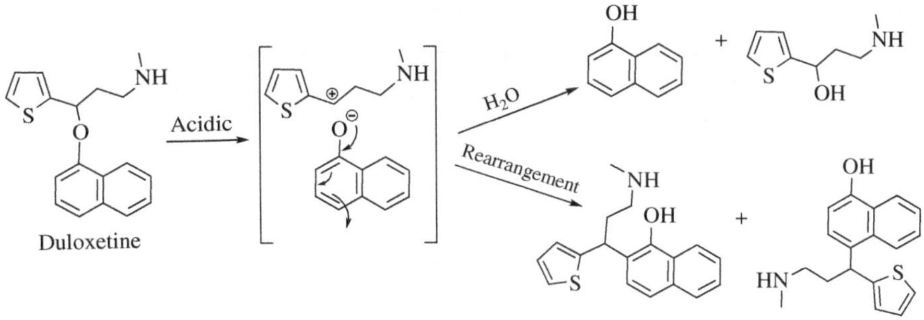

Scheme 11 Degradation of duloxetine under acidic conditions.

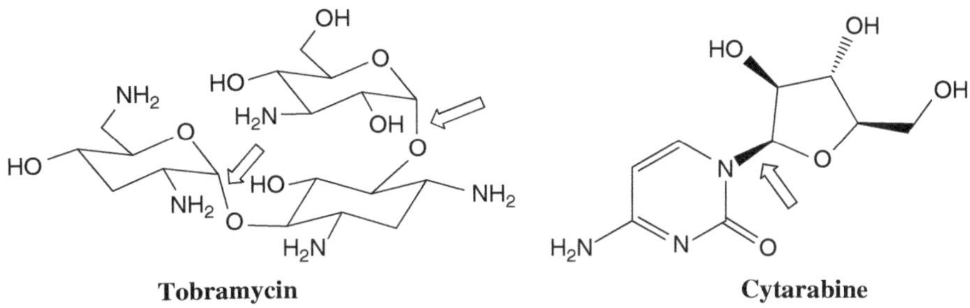

Figure 1.1 Structures of tobramycin and cytarabine. While the two glycosidic bonds in tobramycin may be susceptible to hydrolysis, the glycosyl bond in cytarabine is more resistant than the amino group toward hydrolysis.

Oligosaccharides/Polysaccharides and Nucleoside Analogs

The key linkage in oligosaccharides/polysaccharides and nucleosides is the glycosidic and glycosyl bond, respectively, with the glycosidic bond connecting the monosaccharide units of polysaccharides and glycosyl bond connecting the nucleobase (or its analog) and sugar moiety (Figure 1.1). The glycosidic and glycosyl bonds are usually quite stable hydrolytically; these bonds become less stable under alkaline conditions. For example, the hydrolysis activation energies for the glycosyl bond of tobramycin were reported to be 32 and 15 kcal/mol in 1 N HCl and 1 N NaOH solutions, respectively [36]. It needs to be pointed out that the glycosyl bonds of nucleosides are sometimes incorrectly referred to as glycosidic bonds.

Types of Functional Groups or Structural Moieties That Are Susceptible to Different Types of Oxidative Degradation

Oxidative degradation is another very common degradation pathway. The ultimate "oxidizing reagent" for drug oxidation originates from molecular oxygen, which constitutes ~21% of the atmosphere. The molecular oxygen at ground state is a diradical, which is a triplet species. Usually, the reaction between the triplet molecular oxygen and the vast majority of drug molecules, which are singlet species, is a kinetically forbidden process due to violation of the spin conservation rule [37]. The activation of molecular oxygen is usually mediated by redox reactive transition metal ions, most notably ferrous ions (Scheme 12), during which process three reactive oxidative species (ROS), i.e., superoxide anion radical, hydrogen peroxide, and hydroxyl radical, are generated.

Nucleophilic Oxidation

Tertiary amines and thioethers, particularly the alkyl ones, are capable of reacting with hydroperoxide species (including the prototype hydrogen peroxide; Scheme 13), which are often present in polymeric excipients such as polyethylene glycol (PEG) and polysorbate (e.g., Twin 80), as a result of auto-oxidation [38].

Certain compounds containing electron-rich carbon–carbon double bonds can also undergo analogous nucleophilic oxidation (Scheme 14); this type of compound

$$O_2 + Fe^{2+} \xrightleftharpoons{\text{Electron transfer}} \boxed{O_2^{\bullet -}} + Fe^{3+}$$

$$O_2^{\bullet -} + 2\,H^+ \longrightarrow O_2 + \boxed{H_2O_2}$$

$$H_2O_2 + Fe^{2+} + H^+ \longrightarrow \boxed{HO^{\bullet}} + Fe^{3+}$$

Scheme 12 Activation of molecular oxygen by ferrous ion Fe^{2+}, a redox reactive transition metal ion.

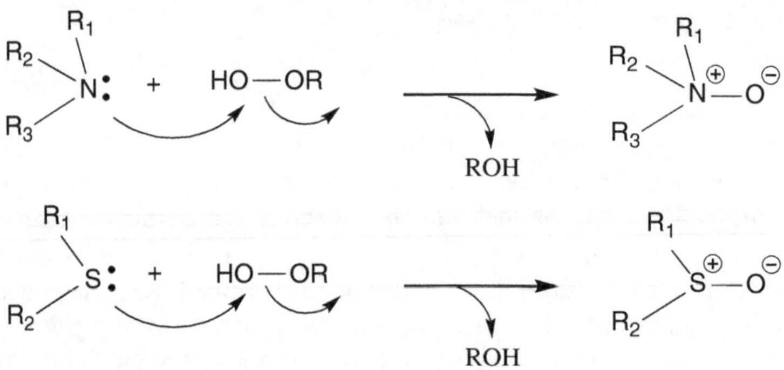

Scheme 13 Nucleophilic oxidation of tertiary amines and thioethers.

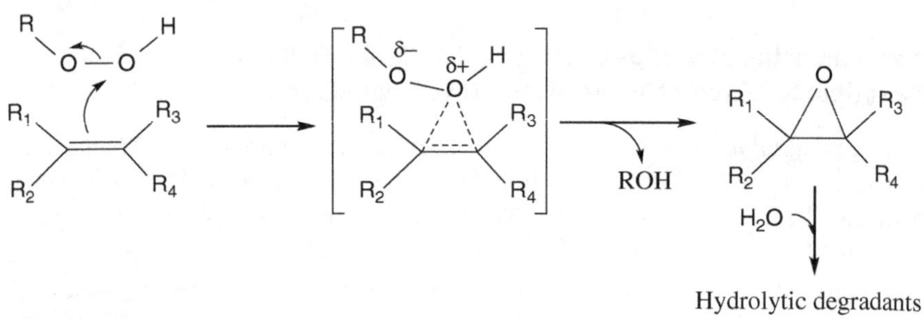

Scheme 14 Nucleophilic oxidation of compounds containing electron-rich carbon–carbon double bonds.

includes those that possess an indole ring, which can be viewed as an enamine (*N*-substituted carbon–carbon double bond).

Free-Radical-Mediated Oxidation

Free-radical-mediated oxidation is probably the most common and yet most complicated type of oxidation. The structural moieties that are susceptible to this type of oxidation include the CH or CH_2 positions that are alpha to carbon–carbon double bonds (particularly to conjugated carbon–carbon double bonds), CH or CH_2 positions that are alpha to a carbonyl group, and benzylic positions. An example is given for the facile free-radical-mediated oxidation of the butadiene moiety that is embedded in the core structures of lovastatin and simvastatin (Scheme 15) [39].

Base-Catalyzed Oxidation

This type of oxidation, also known as carbanion/enolate-mediated auto-oxidation, typically occurs in a group of compounds that contain somewhat "acidic" CH or CH_2 moieties. A classic example is the oxidation of rofecoxib (Scheme 16). The mechanism of the oxidation was originally proposed as free-radical mediated, but was later found to be mediated by carbanion/enolate [40, 41]. The rate of the base-catalyzed oxidation would be two orders of magnitude higher than that mediated by free radicals.

Oxidation Via the Udenfriend Reaction

The Udenfriend reaction is an oxidative degradation pathway that can occur in formulated drug products containing both a metal ion chelator and a reducing agent [38]. The role of the chelator, which is typically EDTA or citrate, is to form a chelating complex with the reduced form of a redox-reactive transition metal ion, most commonly ferrous ion (Fe^{2+}). The resulting complex, Fe^{2+}{EDTA},

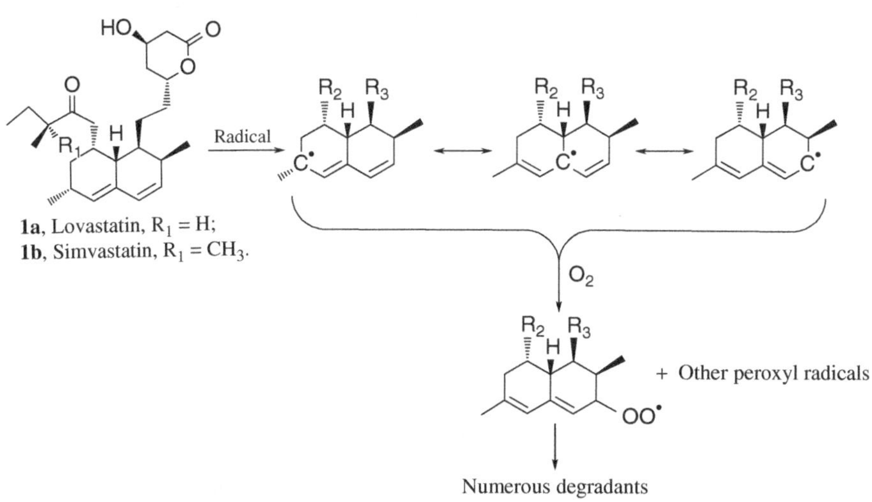

Scheme 15 Free-radical-mediated oxidation of the butadiene moiety embedded in the core structures of lovastatin and simvastatin.

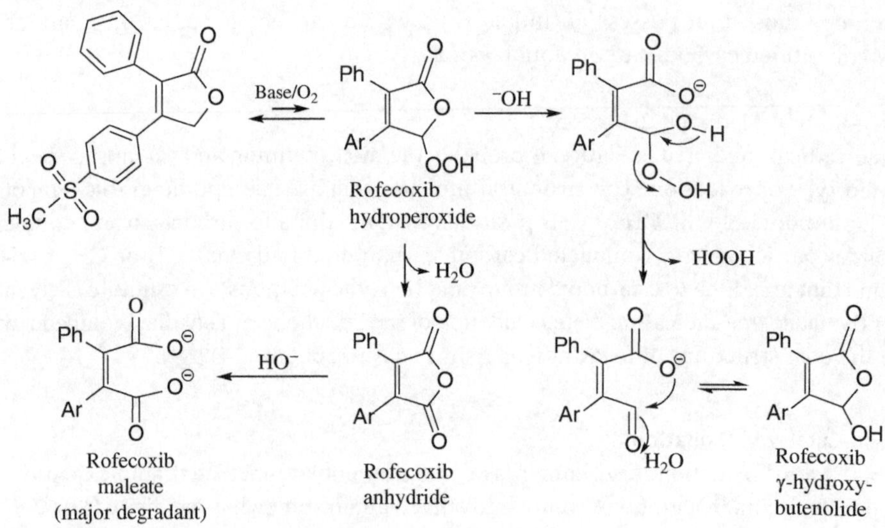

Scheme 16 Base-catalyzed oxidation of rofecoxib.

would be easier to transfer an electron to molecular oxygen than the unchelated Fe^{2+}, resulting in the formation of superoxide anion radical. The latter species can form hydrogen peroxide via disproportionation, and hydrogen peroxide can then, in turn, be transformed to hydroxyl radical via the Fenton reaction. During this process, which is the activation of molecular oxygen, two equivalents of Fe^{2+}\{EDTA\} are converted into the oxidized form of Fe^{3+}\{EDTA\}. In the presence of a reducing agent, such as ascorbic acid or a phenolic compound, Fe^{3+}\{EDTA\} can be reduced back to Fe^{2+}\{EDTA\}, causing the activation of molecular oxygen and subsequent oxidative degradation to continue until all the reducing agent is consumed. This whole process of the Udenfriend reaction can be illustrated in Scheme 17, which indicates that a reducing agent could actually promote oxidative degradation of a drug product if it also contains a chelating agent in its formulation [42–46].

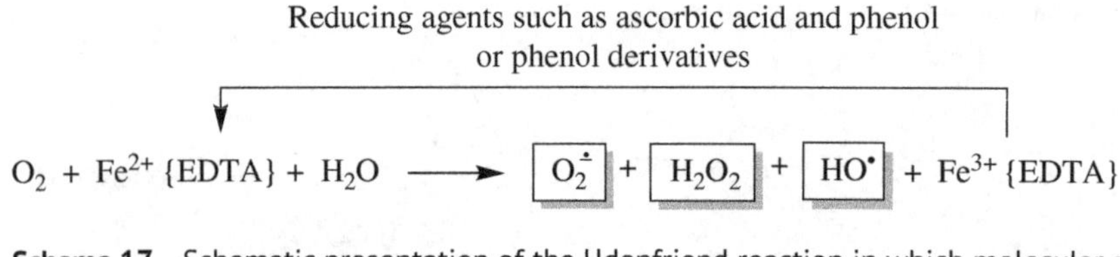

Scheme 17 Schematic presentation of the Udenfriend reaction in which molecular oxygen is activated continuously in the presence of a metal ion chelator and a reducing agent, resulting in the formation of three reactive oxygen species: superoxide anion radical, hydrogen peroxide, and hydroxyl radical.

Types of Functional Groups or Structural Moieties That Are Susceptible to Photochemical Degradation

Photochemical reactions are those that take place at electronically excited states. Hence, the prerequisite for a drug molecule to undergo photochemical degradation is that it needs to have a chromophore that would have adequate absorption under the photolytic conditions, for the molecule to be elevated into electronically excited states. The excited states will return or relax to the ground state via several passages, one of which is a chemical transformation, i.e., a photochemical reaction, while others are physical processes such as light emission and collision-induced relaxation [47]. Some compounds lack appropriate chromophores but nevertheless can form chromogenic complexes with metal ions, thus making themselves photochemically vulnerable [44]. Some of the most common moieties that are susceptible to photochemical degradation include benzophenone and its analogs (any two aryl groups connected by a carbonyl), halogenated aryl compounds, and substituted alkenes (e.g., stilbene and its analogs). Several nonsteroidal anti-inflammatory drug (NSAID) molecules contain benzophenone or its analogous moieties in conjunction with a nearby carboxylic group; the resulting 2-arylpropionic acid moieties can usually undergo photochemical decarboxylation (Scheme 18) [48, 49].

Halogenated aryl drugs, such as carprofen, are susceptible to photochemical dehalogenation (Scheme 19); nevertheless, such dehalogenation tends to occur in solution photolysis and usually would not take place in the solid state, except for iodoaryl compounds. In solution photolysis, if protic solvents such as water and methanol are used, the corresponding hydroxyl- and methoxyl-substituted photodegradants may also be formed. In such cases, the methoxyl-substituted photodegradants are artificial degradants as methanol is usually not a pharmaceutical excipient.

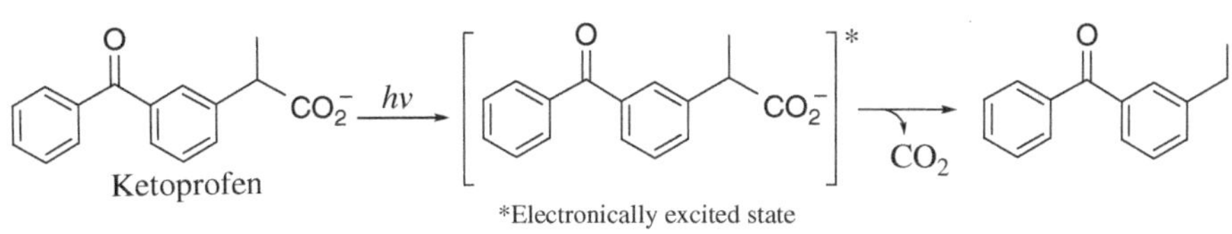

Scheme 18 Photodecarboxylation of ketoprofen, a 2-arylpropionic acid derivative of NSAID.

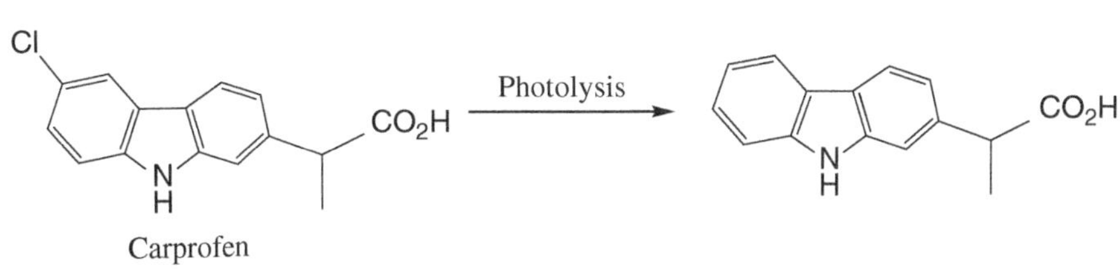

Scheme 19 Photo dehalogenation of carprofen.

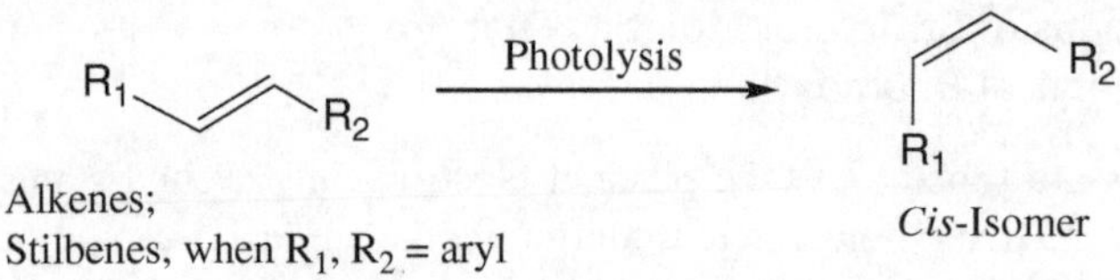

Scheme 20 Photoisomerization of alkenes, including stilbenes and analogs where R_1 and R_2 are aryl groups.

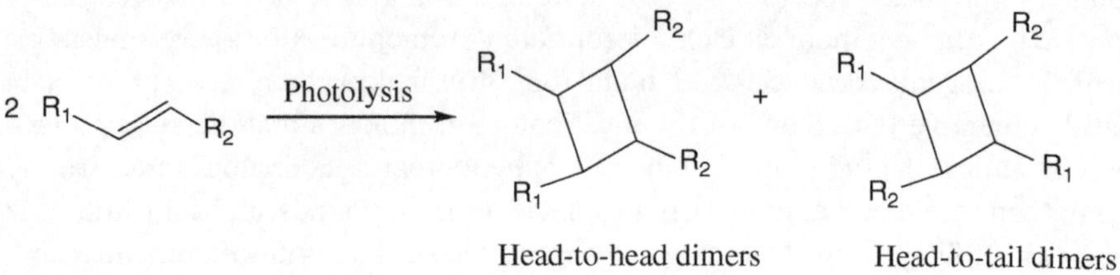

Scheme 21 2 + 2 Photodimerization of alkenes.

The carbon–carbon double bonds in alkenes can undergo photoisomerization (Scheme 20) and such isomerization becomes significant when the substituents of the double bonds render the resulting alkenes to have adequate absorption in UV and even visible regions. Analogous carbon–heteroatom double bonds can also be susceptible to similar photoisomeric degradation. Another important photodegradation pathway for alkenes is the 2 + 2 photodimerization (Scheme 21).

Drug molecules containing multiple aryl groups, either connected or fused, tend to undergo photosensitization during which process molecular oxygen, a triplet molecule, is excited to become very reactive singlet oxygen through collisional energy exchange with the excited state of the aryl rings. The singlet oxygen generated can easily inflict oxidative degradation on the drug molecule itself, such as in the example of an experimental drug containing a substituted thiazole ring (Scheme 22); this photodegradation pathway mediated via singlet oxygen is referred to as type II photosensitized oxidation [50].

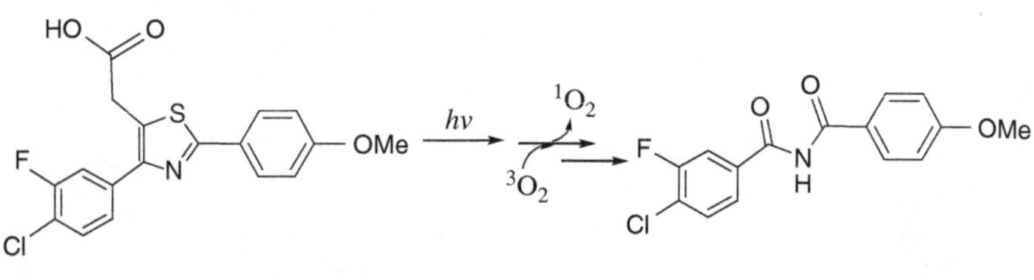

Scheme 22 Photosensitized oxidation of an experimental drug containing an aryl-substituted thiazole ring.

Degradation Due to API-Excipient Interactions

The Maillard Reaction

The most common interaction that occurs between active pharmaceutical ingredients (APIs) and excipients is probably the one between primary and secondary amine API molecules with reducing sugars such as lactose, which is known as the Maillard reaction. The Maillard reaction is a very complicated process consisting of hundreds of reactions and its pharmaceutical relevance is primarily limited to the first few steps leading to the formation of the so-called Amadori rearrangement products (Scheme 23) [51].

Degradation Due to Impurities in Excipients – Formaldehyde, Peroxides, Formic Acid, etc.

Pharmaceutical excipients contain impurities, some of which are of chemical reactivity and thus capable of reacting with API. The frequently seen impurities in excipients are formaldehyde, peroxides (including hydrogen peroxide), and formic acid, which result from auto-oxidative degradation of certain polymeric excipients such as PEG, polysorbate, and PVP (povidone) [51]. Formaldehyde is an electrophile that can react with API capable of nucleophilicity to form hydroxymethyl intermediates and then dimeric degradants, as in the case of two dimeric degradants of ropinirole (Scheme 24) [52]. Peroxides mainly inflict nucleophilic oxidative degradation of tertiary amines and thioethers (refer to Scheme 13). Formic acid can react with primary and secondary amine API to form *N*-formyl degradants, which is a partial source for this type of degradant, with the other source being from further degradation products of the Amadori rearrangement [53]. Formic acid appears to be an agent that may cause rare reductive degradation as in the case of *N*-methylvarenicline, a degradant of varenicline (Scheme 25) [54].

Formation of *N*-nitrosamines Due to the Reaction of Secondary Amine APIs or API-Related Fragments with Residual Nitrites in Excipients

In mid-2018, it was reported to the US FDA and other regulatory agencies that *N*-nitrosodimethylamine (NDMA), an animal carcinogen, was present in valsartan in trace amounts by a manufacturer of the API [55]. Subsequently, NDMA and other low molecular weight *N*-nitrosamines (also referred to as small molecular weight *N*-nitrosamines), e.g., *N*-nitrosodiethylamine (NDEA) and *N*-nitroso-4-(methylamino) butyric acid (NMBA), were found in valsartan and other ARB drugs (angiotensin receptor blockers) [56]. These events prompted the regulatory agencies and the industry to look deeper into the root causes for the occurrence of *N*-nitrosamine impurities in drug products. Starting from 2019, NDMA was found to be present in drug classes other than ARBs, e.g., ranitidine, nizatidine, and metformin [57]. While NDMA is typically a process impurity for ARB drugs, it appears to be a degradant for the latter three drug products. In July 2021, it was announced that varenicline, a secondary amine smoking-cessation drug, contained *N*-nitroso varenicline by the original innovator of the drug [58]. Since that moment, it has become evident that the drugs with secondary amine APIs would be most

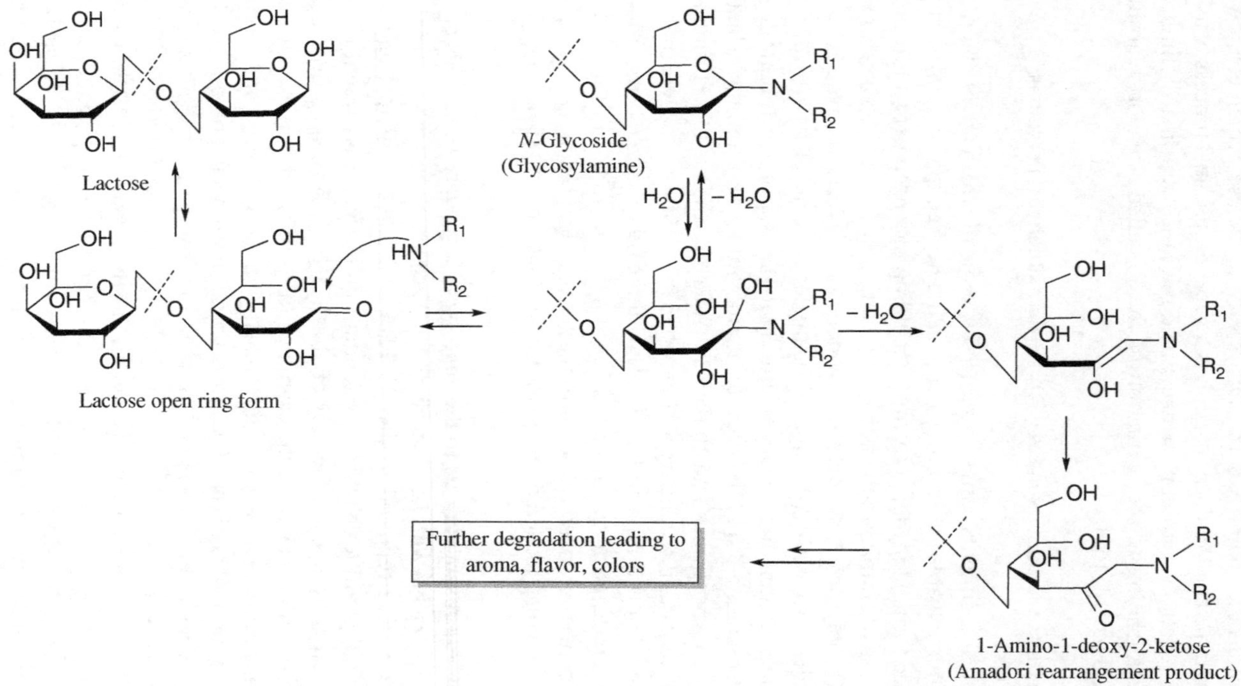

Scheme 23 The Maillard reaction between primary and secondary amine API with lactose, a very common reducing sugar excipient.

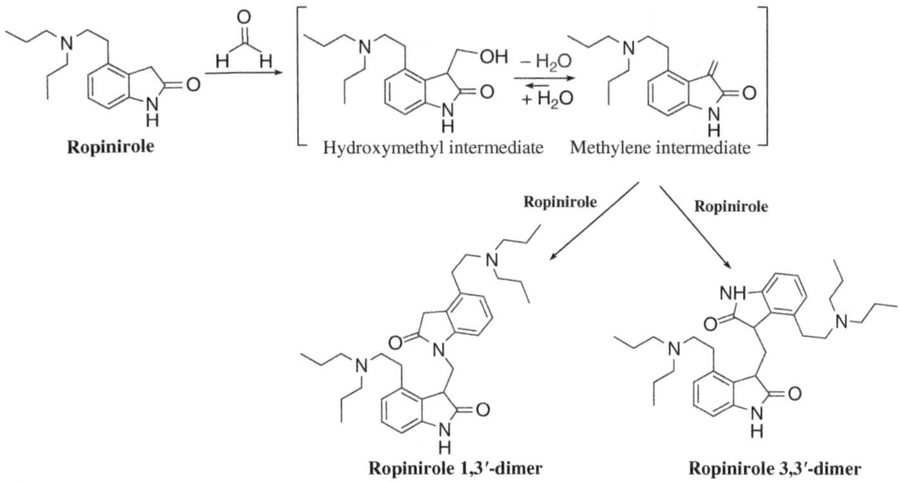

Scheme 24 Proposed mechanism for the formation of two ropinirole dimeric degradants [52].

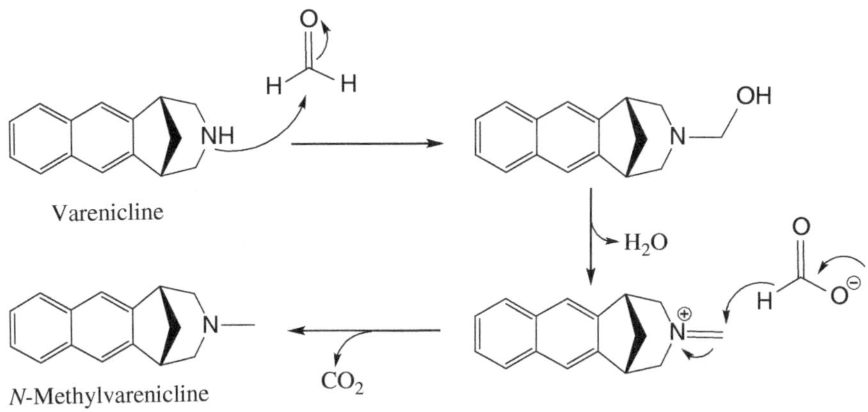

Scheme 25 Formation of *N*-methylvarenicline.

vulnerable towards the formation of *N*-nitrosamine impurities, particularly for the drugs formulated with excipients that contain uncontrolled amounts of residual nitrites. This is because secondary amine APIs can readily react with residual nitrites, particularly in a somewhat acidic microenvironment of the formulations to form API-related *N*-nitrosamines (Scheme 26).

The *N*-nitrosamines of the secondary amine APIs and their secondary amine intermediates are now categorized as part of the "nitrosamine drug substance-related impurities" (NDSRIs), with the remaining part being *N*-nitrosamines of secondary amine fragments of the APIs, which are typically process impurities but could be degradants as well in certain cases. It is obvious that the highest risk for the

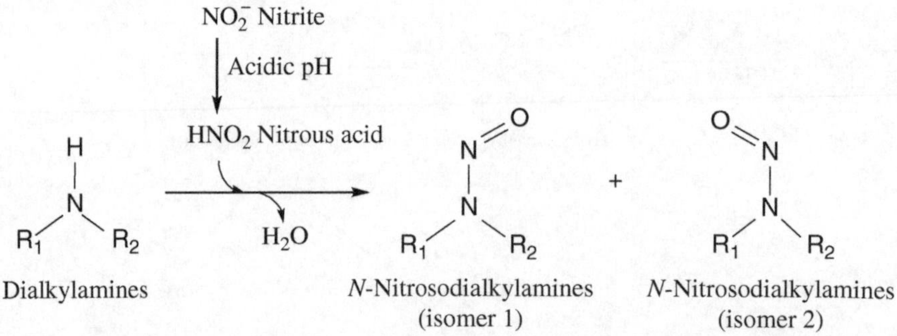

Scheme 26 Reaction of dialkylamines with nitrous acid to form *N*-nitrosodialkylamines. When $R_1 \neq R_2$, *E/Z* isomerism occurs. If R_1 or R_2 is replaced by an aryl substituent, the reactivity toward nitrous acid would be significantly reduced.

formation of NDSRIs comes from the secondary amine APIs, and therefore, as long as the secondary amine API-derived NDSRIs are controlled, the risk from their intermediates and fragments would be negligible since the amounts of the intermediates and fragments are usually three orders of magnitude lower than their APIs and all of them should have similar reactivity toward nitrous acid due to structural similarity. On the other hand, certain reducing agents, such as ascorbic acid and caffeic acid, can inhibit the formation of NDSRIs in drug products with secondary amine APIs [59]. For drug products with tertiary amine APIs, the main risk usually originates from their secondary amine impurities, despite the fact that the tertiary amine APIs could react with nitrous acid to eventually produce *N*-nitrosamines via a few intermediary steps. It seems that such multiplicity in the reaction mechanism would dramatically reduce the efficiency for the tertiary amine APIs to react with nitrous acid, which was reported to be only approximately 1/1,000 of the structurally similar secondary amines [60]. In the laboratories of the present author, we compared the efficiencies of triethylamine and diethylamine to react with nitrous acid, respectively, under the same conditions and found that triethylamine was only able to produce approximately 1/50 to 1/200 quantities of NDEA as compared to those by diethylamine [61]. An exception to the sluggishness of the tertiary amine reactivity toward *N*-nitrosamine formation was observed with those tertiary amines containing an electron-rich benzylic-type substituent [62].

Forced Degradation Study and Prediction of Real-Life Degradation Pathways

Forced degradation is also referred to as stress testing or simply stress. According to the ICH guidance document Q1(R2), *Stability Testing of New Drug Substances and Products*, "Stress testing of the drug substance can help identify the likely degradation products, which can in turn help establish the degradation pathways and the intrinsic stability of the molecule and validate the stability indicating power of the analytical procedures used. The nature of the stress testing will depend on the individual drug substance and the type of drug product involved" [5]. This statement clearly defines the purpose and significance of stress testing. Due to the complexity of the vast variety of drug molecules, ICH guidance only provides general principles for the stress testing of drug substances, except for the photolytic stress in solid state, where the light sources and duration for the exposure are clearly defined in Q1B [5]. At the high level, Q1A(R2) requires that the effect of temperature (typically in 10°C increments above that for the accelerated stability study), humidity (typically at 75%RH or higher), oxidation, and photolysis be evaluated on a drug substance. Furthermore, the susceptibility of the drug substance to hydrolysis under a wide range of pH also needs to be assessed.

In industry practice, there appears to be a wide variety of ways for performing stress testing or forced degradation [63, 64]. For the selection of hydrolytic conditions, although hydrochloric acid (HCl) and sodium hydroxide (NaOH) are usually the choices of selection for acidic and alkaline stress, respectively, the strengths of the acid and base, temperature, and duration of the hydrolysis can vary significantly without clear justification. With regard to oxidative stress, most studies seem to use hydrogen peroxide as the oxidizing reagent; nevertheless, hydrogen peroxide can usually only help predict nucleophilic oxidation of a drug molecule that contains tertiary amines, thioethers, or other structural moieties that are susceptible to nucleophilic oxidation, as explained above.

There have been efforts in trying to come up with best practices for performing forced degradation studies to maximize the predicting capability of stress testing, while minimizing the production of artificial degradation [65–68]. Such best practices focus on purposefully designed or mechanism-based stress conditions [68], which rely on the perceived degradation mechanism of the target drug molecule, as well as on the consideration of its degradation kinetics that would be correlated to the long-term and accelerated stability studies. Specifically, for the oxidative stress, not only hydrogen peroxide (typically not exceeding 3% concentration) needs to be used, but a free radical initiator such as AIBN or other initiators should also be employed for those drug molecules containing structural moieties or functional groups that are susceptible to free radical-mediated oxidation (e.g., benzylic positions, methylene positions alpha to double bonds and conjugated double bonds) [69]. Reynolds et al. reported the use of *N*-methylpyrrolidone (NMP)-water-air-heat system as a stress tool that can generate degradants from free radical-mediated and nucleophilic oxidation as well as hydrolytic degradation [70]. In the laboratories of the present author, benzoyl peroxide was also found to be capable of predicting oxidative degradants resulting from both free radical-mediated and nucleophilic oxidation [71]. For certain drug molecules, redox-active transition metal ion-mediated oxidation may be a significant pathway, although it was considered not relevant on a routine basis by certain researchers [68]. In those relevant cases, metal ions such as Fe^{2+}/Fe^{3+} may be used. As a matter of fact, metal-ion-mediated stress is a requirement by the Brazilian regulatory authority ANVISA [72]. For certain drug molecules containing "acidic CH or CH_2," it may be necessary to examine their susceptibility to base-catalyzed oxidation, which would be more favored in alkaline solutions prepared with a majority of aprotic solvents. The recommended conditions for oxidative stress studies are

Table 1.5 Recommended conditions for oxidative stress studies.

Type of oxidation	Recommended conditions	Notes
Nucleophilic	0.1%–3% H_2O_2, room temperature, 24 hours, protected from light	Higher temperature or under light may induce irrelevant degradation
Free radical-mediated	AIBN or benzoyl peroxide	For a mostly aqueous system, 4,4′-azobis(4-cyanovaleric acid) (ACVA), a water-soluble analog of AIBN, can be used [63]
Base-catalyzed	Aprotic solvent–water system with 1% of 1 M NaOH solution in methanol	Most common aprotic solvent: acetonitrile
Metal-ion-catalyzed	5 mM $FeSO_4$ and/or $FeCl_3$, 30°C, 2 weeks, protected from light	A combination of Fe^{2+}/Fe^{3+} may be used

Table 1.6 Recommended conditions for hydrolytic stress studies.

Type of stress	Recommended conditions	Notes
Acidic	0.1 N HCl, 60°C, 2 days or 1.0 N HCl, 60°C, 2 hours	H_2SO_4 may be used in place of HCl
Alkaline	0.1 N NaOH, 60°C, 2 days or 1.0 N NaOH, 60°C, 2 hours	KOH may be used in place of NaOH

summarized in Table 1.5, based on the recommendations reported in the literature as well as on the experience in the laboratories of the present author.

For the reagents of hydrolysis, 0.1 N HCl and 0.1 N NaOH solutions are generally used as the starting point [63]. While the temperature and duration of the hydrolytic stress can vary dramatically, the recommended conditions based on the literature as well as the experience at the laboratories of the present author are provided in Table 1.6; these conditions are usually strong enough to hydrolyze a typical amide bond (*Ea* ~20–25 kcal/mol) within the desirable range of ~5–20%.

For the stress of solid drug substances under dry heat or heat/moisture conditions, 60°C for 3 weeks would be approximately equivalent to 40°C for 6 months, assuming the type of degradation follows the Arrhenius equation with an activation energy (*Ea*) of ~25 kcal/mol. For the photo stress of solid drug substances, the photolysis conditions recommended by ICH Q1B are generally followed. In cases where photo stress study in solution state may be needed, the photolysis may be conducted with laboratory UV lighting or sunlight (Table 1.7).

There is no specific guidance for the stress testing of the finished drug products other than the photo stress testing, which is outlined in ICH Q1B. The heat and heat/moisture stress conditions typically used for API (e.g., those listed in

Table 1.7 Recommended conditions for stress studies in solid state.

Type of stress	Recommended conditions	Notes
Dry heat	60°C, ambient moisture, 3 weeks	[a]
Heat/high moisture	60°C, minimum 75% RH, 3 weeks	[a]
Heat/low moisture	60°C, 30% RH, 3 weeks	[a]This stress may be adequate for certain molecules susceptible to dehydration
Photolysis	Per conditions outlined in ICH Q1B	With and without packaging

[a] Assuming *Ea* of ~25 kcal/mol.

Table 1.7) can generally be directly applicable to the finished drug products. On the other hand, stress in solutions is generally not quite relevant to the finished drug products unless the dosage form is in liquid formulation. Preferably, the forced degradation study for solid dosage forms can be combined with the compatibility study that is typically conducted during the formulation development stage. The compatibility study is usually performed on a one-to-one basis between the API and each excipient. Sometimes, compatibility study on a one-to-two or more basis may be warranted.

Overall, the practices in stress or forced degradation studies vary significantly among the practicing scientists, despite the efforts to streamline a best practice that would be applicable in the vast majority cases. Nevertheless, it is now generally agreed that the level of the degradation should be controlled in ~5–20% range in order to avoid excessive degradation, which typically would result in irrelevant degradation profiles. While artificial degradants usually cannot be avoided completely, preferably the degradation profiles obtained under stress conditions will include the real degradation profiles, the ones obtained under the long-term stability conditions. At the same time, the practicing scientists should be able to distinguish the artificial degradants, such as the ones derived from the involvement of the organic cosolvents used to solubilize the drug substances in solution stress studies, from the real degradants.

Concluding Remarks

This introduction is intended to provide a quick overview of the most common drug degradation chemistry and pathways. For an in-depth knowledge in this regard, readers are encouraged to refer to the relevant chapters of *Organic Chemistry of Drug Degradation*. The 300+ drug molecules covered in the monographs of the current book, *Compendium of Drug Degradation Pathways*, would still represent a tiny fraction of all the drug products on the market. The present author hopes that readers can extrapolate from the examples in the monographs to other drug molecules based upon structural similarity and analogy of degradation chemistry.

References

1 Jamrógiewicz, M. and Merchel, M. (2018). A history of the physical and chemical stability of pharmaceuticals – a review. *Acta Pol. Pharm. Drug Res.* 75: 297–304.

2 Connors, K.A., Amidon, G.L., and Stella, V.J. (1986). *Chemical Stability of Pharmaceuticals: A Handbook for Pharmacists*, 2e. New York: Wiley-Interscience.

3 Yasmeen, A. and Sofi, G. (2019). A review of regulatory guidelines on stability studies. *J. Phytopharmacol.* 8: 147–151.

4 ICH website. https://www.ich.org/page/history (accessed 28 November 2022).

5 ICH website https://www.ich.org/page/quality-guidelines (accessed 28 November 2022).

6 WHO guidelines on stability testing (30 September 2018). Annex 10, WHO Technical Report Series 1010, 2018. https://www.who.int/publications/m/item/trs1010-annex10 (accessed 11 June 2025).

7 Li, M. (2012). Hydrolytic degradation. In: *Organic Chemistry of Drug Degradation*. Cambridge: RSC Publishing.

8 Tian, Y., Lin, J., Chen, F. et al. (2020). Structure elucidation and formation mechanistic study of a methylene-bridged pregabalin dimeric degradant in pregabalin extended-release tablets. *Int. J. Pharm.* 575: 118910.

9 Marquez, V.E., Tseng, C.K.-H., Kelley, J.A. et al. (1987). 2′,3′-Dideoxy-2′-fluoro-ara-A. An acid-stable purine nucleoside active against human immunodeficiency virus (HIV). *Biochem. Pharmco.* 36: 2719–2722.

10 Namchuk, M.N., McCarter, J.D., Becalski, A. et al. (2000). The role of sugar substituents in glycoside hydrolysis. *J. Am. Chem. Soc.* 122: 1270–1277.

11 Waterman, K.C., Adami, R.C., Alsante, K.M. et al. (2002). Hydrolysis in pharmaceutical formulations. *Pharm. Dev. Technol.* 7: 113–146.

12 Bruice, T.C. and Mayahi, M.F. (1960). The influence of the leaving tendency of the phenoxy group on the ammonolysis and hydrolysis of substituted phenyl acetates. *J. Am. Chem. Soc.* 82: 3067–3071.

13 Robinson, B.A. and Tester, J.W. (1990). Kinetics of alkaline hydrolysis of organic esters and amides in neutrally-buffered solution. *Int. J. Chem. Kinet.* 22: 431–448.

14 Singer, A.K. (2019). Activation parameter and solvent effect on solvolysis of ethyl benzoate in aquo-organic solvent system. https://ajrconline.org/HTML_Papers/Asian. *J. Res. Chem.* 12: 99. accessed 27 October 2022.

15 Tsujikawa, H. and Inoue, H. (1966). The reaction rate of the alkaline hydrolysis of ethyl acetate. *Bull. Chem. Soc. Jpn.* 39: 187–1842.

16 Kaufman, M.J. (1990). Rate and equilibrium constants for acid-catalyzed lactone hydrolysis of HMG-CoA reductase inhibitors. *Int. J. Pharm.* 66: 97–106.

17 Meresaar, U. and Bratt, L. (1974). Hydrolysis of amides. Alkaline and general acid catalyzed alkaline hydrolysis of some substituted acetamides and benzamides. *Acta Chem. Scand. A.* 28: 715–722.

18 King, A.P. and Krespan, C.G. (1974). Secondary amines from trifluoroacetamides. *J. Organomet. Chem.* 39: 1315–1316.

19 Ghosh, A.K. and Brindisi, M. (2015). Organic carbamates in drug design and medicinal chemistry. *J. Med. Chem.* 58: 2895–2940.

20 Dittert, L.W. and Higuchi, T. (1962). Rates of hydrolysis of carbamate and carbonate esters in alkaline solution. *J. Pharm. Sci.* 52: 852–857.

21 Chapman, T.M. (1989). Models for polyurethane hydrolysis under moderately acidic conditions: a comparative study of hydrolysis rates of urethanes, ureas, and amides. *J. Polymer Sci. A: Polymer Chem.* 27: 1993–2005.

22 Loftsson, T., Olafsdottir, B.J., and Baldvinsdottir, J. (1992). Estramustine: hydrolysis, solubilization, and stabilization in aqueous solutions. *J. Int. J. Pharm.* 79: 107–112.

23 Stroud, N., Richardson, N.E., Davies, D.J.G., and Norton, D.A. (1980). Quality control of prednisolone sodium phosphate. *Analyst* 105: 455–461.

24 Flynn, G.L. and Lamb, D.J. (1970). Factors influencing solvolysis of corticosteroid-21-phosphate esters. *J. Pharm. Sci.* 59: 1433–1438.

25 Garrison, A.W. and Boozer, C.E. (1968). The acid-catalyzed hydrolysis of a series of phosphoramidates. *J. Am. Chem. Soc.* 90: 3486–3494.

26 Skrdla, P., Abrahim, A., and Wu, Y. (2006). An HPLC chromatographic reactor approach for investigating the hydrolytic stability of a pharmaceutical compound. *J. Pharm. Biomed. Anal.* 41: 883–890.

27 Malin, J.J., Suárez, I., Priesner, V. et al. (2021). Remdesivir against COVID-19 and other viral diseases. *Clin. Microbiol. Rev.* 34: e00162–e00120.

28 Searles, S. and Nukina, S. (1959). Cleavage and rearrangement of sulfonamides. *Chem. Rev.* 59: 1077–1103.

29 Shaw, W.H.R. and Bordeaux, J.J. (1955). The decomposition of urea in aqueous media. *J. Am. Chem. Soc.* 77: 4729–4733.

30 Garrett, E.R., Bojarski, J.T., and Yakatan, G.J. (1971). Kinetics of hydrolysis of barbituric acid derivatives. *J. Pharm. Sci.* 60: 1145–1154.

31 Bansal, G., Singh, M., Jindal, K.C., and Singh, S. (2008). Ultraviolet-Photodiode array and high-performance liquid chromatographic/mass spectrometric studies on forced degradation behavior of glibenclamide and development of a validated stability-indicating method. *J. AOAC Inter.* 91: 709–719.

32 Wiseman, E.H., Chiaini, J., and Pinson, R. Jr. (1964). Determination of sulfamylurea hypoglycemic agents and their metabolites in biological fluids. *J. Pharm. Sci.* 53: 766–769.

33 Pritchard, J.G. and Long, F.A. (1956). Kinetics and mechanism of the acid-catalyzed hydrolysis of substituted ethylene oxides. *J. Am. Chem. Soc.* 78: 2667–2670.

34 Koskikallio, J. and Whalley, E. (1959). Effect of pressure on the spontaneous and the base-catalyzed hydrolysis of epoxides. *Can. J. Chem.* 37: 783–787.

35 Arava, V.R., Siripalli, U.B.R., and Bandatmakuru, S.R. (2007). Novel acid catalysed rearrangement of duloxetine. *Indian J. Chem. Sect. B* 46B: 1695–1698.

36 Brandl, M. and Gu, L. (1992). Degradation of tobramycin in aqueous solution. *Drug Dev. Ind. Pharm.* 18: 1423–1436.

37 Miller, D.M., Buettner, G.R., and Aust, S.D. (1990). Transition metals as catalysts of "autoxidation" reactions. *Free Radic. Biol. Med.* 8: 95–108.

38 Li, M. (2012). Oxidative degradation. In: *Organic Chemistry of Drug Degradation*. Cambridge: Royal Society of Chemistry.

39 Smith, G.B., DiMichele, L., Colwell, L.F. Jr. et al. (1993). Autooxidation of simvastatin. *Tetrahedron Lett.* 49: 4447–4462.

40 Mao, B., Abrahim, A., Ge, Z. et al. (2002). Examination of rofecoxib solution decomposition under alkaline and photolytic stress conditions. *J. Pharm. Biomed. Anal.* 28: 1101–1113.

41 Harmon, P.A., Biffar, S., Pitzenberger, S.M., and Reed, R.A. (2005). Mechanism of the solution oxidation of rofecoxib under alkaline conditions. *Pharm. Res.* 22: 1716–1726.

42 Hong, J., Lee, E., Carter, J.C. et al. (2004). Antioxidant-accelerated oxidative degradation: a case study of transition metal ion catalyzed oxidation in formulation. *Pharm. Dev. Technol.* 9: 171–179.

43 Wu, S., Waugh, W., and Stella, V.J. (2000). Degradation pathways of a peptide boronic acid derivative, 2-Pyz-(CO)-Phe-Leu-B$(OH)_2$. *J. Pharm. Sci.* 89: 758–765.

44 Reed, R.A., Harmon, P., Manas, D. et al. (2003). The role of excipients and package components in the photostability of liquid formulations. *PDA J. Pharm. Sci. Technol.* 57: 351–368.

45 Dong, J., Karki, S.B., Parikh, M. et al. (2012). Oxidative degradation studies of an oxazolidinone-derived antibacterial agent, RWJ416457, in aqueous solutions. *Drug Dev. Ind. Pharm.* 38: 1289–1297.

46 Pardo-Andreu, G.L., Delgado, R., Nunez-Selles, A.J., and Vercesi, A.E. (2006). Dual mechanism of mangiferin protection against iron-induced damage to 2-deoxyribose and ascorbate oxidation. *Pharmacol. Res.* 53: 253–260.

47 Li, M. (2012). Photochemical degradation. In: *Organic Chemistry of Drug Degradation*. Cambridge: Royal Society of Chemistry.

48 Borsarelli, C.D., Braslavsky, S.E., Sortino, S. et al. (2000). Photodecarboxylation of ketoprofen in aqueous solution. A time-resolved laser-induced optoacoustic study. *Photochem. Photobiol.* 72: 163–171.

49 Cosa, G.L., Martinez, L., and Scaiano, J.C. (1999). Influence of solvent polarity and base concentration on the photochemistry of ketoprofen: independent singlet and triplet pathways. *Phys. Chem. Chem. Phys.* 1: 3533–3537.

50 Wu, L., Hong, T.Y., and Vogt, F.G. (2007). Structural analysis of photo-degradation in thiazole-containing compounds by LC–MS/MS and NMR. *J. Pharm. Biomed. Anal.* 44: 763–772.

51 Li, M. (2012). Drug–excipient interactions and adduct formation. In: *Organic Chemistry of Drug Degradation*. Cambridge: Royal Society of Chemistry.

52 Kuang, Z., Huang, T., Li, D. et al. (2020). Structure elucidation and mechanistic study of a new dimeric degradant in ropinirole hydrochloride extended-release tablets. *Pharm. Res.* 37: 136. https://doi.org/10.1007/s11095-020-02863-3.

53 Wirth, D.D., Baertschi, S.W., Johnson, R.A. et al. (1998). Maillard reaction of lactose and fluoxetine hydrochloride, a secondary amine. *J. Pharm. Sci.* 87: 31–39.

54 Waterman, K.C., Arikpo, W.B., Fergione, M.B. et al. (2008). N-Methylation and N-formylation of a secondary amine drug (varenicline) in an osmotic tablet. *J. Pharm. Sci.* 97: 1499–1507.

55 US FDA news release on July 13, 2018: FDA announces voluntary recall of several medicines containing valsartan following detection of an impurity.

https://www.fda.gov/news-events/press-announcements/fdaannounces-voluntary-recall-several-medicines-containing-valsartanfollowing-detection-impurity (accessed 12 January 2025).

56 Zhang, J., Selaya, S.D., Shakleya, D. et al. (2023). Rapid quantitation of four nitrosamine impurities in angiotensin receptor blocker drug substances. *J. Pharm. Sci.* 112: 1246–1254.

57 US FDA news release on September 24, 2019: FDA announces voluntary recall of Sandoz ranitidine capsules following detection of an impurity. https://www.fda.gov/news-events/press-announcements/fda-announces-voluntary-recall-sandoz-ranitidine-capsules-following-detection-impurity (accessed 12 January 2025).

58 Company announcement posted by the US FDA on July 19, 2021: Pfizer issues a voluntary nationwide recall for twelve lots of CHANTIX® (Varenicline) tablets due to *N*-nitroso varenicline content. https://www.fda.gov/safety/recalls-market-withdrawals-safety-alerts/pfizer-issues-voluntary-nationwide-recall-twelve-lots-chantixr-varenicline-tablets-due-n-nitroso (accessed 12 January 2025).

59 Shakleya, D., Asmelash, B., Alayoubi, A. et al. (2023). Bumetanide as a model NDSRI substrate: N-nitrosobumetanide impurity formation and its inhibition in bumetanide tablets. *J. Pharm. Sci.* 112: 3075–3087.

60 Ashworth, I.W., Dirat, O., Teasdale, A., and Whiting, M. (2020). Potential for the formation of *N*-nitrosamines during the manufacture of active pharmaceutical ingredients: an assessment of the risk posed by trace nitrite in water. *Org. Process. Res. Dev.* 24: 1629–1646.

61 Dong, P., Li, M. Unpublished results.

62 López-Rodríguez, R., McManus, J.A., Murphy, N.S. et al. (2020). Pathways for *N*-nitroso compound formation: secondary amines and beyond. *Org. Process. Res. Dev.* 24: 1558–1585.

63 Singh, S. and Bakshi, M. (2000). Guidance on conduct of stress tests to determine inherent stability of drugs. Pharm. Tech. On-line. https://api.semanticscholar.org/CorpusID:54644738 (accessed 12 January 2025).

64 Alsante, K.M., Martin, L., and Baertschi, S.W. (2003). A stress testing benchmarking study. *Pharm. Technol.* 27: 60–72.

65 Reynolds, D.W., Facchine, K.L., Mullaney, J.F. et al. (2002). Available guidance and best practices for conducting forced degradation studies. *Pharm. Technol.* 26: 48–56.

66 Nelson, E.D., Harmon, P.A., Szymanik, R.C. et al. (2006). Evaluation of solution oxygenation requirements for azonitrile-based oxidative forced degradation studies of pharmaceutical compounds. *J. Pharm. Sci.* 95: 1527–1539.

67 Alsante, K.M., Ando, A., Brown, R. et al. (2007). The role of degradant profiling in active pharmaceutical ingredients and drug products. *Adv. Drug Deliv. Rev.* 59: 29–37.

68 Maheswaran, R. (2012). Scientific considerations of forced degradation studies in ANDA submissions. *Pharm. Technol.* 36: 73–80.

69 Klick, S., Muijselaar, P.G., Waterval, J. et al. (2005). Toward a generic approach for stress testing of drug substances and drug products. *Pharm. Technol.* 29: 48–66.

70 Reynolds, D.W., Galvani, M., Hicks, S.R. et al. (2012). The use of *N*-methylpyrrolidone as a cosolvent and oxidant in pharmaceutical stress testing. *J. Pharm. Sci.* 101: 761–776.

71 Unpublished results from the present author's laboratories.

72 Tattersall, P., Asawasiripong, S., Takenaka, I., and Castoro, J.A. (2016). Impact from the recent issuance of ANVISA resolution RDC-53/2015 on pharmaceutical small molecule forced degradation study requirements. *Am. Pharm. Rev.* https://www.americanpharmaceuticalreview.com/Featured-Articles/184364-Impact-from-the-Recent-Issuance-of-ANVISA-Resolution-RDC-53-2015-on-Pharmaceutical-Small-Molecule-Forced-Degradation-Study-Requirements/ (accessed 11 June 2025).

2

Abacavir to Azithromycin

1: Abacavir

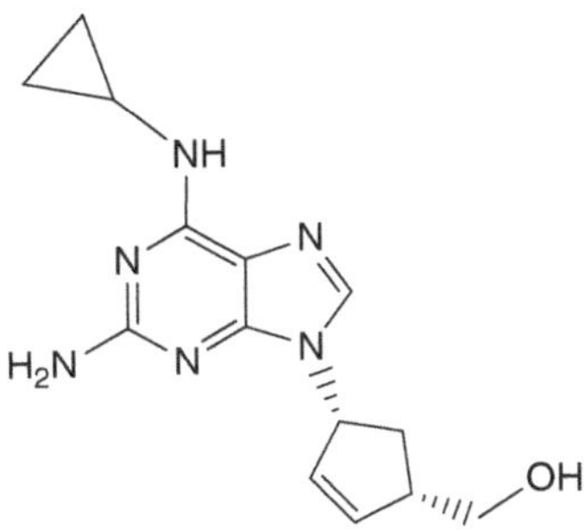

Chemical name: {(1*S*,4*R*)-4-[2-Amino-6-(cyclopropylamino)-9*H*-purin-9-yl] cyclopent-2-en-1-yl}methanol
Brand name: Ziagen
Formula: $C_{14}H_{18}N_6O$
Monoisotopic protonated ion (*m/z*): 287.1615
Molecular weight: 286.34
CAS number: 136470-78-5
Common salt form: Sulfate

Abacavir, a carbocyclic guanosine analog, is available in the form of sulfate salt. Its bicyclic pyrimidine-imidazole core contains three substituents: an amino, cyclopropylamino, and substituted cyclopentenyl group. In the literature, there appears to be no report regarding its degradation chemistry in commercial formulations. On the other hand, a couple of forced degradation studies were reported. In a study reported by Seshachalam et al. [1], abacavir was stable when refluxed in both 0.1 N HCl and 0.1 N NaOH for up to 5 days. In another study, Rao et al. employed 1 N HCl and 1 N NaOH as the acidic and basic stress conditions [2]. After refluxing for 24 hours, abacavir was completely decomposed to form five degradants under the acidic stress, while it remained unchanged under the basic stress. The structures of the

Compendium of Drug Degradation Pathways, First Edition. Min Li.

four major degradants (**1–4**) were proposed based on LC–MS results and MS/MS fragmentation pathways. Two types of degradation, that is, deamination and dealkylation, account for the formation of **1–4**. When a solid sample of abacavir was heated at 100°C for 10 days, no decomposition was observed [2].

Under oxidative stress conditions, abacavir was stable in 3% H_2O_2 at ambient temperature for 7 days [2]. Only at higher concentrations of H_2O_2 (6% and 10%, respectively) [1, 2] was abacavir found to decompose. In the study by Rao et al. [2], abacavir gave rise to degradants **5** and **6**. The same workers also stressed the drug substance using AIBN, and under this radical-initiated stress condition, two degradants, **6** and **7**, were formed. It is not surprising that no oxidation of the hydroxyl group of abacavir was observed, which is consistent with the fact that a typical hydroxyl group is not very susceptible to oxidation under the oxidative stress conditions utilized in the forced degradation studies discussed here [3].

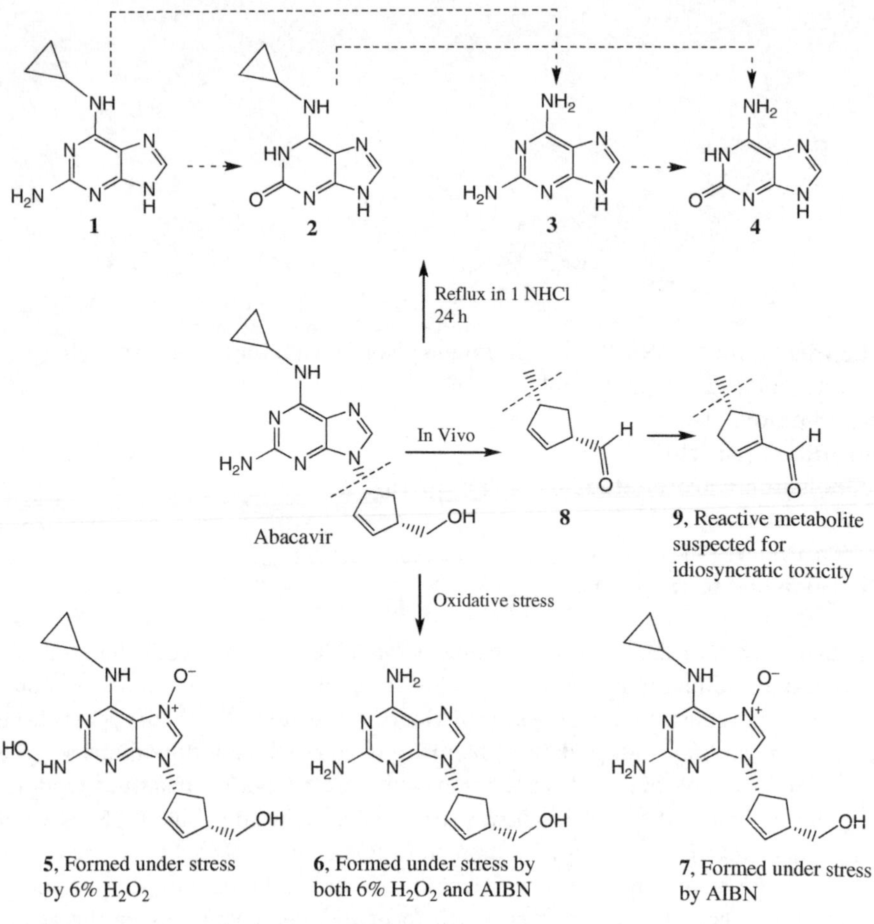

Scheme 1 Degradation pathways of abacavir under stress conditions. Degradant structures were proposed based on only LC–MS results and MS/MS fragmentation pathways. It is likely that all the degradants, except perhaps 6 and 7, may be artificial degradants.

This hydroxyl functionality, however, is susceptible to oxidation in vivo and the aldehyde metabolite formed (**8**) can further degrade to a Michael acceptor (**9**), which is suspected to be responsible for the severe toxicity of the drug in certain patient populations [4, 5].

Under photo stress conditions, abacavir was stable in both solution and solid forms when stressed under UV irradiation at 320 nm for 5 days, although the power of the irradiation source was not disclosed [2]. When stressed under the ICH photo stability condition, however, a low-level degradant was seen, the structure of which was not analyzed [1].

Overall, it appears that abacavir is quite stable. It is likely that the degradants, except perhaps **6** and **7**, observed in the stress studies may be artificial degradants (Scheme 1).

References

1 Seshachalam, U., Haribabu, B., and Chandrasekhar, K.B. (2007). Development and validation of a reverse-phase liquid chromatographic method for assay and related substances of abacavir sulfate. *J. Sep. Sci.* 30: 28–34.

2 Rao, R.N., Vali, R.M., Ramachandra, B., and Raju, S.S. (2011). Separation and characterization of forced degradation products of abacavir sulphate by LC–MS/MS. *J. Pharm. Biomed. Anal.* 54: 279–285.

3 Li, M. (2012). Oxidative degradation. In: *Organic Chemistry of Drug Degradation*, 48–109. Cambridge: Royal Society Chemistry Publishing.

4 Charneira, C., Godinho, A.L.A., Oliveira, M.C. et al. (2011). Reactive aldehyde metabolites from the anti-HIV drug abacavir: amino acid adducts as possible factors in abacavir toxicity. *Chem. Res. Toxicol.* 24: 2129–2141.

5 Charneira, C., Grilo, N.M., Pereira, S.A. et al. (2012). *N*-terminal valine adduct from the anti-HIV drug abacavir in rat haemoglobin as evidence for abacavir metabolism to a reactive aldehyde in vivo. *British J. Pharmacol.* 167: 1353–1361.

2: Acetaminophen (Paracetamol)

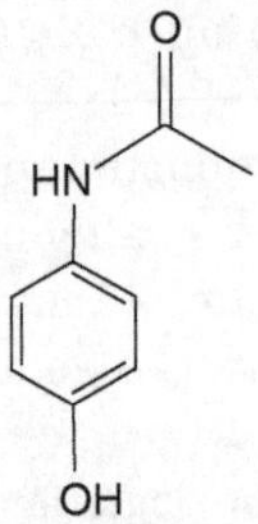

Acetaminophen
(paracetamol)

Chemical name: *N*-Acetyl-4-aminophenol; *N*-(4-hydroxyphenyl)acetamide
Brand name: Tylanol
Formula: $C_8H_9NO_2$
Monoisotopic protonated ion (*m/z*): 152.0706
Molecular weight: 151.17
CAS number: 103-90-2

Acetaminophen is the acetamide of a substituted aniline. The amide linkage of the drug molecule appears to be the only weak spot that may be susceptible to degradation (hydrolysis) in commercial formulations. The activation energy (*Ea*) for the hydrolysis of the amide linkage was determined to be on the order of ~17.5 kcal/mol between pH 2 and 9 [1]. In two more recent studies, the *Ea* for the hydrolysis in 0.5 M HCl was found to be 16.69 and 16.3 kcal/mol, respectively [1, 2]. Hence, although the strength of this particular amide bond is somewhat weaker than a typical aliphatic amide bond, which is on the order of ~20 kcal/mol or higher [3], it is expected that it would still be quite resistant toward hydrolytic degradation. The degradant formed from the hydrolysis is 4-aminophenol (or 4-hydroxyanaline; Scheme 2), which is also a process impurity [4] that has particular toxicological concern [5]. Hence, the pharmacopeia limit for 4-aminophenol in acetaminophen drug substance is a maximum of 0.005% or 50 ppm [6]. In formulated drug products, 4-aminophenol is controlled at maximum 0.15% by major pharmacopeias [7], although internal control by some company was reported to be still at 50 ppm for drug products [8].

One of the critical metabolites of acetaminophen is *N*-acetyl-*p*-benzoquinone imine (NAPQI), which has now been confirmed as the culprit for causing liver toxicity and failure in cases of acetaminophen overdose [9, 10]. There has been no report of this chemical entity in pharmaceutical preparations of acetaminophen. On the other hand, acetaminophen could be oxidized to NAPQI and its subsequent hydrolytic degradant, 1,4-benzoquinone, by residual chlorine in drinking water [11–13].

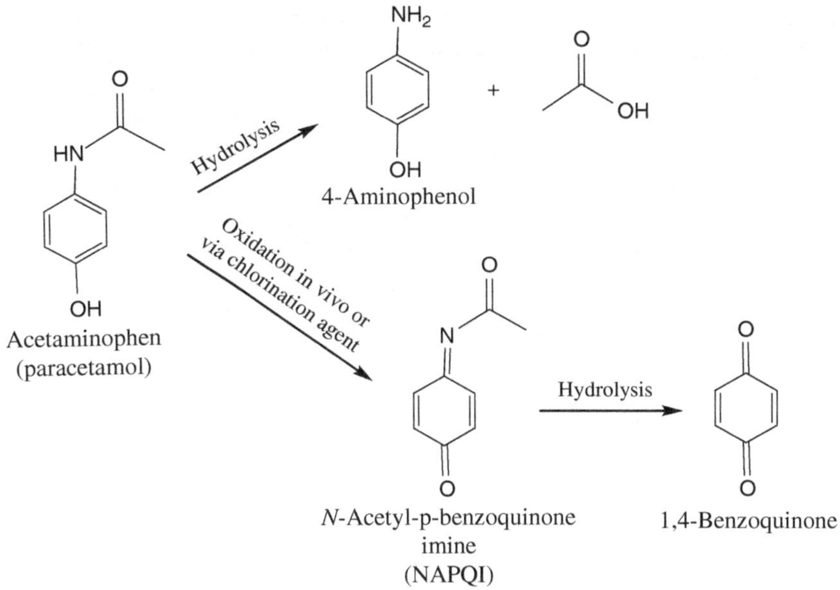

Scheme 2 In vitro and in vivo degradation of acetaminophen.

References

1 Koshy, K.T. and Lach, J.L. (1961). Stability of aqueous solutions of *N*-acetyl-*p*-aminophenol. *J. Pharm. Sci.* 50: 113–118.

2 Chen, G., Ye, J., Bao, H., and Yang, P. (2002). Determination of the rate constants and activation energy of acetaminophen hydrolysis by capillary electrophoresis. *J. Pharm. Biomed. Anal.* 29: 843–850.

3 Li, M. (2012). Hydrolytic degradation. In: *Organic Chemistry of Drug Degradation*, 19. Cambridge, Royal Society Chemistry Publishing.

4 Fairbrother, J.F. (1974). *Analytical profile of drug substances*, vol. 3 (ed. K. Florey), 1–109. New York: Academic Press.

5 Prescott, L.F. (1996). *Paracetamol (Acetaminophen): A Critical Bibliographic Review*. London: Taylor & Francis.

6 USP Monograph Official Acetaminophen. 10.31003/USPNF_M150_04_01 (accessed 1 May 2020).

7 USP Monograph Acetaminophen tablets. 10.31003/USPNF_M200_05_01 (accessed 1 October 2021).

8 Bloomfield, M.S. (2002). A sensitive and rapid assay for 4-aminophenol in paracetamol drug and tablet formulation, by flow injection analysis with spectrophotometric detection. *Talanta* 580: 1301–1310.

9 Dahlin, D.C. and Nelson, S.D. (1982). Synthesis, decomposition kinetics, and preliminary toxicological studies of pure *N*-acetyl-*p*-benzoquinone imine, a proposed toxic metabolite of acetaminophen. *J. Med. Chem.* 25: 885–886.
10 Kalgutkar, A.S., Gardner, I., Obach, R.S. et al. (2005). A comprehensive listing of bioactivation pathways of organic functional groups. *Curr. Drug Metab.* 6: 161–225.
11 Bedner, M. and Maccrehan, W.A. (2006). Transformation of acetaminophen by chlorination produces the toxicants 1,4-benzoquinone and *N*-acetyl-*p*-benzoquinone imine. *Environ. Sci. Technol.* 40: 516–522.
12 Gibs, J., Stackelberg, P.E., Furlong, E.T. et al. (2007). Persistence of pharmaceuticals and other organic compounds in chlorinated drinking water as a function of time. *Sci. Total Environ.* 373: 240–249.
13 Xagoraraki, I., Hullman, R., Song, W. et al. (2008). Effect of pH on degradation of acetaminophen and production of 1,4-benzoquinone in water chlorination. *J. Water Supply: Res. Tech. - AQUA* 57: 381–390.

3: Acetazolamide

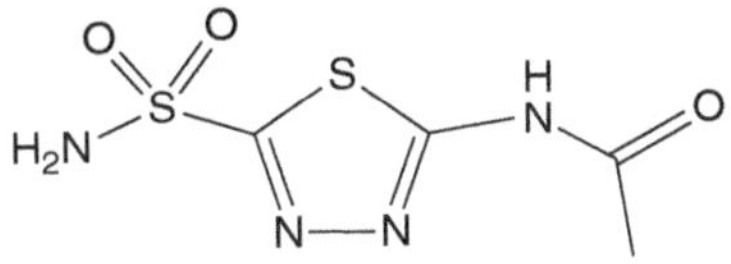

Chemical name: 5-Acetamido-1,3,4-thiadiazo-2-yl-sulfonamide
Brand name: Diamox
Formula: $C_4H_6N_4O_3S_2$
Monoisotopic protonated ion (*m/z*): 222.9954
Molecular weight: 223.24
CAS number: 59-66-5

Acetazolamide, a carbonic anhydrase inhibitor, has a five-membered thiadiazo ring that is connected to a sulfonamide group at the 2-position and an acetamide at the 5-position. According to a few stress studies, the drug molecule is most susceptible to hydrolytic degradation at the acetamide bond to yield degradant **1** (Scheme 3) [1–4]. This is consistent with the fact that a sulfonamide bond is far more stable than an acetamide bond [5]. Parasrampuria and Gupta studied the hydrolytic stability of acetazolamide in phosphate and citrate buffers, respectively, throughout the pH range [1]. It was found that acetazolamide was most stable at pH 4 with an activation energy (*Ea*) of 16.6 kcal/mol. The hydrolytic degradation was determined to be mediated via a specific acid–base catalysis. The rate constants under the acidic and basic catalysis, K_H and K_{OH}, were estimated to be 0.23 and 1.56/d, respectively. The pH stability profile of acetazolamide was confirmed by a study performed by Khamis et al. [2]. In addition, the latter study also determined that the *Ea* for the hydrolysis of acetazolamide in 0.01 N NaOH (pH 12) was 19.6 kcal/mol.[1]

In an oxidative stress of acetazolamide in 30% H_2O_2 at room temperature for 24 hours, an oxidative degradant was found [3]. There was no further structure elucidation by the original workers other than the HPTLC study with UV detection and scanning. However, based on its UV spectrum, the oxidative degradant could be the *S*-oxide (**2**) of acetazolamide, as the UV profile of the latter is dramatically different from that of acetazolamide. The conjugation to the newly formed sulfonyl group should be responsible for the significant shift of UV absorption to longer wavelengths observed. Since the oxidative stress appears to be excessive (in 30% H_2O_2), it is uncertain if this degradant would occur in commercial formulations under long-term stability conditions.

It has been reported that acetazolamide shows phototoxicity clinically. In a photochemical study by Vargas et al. [6], acetazolamide was found to undergo decomposition under both UV-A and UV-B irradiation. Two degradants (**3** and **4**, Scheme 3) were observed under aerobic irradiation, while only **3** was formed under anaerobic irradiation. The two degradants were also produced under a photosensitized

1 The *Ea* was incorrectly reported as 19.6 kJ/mol in Ref. [2]. It should be 19.6 kcal/mol based on the raw data reported in Ref. [2] as calculated by the current author.

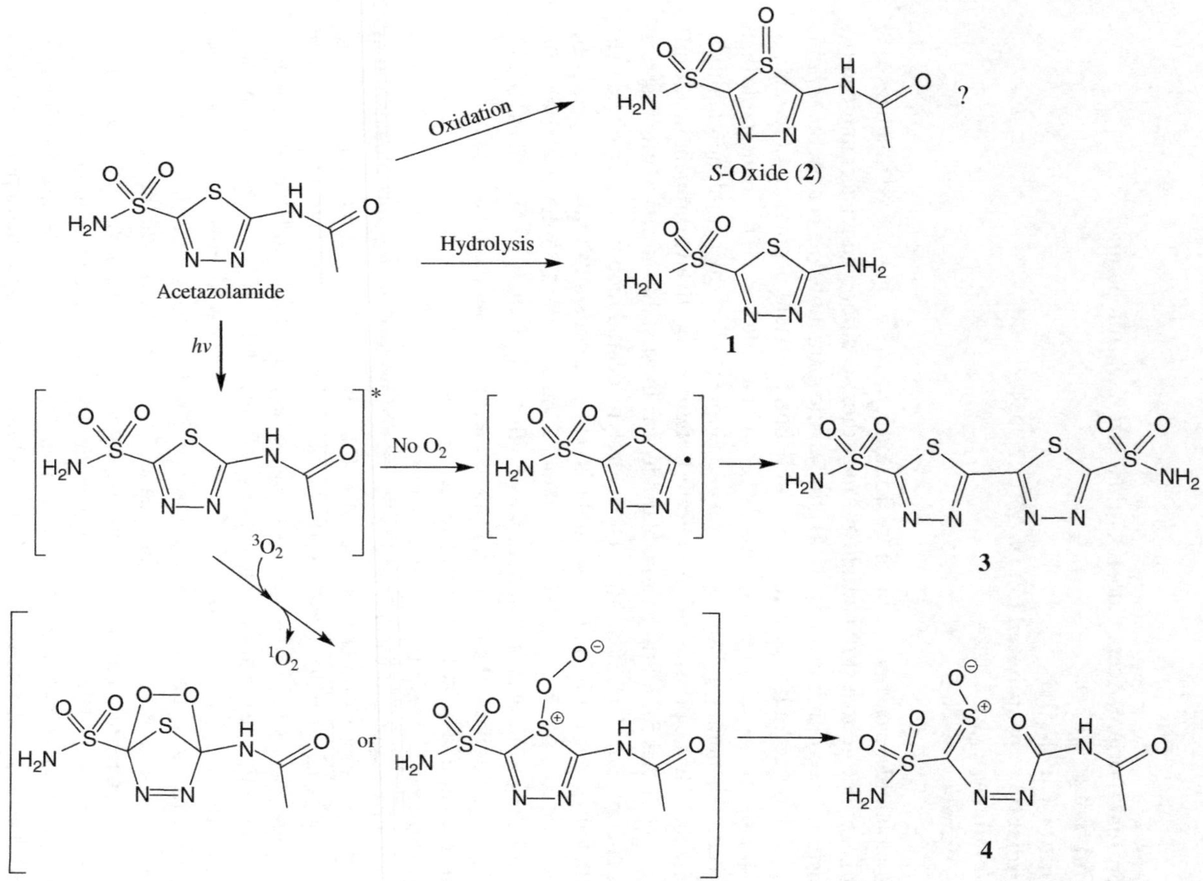

Scheme 3 Degradation pathways of acetazolamide under stress conditions. The structure of degradant 2 was tentative based on its reported UV spectrum in Reference [3].

condition where singlet oxygen ($^1O_2$) was generated via rose Bengal. Therefore, acetazolamide can be both a $^1O_2$ sensitizer and quencher. In a more recent study, Valencia et al. [7] found that acetazolamide is an efficient $^1O_2$ quencher but a poor $^1O_2$ sensitizer (quantum yields 0.015–0.097 in organic protic and aprotic solvents). Based on this finding, it was suggested that the phototoxicity of acetazolamide might involve species other than $^1O_2$ as originally thought.

References

1. Parasrampuria, J. and Gupta, V.D. (1989). Preformulation studies of acetazolamide: effect of pH, two buffer species, ionic strength, and temperature on its stability. *J. Pharm. Sci.* 78: 855–857.
2. Khamis, E.F., Abdel-Hamid, M., Hassan, E.M. et al. (1993). A stability-indicating first-derivative spectrophotometric assay of acetazolamide and its use in dissolution and kinetic studies. *J. Clin. Pharm. Ther.* 18: 97–101.
3. Ramesh, B., Srimannaraya, P., Reddy, A.S., and Sitadevi, P. (2011). Densitometric evaluation of stability-indicating HPTLC method for the analysis of acetazolamide and a study of degradation kinetics. *J. Pharm. Res.* 4: 429–433.
4. Srinivasu, P., Subbarao, D.V., Vegesnaa, R.V.K., and Babu, K.S. (2010). A validated stability-indicating LC method for acetazolamide in the presence of degradation products and its process-related impurities. *J. Pharm. Biomed. Anal.* 52: 142–148.
5. Li, M. (2012). Hydrolytic degradation. In: *Organic Chemistry of Drug Degradation*, 16–47. Cambridge: Royal Society Chemistry Publishing.
6. Vargas, F., Hisbeth, M.V., and Rojas, J.K. (1998). Photolysis and photosensitized degradation of the diuretic drug acetazolamide. *J. Photochem. Photobiol. A Chem.* 118: 19–23.
7. Valencia, C., Tobon, E., and Castano, C. (2013). Acetazolamide as a singlet molecular oxygen quencher. *J. Photochem. Photobiol. A Chem.* 251: 113–117.

4: Acetylcysteine

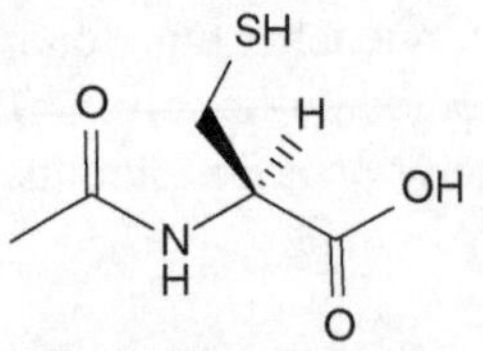

Chemical name: *N*-Acetyl-L-cysteine; 2-Acetamido-3-sulfanylpropanoic acid
Brand name: Mucolysin, Acetadote
Formula: $C_5H_9NO_3S$
Monoisotopic protonated ion (*m/z*): 164.0376
Molecular weight: 163.19
CAS number: 616-91-1

Acetylcysteine is an amino acid derivative. Its pharmacological effects and degradation behavior are primarily attributable to the sulfhydryl or thiol group of the cysteine moiety. The thiol group is a reducing agent as well as a nucleophile. As a reducing agent, acetylcysteine is used as a mucolytic drug because it can liquefy mucus by reducing the disulfide bonds of mucous proteins. As a nucleophile, acetylcysteine is employed as an antidote for liver toxicity caused by acetaminophen overdose. In that capacity, acetylcysteine is capable of intercepting *N*-acetyl-*p*-benzoquinone imine (NAPQI), a toxic, highly electrophilic metabolite of acetaminophen,[2] in a way similar to the action of glutathione, the body's natural defense agent against electrophilic metabolites.

Farquhar et al. studied the solution stability of acetylcysteine in aqueous formulations [1]. Under aerobic conditions and within a pH range of 5–8, *N,N'*-diacetylcystine is the predominant degradant, resulting from the oxidative coupling of the API. The formation of a disulfide bridge, such as the one in this case, is generally catalyzed by trace amounts of redox-active transition metal ions [2]. This catalytic process can be inhibited by the use of a chelating agent, usually EDTA. For this reason, a majority of commercial aqueous formulations of acetylcysteine use EDTA as the stabilizing chelating agent [3]. Since the ultimate oxidizing agent is molecular oxygen or O_2, complete removal of O_2 was shown to be effective in shutting down the oxidative coupling pathway of acetylcysteine. The effectiveness of this approach achieved through deaeration and nitrogen purging has also been demonstrated in kilogram-scale production [3]. Under anaerobic conditions, cysteine was found to be the main degradant within a pH range of 6.5–7 [3]. In the original study of Farquhar et al., cysteine was observed along with *N,S*-diacetylcysteine and *S*-acetylcysteine in pH 2 aqueous solutions under anaerobic conditions. The formation of *N,S*-diacetylcysteine is apparently via an intermolecular transacylation mechanism, during which process cysteine is also formed. On the other hand, the formation of *S*-acetylcysteine would be more likely mediated through an intramolecular transacylation

2 Refer to the monograph of acetaminophen and the relevant references cited therein.

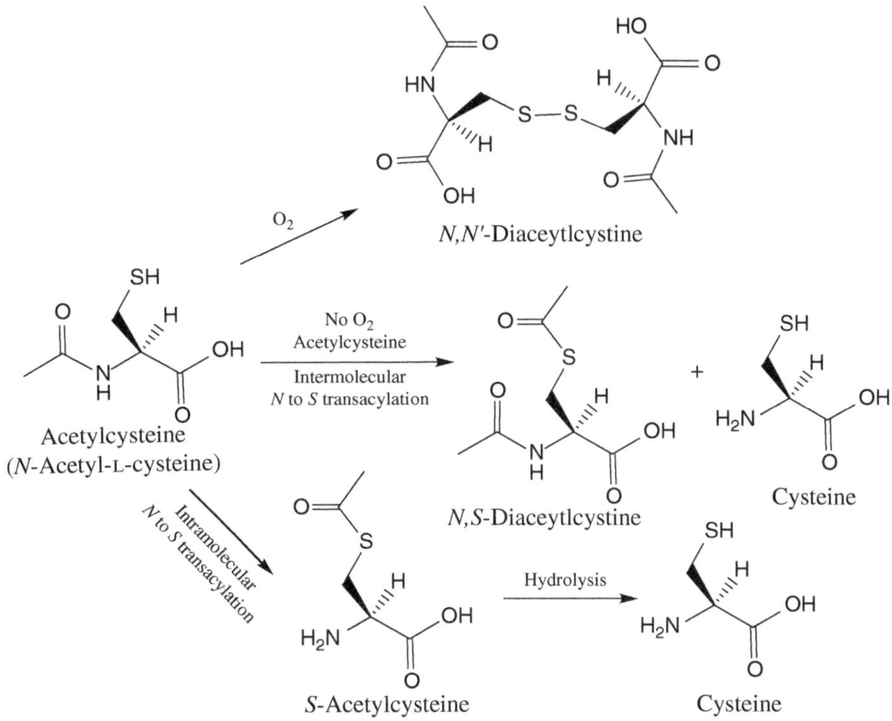

Scheme 4 Degradation pathways of acetylcysteine.

mechanism. The resulting degradant, *S*-acetylcysteine, should be susceptible to hydrolysis to yield cysteine, particularly in the pH range near neutrality and above. The degradation pathways of acetylcysteine are summarized in Scheme 4.

References

1 Farquhar, J., Finlay, G., Ford, P.A., and Martin-Smith, M. (1985). A reversed-phase high-performance liquid chromatographic assay for the determination of *N*-acetylcysteine in aqueous formulations. *J. Pharm. Biomed. Anal.* 3: 279–285.

2 Tomazic, S.J. and Klibanov, A.M. (1988). Mechanisms of irreversible thermal inactivation of Bacillus alpha-amylases. *J. Biol. Chem.* 263: 3086–3091.

3 Pavliv, L. Acetylcysteine composition and uses thereof. United States Patent Application, Pub. No. US2013/0165516 A1. Patent filed 01 February 2013.

5: Acyclovir

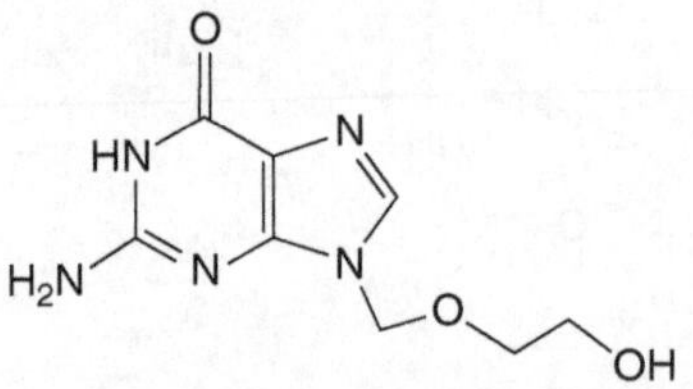

Chemical names: 9-Hyroxyethoxymethylguanine; 2-Amino-9-[(2-hydroxyethoxy) methyl]-1,9-dihydro-6*H*-purin-6-one
Brand name: Zovirax
Formula: $C_8H_{11}N_5O_3$
Monoisotopic protonated ion (*m/z*): 226.0935
Molecular weight: 225.21
CAS number: 59277-89-3

Acyclovir, also spelled as aciclovir, is one of the widely prescribed antiviral drugs for the treatment of many types of herpes virus infections such as cold sores, chickenpox, and shingles. It is a nucleoside (guanosine) analogue with the nucleoside sugar unit being replaced by an acyclic, functionalized alkyl group that is a residual of the sugar unit. In other words, acyclovir can be considered a ^9N-alkylated guanine derivative (**1**, Scheme 5). The removal of this alkyl group constitutes its main degradation pathway under acidic, basic, and photo stress conditions, during which process guanine is formed [1, 2]. Under both the photochemical and non-photochemical stress in solutions, the degradation is more significant when an acid is present. It was suggested that the dealkylation of acyclovir may be facilitated by a nucleophilic attack of the side chain hydroxyl group (Scheme 5) [3]. This mechanism is consistent with the observation that a substituted benzoate prodrug of acyclovir is three times more stable than the parent drug under acidic stress, because the ester oxygen in the prodrug is much less nucleophilic than the hydroxyl oxygen in the parent drug. Despite the fact that guanine, which is also a starting material of acyclovir, can be readily generated, particularly under acidic stress conditions, it appears that it is not a significant degradant in commercial as well as several experimental solid dosage forms [4–6].

Oxidative stress of acyclovir was performed in various concentrations of H_2O_2 and under different temperatures [1, 2], during which processes only a single, unidentified degradant was generated. This stress-generated degradant does not appear to be meaningful in formulated acyclovir products. For the side chain hydroxyl group, it is not expected to be susceptible to oxidation under regular or accelerated stability conditions for the same reason discussed in the monograph of abacavir. Nevertheless, this hydroxyl group is oxidized in vivo to form the major metabolite, carboxyl, via the intermediacy of aldehyde. The intermediary aldehyde was hypothesized to be the culprit for nephrotoxicity observed in some pediatric patients [7].

Since the drug molecule contains a primary amino group, its compatibility with lactose, a reducing sugar, was evaluated by Monajjemzadeh et al. [8].

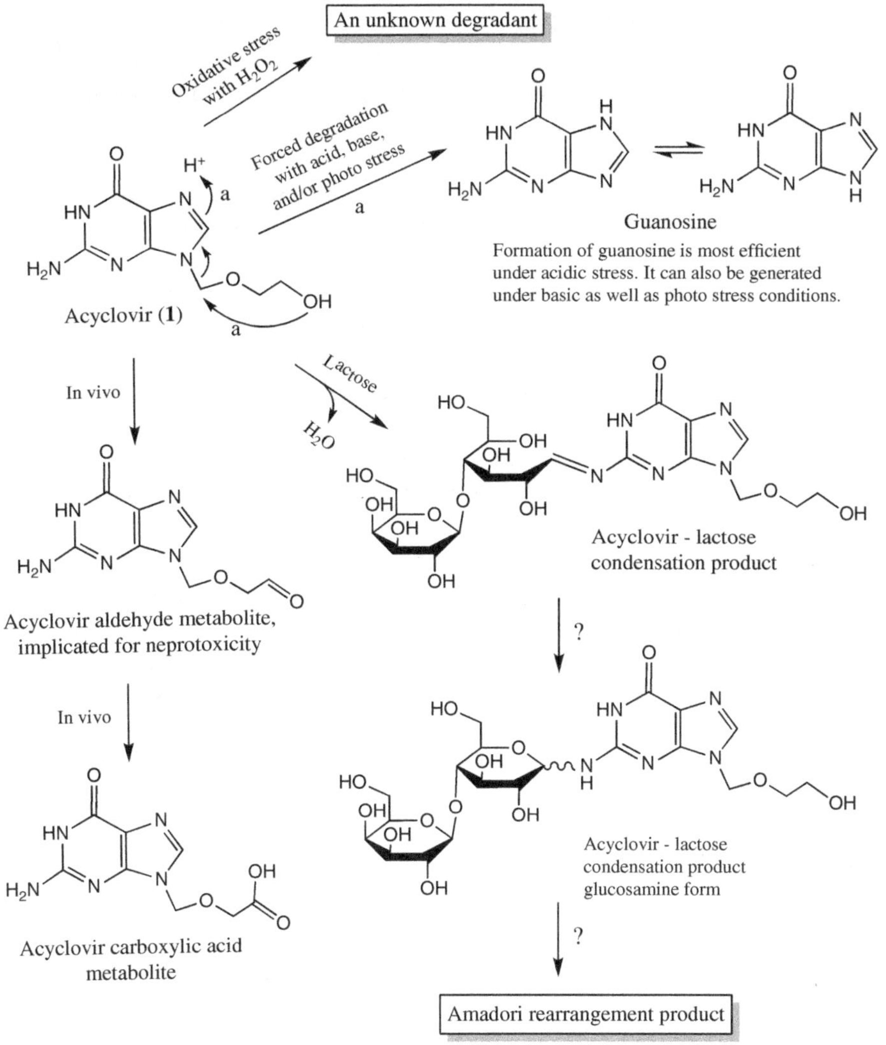

Scheme 5 In vitro and in vivo degradation pathways of acyclovir. None of the in vitro degradants has been reported to be significant in commercial formulations.

A condensation product was formed between the drug molecule and lactose based on LC–MS evidence. This degradant was proposed to be a Schiff base, the first degradant in the Maillard degradation chemistry. Nevertheless, it could also be the isomeric glycosamine or Amadori rearranged product [9]. Monajjemzadeh et al. also analyzed three batches of commercial acyclovir tablets that contain lactose and found no detectable level of the condensation product. Overall, it appears that acyclovir is quite stable in solid dosage forms, and all in vitro degradants discussed in this monograph become significant only when the drug is subjected to forced degradation under various stress conditions.

References

1 Sinha, V.R., Trehan, M.A., Kumar, M. et al. (2007). Stress studies on acyclovir. *J. Chromatogr. Sci.* 45: 319–324.

2 Zacharis, C.K., Verdoukas, A., Tzanavaras, P.D., and Themelis, D.G. (2009). Automated sample preparation coupled to sequential injection chromatography: on-line filtration and dilution protocols prior to separation. *J. Pharm. Biomed. Anal.* 49: 726–732.

3 Jensen, E. and Bundgaard, H. (1991). Kinetics of the acid-catalyzed hydrolysis of acyclovir and an ester prodrug in aqueous solution. *Acta Pharm. Nord.* 3: 147–150.

4 Shafaati, A.R. and Clark, B.J. (2003). Determination of aciclovir and its related substance guanine in bulk drug and tablet preparation by capillary electrophoresis. *Iran. J. Pharm. Res.* 2: 71–75.

5 Vinodbhai, P.K., Gohel, M.C., Parikh, R.K. et al. (2011). Sustained release floating microspheres of acyclovir: formulation, optimization, characterization and in vitro evaluation. *Int. J. Drug Dev. Res.* 3: 242–251.

6 Singh, A.V. and Nath, L.K. (2010). Development and evaluation of acyclovir mucoadhesive tablet using a novel starch based platform. *Lat. Am. J. Pharm.* 29: 279–283.

7 Gunness, P., Aleksa, K., Bend, J., and Koren, G. (2011). Acyclovir-induced nephrotoxicity: the role of the acyclovir aldehyde metabolite. *Transl. Res.* 158: 290–301.

8 Monajjemzadeh, F., Hassanzadeh, D., Valizadeh, H. et al. (2009). Compatibility studies of acyclovir and lactose in physical mixtures and commercial tablets. *Eur. J. Pharm. Biopharm.* 73: 404–413.

9 Li, M. (2012). Drug-excipient interactions and adduct formation. In: *Organic Chemistry of Drug Degradation*, 150–164. Cambridge: Royal Society Chemistry Publishing.

6–8: Albendazole, Fenbendazole, Mebendazole

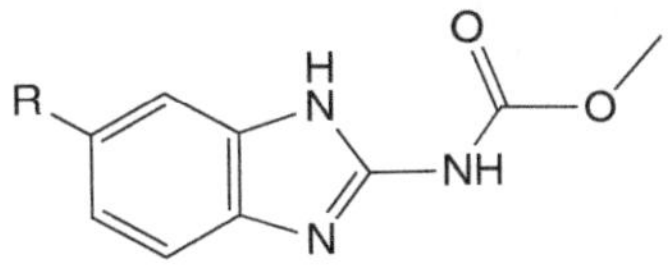

Albendazole (**1**), R = C_3H_7S–;
Fenbendazole (**2**), R = PhS–;
Mebendazole (**3**), R = PhC(O)–.

6: Albendazole

Chemical name: Methyl [5-(propylsulfanyl)-1*H*-benzimidazol-2-yl]carbamate
Brand name: Albenza; Eskazole
Formula: $C_{12}H_{15}N_3O_2S$
Monoisotopic protonated ion (*m/z*): 266.0958
Molecular weight: 265.33
CAS number: 54965-21-8

7: Fenbendazole

Chemical name: Methyl *N*-(6-phenylsulfanyl-1*H*-benzoimidazol-2-yl)carbamate
Brand name: Panacur
Formula: $C_{15}H_{13}N_3O_2S$
Monoisotopic protonated ion (*m/z*): 300.0801
Molecular weight: 299.35
CAS number: 43210-67-9

8: Mebendazole

Chemical name: Methyl (5-benzoyl-1*H*-benzimidazol-2-yl)carbamate
Brand name: Emverm; Vermox
Formula: $C_{16}H_{13}N_3O_3$
Monoisotopic protonated ion (*m/z*): 296.1030
Molecular weight: 295.30
CAS number: 31431-39-7

Albendazole, fenbendazole, and mebendazole are three benzimidazole-based anthelmintic drugs that all contain the same methyl carbamyl group at the two-position and a different substituent at the five-position of the benzimidazole ring. Ragno et al. performed forced degradation studies of the three drugs under thermo as well as photo stress conditions [1]. By heating at 50°C for up to 8 hours, the three drugs were found to be stable in both solid and ethanol solutions. On the other hand, when the ethanol solutions of the three drugs were subjected to photo stress with a xenon lamp under the ICH photo stability conditions, all three drugs were found to decompose in significant quantities with the formation of an intermediary

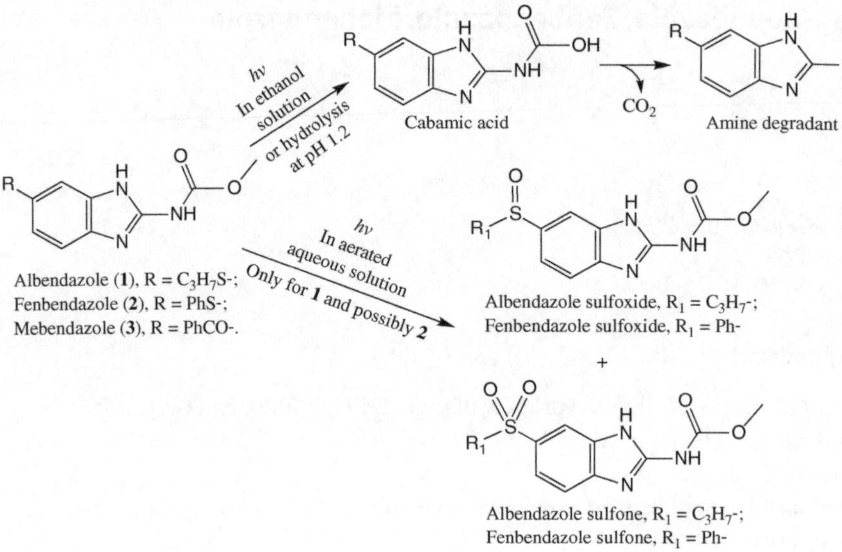

Scheme 6-8 Degradation pathways of albendazole, fenbendazole, and mebendazole.

degradant and the final amine degradant. The intermediary degradant was proposed to be the carbamic acid, resulting from the initial photo-hydrolysis of each drug (Scheme 6–8) based on the GC–MS analysis of the isolated degradant. This is somewhat surprising as carbamic acids can be present in solutions but are usually too unstable to be isolated due to their facile decomposition to the corresponding amines [2].

In another photochemical degradation study, Weerasinghe et al. evaluated the photostability of albendazole in aerated aqueous solutions between pH 7 and 9 [3]. Albendazole was found to undergo stepwise photooxidation, giving rise to albendazole sulfoxide and sulfone, respectively, in addition to a major unknown photodegradant that was found to be very polar in nature. It is worthwhile to mention that this polar degradant is not the amine degradant as the latter was another compound evaluated in the same study. It needs to be pointed out that the Weerasinghe study used all aqueous solutions while the Ragno study used pure ethanol as the solvent. Consequently, the concentrations of the solutions in the Weerasinghe study were much more dilute than those in the Ragno study due to the poor solubility of albendazole in aqueous solutions. These differences are probably the main reason for the different photodegradation patterns observed in the two studies. Since fenbendazole contains a phenylsulfanyl at the 5-position of the benzimidazole ring, it may also be susceptible to the same photo oxidation as albendazole. The latter has a propylsulfanyl in the 5-position.

Torrado et al. conducted a formulation study of albendazole [4]. Over a pH range of 1.2–7.5, the solubility of albendazole was the highest at the lowest pH 1.2. In an experimental aqueous formulation containing 40% transcutol [2-(2-ethoxyethoxy) ethanol)] with a nominal pH of 1.2, albendazole was found to decompose rather quickly. The authors did not specify what degradants were observed. Nevertheless, it appears to be very likely that the degradation of albendazole was due to the hydrolysis of the carbamyl linkage. A carbamyl bond is usually quite stable as the activation energy for its hydrolysis is typically ≥20 kcal/mol [2]. However, the carbamyl linkage of albendazole became quite unstable at pH 1.2 as the activation energy for its hydrolysis is only 9.26 kcal/mol, which is calculated based on the Arrhenius plot determined by the original researchers [4]. Owing to the fact that the same methyl carbamate is also present in both fenbendazole and mebendazole, the latter two drugs should display very similar hydrolytic kinetics.

References

1 Ragno, G., Risoli, A., Ioele, G., and De Luca, M. (2006). Photo- and thermal-stability studies on benzimidazole anthelmintics by HPLC and GC–MS. *Chem. Pharm. Bull.* 54: 802–806.

2 Li, M. (2012). Hydrolytic degradation. In: *Organic Chemistry of Drug Degradation*, 16–47. Cambridge: Royal Society Chemistry Publishing.

3 Weerasinghe, C.A., Lewis, D.O., Mathews, J.M. et al. (1992). Aquatic photodegradation of albendazole and its major metabolites. 1. Photolysis rate and half-life for reactions in a tube. *J. Agric. Food Chem.* 40: 1413–1418.

4 Torrado, S., Torrado, S., Cadorniga, R., and Torrado, J.J. (1996). Formulation parameters of albendazole solution. *Int. J. Pharm.* 140: 45–50.

9: Allopurinol [WHO List of Essential Medicines]

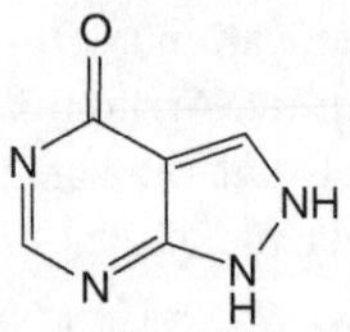

Chemical name: 1*H*-pyrazolo[3,4-*d*]pyrimidin-4(2*H*)-one
Brand name: Zyloprim
Formula: $C_5H_4N_4O$
Monoisotopic protonated ion (*m/z*): 137.0458
Molecular weight: 136.11
CAS number: 315-30-0

Allopurinol is a xanthine oxidase inhibitor, which is used primarily for the treatment of gout. Structurally, it is an isomer of hypoxanthine, a naturally occurring purine compound. Its main degradation is of hydrolytic nature, and the point of the hydrolytic attack is the carbon connecting the two nitrogen atoms in the fused six-membered ring [1]. The initial degradation product should be Impurity B, which would mostly further degrade to Impurity A under strong acid or base catalysis [1–3] (Scheme 9).

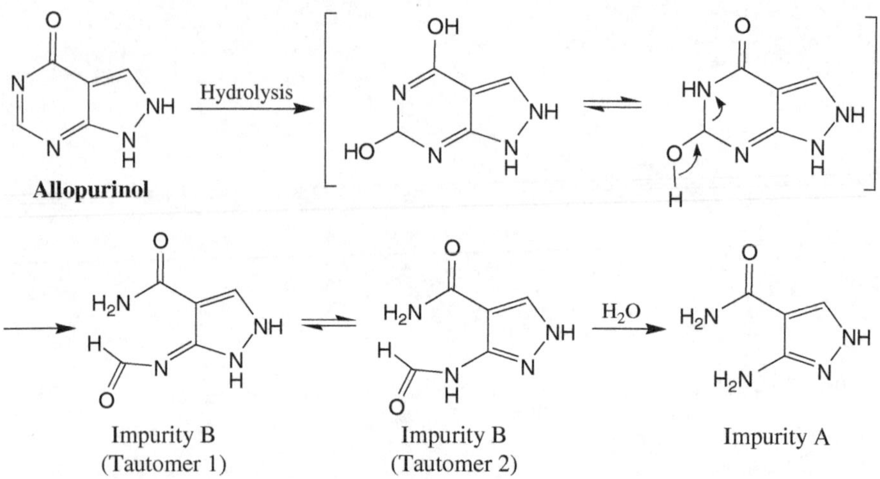

Scheme 9 Hydrolytic degradation of allopurinol.

References

1 Gressel, P.D. and Gallelli, J. (1968). Quantitative analysis and alkaline stability studies of allopurinol. *J. Pharm. Sci.* 57: 335–338.

2 Singh, S. and Gadhawala, Z. (2013). Development of a stability indicating RP-RRLC method for determination of allopurinol and its degradation products in solid oral dosage. *Int. J. PharmTech. Res.* 5: 44–53.

3 Matzko-Hollenbach, K., Jira, T., and Mezey, G. (1987). Zur stabilität von allopurinol und lyophilisiertem allopurinol-natrium. *Arch. Pharm.* 320: 282–284.

10–11: Amantadine, Memantine

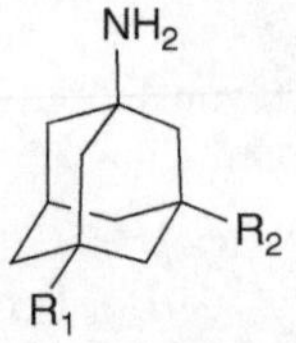

Amantadine: R_1=R_2=H;
Memantine: R_1=R_2=Me

10: Amantadine

Chemical name: Adamantan-1-amine
Brand name: Gocovri; Osmolex
Formula: $C_{10}H_{17}N$
Monoisotopic protonated ion (*m/z*): 152.1434
Molecular weight: 151.25
CAS number: 768-94-5
Common salt form: HCl salt

11: Memantine

Chemical name: 3,5-Dimethyladamantan-1-amine
Brand name: Axura; Ebixa; Marixino
Formula: $C_{12}H_{21}N$
Monoisotopic protonated ion (*m/z*): 180.1747
Molecular weight: 179.31
CAS number: 19982-08-2
Common salt form: HCl salt

Amantadine and memantine are both derivatives of adamantine that contain a primary amino group. Based on their structures, they should be quite stable compounds. In a stress study with amantadine hydrochloride, the drug substance was found to be stable under the stress conditions of boiling, acid, base, and UV irradiation [1]. The photostability is expected as both compounds have no UV chromophore at all. In another forced degradation study where much harsher stress conditions (except base stress) were used, degradation of amantadine was observed [2]. In the latter study where a derivatization assay method was employed, the base stress was milder than the one as reported in Reference [1]; yet, amantadine was found to retain only 77% assay value after being treated in 1 N NaOH solution for 5 minutes at room temperature. In this case, the missing 23% of amantadine could be due to lower derivatization yield, rather than real degradation.

In the development of a stability-indicating GC method for memantine hydrochloride, stress of the drug substance under harsh acid and base conditions, e.g., refluxing in 5 M NaOH solution for 36 hours, did not yield any degradation product [3]. Only in the oxidative stress in 30% hydrogen peroxide at room temperature for 24 hours, an appreciable amount of degradation was observed; such a harsh oxidative condition tends to yield artificial degradants.

In formulation studies, memantine was found to undergo Maillard reaction with reducing end sugars such as lactose under the long-term stability condition of 25C/60%RH [4]. The observed degradation products were sugar adducts resulting from the Amadori rearrangement, after the initial sequential formation of the corresponding Schiff bases and glycosylamines. In addition to the formation of memantine-lactose adduct as expected, memantine also formed adducts in comparable amounts with galactose and glucose, the two components of lactose. Also formed in comparable quantity was an acetic acid adduct (referred to as memantine-dimethylamino glycine or DMAG adduct in Reference [4]). The latter adduct could result from further degradation of all three sugar adducts.

Due to their highly similar structures, it is reasonable to assume that amantadine and memantine would display very similar chemical stability and degradation behaviors (Scheme 10–11).

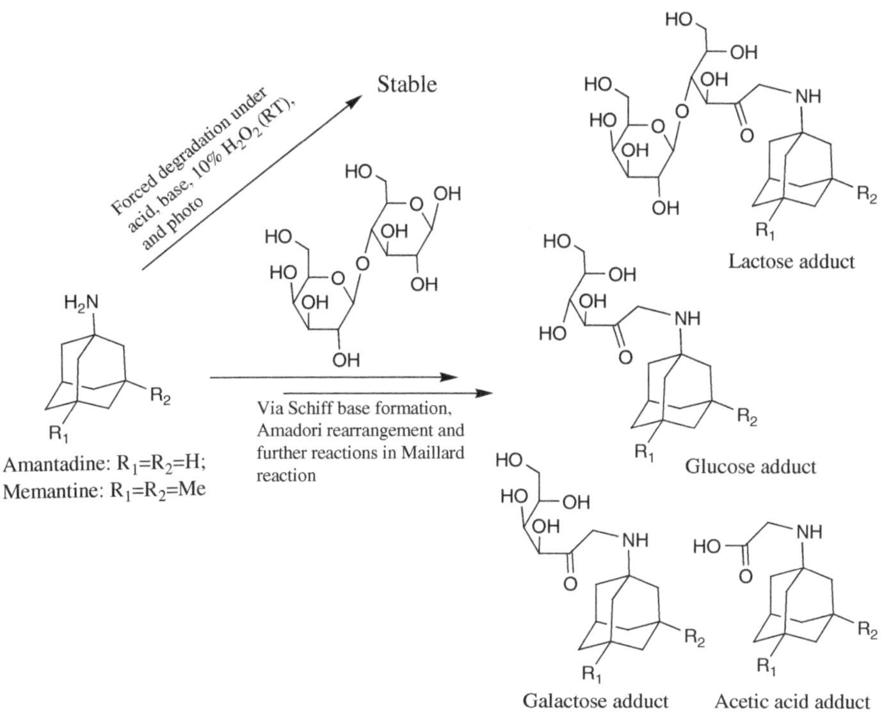

Scheme 10–11 Degradation pathways of amantadine and memantine.

References

1 Askal, H.F., Khedr, A.S., Darwish, I.A., and Mahmoud, R.M. (2008). Quantitative thin-layer chromatographic method for determination of amantadine hydrochloride. *Int. J. Biomed. Sci.* 4: 155–160.

2 Narola, B., Singh, A.S., Santhakumar, P.R., and Chandrashekhar, T.G. (2010). A validated stability-indicating reverse phase HPLC assay method for the determination of memantine hydrochloride drug substance with UV-detection using precolumn derivatization technique. *Anal. Chem. Insights* 5: 37–45. https://doi.org/10.4137/ACI.S3936.

3 Jadhav, S.A., Landge, S.B., Niphade, N.C. et al. (2012). Development and validation of stability-indicating GC-FID method for the quantitation of memantine hydrochloride and its nonchromophoric impurities in bulk and pharmaceutical dosages. *Chromtogr. Res. Inter.* 2012 (1): 806068. https://doi.org/10.1155/2012/806068.

4 Rystov, L., Chadwick, R., Krock, K., and Wang, T. (2011). Simultaneous determination of Maillard reaction impurities in memantine tablets using HPLC with charged aerosol detector. *J. Pharm. Biomed. Anal.* 56: 887–894.

12–13: Ambroxol, Bromhexine

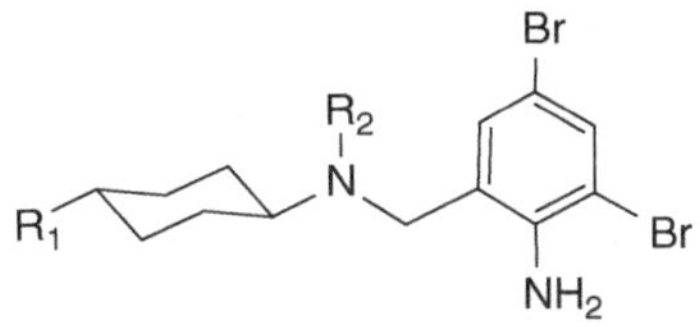

Ambroxol, R_1 = –OH, R_2 = –H;
Bromhexine, R_1 = –H, R_2 = –Me;.

12: Ambroxol

Chemical name: *trans*-4-(2-Amino-3,5-dibrombenzylamino)-cyclohexanol
Brand name: Ambrolex
Formula: $C_{13}H_{18}Br_2N_2O$
Monoisotopic protonated ion (*m/z*): 376.9859
Molecular weight: 378.11
CAS number: 18683-91-5
Common salt form: HCl salt

13: Bromhexine

Chemical name: 2,4-Dibromo-6-{[cyclohexyl(methyl)amino]methyl}aniline
Brand names: Amiorel; many others
Formula: $C_{14}H_{20}Br_2N_2$
Monoisotopic protonated ion (*m/z*): 360.9910
Molecular weight: 362.11
CAS number: 3572-43-8
Common salt form: HCl salt

Bromhexine is a mucolytic drug sold over the counter in many cough remedies. It was introduced by Boehringer Ingelheim in 1963 and is still widely used today. Ambroxol is an active metabolite of bromhexine, resulting from demethylation of the tertiary amine and hydroxylation of the cyclohexane ring. Ambroxol was also developed into a drug in 1978 and is an active ingredient in many drug products with different formulations. In a European patent [1], a stabilization study of the liquid formulation of ambroxol was disclosed. Three degradation products were mentioned in this patent and all of them are listed in the European Pharmacopoeia (EP) [2]. The first degradant is the commonly observed EP Impurity B, which is trans-4-(6,8-dibromo-1,4-dihydroquinazolin-3(2*H*)-yl)cyclohexanol. The methylene bridging the two nitrogen atoms in Impurity B originates from excipients that contain formaldehyde impurity, e.g., PEG. Impurity B is also a metabolite of bromhexine; however, in the latter case, the bridging methylene originates from the methyl group on the tertiary amine. The metabolic formation of Impurity B is an alternative pathway to metabolic demethylation [3].

The other two degradants of ambroxol disclosed in the patent were EP Impurity A and Impurity E, which are (2-amino-3,5-dibromophenyl)methanol and 2-amino-3,5-dibromobenzaldeyde, respectively. According to the patent, both impurities are oxidative degradants and can be controlled by operating under a nitrogen atmosphere. The formation of Impurity E, the benzaldehyde degradant, can be easily rationalized via a radical-mediated auto-oxidation pathway. The benzaldehyde degradant can be generated through oxidation of bromhexine electrochemically [4] or by potassium permanganate [5]. On the other hand, it would be difficult to postulate a reasonable mechanism for the formation of Impurity A from the oxidation of ambroxol.

The same European patent also revealed that unknown degradants were in certain liquid formulations that were projected to increase more than the three known degradants, i.e., Impurities A, B, and E (Scheme 12–13).

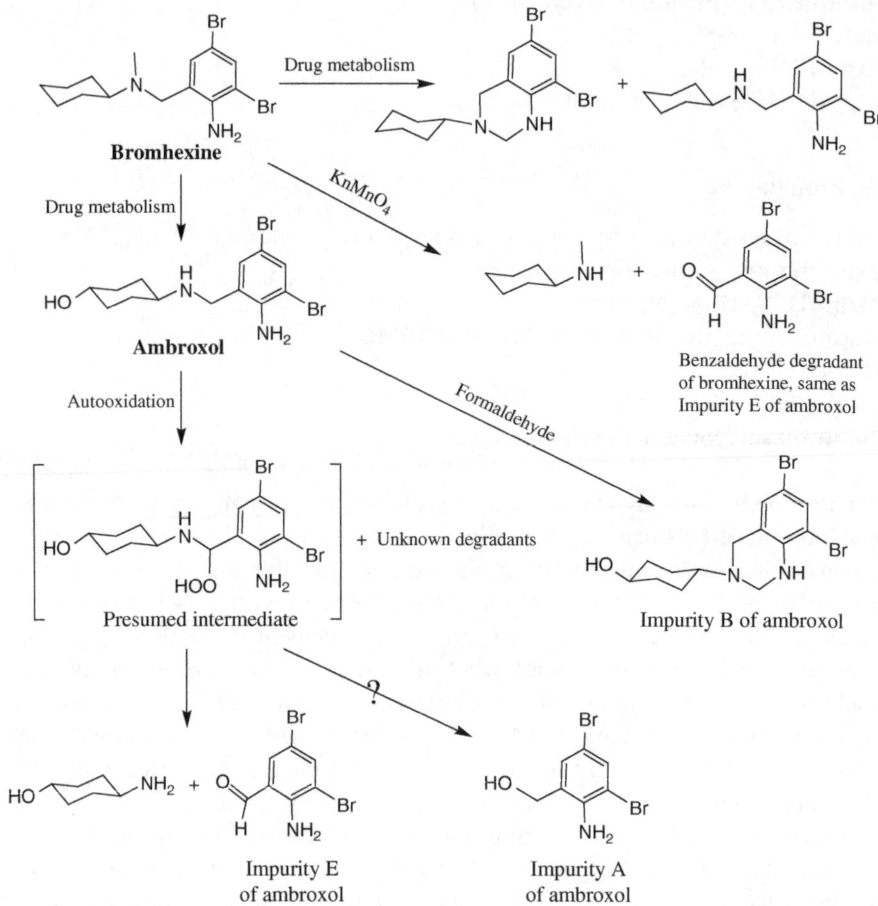

Scheme 12–13 Degradation pathways of ambroxol and metabolism of bromhexine leading to the formation of ambroxol.

References

1 Giorgioc, P. (2003). Concentrated aqueous solution of ambroxol. European Patent 1543826. Patent filed 09 November 1967. Patent granted 25 October 1968.

2 European Pharmacopoeia, 10th ed. (official as of January 2017) Ambroxol hydrochloride. Council of Europe, Strasbourg, France. pp. 1809–1810.

3 Meijer, L.A., Verstegen, J.C.M., Bull, S., and Fink-Gremmels, J. (2004). Metabolism of bromhexine in pig hepatocyte cultures. *J. Vet. Pharm. Col. Therap.* 27: 219–225.

4 Turchan, M., Jara-Ulloa, P., Bollo, S. et al. (2007). Voltammetric behaviour of bromhexine and its determination in pharmaceuticals. *Talanta* 73: 913–919.

5 Susmitha, K., Thirumalachary, M., and Venkateshwarlu, G. (2013). Spectrophotometric determination of bromhexine HCl in pure and pharmaceutical forms. *Int. Sch. Res. Notices* 2013 (1): 861851. http://dx.doi.org/10.1155/2013/861851.

14–16: Aminoglycoside Antibiotics

14: Amikacin [WHO Essential List of Medicines]

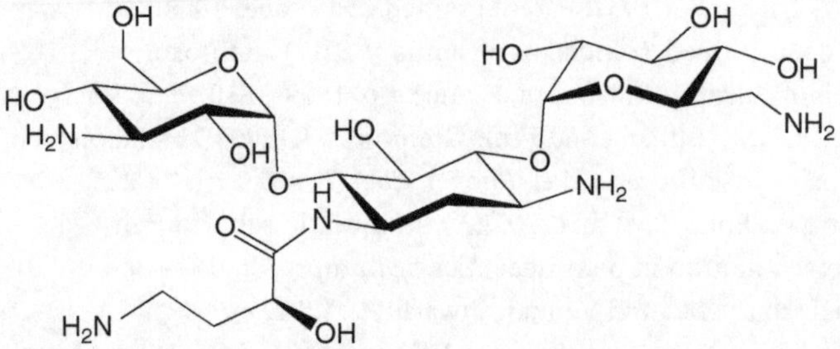

Chemical name: (2*S*)-4-Amino-*N*-{(1*R*,2*S*,3*S*,4*R*,5*S*)-5-amino-2-[(3-amino-3-deoxy-α-D-glucopyranosyl)oxy]-4-[(6-amino-6-deoxy-α-D-glucopyranosyl)oxy]-3-hydroxycyclohexyl}-2 -hydroxybutanamide
Brand name: Amikin, Amiglyde-V, Arikayce
Formula: $C_{22}H_{43}N_5O_{13}$
Monoisotopic protonated ion (*m/z*): 586.2930
Molecular weight: 585.60
CAS number: 37517-28-5
Common salt form: Sulfate

15: Gentamicin

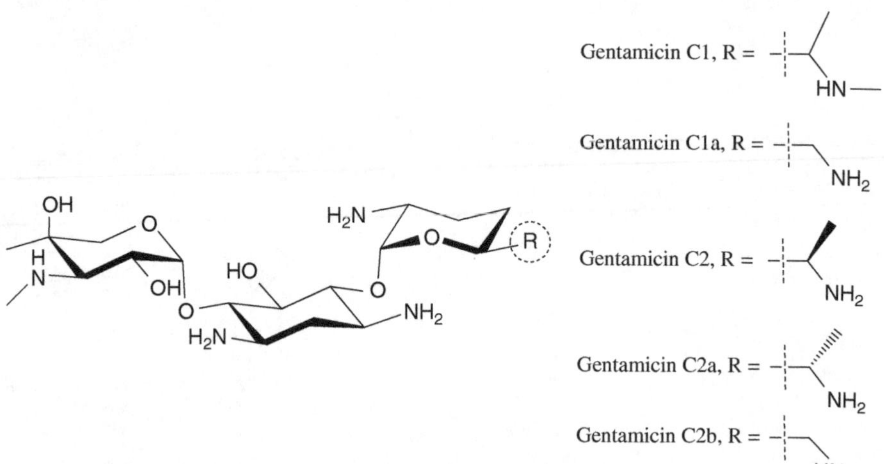

Gentamicin is a mixture of several structural analogs. Its main components include gentamicin C1, C1a, C2, C2a, and C2b. Gentamicin is available in both neutral base form and sulfate salt form.

Brand name: Cidomycin, Genticyn, Garamycin

15a: Gentamicin C1

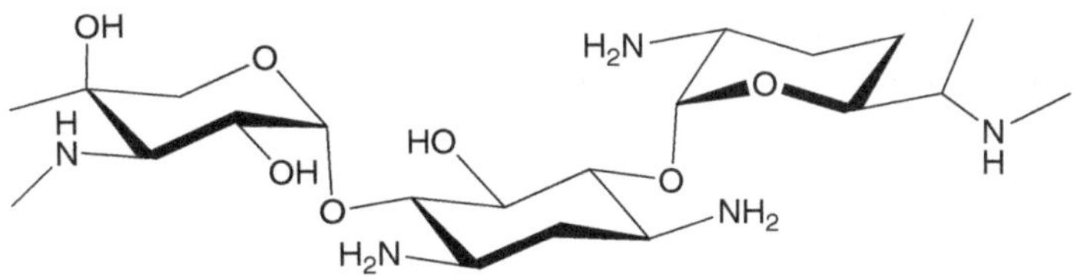

Chemical name: (1*S*,2*R*,3*S*,4*S*,6*S*)-4,6-Diamino-3-({(2*S*,3*S*,6*S*)-3-amino-6-[(1*S*)-1-(methylamino)ethyl]tetrahydro-2*H*-pyran-2-yl}oxy)-2-hydroxycyclohexyl 3-deoxy-4-C-methyl-3-(methylamino)-β-L-lyxopyranoside
Formula: $C_{21}H_{43}N_5O_7$
Monoisotopic protonated ion (*m/z*): 478.3235
Molecular weight: 477.60
CAS number: 1403-66-3

15b: Gentamicin C1a

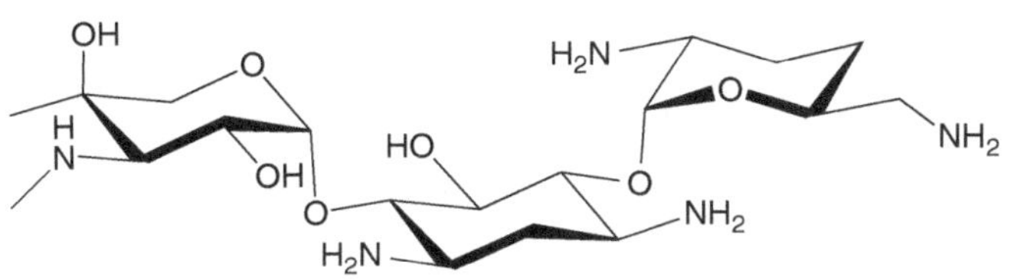

Chemical name: (1*R*,2*S*,3*S*,4*R*,6*S*)-4,6-Diamino-3-{[3-deoxy-4-C-methyl-3-(methylamino)-β-L-arabinopyranosyl]oxy}-2-hydroxycyclohexyl 2,6-diamino-2,3,4,6-tetradeoxy-α-D-erythro-hexopyranoside; (2*R*,3*R*,4*R*,5*R*)-2-[(1*S*,2*S*,3*R*,4*S*,6*R*)-4,6-Diamino-3-[(2*R*,3*R*,6*S*)-3-amino-6-(aminomethyl)oxan-2-yl]oxy-2-hydroxycyclohexyl]oxy-5-methyl-4-(methylamino)oxane-3,5 -diol
Formula: $C_{19}H_{39}N_5O_7$
Monoisotopic protonated ion (*m/z*): 450.2922
Molecular weight: 449.54
CAS number: 26098-04-4

15c: Gentamicin C2

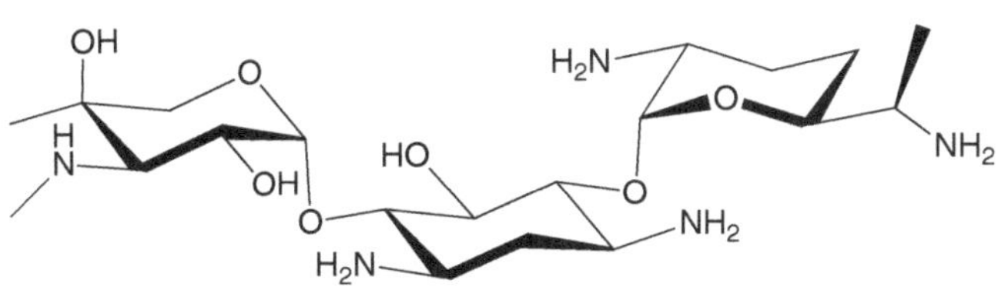

Chemical name: (1*S*,2*S*,3*R*,4*S*,6*R*)-4,6-Diamino-3-({(2*R*,3*R*,6*S*)-3-amino-6-[(1*R*)-1-aminoethyl]tetrahydro-2*H*-pyran-2-yl}oxy)-2-hydroxycyclohexyl-3-deoxy-4-C-methyl-3-(methylamino)-β-L-arabinopyranoside
Formula: $C_{20}H_{41}N_5O_7$
Monoisotopic protonated ion (*m/z*): 464.3079
Molecular weight: 463.57
CAS number: 25876-11-3

15d: Gentamicin C2a

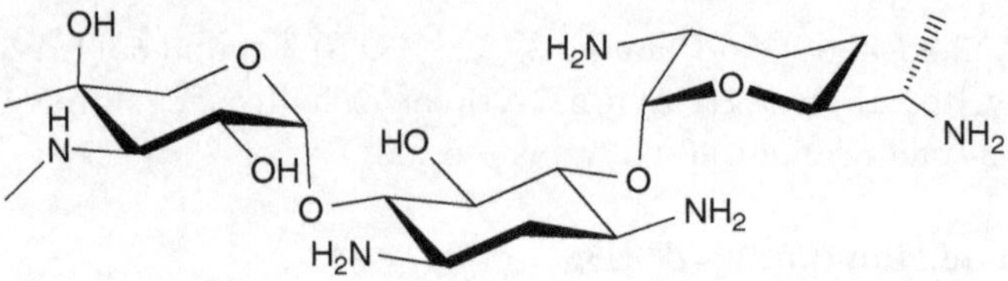

Chemical name: (1*S*,2*S*,3*R*,4*S*,6*R*)-4,6-Diamino-3-{[(2*R*,3*R*,6*S*)-3-amino-6-[(1*S*)-1-aminoethyl]tetrahydro-2*H*-pyran-2-yl]oxy}-2-hydroxycyclohexyl-3-deoxy-4-C-methyl-3-(methylamino)-β-L-arabinopyranoside
Formula: $C_{20}H_{41}N_5O_7$
Monoisotopic protonated ion (*m/z*): 464.3079
Molecular weight: 463.57
CAS number: 59751-72-3

15e: Gentamicin C2b (Micronomicin)

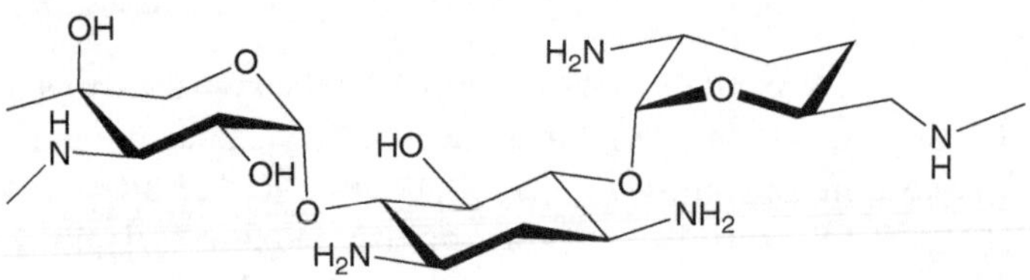

Chemical name: (1*R*,2*S*,3*S*,4*R*,6*S*)-4,6-Diamino-3-{[3-deoxy-4-C-methyl-3-(methylamino)-β-L-arabinopyranosyl]oxy}-2-hydroxycyclohexyl-2-amino-2,3,4,6-tetradeoxy-6-(methylamino)-α-D-erythro-hexopyranoside
Formula: $C_{20}H_{41}N_5O_7$
Monoisotopic protonated ion (*m/z*): 464.3079
Molecular weight: 463.57
CAS number: 52093-21-7

16: Tobramycin [WHO List of Essential Medicines]

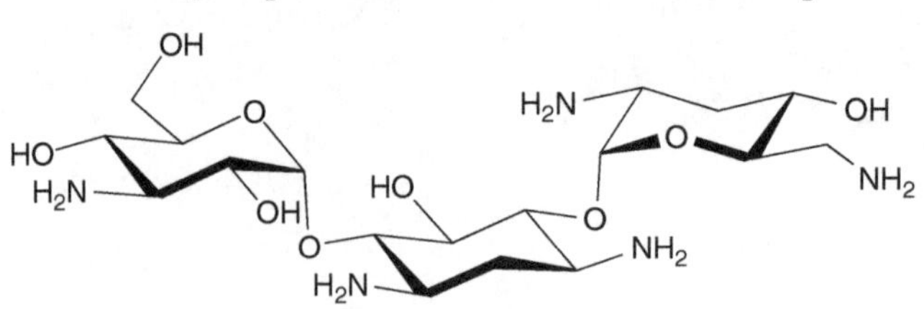

Chemical name: (1*S*,2*S*,3*R*,4*S*,6*R*)-4,6-Diamino-3-[(2,6-diamino-2,3,6-trideoxy-a-D-ribo-hexopyranosyl)oxy]-2-hydroxycyclohexyl 3-amino-3-deoxy-a-D-glucopyranoside
Brand name: Bethkis, Kitabis, Tobi,
Formula: $C_{18}H_{37}N_5O_9$
Monoisotopic protonated ion (*m/z*): 468.2664
Molecular weight: 467.52
CAS number: 32986-56-4
Common salt form: Sulfate

Aminoglycoside antibiotics were first discovered in the 1940s, among which streptomycin is the first medication that can cure tuberculosis [1]. Despite their side effects, they are still among the most prescribed antibiotics worldwide for several diseases caused by infection with Gram-negative pathogens [2, 3]. Aminoglycoside antibiotics are usually quite stable compounds, particularly in their usual sulfate salt forms. The functional groups of aminoglycosides that are potentially susceptible to degradation are the glycosidic linkages and amino groups. The degradation of glycosidic linkages usually would not take place under accelerated or long-term stability conditions. Only under high-temperature and strong acidic or basic conditions would hydrolysis of the glycosidic bonds occur. For the amino moieties, they may undergo the Maillard reaction if their formulations contain excipients possessing reducing sugars [4–6]. In this monograph, we discuss amikacin, gentamicin, and tobramycin as representatives in this important class of drugs.

While amikacin has a somewhat different structure, gentamicin and tobramycin are typical aminoglycoside antibiotics in terms of their structures. Gentamicin has three structural moieties, namely purpurosamine, 2-deoxystreptamine, and garosamine, which are linked together by two glycosidic bonds. The heterogeneity of methylation and/or stereochemistry on the purpurosamine moiety renders the drug substance a heterogeneous mixture that contains five major components: C1, C1a, C2, C2a, and C2b [7]. In addition, numerous minor components or impurities, most notably garamin, sisomicin, and deoxystreptamine, have been identified in gentamicin drug substance, although not all of them may be present at the same time in a single batch [8]. Various long-term stability studies have demonstrated that gentamicin is quite stable and none of the identified impurities have been found to increase over the periods of the studies. For example, a stability study of gentamicin drug substance for at least 5 years was reported [9]. For a number of gentamicin products that were formulated in polymeric excipients such as polymethylmethacrylate (PMMA), the drug was found to be stable under both long-term (25°C/60%RH, 24 months) and accelerated stability conditions (40°C/75%RH, 6 months) [9]. When gentamicin was formulated in dextrose solutions, incompatibility of gentamicin with the reducing sugar was observed as reflected in the development of a yellow color and loss of potency [4]. The compatibility study indicated that several degradation products were quickly formed; although the structures of these degradants were not elucidated, it would be most likely that the degradation was mainly due to the Maillard reaction [6]. Furthermore, when the aminoglycosides were co-formulated with β-lactam antibiotics, the amino groups of the aminoglycosides would be capable of attacking the lactam moieties of the β-lactam

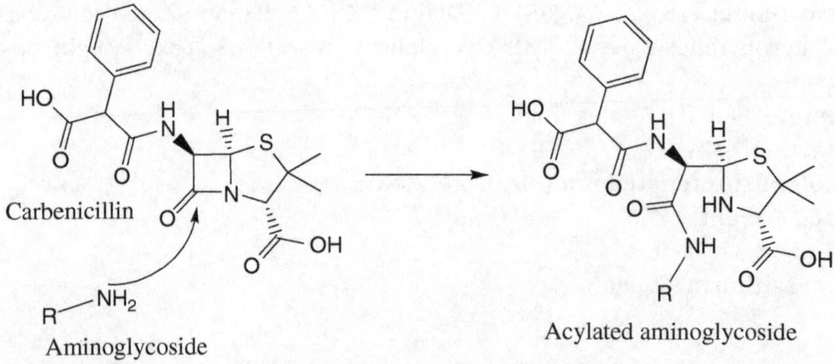

Scheme 14-16-1 Incompatibility between aminoglycosides and carbenicillin, a semi-synthetic penicillin antibiotic, via acylation of the amino groups of aminoglycosides.

antibiotics, resulting in acylation of the amino groups. Such degradation was demonstrated in the case of amikacin acylation by carbenicillin (Scheme 14–16-1) [10].

On the other hand, forced degradation under acidic or basic conditions at elevated temperatures could decompose gentamicin through the cleavage of the glycosidic bonds, which would be expected to be similar to the case of tobramycin as discussed in the next paragraph.

The structure of tobramycin is similar to that of gentamicin, particularly the glycosidic linkages. Brandl and Gu performed a thorough forced degradation study of tobramycin at different pH and temperatures (Scheme 14–16-2) [11]. In 1 N HCl solutions in the temperature range of 60–100°C for up to ~160 hours, only one

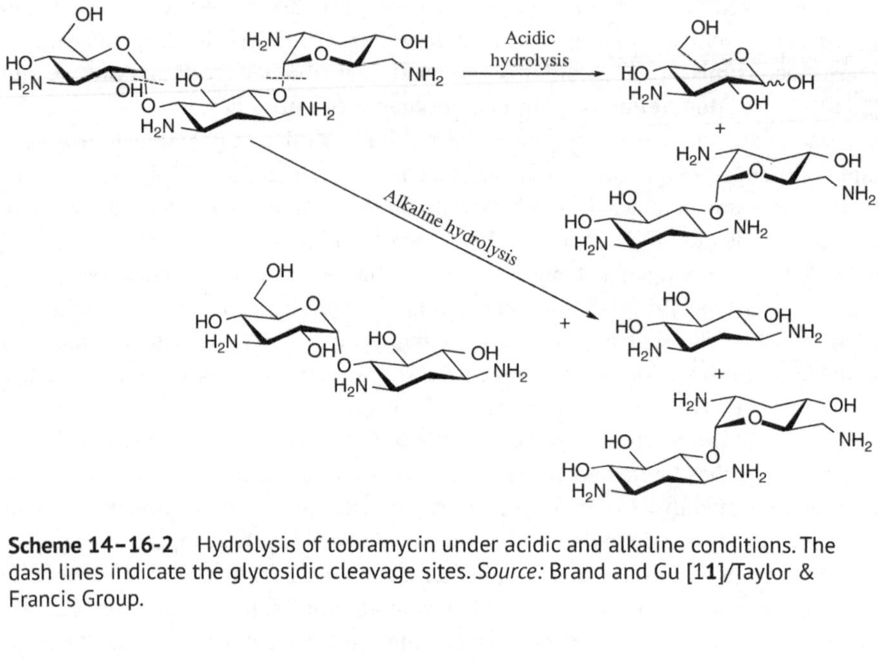

Scheme 14-16-2 Hydrolysis of tobramycin under acidic and alkaline conditions. The dash lines indicate the glycosidic cleavage sites. *Source:* Brand and Gu [11]/Taylor & Francis Group.

glycosidic bond was cleaved, and the activation energy was found to be 30 kcal/mol. At the other end of the extreme pH, both glycosidic bonds were found to be cleaved in a 1 N NaOH in the temperature range of 60–100°C. The activation energy for the hydrolysis under the alkaline conditions was determined to be 15 kcal/mol.

In solutions of neutral pH at temperature range of 60–100°C, cleavage of tobramycin was extremely slow in the absence of oxygen; on the other hand, the degradation became rapid in ambient air, giving similar degradation products under alkaline hydrolytic conditions [11], suggesting the cleavage of the glycosidic bonds may be mediated by auto-oxidation [12]. Nevertheless, the implication of the oxidative degradation pathway toward the stability of tobramycin under long-term (25°C) or accelerated stability (40°C) conditions would be difficult to evaluate. This is because auto-oxidation usually involves free radical intermediates and its degradation mechanism and kinetics may change dramatically as temperature decreases significantly.

In a study of gentamicin and tobramycin in concentrated dextrose solution, tobramycin was found to be less stable than gentamicin [13]. Although the study did not elaborate on the degradation mechanisms, it would be quite likely that the underlying degradation would be due to the Maillard reaction.

Amikacin is somewhat structurally different from gentamicin and tobramycin in that it contains an acyl group in one of the amino groups. There appear to be very few reports on its stability study [14]. Among the few studies reported, one studied the chemical compatibility between amikacin and four β-lactam antibiotics over a period of 3 weeks in temperatures ranging from −70 to 25°C [15]. While amikacin demonstrated no loss of activity when combined in solutions with two penicillin antibiotics, carbenicillin and piperacillin, respectively, gentamicin and tobramycin showed significant loss of activities 8–48 hours after admixing with the two penicillins at both 25°C and 4°C. The inactivation of the aminoglycosides by the penicillins was believed to involve nucleophilic attack by the amino groups of aminoglycosides on the 4-membered betalactam rings. The degradation is similar to the case of amikacin acylation by carbenicillin as mentioned above [10].

References

1 Aminoglycoside. Wikipedia.org (accessed 13 May 2022).

2 Forge, A. and Schacht, J. (2000). Aminoglycoside antibiotics. *Audiol. Neurootol.* 5: 3–22.

3 McWilliam, S.J., Antoine, D.J., Smyth, R.L., and Pirmohamed, M. (2017). Aminoglycoside-induced nephrotoxicity in children. *Pediatr. Nephrol.* 32: 2015–2025.

4 Graham, A.E., Speicher, E., and Williamson, B. (1997). Analysis of gentamicin sulfate and a study of its degradation in dextrose solution. *J. Pharm. Biomed. Anal.* 15: 537–543.

5 Kumar, V. and Banker, G.S. (1998). Maillard reaction and drug stability. In: *Maillard Reactions in Chemistry, Food and Health* (ed. T.P. Labuza, G.A. Reineccius, V.M. Monnier, et al.). Woodhead Publishing.

6 Li, M. (2012). Drug-excipient interactions and adduct formation. In: *Organic chemistry of drug degradation*. Cambridge, UK: RSC Publishing.

7 USP (official as of 1 May 2020) Gentamicin sulfate. https://doi.org/10.310.3/USPNF_M34850_04_01 (accessed 13 May 2022).

8 Grahek, R. and Zupancic-Kralj, L. (2009). Identification of gentamicin impurities by liquid chromatography tandem mass spectrometry. *J. Pharm. Biomed. Anal.* 50: 1037–1043.

9 Kühn, K.-D., Weber, C., Kreis, S., and Holzgrabe, U. (2008). Evaluation of the stability of gentamicin in different antibiotic carriers using a validated MEKC method. *J. Pharm. Biomed. Anal.* 48: 612–618.

10 Iyengar, B.S., Kumar, V., Wunz, T.P., and Remers, W.A. (1986). Aminoglycoside antibiotics. 6. Chemical reactions of aminoglycosides with disodium carbenicillin. *J. Med. Chem.* 29: 611–614.

11 Brand, M. and Gu, L. (1992). Degradation of tobramycin in aqueous solution. *Drug Dev. Ind. Pharm.* 18: 1423–1436.

12 Hodges, N.A. and Singh, J. (1978). Enhancement of neomycin stability using propylene-glycol and sodium metabisulfite. *J. Pharm. Pharmacol.* 30: 737–739.

13 Nance, K.S. and Matzke, G.R. (1984). Stability of gentamicin and tobramycin in concentrate solutions for automated peritoneal dialysis. *Am. J. Nephrol.* 4: 240–243.

14 Zhang, Y. and Trissel, L.A. (2003). Stability of amikacin sulfate in AutoDose Infusion System bags. *Inter. J. Pharma. Comp.* 7: 230–232.

15 Glew, R.H. and Pavuk, R.A. (1983). Stability of gentamicin, tobramycin, and amikacin in combination with four β-lactam antibiotics. *Antimicrob. Agents Chemother.* 24: 474–477.

17: Amiloride

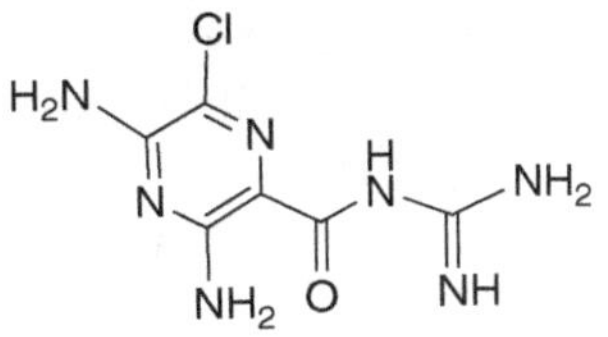

Chemical name: 3,5-Diamino-6-chloro-*N*-(diaminomethylidene)pyrazine-2-carboxamide
Brand name: Midamor
Formula: $C_6H_8ClN_7O$
Monoisotopic protonated ion (*m/z*): 230.0552
Molecular weight: 229.63
CAS number: 2609-46-3
Common salt form: HCl salt

Amiloride is a diuretic with a substituted pyrazine ring and an acylated guanidinyl side chain. In a forced degradation study [1], the drug was found to be unstable under hydrolytic degradations at elevated temperatures. The degradation products

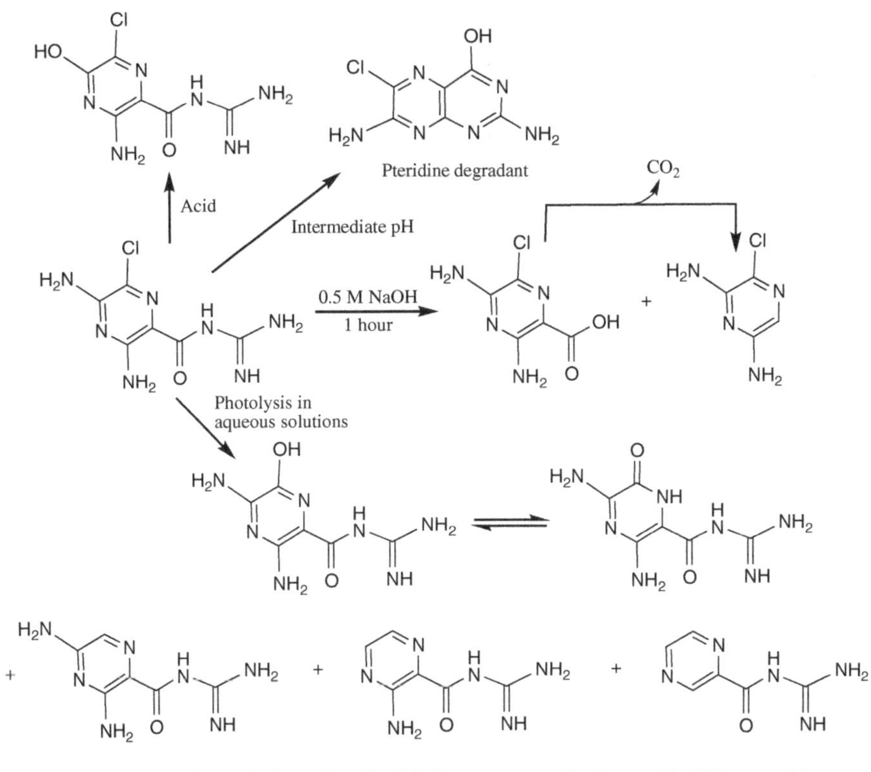

Scheme 17 Photodegradation of amiloride in aqueous solutions with different pH.

were found to be dependent on the pH of the stress conditions. Under acidic stress, one of the amino groups on the pyrazine ring was displaced by a hydroxyl group. In the intermediate pH range, the guanidine group underwent cyclization to form a pteridine ring. In 0.5 N NaOH solution at 100°C for 1 hour, ~50% of amiloride was decomposed to form two major degradants (A and B) [1]. Degradant A apparently results from the cleavage of the acyl-guanidine linkage, while Degradant B appears to stem from decarboxylation of Degradant A.

Amiloride is a known photosensitizer and its photodegradation chemistry has been studied [2–4]. The chromophore of amiloride is pH dependent, and as such its photochemical stability is also pH dependent. In a photodegradation study in aqueous anaerobic solutions of pH 4.5–11, the only photodegradant of amiloride isolated with a quantum yield of 0.23 was a photosubstitution product in which the chlorine is replaced by hydroxyl. At high pH range, the neutral molecule is ~3 times more photochemically reactive than its protonated form at low pH [2]. In another pH-dependent photodegradation study, four photodegradants were identified with LC–MS (Scheme 17).

References

1 Lakshmi Narasimham, Y.S. and Barhate, V.D. (2010). Development and validation of stability indicating UPLC method for the simultaneous determination of beta-blockers and diuretic drugs in pharmaceutical dosage forms. *J. Chem. Metrol.* 4: 1–20.

2 Nian, Y., Li, B., Moore, D.E., and Tattam, B.N. (1999). Photodegradation of amiloride in aqueous solution. *Int. J. Pharm.* 109–116.

3 Calza, P., Massolino, C., Monaco, G. et al. (2008). Study of the photolytic and photocatalytic transformation of amiloride in water. *J. Pharm. Biomed. Anal.* 48: 315–320.

4 De Luca, M., Ioele, G., Mas, S. et al. (2012). A study of pH-dependent photodegradation of amiloride by a multivariate curve resolution approach to combined kinetic and acid–base titration UV data. *Analyst* 137: 5428–5435.

18–19: 4-Aminosalicylic Acid, 5-Aminosalicyclic Acid

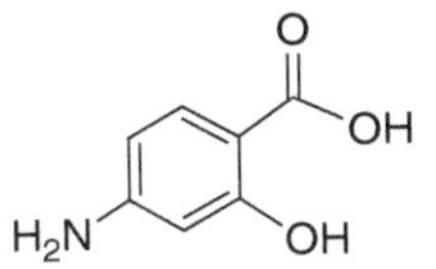

4-Aminosalicylic acid

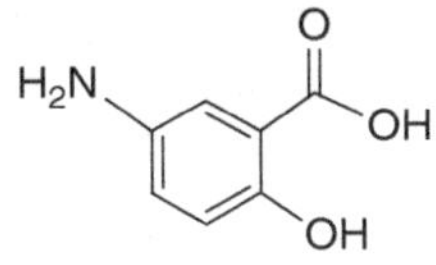

5-Aminosalicylic acid

18: 4-Aminosalicylic Acid

Chemical name: 4-Amino-2-hydroxybenzoic acid
Brand name: Granupas
Formula: $C_7H_7NO_3$
Monoisotopic protonated ion (*m/z*): 154.0499
Molecular weight: 153.14
CAS number: 65-49-6
Common salt form: Calcium and sodium salts

19: 5-Aminosalicyclic Acid (Mesalamine)

Chemical name: 5-Amino-2-hydroxybenzoic acid
Brand name: Asacol, Pentasa
Formula: $C_7H_7NO_3$
Monoisotopic protonated ion (*m/z*): 154.0499
Molecular weight: 153.14
CAS number: 89-57-6

4-Aminosalicylic acid is an antitubercular drug, while 5-aminosalicylic acid (more commonly known as mesalamine) is an anti-inflammatory drug. The main degradation pathways of both drugs are decarboxylation and oxidation, respectively. Sodium, potassium, and calcium salts of 4-aminosalicylic acid are also employed therapeutically.

The decarboxylated degradant of 4-aminosalicylic acid, 3-aminophenol, is of particular toxicological concern, due to the very high therapeutic doses of the drug, which can be as high as 10–12 g/d. The decarboxylation takes place much more easily in acidic solutions than in neutral ones [1]. Hence, it would not be surprising that 4-aminosalicylic acid underwent appreciable decarboxylation in simulated gastric fluid according to the original research by Matsunaka, which was cited by Wesolowski[1]. To overcome the problem, enteric-coated tablets of the drug were developed [2]. In a formulation study, surfactants were found to significantly reduce the rate of decarboxylation [3].

The decarboxylated degradant, 3-aminophenol, is a highly electron-rich species and thus is susceptible to auto-oxidation, giving rise to numerous further degradants including polymeric degradants. Of the many further degradants mentioned in the

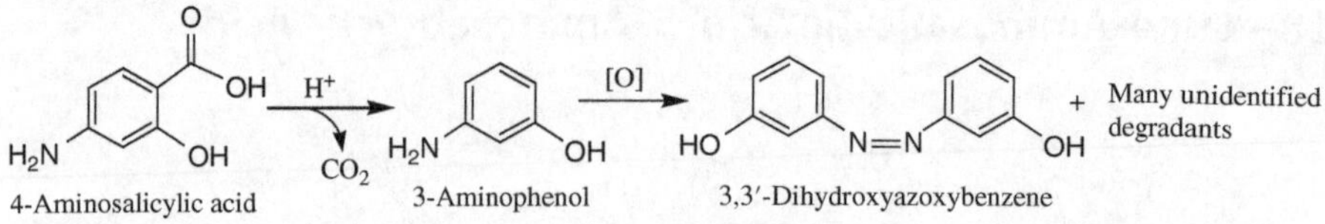

Scheme 18-19-1 Degradation pathway of 4-aminosalicylic acid.

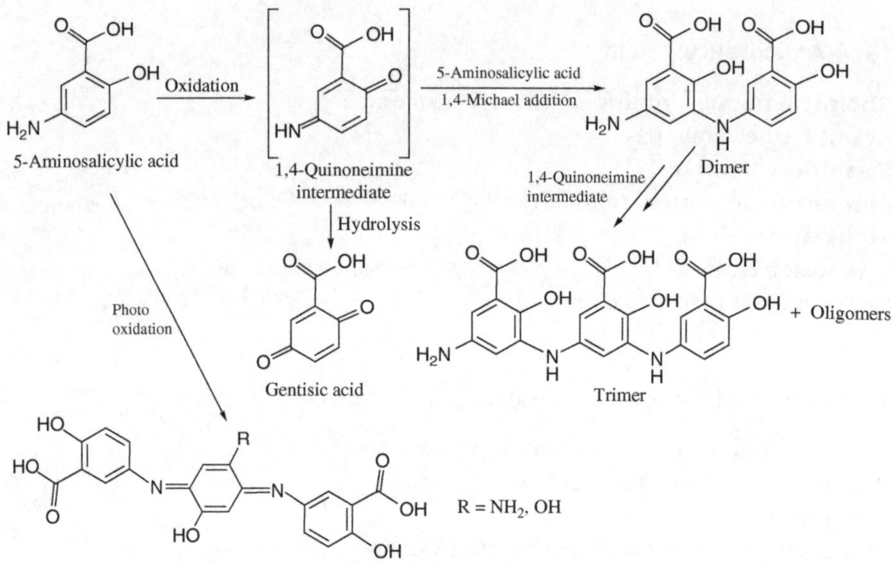

Scheme 18-19-2 Degradation pathways of 5-aminosalicylic acid.

literature, only one was structurally identified: 3,3′-dihydroxyazoxybenzene, which is a degradant resulting from oxidative coupling of 3-aminophenol (Scheme 18–19-1) [4].

According to a study by Palsmeier et al. [5], although 5-aminosalicylic acid can undergo decarboxylation, its degradation in aqueous solutions is mostly of oxidative nature. The oxidation of 5-aminosalicylic acid is mediated through a 1,4-quinoneimine intermediate (Scheme 18–19-2). Hydrolysis of this intermediate leads to the formation of gentisic acid. On the other hand, the reaction of the intermediate with 5-aminosalicylic acid itself causes the formation of dimeric and oligomeric degradants. 5-Aminosalicylic acid can also undergo photooxidation; some trimers were among the photodegradants identified [6].

References

1 Wesolowski, M. (1977). The decarboxylation and thermal stability of *p*-aminosalicylic acid and its salts. *Thermochim. Acta* 21: 243–253.

2 Verma, A., Islam, S., Mishra, A.K., and Pandit, J.K. (2012). Formulation, optimization and evaluation of enteric coated tablets of para-amino salicylate sodium. *J. Sci. Ind. Res.* 71: 667–677.

3 Plamondon, J.E. and Nairn, J.G. (1997). The effect of surfactants on the rate of decarboxylation of *p*-aminosalicylic acid in acidic aqueous solutions. *J. Pharm. Sci.* 86: 205–208.

4 Shih, I.K. (1971). Aminosalicylic acid, identity of a major chromogen in *p*-aminosalicylic acid and sodium *p*-aminosalicylate dosage forms. *J. Pharm. Sci.* 60: 1886–1887.

5 Palsmeier, R.K., Radzik, D.M., and Lunte, C.E. (1992). Investigation of the degradation mechanism of 5-aminosalicylic acid in aqueous solution. *Pharm. Res.* 9: 933–938.

6 Jensen, J., Cornett, C., Olsen, C.E. et al. (1992). Identification of major degradation products of 5-aminosalicylic acid formed in aqueous solutions and in pharmaceuticals. *Int. J. Pharm.* 88: 177–187.

20: Amiodarone

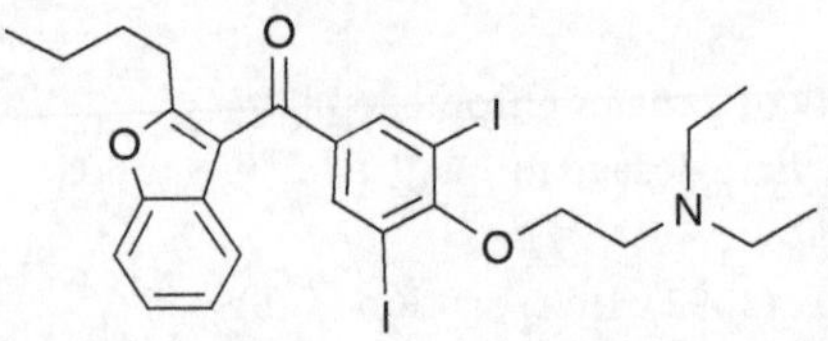

Chemical name: (2-Butyl-1-benzofuran-3-yl)-[4-[2-(diethylamino)ethoxy]-3,5-diiodophenyl]methanone
Brand name: Nexterone, Pacerone
Formula: $C_{25}H_{29}I_2NO_3$
Monoisotopic protonated ion m/z: 646.0310
Molecular weight: 645.31
Common salt form: HCl salt
CAS number: 1951-25-3

Amiodarone is a class III antiarrhythmic drug. Its di-arylketone moiety is susceptible to photochemical degradation. According to the study by Martine and Verrierhe [1], photo-excited triplet state of the drug can sensitize oxygen, leading to the formation of reactive oxygen species. This process was suspected to be a contributing factor to the phototoxicity of the drug [2]. The photo-excited state of the drug can also lead to dehalogenation degradation products (Impurity A and C, Scheme 20) [1]. Further photodegradants include Impurity D, resulting from the

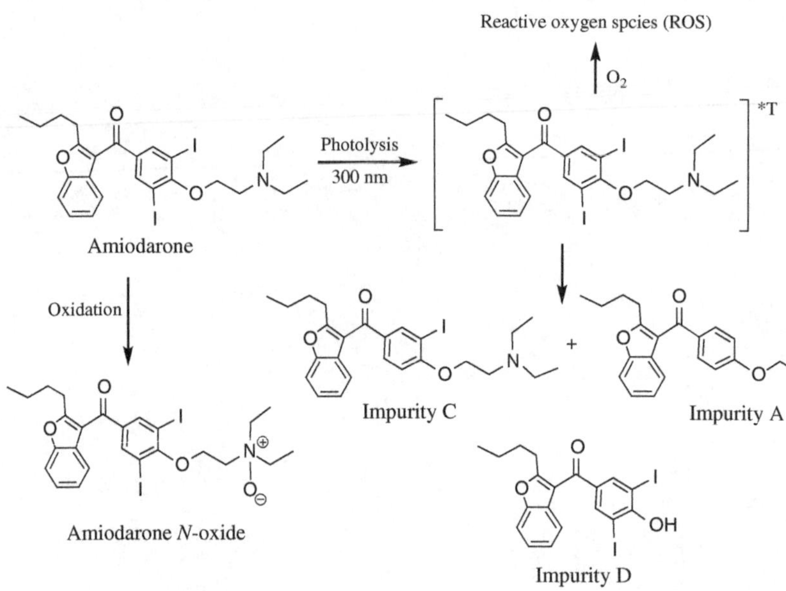

Scheme 20 Degradation pathways of amiodarone.

cleavage of the tertiary amine side chain. Since the drug also contains a tertiary amine moiety, it could potentially form the *N*-oxide degradant; nevertheless, the *N*-oxide would be difficult to form in its HCl salt form [3], as the nucleophilicity of the tertiary amine is blocked by protonation.

References

1 Paillous, N. and Verrier, M. (1988). Photolysis of amiodarone, an antiarrhythmic drug. *Photochem. Photobiol.* 47: 337–343.

2 Henry, B., Foti, C., and Alsante, K. (2009). Can light absorption and photostability data be used to assess the photosafety risks in patients for a new drug molecule? *J. Photochem. Photobio. B: Biol.* 96: 57–62.

3 Khan, M.A., Kumar, S., Jayachandran, J. et al. (2005). Validation of a stability indicating LC method for amiodarone HCL and related substances. *Chromatographia* 61: 599–607.

21: Aripiprazole

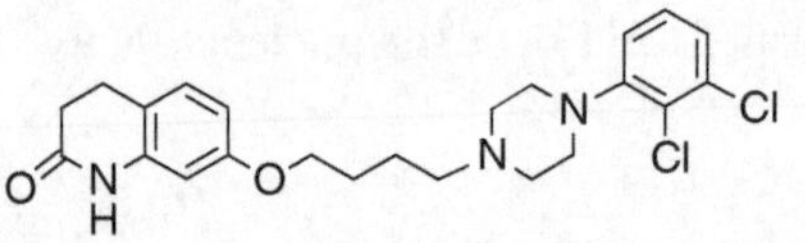

Chemical name: 7-[4-[4-(2,3-Dichlorophenyl)piperazin-1-yl]butoxy]-3,4-dihydro-1*H*-quinolin-2-one
Brand name: Abilify, Aripiprex
Formula: $C_{23}H_{27}Cl_2N_3O_2$
Monoisotopic protonated ion (*m/z*): 448.1553
Molecular weight: 448.39
CAS number: 129722-12-9

Aripiprazole is a novel drug used for the treatment of schizophrenia. Reddy et al. performed a forced degradation study on aripiprazole tablets [1]. After powdered aripiprazole tablets were stressed under dry heat at 105°C for 3 days, three major degradants were formed. Structure elucidation by LC–MS suggested that one of the degradants is the *N*-oxide, while the other two would result from the O–C and N–C cleavage, respectively (Scheme 21). The formation of the O–C cleavage degradant should proceed through the elimination of 7-HDQ, which is rather unusual. Cleavage of an O–C linkage of this nature would usually be preceded by oxidation. In a similar vein, the cleavage of the N–C linkage, leading to the formation of the N–C cleavage degradant, would likely proceed through the *N*-oxide intermediate, which would be followed by the Cope elimination [2]. In the latter scenario, the other cleavage product would be *N*-hydroxyl DCPP, which may be reduced to DCPP by reductive components of the formulation. The *N*-oxide was also generated via stress with hydrogen peroxide in the same forced degradation study by Reddy et al. [1].

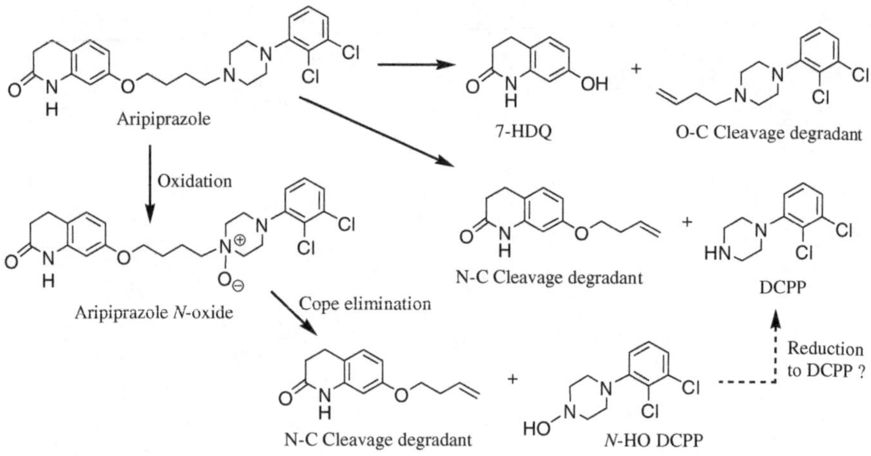

Scheme 21 Degradation pathways of aripiprazole.

References

1 Reddy, G.V.R., Kumarb, A.P., Reddy, B.V. et al. (2010). Identification of degradation products in Aripiprazole tablets by LC-QToF mass spectrometry. *Eur. J. Chem.* 1: 20–27.

2 Li, M. (2012). Oxidative degradation. In: *Organic Chemistry of Drug Degradation*, 73–75. Cambridge: Royal Society Chemistry Publishing.

22: Atazanavir

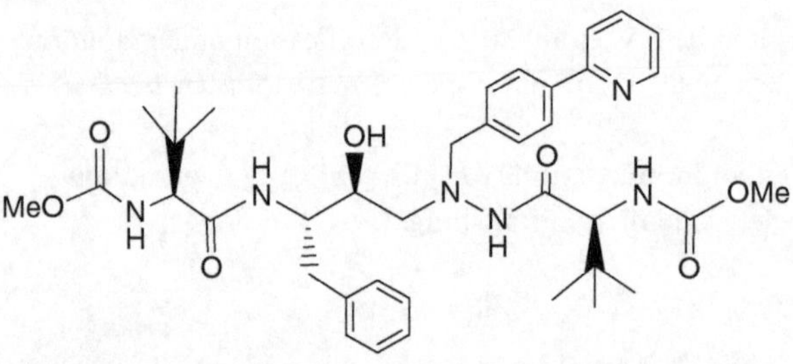

Chemical name: Methyl *N*-[(2*S*)-1-[2-[(2*S*,3*S*)-2-hydroxy-3-[[(2*S*)-2-(methoxycarbonylamino)-3,3-dimethylbutanoyl]amino]-4-phenylbutyl]-2-[(4-pyridin-2-ylphenyl)methyl]hydrazinyl]-3,3-dimethyl-1-oxobutan-2-yl]carbamate
Brand name: Reyataz
Formula: $C_{38}H_{52}N_6O_7$
Monoisotopic protonated ion (*m/z*): 705.3970
Molecular weight: 704.86
CAS number: 198904-31-3
Common salt form: Sulfate

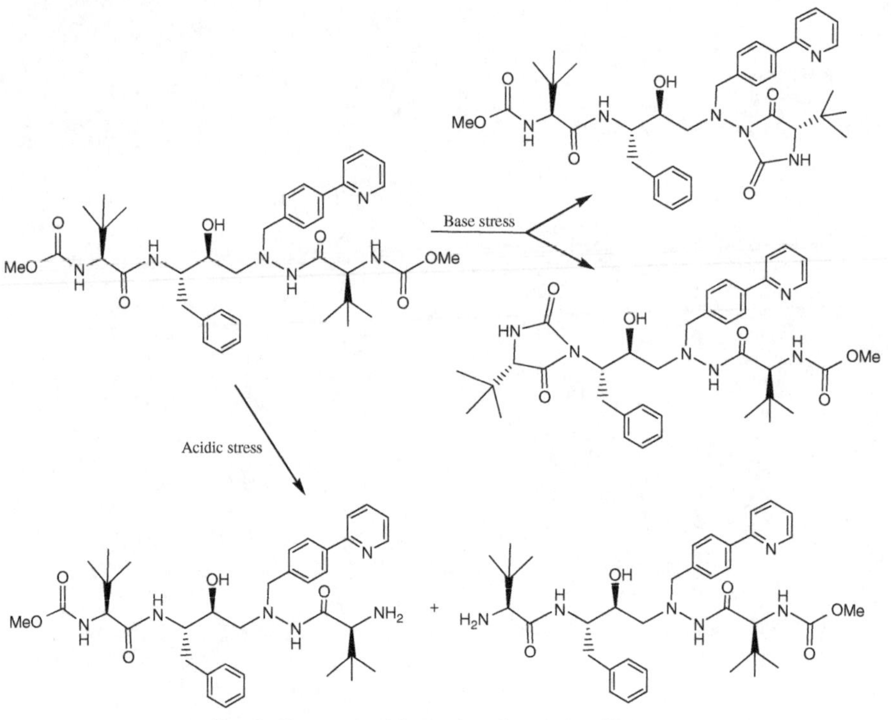

Scheme 22 Degradation pathways of atazanavir under acidic and basic stress conditions.

Atazanavir is an HIV protease inhibitor. Its structure is somewhat symmetric with both ends of the molecule containing a methyl carbamate moiety. Chitturi et al. carried out a comprehensive forced degradation study [1]. Two major base-stress degradants were identified via LC–MS/MS, NMR, and IR. Both degradants stem from the cyclization of the carbamate moieties (Scheme 22). Acidic stress yielded two degradants at 2.91% and 0.68%, but their structures were not identified in the study. Based on the structure of the drug molecule, the two acid-stress degradants could be due to the hydrolysis of the two carbamate moieties, respectively.

Reference

1 Chitturi, S.R., Somannavar, Y.S., Peruri, B.G. et al. (2011). Gradient RP-HPLC method for the determination of potential impurities in atazanavir sulfate. *J. Pharm. Biomed. Anal.* 55: 31–47.

23: Atenolol

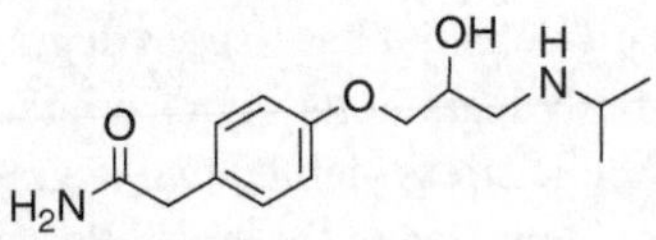

Chemical name: 2-[4-[2-Hydroxy-3-(propan-2-ylamino)propoxy]phenyl]acetamide
Brand names: Tenoretic, Tenormin
Formula: $C_{14}H_{22}N_2O_3$
Monoisotopic protonated ion (*m/z*): 267.1703
Molecular weight: 266.34
CAS number: 29122-68-7

Atenolol is a beta-blocker used as a treatment for hypertension and coronary heart disease. It contains a primary amide and a secondary amine moiety. As a secondary amine, this drug may not be compatible with excipients containing reducing-end sugar such as lactose. In addition, this moiety can interact with impurities of certain excipients that contain reactive keto functionalities. Vijay Kumar et al. performed a series of compatibility studies of atenolol with citric acid, ascorbic acid, and butylated hydroxyanisole (BHA) [1]; solid atenolol was mixed with equal weights of three excipients, respectively and then stressed under the ICH accelerated condition for 1 month. Various degradants were formed and the structures of the degradants were elucidated by LC–MS. In the compatibility study with citric acid, the degradant was the hydrolytic degradant, EP impurity E, resulting from the hydrolysis of the amide linkage. In the compatibility study with ascorbic acid, EP impurity G was also formed. Furthermore, two unknown degradants were observed as well. Accurate MS and MS/MS analysis suggested that these two unknown degradants are *N*-formyl and *N*-formyloxyformyl adducts of atenolol (the latter degradant was referred to as carbamic formic anhydride of atenolol by the original researchers). The *N*-formyl and *N*-formyloxyformyl groups would most likely originate from ascorbic acid, since the latter is not very stable and decomposes to many degradation products under elevated temperature and moisture [2]. Although the structure assignments in the paper seem reasonable, it appears that the majority of the proposed formation mechanisms need additional supporting evidence. In the compatibility study with BHA, four degradants were produced: *N*-formyl atenolol, *N*-acetyl atenolol, 2-(4-hydroxyphenyl)acetamide, and 2-(4-[2-amino-2-oxoethyl]phenoxy)acetic acid. The formation of the latter two degradants apparently resulted from oxidative cleavage of the main backbone of the drug molecule. On the other hand, the formation of the first two degradants was proposed to stem from the reactions of the drug molecule with formyl and acetyl donors, which themselves are products of the oxidative cleavage, respectively. It seems to be paradoxical that the antioxidant, BHA, would cause oxidation of atenolol in this compatibility study when it was present in overwhelming amount as compared to a typical amount that is usually in the range of single digit percentage or less.

In an acidic hydrolytic stress study with 0.1 M HCl at 90°C for 1 hour, in addition to the degradant 2-(4-hydroxyphenyl)acetamide, which was also observed in the oxidative cleavage described above, a dehydrated degradant, 4-([3-isopropyloamino]-2-propenoxy)phenylacetic acid, was also produced [3]. The position of the double bond resulting from the dehydration was confirmed by ^{1}H NMR. The UV spectrum of the dehydrated degradant is very similar to that of atenolol, which would be consistent with the double bond position assignment.

In a photodegradation study of atenolol with UVA–UVB radiation, two photodegradants were found [4]. At pH 7.4, the main degradant was identified as 2-(4-hydroxyphenyl)acetamide based on ^{1}H NMR and UV data and the minor one was identified as a dehydrated degradant based on GC–MS and UV results. However, this dehydrated degradant is different from the dehydrated degradant that was formed in the acidic stress at 90°C as described above. The difference lies in the double bond position of the two degradants.

The degradation products from the aforementioned studies [1–4] are summarized in Scheme 23.

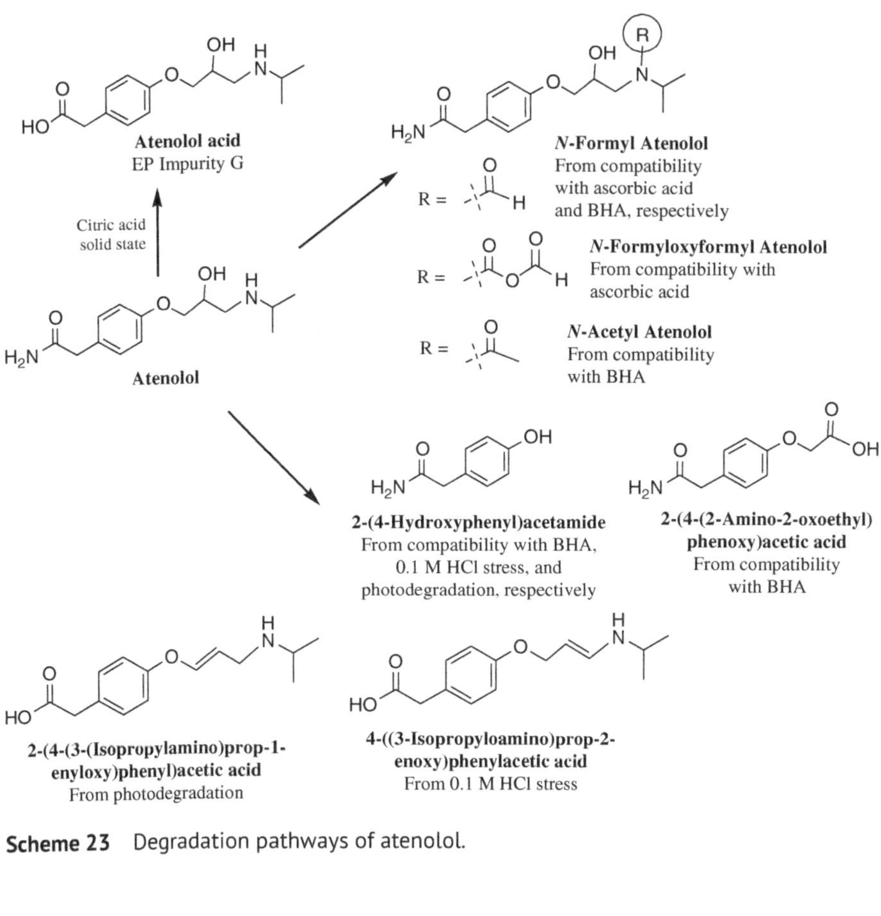

Scheme 23 Degradation pathways of atenolol.

References

1 Kumar, V., Shah, R.P., Malik, S., and Singh, S. (2009). Compatibility of atenolol with excipients: LC–MS/TOF characterization of degradation/interaction products, and mechanisms of their formation. *J. Pharm. Biomed. Anal.* 49: 880–888.

2 Shephard, A.B., Nichols, S.C., and Braithwaite, A. (1999). Moisture induced solid phase degradation of l-ascorbic acid. Part 2, separation and characterization of the major degradation products. *Talanta* 48: 595–606.

3 Krzek, J. and Kwiecien, A. (2006). Stability of atenolol, acebutolol and propranolol in acidic environment depending on its diversified polarity. *Pharm. Dev. Technol.* 11: 409–416.

4 Andrisano, V., Gotti, R., Leoni, A., and Cavrini, V. (1999). Photodegradation studies on atenolol by liquid chromatography. *J. Pharm. Biomed. Anal.* 21: 851–857.

24–25: Amitriptyline, Nortriptyline

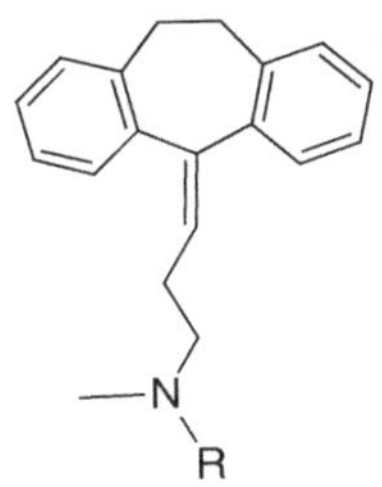

Amitriptyline R=Me;
Nortriptyline R=H.

24: Amitriptyline

Chemical name: *N,N*-Dimethyl-3-(2-tricyclo[9.4.0.0^{3,8}]pentadeca-1(15),3,5,7,11,13-hexaenylidene)propan-1-amine
Brand names: Elavil
Formula: $C_{20}H_{23}N$
Monoisotopic protonated ion (*m/z*): 278.1903
Molecular weight: 277.4041
CAS number: 50-48-6
Common salt form: HCl salt

25: Nortriptyline

Chemical name: *N*-Methyl-3-(2-tricyclo[9.4.0.0^{3,8}]pentadeca-1(15),3,5,7,11,13-hexaenylidene) propan-1-amine
Brand names: Aventyl, Pamelor
Formula: $C_{19}H_{21}N$
Monoisotopic protonated ion (*m/z*): 264.1747
Molecular weight: 263.38
CAS number: 72-69-5
Common salt form: HCl salt

Amitriptyline is a classic tricyclic antidepressant. Nortriptyline is an active metabolite of amitriptyline, which itself is also a drug. In a stability study of amitriptyline in pH 6.8 buffered solution under autoclaving conditions, three degradants, "diene," "aldehyde," and dibenzosubrone, were formed (Scheme 24–25) [1]. The formation of the "diene" can be rationalized by the Cope elimination of the *N*-oxide intermediate. The *N*-oxide, a metabolite of amitriptyline, was not detected in this particular study. It is possible that once formed, the *N*-oxide would immediately undergo the Cope elimination under the autoclaving temperature of 115–116°C. Dibenzosubrone appears to be a terminal degradant, which is also an oxidative degradant of nortriptyline [2].

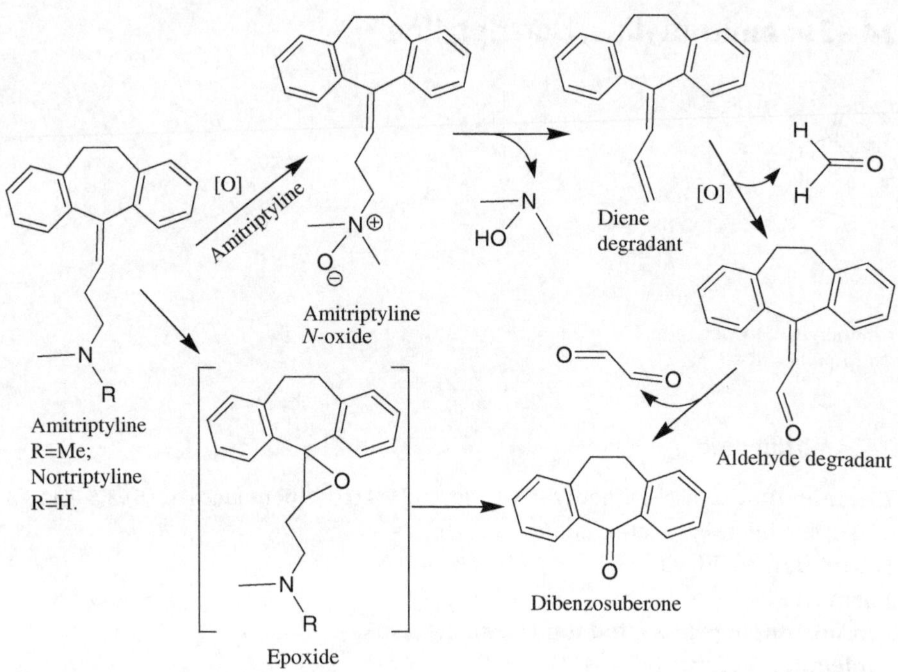

Scheme 24-25 Oxidative degradation pathways of amitriptyline.

References

1 Enever, R.P., Po, A.L.W., Millard, B.J., and Shotton, E. (1975). Decomposition of amitriptyline hydrochloride in aqueous solution: identification of decomposition products. *J. Pharm. Sci.* 64: 1497–1499.

2 Misiuk, W. and Tykocka, A. (2007). Sensitive extractive spectrophotometric methods for the determination of nortriptyline hydrochloride in pharmaceutical formulations. *Chem. Pharm. Bull. (Tokyo)* 55: 1655–1661.

26: Amlodipine

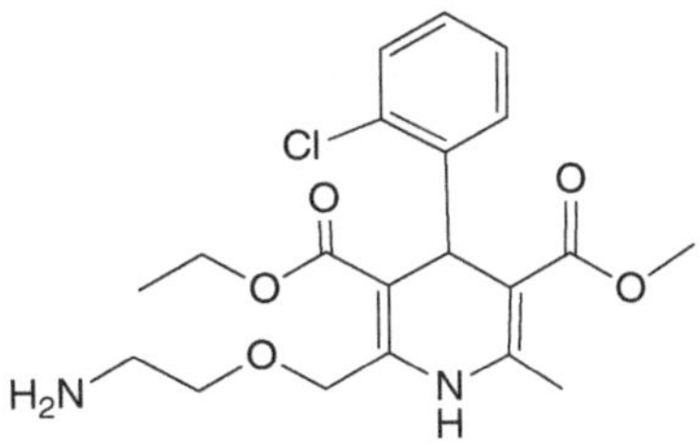

Chemical name: 3-*O*-ethyl 5-*O*-methyl 2-(2-aminoethoxymethyl)-4-(2-chlorophenyl) -6-methyl-1,4-dihydropyridine-3,5-dicarboxylate
Brand names: Norvasc (for Amlodipine Besylate)
Formula: $C_{20}H_{25}ClN_2O_5$
Monoisotopic protonated ion (*m/z*): 409.1525
Molecular weight: 408.88
CAS number: 88150-42-9
Common salt form: Besylate, maleate

Amlodipine is a calcium channel blocker used for the treatment of hypertension and angina. Amlodipine besylate, which was originally introduced by Pfizer, is one of the most widely prescribed cardiovascular drugs. Amlodipine has a 1,4-dihydropyridine core structure that is very susceptible to oxidation, leading to the formation of a more stable, aromatic pyridine ring (Scheme 26). This oxidative aromatization can take place both photochemically and non-photochemically. Photochemical aromatization of amlodipine can occur in both UV and natural light environment; in the UV region, irradiation at 366 nm was found to be more effective than at 254 nm [1]. Other drugs in the 1,4-dihydropyridine family, such as nifedipine and nisoldipine, suffer from the same photooxidative degradation pathway [1–3]. To prevent or minimize this main degradation pathway, Ragno et al. performed formulation studies with supramolecular systems such as liposomes, cyclodextrins, and microspheres and found that these matrices could significantly reduce the photochemical oxidation process through physical and chemical barriers [4].

Amlodipine also contains two ester functional groups, which are susceptible to hydrolytic stress, particularly under base stress conditions [5–7].

Lastly, amlodipine has a primary amine group that has been found to be incompatible toward excipients containing reducing sugar components due to the Maillard reaction [8, 9]. The presence of basic excipients like magnesium stearate would accelerate this degradation process due to its ability to neutralize the acid component of the drug. For amlodipine maleate, its amine moiety was found to react with maleate via Michael addition in commercial formulations [9].

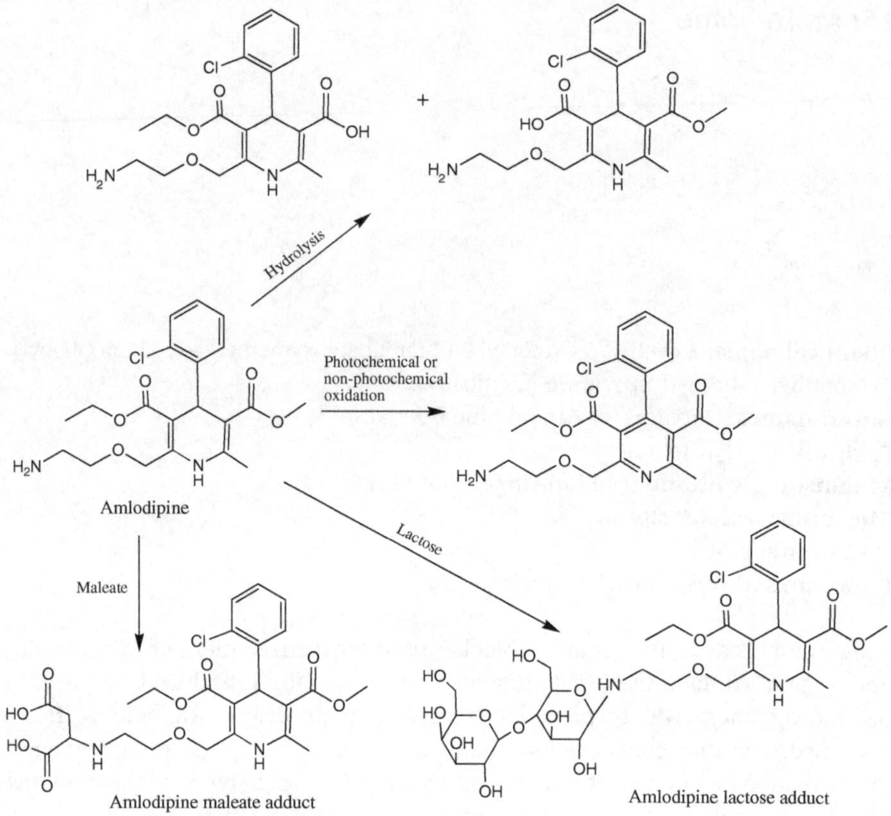

Scheme 26 Degradation pathways of amlodipine.

References

1 Fasani, E., Albini, A., and Gemme, S. (2008). Mechanism of the photochemical degradation of amlodipine. *Int. J. Pharm.* 352: 197–201.

2 Ragno, G., Garofalo, A., and Vetuschi, C. (2002). Photodegradation monitoring of amlodipine by derivative spectrophotometry. *J. Pharm. Biomed. Anal.* 27: 19–24.

3 Fasani, E., Dondi, D., Ricci, A., and Albini, A. (2006). Photochemistry of 4-(2-nitrophenyl)-1,4-dihydropyridines. Evidence for electron transfer and formation of an intermediate. *Photochem. Photobiol.* 82: 225–230.

4 Ragno, G., Cione, E., Garofalo, A. et al. (2003). Design and monitoring of photostability systems for amlodipine dosage forms. *Int. J. Pharm.* 265: 125–132.

5 Mohammadi, A., Rezanour, N., Ansari, D.M. et al. (2007). RP-LC Simultaneous determination of nebivolol hydrochloride and amlodipine besilate in Bi-Layer Tablets. *J. Chromatogr. B.* 846: 215–221.

6 Naidu, K.R., Kale, U.N., and Shingare, M.S. (2005). Stability indicating RP-HPLC method for simultaneous determination of amlodipine and benezepril hydrochloride from their combination drug product. *J. Pharm. Biomed. Anal.* 39: 147–155.

7 Chaudhari, B.G., Patel, N.M., and Shah, P.B. (2007). Stability indicating RP-HPLC method for simultaneous determination of atorvastatin and amlodipine from their combination drug products. *Chem. Pharm. Bull.* 55: 241–246.

8 Abdoh, A., Al-Omari, M.M., Badwan, A.A., and Jaber, A.M.Y. (2004). Amlodipine besylate-excipients interaction in solid dosage form. *Pharm. Dev. Technol.* 9: 15–24.

9 Murakami, T., Fukutsu, N., Kondo, J. et al. (2008). Application of liquid chromatography-two-dimensional nuclear magnetic resonance spectroscopy using pre-concentration column trapping and liquid chromatography–mass spectrometry for the identification of degradation products in stressed commercial amlodipine maleate tablets. *J. Chromatogr. A.* 1181: 67–76.

27–29: Artemisinin, Artemether, Artesunate

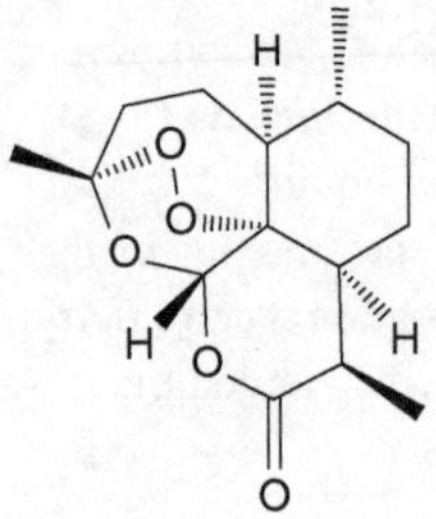

Artemisinin

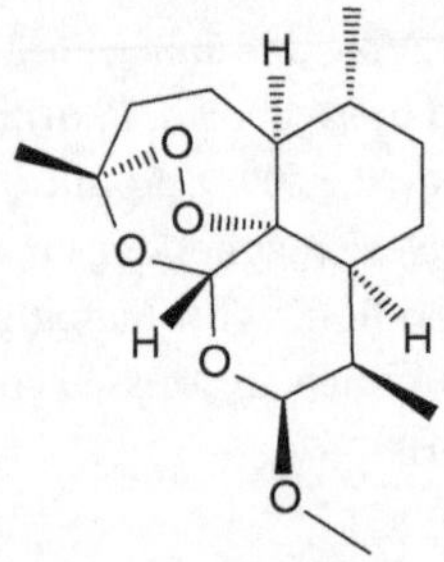

Artemether

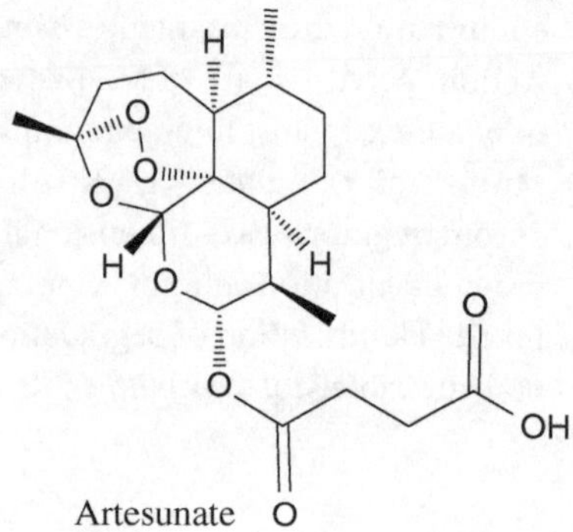

Artesunate

27: Artemisinin

Chemical name: (3*R*,5a*S*,6*R*,8a*S*,9*R*,12*S*,12a*R*)-Octahydro-3,6,9-trimethyl-3,12-epoxy-12*H*-pyrano[4,3-j]-1,2-benzodioxepin-10(3*H*)-one; (1*R*,4*S*,5*R*,8*S*,9*R*,12*S*,13*R*)-1,5,9-trimethyl-11,14,15,16-tetraoxatetracyclo[10.3.1.$0^{4,13}$.$0^{8,13}$]hexadecan-10-one
Brand names: N/A
Formula: $C_{15}H_{22}O_5$
Monoisotopic protonated ion (*m/z*): 283.1540
Molecular weight: 282.33
CAS number: 63968-64-9

28: Artemether

Chemical name: (1*R*,4*S*,5*R*,8*S*,9*R*,10*S*,12*R*,13*R*)-10-Methoxy-1,5,9-trimethyl-11,14,15,16-tetraoxatetracyclo[10.3.1.$0^{4,13}$.$0^{8,13}$]hexadecane
Brand names: Coartem (in combination with Lumefantrine)
Formula: $C_{16}H_{26}O_5$
Monoisotopic protonated ion (*m/z*): 299.1853
Molecular weight: 298.38
CAS number: 71963-77-4

29: Artesunate

Chemical name: 4-oxo-4-[[(1*R*,4*S*,5*R*,8*S*,9*R*,10*S*,12*R*,13*R*)-1,5,9-Trimethyl-11,14,15,16-tetraoxatetracyclo[10.3.1.^{04,13}.$0^{8,13}$]hexadecan-10-yl]oxy]butanoic acid
Brand names: Artesunate-Amivas
Formula: $C_{19}H_{28}O_8$
Monoisotopic protonated ion (*m/z*): 385.1857
Molecular weight: 384.42
CAS number: 88495-63-0
Common salt form: Sodium salt

Artemisinin is a potent antimalarial drug that was isolated in the 1970s by Chinese scientists from *Artemisia annua*, a plant native to China. This plant is called Qinhao in Chinese, and artemisinin is known as Qinhaosu in China, meaning the element or essence of Qinhao. Artemisinin has a unique sesquiterpine endoperoxide moiety which is essential to the activity of the drug. Upon chelation with reduced iron ion (Fe^{2+}), the endoperoxide linkage breaks down via reductive scission to form oxy free radicals. The latter can rearrange to form a more stable carbon-based radical [1]. Under in vivo conditions where the reduced iron is mainly in heme, the carbon-based radical is believed to be the agent that kills the malaria parasite through alkylation [2–5]. This mechanism of action also constitutes a major degradation pathway for artemisinin in the presence of reduced iron ion as illustrated in Scheme 27–29-1 [6]. There were a number of minor degradants whose structures were not identified.

Because of its poor bioavailability, artemisinin itself has become much less utilized clinically. Instead, its two derivatives, artemether and artesunate, are now the main therapeutic forms. In addition, due to the concern of drug resistance, WHO mandates that artemisinin derivatives be used in combination with other antimalaria drugs, such as lumefantrine, amodiaquine, and mefloquine. The water-soluble artesunate has additional degradation pathways owing to the introduction of the succinate moiety. Analysis of artesunate rectal capsules stored under long-term stability conditions revealed the presence of several minor degradants due to the succinate moiety, in addition to the proxyhemiacetal degradant, the major degradant, and 2-deoxyartemisinin (Scheme 27–29-2) [7].

In a fixed dose combination of artesunate and amodiaquine formulated as bi-layer tablets, a glycal version of degradant 2 was observed after the tablets were heated under elevated temperatures [8]. In a triple combination of artesunate, chlorproguanil, and dapsone, a degradant that resulted from the transfer of the succinate moiety from artesunate to dapsone was found in long-term stability samples (Scheme 27–29-3) [9].

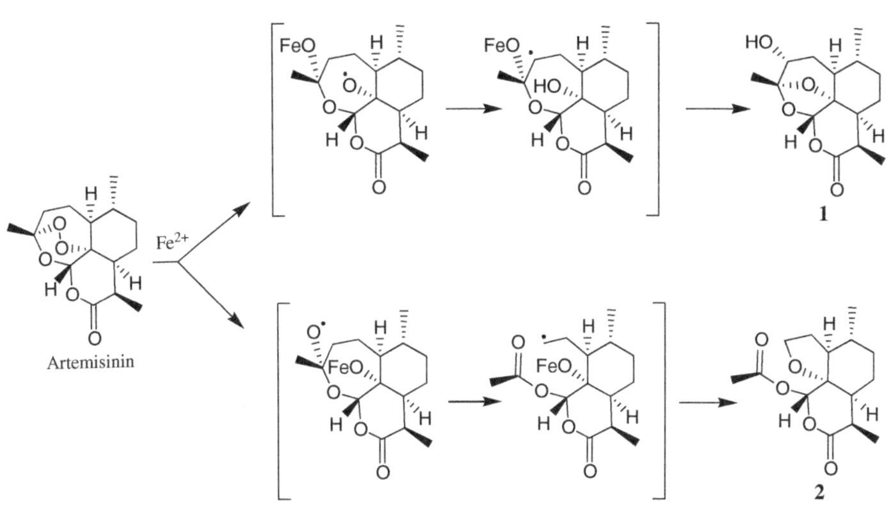

Scheme 27–29-1 Degradation of artemisinin in solutions containing spiked Fe^{2+}. A number of unidentified minor degradants were also present.

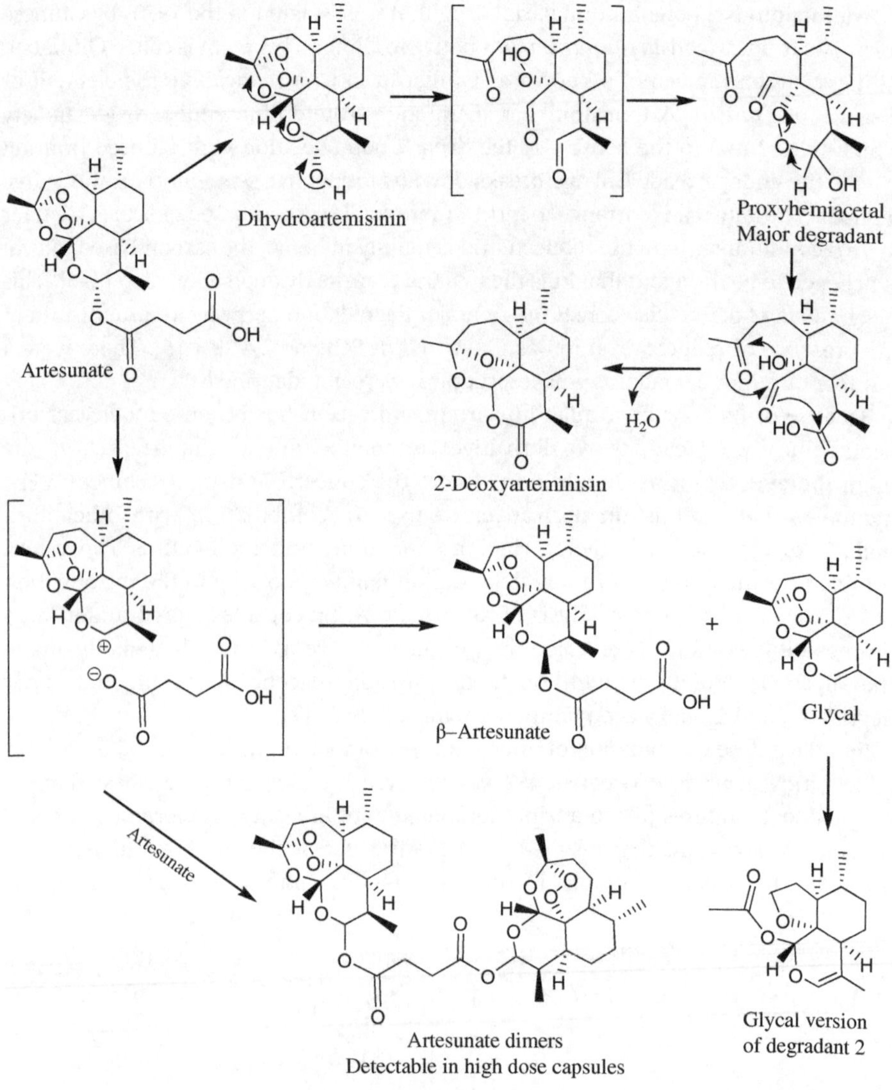

Scheme 27–29-2 Degradation of artesunate in capsules and combination tablets.

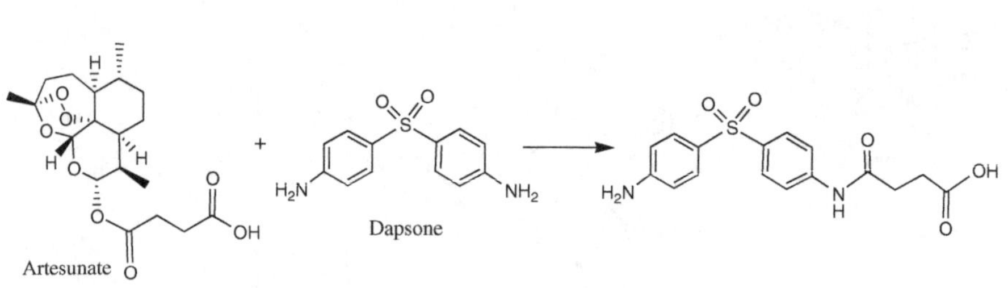

Scheme 27–29-3 Degradation of artesunate in a triple combination product containing dapsone.

References

1 O'Neill, P.M. and Posner, G.H. (2004). A medicinal chemistry perspective on artemisinin and related endo-peroxides. *J. Med. Chem.* 47: 2945–2964.

2 Robert, A., Coppel, Y., and Meunier, B. (2002). Alkylation of heme by the antimalarial drug artemisinin. *Chem. Commun.* 5: 414–415.

3 Wang, D.Y. and Wu, Y.L. (2000). A possible antimalarial action mode of qinghaosu (artemisinin) series compounds. Alkylation of reduced glutathione by C-centered primary radicals produced from anti-malarial compound qinghaosu and 12-(2,4-dime-thoxyphenyl)-12-deoxoqinghaosu. *Chem. Commun.* 22: 2193–2194.

4 Wu, W.M., Yao, Z.J., Wu, Y.L. et al. (1996). Ferrous ion induced cleavage of the peroxy bond in qinghaosu and its derivatives and the DNA damage associated with this process. *Chem. Commun.* 18: 2213–2214.

5 Asawamahasakda, W., Ittarat, I., Pu, Y. et al. (1994). Reaction of antimalarial endo-peroxides with specific parasite proteins. *Antimicrob. Agents Chemother.* 38: 1854–1858.

6 Creek, D.J., Chiu Francis, C.K., Prankerd, R.J. et al. (2005). Kinetics of iron-mediated artemisinin degradation: effect of solvent composition and iron salt. *J. Pharm. Sci.* 94: 1820–1829.

7 Haynes, R.K., Chan, H.-W., Lung, C.-M. et al. (2007). Artesunate and dihydroartemisinin (DHA): unusual decomposition products formed under mild conditions and comments on the fitness of DHA as an antimalarial drug. *ChemMedChem.* 2: 1448–1463.

8 Charrier, C., Bertho, G., Petigny, O. et al. (2013). A new derivative detected in accelerated ageing of artesunate-amodiaquine fixed dose combination tablets. *J. Pharm. Biomed. Anal.* 81–82: 20–26.

9 Bardsley, B., Barry, S.J., Drozdz, M.A. et al. (2011). Formation and identification of a degradant in chlorproguanil–dapsone–artesunate (Dacart™) tablets. *J. Pharm. Biomed. Anal.* 54: 610–613.

30: Ascorbic Acid [Vitamin C]

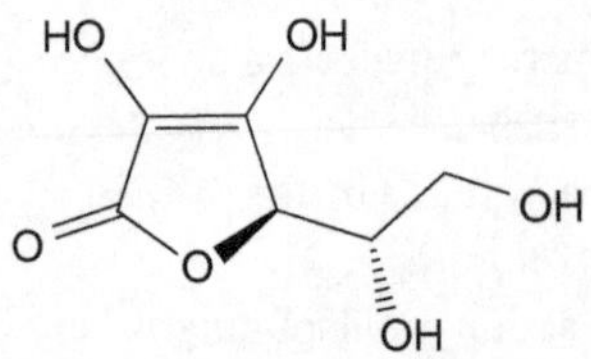

Chemical name: L-*threo*-Hex-2-enono-1,4-lactone; (*R*)-3,4-Dihydroxy-5-((*S*)-1,2-dihydroxyethyl)furan-2(5*H*)-one
Brand names: Ascor and many others
Formula: $C_6H_8O_6$
Monoisotopic protonated ion (*m/z*): 177.0394
Molecular weight: 176.12
CAS number: 50-81-7

Ascorbic acid, better known as vitamin C, is a water-soluble essential vitamin. Clinically, it is used to treat vitamin C deficiency and help iron absorption. More commonly it is marketed as a general antioxidant supplement. Furthermore, ascorbic acid is also employed as an antioxidative excipient in the formulation of certain drugs that are labile to oxidative degradation [1].

Ascorbic acid (AH_2) displays different degradation pathways in solution and solid states [2, 3]. In solution, its main oxidative degradant is dehydroascorbic acid (DHA) upon exposure to light, transition metal ions (especially cupric and ferric ions), and alkaline pH (Scheme 30-1) [2, 4, 5]. According to a recent paper by Njus et al. [6], the real structure of dehydroascorbic acid is not the commonly cited tricarbonyl structure (A), but rather a bicyclic hemiketal structure (AH_2O). For

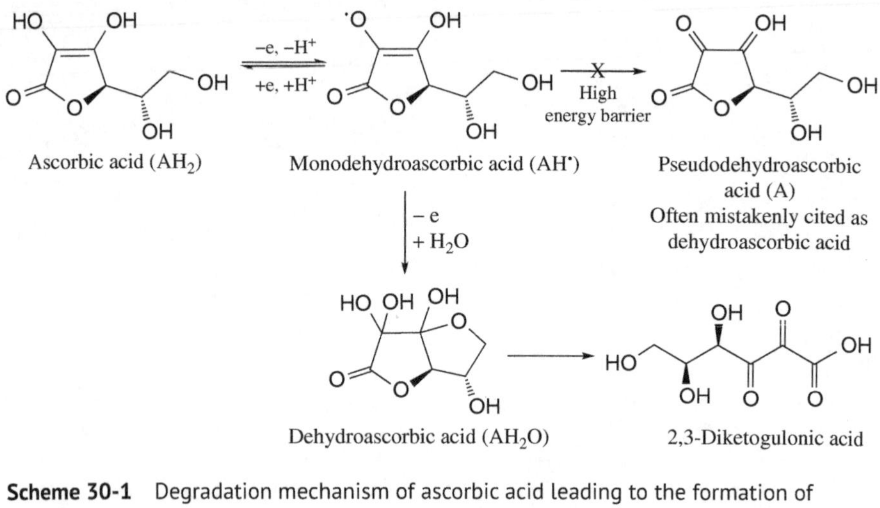

Scheme 30-1 Degradation mechanism of ascorbic acid leading to the formation of 2,3-diketogulonic acid.

this reason, Njus et al. suggested that the tricarbonyl structure be called pseudo-dehydroascorbic acid. Ascorbic acid exerts its antioxidative function by affording an equivalent of H (or $H^+ + e^-$) to form monodehydroascorbic acid, which is a free radical and can react with other radical species, hence becoming a terminator of free radical chain reactions. In biological systems, ascorbic acid cycles back and forth with monodehydroascorbic acid. In chemical degradation, monodehydroascorbic acid can further oxidize to dehydroascorbic acid, which can hydrolyze to produce 2,3-diketogulonic acid [4].

Other than the aforementioned degradation pathways, ascorbic acid can also yield furan degradants [7] including hydroxymethylfurfural (HMF), which is a common degradant for several carbohydrates containing six carbons. In a study of ascorbic acid stability in citrus juice concentrates, the accumulation of HMF was highly correlated with the degradation of ascorbic acid [8]. In a study of examining ascorbic acid's decomposition into furan, a Class 2B carcinogen by IARC classification, using model systems to mimic food cooking [7], a number of known ascorbic acid degradants were proposed as the intermediates leading to the formation of furan. These intermediates are 4-deoxyascorbic acid, 3-dexoypentosulose, 3,4-dideoxypentos-3-en-ulose, 2-furfural, 2-furoic acid, and 2-deoxyaldotetrose (Scheme 30-2). During the decomposition process, oxalic acid, formic acid, and carbon dioxide were also formed as elimination products. In a solid-state degradation study of ascorbic acid, carbon dioxide was found to be a degradant formed in 1:1 stoichiometric quantity from ascorbic acid [3].

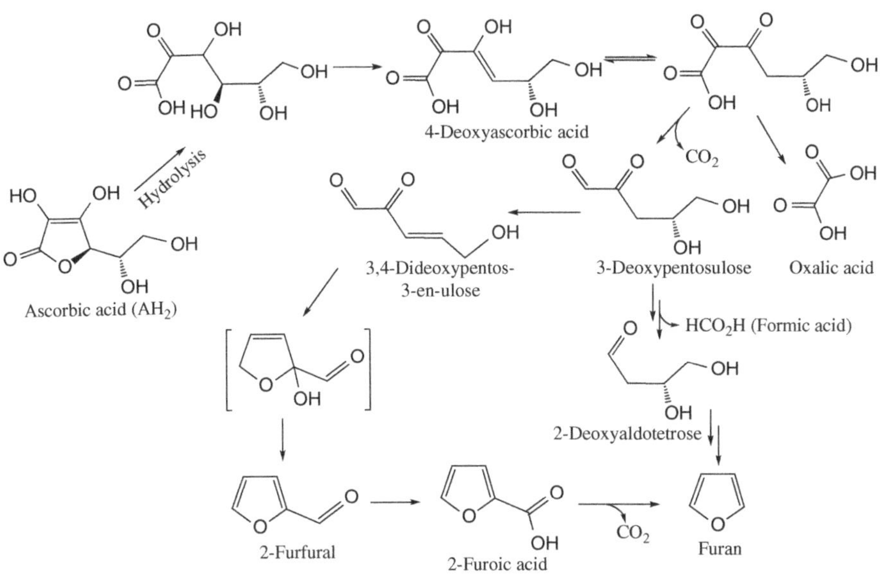

Scheme 30-2 Degradation of ascorbic acid leading to various degradants.

References

1 Kroselj V., Lakse R., Toporisic R., Kastelic J. Pharmaceutical composition comprising simvastatin and ezetimibe. US Patent 8,921,352 B2. filed 30 June 2006. granted 30 December 2014.

2 Shen, J., Griffiths, P.T., Campbell, S.J. et al. (2021). Ascorbate oxidation by iron, copper and reactive oxygen species: review, model development, and derivation of key rate constants. *Sci. Rep.* 11: 7417. https://doi.org/10.1038/s41598-021-86477-8.

3 Shephard, A.B., Nichols, S.C., and Braithwaite, A. (1999). Moisture induced solid phase degradation of l-ascorbic acid part 3, structural characterisation of the degradation products. *Talanta* 48: 607–622.

4 Yin, X., Chen, K., Cheng, H. et al. (2022). Chemical stability of ascorbic acid integrated into commercial products: a review on bioactivity and delivery technology. *Antioxidants* 11: 153. https://doi.org/10.3390/antiox11010153.

5 Jansson, P.J., Jung, H.R., Lindqvist, C., and Nordström, T. (2004). Oxidative decomposition of vitamin C in drinking water. *Free Radic. Res.* 38: 855–860. https://doi.org/10.1080/10.15760410001700497.

6 Njus, D., Kelley, P.M., Tu, Y.-J., and Schlegel, H.B. (2020). Ascorbic acid: the chemistry underlying its antioxidant properties. *Free Radic. Biol. Med.* 159: 37–43. https://doi.org/10.1016/j.freeradbiomed.2020.07.013.

7 Limacher, A., Kerler, J., Conde-Petit, B., and Blank, I. (2007). Formation of furan and methylfuran from ascorbicacid in model systems and food. *Food Addit. Contam.* 24 (Suppl 1): 122–135. https://doi.org/10.1080/02652030701393112.

8 Burdurlu, H.S., Koca, N., and Karadeniz, F. (2006). Degradation of vitamin C in citrus juice concentrates during storage. *J. Food Eng.* 74: 211–216.

31: Aspirin

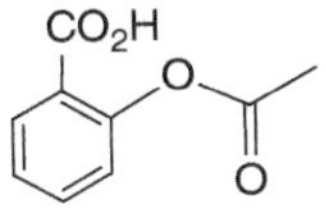

Chemical name: Acetylsalicylic acid; 2-Acetyloxybenzoic acid
Brand name: Aspirin
Formula: $C_9H_8O_4$
Monoisotopic protonated ion (*m/z*): 181.0495
Molecular weight: 180.16
CAS number: 50-78-2

Aspirin, one of the oldest synthetic drugs, is chemically known as acetylsalicylic acid. It remains as one of the most widely used medications. Its ester linkage is quite susceptible to hydrolytic degradation (Scheme 31) [1]. Since aspirin is also co-formulated with other drugs, its acetyl group can transfer to those drugs that contain nucleophilic groups, particularly the amino group. For example, in a tablet formulation of aspirin with phenylephrine, a secondary amine drug molecule, the main degradant was *N*-acetyl phenylephrine, resulting from trans-acetylation of aspirin onto the secondary amino group of phenylephrine [2]. Further degradation produced minor degradants containing two or even three acetyl groups. In the latter case, since the transfer of the second and third acetyl group results in the formation of ester bonds, it can also be considered a type of degradation via trans-esterification. Another case of degradation via trans-esterification was reported in a co-formulation study of aspirin with codeine; acetyl codeine was found to be a detectable degradant even at a low moisture level [3].

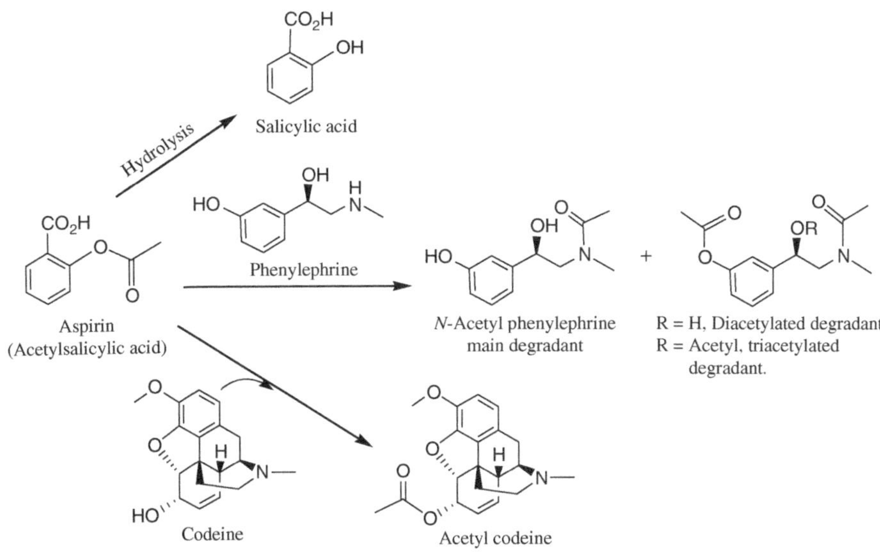

Scheme 31 Degradation pathways of aspirin.

References

1 Garrett, E.R. (1957). The kinetics of solvolysis of acyl esters of salicyclic acid. *J. Am. Chem. Soc.* 79: 3401–3408.

2 Troup, A.E. and Mitchner, H. (1964). Degradation of phenylephrine hydrochloride in tablet formulations containing aspirin. *J. Pharm. Sci.* 53: 375–379.

3 Galante, R.N., Visalli, A.J., and Patel, D.M. (1979). Solid-state acetylation of codeine phosphate by aspirin. *J. Pharm. Sci.* 68: 1494–1498.

32: Atorvastatin

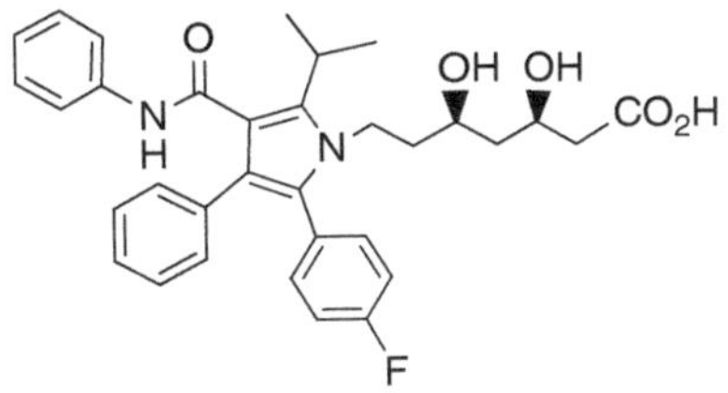

Chemical name: (3*R*,5*R*)-7-[2-(4-Flluorophenyl)-3-phenyl-4-(phenylcarbamoyl)-5-propan-2-ylpyrrol-1-yl]-3,5-dihydroxyheptanoic acid
Brand names: Lipitor
Formula: $C_{33}H_{35}FN_2O_5$
Monoisotopic protonated ion (*m/z*): 559.2603
Molecular weight: 558.64
CAS number: 134523-00-5
Common salt form: Calcium salt

Atorvastatin, a top-selling cholesterol-lowering drug, has a fully substituted pyrrole core with a characteristic side chain of 3,5-dihydroxyheptanoic acid. It is usually used as a calcium salt. The most significant degradation pathway of atorvastatin is the cyclization of the 5-hydroxyl group with the carboxyl group of the heptanoic acid side chain to form a six-membered lactone [1]. Oxidative and photodegradation studies showed that the pyrrole ring is susceptible to oxidative degradation, resulting in several degradants [2]. On the other hand, amorphous atorvastatin calcium is much more oxidizable than the crystalline form. In a patent application, antioxidants were added into amorphous atorvastatin calcium to stabilize the drug substance [3]. In various experimental tablet formulations, a total of five oxidative degradants (I to V, Scheme 32) were found. In another patent application, it was found that amorphous atorvastatin calcium might be stabilized by using appropriate antioxidants [3].

In a mechanistic study with regard to the reported phototoxicity of atorvastatin, Montanaro et al. found that it is atorvastatin phenanthrene, rather than atorvastatin itself, that photosensitizes triplet oxygen to produce singlet oxygen [4]. The latter then oxidizes the pyrrole ring via perepoxide [4] and/or endoepoxide [5] intermediates to yield a number of oxidative degradants (Scheme 32). Some of these degradants are not very stable, and their occurrence may vary depending on the conditions of oxidation. The oxidation of the pyrrole ring can also occur non-photochemically. In an oxidative degradation study of atorvastatin, an oxidative degradant isomeric to IIa was generated by oxidation of atorvastatin with alkaline hydrogen peroxide [5].

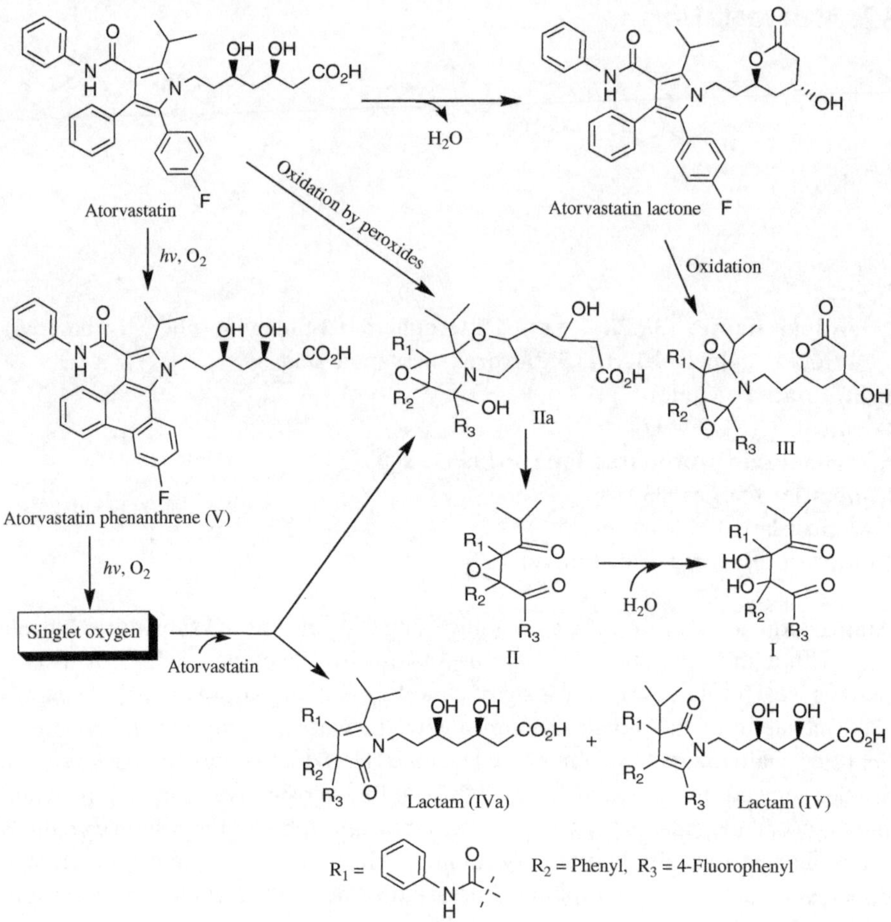

Scheme 32 Degradation pathways of atorvastatin.

References

1 Stach, J., Havlíček, J., Plaček, L., and Rádl, S. (2008). Synthesis of some impurities and/or degradation products of atorvastatin. *Collect. Czechoslov. Chem. Commun.* 73: 229–246.

2 Kracun, M., Kocijan, A., Bastarda, A. et al. (2009). Isolation and structure determination of oxidative degradation products of atorvastatin. *J. Pharm. Biomed. Anal.* 50: 729–736.

3 Kumar Y., Kumar S.M.D., and Sathyanarayana S. (2008). Process for the production of atorvastatin calcium in amorphous form. US patent application 20090216029, filed 21 February 2008.

4 Montanaro, S., Lhiaubet-Vallet, V., Iesce, M. et al. (2009). A mechanistic study on the phototoxicity of atorvastatin: singlet oxygen generation by a phenanthrene-like photoproduct. *Chem. Res. Toxicol.* 22: 173–178.

5 Cermola, F., DellaGreca, M., Iesce, M.R. et al. (2006). Photochemical behavior of the drug atorvastatin in water. *Tetrahedron* 62: 7390–7395.

33: Atracurium

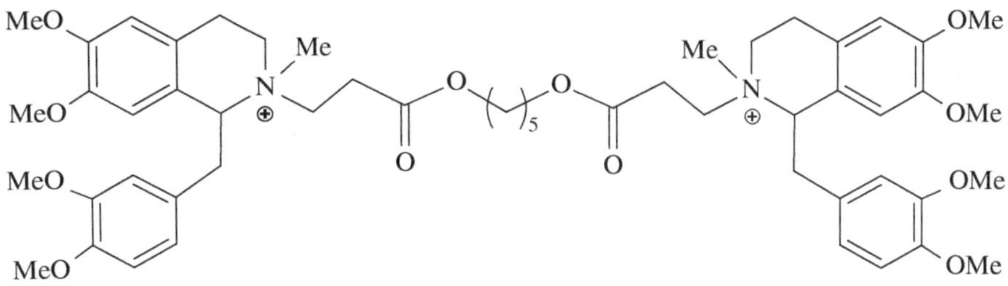

Chemical name: 5-[3-[1-[(3,4-Dimethoxyphenyl)methyl]-6,7-dimethoxy-2-methyl-3,4-dihydro-1*H*-isoquinolin-2-ium-2-yl]propanoyloxy]pentyl-3-[1-[(3,4-dimethoxy phenyl)methyl]-6,7-dimethoxy-2-methyl-3,4-dihydro-1*H*-isoquinolin-2-ium-2-yl] propanoate
Brand names: N/A
Formula: $C_{53}H_{72}N_2O_{12}{}^{+2}$
Monoisotopic cation (*m/z*): 464.2537 (doubly charged ion)
Molecular weight: 1243.5 (as Atracurium Besylate)
CAS number: 64228-79-1
Common salt form: Besylate

Atracurium is a bis-quaternary ion, which is in the form of besylate salt. It has multiple chiral centers, and thus is a mixture of stereoisomers. Cisatracurium is a purified stereoisomer, which accounts for ~15% of the total stereoisomers [1]. Both atracurium and cisatracurium are clinically widely used skeletal muscle relaxants. This molecule is a prodrug made of two identical quaternary salt units that are linked together by a diester spacer. The active drug, laudanosine, is released in vivo from the quaternary salt unit via Hofmann elimination (Scheme 33). Nevertheless, this mechanism of action also constitutes an in vitro degradation pathway of the drug [2]. Since the drug molecule contains two ester bonds, it is also susceptible to hydrolytic degradation. A study in phosphate buffer at the physiological pH of 7.4 indicated that the Hofmann elimination pathway is the rate-determining step in the breakdown of atracurium [3].

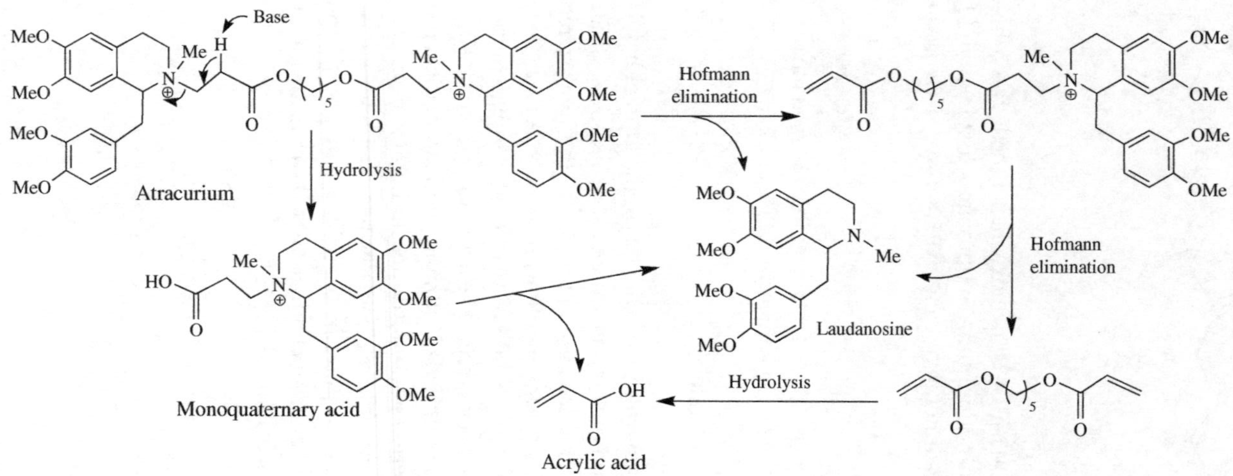

Scheme 33 Degradation pathways of atracurium.

References

1 Mellinghoff, H., Radbruch, L., Diefenbach, C., and Buzello, W. (1996). A Comparison of cisatracurium and atracurium. *Anesth. Analg.* 83: 1072–1075.

2 Welch, R.M., Brown, A., Ravitch, J., and Dahl, R. (1995). The in vitro degradation of cisatracurium, the R, cis-R′-isomer of atracurium, in human and rat plasma. *Clin. Pharmacol. Ther.* 58: 132–142.

3 Weindlmayr-Goettel, M., Gilly, H., and Kress, H.G. (2002). Does ester hydrolysis change the in vitro degradation rate of cisatracurium and atracurium? *Br. J. Anaesth.* 88: 555–562.

34–35: Atropine, Scopolamine

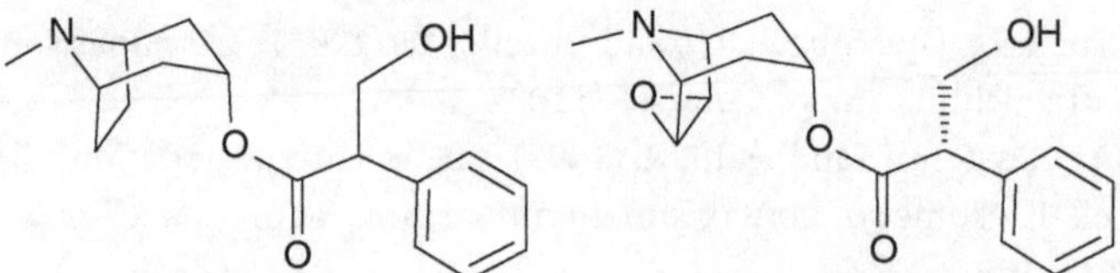

Atropine   Scopolamine

34: Atropine

Chemical: D,L-Hyoscyamine; [(1*S*,5*R*)-8-methyl-8-azabicyclo[3.2.1]octan-3-yl] 3-hydroxy-2-phenylpropanoate
Brand names: Atnaa, Atropen, Busulfex
Formula: $C_{17}H_{23}NO_3$
Monoisotopic protonated ion (*m/z*): 290.1751
Molecular weight: 289.37
CAS number: 51-55-8
Common salt form: Sulfate

35: Scopolamine

Chemical name: Hyoscine; [(1*R*,2*R*,4*S*,5*S*)-9-methyl-3-oxa-9-azatricyclo[3.3.1.0^{2,4}] nonan-7-yl]-(2*S*)-3-hydroxy-2-phenylpropanoate
Brand names: Donnatal, Phenohytro, Transderm Scop
Formula: $C_{17}H_{21}NO_4$
Monoisotopic protonated ion (*m/z*): 304.1543
Molecular weight: 303.35
CAS number: 114-49-8
Common salt form: HBr salt

Atropine is a tropane alkaloid that is present in a number of plants of the family *Solanaceae*. Some of these plants have been used medicinally for more than two millennia. Atropine has an ester linkage and, as such, one of its main degradation pathways is hydrolysis, producing tropic acid and tropine (3-tropanol, Scheme 34–35) [1]. Its tropic acid moiety also contains a hydroxyl group that is β to both the carbonyl and phenyl groups. This structural feature renders atropine susceptible to dehydration, yielding apoatropine [2]. The latter can be hydrolyzed to form atropic acid (also known as 2-phenylacrylic acid). Alternately, atropic acid can be formed from the dehydration of tropic acid as well. Hence, the two degradation pathways, i.e., hydrolysis and dehydration (a form of elimination) produced these four main degradants, tropic acid, apoatropine, atropic acid, and tropine, although tropine is not detectable by a regular HPLC UV detector [3]. On the other hand, a GC method with pentafluoropropionic anhydride as a derivatizing reagent can detect tropine [4]. Lund and Waaler performed a thorough kinetic study of atropine degradation in a pH range of

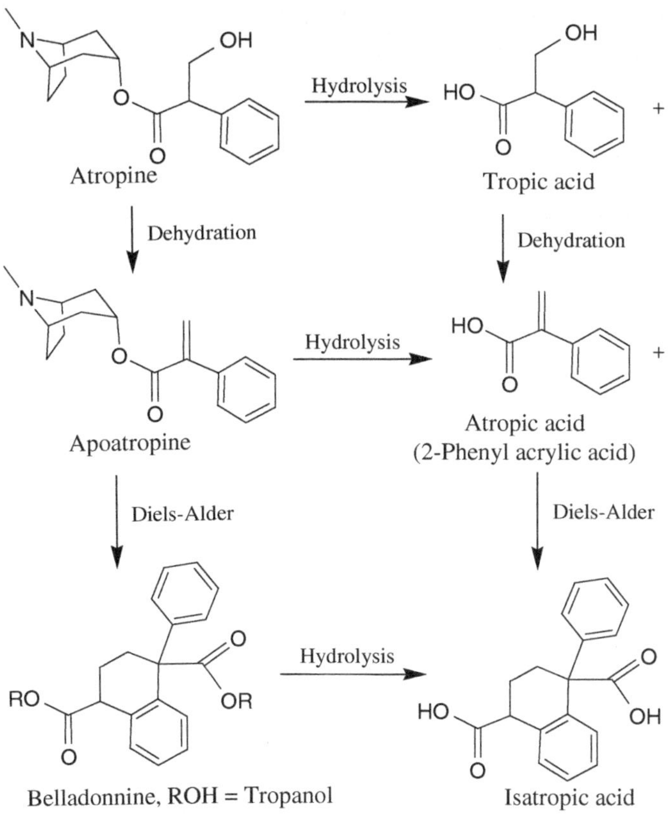

Scheme 34-35 Degradation pathways of atropine.

~1–6 [5]. Atropine displayed a U-shape pH stability curve and at pH 3, it appeared most stable as the overall rate of degradation is the smallest. And at this pH, the rate of dehydration was approximately the same as that of hydrolysis.

The degradant resulting from dehydration, namely apoatropine, could also react with atropine via Diels-Alder reaction to produce belladonnine and then isotropic acid upon further degradation by hydrolysis [2]. These secondary and tertiary degradants would most likely be minor ones. Atropine *N*-oxide is a metabolite of the drug; however, no report has been found to suggest that it would be a meaningful degradant in pharmaceutical preparations.

Scopolamine is another naturally occurring tropane alkaloid that is highly similar to atropine. As such, it may likely have similar degradation pathways.

References

1 Zvirblis, P., Socholitsky, I., and Kondritzer, A.A. (1956). The kinetics of the hydrolysis of atropine. *J. Am. Pharm. Assoc. Sci. Ed.* 45: 450–454.

2 Voigtlander, H.-W. and Rosenberg, W. (1959). The constitution of isatropasic esters belladonnin and scopadonnin (Translation from German). *Arch. Pharm.* 292: 632–642.

3 Zimmermann, T., Dimmel, A., Jüttemeyer, S. et al. (2012). Rapid resolution liquid chromatography for monitoring the quality of stockpiled atropine preparations for injection. *Drug Test. Anal.* 4: 222–228.

4 Schier, J.G., Ravikumar, P.R., Nelson, L.S. et al. (2004). Preparing for chemical terrorism: stability of injectable atropine sulfate. *Acad. Emerg. Med.* 11: 329–334.

5 Lund, W. and Waaler, T. (1968). The kinetics of atropine and apoatropine in aqueous solutions. *Acta Chem. Scand.* 22: 3085–3097.

36: Azathioprine

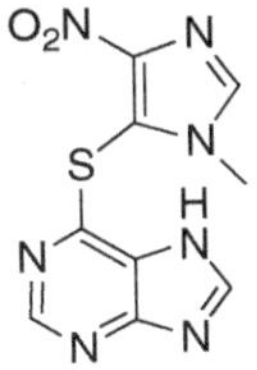

Chemical name: 6-(3-Methyl-5-nitroimidazol-4-yl)sulfanyl-7*H*-purine
Brand names: Azasan, Imuran
Formula: $C_9H_7N_7O_2S$
Monoisotopic protonated ion (*m/z*): 278.0455
Molecular weight: 277.26
CAS number: 446-86-6
Common salt form: Sodium salt

Azathioprine was originally developed as an anticancer drug but it is now mainly used against organ rejection after transplant. It is a prodrug of 6-mercaptopurine (thiopurine). Its main degradation pathway is hydrolysis, and its pH stability profile displayed a U-shape, with it being most stable at ~pH 6 [1]. According to a study by Singh and Gupta [2], azathioprine decomposed to 6-mercaptopurine and 1-methyl-4-nitro-5-hydroxyimidazole at pH greater than 7 (Scheme 36). In acidic pH, it degraded to hypoxanthine and 1-methyl-4-nitro-5-mercaptoimidazole. In either case, the hydrolysis seemed to proceed via a direct nucleophilic attack by either hydroxide ion or water molecule [1].

It was reported that long-term use of azathioprine may cause skin cancer, implying that the drug is photolabile. Photolysis of an azathioprine solution yielded 6-mercaptopurine, 1-methyl-4-nitro-5-hydroxyimidazole, and a cyclized photodegradant [3]. Photolysis of 6-mercaptopurine and its photodegradants, purine-6-sulfinic acid and purine-6-sulfonic acid caused the formation of a variety of free radicals that may be responsible for the phototoxicity of the drug [4].

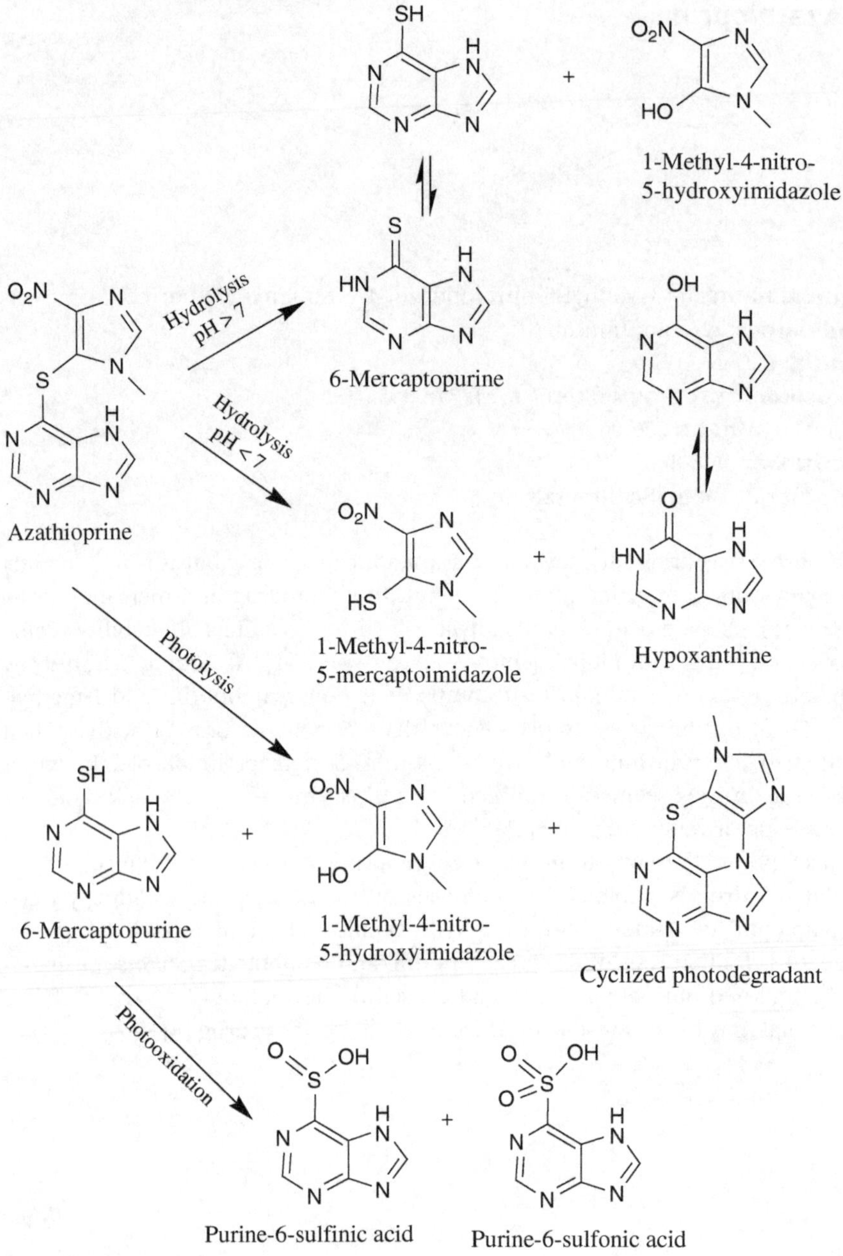

Scheme 36 Degradation pathways of azathioprine.

References

1 Mitra, A.K. and Narurkar, M.M. (1986). Kinetics of azathioprine degradation in aqueous solution. *Int. J. Pharm.* 35: 165–171.

2 Singh, S. and Gupta, R.L. (1988). A critical study on degradation of azathioprine in aqueous solutions. *Int. J. Pharm.* 42: 263–266.

3 Hemmens, V.J. and Moore, D.E. (1986). Photochemical sensitization by azathioprine and its metabolites – II. Azathioprine and nitroimidazole metabolites. *Photochem. Photobiol.* 43: 257–262.

4 Moore, D.E., Sik, R.H., Bilski, P. et al. (1994). Photochemical sensitization by azathioprine and its metabolites. Part 3. A direct EPR and spin-trapping study of light-induced free radicals from 6-mercaptopurine and its oxidation products. *Photochem. Photobiol.* 60: 574–581.

37: Azithromycin

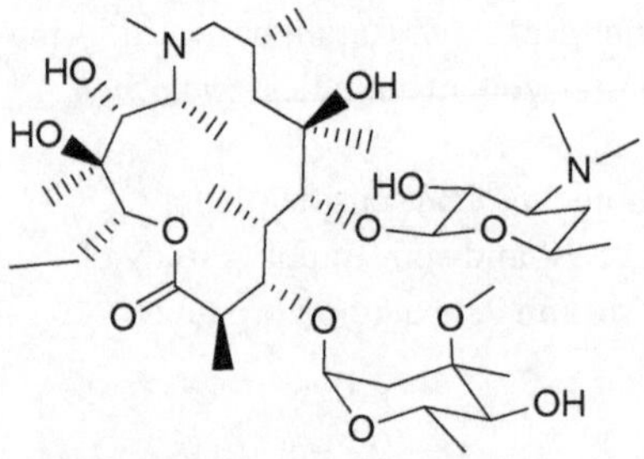

Chemical name: (*2R,3S,4R,5R,8R,10R,11R,12S,13S,14R*)-11-[(*2S,3R,4S,6R*)-4-(Dimethylamino)-3-hydroxy-6-methyloxan-2-yl]oxy-2-ethyl-3,4,10-trihydroxy-13-[(*2R,4R,5S,6S*)-5-hydroxy-4-methoxy-4,6-dimethyloxan-2-yl]oxy-3,5,6,8,10,12,14-heptamethyl-1-oxa-6-azacyclopentadecan-15-one
Brand names: Azasite, Zithromax, Zmax
Formula: $C_{38}H_{72}N_2O_{12}$
Monoisotopic protonated ion (*m/z*): 749.5158
Molecular weight: 748.99
CAS number: 83905-01-5

Azithromycin is an antibiotic that is structurally modified based on erythromycin. The 9-keto functionality of erythromycin is replaced by a methyl-substituted nitrogen moiety in azithromycin. This modification makes azithromycin no longer undergo dehydration as opposed to erythromycin. The main degradation pathway of azithromycin, nevertheless, is the hydrolysis of the glycosidic bond that connects to the cladinose sugar unit at the 3-position of the macrocyclic ring (Scheme 37). The pH stability profile of azithromycin showed a V shape with the most stable pH at 6.3. At this pH, the activation energy for the ether hydrolysis was found to be 96.8 kJ/mol (23.1 kcal/mol), and the degradation product is desosaminylazithromycin (des-cladinose azithromycin, DCA) [1]. On the other hand, azithromycin quickly decomposed in pH 1.5 boiled human gastric fluid with a half-life of ~25 minutes [2].[3]

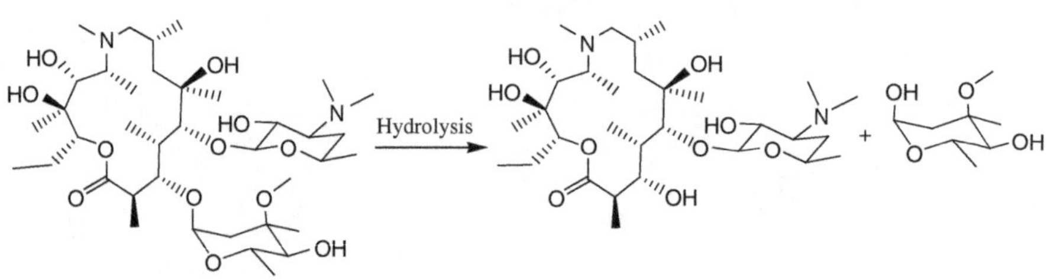

Scheme 37 Hydrolytic degradation pathway of azithromycin.

3 The author of reference 2 did not specify the temperature of the gastric fluid study but it appears the temperature should be 37 °C.

Nonetheless, the overall stability of azithromycin in acidic media is greatly improved as compared to erythromycin. For example, the time for 10% decomposition ($T_{1/10}$) for azithromycin at pH 2 was found to be 20.1 minutes, which compares favorably with erythromycin that had a $T_{1/10}$ of only 3.7 seconds at the same pH [3].

Azithromycin does not have a chromophore that absorbs significantly in UV A and UV B regions. Therefore, it would not be expected to be photochemically active. There are some studies on azithromycin and analogous drugs regarding their photochemical degradation in the environment [4, 5]. However, the relevance of the studies with respect to the photostability of these drugs in pharmaceutical dosage forms is unknown.

References

1 Zhang, Y., Liu, X., Cui, Y. et al. (2009). Aspects of degradation kinetics of azithromycin in aqueous solution. *Chromatographia* 70: 67–73.

2 Curatolo, W. (2011). Interdisciplinary science and the design of a single-dose antibiotic therapy. *Pharm. Res.* 28: 2059–2071.

3 Fiese, E.F. and Steffen, S.H. (1990). Comparison of the acidic stability of azithromycin and erythromycin A. *J. Antimicrob. Chemother.* 25: 39–47.

4 Tong, L., Eichhorn, P., Pérez, S. et al. (2011). Photodegradation of azithromycin in various aqueous systems under simulated and natural solar radiation: kinetics and identification of photoproducts. *Chemosphere* 83: 340–348.

5 Vione, D., Feitosa-Felizzola, J., Minero, C., and Chiron, S. (2009). Phototransformation of selected human-used macrolides in surface water: kinetics, model predictions and degradation pathways. *Water Res.* 43: 1959–1967.

3

Baclofen to Butalbital and Related Barbiturates

38: Baclofen

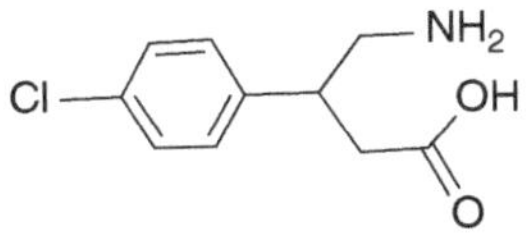

Chemical name: (*R,S*)-4-Amino-3-(4-chlorophenyl)butanoic acid
Brand name: Lioresal
Formula: $C_{10}H_{12}ClNO_2$
Monoisotopic protonated ion (*m/z*): 214.0629
Molecular weight: 213.66
CAS number: 1134-47-0

Baclofen is a derivative of γ-aminobutyric acid (GABA). It is a GABA receptor antagonist and used as a treatment for spasticity. As a γ-amino acid, it is prone to intramolecular dehydration, leading to the formation of a five-membered ring lactam, 4-(4-chlorophenyl)-2-pyrrolidone (CPP), which is its main degradation product (Scheme 38) [1, 2]. In contrast, as it contains a primary amino group, it may also undergo the Maillard reaction. A study between baclofen and lactose indeed showed that early-stage Maillard reaction products were formed, the major one of which was proposed as the Schiff base form of the adduct based on IR analysis of a solid mixture of baclofen and lactose [3]. However, in the solution of baclofen and lactose from which the LC-MS analysis was performed, the glucosamine form or even the Amadori rearrangement product may be possible; all three forms have an identical molecular formula [4].

Compendium of Drug Degradation Pathways, First Edition. Min Li.

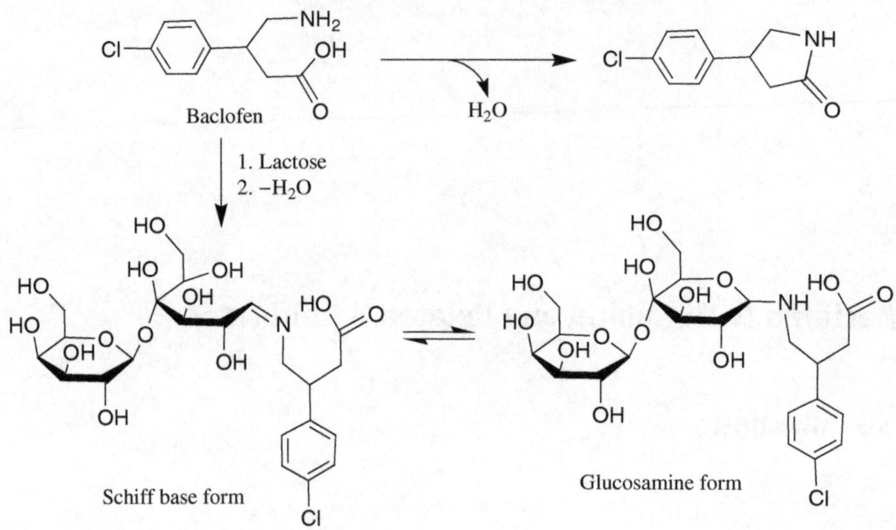

Scheme 38 Degradation of baclofen.

References

1 Ahuja, S. (1985). Baclofen. In: *Analytical Profiles of Drug Substances*, vol. 14 (ed. K. Florey), 527–548. New York, NY: Academic Press Inc.

2 Cutrignelli, A., Denora, N., Lopedota, A. et al. (2007). Comparative effects of some hydrophilic excipients on the rate of gabapentin and baclofen lactamization in lyophilized formulations. *Int. J. Pharm.* 332: 98–106.

3 Monajjemzadeh, F., Hassanzadeh, D., Valizadeh, H. et al. (2009). Assessment of feasibility of Maillard Reaction between baclofen and lactose by liquid chromatography and tandem mass spectrometry, application to pre formulation studies. *AAPS PharmSciTech* 10: 649–659.

4 Li, M. (2012). Drug–excipient interactions and adduct formation. In: *Organic Chemistry of Drug Degradation*, 150–164. Cambridge: RSC Publishing.

39: Benznidazole [WHO List of Essential Medicines]

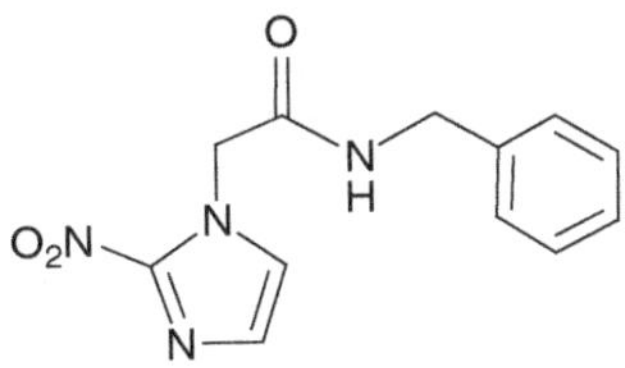

Chemical name: *N*-Benzyl-2-(2-nitro-1*H*-imidazol-1-yl)acetamide
Brand name: Rochagan
Formula: $C_{12}H_{12}N_4O_3$
Monoisotopic protonated ion (*m/z*): 261.0982
Molecular weight: 260.2530
CAS number: 22994-85-0

Benznidazole is used for the treatment of Chagas disease, a tropical disease that is caused by the parasite *Trypanosoma cruzi*. According to the study by Kawaguchi et al., the main degradation pathway of the drug is the substitution of the nitro group by either hydroxide under alkaline conditions or chloride under acidic conditions (in 2 M HCl at 60°C for 96 hours), rather than the hydrolysis of the amide linkage (Scheme 39) [1].

The hydroxyl-substituted degradant was present in "enol and ketone" forms, based on the LC-MS analysis that showed two isomeric peaks [1]. It should be noted that the chloro-substituted degradant is certainly an artificial degradant under the forced degradation conditions. On the other hand, the drug appeared to be stable under other stress conditions: oxidative (3% H_2O_2 at room temperature for 48 hours), photostress (in solution), and heat and moisture (70°C, 65% RH, 60 days) conditions.

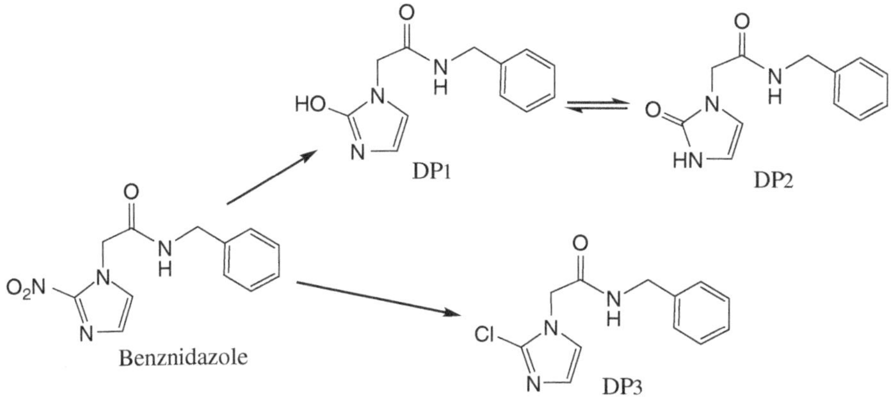

Scheme 39 Degradation of benznidazole under acidic stress conditions.

Reference

1 Kawaguchi, W.H., Fachi, M.M., Cerqueira, L.B. et al. (2020). Stability indicating method for determination of benznidazole and its degradation products in active pharmaceutical ingredient. *J. Braz. Chem. Soc.* 31: 1194–1202.

40: Benzocaine

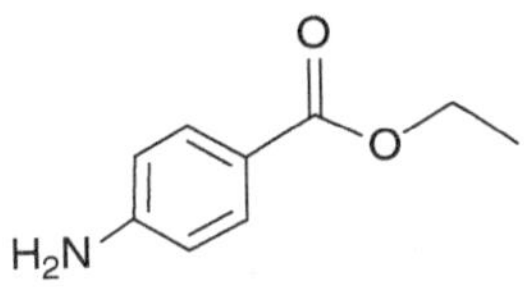

Chemical name: Ethyl 4-aminobenzoate
Brand name: Anbesol, Lanacane, Orajel
Formula: $C_9H_{11}NO_2$
Monoisotopic protonated ion (*m/z*): 166.0863
Molecular weight: 165.19
CAS number: 94-09-7
Common salt form: HCl salt

Benzocaine, chemically known as ethyl 4-aminobenzoate, is a local anesthetic. Its ester functionality is susceptible to hydrolytic degradation (Scheme 40); the activation energy (*E*a) of acid-catalyzed hydrolysis was reported to be ~18.6 kcal/mol [1]. However, the amino group of the aniline moiety is incompatible with excipients containing carbonyl functional groups. For example, *N*-formylbenzocaine was observed in a liquid formulation containing polyvinylpyrrolidone [2] and an orally disintegrating tablet formulation containing silicified microcrystalline cellulose [3]. In a forced degradation study with H_2O_2, the aniline amino group was oxidized to

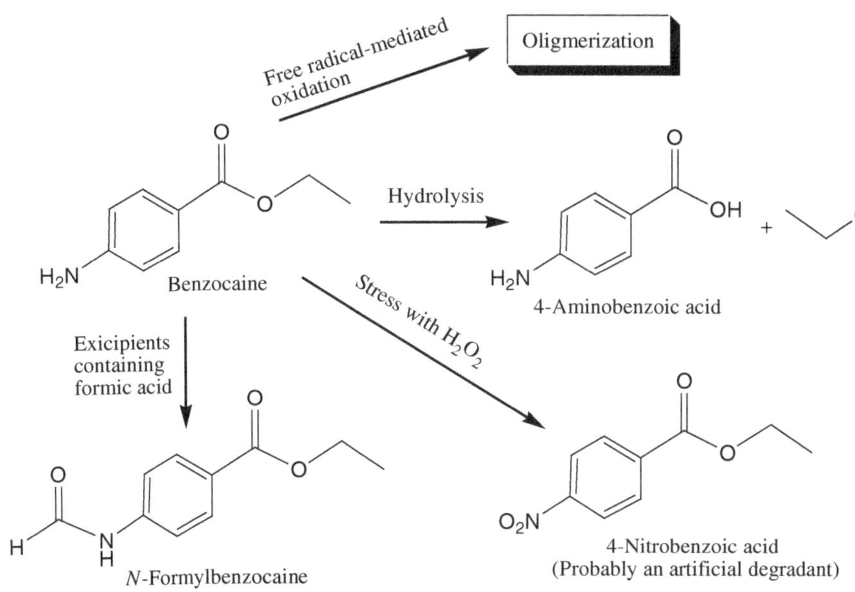

Scheme 40 Degradation pathways of benzocaine under various conditions including forced degradation.

nitro group [4]. Such oxidation probably would not occur under long-term stability storage conditions. On the other hand, free radical-mediated oxidation of the aniline moiety may cause the oligomerization of the drug [5].

References

1 Marcus, A.D. and Baron, S. (1959). A comparison of the kinetics of the acid catalyzed hydrolyses of procainamide, procaine, and benzocaine. *J. Am. Pharm. Assoc.* 48: 85–90.

2 Prasad, A. and Langley, N. *The Effect of Secondary Oxidation Impurities in PVP on API Stability*. Tarrytown, NY: BASF Corporation https://www.scribd.com/document/378037462/The-Effect-of-Secondary-Oxidation-Impurities (accessed 26 February 2022).

3 Köllmer, M., Popescu, C., Manda, P. et al. (2013). Stability of benzocaine formulated in commercial oral disintegrating tablet platforms. *AAPS PharmSciTech* 14: 1333–1340.

4 Perez-Lozano, P., Garcia-Montoya, E., Orriols, A. et al. (2005). A new validated method for the simultaneous determination of benzocaine, propylparaben and benzyl alcohol in a bioadhesive gel by HPLC. *J. Pharm. Biomed. Anal.* 39: 920–927.

5 Marjanović, B., Juranić, I., Ćirić-Marjanović, G. et al. (2011). Chemical oxidative polymerization of benzocaine. *React. Funct. Polym.* 71: 704–712.

41–50: Benzodiazepines

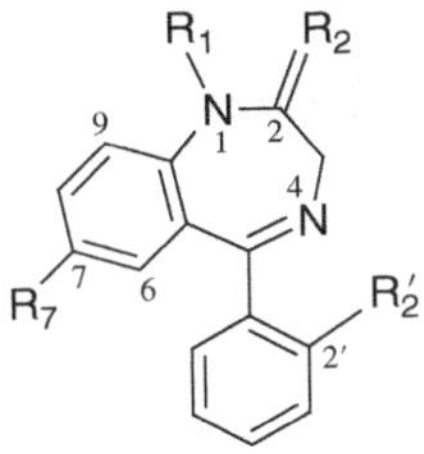

General structures of 1,4-benzodiazepines

41: Alprazolam

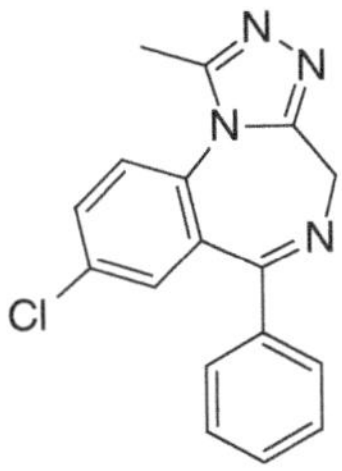

Chemical name: 8-Chloro-1-methyl-6-phenyl-4*H*-[1,2,4]triazolo[4,3-a][1,4] benzodiazepine
Brand name: Xanax, Xanor, Niravam
Formula: $C_{17}H_{13}ClN_4$
Monoisotopic protonated ion (*m/z*): 309.0902
Molecular weight: 308.77
CAS number: 28981-97-7

42: Clonazepam

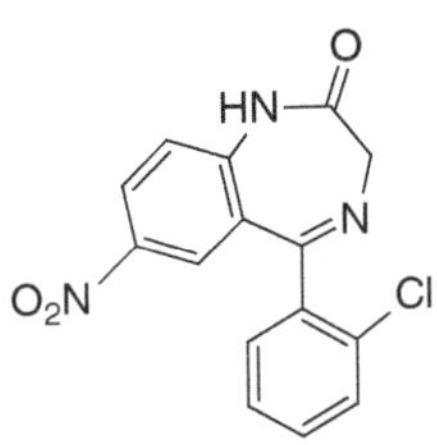

Chemical name: 5-(2-Chlorophenyl)-7-nitro-1,3-dihydro-1,4-benzodiazepin-2-one
Brand name: Klonopin, Rivotril
Formula: $C_{15}H_{10}ClN_3O_3$
Monoisotopic protonated ion (*m/z*): 316.0483
Molecular weight: 315.71
CAS number: 1622-61-3

43: Diazepam [WHO Lists of Essential Medicines]

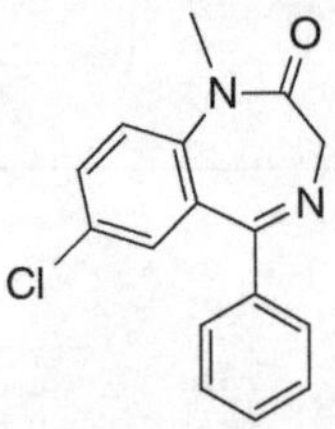

Chemical name: 7-Chloro-1,3-dihydro-1-methyl-5-phenyl-3*H*-1,4-benzodiazepin-2-one
Brand name: Valium, Vazepam, Valtoco
Formula: $C_{16}H_{13}ClN_2O$
Monoisotopic protonated ion (*m/z*): 285.0789
Molecular weight: 284.74
CAS number: 439-14-5

44: Flunitrazepam

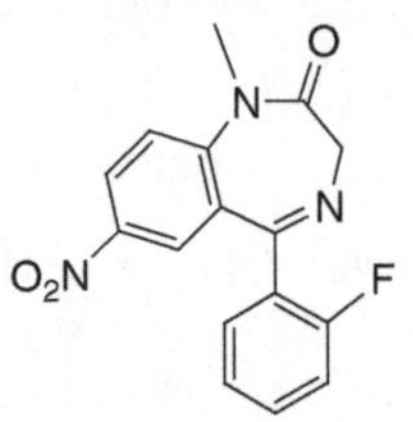

Chemical name: 5-(2-Fluorophenyl)-1-methyl-7-nitro-1*H*-benzo[e][1,4]diazepin-2(3*H*)-one
Brand name: Rohypnol
Formula: $C_{16}H_{12}FN_3O_3$
Monoisotopic protonated ion (*m/z*): 314.0935
Molecular weight: 313.29
CAS number: 1622-62-4

45: Lorazepam

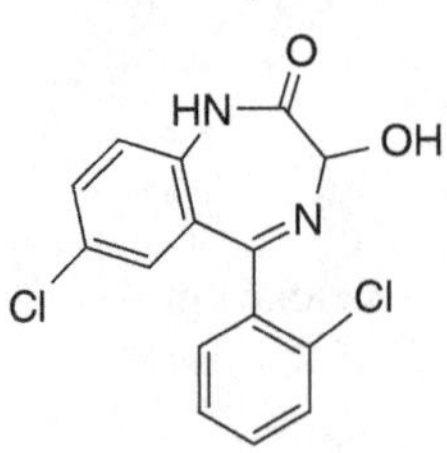

Chemical name: 7-Chloro-5-(2-chlorophenyl)-3-hydroxy-1,3-dihydro-1,4-benzodiazepin-2-one
Brand name: Ativan, Tavor, Temesta
Formula: $C_{15}H_{10}Cl_2N_2O_2$
Monoisotopic protonated ion (*m/z*): 321.0192
Molecular weight: 321.16
CAS number: 846-49-1

46: Midazolam [WHO Lists of Essential Medicines]

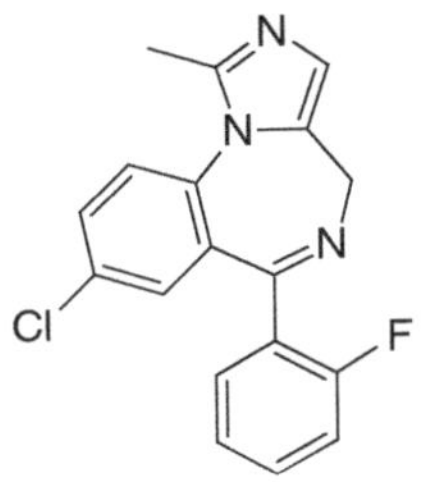

Chemical name: 8-Chloro-6-(2-fluorophenyl)-1-methyl-4*H*-imidazo[1,5-a][1,4] benzodiazepine
Brand name: Dormicum, Hypnovel, Versed
Formula: $C_{18}H_{13}ClFN_3$
Monoisotopic protonated ion (*m/z*): 326.0855
Molecular weight: 325.77
CAS number: 59467-70-8

47: Nitrazepam

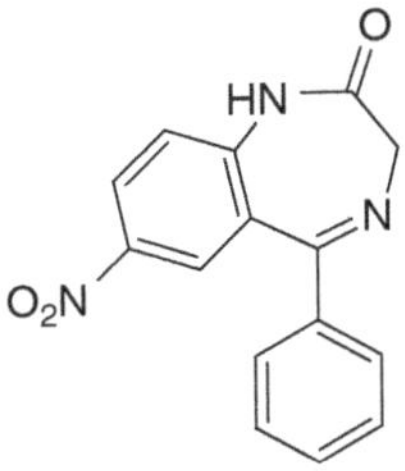

Chemical name: 7-Nitro-5-phenyl-1*H*-benzo[e][1,4]diazepin-2(3*H*)-one
Brand name: Alodorm, Apodorm, Arem
Formula: $C_{15}H_{11}N_3O_3$
Monoisotopic protonated ion (*m/z*): 282.0873
Molecular weight: 281.27
CAS number: 146-22-5

48: Oxazepam

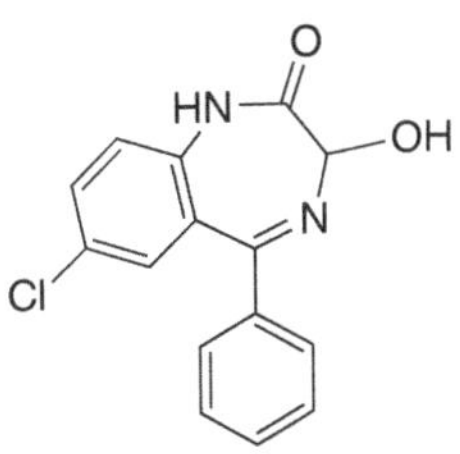

Chemical name: 7-Chloro-3-hydroxy-5-phenyl-1,3-dihydro-1,4-benzodiazepin-2-one
Brand name: Serax, Alepam
Formula: $C_{15}H_{11}ClN_2O_2$
Monoisotopic protonated ion (*m/z*): 287.0582
Molecular weight: 286.72
CAS number: 604-75-1

49: Temazepam

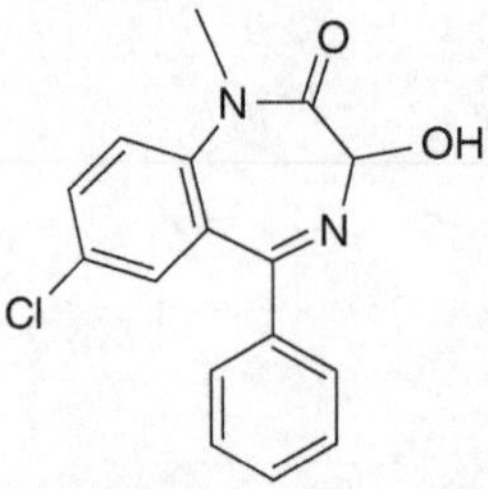

Chemical name: 7-Chloro-1,3-dihydro-3-hydroxy-1-methyl-5-phenyl-1,4-benzodiazepin-2-one
Brand name: Restoril, Normison, Nortem
Formula: $C_{16}H_{13}ClN_2O_2$
Monoisotopic protonated ion (*m/z*): 301.0738
Molecular weight: 300.74
CAS number: 846-50-4

50: Triazolam

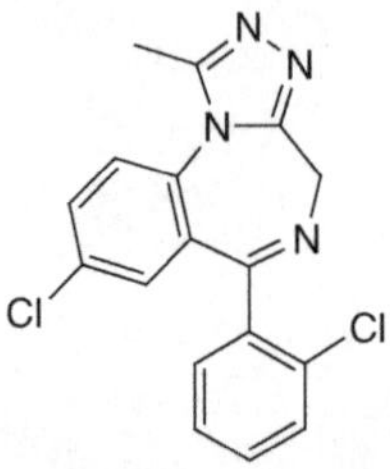

Chemical name: 8-Chloro-6-(2-chlorophenyl)-1-methyl-4*H*-[1,2,4]triazolo[4,3-*a*][1,4]benzodiazepine
Brand name: Halcion
Formula: $C_{17}H_{12}Cl_2N_4$
Monoisotopic protonated ion (*m/z*): 343.0512
Molecular weight: 343.21
CAS number: 28911-01-5

Benzodiazepines, or more specifically 1,4-benzodiazepines, are a class of psychoactive drugs widely used for the treatment of anxiety, insomnia, seizure, and muscle spasms, as well as for inducing sedative effects. The family of benzodiazepines consists of more than a dozen drugs, and the first one, chlordiazepoxide, was synthesized by Leo Sternbach serendipitously in 1955 while working for Hoffmann-La Roch, and chlordiazepoxide was approved for clinical use in 1960 [1]. Generally, the core structure of benzodiazepines can be viewed as deriving from 2-aminobenzophenone upon which a glycine first forms an amide linkage with the amine group and then condenses with the carbonyl group of benzophenone, resulting in the formation of the seven-membered diazepine ring via an intramolecular imine bond. Both the amide and imine bonds are susceptible to hydrolytic degradation, with the imine being more vulnerable, particularly under acidic conditions.

The embedded imine moiety in the diazepine part of the benzodiazepine core structure is sensitive to pH. A study of alprazolam, midazolam, and triazolam showed that at pH below 2, the open ring form was favored, while at pH above 4, the ring-closed form was favored (Scheme 41–50-1) [2]. Although the ring-open form may be considered a hydrolytic degradant, the ring opening and closure are reversible in solution. Hence, for an IV formulation of midazolam, the open form can be considered a prodrug as the ring would be completely closed at the physiological pH of 7.4 [2, 3]. On the contrary, as the ring-open form is a primary amine, it can undergo the Maillard reaction if the excipients contain reducing sugars. In a tablet formulation of alprazolam containing lactose, the Maillard degradant was found to be quickly increasing under the accelerated stability conditions (Scheme 41–50-1) [4].

Other benzodiazepine drugs should follow similar ring open-closure pathways as outlined in Scheme 41–50-1, except for the following differences. First, under very strong acidic conditions such as acid-catalyzed forced degradation, the open form of other benzodiazepines, e.g., diazepam and temazepam, that possess a 2,3-amide bond can be further hydrolyzed to form 2-aminobenzophenone degradants (Pathway a, Scheme 41–50-2) [5, 6]. Second, those benzodiazepines containing a 3-hydroxyl group on the diazepine ring behave differently under acidic and basic conditions; such drugs include lorazepam and temazepam. Under acidic conditions, lorazepam degraded to 2-aminobenzophenone derivative and quinazoline carboxaldehyde (Pathways a and b, Scheme 41–50-2) [7, 8]. The latter aldehyde could disproportionate to yield the corresponding alcohol and carboxylic acid [8]. Under basic conditions, lorazepam

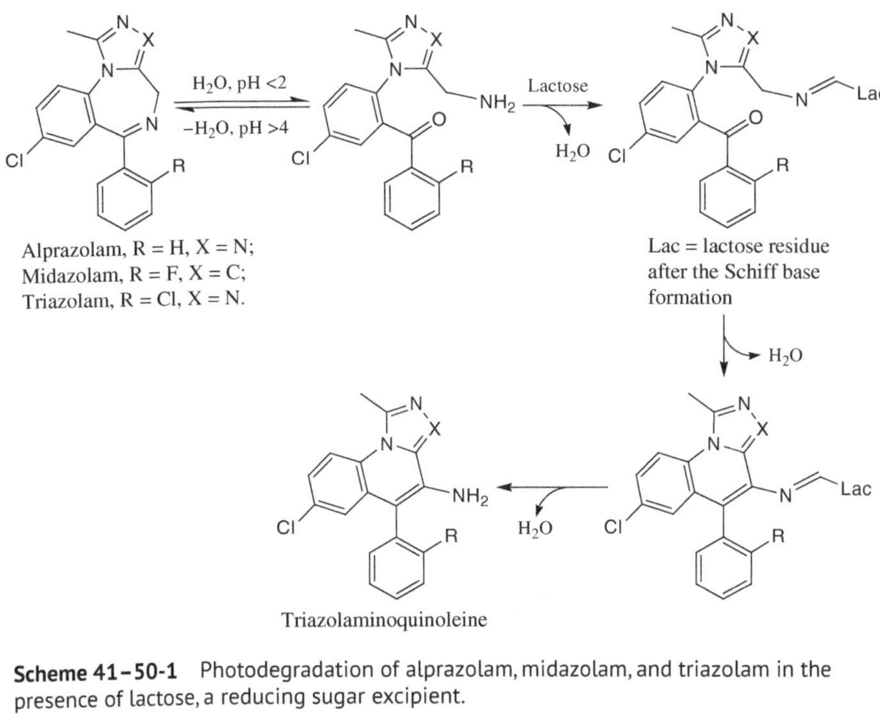

Scheme 41–50-1 Photodegradation of alprazolam, midazolam, and triazolam in the presence of lactose, a reducing sugar excipient.

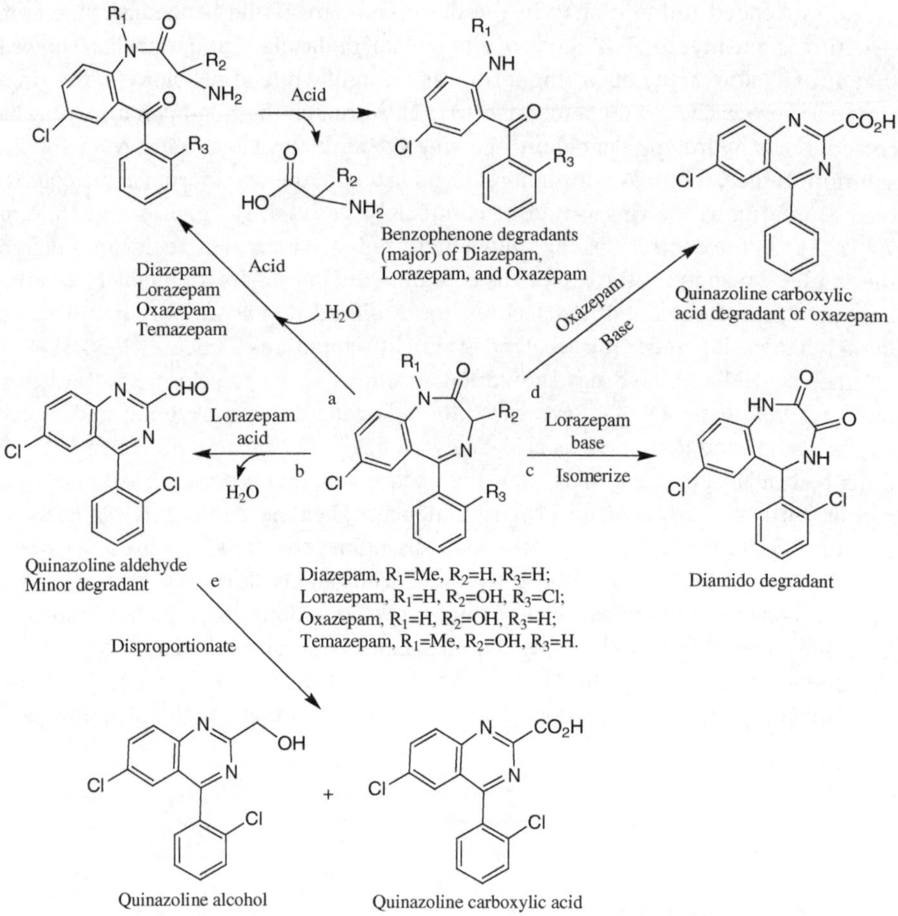

Scheme 41–50-2 Degradation pathways of diazepam, lorazepam, oxazepam, and temazepam.

rearranged to an isomeric diamido degradant (Pathway c, Scheme 41–50-2). Surprisingly, the degradation pathways of oxazepam, which is structurally very similar to lorazepam, are somewhat different from those of lorazepam (Pathway d, Scheme 41–50-2).

Benzodiazepines can also undergo photochemical degradation. For those that do not have a nitro group, the photochemical transformation occurs mainly on the diazepine ring (Scheme 41–50-3). For example, photostress of diazepam under UV and simulated sunlight produced a number of degradants, the vast majority of which are derivatives of benzophenone, acridinone and quinazoline [5, 9]. The first step of the photodegradation is most likely the dissociation of the imine bond [5], which also occurs under acidic stress.

The photochemical degradation of alprazolam (Scheme 41–50-4) is somewhat different from that of simpler benzodiazepines, such as diazepam (Scheme 41–50-3),

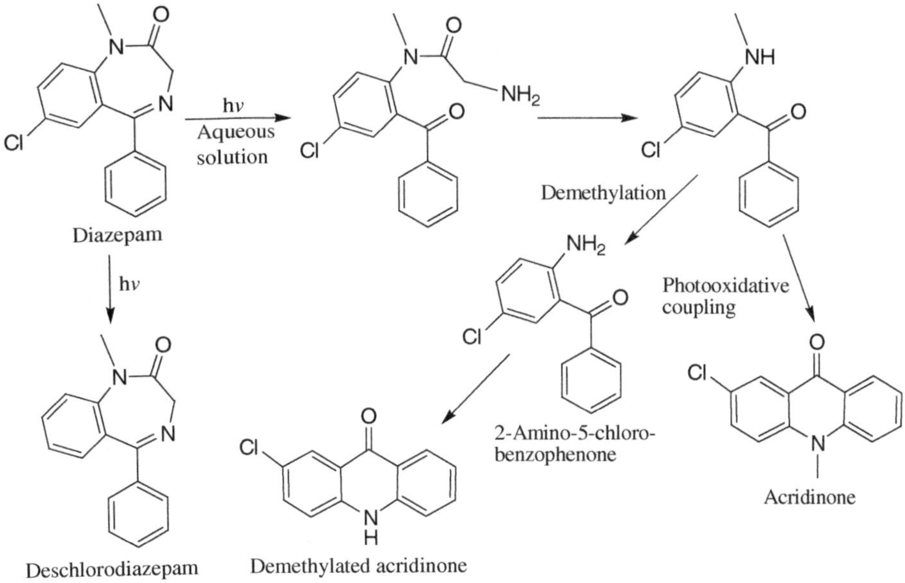

Scheme 41–50-3 Photochemical degradation of diazepam.

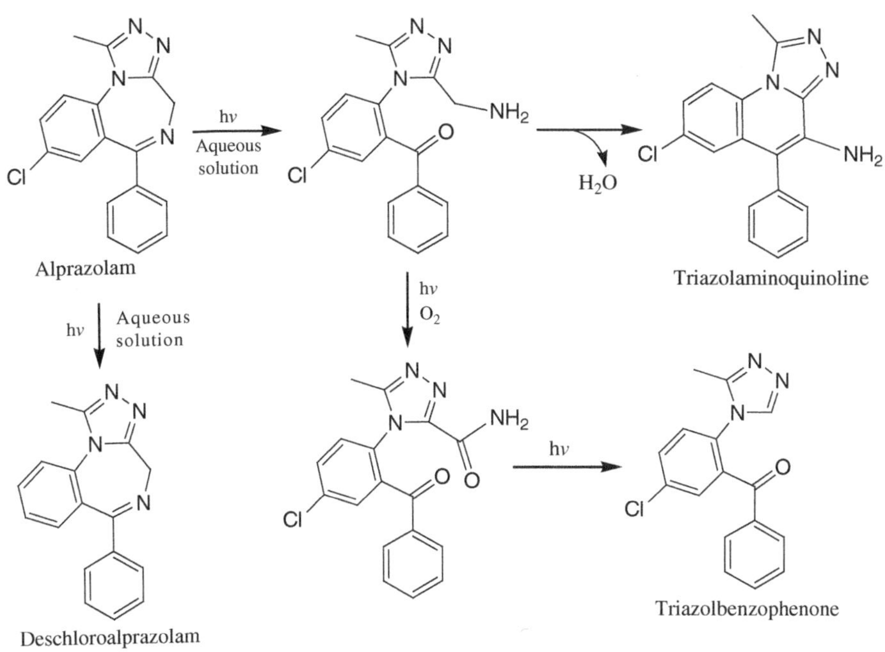

Scheme 41–50-4 Photochemical degradation of alprazolam.

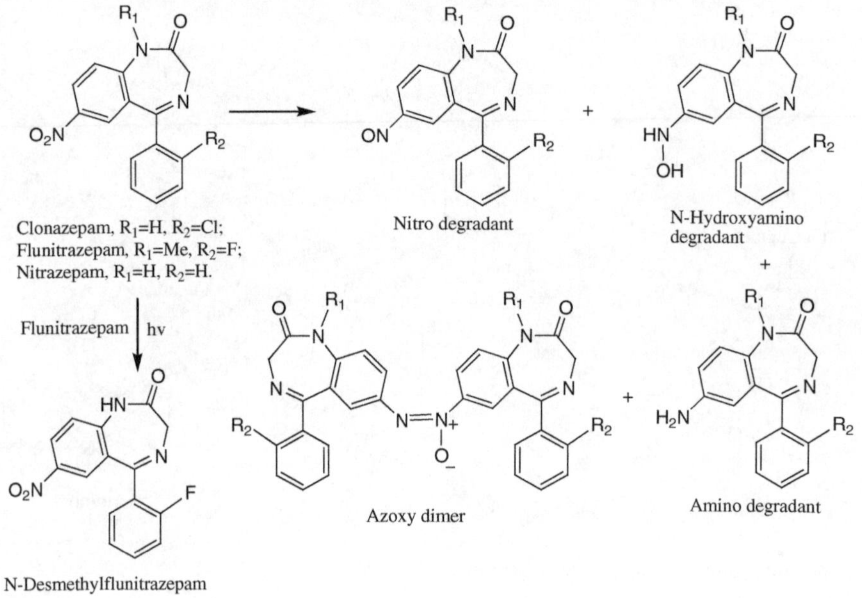

Scheme 41-50-5 Photodegradation pathways of clonazepam, flunitrzepam, and nitrazepam.

due to the presence of the triazilo group, which prevents the formation of an acridinone-type degradant.

For benzodiazepines with a nitro substituent, such as clonazepam, flunitrazepam, and nitrazepam, their photochemical degradation under anaerobic conditions occurs on the nitro group, which is reduced to 7-nitroso, hydroxylamino, and amino groups sequentially (Scheme 41–50-5) [10, 11]. Moreover, during the photoreductive process, an azoxy dimer was also observed [11]. Under aerobic conditions, these nitrobenzodiazepines appeared relatively more stable, probably because their photochemically excited states are quenched by molecular oxygen to produce singlet oxygen. The latter is believed to be the most responsible for the phototoxicity of these drugs. For flunitrazepam containing a 1-methyl group on the diazepine ring, *N*-desmethylflunitrazepam, the demethylated degradant, was found to be the major photodegradant under the aerobic condition [12].

References

1 Benzodiazepines, Wikipedia, https://en.wikipedia.org/wiki/Benzodiazepine (accessed 16 April 2022).

2 Loftsson, T., Guðmundsdottir, H., Sigurjonsdottir, J.F. et al. (2001). Cyclodextrin solubilization of benzodiazepines: formulation of midazolam nasal spray. *Int. J. Pharm.* 212: 29–40.

3 Gerecke, M. (1983). Chemical structure and properties of mida-zolam compared with other benzodiazepines. *Br. J. Clin. Pharmacol.* 16 (Suppl. 1): 11S–16S.

4 Huidobro, A.L. and Barbas, C. (2009). Analytical study proving alprazolam degradation to its main impurity triazolaminoquinoleine through Maillard reaction. *Anal. Bioanal. Chem.* 394: 1349–1359.

5 Cabrera, C.G., Goldberg de Waisbaum, R., and Nudelman, N.S. (2005). Kinetic and mechanistic studies on the hydrolysis and photodegradation of diazepam and alprazolam. *J. Phys. Org. Chem.* 18: 156–161.

6 Fatmi, A.A. and Hickson, E.A. (1988). Determination of temazepam and related compounds in capsules by high-performance liquid chromatography. *J. Pharm. Sci.* 77: 87–89.

7 Siddegowda, M.S., Yathirajan, H.S., and Ramakrishna, R.A. (2012). Facile thermal rearrangement of lorazepam and oxazepam. *Indian J. Chem.* 51B: 1628–1632.

8 Rutgers, J.G. and Shearer, C.M. (1981). Lorazepam. *Anal. Prof. Drug Sub.* 9: 397–426.

9 West, C.E. and Rowland, S.J. (2012). Aqueous phototransformation of diazepam and related human metabolites under simulated sunlight. *Environ. Sci. Technol.* 46: 4749–4756.

10 Cornelissen, P.J.G., Beijersbergen van Henegouwen, G.M.J., and Mohn, G.R. (1981). Relationship between structure and photobiological activity of 7-nitro-1,4-benzodiazepines. *Photochem. Photobiol.* 34: 345–350.

11 Cornelissen, P.J.G. and Beijersbergen Van Henegouwen, G.M.J. (1981). Photochemical activity of 7-nitro-1,4-benzodiazepines. *Pharm. Weekbl.* 3: 800–809.

12 Busker, R.W., Beijersbergen van Henegouwen, G.M.J., Kwee, B.M.C., and Winkens, J.H.M. (1987). Photobinding of flunitrazepam and its major photo-decomposition product N-desmethylflunitrazepam. *Int. J. Pharm.* 36: 113–120.

51: Benzoyl Peroxide

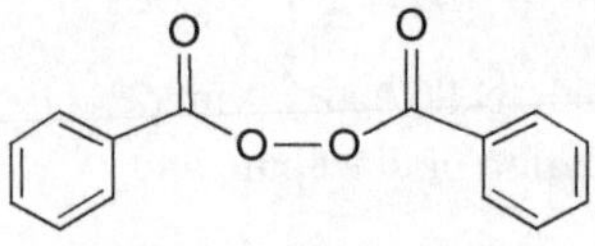

Chemical name: Benzoic peroxyanhydride
Brand name: Benzac
Formula: $C_{14}H_{10}O_4$
Monoisotopic protonated ion (*m/z*): 243.0652
Molecular weight: 242.23
CAS number: 94-36-0

Benzoyl peroxide is widely used in topical formulations for treating acne. Its peroxyl O-O bond is weak, as evidenced by a relatively small bond dissociation energy of 30 kcal/mol [1]. Hence, it can dissociate homolytically into two benzoate radicals under somewhat elevated temperature, particularly in solution (Scheme 51). The homolytic dissociation can also proceed photochemically. The benzoate radical is capable of abstracting a proton from a weak C—H bond of either the solvent of the solution or a substance in the solution to produce benzoic acid, which is the major degradant of benzoyl peroxide [2]. Benzoate radical can also undergo decarboxylation to form phenyl radical, which can combine with another benzoate radical to yield phenyl benzoate.

The stability of benzoyl peroxide in pharmaceutical formulations depends on the nature of solvents used and the solubility of benzoyl peroxide in these solvents [3]. In general, benzoyl peroxide tends to be more stable in media where it may exist as a suspension or emulsion rather than a true solution [4].

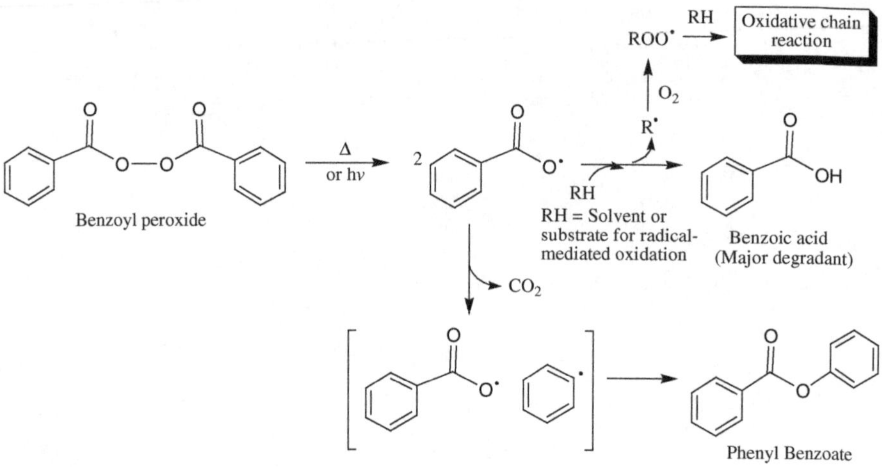

Scheme 51 Degradation of benzoyl peroxide.

References

1 Walling, C. (1995). *Fifty Years of Free Radicals*. New York, NY: Wiley.
2 Cass, W.E. (1947). Products of the reaction of benzoyl peroxide with certain ethers. *J. Am. Chem. Soc.* 69: 500–503.
3 Nielloud, F., Mestres, J.P., and Mestres, G.M. (2002). Consideration on the formulation of benzoyl peroxide at ambient temperature: choice of non-polar solvent and preparation of submicron emulsion gels. *Drug Dev. Ind. Pharm.* 28: 863–870.
4 Chellquist, E.M. and Gorman, W.G. (1992). Benzoyl peroxide solubility and stability in hydric solvents. *Pharm. Res.* 9: 1341–1346.

52–54: Bisoprolol, Metoprolol, Propranolol

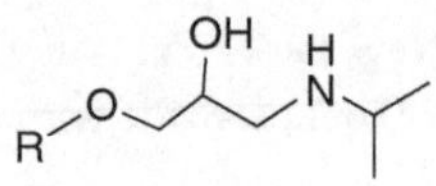

Bisoprolol, R = 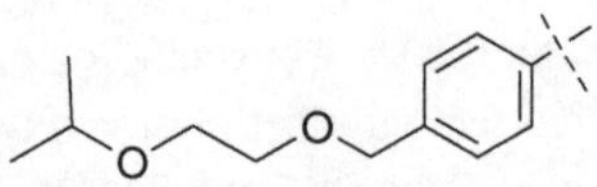

Metoprolol, R = 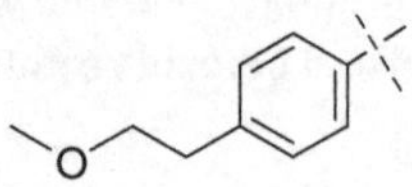

Propranolol, R = 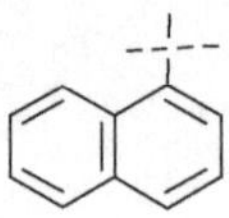

52: Bisoprolol

Chemical name: (*R,S*)-1-{4-[(2-Isopropoxyethoxy)methyl]phenoxy}-3-(isopropylamino)propan-2-ol
Brand name: Zebeta, Concor, Selecta
Formula: $C_{18}H_{31}NO_4$
Monoisotopic protonated ion (*m/z*): 326.2326
Molecular weight: 325.449
CAS number: 66722-44-9
Common salt form: Fumarate

53: Metoprolol

Chemical name: (*R,S*)-1-[4-(2-Methoxyethyl)phenoxy]-3-[(propan-2-yl)amino]propan-2-ol
Brand name: Lopressor, Metolar XR, Toprol XL
Formula: $C_{15}H_{25}NO_3$
Monoisotopic protonated ion (*m/z*): 268.1907
Molecular weight: 267.37
CAS number: 51384-51-1
Common salt form: Succinate

54: Propranolol [WHO List of Essential Medicines]

Chemical name: (*R,S*)-1-(1-methylethylamino)-3-(1-naphthyloxy)propan-2-ol
Brand name: Inderal
Formula: $C_{16}H_{21}NO_2$
Monoisotopic protonated ion (*m/z*): 260.1645
Molecular weight: 259.35
CAS number: 525-66-6
Common salt form: HCl salt

Bisoprolol, metoprolol, and propranolol all belong to the family of β-blockers, which are a group of therapeutic agents widely used for the treatment of hypertension. All these three drugs contain a 3-(isopropylamino)propylene glycol moiety. The latter moiety is also present in atenolol, another β-blocker whose degradation chemistry is discussed previously in this book. Therefore, for the degradation chemistry resulting from this moiety, please also refer to that monograph.

Under acidic and basic stress conditions, the main degradant for bisoprolol is the hydrolysis of the benzylic ether bond, one of its two ether bonds (Scheme 52–54-1) [1]. A formulation study with a solid dosage form of bisoprolol indicated that bisoprolol is not stable in a low pH environment [2]. A kinetic study showed that the half-life of bisoprolol under stress by 1 M HCl at 333.2K was only 0.12 hour [3]. Under the same conditions, the half-life of propranolol was found to be 385 hours. A forced degradation study of metoprolol was performed, and it was found that extensive degradation occurred under basic and thermal stress conditions [4]. The degradation probably took place in the 3-(isopropylamino)propylene glycol side chain of the drug molecule.

For propranolol, as it contains a naphthalene group, it would be expected to be photochemically active (Scheme 52–54-2). A laser-flash photolysis study of propranolol found that it is a good photosensitizer with a quantum yield of 0.24 for sensitizing singlet oxygen formation [5]. Depending on the condition of the photolysis, the initial photodegradant can be naphthol [6]. In the presence of oxygen, propranolol can be oxidized by singlet oxygen, which is formed during the photosensitization process, to give 6-hydroxy-1,4-naphthoquinone via the highly photolabile intermediate 1,4-naphthoquinone [5]. A liquid formulation study with β-cyclodextrin indicated that the rate of photodecomposition of propranolol could be reduced by ~50% [7].

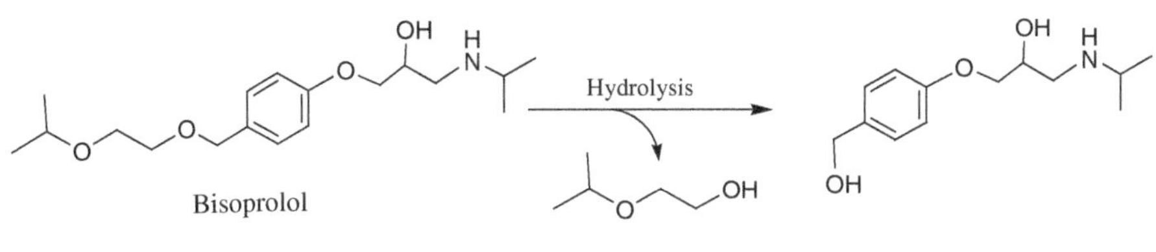

Scheme 52–54-1 Hydrolytic degradation of bisoprolol.

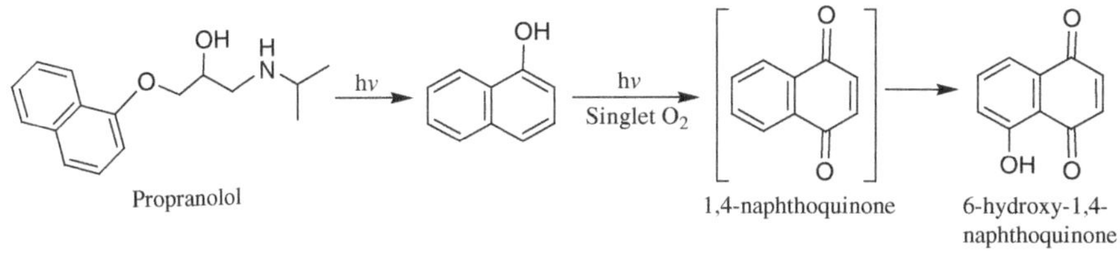

Scheme 52–54-2 Photochemical degradation of propranolol.

References

1 Joshi, S.J., Karbhari, P.A., Bhoir, S.I. et al. (2010). RP-HPLC method for simultaneous estimation of bisoprolol fumarate and hydrochlorothiazide in tablet formulation. *J. Pharm. Biomed. Anal.* 52: 362–371.

2 Dulin, W.A. (1995). Degradation of bisoprolol fumarate in tablets formulated with dicalcium phosphate. *Drug Dev. Ind. Pharm.* 21: 393–409.

3 Szalka, M., Rokaszewski, E., and Kaczmarski, K. (2013). Kinetics of hydrolysis of bisoprolol hemifumarate in aqueous acidic solutions. *Int. J. Chem. Kinet.* 45: 744–754.

4 Phale, M.D. and Hamrapurkar, P.D. (2010). Optimization and establishment of a validated stability-indicating HPLC method for study of the stress degradation behavior of metoprolol succinate. *J. AOAC Int.* 93: 911–916.

5 Sortino, S., Petralia, S., Boscà, F., and Miranda, M.A. (2002). Irreversible photo-oxidation of propranolol triggered by self-photogenerated singlet molecular oxygen. *Photochem. Photobiol. Sci.* 1: 136–140.

6 Uwai, K., Tani, M., Ohtake, Y. et al. (2005). Photodegradation products of propranolol: the structures and pharmacological studies. *Life Sci.* 78: 357–365.

7 Ansolin, T.L., Bariccatti, R.A., Ferreira da Rosa, M., and Lindino, C.A. (2014). The effect of β-cyclodextrin in the photochemical stability of propranolol hydrochloride in aqueous solution. *Acta Sci. Technol.* 36: 337–340.

55: Bleomycin [WHO List of Essential Medicines]

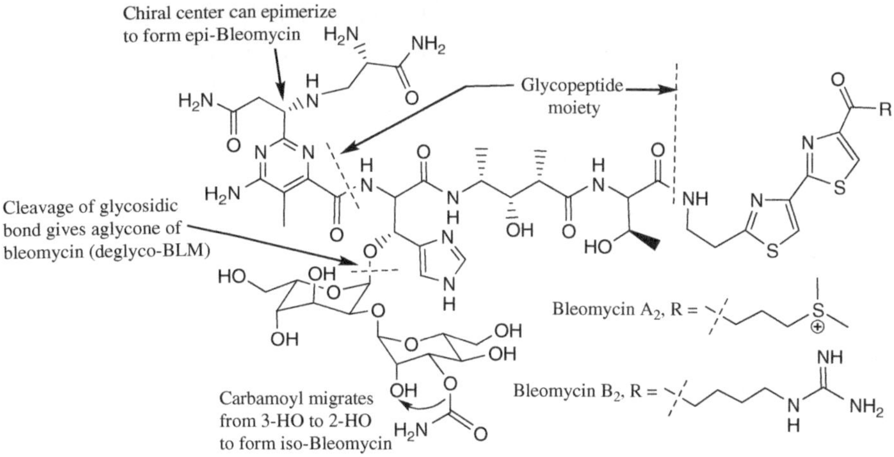

55a: Bleomycin A_2

Chemical name: 3-[[2-[2-[2-[[(2*S*,3*R*)-2-[[(2*S*,3*S*,4*R*)-4-[[(2*S*,3*R*)-2-[[6-Amino-2-[(1*S*)-3-amino-1-[[(2*S*)-2,3-diamino-3-oxopropyl]amino]-3-oxopropyl]-5-methylpyrimidine-4-carbonyl]amino]-3-[(2*R*,3*S*,4*S*,5*S*,6*S*)-3-[(2*R*,3*S*,4*S*,5*R*,6*R*)-4-carbamoyloxy-3,5-dihydroxy-6-(hydroxymethyl)oxan-2-yl]oxy-4,5-dihydroxy-6-(hydroxymethyl)oxan-2-yl]oxy-3-(1*H*-imidazol-5-yl)propanoyl]amino]-3-hydroxy-2-methylpentanoyl]amino]-3-hydroxybutanoyl]amino]ethyl]-1,3-thiazol-4-yl]-1,3-thiazole-4-carbonyl]amino]propyl-dimethylsulfanium

Brand name: N/A

Formula: $C_{55}H_{84}N_{17}O_{21}S_3$

Monoisotopic protonated ion (*m/z*): 1414.5184

Molecular weight: 1415.55

CAS number: 11116-31-7; 11056-06-7 (Sulfate salt)

55b: Bleomycin B_2

Chemical name: [(2*R*,3*S*,4*S*,5*R*,6*R*)-2-[(2*R*,3*S*,4*S*,5*S*,6*S*)-2-[(1*R*,2*S*)-2-[[6-Amino-2-[(1*S*)-3-amino-1-[[(2*S*)-2,3-diamino-3-oxopropyl]amino]-3-oxopropyl]-5-methylpyrimidine-4-carbonyl]amino]-3-[[(2*R*,3*S*,4*S*)-5-[[(2*S*,3*R*)-1-[2-[4-[4-[4-(diaminomethylideneamino)butylcarbamoyl]-1,3-thiazol-2-yl]-1,3-thiazol-2-yl]ethylamino]-3-hydroxy-1-oxobutan-2-yl]amino]-3-hydroxy-4-methyl-5-oxopentan-2-yl]amino]-1-(1*H*-imidazol-5-yl)-3-oxopropoxy]-4,5-dihydroxy-6-(hydroxymethyl)oxan-3-yl]oxy-3,5-dihydroxy-6-(hydroxymethyl)oxan-4-yl] carbamate

Formula: $C_{55}H_{84}N_{20}O_{21}S_2$

Monoisotopic protonated ion (*m/z*): 1424.5562

Molecular weight: 1425.51

CAS number: 9060-10-0

Bleomycin is a widely used anticancer antibiotic containing at least 10 components, and the major ones are bleomycin A_2 (*N1*-[3-(dimethylsulfonio)propyl]-bleomy cinamide) and B_2 (*N1*-[4-[(aminoiminomethyl)amino]butyl]-bleomycinamide) [1]. Structurally, the main skeleton of the molecule is an unusual glycopeptide, which is connected to a substituted pyrimidine moiety on the left and to a substituted bithiazole moiety on the right. The pyrimidine moiety is capable of chelating with transition metal ions with five nitrogens in the pyrimidine moiety as the ligands, while the positively charged bithiazole moiety can intercalate into DNA [2]. When the chelating metal ion is ferrous ion, bleomycin is activated upon binding by molecular oxygen onto the sixth coordination site to the chelated metal ion. This form of bleomycin is referred to as activated bleomycin, which is capable of cleaving DNA, and in the absence of DNA, it undergoes auto-oxidation. The latter oxidation occurs on the β-aminoalanine side chain of the pyrimidine moiety, resulting in the formation of a deaminated bleomycin as the major degradant and an *N*-dealkylated one as a minor degradant in aqueous solutions (Scheme 55-1) [3].

The research group of Umezawa, who discovered bleomycin [4, 5], performed a series of forced degradation studies. In aqueous alcoholic solutions containing weak bases such as triethylamine, the carbamoyl group of metal-free bleomycin can migrate from the 3-hydroxyl to 2-hydroxyl of the mannose moiety to form iso-bleomycin (refer to the structure of bleomycin above) [6]. Under the same conditions, copper-chelated bleomycin undergoes epimerization at the chiral position α to the pyrimidine ring to produce epi-bleomycin.

Under acidic hydrolytic conditions, all the amide, the glycosidic, and carbamoyl linkages are susceptible to hydrolysis; nevertheless, the glycosidic linkage is among the most acid-sensitive bonds in bleomycin, and its cleavage leads to the formation of aglycone of bleomycin (also known as deglyco-BLM) [7]. Treatment under mild alkaline conditions can cause the cleavage of the sugar moiety via β-elimination, forming dehydrohistidine-bleomycin (Scheme 55-2) [8]. Under elevated temperature, the β-aminoalanine can be eliminated via β-elimination as well.

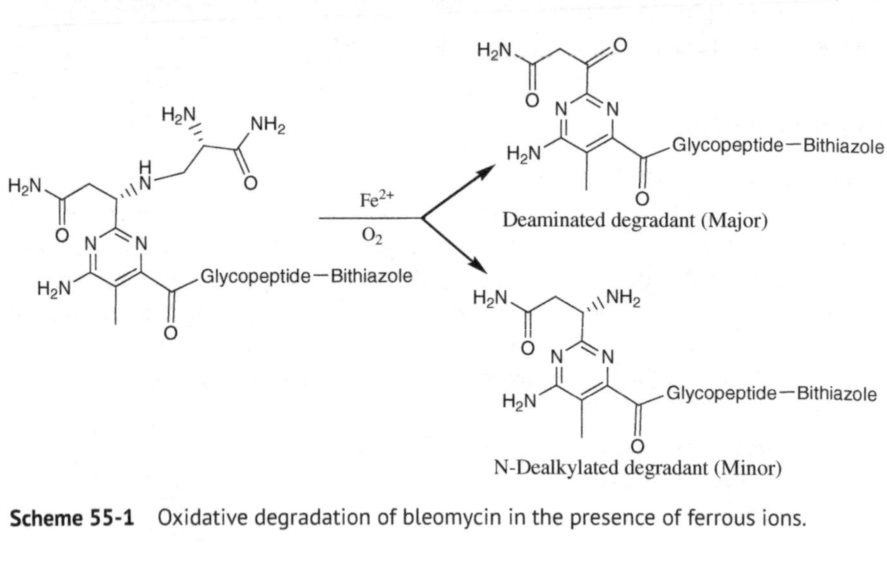

Scheme 55-1 Oxidative degradation of bleomycin in the presence of ferrous ions.

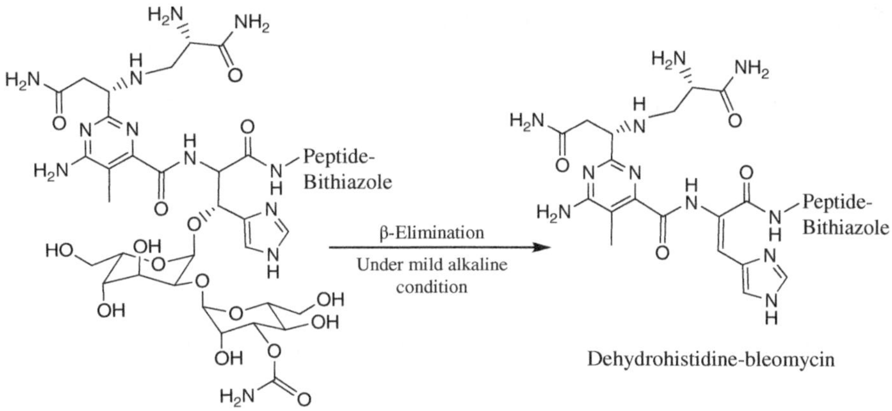

Scheme 55-2 Degradation of bleomycin under mild alkaline condition.

References

1 Hecht, S.M. (2000). Bleomycin: new perspectives on the mechanism of action. *J. Nat. Prod.* 63: 158–168.

2 Stubbe, J. and Kozarich, J.W. (1987). Mechanisms of bleomycin-induced DNA degradation. *Chem. Rev.* 87: 1107–1136.

3 Sam, J.W., Tang, X.-J., Magliozzo, R.S., and Peisachts, J. (1995). Electrospray mass spectrometry of iron bleomycin II: investigation of the reaction of Fe(III)-bleomycin with iodosylbenzene. *J. Am. Chem. Soc.* 117: 1012–1018.

4 Umezawa, H., Maeda, K., Takeuchi, T., and Okami, Y. (1966). New antibiotics, bleomycin A and B. *J. Antibiot.* 19: 200–209.

5 Takita, T., Muraoka, Y., Nakatani, T. et al. (1978). Chemistry of bleomycin. XIX. Revised structures of bleomycin and phleomycin. *J. Antibiot.* 31: 801–804.

6 Muraoka, Y., Kobayashi, H., Fujii, A. et al. (1976). Chemistry of bleomycin. XVI Epi-Bleomycin. *J. Antibiot.* 29: 853–856.

7 Muraoka, Y., Suzuki, M., Fujii, A. et al. (1981). Chemistry of bleomycin. XXVIII preparation of deglycobleomycin by mild acid hydrolysis of bleomycin. *J. Antibiot.* 34: 353–357.

8 Omoto, S., Takita, T., Maeda, K. et al. (1972). The chemistry of bleomycin. VIII The structure of the sugar moiety of bleomycin A_2. *J. Antibiot.* 25: 752–754.

56: Beclometasone Dipropionate

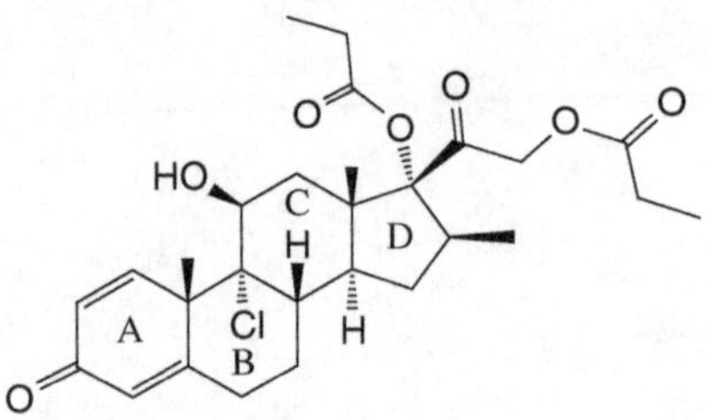

Chemical name: (11β,16β)-9-Chloro-11-hydroxy-16-methyl-3,20-dioxopregna-1,4-diene-17,21-diyl dipropanoate
Brand name: Qvar
Formula: $C_{28}H_{37}ClO_7$
Monoisotopic protonated ion (*m/z*): 521.2301
Molecular weight: 521.05
CAS number: 5534-09-8

Beclometasone dipropionate is a second-generation synthetic corticosteroid, which is a prodrug and its active metabolites, beclometasone 17-dipropionate and beclomethasone, are effective anti-inflammatory agents. Clinically, beclometasone dipropionate is primarily used for the treatment of asthma, allergic rhinitis, dermatitis, and psoriasis, typically in the dosage forms of a pressurized metered-dose inhaler, dry powder inhaler, aqueous nasal spray, and cream. Structurally, beclometasone dipropionate can be regarded as a chloro analog of betamethasone dipropionate in which the 9-fluoro of the latter is replaced by a chloro substituent. Thus, the degradation chemistry of beclometasone 17-propionate would be very similar to that of betamethasone dipropionate, which will be described in the Monograph of Betamethasone 21-acetate, Betamethasone 17-valerate, Betamethasone 17, 21-dipropionate, and Betamethasone 21-phosphate. Indeed, in an accelerated stability study of beclometasone dipropionate, both hydrolytic degradants, i.e., beclometasone 17-propionate and beclometasone 21-propionate, were observed (Scheme 56) [1]. The difference in their degradation behavior between beclometasone dipropionate and betamethasone dipropionate appears to lie in the presence of the 9-chloro substituent in the former drug molecule versus the 9-fluoro one in the latter. For betamethasone and its analogs, the A-ring is susceptible to isomerization under normal laboratory fluorescent lighting. In contrast, beclometasone dipropionate and its analogs (i.e., those containing the 9-chloro substituent) appear to be resistant toward this type of photodegradation, due to the "heavy atom effect" of the 9-chloro substituent that shortens the triplet lifetime of the photochemically excited state [2]. In contrast, beclometasone dipropionate and its analogs tend to form the corresponding 9,11-epoxide degradants via elimination of HCl more easily than betamethasone dipropionate and its analogs via elimination of HF. This is probably because the chloro is a better leaving group than the fluoro. In the beclometasone dipropionate cream accelerated stability study, beclomethasone 9,11-epoxide

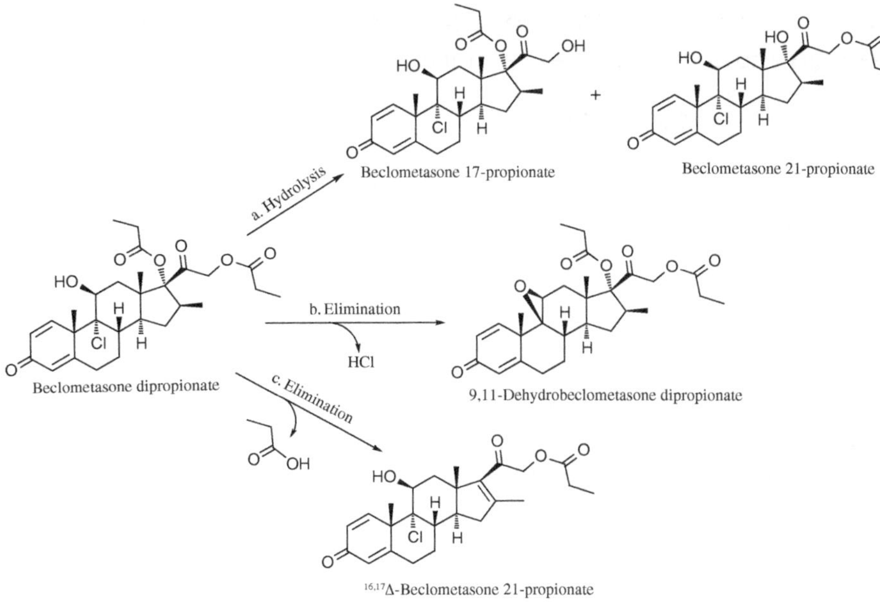

Scheme 56 Degradation pathways of beclomethasone dipropionate.

dipropionate was observed among the degradants [1]. In the same stability study, the major degradant was found to be another elimination degradant, $^{16,17}\Delta$-beclometasone 21-propionate; for comparison purpose, it should be mentioned that $^{16,17}\Delta$-betamethasone (16,17-dehydrobetamethasone) was observed in the forced degradation of betamethasone with HCl (refer to the next Monograph).

References

1 Sambandan, E., Kathavarayan, T., Sellappan, S. et al. (2019). Identification and characterization of unknown degradation impurities in beclomethasone dipropionate cream formulation using HPLC, ESI-MS and NMR. *J. Pharm. Biomed. Anal.* 167: 123–131.

2 Ricci, A., Fasani, E., Mella, M., and Albini, A. (2003). General patterns in the photochemistry of pregna-1,4-dien-3,20-diones. *J. Org. Chem.* 68: 4361–4366.

57–61: Betamethasone, Dexamethasone, Prednisolone, Methylprednisolone, Prednisone

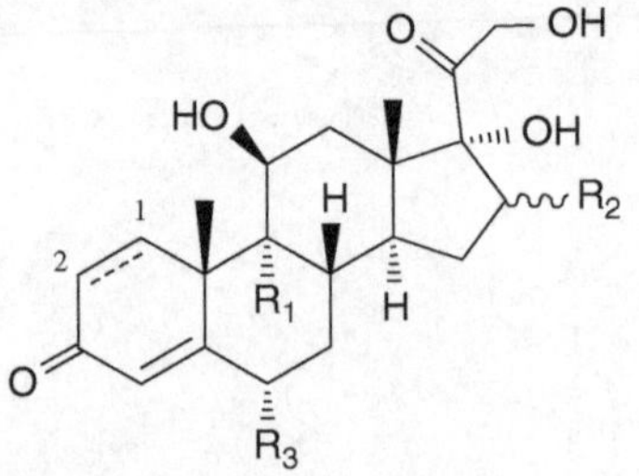

Betamethasone, 1-2 double bond,
R_1 = F, R_2 = CH_3 at β-position, R_3 = H;
Dexamethasone, 1-2 double bond,
R_1 = F, R_2 = CH_3 at α-position, R_3 = H;
Prednisolone, 1-2 double bond,
R_1 = H, R_2 = H, R_3 = H;
Methylprednisolone, 1-2 single double bond,
R_1 = H, R_2 = H, R_3 = CH_3.

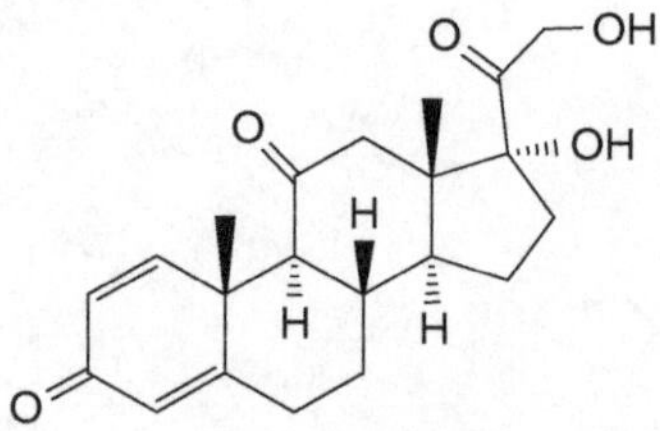

Prednisone

57: Betamethasone

Chemical name: (11β,16β)-9-Fluoro-11,17,21-trihydroxy-16-methylpregna-1,4-diene-3,20-dione
Brand name: Celestone, Eleuphrat
Formula: $C_{22}HFO_5$
Monoisotopic protonated ion (*m/z*): 393.2072
Molecular weight: 392.46
CAS number: 378-44-9

58: Dexamethasone

Chemical name: (11β,16α)-9-Fluoro-11,17,21-trihydroxy-16-methylpregna-1,4-diene-3,20-dione
Brand name: Decadron, Ozurdex, Dexycu
Formula: $C_{22}HFO_5$
Monoisotopic protonated ion (*m/z*): 393.2072
Molecular weight: 392.46
CAS number: 50-02-2

59: Prednisolone

Chemical name: (11β)-11,17,21-Trihydroxypregna-1,4-diene-3,20-dione
Brand name: Orapred, PediaPred, Millipred
Formula: $C_{21}H_{28}O_5$

Monoisotopic protonated ion (*m/z*): 361.2010
Molecular weight: 360.45
CAS number: 50-24-8

60: Methylprednisolone

Chemical name: (6α)-11,17,21-Trihydroxy-6-methylpregna-1,4-diene-3,20-dione
Brand name: Medrol
Formula: $C_{22}H_{30}O_5$
Monoisotopic protonated ion (*m/z*): 375.2166
Molecular weight: 374.48
CAS number: 83-43-2

61: Prednisone

Chemical name: 17,21-Dihydroxypregna-1,4-diene-3,11,20-trione
Brand name: Deltasone
Formula: $C_{21}H_{26}O_5$
Monoisotopic protonated ion (*m/z*): 359.1853
Molecular weight: 358.43
CAS number: 53-03-2

Betamethasone and related corticosteroids such as dexamethasone, prednisolone, methylprednisolone, and prednisone are effective anti-inflammatory therapeutic agents that have been used for many decades. The structural moieties that are susceptible to degradation are mainly the A and D rings of the steroid core.

As a derivative of cyclohexa-2,5-dienone, the A ring undergoes facile photoisomerization particularly in solutions (Scheme 57–61-1), giving rise to lumi-isomers of the steroids; in the case of betamethasone and its esters, lumi-betamethasone and its esters are formed [1, 2]. The lumi-degradants can be observed in solutions of the corticosteroids when exposed to the fluorescent lighting in regular analytical laboratories [2]. Further photolysis can lead to many more photodegradants including photolumi degradants [3, 4].

However, the degradation chemistry on the D ring and its 1,3-dihydroxyacetone side chain is much more abundant than the A ring chemistry. For simplicity, we will use betamethasone as the prototype for our discussion and illustration of the degradation pathways. The degradation chemistry of the D ring comprises primarily two types: dehydration by the 17-hydroxyl group and oxidation of the 21-hydroxyl group. The dehydration can take place via two routes: one proceeds through the enolization of the 20-carbonyl group, which is referred to as the Mattox rearrangement [5], and the other can take place directly along with a hydrogen on the 16-position (Scheme 57–61-2). In formulated drug products, the former dehydration, which yields the *E*- and *Z*-isomers of enol aldehyde, is more frequently observed than the latter one. Both acidic stress and basic stress can catalyze the dehydration. When stressed by diluted sulfuric acid, only the enol aldehydes were

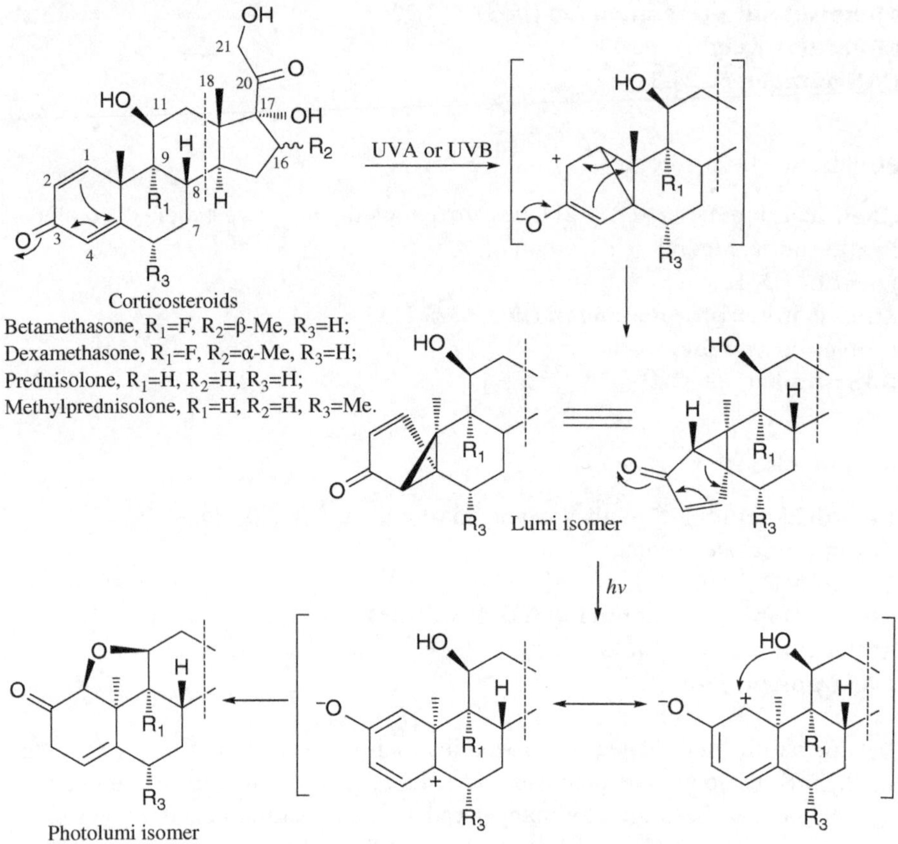

Scheme 57–61-1 Photochemical degradation pathways of betamethasone and related corticosteroids.

observed among the dehydrated products of betamethasone [6]. On the other hand, when stressed by hydrochloric acid, dehydration of betamethasone produces a significant amount of the 16,17-dehydro degradant in addition to the two isomers of enol aldehyde [1].

The enol aldehydes formed can undergo hydration on the aldehyde group, thanks to the activation by the neighboring keto group rendering the aldehyde group much more electrophilic. The hydrated intermediate, observable by LC-MS, can then isomerize to four acidic degradants of 17*H*-20-hydroxy-21-acids via intramolecular Cannizzaro rearrangement (Scheme 57–61-3) [7, 8].

The methylene at the 21-position is connected to both a keto and a hydroxyl group. Structure moiety of this type may be susceptible to oxidation via the carbanion/enolate auto-oxidation mechanism, especially in liquid formulations [8, 9]. The feasibility for such a mechanism (Scheme 57–61-4) is demonstrated in a base stress study of betamethasone [8]. The key intermediate is the Cregee type peroxide anion, which can have several routes of rearrangement, resulting in the formation of various degradants. Route a of Scheme 57–61-4 yields 21-aldehyde degradant, which is also

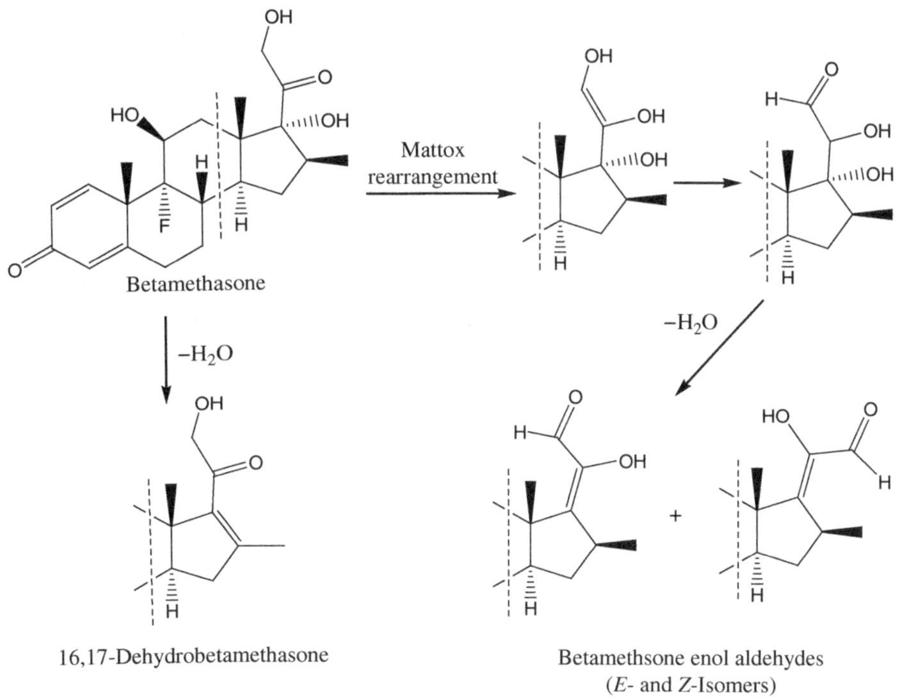

Scheme 57–61-2 Degradation of betamethasone via two different routes of dehydration.

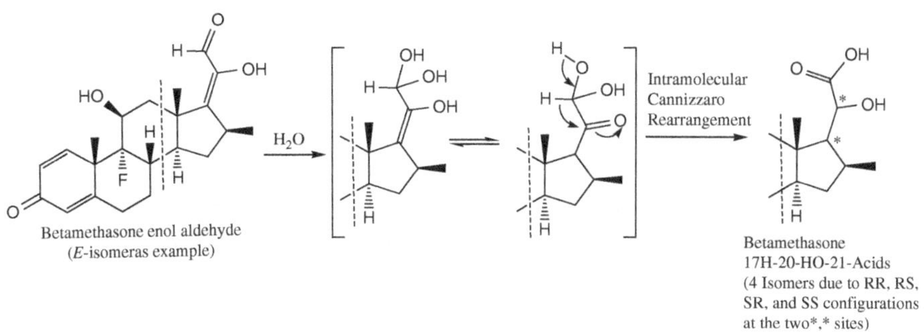

Scheme 57–61-3 Further degradation of betamethasone enol aldehyde, a relatively stable intermediate degradant of betamethasone, leading to the formation of four acidic degradants of betamethasone.

an α-keto aldehyde just like the enol aldehydes. Hence, the aldehyde group is susceptible to hydration. In reality, 21-aldehyde is more readily hydrated than the enol aldehydes, which is evidenced by the fact that it is exclusively present as the hydrated form once dissolved in aqueous solutions based on LC-MS data [8]. The hydrated form can also undergo intramolecular Cannizzaro rearrangement to produce 20-hydroxy-21-acid (Route b), which should be prone to retro-aldo to give 17-ketone degradant. However, it would also be possible that 17-ketone is formed via a direct

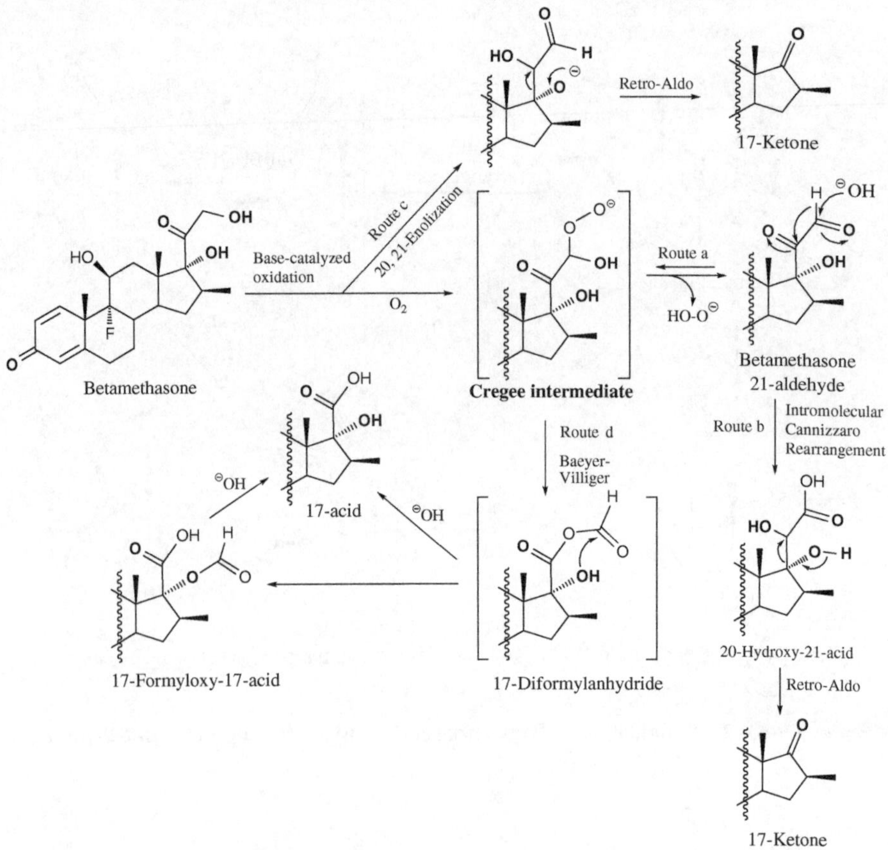

Scheme 57–61-4 Base-catalyzed oxidative degradation of betamethasone.

retro-aldo of the rearranged anion (Route c). In such a case, 17-ketone is not a result of oxidative degradation. Route d of Scheme 57–61-4 produces the diformylanhydride intermediate, which readily isomerizes to 17-formyloxy-17-acid. The latter is an ester of formic acid, which is not very stable and is eventually hydrolyzed to 17-acid. 17-Acid can also be formed directly from the 17-diformylanhydride intermediate.

It should be noted that the degradation pathways elucidated from this study may also be feasible for a free radical version of auto-oxidation in which the Cregee intermediate would be a peroxy free radical. The free radical mechanism may be more suitable for the steroids formulated in solid dosage forms.

The degradants discussed in this monograph have been observed in various drug products containing betamethasone or analogous corticosteroids that were stored under either ICH stability or ambient conditions [10–12], increasing the validity of the proposed carbanion/enolate auto-oxidation mechanism. Obviously not all the degradants are observed all the times. The occurrence or distribution of these degradants depends on the formulation and the age of a drug product sample. Certain degradants, e.g., 17-acid, 17-ketone, and enol aldehydes, are more frequently observed than others.

References

1 Hidaka, T., Huruumi, S., Tamaki, S. et al. (1980). Studies on betamethasone: behavior of betamethasone in acid or alkaline medium, photolysis and oxidation (Original in Japanese). *Yakugaku Zasshi* 100: 72–80.

2 Lin, M., Li, M., Buevich, A.V. et al. (2009). Rapid structure elucidation of drug degradation products using mechanism-based stress studies in conjunction with LC–MS^n and NMR spectroscopy: identification of a photodegradation product of betamethasone dipropionate. *J. Pharm. Biomed. Anal.* 50: 275–280.

3 Ricci, A., Fasani, E., Mella, M., and Albini, A. (2003). General patterns in the photochemistry of pregna-1,4-dien-3,20-diones. *J. Org. Chem.* 68: 4361–4366.

4 Miolo, G., Gallocchio, F., Levorato, L. et al. (2009). UVB photolysis of betamethasone and its esters: characterization of photoproducts in solution, in pig skin and in drug formulations. *J. Photochem. Photobiol. B Biol.* 96: 75–81.

5 Mattox, V.R. (1952). Steroids derived from bile acids. XV. The formation of a glyoxal side chain at C-17 from steroids with dihydroxyacetone and Δ^{16}-ketol side chains. *J. Am. Chem. Soc.* 74: 4340–4347.

6 Li, M., Chen, B., Lin, M. et al. (2007). A variation of Mattox rearrangement mechanism under alkaline condition. *Tetrahedron Lett.* 48: 3901–3905.

7 Görög, S. and Horváth, P. (1978). Analysis of steroids. Part XXXI. Mechanism of the tetrazolium reaction of corticosteroids. *Analyst* 103: 305–316.

8 Li, M., Chen, B., Monteiro, S., and Rustum, A.M. (2009). Mechanism of base-catalyzed autooxidation of corticosteroids containing 20-keto-21-hydroxyl side chain. *Tetrahedron Lett.* 50: 4575–4581.

9 Li, M. (2012). Carbanion/enolate-mediated autooxidation (base-catalyzed autooxidation). In: *Organic Chemistry of Drug Degradation*, 61–62. Cambridge: RSC Publishing.

10 Shou, M., Galinada, W.A., Wei, Y.-C. et al. (2009). Development and validation of a stability-indicating HPLC method for simultaneous determination of salicylic acid, betamethasone dipropionate and their related compounds in Diprosalic Lotion®. *J. Pharm. Biomed. Anal.* 50: 356–361.

11 Johnston, S.E., Gill, N.L., Wei, Y.-C. et al. (2010). Development and validation of a stability-indicating RP-HPLC method for simultaneous assay of betamethasone dipropionate, chlorocresol, and for the estimation of betamethasone dipropionate related compounds in a pharmaceutical cream and ointment. *J. Chromatogr. Sci.* 48: 733–741.

12 Matter, B., Ghaffari, A., Bourne, D. et al. (2019). Dexamethasone degradation in aqueous medium and implications for correction of in vitro release from sustained release delivery systems. *AAPS PharmSciTech* 20: 320. https://doi.org/10.1208/s12249-019-1508-7.

62–65: Betamethasone 21-Acetate, Betamethasone 17-Valerate, Betamethasone 17,21-Dipropionate, Betamethasone 21-Phosphate

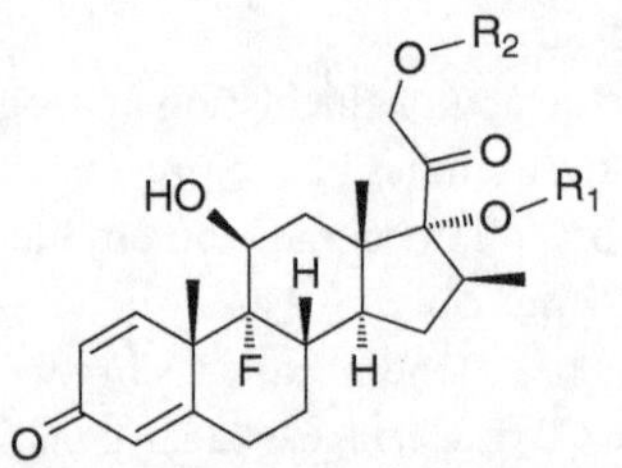

Betamethasone 21-acetate, R_1=H, R_2=Acetyl;
Betamethasone 17-valerate, R_1=Valeryl, R_2=H;
Betamethasone 17,21-dipropionate, R_1=R_2=Propionyl;
Betamethasone 21-Phosphate, R_1=H, R_2=Phosphoryl.

Esters of Betamethasone

62: Betamethasone 21-Acetate (also Known as Betamethasone Acetate)

Chemical name: (11β,16β)-9-Fluoro-11,17-dihydroxy-16-methyl-3,20-dioxopregna-1,4-dien-21-yl acetate
Brand name: Celestone Chronodose (injection combo with betamethasone sodium phosphate), Celestone Soluspan (injection combo with betamethasone sodium phosphate)
Formula: $C_{24}H_{31}FO_6$
Monoisotopic protonated ion (*m/z*): 435.2177
Molecular weight: 434.50
CAS number: 987-24-6

63: Betamethasone 17-Valerate (also Known as Betamethasone Valerate)

Chemical name: (11β,16β)-9-Fluoro-11,17-dihydroxy-16-methyl-3,20-dioxopregna-1,4-dien-21-yl valerate
Brand name: Betnovate, Celestoderm
Formula: $C_{27}H_{37}FO_6$
Monoisotopic protonated ion (*m/z*): 477.2647
Molecular weight: 476.59
CAS number: 2152-44-5

64: Betamethasone 17,21-Dipropionate (also Known as Betamethasone Dipropionate)

Chemical name: (11β,16β)-9-Fluoro-11-hydroxy-16-methyl-3,20-dioxopregna-1,4-diene-17,21-diyl dipropanoate

Brand name: Diprolene, Diprosone, Sernivo
Formula: $C_{28}H_{37}FO_7$
Monoisotopic protonated ion (*m/z*): 505.2596
Molecular weight: 504.60
CAS number: 5593-20-4

65: Betamethasone 21-Phosphate (also Known as Betamethasone Phosphate)

Chemical name: Disodium (11β,16β)-9-fluoro-11,17-dihydroxy-16-methyl-3,20-dioxopregna-1,4-dien-21-yl phosphate
Brand name: Celestone Chronodose (injection combo with betamethasone acetate), Celestone Soluspan (injection combo with betamethasone acetate)
Formula: $C_{22}H_{28}FNa_2O_8P$
Monoisotopic protonated ion (*m/z*): 473.1735
Molecular weight: 516.41
CAS number: 360-63-4

The esters of corticosteroids are also widely used anti-inflammatory therapeutic agents, and they are generally susceptible to hydrolytic degradation. For simplicity, we will use the esters of betamethasone as the prototype for our discussion and illustration of the degradation pathways of analogous corticosteroid esters. Among the esters, 21-acetate would be most susceptible to hydrolysis to yield betamethasone, while 21-phosphate is the least susceptible. This prediction can be made based on the activation energies for hydrolysis of the esters that can be used as model compounds for the ester drugs discussed in this monograph [1]. For example, the

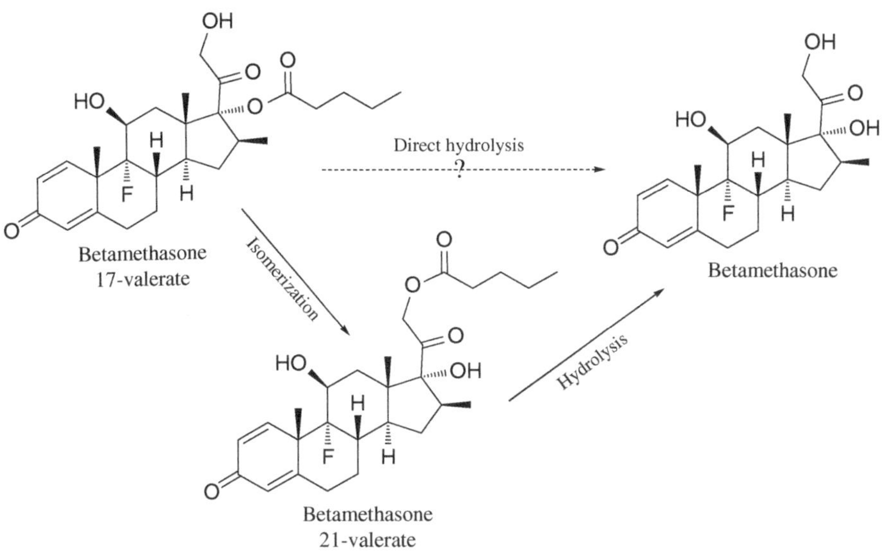

Scheme 62–65-1 Degradation of betamethasone 17-valerate.

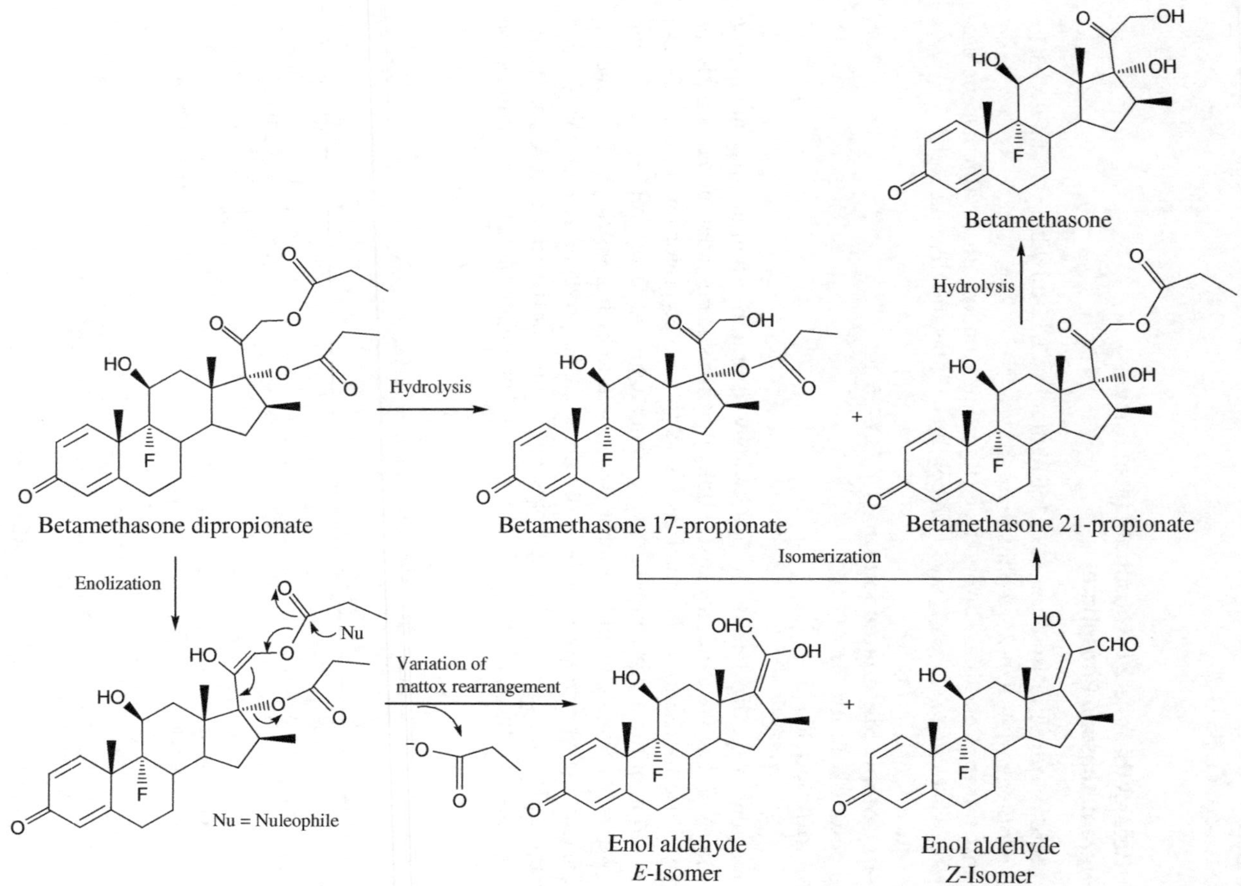

Scheme 62–65-2 Degradation pathways of betamethasone dipropionate.

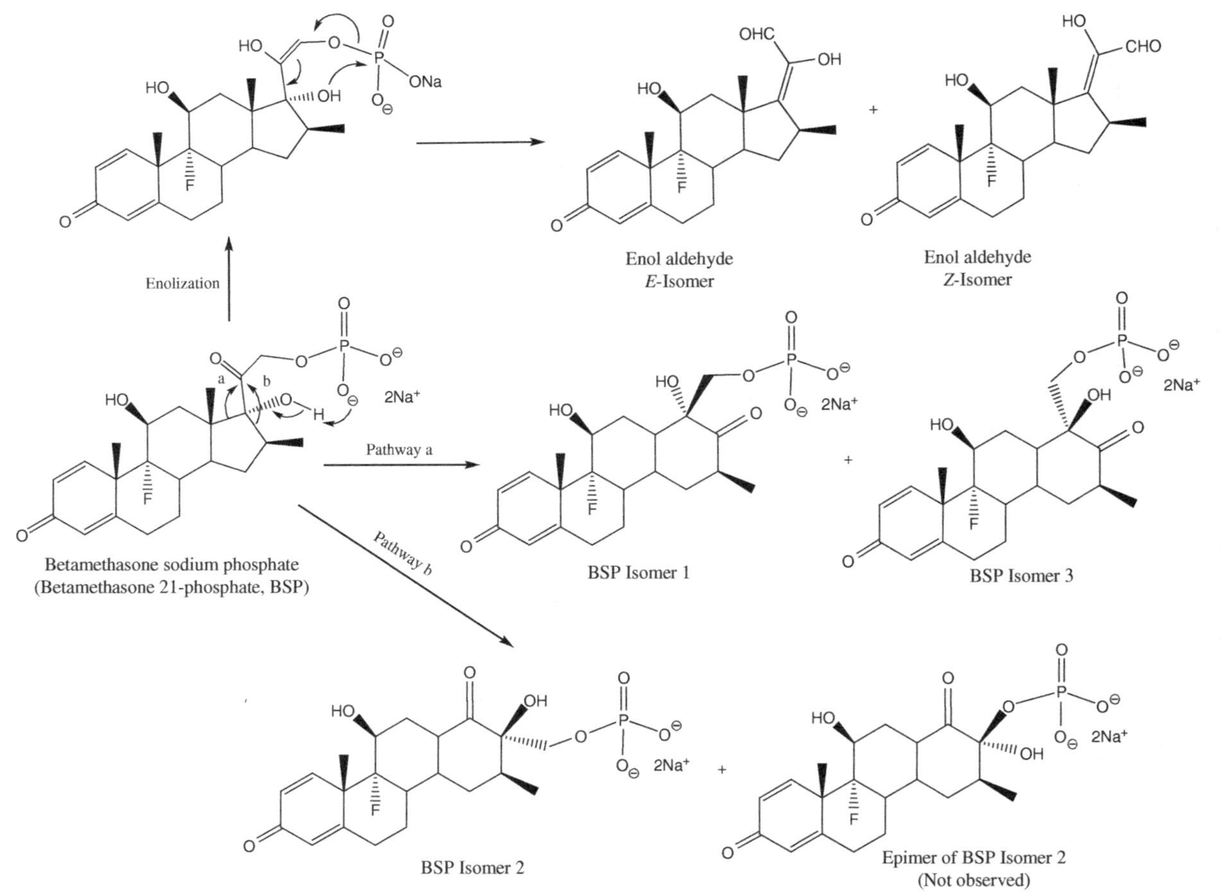

Scheme 62–65-3 Degradation pathways of betamethasone sodium phosphate.

activation energy for hydrolysis of a typical acetate is usually no more than 15 kcal/mol. On the other hand, the activation energy for hydrolysis of a typical phosphate is usually ca. 30 kcal/mol. Such a high energy barrier implies that hydrolysis would be an insignificant degradation pathway for betamethasone 21-phosphate and its formulated products that are stored under long-term stability conditions.

According to a study by Hansen and Bundgaard, betamethasone 17-varylate does not undergo hydrolysis directly [2]. Instead, it isomerizes to 21-varylate, which is then followed by the hydrolysis of the latter (Scheme 62–65-1).

The hydrolysis of betamethasone 17,21-dipropionate produces 17-mono ester and 21-mono ester, respectively. The mono esters can further degrade to betamethasone; the degradation of 17-mono ester would likely proceed through 21-mono ester, in the same fashion as betamethasone 17-valerate. Additionally, betamethasone 17,21-dipropionate can undergo the variation of the Mattox rearrangement to yield the two enol aldehydes directly (Scheme 62–65-2) [3, 4].

Betamethasone 21-phosphate can also degrade to the two enol aldehydes directly, through a mechanism (Scheme 62–65-3) that is somewhat different from the variation of the Mattox rearrangement [5]. The enol aldehydes formed can further degrade to give four 17*H*-20-hydroxy-21-acids as discussed previously. In a solid state, the D-ring of betamethasone 21-phosphate can undergo isomeric ring expansion, a process referred to as D-homoannular ring expansion [5]. During the latter process, four isomeric degradants could be produced; however, in the case of betamethasone sodium phosphate, only three such degradants were observed. The missing one (epimer of BSP Isomer 2, Scheme 62–65-3) could be too difficult to form, probably due to steric hindrance. In all the esters, the A-rings remain intact, and hence they are still susceptible to the photochemical rearrangement as discussed in the previous monograph.

References

1 Li, M. (2012). Hydrolytic degradation. In: *Organic Chemistry of Drug Degradation*, 16–47. Cambridge: RSC Publishing.

2 Bundgaard, H. and Hansen, J. (1981). Studies on the stability of corticosteroids VI. Kinetics of the rearrangement of betamethasone-17-valerate to the 21-valerate ester in aqueous solution. *Int. J. Pharm.* 7: 197–203.

3 Li, M., Chen, B., Lin, M. et al. (2007). A variation of Mattox rearrangement mechanism under alkaline condition. *Tetrahedron Lett.* 48: 3901–3905.

4 Chen, B., Li, M., Lin, M. et al. (2009). A comparative study of enol aldehyde formation from betamethasone, dexamethasone, beclomethasone and related compounds under acidic and alkaline conditions. *Steroids* 74: 30–41.

5 Li, M., Wang, X., Chen, B. et al. (2009). Forced degradation of betamethasone sodium phosphate under solid state: formation, characterization, and mechanistic study of all four 17,20-diastereomers of betamethasone 17-deoxy-20-hydroxy-21-oic acid. *J. Pharm. Sci.* 98: 894–904.

66–69: Brompheniramine, Dexbrompheniramine, Chlorpheniramine, Pheniramine

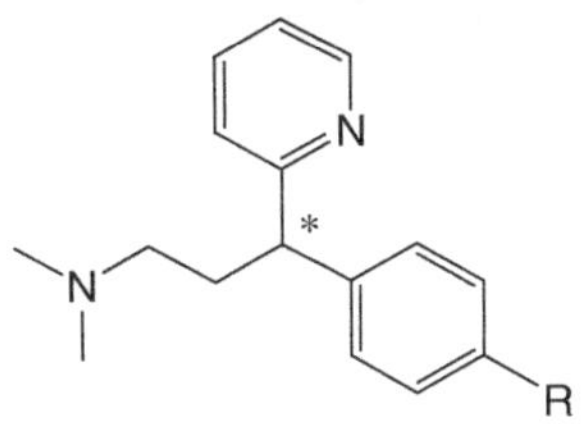

Brompheniramine, R = Br, * = racemic,
Dexbrompheniramine, R = Br, * = *S*-isomer,
Chlorpheniramine, R = Cl, * = Racemic
Pheniramine, R = H, * = Racemic.

66: Brompheniramine

Chemical name: (*R,S*)-3-(4-Bromophenyl)-*N,N*-dimethyl-3-pyridin-2-yl-propan-1-amine
Brand name: Bromfed, Dimetapp, Bromfenex
Formula: $C_{16}H_{19}BrN_2$
Monoisotopic protonated ion (*m/z*): 319.0804
Molecular weight: 319.25
CAS number: 86-22-6
Common salt form: Maleate

67: Dexbrompheniramine

Chemical name: (3*S*)-3-(4-Bromophenyl)-*N,N*-dimethyl-3-pyridin-2-yl-propan-1-amine
Brand name: Drixoril, Conex, Dixaphedrine
Formula: $C_{16}H_{19}BrN_2$
Monoisotopic protonated ion (*m/z*): 319.0804
Molecular weight: 319.25
CAS number: 132-21-8
Common salt form: Maleate

68: Chlorpheniramine

Chemical name: 3-(4-Chlorophenyl)-*N,N*-dimethyl-3-(pyridin-2-yl)-propan-1-amine
Brand name: Demazin
Formula: $C_{16}H_{19}ClN_2$
Monoisotopic protonated ion (*m/z*): 275.1310
Molecular weight: 274.79
CAS number: 132-22-9
Common salt form: Maleate, HCl salt

69: Pheniramine

Chemical name: *N,N*-Dimethyl-3-phenyl-3-pyridin-2-yl-propan-1-amine
Brand name: Under various brand names in many OTC combo medicines
Formula: $C_{16}H_{20}N_2$
Monoisotopic protonated ion (*m/z*): 241.1699
Molecular weight: 240.35
CAS number: 86-21-5
Common salt form: Maleate

Brompheniramine, dexbrompheniramine, chlorpheniramine, and pheniramine are first-generation antihistamine medicines, and their therapeutic use can date back to the 1950s. Nowadays, they are mostly utilized in combination with other therapeutic agents as OTC drugs. Despite their long history of medical use, the literature on their degradation chemistry is scarce. On the contrary, based on their structures, it is reasonable to assume that the tertiary amine moiety of this drug family would be susceptible to nucleophilic oxidation to produce *N*-oxide degradants under long-term stability and forced degradation conditions. During a forced degradation study of a cough-cold medicine containing chlorpheniramine, an oxidative degradant was observed, which contains an extra oxygen as compared to chlorpheniramine [1]. Although the original paper did not identify the exact structure of this degradant, it is quite likely the tertiary amine *N*-oxide (Scheme 66–69-1). Chlorpheniramine also has a pyridine group that could form the pyridine *N*-oxide. Nevertheless, the formation of pyridine *N*-oxide would be less likely than the formation of the tertiary amine *N*-oxide, due to the poor nucleophilicity of the pyridine ring.

Chlorpheniramine and dexbrompheniramine are typically present in their maleate salt forms for use as active pharmaceutical ingredients. In several OTC combo drugs, chlorpheniramine maleate and dexbrompheniramine maleate are used in combination with pseudoephedrine, a very common nasal decongestant. In these formulated products, maleate may react with pseudoephedrine, a secondary amine, to form a degradant via Michael addition (Scheme 66–69-2) [2, 3].

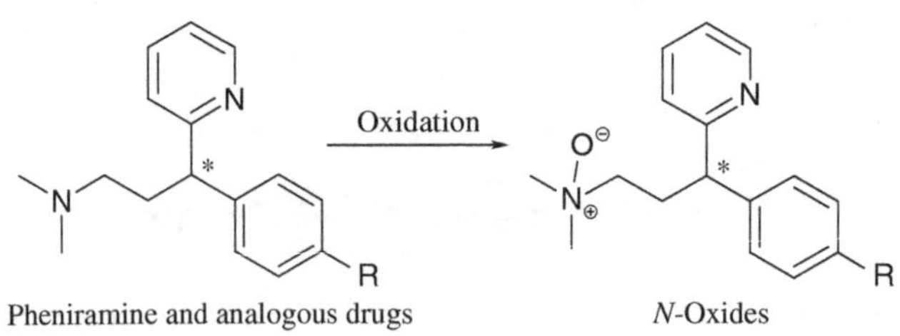

Scheme 66–69-1 Oxidative degradation of pheniramine and analogous drugs.

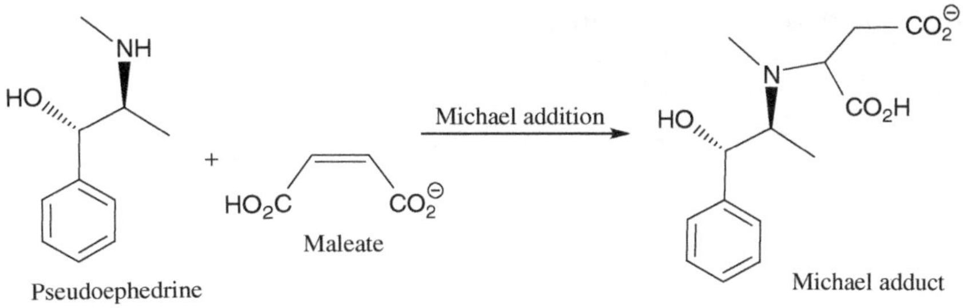

Scheme 66-69-2 Degradation of pseudoephedrine via Michael addition.

References

1 Marin, A. and Barbas, C. (2004). LC/MS for the degradation profiling of cough–cold products under forced conditions. *J. Pharm. Biomed. Anal.* 35: 1035–1045.
2 Wong, J., Wiseman, L., Al-Mamoon, S. et al. (2006). Major degradation product identified in several pharmaceutical formulations against the common cold. *Anal. Chem.* 78: 7891–7895.
3 Chan, T.-M., Kong, J., McNamara, P., and Wong, J.K. (2008). Syntheses of potential degradation products of phenylephrine in OTC products. *Synth. Commun.* 38: 2252–2260.

70–71: Budesonide, Desonide

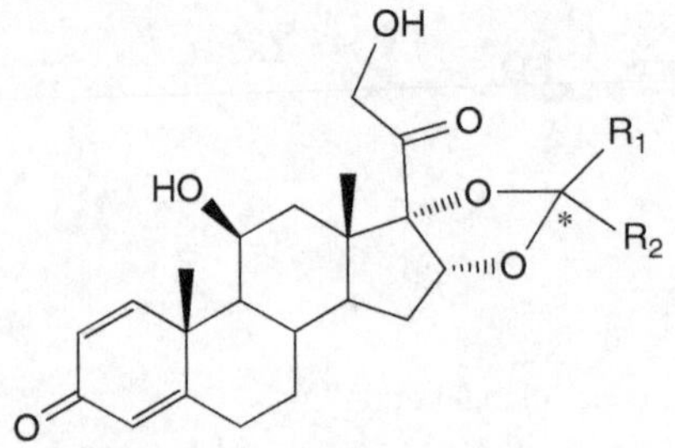

Budesonide, R_1 = Propyl, R_2 = H, * = racemic;
Desonide, R_1 = R_2 = Methyl.

70: Budesonide

Chemical name: 11β,21-Dihydroxy-16α,17α-[butane-1,1-diylbis(oxy)]pregna-1,4-diene-3,20-dione
Brand name: Pulmicort, Rhinocort, Entocort
Formula: $C_{25}H_{34}O_6$
Monoisotopic protonated ion (*m/z*): 431.2428
Molecular weight: 430.54
CAS number: 51333-22-3

71: Desonide

Chemical name: 11β,21-Dihydroxy-16α,17α-[(1-methylethylidene)bis(oxy)]pregna-1,4-diene-3,20-dione
Brand name: Desowen
Formula: $C_{24}H_{32}O_6$
Monoisotopic protonated ion (*m/z*): 417.2272
Molecular weight: 416.51
CAS number: 638-94-8

Budesonide and desonide are acetal derivatives of nonhalogenated glucocorticosteroids that are widely used in the treatment of inflammatory diseases. Budesonide is a condensation product between the steroid core and butyraldehyde; the formation of the five-membered acetal ring creates a new chiral center. Hence, budesonide is a mixture of two epimers, and major pharmacopoeias require a specific range of ratios between the two epimers [1]. On the other hand, desonide is a condensation product between the steroid core and acetone; the formation of its acetal ring does not create any new chiral center due to the presence of the two identical methyl groups.

Ferraboschi et al. performed a detailed study of budesonide tablets that were put on stability at 25°C/60%RH for 3 years [2]. Five major degradants were identified based on structural characterization data and/or synthesis of the degradants that were not included in the European Pharmacopoeia. Although the authors described the general pathways for each major degradant, detailed mechanisms were lacking particularly for the formation of the D-homoannual ring expansion product (EP impurity C) and budesonide 16-butyroyl-17-ketone. The current author attempted to propose possible mechanisms for the degradants (Scheme 70–71-1). Apparently, budesonide 11-ketone results from the oxidation of the 11-hydroxyl group. Notably, the oxidation of the 11-hydroxyl group is usually not a significant degradation pathway for fluoro-containing corticosteroids such as betamethasone and its derivatives under long-term stability conditions. The D-ring annual ring expansion likely involves a concerted rearrangement as depicted in Scheme 70–71-1. The formation of budesonide 16-butyroyl-17-ketone probably starts with the oxidation on the acetal carbon: a free-radical mediated oxidation would first give a peroxyl intermediate, which can decompose to the hydroxyl intermediate with concurrent enolization of the hydroxyacetone side chain (Scheme 70–71-1) [3]. A retro-aldo process of the hydroxyl intermediate should produce the budesonide 16-butyroyl-17-ketone [4]. The oxidation on the 21-carbon (of the hydroxyacetone side chain) should yield the budesonide 21-aldehyde (EP impurity D) and 17-acid. The mechanism for the oxidation should be analogous to that of betamethasone [4]. In the solid state such as the stability study of budesonide tablets, the oxidation is likely mediated via free radicals. However, in liquid formulations, the oxidation would more likely proceed through the carbanion/enolate auto-oxidation pathway [5]. In a stability study of budesonide in propylene glycol that was continuously being oxygenated and stored at 40°C for 3 months, more than 10% of the API was found to be decomposed and approximately 5 to 7 degradants were observed, of which only the two 21-aldehyde epimers were identified based on MS data [6]. In two patents, it was revealed that ethanolic solutions of budesonide can be stabilized by the addition of organic acids [7, 8]. These reports suggest that this steroid is also likely susceptible to base-catalyzed carbanion/enolate auto-oxidation in liquid formulations [9], similar to the base-catalyzed oxidation of betamethasone [4].

The steroid core of desonide is identical to that of budesonide. Hence, desonide should undergo similar degradation pathways with regard to its steroid core. Indeed, desonide was found to be photochemically unstable primarily due to its A-ring [10],[1] which can be problematic because the drug is frequently utilized in topical applications. Mechanistic study revealed that under irradiation at 254 nm, the main photodegradant was the lumi-degradant of desonide (Scheme 70–71-2). When irradiated at 310 nm under the aerated and nonaerated conditions, the main photodegradants were 17-peroxyl desonide and 17*H*-desonide, respectively. The peroxyl degradant can further degrade to desonide 17-ketone. The photochemical stability of formulated desonide can be enhanced by the use of benzophenone-3 [11]. In a study by Nguyen et al., desonide 17-acid and 17*H*-desonide were observed as degradants in a desonide ointment product [12]. The two degradants were synthesized by treating desonide with alkaline hydrogen peroxide.

1 Also refer to the Monograph for betamethasone and its analogous drugs in this book.

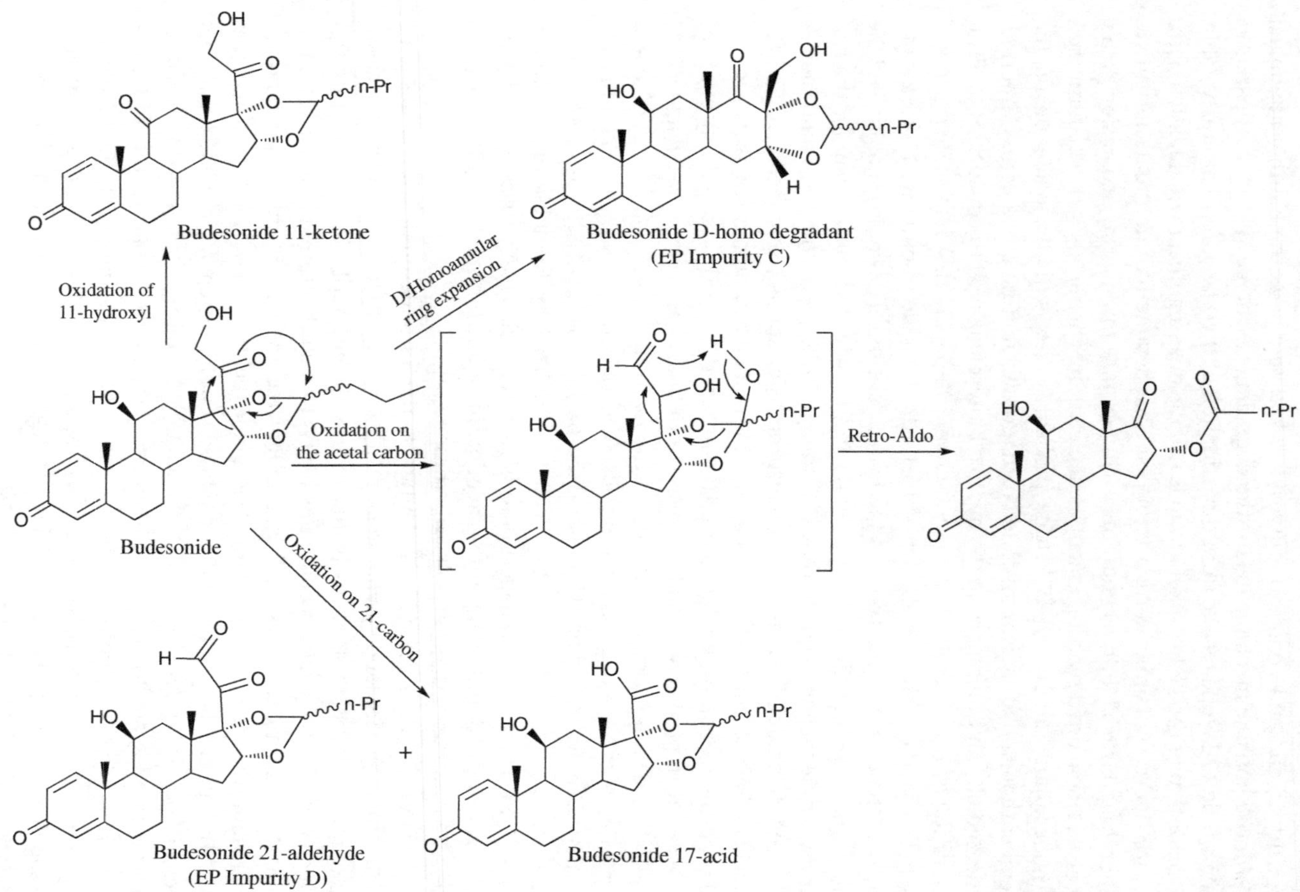

Scheme 70–71-1 Degradation pathways of budesonide.

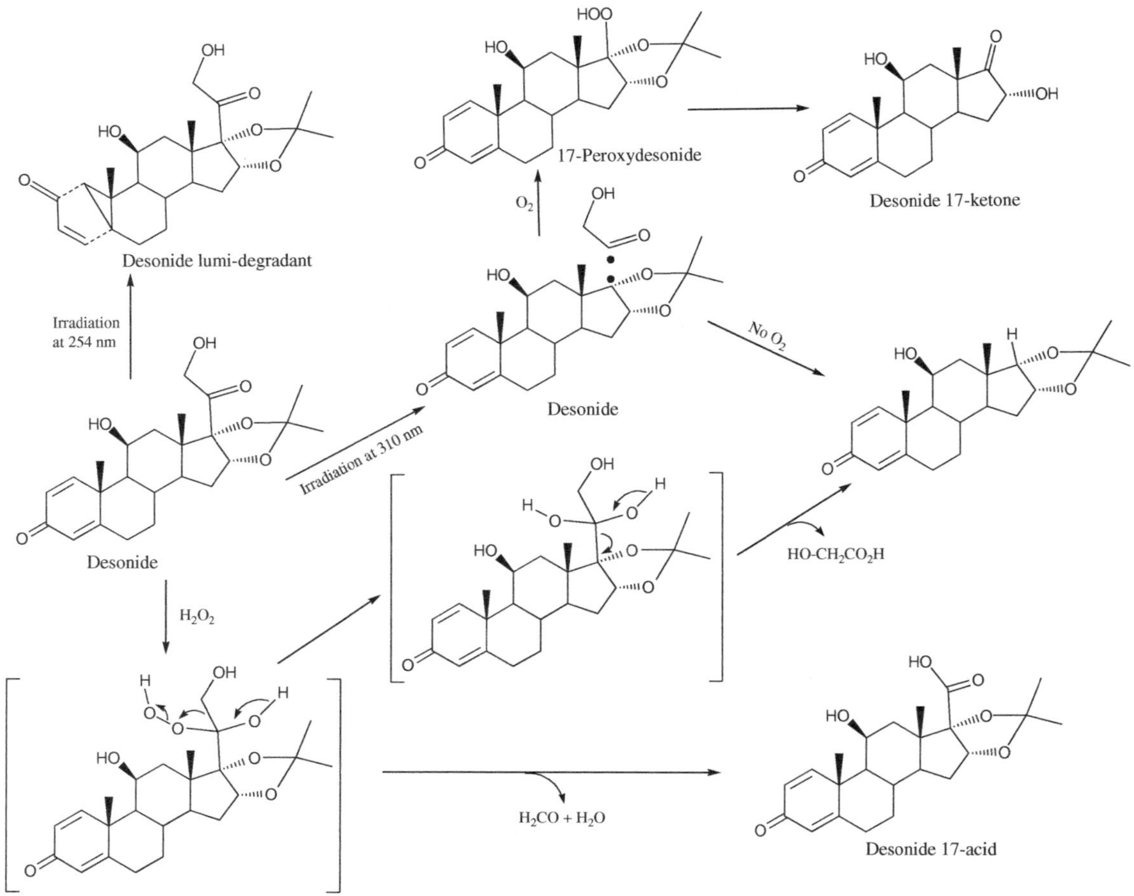

Scheme 70–71-2 Degradation pathways of desonide.

References

1 Budesonide. United States Pharmacopoeia, Official as of 1 July 2021, DOI: 10.31003/USPNF_M10458_04_01.

2 Ferraboschi, P., Bertacche, V., Maccone, I. et al. (2008). Estimation and characterisation of budesonide tablets impurities. *J. Pharm. Biomed. Anal.* 47: 636–640.

3 Li, M., Wang, X., Chen, B. et al. (2009). Forced degradation of betamethasone sodium phosphate under solid state: formation, characterization, and mechanistic study of all four 17,20-diastereomers of betamethasone 17-deoxy-20-hydroxy-21-oic acid. *J. Pharm. Sci.* 98: 894–904.

4 Li, M., Chen, B., Monteiro, S., and Rustum, A.M. (2009). Mechanism of base-catalyzed autooxidation of corticosteroids containing 20-keto-21-hydroxyl side chain. *Tetrahedron Lett.* 50: 4575–4581.

5 Li, M. (2012). Carbanion/enolate-mediated autooxidation (Base-catalyzed autooxidation). In: *Organic Chemistry of Drug Degradation*, 61–62. Cambridge: RSC Publishing.

6 Hou, S., Hindle, M., and Byron, P.R. (2001). A stability-indicating HPLC assay method for budesonide. *J. Pharm. Biomed. Anal.* 24: 371–380.

7 Freund B, Krueger M, Zierenberg B, (2002) Stable pharmaceutical budesonide preparation for producing propellant-free aerosols. US Patent No. 6,491,897 B1.

8 Freund B, Krueger M, Zierenberg B, (2011) Stable pharmaceutical budesonide preparation for producing propellant-free aerosols. US Patent No US 8,062,626 B2.

9 Slavíková, B., Kasal, A., and Buděšínský, M. (1999). Autoxidation vs hydrolysis in 16α-acyloxy steroids. *Collect. Czechoslov. Chem. Commun.* 64: 1125–1134.

10 Iqbal, J., Husain, A., and Gupta, A. (2006). Photochemistry of desonide, a non-fluorinated steroidal anti-inflammatory. *Drug Chem. Pharm. Bull.* 54: 836–838.

11 Nguyen, T.T., Kringstad, R., Aasen, A.J., and Rasmussen, K.E. (1988). Identification and analysis of a degradation product of the glucocorticoid desonide in ointment. *Acta Chem. Scand. B* 42: 403–407.

12 Rosa, P., Salla, A.P.S., de Bona da Silva, C. et al. (2014). Investigation of the stabilizing effects of antioxidants and benzophenone-3 on desonide photostability. *AAPS PharmSciTech* 15: 1155–1162.

72: Bupropion

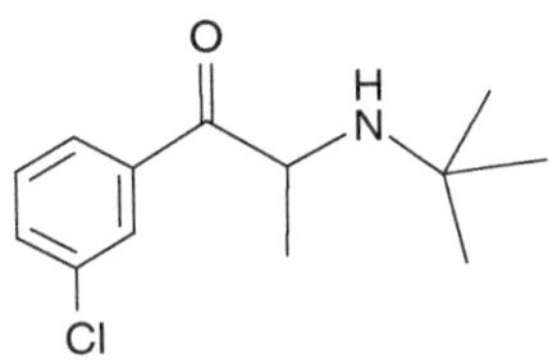

Chemical name: (*R,S*)-2-(*tert*-Butylamino)-1-(3-chlorophenyl)propan-1-one
Brand name: Wellbutrin, Zyban, Elontril
Formula: $C_{13}H_{18}ClNO$
Monoisotopic protonated ion (*m/z*): 240.1150
Molecular weight: 239.74
CAS number: 34911-55-2
Common salt form: HCl salt

Bupropion is an aminoketone that has many therapeutic indications, including antidepression and smoking cessation; it is one of the most prescribed antidepressants in the United States. O'Byrne et al. studied the stability of bupropion in aqueous solutions over a pH range of 2–13 [1]. No bupropion decomposition was found at 60°C in a solution of pH ≤ 5 in which the drug molecule is mostly in its protonated form (the pKa of bupropion is 7.9). In neutral to alkaline solutions, bupropion started to degrade, and the first degradant is the hydroxyketone, which is formed via nucleophilic substitution of the amino moiety by water/hydroxide ion (Scheme 72). The hydroxyketone then undergoes base-catalyzed auto-oxidation as well as enolization to yield other three degradants, i.e., 3-chlorobenzoic acid, the β-keto degradant, and the diketo degradant. There are a number of bupropion formulation patents in which both organic and inorganic acids were used to stabilize the drug products [2–4]. This fact is consistent with the base-catalyzed oxidation mechanism as mentioned above.

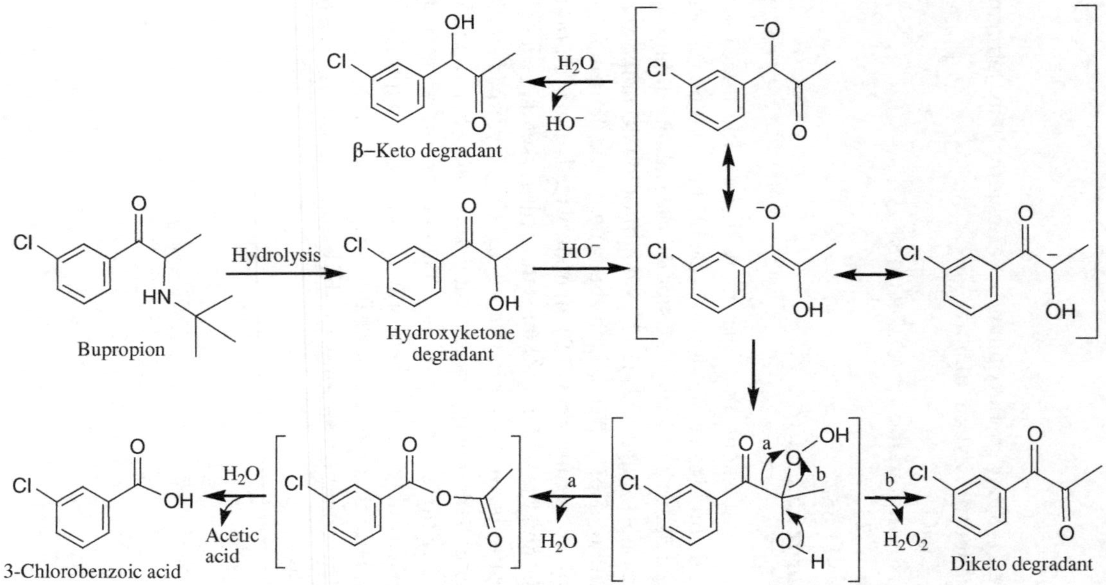

Scheme 72 Hydrolytic degradation of bupropion in neutral to alkaline solutions.

References

1 O'Byrne, P.M., Williams, R., Walsh, J.J., and Gilmer, J.F. (2010). The aqueous stability of bupropion. *J. Pharm. Biomed. Anal.* 53: 376–381.

2 M.D. Ruff, S.R. Kalidindi, J.E. Sutton Jr., Pharmaceutical composition containing bupropion hydrochloride and a stabilizer, US Patent No. 5,358,970 (1994).

3 A. Maitra, P.S. Kulkarni, B.B. Shah, J.M. Devito, Pharmaceutical composition containing bupropion hydrochloride and an inorganic acid stabilizer, US Patent No. 5,968,553 (1999).

4 C.-H. Han, G. Liaw, Stabilized pharmaceutical compositions containing bupropion hydrochloride, US Patent No. 6,333,332 (2001).

73: Buspirone

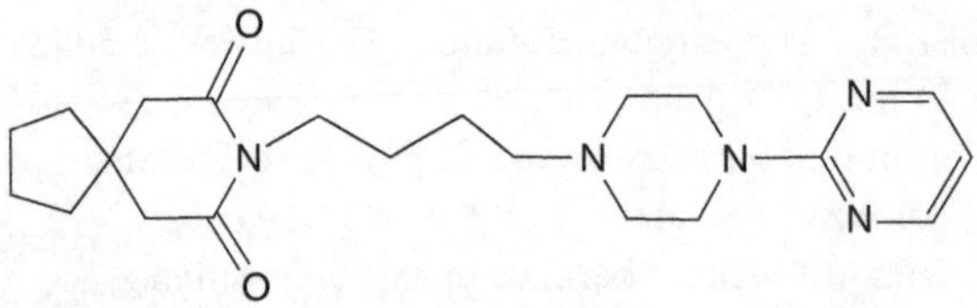

Chemical name: 8-{4-[4-(Pyrimidin-2-yl)piperazin-1-yl]butyl}-8-azaspiro[4.5] decane-7,9-dione
Brand name: Buspar
Formula: $C_{21}H_{31}N_5O_2$
Monoisotopic protonated ion (*m/z*): 386.2551
Molecular weight: 385.51
CAS number: 36505-84-7
Common salt form: HCl salt

Buspirone is a serotonin $5HT_{1A}$ receptor agonist of the arylpiperazine class that is clinically utilized for the treatment of general anxiety [1]. Its structural moieties that are susceptible to degradation include the cyclic imide, piperazine, and pyrimidine ring. The two symmetric imide bonds are equally vulnerable to hydrolytic attack, giving rise to the buspirone acid [2]. The nitrogens of the piperazine ring, particularly the piperazine nitrogen connected to the butyl group, and those of the pyrimidine ring can be oxidized by H_2O_2 to form the piperazine and pyrimidine *N*-oxides, with the piperazine *N*-oxide much preferred over the pyrimidine *N*-oxide [3]. The degradation pathways of buspirone are summarized in Scheme 73.

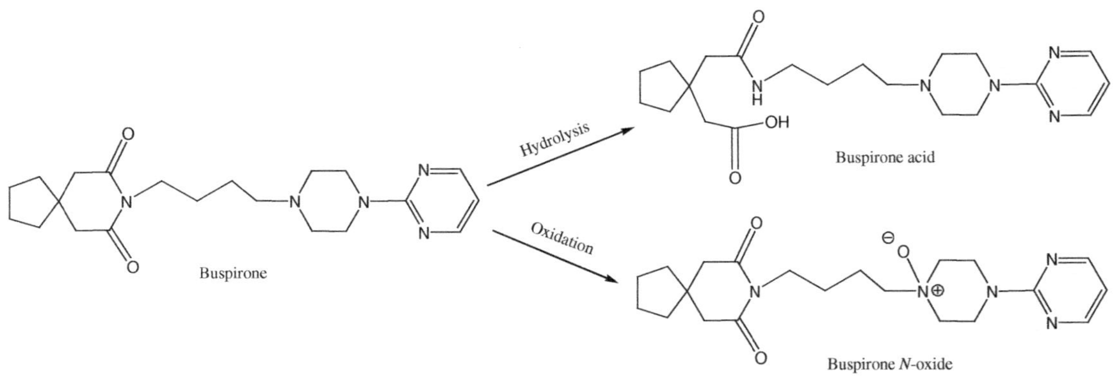

Scheme 73 Degradation pathways of buspirone.

References

1 Bockaert, J., Dumuis, A., Bouhelal, R. et al. (1987). Piperazine derivatives including the putative anxiolytic drugs, buspirone and ipsapirone, are agonists at 5-HT_{1A} receptors negatively coupled with adenylate cyclase in hippocampal neurons. *Naunyn Schmiedeberg's Arch. Pharmacol.* 335: 588–592.

2 Khedr, A. and Sakr, A. (1999). Stability-indicating high-performance liquid chromatographic assay of buspirone HCl. *J. Chromatogr. Sci.* 37: 462–468.

3 Chen, B.-C. and Stark, D.R. (1996). A facile preparation of buspirone N-oxide using Davis' reagent. *Org. Prep. Proced. Int.* 28: 115–117.

74–78: Butalbital and Related Barbiturates

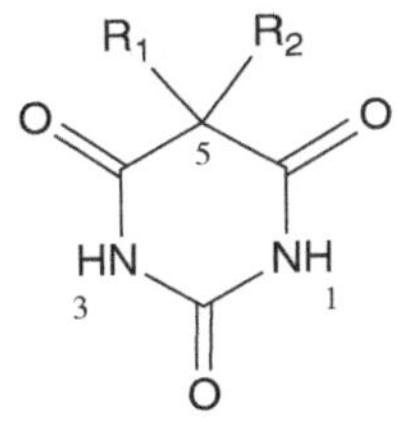

Butalbital, R_1=1-Isobutyl, R_2=2-Propenyl;
Pentobarbital, R_1=Ethyl, R_2=2-Pentyl;
Phenobarbital, R_1=Ethyl, R_2=Phenyl;
Secobarbital, R_1=2-Pentyl, R_2=2-Propenyl;
Talbital, R_1=1-sec-Butyl, R_2=2-Propenyl.

74: Butalbital

Chemical name: 5-(2-Methylpropyl)-5-(2-propenyl)-2,4,6(1*H*,3*H*,5*H*)-pyrimidin etrione
Brand name: Various brand names for its combination with other drugs.
Formula: $C_{11}H_{16}N_2O_3$
Monoisotopic protonated ion (*m/z*): 225.1234
Molecular weight: 224.26
CAS number: 77-26-9

75: Pentobarbital

Chemical name: 5-Ethyl-5-(1-methylbutyl)-2,4,6(1*H*,3*H*,5*H*)-pyrimidinetrione
Brand name: Nembutal
Formula: $C_{11}H_{18}N_2O_3$
Monoisotopic protonated ion (*m/z*): 227.1390
Molecular weight: 226.28
CAS number: 76-74-4

76: Phenobarbital [WHO List of Essential Medicines]

Chemical name: 5-Ethyl-5-phenyl-2,4,6(1*H*,3*H*,5*H*)-pyrimidinetrione
Brand name: Luminal
Formula: $C_{12}H_{12}N_2O_3$
Monoisotopic protonated ion (*m/z*): 233.0921
Molecular weight: 232.24
CAS number: 50-06-6

77: Secobarbital

Chemical name: (*R,S*)-5-(1-Methylbutyl)-5-(2-propen-1-yl)-2,4,6(1*H*,3*H*,5*H*)-pyrimidinetrione
Brand name: Seconal

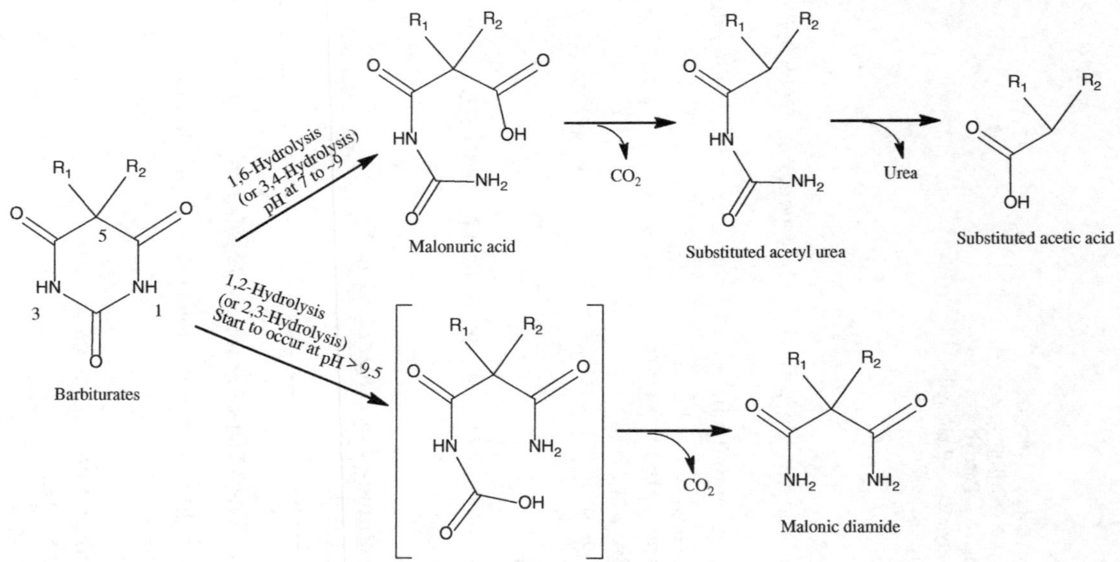

Scheme 74–78 Hydrolytic degradation of barbiturates under different pHs.

Formula: $C_{12}H_{18}N_2O_3$
Monoisotopic protonated ion (*m/z*): 239.1390
Molecular weight: 238.28
CAS number: 76-73-3

78: Talbutal

Chemical name: (*R,S*)-5-Allyl-5-sec-butyl-2,4,6(1*H*,3*H*,5*H*)-pyrimidinetrione
Brand name: Lotusate
Formula: $C_{11}H_{16}N_2O_3$
Monoisotopic protonated ion (*m/z*): 225.1234
Molecular weight: 224.26
CAS number: 115-44-6

Barbiturates are derivatives of barbituric acid with different substituents at the 4-position of the barbituric acid ring. Barbiturates have a long history of medical use as sedatives, hypnotics, and antiepileptic, which dates back to as early as the 1910s [1]. The molecular weights of bartiturates for medical use are relatively small. Their modes of action had not been known until quite recently: they act as positive modulators of $GABA_A$ receptors, and at high concentrations, they become agonists of the receptors [2]. Among the $GABA_A$ receptor-binding sites, barbiturates bind at sites different from those by benzodiazepines, and their use in routine medical practice has largely been replaced by benzodiazepines and other drugs with less side effects or less tendency for dependency. Nevertheless, barbiturates are still used in general anesthesia, epilepsy, and other cases. In pharmaceutical dosage forms, both the free acid form and salt form (such as sodium, calcium, or other salts) are employed. Among the barbiturates, butalbital is usually used in conjunction with other therapeutic agents such as acetaminophen, aspirin, caffeine, and codeine in binary or tertiary combination for the treatment of pain and headache.

The main degradation pathway of the barbiturates is the hydrolysis of the barbituric acid ring, followed by decarboxylation and further hydrolysis (Scheme 74–78) [3–5]. According to a study by Maulding et al., the hydrolytic cleavage site depends on the pH of the media [3]. At neutral pH and a few pH units above, the 1,6-(or 3,4-) cleavage is the predominant pathway. At pH greater than 9.5, the 1,2-(or 2,3-) cleavage plays a role in the overall degradation of the drug. The activation energy for the hydrolytic degradation of phenobarbital was reported to be ~19 kcal/mol [6, 7].

References

1 Barbiturate, Wikipedia. https://en.wikipedia.org/wiki/Barbiturate (accessed 2 May 2022).
2 Löscher, W. and Rogawski, M.A. (2012). How theories evolved concerning the mechanism of action of barbiturates. *Epilepsia* 53: 12–25.

3 Maulding, H.V., Nazareno, J., Polesuk, J., and Michaelis, A. (1972). Investigations of hydrolytic products of butalbital. *J. Pharm. Sci.* 61: 1389–1393.
4 Scott, E.P. (1983). Application of postcolumn ionization in the high-performance liquid chromatographic analysis of butabarbital sodium elixir. *J. Pharm. Sci.* 72: 1089–1091.
5 Gannett, P.M., Daft, J.R., James, D. et al. (2001). In vitro reaction of barbiturates with formaldehyde. *J. Anal. Toxicol.* 25: 443–449.
6 Garrett, E.R., Bojarski, J.T., and Yakatan, G.J. (1971). Kinetics of hydrolysis of barbituric acid derivatives. *J. Pharm. Sci.* 60: 1145–1154.
7 Tarsa, M., Zuchowski, G., Stasiewicz-Urban, A., and Bojarski, J. (2009). Hydrolysis of 2,4-dithiophenobarbital. *Acta Pol. Pharm.-Drug Res.* 66: 123–128.

4

Carbamazepine to Decitabine

79: Carbamazepine [WHO List of Essential Medicines]

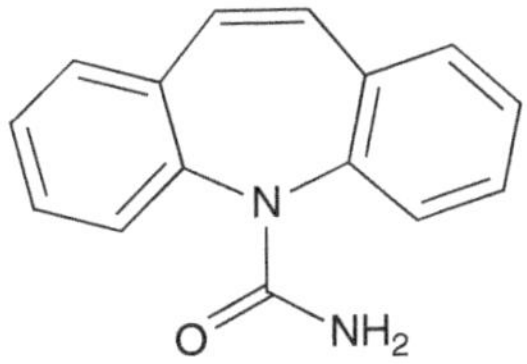

Chemical name: Benzo[b][1]benzazepine-11-carboxamide
Brand name: Tegretol
Formula: $C_{15}H_{12}N_2O$
Monoisotopic protonated ion (*m/z*): 237.1022
Molecular weight: 236.27
CAS number: 298-46-4

Carbamazepine is one of the most prescribed CNS drugs, which is most commonly used for the treatment of epilepsy. Due to its widespread use, there have been a large number of papers describing its disposal or decomposition in sewage [1–3]. On the other hand, few studies in the literature have reported its degradation under pharmaceutically relevant conditions. Hence, an effort is made here to predict its degradation under pharmaceutically relevant conditions, considering the results reported from the papers regarding its disposal.

Based on the structure of carbamazepine, this drug molecule would appear to be susceptible to oxidative, hydrolytic, and photochemical degradation, due to its carbon–carbon double bond, carbamate, and tricyclic moieties, respectively. First, the oxidation of the carbon–carbon double bond could produce carbamazepine 10,11-epoxide as the initial degradant, which is also a metabolite of carbamazepine [4]. The analogous epoxide was reported in the case of cyclobenzaprine, which has a similar carbon-carbon double bond that is also embedded in a tricyclic conjugated ring system (refer to the Monograph of Cyclobenzaprine and Cyproheptadine of this book). The initially formed epoxide would decompose to produce the

Compendium of Drug Degradation Pathways, First Edition. Min Li.

di-hydroxyl degradant, carbamazepine-diol, upon attack by water. Second, the carbamate moiety is usually stable enough to resist significant hydrolytic degradation under pharmaceutically relevant conditions. Nevertheless, under harsh conditions such as refluxing in 1N HCl or 1N NaOH, the carbamate can be hydrolyzed to yield iminostilbene [5], which was reported to be a metabolite of carbamazepine [6]. The above degradation pathways are illustrated in Scheme 79-1.

Lastly, the tricyclic ring would be likely susceptible to photochemical degradation, including photosensitized oxidation [7]. Indeed, under UV irradiation at 254 nm, iminostilbene was also formed and its formation was proposed via either a Chugaev-type elimination or a photochemical hydrolysis. The mechanisms presented in Scheme 79-2 were based on those proposed by Ali et al. [1], with modification.

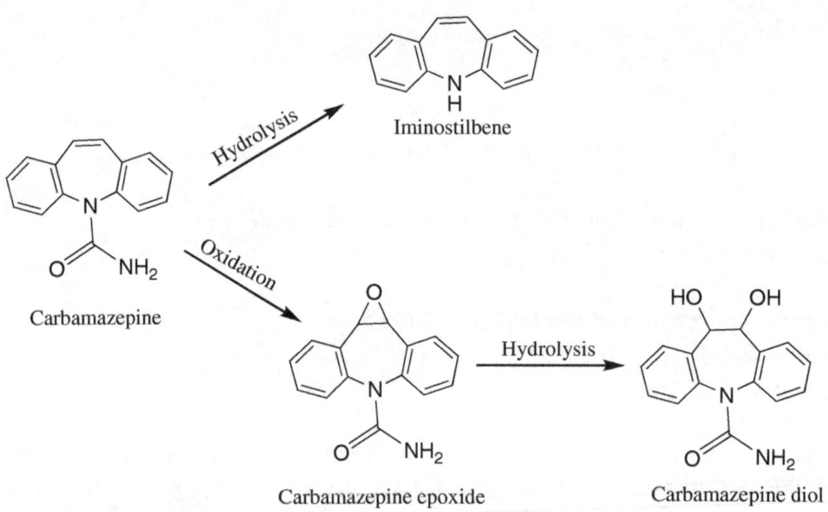

Scheme 79-1 Degradation pathways of carbamazepine.

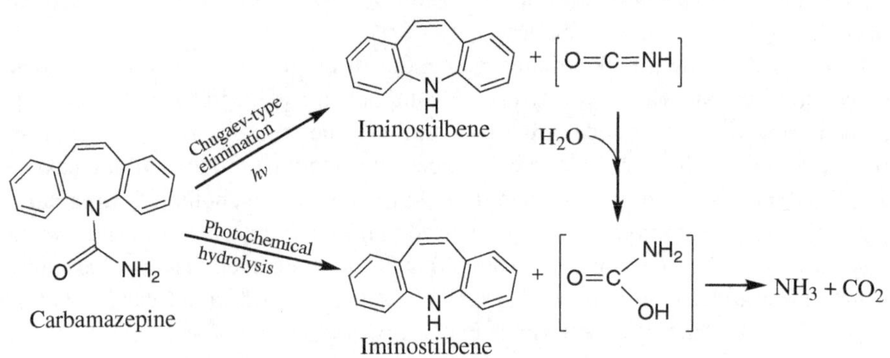

Scheme 79-2 Proposed mechanisms for photochemical degradation of carbamazepine.

References

1 Ali, F., Khan, J.A., Shah, N.S. et al. (2018). Carbamazepine degradation by UV and UV-assisted AOPs: kinetics, mechanism and toxicity investigations. *Proc. Safety Environ. Protec.* 117: 307–314.

2 Dai, C.-M., Zhou, X.-F., Zhang, Y.-L. et al. (2012). Comparative study of the degradation of carbamazepine in water by advanced oxidation processes. *Environ. Technol.* 33: 1101–1109.

3 Bu, L., Zhou, S., Zhu, S. et al. (2018). Insight into carbamazepine degradation by UV/monochloramine: reaction mechanism, oxidation products, and DBPs formation. *Water Res.* 146: 288–297.

4 Bertilsson, L. (1978). Clinical pharmacokinetics of carbamazepine. *Clin. Pharmacokinet.* 3: 128–143.

5 Naguib, I.A., Elyazeed, N.A., Elroby, F.A., and El-Ghobashy, M.R. (2019). Stability indicating spectrophotometric methods for quantitative determination of carbamazepine and its degradation product, iminostilbene, in pure form and pharmaceutical formulations. *Spectrochim. Acta Part A: Mole. Biomol. Spect.* 214: 21–31.

6 Baselt, R.C. (2004). *Disposition of Toxic Drugs and Chemicals in Man*, 7e, 164–167. Foster City, CA: Biomedical.

7 Li, M. (2012). Photochemical degradation. In: *Organic Chemistry of Drug Degradation*, 165–297. Cambridge: Royal Society Chemistry Publishing.

80: Cabergoline

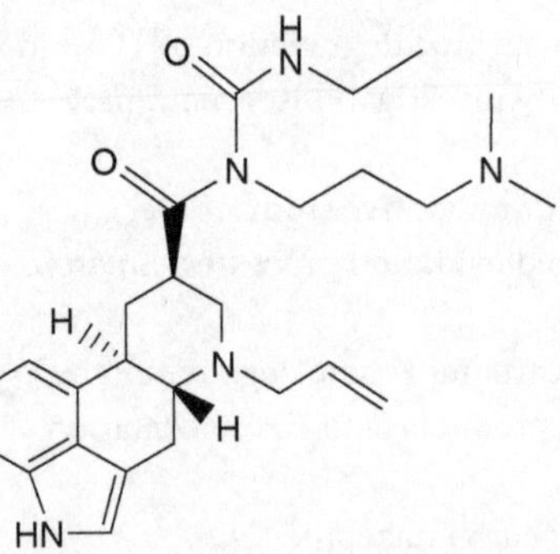

Chemical name: (6*aR*,9*R*,10*aR*)-*N*-[3-(Dimethylamino)propyl]-*N*-(ethylcarbamoyl)-7-prop-2-enyl-6,6*a*,8,9,10,10*a*-hexahydro-4*H*-indolo[4,3-fg]quinoline-9-carboxamide
Brand name: Cabaser, Dostinex
Formula: $C_{26}H_{37}N_5O_2$
Monoisotopic protonated ion (*m/z*): 452.3020
Molecular weight: 451.61
CAS number: 81409-90-7

Cabergoline is a natural product-based therapeutic agent. As a dopamine agonist, it is utilized clinically for the treatment of hyperprolactinemic disorders and Parkinson's disease. The functional groups susceptible to degradation include the

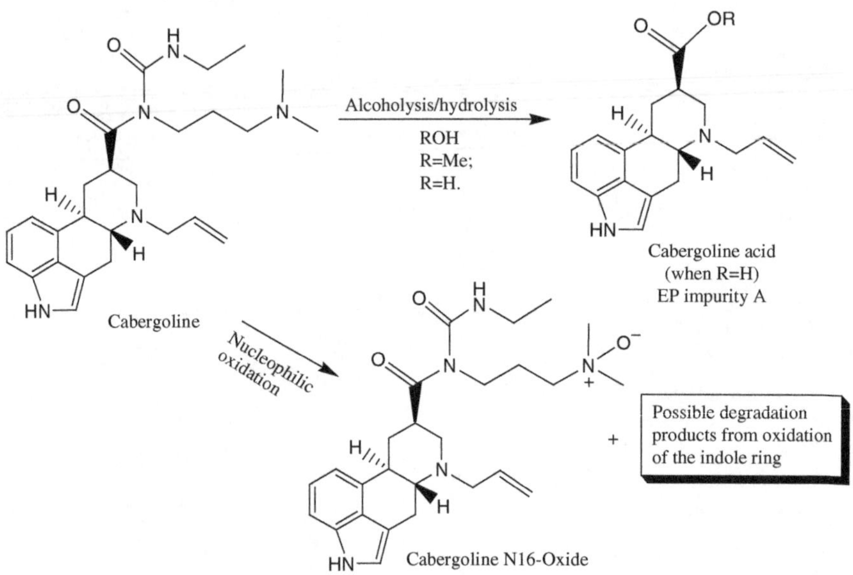

Scheme 80 Degradation pathways of cabergoline.

acylated urea functionality, two tertiary amine groups, and the indole ring. There appear to be very few reports on the degradation chemistry of cabergoline. In a study by Paul et al. [1], cabergoline underwent alcoholysis upon extended storage in methanol solution at room temperature, resulting in the substitution of the urea moiety by methanol (Scheme 80). This observation suggests that cabergoline should also be susceptible to hydrolysis, giving rise to cabergoline acid, which is EP impurity A [2, 3] and also a major metabolite of the drug [4]. The two tertiary amine moieties and the indole ring should be prone to oxidation; indeed, "cabergoline *N*-oxide", the *N*-oxide with oxygen on the dimethylamino side chain, is included in the PubChem database [5] and currently available as an impurity reference material per information on many websites [6]. Recently, Lechel et al. found the second *N*-oxide degradant of cabergoline, i.e., the one with oxygen on the *N*-allyl nitrogen, in an injectable formulation product of cabergoline [7]. Önal et al. performed stress studies under hydrolytic, oxidative, thermal, and photochemical conditions [8]. The results showed that the drug was less stable in 0.1 M HCl than in 0.1 M NaOH solution. However, no elucidation of degradants was done.

References

1 Paul, G., Winnik, W., Hughes, N. et al. (2003). Accurate mass measurement at enhanced mass-resolution on a triple quadrupole mass-spectrometer for the identification of a reaction impurity and collisionally-induced fragment ions of cabergoline. *Rapid Commun. Mass Spectrom.* 17: 561–568.

2 (2012). *European Pharmacopoeia, Monograph C*, 7e (7.5), 1535. Strasbourg, France: Council of Europe.

3 Wagger, J., Pozesb, A., and Pozgan, F. (2013). Synthesis of European pharmacopoeial impurities A, B, C, and D of cabergoline. *RSC Adv.* 3: 23146–23156.

4 Wiseman, L.R. and Fitton, A. (1999). Cabergoline, a review of its efficacy in the treatment of Parkinson's disease. *CNS Drugs* 12: 485–497.

5 Cabergoline N-Oxide. PubChem. https://pubchem.ncbi.nlm.nih.gov/compound/71314472. (accessed 4 July 2025).

6 Cabergoline N-Oxide. Chemical Book. https://www.chemicalbook.com/ChemicalProductProperty_EN_CB12558018.htm. (accessed 4 July 2025).

7 Lechel, T., Otero, R.S., Springer, A. et al. (2024). Identification and structural elucidation of an oxidation product generated during stability studies of cabergoline drug product. *J. Pharm. Biomed. Anal.* 248: 116282.

8 Önal, A., Sağırlı, O., and Sensoy, D. (2007). Selective LC determination of cabergoline in the bulk drug and in tablets: In vitro dissolution studies. *Chromatographia* 65: 561–567.

81: Captopril

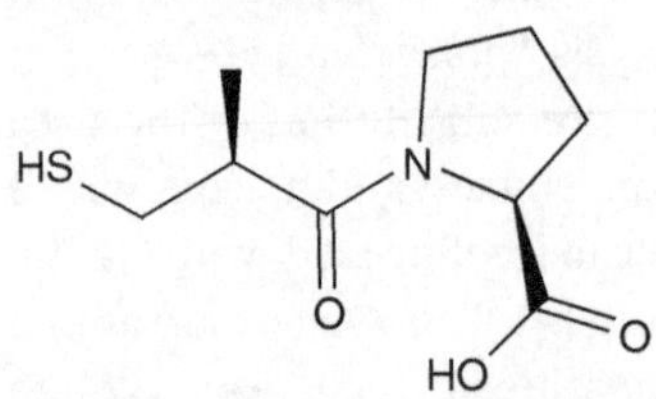

Chemical name: 1-[2(*S*)-3-mercapto-2-methyl-1-oxopropyl]-L-proline; (2*S*)-1-[(2*S*)-2-Methyl-3-sulfanylpropanoyl]pyrrolidine-2-carboxylic acid
Brand name: Capoten
Formula: $C_9H_{15}NO_3S$
Monoisotopic protonated ion (*m/z*): 218.0845
Molecular weight: 217.29
CAS number: 62571-86-2

Captopril is the first angiotensin-converting enzyme (ACE) inhibitor developed by Bristol-Myers Squibb for the treatment of hypertension. It is an acylated L-proline with the acylated side chain containing a thiol (sulfhydryl) group. Since the thiol group is very susceptible to auto-oxidation, particularly in aqueous solutions, the predominant degradation pathway of captopril is the oxidation of its thiol group to form captopril disulfide. The oxidation is generally believed to proceed through the ionized thiol group, from which an electron is removed by either a redox transition metal ion (most likely iron(III) ion in vast majority cases) or molecular oxygen directly. According to a study by Lee and Notari using cupric ion as the model redox transition metal ion [1], both pathways should be operative during the auto-oxidation of captopril in the pH range of 6.6–8.0, because the degradation kinetics cannot be well explained by either of the pathways alone. The auto-oxidative degradation mechanism is shown in Scheme 81-1, which is mostly based on the work by Lee and Notari [1] with some modification.

According to the oxidative degradation mechanism shown in Scheme 81-1, control of oxygen content and redox transition metal ions in conjunction with low pH should promote the stability of formulated captopril products. From the formulation perspective, control of residual metal ions (from the API, excipients and/or packaging materials) is usually achieved by the addition of metal chelators such as EDTA. Since the pKa of captopril thiol group is around 9.8 [2], liquid formulations with pH well below the pKa value should also stabilize the formulated drug products. Indeed, many formulation studies demonstrated that the shelf lives were extended significantly by the addition of EDTA as well as by keeping the pH of the formulation in the acidic range [2–4]. The studies by Timmins et al. [2] and Huang et al. [3] showed that the rates of oxidative degradation increased with ascending pH, particularly at pH above 4, from which point the rates of oxidation dramatically accelerated. On the other hand, addition of antioxidants either had no effect (e.g., ascorbic acid and sodium bisulfite) [3] or unexpectedly promoted oxidative degradation of captopril (e.g., propyl gallate and sodium

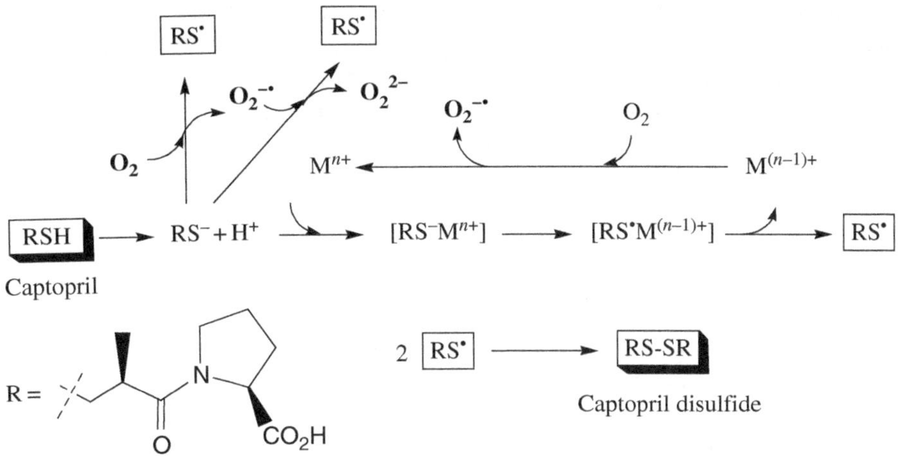

Scheme 81-1 Proposed mechanism for the auto-oxidation of captopril to form captopril disulfide. The mechanism was based on the work of Lee and Notari with some modification; the oxidation study was performed in the pH range of 6.6–8.0.

metabisulfite) [2]. The latter results were attributed to the residual metal ions in the two antioxidative excipients by the original researchers. In light of the elucidation of the Udenfriend chemistry and its role in the auto-oxidation of drugs [5], both seemingly paradoxical results may be explained by the Udenfriend chemistry. As illustrated in Scheme 81-2, the role of propyl gallate and sodium metabisulfite would likely be to reduce [RSH M^{n+}] to [RSH $M^{(n-1)+}$] so that the latter can continue the cycle during which molecular oxygen is activated (or reduced) to form superoxide anion radical ($O_2^{-•}$). The latter could be further converted to hydroxyl free radical via the intermediacy of hydrogen peroxide; the hydroxyl free radical would be more damaging toward captopril than superoxide anion radical.

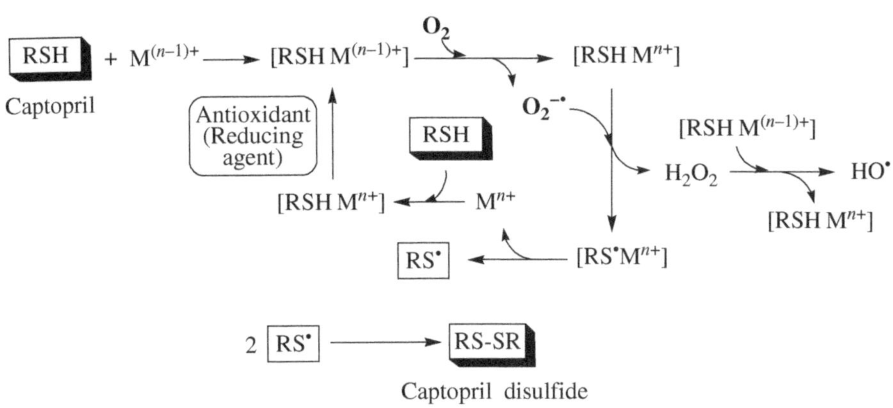

Scheme 81-2 Proposed mechanism for antioxidant-promoted oxidation of captopril based on the Udenfriend chemistry [5]. The original work was performed by Timmins et al. [2] and the study was conducted in aqueous solutions with pH of 4. *Source:* Timmins et al. [2] and Li [5].

As an acylated amino acid, the amide linkage of captopril is technically susceptible to hydrolytic degradation. The activation energy for the hydrolysis was found to be 21.4 kcal/mol when the hydrolysis was performed in 0.5 N HCl [2]. This result suggests that hydrolysis would not be a meaningful degradation pathway for captopril unless under harsh forced degradation conditions. Various formulation studies found captopril disulfide as either the sole or predominant degradant [1–4].

References

1 Lee, T.-Y. and Notari, R.E. (1987). Kinetics and mechanism of captopril oxidation in aqueous solution under controlled oxygen partial pressure. *Pharm. Res.* 4: 98–103.

2 Timmins, P., Jackson, I.M., and Wang, Y.J. (1982). Factors affecting captopril stability in aqueous solution. *Int. J. Pharm.* 11: 329–336.

3 Huang, Y.-B., Tsai, Y.-H., Chang, J.-S. et al. (2002). Effect of antioxidants and anti-irritants on the stability, skin irritation and penetration capacity of captopril gel. *Int. J. Pharm.* 241: 345–351.

4 Berger-Gryllaki, M., Podilsky, G., Widmer, N. et al. (2007). The development of a stable oral solution of captopril for paediatric patients. *EJHP Sci.* 13: 27–32.

5 Li, M. (2012). Oxidative degradation. In: *Organic Chemistry of Drug Degradation*, 49–53. Cambridge: Royal Society Chemistry Publishing.

82–83: Carboplatin, Cisplatin

82: Carboplatin

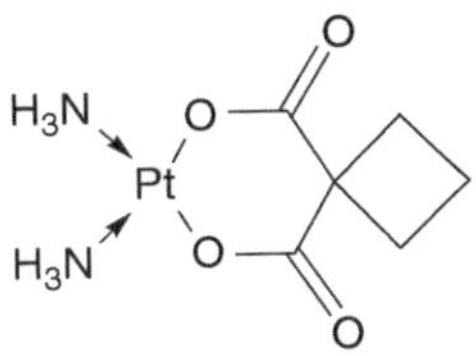

Chemical name: cis-Diammine(cyclobutane-1,1-dicarboxylate-O,O′)platinum (II); [1,1-Cyclobutanedicarboxylato(2-)-κ^2O^1,O^1]platinum diammoniate
Brand name: Paraplatin
Formula: $C_6H_{12}N_2O_4Pt$
Monoisotopic protonated ion (*m/z*): 372.0518 (for the most abundant ^{195}Pt isotope)
Molecular weight: 371.25
CAS number: 41575-94-4

83: Cisplatin

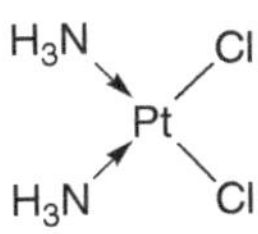

Chemical name: cis-Diamminedichloroplatinum (II)
Brand name: Transplatin
Formula: $Cl_2H_6N_2Pt$
Monoisotopic protonated ion (*m/z*): 299.9629 (for the most abundant ^{195}Pt isotope)
Molecular weight: 300.04
CAS number: 14913-33-8

The degradation pathways of both carboplatin and cisplatin are mostly of hydrolytic nature. Carboplatin is a second generation of this widely used chemotherapeutic family, which was developed to reduce the side effects of cisplatin, the original platinum-containing anticancer drug. Carboplatin is about 100 times more stable in aqueous solutions at 37°C than cisplatin [1] and hence, it can be formulated in liquid dosage forms. The hydrolytic degradation of carboplatin was reported to be catalyzed by acid [2, 3]. However, it was revealed in a patent that the real degradation behavior of carboplatin in acidic pH range may be more complicated: across the pH range between 2.5 and 7.5 achieved by phosphate buffers, carboplatin was found to be more stable in the pH range of 3.5–4.5 than in pH 6.5–7.5 at a concentration of 10 mM [4]. In addition, at pH 3.5 the stability of carboplatin versus its concentration

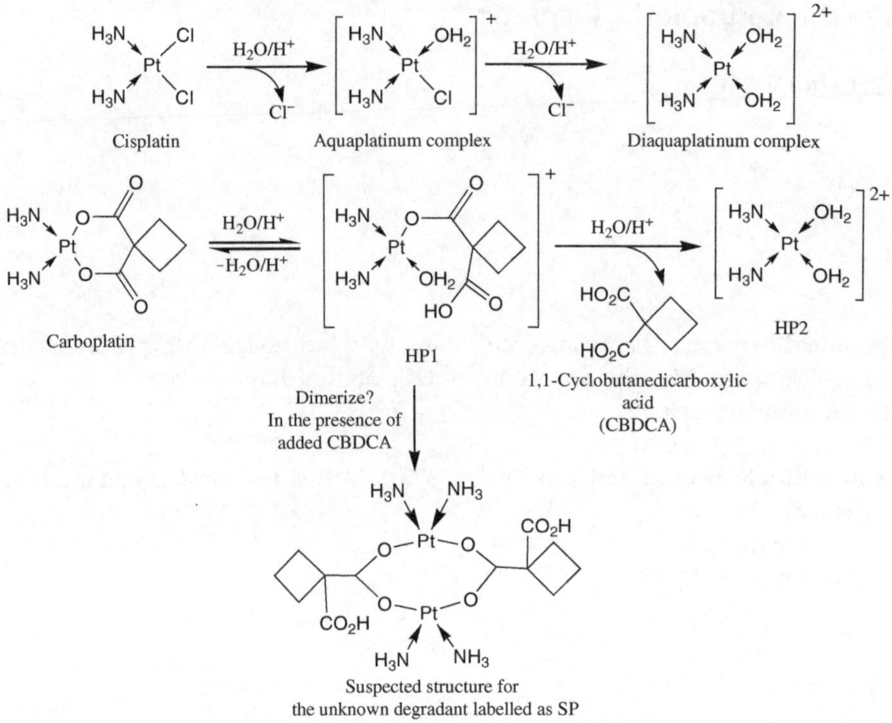

Scheme 82–83 Plausible mechanisms for the hydrolytic degradation of cisplatin and carboplatin, respectively. The nomenclature for the degradation products of carboplatin follows that of Schnurr et al. [1]. *Source:* Schnurr et al. [1]/Springer Nature.

was studied. It was found that carboplatin became more unstable as its concentration increased. Therefore, the actual degradation mechanism of carboplatin may very well be more complex than what is displayed in Scheme 82–83. In a study by Jin et al., light and UV irradiation were found to accelerate the degradation of carboplatin [5]. Furthermore, in the presence of NaCl, the degradation rate increased, and the main degradant was cisplatin. The latter observation was consistent with that reported by Cheng et al. [6].

References

1 Schnurr, B., Heinrich, H., and Gust, R. (2002). Investigations on the decomposition of carboplatin in infusion solutions II. Effect of 1,1-cyclobutanedicarboxylic acid admixture. *Microchim. Acta* 140: 141–148.

2 Canovese, L., Cattalini, L., Chessa, G., and Tobe, M.L. (1988). Kinetics of the displacement of cyclobutane-1,1-dicarboxylate from diammine(cyclobutane-1,1-dicarboxylato)-platinum(II) in aqueous solution. *J. Chem. Soc. Dalton Trans.* 8: 2135–2140.

3 Allsopp, M.A., Sewell, G.J., Rowland, C.G. et al. (1991). The degradation of carboplatin in aqueous solutions containing chloride or other selected nucleophiles. *Int. J. Pharm.* 69: 197–210.

4 Nijker, A.J., Vermeer, J.M.P. Carboplatin composition. European patent application 0401896A1. Filed 28 May 1990. Granted 11 August 1993.

5 Jin, Y.-R., Huang, T.-M., Yang, Y.-K. et al. (1994). Stability study of carboplatin by HPLC (original in Chinese). *Chinese J. Pharm.* 25: 69–72.

6 Cheng, Y., Cheung, Y.-W., Cradock, J.C. et al. (1987). Stability of cisplatin, iproplatin, carboplatin, and tetraplatin in commonly used intravenous solutions. *Am. J. Hosp. Pharm.* 44: 124–130.

84: Carisoprodol

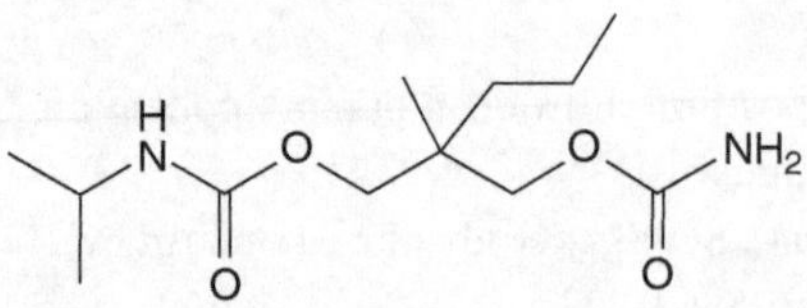

Chemical name: [2-(carbamoyloxymethyl)-2-methylpentyl] *N*-propan-2-ylcarbamate
Brand name: Soma, Vanadom
Formula: $C_{12}H_{24}N_2O_4$
Monoisotopic protonated ion (*m/z*): 261.1809
Molecular weight: 260.33
CAS number: 78-44-4

Carisoprodol and meprobamate are muscle relaxants, and the latter is a metabolite of carisoprodol, upon which carisoprodol was developed according to Wikipedia [1].

There appear to be very few reported studies for the degradation study of both drugs. The only functional groups that are potentially susceptible to degradation are the two carbamates present in both drugs. For carisoprodol, the unsubstituted carbamate moiety would be somewhat less stable than the substituted one because it is less sterically hindered than the latter. Since the activation energy for the hydrolysis of carbamates is typically greater than 20 kcal/mol [2], both drugs are expected to be reasonably stable, particularly in solid dosage forms. In what appear to be the only two reported forced degradation studies of carisoprodol, this drug molecule was found to be quite stable under very harsh acidic conditions, e.g., refluxing in 3N or 5N HCl solutions [3, 4]. Nevertheless, the results in alkaline stress conditions reported by the two research groups were quite different; while Murali and Rambabu reported that carisoprodol was stable when refluxed in 5 N NaOH solution at 80°C for 8 hours [3], the results of Acharya et al. indicated that when refluxed in 0.1 N NaOH for 90 minutes, sufficient degradation of carisoprodol was produced. Among the degradation products whose structures were proposed by Acharya et al. were those resulting from hydrolytic cleavage of the two carbamate bonds (Scheme 84); it needs to point out that the structure for one of the degradants (referred to as Degradant II) appears to be erroneous in the original paper [4].

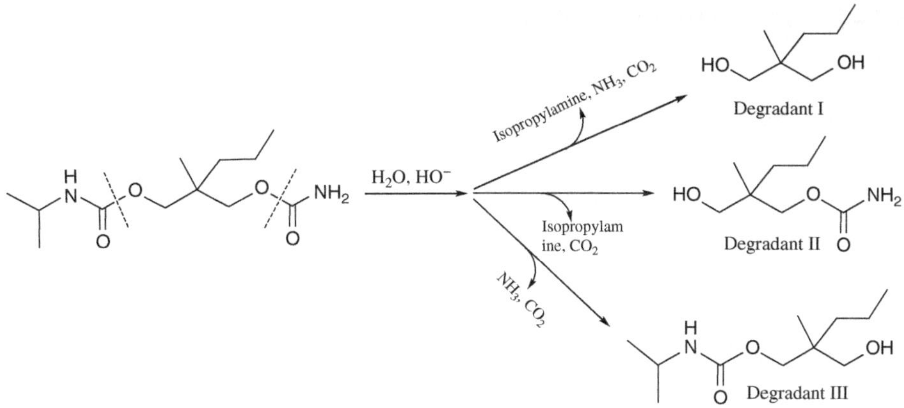

Scheme 84 Hydrolytic degradation of carisoprodol under alkaline conditions. Please note that the structure of Degradant II was erroneous in the original paper by Acharya et al. [4]. *Source:* Acharya et al. [4]/with permission of Elsevier.

References

1 Carisoprodol. https://en.wikipedia.org/wiki/Carisoprodol (accessed 29 November 2015).

2 Li, M. (2012). Hydrolytic degradation. In: *Organic Chemistry of Drug Degradation*, 16–47. Cambridge: Royal Society Chemistry Publishing.

3 Murali, D. and Rambabu, C. (2016). Stability indicating high performance liquid chromatographic method for the estimation of carisoprodol in bulk and in tablet dosage form. *Int. J. PharmTech Res.* 9: 171–180.

4 Acharya, P.C., Vasi, R., and Suares, D. (2016). FTIR assay method for UV inactive drug carisoprodol and identification of degradants by RP-HPLC and ESI-MS. *J. Chromatogr. B* 1029–1030: 16–21.

85–93: Cephalosporin Antibiotics

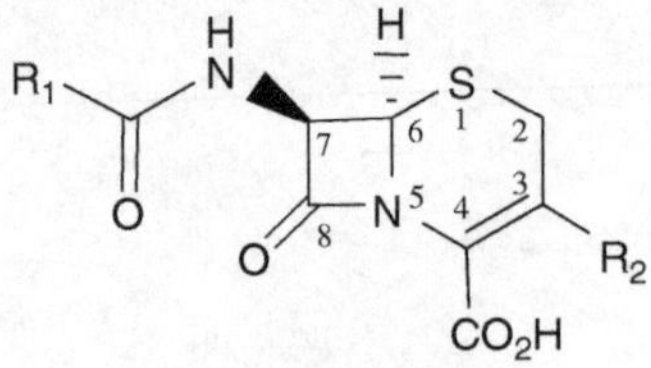

Cephalosporins

85: Cefalexin [Cephalexin]

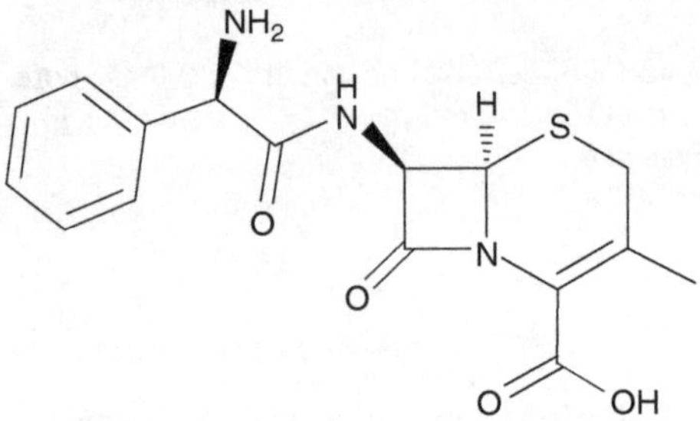

Chemical name: (6*R*,7*R*)-7-[[(2*R*)-2-Amino-2-phenylacetyl]amino]-3-methyl-8-oxo-5-thia-1-azabicyclo[4.2.0]oct-2-ene-2-carboxylic acid
Brand name: Keflex
Formula: $C_{16}H_{17}N_3O_4S$
Monoisotopic protonated ion (*m/z*): 348.1013
Molecular weight: 347.39
CAS number: 15686-71-2

86: Cefadroxil [Cephadroxil]

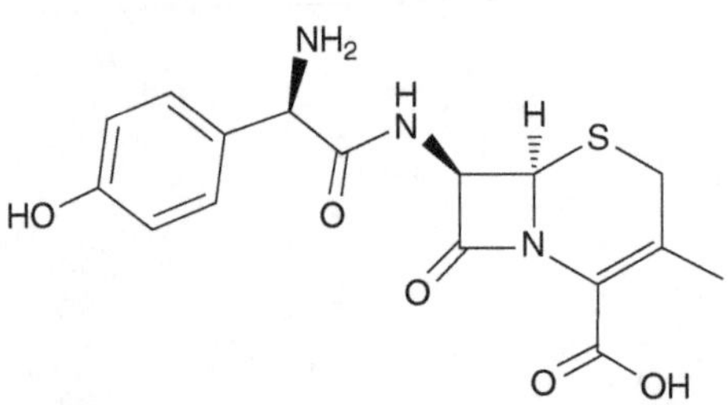

Chemical name: (6*R*,7*R*)-7-[[(2*R*)-2-Amino-2-(4-hydroxyphenyl)acetyl]amino]-3-methyl-8-oxo-5-thia-1-azabicyclo[4.2.0]oct-2-ene-2-carboxylic acid
Brand name: Baxan, Bidocef, Cefamox
Formula: $C_{16}H_{17}N_3O_5S$

Monoisotopic protonated ion (*m/z*): 364.0962
Molecular weight: 363.39
CAS number: 50370-12-2

87: Cefaclor

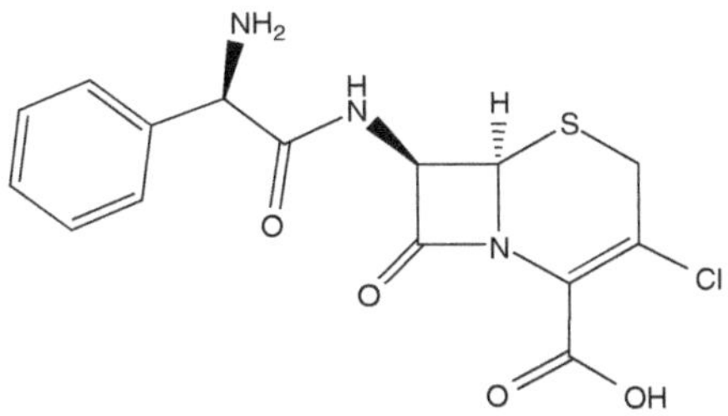

Chemical name: (6*R*,7*R*)-7-[[(2*R*)-2-Amino-2-phenylacetyl]amino]-3-chloro-8-oxo-5-thia-1-azabicyclo[4.2.0]oct-2-ene-2-carboxylic acid
Brand name: Ceclor
Formula: $C_{15}H_{14}ClN_3O_4S$
Monoisotopic protonated ion (*m/z*): 368.0466
Molecular weight: 367.81
CAS number: 53994-73-3

88: Cephazolin [Cefazolin]

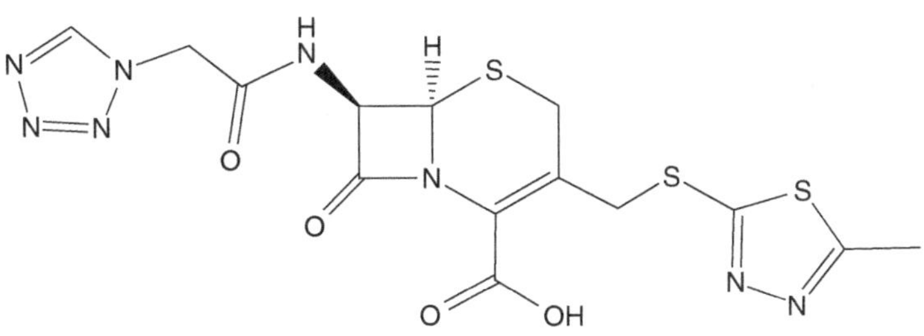

Chemical name: (6*R*,7*R*)-3-[(5-Methyl-1,3,4-thiadiazol-2-yl)sulfanylmethyl]-8-oxo-7-[[2-(tetrazol-1-yl)acetyl]amino]-5-thia-1-azabicyclo[4.2.0]oct-2-ene-2-carboxylic acid
Brand name: Elzogram, Zolicef
Formula: $C_{14}H_{14}N_8O_4S_3$
Monoisotopic protonated ion (*m/z*): 455.0373
Molecular weight: 454.51
CAS number: 25953-19-9

89: Cefuroxime

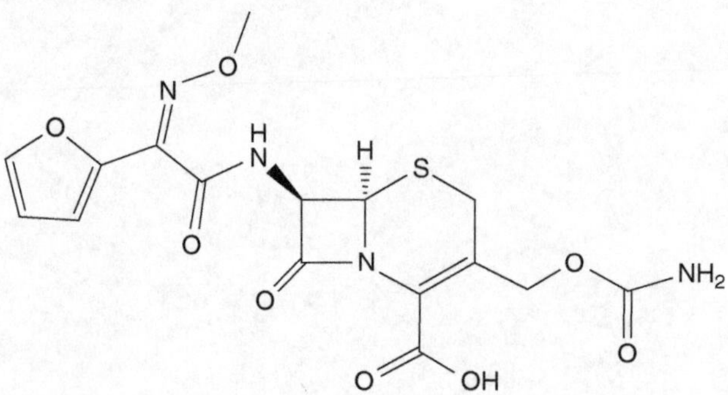

Chemical name: (6*R*,7*R*)-3-(Carbamoyloxymethyl)-7-[[(2*Z*)-2-(furan-2-yl)-2-methoxyiminoacetyl]amino]-8-oxo-5-thia-1-azabicyclo[4.2.0]oct-2-ene-2-carboxylic acid
Brand name: Ceftin
Formula: $C_{16}H_{16}N_4O_8S$
Monoisotopic protonated ion (*m/z*): 425.0762
Molecular weight: 424.39
CAS number: 55268-75-2

90: Cefotaxime

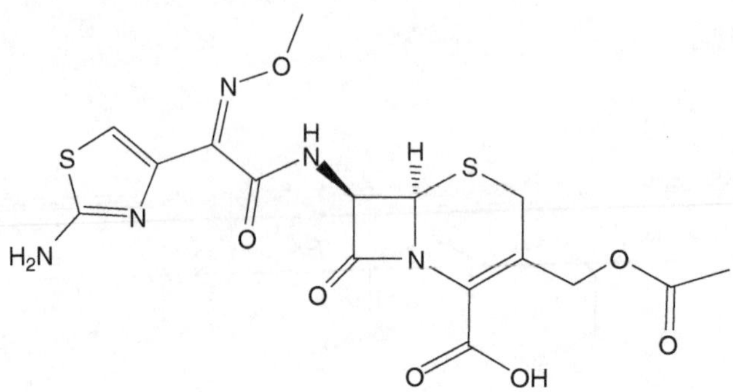

Chemical name: (6*R*,7*R*)-3-(Acetyloxymethyl)-7-[[(2*Z*)-2-(2-amino-1,3-thiazol-4-yl)-2-methoxyiminoacetyl]amino]-8-oxo-5-thia-1-azabicyclo[4.2.0]oct-2-ene-2-carboxylic acid
Brand name: Claforan
Formula: $C_{16}H_{17}N_5O_7S_2$
Monoisotopic protonated ion (*m/z*): 456.0642
Molecular weight: 455.47
CAS number: 63527-52-6

91: Ceftriaxone

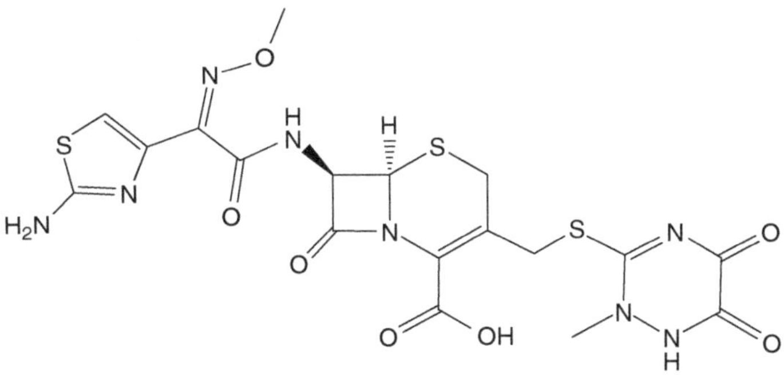

Chemical name: (6*R*,7*R*)-7-[[(2*Z*)-2-(2-Amino-1,3-thiazol-4-yl)-2-methoxyiminoacetyl] amino]-3-[(2-methyl-5,6-dioxo-1*H*-1,2,4-triazin-3-yl)sulfanylmethyl]-8-oxo-5-thia-1-azabicyclo[4.2.0]oct-2-ene-2-carboxylic acid
Brand name: Rocephin
Formula: $C_{18}H_{18}N_8O_7S_3$
Monoisotopic protonated ion (*m/z*): 555.0533
Molecular weight: 554.58
CAS number: 73384-59-5

92: Cefixime

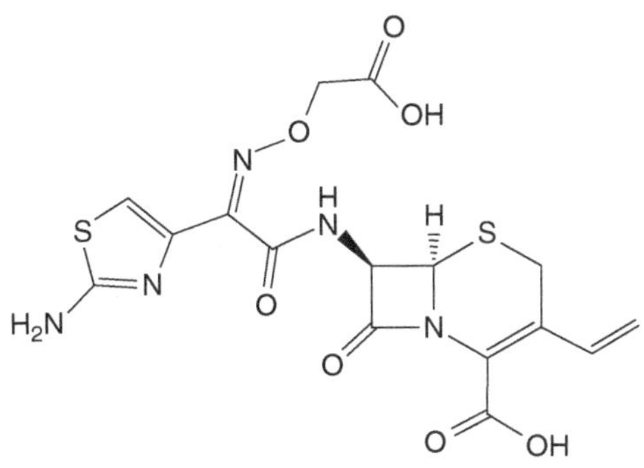

Chemical name: (6*R*,7*R*)-7-[[(2*Z*)-2-(2-Amino-1,3-thiazol-4-yl)-2-(carboxymetho xyimino)acetyl]amino]-3-ethenyl-8-oxo-5-thia-1-azabicyclo[4.2.0]oct-2-ene-2-carboxylic acid
Brand name: Suprax
Formula: $C_{16}H_{15}N_5O_7S_2$
Monoisotopic protonated ion (*m/z*): 454.0486
Molecular weight: 453.45
CAS number: 79350-37-1

93: Ceftazidime

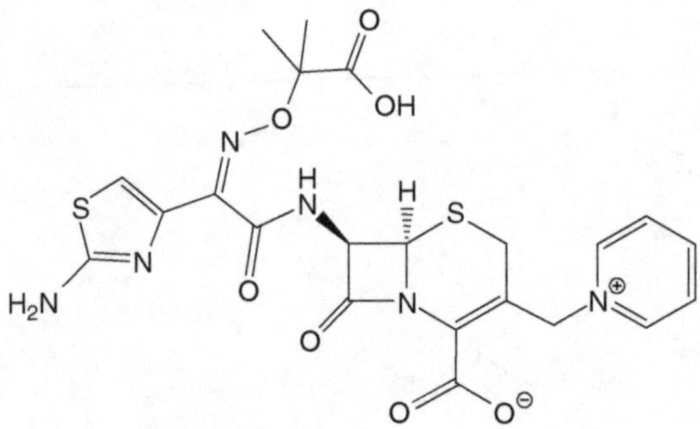

Chemical name: (6*R*,7*R*)-7-[[(2*Z*)-2-(2-Amino-1,3-thiazol-4-yl)-2-(2-carboxypropan-2-yloxyimino)acetyl]amino]-8-oxo-3-(pyridin-1-ium-1-ylmethyl)-5-thia-1-azabicyclo [4.2.0]oct-2-ene-2-carboxylate
Brand name: Fortaz, Tazicef
Formula: $C_{22}H_{22}N_6O_7S_2$
Monoisotopic protonated ion (*m/z*): 547.1064
Molecular weight: 546.58
CAS number: 72558-82-8

Cephalosporins are one of the most important classes of antibacterial drugs. A great number of them [1] were brought to market in different generations. Discussed in this monograph are the representative ones, including those in the first-generation (cefalexin, cefadroxil, cefaclor, and cefazolin), second-generation (cefuroxime), and third-generation cephalosporin families (cefotaxime, ceftriaxone, cefixime, and ceftazidime).

Cephalosporins contain an eight-membered nucleus consisting of a four-membered β-lactam ring fused to a dihydrothiazine ring. The main degradation pathway of cephalosporins is the hydrolysis of the β-lactam ring, giving rise to the open ring form (Pathway a, Scheme 85–93-1) [1]. The susceptibility of the β-lactam to hydrolysis depends upon the nature of the substituents on the core, pH, and temperature. The pH profiles of cephalosporins are typically of a U-shape, with their most stable pH range usually in the acidic range [2]. For example, cefaclor was shown to be most stable in the pH range of 2–4 [3], cefepime pH 3–6 [4], cefixime pH 4–7 [5], cefotaxime pH 3–7 [6], and cefuroxime pH 4–7 [7]. The open ring form can undergo decarboxylation (Pathway a1, Scheme 85–93-1) [5, 8]. Upon further hydrolysis of the dihydrothiazine ring, the decarboxylated degradant gives rise to the aldehyde degradant that is usually present in its hydrated form (Pathway a2, Scheme 85–93-1), which is not stable and tends to polymerize [5].

Under alkaline conditions, cephalosporins can undergo two additional decomposition pathways: epimerization at the 7-position to produce 7-epi-cephalosporins (Pathway b, Scheme 85–93-1) and rearrangement to give Δ^2-cephalosporins (Pathway c, Scheme 85–93-1) [5].

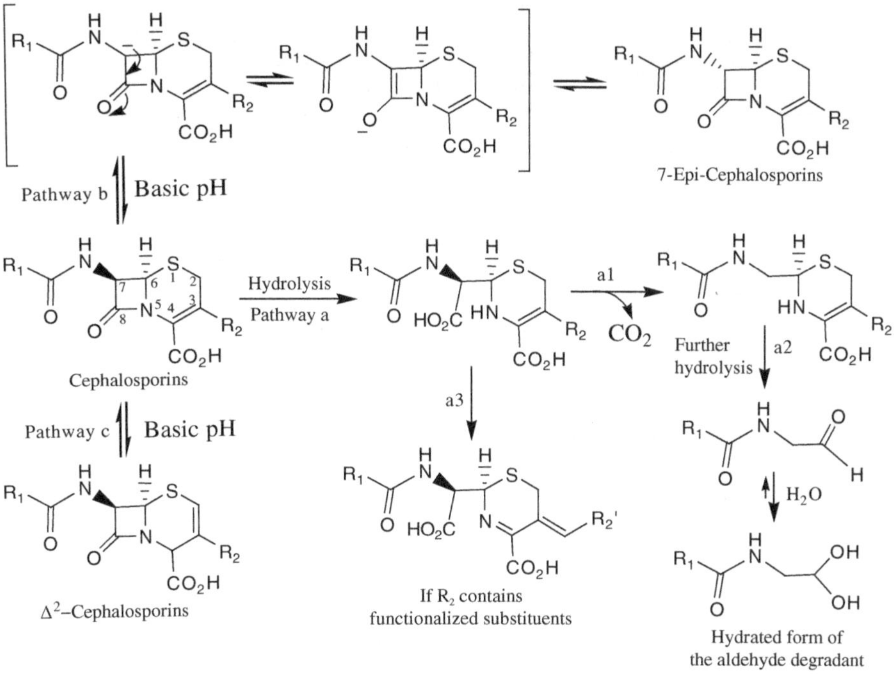

Scheme 85–93-1 Degradation pathways of the cephalosporin core.

On the other hand, if the R_2 group (in Schemes 85–93-1 and 85–93-2) is an alkyl group that is functionalized at the position α to the dihydrothiazine ring (e.g., a vinyl in cefixime and acetoxylmethyl in cefotaxime), the enamine form of the open ring degradant can rearrange to the conjugated imine form with concurrent formation of a simple or substituted methylene group at the 3-position (Pathway a1, Scheme 85–93-2). For cefixime, R_2' (in Scheme 85–93-2) is a methyl group, while in many others, such as cefotaxime, ceftriaxone, ceftazidime, cefepime, cefuroxime, and cefazolin, it is a hydrogen [7, 8]. The conjugated imines thus formed are electrophilic species and are susceptible to nucleophilic attack by the carboxyl group to yield the lactone degradants (Pathway a2, Scheme 85–93-2), particularly under acidic conditions [5, 8]. Many cephalosporins (cefotaxime, ceftriaxone, ceftazidime, cefepime, cefuroxime, and cefazolin) can also directly degrade to the corresponding lactones (Pathway b1, Scheme 85–93-2) or via the intermediacy of 3-hydroxymethyl degradants (Pathway b2 and b3, Scheme 85–93-2) [9].

For those cephalosporins containing an amino group on the 7-position side chain, such as cefalexin, cefadroxil, and cefaclor, this amino group can attack the β-lactam, a process referred to as intramolecular aminolysis, to form diketopiperazine (DKP) degradants (Scheme 85–93-3). The DKP degradants, now containing a dihydrothiazine ring that is no longer fused to the lactam ring, are capable of further degradation, such as rearrangement and lactonization, as illustrated in Scheme 85–93-2.

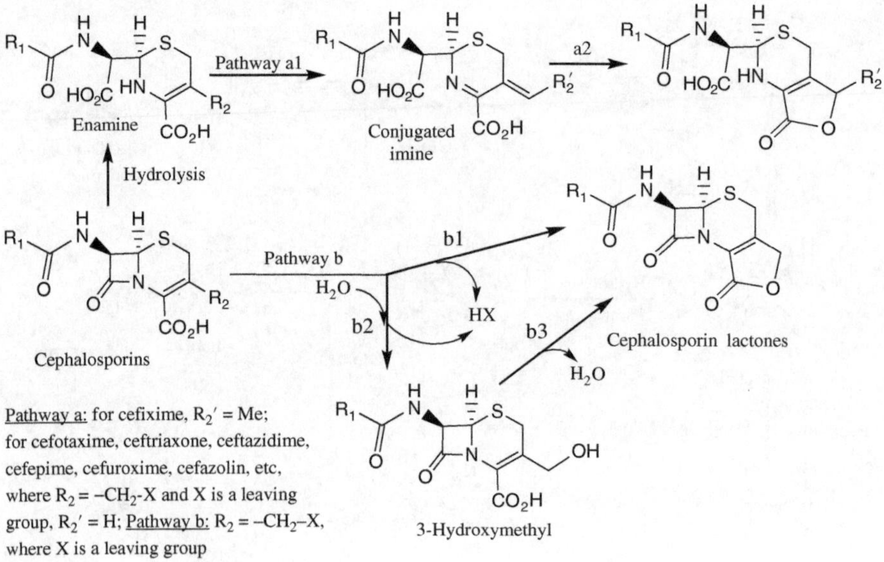

Scheme 85-93-2 Rearrangement to the enamine/imine forms and lactone formation.

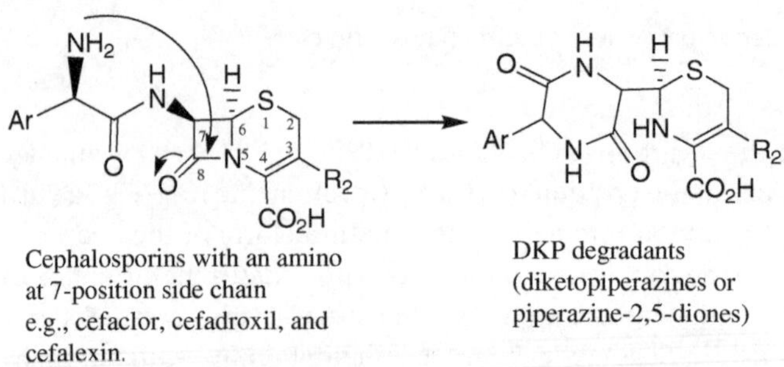

Scheme 85-93-3 Formation of the diketopiperazine (DKP) degradants.

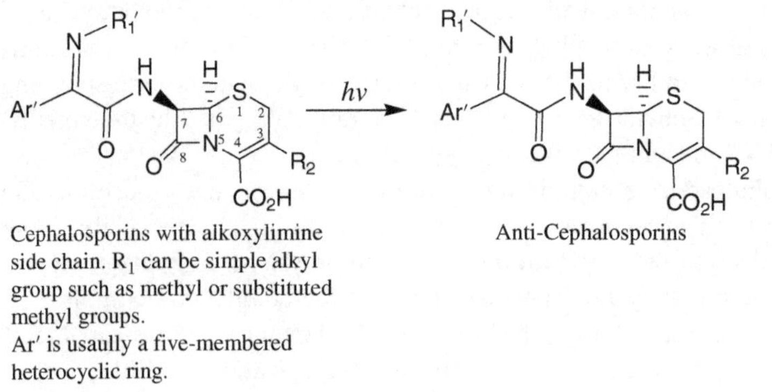

Scheme 85-93-4 Photoisomerization of cephalosporins with an alkoxylimine side chain.

Many newer generations of cephalosporins, such as cefotaxime, ceftriaxone, cefixime, and ceftazidime, contain an alkoxylimine group on the 7-position side chain; the alkoxylimine double bond is susceptible to photochemical isomerization to yield the corresponding anti-cephalosporins (Scheme 85–93-4) [10].

References

1 Alekseev, V.G. (2010). Drug synthesis methods and manufacturing technology–acid-base properties of penicillins and cephalosporings (a review). *Pharm. Chem. J.* 44: 14–24.

2 Llinas, A., Vilanova, B., Frau, J. et al. (1998). Chemical reactivity of penicillins and cephalosporins. Intramolecular involvement of the acyl-amido side chain. *J. Organomet. Chem.* 63: 9052–9060.

3 Dimitrovska, A., Stojanoski, K., and Dorevski, K. (1995). Kinetics of degradation of cefaclor: I. Effects of temperature, phosphate buffer, pH, and copper (II) ion. *Int. J. Pharm.* 115: 175–182.

4 Fubara, J.O. and Notari, R.E. (1998). Influence of pH, temperature and buffers on cefepime degradation kinetics and stability predictions in aqueous solutions. *J. Pharm. Sci.* 87: 1572–1576.

5 Namiki, Y., Tanabe, T., Kobayashi, T. et al. (1987). Degradation kinetics and mechanism of a new cephalosporin, cefixime, in aqueous solution. *J. Pharm. Sci.* 76: 208–214.

6 Fabre, H., Eddine, N.H., and Berge, G. (1984). Degradation kinetics in aqueous solution of cefotaxime sodium, a third-generation cephalosporin. *J. Pharm. Sci.* 73: 611–618.

7 Wang, D. and Notari, R.E. (1994). Cefuroxime hydrolysis kinetics and stability predictions in aqueous solution. *J. Pharm. Sci.* 83: 577–581.

8 Sivakumar, B., Parthasarathy, K., Murugan, R. et al. (2013). Isolation and characterisation of degradation impurities in the cefazolin sodium drug substance. *Sci. Pharm.* 81: 933–950.

9 Kukolja, S. (1968). Chemistry of cephalosporin antibiotics. XI. Preparation and properties of desacetylcephaloglycin and its lactone. *J. Med. Chem.* 11: 1067–1069.

10 Nguyen, N.A. (1991). The relationship of diastereomer hydrolysis kinetics to shelf-life predictions for cefuroxime axetil. *Pharm. Res.* 8: 893–898.

94: Celecoxib

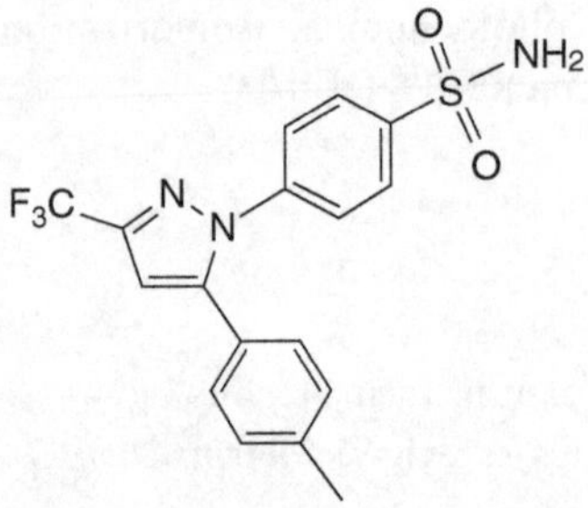

Chemical name: 4-[5-(4-Methylphenyl)-3-(trifluoromethyl)pyrazol-1-yl]benzenesulfonamide
Brand name: Celebrex
Formula: $C_{17}H_{14}F_3N_3O_2S$
Monoisotopic protonated ion (*m/z*): 382.0832
Molecular weight: 381.37
CAS number: 169590-42-5

Celecoxib is a selective COX-2 inhibitor, which differentiates itself from the vast majority of NSAIDs that display no selectivity in inhibiting COX-1 and COX-2. Other than a few forced degradation studies in which none of the degradation products was fully characterized [1–4], there appear to be very few reports regarding the degradation chemistry of celecoxib. In a forced degradation study by Hashem et al. [1], celecoxib was found to be stable under all stress conditions except UV irradiation (in an acetonitrile solution). In another forced degradation study, hydrolysis under very harsh conditions did generate a degradant, which most likely should result from the hydrolysis of the sulfonamide functional group [2]. The hydrolytic degradant was proposed by the original authors as a desulfonyl derivative of the drug based on infrared spectrometric evidence. Nevertheless, it would be more reasonable that the hydrolytic degradant is the sulfonic acid derivative of the drug, which was consistent with the TLC behavior of the degradant (a very polar compound that did not move on a normal phase TLC plate) as presented by the authors. Since the activation energy for the hydrolysis of a sulfonamide bond is usually about 30 kcal/mol or greater [5], this hydrolysis would not be expected to occur under a long-term stability or even the accelerated stability condition of 40°C/75%RH. In other words, the hydrolysis is unlikely to be a real degradation pathway for celecoxib. In two oxidative stress studies using H_2O_2 [3, 4], significant degradation was observed, although the structure of the oxidative degradant was not determined.

On the other hand, a real issue with the degradation of celecoxib in formulated products may be related to the fact that its sulfonylamino group is still nucleophilic [6] and as such, it is capable of reacting with excipients containing electrophilic moiety and/or electrophilic impurities such as formaldehyde. In a liquid formulation of celecoxib in PEG-400 [7], it was revealed that an impurity from PEG-400, presumably ethylene epoxide, reacted with the nucleophilic sulfonylamino group of celecoxib to produce

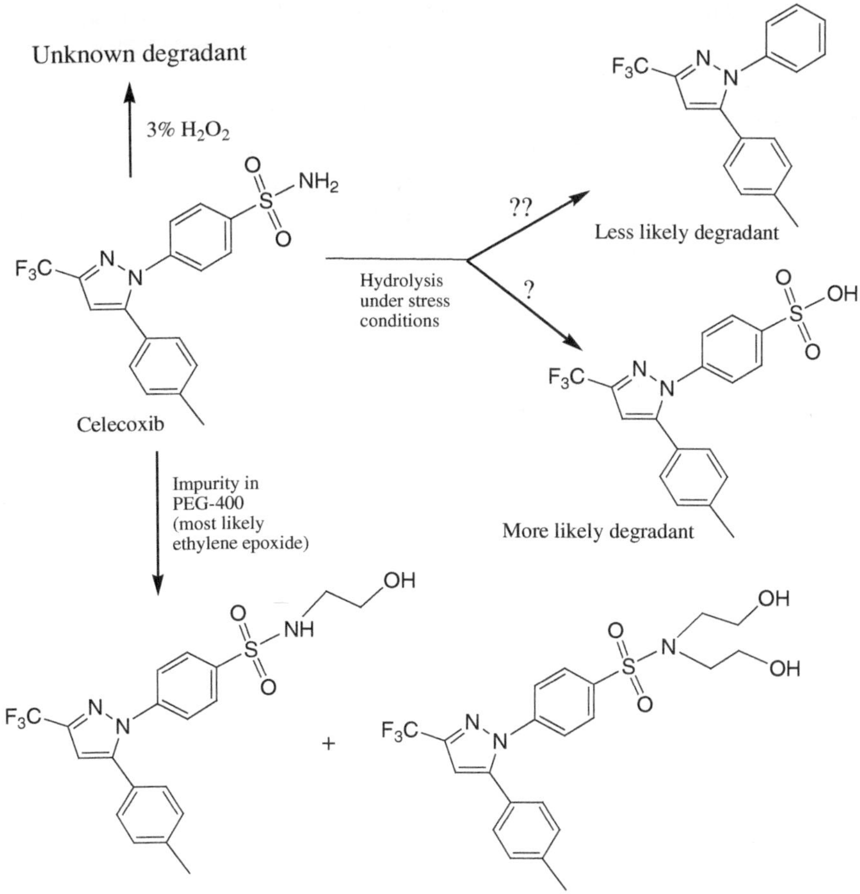

Scheme 94 Degradation of celecoxib under forced degradation conditions and in an experimental formulation.

two adducts (Scheme 94); use of antioxidant such as propyl gallate was shown to be effective in suppressing the formation of these two degradants. The latter result suggests that the presumed ethylene epoxide is an oxidative impurity of PEG-400. The nucleophilic sulfonylamino group is also capable of the Maillard reaction if a formulation contains reducing-end sugars such as lactose.

References

1 Hashem, H., Trundelberg, C., and Jira, T. (2010). Chromatographic application on calixarene bonded stationary phases: a stability indicating LC-method for determination of celecoxib in tablet formulation. *Chromatographia* 71: 91–94.

2 Bebawy, L.I., Moustafa, A.A., and Abo-Talib, N.F. (2002). Stability-indicating methods for the determination of doxazosin mezylate and celecoxib. *J. Pharm. Biomed. Anal.* 27: 779–793.

3 Baboota, S., Faiyaz, S., Ahuja, A. et al. (2007). Development and validation of a stability-indicating HPLC method for analysis of celecoxib (CXB) in bulk drug and microemulsion formulations. *Acta Chromatogr.* 18: 116–129.

4 Donnelly, R.F., Pascuet, E., Ma, C., and Vaillancourt, R. (2009). Stability of celecoxib oral suspension. *Can. J. Hosp. Pharm.* 62: 464–468.

5 Li, M. (2012). Hydrolytic degradation. In: *Organic Chemistry of Drug Degradation*, 30–32. Cambridge: Royal Society Chemistry Publishing.

6 Li, M. (2012). Drug–excipient interactions and adduct formation. In: *Organic Chemistry of Drug Degradation*. Cambridge: Royal Society Chemistry Publishing.

7 Gao, P., Huang, T., Robins, R.H. et al. (2002). Orally deliverable pharmaceutical composition comprising an active compound having an aminosulfonyl group (cox-2 inhibitor), a polyethylene glycol and a free-radical scavenging antioxidant. European patent: EP1494666. Filing date: 12 April 2002.

95–100: Cetirizine and Structurally Related Drugs

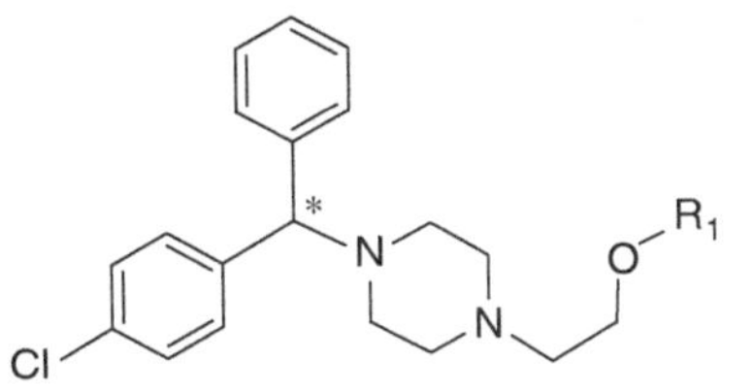

Ceritizine, * = Racemic; R = $-CH_2CO_2H$
Levoceritizine, * = *R*-configuration; R = $-CH_2CO_2H$
Hydroxyzine, * = Racemic; R = $-CH_2CH_2OH$

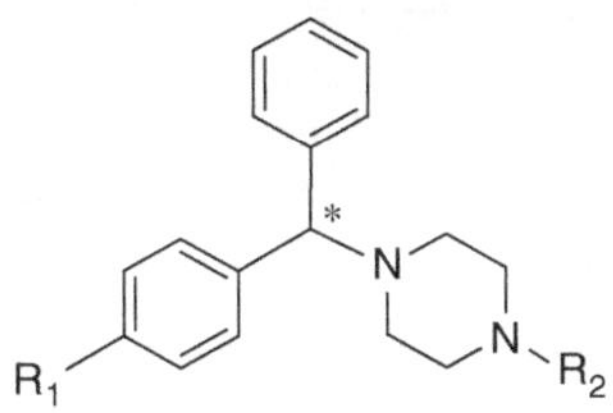

Cyclozine, R_1 = H; R_2 = $-CH_3$;
Chlorcyclizine, R_1 = Cl; R_2 = $-CH_3$;
Meclizine, * = Racemic; R_1= Cl; R_2 = $-CH_2Ph$

95: Cetirizine

Chemical name: 2-[2-[4-[(4-Chlorophenyl)-phenylmethyl]piperazin-1-yl]ethoxy] acetic acid
Brand name: Zyrtec
Formula: $C_{21}H_{25}ClN_2O_3$
Monoisotopic protonated ion (*m/z*): 389.1626
Molecular weight: 388.89
CAS number: 83881-51-0
Common salt form: HCl

96: Levocetirizine

Chemical name: 2-[2-[4-[(*R*)-(4-Chlorophenyl)-phenylmethyl]piperazin-1-yl]ethoxy]acetic acid
Brand name: Xyzal
Formula: $C_{21}H_{25}ClN_2O_3$
Monoisotopic protonated ion (*m/z*): 389.1626
Molecular weight: 388.89
CAS number: 130018-77-8
Common salt form: 2(HCl)

97: Hydroxyzine

Chemical name: 2-[2-[4-[(4-Chlorophenyl)-phenylmethyl]piperazin-1-yl]ethoxy] ethanol
Brand name: Atarax, Vistaril
Formula: $C_{21}H_{27}ClN_2O_2$
Monoisotopic protonated ion (*m/z*): 375.1834
Molecular weight: 374.90
CAS number: 68-88-2
Common salt form: HCl

98: Cyclizine

Chemical name: 1-Benzhydryl-4-methylpiperazine
Brand name: Nausicalm
Formula: $C_{18}H_{22}N_2$
Monoisotopic protonated ion (*m/z*): 267.1856
Molecular weight: 266.38
CAS number: 82-92-8
Common salt form: HCl

99: Chlorcyclizine

Chemical name: 1-[(4-Chlorophenyl)-phenylmethyl]-4-methylpiperazine
Brand name: Ahist, Biclora
Formula: $C_{18}H_{21}ClN_2$
Monoisotopic protonated ion (*m/z*): 301.1466
Molecular weight: 300.83
CAS number: 82-93-9
Common salt form: HCl

100: Meclizine

Chemical name: 1-[(4-Chlorophenyl)-phenylmethyl]-4-[(3-methylphenyl)methyl] piperazine
Brand name: Antivert, Diphen
Formula: $C_{25}H_{27}ClN_2$
Monoisotopic protonated ion (*m/z*): 391.1936
Molecular weight: 390.9488
CAS number: 163837-49-8
Common salt form: HCl

Cetirizine is a metabolite of hydroxyzine, a first-generation antihistamine, and is categorized as a second-generation antihistamine. It is a racemic mixture, and its *R*-enantiomer is levocetirizine, which is a third-generation antihistamine. All of them are members of the phenylpiperazine family of antihistamine drugs. As tertiary amines, cetirizine, levocetirizine, and hydroxyzine should be susceptible to nucleophilic oxidation to form the corresponding *N*-oxide degradants. Indeed, a study on a formulation containing PEG, which is known to contain residual peroxides, showed that *N*-oxidation occurs on the less sterically hindered nitrogen of the piperazine ring (Scheme 95–100-1) [1]. On the other hand, the carboxyl group of cetirizine and levocetirizine can be reactive toward hydroxyl-containing excipients by forming ester-linked adducts [2]. A study by Yu et al. found that two marketed preparations of cetirizine contained between 0.1% and 0.3% of esters that were apparently formed between the drug substance and polyols excipients such as sorbitol and glycerol [3]. For levocetirizine, its stereochemical stability was evaluated by

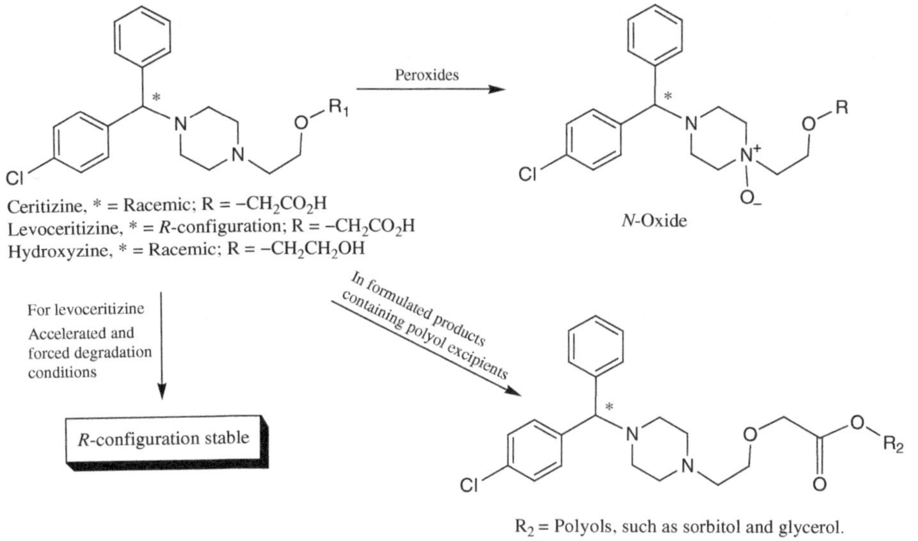

Scheme 95–100-1 Degradation pathways of cetirizine and structurally related drugs.

Hommoss et al.; no racemization was observed under both accelerated stability as well as forced degradation conditions (acid, base, hydrogen peroxide, thermal, and photolysis) [4]. Since cetirizine does not contain a strong chromophore, it should be photochemically stable. However, under extensive photo stress conditions particularly in solutions in the UV-A region, cetirizine was found to produce two photo degradants [5]; The structures of the two photo degradants were not elucidated; it would be likely that the photo degradation such as dichlorination, may occur with the chlorophenyl moiety [6].

Other members of the phenylpiperazine antihistamines include cyclizine and meclizine. Similarly, the distal tertiary amine moieties of both molecules, which are more reactive due to less steric hindrance and more alkalinity (the distal amines having pKa of 8.05 and 7.65 for cyclizine and meclizine, respectively, versus 2.16 and 2.12 for their proximal amine moieties), should be susceptible to nucleophilic oxidation to give the corresponding *N*-oxides (Scheme 95–100-2) [7]. In addition to the

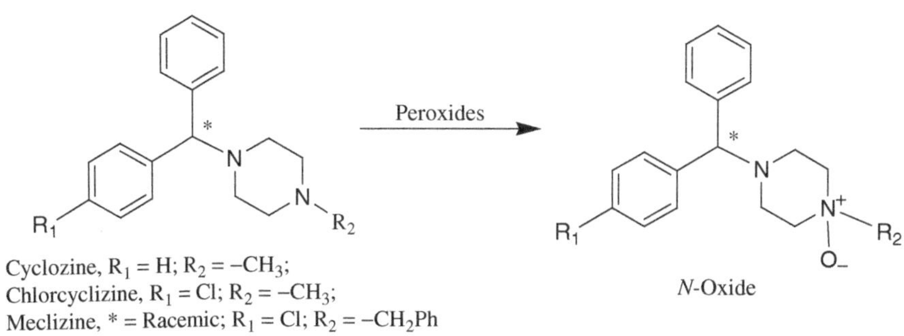

Scheme 95–100-2 *N*-Oxidation of cyclozine, chlorcyclizine, and meclizine.

N-oxide, other forced degradation products of meclizine were reported; however, the identities of these degradants were established only based upon retention time matching, and the proposed structures of certain degradants do not seem reasonable[1].

References

1 Dyakonov, T., Muir, A., Nasri, H. et al. (2010). Isolation and characterization of cetirizine degradation product: mechanism of cetirizine oxidation. *Pharm. Res.* 27: 1318–1324.

2 Li, M. (2012). Drug–excipient interactions and adduct formation. In: *Organic Chemistry of Drug Degradation*, 150–164. Cambridge: Royal Society Chemistry Publishing.

3 Yu, H., Cornett, C., Larsen, J., and Hansen, S.H. (2010). Reaction between drug substances and pharmaceutical excipients: formation of esters between cetirizine and polyols. *J. Pharm. Biomed. Anal.* 53: 745–750.

4 Hommoss, R., Elzein, H., and Haidar, S. (2011). Determination of levocetirizine configurational stability in tablets using chiral HPLC method. *Int. J. Pharm. Pharm. Sci.* 3: 103–107.

5 Jaber, A.M.Y., Al Sherife, H.A., Al Omari, M.M., and Badwan, A.A. (2004). Determination of cetirizine dihydrochloride, related impurities and preservatives in oral solution and tablet dosage forms using HPLC. *J. Pharm. Biomed. Anal.* 36: 341–350.

6 Li, M. (2012). Photochemical degradation. In: *Organic Chemistry of Drug Degradation*, 165–197. Cambridge: Royal Society of Chemistry.

7 Peraman, R., Manikala, M., Kondreddy, V.K., and Yiragamreddy, P.R. (2015). A stability-indicating RP-HPLC method for the quantitative analysis of meclizine hydrochloride in tablet dosage form. *J. Chromatogr. Sci.* 53: 793–799.

---

1 For example, in Reference 7, it was reported that under acidic stress, the only degradant observed was 3-(hydroxymethyl)toluene. One would assume that this degradant should be a product resulting from the breakup of meclizine at the piperazine ring; such a breakup should also result in, concomitantly, the formation of at least another degradant bearing the phenylpiperazine core, such as desmethylchlorcyclizine. Nevertheless, no additional degradant was observed.

101: Chlorhexidine [WHO Essential List of Medicines]

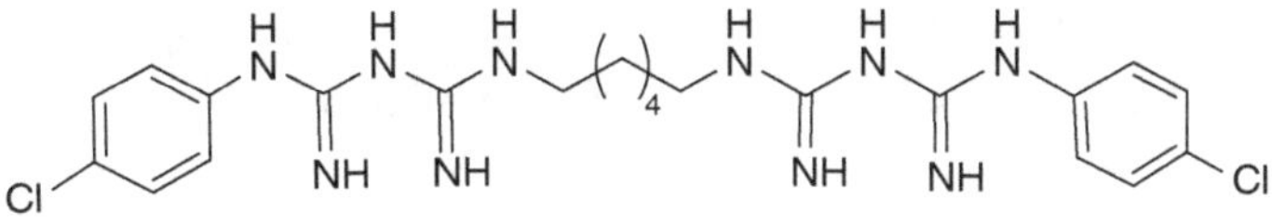

Chemical name: (1*E*)-2-[6-[[Amino-[(*E*)-[amino-(4-chloroanilino)methylidene] amino]methylidene]amino]hexyl]-1-[amino-(4-chloroanilino)methylidene] guanidine
Brand name: Betasept, Hibiclens, Instillagel, etc.
Formula: $C_{22}H_{30}Cl_2N_{10}$
Monoisotopic protonated ion (*m/z*): 505.2105
Molecular weight: 505.45
CAS number: 55-56-1
Common salt form: Diacetate, digluconate, 2(HCl)

Chlorhexidine is a biguanide, which is a very strong base, and present as a positively charged ion at pH < 11 [1]. It is widely used as an antibacterial agent in pharmaceutical preparations as well as in consumer products such as mouthwash. In aqueous solutions, it is most stable in the pH range of 5–8 [2]. In a forced degradation study of a commercial chlorhexidine digluconate solution, the main degradation pathway was proposed to be a retro-ene mechanism by Revelle et al. [3] (Scheme 101-1), to

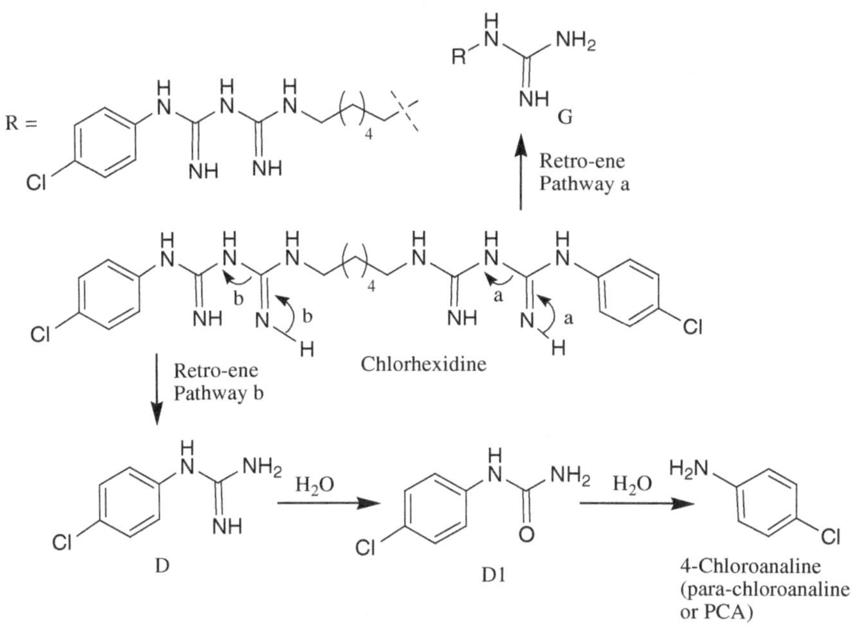

Scheme 101-1 Forced degradation of a commercial chlorhexidine digluconate solution via the proposed retro-ene mechanism. Nomenclature of Revelle et al. for the impurities is adopted [3]. *Source:* Revelle et al. [3]/Springer Nature.

give rise to impurity D and G as the main degradants. It should be noted that these two degradants can also be formed via hydrolytic mechanisms. The degradant D can further degrade to 4-chloroanaline (also referred to as para-chloroanaline or PCA).

Under both acidic and alkaline stress conditions, the main degradation pathway seems to be of hydrolytic nature. At pH 1.0 at 90°C for 5 hours, chlorhexidine was found to decompose to A1, A2, A4, A5, and 4-chloroanaline (Scheme 101-2) [3]. On the other hand, when stressed at pH 8.6 at 90°C for 24 hours, chlorhexidine degraded to G, D1, and 4-chloroanaline (Scheme 101-3). Among all the degradants observed, 4-chloroaniline (PCA) is of most toxicological concern because it is a known carcinogen, and the USP limit on this impurity is set at 3 ppm [4].

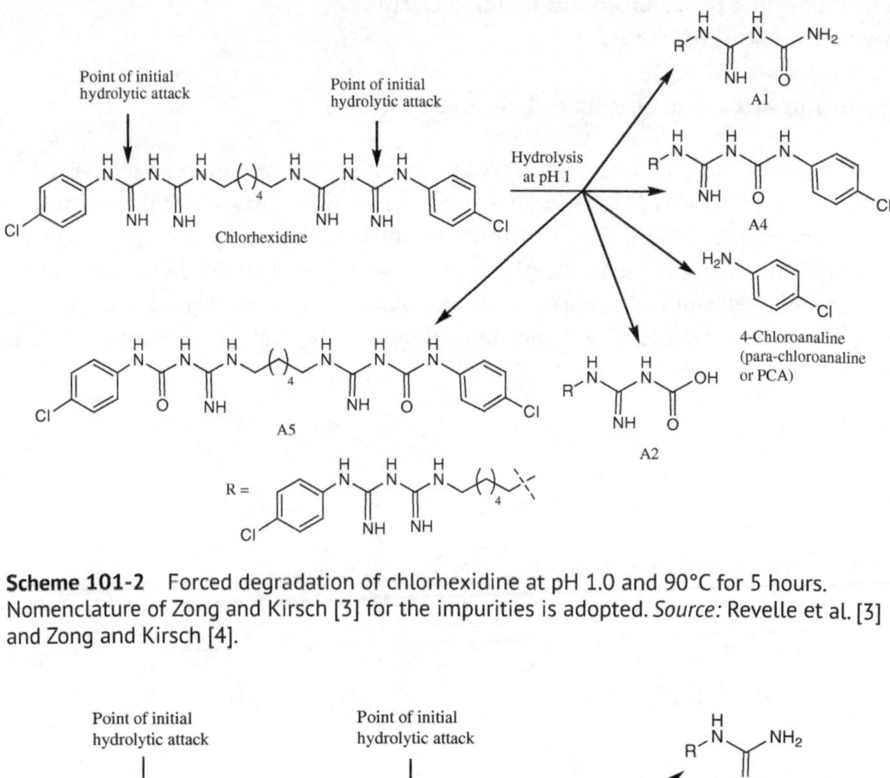

Scheme 101-2 Forced degradation of chlorhexidine at pH 1.0 and 90°C for 5 hours. Nomenclature of Zong and Kirsch [3] for the impurities is adopted. *Source:* Revelle et al. [3] and Zong and Kirsch [4].

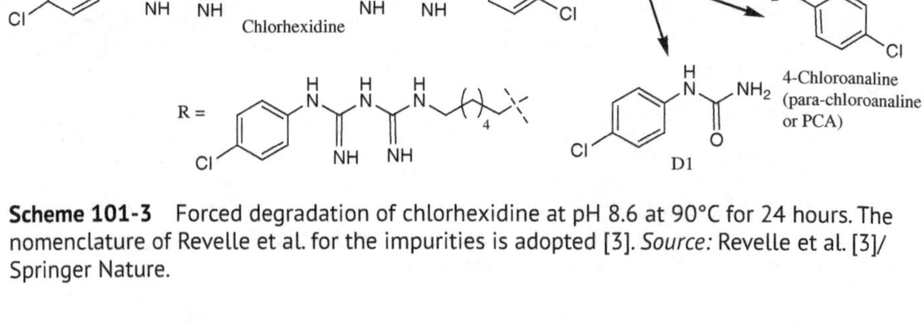

Scheme 101-3 Forced degradation of chlorhexidine at pH 8.6 at 90°C for 24 hours. The nomenclature of Revelle et al. for the impurities is adopted [3]. *Source:* Revelle et al. [3]/ Springer Nature.

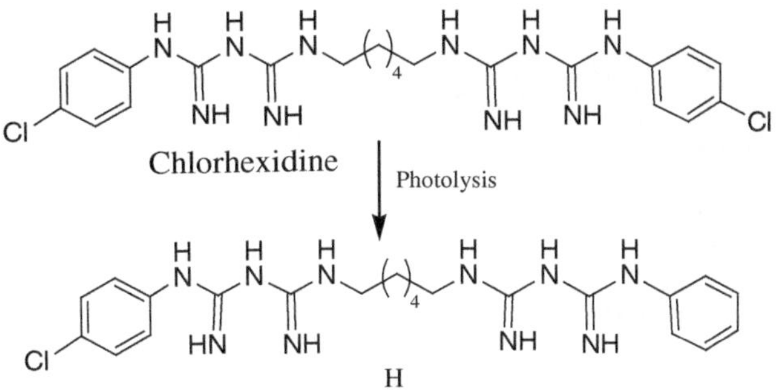

Scheme 101-4 Photochemical dechlorination of chlorhexidine. Nomenclature of Revelle et al. for the impurities is adopted [3]. *Source:* Revelle et al. [3]/Springer Nature.

When stressed under sunlight for 69 hours, impurity H was observed, which apparently results from photodechlorination of chlorhexidine (Scheme 101-4).

References

1 Elpern, B. (1968). Chemistry of the biguanides. *Ann. N. Y. Acad. Sci.* 148: 577–586.

2 Nicolay, A., Wolff, E., Vergnes, M.-F. et al. (2011). Rapid HPLC method for determination of parachloroaniline in chlorhexidine antiseptic agent in mouthrinses, ophthalmic and skin solution. *Am. J. Anal. Chem.* 2: 422–428.

3 Revelle, L.K., Doub, W.H., Wilson, R.T. et al. (1993). Identification and isolation of chlorhexidine digluconate impurities. *Pharm. Res.* 10: 1777–1784.

4 Zong, Z. and Kirsch, L.E. (2012). Studies on the instability of chlorhexidine, part I: kinetics and mechanisms. *J. Pharm. Sci.* 101: 2417–2427.

102: Chloramubcil

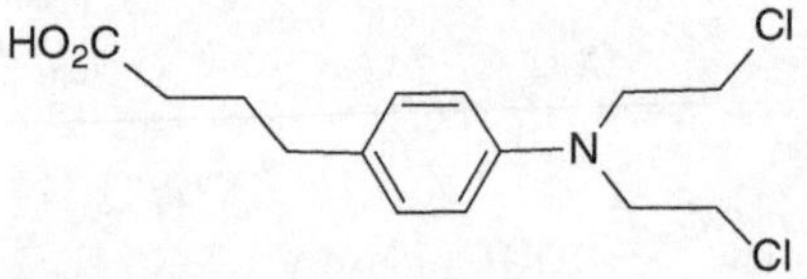

Chemical name: 4-[4-[Bis(2-chloroethyl)amino]phenyl]butanoic acid
Brand name: Leukeran
Formula: $C_{14}H_{19}Cl_2NO_2$
Monoisotopic protonated ion (*m/z*): 304.0866
Molecular weight: 304.21
CAS number: 305-03-3

Chlorambucil is an analine-based nitrogen mustard, which was introduced in the 1940s as an antitumor agent. Its main degradation pathway is hydrolysis via the aziridium intermediate (Scheme 102); this mechanism is consistent with the fact that chloride stabilizes the drug due to its reacting with the intermediate to regenerate the drug [1]. The aziridium intermediate is also supported by density functional theory (DFT) calculation [2]. Using this calculation, the activation energy for the hydrolysis was determined to be 24.8 kcal/mol, which is consistent with the experimental result [3]. The pKa of chlorambucil was determined to be approximately 3.5 and in the pH range of 3.5–8, the rate of chlorambucil hydrolysis was independent of pH. At pH below 3.5, the hydrolysis rate decreased rapidly, apparently due to the protonation of the aniline nitrogen, which should slow down the formation of the aziridium intermediate. In another study for the stability of chlorambucil in aqueous solution, it was found that the stability of chlorambucil was unaffected in the pH range between 5 and 10, while it became more stable at lower pH [4]. The general trends for the stability of chlorambucil in these studies were consistent, despite some minor differences in the pH range.

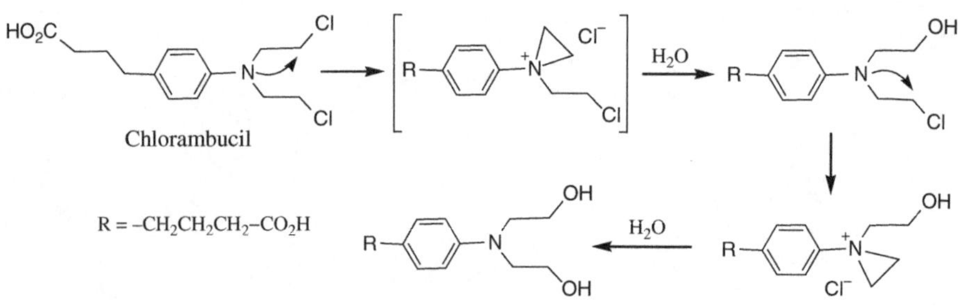

Scheme 102 Hydrolysis of chlorambucil.

References

1 Cullis, P.M., Green, R.E., and Malone, M.E. (1995). Mechanism and reactivity of chlorambucil and chlorambucil-spermidine conjugate. *J. Chem. Soc. Perkin Trans. 2*: 1503–1511.

2 Pineda, F.P., Ortega-Castro, J., Alvarez-Idaboy, J.R. et al. (2011). Hydrolysis of a Chlorambucil analogue. A DFT study. *J. Phys. Chem. A* 115: 2359–2366.

3 Gamcsik, M.P., Millis, K.K., and Hamill, T.G. (1997). Kinetics of the conjugation of aniline mustards with glutathione and thiosulfate. *Chem-Bio. Interac.* 105: 35–52.

4 Ehrsson, H., Eksborg, S., Wallin, I., and Nilsson, S.O. (1980). Degradation of chlorambucil in aqueous solution. *J. Pharm. Sci.* 69: 1091–1094.

103: Citicoline

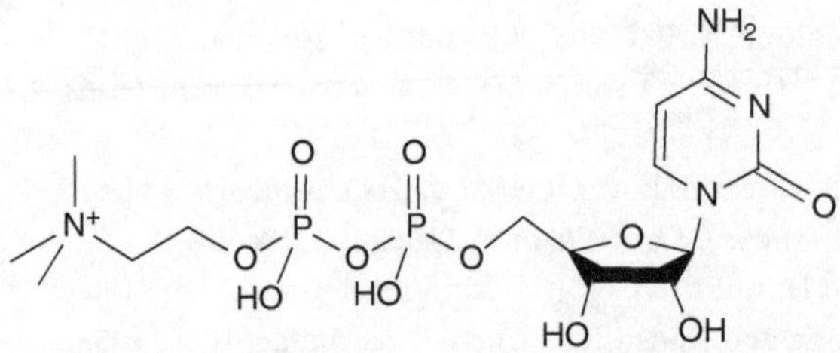

Chemical name: [[(2*R*,3*S*,4*R*,5*R*)-5-(4-Amino-2-oxopyrimidin-1-yl)-3,4-dihydroxyoxolan-2-yl]methoxy-hydroxyphosphoryl] 2-(trimethylazaniumyl)ethyl phosphate
Brand name: CerAxon
Formula: $C_{14}H_{26}N_4NaO_{11}P_2$
Monoisotopic ion (*m/z*): 488.1068 (citicoline is a quaternary ammonium ion)
Molecular weight: 510.31 (for citicoline sodium, $C_{14}H_{25}N_4NaO_{11}P_2$)
CAS number: 987-78-0
Common salt form: Sodium salt

Citicoline is a psychostimulant, and structurally it is a conjugate of cytidine and choline that are linked together by a diphosphate. There are a couple of forced degradation studies of citicoline reported in the literature [1, 2]. Under acidic and alkaline conditions, approximately two to three degradants were observed depending on the harshness of the stress conditions. The main degradant was

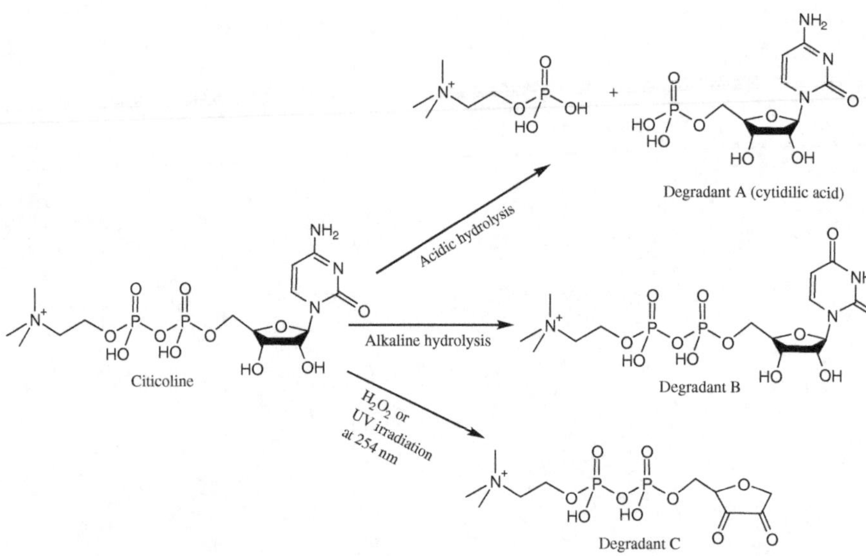

Scheme 103 Degradation pathways of citicoline under acidic, alkaline, oxidative, and photo-forced degradation conditions.

identified via LC-MS as cytidilic acid, which is formed via the cleavage of the phosphoryl bond of diphosphate (Scheme 103) [2]. Holbrook and Ouellet reported in 1957 that the activation energy (*E*a) for hydrolysis of adenosine diphosphate, which can be considered an analog of citicoline for the sake of the phosphoryl bond hydrolysis, was about 24 kcal/mol at pH 8 [3]. The *E*a value of this scale suggests that citicoline should be reasonably stable toward hydrolytic degradation under long-term stability conditions at a similar pH range. In two forced degradation studies, when citicoline was stressed in H_2O_2 solutions under elevated temperature [1, 2], one major degradant was observed; however, its structure was not elucidated. In a later forced degradation study performed by Derbouz et al. [4], citicoline was found to produce three degradants (Degradants A, B, and C) under acidic, alkaline, and oxidative (H_2O_2) conditions, respectively. When the solid citicoline sodium was irradiated at 254 nm, Degradant C was also formed.

The formation of Degradants A and B is relatively straightforward: under acidic stress, the diphosphate linkage was cleaved, while under alkaline stress, the cytosine moiety was converted to the uracil moiety in Degradant B via deamination. On the other hand, the formation of Degradant C appears to be more complicated; the formation mechanism provided by Derbouz et al. [4] seems to be vague.

References

1 Ganduri, R.B., Peddareddigari, J.R., Dasari, N.R., and Saiempu, R.K. (2010). Stability indicating LC method for the determination of citicoline sodium in injection formulation. *Int. J. PharmTech Res.* 2: 427–433.

2 Patel, J.A., Panigrahi, B., Patel, C.N., and Ramalingan, B. (2011). Stress degradation studies on citicoline sodium and development of a validated stability-indicating HPLC assay. *Chron. Young Sci.* 2: 150–154.

3 Holbrook, K.A. and Ouellet, L. (1957). The non-enzymatic hydrolysis of adenosine diphosphate. *Can. J. Chem.* 35: 1496–1503.

4 Derbouz, S., Guermouche, M.H., and Guermouche, S. (2017). Stability-indicating HILIC method for the determination of citicoline and characterization of its degradation products by LC–MS/TOF, 1H and 13C NMR. *Chromatographia* 80: 265–274.

104–116: Chlorpromazine and Structurally Analogous Drugs

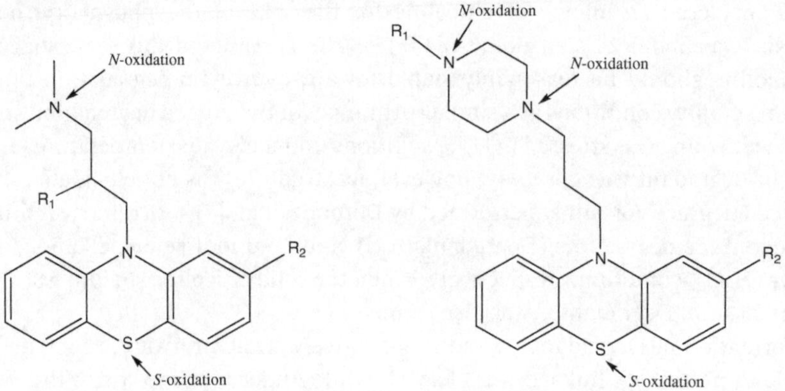

Chlorpromazine, R_1 = hydrogen, R_2 = chloro;
Acepromazine, R_1 = hydrogen, R_2 = acetyl;
Methotrimeparzine, R_1 = (*R*)-methyl, R_2 = methoxyl;
Promazine, R_1 = hydrogen, R_2 = hydrogen;
Triflupromazine, R_1 = hydrogen, R_2 = trifluoromethyl;
Trimeprazine, R_1 = methyl, R_2 = hydrogen.

Perphenazine, R_1 = 2-hydroxyethyl, R_2 = chloro;
Carphenazine, R_1 = 2-hydroxyethyl, R_2 = propionyl;
Fluphenazine, R_1 = 2-hydroxyethyl, R_2 = trifluoromethyl;
Prochlorperazine, R_1 = methyl, R_2 = chloro;
Thiethylperazine, R_1 = methyl, R_2 = ethylmercapto;
Trifluoperazine, R_1 = methyl, R_2 = trifluoromethyl;
Thioproperazine, R_1 = methyl, R_2 = dimethylsulfamyl.

Structures of chlorpromazine and its analogous phenothiazine-based drugs

104: Chlorpromazine [List of WHO Essential Medicines]

Chemical name: 3-(2-Chlorophenothiazin-10-yl)-*N*,*N*-dimethylpropan-1-amine
Brand name: Aminazine, Chlorprom
Formula: $C_{17}H_{19}ClN_2S$
Monoisotopic protonated ion (*m/z*): 319.1030
Molecular weight: 318.87
CAS number: 50-53-3
Common salt form: HCl

105: Acepromazine

Chemical name: 1-[10-[3-(Dimethylamino)propyl]phenothiazin-2-yl]ethanone
Brand name: Concentrat vo34
Formula: $C_{19}H_{22}N_2OS$
Monoisotopic protonated ion (*m/z*): 327.1526
Molecular weight: 326.46
CAS number: 61-00-7
Common salt form: Maleate

106: Carphenazine (Carfenazine)

Chemical name: 1-[10-[3-[4-(2-Hydroxyethyl)piperazin-1-yl]propyl]phenothiazin-2-yl]propan-1-one
Brand name: Proketazin, Proketazine
Formula: $C_{24}H_{31}N_3O_2S$
Monoisotopic protonated ion (*m/z*): 426.22097
Molecular weight: 425.59
CAS number: 2622-30-2
Common salt form: Maleate

107: Fluphenazine

Chemical name: 2-[4-[3-[2-(Trifluoromethyl)phenothiazin-10-yl]propyl]piperazin-1-yl]ethanol
Brand name: Permitil, Prolixin
Formula: $C_{22}H_{26}F_3N_3OS$
Monoisotopic protonated ion (*m/z*): 438.1821
Molecular weight: 437.52
CAS number: 5002-47-1
Common salt form: HCl

108: Levomepromazine (Methotrimeprazine)

Chemical name: (2*R*)-3-(2-Methoxyphenothiazin-10-yl)-*N*,*N*,2-trimethylpropan-1-amine
Brand name: Levoprome, Nozinan
Formula: $C_{19}H_{24}N_2OS$
Monoisotopic protonated ion (*m/z*): 329.1682
Molecular weight: 328.47
CAS number: 60-99-1
Common salt form: HCl, Maleate

109: Perphenazine

Chemical name: 2-[4-[3-(2-Chlorophenothiazin-10-yl)propyl]piperazin-1-yl]ethanol
Brand name: Trilafon
Formula: $C_{21}H_{26}ClN_3OS$
Monoisotopic protonated ion (*m/z*): 404.1558
Molecular weight: 403.97
CAS number: 58-39-9

110: Prochlorperazine

Chemical name: 2-Chloro-10-[3-(4-methylpiperazin-1-yl)propyl]phenothiazine
Brand name: Compazine, Compro
Formula: $C_{20}H_{24}ClN_3S$
Monoisotopic protonated ion (*m/z*): 374.1452
Molecular weight: 373.94
CAS number: 58-38-8

111: Promazine

Chemical name: *N,N*-Dimethyl-3-phenothiazin-10-ylpropan-1-amine
Brand name: Pliva, Talofen
Formula: $C_{17}H_{20}N_2S$
Monoisotopic protonated ion (*m/z*): 285.1420
Molecular weight: 284.42
CAS number: 58-40-2
Common salt form: HCl

112: Thiethylperazine

Chemical name: 2-Ethylsulfanyl-10-[3-(4-methylpiperazin-1-yl)propyl]phenothiazine
Brand name: Torecan
Formula: $C_{22}H_{29}N_3S_2$
Monoisotopic protonated ion (*m/z*): 400.1876
Molecular weight: 399.62
CAS number: 1420-55-9
Common salt form: Malate, Maleate

113: Thioproperazine

Chemical name: *N,N*-Dimethyl-10-[3-(4-methylpiperazin-1-yl)propyl]phenothiazine-2-sulfonamide
Brand name: Majeptil
Formula: $C_{22}H_{30}N_4O_2S_2$
Monoisotopic protonated ion (*m/z*): 447.1883
Molecular weight: 446.63
CAS number: 316-81-4

114: Trifluoperazine

Chemical name: 10-[3-(4-Methylpiperazin-1-yl)propyl]-2-(trifluoromethyl)phenothiazine
Brand name: Stelazine

Formula: $C_{21}H_{24}F_3N_3S$
Monoisotopic protonated ion (*m/z*): 408.1716
Molecular weight: 407.50
CAS number: 117-89-5
Common salt form: HCl

115: Triflupromazine

Chemical name: *N,N*-Dimethyl-3-[2-(trifluoromethyl)phenothiazin-10-yl]propan-1-amine
Brand name: Vesprin
Formula: $C_{18}H_{19}F_3N_2S$
Monoisotopic protonated ion (*m/z*): 353.1294
Molecular weight: 352.42
CAS number: 146-54-3
Common salt form: HCl

116: Trimeprazine

Chemical name: *N,N*,2-Trimethyl-3-phenothiazin-10-ylpropan-1-amine
Brand name: Panectyl
Formula: $C_{18}H_{22}N_2S$
Monoisotopic protonated ion (*m/z*): 299.1576
Molecular weight: 298.45
CAS number: 84-96-8
Common salt form: Tartrate

Chlorpromazine and its analogous drugs are used as antipsychotic medications, except in rare cases, such as trimeprazine (which is an antihistamine). This family of drugs contains a phenothiazine core with one or two substituents on the core (refer to the structures of chlorpromazine and its analogous drugs above). Based on the structure of the *N*-substituent, this family of drugs can be further divided into two subtypes: one has a simple tertiary amine group in the side chain (as represented by chlorpromazine), and the other contains a piperazine ring (as represented by perphenazine). All the tertiary amine functionalities in the side chain, including both tertiary amines in the piperazine ring, are susceptible to nucleophilic oxidation to form the corresponding *N*-oxides [1]. The sulfur in the phenothiazine ring is susceptible to both nucleophilic [2, 3], free-radical mediated oxidation, and photochemical oxidation [4, 5]; usually, the degradation product is limited to sulfoxide [5], but sulfone could be formed [6]. In a forced degradation study of perphenazine in methanol solution containing H_2O_2, the ^{17}N-oxide was the main degradant in the early stage of the stress, in addition to ^{14}N-oxide, perphenazine sulfoxide, and a low level of ^{14}N,^{17}N-dioxide [2]. The solution stress study was in good agreement with a study on solid dosage form of perphenazine in which ^{17}N-oxide was also found as the main oxidative degradant.

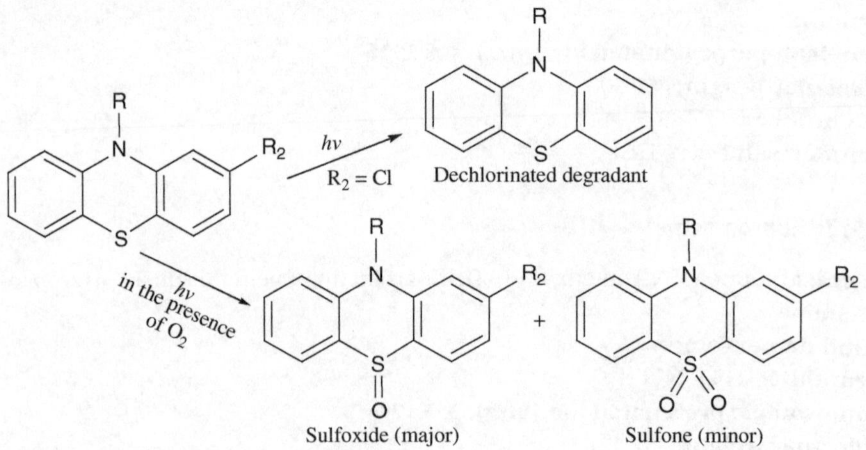

Scheme 104–116 Photochemical degradation of phenothiazine-based drugs.

The phenothiazine ring is quite susceptible to photochemical activation, and as such, this class of drugs displays certain levels of phototoxicity [4]. For those phenothiazine-derived drugs that contain a chlorine on the phenothiazine ring, e.g., chlorpromazine and perphenazine, photo dechlorination is a common degradation pathway (Scheme 104–116) [3, 7]. *S*-Oxidation could also occur in the photochemical degradation process.

For fluphenazine, a drug molecule in the phenothiazine-based family that contains a 2-hydroxyethyl side chain, it is also available in the forms of decanoate and enanthate esters with the ester linkages with the 2-hydroxyethyl side chain. The ester APIs are employed in intramuscular or subcutaneous injections for an extended duration of pharmacological effect. The ester linkages would be susceptible to hydrolytic degradation,

References

1 Li, M. (2012). Oxidative degradation. In: *Organic Chemistry of Drug Degradation*, 71–73. Cambridge: Royal Society Chemistry Publishing.
2 Wang, X., Li, M., and Rustum, A.M. (2010). Thermally induced intramolecular oxygen migration of *N*-oxides in atmospheric pressure chemical ionization mass spectrometry. *Rapid Commun. Mass Spectrom.* 24: 2805–2811.
3 Li, X., Blondino, F.E., Hindle, M. et al. (2005). Stability and characterization of perphenazine aerosols generated using the capillary aerosol generator. *Int. J. Pharm.* 303: 113–124.

4 Chagonda, L.F.S. and Millership, J.S. (1989). The determination of chlorpromazine, related impurities and degradation products in pharmaceutical dosage forms. *J. Pharm. Biomed. Anal.* 7: 271–278.

5 Chagonda, L.F.S. and Millership, J.S. (1988). High-performance liquid chromatographic determination of chlorpromazine and its degradation products in pharmaceutical dosage forms: a stability-indicating assay. *Analyst* 113: 233–237.

6 Takahashi, D.M. (1980). Rapid determination of chlorpromazine hydrochloride and two oxidation products in various pharmaceutical samples using high-performance liquid chromatography and fluorescence detection. *J. Pharrn. Sci.* 69: 184–187.

7 Li, M. (2012). Photochemical degradation. In: *Organic chemistry of drug degradation*, 178–179. Cambridge: Royal Society Chemistry Publishing.

117: Chlorzoxazone

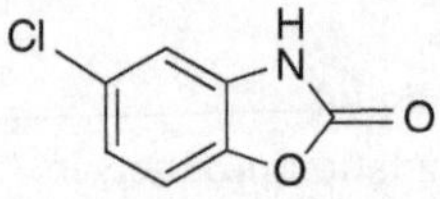

Chemical name: 5-Chloro-3*H*-1,3-benzoxazol-2-one
Brand name: Lorzone, Parafon Forte
Formula: $C_7H_4ClNO_2$
Monoisotopic protonated ion (*m/z*): 168.9931
Molecular weight: 169.57
CAS number: 95-25-0

Chlorzoxazone is a skeletal muscle relaxant. Structurally, it can be viewed as a cyclic carbamate fused into the chlorobenzene ring. Hence, its main degradation pathway is the hydrolysis of the carbamate [1] in which 2-amino-4-chlorophenol is produced (Scheme 117).

The hydrolysis is greatly enhanced under the catalysis of a strong base [2]; according to a forced degradation study, the activation energies for the hydrolysis were found to be 62.8 and 54.4 kJ/mol in 3 and 8 M sodium hydroxide solutions, respectively[2]. This impurity is also a process impurity that has some toxicological concerns [3]. There is no evidence that this impurity is a metabolite of the drug [4].

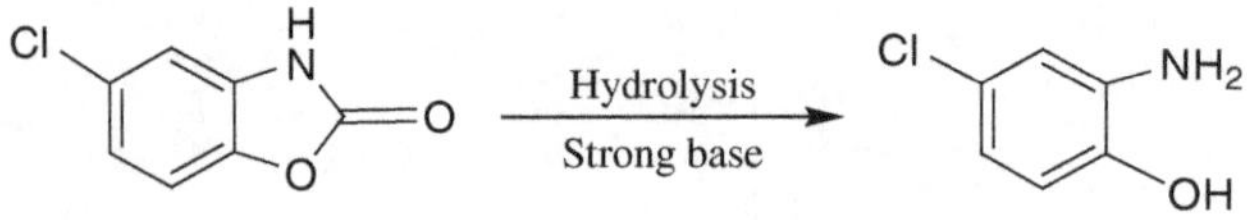

Scheme 117 Hydrolysis of chlorzoxane under strong basic conditions.

References

1 Li, M. (2012). Hydrolytic degradation. In: *Organic Chemistry of Drug Degradation*, 30–32. Cambridge: Royal Society Chemistry Publishing.
2 Stewart, J.T. (1987). *Analytical Profiles of Drug Substances and Excipients*, vol. 16, 119–144. Academic Press.
3 Hazardous Chemicals Fact Sheet, New Jersey Department of Health and Senior Services, 1996.
4 Twele, R. and Spiteller, G. (1982). Identification of chlorzoxazone metabolites in human urine (English translation from German). *Arzneimittelforschung* 32: 759–763.

2 In the original paper by Ellaithy et al. (*Il Farmaco*, 2003, 58, 337–342.), the activation energies in 3 M and 8 M sodium hydroxide solutions were reported to be 62 829.43 and 54 393.2 kcal/mol, respectively, which are apparently erroneous.

118: Cholecalciferol

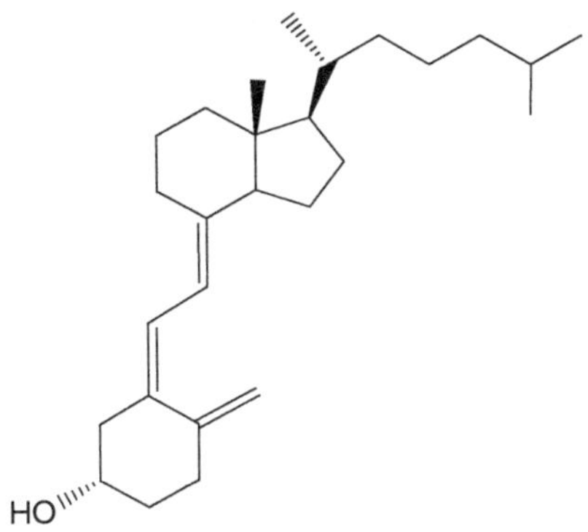

Cholecalciferol (Vitamin D_3)
Chemical name: (1*S*,3*Z*)-3-[(2*E*)-2-[(1*R*,3a*S*,7a*R*)-7*a*-Methyl-1-[(2*R*)-6-methylheptan-2-yl]-2,3,3*a*,5,6,7-hexahydro-1*H*-inden-4-ylidene]ethylidene]-4-methylidene cyclohexan-1-ol
Formula: $C_{27}H_{44}O$
Monoisotopic protonated ion (*m/z*): 385.3465
Molecular weight: 384.64
CAS number: 67-97-0

Cholecalciferol, more commonly known as vitamin D_3, is a fat-soluble vitamin, which is used as a supplement as well as a therapeutic agent particularly for helping calcium absorption. In human body, vitamin D_3 is produced photochemically in skin from 7-dehydrocholesterol via the intermediacy of previtamin D_3 by exposure to sunlight [1]. Previtamin D_3 is thermodynamically less stable than vitamin D_3 and upon heating in solution (but not in solid state) [2], previtamin D_3 is converted to vitamin D_3 [3]. Since this isomerization process is reversible, vitamin D_3 can also convert back to previtamin D_3 through 1,7-H shift [4]; the ratio of vitamin D_3 versus D_2 is a function of temperature, and at 20°C, this ratio is 93:7 [5]. Upon sunlight irradiation, vitamin D_3 can degrade to 5,6-transvitamin D_3, suprasterol I and suprasterol II as its main photo degradants (Scheme 118) [6]. Vitamin D_3 is stable under alkaline conditions but undergoes isomerization to give 5,6-transvitamin D_3 and isotachysterol under mild acidic condition [7]. On the other hand, the hydroxyl group of vitamin D_3 and previtamin D_3 can undergo transesterification if the excipients of a finished product contain an ester component such as triglyceride [8]. Ballard et al. reported four degradants resulting from transesterification between vitamin D_3 and previtamin D_3 with the octanoate and decanoate components of triglycerides in an experimental tablet formulation [9].

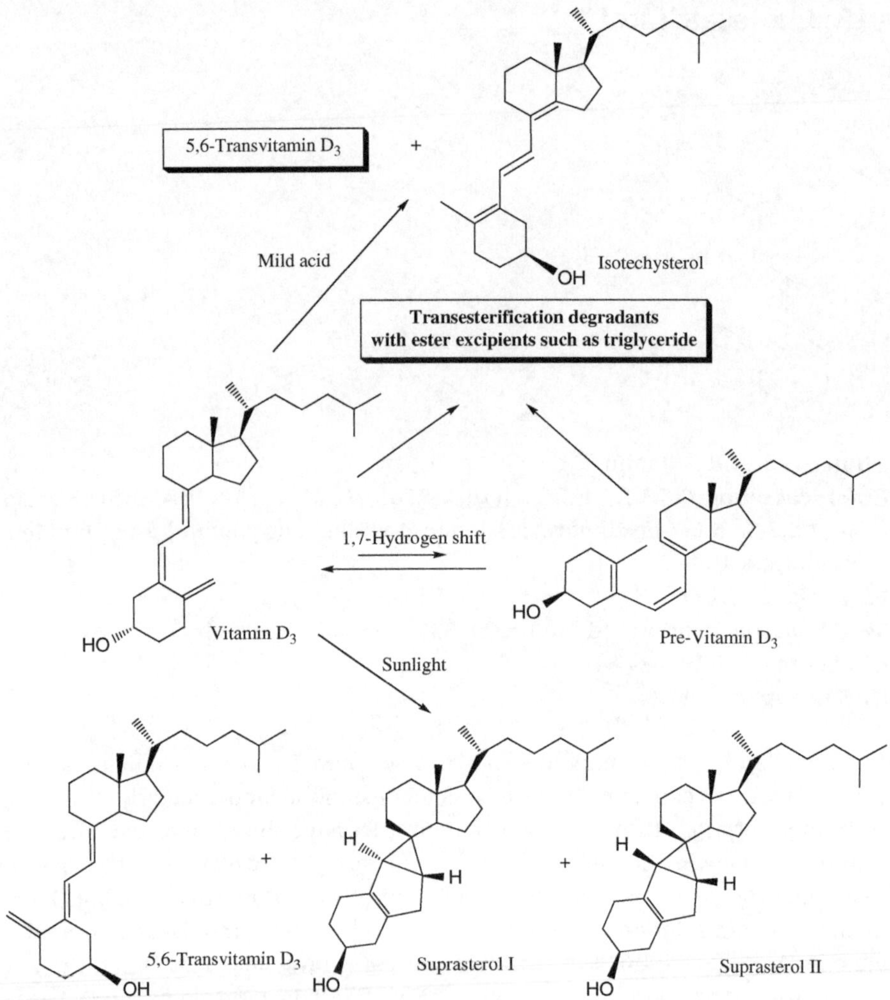

Scheme 118 Degradation pathways of cholecalciferol.

References

1 Holick, M.F. (1989). *Vertebrate Endocrinology: Fundamentals and Implications*, vol. 3 (ed. P.K.T. Pang and M., F. Holick), 7–34. Orlando, FL: Academic.

2 Ball, G.F.M. (2006). Vitamin D. In: *Vitamins in Foods, Analysis, Bioavailability and Stability*, 107–118. Boca Raton, FL: CRC Press.

3 Holick, M.F., MacLaughlin, J.A., Clark, M.B. et al. (1980). Photosynthesis of previtamin D_3 in human skin and the physiologic consequences. *Science* 210: 203–205.

4 Brandl, M., Wu, X.Y., Liu, Y. et al. (2003). Chemical reactivity of Ro-26-9228, 1α-fluoro-25-hydroxy-16,23E-diene -26,27-bishomo-20-epi-cholecalciferol in aqueous solution. *J. Pharm. Sci.* 92: 1981–1989.

5 Mulder, F.J., de Vries, E.J., and Borsje, B. (1971). Chemical analysis of vitamin D in concentrates and its problems. XII. Analysis of fat-soluble vitamins. *J. Assoc. Off. Anal. Chem.* 54: 1168–1178.

6 Webb, A.R., DeCosta, B.R., and Holick, M.F. (1989). Sunlight regulates the cutaneous production of vitamin D_3 by causing its photodegradation. *J. Clin. Endocrinol. Metab.* 68: 882–887.

7 DeLuca, H.F. (1978). Vitamin D, in vitamin D. In: *Handbook of Lipid Research*, vol. 2 (ed. H.F. DeLuca), 69. New York: Plenum Press.

8 Li, M. (2012). Drug-excipient interactions and adduct formation. In: *Organic Chemistry of Drug Degradation*, 150–164. Cambridge: RSC Publishing.

9 Ballard, J.M., Zhu, L., Nelson, E.D., and Seburg, R.A. (2007). Degradation of vitamin D_3 in a stressed formulation: the identification of esters of vitamin D3 formed by a transesterification with triglycerides. *J. Pharm. Biomed. Anal.* 43: 142–150.

119: Cyclosporin A [Ciclosporin]

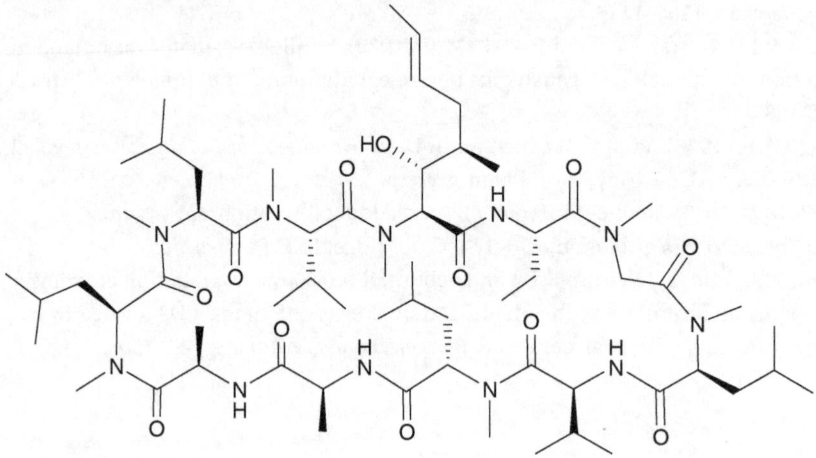

Chemical name: Cyclo[((2*S*)-2-aminobutyryl)-sarcosyl-*N*-methyl-L-leucyl-L-valyl-*N*-methyl-L-leucyl-L-alanyl-D-alanyl-*N*-methyl-L-leucyl-*N*-methyl-L-leucyl-*N*-methyl-L-valyl-*N*-methyl-(4*R*)-4-[(*E*)-but-2-enyl]-4-methyl-L-threonyl]
Brand name: Cequa, Gengraf, Neoral, SandIMMUNE
Formula: $C_{62}H_{111}N_{11}O_{12}$
Monoisotopic protonated ion (*m/z*): 1202.8486
Molecular weight: 1202.61
CAS number: 59865-13-3

Cyclosporin A, an immunosuppressive drug, is a cyclic undecapeptide, which consists of residues of many methylated amino acids, a D-alanine, and two unusual amino acids. The major degradation of cyclosporin A starts from the N—O peptidyl shift in which the hydroxyl side chain of the unusual amino acid residue attacks the preceding amide bond, forming the oxazoline-type intermediate (Pathway a, Scheme 119) [1]. In a study of degradation kinetics in aqueous solutions of pH 0.1–4.0, isocyclosporin A was found to be the predominant degradant at pH < 2 (Pathway a1) [2]. In a solid state, according to a study by Bonifacio et al. [1], the oxazoline-type intermediate seems to prefer the enolization pathway, which leads to the racemization of the $MeVal^{11}$ residue and produces cyclosporin H (Pathway a2). Bonifacio et al. analyzed commercial batches of cyclosporin A capsules from several manufacturers, which showed that only cyclosporin H may be present as the predominant degradant dependent upon the formulation, storage condition, and age of the sample analyzed [1]. The oxazoline-type intermediate should be susceptible to hydrolysis, yielding the open-chain degradant, which was mentioned as one of the principal degradants in the same study by Bonifacio et al., although it was not detected in commercial samples of cyclosporin A capsules [1]. In a hydrolytic stress study with 3 M HCl at 60°C, two open-chain peptides (a deca- and nona-peptide) were identified among the major degradation products [3]. This forced degradation study seems

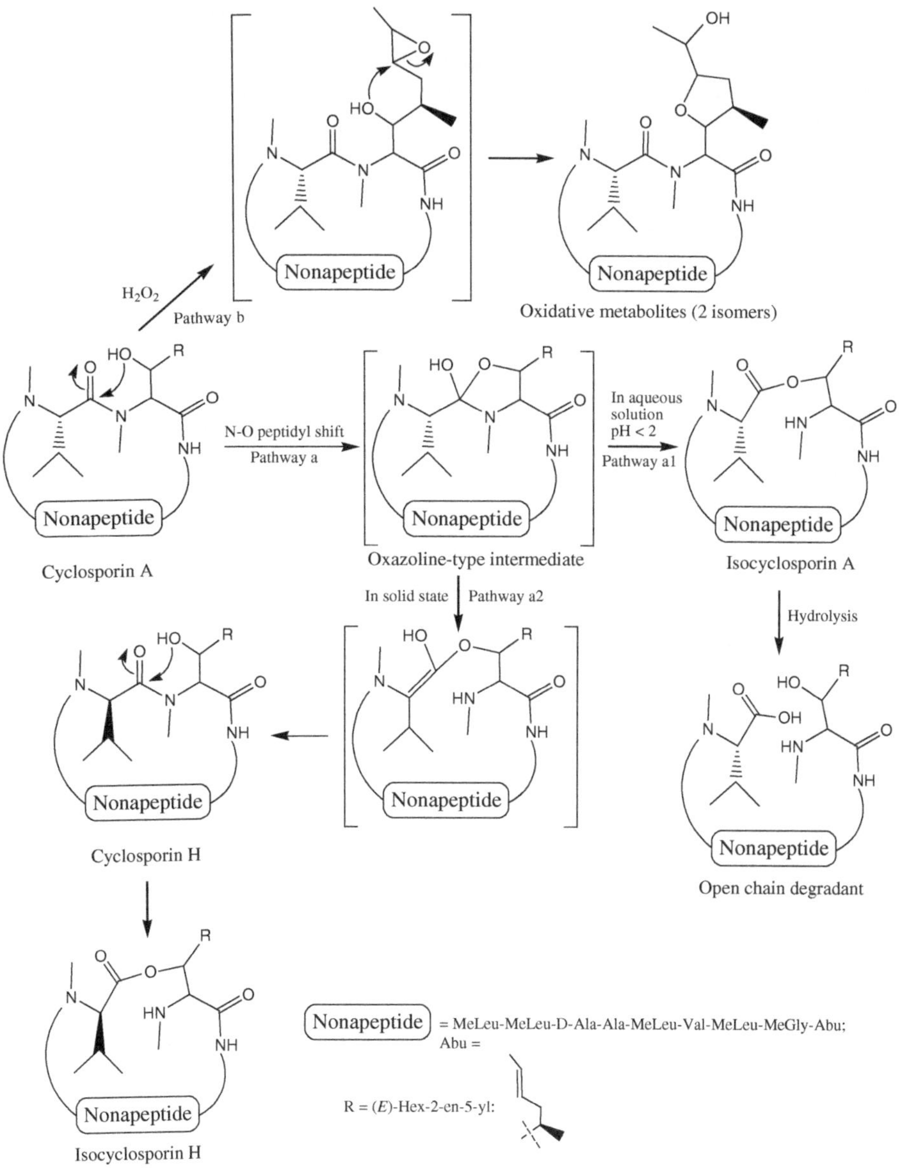

Scheme 119 Degradation pathways of cyclosporine A.

to be too harsh to represent the real degradation chemistry of the drug product. In an oxidative degradation study, Liu et al. were able to generate two oxidative metabolites by oxidizing an ether solution of cyclosporin A with 30% hydrogen peroxide through heterogeneous vortexing at 37°C [4]. It is apparent that the cyclized side chain is formed via an epoxide intermediate (Pathway b, Scheme 119). There has been no report to indicate that these two metabolites are also degradation products in finished dosage forms of cyclosporine.

References

1 Nunes Bonifacio, D.F., Giocanti, M., Reynier, J.P. et al. (2009). Development and validation of HPLC method for the determination of Cyclosporin A and its impurities in Neoral® capsules and its generic versions. *J. Pharm. Biomed. Anal.* 49: 540–546.

2 Friis, G.J. and Bundgaad, H. (1992). Kinetics of degradation of cyclosporin A in acidic aqueous solution and its implication in its oral absorption. *Int. J. Pharm.* 82: 79–83.

3 Magni, F., Arcelloni, C., Paroni, R. et al. (1994). Open-chain peptides obtained by acidic hydrolytic cleavage of cyclosporin A. *Biol. Mass Spec.* 23: 514–518.

4 Liu, W.T., Marat, K., Ren, Y. et al. (1998). Structural characterization of two novel oxidative derivatives of cyclosporine generated by a chemical method. *Clin. Biochem.* 31: 173–180.

120–123: Ciprofloxacin, Norfloxacin, Ofloxacin, Levofloxacin

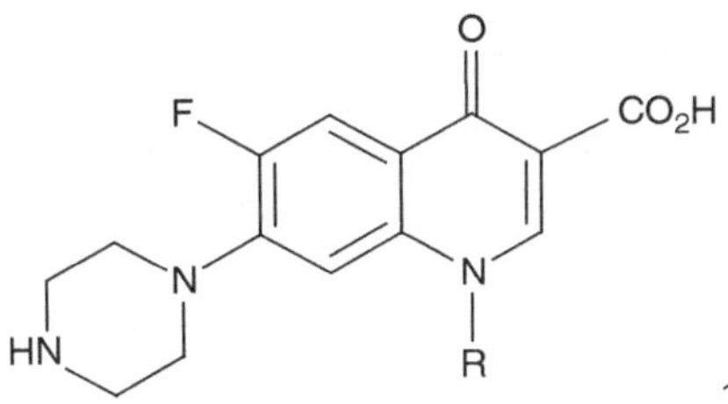

Ciprofloxacin, R = Cyclopropyl.
(Norfloxacin, R = Ethyl)

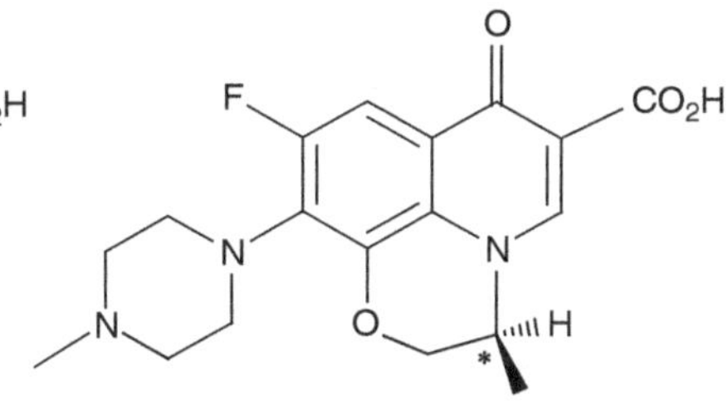

Levofloxacin
(Norfloxacin, * = racemic)

120: Ciprofloxacin

Chemical name: 1-Cyclopropyl-6-fluoro-4-oxo-7-piperazin-1-ylquinoline-3-carbo xylic acid
Brand name: Cipro
Formula: $C_{17}H_{18}FN_3O_3$
Monoisotopic protonated ion (*m/z*): 332.1405
Molecular weight: 331.34
CAS number: 85721-33-1

121: Levofloxacin

Chemical name: (2*S*)-7-Fluoro-2-methyl-6-(4-methylpiperazin-1-yl)-10-oxo-4-oxa-1-azatricyclo[7.3.1.$0^{5,13}$]trideca-5(13),6,8,11-tetraene-11-carboxylic acid
Brand name: Levaquin, Quinsair
Formula: $C_{18}H_{20}FN_3O_4$
Monoisotopic protonated ion (*m/z*): 362.1511
Molecular weight: 361.37
CAS number: 100986-85-4

122: Ofloxacin

Chemical name: 7-Fluoro-2-methyl-6-(4-methylpiperazin-1-yl)-10-oxo-4-oxa-1-azatricyclo[7.3.1.$0^{5,13}$]trideca-5(13),6,8,11-tetraene-11-carboxylic acid
Brand name: Ocuflox
Formula: $C_{18}H_{20}FN_3O_4$
Monoisotopic protonated ion (*m/z*): 362.1511
Molecular weight: 361.37
CAS number: 82419-36-1

123: Norfloxacin

Chemical name: 1-Ethyl-6-fluoro-4-oxo-7-piperazin-1-ylquinoline-3-carboxylic acid
Brand name: Noroxin
Formula: $C_{16}H_{18}FN_3O_3$

Monoisotopic protonated ion (*m/z*): 320.1405
Molecular weight: 319.33
CAS number: 70458-96-7

Ciprofloxacin and analogous fluoroquinolones are widely used antibiotics. The fluoroquinolone core is usually quite stable thermally. On the other hand, the substituted amino group on the fluoroquinolone core can undergo elimination under photochemical stress conditions; e.g., the piperazine ring of ciprofloxacin was destroyed under both artificial and sunlight exposure to yield the ethylenediamine and amine degradants, sequentially, in acidic solutions (Scheme 120–123-1) [1, 2]. With the increase of pH, the rate of photo degradation was found to be accelerated [2]. Another photo degradation study conducted at pH 9 indicated that ciprofloxacin produced three additional degradants [3]. Norfloxacin, which is structurally very similar to ciprofloxacin, also underwent this photo elimination process [4].

For ofloxacin and its *S*-enantiomer, levofloxacin, they have a methyl substituent on the piperazine ring, and the resulting tertiary amine is susceptible to nucleophilic oxidation, yielding the *N*-oxide degradant (Scheme 120–123-2) [5]. On the

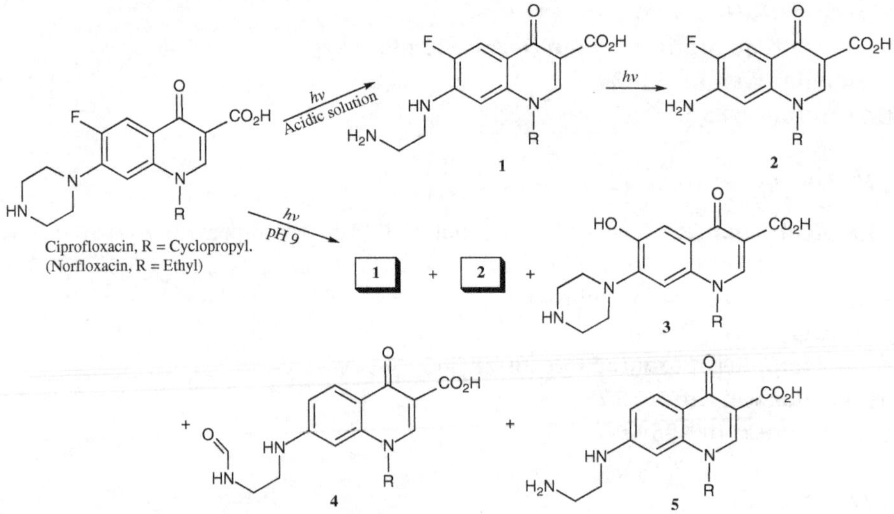

Scheme 120–123-1 Photo degradation pathways of ciprofloxacin. The structures of **3–5** were proposed based on LC-MS data.

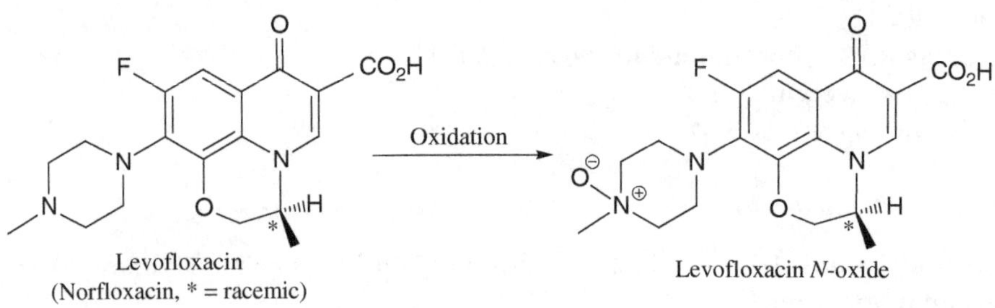

Scheme 120–123-2 Oxidation of levofloxacin.

contrary, the second amine functional group of ciprofloxacin and ofloxacin would be reluctant to undergo this kind of nucleophilic oxidation.

References

1 Torniainen, K., Mattinen, J., Askolin, C.-P., and Tammilehto, S. (1997). Structure elucidation of a photodegradation product of ciprofloxacin. *J. Pharm. Biomed. Anal.* 15: 887–894.
2 Torniainen, K., Askolin, C.-P., and Mattinen, J. (1997). Isolation and structure elucidation of an intermediate in the photodegradation of ciprofloxacin. *J. Pharm. Biomed. Anal.* 16: 439–445.
3 Vasconcelos, T.G., Henriques, D.M., König, A. et al. (2009). Photo-degradation of the antimicrobial ciprofloxacin at high pH: identification and biodegradability assessment of the primary by-products. *Chemosphere* 76: 487–493.
4 Córdoba-Borrego, M. and M. Córdoba-Díaz, D. Córdoba-Díaz. (1999). Validation of a high-performance liquid chromatographic method for the determination of norfloxacin and its application to stability studies (photo-stability study of norfloxacin). *J. Pharm. Biomed. Anal.* 18: 919–926.
5 Devi, M.L. and Chandrasekhar, K.B. (2009). A validated stability-indicating RP-HPLC method for levofloxacin in the presence of degradation products, its process related impurities and identification of oxidative degradant. *J. Pharm. Biomed. Anal.* 50: 710–717.

124–125: Citalopram, Escitalopram

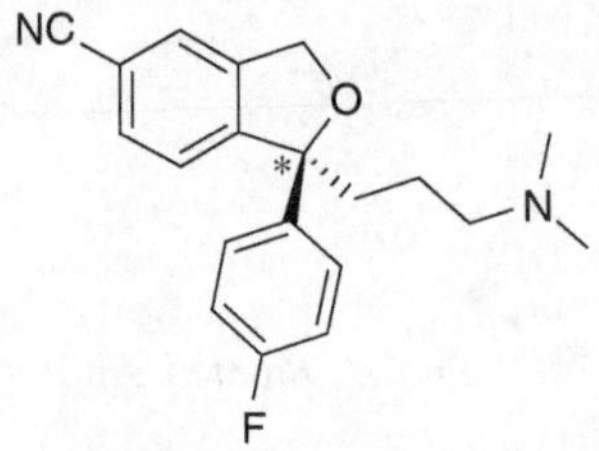

Escitalopram
(Citalopram, * = racemic)

124: Citalopram

Chemical name: 1-[3-(Dimethylamino)propyl]-1-(4-fluorophenyl)-3*H*-2-benzofuran-5-carbonitrile
Brand name: Celexa
Formula: $C_{20}H_{21}FN_2O$
Monoisotopic protonated ion (*m/z*): 325.1711
Molecular weight: 324.39
CAS number: 59729-33-8
Common salt form: HBr, Oxalate

125: Escitalopram

Chemical name: (1*S*)-1-[3-(Dimethylamino)propyl]-1-(4-fluorophenyl)-3*H*-2-benzofuran-5 -carbonitrile
Brand name: Cipralex, Lexapro
Formula: $C_{20}H_{21}FN_2O$
Monoisotopic protonated ion (*m/z*): 325.1711
Molecular weight: 324.39
CAS number: 128196-01-0
Common salt form: Oxalate

Citalopram and its *S*-enantiomer escitalopram are selective serotonin reuptake inhibitors (SSRI), which are used clinically for the treatment of depression and anxiety. The drug molecule contains three degradable functional groups: the cyano, benzylic position, and tertiary amine groups. Hydrolysis of the cyano group should first yield the carboxamide degradant (**1**) [1, 2], which upon further hydrolysis should give the carboxylic acid degradant (**2**), especially under forced degradation conditions (Scheme 124–125-1) [2]. The benzylic position is susceptible to oxidation, particularly free radical-mediated oxidation, in which process the peroxide would be a plausible intermediate. Decomposition of the latter intermediate should give rise to the hydroxyl and keto degradants (**3** and **4**), respectively. Under a nucleophilic oxidative condition, the tertiary amine group can be oxidized to produce the *N*-oxide real degradant (**5**) [2].

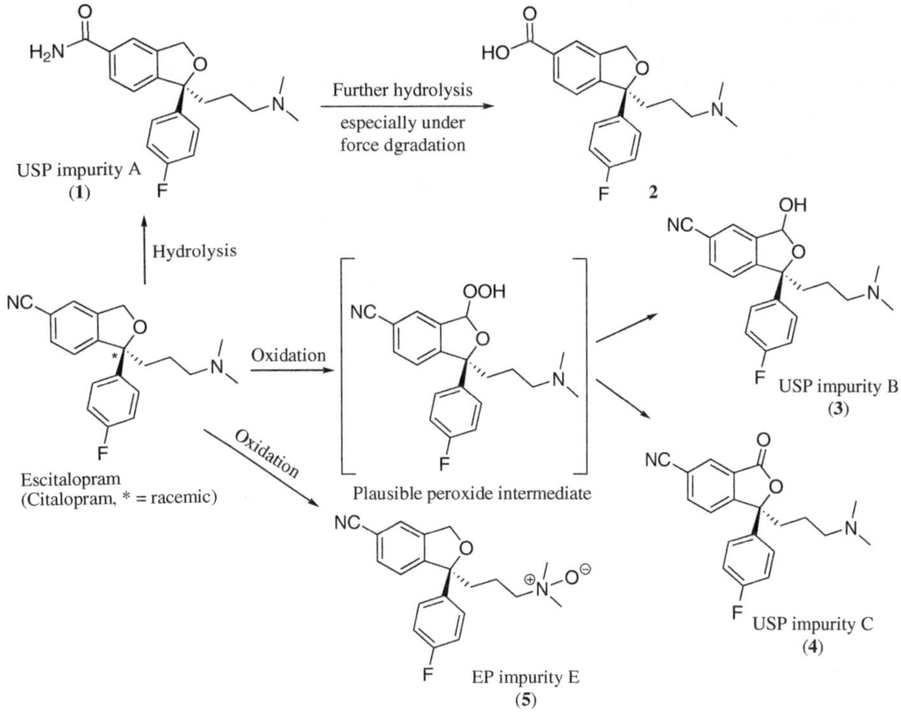

Scheme 124–125-1 Degradation pathways of escitalopram.

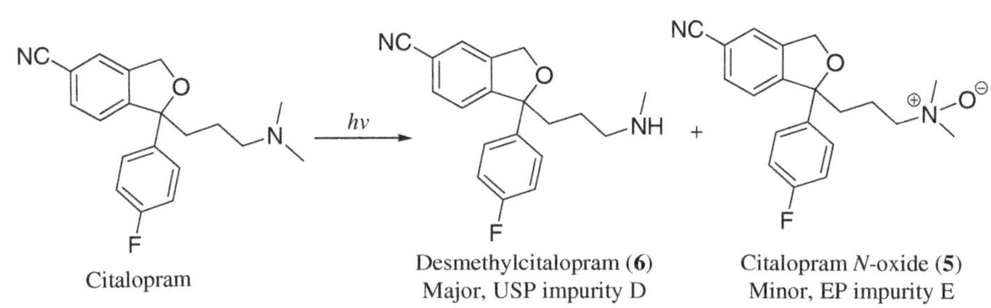

Scheme 124–125-2 Photochemical degradation of citalopram.

In a photo degradation study of citalopram HBr in water with simulated sunlight over a period of 30 days, *N*-desmethylcitalopram (**6**) was observed as the major degradant, while citalopram *N*-oxide (**5**) was formed as the minor degradant (Scheme 124–125-2) [3].

References

1 Sun, C., Xu, H., Pan, Y. et al. (2007). Characterization of the trace-level impurities in the bulk drug citalopram by high-performance liquid chromatography/tandem multistage mass spectrometry. *Rapid Commun. Mass Spectrom.* 21: 2889–2894.
2 Rao, R.N., Raju, A.N., and Narsimha, R. (2008). Isolation and characterization of degradation products of citalopram and process-related impurities using RP-HPLC. *J. Sep. Sci.* 31: 1729–1738.
3 Kwon, J.-W. and Armbrust, K.L. (2005). Degradation of citalopram by simulated sunlight. *Environ. Toxicol. Chem.* 24: 1618–1623.

126–127: Clindamycin, Clindamycin Phosphate

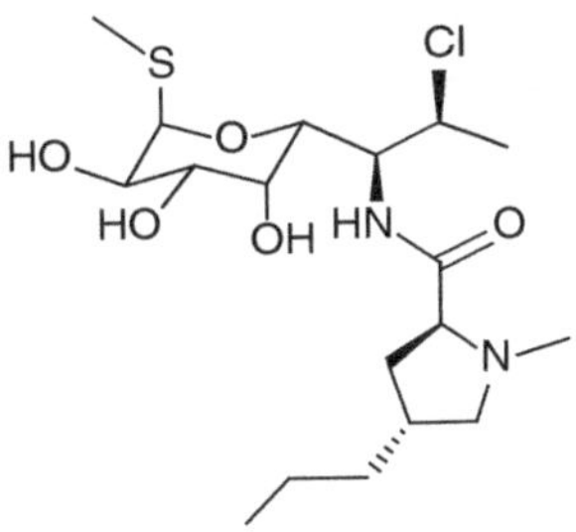

Clindamycin

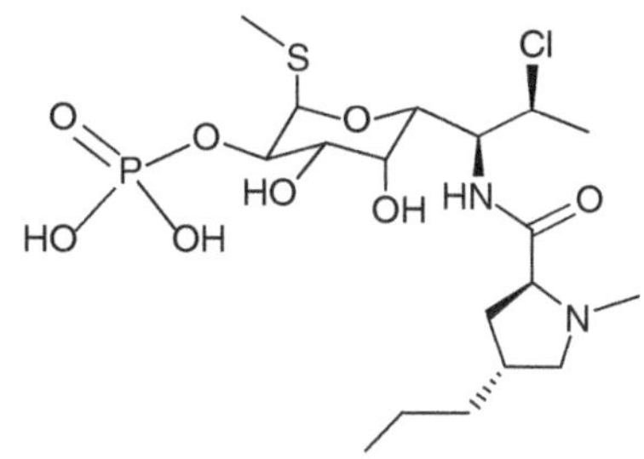

Clindamycin-2-phosphate

126: Clindamycin

Chemical name: (2*S*,4*R*)-*N*-[(1*S*,2*S*)-2-Chloro-1-[(2*R*,3*R*,4*S*,5*R*,6*R*)-3,4,5-trihydroxy-6-methylsulfanyloxan-2-yl]propyl]-1-methyl-4-propylpyrrolidine-2-carboxamide
Brand name: Acanya, Benzaclin, Cleocin
Formula: $C_{18}H_{33}ClN_2O_5S$
Common salt form: HCl
Monoisotopic protonated ion (*m/z*): 425.1871
Molecular weight: 424.98
CAS number: 18323-44-9

127: Clindamycin Phosphate (Clindamycin-2-Phosphate; Injectable Version of Clindamycin)

Chemical name: [(2*R*,3*R*,4*S*,5*R*,6*R*)-6-[(1*S*,2*S*)-2-Chloro-1-[[(2*S*,4*R*)-1-methyl-4-propylpyrrolidine-2-carbonyl]amino]propyl]-4,5-dihydroxy-2-methylsulfany loxan-3-yl] dihydrogen phosphate
Brand name: Cleocin Phosphate
Formula: $C_{18}H_{34}ClN_2O_8PS$
Monoisotopic protonated ion (*m/z*): 505.1535
Molecular weight: 504.96
CAS number: 24729-96-2

Clindamycin is a lincosamide family antibiotics, which was developed based on lincomycin, a natural product and first member of the lincosamide family antibiotics.

Oesterling performed a comprehensive stability study of clindamycin solution in the pH range of 0.4–12 [1]; it showed that clindamycin is most stable at pH 3–5. In the acidic pH range of 0.4–4, the degradation of clindamycin is mainly the hydrolysis at the thioglycosidic linkage to produce 1-dethiomethyl-1-hydroxy clindamycin (**1**) and mercaptomethane (Scheme 126–127-1). The activation energy for the hydrolysis was found to be 38 kcal/mol. At pH 5–10, the displacement of the chloro substituent via hydrolysis became the main degradation pathway, producing

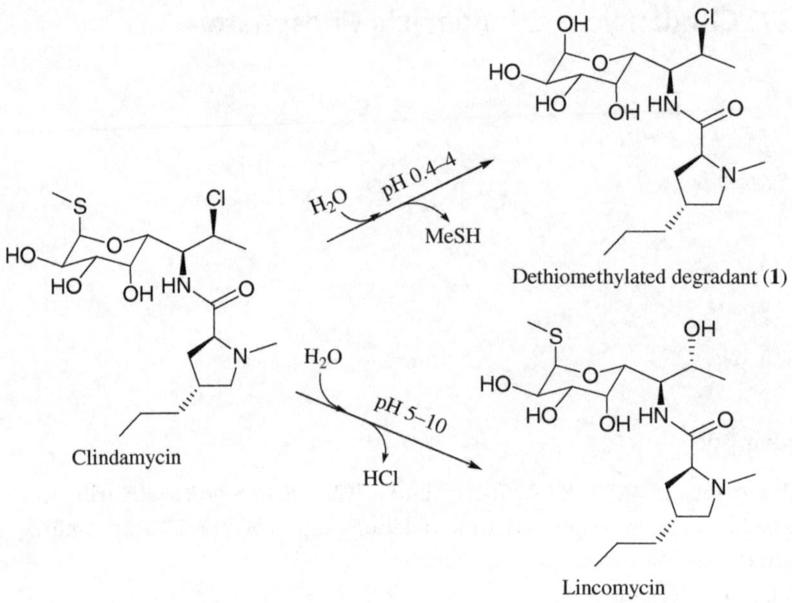

Scheme 126–127-1 Hydrolytic degradation of clindamycin.

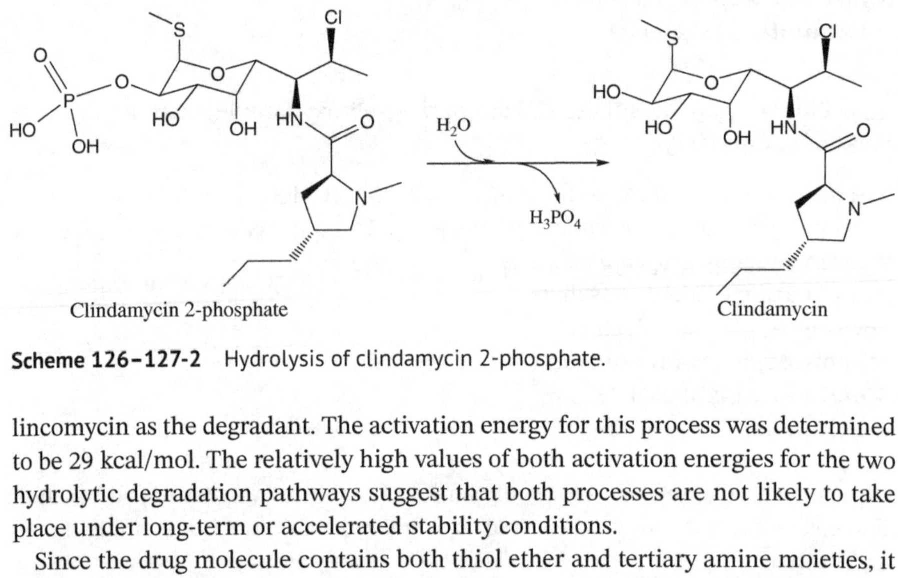

Scheme 126–127-2 Hydrolysis of clindamycin 2-phosphate.

lincomycin as the degradant. The activation energy for this process was determined to be 29 kcal/mol. The relatively high values of both activation energies for the two hydrolytic degradation pathways suggest that both processes are not likely to take place under long-term or accelerated stability conditions.

Since the drug molecule contains both thiol ether and tertiary amine moieties, it should be susceptible to nucleophilic oxidation on both functional groups; nevertheless, no report has been found to suggest the presence of the sulfoxide and *N*-oxide degradants, even though the sulfoxide is a metabolite of clindamycin [2, 3].

Oesterling and Rowe conducted a kinetic study for the hydrolysis of clindamycin-2-phosphate, the injectable version of clindamycin, and found the activation energy was 32.9 kcal/mol [4], which is typical for the hydrolysis of an alkyl phosphate ester (Scheme 126–127-2) [5].

References

1 Oesterling, T.O. (1970). Aqueous stability of clindamycin. *J. Pharm. Sci.* 59: 63–67.
2 Sun, F.F. (1973). Metabolism of clindamycin II: urinary excretion products of clindamycin in rat and dog. *J. Pharm. Sci.* 62: 1657–1662.
3 Wynalda, M.A., Hutzler, J.M., Koets, M.D. et al. (2003). In vitro metabolism of clindamycin in human liver and intestinal microsomes. *Drug Metab. Dispos.* 31: 878–887.
4 Oesterling, T.O. and Rowe, E.L. (1970). Hydrolysis of lincomycin-2-phosphate and clindamycin-2-phosphate. *J. Pharm. Sci.* 59: 175–179.
5 Li, M. (2012). Hydrolytic degradation. In: *Organic Chemistry of Drug Degradation*, 16–47. Cambridge: Royal Society of Chemistry.

128–129: Clobetasol Propionate, Halobetasol Propionate

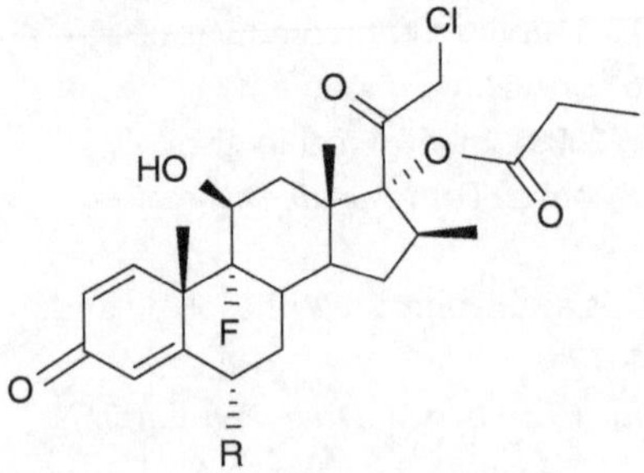

Clobetasol 17-propionate, R = H;
Halobetasol 17-propionate, R = F.

128: Clobetasol Propionate (Clobetasol 17-Propionate)

Chemical name: [(8*S*,9*R*,10*S*,11*S*,13*S*,14*S*,16*S*,17*R*)-17-(2-Chloroacetyl)-9-fluoro-11-hydroxy-10,13,16-trimethyl-3-oxo-6,7,8,11,12,14,15,16-octahydrocyclopenta[a]phenanthren-17-yl] propanoate
Brand name: Clobex, Clodan, Dermovate
Formula: $C_{25}H_{32}ClFO_5$
Monoisotopic protonated ion (*m/z*): 467.1995
Molecular weight: 466.97
CAS number: 25122-46-7

129: Halobetasol Propionate

Chemical name: [(6*S*,8*S*,9*R*,10*S*,11*S*,13*S*,14*S*,16*S*,17*R*)-17-(2-Chloroacetyl)-6,9-difluoro-11-hydroxy-10,13,16-trimethyl-3-oxo-6,7,8,11,12,14,15,16-octahydrocyclopenta[a]phenanthren-17-yl] propanoate
Brand name: Bryhali, Duobrii, Lexette, Ultravate
Formula: $C_{25}H_{31}ClF_2O_5$
Monoisotopic protonated ion (*m/z*): 485.1901
Molecular weight: 484.96
CAS number: 66852-54-8

Clobetasol propionate, or more specifically clobetasol 17-propionate, is a synthetic corticosteroid used for the topic treatment of various dermatologic disorders such as eczema and psoriasis. Despite being a drug product that has been on the market since the 1970s, there have been very few studies published on the degradation chemistry of clobetasol propionate. For the two forced degradation studies reported in the literature [1, 2], either no characterization of the degradants was performed [1] or only partial characterization was done [2]. The latter study, performed by Damle and Polawar, reported two base-catalyzed degradants, which appeared to be

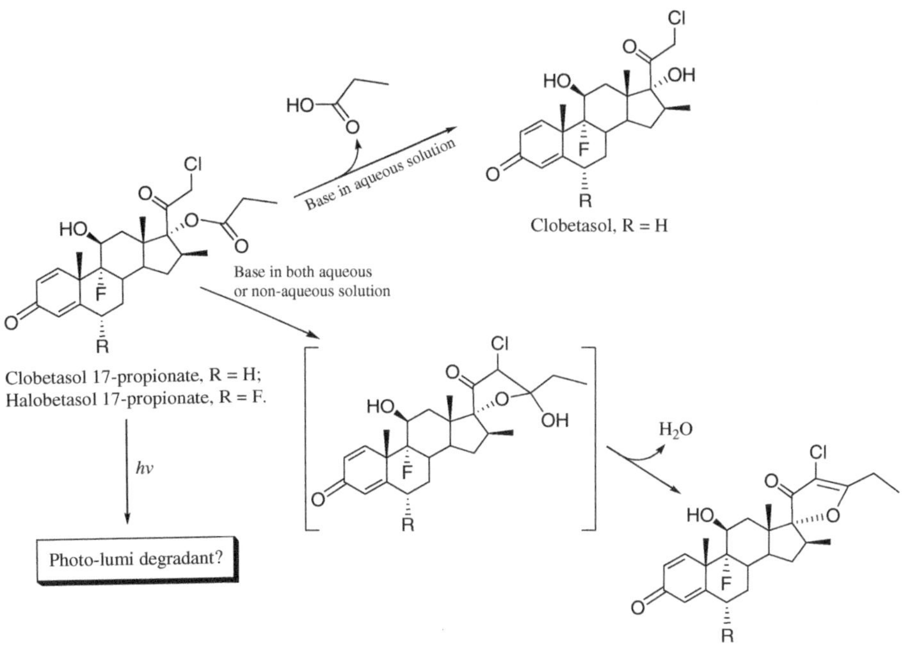

Scheme 128–129-1 Degradation pathways of clobetasol 17-propionate and halobetasol 17-propionate.

clobetasol (via cleavage of the ester linkage) and an unknown dehydrated product, respectively, based on LC-MS evidence. On the other hand, Cravotto et al. reported complete structure characterization of a dehydrated impurity (**1b**, Scheme 128–129-1) of halobetasol propionate (a structural analog of clobetasol propionate) that was also formed under base catalysis [3]. This forced degradation product (**1b**) resulted from condensation (via dehydration) between the two 17-substituents on the D-ring of the halobetasol steroid core (Scheme 128–129-1). Since clobetasol propionate and halobetasol propionate have identical substituents on the 17-position of the steroid cores, it would be reasonable to assume that the dehydrated degradant (**1a**), as observed by Damle and Polawar [2], was formed via the same Claisen-type condensation (Scheme 128–129-1). This type of dehydrative condensation was also observed with mometasone furoate, which has similar 17-position substituents [4].

It is noted that the base-catalyzed degradation of clobetasol propionate yielded two degradants [2], while that of halobetasol propionate produced only one degradant [3]. This difference could probably be attributed to the fact that different solvent systems were used in the base-catalyzed degradation studies: the study of clobetasol propionate used a mixture of methanol/0.1 N aqueous NaOH (10/1, v/v), while that of halobetasol propionate employed ethanol solution of K_2CO_3 in the absence of water.

In addition to the two 17-position substituents, the A-ring of the clobetasol propionate could also be susceptible to photochemical rearrangement, which is not

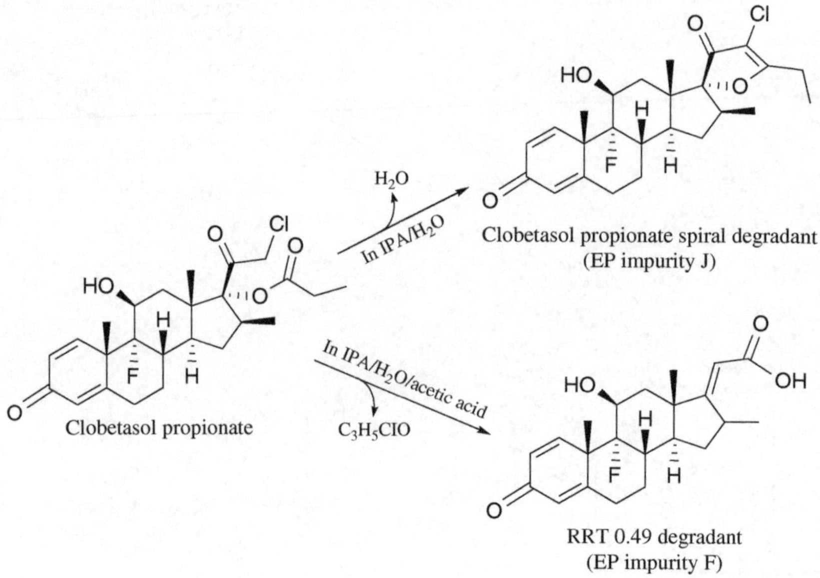

Scheme 128–129-2 Degradation pathways of clobetasol propionate leading to the formation of EP Impurity J and RRT 0.49 impurity (EP Impurity F) as the major degradant under each respective condition [7]. IPA, isopropanol. *Source:* Adapted from Tian et al. [7].

uncommon for corticosteroids containing the same A-ring such as betamethasone and its derivatives [5, 6].

In a stability study of clobetasol 17-propionate topical solution by Tian et al. [7], a major degradant was found, which was determined to be EP Impurity F through the use of LC-MS and NMR. The formation of this degradant was proposed to proceed through a Favorskii-like rearrangement mechanism under weakly acidic condition (Scheme 128–129-2). Even though this impurity is listed in the EP, there has been no clear indication that this impurity is a degradant of clobetasol 17-propionate in the literature.

References

1 Fauzee, A.F.B. and Walker, R.B. (2013). Forced degradation studies of clobetasol 17-propionate in methanol, propylene glycol, as bulk drug and cream formulations by RP-HPLC. *J. Sep. Sci.* 36: 849–856.

2 Damle, M.C. and Polawar, A.R. (2014). Stability indicating Hptlc method for the estimation of clobetasol propionate in presence of alkali induced degradation product. *Int. J. PharmTech Res.* 6: 1914–1925.

3 Cravotto, G., Giovenzana, G.B., Masciocchi, N. et al. (2007). A degradation product of halobetasol propionate – characterization and structure. *Steroids* 72: 787–791.

4 Sahasranaman, S., Issar, M., Toth, G. et al. (2004). Characterization of degradation products of mometasone furoate. *Pharmazie* 59: 367–373.

5 Hidaka, T., Huruumi, S., Tamaki, S. et al. (1980). *Yakugaku Zasshi* 100: 72–80.

6 Lin, M., Li, M., Buevich, A.V. et al. (2009). Rapid structure elucidation of drug degradation products using mechanism-based stress studies in conjunction with LC–MSn and NMR spectroscopy: identification of a photodegradation product of betamethasone dipropionate. *J. Pharm. Biomed. Anal.* 50: 275–280.

7 Tian, Y., Zhang, H., Zhu, Z. et al. A variation of Favorskii rearrangement mechanism under weakly acidic conditions: the case of clobetasol propionate degradation in solution. *Chemrxiv* https://doi.org/10.26434/chemrxiv.11809077.v1.

130: Clomipramine

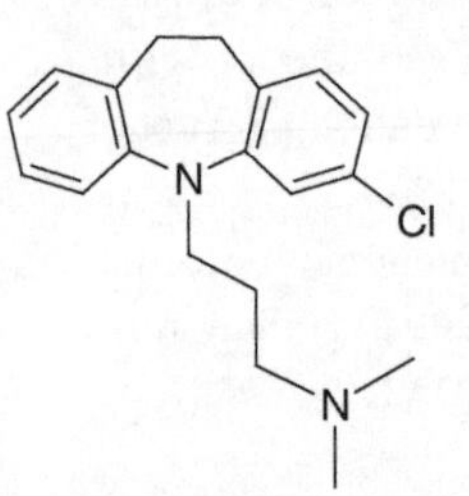

Chemical name: 3-(2-Chloro-5,6-dihydrobenzo[b][1]benzazepin-11-yl)-*N,N*-dime thylpropan-1-amine
Brand name: Anafranil
Formula: $C_{19}H_{23}ClN_2$
Monoisotopic protonated ion (*m/z*): 315.1623
Molecular weight: 314.85
CAS number: 303-49-1
Common salt form: HCl salt

Clomipramine is a tricyclic antidepressant. It contains two likely degradable moieties: the aryl chloride and the tertiary amine. In a photo degradation study carried

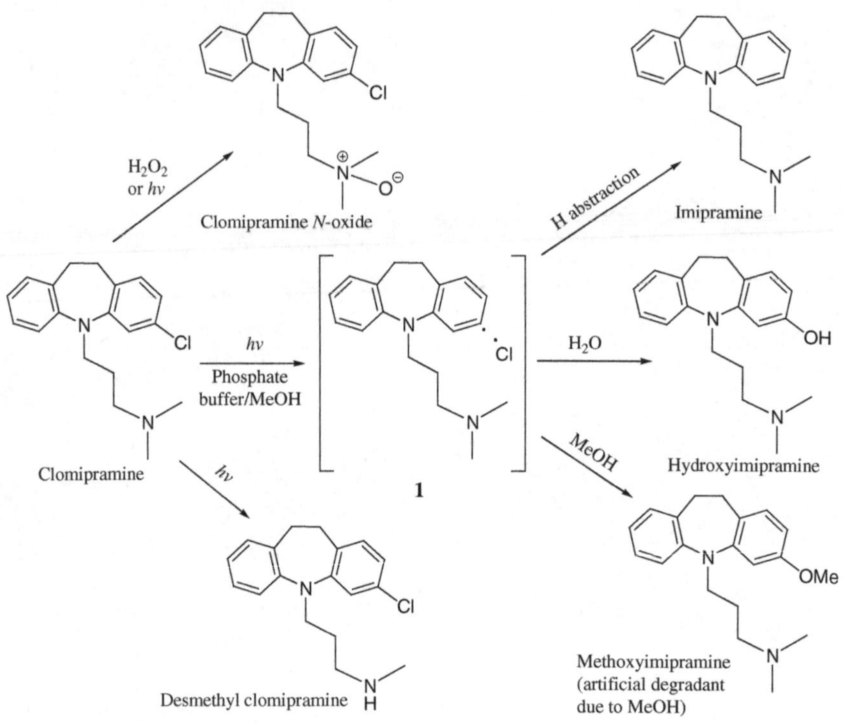

Scheme 130 Degradation pathways of clomipramine.

out in a mixture of phosphate buffer and methanol, several degradants were isolated and identified [1]. It was proposed that the aryl chloride underwent homolytic cleavage to give the carbon-based radical intermediate (**1**), from which deschloro, hydroxyl, and methoxyl degradants were formed (Scheme 130). Clomipramine *N*-oxide was also found among the photo degradants, although this degradant can also be formed under non-photo degradation conditions. Due to the difference in irradiation sources, solvents, and length of the irradiation, the formation and distribution of photo degradants may be different. In a photo degradation study of clomipramine performed in pure water as the solvent, only the hydroxyl degradant was observed [2], while in another photo degradation study, desmethyl clomipramine was detected [3].

References

1 Canudas, N. and Contreras, C. (2002). Isolation and identification of the photodegradation products of the photosensitizing antidepressant drug clomipramine. Phototoxicity studies on erythrocytes. *Pharmazie* 57: 405–408.
2 Nassar, R., Trivella, A., Mokh, S. et al. (2017). Photodegradation of sulfamethazine, sulfamethoxypiridazine, amitriptyline, and clomipramine drugs in aqueous media. *J. Photochem. Photobiol. A Chem.* 336: 176–182.
3 Székely, P., Gyéresi, A., Hancu, G. et al. (2011). Study regarding the phototoxicity of some tricyclic antidepressants. *Acta Med. Marisiensis* 57: 745–747.

131: Clopidogrel

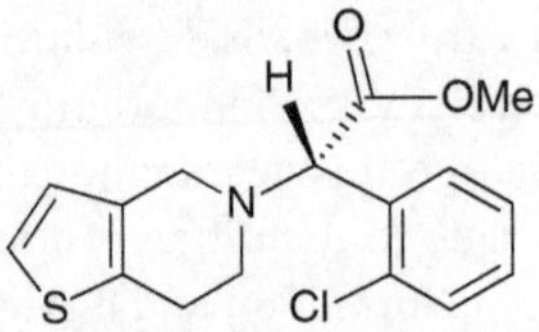

Chemical name: methyl (2*S*)-2-(2-Chlorophenyl)-2-(6,7-dihydro-4*H*-thieno[3,2-c]pyridin-5-yl)acetate
Brand name: Plavix
Formula: $C_{16}H_{16}ClNO_2S$
Monoisotopic protonated ion (*m/z*): 322.0663
Molecular weight: 321.82
CAS number: 113665-84-2
Common salt form: Bisulfate

Clopidogrel is an antiplatelet drug, which was marketed by Sanofi–Synthelabo and Bristol-Myers Squibb worldwide under the brand names PLAVIX® and ISCOVER® and was the second best-selling drug prior to its patent expiration. Structurally, clopidogrel is a chiral drug containing a single chiral center and a methyl ester functionality. As such, clopidogrel would be expected to undergo hydrolysis and may be susceptible to racemization, leading to the formation of clopidogrel acid (**1**) and the R-isomeric degradant (**2**, Scheme 131); indeed, in an accelerated stability study of the original brand name drug product versus several generic products on the market, **1** and **2** were observed as the main degradants [1]. A degradation kinetic study of clopidogrel in methanolic phosphate buffers within the pH range of 3.0–9.0 indicated that the chiral conversion was slowest at pH 3.0, while the hydrolytic degradation was slowest at pH 5.6 [2]. In an investigation performed by Mohan et al., an oxidative degradant present in clopidogrel drug substance and product was characterized (**3**) [3, 4]. This degradant (**3**) could undergo further oxidation to form degradant **4a**; the hydrolyzed version of **4a**, namely **4b**, appeared to be present in an accelerated stability study of amorphous clopidogrel [5]. In a forced degradation study of clopidogrel with H_2O_2, the presence of clopidogrel *N*-oxide was reported, but its structure was not fully characterized due to its instability [6].

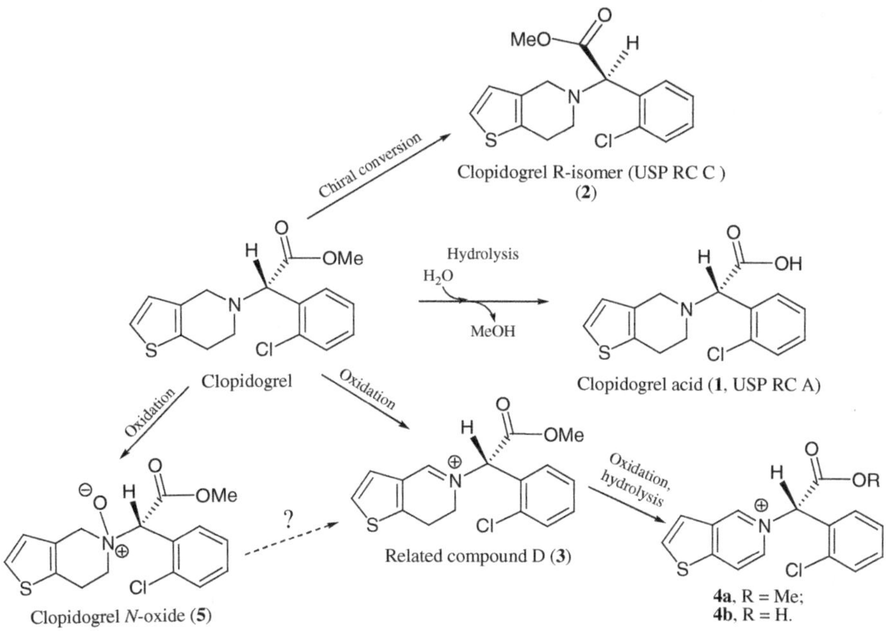

Scheme 131 Degradation pathways of clopidogrel.

References

1 Gomez, Y., Adams, E., and Hoogmartens, J. (2004). Analysis of purity in 19 drug product tablets containing clopidogrel: 18 copies versus the original brand. *J. Pharm. Biomed. Anal.* 34: 341–348.

2 Reist, M., Roy-de Vos, M., Montseny, J.-P. et al. (2000). Very slow chiral inversion of clopidogrel in rats: a pharmacokinetic and mechanistic investigation. *Drug Metab. Dispos.* 28: 1405–1410.

3 Mohan, A., Hariharan, M., Vikraman, E. et al. (2008). Identification and characterization of a principal oxidation impurity in clopidogrel drug substance and drug product. *J. Pharm. Biomed. Anal.* 47: 183–189.

4 Li, M. (2012). *Organic Chemistry of Drug Degradation*, 64–65. Cambridge: Royal Society of Chemistry.

5 Raijada, D.K., Prasad, B., Paudel, A. et al. (2010). Characterization of degradation products of amorphous and polymorphic forms of clopidogrel bisulphate under solid state stress conditions. *J. Pharm. Biomed. Anal.* 52: 332–344.

6 Mashelkar, U.C. and Renapurkar, S.D. (2010). A LCMS compatible stability-indicating HPLC assay method for clopidogrel bisulphate. *Int. J. ChemTech Res.* 2: 822–829.

132: Clotrimazole

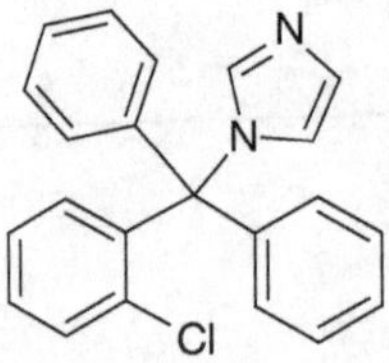

Chemical name: 1-[(2-Chlorophenyl)diphenylmethyl]imidazole
Brand name: Lotrimin
Formula: $C_{22}H_{17}ClN_2$
Monoisotopic protonated ion (*m/z*): 345.1153
Molecular weight: 344.84
CAS number: 23593-75-1

Clotrimazole is an antifungal drug. Its main degradation pathway is of hydrolytic nature to give (2-chlorophenyl)diphenylmethanol (**1**) and imidazole (**2**, Scheme 132); this degradation is more significant at low pH [1, 2], at which the solubility of the drug is greatly enhanced as the pK of the drug is 4.7. On the other hand, clotrimazole appeared to be stable under basic conditions; in a clotrimazole solution in 0.5 N NaOH, no degradation was observed after the solution was heated at 80°C for 8 hours [3]. In an ionizing irradiation study, an intramolecular crosslinked impurity (**3**) was found as the only degradant [4].

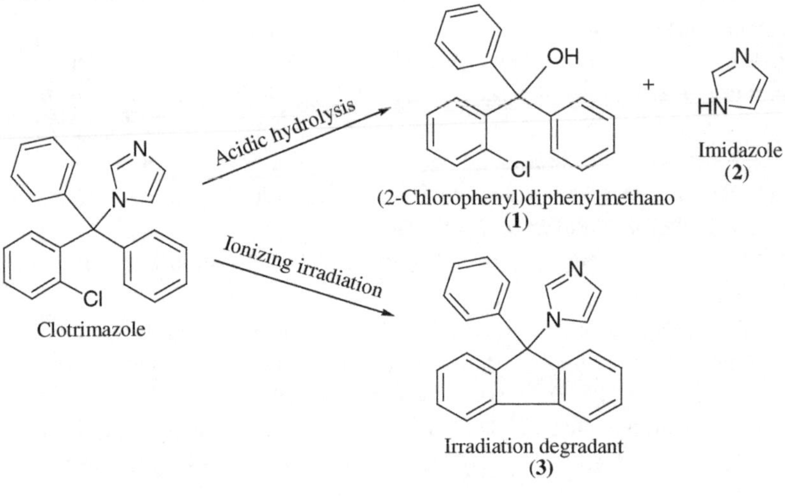

Scheme 132 Degradation pathways of clotrimazole.

References

1 Solich, P., Hajkova, R., Pospisilov, M., and Sicha, J. (2002). Determination of methylparaben, propylparaben, clotrimazole and its degradation products in topical cream by RP-HPLC. *Chromatographia* 56 (Suppl): S181–S184.

2 Hajkova, R., Sklenarova, H., Matysova, L. et al. (2007). Development and validation of HPLC method for determination of clotrimazole and its two degradation products in spray formulation. *Talanta* 73: 483–489.

3 Borhade, V., Pathak, S., Sharma, S., and Patravale, V. (2012). Clotrimazole nanoemulsion for malaria chemotherapy. Part II: stability assessment, in vivo pharmacodynamic evaluations and toxicological studies. *Int. J. Pharm.* 431: 149–160.

4 Marciniec, B., Dettlaff, K., and Naskrent, M. (2009). Influence of ionising irradiation on clotrimazole in the solid state. *J. Pharm. Biomed. Anal.* 50: 675–678.

133: Cocaine

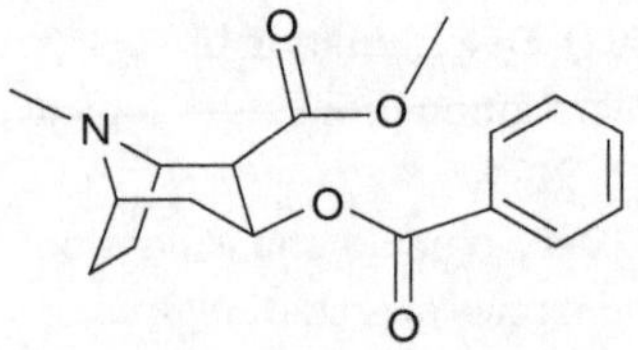

Chemical name: Methyl (1*R*,2*R*,3*S*,5*S*)-3-benzoyloxy-8-methyl-8-azabicyclo[3.2.1]octane-2-carboxylate
Brand name: Goprelto, Numbrino
Formula: $C_{17}H_{21}NO_4$
Monoisotopic protonated ion (*m/z*): 304.1543
Molecular weight: 303.35
CAS number: 50-36-2
Common salt form: HCl

Despite being the second most abused controlled substance on a global scale, only next to cannabis, cocaine does have a few legitimate medical applications. It contains two ester groups and one tertiary amine group, which should be susceptible to hydrolytic and oxidative degradation, respectively. In a study assessing hydrolytic stability of cocaine in aqueous solutions of pH 2.2, 5.0, and 6.8, the activation energies for the hydrolysis were found to be 97.9, 107.3, and 95.9 kJ/mol (or 23.4, 25.6, and 22.9 kcal/mol), respectively, at each pH [1]. Hence, in the pH range studied, cocaine was most hydrolytically stable at pH 5.0. From a structural perspective, one may expect that the methyl ester functionality would be much more susceptible to

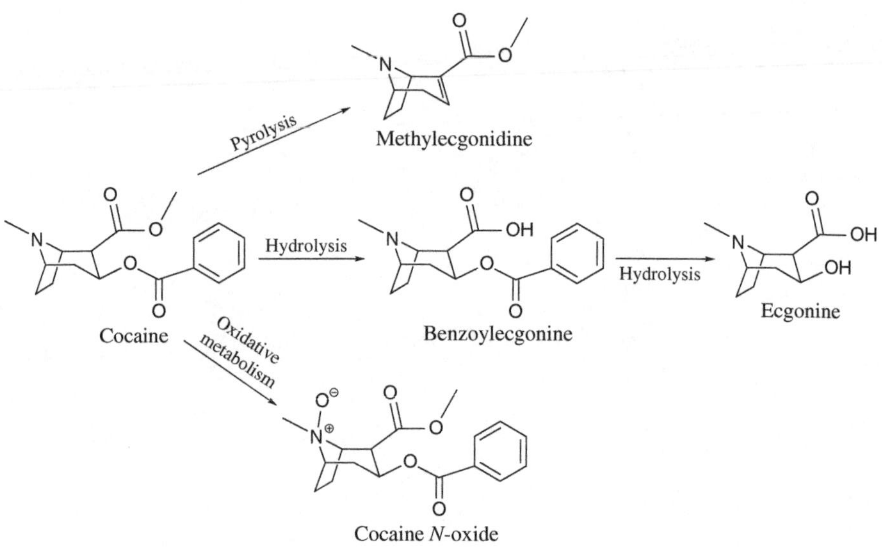

Scheme 133 Degradation pathways of cocaine.

hydrolysis than the benzoyl ester functionality [2]. Indeed, a study of cocaine stability in blood samples demonstrated that only the methyl ester was hydrolyzed to produce benzoylecgonine under non-enzymatic conditions [3]. Interestingly, under pyrolysis conditions where the solid cocaine was subjected to temperatures in the order of several hundred °C, elimination became a main route of degradation, in which methylecgonidine was formed (Scheme 133) [4]. Obviously, such a high temperature employed in pyrolysis is usually not encountered in drug degradation studies. Technically, the tertiary amine moiety of cocaine could be oxidized to give cocaine *N*-oxide, a metabolite of the drug [5]; nevertheless, it has not been reported as a degradant in a pharmaceutical preparation.

References

1 Murray, J.B. and Al-Shora, H.I. (1978). Stability of cocaine in aqueous solution. *J. Clin. Pharm. Therap.* 3: 1–6.

2 Li, M. (2012). Hydrolytic degradation. In: *Organic Chemistry of Drug Degradation*, 16–47. Cambridge: Royal Society of Chemistry.

3 Isenschmid, D.S., Levine, B.S., and Caplan, Y.H. (1989). A Comprehensive study of the stability of cocaine and its metabolites. *J. Anal. Toxicol.* 13: 250–256.

4 Cook, C.E. and Jeffcoat, A.R. National institute on drug abuse. In: *Research Monograph*, vol. 99, 97–120.

5 Wang, P.P. and Bartlett, M.G. (1999). Identification and quantitation of cocaine *N*-oxide: a thermally labile metabolite of cocaine. *J. Anal. Toxicol.* 23: 62–66.

134: Colchicine

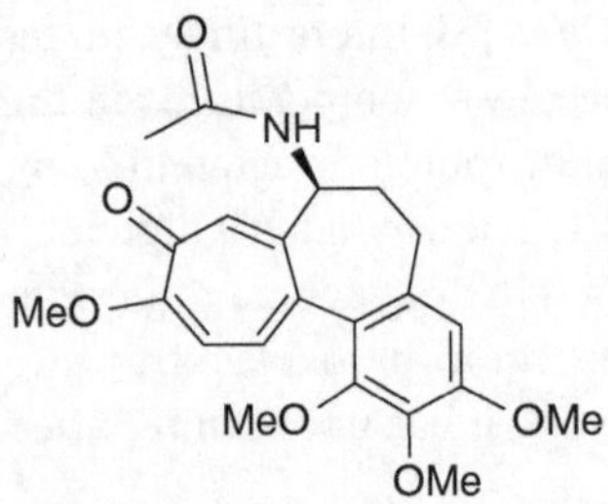

Chemical name: *N*-[(7*S*)-1,2,3,10-Tetramethoxy-9-oxo-6,7-dihydro-5*H*-benzo[a]heptalen-7-yl]acetamide
Brand name: Colcrys, Gloperba, Mitigare
Formula: $C_{22}H_{25}NO_6$
Monoisotopic protonated ion (*m/z*): 400.1755
Molecular weight: 399.44
CAS number: 64-86-8

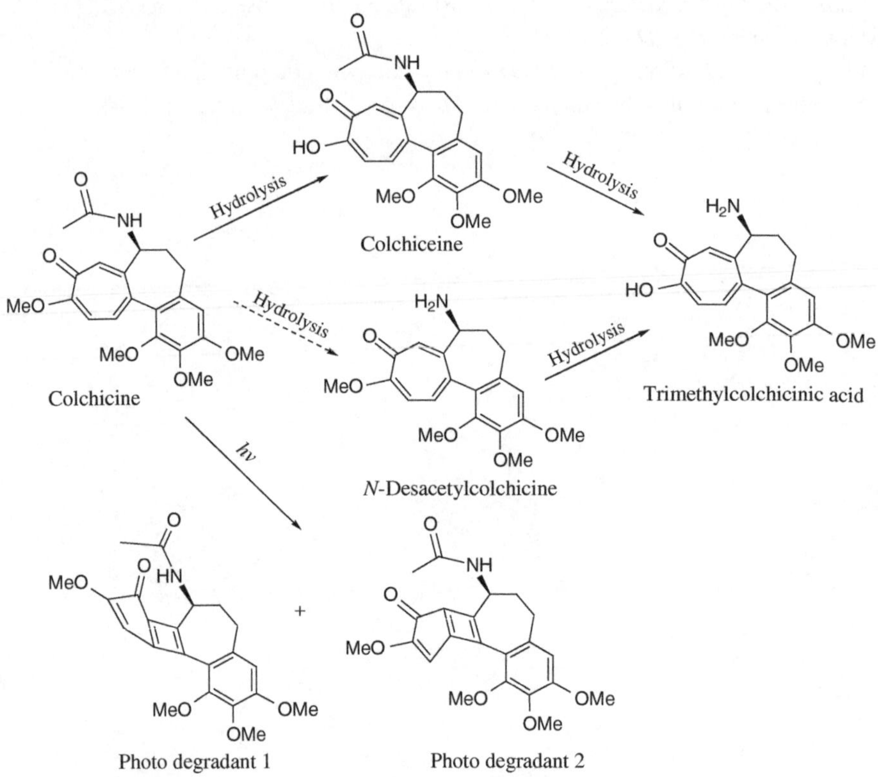

Scheme 134 Degradation pathways of colchicine.

Colchicine, a natural product originally isolated from plants of the genus *Colchicum*, is usually used for treating gout; the description of *Colchicum* extract for medicinal usage dates back as early as 1500 BC [1]. The first effort to elucidate the structure of colchicine appears to have occurred in 1924 [2]. Its degradation chemistry consists of hydrolytic and photochemical degradation pathways (Scheme 134), which were well summarized by Lacey and Brady [3]. When heated in 0.1 N HCl at 100°C for 1 hour, the tropolone methyl ether linkage can be mostly cleaved to produce colchiceine [4] and possibly other degradants, such as trimethylcolchicinic acid. Under photo irradiation, colchicine undergoes [2+2] photo cyclization to yield two isomeric degradants.

References

1 Colchicine. https://en.wikipedia.org/wiki/Colchicine (accessed 26 February 2017).

2 Dewar, M.J.S. (1945). The structure of colchicine. *Nature* 155: 141–142.

3 Lacey, E. and Brady, R.L. (1984). Separation of colchicine and related hydrolysis and photodecomposition products by high-performance liquid chromatography, using copper ion complexation. *J. Chrom. A.* 315: 233–241.

4 Borisy, G.G. and Taylor, E.W. (1967). The mechanism of action of colchicine. *J. Cell Biol.* 34: 525–533.

135–136: Cyclobenzaprine, Cyproheptadine

135: Cyclobenzaprine

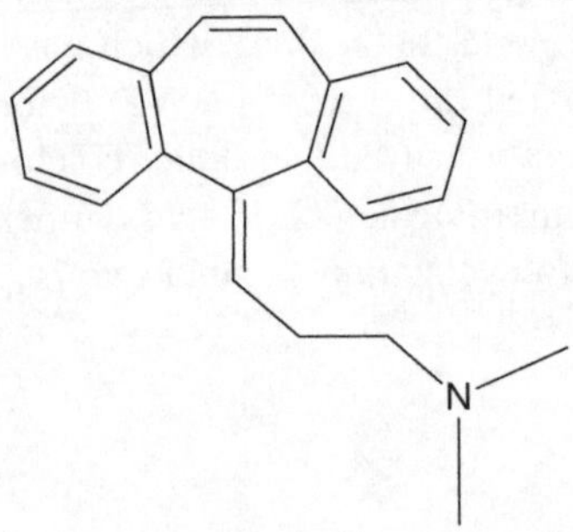

Chemical name: *N,N*-Dimethyl-3-(2-tricyclo[9.4.0.$0^{3,8}$]pentadeca-1(15),3,5,7,9,11,13-heptaenylidene)-propan-1-amine
Brand name: Amrix, Fexmid, Flexeril
Formula: $C_{20}H_{21}N$
Common salt form: HCl
Monoisotopic protonated ion (*m/z*): 276.1747
Molecular weight: 275.39
CAS number: 303-53-7

136: Cyproheptadine

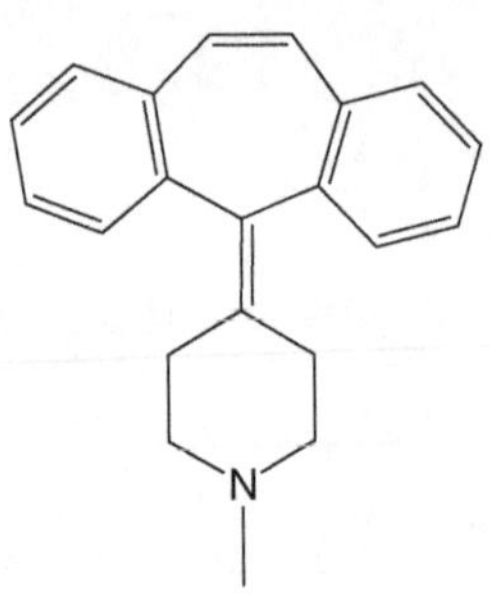

Chemical name: 1-Methyl-4-(2-tricyclo[9.4.0.$0^{3,8}$]pentadeca-1(15),3,5,7,9,11,13-heptaenylidene)piperidine
Brand name: Periactin
Formula: $C_{21}H_{21}N$
Common salt form: HCl
Monoisotopic protonated ion (*m/z*): 288.1747
Molecular weight: 287.40
CAS number: 129-03-3

Cyclobenzaprine is used to relieve skeletal muscle spasm of local origin [1]. At the core of the molecule is a highly conjugated tricyclic ring with two phenyl groups fused together by an endocyclic and exocyclic carbon–carbon double bond. Being an electron-rich molecule, the two carbon–carbon double bonds, as well as the tertiary amine moiety of cyclobenzaprine are susceptible to oxidative degradation. Liu et al. performed a comprehensive forced degradation study of cyprobenzaprine [2]; more than a dozen oxidative degradants were observed (Scheme 135–136-1). Many of these degradants may not be very stable; nevertheless, cyclobenzaprine *N*-oxide, dibenzosubrenone, and 9,10-anthraquinone seem to be more stable than others and their presence in a topic formulation of cyclobenzaprine HCl, piroxicam, and

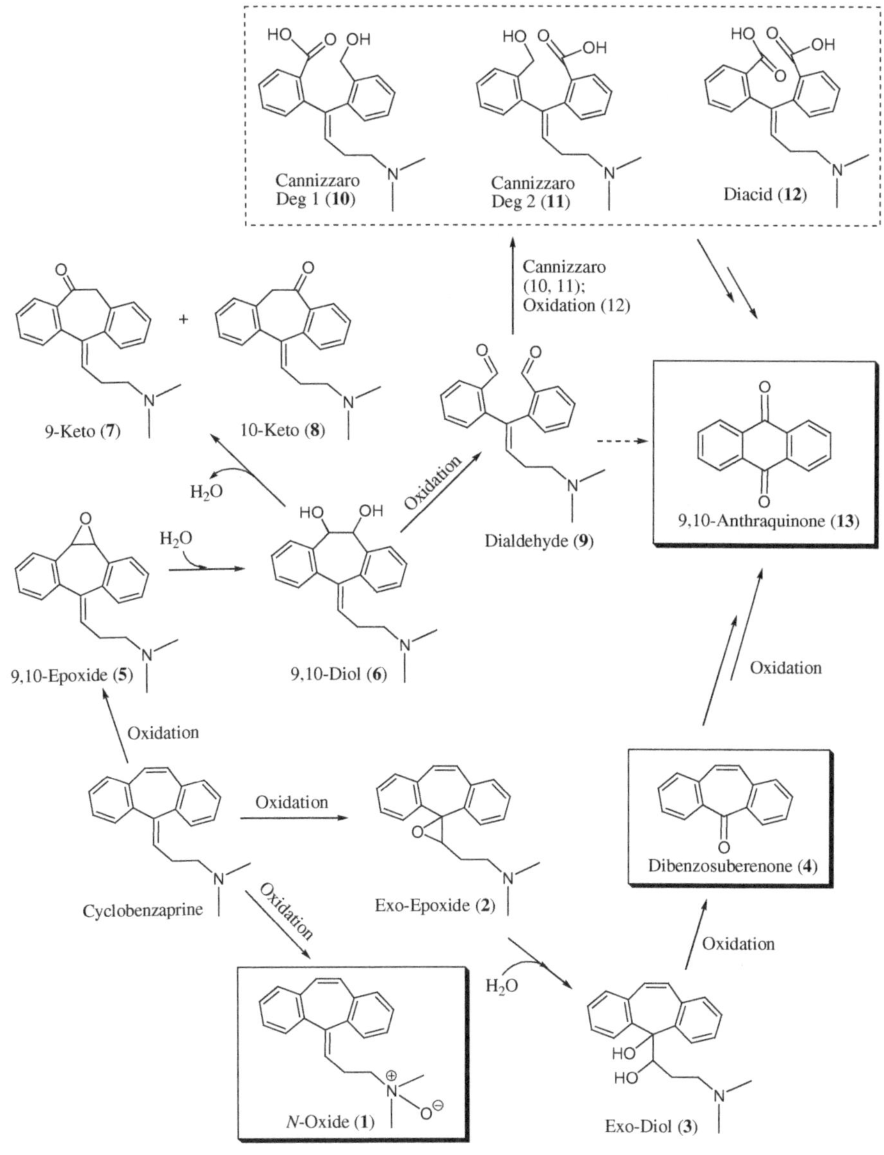

Scheme 135–136-1 Degradation pathways of cyclobenzaprine.

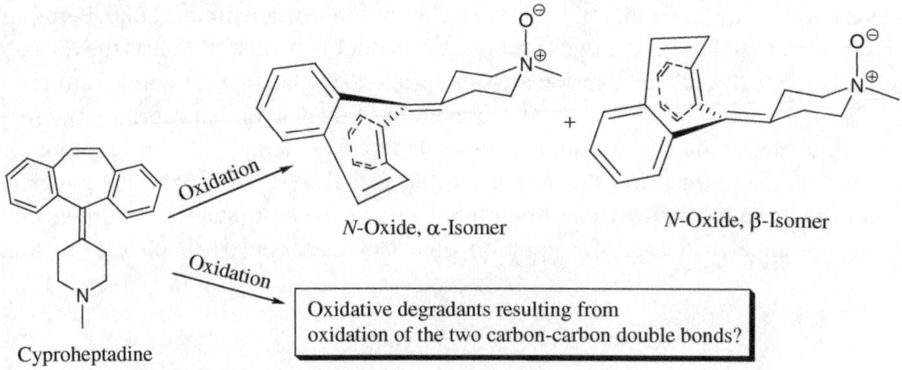

Scheme 135–136-2 Degradation pathways of cyproheptadine.

lidocaine was reported [3]. In an experimental tablet formulation, anthraquinone was formed as the primary degradant, although trace amounts of other oxidative degradants were also seen [4]. The oxidation could be suppressed by adding antioxidant, BHA, into the formulation.

Cyproheptadine, a first-generation antihistamine, has a structure very similar to that of cyclobenzaprine. Consequently, it would be reasonable to assume that it has very similar degradation pathways. Nevertheless, there have been very few reports regarding its degradation chemistry. In a study by Christy et al., it was reported that cyproheptadine can be oxidized by hydrogen peroxide, and the resulting *N*-oxide has two stereoisomers, the α and β-isomers (Scheme 135–136-2) [5].

On the other hand, amitriptyline, a tricyclic antidepressant, is structurally identical to cyprobenzaprine, except that it has a saturated 9,10-double bond. For the discussion of its degradation chemistry, refer to the monograph for amitriptyline and nortriptyline.

References

1 FDA Approval package for Flexeril. 2003. http://www.accessdata.fda.gov/drugsatfda_docs/nda/2003/017821_S045_FLEXERIL.pdf (accessed 3 March 2017).

2 Liu, Y., Zhao, D., and Zhao, Z.Z. (2014). Study of degradation mechanisms of cyclobenzaprine by LC-MS/MS. *Anal. Methods* 6: 2320–2330.

3 Cioroiu, B.I., Grigoriu, I.C., Cioroiu, M.E. et al. (2016). Determination of degradation products of cyclobenzaprine hydrochloride, lidocaine and piroxicam in a semi-topical formulation: MS–MS confirmation of unknown impurities. *J. Chromatogr. Sci.* 54: 902–911.

4 Cotton, M.L. and Down, G.R.B. Cyclobenzaprine. In: *Anal. Prof. Drug Subs.*, vol. 17 (ed. K. Florey), 41–72. Elsevier.

5 Christy, M.E., Anderson, P.S., Arison, B.H. et al. (1977). Stereoisomerism of cyproheptadine *N*-oxide. *J. Organomet. Chem.* 42: 378–379.

137: Cyclophosphamide [WHO List of Essential Medicines]

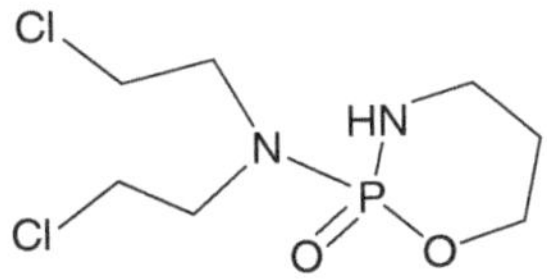

Chemical name: (*RS*)-*N*,*N*-Bis(2-chloroethyl)-1,3,2-oxazaphosphinan-2-amine 2-oxide
Brand name: Procytox
Formula: $C_7H_{15}Cl_2N_2O_2P$
Monoisotopic protonated ion (*m/z*): 261.0321
Molecular weight: 261.09
CAS number: 50-18-0

Cyclophosphamide, a pro-drug of nitrogen mustard, is the prototype of the drug family of oxazaphosphorine cytostatics. It was developed by Brock and his research team at ASTA Pharma in the 1950s [1]. Its clinical use includes the treatment of cancers and autoimmune diseases. The degradation chemistry of cyclophosphamide in aqueous solutions has been studied a few times, and those performed by ASTA Pharma are illustrated in Scheme 137-1 [2, 3]. In general, the degradation of cyclophosphamide is of hydrolytic nature, and the formation and distribution of the degradants depends upon the nature of solvents, pH, temperature, and duration of

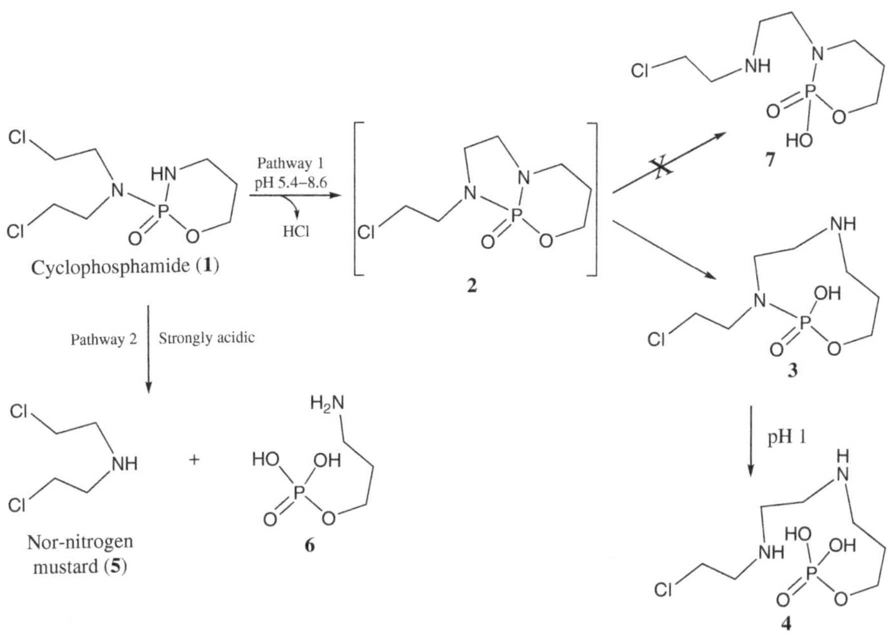

Scheme 137-1 Degradation pathways of cyclophosphamide based on the results from references 2 and 3.

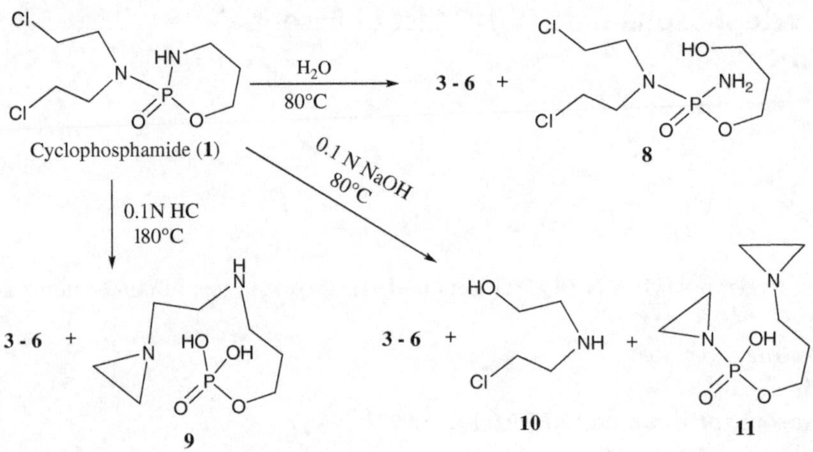

Scheme 137-2 Degradation pathways of cyclophosphamide in solution of water, 0.1 N HCl and 0.1 N NaOH at 80°C for 24 hours.

the degradation studies. In solutions with pH between 5.4 and 8.6, cyclophosphamide underwent intramolecular alkylation to form a transient intermediate (**2**), which immediately decomposed to produce the degradant with a 9-membered ring (**3**, Pathway 1). Upon further treatment in pH 1 solution, the latter degradant decomposed to yield phosphoric acid monoester **4**. In strongly acidic solutions, cyclophosphamide degraded to give *nor*-nitrogen mustard (**5**) and phosphoric acid monoester **6** (Pathway 2); at pH 1.2 and 37°C, the half-life of the degradation was 1.4 days. In the pH range of 2.2–3.4, both Pathways 1 and 2 coexist; in solid-state degradation, where the samples were heated to the melting temperature, degradation proceeded predominantly via Pathway 1 over Pathway 2 [3].

In a more recent forced degradation study [4], cyclophosphamide was subjected to solutions of 0.1 N HCl, 0.1 N NaOH, and water, respectively, at 80°C for 24 hours. While Degradant **8** was observed under all forced degradation conditions (Scheme 137-2), Degradant **9** was in 0.1 N HCl solution, and Degradants **10** and **11** were observed in 0.1 N NaOH solution. The structural elucidation was based on accurate MS and MS/MS.

References

1 Brock, N. (1989). Oxazaphosphorine cytostatics: past-present-future. Seventh Cain memorial award lecture. *Cancer Res.* 49: 1–7.

2 Gilard, V., Martino, R., Malet-Martino, M.-C. et al. (1994). Chemical and biological evaluation of hydrolysis products of cyclophosphamide. *J. Med. Chem.* 37: 3986–3993.

3 Niemeyer, U., Kutscher, B., Engel, J. et al. Degradation products of cyclophosphamide synthesis and structural studies. *Phosphorus Sulfur Silicon Relat. Elem.* 109: 473–476.

4 Shivakumar, G. and Dwivedi, J. (2015). Identification of degradation products in cyclophosphamide API by LC-QTOF mass spectrometry. *J. Liq. Chromatogr. Relat. Technol.* 38: 190–195.

138–140: Cytarabine, Azacitadine, Decitabine

138: Cytarabine [WHO List of Essential Medicines]

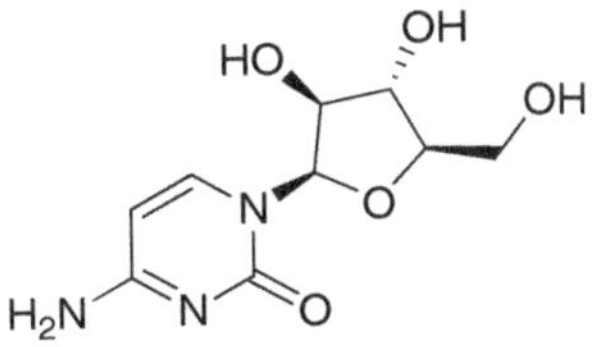

Chemical name: 1β-Arabinofuranosylcytosine; 4-Amino-1-[(2*R*,3*S*,4*S*,5*R*)-3,4-dihydroxy-5-(hydroxymethyl)oxolan-2-yl]pyrimidin-2-one;
Brand name: Cytosar, Vyxeos
Formula: $C_9H_{13}N_3O_5$
Monoisotopic protonated ion (*m/z*): 244.0928
Molecular weight: 243.22
CAS number: 147-94-4
Common salt form: HCl salt

139: Azacitidine (Azacytidine)

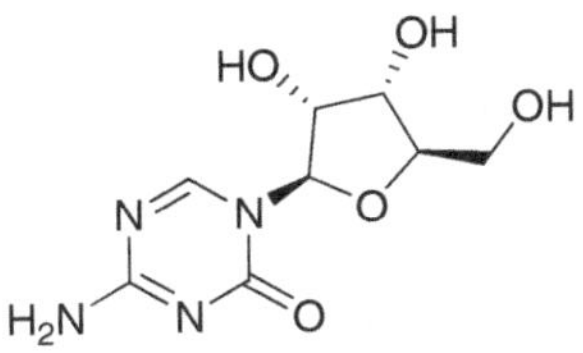

Chemical name: 4-Amino-1-[(2*R*,3*R*,4*S*,5*R*)-3,4-dihydroxy-5-(hydroxymethyl)oxolan-2-yl]-1,3,5-triazin-2-one
Brand name: Onureg, Vidaza
Formula: $C_8H_{12}N_4O_5$
Monoisotopic protonated ion (*m/z*): 245.0880
Molecular weight: 244.21
CAS number: 320-67-2

140: Decitabine

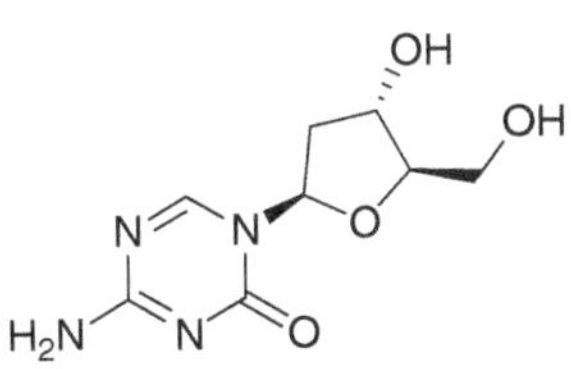

Chemical name: 4-Amino-1-[(2*R*,4*S*,5*R*)-4-hydroxy-5-(hydroxymethyl)oxolan-2-yl]-1,3,5-triazin-2- one

Brand name: Dacogen
Formula: $C_8H_{12}N_4O_4$
Monoisotopic protonated ion (*m/z*): 229.0931
Molecular weight: 228.21
CAS number: 2353-33-5

Cytarabine, azacytadine, and decitabine are analogs of the nucleoside cytidine and are clinically used for the treatment of certain types of cancers, particularly leukemia. Cytarabine was first synthesized in 1959, as a result of the research following the discovery that the structurally similar spongonucleosides, isolated from the sponge, *Cryptotethia crypta*, possess anticancer activity [1]; it was approved for clinical use in 1969 [2]. Azacitidine and its 2′-deoxy derivative, decitabine, which were discovered in the mid-1970s [3], were approved by the FDA in 2004 and 2006, respectively, for the treatment of myelodysplastic syndromes (MDS) [4, 5], conditions that can progress to acute myeloid leukemia (AML).

The main degradation pathways of cytarabine, azacytadine, and decitabine are of hydrolytic nature. In aqueous buffered solutions with acidic to neutral pH, cytarabine primarily underwent hydrolytic deamination to produce uracil arabinoside (Scheme 138–140-1) [6]. The degradant, uracil arabinoside, was present in a sterile powder of cyrarabine in low level (0.05%) [7], which was lower than the USP specification of 0.30% for this impurity [8]. Cytarabine was found to be most stable at pH 6.9 [9].

In neutral to alkaline solutions, the triazine of azacytadine would also be susceptible to hydrolytic degradation, resulting in the formation of *N*-(formylaminine)-*N*′-β-D-ribofuranosylurea (**1**) and its further des-formyl degradant (**2**, Scheme 138–140-2) [7]. This degradation pathway was also observed in a stability study of reconstituted azacitidine suspensions for injection [10].

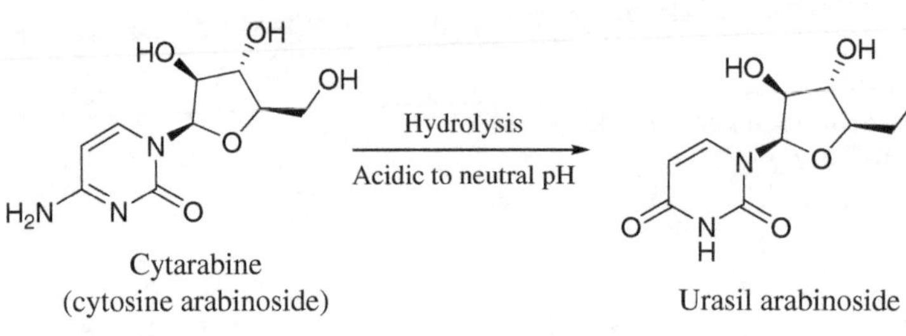

Scheme 138–140-1 Hydrolysis of cytarabine in acidic to neutral solutions.

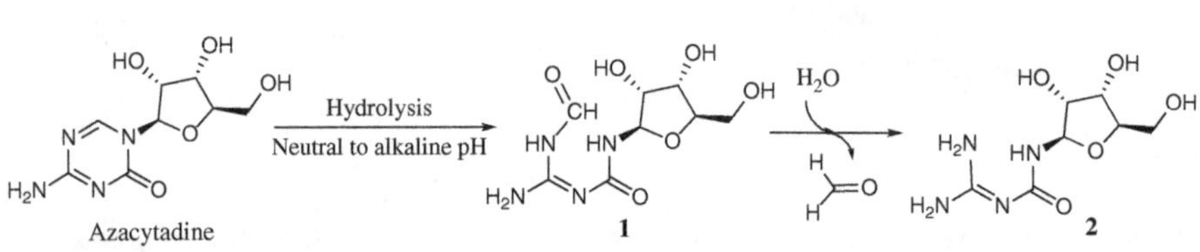

Scheme 138–140-2 Hydrolysis of azacytadine in neutral to alkaline solutions.

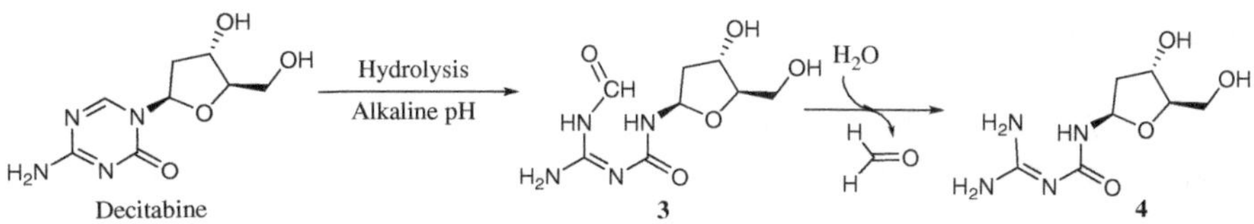

Scheme 138–140-3 Hydrolysis of decitabine in alkaline solutions.

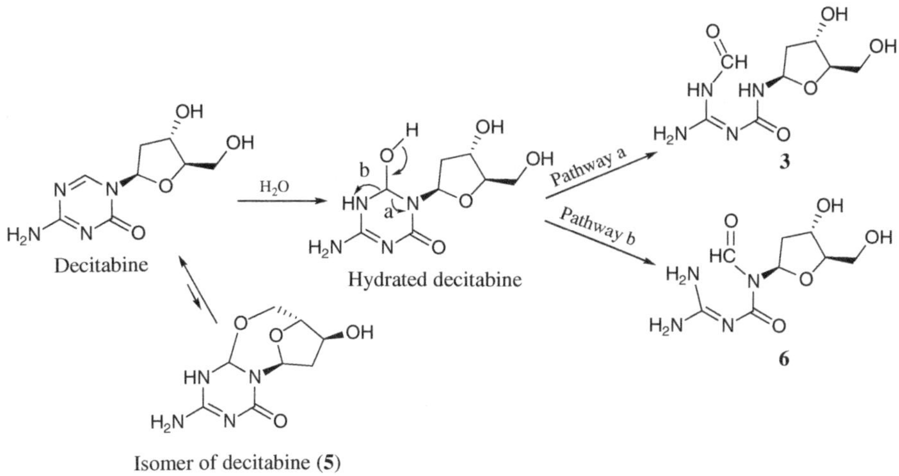

Scheme 138–140-4 Stability of decitabine in water solution. Illustrated above are the more reasonable structures (**5** and **6**) among those proposed by Liu et al. [13] for the respective isomers of decitabine and **3**. *Source:* Liu et al. [13]/John Wiley & Sons.

Decitabine, the 2′-deoxyl derivative of azacytadine, would be expected to behave similarly in terms of hydrolytic stability with regard to the triazine moiety as compared to azacytadine. Indeed, in a study performed by Lin et al. [11], decitabine was found to undergo the same sequential decomposition of the triazine moiety in alkaline solution (Scheme 138–140-3).

Nevertheless, in a pH 7.4 phosphate buffer or water, the degradation profile of decitabine was quite different from that of azacytadine: in addition to the two degradants due to the triazine ring-opening, additional unidentified degradant peaks were observed [11]. During a stability study of reconstituted decitabine solutions [12], a degradation profile similar to that observed in the pH 7.4 solution [11] was seen. In a 2006 study of decitabine stability in water by Liu et al. [13], two of the previously unidentified major degradants were found to be isomers of decitabine and **3**, respectively. Among the structures proposed by Liu et al., it appears that the following two structures (**5** and **6**) are more reasonable for the respective isomers of decitabine and **3** (Scheme 138–140-4) than the others.

References

1 Lichtman, M.A. (2013). A historical perspective on the development of the cytarabine (7 days) and daunorubicin (3 days) treatment regimen for acute myelogenous leukemia: 2013 the 40th anniversary of 7+3. *Blood Cells Mol. Dis.* 50: 119–130.
2 Cytarabine. https://en.wikipedia.org/wiki/Cytarabine (accessed 30 August 2022).
3 Cihák, A. (1974). Biological effects of 5-azacytidine in eukaryotes. *Oncology* 30: 405–422.
4 Azacitidine. https://en.wikipedia.org/wiki/Azacitidine (accessed 30 August 2022).
5 Decitabine. https://go.drugbank.com/drugs/DB01262 (accessed 30 August 2022).
6 Notari, R.E. (1967). A mechanism for the hydrolytic deamination of cytosine arabinoside in aqueous buffer. *J. Pharm. Sci.* 56: 804–809.
7 Kissinger, L.D. and Stemm, N.L. (1986). Determination of the antileukemia agents cytarabine and azacitidine and their respective degradation products by high-performance liquid chromatography. *J. Chromatogr.* 353: 309–318.
8 Cytarabine (2020). USP Monograph, official as of 01 May 2020. https://doi.org/10.31003/USPNF_M21810_05_01 (accessed 30 August 2022).
9 Notari, R.E., Chin, M.L., and Wittebort, R. (1972). Arabinosylcytosine stability in aqueous solutions: pH profile and shelflife predictions. *J. Pharm. Sci.* 61: 1189–1196.
10 Legeron, R., Xuereb, F., Djabarouti, S. et al. (2013). Chemical stability of azacitidine suspensions for injection after cold-chain reconstitution of powder and storage. *Am. J. Health-Sys. Pharm.* 70: 2137–3142.
11 Lin, K.-T., Momparler, R.L., and Rivard, G.E. (1981). High-performance liquid chromatographic analysis of chemical stability of 5-aza-2′-deoxycytidine. *J. Pharm. Sci.* 70: 1228–1232.
12 Kim, S.H., Heeb, R.M., and Krämer, I. (2017). Physicochemical stability of reconstituted Decitabine (Dacogen®) solutions and ready-to-administer infusion bags when stored refrigerated or frozen. *Pharm. Tech. Hosp. Pharm.* 2: 145–157. https://doi.org/10.1515/pthp-2017-0025.
13 Liu, Z., Marcucci, G., Byrd, J.C. et al. (2006). Characterization of decomposition products and preclinical and low dose clinical pharmacokinetics of decitabine (5-aza-2′-deoxycytidine) by a new liquid chromatography/tandem mass spectrometry quantification method. *Rapid Commun. Mass Spectrom.* 20: 1117–1126.

5

Dabigatran Etexilate to Duloxetine

141: Dabigatran Etexilate

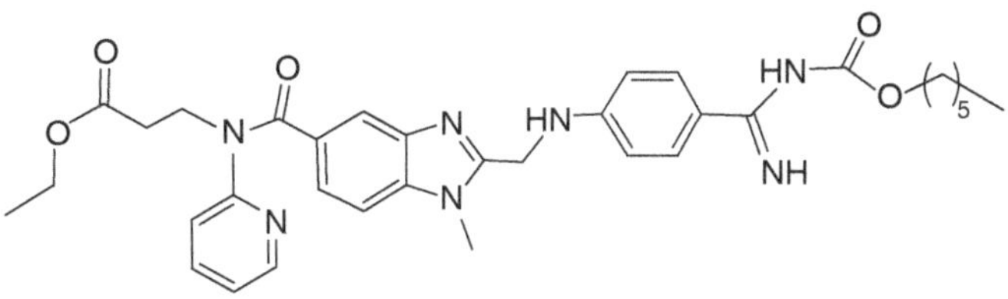

Chemical name: Ethyl 3-[[2-[[4-(*N*-hexoxycarbonylcarbamimidoyl)anilino] methyl]-1-methylbenzimidazole-5-carbonyl]-pyridin-2-ylamino]propanoate
Brand name: Pradaxa
Formula: $C_{34}H_{41}N_7O_5$
Monoisotopic protonated ion (*m/z*): 628.3242
Molecular weight: 627.73
CAS number: 1610666-09-5
Common salt form: Mesylate

Dabigatran etexilate is the prodrug of dabigatran, which itself is a direct thrombin inhibitor, used as an oral anticoagulant. The function of the prodrug is achieved via the two functionalities: the ethyl ester and carbamate, both of which would be susceptible to hydrolytic degradation. As the other parts of the drug molecule seem to be reasonably stable, the main degradation pathways of dabigatran etexilate would be hydrolytic. In a patent application [1], Miralles et al. described a few hydrolytic degradation products observed from stability studies of several tablet formulations of dabigatran etexilate (Scheme 141-1). The major one, the imide degradant (**B3**), was not reported before. Such a degradation profile in the solid state is apparently quite different from the in vivo study, where the active drug, dabigatran, is the main product from the hydrolysis of dabigatran etexilate [2]. It seems that the imide degradant **B3** tended to form rather easily, apparently more easily than **C2**, which results from hydrolysis of the ethyl ester moiety, in the solid formulations as reported

Compendium of Drug Degradation Pathways, First Edition. Min Li.

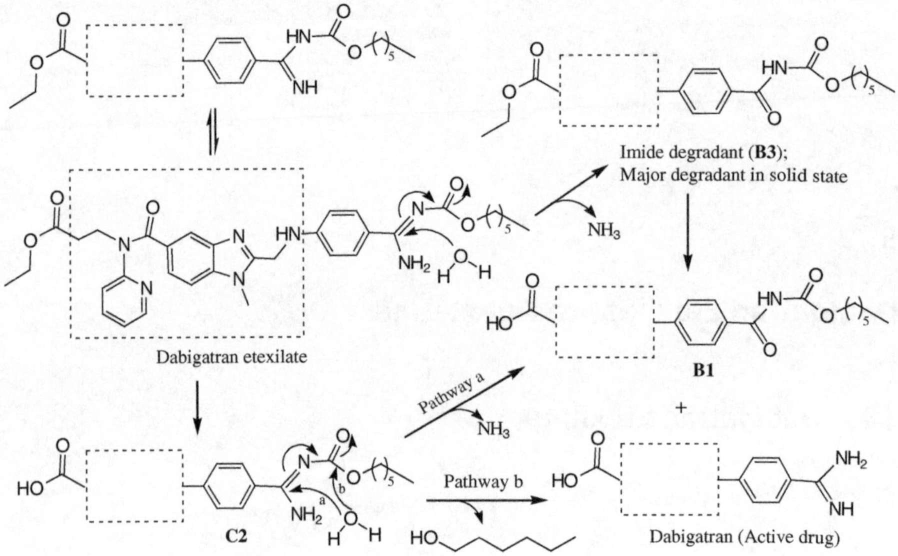

Scheme 141-1 Hydrolytic degradation pathways of formulated dabigatran etexilate in solid dosage. The nomenclature of the degradants is that used by Miralles et al. [1].

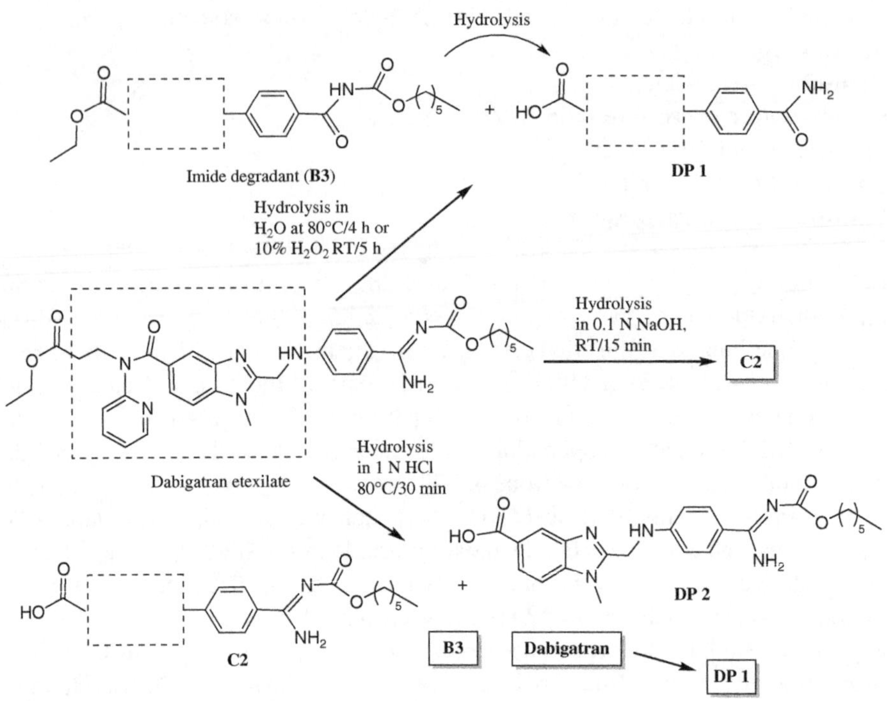

Scheme 141-2 Forced degradation of dabigatran etexilate solutions under different conditions.

in the patent application, particularly if the formulated product lacked proper protection from moisture by its packaging.

On the other hand, in certain solution forced degradation studies [3, 4], the degradation profiles (Scheme 141-2) were different from that obtained in the solid state [1]. Under the harsh conditions of these forced degradation studies, another two hydrolytic degradants, i.e., DP 1 and DP 2, were also observed in addition to B3 and C2. DP 1 could be a direct degradant from dabigatran etexilate or a secondary degradant from the imide degradant (B3), while DP 2 results from the hydrolysis of the amide linkage of the drug molecule.

References

1 Miralles, J.S., Lépez, E.D., and Rao, L. US Patent Application, US 2014/0076036 A1.
2 Laizure, S.C., Parker, R.B., Herring, V.L., and Hu, Z.-Y. (2014). Identification of carboxylesterase-dependent dabigatran etexilate hydrolysis. *Drug Metab. Dispos.* 42: 201–206.
3 Swain, D., Patel, P.N., Nagaraj, G. et al. Liquid chromatographic method development for forced degradation products of dabigatran etexilate: characterisation and in silico toxicity evaluation. *Chromatographia* https://doi.org/10.1007/s10337-015-3009-5.
4 Amrani, F., Secrétan, P.-H., Sadou-Yayé, H. et al. (2015). Identification of dabigatran etexilate major degradation pathways by liquid chromatography coupled to multi stage higher resolution mass spectrometry. *RSC Adv.* 5: 45068–45081. https://doi.org/10.1039/C5RA04251H.

142: Dacarbazine [WHO List of Essential Medicines]

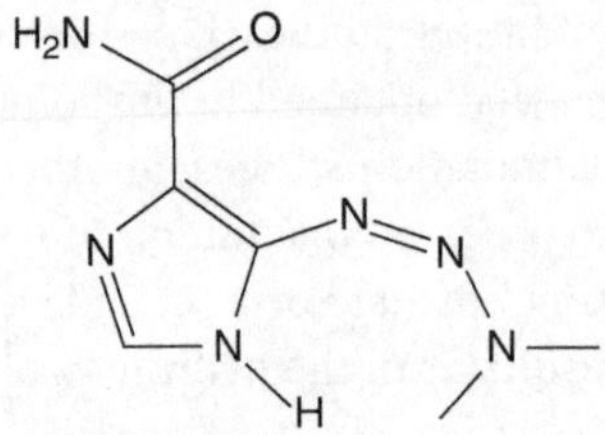

Chemical name: 4-[(*E*)-Dimethylaminodiazenyl]-1*H*-imidazole-5-carboxamide
Brand name: DTIC, DTIC-Dome
Formula: $C_6H_{10}N_6O$
Monoisotopic protonated ion (*m/z*): 183.0989
Molecular weight: 182.18
CAS number: 4342-03-4

Dacarbazine is a triazene anticancer drug with its cytotoxicity deriving from its DNA alkylating capability. Its main degradant is 2-azahypoxanthine (**2**, Scheme 142) [1],

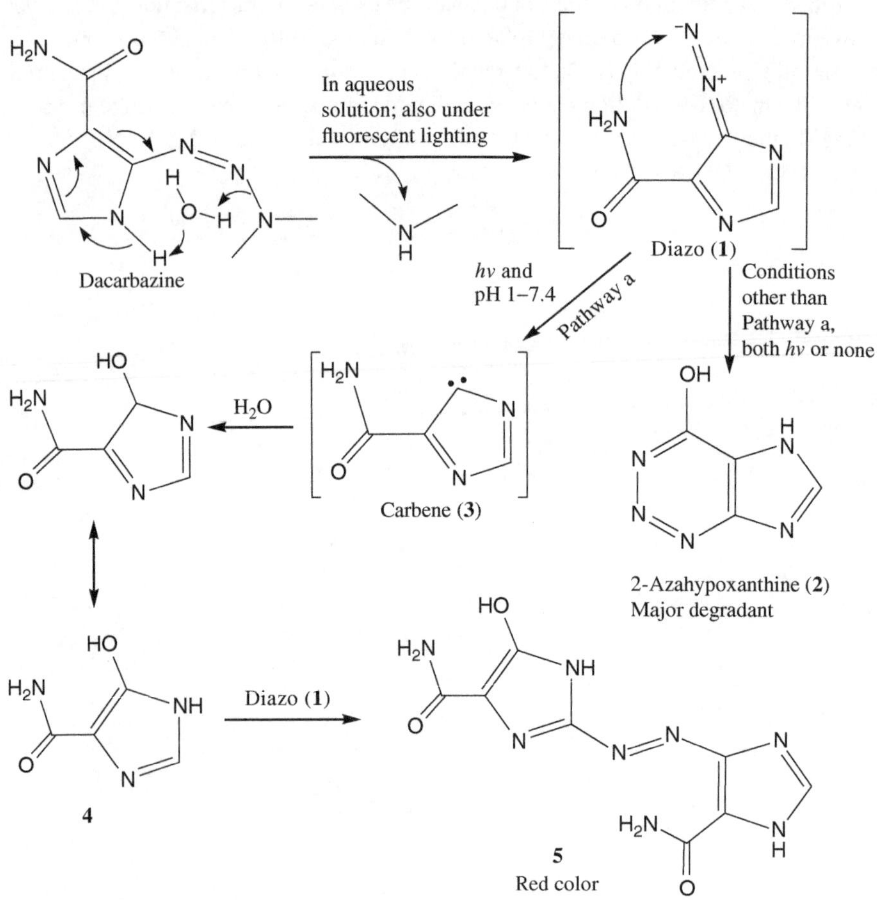

Scheme 142 Degradation of dacarbazine.

which is a photodegradant; the overall degradation was enhanced between ~15 and 100 times at different pH under fluorescent lighting [2]. A stability study of dacarbazine in commercial amber glass vials and polyvinyl chloride bags was conducted, and the results indicated that "dacarbazine must be carefully protected from light, administered using opaque infusion tubing" [3]. When the photolysis of dacarbazine solutions was performed in pH 1–7.4, either by UV or by fluorescent lighting, the diazo intermediate (**1**) can produce a red-colored degradant (**5**) via the intermediacy of degradant **4** [4].

References

1 Fiore, D., Jackson, A.J., Didolkar, M.S., and Dandu, V.R. (1985). Simultaneous determination of dacarbazine, its photolytic degradation product, 2-Azahypoxanthine, and the metabolite 5-Aminoimidazole-4-Carboxamide in plasma and urine by high-pressure liquid chromatography. *Antimicrob. Agents Chemother.* 27: 977–979.

2 Shetty, B.V., Schowen, R.L., Slavik, M., and Riley, C.M. (1992). Degradation of dacarbazine in aqueous solution. *J. Pharm. Biomed. Anal.* 10: 675–683.

3 El Aatmani, M., Poujol, S., Astre, C. et al. (2002). Stability of dacarbazine in amber glass vials and polyvinyl chloride bags. *Am. J. Health-Sys. Pharm.* 59: 1351–1356.

4 Asahi, M., Matsushita, R., Kawahara, M. et al. (2002). Causative agent of vascular pain among photodegradation products of dacarbazine. *J. Pharm. Pharmacol.* 54: 1117–1122.

143: Dapsone [WHO List of Essential Medicines]

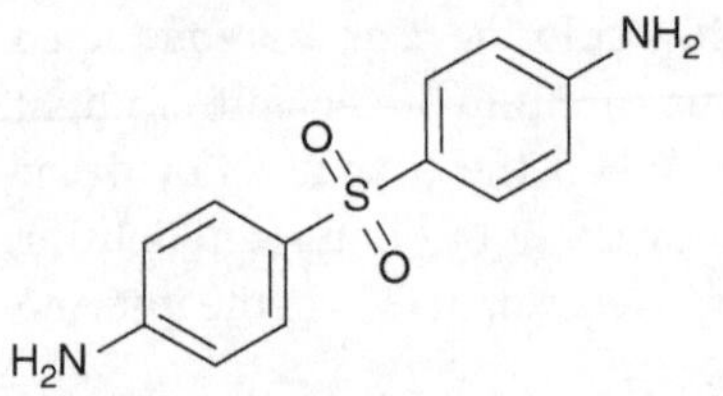

Chemical name: 4-(4-Aminophenyl)sulfonylaniline
Brand name: Aczone
Formula: $C_{12}H_{12}N_2O_2S$
Monoisotopic protonated ion (*m/z*): 249.0692
Molecular weight: 248.30
CAS number: 80-08-0

Dapsone, also known as diaminodiphenylsulfone (or DDS), is an old antibiotic for the treatment of leprosy as well as other bacterial or parasitic infections, which dates back to the 1930s [1]. Its potentially degradable functional group is the aromatic amino group, which is nucleophilic as well as oxidizable as an aniline analog. In the study of a triple combination drug for malaria, it was found that dapsone reacted with the succinic moiety of artesunate, one of the components in the triple combination drug, to form succinyl dapsone [2].

The formation of the degradant, succinyldapsone, could proceed through a direct interaction between dapsone and artesunate (Pathway a, Scheme 143-1). Nevertheless, the original authors preferred a step-wise mechanism (Pathway b, Scheme 143-1), in which dapsone is coupled with the succinic acid that is formed from the hydrolytic degradation of artesunate. Such a step-wise mechanism would not only require a quite significant hydrolytic degradation of artesunate but also a quite efficient coupling reaction between dapsone and the succinic acid formed in the hydrolysis. Unfortunately, the latter coupling reaction between a weak organic base and nonactivated acid would not be expected to be efficient, suggesting that the indirect mechanism lacks a reasonable mechanistic basis. Although the direct mechanism would be more reasonable than the step-wise one, it may suffer from a potential drawback that the point of the direct attacking may have too much steric hindrance. Another possible and perhaps more reasonable mechanism would be for artesunate to undergo an intramolecular decomposition in which succinic anhydride is formed. The anhydride is quite reactive and can react with dapsone much more efficiently (Pathway c, Scheme 143-1).

The aniline-type amino groups of dapsone would certainly be susceptible to oxidative degradation. In a forced degradation study with hydrogen peroxide, two degradants were formed (Scheme 143-2): dihydroxydapsone (*N,N'*-[sulfonyl(bis(4,1-phenylene))]-bis(hydroxylamine); Deg 1) and sulfanilic acid (Deg 2) [3]. The formation of Deg 1, dihydroxydapsone, is apparently via a nucleophilic oxidation pathway. The monohydroxydapsone, an intermediate of the nucleophilic oxidation, is also a metabolite of dapsone [4]; however, it was not reported in this forced degradation study.

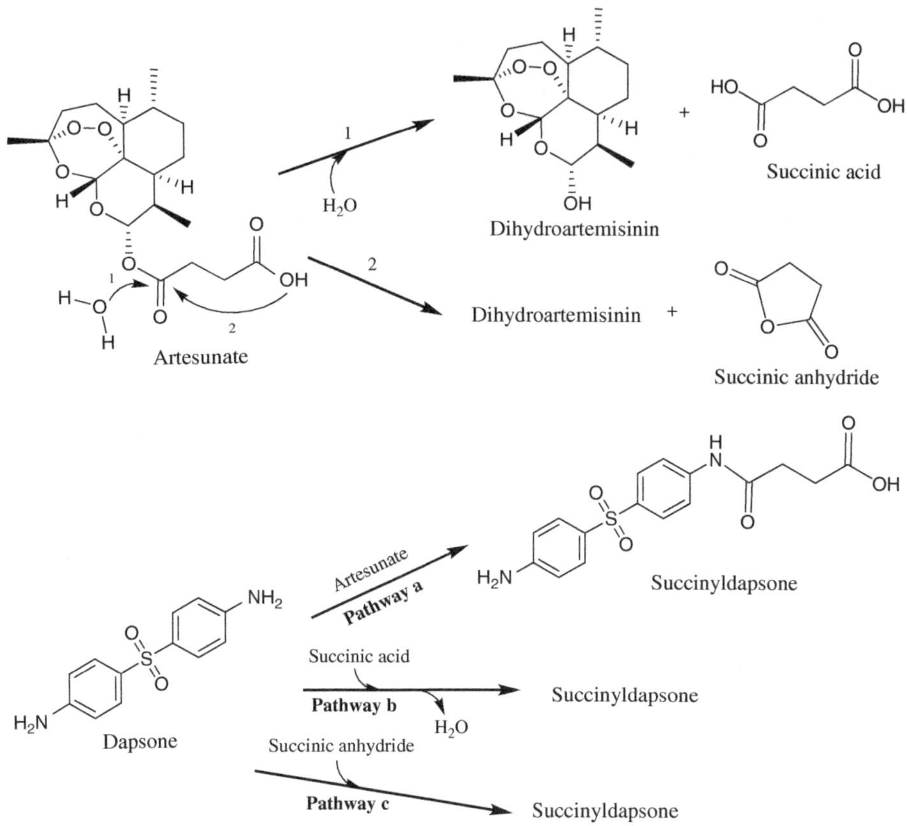

Scheme 143-1 Degradation pathways of dapsone and artesunate in their combination formulation.

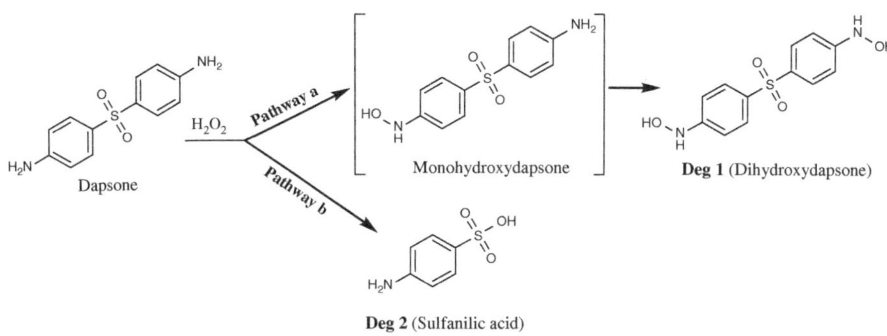

Scheme 143-2 Oxidative degradation of dapsone under forced degradation by hydrogen peroxide.

References

1 Dapsone. https://en.wikipedia.org/wiki/Dapsone (accessed 17 February 2018).

2 Bardsley, B., Barry, S.J., Drozdz, M.A. et al. (2011). Formation and identification of a degradant in chlorproguanil–dapsone– artesunate (Dacart™) tablets. *J. Pharm. Biomed. Anal.* 54 (3): 610–613.

3 Salama, N.N.E.-D.A., El Ries, M.A., Toubar, S. et al. (2012). Validated TLC and HPLC stability-indicating methods for the quantitative determination of dapsone. *J. Planar Chromatogr.* 25: 65–71.

4 Aberg, A.K.G., Zolotoy, A., Bain, A.I. (2000). Galenical preparations of dapsone and related sulphones, and method of therapeutic and preventative treatment of disease. WO2001041772 A1.

144–146: Daunorubicin, Doxorubicin, Idarubicin

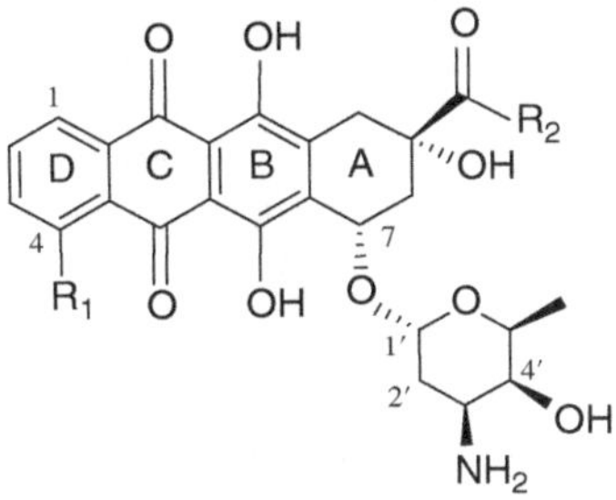

Daunorubicin: R_1 = –OMe; R_2 = –Me.
Doxorubicin: R_1 = –OMe; R_2 = –CH_2OH.
Idarubicin: R_1 = –H; R_2 = –Me.

144: Daunorubicin

Chemical name: (7*S*,9*S*)-9-Acetyl-7-[(2*R*,4*S*,5*S*,6*S*)-4-amino-5-hydroxy-6-methyl oxan-2-yl]oxy-6,9,11-trihydroxy-4-methoxy-8,10-dihydro-7*H*-tetracene-5,12-dione
Brand name: Cerubidine, Vyxeos
Formula: $C_{27}H_{29}NO_{10}$
Monoisotopic protonated ion (*m/z*): 528.1864
Molecular weight: 527.52
Common salt form: HCl salt
CAS number: 20830-81-3

145: Doxorubicin [WHO List of Essential Medicines]

Chemical name: (7*S*,9*S*)-7-[(2*R*,4*S*,5*S*,6*S*)-4-Amino-5-hydroxy-6-methyloxan-2-yl]oxy-6,9,11-trihydroxy-9-(2-hydroxyacetyl)-4-methoxy-8,10-dihydro-7*H*-tetracene-5,12-dione
Brand name: Adriamycin, Caelyx, Myocet
Formula: $C_{27}H_{29}NO_{11}$
Monoisotopic protonated ion (*m/z*): 544.1813
Molecular weight: 543.52
Common salt form: HCl salt
CAS number: 23214-92-8

146: Idarubicin

Chemicalname:(7*S*,9*S*)-9-Acetyl-7-[(2*R*,4*S*,5*S*,6*S*)-4-amino-5-hydroxy-6-methyloxan-2-yl]oxy-6,9,11-trihydroxy-8,10-dihydro-7*H*-tetracene-5,12-dione
Brand name: Idamycin
Formula: $C_{26}H_{27}NO_9$
Monoisotopic protonated ion (*m/z*): 498.1759

Molecular weight: 497.49
Common salt form: HCl salt
CAS number: 58957-92-9

In this monograph, the degradation chemistry of daunorubicin (also known as daunomycin) and its closely related analogs, doxorubicin and idarubicin, is examined. All three drugs are anthracycline antibiotics used as chemotherapeutic agents for treating cancers. The degradation chemistry of these drugs seems to occur mainly on the glycosidic linkage, which could be cleaved under both acidic and alkaline conditions, while it was most stable in the pH range of 4.5–6.0 [1]. The degradation profile is dependent on the pH of the solution: at pH < 3.5, daunorubicin seemed to be exclusively hydrolyzed to give daunorubicinone (I, Scheme 144–146-1) and the amino sugar, if the degradation conditions were somewhat moderate [1, 2]. Notably, the configuration of the chiral center in I remained unchanged, indicating that the nucleophilic attack by water would occur on the glycosidic carbon (C1′), rather than on the benzilic-like position (C7) in the anthracycline D ring [3]. At pH > 4.0, daunorubicin primarily decomposed to yield the other two aglycone degradants (II and III, Scheme 144–146-1) [1]. Under harsher forced degradation conditions, daunorubicin produced a few more degradants [3]; nevertheless, they may not be relevant under pharmaceutical storage conditions.

In a forced degradation study of idarubicin, which is 4-desmethoxyldaunorubicin and reported to have less cardio toxicity and better efficacy, it was found to yield two oxidative degradants through the Baeyer-Villiger oxidation of the keto group (Scheme 144–146-2) [4]. However, the oxidation condition used was quite harsh (30% H_2O_2), which resulted in excessive degradation. Hence, the validity of this forced degradation study in terms of predicting the degradation chemistry of idarubicin (and its analogs) may need to be further explored.

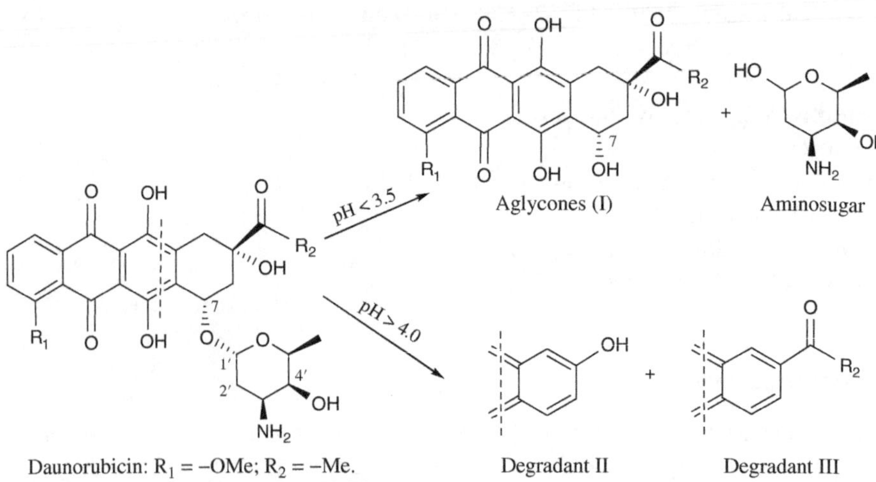

Scheme 144–146-1 Degradation of anthracycline antibiotics in solutions.

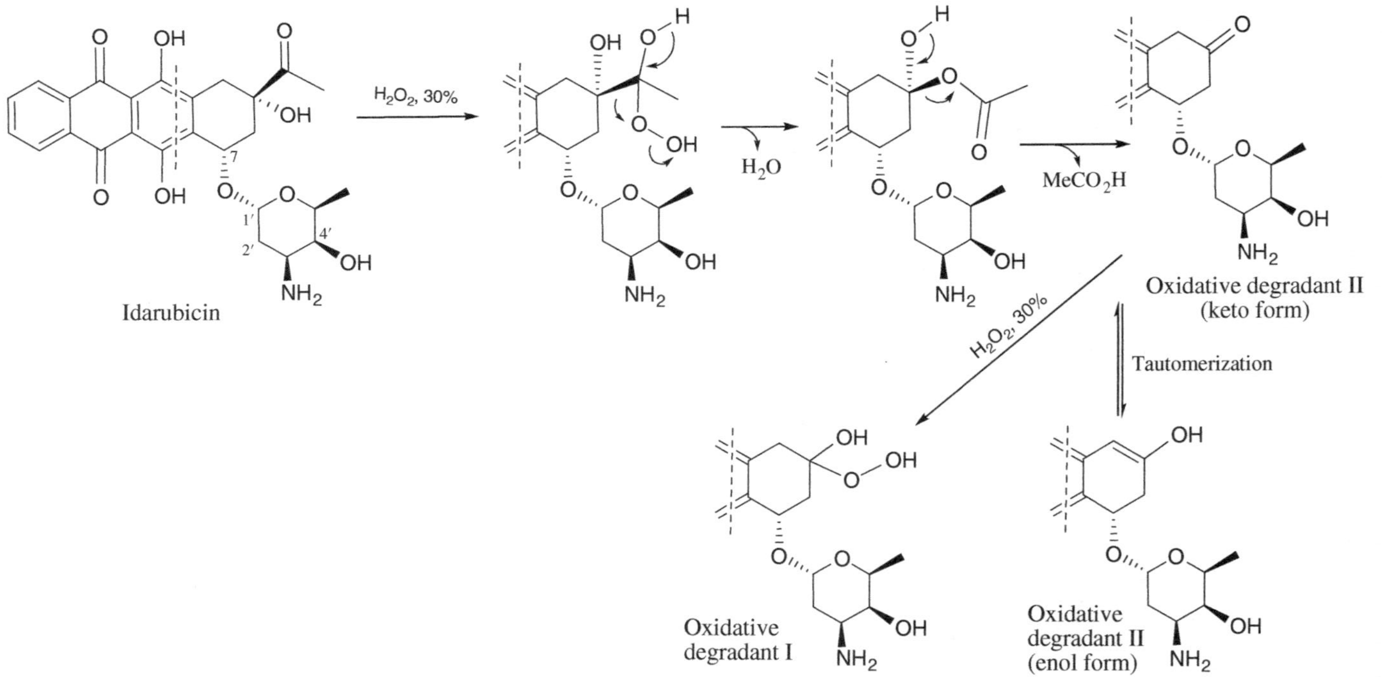

Scheme 144–146-2 Oxidative degradation of idarubicin in 30% H_2O_2. The mechanism presented is modified from that proposed by the original authors. *Source:* Kaushik and Bansal [4]/with permission of Elsevier.

The degradation kinetic parameters, activation energies, and frequency factors of daunorubicin were determined to be 114 kJ/mol and 5.5×10^{13}/s at pH 1.5 and 79 kJ/mol and 5.5×10^{6}/s at pH 8, respectively, over the temperature range of 40–60°C [1]. For a degradation kinetic study of doxorubicin, the activation energy for degradation in a 0.5 M HCl solution was found to be 92.0 kJ/mol over the temperature range 22–50°C [5].

Photochemical degradation of daunorubicin, doxorubicin, and epirubicin (4′-epimer of doxorubicin) conducted by Wood et al. [6] indicated that under intense sunlight and fluorescent lighting, these drugs quickly decomposed, particularly in solutions of low concentrations. However, in high concentration solutions, e.g., those used in typical anticancer treatment (≥500 mg/l), it was concluded "no special precautions are necessary to protect freshly prepared solutions of these agents from light during intravenous administration." The major photodegradant was reported to be degradant II (Scheme 144–146-1) [7]. In a more recent photochemical study, doxorubicin exhibits unusual photoreactivity in aqueous solutions via two pathways [8]. The first pathway is a two-step process that leads to the formation of 3-methoxysalicylic acid through an unstable intermediate (Scheme 144–146-3). The second pathway is a photocatalyzed process in which doxorubicin is recycled back and forth with its reduced form, dihydrodoxorubicin, via photoreduction of doxorubicin and subsequent spontaneous oxidation of dihydrodoxorubicin. The latter process is mediated by dissolved oxygen, which is reduced to H_2O_2 (Pathway b, Scheme 144–146-3). During this redox cycling, the amount of H_2O_2 may accumulate

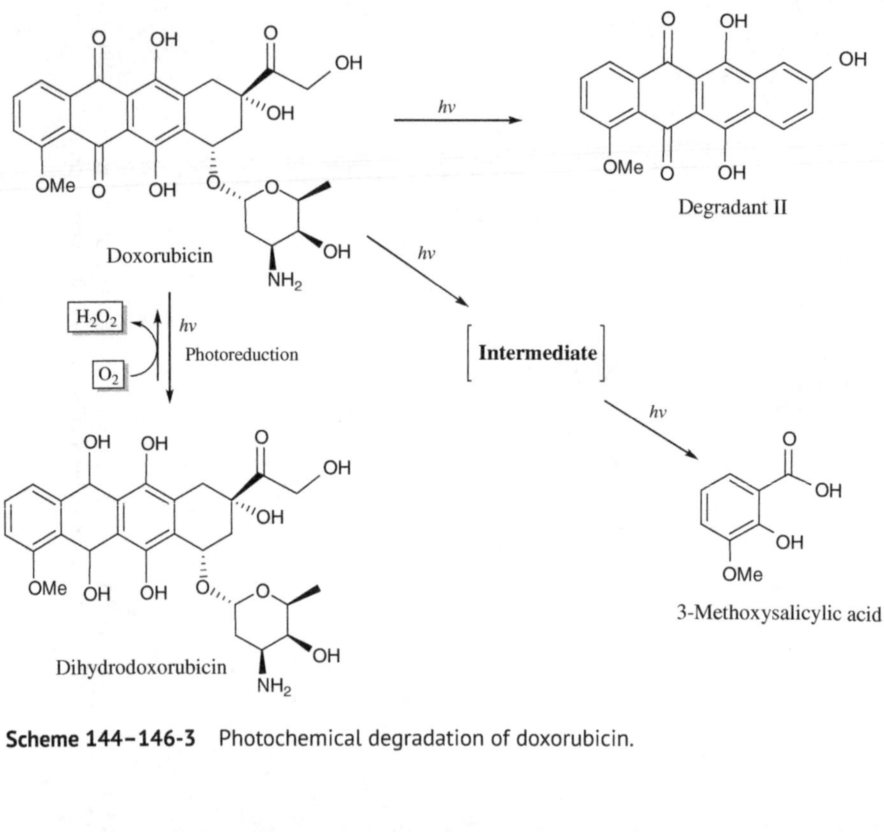

Scheme 144–146-3 Photochemical degradation of doxorubicin.

to toxic levels in pharmaceutical preparations, according to the evidence presented by the same researchers [9].

These anthracycline antitumor antibiotics can form stable complexes with metal ions such as Fe(III) and Cu(II), and under slightly acidic conditions, these metal ions can induce an appreciable amount of degradant II [10, 11]. All the identified degradants of daunorubicin do not have the aminosugar ring, commonly referred to as aglycones, and they are devoid of anticancer activity for which the aminosugar moiety is required [12].

References

1 Beijnen, J.H., van der Houwen, O.A.G.J., Voskuilen, M.C.H., and Underberg, W.J.M. (1986). Aspects of the degradation kinetics of daunorubicin in aqueous solution. *Int. J. Pharm.* 31: 75–82.

2 Arcamone, F., Franceschi, G., Orezzi, P. et al. (1964). The structure of daunomycinone. *J. Am. Chem. Soc.* 86: 5334–5335.

3 Beijnen, J.H., Potman, R.P., van Ooijen, R.D. et al. (1987). Structure elucidation and characterization of daunorubicin degradation products. *Int. J. Pharm.* 34: 247–251.

4 Kaushik, D. and Bansal, G. (2013). Characterization of degradation products of idarubicin through LC-UV, MS^n and LC–MS-TOF studies. *J. Pharm. Biomed. Anal.* 85: 123–131.

5 Wassermann, K. and Bundgaard, H. (1983). Kinetics of the acid-catalyzed doxorubicin. *Int. J. Pharm.* 14: 73–78.

6 Wood, M.J., Irwin, W.J., and Scott, D.K. (1990). Photodegradation of doxorubicin, daunorubicin and epirubicin measured by high-performance liquid chromatography. *J. Clin. Pharm. Therap.* 15: 291–300.

7 Gray, P.J. and Phillips, D.R. (1981). Ultraviolet photoirradiation of daunomycin and DNA-daunomycin complexes. *Photochem. Photobiol.* 33: 297–303.

8 Nawara, K., Krysinski, P., and Blanchard, G.J. (2012). Photoinduced reactivity of doxorubicin: catalysis and degradation. *J. Phys. Chem. A* 116: 4330–4337.

9 Nawara, K., Krysinski, P., and Blanchard, G.J. (2012). Doxorubicin is a photocatalyst for the generation of H_2O_2. *RSC Adv.* 2: 4059–4061.

10 Fiallo, M.M.L., Garnier-Suillerot, A., Matzanke, B., and Kozlowski, H. (1999). How Fe^{3+} binds anthracycline antitumour compounds: the myth and the reality of a chemical sphinx. *J. Inorg. Biochem.* 75: 105–115.

11 Fiallo, M.M.L., Haj, H.T.-B., and Garnier-Suillerot, A. (1994). Degradation of anthracycline antitumor compounds catalysed by metal ions. *Metal-Based Drugs* 1: 183–193.

12 Henry, D.W. (1979). Structure-activity relationships among daunorubicin and adriamycin analogs. *Cancer Treat. Rep.* 63: 845–854.

147–148: Desvenlafaxine, Venlafaxine

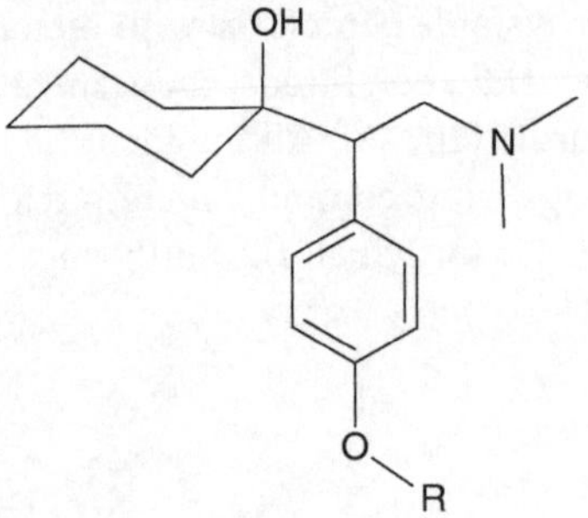

Desvenlafaxine (active metabolite of venlafaxine), R = H;
Venlafaxine, R = Me.

147: Desvenlafaxine

Chemical name: 4-[2-(Dimethylamino)-1-(1-hydroxycyclohexyl)ethyl]phenol
Brand name: Pristiq
Formula: $C_{16}H_{25}NO_2$
Monoisotopic protonated ion (*m/z*): 264.1958
Molecular weight: 263.38
CAS number: 93413-62-8
Common salt form: HCl salt, fumarate, succinate

148: Venlafaxine

Chemical name: 1-[2-(Dimethylamino)-1-(4-methoxyphenyl)ethyl]cyclohexan-1-ol
Brand name: Effexor
Formula: $C_{17}H_{27}NO_2$
Monoisotopic protonated ion (*m/z*): 278.2115
Molecular weight: 277.40
CAS number: 93413-69-5
Common salt form: HCl salt

The two functional groups in both desvenlafaxine and venlafaxine, which may be susceptible to degradation, are the hydroxyl and tertiary amino groups. Indeed, under acidic-forced degradation conditions, both desvenlafaxine and venlafaxine were found to undergo dehydration. Interestingly, the double bond formed as a result of the dehydration is in the cyclohexane ring (Scheme 147–148), which is supported by the presence of a single proton signal in the olefinic region [1, 2]. It needs to be pointed out that in a 2012 paper published by Pawar et al. [3], the structure proposed for dehydrated desvenlafaxine was incorrect with regard to the position of the double bond formed.

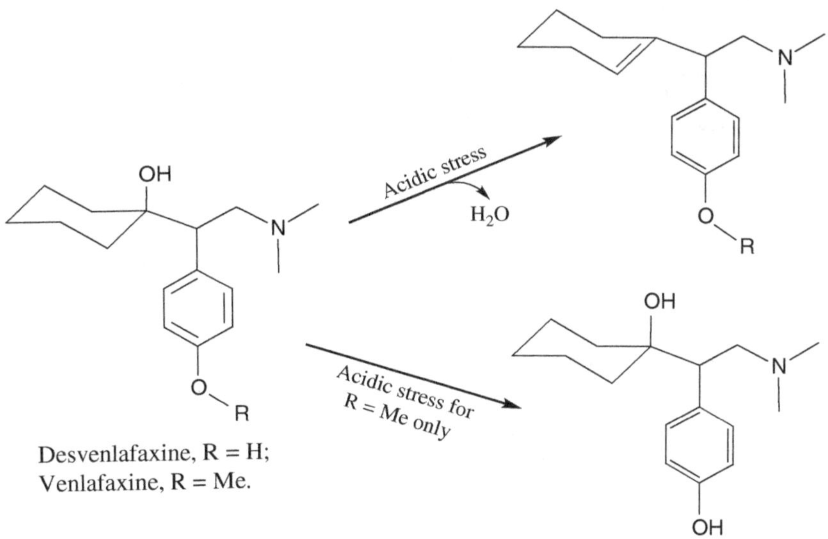

Desvenlafaxine, R = H;
Venlafaxine, R = Me.

Scheme 147–148 Degradation of desvenlafaxine and venlafaxine under acidic forced degradation conditions.

A later forced degradation study of venlafaxine by Carneiro et al. found that venlafaxine not only yielded the dehydrated venlafaxine but also gave desvenlafaxine as a minor degradant under the acidic stress condition in 1 N HCl at 60°C for 24 hours [4]. Nevertheless, the minor degradant peak of desvenlafaxine disappeared upon further stress by the time point of 50 hours. The results obtained by Carneiro et al. were somewhat different from those of Asafu-Adjaye et al. [5] and Makhija and Vavia [6]: in the case of Asafu-Adjaye et al., only dehydrated venlafaxine was observed, while in the case of Makhija and Vavia, only desvenlafaxine was seen. Such differences could be attributed to the different forced degradation conditions or different chromatographic conditions.

In all the forced degradation studies cited above, both desvenlafaxine and venlafaxine were found to be stable under conditions of basic (e.g., 1 M NaOH) and hydrogen peroxide. As both desvenlafaxine and venlafaxine are aliphatic tertiary amines and, as such, they would be expected to form the *N*-oxide degradants, the observation that they were stable under the stress by hydrogen peroxide would appear to be contradictory to the prediction. Nevertheless, the observation may be explained by the fact that in the stress studies, both APIs were utilized in their salt forms, and thus, the nucleophilicity of the tertiary amine functionality was suppressed by protonation. On the other hand, with regard to the degradation behaviors of the formulated products of desvenlafaxine and venlafaxine, there seems to be no evidence to indicate that the dehydrated products are the real degradants for both products, respectively. Furthermore, the degradation of venlafaxine to desvenlafaxine would also seem to be a remote possibility for a formulated product of venlafaxine, particularly in solid dosage form.

References

1 Moussa, B.A., El-Bagary, R.I., and Al-Eryani, Y.A. (2011). Development and validation of a stability-indicating HPLC method for the analysis of desvenlafaxine succinate in the presence of its acidic induced degradation product in bulk and pharmaceutical preparation. *J. Chem. Pharm. Res.* 3: 425–437.

2 Rao, R.N. and Raju, A.N. (2006). Simultaneous separation and determination of process-related substances and degradation products of venlafaxine by reversed-phase HPLC. *J. Sep. Sci.* 29: 2733–2744.

3 Pawar, S.M., Khatal, L.D., Gabhe, S.Y., and Dhaneshwar, S.R. (2012). LC–UV and LC–MS evaluation of stress degradation behavior of desvenlafaxine. *J. Pharm. Anal.* 2: 264–271.

4 Carneiro, W.J., Andrade, C.H., Braga, R.C., and de Oliveira, V. (2010). Identification of desvenlafaxine, the major active metabolite of venlafaxine, in extended-release capsules. *Rev. Eletr. Farm.* 7: 39–53.

5 Asafu-Adjaye, E.B., Faustino, P.J., Tawakkul, M.A. et al. (2007). Validation and application of a stability-indicating HPLC method for the in vitro determination of gastric and intestinal stability of venlafaxine. *J. Pharm. Biomed. Anal.* 43: 1854–1859.

6 Makhija, S.N. and Vavia, P.R. (2002). Stability indicating LC method for the estimation of venlafaxine in pharmaceutical formulations. *J. Pharm. Biomed. Anal.* 28: 1055–1059.

149: Dextromethorphan

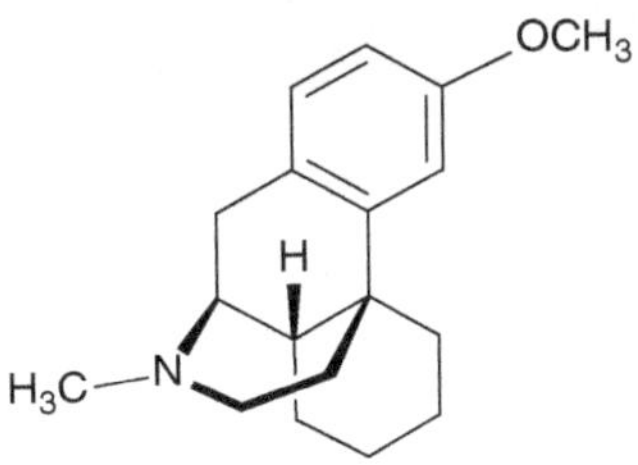

Chemical name: (1*S*,9*S*,10*S*)-4-Methoxy-17-methyl-17-azatetracyclo[7.5.3.$0^{1,10}$.$0^{2,7}$] heptadeca-2(7),3,5-triene
Brand name: Robitussin, Delsym, Others
Formula: $C_{18}H_{25}NO$
Monoisotopic protonated ion (*m/z*): 272.2009
Molecular weight: 271.40
CAS number: 125-71-3
Common salt form: HBr salt

Dextromethorphan (DM) is a widely used antitussive (cough suppressant) drug, which is typically coformulated with other active pharmaceutical ingredients in a great number of over-the-counter cold and cough medicines. The stereoisomer of dextromethorphan, levomethorphan, on the other hand, is a potent opioid analgesic but has never been developed into a clinically approved drug.

Due to its structural similarity to opioid analgesics, such as morphine, it can be reasonably expected that the benzylic position (10-position) is susceptible to free-radical mediated oxidation; this oxidation process can be initiated both photochemically [1] and nonphotochemically [2], which nonetheless should both produce the hydroperoxyl intermediate (Scheme 149-1).

The formation of the free radical should be catalyzed by redox-active transition metal ions, most likely iron ions. In a photochemical degradation study of dextromethorphan in 1 M HCl solution, Boccardi et al. found that Fe(III) was required for the photochemical oxidation to occur, in which both the 10β-hydroxyl and 10-keto degradants were formed [3]. From the hydroperoxyl intermediate, the dehydration process would lead to the formation of the 10-keto degradant, while a reductive process of the hydroperoxyl intermediate should give the 10β-hydroxyl degradant. Nevertheless, according to Boccardi et al. [3], this reductive process would be a minor process. The major process should be the one illustrated in Pathway b, Scheme 149-1: the formation of a carbocation via the elimination of H_2O_2 from the hydroperoxyl intermediate, followed by quenching of the carbocation by a water molecule. This elimination/quenching mechanism was supported by an experiment in which the corresponding 10β-methoxyl degradant was observed in good yield when the oxidation was carried out in a solution of 2 M HCl/methanol (1/1, v/v) [3].

Boccardi et al. also found that the photochemical degradation of dextromethorphan performed in a HCl solution gave different degradation profile than that

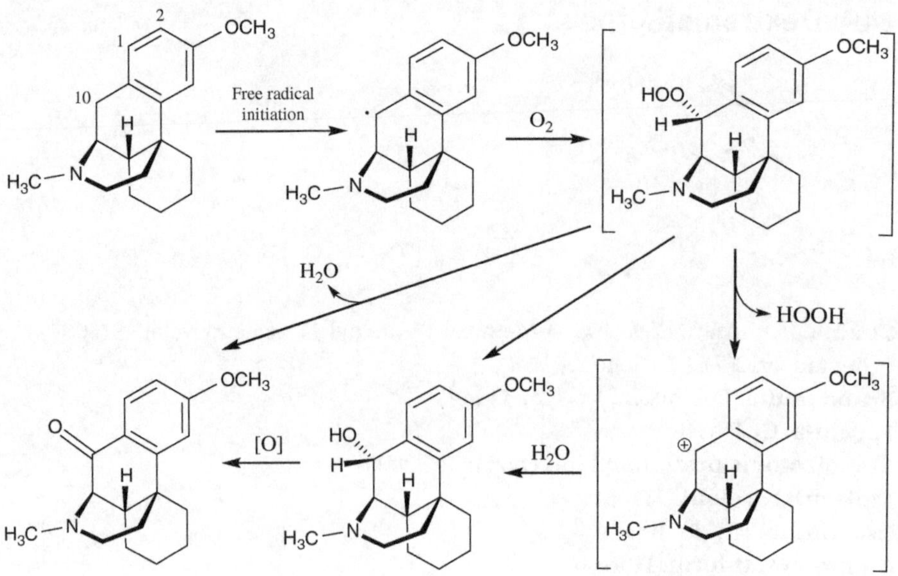

Scheme 149-1 Plausible mechanism for the oxidation on the 10-position of dextromethorphan.

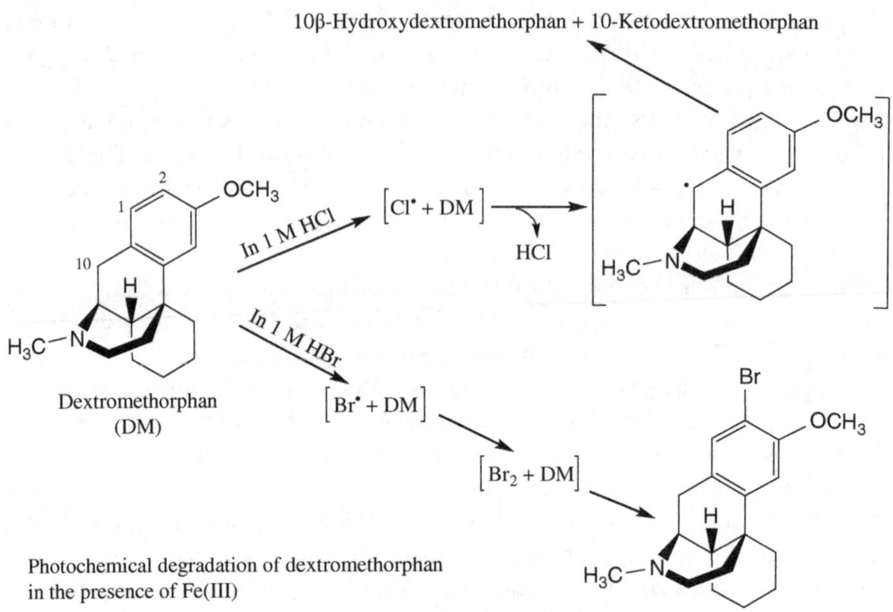

Scheme 149-2 Photochemical degradation of dextromethorphan in solution in the presence of Fe(III).

performed in a HBr solution: while the former produced the 10β-hydroxyl and 10-keto degradants, the latter yielded only the 2-bromo degradant (Scheme 149-2) [3]. The difference was explained based on the reactivity of the chlorine and bromine radicals ($Cl^{\cdot}$ and $Br^{\cdot}$). The $Cl^{\cdot}$ would be capable of initiating the free radical process

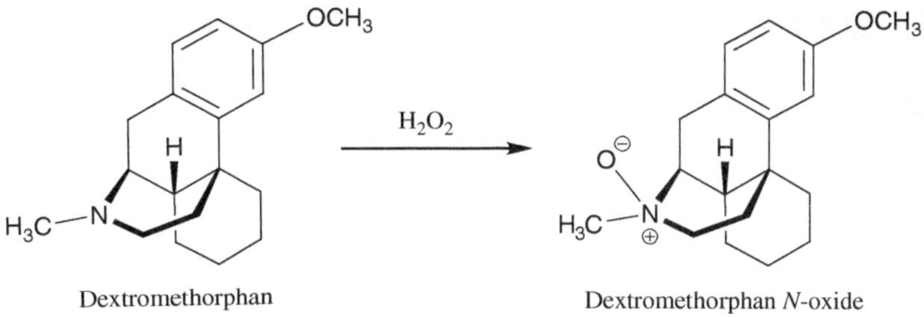

Scheme 149-3 Nucleophilic oxidation of dextromethorphan.

by abstracting the 10β-H. On the contrary, Br˙ would not be capable of the H abstraction due to its lower energy as compared to that of Cl˙. Thus, Br˙ could only react with each other to form Br_2, and the latter would then brominate dextromethorphan at the 2-position via electrophilic substitution.

Under the oxidative stress condition using 3.0% hydrogen peroxide at 60°C for 3 hours, dextromethorphan *N*-oxide seemed to be the only significant degradant with 8% yield (Scheme 149-3) [4]. Another similar oxidative stress study of a formulated combination product containing guaifenesin and dextromethorphan also indicated that dextromethorphan underwent significant degradation [5].

References

1 Haefiger, Q., Brossi, A., Chopard-dit-Jean, L.H. et al. (1956). Mitteilung. Über ein photooxydationsprodukt von (+)-3-Methoxy-N-methyl-morphinan. *Helv. Chim. Acta* 39: 2053–2062.

2 Proska, B., Voticky, Z., Molanar, L. et al. (1978). 10-Oxomorphine, a decomposition product of morphine. *Chem. Zvestii* 32: 710–714.

3 Boccardi, G., Mezzanzanica, P., Guzzi, U. et al. (1989). Photochemical Iron(III)-meidated autoxidation of dextromethorphan. *Chem. Pharm. Bull.* 37: 308–310.

4 Puppala, U., Srinivas, K.S.V., Reddy, K.V., and Jogi, B.R. (2018). Development and validation of a stability indicating and LCMS compatible UPLC method for estimation of assay, related substances and degradants in dextromethorphan hydrobromide active pharmaceutical ingredient. *Euro. J. Biomed. Pharm. Sci.* 5: 757–765.

5 Raju, T.V.R., Kumar, N.A., Kumar, S.R. et al. Development and validation of a stability-indicating RP-HPLC method for the simultaneous estimation of guaifenesin and dextromethorphan impurities in pharmaceutical formulations. *Chrom. Res. Int.* https://doi.org/10.1155/2013/315145.

150–152: Diclofenac, Aceclofenac, Meclofenamic Acid

150: Diclofenac

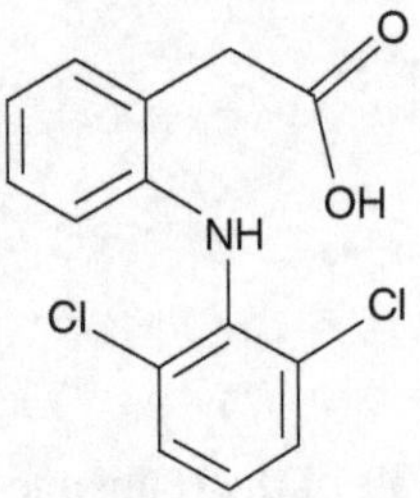

Chemical name: 2-[2-(2,6-Dichloroanilino)phenyl]acetic acid
Brand name: Arthrotec, Cataflam, Voltaren, Zipsor
Formula: $C_{14}H_{11}Cl_2NO_2$
Monoisotopic protonated ion (*m/z*): 296.0240
Molecular weight: 296.15
Common salt form: Sodium salt
CAS number: 15307-86-5

151: Aceclofenac

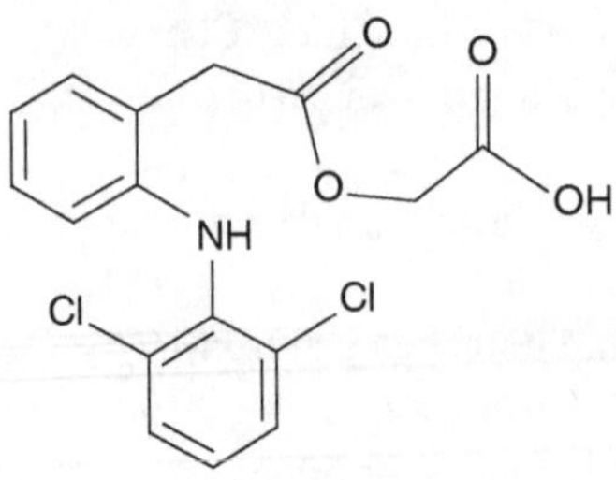

Chemical name: 2-[2-[2-(2,6-Dichloroanilino)phenyl]acetyl]oxyacetic acid
Brand name: Cincofen, Clanza, Hifenac
Formula: $C_{16}H_{13}Cl_2NO_4$
Monoisotopic protonated ion (*m/z*): 354.0294
Molecular weight: 354.18
CAS number: 89796-99-6

152: Meclofenamic Acid

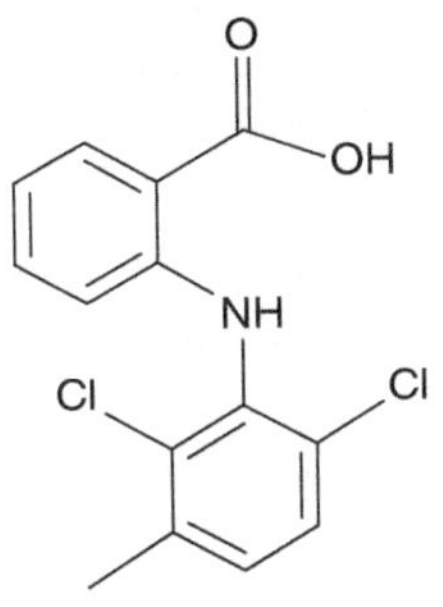

Chemical name: 2-(2,6-Dichloro-3-methylanilino)benzoic acid
Brand name: Meclomen
Formula: $C_{14}H_{11}Cl_2NO_2$
Monoisotopic protonated ion (*m/z*): 296.0240
Molecular weight: 296.15
Common salt form: Sodium salt
CAS number: 644-62-2

Diclofenac is a nonsteroidal anti-inflammatory drug, which is a substituted *N*-(2,6-dichlorophenyl)aniline containing a methylenecarboxyl group ortho to the amine group. The carboxyl group could be cyclized with the amino group due to the particular spatial distance between the two groups. Indeed, during an accelerated stability study of diclofenac in liquid formulation, diclofenac underwent the cyclization between the carboxyl and amine groups, yielding the five-membered lactam degradant (Scheme 150–152-1) [1].

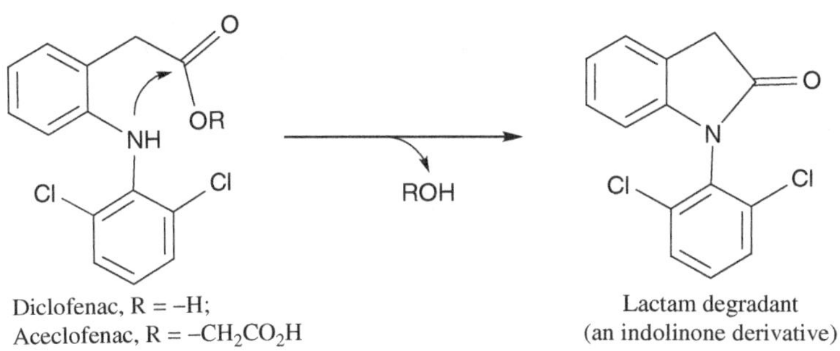

Scheme 150–152-1 Degradation of diclofenac and aceclofenac due to intramolecular cyclization.

Although diclofenac was photochemically stable in the solid state, its two substituted phenyl rings were found to undergo photochemical cyclization to produce the tricyclic chlorocarbazole ring, accompanied by the elimination of one chlorine, upon UV irradiation of its solution [2]. It was believed that the formation of the carbazole photodegradant is not via a free radical pathway (Pathway a, Scheme 150–152-2) but rather a 6π-electrocyclization pathway (Pathway b, Scheme 150–152-2). A study performed by Encinas and Miranda suggested that the tricyclic chlorocarbazole degradant may be responsible for the phototoxicity of diclofenac [3]. Further photolysis of the chlorocarbazole degradant affected decarboxylation, producing the decarboxylated degradant (Scheme 150–152-2).

Aceclofenac is a prodrug of diclofenac in which the original carboxyl group is esterified by hydroxyacetic acid to form the glycolic acid ester of diclofenac. Based on its structural feature and similarity with diclofenac, aceclofenac would also be expected to undergo the same cyclization to form the same lactam degradant (Scheme 150–152-1) and the same photochemical degradation to yield the analogous tricyclic chlorocarbazole degradant (with R_1 being $-CH_2CO_2CH_2CO_2H$, Scheme 150–152-2). On the other hand, due to the blockage of the methylenecarboxyl group attached to the phenyl ring by the formation of the ester, aceclofenac would not be expected to undergo decarboxylation, unless the ester is hydrolyzed. In a compatibility study with a number of excipients, aceclofenac was found to be incompatible with magnesium stearate due to the formation of aceclofenac magnesium salt [4].

Meclofenamic acid is a desmethylene analog of diclofenac. Due to the absence of the methylene group between the carboxyl and the phenyl groups, meclofenamic acid would not be expected to undergo facile cyclization to form the analogous degradant with a four-membered ring, owing to the high tension of the four-membered ring. Nevertheless, its *N*-(2,6-dichlorophenyl)aniline core is still very similar to those of diclofenac and aceclofenac; as such, the substituted aniline core can still photocyclize to yield two analogous tricyclic chlorocarbazole degradants due to its asymmetric 2,6-dichloro-4-methylphenyl ring (Scheme 150–152-2). It also needs to point out that the chlorocarbazole degradants would not be expected to undergo the subsequent photochemical decarboxylation, because the carboxyl group in the benzoic acid moiety of meclofenamic acid is no longer "activated" for decarboxylation [5]. The chlorocarbazole degradants, nevertheless, could undergo dechlorination to give two carbazole degradants (Scheme 150–152-2).

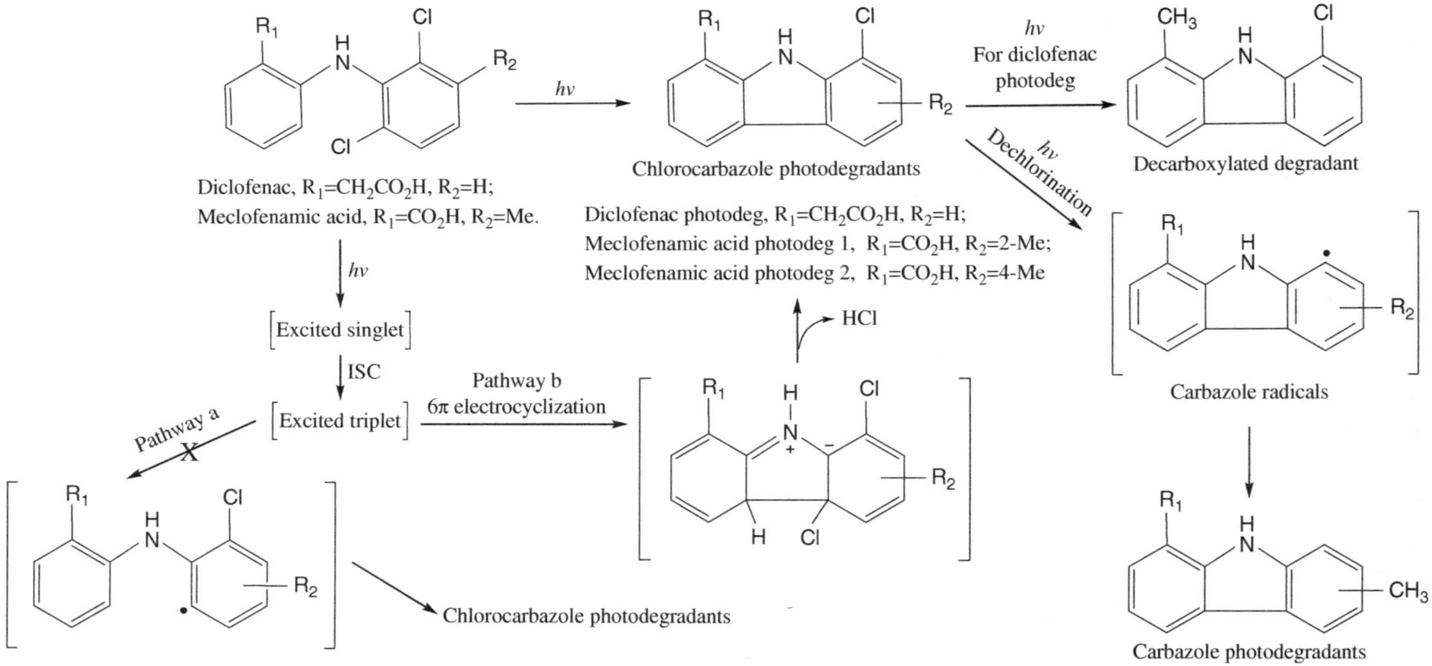

Scheme 150–152-2 Photochemical degradation of diclofenac and structurally similar meclofenamic acid. Aceclofenac is expected to undergo the same photochemical pathway.

References

1 Galmier, M.-J., Bouchona, B., Madelmont, J.-C. et al. (2005). Identification of degradation products of diclofenac by electrospray ion trap mass spectrometry. *J. Pharm. Biomed. Anal.* 38: 790–796.

2 Encinas, S., Bosca, F., and Miranda, M.A. (1998). Photochemistry of 2,6-dichlorodiphenylamine and 1-chlorocarbazole, the photoactive chromophores of diclofenac, meclofenamic acid and their major photoproducts. *Photochem. Photobiol.* 68: 640–645.

3 Encinas, S., Bosca, F., and Miranda, M.A. (1998). Phototoxicity associated with diclofenac: a photophysical, photochemical, and photobiological study on the drug and its photoproducts. *Chem. Res. Toxicol.* 11: 946–952.

4 Aigner, Z., Heinrich, R., Sipos, E. et al. (2011). Compatibility studies of aceclofenac with retard tablet excipients by means of thermal and FT-IR spectroscopic methods. *J. Therm. Anal. Calorim.* 104: 265–271.

5 Li, M. (2012). Chapter 4, section 4.2 decarboxylation. In: *Organic Chemistry of Drug Degradation*, 118–121. Cambridge: RSC Publishing.

153: Dicyclomine (Dicycloverine)

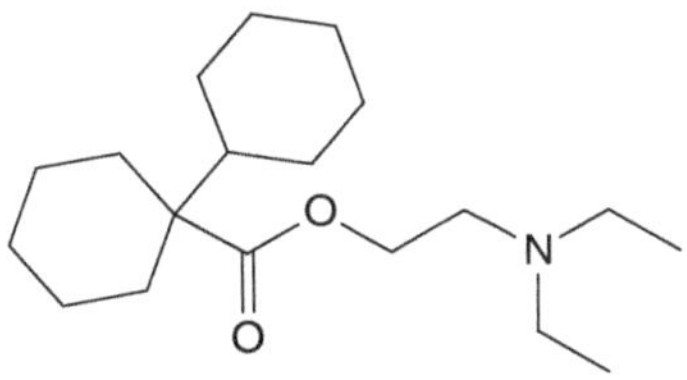

Chemical name: 2-(Diethylamino)ethyl 1-cyclohexylcyclohexane-1-carboxylate
Brand name: Bentyl
Formula: $C_{19}H_{35}NO_2$
Monoisotopic protonated ion (*m/z*): 310.2741
Molecular weight: 309.49
Common salt form: HCl salt
CAS number: 77-19-0

Dicyclomine is an anticholinergic agent that is clinically used as antispasmodic for patients with irritable bowel syndrome (IBS). Despite its long and widespread clinical use since 1950, it appears that no serious study has been conducted on the stability of this drug product. Based on its structure, dicyclomine has two functional groups that may be susceptible to degradation: the ester and tertiary amine groups (Scheme 153). In one study aimed at developing a stability-indicating capillary gas-liquid chromatography–FID method for a few formulated products of dicyclomine hydrochloride, the authors demonstrated that the drug substance underwent degradation when stressed under both acidic (9 M H_2SO_4) and alkaline (1 M NaOH) conditions at 80°C for up to 45 minutes [1]. The authors assumed that the degradation was hydrolysis, and it appeared to be confined in scope. This finding is consistent with the pharmacokinetic result that the in vivo half-life of dicyclomine was reported to be 5 hours. The resistance to hydrolytic degradation, either in vitro or in vivo, may be rationalized by the fact that the ester functionality of dicyclomine is very sterically hindered due to the presence of the two nearby cyclohexane rings.

As a tertiary alkyl amine, it would be quite likely that dicyclomine could be susceptible to nucleophilic oxidation to form the corresponding *N*-oxide [2], particularly in a somewhat neutral to alkaline formulation.

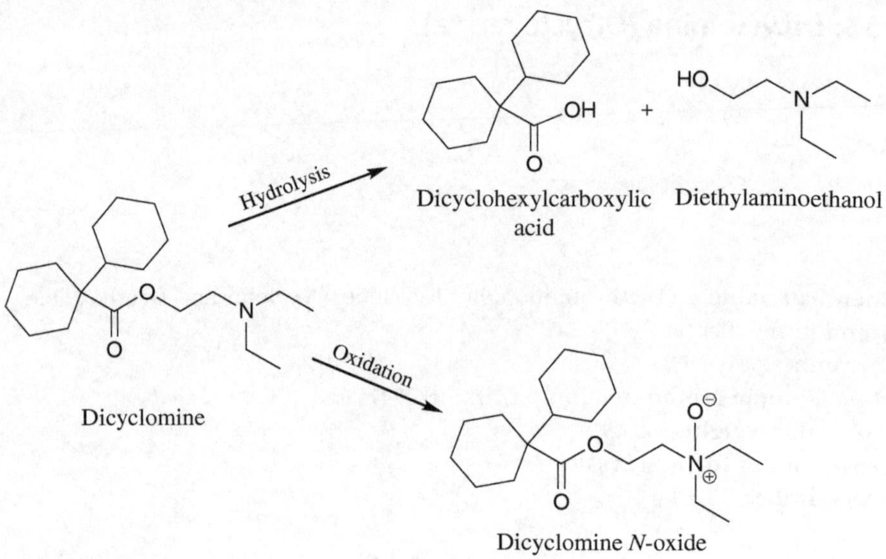

Scheme 153 Plausible degradation pathways of dicyclomine.

Reference

1 Tan, H.S.I. and Yan, Y.-P. (1989). Stability-indicating capillary gas-liquid chromatographic assay of dicyclomine hydrochloride in some pharmaceutical formulations. *J. Chromatogr.* 475: 381–389.

154: Didanosine [WHO List of Essential Medicines]

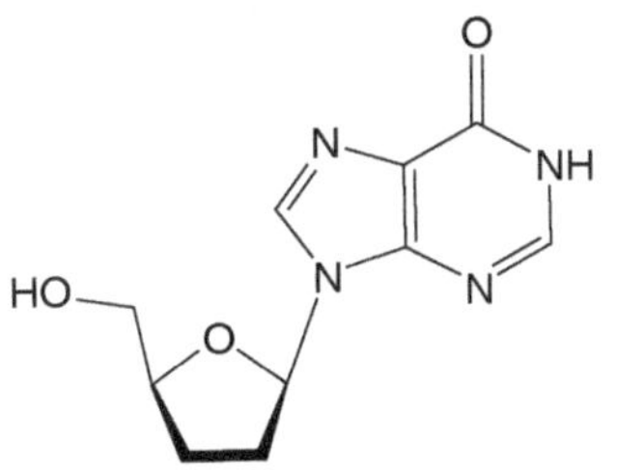

Chemical name: 9-[(2*R*,5*S*)-5-(Hydroxymethyl)oxolan-2-yl]-1*H*-purin-6-one
Brand name: Videx
Formula: $C_{10}H_{12}N_4O_3$
Monoisotopic protonated ion (*m/z*): 237.0982
Molecular weight: 236.23
CAS number: 69655-05-6

Didanosine, also known as 2′,3′-dideoxyinosine (ddI or DDI), is an HIV reverse transcriptase inhibitor. It is most commonly used in combination with HIV protease inhibitors for the treatment of AIDS. It is a dideoxypurinenucleoside analog, and its main degradation pathway is hydrolytic. Didanosine is quite unstable under acidic conditions but stable under alkaline conditions. For example, it decomposed by 30% in 6 hours in an acidic mobile phase consisting of a pH 4.0 buffer solution and methanol, giving hypoxanthine as the degradant [1]. In an earlier study, the half-life of didanosine was found to be only a few minutes in a pH 2 solution [2]. Berkers et al. studied the stability of didanosine in aqueous solutions over a pH range of 0–12 with a stability-indicating HPLC method [3]. At pH 3.2 and 3.6, the activation energies for the hydrolytic degradation were found to be 82.0 and 87.3 kJ/mol, while the frequency factors were 2.9×10^{10} and 8.8×10^{10}/s, respectively. The study also examined the stability of didanosine in the presence of several cyclodextrins. In the presence of 6×10^{-2} M hydroxypropyl-β-cyclodextrin (HP-β-CyD), the stability of didanosine was increased 5.1 times, as evidenced by the degradation rate constant of didanosine-HP-β-CyD complex (3.34×10^{-4}) versus that of uncomplexed didanosine (1.73×10^{-3}).

With regard to the mechanism of the acidic hydrolysis of didanosine, it appears that an A-1 mechanism would be applicable (Scheme 154), based on the study on deoxyuridine and related nucleosides by Shapiro and Kang [4].

The main acidic degradants are hypoxanthine and presumably 2′,3′-deoxyribose although it appears that the latter degradant has not been mentioned or verified in the stability studies of didanosine reported. Structurewise, didanosine is 2′,3′-deoxyinosine; nevertheless, inosine is much more stable than didanosine, i.e., 2′,3′-deoxyinosine. The activation energy (*Ea*) for acidic hydrolysis of inosine was determined to be 27.9 kcal/mol (or 117 kJ/mol) [5], which is significantly higher than that for didanosine (*Ea*: 82.0–87.3 kJ/mol) [3]. The enormous difference in the susceptibility toward acidic hydrolysis between didanosine and inosine is apparently because the 2′- and 3′-hydroxyl groups are missing in didanosine.

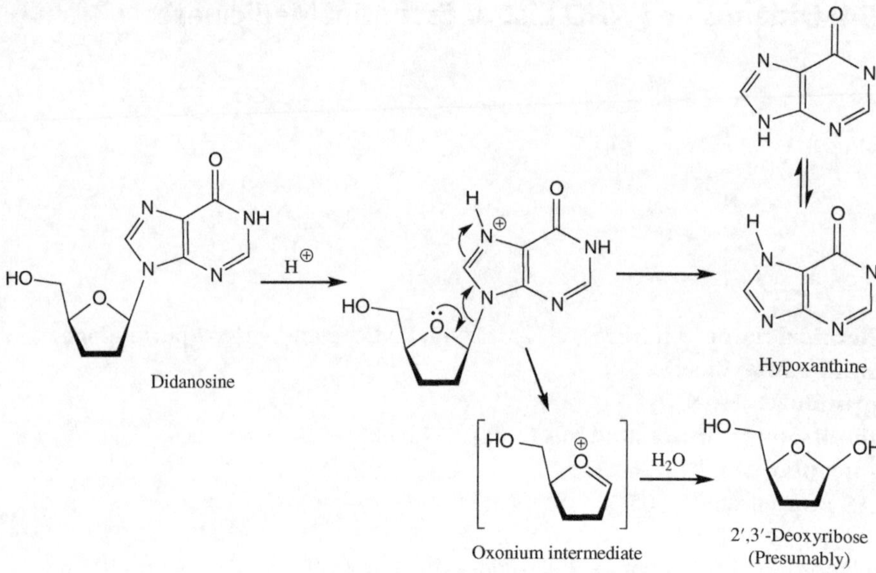

Scheme 154 Acidic hydrolysis of didanosine via an A-1 mechanism.

The presence of the hydroxyl groups, in particular the 2′-hydroxyl, can retard the solvolysis in neutral and acidic solution, which could explain the instability of didanosine according to Shapiro and Kang [4]. A similar example can be seen in the case of deoxyuridine and uridine, where deoxyuridine is much more hydrolyzable in acid than uridine [6]. On the other hand, in a search for acid-stable analogs of dideoxyadenosine (ddA), which is structurally similar to didanosine (ddI), introduction of a fluorine at 2′-position (2′-F-ddA) completely suppressed the acidic lability of ddA [7]. While ddA was almost immediately hydrolyzed in pH 1.0 solution at 37°C with a half-life of only 35 seconds, both epimers of 2′-F-ddA remained stable for 24 hours under the same conditions. The much-improved stability was attributed to the destabilization of the oxonium intermediate (refer to Scheme 154) by the fluorine substituent. Apparently, 2′-hydroxyl substituent can play the same role, thus explaining the instability of didanosine as compared to inosine in acidic solutions.

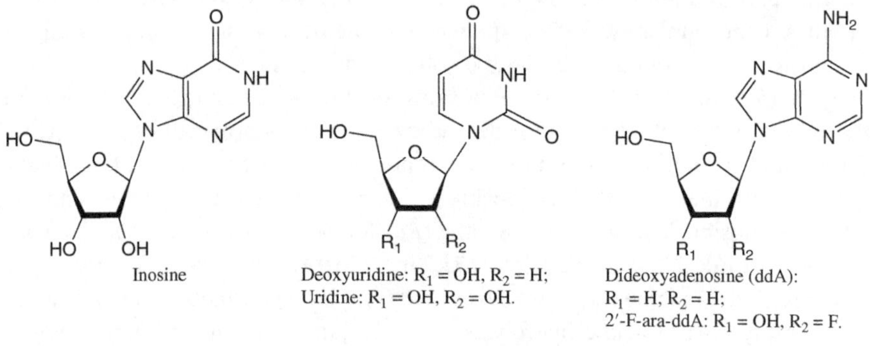

Structures of nucleosides also discussed in this monograph.

References

1 Rebiere, H., Mazel, B., Civade, C., and Bonnet, P.-A. (2007). Determination of 19 antiretroviral agents in pharmaceuticals or suspected products with two methods using high-performance liquid chromatography. *J. Chromatogr. B* 850: 376–383.

2 Marquez, V.E., Tseng, C.K.H., Kelley, J.A. et al. (1987). 2′,3′-Dideoxy-2′-fluoro-ara-A. An acid-stable purine nucleoside active against human immunodeficiency virus (HIV). *Biochem. Pharmacol.* 36: 2719–2722.

3 Berkers, O., Beijnen, J.H., Tank, M.J.T.K. et al. (1993). 2′,3′-Dideoxyinosine (dd1): its chemical stability and cyclodextrin complexation in aqueous media. *J. Pharm. Biomed. Anal.* 11: 489–493.

4 Shapiro, R. and Kang, S. (1969). Uncatalyzed hydrolysis of deoxyuridine, thymidine, and 5-bromodeoxyuridine. *Biochemist* 8: 1806–1810.

5 Suzuki, Y. (1974). The stability of inosine in acid and in alkali. *Bull. Chem. Soc. Jap.* 47: 2469–2472.

6 Pfitzner, K.E. and Moffatt, J.G. (1964). The synthesis and hydrolysis of 2′,3′-Dideoxyuridine. *J. Organomet. Chem.* 29: 1508–1511.

7 Marquez, V.E., Tseng, C.K.-H., Kelley, J.A. et al. (1987). 2′,3′-Dideoxy-2′-fluoro-ara-A. An acid-stable purine nucleoside active against human immunodeficiency virus (HIV). *Biochem. Pharmacol.* 36: 2719–2722.

155–157: Digoxin, Digitoxin, Deslanoside

155: Digoxin [WHO List of Essential Medicines]

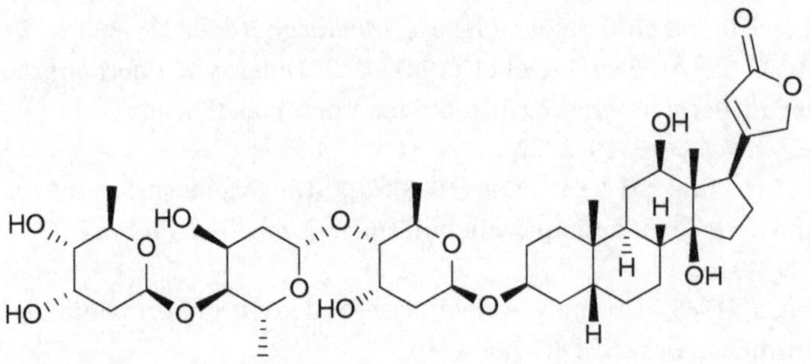

Chemical name: 3-[(3*S*,5*R*,8*R*,9*S*,10*S*,12*R*,13*S*,14*S*,17*R*)-3-[(2*R*,4*S*,5*S*,6*R*)-5-[(2*S*,4*S*,5*S*,6*R*)-5-[(2*S*,4*S*,5*S*,6*R*)-4,5-Dihydroxy-6-methyloxan-2-yl]oxy-4-hydroxy-6-methyloxan-2-yl]oxy-4-hydroxy-6-methyloxan-2-yl]oxy-12,14-dihydroxy-10,13-dimethyl-1,2,3,4,5,6,7,8,9,11,12,15,16,17-tetradecahydrocyclopenta[a]phenanthren-17-yl]-2*H*-furan-5-one
Brand name: Digitek, Digox, Lanoxin, Toloxin
Formula: $C_{41}H_{64}O_{14}$
Monoisotopic protonated ion (*m/z*): 781.4369
Molecular weight: 780.94
CAS number: 20830-75-5

156: Digitoxin

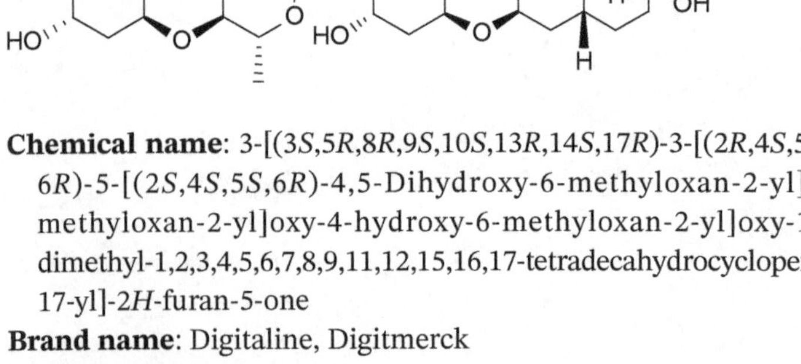

Chemical name: 3-[(3*S*,5*R*,8*R*,9*S*,10*S*,13*R*,14*S*,17*R*)-3-[(2*R*,4*S*,5*S*,6*R*)-5-[(2*S*,4*S*,5*S*,6*R*)-5-[(2*S*,4*S*,5*S*,6*R*)-4,5-Dihydroxy-6-methyloxan-2-yl]oxy-4-hydroxy-6-methyloxan-2-yl]oxy-4-hydroxy-6-methyloxan-2-yl]oxy-14-hydroxy-10,13-dimethyl-1,2,3,4,5,6,7,8,9,11,12,15,16,17-tetradecahydrocyclopenta[a]phenanthren-17-yl]-2*H*-furan-5-one
Brand name: Digitaline, Digitmerck
Formula: $C_{41}H_{64}O_{13}$
Monoisotopic protonated ion (*m/z*): 765.4420
Molecular weight: 764.94
CAS number: 71-63-6

157: Deslanoside

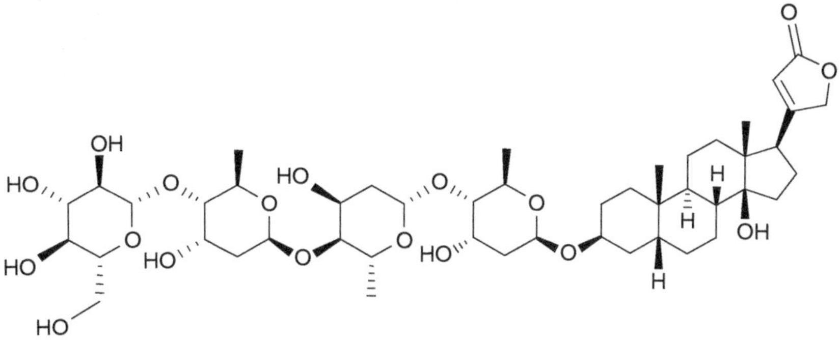

Chemical name: 3-[(3*S*,5*R*,8*R*,9*S*,10*S*,12*R*,13*S*,14*S*,17*R*)-12,14-Dihydroxy-3-[(2*R*,4*S*,5*S*,6*R*)-4-hydroxy-5-[(2*S*,4*S*,5*S*,6*R*)-4-hydroxy-5-[(2*S*,4*S*,5*S*,6*R*)-4-hydroxy-6-methyl-5-[(2*S*,3*R*,4*S*,5*S*,6*R*)-3,4,5-trihydroxy-6-(hydroxymethyl)oxan-2-yl]oxyoxan-2-yl]oxy-6-methyloxan-2-yl]oxy-6-methyloxan-2-yl]oxy-10,13-dimethyl-1,2,3,4,5,6,7,8,9,11,12,15,16,17-tetradecahydrocyclopenta[a]phenanthren-17-yl]-2*H*-furan-5-one
Brand name: Cedilanide
Formula: $C_{47}H_{74}O_{19}$
Monoisotopic protonated ion (*m/z*): 943.4897
Molecular weight: 943.08
CAS number: 17598-65-1

Digoxin is one of the oldest medicines used for the treatment of congestive heart failure and certain types of arrhythmia such as atrial fibrillation. It was isolated from the medical herb of *Digitalis lanata*, and the medical use of the latter dates back to the 18th century. The molecule of digoxin consists of two parts: the steroid and trisaccharide parts; without the saccharide part, the steroid itself (i.e., digoxigenin) is only about 1/10 as active [1].

The trisaccharide part is made of three 2-dexoyhexose units (referred to as digitoxosides), and thus the glycosidic linkages between the digitoxosides and between the tridigitoxoside and steroid would be susceptible to hydrolytic degradation under acidic conditions [2]. Therefore, the main degradation pathway of digoxin is hydrolytic. The three glycosidic bonds can be hydrolyzed, respectively, to give digoxigenin-bisdigitoxoside and digoxigenin-monodigitoxoside and digoxigenin (Scheme 155–157) [3, 4]. Furthermore, the intermediary digoxigenin-bisdigitoxoside and digoxigenin-monodigitoxoside can be further hydrolyzed to produce digoxigenin-monodigitoxoside and digoxigenin. Sternson et al. determined the rates of hydrolysis at pH 1.0 and 37°C, and the results showed that the rates of hydrolysis were generally in the same magnitude, with the rate of digoxin to digoxigenin-bisdigitoxoside being the fastest (0.0263/min) and the rate of digoxin to digoxigenin the slowest (0.0089/min) [5]. This trend for the rates of hydrolysis can be readily understood from the perspective of the stereochemistry of digoxin as the steric hindrance of the steroid moiety should be the largest.

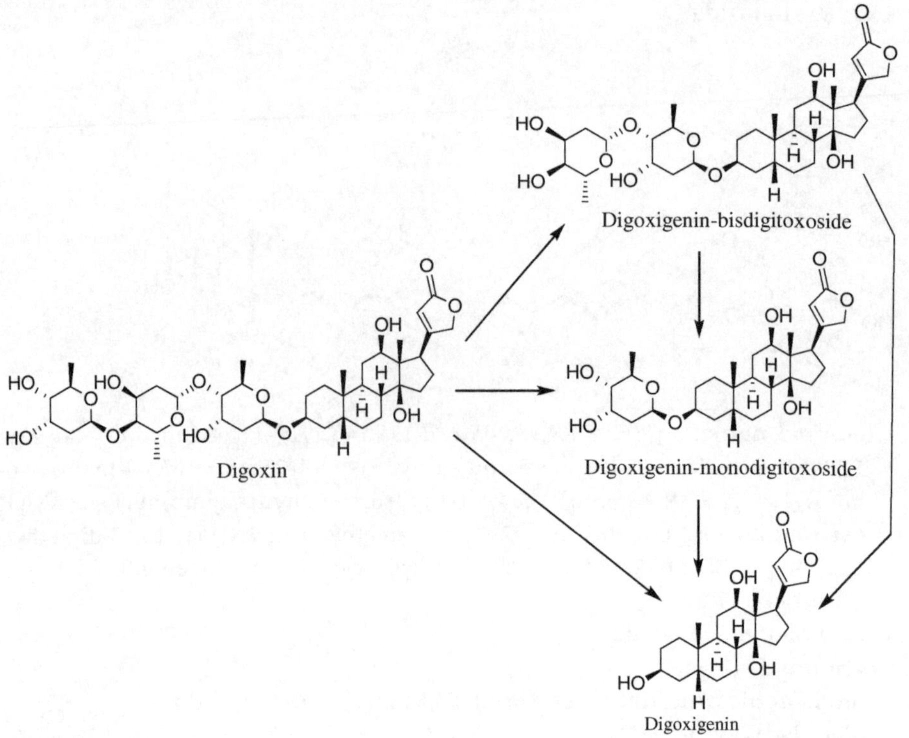

Scheme 155–157 Hydrolytic degradation of digoxin.

Digitoxin is a close analog of digoxin, and structurewise, it is 12-deoxydigoxin. The loss of the 12-hydroxyl group from the steroid core causes a dramatic difference in the ADME behavior of digitoxin, rendering its significant elimination via feces as opposed to digoxin's elimination via urine. Nevertheless, the chemical degradation pathway of digitoxin should be the same as that of digoxin [6]. Thus, the pathway illustrated in Scheme 155–157 would be readily applicable to digotoxin.

Deslanoside has an additional glucose unit than digoxin and digitoxin; it can be considered D-glucopyranosyldigoxin, i.e., the glucose unit is attached to the terminal digitoxose of digoxin. As the additional glucose unit has a hydroxyl group at its 2-position, the glycosidic bond between the first sugar unit (D-glucopyranosyl) and the second sugar unit is expected to be stable under acidic conditions. Hence, the acidic degradants of deslanoside would be expected to be the same as those of digoxin.

References

1 Kroneberg, G. (1959). Zuckergehalt und Wirksamkeit der Digitoxigenin- und Digoxigenindigitoxose. *Naunyn-Schmiedeberg's Areh. exp. Path. Pharmak* 287: 222–228.

2 Namchuk, M.N., McCarter, J.D., Becalski, A. et al. (2000). The role of sugar substituents in glycoside hydrolysis. *J. Am. Chem. Soc.* 122: 1270–1277.

3 Kuhlmann, J., Abshagen, U., and Rietbrock, N. (1973). Cleavage of glycosidic bonds of digoxin and derivatives as function of pH and time. *Naunyn Schmiedeberg's Arch. Pharmacol.* 276: 149–156.

4 Jedlicka, A., Grafnetterova, T., and Miller, V. (2003). HPLC method with UV detection for evaluation of digoxin tablet dissolution in acidic medium after solid-phase extraction. *J. Pharm. Biomed. Anal.* 33: 109–115.

5 Sternson, L.A. and Shaffer, R.D. (1978). Kinetics of gigoxin stability in aqueous solution. *J. Pharm. Sci.* 67: 327–330.

6 Desta, B., Kwong, E., and McErlane, K.M. (1982). Separation of digoxin, digitoxin and their potential metabolites, impurities or degradation products by high performance liquid chromatography. *J. Chromatogr.* 240: 137–143.

158: Diltiazem

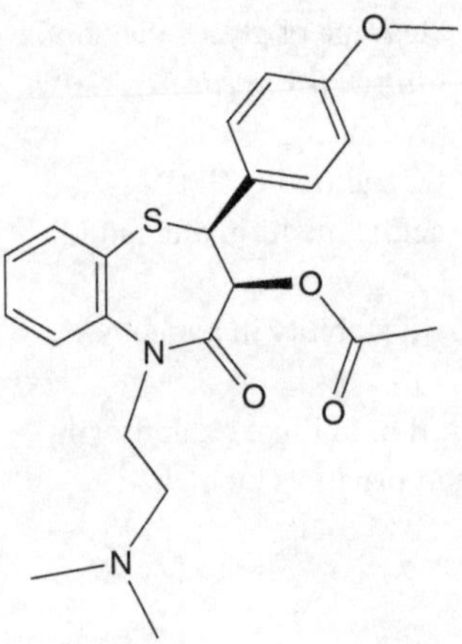

Chemical name: [(2*S*,3*S*)-5-[2-(Dimethylamino)ethyl]-2-(4-methoxyphenyl)-4-oxo-2,3-dihydro-1,5- benzothiazepin-3-yl] acetate
Trade name: Cardizem, Dilacorxr
Formula: $C_{22}H_{26}N_2O_4S$
Monoisotopic protonated ion (*m/z*): 415.1686
Molecular weight: 414.52
CAS number: 42399-41-7
Common salt form: HCl salt

Diltiazem is a calcium channel blocker that is clinically used to treat hypertension and certain type of arrhythmia such as atrial fibrillation. It is a benzothiazepine derivative that contains the following functional groups or moieties that may be susceptible to various types of degradation: an acetyl ester, a tertiary amine, and a thioether. The acetyl ester functionality is particularly susceptible to hydrolytic degradation, under both acidic and alkaline conditions, leading to the formation of the deacetylated degradant, which is both a process impurity and metabolite (Scheme 158) [1, 2]. Won and Iula performed a kinetic study for the hydrolysis of diltiazem, in which the Log *k*-pH curves were determined at different temperatures; the results indicated that diltiazem is most stable at pH near 4 [3], which is consistent with a prior study [4]. Furthermore, at both ends of the pH range (1–8), the rates of the hydrolysis were similar [3]. The tertiary amine moiety of the drug molecule could be likely oxidized to produce *N*-oxide degradant, which is a known metabolite [5], although its existence in drug products has not been reported. On the other hand, desacetyldiltiazem *N*-oxide is also a metabolite [2].

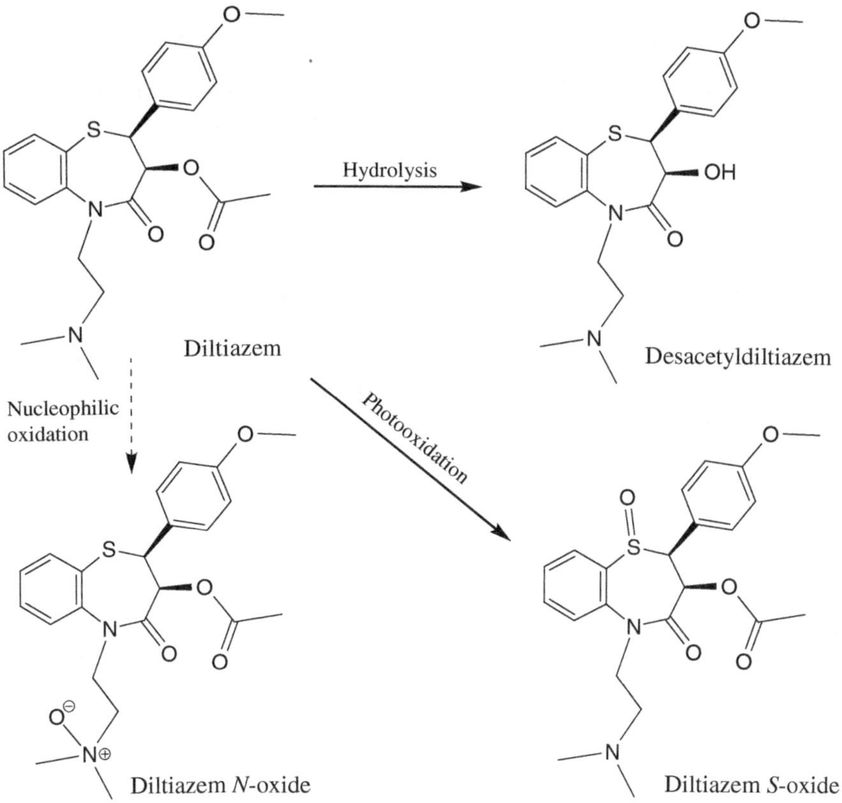

Scheme 158 Degradation pathways of diltiazem. The dashed arrow indicates a likely scenario in which the *N*-oxide would be expected to form under nucleophilic oxidative conditions.

Reports of skin disorders after taking this drug prompted some researchers to study its phototoxic mechanism. According to a study by Andrisano et al., diltiazem *S*-oxide was the main photodegradant in acidic and near-neutral solutions that were exposed to UVB–UVA radiations (xenon arc lamp) [6]. In a solution of pH 9.0, photodegradation produced a few more degradants, among which the only identified degradant was the deacetylated degradant.

Muszalska and Siakowski studied the degradation of diltiazem in the solid state, and they found that under an atmosphere of moisture of 76.4%, diltiazem decomposed only to deacetylated diltiazem with the following thermodynamic parameters: *Ea* 134.5 kJ/mol, $\Delta H^{\neq}$ 132.0 kJ/mol, and $\Delta S^{\neq}$ 9.31 J/K mol [7]. The activation energy, 134.5 kJ/mol, determined under the solid state was significantly higher than the one (71.0 kJ/mol) in solution measured by Davis et al. [8].

References

1 Mahajan, N., Deshmukh, S., and Farooqui, M. (2021). A novel stability-indicating method for known and unknown impurities profiling for diltiazem hydrochloride pharmaceutical dosage form (tablets). *Future J. Pharm. Sci.* https://doi.org/10.1186/s43094-021-00352-x.

2 Miyazaki, M., Iwakuma, T., and Tanaka, T. (1978). Synthesis of the metabolites and related compounds of diltiazem. *Chem. Pharm. Bull.* 26: 2889–2893.

3 Won, C.M. and Iula, A.K. (1992). Kinetics of hydrolysis of diltiazem. *Int. J. Pharm.* 79: 183–190.

4 Davis, S.S., Illum, L., Triccas, I.M., and Winchcomb, K.N. (1986). The simultaneous degradation and sorption of diltiazem in aqueous solution. *Int. J. Pharm.* 30: 29–33.

5 Li, R., Farmer, P.S., Xie, M. et al. (1992). Synthesis, characterization, and Ca^{2+} antagonistic activity of diltiazem metabolites. *J. Med. Chem.* 35: 3246–3253.

6 Andrisano, V., Hrelia, P., Gotti, R. et al. (2001). Photostability and phototoxicity studies on diltiazem. *J. Pharm. Biomed. Anal.* 25: 589–597.

7 Muszalska, I. and Siakowski, M. (2003). Kinetics of diltiazem hydrochloride in solid phase. *Act Polon. Pharm.–Drug Res.* 60: 27–30.

8 Davis, S.S., Illum, L., Triccas, I.M., and Winchcomb, K.N. (1986). The simultaneous degradation and sorption of diltiazem in aqueous solution. *Int. J. Pharm.* 30: 29–33.

159: Diphenhydramine

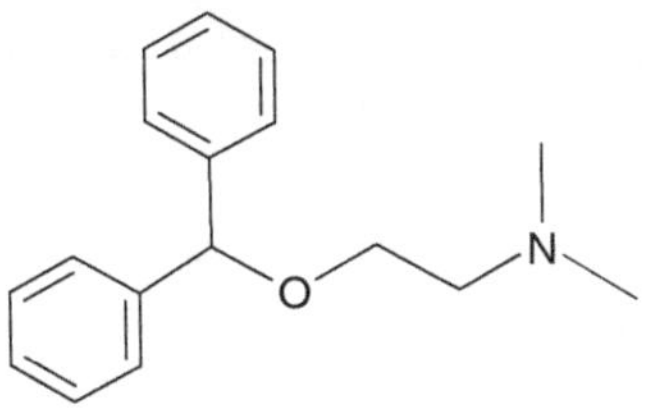

Chemical name: 2-Benzhydryloxy-*N,N*-dimethylethanamine
Trade name: Benadryl
Formula: $C_{17}H_{21}NO$
Monoisotopic protonated ion (*m/z*): 256.1696
Molecular weight: 255.36
CAS number: 58-73-1
Common salt form: HCl salt, Citrate

Diphenhydramine is a first-generation antihistamine medication that was discovered and commercialized in the 1940s. In addition to its indication for treatment of allergic symptoms, diphenhydramine also has sedative properties, due to which it is widely used as an over-the-counter sleeping aid, as well as a local anesthetic. There had been some studies regarding the stability of diphenhydramine between the 1950s and the 1960s, mostly by Japanese research groups, which were summarized in a monograph by Holcomb and Fusari [1]. According to the monograph, diphenhydramine was susceptible to decomposition in an acidic medium due to the hydrolytic cleavage of the ether linkage, producing benzhydrol and 2-(dimethylamino) ethanol (Scheme 159); a separate UV irradiation study of diphenhydramine also

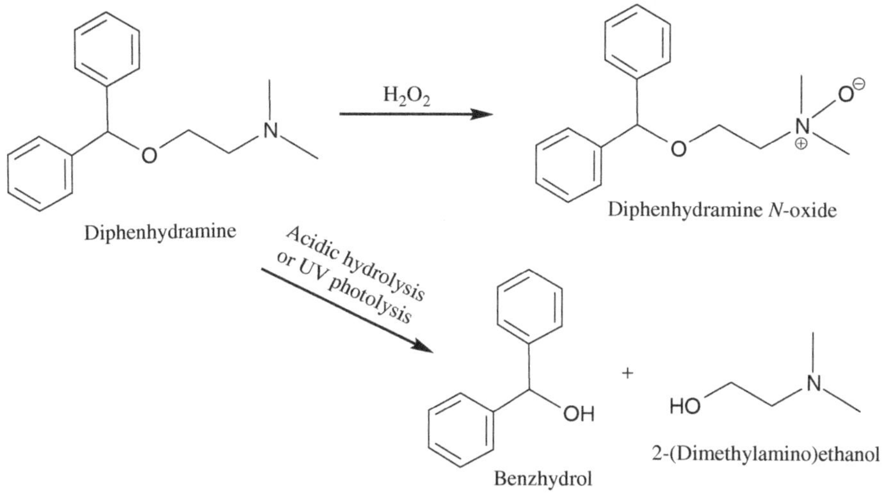

Scheme 159 Degradation pathways of diphenhydramine.

yielded the two main degradants. In a more recent study, combination tablets of ibuprofen and diphenhydramine citrate were subjected to acidic degradation in 2 N HCl solution at 25°C for 2 hours, and no degradation product was observed, which may be due to the milder stress condition used [2]. On the other hand, diphenhydramine was fairly stable in an alkaline solution [1, 2]. In an oxidative stress study of diphenhydramine hydrochloride, it appeared that it was substantially degraded, and the degradation products included toluene, benzophenone, benzyl alcohol, benzoic acid, and several phenolic substances, in addition to 2-(dimethylamino) ethanol [1]. The results of this early oxidative stress study were quite different from those conducted in much later times, where only diphenhydramine *N*-oxide was observed as the primary degradant [2, 3]. The discrepancy is most likely due to the much harsher conditions used in the early study, as the initial degradants underwent substantial subsequent degradation. It is somewhat surprising to note that there have been very few reports for a systematic forced degradation study on diphenhydramine with emphasis on structural elucidation of degradation products [4], since those conducted in early times [1], despite its long and widespread medical use.

References

1 Holcomb, I.J. and Fusari, S.A. (1974). Diphenhydramine hydrochloride. *Ana. Prof. Drug subs.* 3: 173–232.

2 Rao, D.D., Sait, S.S., and Mukkanti, K. (2011). Development and validation of an UPLC method for rapid determination of ibuprofen and diphenhydramine citrate in the presence of impurities in combined dosage form. *J. Chromatogr. Sci.* 49: 281–286.

3 Ali, N.W., Abdelkawy, M., and Abdelhamid, N.S. (2013). Simultaneous determination of paracetamol and diphenhydramine hydrochloride mixture in the presence of their degradation products. *IOSR J. Pharm. Bio. Sci.* 6: 44–52.

4 Barbas, C., García, A., Saavedra, L., and Castro, M. (2000). Optimization and validation of a method for the determination of caffeine, 8-chlorotheophylline and diphenhydramine by isocratic high-performance liquid chromatography: stress test for stability evaluation. *J. Chrom. A* 870: 97–103.

160: Dipyridamole

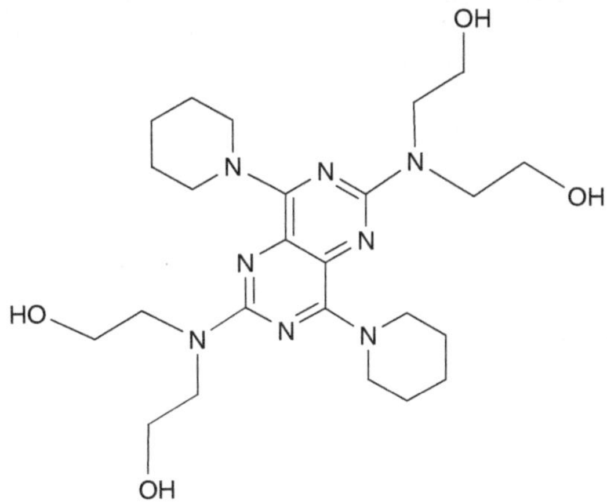

Chemical name: 2-[[2-[Bis(2-hydroxyethyl)amino]-4,8-di(piperidin-1-yl)pyrimido [5,4-d]pyrimidin-6- yl]-(2-hydroxyethyl)amino]ethanol
Trade name: Persantin
Formula: $C_{24}H_{40}N_8O_4$
Monoisotopic protonated ion (*m/z*): 505.3245
Molecular weight: 504.64
CAS number: 58-32-2

Dipyridamole is a fused pyrimidine compound with certain level of structural symmetry. It was introduced as a therapeutic agent in 1959 due to its coronary vasodilator properties and later demonstrated to have antiplatelet effects by being an inhibitor of cAMP-platelet phosphodiesterase [1, 2]. Despite some controversy regarding some of its therapeutic uses, it has been widely prescribed [1, 3].

In a stability batch of a formulated dipyridamole drug product, Subbaiah et al. reported two degradation products: a hydrolytic degradant and an adduct resulting from monoesterification of the API with tartaric acid, an excipient in the formulation [4]. The latter monoester degradant was also observed in another formulated product of dipyridamole during its shelf-life stability study; furthermore, dipyridamole di-tartaric ester was also formed in the study [5]. Both the mono- and di-ester degradants, which are listed in the British Pharmacopeia [6], were greatly enriched through an accelerated stress of a 1:1 solid mixture of dipyridamole and tartaric acid at 40°C/75%RH [5]. The degradation pathway of dipyridamole as formulated in the presence of tartaric acid is summarized in Scheme 160.

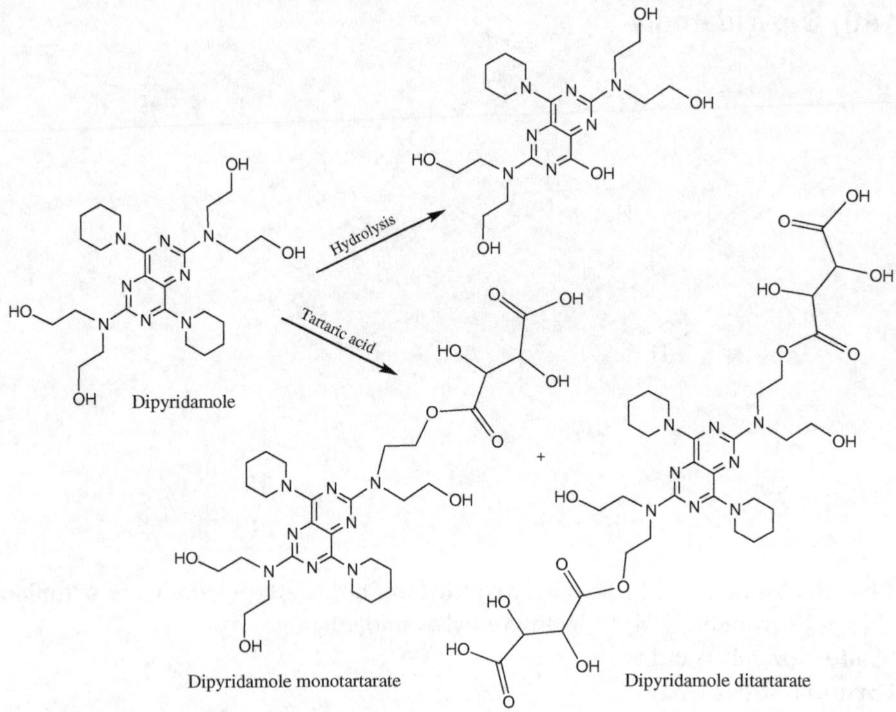

Scheme 160 Degradation pathways of dipyridamole formulated with tartaric acid.

References

1 Bibbs, C.R. and Lip, G.Y.H. (1998). Do we still need dipyridamole. *Br. J. Clin. Pharmacol.* 45: 323–328.

2 Harker, L.A. and Kadatz, R.A. (1983). Mechanism of action of dipyridamole. *Thromb. Res.* 29: 39–46.

3 Spano, D., Marshall, J.-C., Marino, N. et al. (2013). Dipyridamole prevents triple-negative breast-cancer progression. *Clin. Exp. Metastasis* 30: 47–68.

4 Subbaiah, B.V., Ganesh, K.K.S., Krishna, G.V. et al. (2012). Isolation and characterisation of degradant impurities in dipyridamole formulation. *J. Pharm. Biomed. Anal.* 61: 256–264.

5 Nagadeep, J., Kamaraj, P., Arthanareeswar, M., and Vivekanand, P.A. (2021). Identification of tartaric acid adduct impurities in dipyridamole capsule formulation related substances method. *Asian J. Chem.* 33: 307–313.

161–164: Dopamine and related catecholamines, Dobutamine, Norepinephrine, Epinephrine

161: Dopamine [WHO List of Essential Medicines]

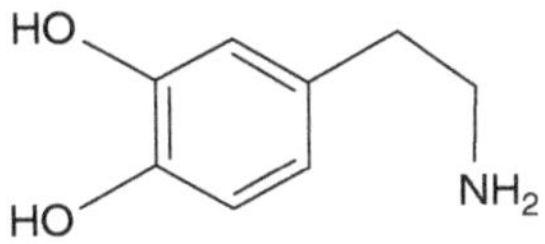

Chemical name: 4-(2-Aminoethyl)benzene-1,2-diol
Trade name: Intropin
Formula: $C_8H_{11}NO_2$
Monoisotopic protonated ion (*m/z*): 154.0863
Molecular weight: 153.18
CAS number: 51-61-6
Common salt form: HCl salt

162: Dobutamine

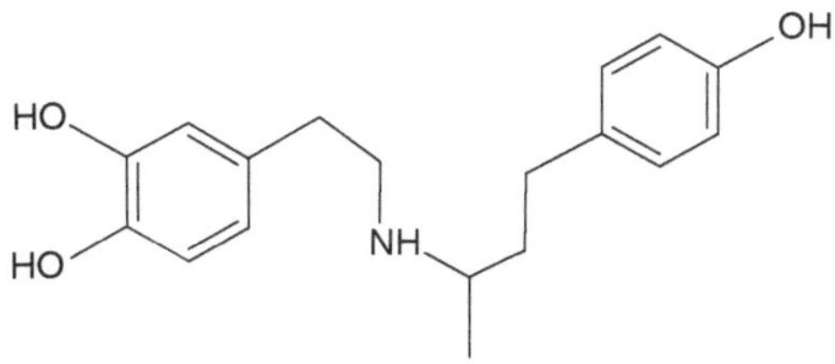

Chemical name: 4-[2-[4-(4-Hydroxyphenyl)butan-2-ylamino]ethyl]benzene-1,2-diol
Trade name: Dobuject
Formula: $C_{18}H_{23}NO_3$
Monoisotopic protonated ion (*m/z*): 302.1751
Molecular weight: 301.38
CAS number: 34368-04-2
Common salt form: HCl salt

163: Norepinephrine

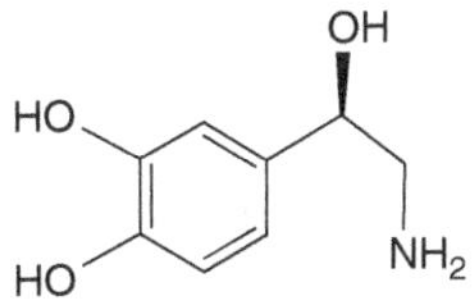

Chemical name: 4-[(1*R*)-2-Amino-1-hydroxyethyl]benzene-1,2-diol
Trade name: Levophed

Formula: $C_8H_{11}NO_2$
Monoisotopic protonated ion (*m/z*): 170.0812
Molecular weight: 169.18
CAS number: 51-41-2
Common salt form: Bitartrate

164: Epinephrine

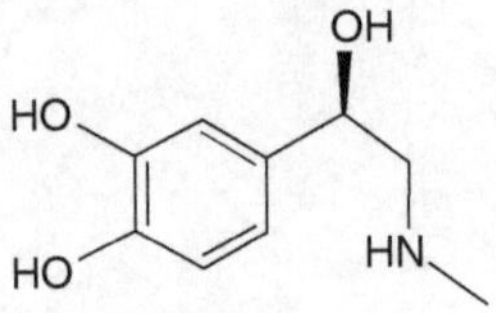

Chemical name: 4-[(1*R*)-1-Hydroxy-2-(methylamino)ethyl]benzene-1,2-diol
Trade name: EpiPen, Adrenaclick, and many others
Formula: $C_9H_{13}NO_3$
Monoisotopic protonated ion (*m/z*): 184.0968
Molecular weight: 183.21
CAS number: 51-43-4
Common salt form: HCl salt, bitartrate

Dopamine is a neurotransmitter when secreted by neurons in the brain, and it is biosynthesized from L-DOPA via decarboxylation. It belongs to the family of catecholamines and is the simplest form of catecholamine. When used for therapeutic purposes, dopamine is given via intravenous injection for the treatment of very low blood pressure, particularly for neonatal use; as dopamine cannot penetrate blood-brain barrier, it does not have CNS effect when administered in this way.

The catechol part of dopamine, as well as other members of the catecholamine family, is the very electron-rich 3,4-dihydroxyl-form of catechol (or 3,4-catechol; the other form of catechol being the 2,3-dihydroxyl isomer, i.e., 2,3-catechol); thus, it is very susceptible to oxidation, particularly its deprotonated forms. The pH-dependent oxidation of catechol has been well recognized, and the higher the pH, the quicker the oxidation [1, 2]. In a systematic study of catechol oxidation by molecular oxygen (O_2) in relation to pH, Maier et al. studied the oxidation rates of several catechols [3]. They found that the oxidation rates of fully deprotonated catechols (di-anions) were several orders of magnitude higher than those of mono-deprotonated anions (mono-anions). For example, for 4-ethylcatechol, which is a close analog of dopamine [4-(2′-aminoethyl)catechol], its di-anion form would be oxidized almost 4,000-folds faster than its mono-anion form (Scheme 161–164-1). This indicates that there will be an exponential increase in the oxidation rates of catechol from neutral to alkaline pH range.

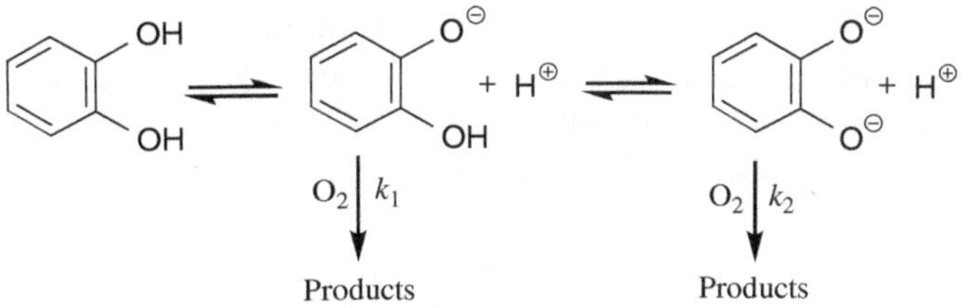

Scheme 161–164-1 Mechanism for the pH-dependent oxidation of catechol by O_2 proposed by Maier et al. [3]. *Source:* Maier et al. [3]/Royal Society of Chemistry.

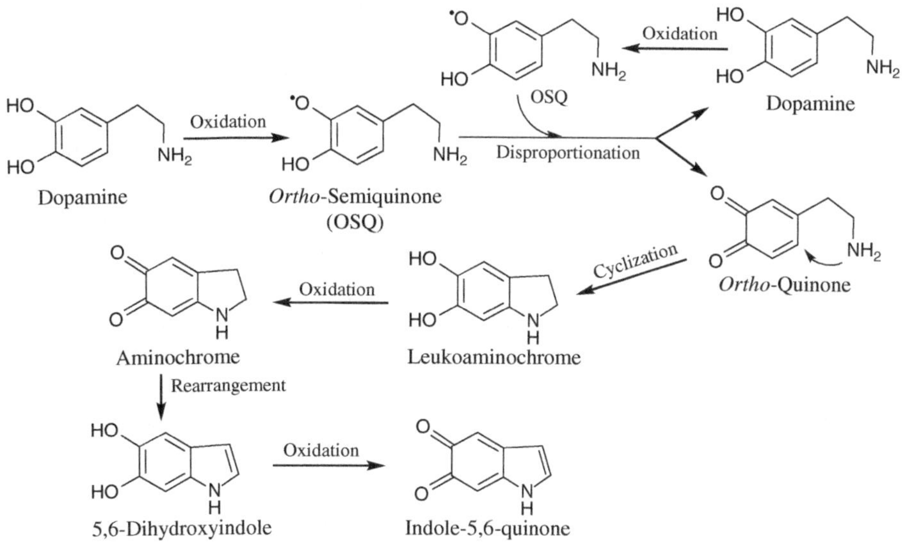

Scheme 161–164-2 Oxidative degradation pathway of dopamine. The pathway displayed is based on the mechanisms proposed by Maier et al. [3] as well as Bindoli et al. [4]. *Source:* Maier et al. [3] and Bindoli et al. [4].

The product from the initial oxidation of the catechol moiety of dopamine is *ortho*-quinone via the intermediacy of *ortho*-semiquinone disproportionation (Scheme 161–164-2) [4]. According to the mechanism illustrated in Scheme 161–164-1, the initial oxidation should occur on either of the anion forms. The *ortho*-quinone formed undergoes intramolecular cyclization to produce an unstable leucoaminochrome intermediate through a Michael addition by the amino group onto the *ortho*-quinone moiety. The leucoaminochrome intermediate is then rapidly oxidized to aminochrome. The latter can rearrange to 5,6-dihydroxyindole with an activation energy of 69 kJ/mol, which can be further oxidized to yield indole-5,6-quinone [5].

Dobutamine is a structural analog of dopamine with the amino group derivatized and the catechol moiety remained unchanged. Hence, its oxidative degradation pathway would be expected to be similar to that of dopamine (Scheme 161–164-2). The other naturally occurring members of the catecholamine family, which are also utilized therapeutically, include norepinephrine and epinephrine. The latter two molecules are biosynthesized from dopamine via hydroxylation and subsequent methylation. Due to the presence of the identical catechol moiety, norepinephrine and epinephrine would be expected to undergo the same or very similar oxidative degradation pathway as shown in Scheme 161–164-2. The hydroxyl groups on the benzylic positions of both molecules are usually very stable and resistant to oxidation as compared with the catechol moiety.

References

1 Joslyn, M.A. and Branch, G.E.K. (1935). The kinetics of absorption of oxygen by catechol. *J. Am. Chem. Soc.* 57: 1779–1785.

2 Roberts, N.B., Higgins, G., and Sargazi, M. (2010). A study on the stability of urinary free catecholamines and free methyl-derivatives at different pH, temperature and time of storage. *Clin. Chem. Lab. Med.* 48: 81–87.

3 Maier, G.P., Bernt, C.M., and Butler, A. (2017). Catechol oxidation: considerations in the design of wet adhesive materials. *Biomater. Sci.* https://doi.org/10.1039/c7bm00884h.

4 Bindoli, A., Rigobello, M.P., and Deeble, D.J. (1992). Biochemical and toxicological properties of the oxidation products of catecholamines. *Free Radic. Biol. Med.* 13: 391–405.

5 Bisaglia, M., Mammi, S., and Bubacco, L. (2007). Kinetic and structural analysis of the early oxidation products of dopamine. *J. Biol. Chem.* 282: 15597–15605.

165: Domperidone

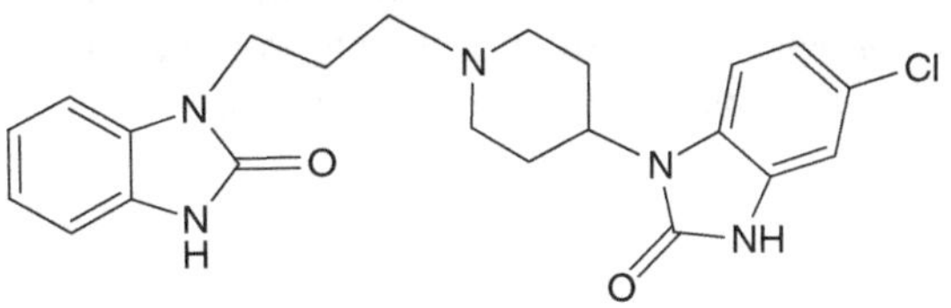

Chemical name: 6-Chloro-3-[1-[3-(2-oxo-3*H*-benzimidazol-1-yl)propyl] piperidin-4-yl]-1*H*-benzimidazol-2-one
Trade name: Motilium, Euciton, Moperidona, Nauzelin
Formula: $C_{22}H_{24}ClN_5O_2$
Monoisotopic protonated ion (*m/z*): 426.1691
Molecular weight: 425.91
CAS number: 57808-66-9
Common salt form: Maleate

Domperidone is a peripheral dopamine receptor antagonist, which has been marketed outside of the United States since 1978 [1]. Its clinical use includes the treatment of vomiting and nausea, as well as gastrointestinal motility disorders.

There are a few studies for the development of stability-indicating methods in which forced degradation was performed [2, 3]. Nevertheless, there seems to be only one study performed by Mutha et al., who conducted a full structural elucidation of the degradation products generated [4]. In the study by Mutha et al., three degradants were observed, and their structures elucidated by mass spectrometry and 1D and 2D NMR. The first degradant is the *N*-oxide of domperidone, which was generated by stress with 5% hydrogen peroxide, while the second and third degradants are a pair of isomeric chloro-substituted domperidone, which were produced by stress in 1 N HCl at 60°C for 12 hours (Scheme 165). While domperidone *N*-oxide could be

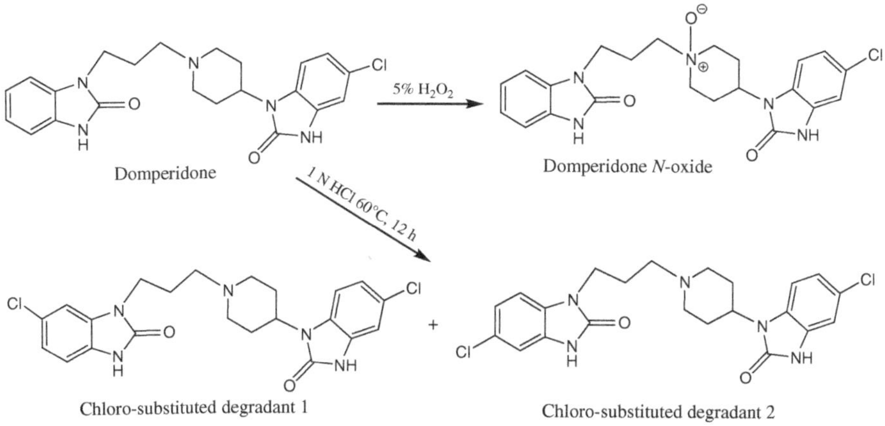

Scheme 165 Degradation pathway of domperidone under oxidative and acidic stress conditions.

a real degradant, the two chloro-substituted isomers are most likely solution degradation products due to the use of 1 N HCl in the acidic stress condition. The formation of these two artifact degradants appears to be rather unusual; nonetheless, no further study was performed with regard to their formation mechanism.

References

1 Barone, J.A. (1999). Domperidone: a peripherally acting dopamine2-receptor antagonist. *Ann. Pharmacother.* 33: 429–440.

2 Thanikachalam, S., Rajappan, M., and Kannappan, V. (2008). Stability-indicating HPLC method for simultaneous determination of pantoprazole and domperidone from their combination drug product. *Chromatographia* 67: 41–47.

3 Seshadri, R.K., Raghavaraju, T.V., and Chakravarthy, I.E. (2013). A single gradient stability-indicating reversed-phase LC method for the estimation of impurities in omeprazole and domperidone capsules. *Sci. Pharm.* 81: 437–458.

4 Mutha, A.K., Chandra, C., Vijayabhaskar, B. et al. (2019). Separation of unprecedented degradants of domperidone by ultra-performance convergence chromatography and their structure elucidation. *J. Chromatogr. Sci.* 57: 806–814.

166: Donepezil

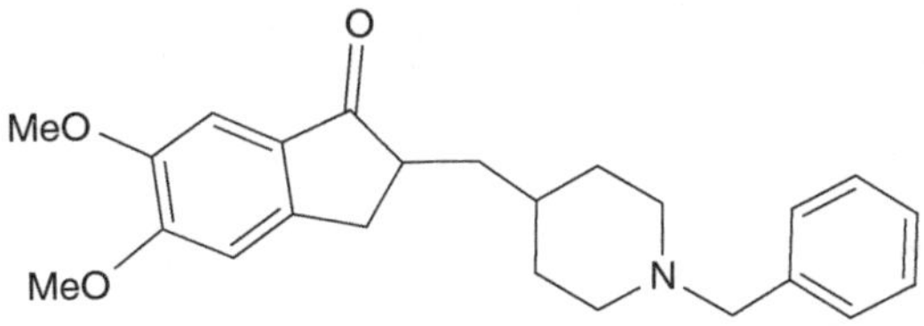

Chemical name: 2-[(1-Benzylpiperidin-4-yl)methyl]-5,6-dimethoxy-2,3-dihydroinden-1-one
Trade name: Aricept
Formula: $C_{24}H_{29}NO_3$
Monoisotopic protonated ion (*m/z*): 380.2220.
Molecular weight: 379.50
CAS number: 120014-06-4
Common salt form: HCl salt

Donepezil is an acetylcholinesterase inhibitor and clinically used for the treatment of Alzheimer's disease. The structural moiety in this drug molecule that is most obviously susceptible to degradation is the benzyl-substituted piperidinyl tertiary amine group, which could be degraded into the corresponding *N*-oxide under nucleophilic oxidative conditions. There have been a number of papers reporting stress studies of donepezil HCl and its formulated products under different stress conditions, including oxidative stress condition. The latter stress used hydrogen peroxide in different concentrations (from 3% to 30%) and temperatures (from room temperature to 80°C). Most of these studies reported oxidative degradation products [1–6], except few that did not observe oxidative degradation [3]. The main reason for not observing the oxidative degradant is most likely due to one or more of the following factors: (i) donepezil HCl was not neutralized with base prior to the oxidation, (ii) the concentration of hydrogen peroxide used was relatively low, (iii) the stress temperature was low, e.g., at room temperature, and (iv) the method used could not elute the degradants. For those studies that reported oxidative degradation products of donepezil under hydrogen peroxide stress, no efforts were demonstrated to elucidate the structures of the degradants, except the one by Mladenovic et al., who performed high-resolution LC-MS analysis of the degradants [4], and the one by Sreelatha et al. [5], who performed both LC-MS and ^{1}H NMR analysis. In the oxidative stress study performed by Mladenovic et al. using 3.0% hydrogen peroxide at 80°C for 30 minutes, a major oxidative degradant was observed at a relative retention time of 1.79 [4]. This degradant has an additional oxygen than donepezil and thus would be most likely the *N*-oxide of donepezil, which is consistent with the results of a later study performed by Sreelatha et al., where the structure of the *N*-oxide was confirmed by ^{1}H NMR measurement [5]. The plausible mechanism for the formation of the *N*-oxide is illustrated in Scheme 166 (Pathway a).

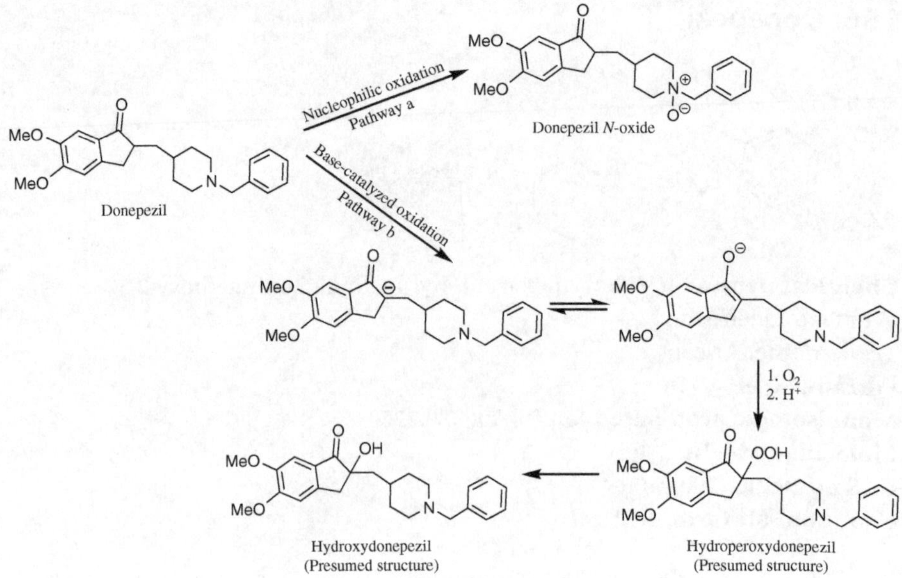

Scheme 166 Oxidative degradation pathways of donepezil under forced degradation conditions.

In the stress studies mentioned above, nearly all of them also reported degradation products under alkaline stress conditions. Nevertheless, no structural elucidation was performed in these studies, except the ones performed by Ruela et al. [3] and Mladenovic et al. [4] in which unit-resolution LC-MS and high-resolution LC-MS were utilized, respectively, to analyze the degradants. Three major degradants were observed eluting before donepezil by both research groups [3, 4], and their formulas were found to be [donepezil + O_2–H_2], [donepezil + O], and [donepezil + O_2], respectively [4]. It was found that the [donepezil + O_2] degradant was increasing along with the time of alkaline stress and became the major degradant in 6 hours [3]. Based on these results obtained by the two groups, one can reasonably speculate that donepezil underwent base-catalyzed oxidation [7], and the [donepezil + O] and [donepezil + O_2] degradants would likely be the hydroxydonepezil and hydroperoxydonepezil. The plausible mechanism for the formation of the two base-catalyzed oxidative degradants is displayed in Scheme 166 (Pathway b).

References

1 Pappa, H., Farru, R., Vilanova, P.O. et al. (2002). A new HPLC method to determine donepezil hydrochloride in tablets. *J. Pharm. Biomed. Anal.* 27: 177–182.

2 Kafkala, S., Matthaiou, S., Alexaki, P. et al. (2008). New gradient high-performance liquid chromatography method for determination of donepezil hydrochloride assay and impurities content in oral pharmaceutical formulation. *J. Chrom. A* 1189: 392–397.

3 Ruela, A.L.M., Santos, M.G., Figueiredo, E.C., and Pereira, G.R. (2014). LC-PDA and LC-MS studies of donepezil hydrochloride degradation behaviour in forced stress conditions. *J. Braz. Chem. Soc.* 25: 2094–2101.

4 Mladenovic, A.R., Jadranin, M.B., Pavlovic, A.D. et al. (2015). Liquid chromatography and liquid chromatography-mass spectrometry analysis of donepezil degradation products. *Chem. Ind. Chem. Eng. Q.* 21: 447–455.

5 Sreelatha, P., Reddy, D.V., Palavai, S.R., and Devi, B.R. (2016). Impurity profiling for donepezil hydrochloride tablet formulations and characterisation of potential degradant. *Am. J. PharmTech Res.* 6: 213–229.

6 Korany, M.A., Mahgoub, H., Haggag, R.S. et al. (2017). Development of a green stability-indicating HPLC–DAD method for the determination of donepezil hydrochloride in the presence of its related substance and degradation products. *J. Liq. Chromatogr. Relat. Technol.* 40: 930–942.

7 Li, M. (2012). Section 3.4, Carbanion/enolate-mediated autooxidation (base-catalyzed autooxidation). In: *Organic Chemistry of Drug Degradation*, 61–62. Cambridge: RSC.

167–168: L-Dopa (Levodopa), Methyldopa

167: L-Dopa (Levodopa)

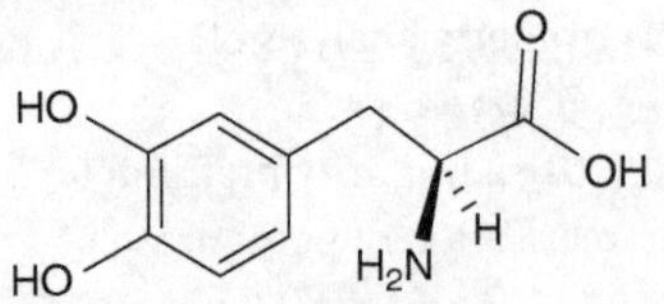

Chemical name: (2*S*)-2-Amino-3-(3,4-dihydroxyphenyl)propanoic acid
Trade name: Sinemet, Larodopa, Dopar, Inbrija
Formula: $C_9H_{11}NO_4$
Monoisotopic protonated ion (*m/z*): 198.0761
Molecular weight: 197.19
CAS number: 59-92-7

168: Methyldopa

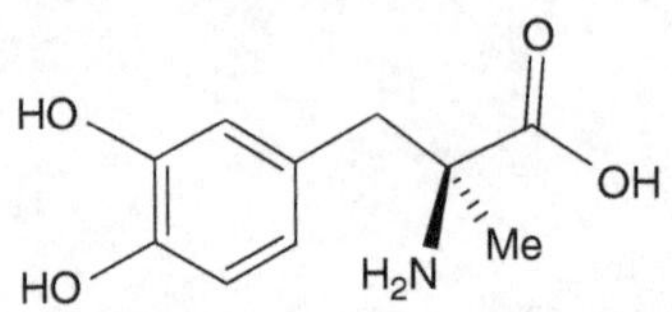

Chemical name: (2*S*)-2-Amino-3-(3,4-dihydroxyphenyl)-2-methylpropanoic acid
Trade name: Aldomet, Aldoril, Dopamet
Formula: $C_{10}H_{13}NO_4$
Monoisotopic protonated ion (*m/z*): 212.0917
Molecular weight: 211.21
CAS number: 555-30-6
Common salt form: HCl salt

Levodopa, also known as L-DOPA, is L-3,4-dihydroxyphenylalanine. Biosynthetically, it is the precursor of dopamine, and thus, it is used as a treatment for Parkinson's disease, which is characterized by a decrease in the production of dopamine in the brain. Conversion of levodopa into dopamine is catalyzed by the enzyme levodopa decarboxylase, and as the enzyme is not only present in the brain but also in the peripheral tissues, levodopa is typically coadministered with an inhibitor of the decarboxylase, such as carbidopa, to prevent levodopa decarboxylation before it reaches the brain.

Compared to dopamine, the structure of levodopa has an extra carboxyl group, and the remaining parts are the same. Thus, the catecholamine moiety of levodopa undergoes a similar degradation pathway as that of dopamine (Scheme 167–168; refer to Scheme 161–164-2 for the oxidative degradation pathway of dopamine) [1].

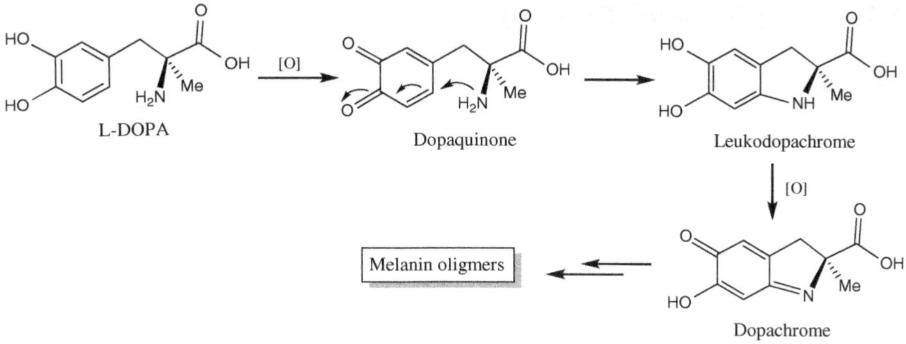

Scheme 167–168 Oxidative degradation pathway of L-DOPA. The intermediates are unstable, which were not detected by Zhou et al. in their stress study [1]. *Source:* Zhou et al. [1]/Oxford University Press.

Methyldopa is a medication for treating hypertension, and it is a methyl derivative of levodopa, with its methyl group replacing the alpha-H of levodopa. Due to the high similarity of its structure with that of L-DOPA, the degradation chemistry of methyldopa should be very similar to that of levodopa.

Reference

1 Zhou, Y.Z., Alany, R.G., Chuang, V., and Wen, J. (2012). Studies of the rate constant of L-DOPA oxidation and decarboxylation by HPLC. *Chromatographia* 75: 597–606.

169–170: Doxepin, Dothiepin

169: Doxepin

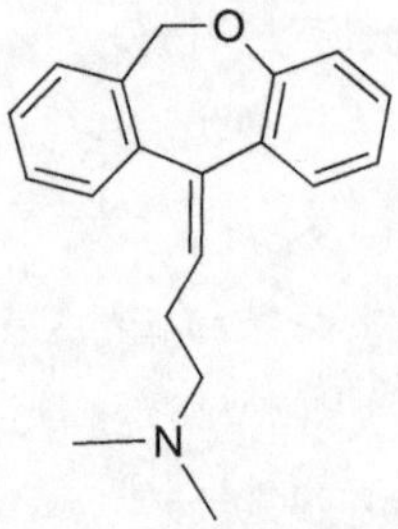

Chemical name: (3*E*)-3-(6*H*-Benzo[c][1]benzoxepin-11-ylidene)-*N,N*-dimethylpropan-1-amine
Trade name: Sinequan
Formula: $C_{19}H_{21}NO$
Monoisotopic protonated ion (*m/z*): 280.1696
Molecular weight: 279.38
Common salt form: HCl salt
CAS number: 1668-19-5

170: Dothiepin (Dosulepin)

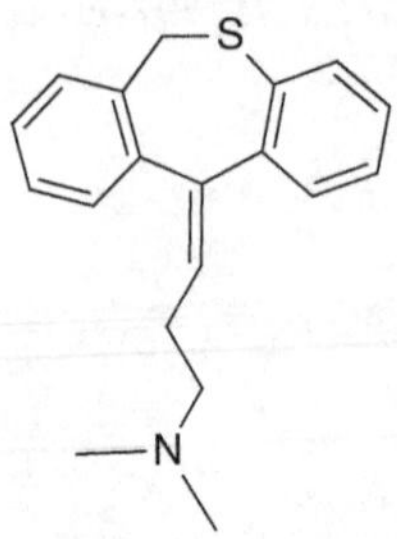

Chemical name: (3*Z*)-3-(6*H*-Benzo[c][1]benzothiepin-11-ylidene)-*N,N*-dimethylpropan-1-amine
Trade name: Prothiaden, Dopress
Formula: $C_{19}H_{21}NS$
Monoisotopic protonated ion (*m/z*): 296.1467
Molecular weight: 295.44
CAS number: 113-53-1
Common salt form: HCl salt

Doxepin is used clinically for the treatment of depression and anxiety disorders. Structurally, it is similar to the drugs in the tricyclic antidepressant family, such as amitriptyline, particularly in the part of the tertiary amine moiety that contains a

C=C double bond linked to the tricyclic ring. Hence, the degradation chemistry of the tertiary amine moiety would be expected to be similar to that of amitriptyline and related drugs. Specifically, the tertiary amine group could undergo nucleophilic oxidation to yield the corresponding *N*-oxide, while the C=C double bond might be susceptible to epoxidation (particularly under photo-oxidative conditions) and photoisomerization, respectively (Pathway a, Scheme 169–170). In a study by Po and Irwin [1], irradiation of a doxepin aqueous solution with a mercury lamp led to the increase of the cis-isomer (Pathway b, Scheme 169–170). In the presence of nitrogen, there was another photo-rearranged degradant, while in the presence of air, the degradation products were much more complex. There was a study in which doxepin was found to slightly degrade to give one degradant after being subjected to hydrolysis in 2 M HCl at 90°C for 24 hours [2]. Nevertheless, the structure of the acidic degradant proposed does not seem to be reasonable.

Dothiepin, also known as dosulepin, is the sulfur analog of doxepin in that the oxygen atom of doxepin is replaced by a sulfur atom. It is also employed as a treatment for depression. Due to its structure being highly similar to that of doxepin, the degradation chemistry of dothiepin would be expected to follow that of doxepin, except for the moiety containing the sulfur atom. In a photostability study of dothiepin in aqueous solutions, when irradiation was carried out under high pressure mercury lamp, dothiepin was found to undergo another isomerization in which the

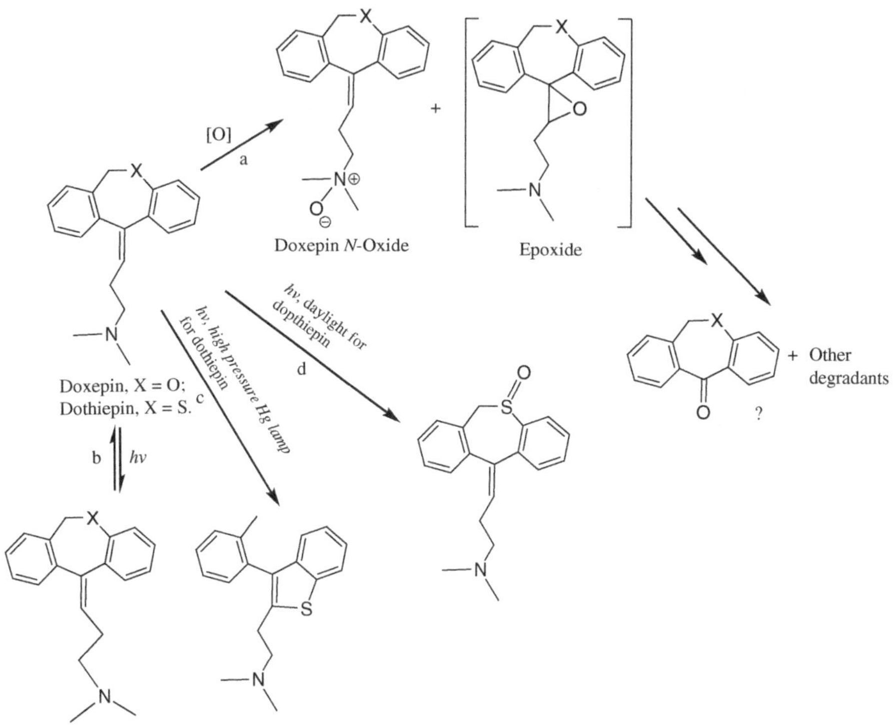

Scheme 169–170 Probable degradation pathways of doxepine and dothiepin; the presumed oxidative degradation is based on the degradation chemistry of amitriptyline.

7-membered ring was opened and the sulfur attached to the carbon-carbon double bond (Pathway c, Scheme 169–170) [3]. When the irradiation of dothiepin was performed under daylight, the main degradation was the oxidation of the sulfur moiety, leading to the formation of *S*-oxide (Pathway d, Scheme 169–170).

References

1 Po, A.L.W. and Irwin, W.J. (1980). The photochemical stability of cis- and trans-isomers of tricyclic neuroleptic drugs. *J. Pharm. Pharmacol.* 32: 25–29.

2 Maślanka, A., Krzek, J., and Stolarczyk, M. (2011). Stability studies of clonazepam, diazepam, haloperidol, and doxepin with diverse polarities in an acidic environment. *J. AOAC Inter.* 94: 1791–1799.

3 Tammilehto, S. and Torniainen, K. (1989). Photochemical stability of dothiepin in aqueous solutions. *Int. J. Pharm.* 52: 123–128.

171–175: Doxycycline, Chlortetracycline, Oxytetracycline, Tetracycline, Minocycline

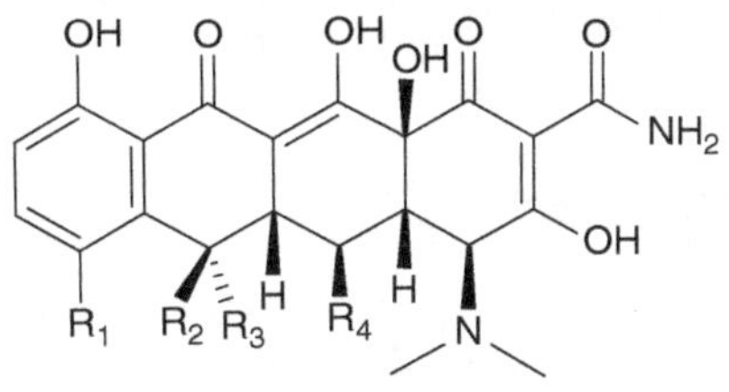

Doxycycline: R_1=H, R_2=Me, R_3=H, R_4=OH;
Chlortetracycline: R_1=Cl, R_2=Me, R_3=OH, R_4=H;
Oxytetracycline: R_1=H, R_2=Me, R_3=OH, R_4=OH;
Tetracycline: R_1=H, R_2=Me, R_3=H, R_4=H;
Minocycline: R_1=$N(Me)_2$, R_2=H, R_3=H, R_4=H.

171: Doxycycline [WHO List of Essential Medicines]

Chemical name: (4*S*,4*aR*,5*S*,5*aR*,6*R*,12*aR*)-4-(Dimethylamino)-1,5,10,11,12a-pentahydroxy-6-methyl-3,12-dioxo-4*a*,5,5*a*,6-tetrahydro-4*H*-tetracene-2-carboxamide
Trade name: Doxy, Doryx, Vibramycin
Formula: $C_{22}H_{24}N_2O_8$
Monoisotopic protonated ion (*m/z*): 445.1605
Molecular weight: 444.44
CAS number: 564-25-0
Common salt form: HCl salt

172: Chlortetracycline [WHO List of Essential Medicines]

Chemical name: (4*S*,4*aS*,5*aS*,6*S*,12*aR*)-7-Chloro-4-(dimethylamino)-1,6, 10,11,12*a*-pentahydroxy-6- methyl-3,12-dioxo-4,4*a*,5,5*a*-tetrahydrotetracene-2-carboxamide
Trade name: Aueromycin
Formula: $C_{22}H_{23}ClN_2O_8$
Monoisotopic protonated ion (*m/z*): 479.1216
Molecular weight: 478.88
CAS number: 57-62-5
Common salt form: HCl salt

173: Oxytetracycline [WHO List of Essential Medicines]

Chemical name: (4*S*,4*aR*,5*S*,5*aR*,6*S*,12*aR*)-4-(Dimethylamino)-1,5,6,10,11,12a-hexahydroxy-6-methyl-3,12-dioxo-4,4a,5,5a-tetrahydrotetracene-2-carboxamide
Trade name: Terramycin, Urobiotic
Formula: $C_{22}H_{24}N_2O_9$

Monoisotopic protonated ion (*m/z*): 461.1555
Molecular weight: 460.43
CAS number: 79-57-2
Common salt form: HCl salt

174: Tetracycline [WHO List of Essential Medicines]

Chemical name: (4*S*,4a*S*,5a*S*,6*S*,12a*R*)-4-(Dimethylamino)-1,6,10,11,12a-pentahydroxy-6-methyl-3,12-dioxo-4,4a,5,5a-tetrahydrotetracene-2-carboxamide
Trade name: Achromycin, Pylera, Sumycin
Formula: $C_{22}H_{24}N_2O_8$
Monoisotopic protonated ion (*m/z*): 445.1605
Molecular weight: 444.44
CAS number: 60-54-8
Common salt form: HCl salt

175: Minocycline

Chemical name: (4*S*,4a*S*,5a*R*,12a*R*)-4,7-Bis(dimethylamino)-1,10,11,12*a*-tetrahydroxy-3,12-dioxo-4*a*,5,5*a*,6-tetrahydro-4*H*-tetracene-2-carboxamide
Trade name: Minocin, Minomycin, Akamin, Amzeeq
Formula: $C_{23}H_{27}N_3O_7$
Monoisotopic protonated ion (*m/z*): 458.1922
Molecular weight: 457.48
CAS number: 10118-90-8
Common salt form: HCl salt

Doxycycline is the most prescribed member in the tetracycline family of antibiotics that have broad antibacterial activity. The other members of the family include chlortetracycline, oxytetracycline, tetracycline, and minocycline. They all have the same core structure of a linear fused tetracyclic nucleus, with each of the fused rings named as A, B, C, and D rings, respectively.

In general, the family of tetracycline drugs displays poor stability, and their main degradation pathway is the epimerization at either the 4- or the 6-position of the core structure [1, 2]. Other degradation types may also occur depending on the specific structure of a tetracycline, pH of the environment, and other factors such as excipients used in their formulated products.

Injac et al. reported two degradation products, 6-epidoxycycline and metacycline, after subjecting doxycycline capsules to 70°C for 90 days (Scheme 171–175-1) [1]. Metacycline, which is also a synthetic intermediate of doxycycline, and 6-epidoxycycline are both controlled at no more than 2% by USP and EP.

Pena et al. performed acidic stress of tetracycline (TC) and found it underwent epimerization at the 4-position to form 4-epitetracycline (4-ETC) in a weak pH range of 3–4.5 (Scheme 171–175-2) [3]. In solution of $pH < 2$, anhydrotetracycline was produced, while at the elevated temperature of 80°C, both anhydrotetracycline (ATC) and 4-epianhydrotetracycline (4-EATC) were formed.

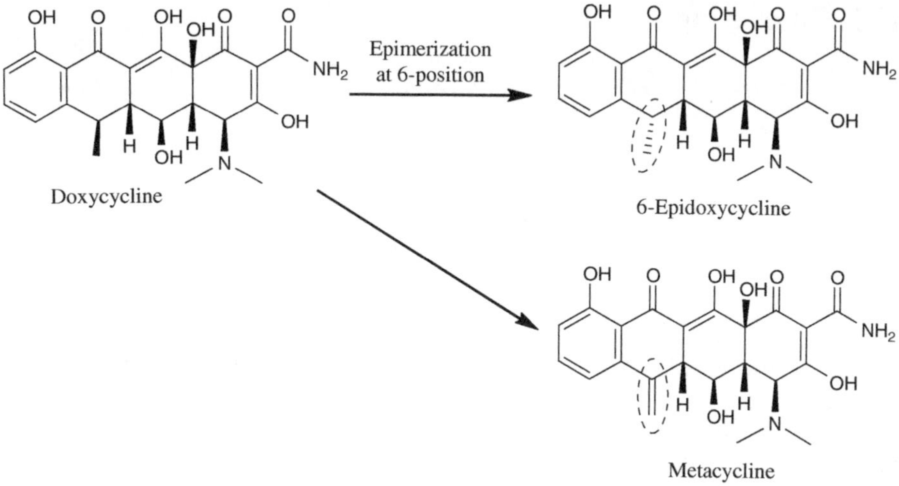

Scheme 171–175-1 Degradation of doxycycline capsules under stress at 70°C for 90 days [1].

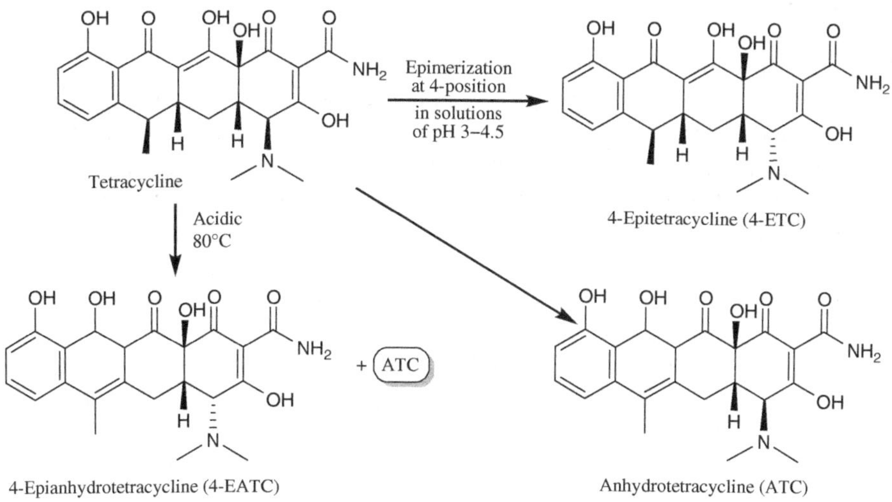

Scheme 171–175-2 Degradation of tetracycline under conditions of acidic stress and elevated temperature of 80°C.

References

1 Injac, R., Djordjevic-Milic, V., and Srdjenovic, B. (2007). Thermostability testing and degradation profiles of doxycycline in bulk, tablets, and capsules by HPLC. *J. Chromatogr. Sci.* 45: 623–628.

2 Skulason, S., Ingolfsson, E., and Kristmundsdottir, T. (2003). Development of a simple HPLC method for separation of doxycycline and its degradation products. *J. Pharm. Biomed. Anal.* 33: 667–672.

3 Pena, A., Palilis, L.P., Lino, C.M. et al. (2000). Determination of tetracycline and its major degradation products by chemiluminescence. *Anal. Chim. Acta* 405: 51–56.

176: Duloxetine

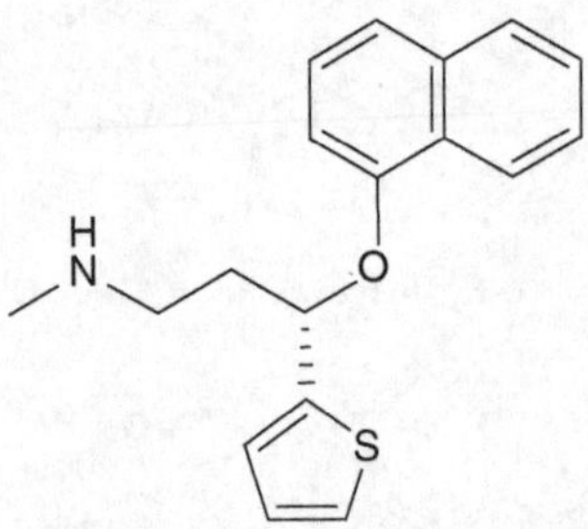

Chemical name: (3*S*)-*N*-Methyl-3-naphthalen-1-yloxy-3-thiophen-2-ylpropan-1-amine
Trade name: Cymbalta
Formula: $C_{18}H_{19}NOS$
Monoisotopic protonated ion (*m/z*): 298.1260
Molecular weight: 297.42
CAS number: 116539-59-4
Common salt form: HCl salt

Duloxetine is an inhibitor for the reuptake of both serotonin and norepinephrine. It is clinically used for the treatment of major depression, as well as neuropathic pain and other indications. The 1-naphthyl ether linkage is the primary weak point of the drug molecule that is readily degradable under acidic conditions; various forced degradation studies have demonstrated its acidic lability [1–4]. In these studies, the structures of certain degradation products were identified or specified based on solid evidence such as NMR and X-ray crystallography [1] or authentic reference compounds [2]. Based on the results of these studies, the degradation pathway of duloxetine is summarized in Scheme 176.

Under acidic conditions, the 1-naphthyl ether moiety is first protonated, which is followed by the cleavage of the ether linkage to generate 1-naphthol and the carbocation intermediate. The latter intermediate, which can be stabilized by conjugation with the thiophene ring, is then intercepted by either 1-naphthol or water to yield 2- and 4-alkylated derivatives of 1-naphthol, as well as *N,N*-dimethyl-3-hydroxy-(2-thienyl)-propanamine (DHTP).

Yet, in another forced degradation study, the apparent overstress generated 17 degradation products under various stress conditions, and the structures of 15 of the 17 degradants were proposed solely based on LC-MS and MS-MS analyses [5]. The structures proposed in this study do not seem to be relevant for the degradation of duloxetine under real-life scenarios.

In summary, duloxetine is reasonably stable in conditions other than acidic ones. Under photolytic stress conditions, duloxetine HCl API was found to be stable in solid but unstable in solution. On the other hand, the secondary amine moiety of duloxetine may be incompatible with aldehyde-containing excipients due to potential Maillard reaction between the secondary amine and aldehyde functionalities [6].

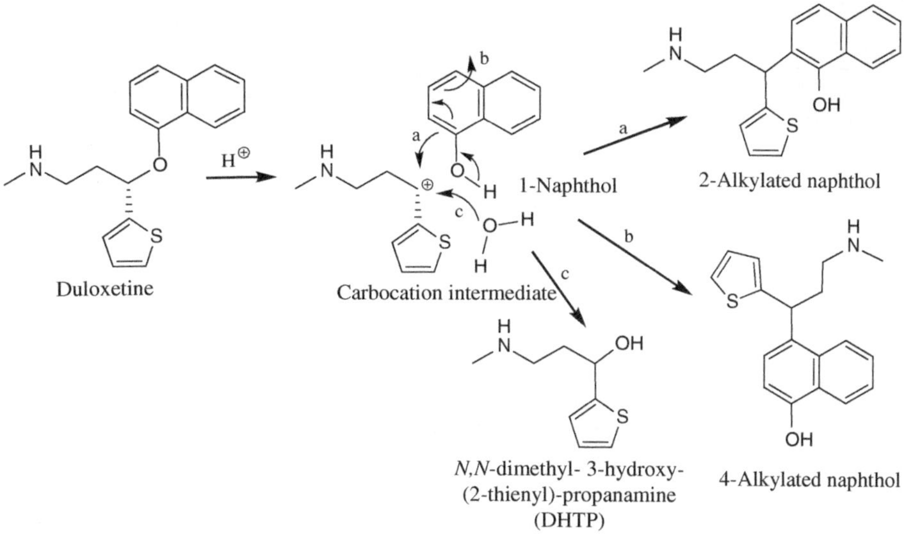

Scheme 176 Degradation pathway of duloxetine under acidic conditions. The chiral centers in the three degradants, 2-alkyldated naphthol, 4-alkylated naphthol, and DHTP, are most likely racemized.

References

1 Arava, V.R., Siripalli, U.B.R., and Bandatmakuru, S.R. (2007). Novel acid catalysed rearrangement of duloxetine. *Indian J. Chem.* 46B: 1695–1698.

2 Sinha, V.R., Anamika, R., and Kumria, J.R.B. (2009). Stress degradation studies on duloxetine hydrochloride and development of an RP-HPLC method for its determination in capsule formulation. *J. Chromatogr. Sci.* 47: 589–593.

3 Raman, N.V.V.S.S., Harikrishna, K.A., Prasad, A.V.S.S. et al. (2010). Determination of duloxetine hydrochloride in the presence of process and degradation impurities by a validated stability-indicating RP-LC method. *J. Pharm. Biomed. Anal.* 51: 994–997.

4 Reddy, P.R.M., Sreeramulu, J., Naidu, P.Y., and Reddy, A.R. (2010). Stability indicating fast LC for the simultaneous estimation of intermediates and degradants of duloxetine hydrochloride. *Chromatographia* 71: 95–100.

5 Chadha, R., Bali, A., and Bansal, G. (2016). Characterization of stress degradation products of duloxetine hydrochloride employing LC–UV/PDA and LC–MS/TOF studies. *J. Pharm. Biomed. Anal.* 121: 39–55.

6 Wirth, D.D., Baertschi, S.W., Johnson, R.A. et al. (1998). Maillard reaction of lactose and fluoxetine hydrochloride, a secondary amine. *J. Pharm. Sci.* 87: 31–39.

6

Efavirenz to Hydroxychloroquine

177: Efavirenz

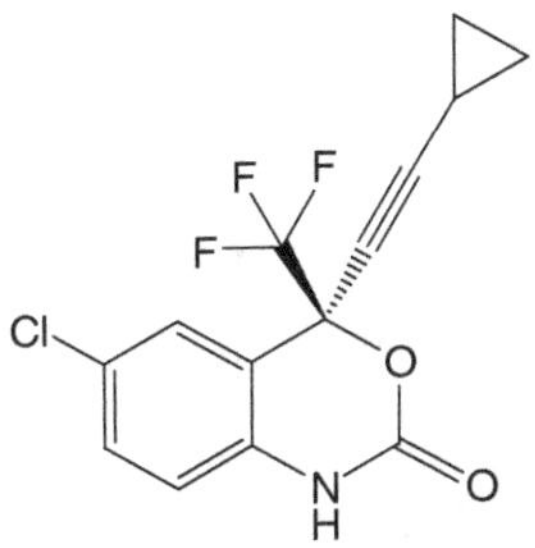

Chemical name: (4*S*)-6-Chloro-4-(2-cyclopropylethynyl)-4-(trifluoromethyl)-1*H*-3,1-benzoxazin-2-one
Trade name: Stoctin, Sustiva
Formula: $C_{14}H_9ClF_3NO_2$
Monoisotopic protonated ion (*m/z*): 316.0347
Molecular weight: 315.67
CAS number: 154598-52-4

Efavirenz is a non-nucleoside inhibitor of the reverse transcriptase of human immunodeficiency virus type-1 (HIV-1), the virus that causes AIDS (acquired immune deficiency syndrome). It contains a fused cyclic carbamate moiety, which is susceptible to hydrolytic degradation under either acidic or alkaline condition [1, 2]. The initial degradation product is an amino alcohol (code name SD573) [3], which can further degrade to a quinoline derivative (code name SM097) [3] via Rupe or Meyer–Shuster rearrangement depending on the catalysis by either acidic or alkaline conditions, respectively (Scheme 177) [1]. Maurin et al. studied the kinetics of efavirenz across a wide range of pH and found that it displayed a V-shape pH-rate profile with the most stable pH range at ~3–5 [1].

Compendium of Drug Degradation Pathways, First Edition. Min Li.

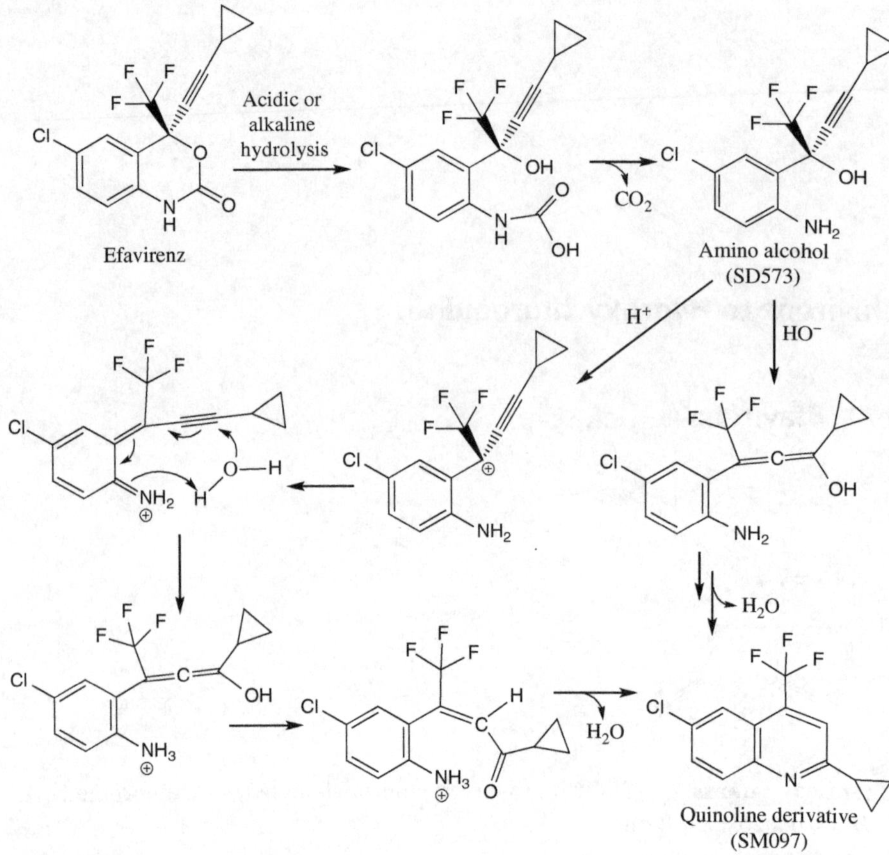

Scheme 177 Degradation of efavirenz under acidic or alkaline condition.

References

1 Maurin, M.B., Rowe, S.M., Blom, K., and Pierce, M.E. (2002). Kinetics and mechanism of hydrolysis of efavirenz. *Pharm. Res.* 19: 517–521.

2 de Aquino Ribeiro, J.A., Moreira de Campos, L.M., Alves, R.J. et al. (2007). Efavirenz related compounds preparation by hydrolysis procedure: setting reference standards for chromatographic purity analysis. *J. Pharm. Biomed. Anal.* 43: 298–303.

3 Alexander, A.J., Zhang, L., Hooker, T.F., and Tomasella, F.P. (2013). Comparison of supercritical fluid chromatography and reverse phase liquid chromatography for the impurity profiling of the antiretroviral drugs lamivudine/BMS-986001/efavirenz in a combination tablet. *J. Pharm. Biomed. Anal.* 78–79: 243–251.

178–179: Emtricitabine (FTC), Lamivudine (3TC)

178: Emtricitabine (FTC)

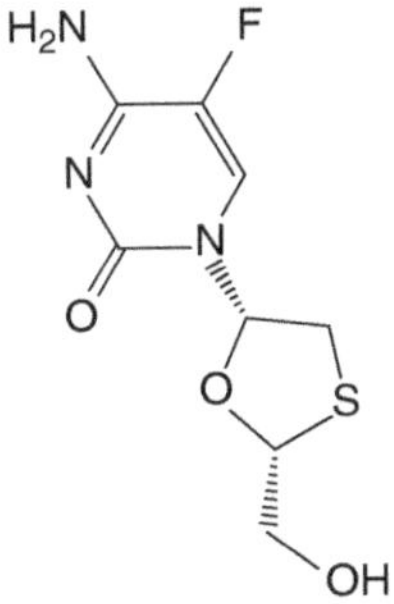

Chemical name: 4-Amino-5-fluoro-1-[(2*R*,5*S*)-2-(hydroxymethyl)-1,3-oxathiolan-5-yl]pyrimidin-2-one
Trade name: Emtriva
Formula: $C_8H_{10}FN_3O_3S$
Monoisotopic protonated ion (*m/z*): 248.0500
Molecular weight: 247.24
CAS number: 143491-57-0

179: Lamivudine

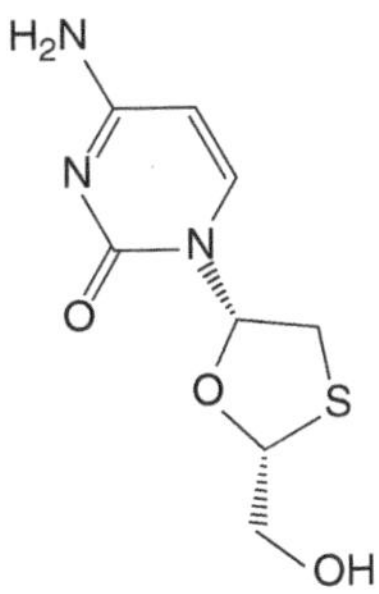

Chemical name: 4-Amino-1-[(2*R*,5*S*)-2-(hydroxymethyl)-1,3-oxathiolan-5-yl]pyrimidin-2-one
Trade name: Epivir, Zeffix, Heptovir
Formula: $C_8H_{11}N_3O_3S$
Monoisotopic protonated ion (*m/z*): 230.0594
Molecular weight: 229.25
CAS number: 134678-17-4

Emtricitabine, also referred to as FTC, is a nucleoside analog that inhibits the HIV-1 reverse transcriptase, and it is typically utilized in combination with other anti-AIDS medications. FTC consists of three functional groups or structural moieties

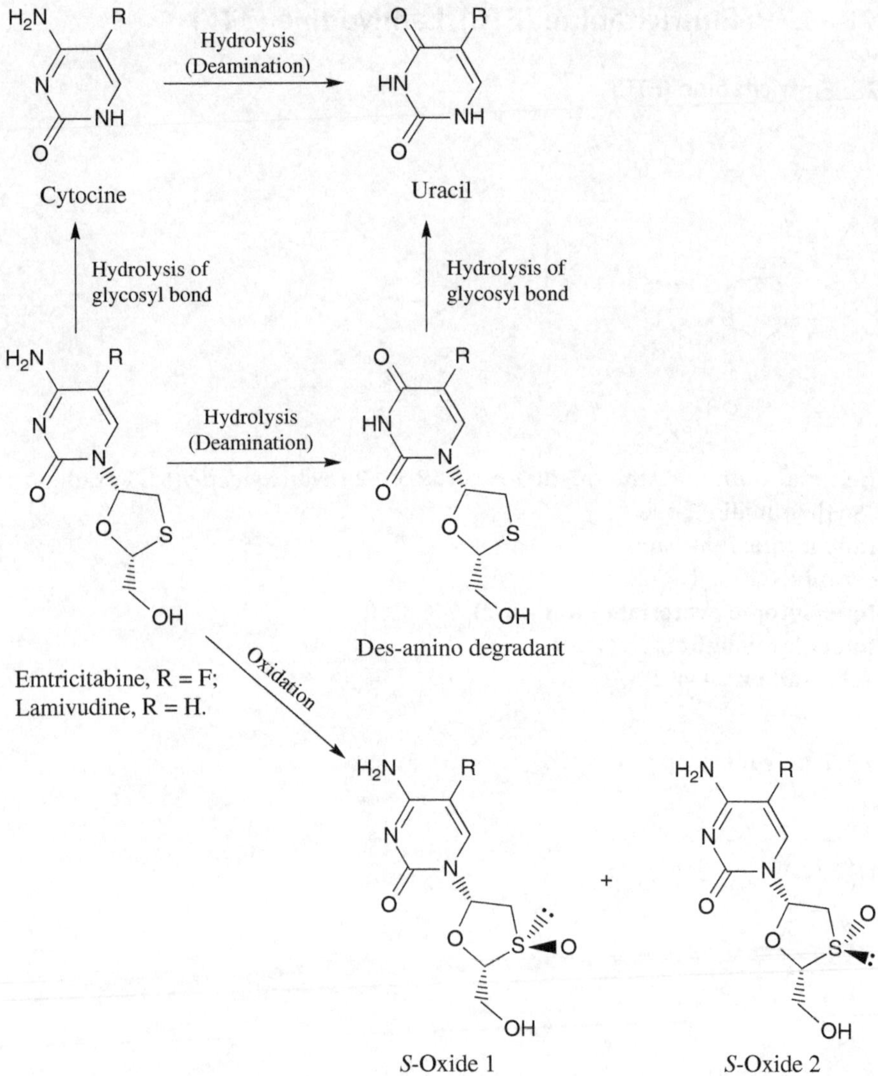

Scheme 178–179 Degradation pathways of emtricitabine and lamivudine.

that are susceptible to degradation: the amino group can be hydrolyzed to produce the des-amino degradant, the sulfur in the modified sugar ring can be subjected to nucleophilic oxidation to give the sulfoxide degradant (*S*-oxide), and the glycosyl linkage could be cleaved to yield 5-fluorocytocine [1]. In an oral liquid formulation, the des-amino and sulfoxide degradants were observed, while in solid dosage forms such as capsules and tablets, only the sulfoxide degradant was detected [2]. Both the des-amino and sulfoxide degradants can be generated under the conditions of acidic hydrolysis and hydrogen peroxide oxidation, respectively, as reported by Seshachalam et al. and Ashenafi et al. [3, 4]. It needs to be noted that the structure of the des-amino degradant presented in the papers of Seshachalam et al. and Ashenafi et al. was incorrect.

Lamivudine, also known as 3TC, is the des-fluoro analog of emtricitabine, which is also an inhibitor of the HIV-1 reverse transcriptase. Due to the very high similarity between the two drug molecules, their degradation pathways are expected to be the same or very similar, which turned out to be true according to a forced degradation study performed by Bedse et al. [5]. The degradation pathways for Emtricitabine and lamivudine are shown in Scheme 178–179. It is interesting to note that the two oxidative degradants of lamivudine, i.e., the two diastereomeric lamivudine *S*-oxides, were separated in the study by Bedse et al., while their counterparts of emtricitabine were not resolved by the HPLC analyses in the previous studies [3, 4]. Further looking into the HPLC methods of emtricitabine shows that the peak of *S*-oxides eluted quite early, suggesting the methods lack proper resolving power to separate the two structurally very similar diastereomers.

References

1 Hamarapurkar, P.D. and Parate, A.N. (2013). HPLC method for the determination of emtricitabine and related degradation substances. *J. Chromatogr. Sci.* 51: 419–424.

2 Ashenafi, D., Ungerbock, M., Hoogmartens, J., and Adams, E. (2013). Liquid chromatographic analysis of various formulations containing emtricitabine. *Chromatographia* 76: 1495–1503.

3 Seshachalam, U., Haribabu, B., and Chandrasekhar, K.B. (2007). Development and validation of a stability-indicating liquid chromatographic method for determination of emtricitabine and related impurities in drug substance. *J. Sep. Sci.* 30: 999–1004.

4 Ashenafi, D., Verbeek, A., Hoogmartens, J., and Adams, E. (2009). Development and validation of an LC method for the determination of emtricitabine and related compounds in the drug substance. *J. Sep. Sci.* 32: 1823–1830.

5 Bedse, G., Kumar, V., and Singh, S. (2009). Study of forced decomposition behavior of lamivudine using LC, LC–MS/TOF and MS^n. *J. Pharm. Biomed. Anal.* 49: 55–63.

180: Enalapril

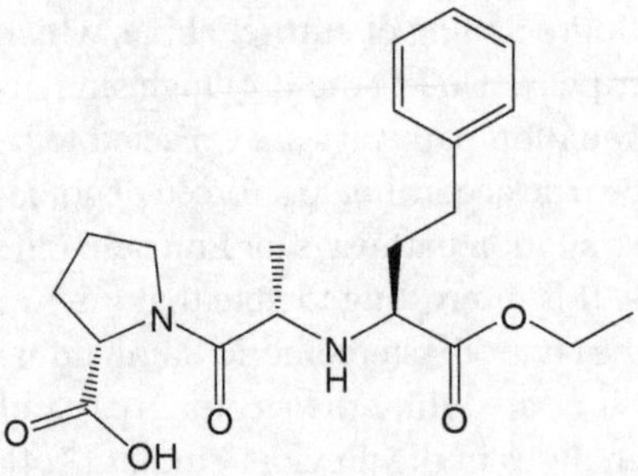

Chemical name: (2*S*)-1-[(2*S*)-2-[[(2*S*)-1-Ethoxy-1-oxo-4-phenylbutan-2-yl]amino] propanoyl]pyrrolidine-2-carboxylic acid
Trade name: Vasotec
Formula: $C_{20}H_{28}N_2O_5$
Monoisotopic protonated ion (*m/z*): 377.2071
Molecular weight: 376.45
CAS number: 75847-73-3
Common salt form: Maleate

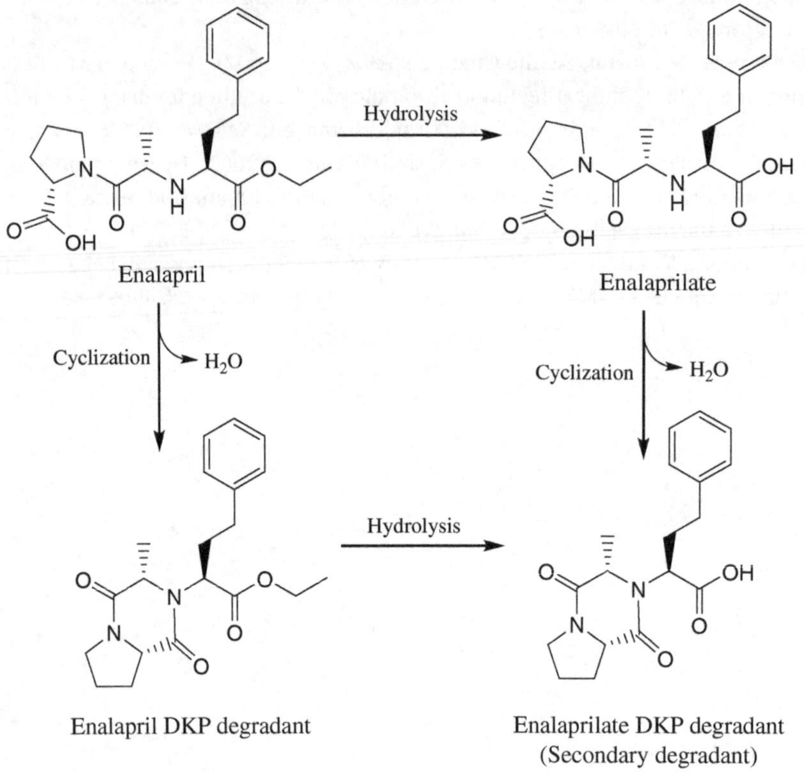

Scheme 180 Degradation pathways of enalapril.

Enalapril is clinically used for the treatment of hypertension. It is a prodrug that is hydrolyzed in vivo by esterases into its active metabolite, enalaprilate, which is an angiotensin-converting enzyme (ACE) inhibitor. The ethyl ester functionality of enalapril can also be hydrolyzed to form enalaprilate, particularly under alkaline conditions [1, 2]. Under acidic conditions, the main degradation pathway seems to be the formation of the so-called diketopiperazine (DKP) degradant, which is a common degradation pathway for many ACE inhibitors that possess a free carboxyl and a secondary amine group [3]. In formulated enalapril products, both enalaprilate and the DKP degradants were observed in meaningful amounts [4, 5]. In forced degradation studies in acidic and alkaline solutions, secondary degradation products could be formed in significant amounts [6]. Like some of the other ACE inhibitors, enalapril could be present in two rotamers due to the hindered rotation around the amide bond; the two rotamers would usually appear as two split peaks in HPLC analysis, which could be converted into a single peak with the control of several factors of the HPLC methods such as column temperature, pH of the mobile phase, and choice of the organic modifier [7]. In a study of enalapril maleate in solid, the formation of the DKP degradant was found to be independent of the moisture [8]. The degradation pathways of enalapril are shown in Scheme 180.

References

1 Shiromani, P.K. and Bavitz, J.F. (1986). Effect of moisture on the physical and chemical stability of granulations and tablets of the angiotensin converting enzyme inhibitor, enalapril maleate. *Drug Dev. Ind. Pharm.* 12: 2467–2480.

2 Ip, D.P. and Brenner, G.S. (1987). Enalapril Maleate. In: *Analytical Profiles of Drug Substances*, vol. 16 (ed. K. Florey), 207–243. London: Academic Press.

3 Li, M. (2012). Various types and mechanisms of degradation reactions. In: *Organic Chemistry of Drug Degradation*, 110–146. Cambridge, UK: RSC Publishing.

4 Qin, X.-Z., DeMarco, J., and Ip, D.P. (1995). Simultaneous determination of enalapril, felodipine and their degradation products in the dosage formulation by reversed phase high-performance liquid chromatography using a Spherisorb C_8 column. *J. Chromatogr. A* 707: 245–254.

5 Al-Omari, M.M., Abdelah, M.K., Badwan, A.A., and Jaber, A.M.Y. (2001). Effect of the drug-matrix on the stability of enalapril maleate in tablet formulations. *J. Pharm. Biomed. Anal.* 25: 893–902.

6 Bhardwaj, S.P. and Singh, S. (2008). Study of forced degradation behavior of enalapril maleate by LC and LC–MS and development of a validated stability-indicating assay method. *J. Pharm. Biomed. Anal.* 46: 113–120.

7 Trabelsi, H., Bouabdallah, S., Sabbah, S. et al. (2000). Study of the cis–trans isomerization of enalapril by reversed-phase liquid chromatography. *J. Chromatogr. A* 871: 189–199.

8 Stanisz, B. (2003). Evaluation of stability of enalapril maleate in solid phase. *J. Pharm. Biomed. Anal.* 31: 375–380.

181–183: Ergometrine, Methylergometrine, Ergotamine

181: Ergometrine (Ergonovine) [WHO List of Essential Medicines]

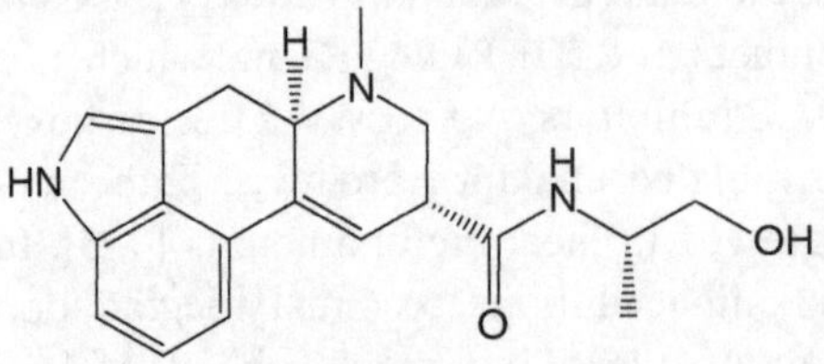

Chemicalname: (6a*R*,9*R*)-*N*-[(2*S*)-1-Hydroxypropan-2-yl]-7-methyl-6,6a,8,9-tetrahydro-4*H*-indolo[4,3-fg]quinoline-9-carboxamide
Trade name: Ergotrate, Ergostat, Syntometrine
Formula: $C_{19}H_{23}N_3O_2$
Monoisotopic protonated ion (*m/z*): 326.1863
Molecular weight: 325.41
CAS number: 60-79-7
Common salt form: Maleate

182: Methylergometrine [WHO List of Essential Medicines]

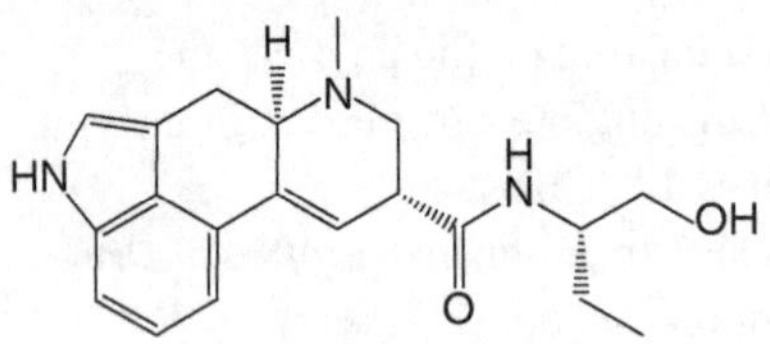

Chemical name: (6a*R*,9*R*)-*N*-[(2*S*)-1-Hydroxybutan-2-yl]-7-methyl-6,6a,8,9-tetrahydro-4*H*-indolo[4,3-fg]quinoline-9-carboxamide
Trade name: Methergine
Formula: $C_{20}H_{25}N_3O_2$
Monoisotopic protonated ion (*m/z*): 340.2020
Molecular weight: 339.44
CAS number: 113-42-8
Common salt form: Maleate

183: Ergotamine

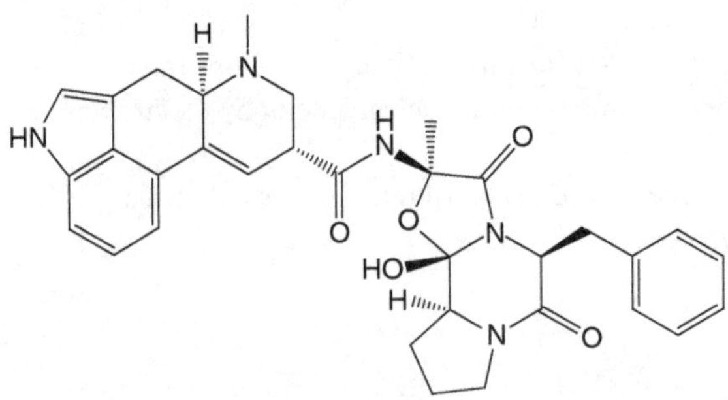

Chemical name: (6a*R*,9*R*)-*N*-[(1*S*,2*S*,4*R*,7*S*)-7-Benzyl-2-hydroxy-4-methyl-5,8-dioxo-3-oxa-6,9-diazatricyclo[7.3.0.0^{2,6}]dodecan-4-yl]-7-methyl-6,6a,8,9-tetrahydro-4*H*-indolo[4,3-fg]quinoline-9-carboxamide
Trade name: Cafergot, Ergomar, Migergot
Formula: $C_{33}H_{35}N_5O_5$
Monoisotopic protonated ion (*m/z*): 582.2711
Molecular weight: 581.67
CAS number: 379-79-3
Common salt form: Tartrate

Ergometrine, also known as ergonovine, is a member of the ergot alkaloids, which are produced by fungi of the *Claviceps* species, especially *Claviceps purpurea* on rye and other cereal-producing plants. Although the medical use of ergot, as an aid in childbirth, dates back to the middle ages, it was not until 1935 when Dudley and Moir identified ergometrine as the component responsible for the traditional clinical effect of ergot [1]. Today, ergometrine is used clinically for treating heavy bleeding after childbirth through contraction of the uterus, and it is commonly used in combination with oxytocin [2]. Although methylergometrine is a methyl analog of ergometrine, it originates from being an active metabolite of methysergide, a medication for migraine, via demethylation of the latter [3]. Clinically, methylergometrine is also employed for the treatment of migraine in addition to its use in childbirth [4].

Both ergometrine and methylergometrine are amides of lysergic acid, and the latter is a precursor for LSD. Hydrolysis of ergometrine and methylergometrine can produce lysergic acid [5, 6]. For this reason, ergometrine is a controlled substance. The lysergic part, which is a common structure moiety in both ergometrine and methylergometrine, contains a fused indole ring, a tertiary amine functional group, and a carbon-carbon double bond conjugated to the indole ring. The α-proton of the conjugated double bond is activated by the amide carbonyl group and, thus, can undergo enolization, particularly under alkaline or acidic conditions, which leads to the formation of the epimeric degradants for both ergometrine and methylergometrine [7]. Furthermore, the conjugated double bond is susceptible to photochemical hydration under somewhat acidic conditions as well [8]. These two types of degradation pathways, i.e., epimerization and photochemical hydration, can also be operative in the derivatives (including their degradants) of ergometrine and methylergometrine that contain the same structural moieties. The degradation pathways of ergometrine and methylergometrine are summarized in Scheme 181–182-1.

Ergotamine is also a member of the ergot alkaloids and has been used medicinally since the 17th century with similar indications as some other members of the alkaloid family. Nowadays, it is most widely used in combination with caffeine for treating migraine (e.g., Cafergot) [9]. It shares the same lysergic moiety as those of ergometrine and methylergometrine. Nevertheless, the amine part of ergotamine is much more structurally complex than those of ergometrine and methylergometrine. Thus, on one hand, it is expected that ergotamine undergoes the same degradation pathways as those of ergometrine and methylergometrine (refer to Scheme 181–183-1) with regard to its lysergic moiety [10]. On the other hand, the amine part of ergometrine contains one cyclic amide bond and one imide bond, with the latter forming a hemi-acetal linkage with the nearby hydroxyl group. The hemi-acetal linkage could be susceptible to degradation under acidic or alkaline catalysis, which leads to the formation of lysergic amide (Scheme 181–183-2) [10]. This dissociative degradation pathway is different from that of ergometrine and methylergometrine, in that the latter two would produce lysergic acid rather than lysergic amide.

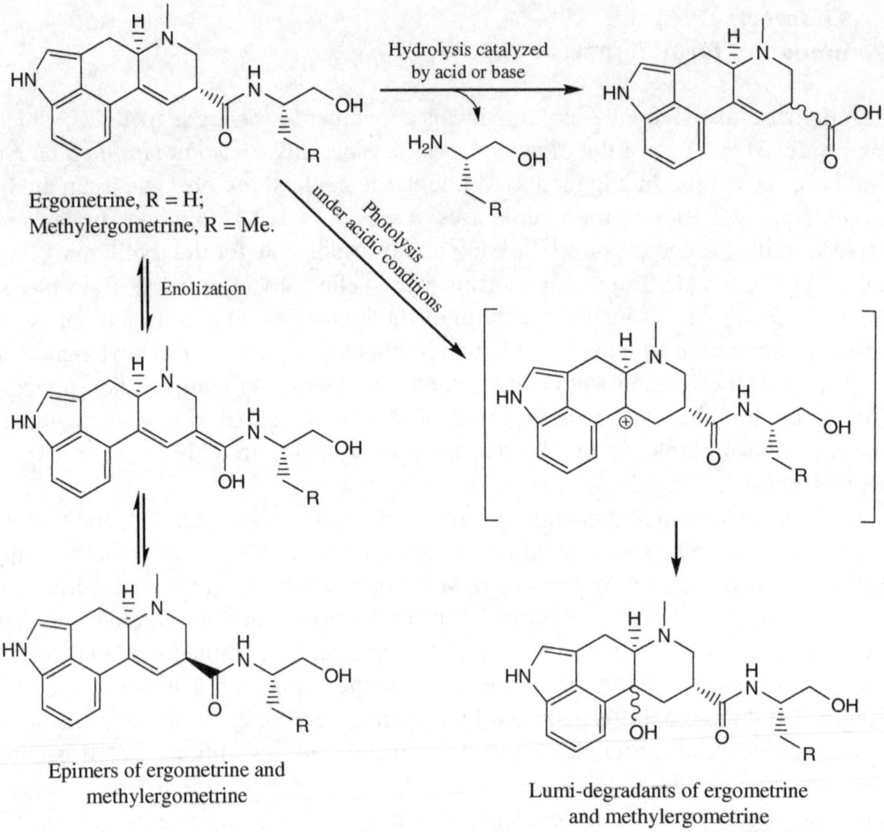

Scheme 181–183-1 Degradation pathways of ergometrine and methylergometrine.

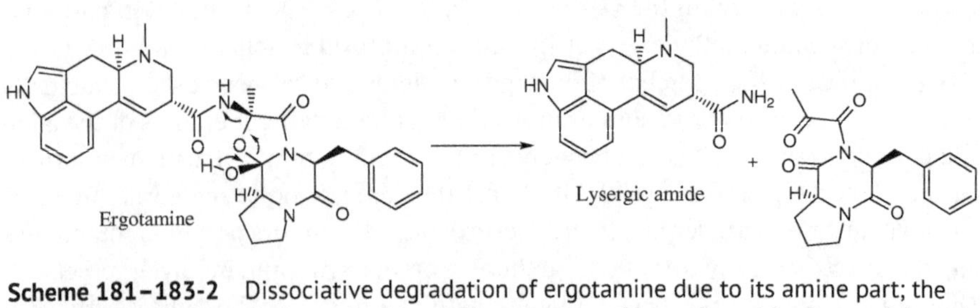

Scheme 181–183-2 Dissociative degradation of ergotamine due to its amine part; the paper citing this degradation only showed the structure of lysergic amide. *Source:* Adapted from Bethke et al. [10].

References

1 Dudley, H.W. and Moir, C. (1935). The substance responsible for the traditional clinical effect of ergot. *BMJ* 1: 520–523.
2 Ergometrine – Wikipedia. https://en.wikipedia.org/wiki/Ergometrine (accessed 2 February 2023).
3 Müller-Schweinitzer, E. and Tapparelli, C. (1986). Methylergometrine, an active metabolite of methysergide. *Cephalalgia* 6: 35–41.
4 Methylergometrine – Wikipedia. https://en.wikipedia.org/wiki/Methylergometrine (accessed 2 February 2023).
5 Reif, V.D. (1982). Ergonovine maleate. In: *Analytical Profiles of Drug Substances*, vol. 11 (ed. K. Florey), 273–312. London: Academic Press.
6 Marigny, K., Lohezic-Ledevehat, F., Aubin, F., and Burgot, G. (2006). Stability of oral liquid preparations of methylergometrine. *Pharmazie* 61: 701–705.
7 Hafner, M., Sulyok, M., Schuhmacher, R. et al. (2008). Stability and epimerisation behaviour of ergot alkaloids in various solvents. *World Mycotoxin J.* 1: 67–78.
8 Merkel, S., Köppen, R., Koch, M. et al. (2012). Lumi-ergometrine – structural identification and occurrence in sclerotia. *Mycotoxin Res.* 28: 59–66.
9 Ergotamine – Wikipedia. https://en.wikipedia.org/wiki/Ergotamine (accessed 2 February 2023).
10 Bethke, H., Delz, B., and Stich, K. (1976). Determination of the content and purity of ergotamine preparations by means of high-pressure liquid chromatography. *J. Chromatogr.* 123: 193–203.

184–186: Epinephrine, Norepinephrine, Isoprenaline

184: Epinephrine [L-Adrenaline or R-Adrenaline]

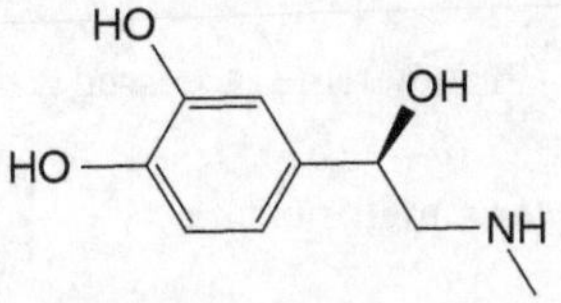

Chemical name: 4-[(1*R*)-1-Hydroxy-2-(methylamino)ethyl]benzene-1,2-diol
Formula: $C_9H_{13}NO_3$
Monoisotopic protonated ion (*m/z*): 184.0968
Molecular weight: 183.207
CAS number: 51-43-4
Common salt form: HCl salt, Acetate, Bitartrate

185: Norepinephrine [Noradrenaline]

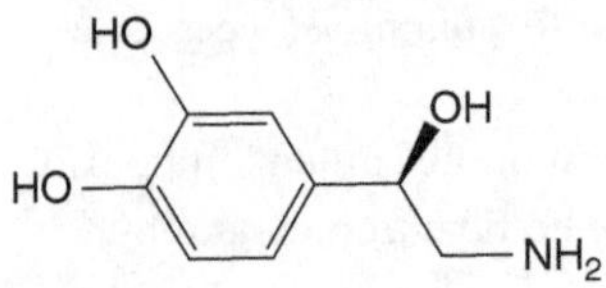

Chemical name: 4-[(1*R*)-2-Amino-1-hydroxyethyl]benzene-1,2-diol
Trade name: Levophed
Formula: $C_8H_{11}NO_3$
Monoisotopic protonated ion (*m/z*): 170.0812
Molecular weight: 169.18
CAS number: 51-41-2
Common salt form: Bitartrate

186: Isoprenaline [Racemate]

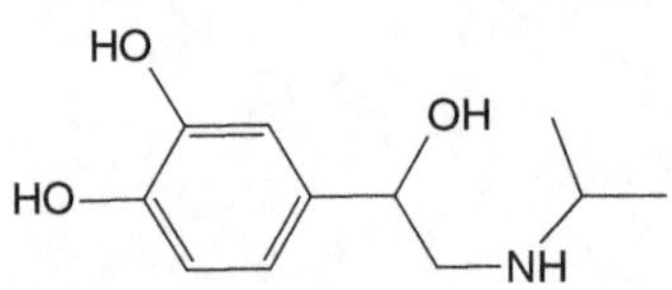

Chemical name: 4-[1-Hydroxy-2-(propan-2-ylamino)ethyl]benzene-1,2-diol
Trade name: Isuprel
Formula: $C_{11}H_{17}NO_3$
Monoisotopic protonated ion (*m/z*): 212.1281
Molecular weight: 211.26
CAS number: 7683-59-2
Common salt form: HCl salt

Epinephrine and norepinephrine are both neurotransmitters and hormones secreted by the adrenal glands to help deal with body's acute stress. Clinically, they have been used to treat allergic reactions and restore cardiac rhythm; the latter application typically occurs in emergency rooms. Isoprenaline, the isopropyl analog of epinephrine, is a nonselective beta-adrenergic receptor agonist that is clinically used for treating bradycardia, heart block, and other symptoms such as congestive heart failure and septic shock.

Chemically, epinephrine, norepinephrine, and isoprenaline are all catecholamines. Their catechol moieties are quite electron rich and, thus, are very susceptible to oxidative degradation, particularly under catalysis by redox reactive transition metal ions, exposure to light, and increased pH [1]. These drug substances are usually formulated in aqueous solutions as injectable formats. The most common antioxidative measure to preserve the active ingredients in solutions is the use of antioxidants, typically sulfite or bisulfite salts, in addition to measures such as reduction of oxygen content (by nitrogen), buffered pH, and protection from light. Although bisulfite can effectively suppress the oxidative degradation [2], a drawback of this protective measure is that it can cause the interaction between the drug substances and sulfites or bisulfites, resulting in the formation of the corresponding sulfonates [3–5]. Racemization of the chiral center could occur in the absence of bisulfite, and the presence of bisulfite did not seem to accelerate the racemization of epinephrine [3]. The formation of the sulfonate was found to be significant in a study where the formulated product was subjected to simulated environmental stress during storage and transportation [1]. The degradation pathways of epinephrine, as summarized by Stepensky et al. [6], are shown in Scheme 184–186-1 with some modification according to the original studies by Schroeter et al. [3] and by Grunert and Wollmann [7]. The initial oxidation of the catechol moiety of epinephrine leads to the formation of 3,4-benzoquinone, which is followed by the Michael addition of the amino group to result in leucoadrenochrome. Further oxidation would produce adrenochrome and adrenolotin, respectively. On the other hand, dehydration of adrenochrome would yield adrenolutin [7].

In a comprehensive review on epinephrine and norepinephrine stability in injectable solutions [8], it was indicated that the use of metal chelates such as diethylenetriaminepentaacetic acid (DTPA) and 8-hydroxyquinoline sulfate could significantly enhance the stability of epinephrine solutions. Interestingly, EDTA did not show any stabilization effect and even promoted degradation at pH 5.5. On the other hand, the combination of different control strategies, such as the use of antioxidants and DTPA, may yield a synergistic protective effect.

In a study aiming to identify the role of sulfites in promoting photochemical degradation of epinephrine [9], Brustugun et al. found that bisulfite could react with adrenochrome, one of the intermediate oxidative degradants of epinephrine, to form adrenochrome sulfonate [10]. Upon UV irradiation at 350 nm, the excited state of adrenochrome sulfonate collides with triplet molecular oxygen to generate singlet oxygen, the presence of which was supported by the characteristic luminescence signal detected at 1270 nm. The singlet oxygen would inflict oxidative degradation of epinephrine, resulting in the formation of more adrenochrome, which could further react with bisulfite to produce another batch of adrenochrome sulfonate, forming a catalytic cycle of degradation (Scheme 184–186-2).

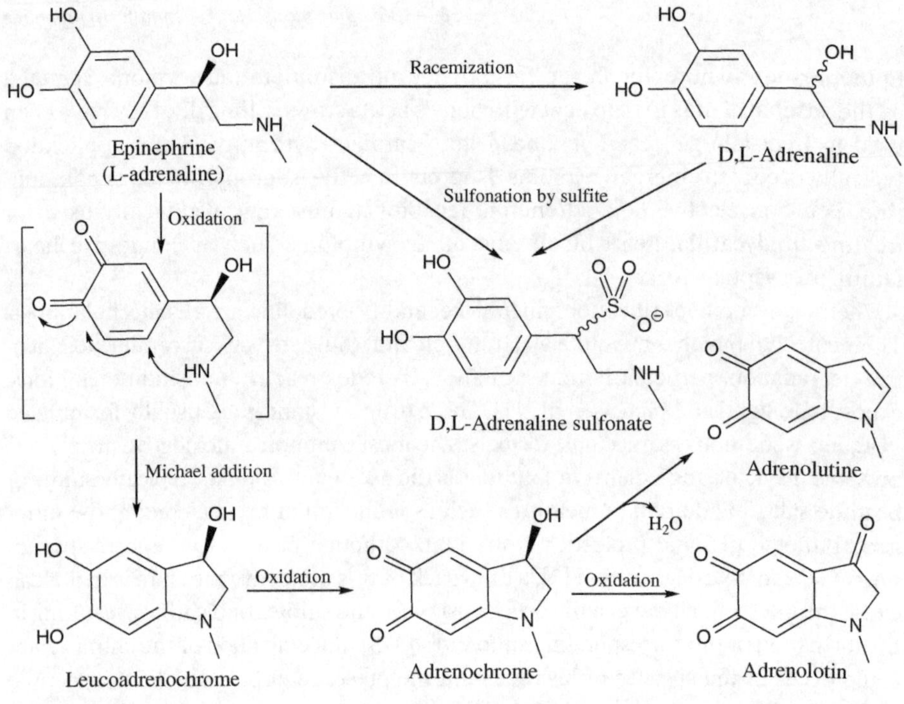

Scheme 184–186-1 Degradation pathways of epinephrine (L-adrenaline) via oxidation and racemization. The degradation pathways were adapted from Stepensky et al. [6] with modifications according to the original studies by Schroeter et al. [3] and by Grunert and Wollmann [7]. *Source:* Stepensky et al. [6]; Schroeter et al. [3]; Grunert and Wollmann [7].

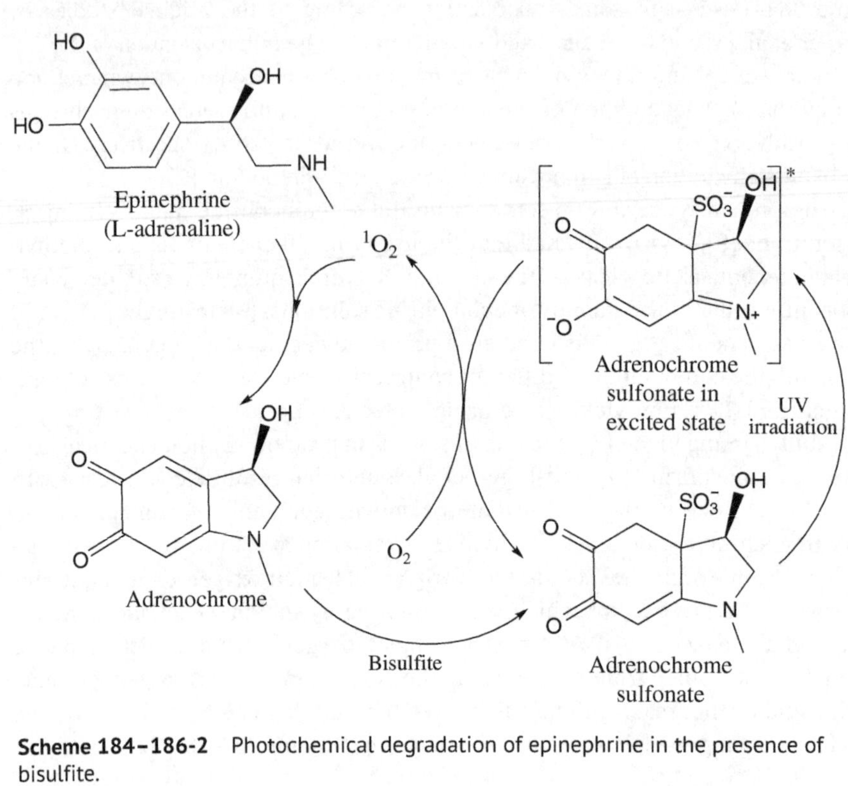

Scheme 184–186-2 Photochemical degradation of epinephrine in the presence of bisulfite.

References

1 Church, W.H., Hu, S.S., and Henry, A.J. (1994). Thermal degradation of injectable epinephrine. *Am. J. Emerg. Med.* 12: 306–309.

2 Wollman, H. and Raether, G. (1993). Efficacy testing of stabilizing agents in epinephrine model solutions: 19: stability of drugs and preparations. *Pharmazie* 38: 37–42.

3 Schroeter, L.C., Higuchi, T., and Schuler, E.E. (1958). Degradation of epinephrine induced by bisulfite. *J. Am. Pharm. Assoc.* 47: 723–728.

4 Higuchi, T. and Schroeter, L.C. (1959). Reactivity of bisulfite with a number of pharmaceuticals. *J. Am. Pharm. Assoc.* 48: 535–540.

5 Higuchi, T. and Schroeter, L.C. (1960). Kinetics and mechanism of formation of sulfonate from epinephrine and bisulfite. *J. Am. Chem. Soc.* 82: 1904–1907.

6 Stepensky, D., Chorny, M., Dabour, Z., and Schumacher, I. (2004). Long-term stability study of l-adrenaline injections: kinetics of sulfonation and racemization pathways of drug degradation. *J. Pharm. Sci.* 93: 969–980.

7 Grunert, R. and Wollmann, H. (1982). Effect of ultraviolet, visible light on drugs of the phenylalkylamine series with a view toward their stability in plastic containers. 17. Stability of drugs, preparations. 83. Problems of the use of plastic containers for liquid pharmaceuticals. *Pharmazie* 37: 798–799.

8 Hoellein, L. and Holzgrabe, U. (2012). Ficts and facts of epinephrine and norepinephrine stability in injectable solutions. *Int. J. Pharm.* 434: 468–480.

9 Brustugun, J., Tønnesen, H.H., Klem, W., and Kjønniksen, I. (2000). Photodestabilization of epinephrine by sodium metabisulfite. *PDA J. Pharm. Sci. Technol.* 54: 136–143.

10 Brustugun, J., Kristensen, S., and Tønnesen, H.H. (2004). Photostability of epinephrine – the influence of bisulfite and degradation products. *Pharmazie* 59: 457–463.

187: Ethambutol

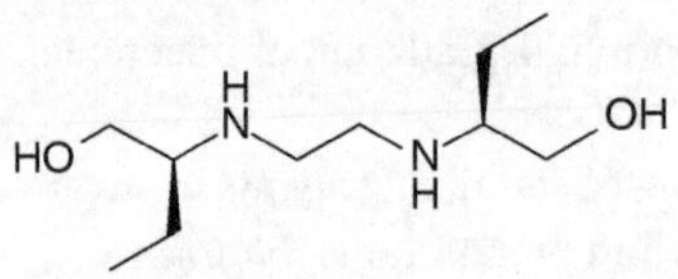

Chemical name: (2*S*)-2-[2-[[(2*S*)-1-Hydroxybutan-2-yl]amino]ethylamino] butan-1-ol
Trade name: Etibi, Myambutol
Formula: $C_{10}H_{24}N_2O_2$
Monoisotopic protonated ion (*m/z*): 205.1911
Molecular weight: 204.31
CAS number: 74-55-5
Common salt form: HCl salt

Ethambutol is an aminoalcohol clinically employed for treating tuberculosis (TB) since the 1960s, which remains to be a major global health threat causing 1.5 million deaths each year and 1.7 billion infections as of 2018 [1]. 2-Amino-1-butanol was reported to be a degradant of ethambutol; nevertheless, no mechanistic explanation was provided in these studies [2, 3]. It could be formed under hydrolytic degradation conditions, either via direct hydrolysis or through a tricyclic aziridine intermediate (Scheme 187). Under either mechanism, the second degradant, (*S*)-2-((2-hydroxyethyl)amino)butan-1-ol, would be expected to form; however, no such degradant has been reported. Ethambutol is usually used in combination with other anti-TB drugs, such as isoniazid, rifampicin, and pyrazinamide, and a number of studies found that the presence of ethambutol promoted the degradation occurred between isoniazid and rifampicin, which was attributed to the hygroscopicity and acidic nature of ethambutol hydrochloride [4–6].

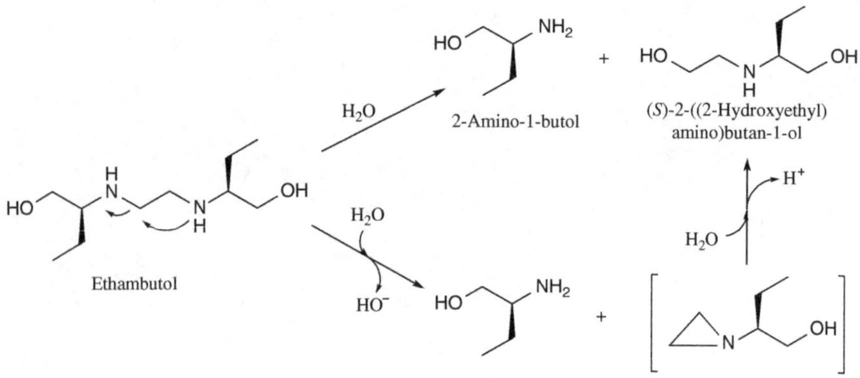

Scheme 187 Possible degradation pathway of ethambutol.

References

1 Tuberculosis. CDC archive, last reviewed: April 6, 2020. https://archive.cdc.gov/#/details?q=tuberculosis&start=0&rows=10&url=https://www.cdc.gov/globalhealth/newsroom/topics/tb/index.html (accessed 26 June 2025).
2 Ragonese, R., Macka, M., Hughes, J., and Petocz, P. (2002). The use of the Box–Behnken experimental design in the optimisation and robustness testing of a capillary electrophoresis method for the analysis of ethambutol hydrochloride in a pharmaceutical formulation. *J. Pharm. Biomed. Anal.* 27: 995–1007.
3 da Silva, J.A.F., de Castro, N.V., de Jesus, D.P. et al. (2010). Fast determination of ethambutol in pharmaceutical formulations using capillary electrophoresis with capacitively coupled contactless conductivity detection. *Electrophoresis* 31: 570–574.
4 Bhutani, H., Singh, S., and Jindal, K.C. (2005). Drug-drug interaction studies on first-line anti-tuberculosis drugs. *Pharm. Dev. Technol.* 10: 517–524.
5 Bhutani, H., Mariappan, T.T., and Singh, S. (2004). The physical and chemical stability of anti-tuberculosis fixed-dose combination products under accelerated climatic conditions. *Int. J. Tuberc. Lung Dis.* 8: 1073–1080.
6 Bhutani, H., Singh, S., Jindal, K.C., and Chakraborti, A.K. (2005). Mechanistic explanation to the catalysis by pyrazinamide and ethambutol of reaction between rifampicin and isoniazid in anti-TB FDCs. *J. Pharm. Biomed. Anal.* 39: 892–899.

188–189: Ethinylestradiol, Mestranol

188: Ethinylestradiol

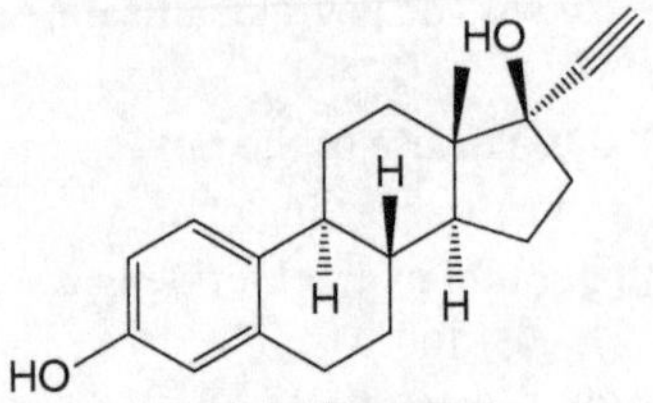

Chemical name: (8*R*,9*S*,13*S*,14*S*,17*R*)-17-Ethynyl-13-methyl-7,8,9,11,12,14,15,16-octahydro-6*H*-cyclopenta[a]phenanthrene-3,17-diol
Formula: $C_{20}H_{24}O_2$
Monoisotopic protonated ion (*m/z*): 297.1849
Molecular weight: 296.41
CAS number: 57-63-6

189: Mestranol

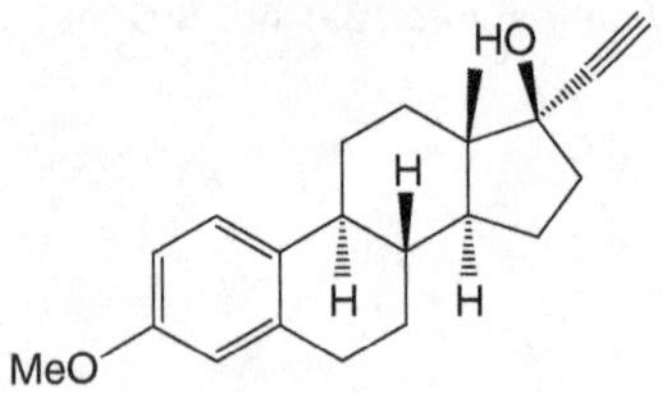

Chemical name:
Formula: $C_{21}H_{26}O_2$
Monoisotopic protonated ion (*m/z*): 311.2006
Molecular weight: 310.44
CAS number: 72-33-3

Ethinylestradiol and its methyl derivative, mestranol, have long been utilized as oral contraceptive agents.[1] Nevertheless, there seem to be only a limited number of studies looking into the degradation behaviors of both drug molecules. In a thermal degradation study of ethinylestradiol and mestranol [1], both drug molecules underwent oxidative degradation at the 6-benzylic position of the steroid core after the solid samples were subjected to heating at 178°C and 147°C, respectively, for a few hours while exposing to air, yielding the corresponding 6α-hydroxy- and 6β-hydroxy-derivatives as the major degradants and 6-keto- and $\Delta^{6,7}$-derivatives as the minor

1 Ethinylestradiol has been known since 1938: H.H. Inhoffen, W. Logemann, W. Hohlweg, and A. Serini, Chem. Ber., 71: 1024 (1938), while mestranol was disclosed in a 1954 patent: F.B. Colton, U.S. Patent 2,666,769. January 19, 1954.

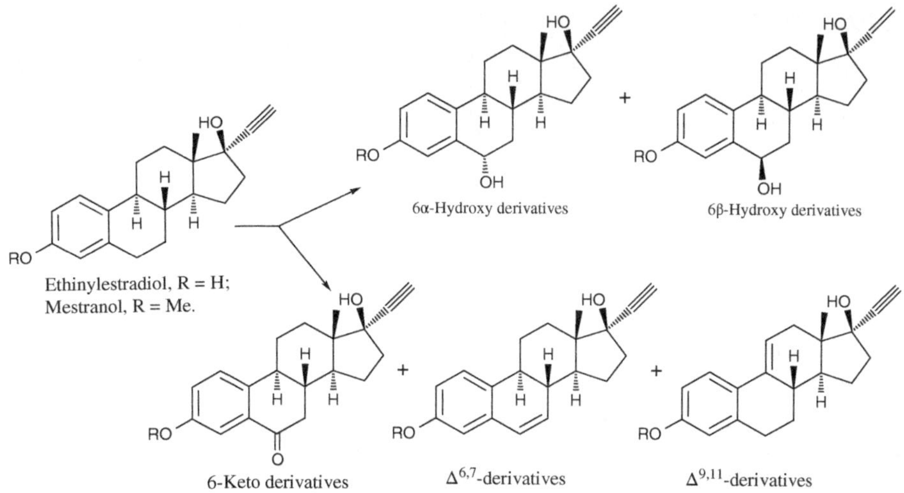

Scheme 188–189 Oxidative degradation pathways of ethinylestradiol and mestranol.

degradants. According to a paper by Ekhato et al. [2], the above four degradants were observed in formulated ethinylestradiol products stored under ambient laboratory conditions for 3 years. In addition, $\Delta^{9,11}$-degradant was also observed. The oxidative degradation pathways of ethinylestradiol and mestranol are illustrated in Scheme 188–189. Li et al. reported improved syntheses of these oxidative degradants of ethinylestradiol and mestranol [3].

References

1 Cotter, M.L., Levine, S.D., Mallory, R., and Shaw, C. (1978). Thermal degradation of mestranol and ethinyl estradiol. *Tetrahedron Lett.* 22: 1939–1942.

2 Ekhato, I.V., Hurley, T., Lovdahl, M. et al. (2002). Norethindrone acetate (NA) and ethinyl estradiol (EE) related oxidative transformation products in stability samples of formulated drug product: synthesis of authentic references. *Steroids* 67: 165–174.

3 Li, H., Song, Y., and Peng, X. (2008). Improved synthesis of mestranol and ethinyl estradiol (EE) related degradation products as authentic references. *Steroids* 73: 488–494.

190: Etodolac

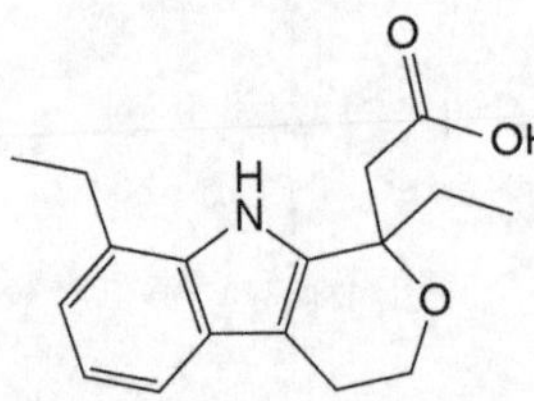

Chemical name: 2-(1,8-Diethyl-4,9-dihydro-3*H*-pyrano[3,4-b]indol-1-yl)acetic acid
Trade name: Lodine
Formula: $C_{17}H_{21}NO_3$
Monoisotopic protonated ion (*m/z*): 288.1594
Molecular weight: 287.36
CAS number: 41340-25-4

Etodolac is a selective cyclooxygenase inhibitor clinically used for treating pain, fever, etc. Lee et al. studied its degradation pathway and kinetics in acidic aqueous solutions [1]. The degradation started with the decarboxylation of the drug molecule (**1**) via a concerted mechanism through a six-membered intermediate. The concerted mechanism was consistent with the small positive entropy of the activation ($\Delta S^{\neq}$) as determined in the study. The initial degradation product from the decarboxylation was degradant **2A**, which was partially converted to degradant **2B** at pH below 3. Both **2A** and **2B** appeared to be intermediary degradants as **3** was observed as the major degradant, which further degraded into degradant **4** through a retro-aldo-like process [2] in which butanone was eliminated. The retro-aldo-like process was probably made possible via the enolization of the open ring form of **3**. The degradation pathway of etodolac under acidic conditions is summarized in Scheme 190.

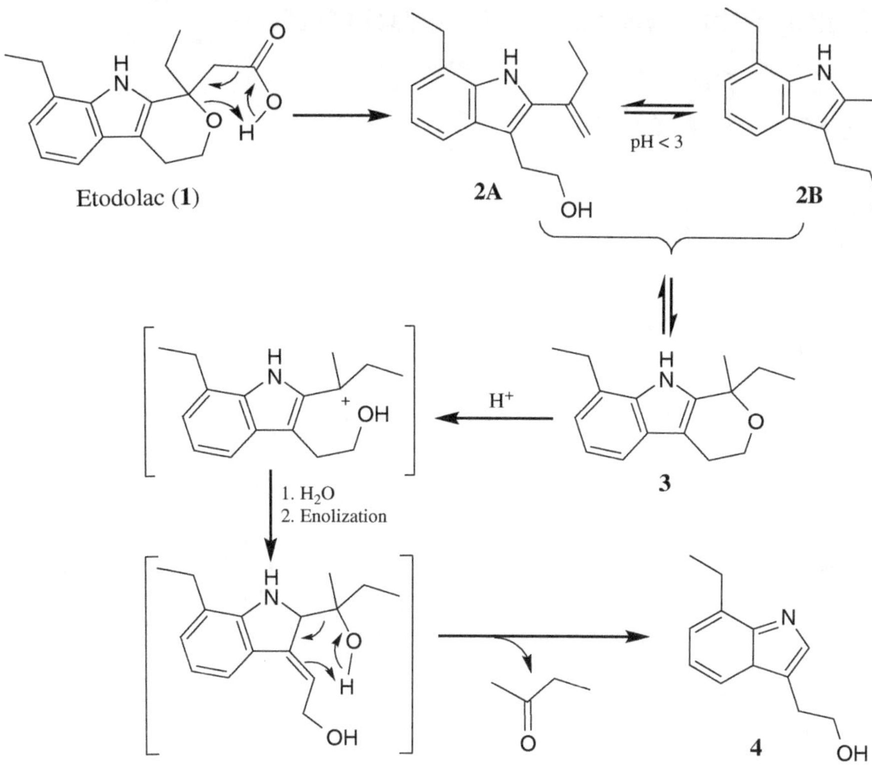

Scheme 190 Degradation pathway of etodolac under acidic conditions.

References

1 Lee, Y.J., Padula, J., and Lee, H.-K. (1988). Kinetics and mechanisms of etodolac degradation in aqueous solutions. *J. Pharm. Sci.* 77: 81–86.

2 Li, M. (2012). Aldol condensation and retro-aldol. In: *Organic Chemistry of Drug Degradation*, 124–127. Cambridge, UK: RSC Publishing.

191: Etoposide [WHO List of Essential Medicines]

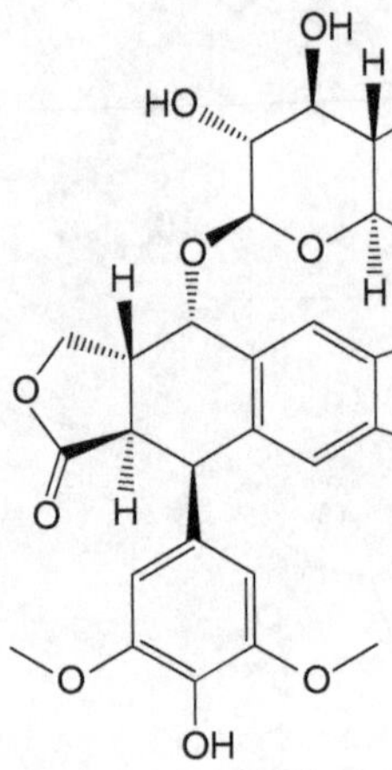

Chemical name: (5*S*,5a*R*,8a*R*,9*R*)-5-[[(2*R*,4a*R*,6*R*,7*R*,8*R*,8a*S*)-7,8-Dihydroxy-2-methyl-4,4a,6,7,8,8a-hexahydropyrano[3,2-d][1,3]dioxin-6-yl]oxy]-9-(4-hydroxy-3,5-dimethoxyphenyl)-5a,6,8a,9-tetrahydro-5*H*-[2]benzofuro[6,5-f][1,3]benzodioxol-8-one
Trade name: Vepesid, Etopophos, Toposar
Formula: $C_{29}H_{32}O_{13}$
Monoisotopic protonated ion (*m/z*): 589.1916
Molecular weight: 588.56
CAS number: 33419-42-0

Etoposide is a chemotherapeutic agent for the treatment of a variety of cancers, including lung cancer, lymphoma, leukemia, testicular cancer, and ovarian cancer [1]. It is a semi-synthetic epipodophyllotoxin from the wild mandrake; structurally, it is a glycoside that contains a modified D-glucose unit. Both the sugar unit and the lactone moiety of etoposide are susceptible to hydrolytic degradation, and its degradation pathways mostly result from the hydrolysis of these two structural moieties (Scheme 191) [2]. In acidic solutions with pH below 5.0, the glycosidic linkage was cleaved to give the aglycone 4′-demethylepipodophyllotoxin (I), which could further degrade to form the corresponding *trans*-hydroxyl carboxylic acid via the opening of the lactone ring (II). In solutions with pH greater than 5.0, the epimerization of the *trans*-fused lactone ring became the most significant degradation pathway, in which the isomeric degradant with the *cis*-fused lactone ring (IV) was formed. The lactone moiety of IV could further decompose to give the corresponding *cis*-hydroxyl carboxylic degradant (V) [2, 3]. Overall, etoposide displayed a V-shaped pH stability profile with the most stable region near pH 5 [2]. In a study by Tian et al., etoposide in liquid emulsion and aqueous solution was found to be most stable at pH 5.0 with half-lives of 54.7 hour and 38.6 minutes, respectively, at 80°C [4]. Joel et al. studied the stability of both i.v. and oral-formulated etoposide solutions and found significant concentration-dependent degradation, with the more concentrated solutions degraded faster than less concentrated ones [5].

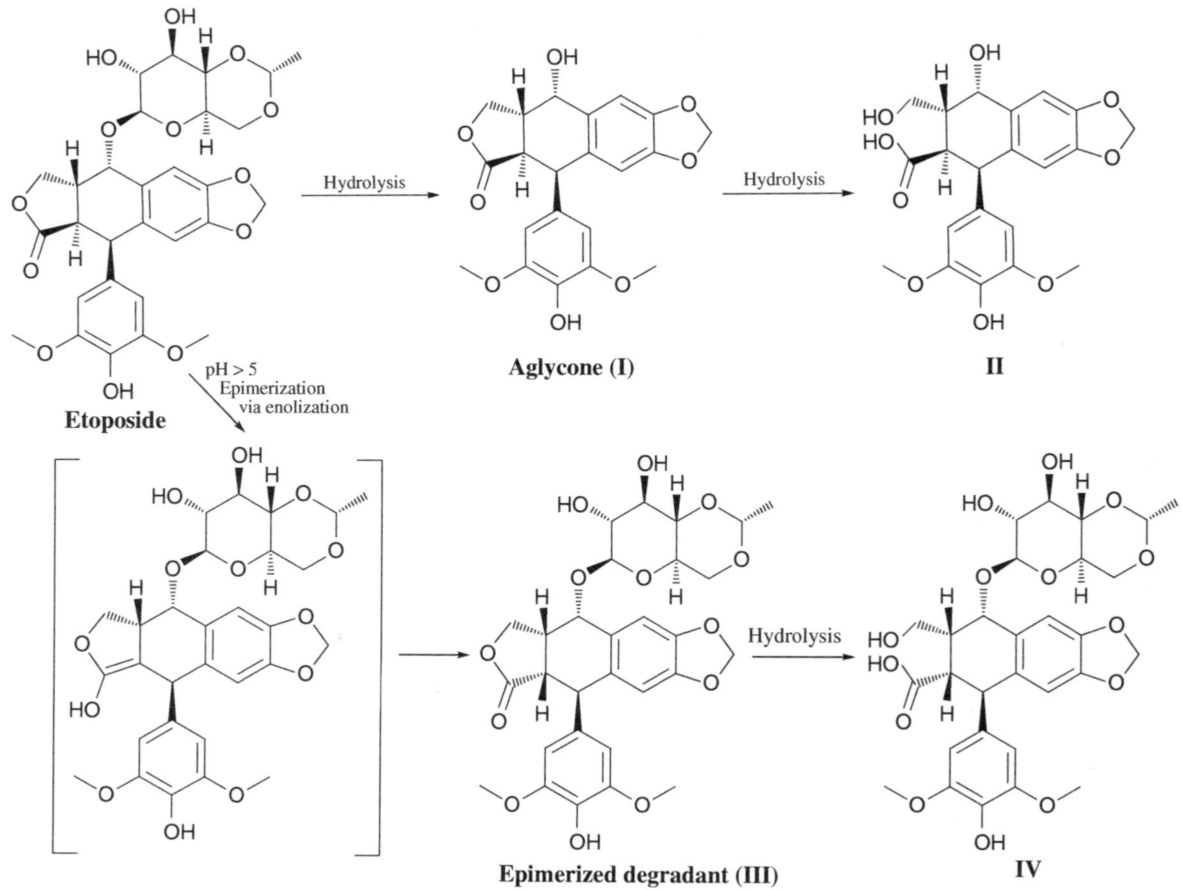

Scheme 191 Degradation pathways of etoposide under acidic and alkaline solutions.

References

1 Etoposide. https://en.wikipedia.org/wiki/Etoposide (accessed 5 April 2023).
2 Beijnen, J.H., Holthuis, J.J.M., Kerkdijk, H.G. et al. (1988). Degradation kinetics of etoposide in aqueous solution. *Int. J. Pharm.* 41: 169–178.
3 Strife, R.J., Jardine, I., and Colvin, M. (1980). Analysis of the anticancer drugs VP16-213 and VM 26 and their metabolites by high performance liquid chromatography. *J. Chromatogr.* 182: 211–220.
4 Tian, L., He, H., and Tang, X. (2007). Stability and degradation kinetics of etoposide-loaded parenteral lipid emulsion. *J. Pharm. Sci.* 96: 1719–1728.
5 Joel, S.P., Clark, P.I., and Slevin, M.L. (1995). Stability of the i.v. and oral formulations of etoposide in solution. *Cancer Chemother. Pharmacol.* 37: 117–124.

192: Famotidine

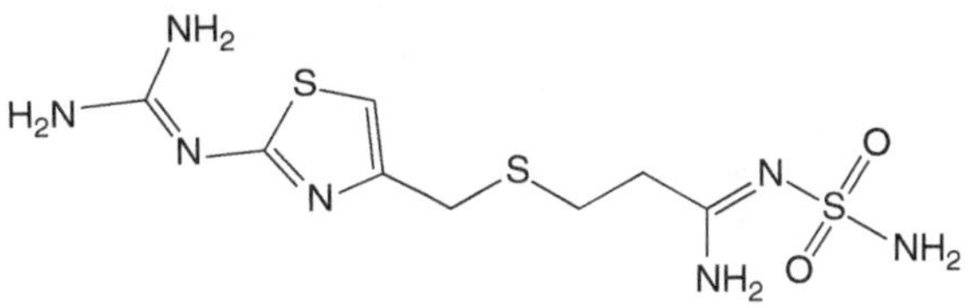

Chemical name: 3-[[2-(Diaminomethylideneamino)-1,3-thiazol-4-yl] methylsulfanyl]-*N*′-sulfamoylpropanimidamide
Trade name: Pepcid
Formula: $C_8H_{15}N_7O_2S_3$
Monoisotopic protonated ion (*m/z*): 338.0522
Molecular weight: 337.44
CAS number: 76824-35-6

Famotidine is a histidine-2 (H_2) receptor antagonist that pharmacologically suppresses gastric secretion. It is clinically used for the treatment of duodenal ulcers, gastric ulcers, gastritis, and gastroesophageal reflux. Its main degradation is hydrolytic, resulting from the hydrolysis of its *N*'-sulfamoylpropanimidamide moiety under both acidic and alkaline conditions, yielding famotidone and several other degradants [1]. According to a study by Islam and Narurkara, the maximum stability of famotidine occurred at pH 6.3 [2]. A later study reported that in a 2M NaOH solution, famotidine was also found to undergo retro-Michael addition, resulting in the breakage of the thiol ether linkage [3]. Nevertheless, the acidic hydrolysis appears to be more pharmacologically relevant, as the acidic degradation is believed to be mostly responsible for the poor bioavailability of famotidine. The pseudo-first-order reaction rate constants were found to be 0.392/hour (or 6.53×10^{-3}/minutes; $t_{1/2}$ 1.77 hour) and 8.1×10^{-3}/minutes in 0.1 N HCl and simulated gastric fluid at 37°C, respectively [4, 5]. In another kinetic study, the activation energy for famotidine degradation was reported to be 16.000–17.563 kcal/mol within the pH range of 1.71–4.00 [1].

When treated with hydrogen peroxide, famotidine can be converted into its *S*-oxide [6], which is also a metabolite of famotidine [7]. The overall degradation pathways of famotidine are illustrated in Scheme 192.

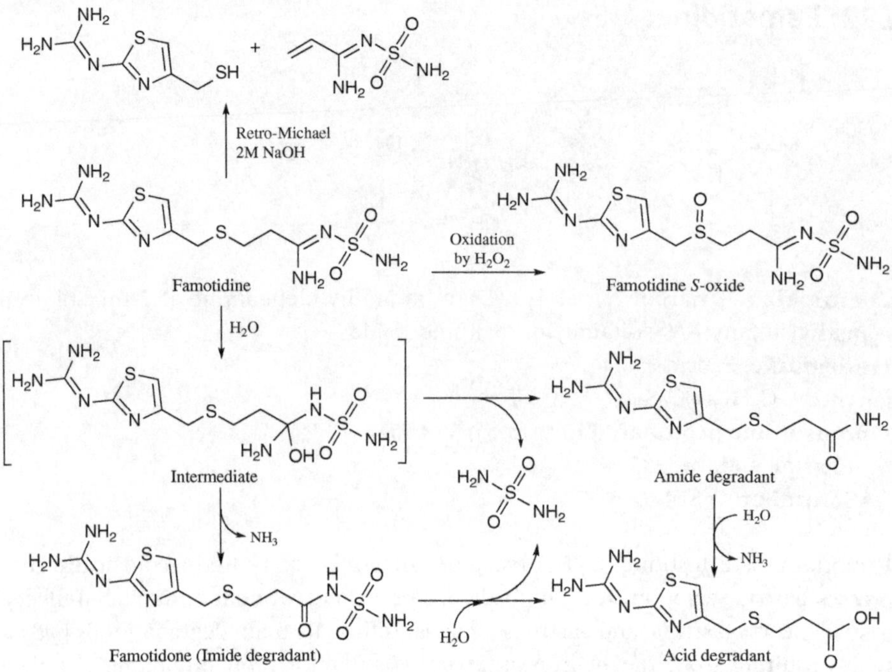

Scheme 192 Degradation pathways of famotidine.

References

1 Junnarkar, G.H. and Stavchansky, S. (1995). Isothermal and nonisothermal decomposition of famotidine in aqueous solution. *Pharm. Res.* 12: 599–604.

2 Islam, M.S. and Narurkar, M.M. (1993). Solubility, stability and ionization behaviour of famotidine. *J. Pharm. Pharmacol.* 45: 682–686.

3 Singh, S., Kumar, S., Sharda, N., and Chakrabort, A.K. (2001). New findings on degradation of famotidine under basic conditions: Identification of a hitherto unknown degradation product and the condition for obtaining the propionamide intermediate in pure form. *J. Pharm. Sci.* 91: 253–257.

4 Suleiman, M.S., Muti, H.Y., Abdel-Hamid, M.E. et al. (1989). A stability-indicating HPLC analysis of famotidine and its application to kinetic studies. *Anal. Lett.* 22: 1499–1512.

5 El-Shaheny, R., Radwan, M.O., Belal, F., and Yamadaf, K. (2020). Pentabromobenzyl-RP versus triazole-HILIC columns for separation of the polar basic analytes famotidine and famotidone: LC method development combined with in silico tools to follow the potential consequences of famotidine gastric instability. *J. Pharm. Biomed. Anal.* 186: 113305.

6 Kelani, K.M., Aziz, A.M., Hegazy, M.A., and Fattah, L.A. (2002). UV-spectrophotometric stability indicating methods for the quantitative determination of cimetidine, famotidine, and ranitidine hydrochloride in the presence of their oxidative derivatives. *Anal. Lett.* 35: 1055–1073.

7 Al-Omar, M.A. and Al-Mohizea, A.M. (2009). Famotidine. In: *Profiles of Drug Substances, Excipients and Related Methodology*, vol. 34 (ed. H.G. Brittain), 115–151. Elsevier.

193: Fentanyl [WHO List of Essential Medicines]

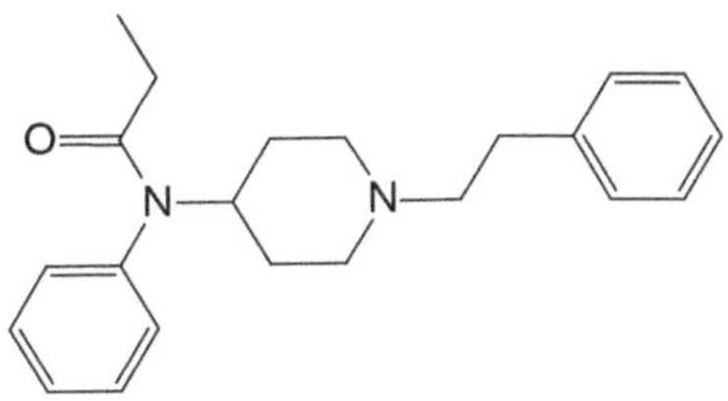

Chemical name: *N*-Phenyl-*N*-[1-(2-phenylethyl)piperidin-4-yl]propanamide
Trade name: Sublimaze, Abstral, Durogesic, Fentanest
Formula: $C_{22}H_{28}N_2O$
Monoisotopic protonated ion (*m/z*): 337.2274
Molecular weight: 336.48
CAS number: 437-38-7
Common salt form: Citrate

Fentanyl, first synthesized by Paul Janssen in 1959, is a highly potent synthetic narcotic, used primarily as a pain reliever for cancer patients and those after painful surgical operations [1]. It is also used as a sedative in combination with a benzodiazepine such as midazolam [2]. On the other hand, abuse of fentanyl has become a major issue in the United States, particularly in recent years. It is a Schedule II-controlled substance by the US DEA. Fentanyl contains an acyl group and a tertiary amine moiety, which are susceptible to hydrolytic and oxidative degradation, respectively. Garg et al. performed a comprehensive forced degradation study of fentanyl under acidic, alkaline, oxidative, thermal, and photolytic conditions [3]. Acidic stress of fentanyl in 5N HCl/acetonitrile (3/6, v/v) at 70°C for 24 hours generated ~35% yield of *N*-phenyl-1-(2-phenylethyl)-piperidin-4-amine (PPA), while it was stable in alkaline stress in 5N NaOH/acetonitrile (3/6, v/v) at the same temperature and duration. Oxidative stress of fentanyl (neutral base) in 0.3% hydrogen peroxide/acetonitrile (1/10, v/v) at ambient temperature produced two diastereomeric *N*-oxide degradants, with the major β-diastereomer formed in 6.51%. On the contrary, a previous oxidative stress of fentanyl citrate in 3% hydrogen peroxide by Lambropoulos et al. did not induce any degradation [4]. The inactivity of fentanyl citrate toward the nucleophilic oxidation under the apparently stronger oxidative stress condition employed by Lambropoulos et al. is most likely because the nucleophilicity of the tertiary amine functionality is inhibited by protonation via citric acid. Heating of fentanyl at 350°C for 5 minutes yielded five major degradants, three of which were identified in a previous thermal stress study [5]. The degradation pathways of fentanyl are shown in Scheme 193. The degradation that occurred under thermal stress (350°C, 5 minutes) seems to start from two pathways of C—N cleavage via β-elimination, followed by several steps of oxidation.

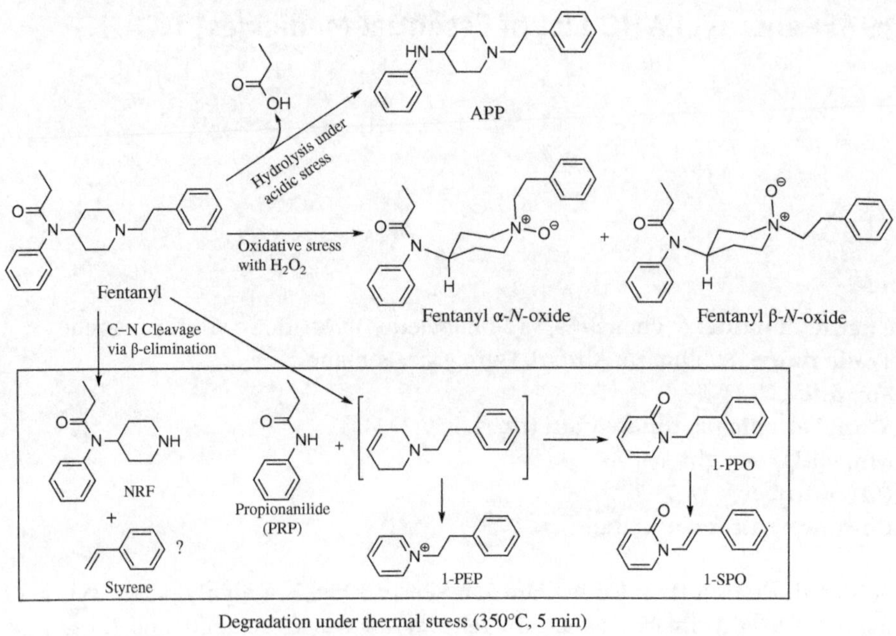

Scheme 193 Degradation pathways of fentanyl. The proposed pathways for the degradation that occurred under thermal stress are somewhat different from those described by Garg et al. [3]. *Source:* Garg et al. [3]/with permission of Elsevier.

References

1 Fentanyl. https://en.wikipedia.org/wiki/Fentanyl (accessed 19 May 2023).

2 Lobb, D., Clarke, A., and Lai, H. (2018). Administration order of midazolam/fentanyl for moderate dental sedation. *J. Dent. Anesth. Pain Med.* 18: 47–56.

3 Garg, A., Solas, D.W., Takahashi, L.H., and Cassella, J.V. (2010). Forced degradation of fentanyl: identification and analysis of impurities and degradants. *J. Pharm. Biomed. Anal.* 53: 325–334.

4 Lambropoulos, J., Spanos, G.A., Lazaridis, N.V. et al. (1999). Development and validation of an HPLC assay for fentanyl and related substances in fentanyl citrate injection, USP. *J. Pharm. Biomed. Anal.* 20: 705–716.

5 Rabinowitz, J.D., Wensley, M., Lloyd, P. et al. (2004). Fast onset medications through thermally generated aerosols. *J. Pharmacol. Exp. Ther.* 309: 769–775.

194–195: Fexofenadine, Terfenadine

194: Fexofenadine

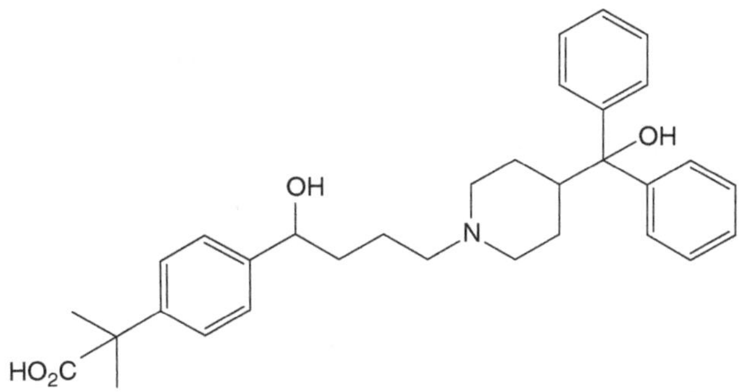

Chemical name: 2-[4-[1-Hydroxy-4-[4-[hydroxy(diphenyl)methyl]piperidin-1-yl]butyl]phenyl]-2-methylpropanoic acid
Trade name: Allegra
Formula: $C_{32}H_{39}NO_4$
Monoisotopic protonated ion (*m/z*): 502.2952
Molecular weight: 501.67
CAS number: 153439-40-8
Common salt form: HCl salt

195: Terfenadine

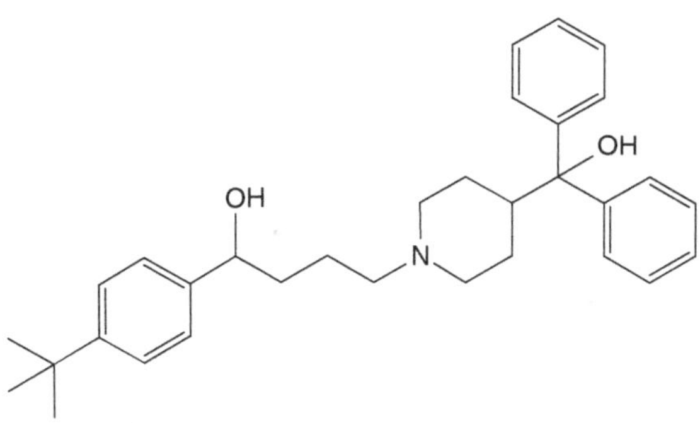

Chemical name: 1-(4-tert-Butylphenyl)-4-[4-[hydroxy(diphenyl)methyl]piperidin-1-yl]butan-1-ol
Formula: $C_{32}H_{41}NO_2$
Monoisotopic protonated ion (*m/z*): 472.3210
Molecular weight: 471.67
CAS number: 50679-08-8

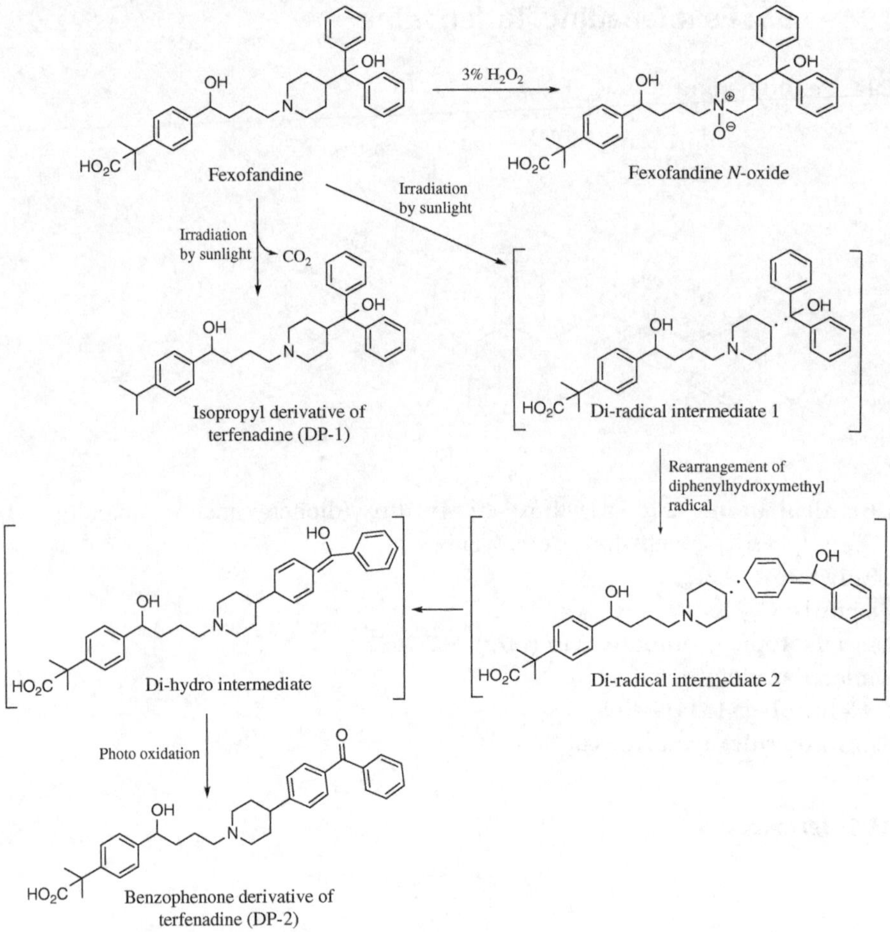

Scheme 194–195-1 Degradation pathways of fexofenadine.

Terfenadine was originally intended to be developed as a potential tranquilizer, but, instead, it was brought to market in 1985 as the first of so-called second-generation antihistamines, which have minimal sedative effect due to their inability to cross the blood-brain barrier [1]. Terfenadine was withdrawn from the US market in 1998 due to the cardiac risk at high doses, and it was superseded by its carboxylic metabolite, fexofenadine. Both drugs contain a piperidine moiety, a tertiary amine functionality, and thus are susceptible to oxidation to form the corresponding *N*-oxide degradants. In a forced degradation study of fexofenadine by Vaghela et al., its *N*-oxide degradant was formed in ~2% yield after being subjected to stress in 3% hydrogen peroxide at 60°C for 5 hours [2]. In a photostability study of fexofenadine drug substance in methanolic solutions with UV irradiation at 254 nm, two major photodegradants (DP-1 and DP-2) were observed. While DP-1 apparently resulted from the photodecarboxylation, the formation of DP-2 does not appear to be straightforward. Although the original authors did not provide a probable formation mechanism for DP-2, it would likely be mediated by a free radical process, in which the initially formed diphenylhydroxymethyl radical would undergo a rearrangement followed by radical coupling and

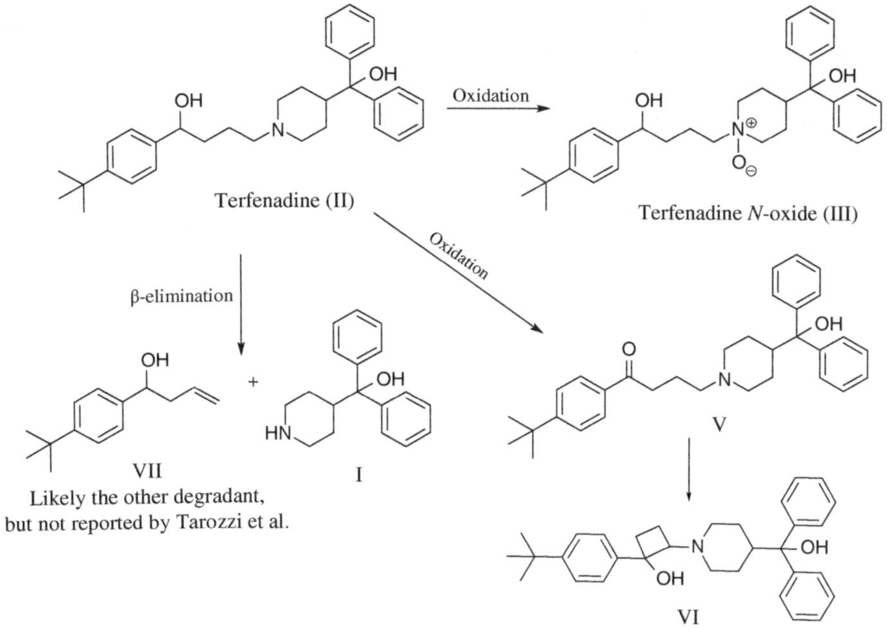

Scheme 194–195-2 Photodegradation pathways of terfenadine. The nomenclature of the degradants is that of Tarozzi et al., except for VII, which was not reported in the original paper. *Source:* Tarozzi et al. [5]/John Wiley & Sons.

subsequent photooxidation (Scheme 194–195-1) [3]. This photo-initiated free radical process would be similar to a photo-Fries rearrangement process [4]. These two photodegradants were also present in fexofenadine tablets after being subjected to the same UV irradiation at 254 nm, indicating that the formulated fexofenadine-formulated products should be protected from light.

In another photostability study of terfenadine by Tarozzi et al., terfenadine was exposed to both UV and natural sunlight, giving rise to different and complex degradation product distributions. The photodegradants formed under natural sunlight irradiation were analyzed by LC-MS, and their structures were proposed (Scheme 194–195-2) [5]. It needs to be noted that Degradant I, reported by Tarozzi et al., would likely result from a β-elimination process, and the other degradant stemming from this process would be Degradant VII or its analog, which was not reported in the original paper.

References

1 Terfenadine. https://en.wikipedia.org/wiki/Terfenadine (accessed 20 May 2023).
2 Vaghela, B., Rao, S.S., Reddy, A.M. et al. (2012). Identification and characterization of an oxidative degradation product of fexofenadine, development and validation of a stability-indicating RP-UPLC method for the estimation of process related impurities and degradation products of fexofenadine in pharmaceutical formulations. *Sci. Pharm.* 80: 295–309.

3 Breier, A.R., Nudelman, N.S., Steppe, M., and Schapoval, E.E.S. (2008). Isolation and structure elucidation of photodegradation products of fexofenadine. *J. Pharm. Biomed. Anal.* 46: 250–257.
4 Li, M. (2012). Photo-Fries rearrangement. In: *Organic Chemistry of Drug Degradation*. 173–174. Cambridge, UK: RSC Publishing.
5 Tarozzi, A., Andrisano, V., Fiori, J. et al. (2003). Photomutagenic properties of terfenadine as revealed by a stepwise photostability, phototoxicity and photomutagenicity testing approach. *Photochem. Photobiol.* 77: 356–361.

196–197: Fluconazole, Voriconazole

196: Fluconazole [WHO List of Essential Medicines]

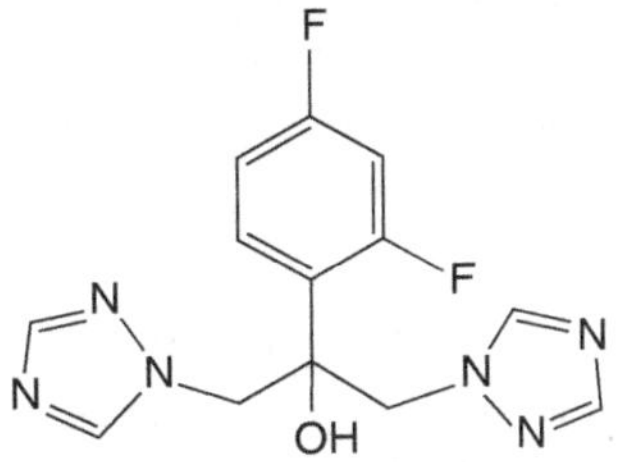

Chemical name: 2-(2,4-Difluorophenyl)-1,3-bis(1,2,4-triazol-1-yl)propan-2-ol
Trade name: Diflucan
Formula: $C_{13}H_{12}F_2N_6O$
Monoisotopic protonated ion (*m/z*): 307.1113
Molecular weight: 306.28
CAS number: 86386-73-4

197: Voriconazole

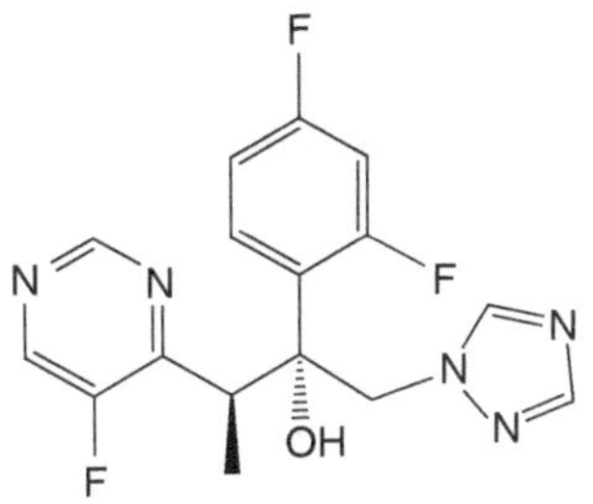

Chemical name: (2*R*,3*S*)-2-(2,4-Difluorophenyl)-3-(5-fluoropyrimidin-4-yl)-1-(1,2,4-triazol-1-yl)butan-2-ol
Trade name: Vfend
Formula: $C_{16}H_{14}F_3N_5O$
Monoisotopic protonated ion (*m/z*): 350.1223
Molecular weight: 349.32
CAS number: 188416-29-7

Fluconazole is a first-generation antifungal drug based on a triazole structural moiety. Despite its long history since its commercialization in 1998, there have been very limited studies with regard to its stability or degradation chemistry. Among those few studies, there seems to be no reporting of structures for the potential degradants observed in the forced degradation, except for one photostability study [1]. In this study, a number of degradants were observed in the photostress of fluconazole in aqueous solutions with UV irradiation at 254 nm, two of which were determined to be monohydroxylated and dihydroxylated defluorination products, respectively. As the photostability experiments were performed in pure aqueous environment, the corresponding simple defluorinated degradation products were not seen, apparently due to a lack of a proper proton radical donor, which would have been available if an organic cosolvent is present (Scheme 196–197-1) [2]. In a forced degradation study by Corrêa et al., fluconazole was found to be mostly stable under the conditions of acidic (0.1 M HCl) and alkaline (0.1 M NaOH) stress, either at room temperature or 90°C for up to 6 hours [3]. While no degradation was observed after being stressed in 3% hydrogen peroxide at room temperature for 1 hour, heating the 3% hydrogen peroxide stress solution at 90°C produced a possible degradant peak; nevertheless, the structure of this possible degradant was not elucidated. In another oxidative degradation study, fluconazole was found to produce a degradant in ~32% yield after being stressed in 35% hydrogen peroxide solution at 150°C for 30 minutes [4]. Nevertheless, the degradant formed under such an extreme condition would most likely be artificial.

On the other hand, voriconazole produced one degradant (DFH impurity) under an alkaline condition (0.1 M NaOH, room temperature, 30 minutes) (Scheme 196–197-2) [5]. The formation of this particular degradant may be due to the introduction of 5-fluoropyrimidinyl group in voriconazole, which made the breakage of the C—C bond possible via a concerted rearrangement mechanism. The other degradant 4-ethyl-5-fluoropyrimidine, which would be expected to form from this degradation process, was not reported in the study, which may be attributed to its volatility. A later study indicated that voriconazole underwent the same degradation under thermal stress conditions at 60°C for 24 hours [6].

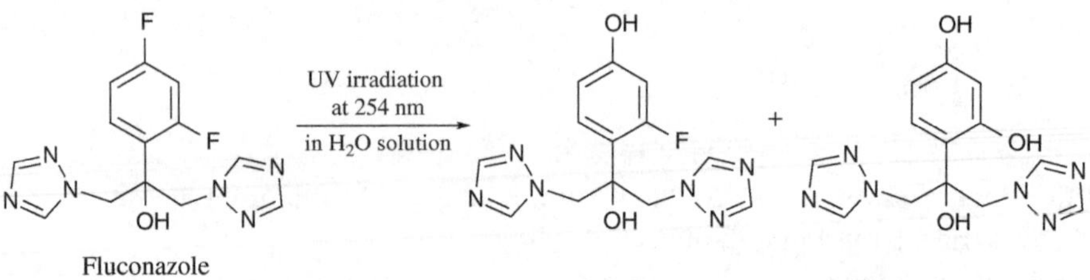

Scheme 196–197-1 Photodegradation pathway of fluconazole in pure aqueous solution.

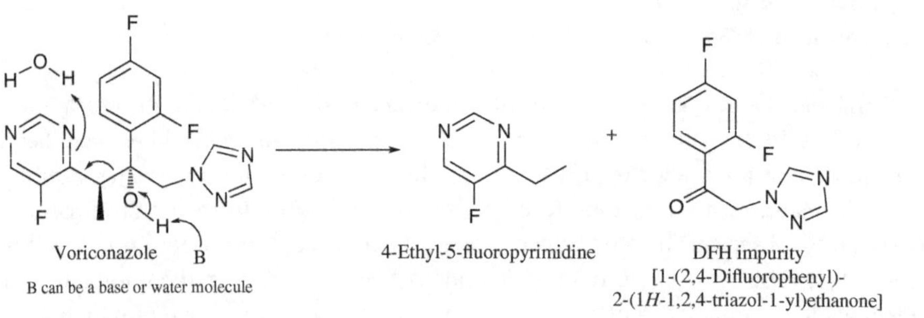

Scheme 196–197-2 Degradation of voriconazole under alkaline and thermal stress conditions.

References

1 Corrêa, J.C.R., Reichman, C., Vianna-Soares, C.D., and Salgado, H.R.N. (2011). Stability study of fluconazole applying validated bioassay and stability-indicating LC methods. *J. Anal. Bioanal. Tech.* 2: 1000126.

2 Li, M. (2012). Photochemical Degradation. In: *Organic Chemistry of Drug Degradation*, 165–197. Cambridge, UK: RSC Publishing.

3 Chen, Z.-F., Ying, G.G., Jiang, Y.-X. et al. (2014). Photodegradation of the azole fungicide fluconazole in aqueous solution under UV-254: kinetics, mechanistic investigations and toxicity evaluation. *Water Res.* 52: 83–91.

4 Lotfy, H.M., Abdel-Aleem, A.-A.A.-B., and Monir, H.H. (2013). Stability-indicating high performance liquid chromatography determination of fluconazole in the presence of its oxidative degradation product – kinetic and stress study. *J. Liq. Chromatogr. Relat. Technol.* 36: 1013–1029.

5 Adams, A.I.H., Gosmann, G., Schneider, P.H., and Bergold, A.M. (2009). LC stability studies of voriconazole and structural elucidation of its major degradation product. *Chromatographia* 69: 115–122.

6 Shaikh, K.A. and Patil, A.T. (2012). A validated stability-indicating liquid chromatographic method for determination of degradation impurities and diastereomers in voriconazole tablets. *Sci. Pharm.* 80: 879–888.

198: Fludrocortisone Acetate [WHO List of Essential Medicines]

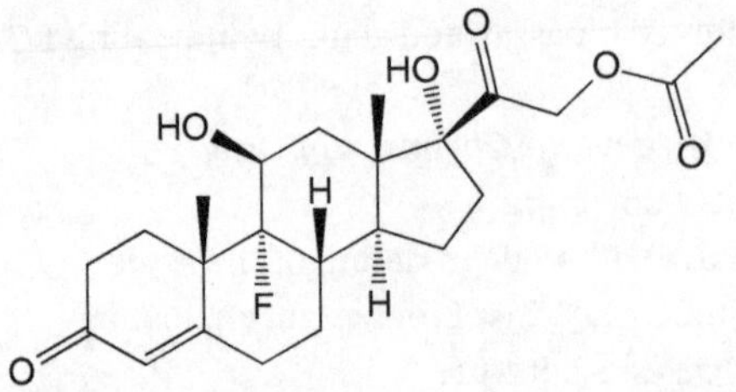

Chemical name: (11β)-9-Fluoro-11,17-dihydroxy-3,20-dioxopregn-4-en-21-yl acetate
Trade name: Florinef
Formula: $C_{23}H_{31}FO_6$
Monoisotopic protonated ion (*m/z*): 423.2177
Molecular weight: 422.49
CAS number: 514-36-3

Fludrocortisone is an adrenal steroid clinically used to treat primary and secondary adrenocortical insufficiency [1]. It was the first synthetic corticosteroid drug and the first fluorine-containing drug [2], which is usually administered in its 21-acetate prodrug form. Hence, hydrolysis is its primary degradation pathway, in which fludrocortisone is generated (Scheme 198). On the other hand, oxidative insertion between 3,4-positions of the A-ring via the Baeyer–Villiger mechanism could be a potential degradation pathway, as such a pathway was reported for other corticosteroids containing the 1,2-dihydro A ring moiety [3, 4].

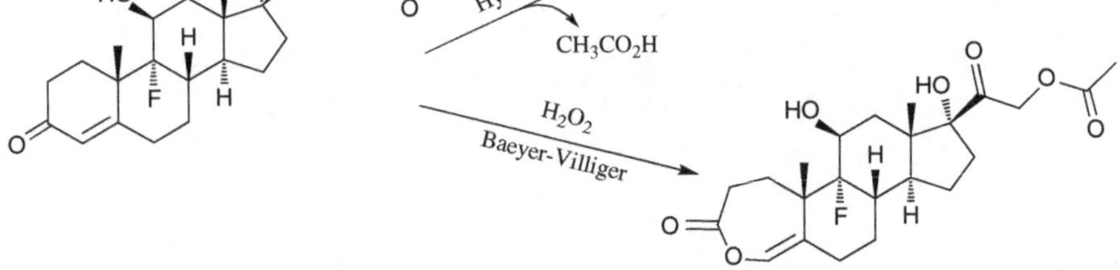

Scheme 198 Hydrolytic and potential oxidative degradation pathways of fludrocortisone.

References

1 Rahman, M. and Anjum, F. (2023). *Fludrocortisone*. StatPearls Publishing https://www.ncbi.nlm.nih.gov/books/NBK564331/ (accessed 27 May 2023).

2 Fludrocortisone. https://en.wikipedia.org/wiki/Fludrocortisone (accessed 27 May 2023).

3 Caspi, E., Chang, Y.W., and Dorfman, R.I. (1962). Degradation of corticosteroids. VII. The synthesis of 7-membered ring-A enol-lactones. *J. Med. Pharm. Chem.* 5: 714–719.

4 Pinhey, J.T. and Schaffner, K. (1968). Baeyer-Villiger oxidation of Δ^4-3-oxo-steroids and 4,5-epoxy-3-oxo-steroids. A route to some novel A-nor-steroids. *Aust. J. Chem.* 21: 1873–1889.

199: Flucytosine [WHO List of Essential Medicines]

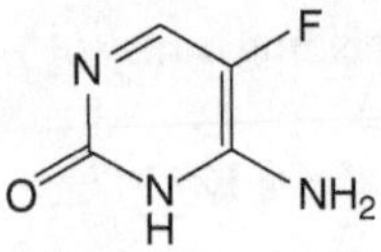

Chemical name: 6-Amino-5-fluoro-1*H*-pyrimidin-2-one
Trade name: Ancobon
Formula: $C_4H_4FN_3O$
Monoisotopic protonated ion (*m/z*): 130.0411
Molecular weight: 129.09
CAS number: 2022-85-7

Flucytosine, introduced into the market in the 1950s, is an antifungal drug specifically for treating serious *Candida* infections and cryptococcosis, and for this indication, it is used in combination with another antifungal drug amphotericin B [1]. It exerts its pharmacological and therapeutic effects by transforming in vivo into fluorouracil, which is its synthetic precursor as well as degradation product due to deamination formed under hydrolytic conditions (Scheme 199) [2]. There are a limited number of forced degradation studies, in which a major degradant was observed under both hydrolytic (acidic) and oxidative (H_2O_2) conditions. Nevertheless, the structure of the degradant was not identified in these studies [3, 4].

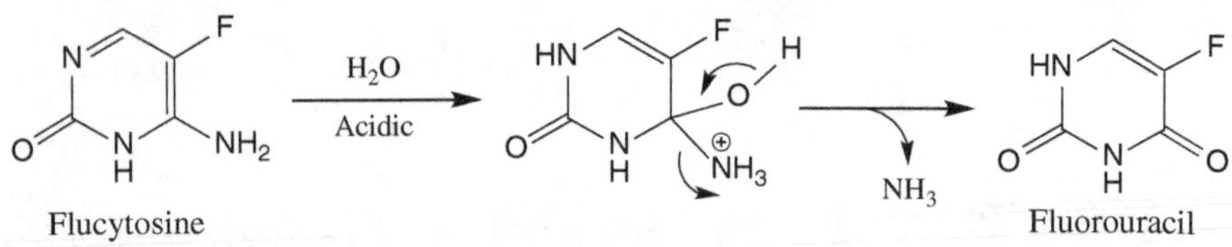

Scheme 199 Deamination of flucytosine under acidic conditions.

References

1 Flucytosine. https://en.wikipedia.org/wiki/Flucytosine (accessed 20 May 2023).

2 Cavrini, V., Bonazzi, D., and Di Pietra, A.M. (1991). Analysis of flucytosine dosage forms by derivative UV spectroscopy and liquid chromatography. *J. Pharm. Biomed. Anal.* 9: 401–407.

3 Annapurna, M.M., Baswani, R., and Pradhan, D.P. (2019). New stability indicating RP-UFLC method for the determination of flucytosine – an anti fungal agent. *Acta Sci. Pharm. Sci.* 3: 38–46.

4 Venkata, N.P. and Seru, G. (2021). A new stability indicating RP-HPLC method for the estimation of flucytosine in presence of an internal standard. *Res. J. Pharm. Technol.* 14: 289–294.

200: Fluorouracil [WHO List of Essential Medicines]

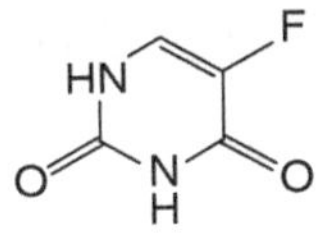

Chemical name: 5-Fluoro-1*H*-pyrimidine-2,4-dione
Formula: $C_4H_3FN_2O_2$
Monoisotopic protonated ion (*m/z*): 131.0251
Molecular weight: 130.08
CAS number: 51-21-8

Fluorouracil, also known as 5-fluorouracil (5-FU or FU), is one of the widely used drugs for the treatment of various cancers. Its mode of action is believed to block thymidylate synthase and thus inhibit the production of DNA [1]. During the process of developing a stability-indicating method, fluorouracil was found to be more unstable under alkaline conditions than acidic one although the forced degradation products were not identified [2]. In a degradation study in aqueous solutions, Legay et al. investigated the hydrolytic pathway of fluorouracil in NaOH and Tris solutions, respectively [3]. In NaOH solutions, the initial degradation product was a pair of 6-hydroxyl intermediates (FUOH), formed via Michael addition of the hydroxide ion onto fluorouracil (Scheme 200). Subsequent degradation proceeded via either defluorination to afford 5-hydroxyuracil or hydrolysis of the imide linkage to give two diastereomeric, open-ring degradants (FHUPA). Cleavage of the C—N bond led to 2-fluoro-3-oxopropanoic acid (FOPA), which would be prone to decarboxylation,

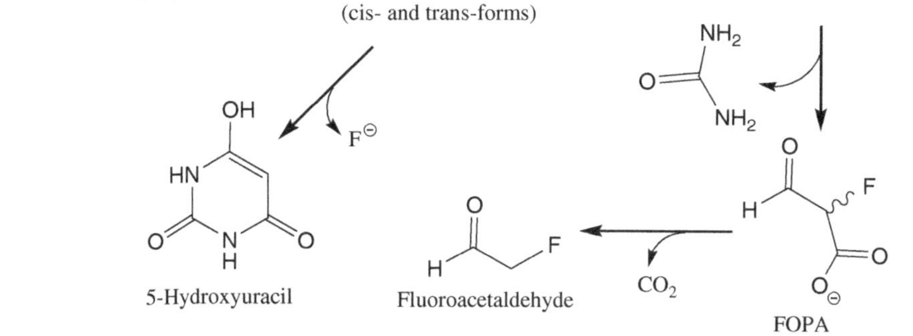

Scheme 200 Hydrolytic degradation pathway of fluorouracil.

resulting in the formation of fluoroacetaldehyde. Both 2-fluoro-3-oxopropanoic acid and fluoroacetaldehyde are cardiotoxic [4]. In Tris solutions, fluoroacetaldehyde was found to form several condensation products between the functional groups of Tris and fluoroacetaldehyde.

References

1 Fluorouracil. https://en.wikipedia.org/wiki/Fluorouracil (accessed 25 May 2023).
2 Alanazi, F.K., Yassin, A.E., El-Badry, M. et al. (2009). Validated high-performance liquid chromatographic technique for determination of 5-fluorouracil: applications to stability studies and simulated colonic media. *J. Chromatogr. Sci.* 47: 558–563.
3 Legay, R., Massou, S., Azéma, J. et al. (2014). Hydrolytic pathway of 5-fluorouracil in aqueous solutions for clinical use. *J. Pharm. Biomed. Anal.* 98: 446–462.
4 Fournet, A., Gilard, V., Malet-Martino, M. et al. (2000). Stability of commercial solutions of 5-fluorouracil for continuous infusion in an ambulatory pump. *Cancer Chemother. Pharmacol.* 46: 501–506.

201: Fluoxetine [WHO List of Essential Medicines]

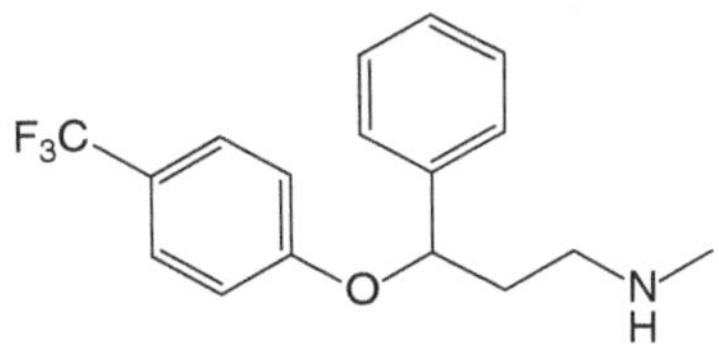

Chemical name: *N*-Methyl-3-phenyl-3-[4-(trifluoromethyl)phenoxy]propan-1-amine
Trade name: Prozac
Formula: $C_{17}H_{18}F_3NO$
Monoisotopic protonated ion (*m/z*): 310.1413
Molecular weight: 309.33
CAS number: 54910-89-3
Common salt form: HCl salt

Fluoxetine is a selective serotonin reuptake inhibitor (SSRI), clinically used for the treatment of major depressive disorder, obsessive-compulsive disorder, eating disorder, and panic disorder, among others. It was discovered in 1970 and received the US FDA's approval in 1987 [1]. Fluoxetine was the third most prescribed SSRI in the USA with approximately 27 million prescriptions in 2018 [2]. It is a secondary amine containing a 4-trifluoromethylphenyl ether moiety. The secondary amine moiety renders fluoxetine to undergo the Maillard reaction with excipients containing reducing sugars, for example, lactose. Wirth et al. reported the glycosylamine degradant and Amadori rearrangement degradant of fluoxetine, two of the degradation products formed in the early stage of the Maillard reaction (Scheme 201-1) [3]. Further degradation of the Amadori rearrangement degradant gave *N*-formylfluoxetine and 2,3-dihydro-3,5-dihydroxy-6-methyl-4*H*-pyran-4-one (DDMP) as the major degradants.

The ether moiety of fluoxetine is susceptible to acidic hydrolysis, resulting in the formation of 4-trifluoromethylphenol and 3-phenyl-3-hydroxypropylamine (also known as MEBM) (Scheme 201-2) [4, 5].

There are a number of papers studying the fate of fluoxetine discharged into the environment via wastewater by conducting photochemical stability studies in aqueous solutions under natural or simulated sunlight. Irradiation of a solution of fluoxetine in water with simulated sunlight caused the *O*-dealkylation as the major photodegradation pathway, in which 4-(difluoromethylene)-2,5-cyclohexadiene-1-one (I, a semiquinone), and 3-phenyl-3-hydroxypropylamine (II) were formed. The minor photodegradation pathway was found to be the photonucleophilic substitution of the trifluoromethyl group, in which the corresponding carboxylic degradant (III) was produced (Scheme 201-3) [6]. The same study suggested that this

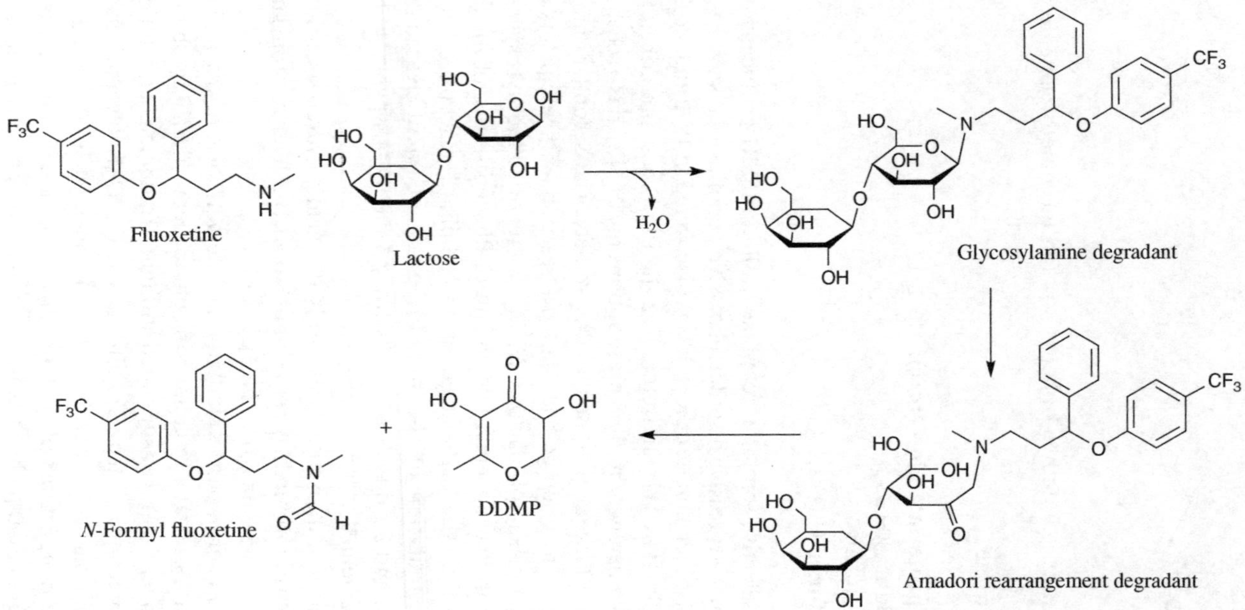

Scheme 201-1 Degradation of fluoxetine in the presence of lactose via the Maillard reaction.

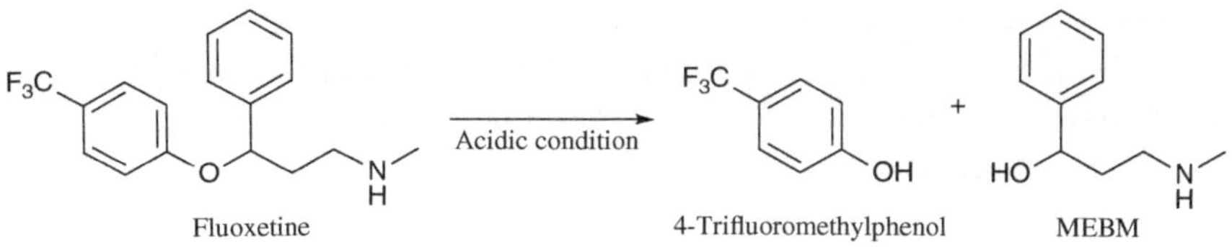

Scheme 201-2 Acidic degradation of fluoxetine.

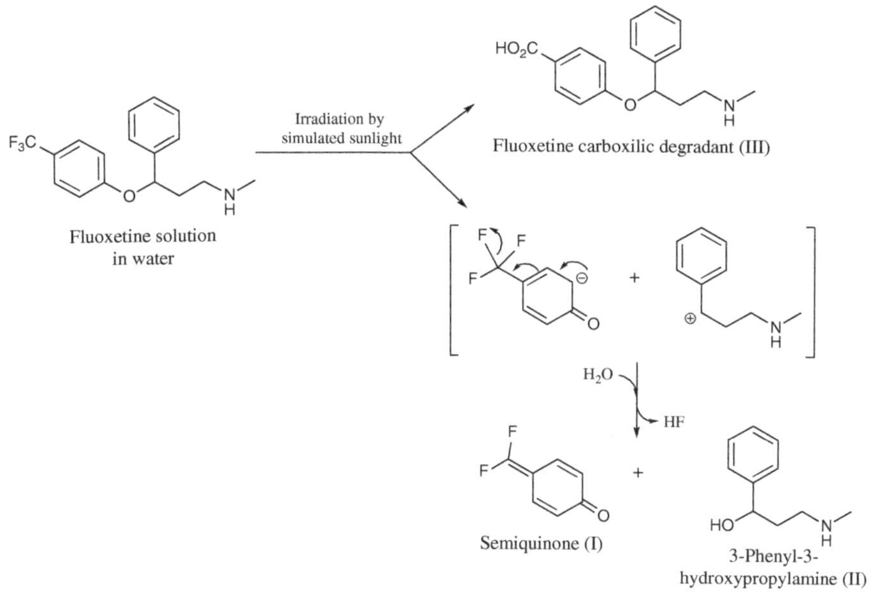

Scheme 201-3 Photodegradation of fluoxetine.

photonucleophilic substitution appeared to be a common photodegradation pathway for molecules containing a trifluoromethyl group, for example, fluometuron and flutalanil [6].

In a photodegradation study of fluoxetine in solid state spiked with metal ions, fluoxetine was found to produce 4-trifluoromethylphenol and 3-phenyl-3-hydroxypropylamine [7], the degradation profile of which appeared to be similar to that under acidic stress conditions (refer to Scheme 201-2).

References

1 Fluoxetine. https://en.wikipedia.org/wiki/Fluoxetine (accessed 27 May 2023).
2 Correia, D., Domingues, I., Faria, M., and Oliveira, M. (2023). Effects of fluoxetine on fish: what do we know and where should we focus our efforts in the future? *Sci. Total Environ.* 857: 159486.
3 Wirth, D.D., Baertschi, S.W., Johnson, R.A. et al. (1998). Maillard reaction of lactose and fluoxetine hydrochloride, a secondary amine. *J. Pharm. Sci.* 87: 31–39.

4 Risley, D.S. and Bopp, R.J. (1990). Fluoxetine. *Anal. Prof. Drug Sub.* 19: 193–219.
5 Pathak, A. and Rajput, S.J. (2009). Development of a stability-indicating HPLC method for simultaneous determination of olanzapine and fluoxetine in combined dosage forms. *J. Chromatogr. Sci.* 47: 605–611.
6 Lam, M.W., Young, C.J., and Mabury, S.A. (2005). Aqueous photochemical reaction kinetics and transformations of fluoxetine. *Environ. Sci. Technol.* 39: 513–522.
7 Maslanka, A., Hubicka, U., Krzek, J. et al. (2013). Determination of fluoxetine in the presence of photodegradation products appearing during UVA irradiation in a solid phase by chromatographic-densitometric method, kinetics and identification of photoproducts. *Acta Chromatogr.* 25: 465–481.

202: Fluticasone [Propionate; Furoate] [WHO List of Essential Medicines]

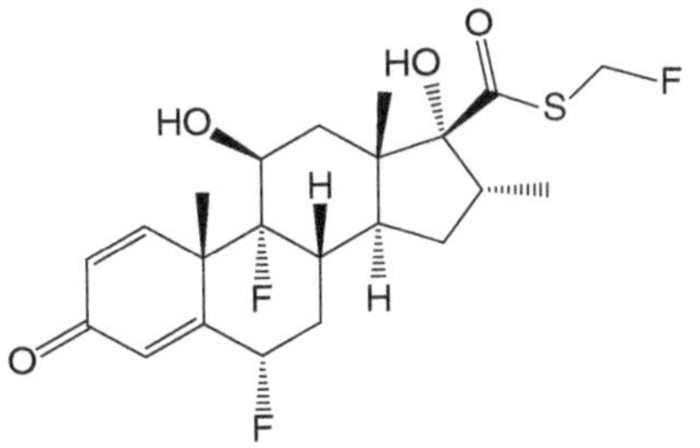

Chemical name: *S*-(Fluoromethyl) (6*S*,8*S*,9*R*,10*S*,11*S*,13*S*,14*S*,16*R*,17*R*)-6,9-difluoro-11,17-dihydroxy-10,13,16-trimethyl-3-oxo-6,7,8,11,12,14,15,16-octahydrocyclopenta[a]phenanthrene-17-carbothioate
Trade name: Flovent, others
Formula: $C_{22}H_{27}F_3O_4S$
Monoisotopic protonated ion (*m/z*): 445.1655
Molecular weight: 444.51
CAS number: 54910-89-3

Fluticasone is a potent anti-inflammatory corticosteroid with low systemic activity [1]. It is much more commonly available in its 17-ester forms, 17-propionate and 17-furoate, for dermal, nasal, and inhalation delivery. When used for the treatment of asthma and chronic obstructive pulmonary disease (COPD), fluticasone 17-esters are typically used in combination with salmeterol, a beta-2 agonist bronchodilator. Both the 17-ester functionality and 21-thiol ester should be susceptible to hydrolytic degradation; nevertheless, it appeared that the 21-thiol ester is much more vulnerable than the 17-ester functionality. In both the forced degradation studies by da Silva Sangoi et al. and Ludvigsson et al. [2, 3], fluticasone 17-propionate produced one major degradant under alkaline and acidic stress conditions, with the alkaline one causing much more degradation. Mass spectrometric analysis of this major degradant indicated that it was the carboxylic degradant of fluticasone 17-propionate (Scheme 202-1) [3]. It should be pointed out that during the hydrolytic degradation, another degradant, fluoromethanethiol, should be formed concurrently; this degradant would not be detected by a typical UV detector.

In its steroid core, fluticasone contains a 2,5-cyclodienone A-ring, which is photochemically reactive and would yield so-called lumi and photo-lumi degradants for this class of corticosteroids [4]. Indeed, in a photodegradation study of fluticasone furoate, the corresponding lumi and photo-lumi degradants were detected (Scheme 202-2) [5], while in a photodegradation study of fluticasone propionate, only the lumi degradant was observed (Scheme 202-3) [6], which could be attributed to insufficient photolysis, as the photo-lumi degradant is a downstream

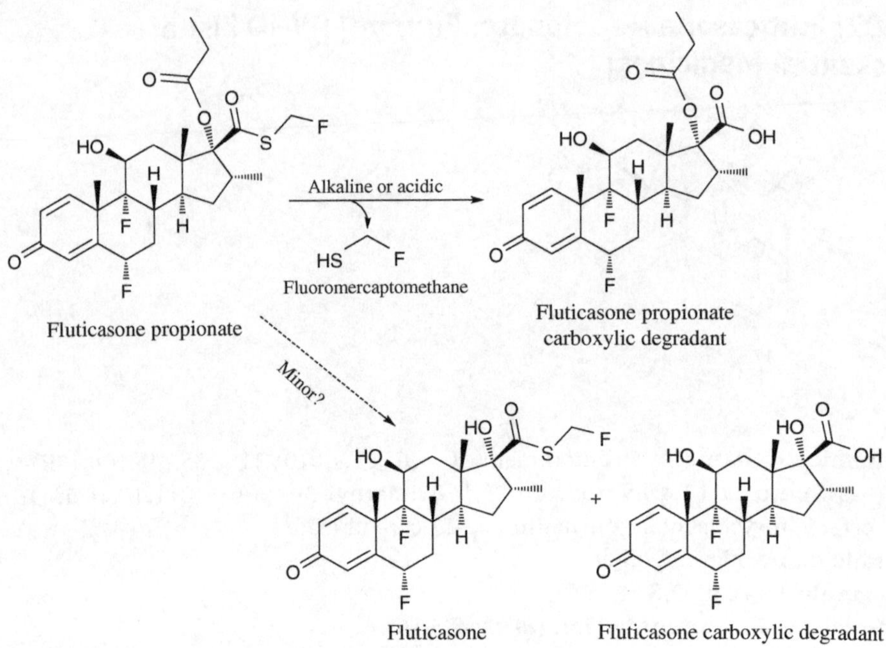

Scheme 202-1 Hydrolytic degradation of fluticasone propionate.

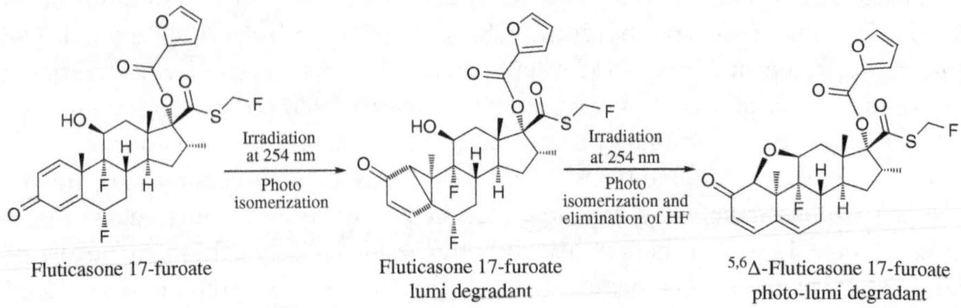

Scheme 202-2 Photodegradation of fluticasone 17-furoate.

degradant upon further photolysis [4]. In the latter study of fluticasone propionate, irradiation at 310 nm caused the formation of other two degradants [5], resulting from Norris type I and II photodegradation mechanisms [4], respectively.

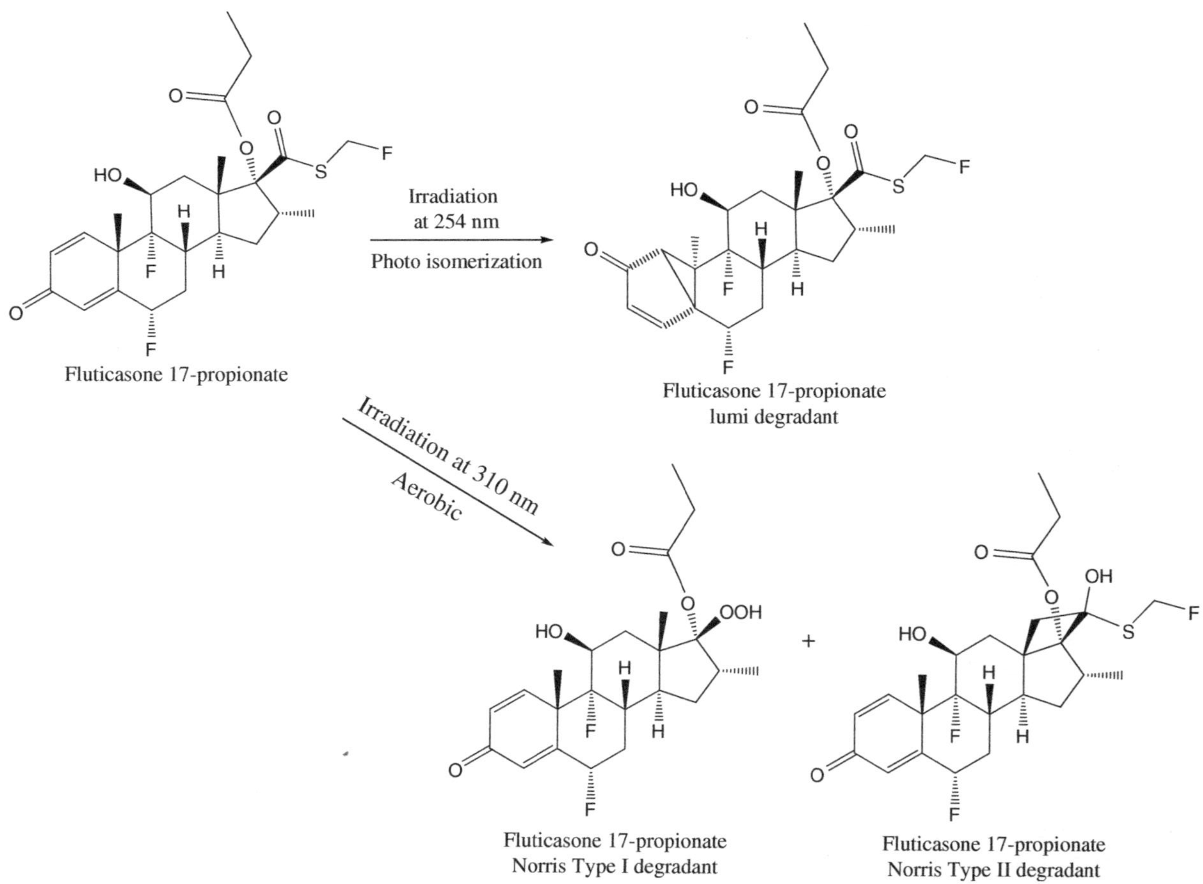

Scheme 202-3 Photodegradation of fluticasone via photoisomerization and Norris Type I and II photoreactions. It is noted that in the original paper [6], the structures of fluticasone and its degradants are erroneous with regard to certain stereo configurations of these molecules. *Source:* Adapted from Ahmad [6].

References

1 Bryson, H.M. and Faulds, D. (1992). Intranasal fluticasone propionate. *Drugs* 43: 760–775.
2 da Silva Sangoi, M., da Silva, L.M., D'Avila, F.B., and Dalmora, S.L. (2010). Determination of fluticasone propionate in nasal sprays by a validated stability-indicating MEKC method. *J. Chromatogr. Sci.* 48: 641–646.
3 Ludvigsson, J.W., Andersson, T., and Kjellberg, V. (2016). A new method to identify hydrolytic degradants in drug substances with UPLC-MS using ^{18}O-labelled water. *J. Pharm. Biomed. Anal.* 122: 9–15.
4 Li, M. (2012). Photochemical Degradation. In: *Organic Chemistry of Drug Degradation*, 165–197. Cambridge, UK: RSC Publishing.
5 Bardsley, B., Smith, M.S., and Gibbon, B.H. (2010). Structure elucidation and spectroscopic analysis of photodegradants of the anti-rhinitis drug fluticasone furoate. *Org. Biomol. Chem.* 8: 1876–1880.
6 Ahmad, W. (2016). Photodegradation behaviour of fluticasone propionate under different irradiation conditions. *Int. J. Pharm. Sci. Res.* 7: 390–395.

203–204: Folic Acid, Folinic Acid

203: Folic Acid [WHO List of Essential Medicines]

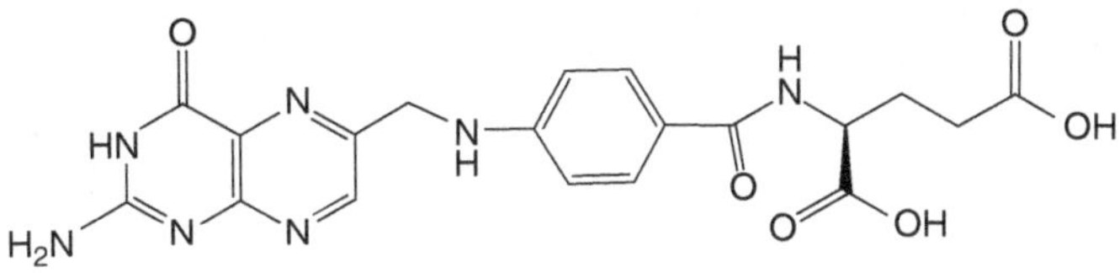

Chemical name: (4-(((2-Amino-4-oxo-3,4-dihydropteridin-6-yl)methyl)amino) benzoyl)-L-glutamic acid
Formula: $C_{19}H_{19}N_7O_6$
Monoisotopic protonated ion (*m/z*): 442.1470
Molecular weight: 441.40
CAS number: 59-30-3
Common salt form: Sodium salt

204: Folinic Acid

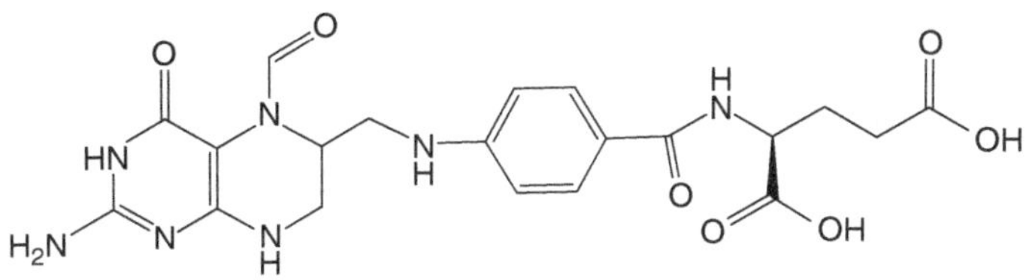

Chemical name: (4-(((2-Amino-5-formyl-4-oxo-3,4,5,6,7,8-hexahydropteridin-6-yl)methyl)amino)benzoyl)-L-glutamic acid
Trade name: Leucovorin, others
Formula: $C_{20}H_{23}N_7O_7$
Monoisotopic protonated ion (*m/z*): 474.1732
Molecular weight: 473.45
CAS number: 58-05-9
Common salt form: Calcium salt; sodium salt

Folic acid, along with its congeners in the folate family, is also known as vitamin B_9.

Usually available in its synthetic form, folic acid is used as a dietary supplement and a therapeutic agent. It is an essential nutrient required for many critical cellular pathways as a one-carbon source, including DNA methylation, synthesis, and maintenance [1].

Also known as pteroylglutamic acid, folic acid is the simplest member of a complex family of closely related compounds that are made of a 4-[(pteridine-6-ylmethyl) amino]benzoic acid unit conjugated with one or more L-glutamate units [2]. As the 4-[(pteridine-6-ylmethyl)amino]benzoic acid unit can be further divided into 2-amino-4-hydroxypteridine and 4-aminobenzoic acid (PABA) subunits, folic acid

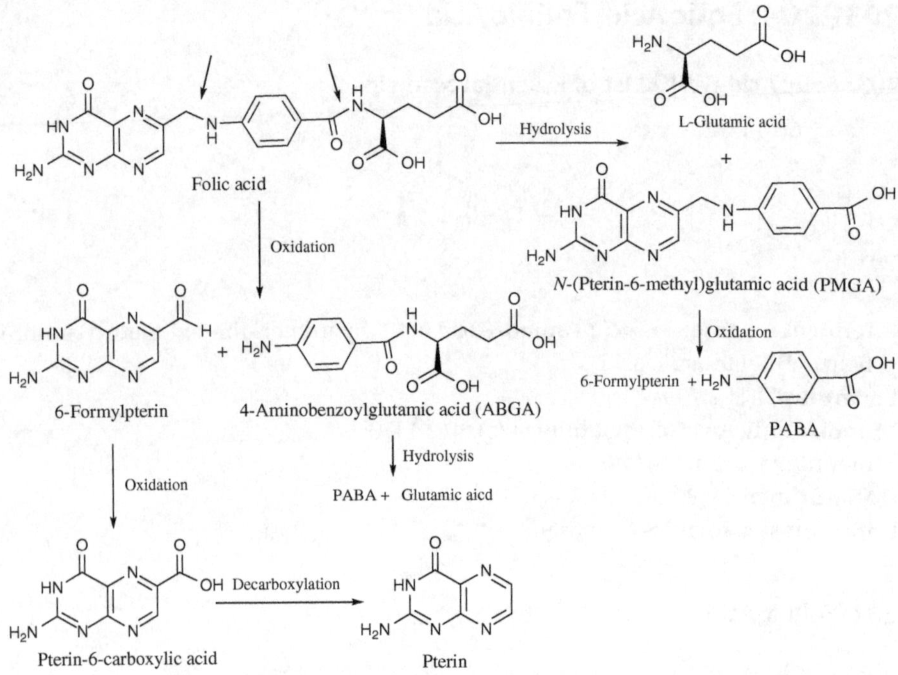

Scheme 203–204-1 Degradation pathways of folic acid under various forced degradation conditions.

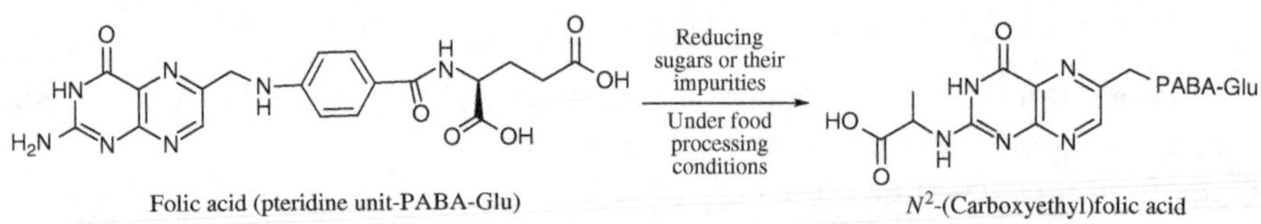

Scheme 203–204-2 Isolated degradation product of folic acid in the presence of reducing sugars under food processing conditions.

can be viewed as consisting of three linearly connected structural components: 2-amino-4-hydroxypteridine (AHP), 4-aminobenzoic acid, and glutamic acid. Its degradation chemistry is largely related to these structural features described above. In forced degradation studies under stress conditions of acidic, alkaline, thermal, oxidative, and photochemical, folic acid was found to degrade into its components, subcomponents, and further degradants such as pterin-6-carboxylic acid and pterin (Scheme 203–204-1) [3–6].

The two arrows associated with folic acid (Scheme 203-204-1) indicate the two bonds vulnerable to the respective oxidative and hydrolytic dissociation, and as such, four component degradants (L-glutamic acid, PMGA, 6-formylpterin, and ABGA) could occur as primary degradants. Further degradation of the primary degradants could lead to the formation of secondary and possibly tertiary

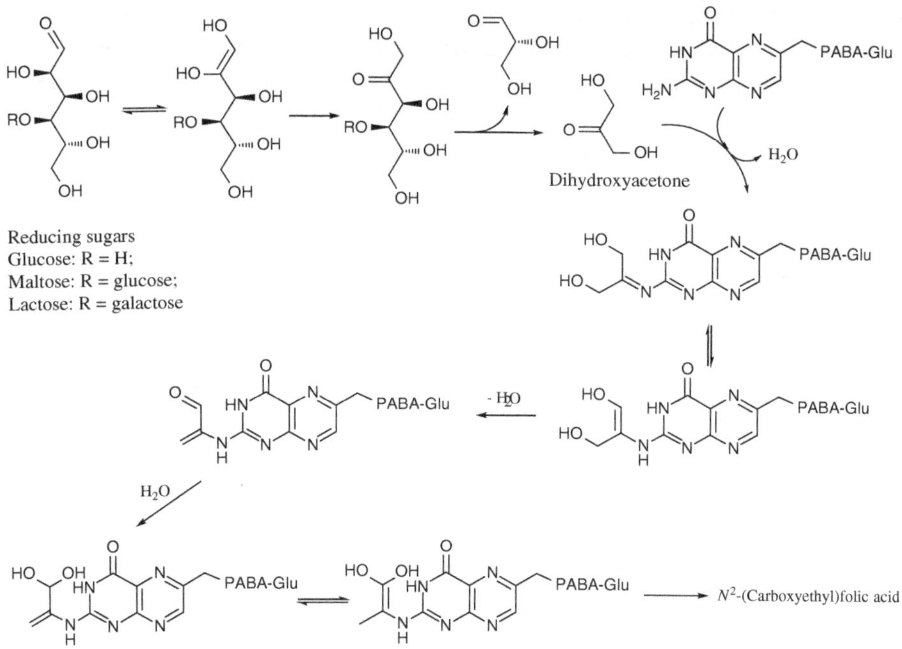

Scheme 203–204-3 Plausible degradation mechanism of folic acid in the presence of reducing sugars under food processing conditions.

degradants, e.g., pterin-6-carboxylic acid and pterin. It needs to be pointed out that not all degradants, primary, secondary, or tertiary, would occur at the same time under each forced degradation condition. Furthermore, oxidative degradation could take place under acidic or alkaline forced degradation conditions as long as oxygen is present in the stress system.

The two degradants, 6-formylpterin and pterin-6-carboxylic acid, could undergo type II photosensitized oxidation in which reactive oxygen species (ROS) can be produced, causing cellular oxidative stress [7].

As folic acid is also used in food fortification, Schneider et al. studied the stability of folic acid under conditions that can occur during food processing and preparation. It was found that folic acid reacted with reducing sugars and sugar impurities, such as maltose, lactose, and dihydroxyacetone (a sugar degradation product), to yield a major glycation degradant, *N*-[1-(carboxyethyl)]folic acid, in addition to PABA (Scheme 203–204-2) [8]. Although the degradation reactions between primary and secondary amines with reducing sugars are generally categorized into the broad scope of the Maillard reaction, which typically proceeds through the glycosylamine and Amadori intermediates [9], the formation of *N*-[1-(carboxyethyl)]folic acid was proposed as a direct reaction between folic acid and the sugar degradant, dihydroxyacetone (Scheme 203–204-3) [8].

Folinic acid or more specifically L-folinic acid, also known as levoleucovorin, is the N^5-formyl derivative of tetrahydrofolic acid. It is used clinically as an antidote to counteract the toxic effects of folic acid antagonists, such as methotrexate, an inhibitor of the enzyme dihydrofolate reductase (DHFR). It is also used in combination with the chemotherapy agent 5-fluorouracil for treating advanced metastatic colorectal cancer [10]. Due to its high similarity to folic acid, its thermal degradation pathway would be expected to be similar to that of folic acid, i.e., it could be degraded into its primary and possibly further degradants such as formyltetrahydropterin, tetrahydropterin, 4-aminobenzoic acid, and glutamic acid.

References

1 Crider, K.S., Yang, T.P., Berry, R.J., and Bailey, L.B. (2012). Folate and DNA methylation: a review of molecular mechanisms and the evidence for folate's role. *Adv. Nutr.* 3: 21–38.

2 Vora, A., Riga, A., Dollimore, D., and Alexander, K. (2004). Thermal stability of folic acid in the solid-state. *J. Therm. Anal. Calorim.* 75: 709–717.

3 Gazzali, A.M., Lobry, M., Colombeau, L. et al. (2016). Stability of folic acid under several parameters. *Eur. J. Pharm. Sci.* 93: 419–430.

4 Akhtar, M.J., Khan, M.A., and Ahmad, I. (2003). Identification of photoproducts of folic acid and its degradation pathways in aqueous solution. *J. Pharm. Biomed. Anal.* 31: 579–588.

5 Akhtar, M.J., Khan, M.A., and Ahmad, I. (1999). Photodegradation of folic acid in aqueous solution. *J. Pharm. Biomed. Anal.* 19: 269–275.

6 Grncharoska, K., Todevska, E.L., Spirevska, I.S. et al. (2022). A stability-indicating HPLC method for determination of folic acid and its related substances in tablets. *Maced. Pharm. Bull.* 68: 27–36. http://bulletin.mfd.org.mk/volumes/Volume%2068_1/68_1_002.pdf (accessed 14 August 2023).

7 Goossens, J.-F., Thuru, X., and Bailly, C. (2021). Properties and reactivity of the folic acid and folate photoproduct 6-formylpterin. *Free Radic. Biol. Med.* 171: 1–10.

8 Schneider, M., Klotzsche, M., Werzinger, C. et al. (2002). Reaction of folic acid with reducing sugars and sugar degradation products. *J. Agric. Food Chem.* 50: 1647–1651.

9 Li, M. (2012). Degradation via the Maillard Reaction. In: *Organic Chemistry of Drug Degradation*, 150–153. Cambridge, UK: RSC Publishing.

10 Levoleucovorin. https://pubchem.ncbi.nlm.nih.gov/compound/135398559 (accessed 29 May 2023).

205: Fosfomycin [WHO List of Essential Medicines]

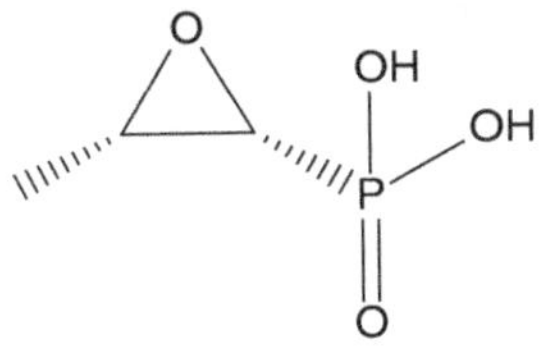

Chemical name: [(2*R*,3*S*)-3-Methyloxiran-2-yl]phosphonic acid
Trade name: Monurol
Formula: $C_3H_7O_4P$
Monoisotopic protonated ion (*m/z*): 139.0155
Molecular weight: 138.06
CAS number: 23155-02-4
Common salt form: Sodium salt

Fosfomycin, originally known as phosphomycin, is an antibiotic used primarily for the treatment of lower urinary tract infections. Discovered in 1969 and approved by the US FDA in 1996, it is a relatively old drug but experiencing a comeback in recent years [1, 2].

Fosfomycin is a phosphinic acid derivative containing an epoxide functional group, which is critical to its antibacterial activity but also susceptible to hydrolytic degradation. Bundgaard studied the degradation kinetics of fosfomycin in acidic solutions with pH ranging between 0.25 and 4.0 [3]. The half-lives of fosfomycin were found to be 27, 37, and 133 minutes at pH 1.0, 1.2, and 2.0, respectively. The hydrolytic degradation product is assumed to be the opening ring dihydroxyl degradant (Scheme 205). The stereochemistry of the degradant could be altered during the hydrolytic degradation, dependent on the hydrolysis conditions.

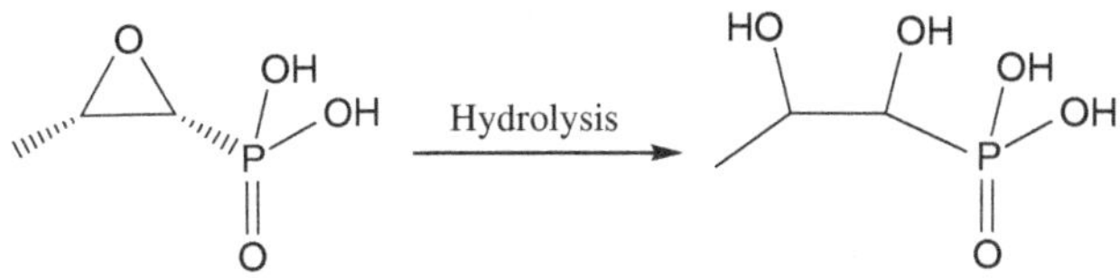

Scheme 205 Hydrolytic degradation of fosfomycin.

References

1 Fosfomycin. https://en.wikipedia.org/wiki/Fosfomycin (accessed 29 May 2023).

2 Fajfr, M., Balik, M., Cermakova, E., and Bostik, P. (2020). Effective treatment for uncomplicated urinary tract infections with oral fosfomycin, single center four year retrospective study. *Antibiotics (Basel).* 9: 511.

3 Bundgaard, H. (1980). Acid-catalyzed hydrolysis of fosfomycin and its implication in oral absorption of the drug. *Int. J. Pharm.* 6: 1–9.

206: Furosemide [WHO List of Essential Medicines]

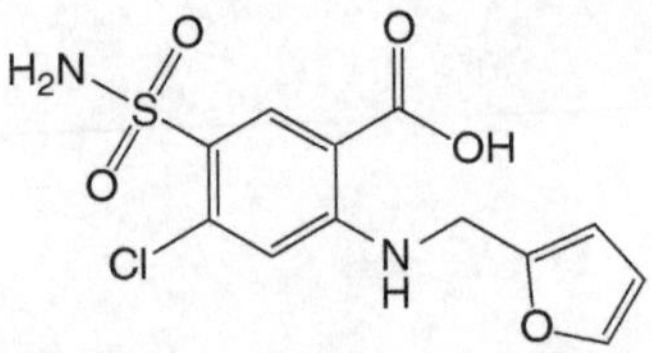

Chemical name: 4-Chloro-2-(furan-2-ylmethylamino)-5-sulfamoylbenzoic acid
Trade name: Furoscix, Lasix
Formula: $C_{12}H_{11}ClN_2O_5S$
Monoisotopic protonated ion (*m/z*): 331.0150
Molecular weight: 330.74
CAS number: 54-31-9
Common salt form: Sodium salt

Furosemide is a diuretic used clinically for treating edema caused by heart failure, liver scarring, or kidney disease. It is also used for the treatment of hypertension in patients with kidney or heart impairment [1]. It was first approved for medical use in the 1960s and continues to be a widely prescribed medication.

Furosemide is a highly substituted aniline derivative, and its C—N linkage of the *N*-substituted furan-2-ylmethyl group is susceptible to dissociation under both thermal and photodegradation conditions [2–4]. In a thermal degradation kinetic study of furosemide in solid state, the main degradant observed was 4-chloro-5-sulfamoylanthranilic acid (CSA; also known as saluamine) [2]. The activation energy for the solid-state thermal degradation of furosemide was 47.7 kcal/mol as measured by TGA, which is meaningfully lower than a typical activation energy for C—N breakage at 59 kcal/mol [3]. This solid-state degradation would be likely of oxidative nature, and thus, the other degradant would be 2-furaldehyde (furfural), although it was not reported in the original paper [2], probably due to the volatile nature of 2-furaldehyde (Scheme 206-1).

In acidic solutions, furosemide was found to decompose to CSA and furfural alcohol (Scheme 206-2), and the degradation was accelerated dramatically as the pH was decreased: the apparent first-order rate constant of hydrolysis was determined to be 67.02×10^{-3}/minute at pH 1.5, which is greater than 500 times versus that at pH 7.17.

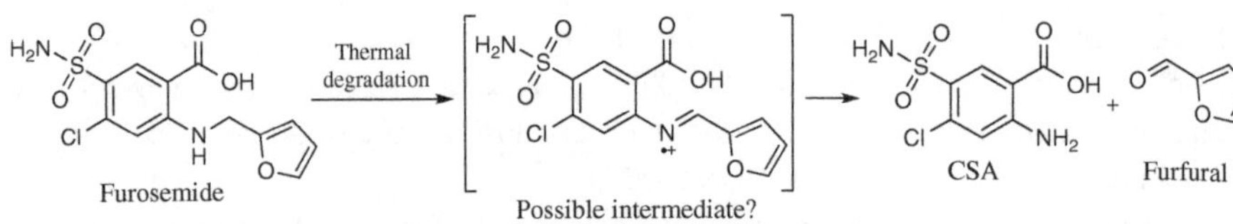

Scheme 206-1 Thermal degradation of furosemide.

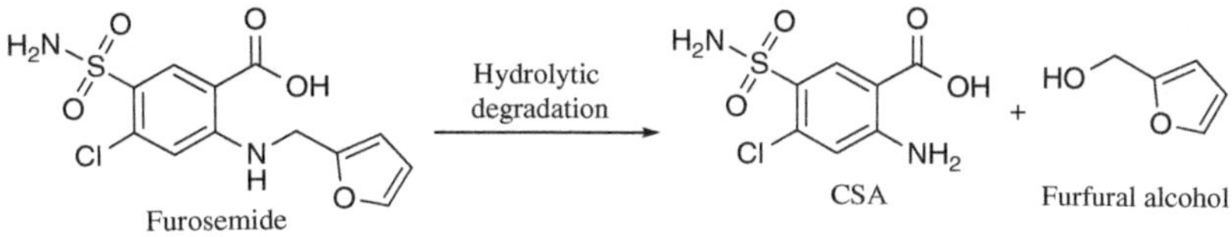

Scheme 206-2 Hydrolytic degradation of furosemide.

With regard to the photodegradation pathways of furosemide, Giannetti et al. provided a good summary in their research article (Scheme 206-3) [5]. Under photolytic conditions in the presence of water, the cleavage products were CSA and furfural alcohol. Furfural alcohol would undergo further degradation: aerial oxidation to form furfural aldehyde and acid-catalyzed oligomerization/polymerization, giving rise to colored degradants [6, 7]. On the other hand, CSA could also undergo photocatalyzed dimerization, trimerization, and oligomerization, in addition to the photo-dechlorinated product and its dimer. In the absence of water, photolysis would lead to the formation of dechlorinated dimer. Photodecarboxylation of furosemide was also observed in a study in aqueous solutions by Bundgaard et al. [8].

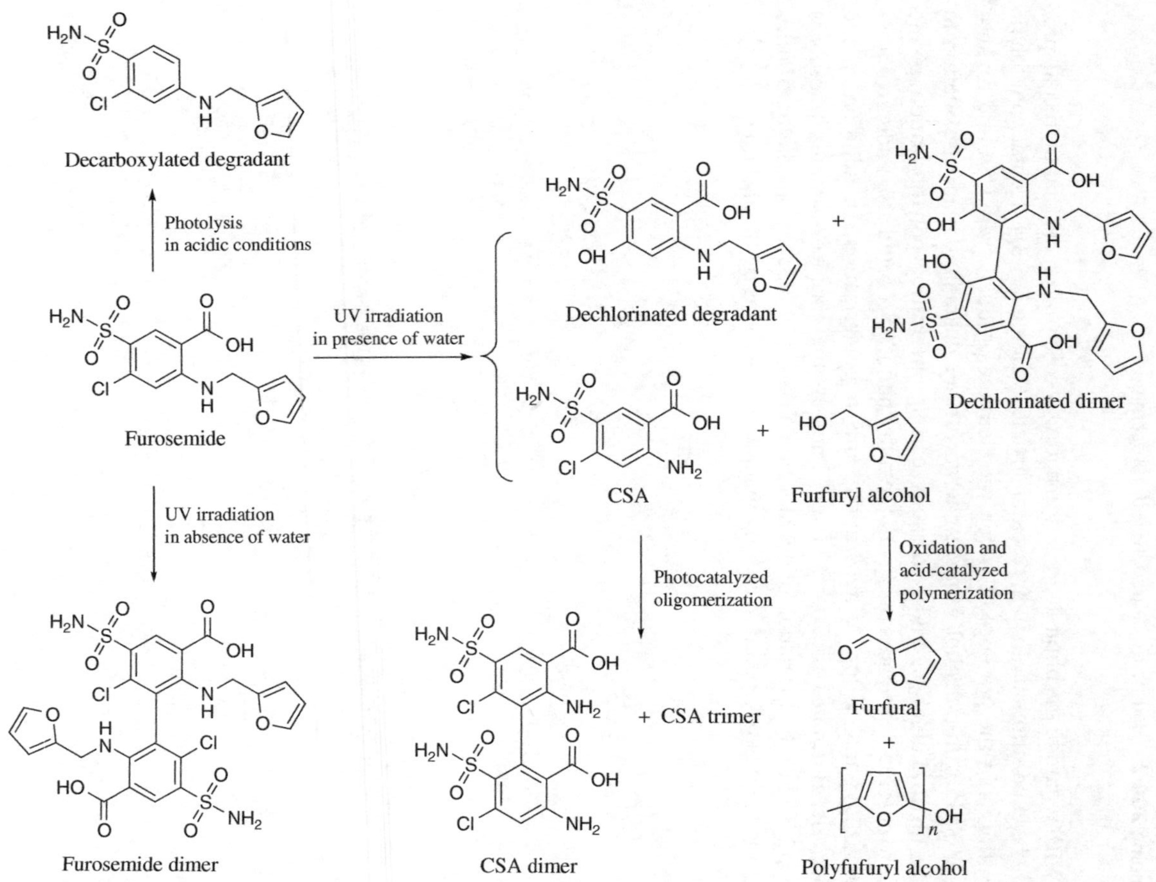

Scheme 206-3 Photodegradation pathways of furosemide.

References

1 Furosemide. https://en.wikipedia.org/wiki/Furosemide (accessed 30 May 2023).

2 Adrjanowicz, K., Kaminski, K., Grzybowska, K. et al. (2011). Effect of cryogrinding on chemical stability of the sparingly water-soluble drug furosemide. *Pharm. Res.* 28: 3220–3236.

3 Beyers, H., Malan, S.F., van der Watt, J.G., and de Villiers, M.M. (2000). Structure-solubility relationship and thermal decomposition of furosemide. *Drug Dev. Ind. Pharm.* 26: 1077–1083.

4 Cruz, J.E., Maness, D.D., and Yakatan, G.J. (1979). Kinetics and mechanism of hydrolysis of furosemide. *Int. J. Pharm.* 2: 275–281.

5 Giannetti, M., Canale, V.C., Micheli, L. et al. (2023). An insight into the degradation processes of the anti-hypertensive drug furosemide. *Molecules* 28: 381.

6 Kim, T., Assary, R.S., Marshall, C.L. et al. (2011). Acid-catalyzed furfuryl alcohol polymerization: characterizations of molecular structure and thermodynamic properties. *Chem. Cat. Chem.* 3: 1451–1458.

7 Shibutani, K., Nakai, J., Aphichartsuphapkhajorn, K. et al. (2021). The activation of furfuryl alcohol polymerization by oxygen and its enhanced mechanical properties. *J. Appl. Polym. Sci.* 138: 50311.

8 Bundgaard, H., Nørgaard, T., and Nielsen, N.M. (1988). Photodegradation and hydrolysis of furosemide and furosemide esters in aqueous solutions. *Int. J. Pharm.* 42: 217–224.

207: Gabapentin

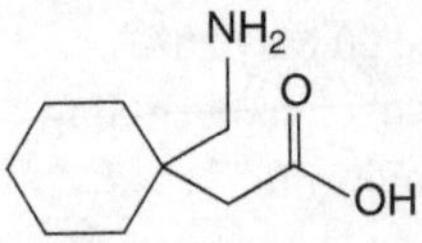

Chemical name: 2-[1-(Aminomethyl)cyclohexyl]acetic acid
Trade name: Neurontin
Formula: $C_9H_{17}NO_2$
Monoisotopic protonated ion (*m/z*): 172.1332
Molecular weight: 171.24
CAS number: 60142-96-3

Gabapentin, originally developed for the treatment of epilepsy, is a structural analogue of the neurotransmitter γ-aminobutyric acid (GABA) [1]. Its main clinical use is to relieve neuropathic pain caused by diabetic neuropathy, postherpetic neuralgia, and central pain [2]. In 2020, it was the 10th most prescribed medication in the US.

Gabapentin is a γ-amino acid, and its main degradation product is the lactam (3,3-pentamethylene-4-butyrolactam, or GBP-L) that is formed via cyclization between the carboxyl and amino groups (Scheme 207) [3]; the lactam degradant was shown to cause seizures in an animal model [4]. It was reported to be most stable at pH around 6, where the zwitterion form dominates [5]. Zong et al. studied the factors impacting lactamization during gabapentin tablets manufacturing process and found that the lactamization rate in the milled gabapentin samples was correlated to increased surface area, milling time, and lower moisture level [6]. The stabilization effect by higher moisture (> 31% relative humidity) was attributed to a moisture-driven competitive annealing process, which would decrease the milling-induced crystal defects. In a forced degradation study, the amino group of gabapentin was found to be oxidized into a nitroso group [7]; this stress-generated degradant is most likely to be an artificial degradant, as the oxidative stress with hydrogen peroxide seemed to be quite harsh.

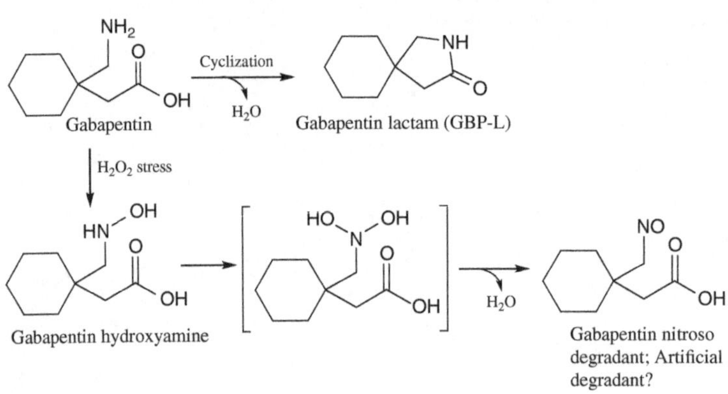

Scheme 207 Degradation pathways of gabapentin.

References

1 Gabapentin. https://en.wikipedia.org/wiki/Gabapentin (accessed 4 June 2023).

2 Goa, K.L. and Sorkin, E.M. (1993). A review of its pharmacological properties and clinical potential in epilepsy. *Drugs* 46: 409–427.

3 Ciavarella, A.B., Gupta, A., Sayeed, V.A. et al. (2007). Development and application of a validated HPLC method for the determination of gabapentin and its major degradation impurity in drug products. *J. Pharm. Biomed. Anal.* 43: 1647–1653.

4 Potschka, H., Feuerstein, T.J., and Löscher, W. (2000). Gabapentin-lactam, a close analogue of the anticonvulsant gabapentin, exerts convulsant activity in amygdala kindled rats. *Naunyn Schmiedeberg's Arch. Pharmacol.* 361: 200–205.

5 Zour, E., Lodhi, S.A., Nesbitt, R.U. et al. (1992). Stability studies of gabapentin in aqueous solutions. *Pharm. Res.* 9: 595–600.

6 Zong, Z., Desai, S.D., Kaushal, A.M. et al. (2011). The stabilizing effect of moisture on the solid-state degradation of gabapentin. *AAPS PharmSciTech* 12: 924–931.

7 Ragham, P.K. and Chandrasekhar, K.B. (2016). Development and validation of a stability-indicating RP-HPL C-CAD method for gabapentin and its related impurities in presence of degradation products. *J. Pharm. Biomed. Anal.* 125: 122–129.

208–209: Glibenclamide, Glipizide

208: Glibenclamide (Glyburide)

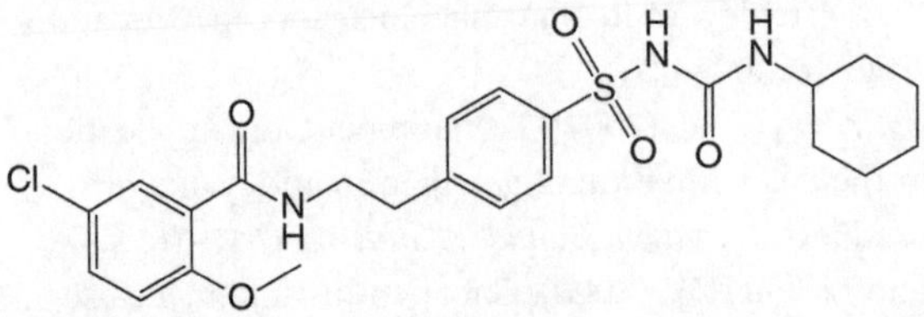

Chemical name: 5-Chloro-*N*-[2-[4-(cyclohexylcarbamoylsulfamoyl)phenyl] ethyl]-2-methoxybenzamide
Trade name: Diabeta, Glucovance, Glynase
Formula: $C_{23}H_{28}ClN_3O_5S$
Monoisotopic protonated ion (*m/z*): 494.1511
Molecular weight: 494.00
CAS number: 10238-21-8

209: Glipizide

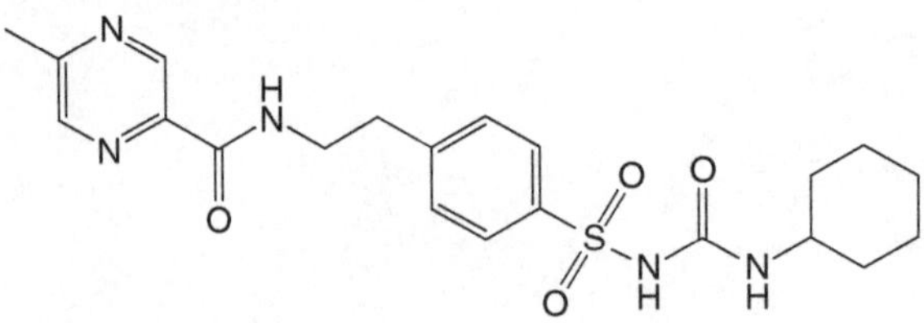

Chemical name: *N*-[2-[4-(Cyclohexylcarbamoylsulfamoyl)phenyl]ethyl]-5-methylpyrazine-2-carboxamide
Trade name: Glucotrol
Formula: $C_{21}H_{27}N_5O_4S$
Monoisotopic protonated ion (*m/z*): 446.1857
Molecular weight: 445.54
CAS number: 29094-61-9

Both glibenclamide and glipizide are antidiabetic drugs of the sulfonylurea family for the treatment of type II diabetes. Glibenclamide was discovered in the 1960s and approved by the US FDA in 1984 [1]. It works by inhibiting the ATP-sensitive K^+ channels, causing depolarization of the cells and subsequent insulin release [2].

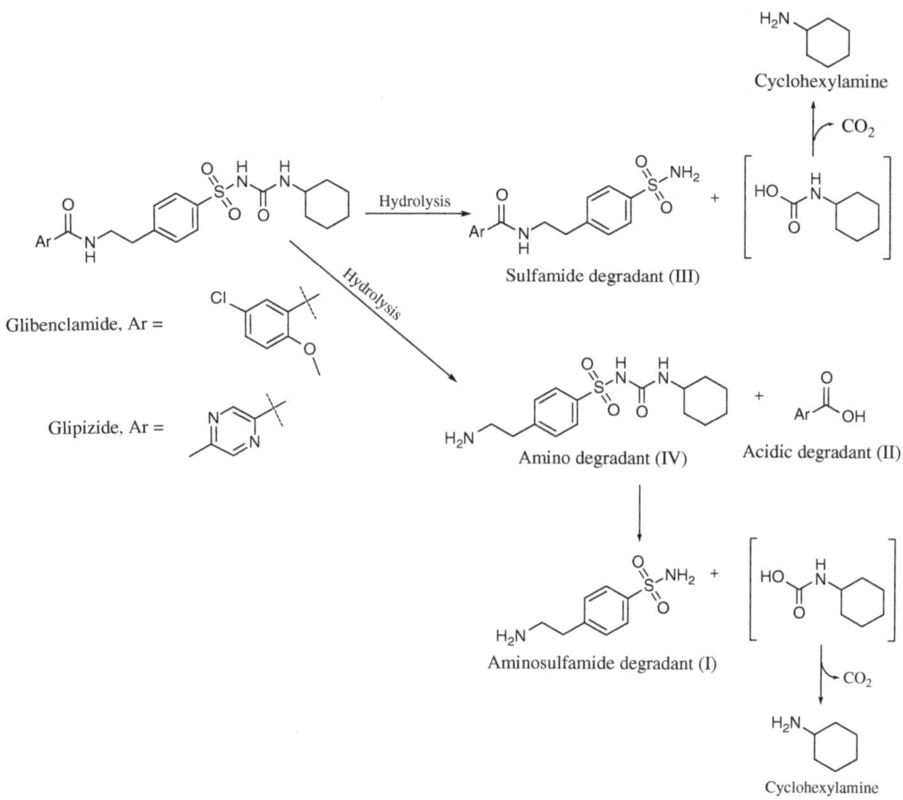

Scheme 208–209-1 Hydrolytic degradation pathways of glibenclamide and glipizide.

The sulfonylurea and benzamide moieties of both drug molecules are susceptible to hydrolytic degradation. The most vulnerable linkage is the sulfonyl-urea bond, and its breakage results in the formation of the sulfamide (III) and cyclohexylamine degradants (Scheme 208–209-1) [3–5]. For glipizide, the rates of hydrolysis under acidic and alkaline conditions were found to be comparable [5, 6], which would be expected to be the same for glibenclamide. Hydrolysis of the benzamide linkage produced degradant IV. In a thermal degradation study of amorphous glibenclamide, it decomposed to 1,3-dicyclohexylurea, probably via the intermediate of cyclohexylisocyanate (Scheme 208–209-2) [7]. Under photochemical stress of glipizide in solution, degradant IV was also formed, and its accelerated formation was proposed through a diradical intermediate produced in the homolysis of the diazole-carbonyl linkage (Scheme 208–209-3) [5, 8].

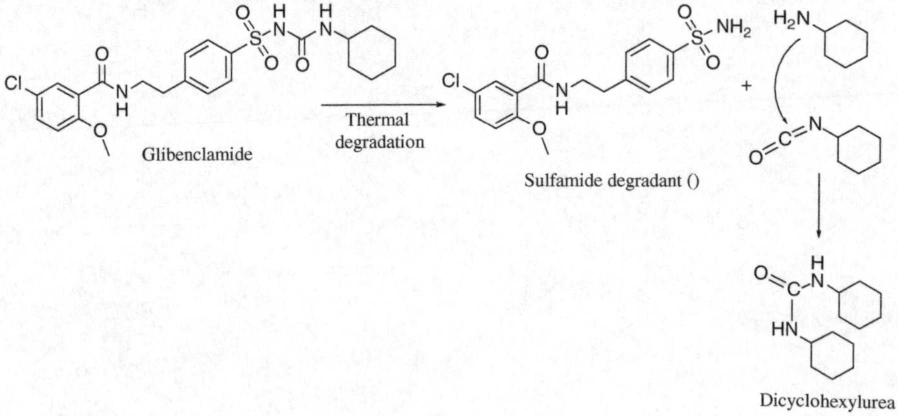

Scheme 208-209-2 Thermal degradation of glibenclamide.

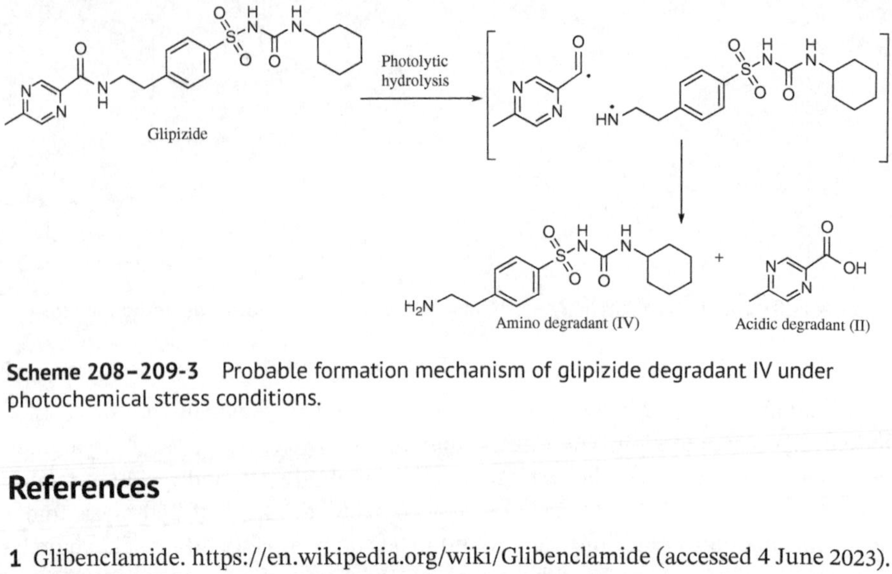

Scheme 208-209-3 Probable formation mechanism of glipizide degradant IV under photochemical stress conditions.

References

1 Glibenclamide. https://en.wikipedia.org/wiki/Glibenclamide (accessed 4 June 2023).

2 Luzi, L. and Pozza, G. (1997). Glibenclamide: an old drug with a novel mechanism of action? *Acta Diabetol.* 34: 239–244.

3 Bansal, G. and Singh, M. (2008). Ultraviolet-photodiode array and high-performance liquid chromatographic/mass spectrometric studies on forced degradation behavior of glibenclamide and development of a validated stability-indicating method. *J. AOAC Int.* 91: 709–719.

4 Gupta, S. and Bansal, G. (2011). Validated stability-indicating HPLC-UV method for simultaneous determination of glipizide and four impurities. *J. AOAC Int.* 94: 523–530.

5 Bansal, G., Singh, M., Jindal, K.C., and Singh, S. (2008). LC and LC-MS study on establishment of degradation pathway of glipizide under forced decomposition conditions. *J. Chromatogr. Sci.* 46: 510–517.

6 Gedawy, A., Al-Salami, H., and Dass, C.R. (2020). Advanced and multifaceted stability profiling of the first-line antidiabetic drugs metformin, gliclazide and glipizide under various controlled stress conditions. *Saudi Pharm. J.* 28: 362–368.

7 Rehder, S., Sakmann, A., Rades, T., and Leopold, C.S. (2012). Thermal degradation of amorphous glibenclamide. *Eur. J. Pharm. Biopharm.* 80: 203–208.

8 Rosenthal, I. (1970). Photochemistry of the amido group. In: *The Chemistry of Amides* (ed. I.J. Zabicky), 289–308. London: Interscience Publisher.

210–211: Glyceryl Trinitrate, Isosorbide Dinitrate

210: Glyceryl Trinitrate (Nitroglycerin)

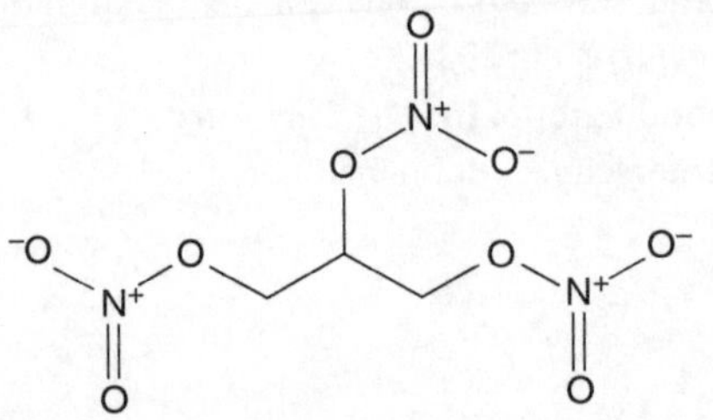

Chemical name: 1,3-Dinitrooxypropan-2-yl nitrate
Formula: $C_3H_5N_3O_9$
Monoisotopic protonated ion (*m/z*): 228.0099
Molecular weight: 227.09
CAS number: 55-63-0

211: Isosorbide Dinitrate [WHO List of Essential Medicines]

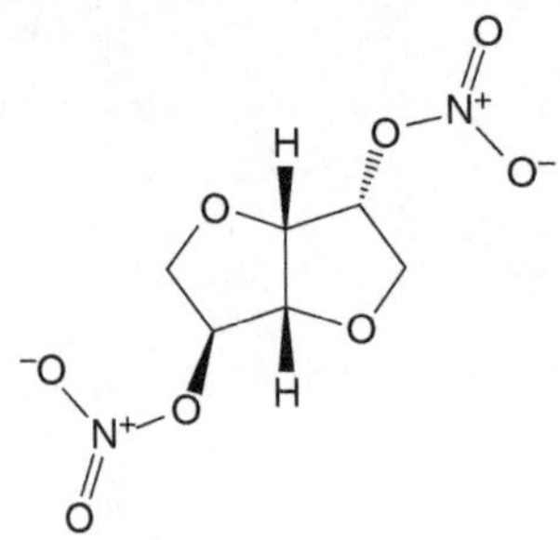

Chemical name: [(3*S*,3a*S*,6*R*,6a*S*)-3-Nitrooxy-2,3,3a,5,6,6a-hexahydrofuro[3,2-b]furan-6-yl] nitrate
Trade name: Bidil, Dilatrate, Isordil
Formula: $C_6H_8N_2O_8$
Monoisotopic protonated ion (*m/z*): 237.0353
Molecular weight: 236.14
CAS number: 87-33-2

Glyceryl trinitrate, commonly known as nitroglycerin, was discovered by the Italian chemist Ascanio Sobrero in 1847, and its first medical use was initiated by the English physician William Murrell in 1878 [1]. It is used for the treatment of angina pectoris, or chest pain, caused by insufficient blood flow to the heart muscle. Glyceryl trinitrate relieves the symptoms of angina via its metabolic release of nitric oxide, NO, a powerful vasodilator [2].

It is a trinitrate ester, and its degradation is mainly hydrolytic. In a degradation kinetic study in calcium hydroxide solutions, degradation of glyceryl trinitrate was

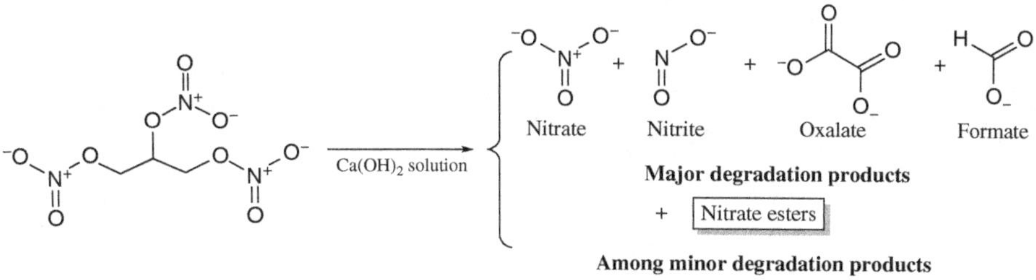

Scheme 210–211-1 Degradation of glyceryl trinitrate in alkaline solutions.

found to proceed via a second-order reaction with an activation energy of 27.53 kcal/mol [3]. The major degradation products were calcium nitrate and calcium nitrite (and the corresponding diesters), with calcium oxalate, nitrate esters, and calcium formate among the minor degradants (Scheme 210–211-1).

Halasz et al. studied the degradation product distribution of glyceryl trinitrate at pH 9 and pH 12, respectively, under heating at 50°C by either conventional way or by microwaving [4]. Although the degradation was much higher under microwave heating, the major degradation products were largely the same. Based on the degradation product distribution, a degradation mechanism was proposed, which involved two critical β-elimination steps to generate two equivalent nitrites (Scheme 210–211-2) [4]. It needs to be pointed out that none of the dinitrate degradants, mononitrate degradants, and glycerin were detected in the study by Halasz et al. [4].

Isosorbide dinitrate is also a member of the nitrate esters, which was originally reported in 1939 [4]. Its mode of action for treating angina is similar to that of

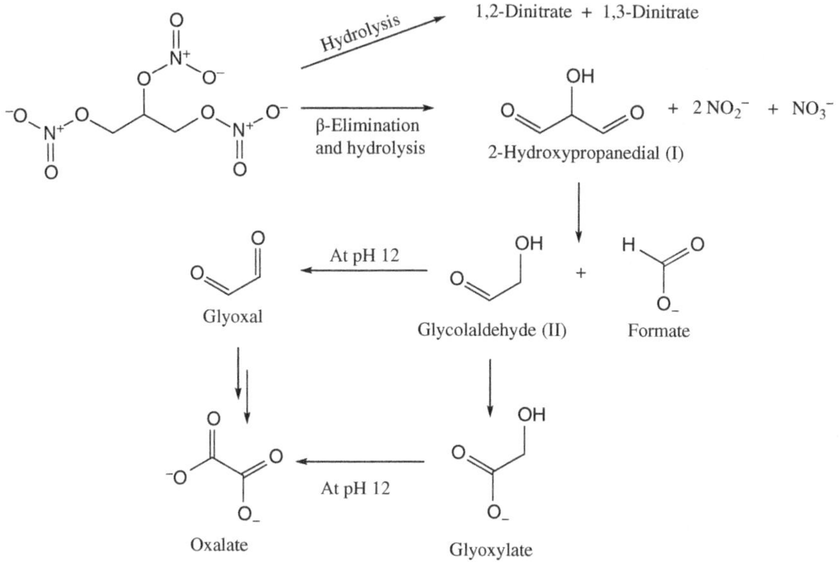

Scheme 210–211-2 Proposed mechanism for degradation of glyceryl trinitrate under alkaline conditions.

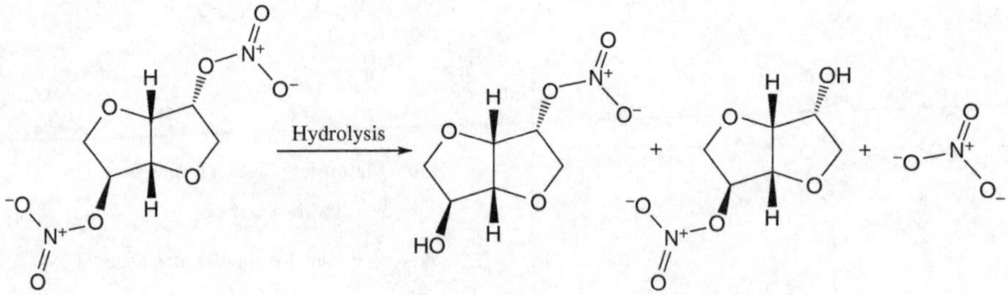

Scheme 210–211-3 Hydrolytic degradation of isosorbide dinitrate.

glyceryl trinitrate. The onset of action for isosorbide dinitrate is 30 minutes [5], which makes it a long-acting nitrate ester drug, whereas glyceryl trinitrate is a short-acting nitrate ester drug [6]. Isosorbide dinitrate is also utilized for the treatment of heart failure due to systolic dysfunction, and for this indication, it is usually used in combination with hydralazine [5]. The nature of its degradation is also hydrolytic, with the two monoesters, i.e., isosorbide-5-mononitrate (5-ISMN) and isosorbide-2-mononitrate (2-ISMN) as the major degradants (Scheme 210–211-3) [7, 8], which is quite different from that of glyceryl trinitrate degradation. The difference in their chemical stability behaviors appears to be consistent with their respective short- and long-acting modes.

References

1 Nitroglycerin. https://en.wikipedia.org/wiki/Nitroglycerin (accessed 10 June 2023).

2 Hashimoto, S. and Kobayashi, A. (2003). Clinical pharmacokinetics and pharmacodynamics of glyceryl trinitrate and its metabolites. *Clin. Pharm.* 42: 205–221.

3 Capellos, C., Fisco, W.J., Ribaudo, C. et al. (1984). Basic hydrolysis of glyceryl nitrate esters. III. Trinitroglycerin. *Int. J. Chem. Kinet.* 16: 1027–1051.

4 Halasz, A., Thiboutot, S., Ampleman, G., and Hawari, J. (2010). Microwave-assisted hydrolysis of nitroglycerin (NG) under mild alkaline conditions: new insight into the degradation pathway. *Chemosphere* 79: 228–232.

5 Isosorbide_dinitrate. https://en.wikipedia.org/wiki/Isosorbide_dinitrate (accessed 10 June 2023).

6 Schaumann, W. (1989). Pharmacokinetics of isosorbide dinitrate and isosorbide-5-mononitrate. *Int. J. Clin. Pharmacol. Ther. Toxicol.* 27: 445–453.

7 Carlson, M., Thompson, R.D., and Snell, R.P. (1988). Determination of isosorbide dinitrate in pharmaceutical products by HPLC. *J. Chromatogr. Sci.* 26: 574–578.

8 Liu, W.Y. (1989). Determination of isosorbide dinitrate and its degradation products in pharmaceuticals by gradient RP-HPLC. *Yao Xue Bao (Acta Pharm. Sinica)* 24: 797–800.

212: Guaifenesin

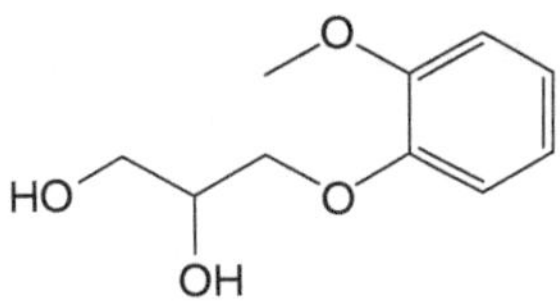

Chemical name: 3-(2-Methoxyphenoxy)propane-1,2-diol
Formula: $C_{10}H_{14}O_4$
Monoisotopic protonated ion (*m/z*): 199.0965
Molecular weight: 198.22
CAS number: 93-14-1

Guaifenesin is an expectorant drug clinically used to help the elimination of sputum from the respiratory tract. It is often used in combination with other types of medications, such as dextromethorphan (cough suppressant), pseudoephedrine (decongestant), codeine, and acetaminophen (pain relievers). Structurally, it is an aryl alkyl ether formed between guaiacol and glycerin. Its degradation pathways are primarily the hydrolytic breakage of the alkyl ether bond and isomerization, resulting in the formation of guaiacol and glycerin and isomeric beta-guaifenesin, respectively (Scheme 212) [1–4]. In a forced degradation study performed by Reddy et al. [2], under acidic condition (1N HCl at 60°C for 12 hours), it appeared that the predominant degradation was isomerization, while under alkaline condition (1N NaOH at 60°C for 12 hours), both isomerization and ether bond cleavage were observed.

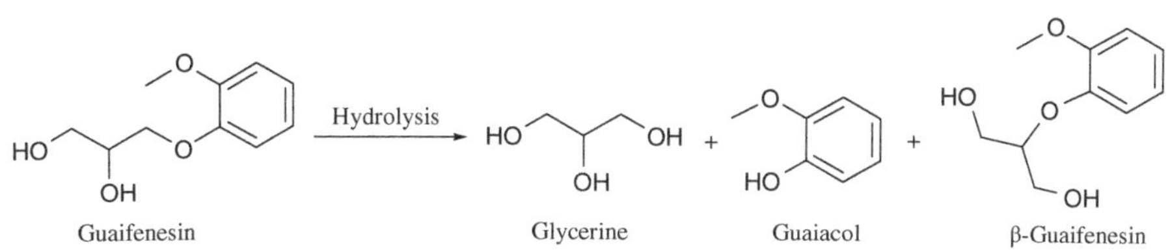

Scheme 212 Hydrolytic degradation of guaifenesin.

References

1 Grosa, G., Grosso, E.D., Russo, R., and Allegrone, G. (2006). Simultaneous, stability indicating, HPLC-DAD determination of guaifenesin and methyl and propyl-parabens in cough syrup. *J. Pharm. Biomed. Anal.* 41: 798–803.

2 Reddy, S.P., Babu, K.S., Kumar, N., and Sekhar, Y.S. (2011). Development and validation of stability indicating the RP-HPLC method for the estimation of related compounds of guaifenesin in pharmaceutical dosage forms. *Pharm. Methods* 2: 229–234.

3 Raju, T.V.R., Kumar, N.A., Kumar, S.R. et al. (2013). Development and validation of a stability-indicating RP-HPLC method for the simultaneous estimation of guaifenesin and dextromethorphan impurities in pharmaceutical formulations. *Chromatogr. Res. Int.* 2013: 315145.

4 Rathnakar, N. and Shanker, D.G. (2020). Stability-indicating simultaneous method development and validation of guaifenesin and dextromethorphan HBr by reverse-phase high-performance liquid chromatography. *Int. J. Pharm. Qual. Assur.* 11: 262–270.

213: Haloperidol [WHO List of Essential Medicines]

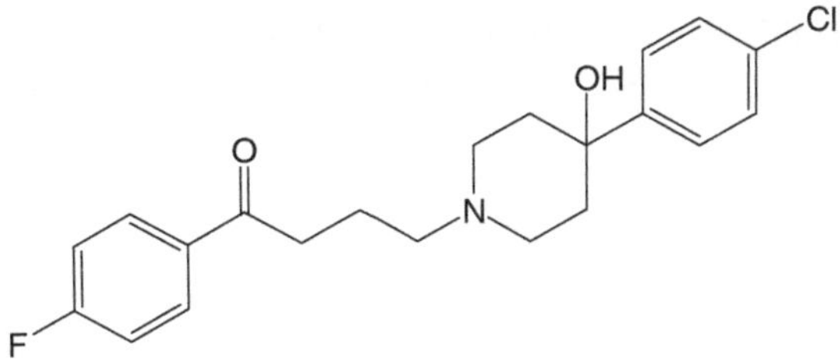

Chemical name: 4-[4-(4-Chlorophenyl)-4-hydroxypiperidin-1-yl]-1-(4-fluorophenyl) butan-1-one
Trade name: Haldol
Formula: $C_{21}H_{23}ClFNO_2$
Monoisotopic protonated ion (*m/z*): 376.1474
Molecular weight: 375.87
CAS number: 52-86-8
Common salt form: Lactate

Discovered by Paul Janssen in 1958, haloperidol remains to be a typical antipsychotic medication used for the treatment of schizophrenia, bipolar disorder, acute psychosis, and agitation, among others [1]. The fluoro and hydroxyl groups are potentially susceptible to hydrolytic degradation, particularly under strong forced alkaline and acidic degradation conditions, to produce the hydroxylated desfluoro and dehydrated degradants [2, 3]. On the other hand, its tertiary amine moiety is susceptible to nucleophilic oxidation to form two diastereomeric *N*-oxide degradants (Scheme 213).

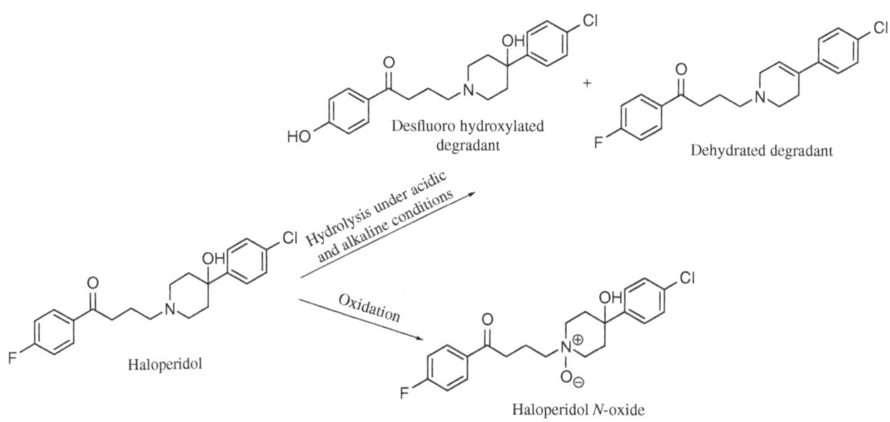

Scheme 213 Hydrolytic and oxidative degradation pathways of haloperidol.

References

1 Haloperidol. https://en.wikipedia.org/wiki/Haloperidol (accessed 14 June 2023).
2 Trabelsi, H., Bouabdallah, S., Bouzouita, K., and Safta, F. (2002). Determination and degradation study of haloperidol by high performance liquid chromatography. *J. Pharm. Biomed. Anal.* 29: 649–657.
3 Monser, L. and Trabels, H. (2003). A rapid LC method for the determination of haloperidol and its degradation products in pharmaceuticals using a porous graphitic carbon column. *J. Liq. Chromatogr. Relat. Technol.* 26: 257–266.

214–215: Heparin, Enoxaparin

214: Heparin [WHO List of Essential Medicines]

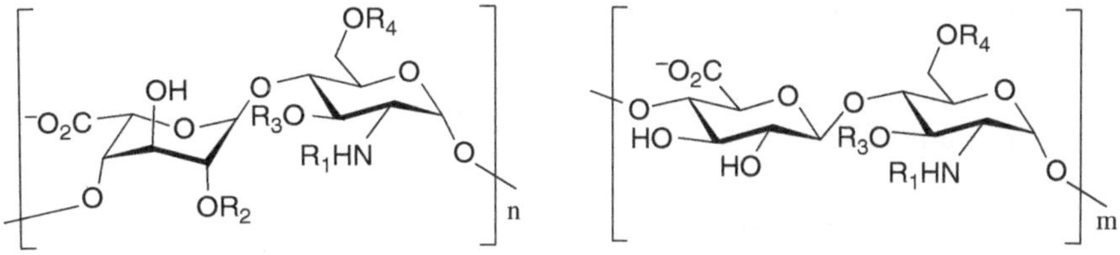

R_1 = H, SO_3^-, or Acetyl; R_2, R_3, R_4 = H or SO_3^-

Major and minor heparin disaccharide repeating units. Left: →4)-L-iduronic acid-(1→4)-αD-glucosamine (1→, which accounts for ~85%. Right: →4)-D-glucuronic acid-(1→4)-D-glucosamine (1→, which accounts for the majority of the remaining ~15%. Each of the monosaccharide units has various degree of *N*- and *O*-sulfation, with occasional *N*-acetylation (~5% in porcine mucosal heparin).

Chemical name: IUPAC name not applicable
Formula: N/A
Monoisotopic protonated ion (*m/z*): N/A
Molecular weight: ~12,000–~15,000 (average); 3,000–30,000 (range)
CAS number: 9005-49-6

215: Enoxaparin

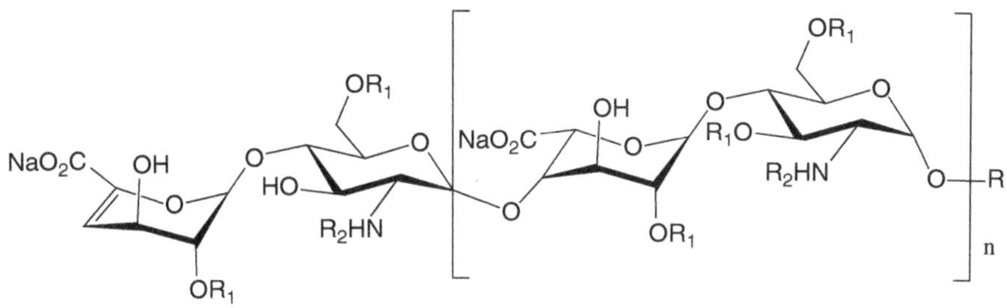

R_1 = H, or SO_3Na; R_2 = SO_3Na or Acetyl.

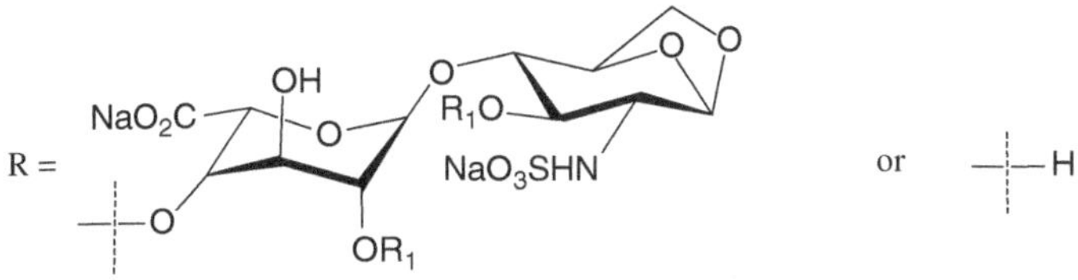

15 – 25%, n = 0 – 20 | Remaining of 15 – 25%, n = 1 – 21

(low-molecular-weight heparin)

Chemical name: IUPAC name not applicable
Trade name: Lovenox
Formula: N/A
Monoisotopic protonated ion (*m/z*): N/A
Molecular weight: ~4500 (average)
CAS number: 679809-58-6 (Sodium salt)
Common salt form: Sodium salt

Heparin, a highly sulfated glycosaminoglycan, is a naturally occurring anticoagulant present in mast cells of mammals. Heparin exerts its anticoagulant activity primarily via inducing or enhancing the inhibition of thrombin (Factor IIa) and Factor Xa by antithrombin III. Clinically, it is used for preventing and treating conditions that involve undesirable formation of blood clots, including deep vein thrombosis, pulmonary embolism, atrial fibrillation, and acute coronary syndrome [1].

The early work conducted by Jay McLean in the laboratory of Prof. William H. Howell at Johns Hopkins University in the 1910s led to the eventual discovery of heparin by Howell's group in 1922 [2]. Although Howell named both substances "heparin," i.e., the first one prepared by McLean in 1918 and the second one by Howell in 1922, due to their originating from dog liver, the one prepared by McLean was not the heparin as we know it today. McLean's "heparin" was likely a lipophilic mixture containing phospholipids and inositol phosphatides, among other components; all possess anticoagulant activity [3]. In 1929, Charles H. Best, the co-discoverer of insulin (along with fellow Canadian Frederick Banting), and his graduate student, Arthur Charles, decided to optimize the purification of heparin. Their effort resulted in a few papers published in 1933, which outlined the purification protocols for isolating heparin from bovine liver [4]. In 1935, Erik Jorpes of Karolinska Institute, who had visited Best in 1929 in Canada [4], purified heparin and determined its structure [5]. The heparin prepared by Best's group and Jorpes' was pure enough to rid the toxic side effects of the previous heparin preparations and thus enabled commercial production of heparin for clinical use in the mid-1930s. Nowadays, heparin is prepared via extraction and subsequent purification from the intestinal mucosa of pigs or cattle.

Heparin is a water-soluble, heterogeneous mixture of congeners consisting of highly sulfated glycosaminoglycan units. The molecular weight distribution of heparin ranges from ~3,000 to ~30,000 Da, with an average at ~12,000–15,000 Da [1]. The level of heterogeneity in heparin varies depending on its sources – either species-wise or different organs of the same species. The sulfated substituents account for ~40% of its average molecular weight. The simplest presentation of heparin in terms of molecular composition may be defined by its predominant (~85%) repeating disaccharide units of 1,4-linked L-iduronate (IdoA)-D-glucosamine (GlcN). The majority of the remaining ~15% repeating disaccharide units are 1,4-linked D-glucuronate (GlcA)-D-glucosamine (GlcN). Each of the monosaccharide units has various degree of *N*- and *O*-sulfation, and overall, for every tetrasaccharide unit, it contains close to five sulfate groups [6]. For porcine mucosal heparin, it contains approximately 20 hexosamine residues per chain, the vast majority of which are *N*-sulfated, with a single *N*-acetylation per chain on average [7].

The degradation pathways of heparin are hydrolytic as several of its structural features are susceptible to hydrolytic degradation with increasing levels of difficulty: *N*-sulfamate, *O*-sulfonate, and glycosidic bonds. In general, *N*-sulfamate linkage is somewhat less stable hydrolytically than *O*-sulfonate linkage; nevertheless, the individual stability of a particular sulfate linkage is also influenced by its surrounding environment as well as pH of the media. On the other hand, the glycosidic linkage is significantly more stable. Nagasawa et al. were able to achieve complete desulfation, with no depolymerization (i.e., no glycosidic cleavage), by heating heparin pyridinium salt solutions in DMSO containing up to 10% water or methanol at 80–100°C [8]. This study also found that the loss of *N*-sulfate preceded that of *O*-sulfate, and that both the *N*-sulfamate and *O*-sulfonate bonds are less stable than the *N*-acetamide bond as no de-*N*-acetylation was observed under the solvolytic conditions. Kozlowski et al. examined heparin hydrolysis in an acidic environment and found that desulfation was responsible for most of the heparin molecular weight reduction (due to hydrolytic degradation at pH 1, 40°C) [9]. Jandik et al. studied the stability of heparin sodium derived from porcine intestinal mucosa, which is probably most pharmaceutically relevant, in 0.1 N HCl, 0.1 NaOH, and deoxygenated neutral phosphate buffer solutions, respectively, at different temperatures [10]. While the degradation mechanisms under acidic and alkaline conditions were quite different, heparin in neutral phosphate buffer was relatively stable upon heating at 100°C for up to 500 hours. Kozlowski et al. also studied the stability of heparin solutions in pH 7–12 and under temperatures of 40°C, 60°C, and 80°C [11]. No significant *N*-desulfation or 6-*O*-desulfation, nor glycosidic bond dissociation, was observed. Nevertheless, stereochemical transformation of α-L-iduronate 2-*O*-sulfate to α-L-galacturonate occurred in 0.1 M phosphate buffer at pH 12 and 80°C (Scheme 214–215).

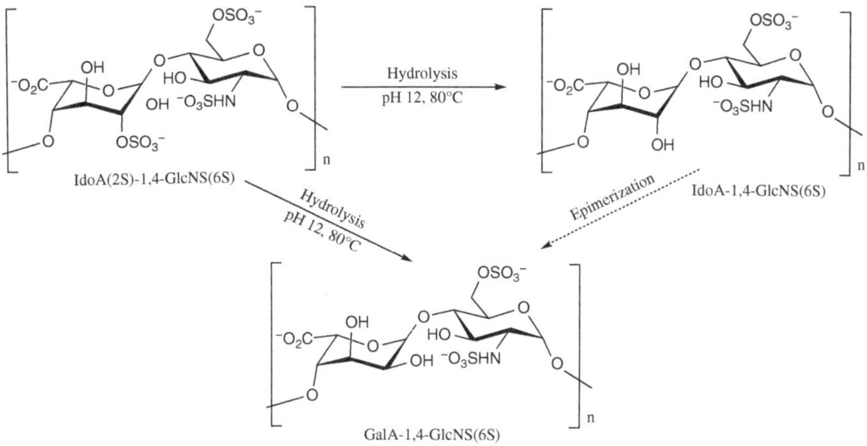

Scheme 214–215 Stereochemical transformation of α-L-iduronate 2-*O*-sulfate to α-L-galacturonate in pH 12 solution at 80°C.

Enoxaparin, available as its sodium salt, is one of the low-molecular-weight heparin preparations or LMWH, which was approved by the US FDA in 1993 for indications similar to those of regular heparin [12], commonly referred to as unfractionated heparin (UFH) when compared to LMWH. The average molecular weight of enoxaparin is ~4,000–5,000, as compared to ~12,000–15,000 for UFH, and it displayed improved pharmacokinetic and pharmacodynamic profiles than UFH [13]. Enoxaparin is prepared from UFH via controlled depolymerization, i.e., controlled hydrolysis of the glycosidic bonds, by using alkaline hydrolysis of the benzylic ester of UFH. Thus, enoxaparin retains the essential major and minor disaccharide units as mentioned above. Due to the high structural similarity between enoxaparin and UFH, the degradation chemistry of enoxaparin would be expected to be similar to that of UFH.

References

1 Heparin. https://en.wikipedia.org/wiki/Heparin (accessed 27 August 2023).

2 Mueller, R.L. and Scheidt, S. (1994). History of drugs for thrombotic disease. Discovery, development, and directions for the future. *Circulation* 89: 432–449.

3 Silver, J.M., Turner, D.L., and Tocantins, L.M. (1959). Lipid anticoagulants. *Prog. Hematol.* 2: 264–281.

4 Lim, G.B. (2017). Discovery and purification of heparin. *Nat. Rev. Cardiol.* https://doi.org/10.1038/nrcardio.2017.171.

5 Jorpes, E. (1936). On heparin, its chemical nature and properties. *Acta Med. Scand.* 88: 427–433.

6 Wolfrom, M.L., Weisblat, D.I., Karabinos, J.V. et al. (1943). Chemical studies on crystalline barium acid heparinate. *J. Am. Chem. Soc.* 65: 2077–2085.

7 Bae, J.H., Desai, U.R., Pervin, A. et al. (1994). Interaction of heparin with synthetic antithrombin III peptide analogues. *Biochem. J.* 301: 121–129.

8 Nagasawa, K., Inoue, Y., and Kamata, T. (1977). Solvolytic desulfation of glycosaminoglycuronan sulfates with dimethyl sulfoxide containing water or methanol. *Carbohydr. Res.* 58: 47–55.

9 Kozlowski, A.M., Yates, E.A., Roubroeks, J.P. et al. (2021). Hydrolytic degradation of heparin in acidic environments: nuclear magnetic resonance reveals details of selective desulfation. *ACS Appl. Mater. Interfaces* 13: 5551–5563.

10 Jandik, K.A., Kruep, D., Cartier, M., and Linhardt, R.J. (1996). Accelerated stability studies of heparin. *J. Pharm. Sci.* 85: 45–51.

11 Kozlowski, A.M., Yates, E.A., Roubroeks, J.P. et al. (2023). Monitoring the stability of heparin: NMR evidence for the rearrangement of sulfated iduronate in phosphate buffer. *Carbohydr. Polym.* 308: 120649.

12 Enoxaparin. https://go.drugbank.com/drugs/DB01225 (accessed 2 September 2023).

13 Iqbal, Z. and Cohen, M. (2011). Enoxaparin: a pharmacologic and clinical review. *Expert. Opin. Pharmacother.* 12: 1157–1170.

216: Hydralazine [WHO List of Essential Medicines]

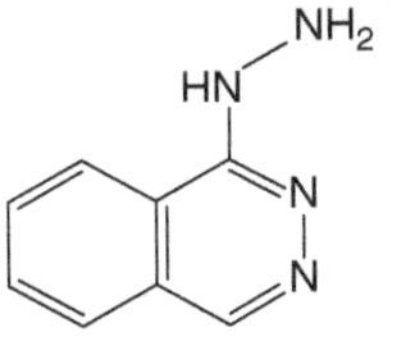

Chemical name: Phthalazin-1-ylhydrazine
Trade name: Apresoline, Bidil
Formula: $C_8H_8N_4$
Monoisotopic protonated ion (*m/z*): 161.0822
Molecular weight: 160.18
CAS number: 86-54-4
Common salt form: HCl salt

Hydralazine is a potent peripheral vasodilator for treating essential hypertension and is available in both oral and injectable dosage forms. Its injectable forms are used during hypertensive crises in emergency rooms, operating rooms, and intensive care units. Originally synthesized by J. Druey at Ciba as an antimalaria treatment in the 1940s, it received US FDA approval for treating hypertension in 1953 [1, 2].

The reactive hydrazino group of hydralazine is the main reason for the instability of the drug as well as for some of its toxicity, such as genotoxicity. The functional group is both fairly strongly nucleophilic and susceptible to oxidative and hydrolytic degradation [1, 3, 4]. The hydrazino group, in tandem with its nearby 2,3-azine moiety, is also capable of chelating metal ions [5]. Hence, hydralazine can readily condense with aldehydes and ketones, which are often present in some of the pharmaceutical excipients or as their impurities. Noda et al. reported that hydralazine rapidly reacted with formaldehyde to form the cyclized triazolophthalazine degradant under weakly acidic conditions (Scheme 216-1) and used this reaction for quantitative HPLC determination of formaldehyde [6], which is a common impurity in certain excipients and other sources. When the source of aldehydes is from reducing sugars or starch, which contains terminal glucoses, hydralazine can react with the reducing sugars to form similar cyclized triazolophthalazine degradants (Scheme 216-1), which have UV–Vis absorbance at longer wavelengths, thus causing discoloration of the drug product as reported by Lesson and Zhao [7].

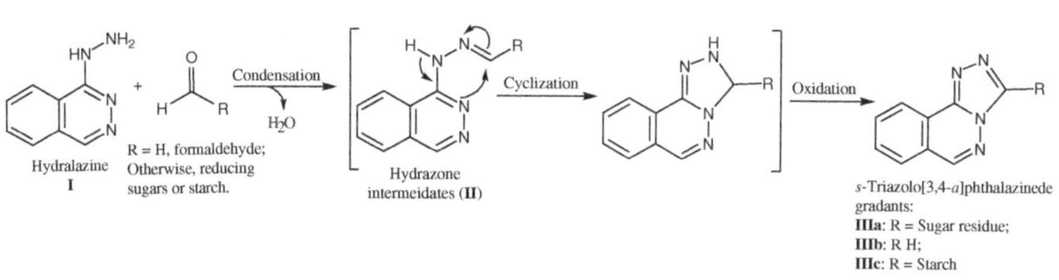

Scheme 216-1 Degradation of hydralazine in the presence of aldehyde-containing excipients or their impurities.

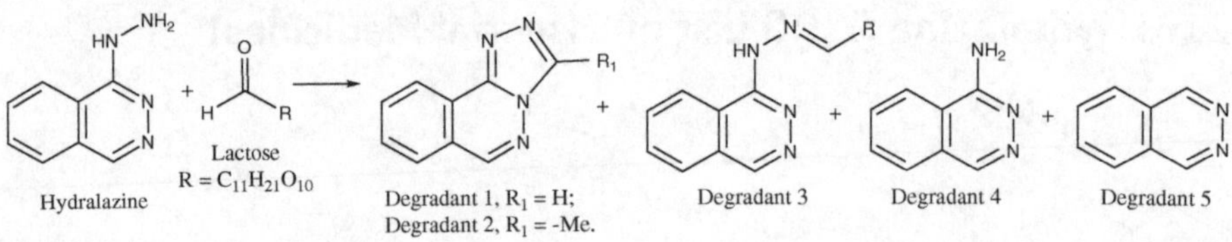

Scheme 216-2 Degradation of hydralazine due to the presence of lactose. *Source:* Adapted from Dong et al. [8].

In a forced degradation study of hydralazine tablets, Dong et al. found four hydralazine degradants formed due to the presence of lactose, an excipient of the formulation [8]. The structures of the degradants were proposed based on LC-MS analysis as well as prior knowledge of hydralazine degradation (Scheme 216-2). However, Dong et al. did not propose probable formation mechanisms for these degradants, and thus, the present author attempted to fill this gap as outlined in Scheme 216-3.

For the formation of Degradant 1, Lesson and Zhao proposed a hydrolytic dissociation mechanism, which was supported by their experiment of converting the sugar-substituted degradant (**IIIa**, Scheme 215-1) into Degradant **IIIb** (same as Degradant 1 in Scheme 215-2) in acidic solution [7]. Nevertheless, in solid phase, it would also be possible for the sugar-substituted degradants, such as Degradants **IIIa**, **IIIb**, and sugar-substituted triazolophthalazine, to undergo retro-aldol reaction to produce not only Degradant 1 but also Degradant 2 (Scheme 215-3). Furthermore, for the formation of Degradant 2, it would be difficult to explain by a hydrolytic process.

Halasi and Nairn studied the hydrolysis kinetics of hydralazine in pH 1–12 at temperatures of 35°C, 50°C, and 70°C and determined that hydralazine has maximum stability near pH 3.5, and its activation energy for hydrolysis at this pH was determined to be 17.9 kcal/mol [9]. In theory, the hydrolysis of hydralazine produces hydrazine [10], which appears to be a common pathway for drugs containing the hydrazine moiety in liquid formulations. In such a hydrolytic mechanism, 1-hydroxyphthalazine would be expected to be the other degradant, rather than phthalazine itself, which is, nevertheless, contradictory to the observation that phthalazine was formed after hydralazine was stressed in pH 12 alkaline solutions (Scheme 216-4). 1-Hydroxyphthalazine can be present in three tautomeric forms, and its synthesis was reported as early as 1893 [11].

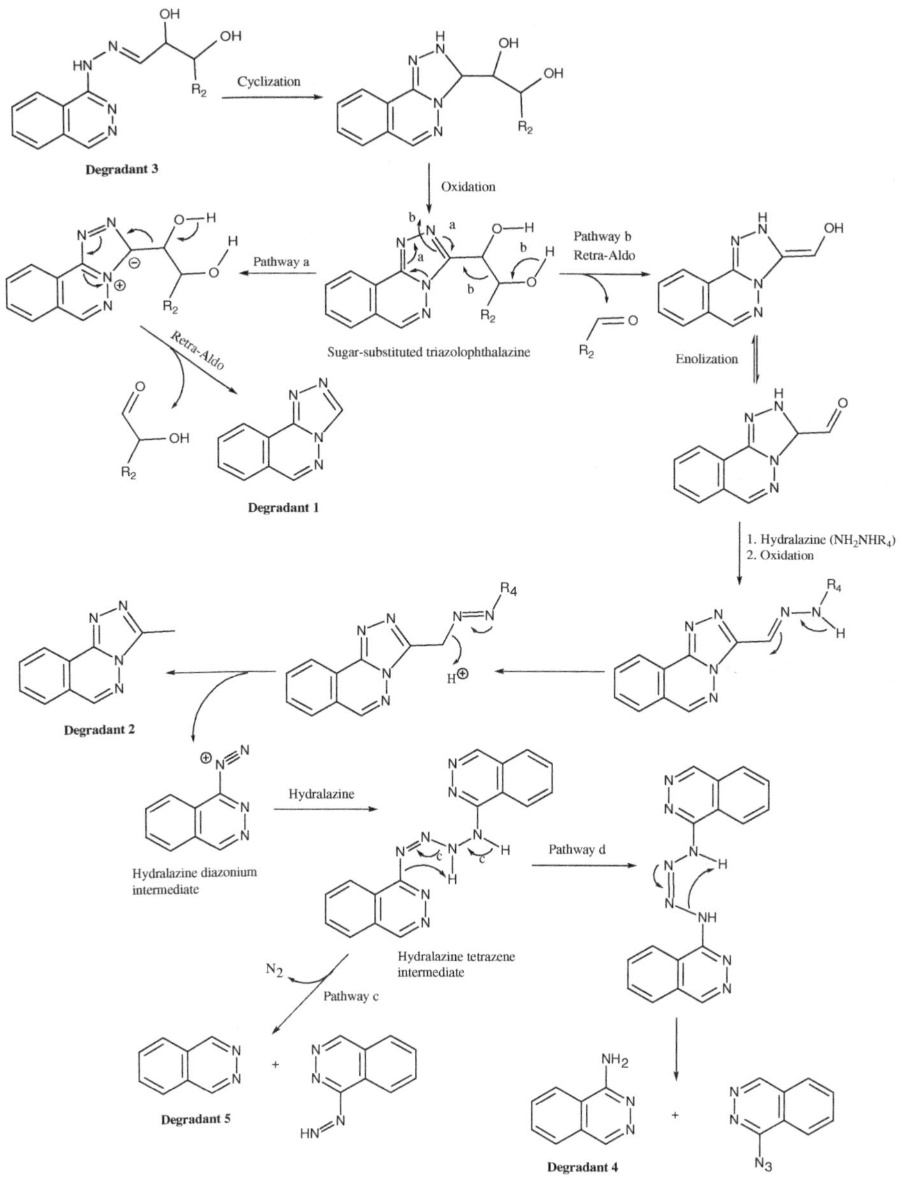

Scheme 216-3 Mechanisms for the formation of hydralazine degradants were proposed by the author of this book; the degradants were observed in the forced degradation study of hydralazine tablets by Dong et al. [8]. *Source:* Adapted from Dong et al. [8].

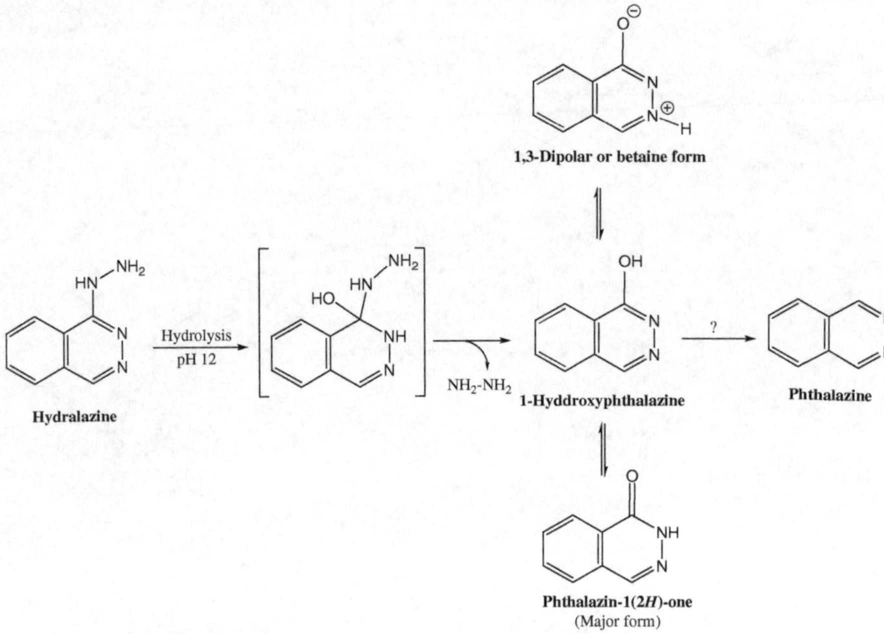

Scheme 216-4 Hydrolytic degradation of hydralazine in pH 12 solutions.

References

1 Israili, Z.H. and Dayton, P.G. (1977). Metabolism of hydralazine. *Drug Metab. Rev.* 6: 283–305.

2 Hydralazine. https://en.wikipedia.org/wiki/Hydralazine (accessed 2 September 2023).

3 Reidenberg, M.M., Drayer, D., deMarco, A.L., and Bello, C.T. (1973). Hydralazine elimination in man. *Clin. Pharmacol. Ther.* 14: 970–977.

4 Barbeau, D. L. (2010). U.S. Patent No. 7,728,133. Washington, DC: U.S. Patent and Trademark Office.

5 Jačić, J.K., Bajuk-Bogdanović, D., Savić, S. et al. (2023). Coordination of hydralazine with Cu^{2+} at acidic pH promotes its oxidative degradation at neutral pH. *J. Inorg. Biochem.* 243: 112181.

6 Noda, H., Minemoto, M., Noda, A., and Ushio, T. (1986). High performance liquid chromatographic determination of formaldehyde accomplished using hydralazine. *Chem. Pharm. Bull.* 34: 3499–3501.

7 Lesson, T. and Zhao, D.-C. (1996). Interactions between drug substances and excipients. 1. Fluorescence and HPLC studies of triazolophthalazine derivatives from hydralazine hydrochloride and starch. *J. Pharm. Sci.* 85: 326–329.

8 Dong, X., Tao, W., Liu, M. et al. (2016). Study on the related substances in hydralazine hydrochloride tablets by LC-MS/MS. *Chinese J. Pharm. Anal.* 36: 155–161. (in Chinese).

9 Halasi, S. and Nairn, J.G. (1990). Stability studies of hydralazine hydrochloride in aqueous solutions. *PDA J. Pharm. Sci. Technol.* 44: 30–34.

10 Gyllenhaal, O., Gronberg, L., and Vessman, J. (1990). Determination of hydrazine in hydralazine by capillary gas chromatography with nitrogen-selective detection after benzaldehyde derivatization. *J. Chromatogr.* 511: 303–315.

11 Dennis, N., Katritzky, A.R., and Ramaiah, M. (1974). 1.3-Dipolar character of six-membered aromatic rings. Part X. Pyridazine and benzopyridazine betaines. *J. Chem. Soc., Perkin Trans. 1* I: 1506–1514.

217–218: Hydrochlorothiazide, Chlorothiazide

217: Hydrochlorothiazide [WHO List of Essential Medicines]

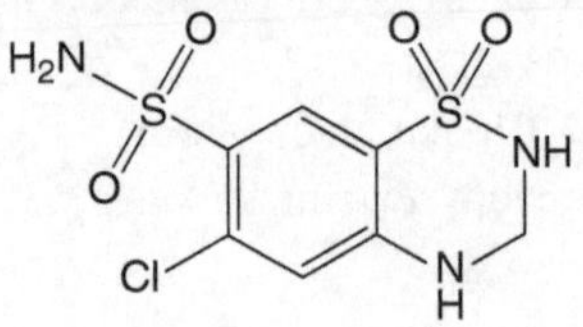

Chemical name: 6-Chloro-3,4-dihydro-2*H*-benzo[e][1,2,4]thiadiazine-7-sulfonamide
Trade name: Hydrodiuril
Formula: $C_7H_8ClN_3O_4S_2$
Monoisotopic protonated ion (*m/z*): 297.9718
Molecular weight: 297.73
CAS number: 58-93-5

218: Chlorothiazide [WHO List of Essential Medicines]

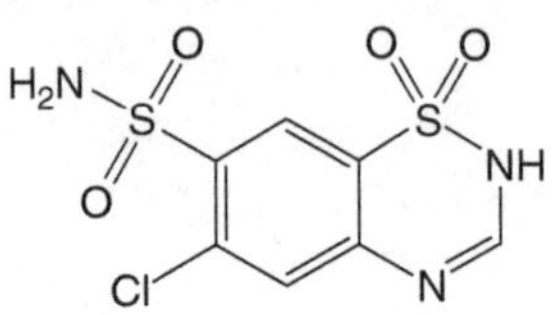

Chemical name: 6-Chloro-2*H*-1,2,4-benzothiadiazine-7-sulfonamide
Trade name: Diuril
Formula: $C_7H_6ClN_3O_4S_2$
Monoisotopic protonated ion (*m/z*): 295.9561
Molecular weight: 295.71
CAS number: 58-94-6

Hydrochlorothiazide and chlorothiazide belong to the thiazide diuretic family, which are used for the treatment of hypertension, often recommended as first-line treatment, and for fluid retention conditions associated with heart failure, nephrotic syndrome, and cirrhosis [1, 2]. Both drugs were approved for clinical use in the 1950s, with chlorothiazide coming onto the market a few years earlier than hydrochlorothiazide [2, 3].

The main degradation pathways of hydrochlorothiazide and chlorothiazide are of hydrolytic and photolytic nature. The hydrolytic degradation gives rise to 4-amino-6-chloro-1,3-benzendisulfonamine (EP impurity B, also known as DSA), which is also a starting material for both hydrochlorothiazide and chlorothiazide (Scheme 217–218-1) [4]. Chlorothiazide was found to be most stable near neutral

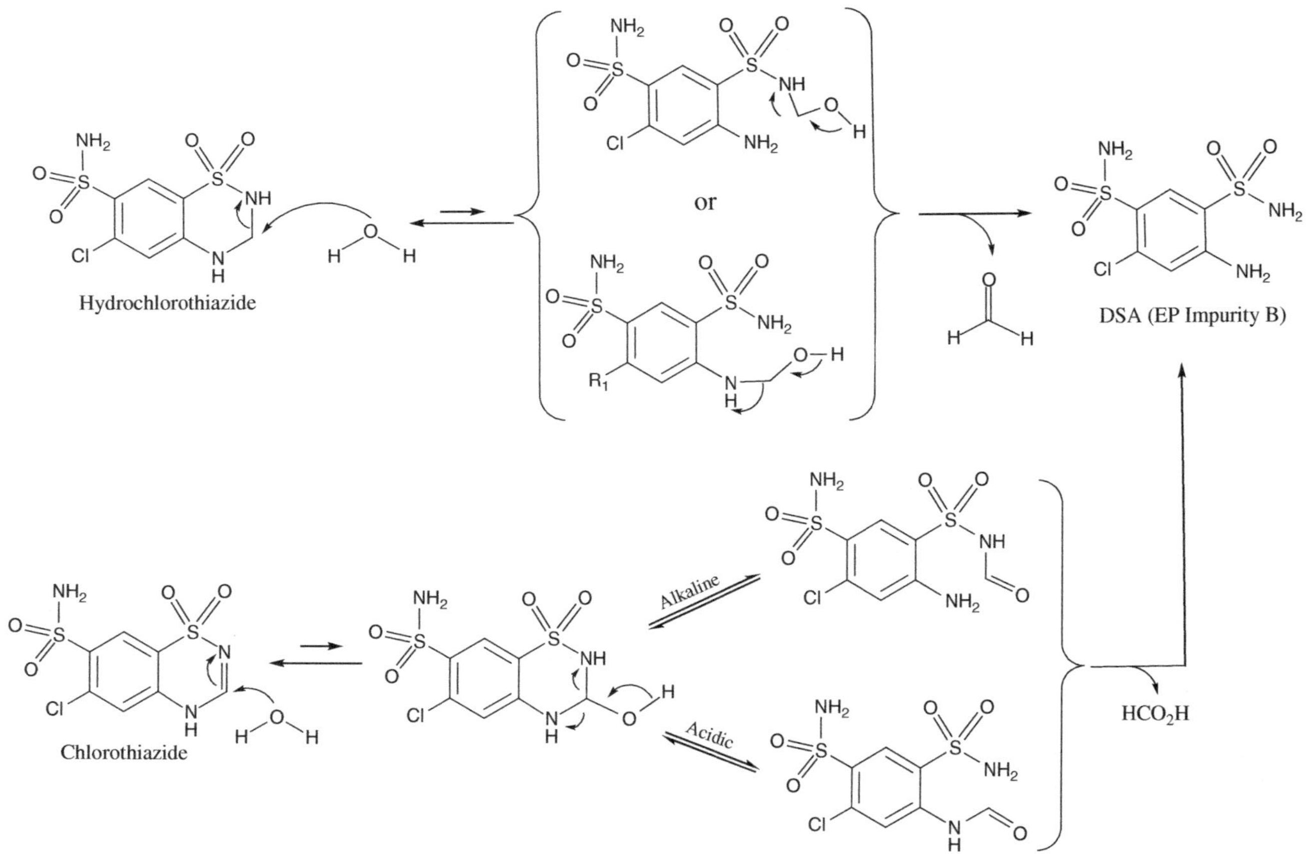

Scheme 217–218-1 Hydrolytic degradation of hydrochlorothiazide and chlorothiazide.

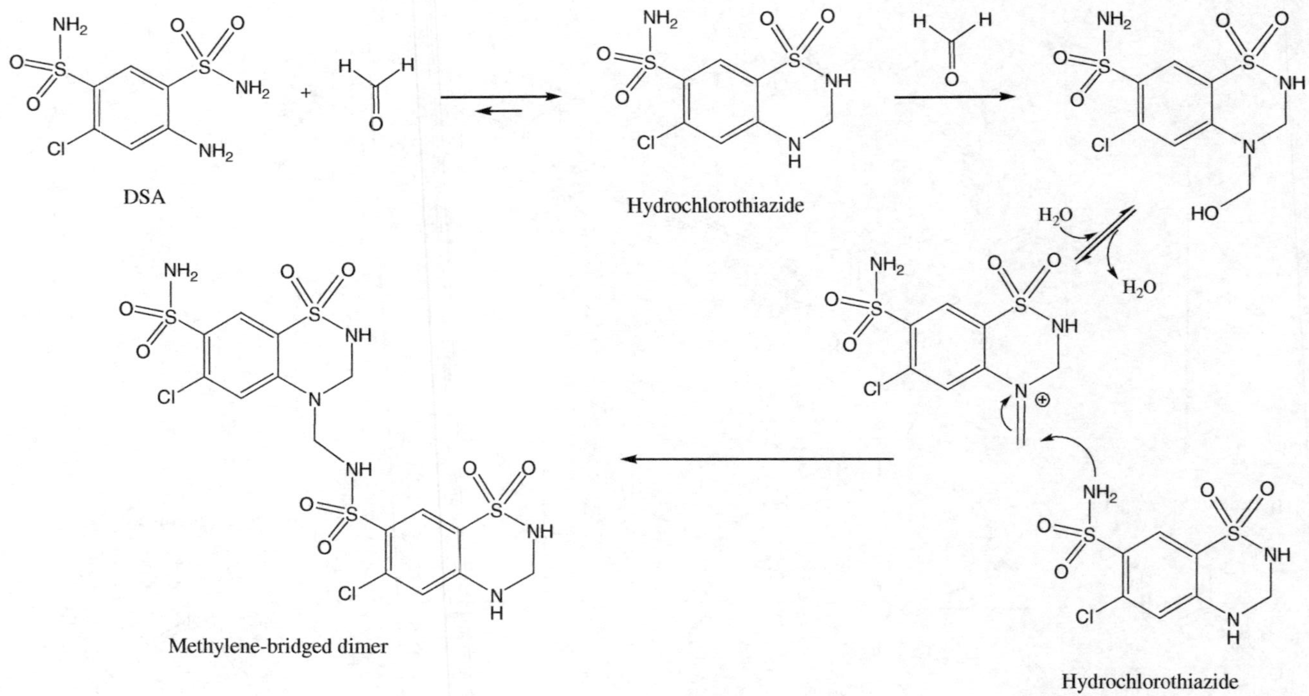

Scheme 217–218-2 Formation of the dimeric degradant of hydrochlorothiazide.

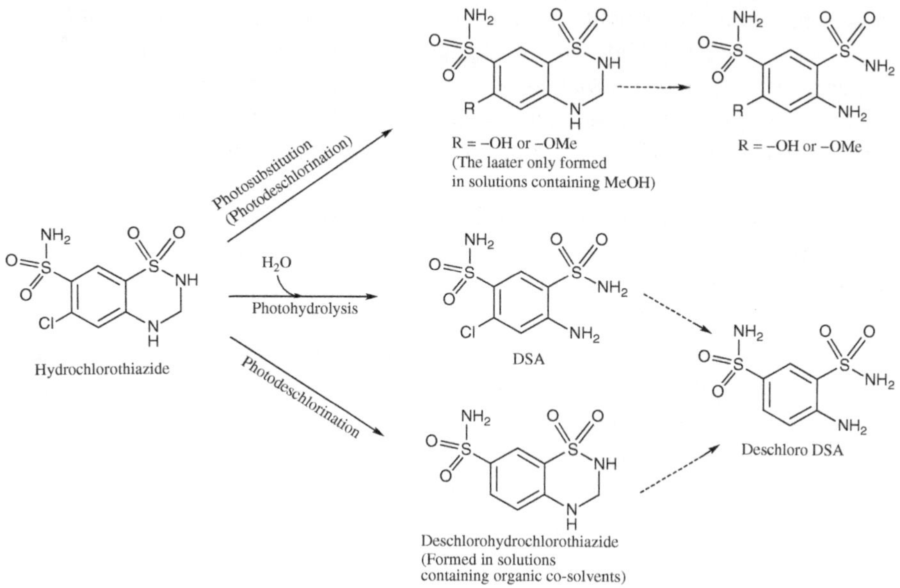

Scheme 217–218-3 Photosubstitution, photohydrolysis, and photodehydrogenation of hydrochlorothiazide. The dotted lines denote the secondary degradation pathways, which may occur under extended photolysis.

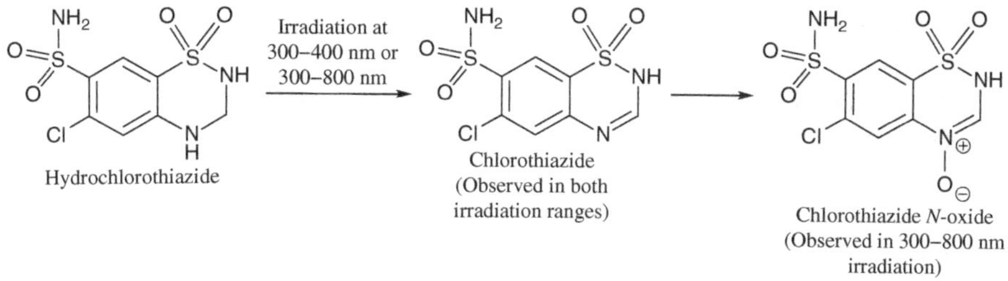

Scheme 217–218-4 Photooxidation of hydrochlorothiazide.

pH [5], and it seems that chlorothiazide may be hydrolytically more stable than hydrochlorothiazide [6].

Hydrochlorothiazide is synthesized by the reaction of DSA with formaldehyde, as such the excess formaldehyde can cause the formation of a methylene-bridged dimeric process impurity [7, 8]. During the hydrolytic degradation of hydrochlorothiazide, formaldehyde is also formed, which might further react with another molecule of hydrochlorothiazide to produce the same methylene-bridged dimeric impurity [7]. In such a case, the dimeric impurity would be a degradant (Scheme 217–218-2).

The photodegradation pathways of hydrochlorothiazide and chlorothiazide appear to be quite similar. The photodegradants and their distribution depend on the conditions of the photolytic degradation, which include wavelength, energy/duration, and solvents (for photodegradation in solutions). In Tamat and Moore's pioneering photodecomposition study of hydrochlorothiazide in methanolic and aqueous solutions [9], hydrochlorothiazide was found to undergo dechlorination (a photosubstitution process), hydrolysis, and reduction (Scheme 217–218-3). The same photodegradants were observed in a photodegradation study by Brigante et al., in which hydrochlorothiazide was irradiated by simulated or natural sunlight in pure water or sewage plant-treated water [10]. The most abundant photodegradant was the hydrolytic degradant, DSA, followed by the photosubstituted degradant, i.e., 6-hydroxydeschloro degradant, and the secondary photodegradant, 6-hydroxydeschloro DSA.

On the other hand, Revelle et al. reported that hydrochlorothiazide underwent photodehydrogenation (a form of photooxidation) in a methanolic solution of hydrochlorothiazide to yield chlorothiazide under irradiation of 300–400 nm [11]. Gumieniczek et al. also reported that chlorothiazide was one of the photooxidative degradants of hydrochlorothiazide, with another one being chlorothiazide *N*-oxide, after hydrochlorothiazide was subjected to UV–Vis photolysis of 300–800 nm (Scheme 217–218-4) [12].

References

1 Sica, D.A., Carter, B., Cushman, W., and Hamm, L. (2011). Thiazide and loop diuretics. *J. Clin. Hypertens.* 13: 639–643.

2 Hydrochlorothiazide. https://en.wikipedia.org/wiki/Hydrochlorothiazide (accessed 15 September 2023).

3 Chlorothiazide. https://en.wikipedia.org/wiki/Chlorothiazide (accessed 15 September 2023).

4 Mollica, J.A., Rehm, C.R., and Smith, J.B. (1969). Hydrolysis of hydrochlorothiazide. *J. Pharm. Sci.* 58: 635–636.

5 Yamana, T., Mizukami, Y., and Tsuji, A. (1968). Studies on the stability of drugs. XVII. Stability of benzothiadiazines. (8). Kinetics of acid-base catalysed hydrolysis of chlorothiazide. *Chem. Pharm. Bull.* 16: 396–404.

6 Huang, G., Smith, R.M., Albishri, H.M., and Lin, J.-M. (2010). Thermal stability of thiazide and related diuretics during superheated water chromatography. *Chromatographia* 72: 1177–1181.

7 Franolic, J.D., Lehr, G.J., Barry, T.L., and Petzinger, G. (2001). Isolation of a 2:1 hydrochlorothiazide–formaldehyde adduct impurity in hydrochlorothiazide drug substance by preparative chromatography and characterization by electrospray ionization LC–MS. *J. Pharm. Biomed. Anal.* 26: 651–663.

8 Fang, X., Bibart, R.T., Mayr, S. et al. (2001). Purification and identification of an impurity in bulk hydrochlorothiazide. *J. Pharm. Sci.* 90: 1800–1809.

9 Tamat, S.R. and Moore, D.E. (1983). Photolytic decomposition of hydrochlorothiazide. *J. Pharm. Sci.* 72: 180–183.

10 Brigante, M., DellaGreca, M., Previtera, L. et al. (2005). Degradation of hydrochlorothiazide in water. *Environ. Chem. Lett.* 2: 195–198.

11 Revelle, L.K., Musser, S.M., Rowe, B.J., and Feldman, I.C. (1997). Identification of chlorothiazide and hydrochlorothiazide UV-A photolytic decomposition products. *J. Pharm. Sci.* 631–634.

12 Gumieniczek, A., Galeza, J., Mroczek, T. et al. (2018). Kinetics and characterization of degradation products of dihydralazine and hydrochlorothiazide in binary mixture by HPLC-UV, LC-DAD and LC-MS methods. *Chromatographia* 81: 1147–1162.

219–220: Hydrocortisone, Cortisone

219: Hydrocortisone [WHO List of Essential Medicines]

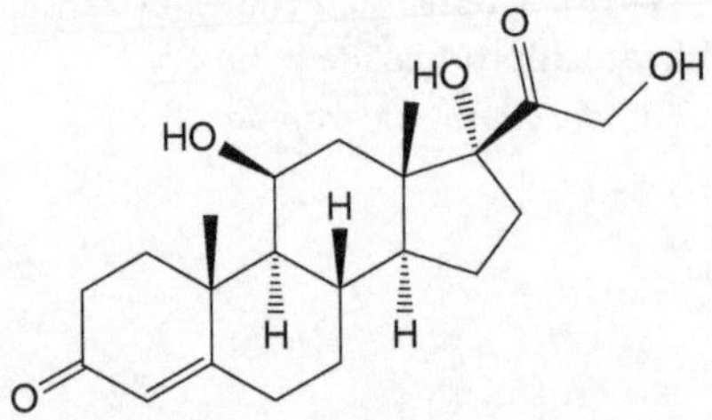

Chemical name: (8*S*,9*S*,10*R*,11*S*,13*S*,14*S*,17*R*)-11,17-Dihydroxy-17-(2-hydroxyacetyl)-10,13-dimethyl-2,6,7,8,9,11,12,14,15,16-decahydro-1*H*-cyclopenta[a]phenanthren-3-one
Formula: $C_{21}H_{30}O_5$
Monoisotopic protonated ion (*m/z*): 363.2166
Molecular weight: 362.47
CAS number: 50-23-7

220: Cortisone

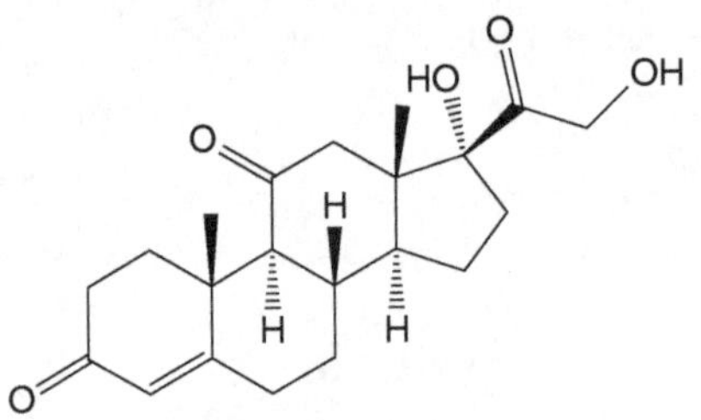

Chemical name: (8*S*,9*S*,10*R*,13*S*,14*S*,17*R*)-17-Hydroxy-17-(2-hydroxyacetyl)-10,13-dimethyl-1,2,6,7,8,9,12,14,15,16-decahydrocyclopenta[a]phenanthrene-3,11-dione
Formula: $C_{21}H_{28}O_5$
Monoisotopic protonated ion (*m/z*): 361.2010
Molecular weight: 360.45
CAS number: 53-06-5

Hydrocortisone is the medicinal name for the endogenous hormone cortisol, which was patented for medicinal use in 1936 and approved in 1941. It is a glucocorticoid steroid, and its pharmacological effects are mainly anti-inflammation and immune suppression [1].

On the other hand, cortisone is an oxidative metabolite of hydrocortisone, and clinically, it is used as a prodrug for hydrocortisone, because cortisone can be converted into hydrocortisone by the enzyme 11β-hydroxysteroid dehydrogenase type 1 in the liver [2].

Like other structurally similar steroids, the degradation of both drugs occurs mostly on their D rings, and the types of degradation include dehydration (referred to as the Mattox rearrangement) [3], oxidation, and subsequent degradation such as intramolecular Cannizzaro rearrangement [4]. Hansen and Bundgaard studied hydrocortisone stability in aqueous solutions of different pHs [5]. In strongly acidic solutions, hydrocortisone underwent dehydration to form two enol aldehyde degradants, a process referred to as the Mattox rearrangement (Scheme 219–220-1). In weakly acidic to near-neutral solutions, hydrocortisone also degraded to oxidative degradants, 21-aldehyde (also called 21-dehydrohydrocortisone) and 17-keto degradants, in addition to the enol aldehyde degradants. In alkaline solutions, the degradation appeared more complicated: the major degradants were 17-hydroxy-21-acid (glycolic acid 1) and 17-hydroxy-17-acid (etienic acid), both of which seemed to stem from further oxidative degradation of 21-aldehyde. The minor degradants observed in the alkaline solutions were 17-deoxy-21-acid and 17-ketone, both of which would appear to result from further degradation of the enol aldehyde degradants according to more recent studies [4, 6].

Timmins and Gary studied hydrocortisone degradation behavior in zinc oxide lotion and found that the degradation was oxidative: the principal degradant was 21-aldehyde, and the minor degradants were 17-hydroxy-17-acid and 17-ketone (Scheme 219–220-2) [7]. The results of this study are consistent with those of Hansen and Bundgaard [5]; the nominal pH of the zinc oxide lotion would be somewhat alkaline.

The 11-hydroxyl group of hydrocortisone is usually quite stable and resistant toward oxidation. Nevertheless, it was reported that in certain crystal forms of hydrocortisone 17-esters, photooxidation of 11-hydroxyl into the corresponding 11-keto was observed (Scheme 219–220-3) [7].

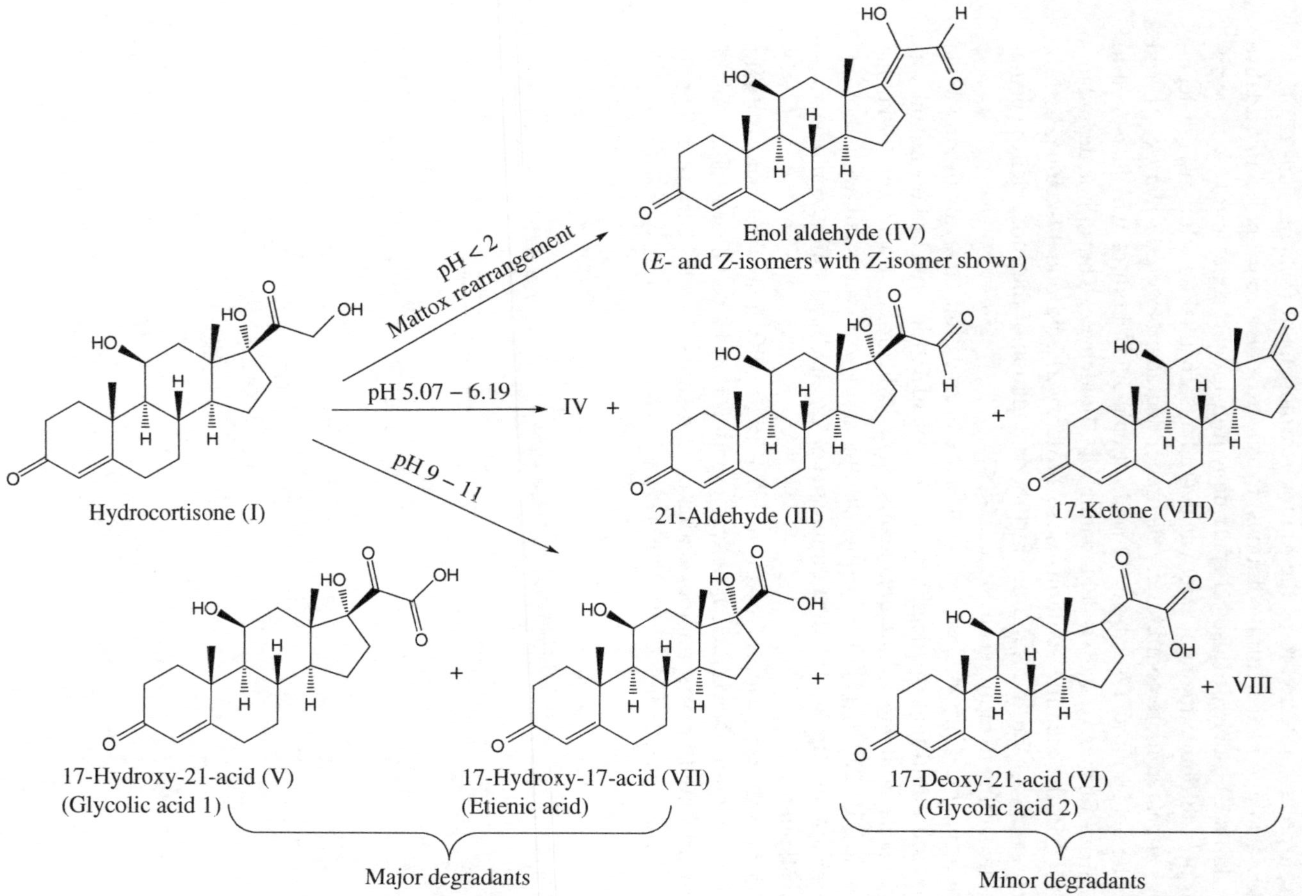

Scheme 219–220-1 Degradation pathways of hydrocortisone in aqueous solutions of different pH [5]. The oxidative degradant, 21-aldehyde (III), is also referred to as 21-dehydrohydrocortisone. The numbering of the compounds follows that of the authors [5]. *Source:* Hansen and Bundgaard [5]/with permission of Elsevier.

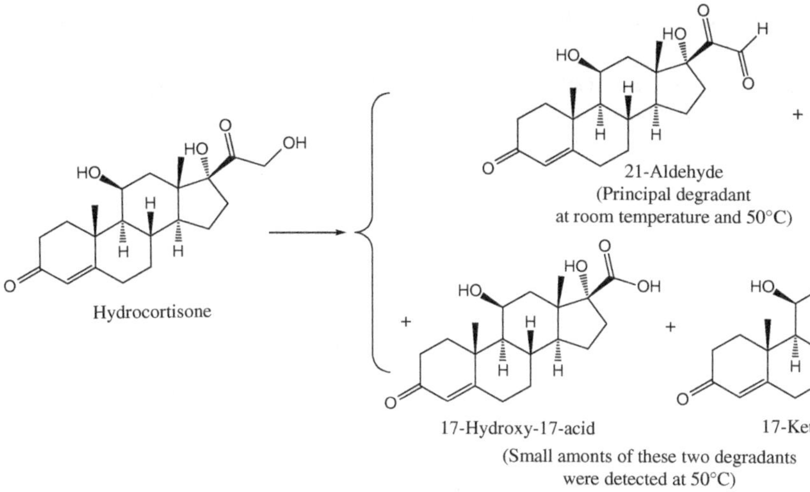

Scheme 219–220-2 Degradation pathways of hydrocortisone in a zinc oxide lotion [7]. *Source:* Timmins and Gary [7]/John Wiley & Sons.

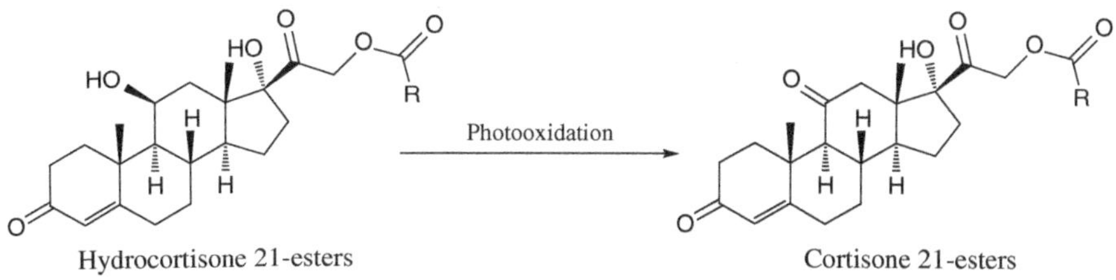

Scheme 219–220-3 Photooxidation of the 11-hydroxyl group of certain crystal forms of hydrocortisone 17-esters in solid state. *Source:* Byrn and Kessler [8]/with permission of Elsevier.

References

1 Hydrocortisone. https://en.wikipedia.org/wiki/Hydrocortisone (accessed 22 November 2023).

2 Cortisone. https://en.wikipedia.org/wiki/Cortisone (accessed 22 November 2023).

3 Mattox, V.R. (1952). Steroids derived from bile acids. XV. The formation of a glyoxal side chain at C-17 from steroids with dihydroxyacetone and Δ16-ketol side chains. *J. Am. Chem. Soc.* 74: 4340–4347.

4 Li, M., Chen, B., Monteiro, S., and Rustum, A.M. (2009). Mechanism of base-catalyzed autooxidation of corticosteroids containing 20-keto-21-hydroxyl side chain. *Tetrahedron Lett.* 50: 4575–4581.

5 Hansen, J. and Bundgaard, H. (1980). Studies on the stability of corticosteroids V. The degradation pattern of hydrocortisone in aqueous solution. *Int. J. Pharm.* 6: 307–319.

6 Li, M., Wang, X., Chen, B. et al. (2009). Forced degradation of betamethasone sodium phosphate under solid state: formation, characterization, and mechanistic study of

all four 17,20-diastereomers of betamethasone 17-deoxy-20-hydroxy-21-oic acid. *J. Pharm. Sci.* 98: 894–904.

7 Timmins, P. and Gary, E.A. (1983). Degradation of hydrocortisone in zinc oxide lotion. *J. Clin. Hosp. Pharm.* 8: 79–85.

8 Byrn, S.R. and Kessler, D.W. (1987). The solid state reactivity of the crystal forms of hydrocortisone esters. *Tetrahedron* 43: 1335–1343.

221–222: Hydroxychloroquine, Chloroquine

221: Hydroxychloroquine [WHO List of Essential Medicines]

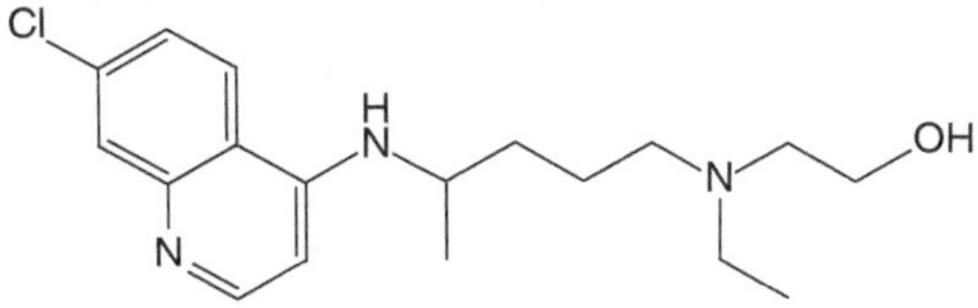

Chemical name: 2-[4-[(7-Chloroquinolin-4-yl)amino]pentyl-ethylamino]ethanol
Trade name: Plaquenil, Sovuna
Formula: $C_{18}H_{26}ClN_3O$
Monoisotopic protonated ion (*m/z*): 336.1837
Molecular weight: 335.88
CAS number: 118-42-3
Common salt form: Sulfate

222: Chloroquine [WHO List of Essential Medicines]

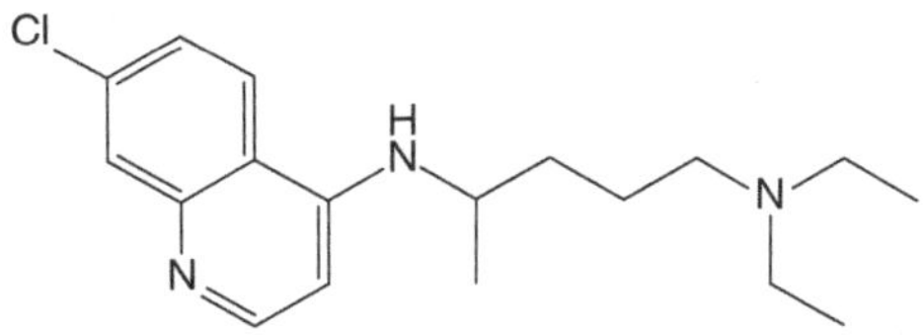

Chemical name: N^4-(7-Chloroquinolin-4-yl)-N^1,N^1-diethylpentane-1,4-diamine
Formula: $C_{18}H_{26}ClN_3$
Monoisotopic protonated ion (*m/z*): 320.1888
Molecular weight: 319.88
CAS number: 54-05-7
Common salt form: HCl salt, Sulfate, Phosphate

Hydroxychloroquine and chloroquine are primarily used for preventing and treating malaria in cases where the particular malaria strains have not developed resistance toward these two drugs. Chloroquine was discovered by Hans Andersag and coworkers at Bayer Laboratories in 1934, followed up on the lead by quinine, the natural product isolated from the bark of a cinchona tree [1]. Hydroxychloroquine was subsequently discovered among the chloroquine analogs [2]. In addition to their use as antimalarial agents, both drugs are utilized for autoimmune diseases such as rheumatoid arthritis and lupus [3].

Both drug molecules have the same moieties that are susceptible to degradation, which are the tertiary amine and chloroquinoline functional groups. The former moiety can be oxidized into the corresponding *N*-oxide under either nucleophilic oxidative or photooxidative conditions [4–6]. For example, Doddaga and Peddakonda reported chloroquine *N*-oxide as the major oxidative degradant of chloroquine [4], while Coelho et al. also identified chloroquine *N*-oxide (code name DP-2) as one of the major oxidative degradants of chloroquine (Scheme 221–222-1) [5]. In addition, Coelho et al. also reported another degradant, code name DP-1, in their oxidative stress of hydroxychloroquine in 3% hydrogen peroxide solution. Likewise, hydroxychloroquine can undergo the same oxidative degradation to give hydroxychloroquine *N*-oxide [6].

It needs to be pointed out that the structure of DP-1 was proposed solely based on mass spectrometric data. If DP-1 is a true degradant, it should be formed via cleavage off the chloroquinoline core, thus causing the simultaneous formation of another degradant that should bear the remaining chloroquinoline core, e.g., a degradant such as DP-3. Nevertheless, no additional degradant was observed in the LC-MS analysis presented by the authors. On the other hand, in the HPLC-UV chromatogram of the forced oxidative degradation solution, DP-1 showed up as a strong peak at the UV detection wavelength of 250 nm, which is contradictory to the fact that DP-1 should have essentially no UV absorption due to its lack of UV chromophore. These data presented by the authors cast considerable doubt on the validity of DP-1.

On the other hand, the chloroquinoline moiety can be susceptible to photochemical degradation. In 1982, Moore and Hemmens investigated photosensitization of a number of antimalarial drugs, including chloroquine and hydroxychloroquine. Under UV irradiation at 365 nm, chloroquine was found to undergo significant dechlorination, particularly with increasing pH; the degradation products resulting from the dechlorination were not studied. Nevertheless, it can be reasonably assumed that the hydroxylated deschloro degradant should be among the major photodegradants (Scheme 221–222-2) [7]. In the photodegradation study of chloroquine in solutions of pH 5.0, 7.0, and 8.0 performed by Karim et al. [8], it was found

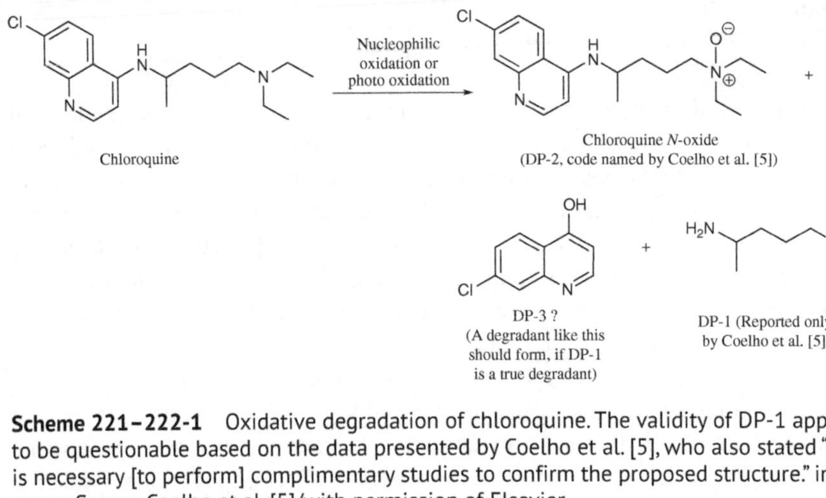

Scheme 221–222-1 Oxidative degradation of chloroquine. The validity of DP-1 appears to be questionable based on the data presented by Coelho et al. [5], who also stated "... it is necessary [to perform] complimentary studies to confirm the proposed structure." in their paper. *Source:* Coelho et al. [5]/with permission of Elsevier.

that chloroquine degraded the fastest at pH 8.0, and desethylchloroquine (II, Scheme 221–222-3) was identified as one of the degradants.

In 1988, Tonnesen et al. studied photolysis of hydroxychloroquine in neutral base form in either isopropanol or water solution with irradiation at 240–600 nm [9]. Three dealkylated degradants (I–III) and one dimeric degradant (IV) were identified based on mass spectrometric data and NMR measurement (Scheme 221–222-3).

In 2013, Saini and Bansal found that in alkaline solutions (acetonitrile/0.1 M NaOH; 2/3, v/v), hydroxychloroquine degraded into six degradants, five of which were structurally elucidated on the basis of mass spectrometric analysis (Scheme 221–222-4) [10]. One of the five degradants, desethyl hydroxychloroquine (3), was previously reported [11], while the other four degradants had not been reported according to the authors.

By combining the above photochemical degradation results, it is clearly that photochemical dealkylation can occur at both the tertiary amine and secondary amine sites.

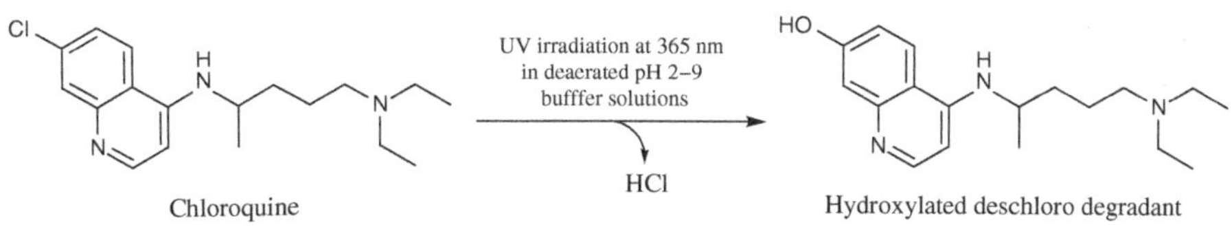

Scheme 221–222-2 Photodechlorination of chloroquine in pH 2–9 buffer solutions.

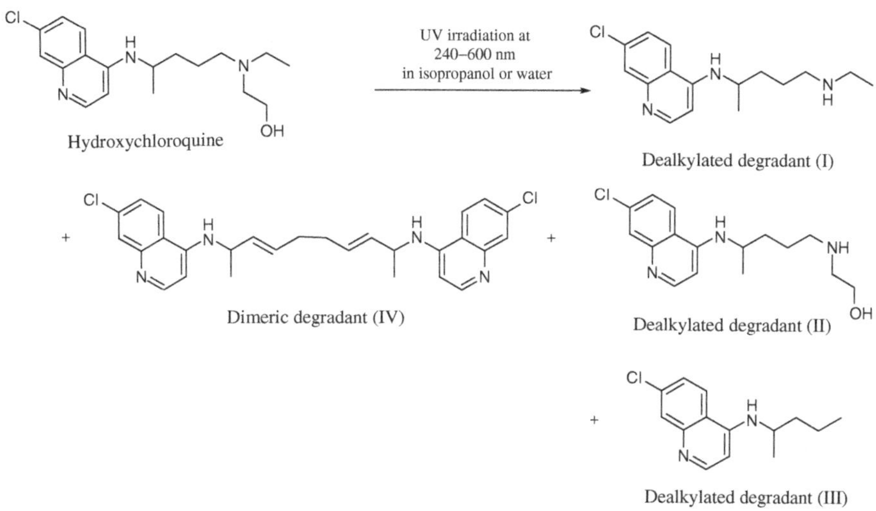

Scheme 221–222-3 Photolysis of hydroxychloroquine in either isopropanol or water solution with irradiation of 240–600 nm.

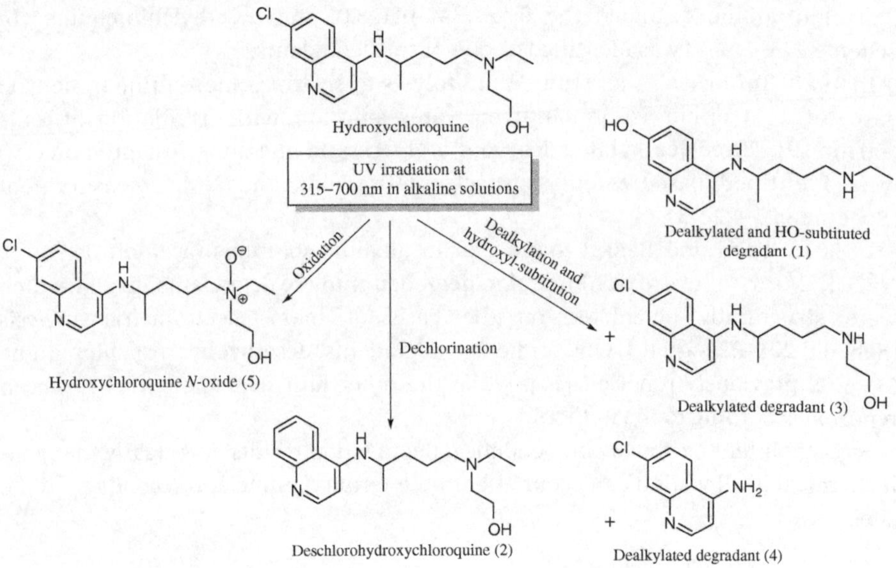

Scheme 221-222-4 Photolysis of hydroxychloroquine in alkaline solutions (acetonitrile/0.1 M NaOH, v/v) with irradiation of 315–700 nm.

References

1 Hydroxychloroquine. https://en.wikipedia.org/wiki/Hydroxychloroquine (accessed 11 November 2023).

2 Chloroquine. https://en.wikipedia.org/wiki/Chloroquine (accessed 11 November 2023).

3 Ben-Zvi, I., Kivity, S., Langevitz, P., and Shoenfeld, Y. (2012). Hydroxychloroquine: from malaria to autoimmunity. *Clin. Rev. Allergy Immunol.* 42: 145–153.

4 Doddaga, S. and Peddakonda, R. (2011). Chloroquine-N-oxide, a major oxidative degradation product of chloroquine: identification, synthesis and characterization. *J. Pharm. Biomed. Anal.* 81–82: 118–125.

5 Coelho, A.S., Chagas, C.E.P., de Padua, R.M. et al. (2017). A comprehensive stability-indicating HPLC method for determination of chloroquine in active pharmaceutical ingredient and tablets: identification of oxidation impurities. *J. Pharm. Biomed. Anal.* 145: 248–254.

6 Dongala, T., Katari, N.K., Palakurthi, A.K. et al. (2020). Stability indicating LC method development for hydroxychloroquine sulfate impurities as available for treatment of COVID-19 and evaluation of risk assessment prior to method validation by quality by design approach. *Chromatographia* 83: 1269–1281.

7 Moore, D.E. and Hemmens, V. (1982). Photosensitization by antimalarial drugs. *Photochem. Photobiol.* 36: 71–77.

8 Karim, E.I.A., Ibrahim, K.E.E., Abdelrahman, A.N., and Fell, A.F. (1994). Photodegradation studies on chloroquine phosphate by high-performance liquid chromatography. *J. Pharm. Biomed. Anal.* 12: 667–674.

9 Tonnesen, H.H., Grislingaas, A.-L., Woo, S.O., and Karlsen, J. (1988). Photochemical stability of antimalarials. I. Hydroxychloroquine. *Int. J. Pharm.* 43: 215–219.

10 Saini, B. and Bansal, G. (2013). Characterization of four new photodegradation products of hydroxychloroquine through LC-PDA, ESI-MS^n and LC-MS-TOF studies. *J. Pharm. Biomed. Anal.* 84: 224–231.

11 Dongre, V.G., Ghugare, P.D., Karmuse, P. et al. (2009). Identification and characterization of process related impurities in chloroquine and hydroxychloroquine by LC/IT/MS, LC/TOF/MS and NMR. *J. Pharm. Biomed. Anal.* 49: 873–879.

7

Ibuprofen to Nitrofurantoin

223: Ibuprofen

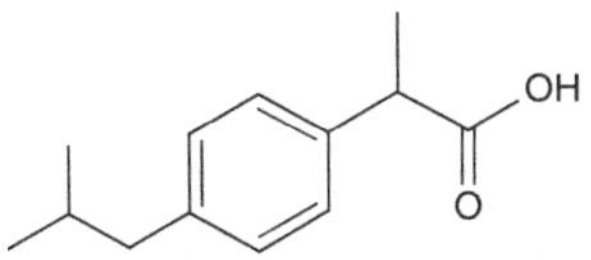

Chemical name: (*R*,*S*)-2-(4-(2-Methylpropyl)phenyl)propanoic acid
Trade name: Motrin, others
Formula: $C_{13}H_{18}O_2$
Monoisotopic protonated ion (*m/z*): 207.1380
Molecular weight: 206.29
CAS number: 154598-52-4

Ibuprofen is one of the most widely used nonsteroidal anti-inflammatory drugs (NSAIDs) and its production is nearly 15,000 tons per year [1]. It is clinically employed as an anti-inflammatory, analgesic, and antipyretic agent. It was discovered by Stewart Adams and John Nicholson at Boots Pure Drug Co., Ltd., in the United Kingdom during the 1950s [2]. Ibuprofen received the first market approval in the United Kingdom in 1969 as a prescription drug and became an over-the-counter (OTC) drug in the 1980s. It inhibits the enzymes cyclooxygenase-1 (COX-1) and cyclooxygenase-2 (COX-2) [3]. Ibuprofen is available in many formulation types, including tablet, hard shell capsule, and soft gel capsule.

Ibuprofen is a 2-aryl-substituted propionic acid, which has two benzylic positions that could be susceptible to free-radical-mediated oxidative degradation. Thus, its main degradation pathway should be of oxidative nature [4]. Indeed, its major oxidative degradants were reported to be 4-isobutylacetophenone (**1**), 2-(4-formylphenyl)propionic acid (**2**), 2-(4-isobutyrylphenyl)propionic acid (**3**), and 2-(4-(1-hydroxy-2-methylpropyl)phenyl)propanoic acid (**4**) [5, 6]. In an accelerated stress study of ibuprofen tablet batches from different manufacturers under 70°C/75%RH, additional oxidative degradants (**5–7**) were also observed in some batches. For those batches that experienced significant degradation, they all contained polyethylene glycols (PEGs), an excipient family known to have varying

Compendium of Drug Degradation Pathways, First Edition. Min Li.

degrees of hydroperoxide impurities resulting from its rather facile autooxidation [7]. In the batch that showed the most degradation also contains polysorbate 80, another excipient that undergoes autooxidation easily. These results are consistent with a most likely scenario that ibuprofen underwent free-radical-mediated oxidative degradation within these formulated formats; a probable degradation mechanism is shown in Scheme 223-1, with some inputs based on the study of Caviglioli et al. [8]

Degradants **2–5** were formed due to the oxidation on the benzylic position of the 2-methylpropyl substituent, degradants **1**, **2′**, and **6′** were produced due to the oxidation on the benzylic position of the propionic moiety, while degradant **6** was generated due to the oxidation on both benzylic positions. The formation of 7 appears rather unusual; on the other hand, its structure was merely inferred from mass spectrometric data [8].

In a long-term stability study of ibuprofen soft gel capsules, four degradants were formed via esterification of the API with the excipients PEG, sorbitol, and sorbitan (Scheme 223-2) [9]. PEG was used as a filler of the soft gel, while sorbitol and sorbitan served as plasticizers for the shell. The same or similar esterification degradants were observed in a forced degradation study of ibuprofen soft gel capsules [10].

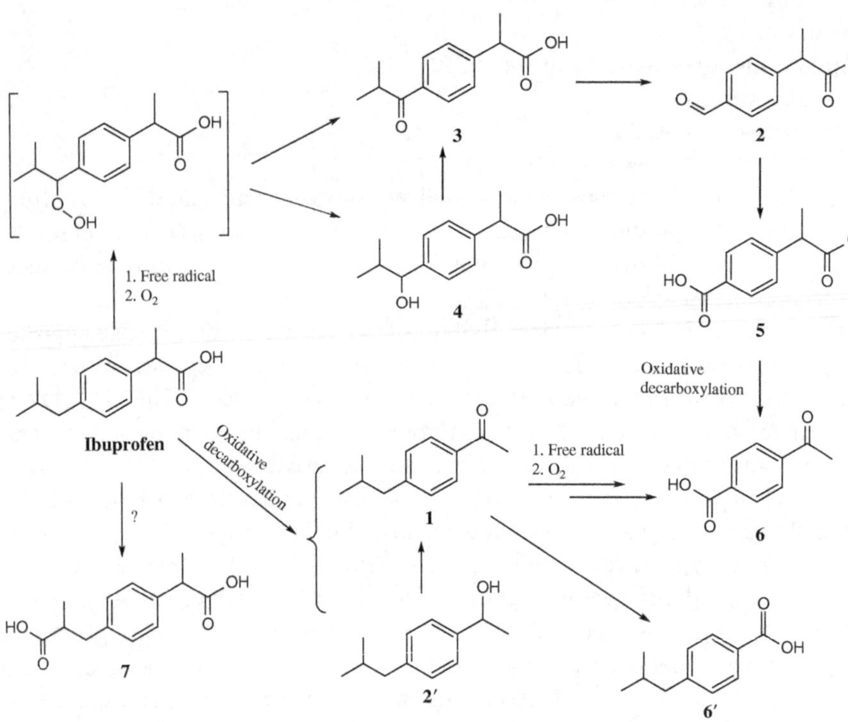

Scheme 223-1 Probable oxidative degradation mechanism of ibuprofen, with some inputs based on the study of Caviglioli et al. [8]. *Source:* Caviglioli et al. [8] / with permission of Elsevier.

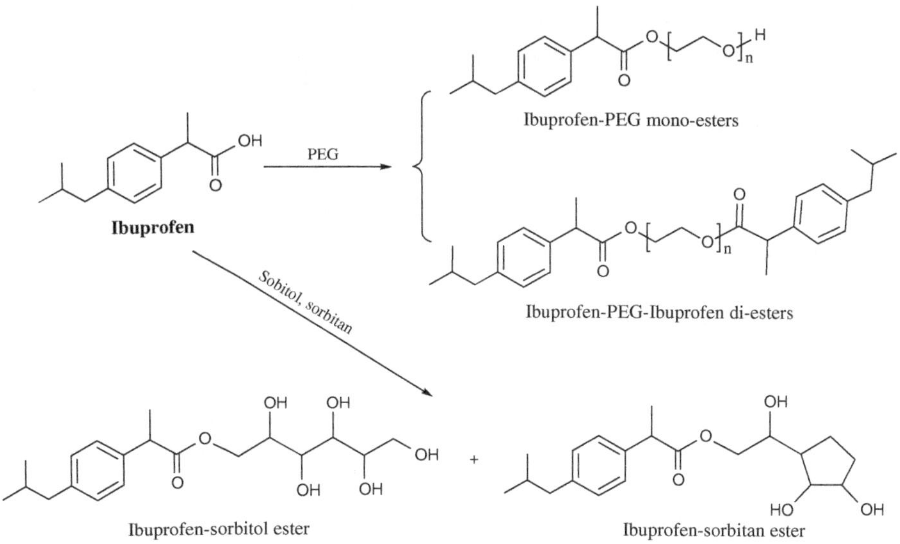

Scheme 223-2 Probable formation mechanism of ibuprofen esters in soft gel capsules containing PEG as the capsule fill and sorbitol and sorbitan as plasticizers for the shell.

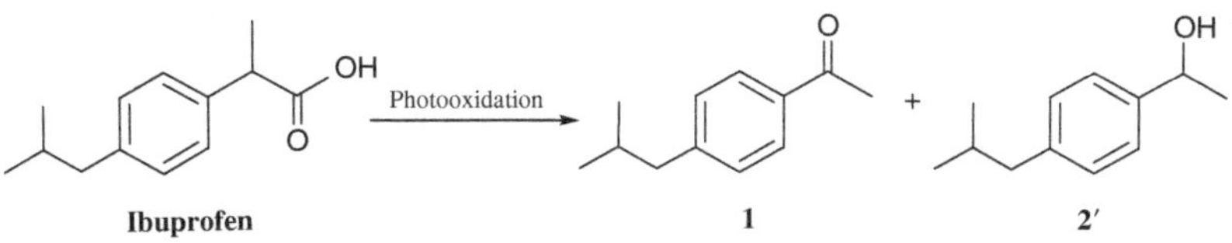

Scheme 223-3 Photooxidation of ibuprofen. Degradant 1 and 2′ were the main degradants in solutions, while 1 was observed as the degradant in the solid state under irradiation by natural sunlight.

Although ibuprofen appeared stable photochemically in the solid state, it is photolabile in solutions. The main photodegradants were reported to be **1** and **2′** (Scheme 223-3) [11, 12]. On the other hand, Farmer et al. showed ibuprofen underwent significant degradation in the solid state under natural sunlight, and the main degradant was **1** [13].

References

1 Myers, R.L. (2007). *The 100 Most Important Chemical Compounds: A Reference Guide*. New York: Bloomsbury Publishing.

2 Varrassi, G., Pergolizzi, J.V., Dowling, P., and Paladini, A. (2020). Ibuprofen safety at the golden anniversary: are all NSAIDs the same? A narrative review. *Adv. Ther.* 37: 61–82. https://doi.org/10.1007/s12325-019-01144-9.

3 Rainsford, K.D. (2009). Ibuprofen: pharmacology, efficacy and safety. *Inflammopharm.* 17: 275–342.

4 Dondoni, A., Dall'Occo, T., Fantin, G. et al. (1986). Studies on the actual and potential impurities in ibuprofen. *Farmaco. Prat.* 41: 237–244.

5 Asmus, P.A. (1985). Determination of 2-(4-isobutylphenyl)propionic acid in bulk and compressed tablets by reversed-phase high performance liquid chromatography. *J. Chromat.* 331: 169–176.

6 Cory, W.C., Harris, C., and Martinez, S. (2010). Accelerated degradation of ibuprofen in tablets. *Pharm. Dev. Tech.* 15: 636–643.

7 Li, M. (2012). Oxidative degradation. In: *Organic Chemistry of Drug Degradation*. Cambridge, UK: RSC Publishing.

8 Caviglioli, G., Valeria, P., Brunella, P. et al. (2002). Identification of degradation products of ibuprofen arising from oxidative and thermal treatments. *J. Pharm. Biomed. Anal.* 30: 499–509.

9 Dousa, M., Meca, L., Gibala, P. et al. (2017). Esterification of ibuprofen in soft gelatin capsules formulations: identification, synthesis and liquid chromatography separation of the degradation products. *J. Chromat. Sci.* 1–8: https://doi.org/10.1093/chromsci/bmx036.

10 Farias, F.F., Martins, V.A.P., Dorr, F.A. et al. (2021). Forced degradation study and characterization of main impurities of ibuprofen soft gelatin capsules by LC–MS-QTOF. *Pharmazie* 76: 138–144.

11 Castell, J.V., Gomez, M.J., Miranda, M.A., and Morera, I.M. (1987). Photolytic degradation of ibuprofen. Toxicity of the isolated photoproducts on fibroblasts and erythrocytes. *Photochem. Photobiol.* 46: 991–996.

12 Iovino, P., Chianese, S., Canzano, S. et al. (2016). Ibuprofen photodegradation in aqueous solutions. *Environ. Sci. Pollut. Res.* https://doi.org/10.1007/s11356-016-7339-0.

13 Farmer, S., Anderson, P., Burns, P., and Velagaleti, R. (2002). Forced degradation of ibuprofen in bulk drug and tablets and determination of specificity, selectivity, and the stability-indicating nature of the USP ibuprofen assay method. *Pharm. Tech.* 28–42. https://www.pharmtech.com.

224: Ifosfamide [WHO List of Essential Medicines]

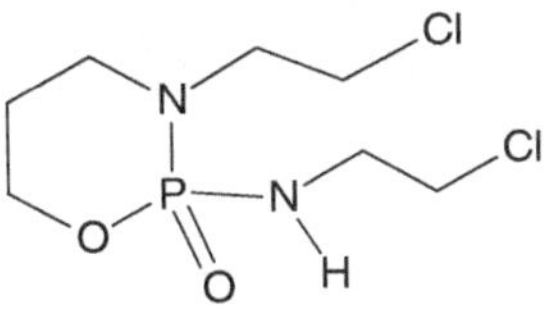

Chemical name: *N*,3-Bis(2-chloroethyl)-1,3,2-oxazaphosphinan-2-amide 2-oxide
Trade name: Ifex, others
Formula: $C_7H_{15}Cl_2N_2O_2P$
Monoisotopic protonated ion (*m/z*): 261.0321
Molecular weight: 261.08
CAS number: 3778-3773-2

Ifosfamide is an oxazaphosphorine anticancer alkylating agent, which is isomeric to cyclophosphamide. It came onto the US market in 1987 [1], which is much later than cyclophosphamide. Like cyclophosphamide, ifosfamide is also a prodrug and differs from cyclophosphamide in that its two chloroethyl groups are attached to the endocyclic and exocyclic nitrogen atoms, respectively, as opposed to both attaching to the same exocyclic nitrogen atom as in the case of cyclophosphamide. Its degradation pathway is essentially hydrolysis, which is also the case for cyclophosphamide, although their degradation products and distributions may be different. The two P–N bonds are more labile toward hydrolysis than the P–O bond. According to the study by Gilard et al., the first step of the hydrolysis started at the endocyclic P–N bond, giving rise to an open-chain phosphoramidic acid monoester (**2**, Scheme 224-1)[2]. Degradant **2** then underwent an intramolecular cyclization to produce the 2-hydroxyoxazaphosphorine degradant (**3**). Last, the endocyclic P–N bond of **3** was hydrolyzed to yield the three phosphoric acid monoesters (**4**). The phosphoric ester linkage is usually more stable than the corresponding phosphoric amide linkage, as evidenced by the fact that the hydrolytic activation energies of phosphoric esters are usually higher than those of phosphoric amides [3]. Thus, it would not be surprising to find that the free-phosphate anion, PO_4^{2-}, was only present at low levels, even after a long period of hydrolysis [2].

Before the aforementioned study of Gilard et al., some studies on the stability of ifosfamide were performed, with the first one reported by Zon et al. [4]. Nevertheless, these studies appeared fragmented, and some of the results may be erroneous [5] because the derivatization method used in the analysis might cause sample degradation, as demonstrated by Gilard et al. [6]. In strongly alkaline conditions, the main degradant was reported to be the aziridine degradant (**5**, Scheme 224-2) [7, 8].

Mufioz et al. studied the degradation kinetics of ifosfamide in aqueous solutions of pH 3.6–9.8 and found that the degradation followed first-order kinetics and the activation energy for the degradation was 118 kJ/mol at pH 5.0 [9]. Furthermore,

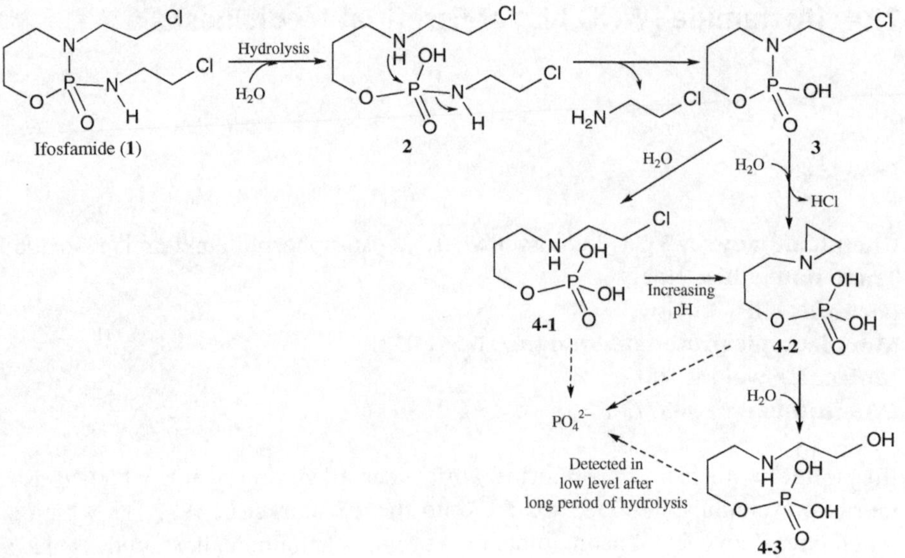

Scheme 224-1 Hydrolytic degradation pathway of ifosfamide in acidic to neutral solutions.

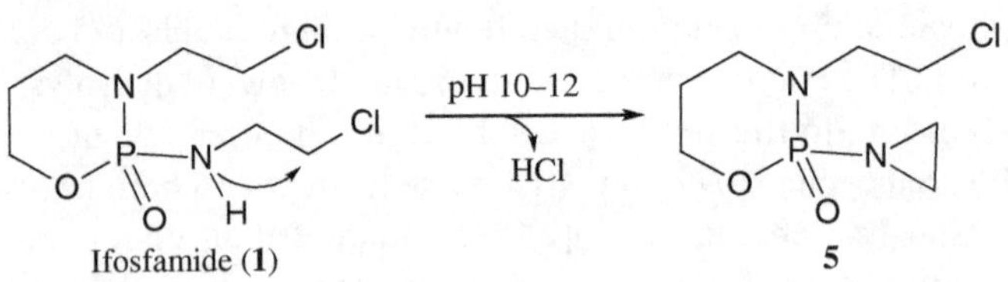

Scheme 224-2 Degradation of ifosfamide in solutions of pH 10–12.

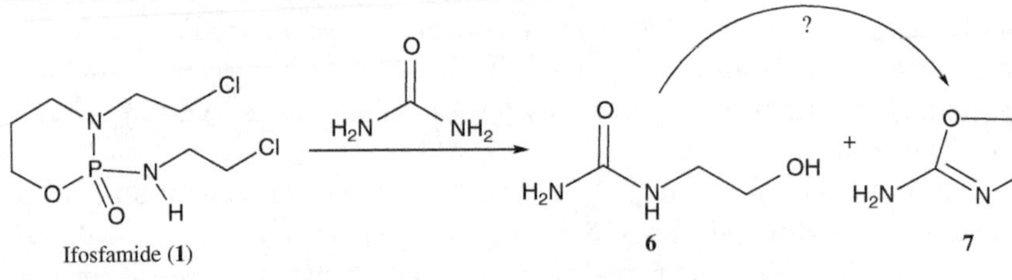

Scheme 224-3 Degradation of ifosfamide in the presence of urea.

ifosfamide was found to be most stable between pH 4 and 7.9; this result was largely consistent with a previous study in which ifosfamide was found to be most stable in the pH region of 4–9 [10].

In a urea-containing formulation of ifosfamide, two degradants were found to result from the interaction between the API and urea (Scheme 224-3) [11].

References

1 Ifosfamide. https://en.wikipedia.org/wiki/Ifosfamide (accessed September 18 2023).

2 Gilard, V., Martino, R., Malet-Martino, M. et al. (1999). Chemical stability and fate of the cytostatic drug ifosfamide and its *N*-dechloroethylated metabolites in acidic aqueous solutions. *J. Med. Chem.* 42: 2542–2560.

3 Li, M. (2012). Hydrolytic degradation. In: *Organic Chemistry of Drug Degradation*, 16–47. Cambridge: Royal Society Chemistry Publishing.

4 Zon, G., Ludeman, S.M., and Egan, W. (1977). High-resolution nuclear magnetic resonance investigations of the chemical stability of cyclophosphamide and related phosphoramidic compounds. *J. Am. Chem. Soc.* 99: 5785–5795.

5 Highley, M.S., Momerency, G., Van Cauwenberghe, K. et al. (1995). Formation of chloroethylamine and 1,3-oxazolidine-2-one following ifosfamide administration in humans. *Drug Metab. Dispos.* 23: 433–437.

6 Gilard, V., Martino, R., Malet-Martino, M., and Niemeyer, U. (1997). Stability of commercial formulations and aqueous solutions of ifosfamide. *Drug Metab. Dispo.* 25: 927–931.

7 Kaijser, G.P., Beijnen, J.H., Bult, A. et al. (1991). A systematic study on the chemical stability of ifosfamide. *J. Pharm. Biomed. Anal.* 9: 1061–1067.

8 Kaijser, G.P., Aalbers, T., Beijnen, J.H. et al. (1996). Chemical stability of cyclophosphamide, trofosfamide, and 2- and 3-dechloroethylifosfamide in aqueous solutions. *J. Oncol. Pharm. Pract.* 2: 15–21.

9 Mufioz, M., Bonjoch, J., Prat, J. et al. (1996). Degradation kinetics of ifosfamide in aqueous solution. *Inter. J. Pharm.* 139: 249–253.

10 Kaijser, G.P., Beijnen, J.H., Bult, A. et al. (1991). A systematic study on the chemical stability of Ifosfamide. *J. Pharm. Biomed. Anal.* 9: 1061–1067.

11 Yüce, I. and Morlock, G.E. (2018). All on one high-performance thin-layer chromatographyplate: solvent-free nanomole-scaled on-surface synthesis, workup and online high-resolution mass spectrometry for elucidation of two new degradation products in anifosfamide formulation. *J. Chromat. A* 1 (572): 145–151.

225: Imatini [WHO List of Essential Medicines]

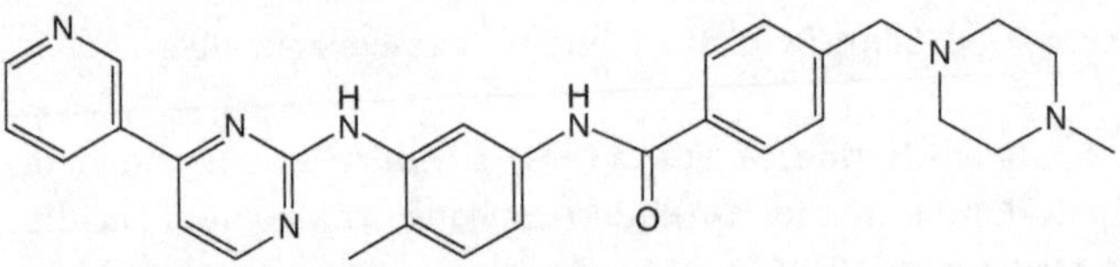

Chemical name: 4-[(4-Methylpiperazin-1-yl)methyl]-*N*-(4-methyl-3-{[4-(pyridin-3-yl)pyrimidin-2-yl]amino}phenyl)benzamide
Trade name: Gleevac, Glivac
Formula: $C_{29}H_{31}N_7O$
Monoisotopic protonated ion (*m/z*): 494.2663
Molecular weight: 493.62
CAS number: 152459-95-5
Salt form: Mesylate

Imatinib is an inhibitor of several nonreceptor tyrosine kinases, including ABL, BCR-ABL, and KIT tyrosine kinases. It is used for the treatment of chronic myelogenous leukemia (CML) and acute lymphoblastic leukemia (ALL) in addition to other tumors [1, 2]. CML is a rare clonal hematopoietic stem cell disorder, manifested by the presence of the so-called Philadelphia chromosome (Ph) [3]. The formation of Ph is due to a reciprocal translocation between the long arms of chromosomes 9 and 22, which causes the generation of a 210k fusion protein that possesses kinase activity (BCR-ABL tyrosine kinase). The fusion protein is present in 95% of CML patients and 15–30% ALL patients. Imatinib was discovered in the late 1990s by scientists at Ciba-Geigy, which was later merged with Sandoz to become Novartis [2, 4]. The clinical use of imatinib for treating CML was spearheaded by Brian Druker of Oregon Health & Science University and it was approved by the US FDA in 2001 as the first selective tyrosine kinase inhibitor for the treatment of a cancer [4], marking a new paradigm for cancer-targeted therapies with various tyrosine kinases as drug targets [5–7].

Imatinib has a central amide linkage, which would be susceptible to hydrolytic degradation to yield the aniline and carboxyl degradants (Scheme 225) [8, 9]. It also possesses a piperazine moiety, which would be susceptible to oxidative degradation. Under harsh oxidative stress conditions, e.g., 30% hydrogen peroxide or 1% hydrogen peroxide at 60°C, the two nitrogen atoms of the piperazine moiety could all undergo nucleophilic oxidation, leading to the formation of two mono-*N*-oxide and one di-*N*-oxide [9]. Under a less harsh oxidative condition, only *N*-oxide 1 was observed [10]. It needs to be pointed out that under oxidative conditions, such as those using hydrogen peroxide as the oxidizing reagent, hydrolytic degradants like the aniline degradant could also be formed, which does not make it an oxidative degradant (as incorrectly noted in Reference [9]). Furthermore, the structural characterization of the imatinib degradants, formed under various stress conditions, was primarily based on mass spectrometric analysis. Thus, some of the degradant structures proposed do not appear to be reasonable [10]; further structural

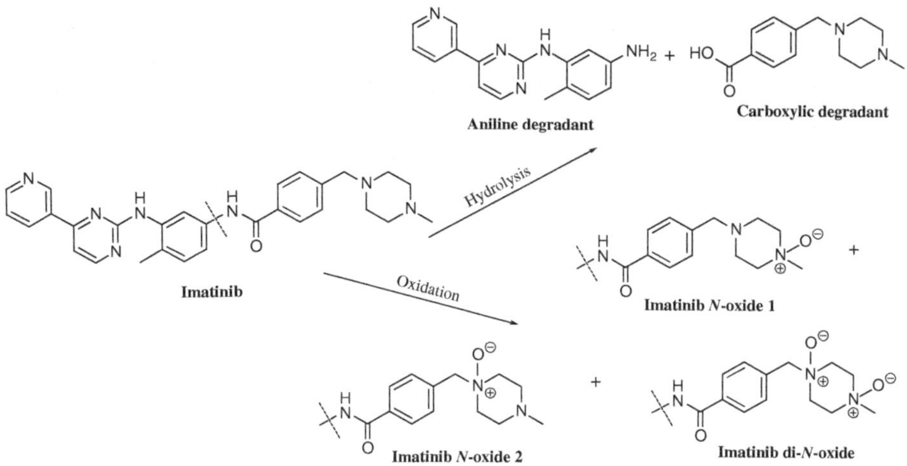

Scheme 225 Hydrolytic and oxidative degradation pathways of Imatinib.

confirmation by complementary techniques such as NMR spectroscopy seems to be necessary.

Although imatinib absorbs light in the visible range, it was shown to be stable under simulated light, and no triplet excited state could be detected, which was attributed to a substructure of imatinib that may act as a self-quencher for the excited state [11].

References

1. Wong, S. and Witte, O.N. (2004). The BCR-ABL Story: bench to bedside and back. *Annu. Rev. Immunol.* 22: 247–306.
2. Imatinib. https://en.wikipedia.org/wiki/Imatinib (accessed December 21 2023).
3. Druker, B.J. (2002). STI571 (GleevecTM) as a paradigm for cancer therapy. *Trends Mol. Med.* 8 (4 Suppl): S14–S18.
4. Capdeville, R., Buchdunger, E., Zimmermann, J., and Matte, A. (2002). Glivec (STI571, imatinib), a rationally developed, targeted anticancer drug. *Nat. Rev. Drug Discov.* 1: 493–502.
5. Druker, B.J. (2004). Imatinib as a paradigm of targeted therapies. *Adv. Cancer Res.* 91: 1–30.
6. Arora, A. and Scholar, E.M. (2005). Role of tyrosine kinase inhibitors in cancer therapy. *J. Pharmacol. Exper. Ther.* 315: 971–979.
7. Fischer, O.M., Streit, S., Hart, S., and Ullrich, A. (2003). Beyond Herceptin and Gleevec. *Curr. Opin. Chem. Bio.* 7: 490–495.
8. Al-Hadiya, B.M.H., Bakheit, A.H.H., and Abd-Elgalil, A.A. (2014). Chapter six - imatinib mesylate. *Prof. Drug Subs. Excip. Rel. Methodol.* 39: 265–297.
9. Ren, Z., Zhang, X., Wang, H., and Jin, X. (2016). Using an innovative quality-by-design approach for the development of a stability-indicating UPLC/

QTOF-ESI-MS/MS method for stressed degradation products of imatinib mesylate. *RSC Adv.* 6: 13050.

10 Nageswaria, A., Reddya, K.V.S.R.K., and Mukkanti, K. (2012). Stability-indicating UPLC method for determination of imatinib mesylate and their degradation products in active pharmaceutical ingredient and pharmaceutical dosage forms. *J. Pharm. Biomed. Anal.* 66: 109–115.

11 Secrétan, P.H., Karoui, M., Sadou-Yaye, H. et al. (2019). Imatinib: major photocatalytic degradation pathways in aqueous media and the relative toxicity of its transformation products. *Sci. Total Environ.* 655: 547–556.

226: Indapamide [WHO List of Essential Medicines]

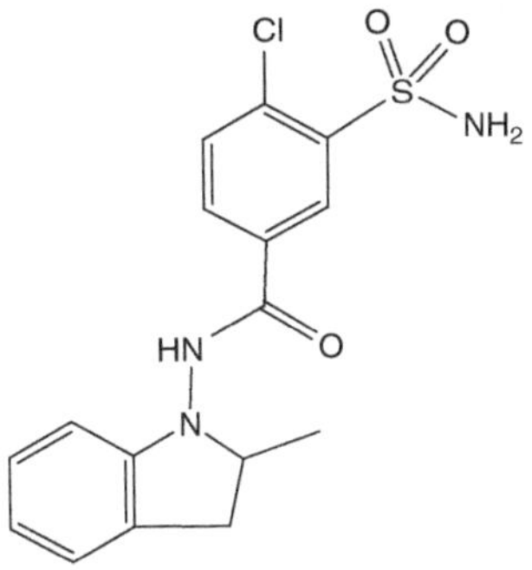

Chemical name: 4-Chloro-*N*-(2-methyl-2,3-dihydroindol-1-yl)-3-sulfamoylbenzamide
Trade name: Arifon, Lozol, Natrilix, others
Formula: $C_{16}H_{16}ClN_3O_3S$
Monoisotopic protonated ion (*m/z*): 366.0674
Molecular weight: 365.83
CAS number: 26807-65-8

Indapamide is a thiazide-like diuretic indicated for the treatment of hypertension and heart failure [1]. It is also used in combination with perindopril, an ACE inhibitor.

Indapamide has an amide-like linkage, which would be labile toward hydrolytic stress and expected to yield the hydrazine degradant (DP4, Impurity C of European Pharmacopeia), and the benzoic acid degradant (DP3) (Scheme 226), which are also the starting materials for the synthesis of the drug [2]. Nevertheless, the hydrazine degradant (DP4) was not observed in the forced degradation studies performed by two research groups [3, 4]. Instead, the oxidized form of the hydrazine degradant (DP2) was produced under either acidic [3] or alkaline [4] conditions. The 2,3-dihydroindole moiety of indapamide is susceptible to oxidative degradation. It seems that the hydrazine degradant (DP4) might be oxidized during the hydrolytic stress to form DP2, or DP2 might be generated from DP1, an oxidative degradant of indapamide that can be formed under either oxidative stress conditions (such as by hydrogen peroxide or $CuCl_2$) or alkaline stress conditions. Under the latter condition, indapamide may undergo a base-catalyzed autooxidation process (Scheme 226) [5]. Palaric et al. also reported a novel degradant (DP5) that apparently stemmed from the expansion of the 1-aminoindole ring into the cinnoline ring [4, 6].

In a test of generic indapamide sustained-release tablets, Yao et al. detected 4-chloro-*N*-(2-methyl-1*H*-indol-1-yl)-3-sulfamoylbenzamide (BP Impurity B) [7], which could be a likely degradant. Such types of amide degradants, due to the cleavage of the hydrazide *N*–*N* linkage, were reported for other hydrazide drugs such as isoniazid [8]. Nevertheless, the exact mechanism for the cleavage remains unclear.

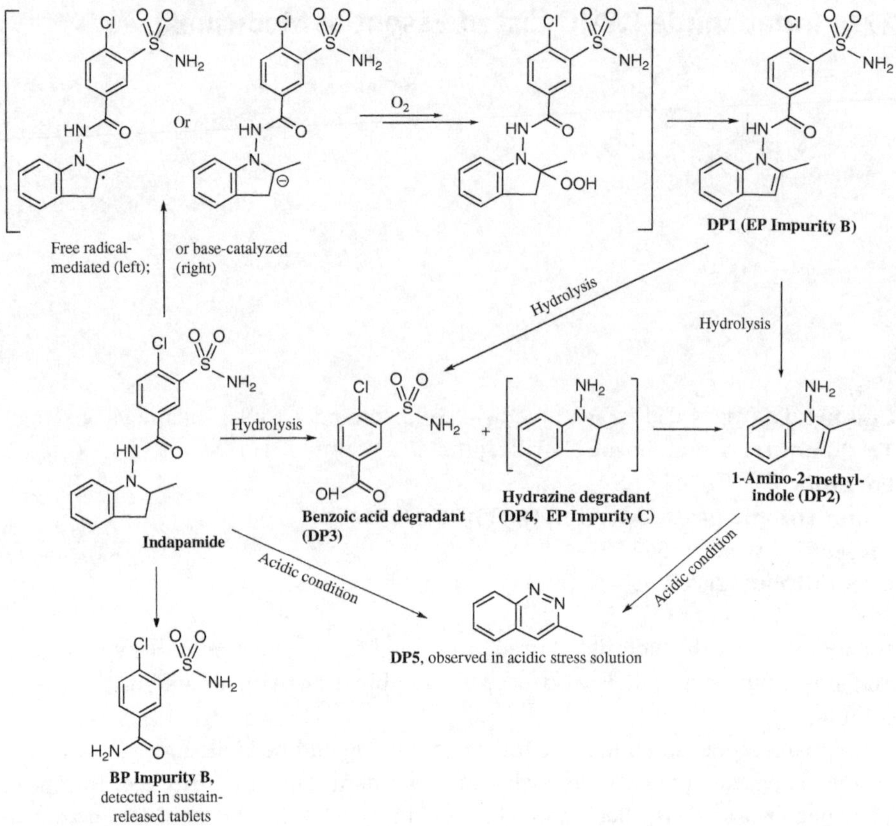

Scheme 226 Degradation pathways of indapamide under forced degradation conditions (all degradants except BP Impurity B) and in sustained-release tablets (BP Impurity B only).

References

1 Indapamide. https://en.wikipedia.org/wiki/Indapamide (accessed September 18 2023).

2 Heghes, S.C., Rus, L.M., Rus, L.-L. et al. (2017). HPLC-UV determination of indapamide in the presence of its main synthesis and degradation impurities. Method validation. *Farmacia* 65: 755–760.

3 Attia, K.A.M., Nassar, M.W.I., El-Din, M.M.K.S. et al. (2016). A stability-indicating QTRAP LC–MS/MS method for identification and structural characterization of degradation products of indapamide. *Anal. Methods* 8: 1836–1851.

4 Palaric, C., Molinié, R., Cailleu, D. et al. (2019). A deeper investigation of drug degradation mixtures using a combination of MS and NMR data: application to indapamide. *Molecules* 24: 1764.

5 Li, M. (2012). Oxidative degradation. In: *Organic Chemistry of Drug Degradation*, 48–109. Cambridge: Royal Society Chemistry Publishing.

6 Somei, M. and Ura, K. (1978). Ring enlargement reaction of 1-aminoindoles to cinnoline derivatives. *Chem. Lett.* 707–708.

7 Yao, W., Zhou, S., and Cheng, Q. (2023). Simultaneous analysis of indapamide and related impurities in sustained-release tablets by a single-run HPLC-PDA method. *Wuhan Univ. J. Nat. Sci.* 28: 333–340.

8 Prajapati, P. and Agrawal, Y.K. (2015). Development of a green method for separation and identification of the degradation impurity of isoniazid by SFC-MS/MS. *Anal. Methods* 7: 7776–7783.

227: Indomethacin (Indometacin) [WHO List of Essential Medicines]

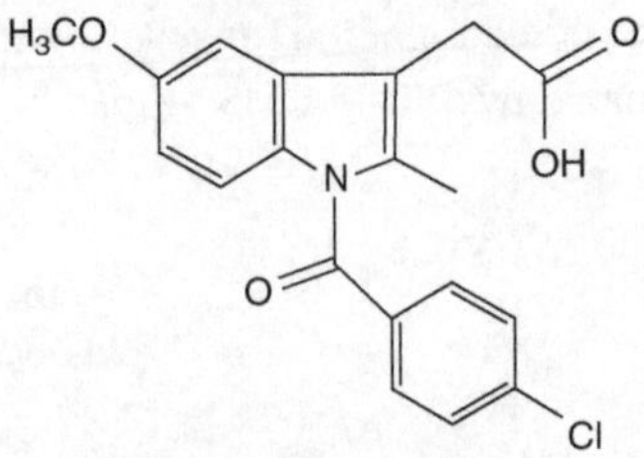

Chemical name: 2-{1-[(4-Chlorophenyl)carbonyl]-5-methoxy-2-methyl-1*H*-indol-3-yl}acetic acid
Trade name: Coversyl, Lozol, others
Formula: $C_{19}H_{16}ClNO_4$
Monoisotopic protonated ion (*m/z*): 358.0841
Molecular weight: 357.79
CAS number: 53-86-1

Indomethacin, also known as indometacin, is one of the nonsteroidal anti-inflammatory drugs (NSAIDs) prescribed for reducing fever, pain, and swelling from inflammation such as that associated with rheumatoid arthritis (RA) [1]. It is a nonselective inhibitor of cyclooxygenases 1 and 2 (COX-1 and 2), enzymes that catalyze the production of prostaglandins.

Indomethacin is an acylated indole derivative and its amide linkage is susceptible to hydrolytic degradation, leading to its two major degradants, 4-chlorobenzoic acid (**1**) and 5-methoxy-2-methylindoleacetic acid (**2**) (Scheme 227-1) [2, 3]. The activation energy for the hydrolysis was reported to be 97.6 kJ/mol in pH 7.4 phosphate buffer [2] and ~42 kJ/mol (~10 kcal/mol) in alkaline solutions [4], indicating it would be much more unstable under alkaline conditions. Degradant **2** could undergo decarboxylation to form degradant **3**. On the other hand, when crystalline indomethacin was heated to prepare its amorphous form, decarboxylated indomethacin (**4**) was produced [5].

In a forced oxidative degradation study by Li et al., two degradants (**5** and **6**) were found after indomethacin was treated with 0.23% hydrogen peroxide at 80°C. The structures of the degradants were elucidated based on MS and NMR measurement and their formation mechanisms were proposed (Scheme 227-2) [6]. The results of this study indicated that indomethacin, which is an acylated indole derivative and can be considered an embedded enamine, can still undergo nucleophilic oxidation.

Indomethacin was reported to be photochemically stable in crystalline form [7] but labile in solutions [8, 9]. In a suspension in benzene under irradiation by a medium-pressure mercury lamp, indomethacin was found to primarily undergo photo decarboxylation, leading to the formation of degradant **4** [8]. In another photodegradation study in organic solvents under irradiation by a low-pressure mercury lamp with a wavelength cutoff below 330 nm [9], degradant **4** and its tautomer

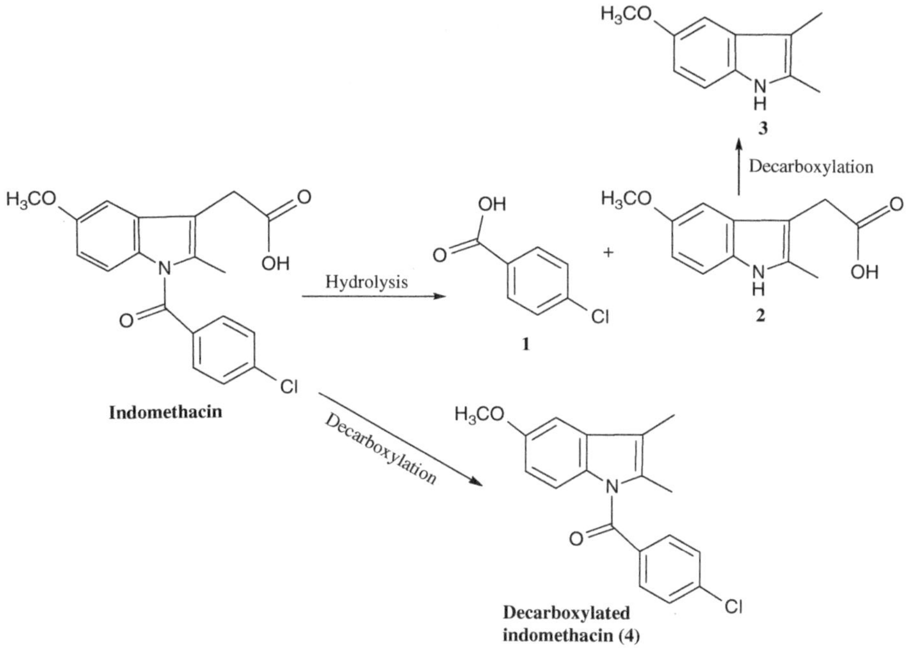

Scheme 227-1 Degradants from the hydrolysis and decarboxylation of indomethacin.

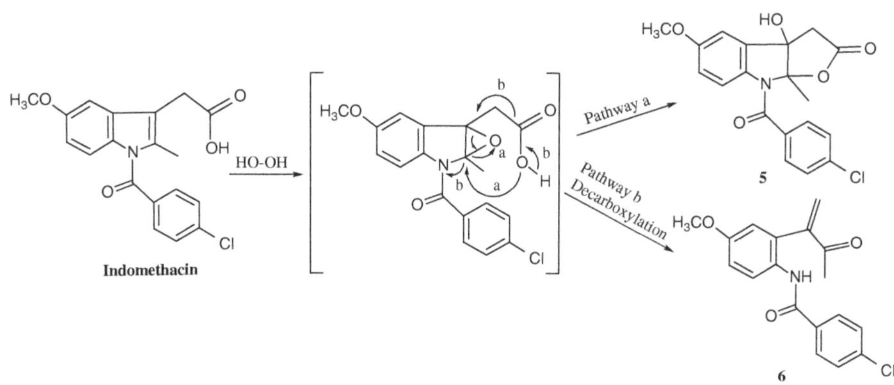

Scheme 227-2 Degradants from the oxidative degradation of indomethacin.

(**4′**) were observed as the major degradants, while a few minor degradants (**7–9**) were also present, among which **8** and **9** are photooxidative degradants (Scheme 227-3). Dabestani et al. also found that the photodegradation proceeded extremely slowly in more polar organic solvents, such as acetonitrile and ethanol [9].

In aqueous solutions, photodegradation products appeared to be more complex. In a photo stress study in aqueous solutions under irradiation at 254 nm and 310 nm, indomethacin was mostly decomposed to yield several photodegradants (**10–15**), among which **10** and **11** are oxidative degradants, **12** and **14** are oxidative and

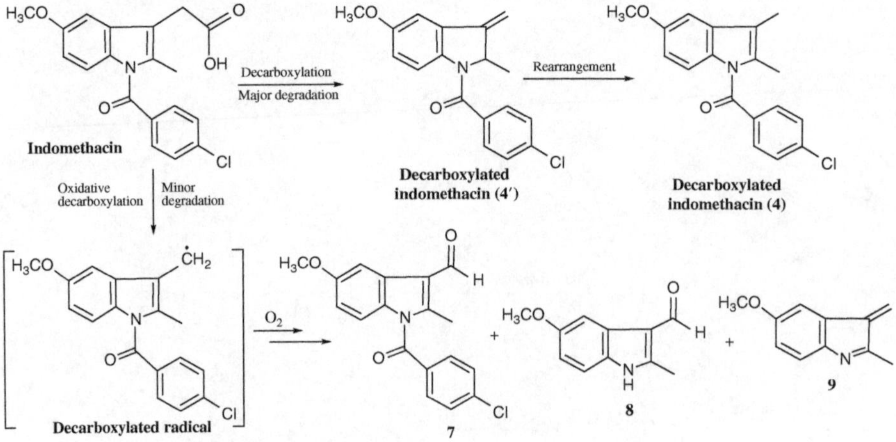

Scheme 227-3 Photodegradation of indomethacin in organic solvents.

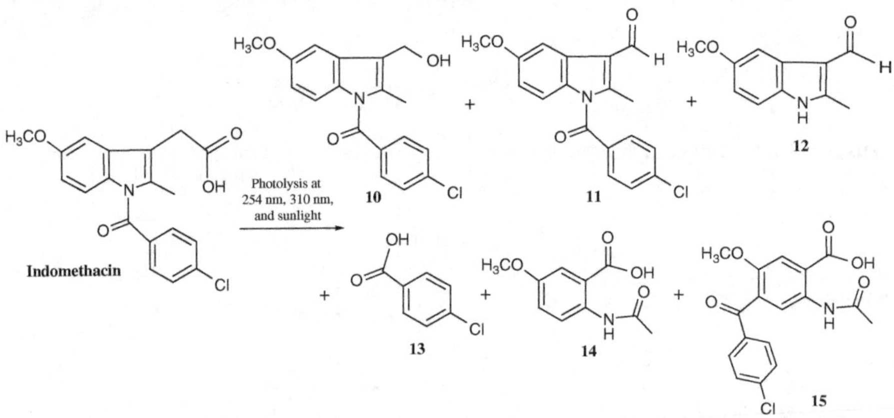

Scheme 227-4 Photodegradation of indomethacin in aqueous solutions.

hydrolytic degradants, **13** is hydrolytic degradant, and **15** is hydrolytic and photo-Fries degradant (Scheme 227-4) [10].

References

1 Indometacin. https://en.wikipedia.org/wiki/Indometacin (accessed December 23 2023).

2 Backensfeld, T., Muller, B.W., Wiese, M., and Seydel, J.K. (1990). Effect of cyclodextrin derivatives on indomethacin stability in aqueous solution. *Pharm. Res.* 7: 484–490.

3 Nováková, L., Matysová, L., Havlíková, L., and Solich, P. (2005). Development and validation of HPLC method for determination of indomethacin and its two degradation products in topical gel. *J. Pharm. Biomed. Anal.* 37: 899–905.

4 Hajratwala, B.R. and Dawson, J.E. (1977). Kinetics of indomethacin degradation I: presence of alkali. *J. Pharm. Sci.* 66: 27–29.

5 Shimada, Y., Komaki, H., Hirai, A. et al. (2018). Decarboxylation of indomethacin induced by heat treatment. *Int. J. Pharm.* 545: 51–56.

6 Li, M., Conrad, B., Maus, R.G. et al. (2005). A novel oxidative degradation pathway of indomethacin under the stressing by hydrogen peroxide. *Tetra. Lett.* 46: 3533–3536.

7 Ekiz-Guecer, N. and Reisch, J. (1991). Photostability of indomethacin in the crystalline state. *Pharm. Acta Helv.* 66: 66–67.

8 Weedon, A.C. and Wong, D.F. (1991). The photochemistry of indomethacin. *J. Photochem. Photobiol. A: Chem.* 61: 27–33.

9 Dabestani, R., Sik, R.H., Dvis, D.G. et al. (1993). Spectroscopic studies of cutaneous photosensitizing agents. XVIII. Indomethacin. *Photochem. Photobiol.* 58: 367–373.

10 Temussi, F., Cermola, F., DellaGreca, M. et al. (2011). Determination of photostability and photodegradation products of indomethacin in aqueous media. *J. Pharm. Biomed. Anal.* 56: 678–683.

228: Isoniazid [WHO List of Essential Medicines]

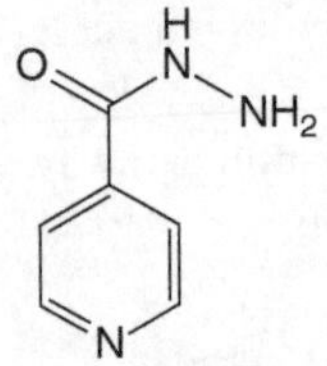

Chemical name: Pyridine-4-carbohydrazide
Trade name: Hydra, Hyzyd, Isovit, Rimifon, others
Formula: $C_6H_7N_3O$
Monoisotopic protonated ion (*m/z*): 138.0662
Molecular weight: 137.14
CAS number: 54-85-3

Isoniazid is an antibacterial drug used as a first-line treatment for both latent and active tuberculosis. For the latter category, it is often utilized in combination with other drugs such as rifampicin, pyrazinamide, and ethambutol [1]. The synthesis of the molecule was reported in 1912 and it was launched to treat tuberculosis in 1952 [2].

Isoniazid is an acylated hydrazine and, as such, it is susceptible to hydrolytic degradation, producing hydrazine (**1**) and isonicotinic acid (**2**) [3–6], both of which were detected in formulated products of isoniazid [3, 4]. In certain hydrolytic stress studies, isonicotinamide (**3**) was also obtained, in addition to isonicotinic acid (**2**) [5, 6]. In the forced degradation study by Bhutani et al. [5], **2** was the only degradant observed under acidic conditions (0.1 M HCl, 80°C), while both **2** (usually the major degradant) and **3** were formed in solutions of water, 0.1 M NaOH, or 30% hydrogen peroxide. In another forced degradation study, both **2** and **3** were reported to be present in comparable amounts under both acidic and alkaline conditions [6]. In solid state under irradiation by UV and fluorescent light, a dimeric degradant, *N′*-(pyridine-4-carbonyl)-hydrazide (**4**), was formed [5]. Degradant **4** was originally observed by Kakemi et al. under nonphotochemical degradation conditions [7], and later by Wollinger et al. in their drug-excipients compatibility study with certain colorants [8]. The degradation chemistry of isoniazid under hydrolytic and solid-state photodegradation conditions is summarized in Scheme 228-1.

The acylated hydrazine moiety of isoniazid is still quite nucleophilic and, as such, it is capable of reacting with carbonyl functionalities, particularly those more reactive ones such as aldehydes [9]. Rifampicin, another first-line anti-tuberculosis drug, contains a hydrazone moiety that can be hydrolyzed to produce the aldehyde degradant, 3-formyl-rifampicin. The latter can react with isoniazid in combination products of isoniazid and rifampicin to yield the isonicotinyl-hydrazone degradant (Scheme 228-2) [10, 11]. Likewise, isoniazid may be incompatible with excipients containing reducing sugar moieties, as it can form the corresponding isoniazid hydrazone degradants with these excipients (in such cases, these excipients would take the role of 3-formyl-rifampicin as shown in Scheme 228-2).

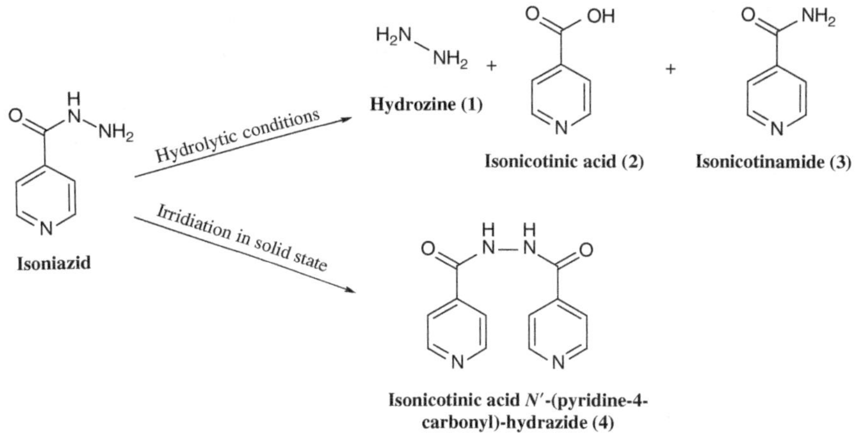

Scheme 228-1 The degradation chemistry of isoniazid under hydrolytic and solid state photodegradation conditions.

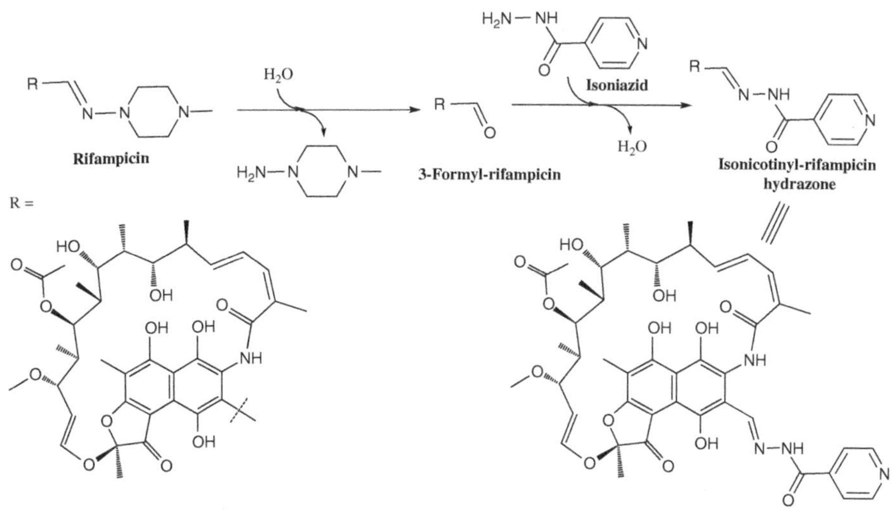

Scheme 228-2 Formation mechanism of isonicotinyl-rifampicin hydrazone degradant.

References

1 Isoniazid. https://en.wikipedia.org/wiki/Isoniazid (accessed September 18 2023).

2 Judge, V., Narasimhan, B., and Ahuja, M. (2012). Isoniazid: the magic molecule. *Med. Chem. Res.* 21: 3940–3957.

3 Matsui, F., Robertson, D.L., and Lovering, E.G. (1983). Determination of hydrazine in pharmaceuticals III: hydralazine and isoniazid using GLC. *J. Pharm. Sci.* 72: 948–951.

4 Faria, A.F., Vasconcelos, J.P., Goncalves, R.S.B. et al. (2012). Simultaneous analysis of isoniazid and its impurities by CZE. *Chromatographia* 75: 1335–1339.

5 Bhutani, H., Singh, S., Vir, S. et al. (2007). LC and LC–MS study of stress decomposition behaviour of isoniazid and establishment of validated stability-indicating assay method. *J. Pharm. Biomed. Anal.* 43: 1213–1220.

6 Prajapati, P. and Agrawal, Y.K. (2015). Development of a green method for separation and identification of the degradation impurity of Isoniazid by SFC-MS/MS. *Anal. Methods* 7: 7776–7783.

7 Kakemi, K., Sezaki, H., and Inoue, S. (1966). Stability of isoniazid and its related compounds. II. Base-catalyzed degradation of isoniazid (article in Japanese). *Yakugaku Zasshi* 86: 163–168.

8 Wollinger, W., da Silva, R.A., da Nóbrega, A.B. et al. (2016). Assessing drug-excipient interactions in the formulation of isoniazid tablets. *J. Braz. Chem. Soc.* 27: 826–833.

9 Devani, M.B., Shishoo, C.J., Doshi, K.J., and Patel, H.B. (1985). Kinetic studies of the interaction between isoniazid and reducing sugars. *J. Pharm. Sci.* 74: 427–432.

10 Singh, S., Mariappan, T.T., Sharda, N. et al. (2000). The reason for an increase in decomposition of rifampicin in the presence of isoniazid under acid conditions. *Pharm. Pharmacol. Commun.* 6: 405–410.

11 Prasad, B. and Singh, S. (2010). LC–MS/TOF and UHPLC–MS/MS study of in vivo fate of rifamycin isonicotinyl hydrazone formed on oral co-administration of rifampicin and isoniazid. *J. Pharm. Biomed. Anal.* 52: 377–383.

229: Ivermectin [WHO List of Essential Medicines]

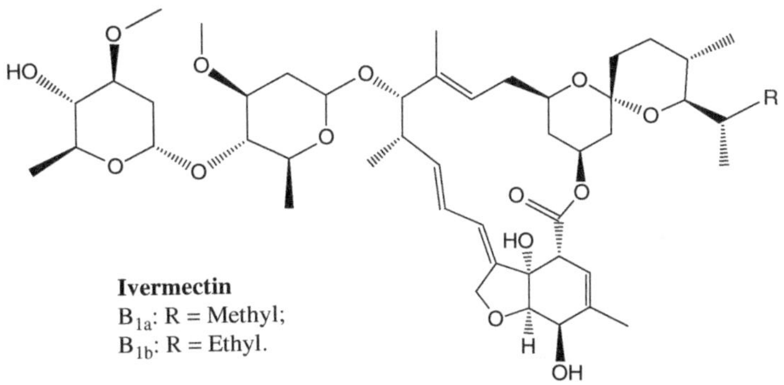

Ivermectin
B_{1a}: R = Methyl;
B_{1b}: R = Ethyl.

Chemical name: 22,23-Dihydroavermectin B_{1a} + 22,23-Dihydroavermectin B_{1b}
Trade name: Stromectol, Sklice, Mectizan, others
Formula: $C_{48}H_{74}O_{14}$; $C_{47}H_{72}O_{14}$
Monoisotopic protonated ion (*m/z*): 875.5151; 861.4995
Molecular weight: 875.11; 861.08
CAS number: 70288-86-7

Ivermectin is an antiparasitic agent, which is a semi-synthetic compound prepared from the natural product, Avermectin. The latter was isolated by William Campbell of Merck Sharp & Dohme Research Laboratories from the sample provided by Satoshi Ōmura of Kitasato University [1]. The scientists at Merck modified the structure of avermectin to improve its pharmacologic property, which resulted in ivermectin, a mixture of two congeners (22,23-dihydroavermectin B_{1a} and 22,23-dihydroavermectin B_{1b}) with at least 80% B_{1a}. Ivermectin was initially approved for veterinary use to prevent and treat heartworm and acariasis, and later approved for human use to treat a number of infestations including river blindness. For the latter indication, Merck & Co., Inc. has established a donation program in Africa for several decades with the product trade name of Mectizan®.

Both components of ivermectin are disaccharide derivatives of a 16-membered macrocyclic lactone. Hence, ivermectin would be susceptible to hydrolytic degradation. Under acidic conditions, it was found to degrade initially as the monosaccharide degradant and then the aglycone degradant (Scheme 229-1) [2, 3]. Under alkaline conditions, the predominant degradation products observed were 2-epi-ivermectin and Δ^2-ivermectin, both of which were formed via base-catalyzed enolization at the position alpha to the lactone carbonyl group (Scheme 229-2) [2, 3]. At pH 6.3, ivermectin was found to be most stable [2].

Ivermectin also contains structural moieties that are labile toward oxidative degradation. The 8a-methylene position, α to the butadiene moiety, and 5-position hydroxyl group would be expected to undergo oxidation under free-radical-mediated oxidative conditions to produce many oxidative degradants (Scheme 229-3) [2, 4, 5]. It would be reasonable to assume that the oxidation proceeds via a peroxide intermediate in both cases [6]. The free-radical-mediated oxidation can also take place

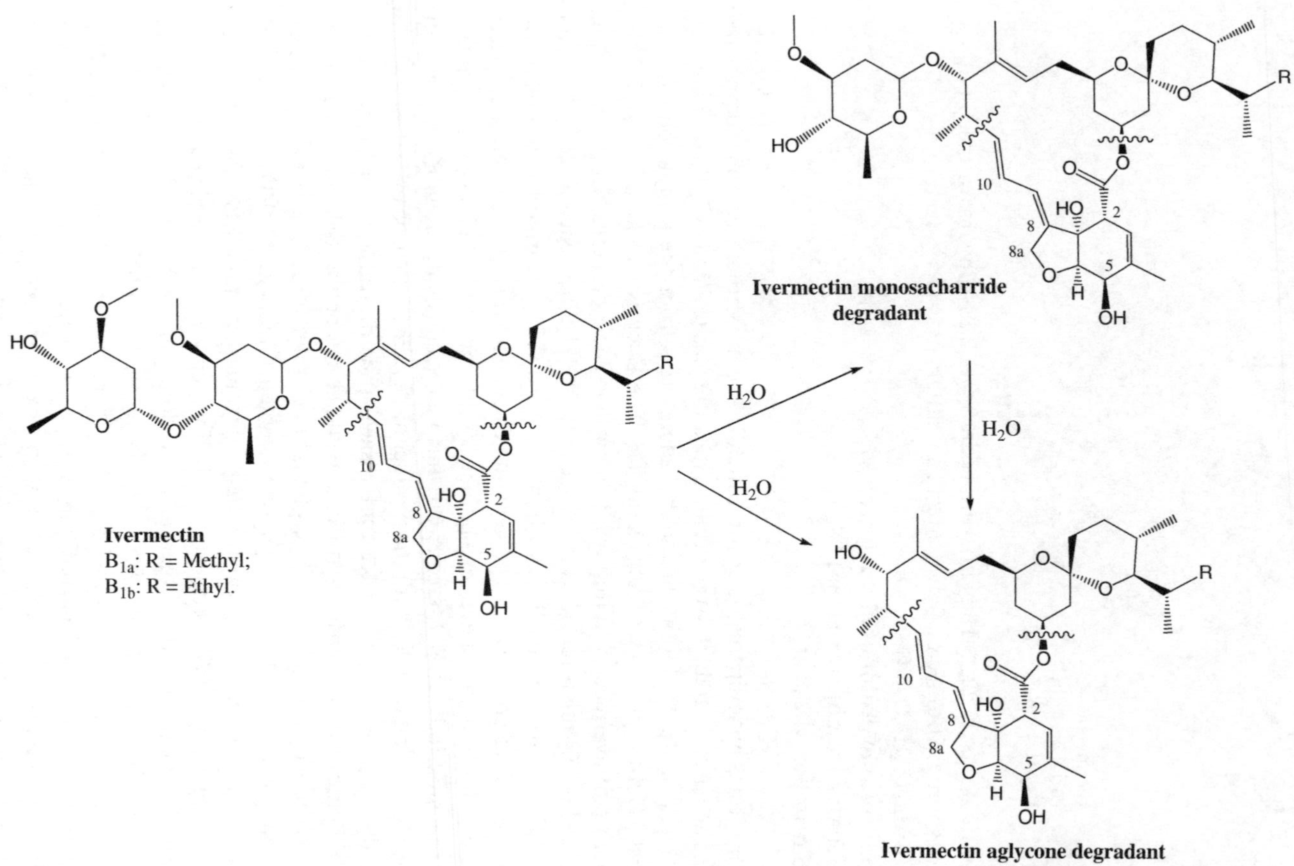

Scheme 229-1 Degradation of ivermectin under acidic conditions.

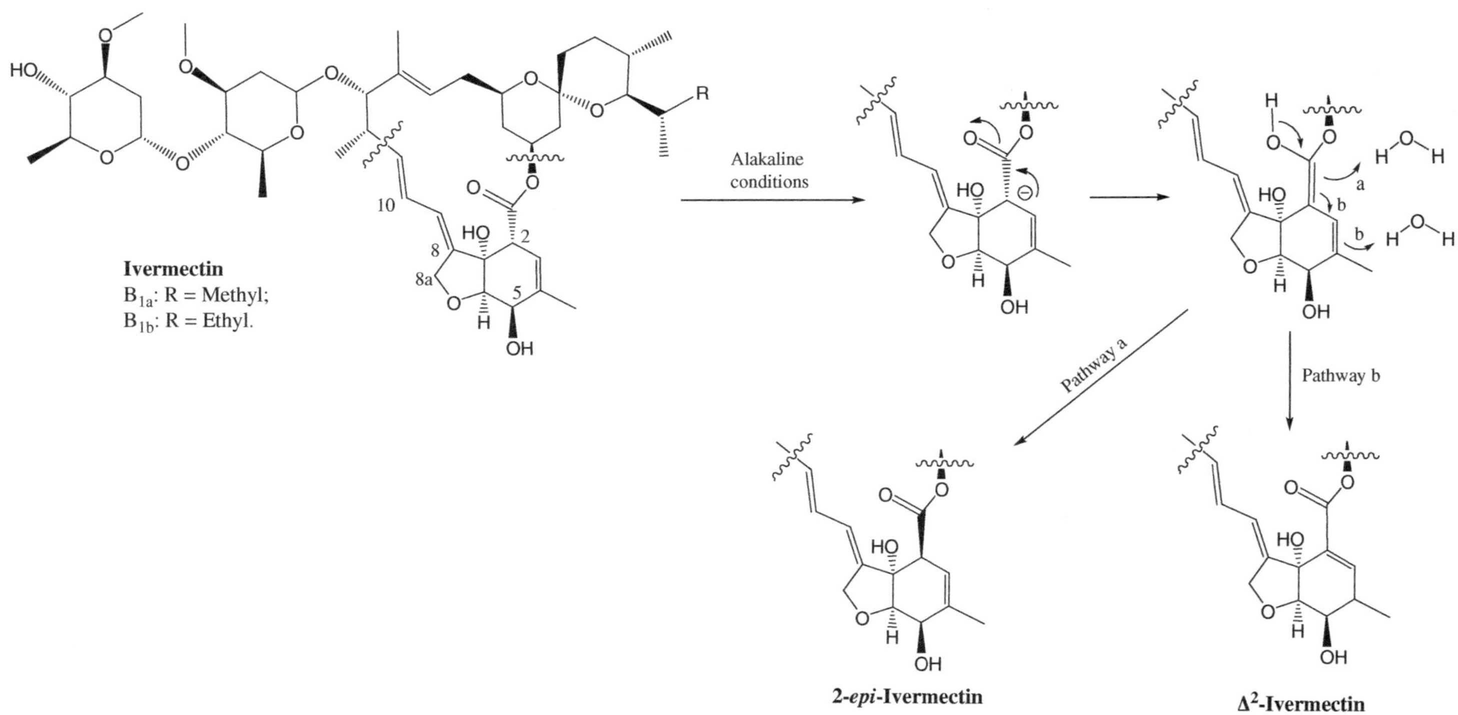

Scheme 229-2 Degradation of ivermectin under alkaline conditions.

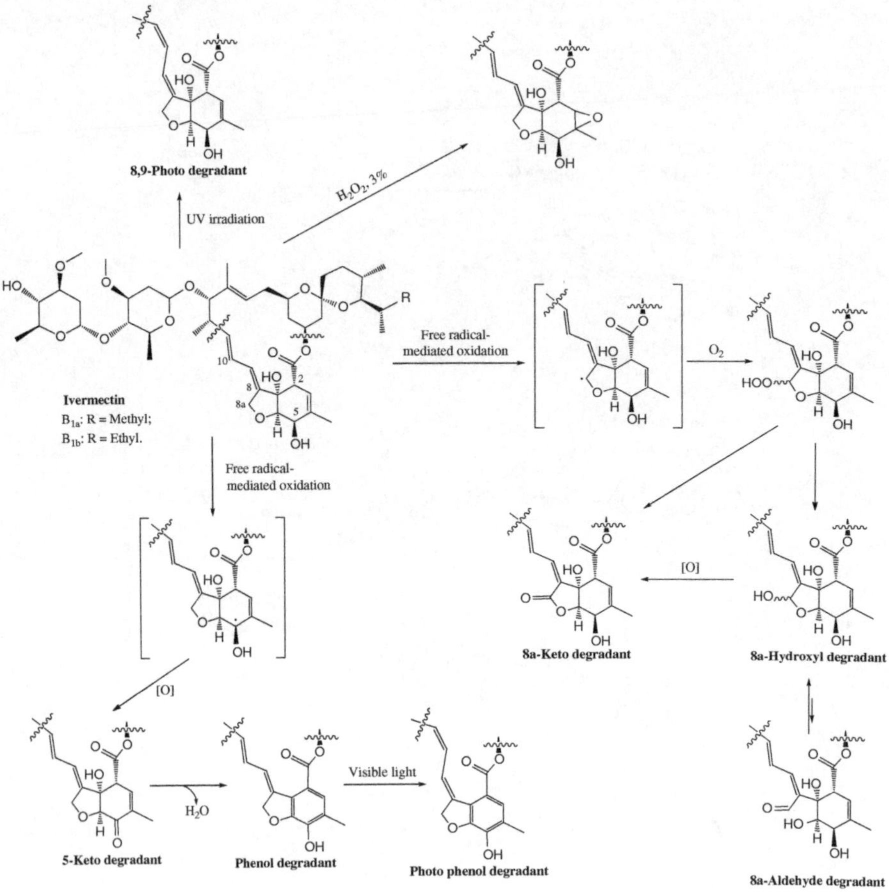

Scheme 229-3 Degradation of ivermectin under free-radical-mediated oxidative and photo stress conditions.

under photo stress of ivermectin in the solid state [3]. On the other hand, the 3,4-olefnic double bond can be oxidized by forced degradation in 3% hydrogen peroxide solution to produce the corresponding peroxide degradant through a nucleophilic oxidation mechanism [3].

The butadiene functionality is also photochemically reactive, giving rise to C_8-C_9 and C_{10}-C_{11} geometric isomers under UV irradiation [2]. For the phenol degradant, the photochemical isomerization can even take place in the visible region due to the formation of the highly conjugated phenol chromophore (Scheme 229-3) [5].

References

1 Ivermectin. https://en.wikipedia.org/wiki/Ivermectin (accessed January 3 2024).

2 Fink, D.W. (1998). Ivermectin. *Anal. Prof. Drug Sub.* 17: 155–184.

3 Adhikari, S. and Rustum, A.M. (2022). A comprehensive study to identify and characterize major degradation products of ivermectin drug substance including its degradation pathways using LC-HRMS and NMR. *J. Pharm. Biomed. Anal.* 214: 114730.

4 Stong, J.D.J., Pivnichny, V., Mrozik, H., and Waksmunski, F.S. (1992). Radical-induced oxidation of oxahydrindene (hexahydrobenzofuran) portion of 22,23-dihydroavermectin B_{1a}: identification of autooxidation products. *J. Pharm. Sci.* 81: 1000–1003.

5 Wang, Q., Stong, J.D., Demontigny, P. et al. (1996). Identification of the secondary degradates of L-648,548 in an animal health formulation. *J. Pharm. Sci.* 85: 446–450.

6 Li, M. (2012). Oxidative degradation. In: *Organic Chemistry of Drug Degradation*, 48–109. Cambridge, UK: RSC Publishing.

230: Lamivudine [WHO List of Essential Medicines]

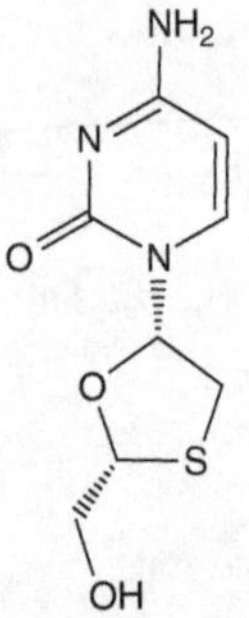

Chemical name: (*2R,5S*)-4-Amino-1-(2-hydroxymethyl-1,3-oxathiolan-5-yl)-(1*H*)-pyrimidin-2-one
Trade name: Epivir, Zeffix, others
Formula: $C_8H_{11}N_3O_3S$
Monoisotopic protonated ion (*m/z*): 230.0594
Molecular weight: 229.25
CAS number: 134678-17-4

Lamivudine, also known as 3TC, is a reverse transcriptase inhibitor typically utilized for the treatment and prevention of AIDS. It is commonly used in combination with other anti-AIDS drugs such as zidovudine, dolutegravir, and abacavir [1]. It is

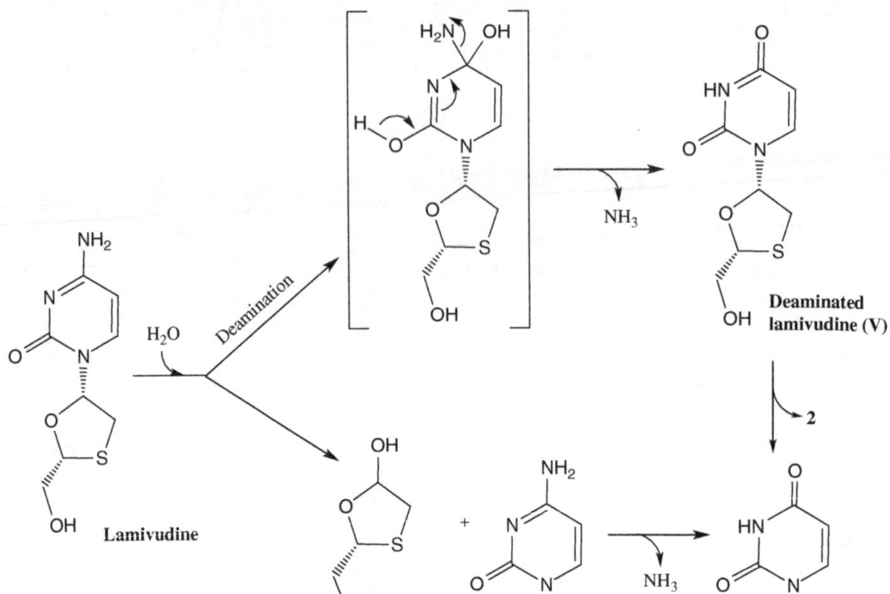

Scheme 230-1 Hydrolytic degradation of lamivudine. The numbering of the degradants follows that of Bedse et al. [2]. *Source:* Bedse et al. [2] / with permission of Elsevier.

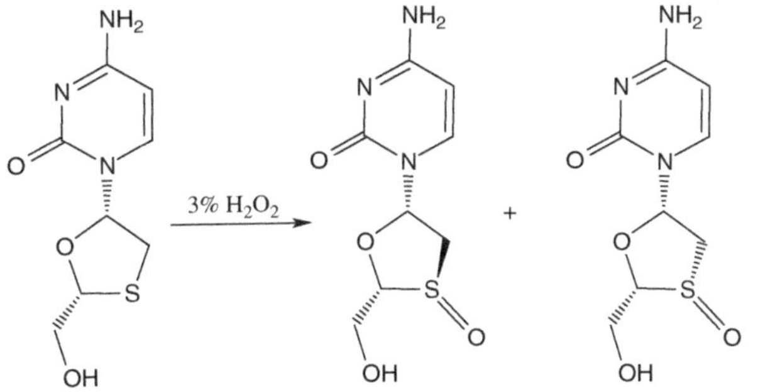

Scheme 230-2 Oxidative degradation of lamivudine. The numbering of the degradants follows that of Bedse et al. [2]. *Source:* Bedse et al. [2] / with permission of Elsevier.

a nucleoside analog with a sulfur-containing pentose analog. Both its nucleobase (cytosine) and the glycosidic-like linkage would be susceptible to hydrolytic degradation to give cytosine (**I**), deaminated cytosine (**II**), deaminated lamivudine (**V**), and pseudopentose degradant (**VI**), as demonstrated in a forced degradation study by Bedse et al. [2]. The hydrolytic removal of the amino group of the cytosine moiety is also called a deamination process (Scheme 230-1). On the other hand, the thioether moiety can be oxidized under nucleophilic oxidative conditions to yield two isomeric degradants (**III** and **IV**, Scheme 230-2).

Wang and You studied thermal degradation of lamivudine in solid state using TG and DSC in both nitrogen and air and found that the activation energies were 126.5 kJ/mol and 133.0 kJ/mol, respectively, indicating lamivudine has good thermal stability [3].

References

1 Lamivudine. https://en.wikipedia.org/wiki/Lamivudine (accessed January 5 2024).

2 Bedse, G., Kumar, V., and Singh, S. (2009). Study of forced decomposition behavior of lamivudine using LC, LC–MS/TOF and MS*n*. *J. Pharm. Biomed. Anal.* 49: 55–63.

3 Wang, X. and You, J. (2010). Thermal degradation and kinetics of lamivudine. *J. Shenyang Pharm. Univ.* 8: http://www.cnki.com.cn/Article/CJFDTotal-SYYD201008005.htm (accessed January 26 2014.

231: Lamotrigine [WHO List of Essential Medicines]

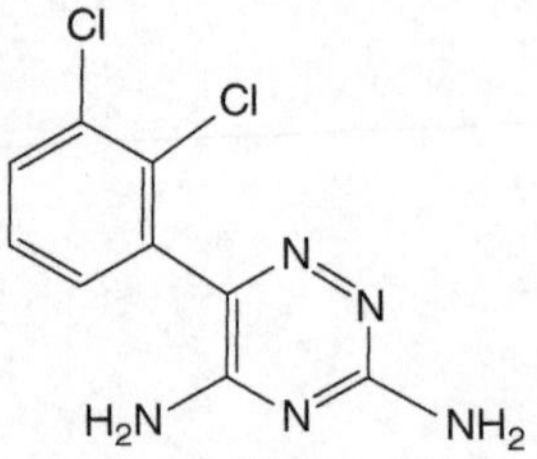

Chemical name: 6-(2,3-Dichlorophenyl)-1,2,4-triazine-3,5-diamine
Trade name: Lamictal, others
Formula: $C_9H_7Cl_2N_5$
Monoisotopic protonated ion (*m/z*): 256.0151
Molecular weight: 256.09
CAS number: 84057-84-1

Lamotrigine is a medicine used to treat epilepsy and stabilize mood in bipolar disorder, which was first approved for medical use in Ireland in 1991 [1]. Chemically, lamotrigine is a phenyltriazine derivative with two amino substituents on the triazine ring. Under forced hydrolytic stress conditions, either acidic (1 N HCl) or alkaline (1 N NaOH) at 80°C, the main degradation pathway is deamination of the 5-amino group, according to the forced degradation study by Mahajan et al. [2]. Under the alkaline stress condition, there was an early-eluting degradant (**I**), whose structure was not identified due to the inability to obtain its mass spectrometric data, in addition to the deaminated degradant (**IV**). Under the forced oxidative stress using 1% hydrogen peroxide at 80°C, two *N*-oxide degradants (**II** and **III**) were produced (Scheme 231-1). On the other hand, lamotrigine was found to be stable under the thermal stress (70°C for 3 days) and photolysis in the solid state [2].

In solution photolysis of lamotrigine with irradiation by either simulated sunlight [3] or UV light by both low- and medium-pressure mercury lamps [4], a number of photodegradants were observed, and the structures for the vast majority of these degradants were proposed based on high-resolution mass spectrometric data (Scheme 231-2) [3, 4]. The photodegradation pathways of lamotrigine can be categorized into four major types. The first one is photoisomerization, from which degradant I was formed. The second type is photo substitution, in which a chloro group was replaced by a hydroxyl group. Hence, further decomposition of degradant I via photo substitution resulted in degradant II. Likewise, direct photo substitution of lamotrigine gave rise to degradants III and IV. The third type is a straightforward dechlorination process, in which degradants V and VI were formed. The last one is a photo cyclization process, which produced degradants VII and VIII; such type of photo cyclization is not uncommon for bi-aryl compounds [5].

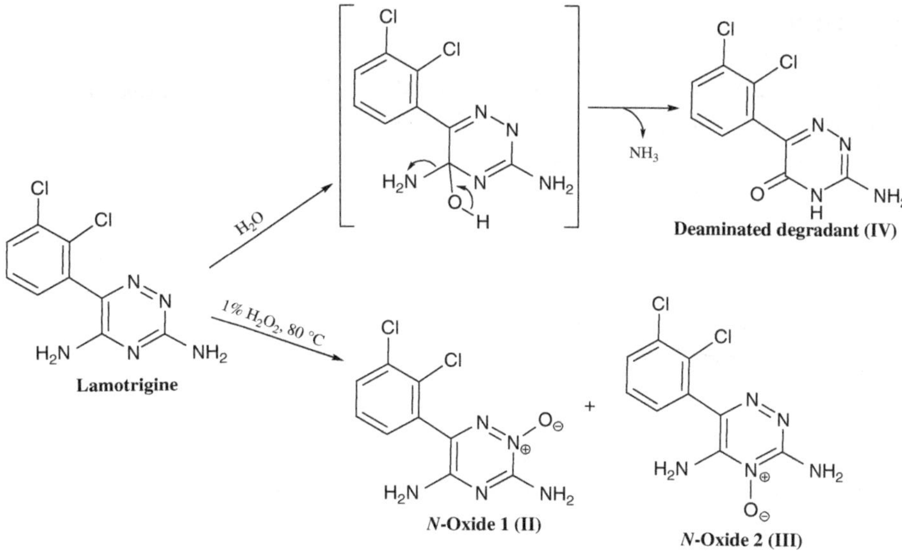

Scheme 231-1 Hydrolytic and oxidative degradation of lamotrigine under forced degradation conditions. The numbering of degradants follows that of Mahajan et al. [2]. *Source:* Mahajan et al. [2] /Taylor & Francis Group.

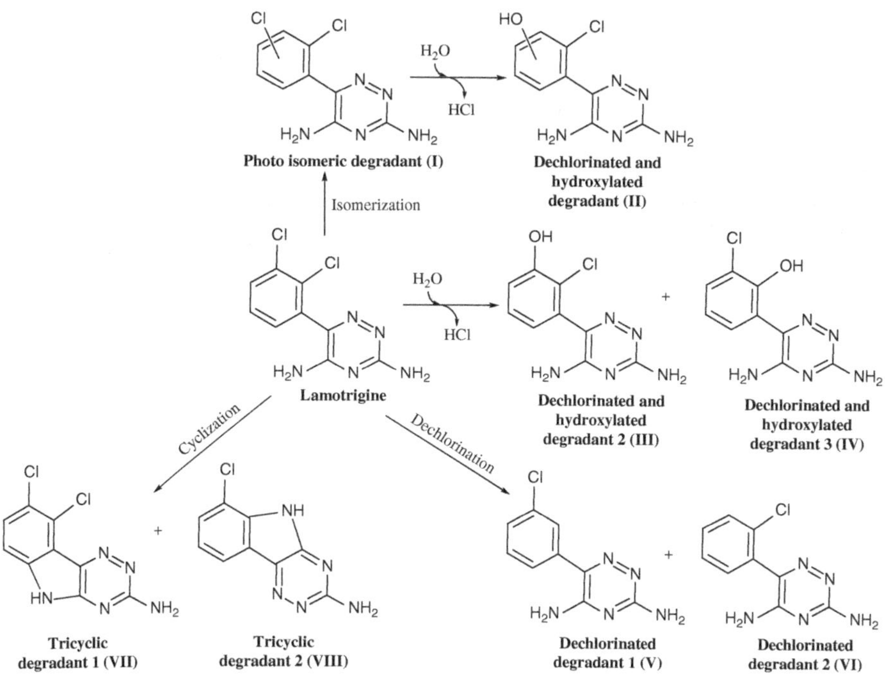

Scheme 231-2 Photochemical degradation pathways of lamotrigine.

References

1 Lamotrigine. https://en.wikipedia.org/wiki/Lamotrigine (accessed January 5, 2024).

2 Mahajan, A.A., Thaker, A.K., Kale, S., and Mohanraj, K. (2012). LC, LC–MS/MS studies for identification and characterization of degradation products of lamotrigine and establishment of mechanistic approach towards degradation. *J. Liq. Chromat. Rel. Tech.* 35: 2255–2271.

3 Young, R.B., Chefetz, B., Liu, A. et al. (2014). Direct photodegradation of lamotrigine (an antiepileptic) in simulated sunlight – pH influenced rates and products. *Environ. Sci. Process. Impacts* 16: 848–857.

4 Keen, O.S., Ferrer, I., Thurman, E.M., and Linden, K.G. (2014). Degradation pathways of lamotrigine under advanced treatment by direct UV photolysis, hydroxyl radicals, and ozone. *Chemosphere* 117: 316–323.

5 Li, M. (2012). Section 6.2.5. Cyclization in polyaromatic ring systems. In: *Organic Chemistry of Drug Degradation*, 180–182. Cambridge, UK: RSC Publishing.

232: Levothyroxine [L-T4] [WHO List of Essential Medicines]

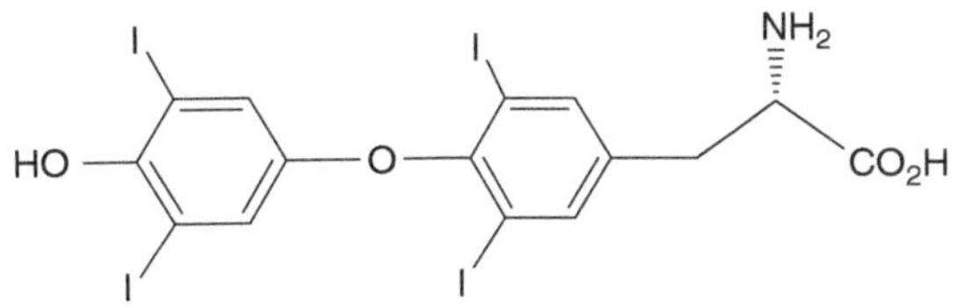

Chemical name: (*S*)-2-Amino-3-[4-(4-hydroxy-3,5-diiodophenoxy)-3,5-diiodophenyl]propanoic acid
Trade name: Synthyroid, others
Formula: $C_{15}H_{11}I_4NO_4$
Monoisotopic protonated ion (*m/z*): 777.6940
Molecular weight: 776.87
CAS number: 51-48-9
Salt form: **Sodium salt**

Levothyroxine, or thyroxine, was isolated by E. C. Kendall from hog thyroid at the Mayo Clinic in 1914 [1, 2]. Its structure was determined by C.R. Harington in 1926, and subsequently it was synthesized by C.R. Harington and G. Barger in 1927 [3, 4]. Levothyroxine is used to treat thyroid hormone deficiency (hypothyroidism) and was the most prescribed medication in the United States in 2016 [5].

Chemically, thyroxine is the tetraiodo derivative of thyronine, and the latter is *O*-4-hydroxyphenyl-substituted tyrosine. The iodo substituents tend to fall off the substituted phenyl rings, i.e., they are prone to deiodination, particularly under photochemical conditions [6, 7]. The photo deiodination produces initially T3 and then a few T2 degradants and T1 degradant, upon further photolysis. Photo irradiation by UVA and UVB light can also cause the breakage of the biaryl ether linkage, leading to the formation of phenylalanine, tyrosine, and their derivatives (Scheme 232-1) [7]. Under irradiation with UV light at wavelength greater than 300 nm, no phenylalanine, tyrosine, or their derivatives were observed, suggesting that the cleavage of the biaryl ether linkage would require higher energy than that of deiodination.

In a forced degradation study of levothyroxine by Won, levothyroxine was reported to undergo nonphotochemical deiodination to produce T3 in 0.01 N NaOH solution under refluxing [8]. In solid state, the main degradation of levothyroxine was elimination of ammonia, followed by oxidative breakage of the carbon–carbon double bond, observed in a sample of levothyroxine sodium that was placed in the open dish at 60°C for 7 weeks (Scheme 232-2, Pathway a).

The stability of formulated levothyroxine products (with its sodium salt as the API), such as the tablets, seems to be a challenge ever since its market approval in the 1950s, as formulated levothyroxine products remain to be the most recalled products, primarily due to the chemical instability of the API [9, 10]. The instability of the formulated products was attributed to the dehydration of the API, in the form

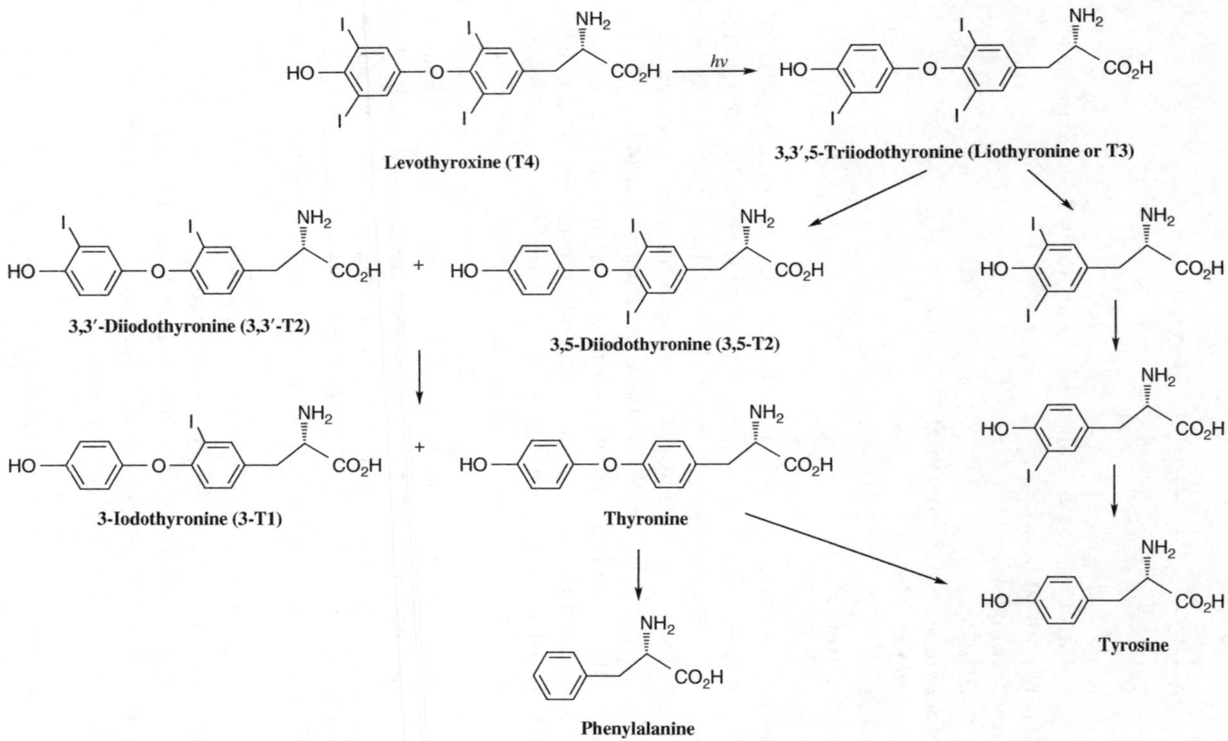

Scheme 232-1 Photochemical degradation of levothyroxine.

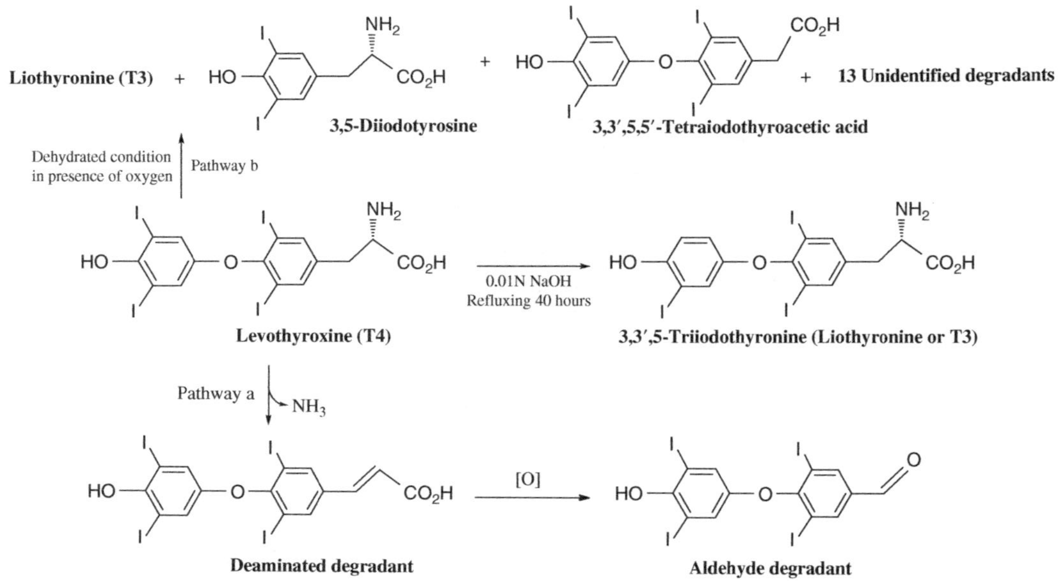

Scheme 232-2 Nonphotochemical degradation of levothyroxine under forced degradation conditions. The identification of the three degradants, formed under dehydrated conditions in the presence of oxygen, was performed only based on retention time comparison with reference materials.

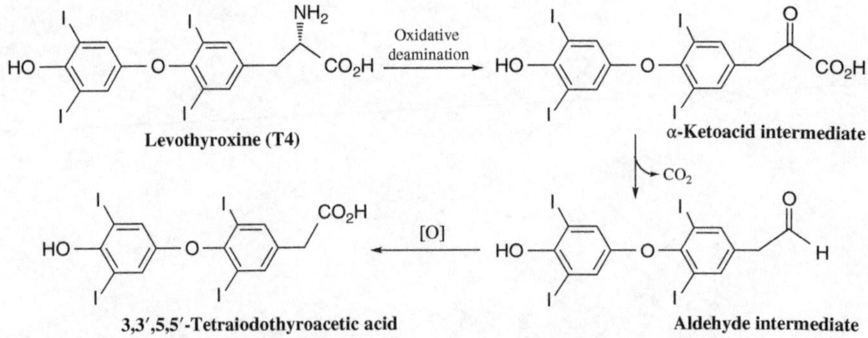

Scheme 232-3 Plausible mechanism for the degradation of levothyroxine to produce 3,3′,5,5′-tetraiodothyroacetic acid.

of pentahydrate, and subsequent oxidation of the dehydrated API [9, 11]. Shah et al. reported that a sample of levothyroxine drug substance, after exposing to a dehydrated condition in the presence of oxygen for a duration of 32 days, produced ~16 degradants, among which only 3 degradants were identified (3,5-diiodo-tyrosine, 3,3′,5-triiodo-thyronine, and 3,3′,5,5′-tetraiodothyroacetic acid) based on comparison to reference materials (Scheme 232-2, Pathway b) [11]. It needs to be pointed out that 3,3′,5,5′-tetraiodothyroacetic acid is also a metabolite of levothyroxine and its formation mechanism under the in vitro degradation conditions, which was not elaborated by Shah et al. [11], could be rationalized by the mechanism illustrated in Scheme 232-3, which is analogous to that proposed by Andre et al. for the thermal degradation of liothyronine sodium (T3 sodium salt) [12]. Most of the 16 degradants eluted after levothyroxine, as analyzed by a reversed-phase HPLC method with a C8 column.

References

1 Simoni, R.D., Hill, R.L., and Vaughan, M. (2002). The isolation of thyroxine and cortisone: the work of Edward C. *Kendall. J. Biol. Chem.* 277: 21–22.

2 Slater, S. (2010). The discovery of thyroid replacement therapy. Part 3: a complete transformation. *J. R. Soc. Med.* 104: 100–106.

3 Harington, C.R. (1926). Chemistry of thyroxine. II: constitution and synthesis of desiodo-thyroxine. *Biochem. J.* 20: 300–313.

4 Harington, C.R. and Barger, G. (1927). Chemistry of thyroxine. III: constitution and synthesis of thyroxine. *Biochem. J.* 21: 169–183.

5 Levothyroxine. https://en.wikipedia.org/wiki/Levothyroxine (accessed January 5 2024).

6 Van Der Walt, B. and Cahnmann, H.J. (1982). Synthesis of thyroid hormone metabolites by photolysis of thyroxine and thyroxine analogs in the near UV Proc. *Natl. Acad. Sci. USA*. 79: 1492–1496.

7 Kazemifard, A.G., Moore, D.E., and Aghazadeh, A. (2001). Identification and quantitation of sodium-thyroxine and its degradation products by LC using electrochemical and MS detection. *J. Pharm. Biomed. Anal.* 25: 697–711.

8 Won, C.M. (1992). Kinetics of degradation of levothyroxine in aqueous solution and in solid state. *Pharm. Res.* 9: 131–137.

9 Hamad, M.L., Engen, W., and Morris, K.R. (2013). Impact of hydration state and molecular oxygen on the chemical stability of levothyroxine sodium. *Pharm. Dev. Technol.* https://doi.org/10.3109/10.37450.2013.862635.

10 Rhodes, C. (1998). Regulatory aspects of the formulation and evaluation of levothyroxine tablets. *Clin. Res. Regul. Aff.* 15: 173–186.

11 Shah, H.S., Chaturvedi, K., Hamad, M. et al. (2019). New insights on solid-state changes in the levothyroxine sodium pentahydrate during dehydration and its relationship to chemical instability. *AAPS PharmSciTech.* 20: 39. https://doi.org/10.1208/s12249-018-1264-0.

12 Andre, M., Domanig, R., Riemer, E. et al. (1996). Identification of the thermal degradation products of G-triiodothyronine sodium (liothyronine sodium) by reversed-phase high-performance liquid chromatography with photodiode-array UV and mass spectrometric detection. *J. Chromat. A* 725: 287–294.

233–234: Loratadine, Desloratadine

233: Loratadine [WHO List of Essential Medicines]

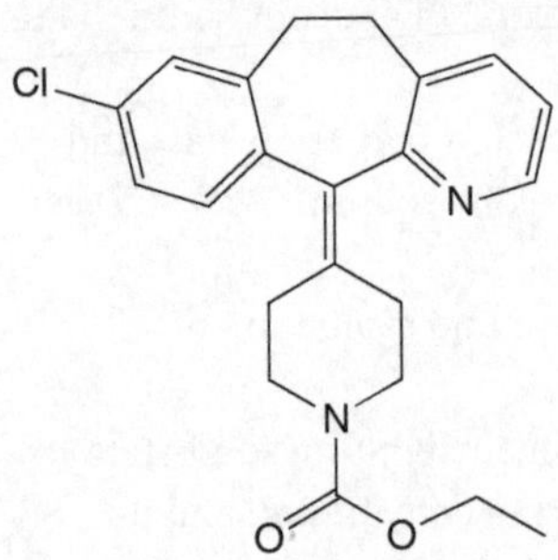

Chemical name: Ethyl 4-(8-chloro-5,6-dihydro-11*H*-benzo[5,6]cyclohepta[1,2-b]pyridin-11-ylidene)-1-piperidinecarboxylate
Trade name: Claritin
Formula: $C_{22}H_{23}ClN_2O_2$
Monoisotopic protonated ion (*m/z*): 383.1521
Molecular weight: 382.89
CAS number: 79794-75-5

234: Desloratadine

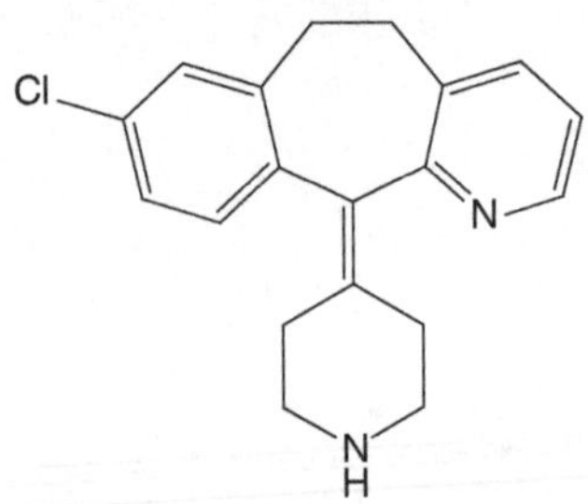

Chemical name: 8-Chloro-11-(4-piperidinylidene)-6,11-dihydro-5*H*-benzo[5,6]cyclohepta[1,2-b]pyridine
Trade name: Clarinex, Aerius
Formula: $C_{19}H_{19}ClN_2$
Monoisotopic protonated ion (*m/z*): 311.1310
Molecular weight: 310.82
CAS number: 100643-71-8

Loratadine is a second-generation antihistamine medication used to treat allergies. It was developed by Schering-Plough Corp. (later merged into Merck & Co., Inc.) and approved in 1988 [1]. It became an over-the-counter (OTC) drug after its patent expiration. Loratadine is a carbamate prodrug with a tricyclic ring that contains a fused pyridine moiety. In solid state, it is quite stable as a drug substance or in solid dosage forms [2]. In liquid formulations such as syrups containing sugar excipients,

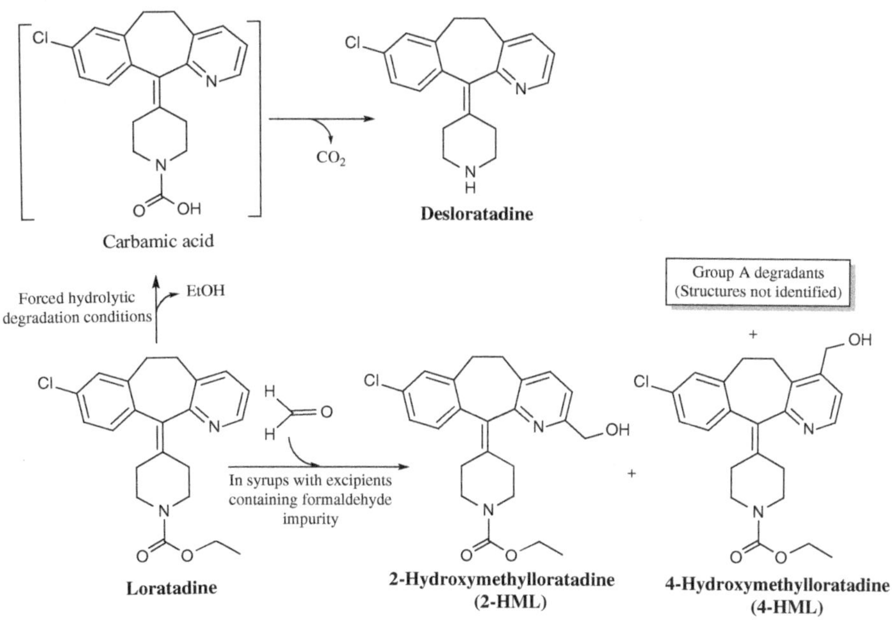

Scheme 233–234-1 Degradation of loratadine in syrups or under forced hydrolytic degradation conditions.

it was revealed that the pyridine moiety could react with certain impurities/degradants of these excipients such as formaldehyde, resulting in the formation of 2-hydroxymethylloratadine (2-HML), 4-hydroxymethylloratadine (4-HML), and other degradants (collectively known as "Group A impurities") (Scheme 233–234-1) [3]. The mechanism for the degradation is of oxidative nature, as many measures using either chelating agents (such as EDAT) [3] or antioxidants (such as ascorbic acid or phenol derivatives) [3, 4] can inhibit the formation of these degradants. Nevertheless, the exact mechanism remains unclear, although it was hypothesized as being mediated through a pyridine *N*-oxide intermediate [5]. Based on the known reactivity of pyridine, it could be a free-radical-mediated addition between loratadine and formaldehyde [6]. On the other hand, hydrolysis of the carbamate linkage does not appear to occur in appreciable amounts even in aqueous liquid formulations [3]. Therefore, desloratadine may not be a real degradant of loratadine, although desloratadine can be generated under forced hydrolytic degradation conditions [7].

Desloratadine is the major metabolite of loratadine formed by hydrolysis of the carbamate group. The hydrolysis exposes the secondary amine functionality of the drug molecule and the secondary amine can react with reducing sugars, such as lactose, to undergo the Maillard reaction, which produces *N*-formyl desloratadine as one of the identified degradants (Scheme 233–234-2) [8, 9]. Yu et al. reported that desloratadine underwent the Maillard degradation with maltose, an impurity in starch, to yield an Amadori degradant during a compatibility study between desloratadine and starch [10]. The Maillard reaction consists of numerous individual

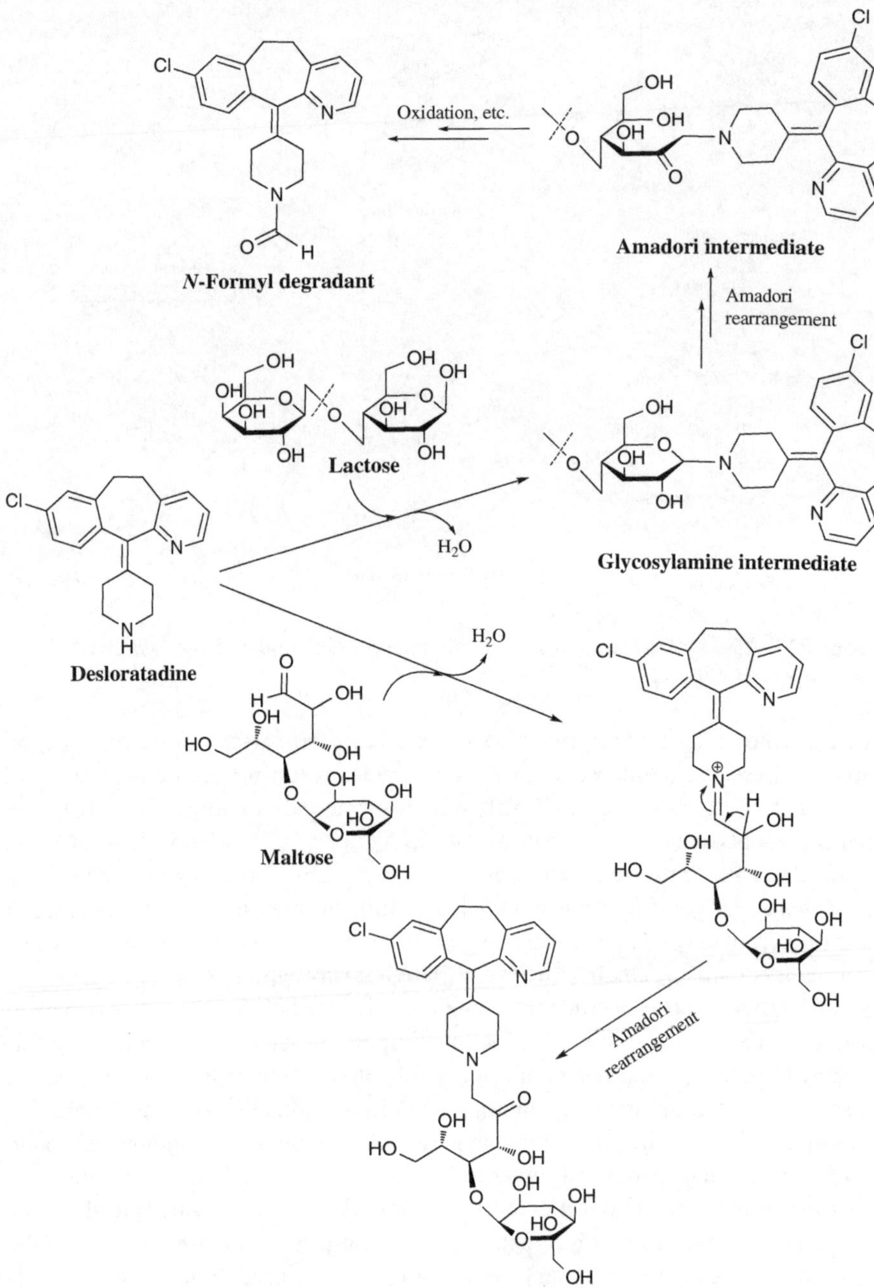

Scheme 233–234-2 Degradation of desloratadine due to the Maillard reaction.

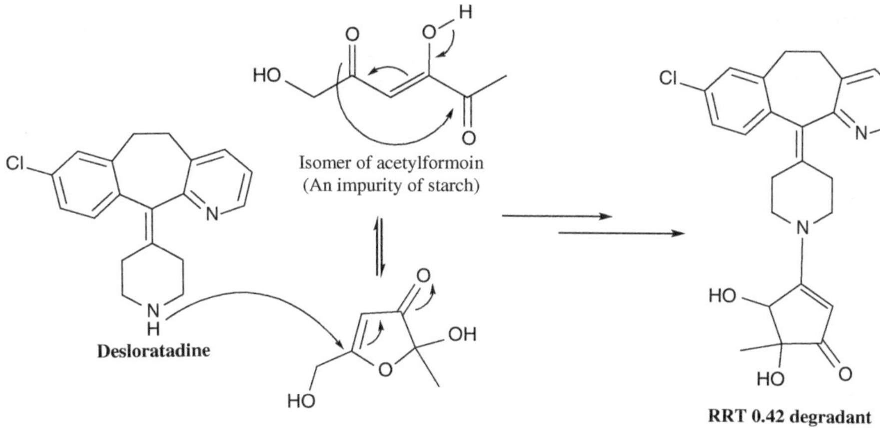

Scheme 233–234-3 Degradation of desloratadine due to reaction with an impurity of starch. *Source:* Yu et al. [10] / with permission of Elsevier.

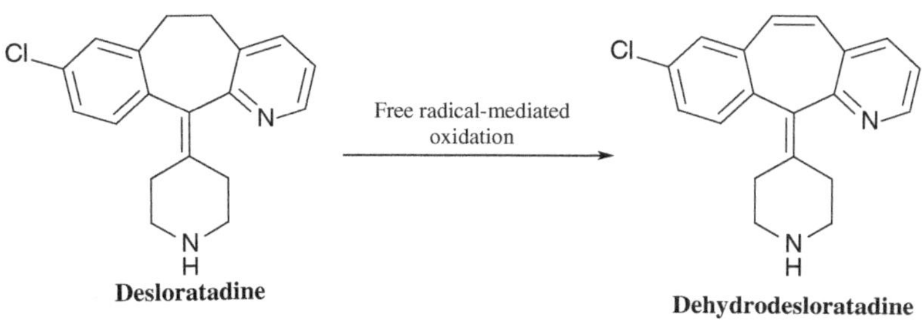

Scheme 233–234-4 Oxidative degradation of desloratadine to form dehydrodesloratadine [8].

reactions, and its early-stage degradants include glycosylamine and Amadori rearrangement products [11]. Further degradation from the Amadori degradants can occur, depending on the extent of a particular stability or forced degradation study.

In the same study, Yu et al. also found that the secondary amine group of desloratadine reacted with an isomer of acetylformoin, also an impurity of polysaccharide origin, to produce a novel degradant (Scheme 233–234-3). On the other hand, it was revealed that dehydrodesloratadine was identified in one of the experimental formulae [8]; this degradant was obviously resulting from the oxidation that occurred at either the 9- or 10-position of the tricyclic ring, both of which are benzylic-type positions and hence would be susceptible to free-radical-mediated oxidation (Scheme 233–234-4). Such an oxidation could also take place with loratadine, although it has not been reported in the literature.

References

1 Loratadine. https://en.wikipedia.org/wiki/Loratadine (accessed January 16 2024).

2 AlMasoud, N., Bakheit, A.H., Alshammari, M.F.M. et al. (2022). Loratadine. *Prof. Drug Sub. Excip. Rel. Methodol.* 47: 55–90.

3 Munayyer, F.J., Guazzo, F., Stupak, E.I. et al. (2003). Stabilized antihistamine syrup. US Patent 6,514,520 B2. Patent filed 05 July 2001. Patent granted 4 February 2003.

4 Dagar, A., Gudi, S., Opawale, F. (2007). Oral liquid loratadine formulations and methods. US Patent Application 2007/0286875 A1. Application filed 6 June 2007.

5 Eyjolfsson, R. (2003). Loratadine: hydroxymethylation in syrup. *Pharmazie* 58: 154.

6 Wang, G., Cao, J., Gao, L. et al. (2017). Metal-free synthesis of C-4 substituted pyridine derivatives using pyridine-boryl radicals via a radical addition/coupling mechanism: a combined computational and experimental study. *J. Am. Chem. Soc.* 139: 3904–3910.

7 El Ragehy, N.A., Badawey, A.M., and Khateeb, S.Z.-El. (2002). Stability indicating methods for the determination of loratadine in the presence of its degradation product. *J. Pharm. Biomed. Anal.* 28: 1041–1053.

8 Rahul, G.S., Subbappa, P.K., Kodipyaka, R. et al. (2007). Stabilized loratadine compositions. US Patent Application 2007/0014855 A1. Application filed 12 July 2006.

9 Mehta, P.R., Bushan, I., Nasare, V.D. et al. (2006). Stabilized desloratadine composition. PCT Patent Application WO 2006/020534 A2. Application filed 8 August 2005.

10 Yu, X., Buevich, A.V., Li, M. et al. (2013). A compatibility study of a secondary amine active pharmaceutical ingredient with starch: identification of a novel degradant formed between desloratadine and a starch impurity using LC–MS^n and NMR spectroscopy. *J. Pharm. Sci.* 102: 717–731.

11 Li, M. (2012). Section 5.1.1. Degradation via the Maillard reaction. In: *Organic Chemistry of Drug Degradation*, 150–153. Cambridge, UK: RSC Publishing.

235: Losartan [WHO List of Essential Medicines]

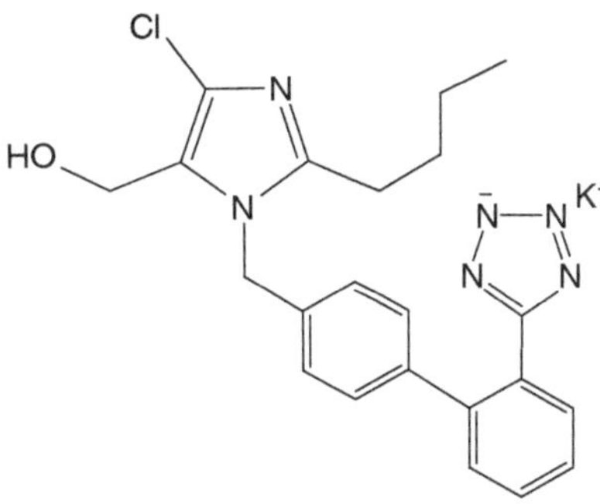

Chemical name: (2-Butyl-4-chloro-1-{[2′-(2*H*-tetrazol-5-yl)biphenyl-4-yl]methyl}-1*H*-imidazol-5-yl)methanol
Trade name: Cosaar
Formula: $C_{22}H_{23}ClN_6O$
Monoisotopic protonated ion (*m/z*): 423.1695
Molecular weight: 422.92
CAS number: 114798-26-4
Salt form: Potassium salt

Losartan is a nonpeptide angiotensin II receptor blocker (ARB) which is utilized, in the form of its potassium salt, for the treatment of hypertension as well as heart

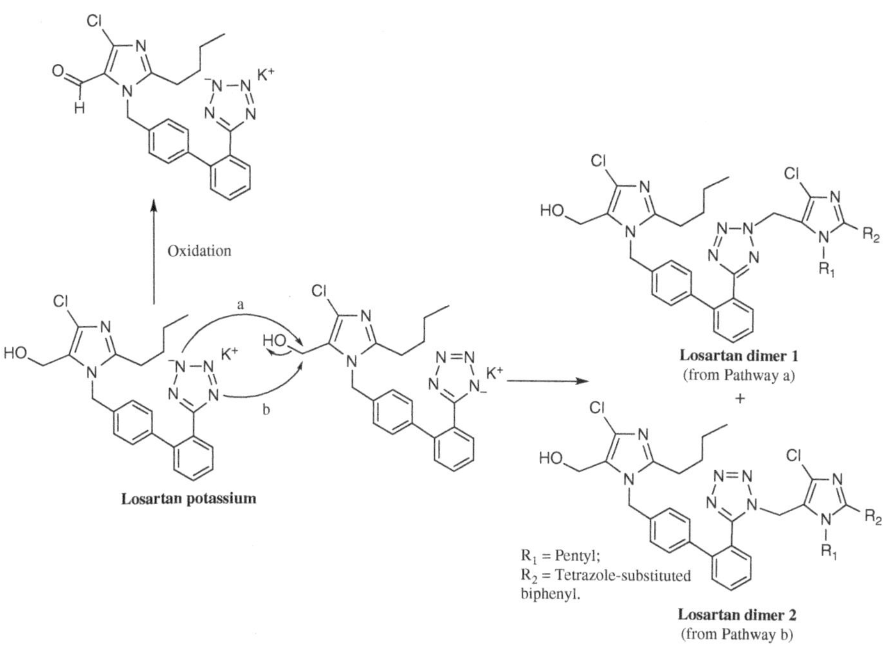

Scheme 235-1 Oxidative and dimeric degradation pathways of losartan observed under long-term and accelerated stability studies [2, 3].

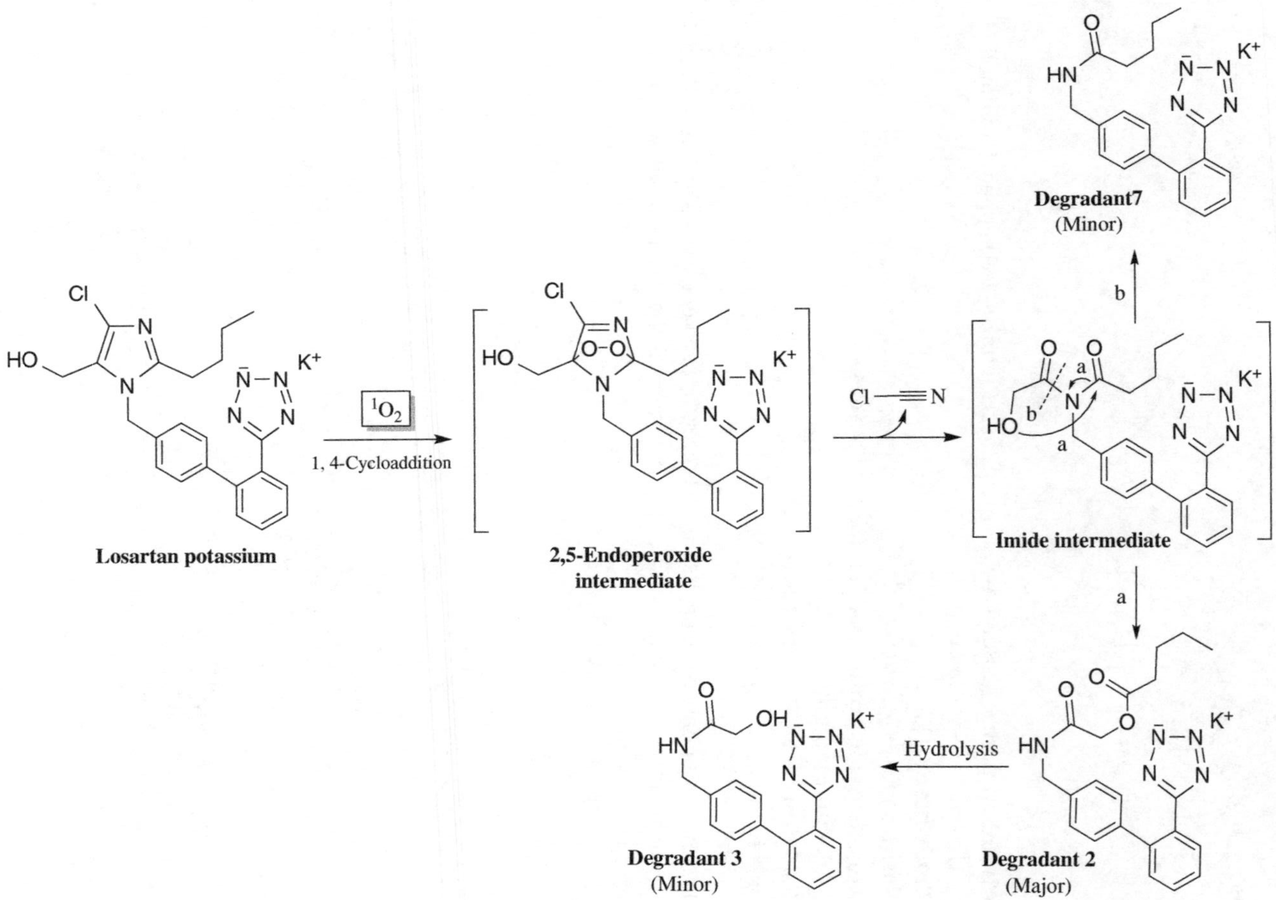

Scheme 235-2 Photo-sensitized oxidation of losartan. The numbering of degradants follows that of Seburg et al. [4].
Source: Seburg et al. [4] / with permission of Elsevier.

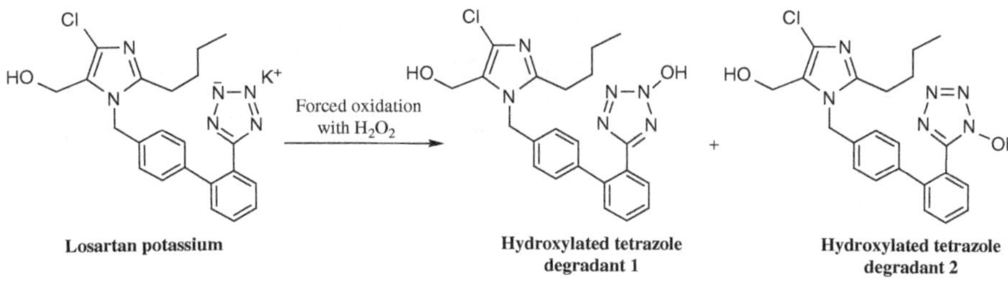

Scheme 235-3 Oxidation of losartan potassium with hydrogen peroxide.
Source: Singamsetti et al. [6] / John Wiley & Sons.

failure and left ventricular enlargement [1]. It was developed by Merck & Co., Inc. and approved in the United States in 1995.

Chemically, losartan is a highly substituted imidazole derivative, among which the 5-hydroxymethyl substituent could be susceptible to degradation. In a study of forced degradation of losartan tablets by Zhao et al., three degradants (≤0.1%) were observed in the severely stressed tablets, among which the aldehyde degradant resulted from oxidation of the hydroxymethyl group, while the two dimeric degradants were formed via SN2 nucleophilic substitution (Scheme 235-1). The two dimeric degradants were also observed during the stability study of losartan tablets at Merck [3].

During the development of a syrup formulation of losartan, a major degradant was observed, which was determined as a photo-sensitized degradant (Scheme 235-2) [4]. Although the species responsible for the photo-sensitization was not identified, it was clearly demonstrated by the authors that the degradation was mediated via singlet oxygen, a process known as "type II photosensitized oxidation" [5], which led to the destruction of the imidazole ring.

More recently, there have been a few forced degradation studies of losartan potassium. The vast majority of the degradants formed in these studies appear to be artificial degradants. For example, Singamsetti et al. reported two novel *N*-hydroxyl degradants produced under oxidative stress conditions (Scheme 235-3) [6]. Pandey et al. observed three novel dimeric degradants, which were formed via double cross-linkage between the two losartan molecules [7].

References

1 Losartan. https://en.wikipedia.org/wiki/Losartan (accessed January 28 2024).

2 Zhao, Z., Wang, Q., Tsai, E.W. et al. (1999). Identification of losartan degradates in stressed tablets by LC-MS and LC-MS/MS. *J. Pharm. Biomed. Anal.* 20: 129–136.

3 McCarthy, K.E., Wang, Q., Tsai, E.W. et al. (1998). Determination of losartan and its degradates in COZAAR® tablets by reversed-phase high-performance thin-layer chromatography. *J. Pharm. Biomed. Anal.* 17: 671–677.

4 Seburg, R.A., Ballard, J.M., Hwang, T.-L., and Sullivan, C.M. (2006). Photosensitized degradation of losartan potassium in an extemporaneous suspension formulation. *J. Pharm. Biomed. Anal.* 42: 411–422.

5 Li, M. (2012). Photochemical degradation. In: *Organic Chemistry of Drug Degradation*. Cambridge, UK: RSC Publishing.

6 Singamsetti, J.M., Gandham, H., Geereddy, M.K.R. et al. (2022). Isolation and structural characterization of two novel oxidative degradation products of losartan potassium active pharmaceutical ingredient using advanced analytical techniques. *Rapid Comm. Mass Spec.* 37: e9432.

7 Pandey, A.K., Rapolu, R., Raju, C.K. et al. (2016). The novel acid degradation products of losartan: isolation and characterization using Q-TOF, 2D-NMR and FTIR. *J. Pharm. Biomed. Anal.* 120: 65–71.

236–237: Meloxicam, Piroxicam

236: Meloxicam

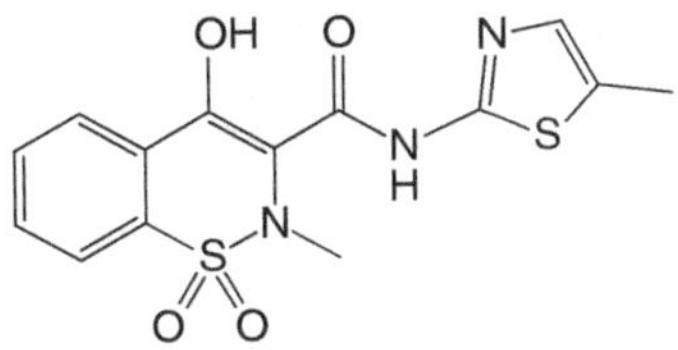

Chemicalname:4-Hydroxy-2-methyl-*N*-(5-methyl-2-thiazolyl)-2*H*-1,2-benzothiazine-3-carboxamide-1,1-dioxide
Trade name: Metacam, Mobic, others
Formula: $C_{14}H_{13}N_3O_4S_2$
Monoisotopic protonated ion (*m/z*): 352.0420
Molecular weight: 351.40
CAS number: 71125-38-7

237: Piroxicam

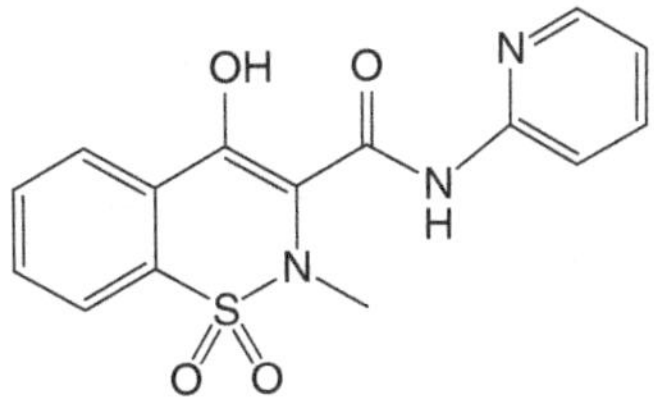

Chemical name: 4-Hydroxy-2-methyl-*N*-(2-pyridinyl)-2*H*-1,2-benzothiazine-3-carboxamide 1,1-dioxide
Trade name: Feldene
Formula: $C_{15}H_{13}N_3O_4S$
Monoisotopic protonated ion (*m/z*): 332.0700
Molecular weight: 331.35
CAS number: 36322-90-4

Meloxicam and piroxicam are nonsteroidal anti-inflammatory drugs (NSAIDs) that inhibit cyclooxygenase (COX) enzymes, particularly COX-2, a member of the prostaglandin family that triggers inflammation. Clinically, they are used to treat pain and inflammation in rheumatic diseases and osteoarthritis [1, 2]. Meloxicam was developed by Boehringer Ingelheim and approved in the United States in 2000, while piroxicam was developed by Pfizer and approved for clinical use in 1982.

Both drug molecules contain a common 4-hydroxybenzothiazine core, which connects to their heterocyclic amine moieties via an amide linkage, respectively. The amide linkage would be susceptible to hydrolytic degradation, leading to the formation of the corresponding amines (**1**) and the benzothiazine-3-carboxylic acid degradant (**2**, Scheme 236–237) [3]. Under forced oxidative and photochemical degradation conditions, both meloxicam and piroxicam produced degradants (**4–6**)

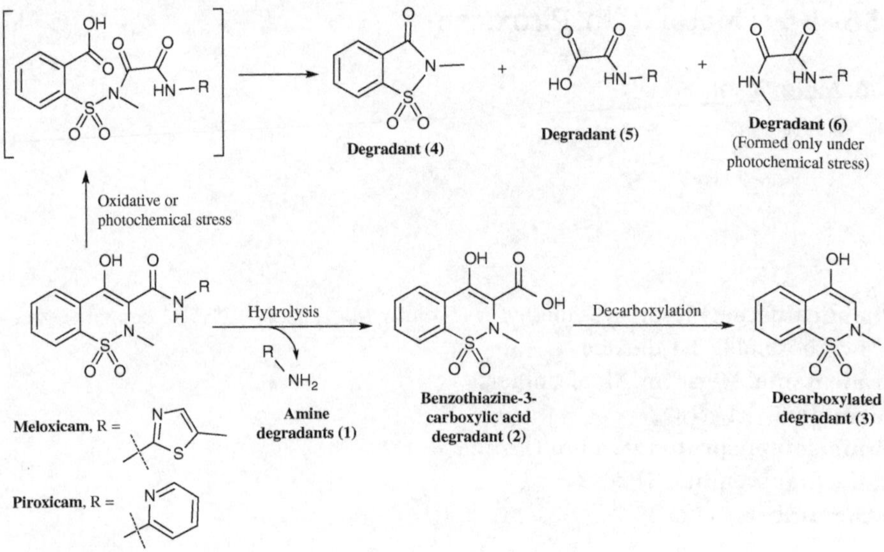

Scheme 236–237 Degradation pathways of meloxicam and piroxicam.

resulting from the destruction of the benzothiazine core, besides the acid degradants (**2**), while the acid degradant (**2**) could undergo decarboxylation in neutral aqueous solution at 80°C [4]. It is worth noting that the sulfonamide bond, which is usually very stable under forced hydrolytic degradation conditions, was cleaved under the photochemical degradation conditions to yield Degradant 6.

References

1 Meloxicam. https://en.wikipedia.org/wiki/Meloxicam (accessed January 30 2024).
2 Piroxicam. https://en.wikipedia.org/wiki/Piroxicam (accessed January 30 2024).
3 Taha, E.A., Salama, N.N., and Fattah, L.S.A. (2006). Stability-indicating methods for determination of meloxicam and tenoxicam in the presence of their degradation products. *Spectr. Lett.: An Int. J. Rapid Comm.* 35: 501–516.
4 Modhave, D.T., Handa, T., Shah, R.P., and Singh, S. (2011). Successful characterization of degradation products of drugs using LC-MS tools: application to piroxicam and meloxicam. *Anal. Methods* 3: 2864–2872.

238: Metformin [WHO List of Essential Medicines]

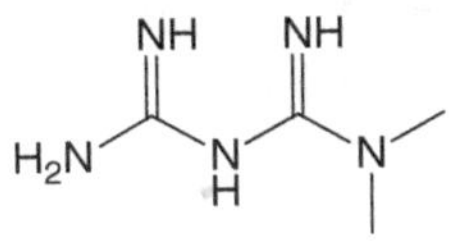

Chemical name: *N,N*-Dimethylimidodicarbonimidic diamide
Trade name: Glucophage
Formula: $C_4H_{11}N_5$
Monoisotopic protonated ion (*m/z*): 130.1087
Molecular weight: 129.17
CAS number: 657-24-9
Salt form: HCl salt

Metformin is the major first-line treatment for type II diabetes, and in 2021, it was the second most prescribed medication in the United States [1]. It was synthesized in the 1920s based on the lead compound, galegine, a natural product present in the medicinal herb *Galega officinalis,* with its usage since medieval times [2]. The medical use of metformin was first introduced in France in 1957, which is long before its introduction into the United States in 1995 [1]. Despite its long clinical history, during which time it has been demonstrated that metformin enhances hepatic insulin sensitivity, its exact molecular mechanism of action remains debatable [2].

Chemically, metformin is a nearly symmetric biguanide consisting of two coupled guanidine moieties. Its main degradation pathway is hydrolytic deamination. In a forced degradation study, metformin was found to produce two degradants under the acidic condition of 2 N HCl/70°C/4 hours (Scheme 238-1), while being stable in the alkaline condition of 2 N NaOH/70°C/2 hours [3].

Under a harsh oxidative stress condition (30% H_2O_2, 70°C, 2 hours), Degradants **1** and **2** were identified by Luo et al.; nevertheless, no formation mechanism was proposed by the original authors [3]. It might be possible that the two degradants were

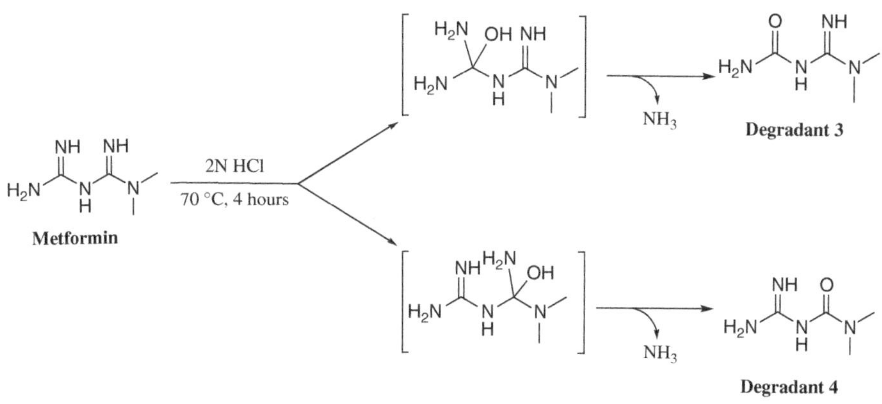

Scheme 238-1 Hydrolytic deamination of metformin. The numbering of the degradants follows that of Luo et al. [3]. *Source:* Luo et al. [3] / with permission of Elsevier.

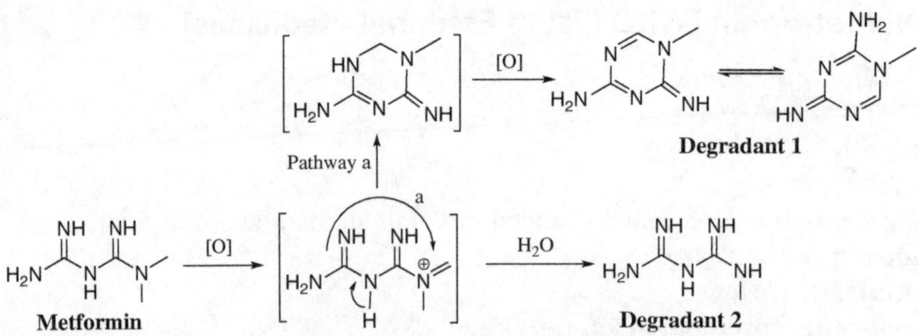

Scheme 238-2 Forced degradation of metformin in 30% H_2O_2 at 70°C for 2 hours. The numbering of the degradants follows that of Luo et al. [3]. *Source:* Luo et al. [3] / with permission of Elsevier.

formed via a common iminium intermediate (Scheme 238-2). Both degradants were previously reported by Collin et al. in their experiments of gamma-irradiation of metformin solutions [4, 5]. In the accelerated stability study of metformin drug substance (40°C/75%RH) performed by Luo et al., Degradants **2** and **3** were produced, indicating they are more likely real degradants [3].

Recently, metformin drug products in extended release formulations were found to contain trace amounts of *N*-nitrosodimethylamine (NDMA) [6], a genotoxic nitrosamine impurity with an allowable intake (AI) of 96 ng/day [7]. Various investigations have indicated that NDMA is both a process impurity and degradant in the extended release formulations [8–12]. With regard to the formation of NDMA in metformin drug products, a study by Schlingemann et al. showed that metformin drug substance was not a significant source for NDMA and that the majority of NDMA was formed during the manufacturing steps that introduced heat and residual nitrite [9]. Their study also showed that the levels of NDMA increased during the storage of the formulated products, but the increase was minor and not shelf-life limiting for the products in their study. In another study of probing NDMA formation in metformin drug products, solutions and film-coated tablets with no obvious source of nitrosating agents were examined for the plausible formation mechanism of NDMA [11]. The results suggested that in solutions, dimethylamine, a degradant of metformin, could be converted into nitrite in the presence of hydrogen peroxide, and the nitrite

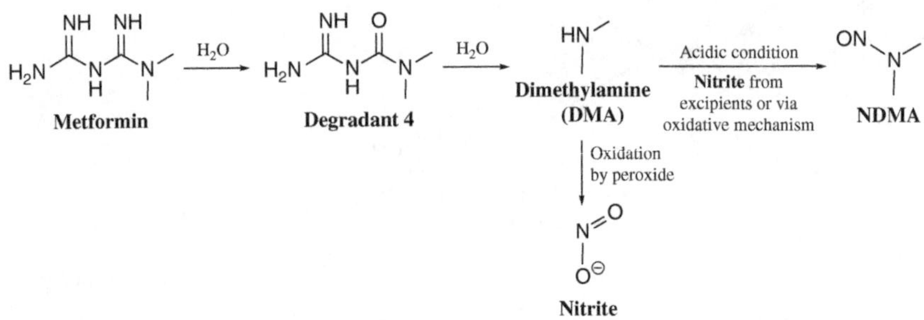

Scheme 238-3 Possible mechanisms for the formation of NDMA in metformin finished drug products.

formed could then turn dimethylamine into NDMA at acidic pH. Although the observation that metformin can be oxidized by hydrogen peroxide to produce ppm-levels of NDMA has been confirmed independently [12], this oxidative mechanism only contributed to less than 10% NDMA formation in the solid dosage of film-coated tablets according to Jires et al. [11]. The latter authors also found that spiking sodium carbonate into the excipients during the tablet manufacturing process could produce tablets with no quantifiable level of NDMA, even in tablet batches where dimethylamine and hydrogen peroxide were intentionally added. The inhibitory effect of an alkaline micro-environment in solid dosage achieved by sodium carbonate was also reported by Hao et al. [12]. The possible mechanisms of the formation of NDMA in metformin drug products are presented in Scheme 238-3.

References

1. Metformin. https://en.wikipedia.org/wiki/Metformin (accessed January 31 2024).
2. Rena, G., Hardie, D., and G., Pearson, E.R. (2017). The mechanisms of action of metformin. *Diabetologia* 60: 1577–1585.
3. Luo, M., Zheng, L., Wang, S. et al. (2021). Characterization of four major degradation products in metformin by 2D LC-QTOF/MS/MS. *J. Pharm. Biomed. Anal.* 192: 113662.
4. Collin, F., Khoury, H., Bonnefont-Rousselot, D. et al. (2004). Liquid chromatographic/electrospray ionisation mass spectrometric identification of the oxidation end-products of metformin in aqueous solutions. *J. Mass. Spectrom.* 39: 890–902.
5. Khouri, H., Collin, F., Bonnefont-Rousselot, D. et al. (2004). Radical-induced oxidation of metformin. *Eur. J. Biochem.* 271: 4745–4752.
6. FDA US FDA's search list of recalled metformin products. https://www.fda.gov/drugs/drug-safety-and-availability/search-list-recalled-metformin-products (accessed February 2 2024).
7. US FDA Guidance Control of nitrosamine impurities in human drugs. https://www.fda.gov/regulatory-information/search-fda.guidancedocuments/control-nitrosamine-impurities-human-drugs (accessed February 2 2024).
8. Nasr, N.E.H., Metwaly, M.G., Ahmed, E.O. et al. (2021). Investigating the root cause of *N*-nitrosodimethylamine formation in metformin pharmaceutical products. *Expert Opin. Drug Saf.* 20: 855–862.
9. Schlingemann, J., Boucley, C., Hickert, S. et al. (2022). Avoiding *N*-nitrosodimethylamine formation in metformin pharmaceuticals by limiting dimethylamine and nitrite. *Int. J. Pharm.* 620: 121740.
10. Jireš, J., Kalášek, S., Gibala, P. et al. (2021). Insight into the formation of *N*-nitrosodimethylamine in metformin products. *J. Pharm. Biomed. Anal.* 195: 113877.
11. Jireš, J., Dousa, M., Gibala, P., and Kubelka, T. (2022). *N*-Nitrosation in the absence of nitrosating agents in pharmaceuticals? *J. Pharm. Biomed. Anal.* 218: 114872.
12. Hao, G., Hu, R., Wang, X. et al. (2022). *N*-Nitrosodimethylamine formation in metforminhydrochloride sustained-release tablets: effects of metformin and hypromellose used in drug product formulation. *J. Pharm. Biomed. Anal.* 222: 115066.

239: Metoclopramide [WHO List of Essential Medicines]

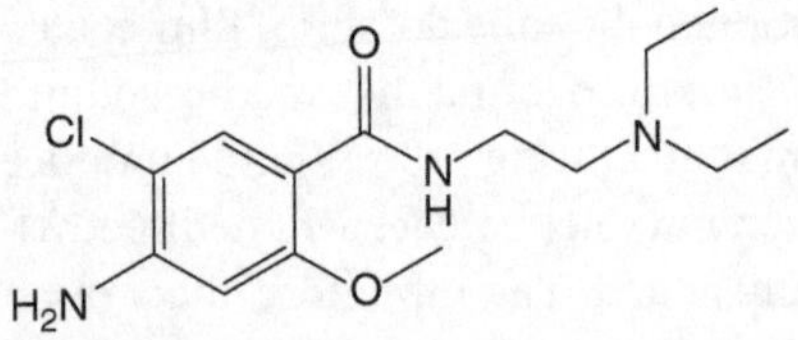

Chemical name: 4-Amino-5-chloro-*N*-[2-(diethylamino)ethyl]-2-methoxybenzamide
Trade name: Primperan, Reglan, others
Formula: $C_{14}H_{22}ClN_3O_2$
Monoisotopic protonated ion (*m/z*): 300.1473
Molecular weight: 299.80
CAS number: 364-62-5
Salt form: HCl salt

Metoclopramide is an antagonist for the dopamine D_2 receptor and 5-HT_3 receptor, and an agonist for the 5-HT_4 receptor. Clinically, it is used as an antiemetic agent to treat and prevent nausea and vomiting, as well as to treat migraine headaches [1]. It was developed and approved for medical use in the 1960s by Laboratoires Delagrange, a company later acquired by SynthoLab and eventually became part of Sanofi.

Metoclopramide contains three functional groups susceptible to degradation: the amide, tertiary amino, and aniline amino group. In a forced degradation study by Sowjanya et al. [2], limited hydrolysis of the amide linkage was observed under both acidic and alkaline stress conditions (1 N HCl or 1 N NaOH solution, 80°C, 5 hours), seemingly with more hydrolysis under the alkaline condition than the acidic one (Scheme 239-1). The results of the hydrolytic forced degradation suggested that metoclopramide would be quite resistant to hydrolytic degradation in its formulated products. Under the oxidative stress condition with hydrogen peroxide (6% H_2O_2, 40°C, 6 hours), numerous degradants were observed, among which two were large

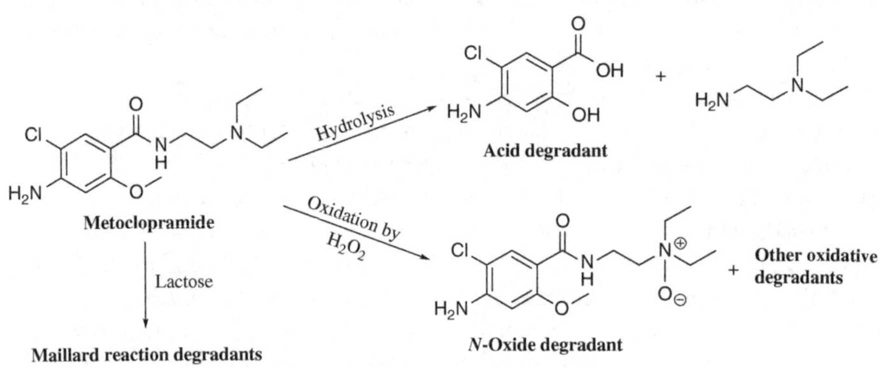

Scheme 239-1 Hydrolytic and oxidative degradation of metoclopramide.

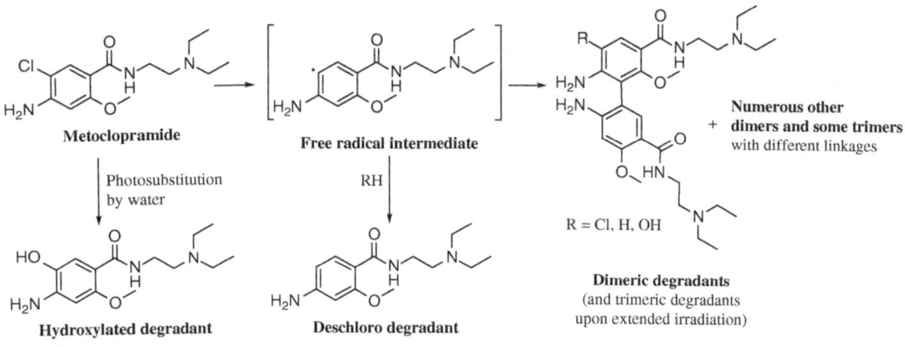

Scheme 239-2 Photochemical degradation of metoclopramide.

degradants and two were medium degradants, while the rest were quite small. Among all the degradants, only one of the two major ones was identified as the *N*-oxide degradant. It could be reasonably assumed that the other oxidative degradants may be largely due to the oxidation of the aniline moiety. On the other hand, the aniline group could undergo the Maillard reaction with reducing sugars. In a compatibility study with lactose by Qiu et al. [3], metoclopramide was degraded in solid mixtures with lactose under heat and moisture. The degradant was presumed to be the corresponding ketosamine (i.e., the glycosylamine degradant) by the authors, which may not be entirely correct, as the initially formed ketosamine degradant could undergo the Amadori rearrangement to produce further degradants [4]. A pseudo-second-order kinetic model was developed for the solid-state degradation, and the activation energy was found to be 53.8 kcal/mol.

Metoclopramide is a substituted benzamide drug that is susceptible to photochemical degradation and thus requires protection from light [5, 6]. In aqueous matrices, metoclopramide was found to decompose rather easily under irradiation of either UV or sunlight [7, 8]. In the photo stress study by Maquille and Jiwan in which UV irradiation at 254 nm was used [7], the initial primary degradation appeared to be a free-radical-mediated dechlorination process, which was followed by abstraction of a hydrogen radical, substitution by a water molecule, and numerous dimerization products (a few trimers were also observed, particularly upon extended irradiation), respectively (Scheme 239-2). The dimers and trimers are linked together through the phenyl group of the drug molecule.

References

1 Metoclopramide. https://en.wikipedia.org/wiki/Metoclopramide (accessed February 4, 2024).

2 Sowjanya, P., Shanmugasundaram, P., Naidu, P., and Singamsetty, S.K. (2013). Novel validated stability-indicating UPLC method for the determination of metoclopramide and its degradation impurities in API and pharmaceutical dosage form. *J. Pharm. Res.* 6: 765–773.

3 Qiu, Z., Stowell, J.G., Morris, K.R. et al. (2005). Kinetic study of the Maillard reaction between metoclopramide hydrochloride and lactose. *Int. J. Pharm.* 303: 20–30.

4 Li, M. (2012). Section 5.1.1. Degradation via the Maillard reaction. In: *Organic Chemistry of Drug Degradation*, 150–153. Cambridge, UK: RSC Publishing.

5 Weclawska, K. and Siminska, E. (1985). The effect of visible and ultraviolet radiation on metoclopramide solutions. *Acta Pol. Pharm.* 42: 555–560.

6 Sánchez-Quiles, I., Nájera-Pérez, M.D., Espuny-Miró, A., and Titos-Arcos, J.C. (2011). Review of the stability of photosensitive medication. *Farm. Hosp.* 35: 204–215.

7 Maquille, A. and Jiwan, J.-L.H. (2009). LC–MS characterization of metoclopramide photolysis products. *J. Photochem. Photobiol. A: Chem.* 205: 197–202.

8 Dabić, D., Hanževački, M., Škorić, I. et al. (2022). Photodegradation, toxicity and density functional theory study of pharmaceutical metoclopramide and its photoproducts. *Sci. Total Environ.* 807: 150694.

240–241: Mexiletine, Tocainide

240: Mexiletine

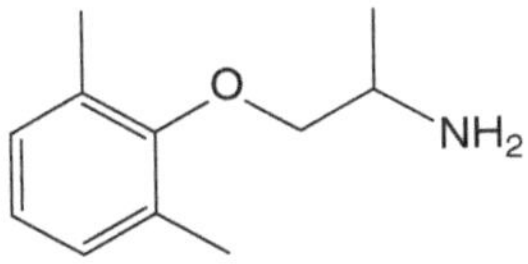

Chemical name: (*R*,*S*)-1-(2,6-dimethylphenoxy)propan-2-amine
Trade name: Mexitil, NaMuscla
Formula: $C_{11}H_{17}NO$
Monoisotopic protonated ion (*m/z*): 180.1383
Molecular weight: 179.26
CAS number: 31828-71-4
Salt form: HCl salt

241: Tocainide

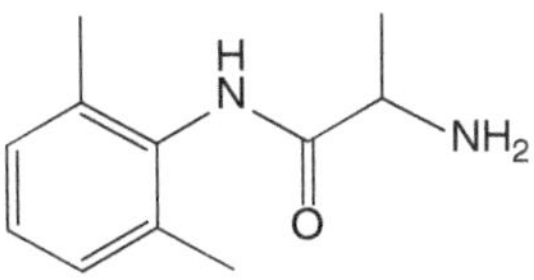

Chemical name: *N*-(2,6-dimethylphenyl)alaninamide
Trade name: Tonocard
Formula: $C_{11}H_{16}N_2O$
Monoisotopic protonated ion (*m/z*): 193.1335
Molecular weight: 192.26
CAS number: 41708-72-9

Mexiletine is a nonselective voltage-gated sodium channel blocker, which is clinically used for treating arrhythmias, chronic pain, and myotonia [1]. Chemically, mexiletine is a phenylether-substituted primary amine. Although there were some forced degradation studies, it appears only one of them performed structural elucidation of degradation products formed under photochemical or oxidative conditions [2]. Irradiation of mexilentine in solid or solution by either sunlight or UV produced a cyclized degradant as the main degradation product (Scheme 240–241). This degradant was proven to be an oxidative degradant, as it could not be produced in the absence of oxygen during the photolysis. Furthermore, it could also be generated under nonphotochemical oxidative conditions, such as using hydrogen peroxide or potassium permanganate. Interestingly, tocainide, a structural analog of mexiletine, was found to be stable under irradiation by sunlight over an extended period [2]. In two forced degradation studies in which mexiletine was subjected to both acidic and alkaline conditions, significant degradation occurred in both cases [3, 4]. In a stability study of extemporaneous solutions of mexiletine, degradation was also observed [5]. Nevertheless, the structural elucidation of these

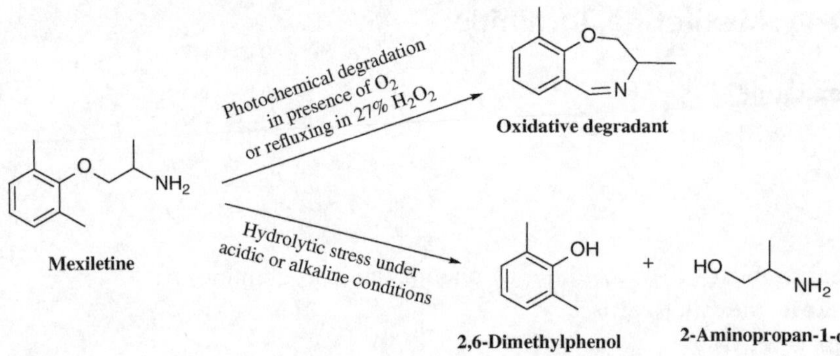

Scheme 240–241 Degradation pathways of mexilentine.

degradants was not attempted [3, 4] or was not successful [5]. It can be surmised, however, that one of the expected degradants would be 2,6-dimethylphenol, a related substance listed in US Pharmacopeia and British Pharmacopeia [4].

References

1 Mexiletine. https://en.wikipedia.org/wiki/Mexiletine (accessed February 4 2024).

2 Takács, M., Vámos, J., Tóth, G., and Mikó-Hideg, Z. (2000). Photochemical and chemical oxidation of mexiletine and tocainide. Structure elucidation of the major products. *Arch. Pharm. Pharm. Med. Chem.* 333: 48–52.

3 Kaushik, S. and Alexander, K.S. (2003). A modified reverse-phase HPLC method for the analysis of mexiletine hydrochloride. *J. Liq. Chromat. Rel. Tech.* 26: 1287–1296.

4 Belal, T.S., Haggag, R.S., and Shaalan, R.A. (2008). Selective and stability-indicating methods for the simultaneous determination of mexiletine hydrochloride and/or its related substance: 2,6-dimethylphenol. *J. AOAC Int.* 91: 720–730.

5 Alexander, K.S. and Kaushik, S. (2004). Extemporaneous formulation and stability testing of mexilentine hydrochloride solution. *Int. J. Pharm. Comp.* 8: 147–152.

242–243: Miconazole, Econazole

242: Miconazole [WHO List of Essential Medicines]

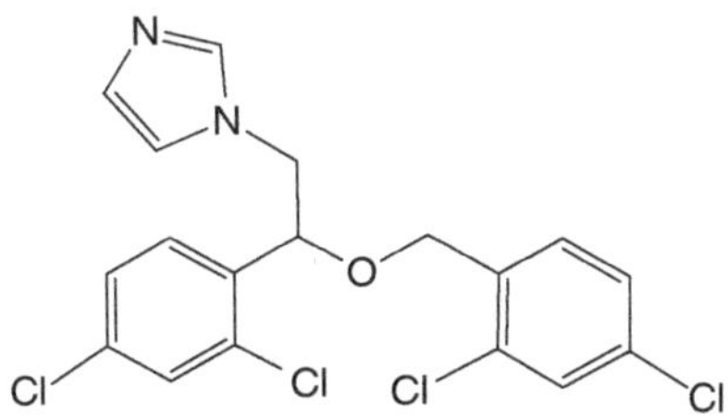

Chemical name: (*R,S*)-1-(2-(2,4-Dichlorobenzyloxy)-2-(2,4-dichlorophenyl)ethyl)-1 *H*-imidazole
Trade name: Monistat, others
Formula: $C_{18}H_{14}Cl_4N_2O$
Monoisotopic protonated ion (*m/z*): 414.9933
Molecular weight: 416.12
CAS number: 22916-47-8
Salt form: Nitrate

243: Econazole

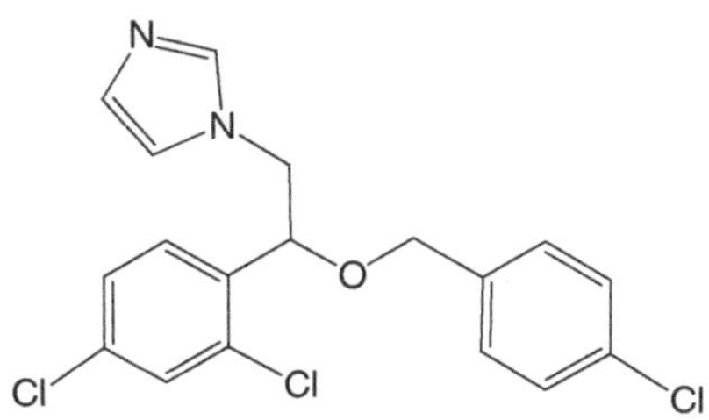

Chemical name: (*R,S*)-1-{2-[(4-Chlorophenyl)methoxy]-2-(2,4-dichlorophenyl)ethyl}-1 *H*-imidazole
Trade name: Spectrazole, others
Formula: $C_{18}H_{15}Cl_3N_2O$
Monoisotopic protonated ion (*m/z*): 381.0323
Molecular weight: 381.68
CAS number: 27220-47-9
Salt form: Nitrate

Miconazole is a first-generation antifungal drug in the imidazole family [1], which is typically in topical formulations and clinically used to treat ring worm, pityriasis versicolor, and yeast infections of skin and vagina [2]. It was developed by Janssen and approved for medical use in 1969 [3]. Chemically, miconazole is a substituted dibenzylic ether with a distinctive imidazole group. Both the ether and imidazole moieties were found to undergo autooxidative degradation in the presence of

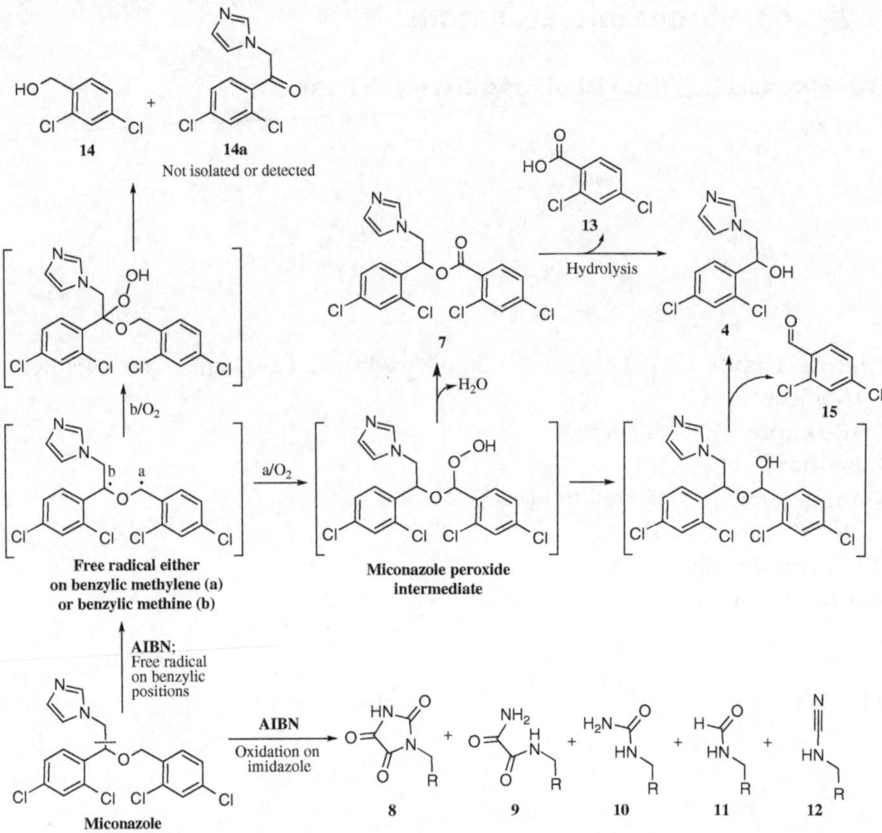

Scheme 242–243-1 Autooxidative degradation of miconazole by catalysis of AIBN, a free-radical initiator. The numbering of the degradants follows that of Oyler et al. [4]. *Source:* Oyler et al. [4] / with permission of Elsevier.

free-radical initiator, AIBN [4]. A great number of degradants were isolated and their structures identified, which resulted from the oxidation of the benzylic positions and imidazole, respectively (Scheme 242–243-1). The radical formation occurred mostly on the benzylic methylene position (a, Scheme 242–243-1), probably due to its less steric hindrance, followed by the formation of the hydroperoxide intermediate, which resulted in the formation of degradants **7**, **13**, and **15**, respectively. The formation of the other hydroperoxide intermediate on the more steric benzylic methine position (b, Scheme 242–243-1) led to the production of degradant **14** . According to this formation pathway (b), degradant **14a** should also be expected to form; nevertheless, this degradant was not reported in the study [4].

Under the autooxidative degradation condition with AIBN, another set of oxidative degradants (**8–12**) was also isolated and identified, which resulted from the oxidation on the imidazole ring. Such degradation usually occurred through reaction with singlet oxygen; a mechanism with triplet oxygen was proposed by the authors [4].

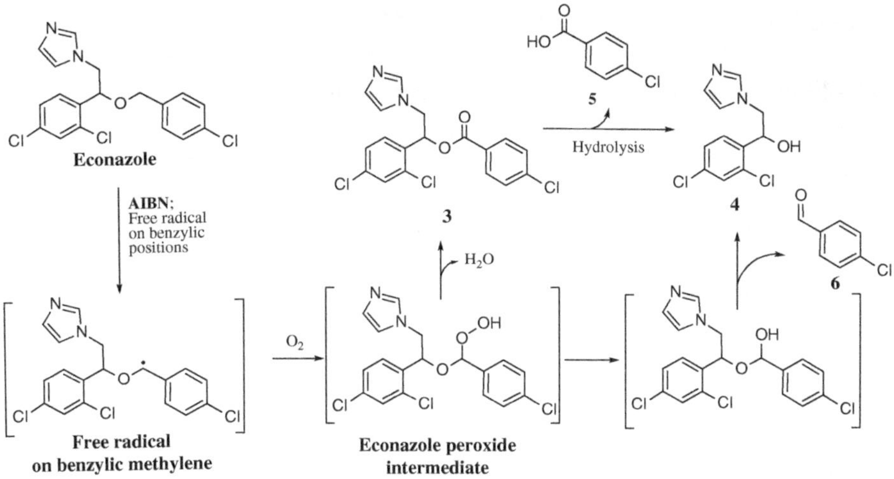

Scheme 242–243-2 Autooxidative degradation of econazole by catalysis of AIBN, a free-radical initiator. The numbering of the degradants follows that of Oyler et al. [4]. *Source:* Shafiei et al. [3] and Oyler et al. [4].

The same autooxidative degradation conditions were applied to econazole, a closely related analogous antifungal drug with one less chloro substituent than miconazole, which was also developed by Janssen and marketed in 1974 [1]. The oxidation that occurred on the benzylic positions was similar to that of miconazole (Scheme 242–243-2). However, no degradants resulting from the oxidation on the imidazole ring were isolated or identified; the authors speculated that some of these analogous degradants might be among the unknown peaks not isolated for identification [4].

Other first-generation imidazole antifungal drugs, such as isoconazole and dapaconazole, would be expected to undergo the same autooxidative degradation.

References

1 Shafieia, M., Peytonb, L., Hashemzadehc, M., and Foroumadi, A. (2020). History of the development of antifungal azoles: a review on structures, SAR, and mechanism of action. *Bioorg. Chem.* 104: 104240.

2 Miconazole. https://en.wikipedia.org/wiki/Miconazole (accessed February 4 2024).

3 Shafiei, M., Peyton, L., Hashemzadeh, M., and Foroumadi, A. (2020). History of the development of antifungal azoles: a review on structures, SAR, and mechanism of action. *Bioorg. Chem.* 104: 104240.

4 Oyler, A.R., Naldi, R.E., Facchine, K.L. et al. (1991). Characterization of autoxidation products of the antifungal compounds econazole nitrate and miconazole nitrate. *Tetrahedron* 47: 6549–6560.

244–245: Mifepristone, Misoprostol

244: Mifepristone [WHO List of Essential Medicines]

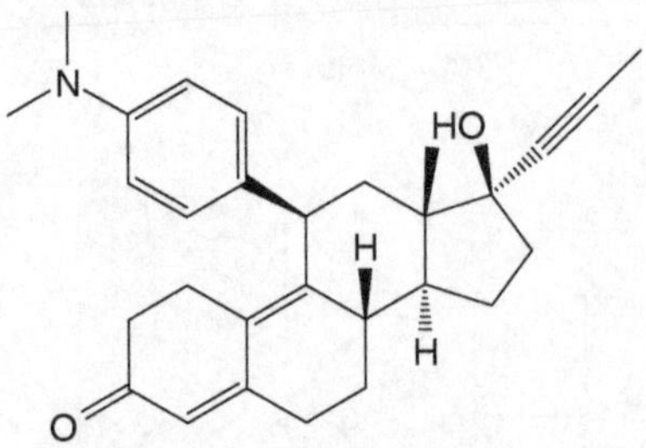

Chemical name: (*8S,11R,13S,14S,17S*)-11-[4-(Dimethylamino)phenyl]-17-hydroxy-13-methyl-17-prop-1-ynyl-1,2,6,7,8,11,12,14,15,16-decahydrocyclopenta[a]phenanthren-3-one
Trade name: Mifegyne, Mifeprex, Korlym, others
Formula: $C_{29}H_{35}NO_2$
Monoisotopic protonated ion (*m/z*): 430.2741
Molecular weight: 429.60
CAS number: 84371-65-3

245: Misoprostol [WHO List of Essential Medicines]

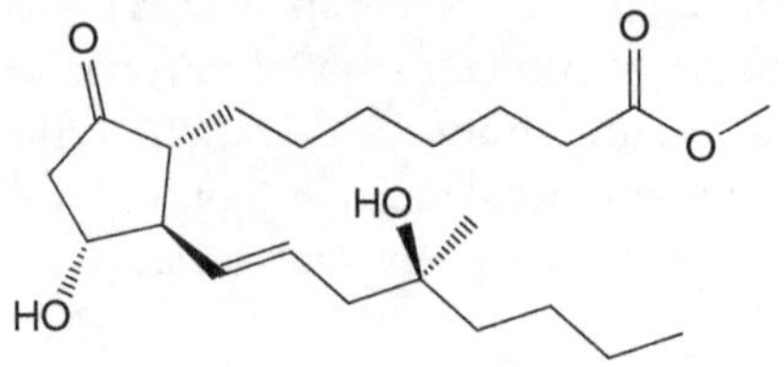

Chemical name: Methyl 7-((*1R,2R,3R*)-3-hydroxy-2-((*S,E*)-4-hydroxy-4-methyloct-1-enyl)-5-oxocyclopentyl)heptanoate
Trade name: Cytotec, Misodel, others
Formula: $C_{22}H_{38}O_5$
Monoisotopic protonated ion (*m/z*): 383.2792
Molecular weight: 382.54
CAS number: 59122-46-2

Mifepristone is also known as RU-486, produced in 1980 at the French pharmaceutical company Roussel-Uclaf with the original compound code RU-38486 [1]. It is an antiprogestogen and is clinically used for inducing medical abortion and managing early-stage miscarriage. Literature search indicated that the major degradation site is the dimethyl-amino moiety, which would be subject to oxidative demethylation to form the *N*-monodemethylated mifepristone as the major degradant and perhaps to a much less degree, the *N,N*-didemethylated mifepristone (Scheme 244–245-1) [2, 3].

Mifepristone is often used in combination with misoprostol, which is a prostaglandin analog, and when utilized as a single therapeutic agent, misoprostol is

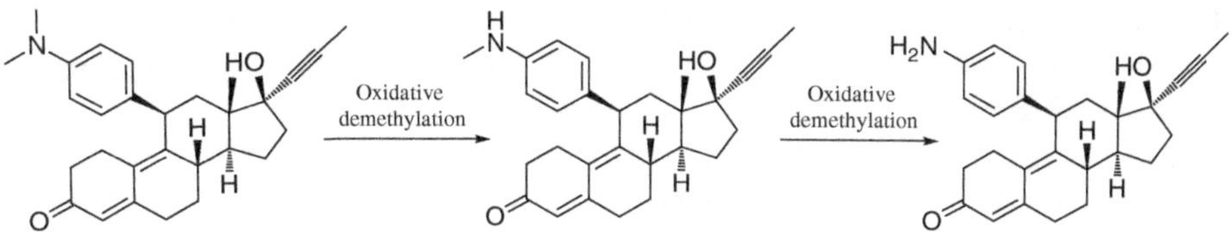

Scheme 244–245-1 Oxidative demethylation of mifepristone. *Source:* Misoprostol [2] and Shi et al. [3].

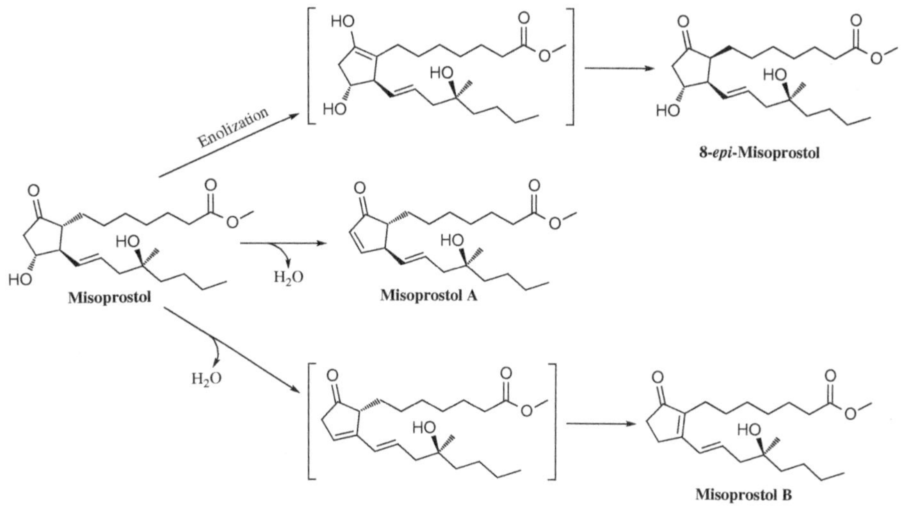

Scheme 244–245-2 Degradation pathways of misoprostol per Hagen et al. [5] with modification. *Source:* Hagen et al. [5] / PLOS / CC BY 4.0.

indicated for treating post-partum hemorrhage (PPH), stomach and duodenal ulcers, among others [4, 5]. Misoprostol can undergo two types of degradation: epimerization and dehydration. The former degradation should lead to the formation of 8-*epi*-misoprostol, most likely mediated via enolization. Dehydration of the hydroxyl group on the 5-membered ring could take place in two directions: one leading to the formation of misoprostol A, while the other one to misoprostol B. In the paper by Hagen et al. [5], misoprostol B was proposed as forming from misoprostol A via isomerization. Nevertheless, it appears that the pathway proposed in Scheme 244–245-2 would be more probable from a mechanistic perspective.

References

1 Mifepristone. https://en.wikipedia.org/wiki/Mifepristone (accessed June 14 2024).
2 Misoprostol. https://en.wikipedia.org/wiki/Misoprostol (accessed June 14 2024).
3 Shi, J., Che, B.-Q., Zheng, Y. et al. (2014). HPLC determination of related substances in mifepristone capsules from different sources. (Original in Chinese). *Chin. J. Pharm. Anal.* 34: 480–484.

4 Iyer, J., Morgan, L.M., Harrison, P. et al. (2023). Applying material science principles to chemical stability: modelling solid state autoxidation in mifepristone containing different degrees of crystal disorder. *J. Pharm. Sci.* 112: 2463–2482.

5 Hagen, N., Bizimana, T., Kayumba, P.C. et al. (2020). Stability of misoprostol tablets collected in Malawi and Rwanda: importance of intact primary packaging. *PLoS ONE* 15 (9): e0238628. https://doi.org/10.1371/journal.pone.0238628.

246: Mirtazapine

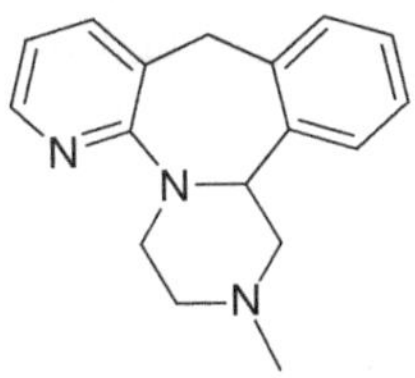

Chemical name: 2-Methyl-1,2,3,4,10,14b-hexahydropyrazino[2,1-a]pyrido[2,3-c][2]benzazepine
Trade name: Remeron, others
Formula: $C_{17}H_{19}N_3$
Monoisotopic protonated ion (*m/z*): 266.1652
Molecular weight: 265.36
CAS number: 85650-52-8
Salt form: HCl salt

Mirtazapine is a tricyclic antidepressant, which is often utilized in major depressive cases that are complicated by anxiety and insomnia [1]. Chemically, it is a racemic mixture of tetracyclic piperazine-azepine that contains a methylpyrazine moiety. The latter is a tertiary amine functionality and thus would be susceptible to nucleophilic oxidative degradation. Indeed, in a forced degradation study by Rao and Raju [2], mirtazapine *N*-oxide was found to form after treating mirtazapine in 3% hydrogen peroxide [2]. Under other forced degradation conditions, such as acidic, alkaline, thermal, and solid phase photolysis, mirtazapine was found to be stable. Nevertheless, in neutral and acidic solutions, photolysis led to the formation of a major and a minor photo-oxidative degradant (Scheme 246) [3]. The major

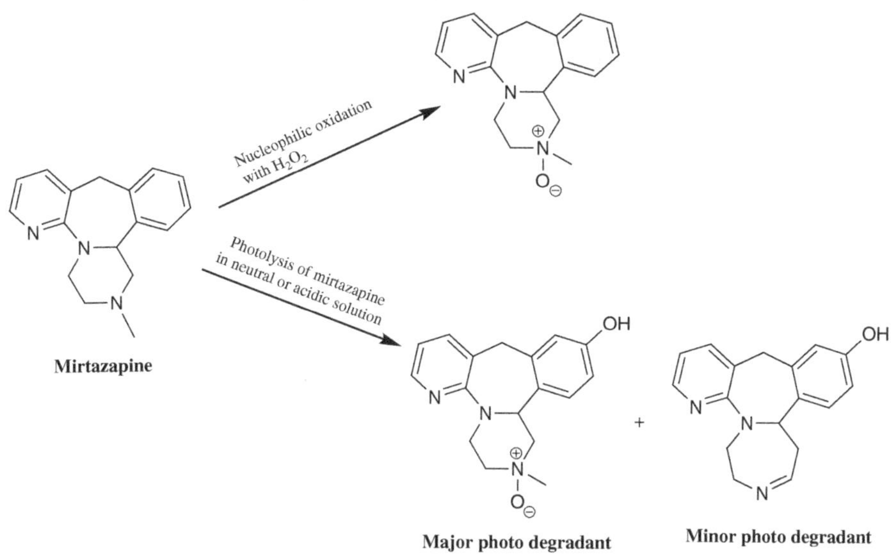

Scheme 246 Nucleophilic and photolytic oxidation of mirtazapine.

degradant was identified as a hydroxylated mirtazapine *N*-oxide, with the hydroxyl group being assigned to the phenyl group based on mass spectrometric data. The minor degradant was proposed to form via *N*-oxidation followed by ring expansion.

References

1 Mirtazapine. https://en.wikipedia.org/wiki/Mirtazapine (accessed February 4 2024).

2 Rao, R.N. and Raju, A.N. (2009). Development and validation of a reversed-phase HPLC method for separation and simultaneous determination of process-related substances of mirtazapine in bulk drugs and formulations. *J. Chromat. Sci.* 47: 223–230.

3 Anand, A. and Bansal, G. (2011). LC-UV and LC-MS studies on mirtazapine to characterize its degradation products. *J. Pharm. Educ. Res.* 2: 66–72.

247: Montelukast

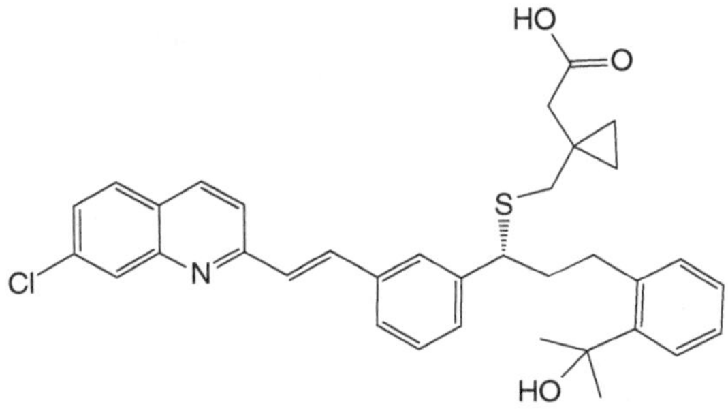

Chemical name: (*E*)-2-(1-((1-(3-(2-(7-Chloroquinolin-2-yl)vinyl)phenyl)-3-(2-(2-hydroxypropan-2-yl)phenyl)propylthio)methyl)cyclopropyl)acetic acid
Trade name: Singulair
Formula: $C_{35}H_{36}ClNO_3S$
Monoisotopic protonated ion (*m/z*): 586.2177
Molecular weight: 586.19
CAS number: 158966-92-8
Salt form: Sodium salt

Montelukast is a medication used in the maintenance treatment of asthma, but is not effective for acute asthma attack [1]. It is a leukotriene receptor antagonist that causes reduced inflammation and relaxation of smooth muscle. Montelukast was developed by the former Merck Frosst Centre for Therapeutic Research in Montreal,

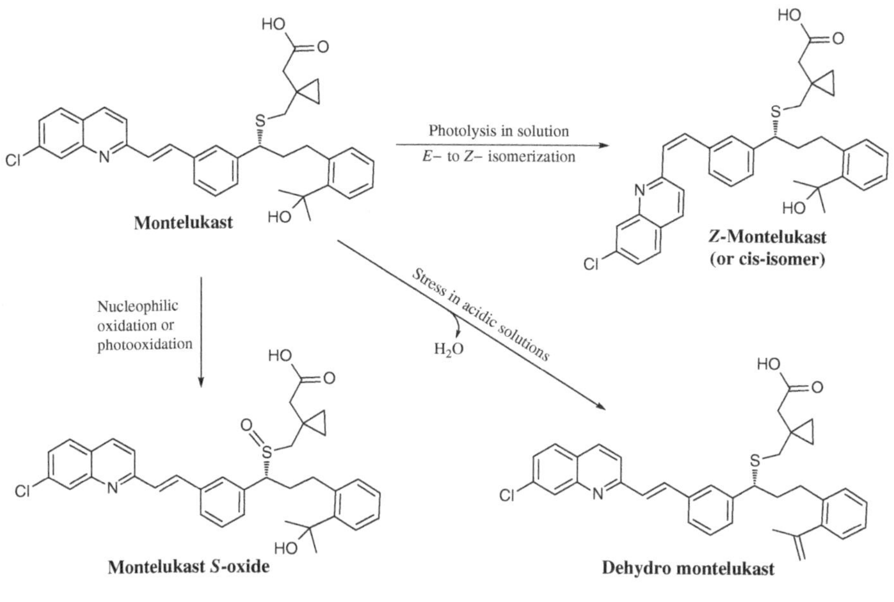

Scheme 247 Degradation pathways of montelukast.

Canada [2], which was part of the Merck Research Laboratories. It was approved for medical use as its sodium salt in the United States in 1998. Chemically, montelukast is a stilbene-like derivative, and hence, its central carbon–carbon double bond would be susceptible to photochemical isomerization to produce the *Z*- or *cis*-isomer. Several papers have reported this isomeric degradation pathway (Scheme 247) [3–5]. Interestingly, it appeared that the *E*- to *Z*-isomerization (or *trans*- to *cis*- isomerization) only occurred in solution. In a photolysis study of uncoated chewable tablets, the major degradant was found to be the sulfoxide degradant, resulting from photooxidation of the thioether moiety of montelukast. The sulfoxide degradant was also the major degradation product observed in an accelerated stability study at 40°C/75%RH for 6 months [3]. Forced degradation studies under other degradation conditions showed that montelukast was stable in alkaline solutions, while in acidic solutions, it seemed to produce dehydro-montelukast (Scheme 247). On the other hand, the stress of montelukast in hydrogen peroxide solutions yielded the *S*-oxide degradant as expected [3, 5]. In principle, two diastereomers of the *S*-oxide degradant should be formed due to the presence of the sulfur lone pair electrons; nevertheless, it appears that only one peak of *S*-oxide was observed in the studies reported [3, 5], which may be attributed to the co-elution of the two isomers.

References

1 Montelukast. https://en.wikipedia.org/wiki/Montelukast (accessed February 4, 2024).

2 Young, R.N. (2001). Discovery of montelukast: a once-a-day oral antagonist of leukotriene D_4 for the treatment of chronic asthma. *Prog. Med. Chem.* 38: 249–277.

3 Al Omari, M.M., Zoubi, R.M., Hasan, E.I. et al. (2007). Effect of light and heat on the stability of montelukast in solution and in its solid state. *J. Pharm. Biomed. Anal.* 45: 465–471.

4 Roman, J., Breier, A.R., and Steppe, M. (2011). Stability indicating LC method to determination of sodium montelukast in pharmaceutical dosage form and its photodegradation kinetics. *J. Chromat. Sci.* 49: 540–546.

5 Tiwari, S.K., Singh, D.K., Ladumor, M.K. et al. (2018). Study of degradation behaviour of montelukast sodium and its marketed formulation in oxidative and accelerated test conditions and prediction of physicochemical and ADMET properties of its degradation products using ADMET Predictor™. *J. Pharm. Biomed. Anal.* 158: 106–118.

248: Mupirocin [WHO List of Essential Medicines]

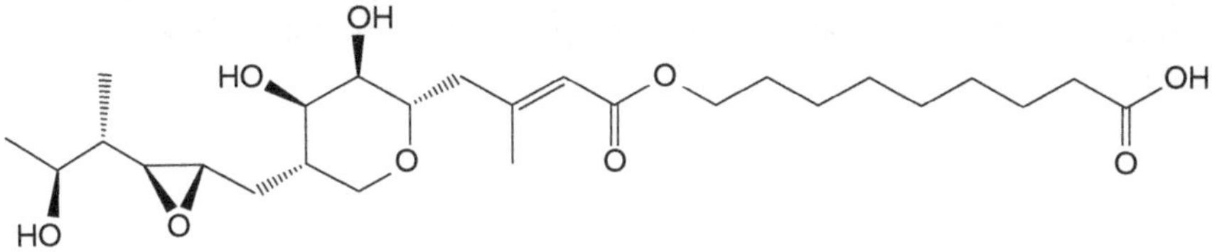

Chemical name: 9-[(*E*)-4-[(2*S*,3*R*,4*R*,5*S*)-3,4-dihydroxy-5-[[(2*S*,3*S*)-3-[(2*S*,3*S*)-3-hydroxybutan-2-yl]oxiran-2-yl]methyl]oxan-2-yl]-3-methylbut-2-enoyl]-oxynonanoic acid
Trade name: Bactroban, others
Formula: $C_{26}H_{44}O_9$
Monoisotopic protonated ion (*m/z*): 501.3058
Molecular weight: 500.63
CAS number: 12650-69-0

Mupirocin is a natural product isolated from *Pseudomonas fluorescens* in 1971. It is used as a topical antibacterial for skin infections, particularly for treating methicillin-resistant *Staphylococcus aureus* (MRSA) in cases of impetigo [1]. Its mode of action is believed to be the inhibition of bacterial protein and RNA synthesis by inhibiting the bacterial isoleucyl-transfer RNA, which leads to cell death [2]. Chemically, it belongs to the pseudomonic acid family, with the predominant component being pseudomonic acid A, which accounts for more than 90% of the mixture. It contains an epoxide group which would react with the adjacent hydroxyl group on the six-membered ring in strongly acidic or alkaline solutions, leading to the formation of two isomeric degradants (**1** and **2**) [3, 4], In strongly alkaline solutions, the ester

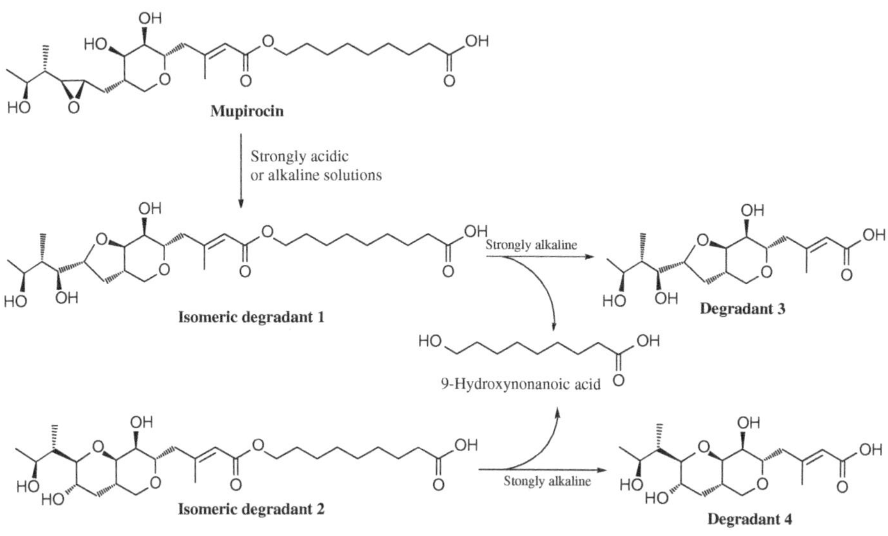

Scheme 248 Degradation of mupirocin in strongly acidic or alkaline solutions.

functionality could be hydrolyzed to produce degradant **3** by eliminating 9-hydroxynonanoic acid. The degradation pathways of mupirocin are summarized in Scheme 248. In the pH range of 4–9, mupirocin was found to be reasonably stable [3].

References

1 Mupirocin. https://en.wikipedia.org/wiki/Mupirocin (accessed February 4 2024).
2 Erwin, D.Z. and Chen, P. Mupirocin. https://www.ncbi.nlm.nih.gov/books/NBK599499/ (accessed February 27 2024).
3 Clayton, J.P., Oliver, R.S., Rogers, N.H., and King, T.J. (1979). The chemistry of pseudomonic acid. Part 3. The rearrangement of pseudomonic acid A in acid and basic solution. *J. Chem. Soc. Perkin Trans.* 1: 838–846.
4 Scott, R.W., Murphy, A.C., Wu, J. et al. (2011). Mupirocin F: structure elucidation, synthesis and rearrangements. *Tetrahedron* 67: 5098–5106.

249: Naproxen

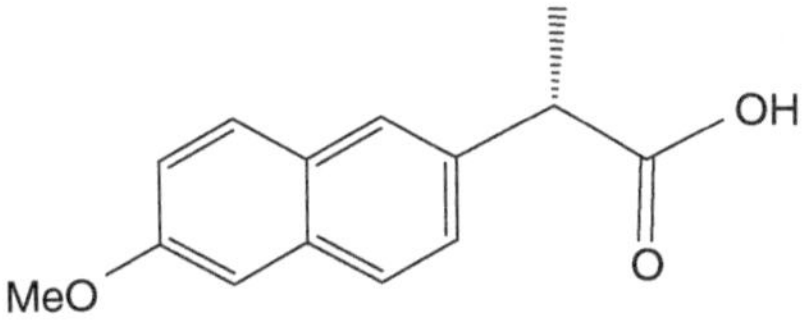

Chemical name: (+)-(*S*)-2-(6-Methoxynaphthalen-2-yl)propanoic acid
Trade name: Naprosyn, Aleve, others
Formula: $C_{14}H_{14}O_3$
Monoisotopic protonated ion (*m/z*): 231.1016
Molecular weight: 230.26
CAS number: 22204-53-1
Salt form: sodium salt

Naproxen is a nonselective COX enzyme inhibitor, which belongs to the family of nonsteroidal anti-inflammatory drugs (NSAIDs), and is indicated for the treatment of pain and inflammatory diseases such as rheumatoid arthritis, gout, and fever [1]. Patented by the pharmaceutical company Sythex in 1967 [2], it was approved for medical use in the United States in 1976 as a prescription drug and became an over-the-counter medication in 1994 [3]. Chemically, naproxen is an 1-substituted aryl propionic acid. A forced degradation study of naproxen soft gel capsules suggested that the major degradation resulted from decarboxylation as well as subsequent oxidation (Scheme 249-1) [4]. Under acidic (5 M HCl, 25°C, 60 minutes), alkaline

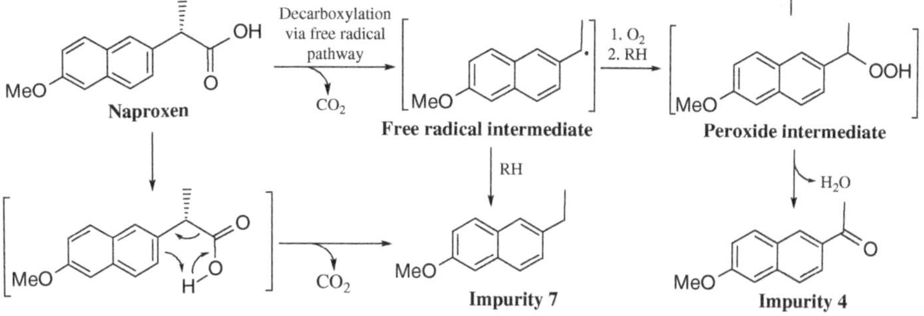

Scheme 249-1 Degradation pathways of naproxen under forced acidic, alkaline, and oxidative degradation conditions. The numbering of impurities follows that of Reddy et al. [4]. There was some contradictory description in the original paper: while in the main text of the paper, one of the major degradants was described as Impurity 2, while in the summary table of the forced degradation study, this major degradant was written as Impurity 3. Based on the mechanistic analysis performed by the current author, Impurity 3 would be a more reasonable major degradant than Impurity 2. *Source:* Reddy et al. [4] / Taylor & Francis Group.

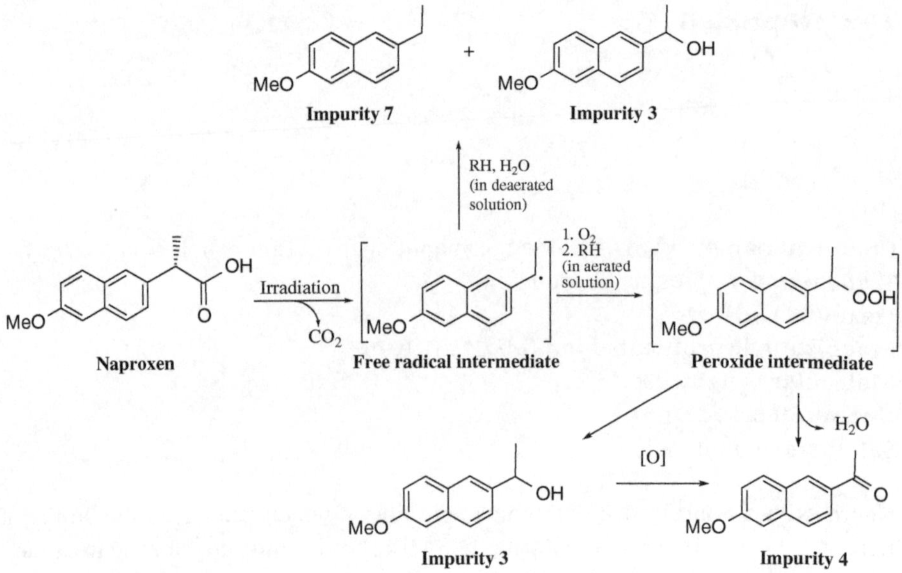

Scheme 249-2 Photochemical degradation pathways of naproxen. For consistency, the numbering of impurities follows that of Reddy et al. [4]. *Source:* Reddy et al. [4] / Taylor & Francis Group.

(5 M NaOH, 85°C, 120 minutes), and oxidative (30% H_2O_2, 25°C, 60 minutes) stress conditions, three degradants (**2**, **4**, and **7**) were observed. Under humid (90% RH, 25°C, 120 minutes) and thermal (60°C, 24 hours) conditions, the observed degradants were **2** and **7**, and **7**, respectively. There was no mechanistic explanation given for the forced degradation results in the original paper; the current author hence provided one as shown in Scheme 249-1 based on the photochemical degradation mechanism proposed by Moore and Chappuis [5].

A photochemical degradation study in solution was performed by Moore and Chappuis [5], naproxen was found to produce **Impurities 3** and **4** in aerated solution, while the degradants were **Impurities 3** and **7** in deaerated solution (Scheme 249-2). It is generally believed that naproxen first underwent intramolecular decarboxylation at an excited triplet state to produce the decarboxylated intermediate, from which the photodegradants were formed [6]. The photodegradation profiles were similar to those under nonphotochemical conditions.

References

1 Naproxen. https://en.wikipedia.org/wiki/Naproxen (accessed February 4 2024).
2 Naproxen sodium. https://www.osc-ortho.com/blog/naproxen-sodium/ (accessed February 29 2024).
3 Naproxen. https://www.ncbi.nlm.nih.gov/books/NBK525965/ (accessed February 29 2024).

4 Reddy, R.S., Krishna, R.M., Vekaria, N.A. et al. (2016). Determination of potential impurities of naproxen sodium in soft gelatin capsules dosage by using ultra performance liquid chromatography. *Anal. Chem. Lett.* 6: 55–69.

5 Moore, D.E. and Chappuis, P.P. (1998). A comparative study of the photochemistry of the non-steroidal anti-inflammatory drugs, naproxen, benoxaprofen and indomethacin. *Photochem. Photobiol.* 47: 173–180.

6 Tu, N., Liu, Y., Li, R. et al. (2019). Experimental and theoretical investigation on photodegradation mechanisms of naproxen and its photoproducts. *Chemosphere* 227: 142–150.

250: Neostigmine Bromide [WHO List of Essential Medicines]

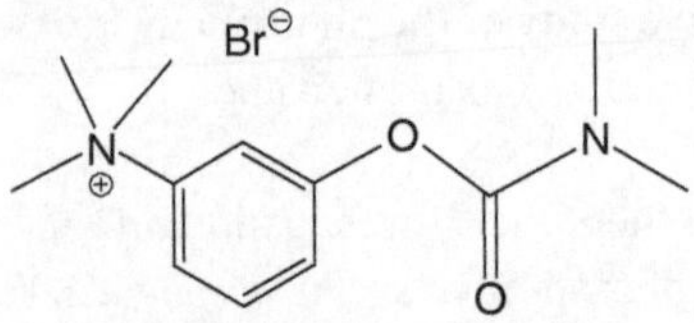

Chemical name: 3-{[(Dimethylamino)carbonyl]oxy}-*N,N,N*-trimethylbenzenaminium
Trade name: Bloxiverz
Formula: $C_{12}H_{19}N_2O_2^+$
Monoisotopic protonated ion (*m/z*): 223.1441 (excluding bromide)
Molecular weight: 303.20
CAS number: 59-99-4
Salt form: Bromide

Neostigmine is a medication for the treatment of myasthenia gravis (an autoimmune disorder manifested by generalized muscle weakness), Ogilvie syndrome (a rare disorder of colonic dilation without anatomic obstruction), and urinary retention without the presence of a blockage [1]. It was synthesized by Aeschlimann and Reinert at Hoffmann La Roche in 1931 and was patented by Aeschlimann in 1933. Chemically, it is a phenylcarbamate as well as a quaternary amine of aniline derivative. Both moieties would be susceptible to degradation reactions, which are of hydrolytic and nucleophilic nature. In a stress study of neostigmine bromide in aqueous solutions at elevated temperature, six degradation products were observed and identified (Scheme 250) [2]. Degradant **2** was due to the carbamate linkage hydrolysis, while further degradation of **2** via demethylation produced degradant **3**.

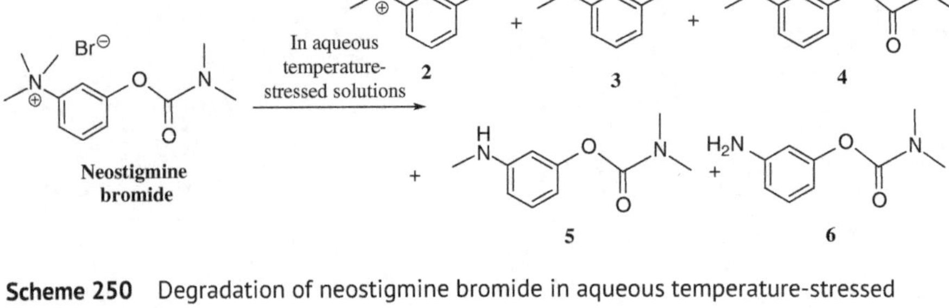

Scheme 250 Degradation of neostigmine bromide in aqueous temperature-stressed solutions.

Degradants **4** to **6** stemmed from different levels of demethylation, while the carbamate linkage remained intact. Bromination of **4** would yield degradant **7**.

References

1 Neostigmine. https://en.wikipedia.org/wiki/Neostigmine (accessed February 4 2024).

2 Porst, H. and Kny, L. (1985). The structure of degradation products of neostigmine bromide. *Pharmazie* 40: 325–328.

251: Nitrofurantoin [WHO List of Essential Medicines]

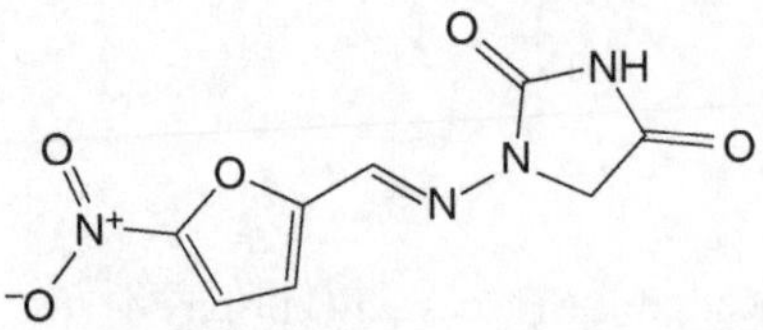

Chemical name: (*E*)-1-[(5-Nitro-2-furyl)methylideneamino]imidazolidine-2,4-dione
Trade name: Macrobid, Macrodantin, Macpac, others
Formula: $C_8H_6N_4O_5$
Monoisotopic protonated ion (*m/z*): 239.0411
Molecular weight: 238.16
CAS number: 67-20-9

Nitrofurantoin is an antibacterial drug in the nitrofuran family and clinically used primarily for the treatment of urinary tract and bladder infections. It was invented by Kenyon J. Hayes of the pharmaceutical company Eaton Laboratories and was approved for medical use in 1953 [1]. Synthetically, it is made by a condensation reaction between 5-nitro-2-furaldehyde (**1**) and 3-aminohydantoin (**2**). Hence, nitrofurantoin would be susceptible to hydrolytic degradation to revert back to the two starting materials. Indeed, in solutions of low pH, these two compounds were observed as the major degradants (Scheme 251-1) [2], while in citrate buffer, the hydrolytic degradation occurred on the hydantoin ring, yielding 3-(5-nitrofurfurylideneamino)hydantoic acid (**3**) as the primary degradant [3]. In alkaline solutions, the hydantoin ring further degraded to produce

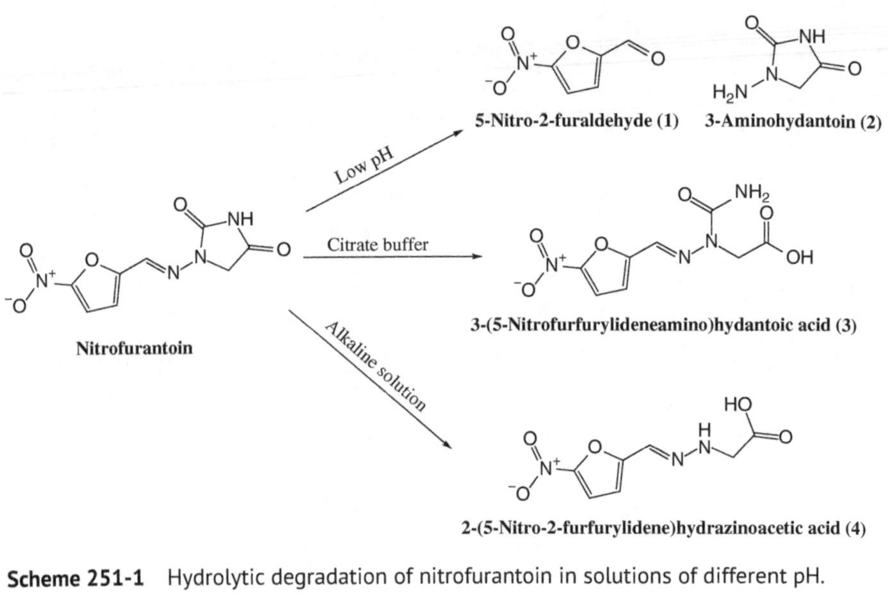

Scheme 251-1 Hydrolytic degradation of nitrofurantoin in solutions of different pH.

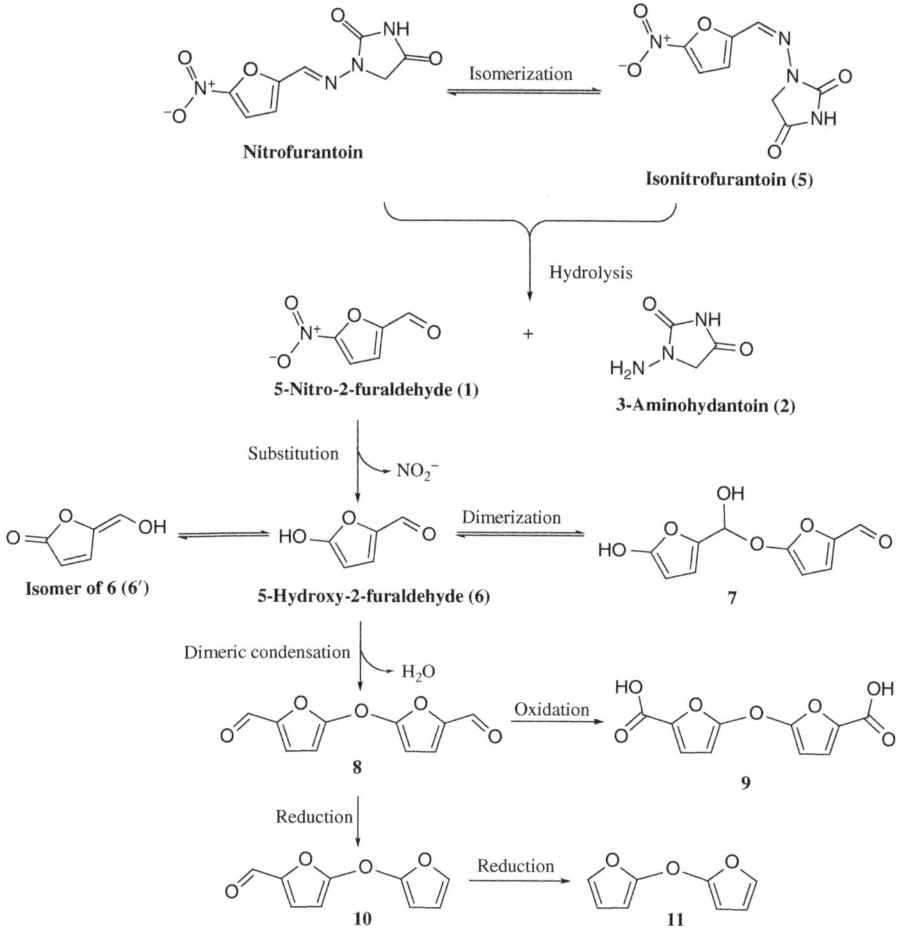

Scheme 251-2 Photochemical degradation pathways of nitrofurantoin in aqueous solution. The structural assignment of the degradants was based upon mass spectrometric data and thus considered tentative. *Source:* Szabó-Bárdos et al. [6] / with permission of Elsevier.

2-(5-nitro-2-furfurylidene)hydrazinoacetic acid (**4**) [4]. Biosic et al. studied the hydrolytic degradation kinetics of nitrofurantoin in solutions of pH 4, 7, and 9 and found that the half-lives at 40°C were 68.0, 10.9, and 4.2 days, respectively, indicating that the hydrolysis rate was the slowest at pH 4 [5].

In a photolysis study of nitrofurantoin in aqueous solutions with irradiation between 375 and 400 nm, which is near the maximum UV absorption of nitrofurantoin, nitrofurantoin first underwent rapid *trans-* to *cis*-transformation to produce isonitrofurantoin (**5**, Scheme 251-2) [6]. Both isomers were subsequently hydrolyzed to yield degradants **1** and **2**, and **1** was further degraded via photosubstitution to generate 5-hydroxy-2-furaldehyde (**6**). The latter degradant (**6**) could further degrade into a few more degradation products (**7–11**).

References

1 Molecule of the week archive, Nitrofurantoin. https://www.acs.org/molecule-of-the-week/archive/n/nitrofurantoin.html (accessed March 1 2024).

2 Paul, H.E., Ells, V.R., Kopko, F., and Bender, R.C. (1960). Metabolic degradation of the nitrofurans. *J. Med. Pharm. Chem.* 2: 563–584.

3 Juenge, E.C., Kreienbaum, M.A., and Gurka, D.F. (1985). Assay of nitrofurantoin oral suspensions contaminated with 3-(5-nitrofurfurylideneamino)hydantoic acid. *J. Pharmacy Sci.* 74: 100–102.

4 Ertan, G., Karasulu, Y., and Guneri, T. (1993). Degradation and gastrointestinal stability of nitrofurantoin in acidic and alkaline media. *Int. J. Pharm.* 96: 243–248.

5 Biosic, M., Skoric, I., Beganovic, J., and Babic, S. (2017). Nitrofurantoin hydrolytic degradation in the environment. *Chemosphere* 186: 660–668.

6 Szabó-Bárdos, E., Cafuta, A., Hegedűs, P. et al. (2020). Photolytic and photocatalytic degradation of nitrofurantoin and its photohydrolytic products. *J. Photochem. Photobiol. A: Chem.* 386: 112093.

8

Omeprazole and Related Azoles to Primaquine

252–257: Omeprazole and Related Azoles

252: Omeprazole

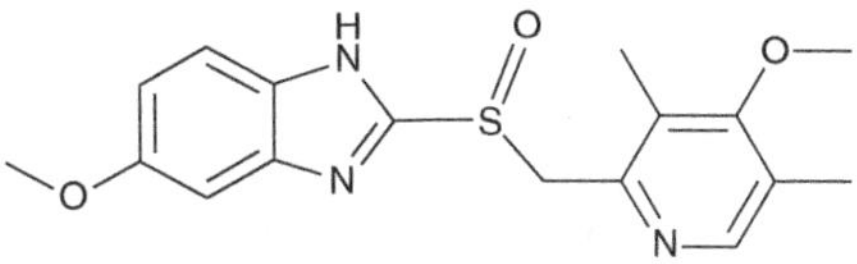

Chemical name: (*R*,*S*)-5-Methoxy-2-[(4-methoxy-3,5-dimethylpyridin-2-yl) methanesulfinyl]-1*H*
-benzimidazole
Trade name: Prilosec, Losec
Formula: $C_{17}H_{19}N_3O_3S$
Monoisotopic protonated ion (*m/z*): 346.1220
Molecular weight: 345.42
CAS number: 73590-58-6
Common salt form: Magnesium, Sodium salts

253: Esomeprazole

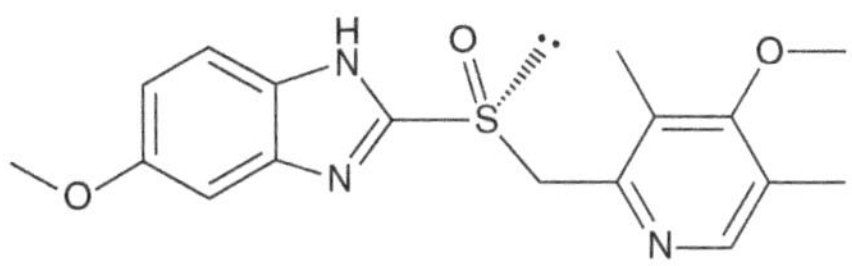

Chemical name: (*S*)-(−)-5-Methoxy-2-[(4-methoxy-3,5-dimethylpyridin-2-yl) methanesulfinyl]-1*H*
-benzimidazole
Trade name: Nexium, Neksium
Formula: $C_{17}H_{19}N_3O_3S$
Monoisotopic protonated ion (*m/z*): 346.1220
Molecular weight: 345.42

Compendium of Drug Degradation Pathways, First Edition. Min Li.

CAS number: 119141-88-7
Common salt form: Magnesium, Sodium, Strontium salts

254: Lansoprazole

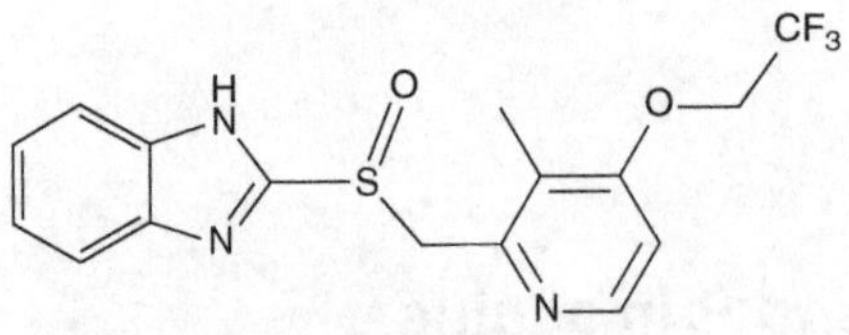

Chemical name: (*R,S*)-2-([3-Methyl-4-(2,2,2-trifluoroethoxy)pyridin-2-yl]methylsulfinyl)-1
H-benzo[d]imidazole
Trade name: Agopton, Ogast, others
Formula: $C_{16}H_{14}F_3N_3O_2S$
Monoisotopic protonated ion (*m/z*): 370.0832
Molecular weight: 369.36
CAS number: 103577-45-3

255: Pantoprazole

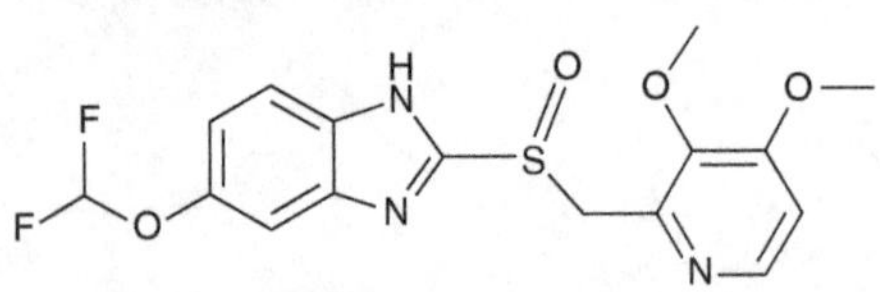

Chemical name: (*R,S*)-6-(Difluoromethoxy)-2-[(3,4-dimethoxypyridin-2-yl)methylsulfinyl]-1
H-benzo[d]imidazole
Trade name: Protonix
Formula: $C_{16}H_{15}F_2N_3O_4S$
Monoisotopic protonated ion (*m/z*): 384.0824
Molecular weight: 383.37
CAS number: 102625-70-7
Common salt form: Magnesium, Sodium salts

256: Rabeprazole

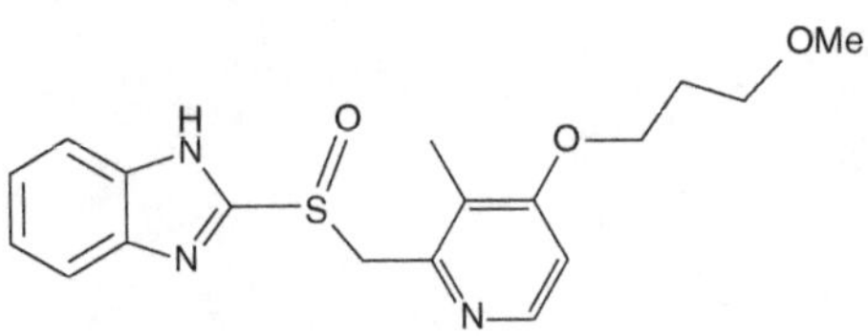

Chemical name: (*R,S*)-2-([4-(3-Methoxypropoxy)-3-methylpyridin-2-yl]methylsulfinyl)-1
H-benzo[d]imidazole
Trade name: Aciphex

Formula: $C_{18}H_{21}N_3O_3S$
Monoisotopic protonated ion (*m/z*): 360.1376
Molecular weight: 359.44
CAS number: 117976-89-3
Common salt form: Sodium salt

257: Ilaprazole

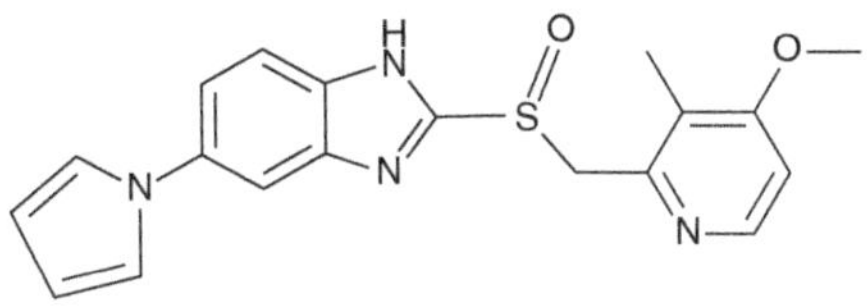

Chemicalname: 2-[(*R,S*)-[(4-methoxy-3-methylpyridin-2-yl)methyl]sulfinyl]-5-(1*H*-pyrrol-1-yl)-1*H*-benzimidazole
Trade name: Noltec
Formula: $C_{19}H_{18}N_4O_2S$
Monoisotopic protonated ion (*m/z*): 367.1223
Molecular weight: 366.44
CAS number: 172152-36-2

Omeprazole is a medication that is used for treating gastroesophageal reflux disease (GERD), peptic ulcer disease, and Zollinger–Ellison syndrome [1]. It is a specific, covalent irreversible inhibitor of H^+/K^+-ATPase, the enzyme controlling the secretion of gastric acid; hence, this class of drugs is also known as the proton pump inhibitors (PPIs). Omeprazole is the first clinically approved PPI [2], and it is a racemic mixture due to the chirality of the sulfoxide moiety. Chemically, it is a benzoimidazole-substituted sulfinylmethyl pyridine derivative. While it is reasonably stable in neutral and alkaline pH, under acidic conditions (particularly at pH 5 or below), omeprazole undergoes rather facile intramolecular rearrangement to form the isomeric sulfenic acid derivative (**2**), which is followed by dehydration to produce the cyclized sulfenamide (**3**) [3–6]. Either the sulfenic acid or the sulfenamide was thought to be the active form of the drug [5]. This transformation process is also its degradation pathway under acidic conditions, under which a dimeric degradant (**4**) would also be formed (Scheme 252–257-1) [5, 6] in addition to numerous other degradants, many of which may be of transient nature [5]. Because of its acidic lability, omeprazole is typically formulated with an enteric coating to prevent its degradation when passing through the stomach.

Roberts et al. studied acid-induced degradation of omeprazole in 18 mM HCl solution at 10°C for up to 48 hours [7]. It was found that the vast majority of omeprazole was decomposed, and ~10 degradants were identified, among which the major one was the sulfhydryl degradant (**6**, Scheme 252–257-2), which was also observed in the original study of Brandstrom et al. [5]. Another monomeric degradant identified was the des-sulfur degradant and its structure was proposed as **7** by Roberts et al. Nevertheless, a more reasonable structure for the des-sulfur degradant might be **7′**, as it can be more rationally accounted for by an elimination mechanism from dimer **4**. Furthermore, **7′** was also identified as an acidic degradant of

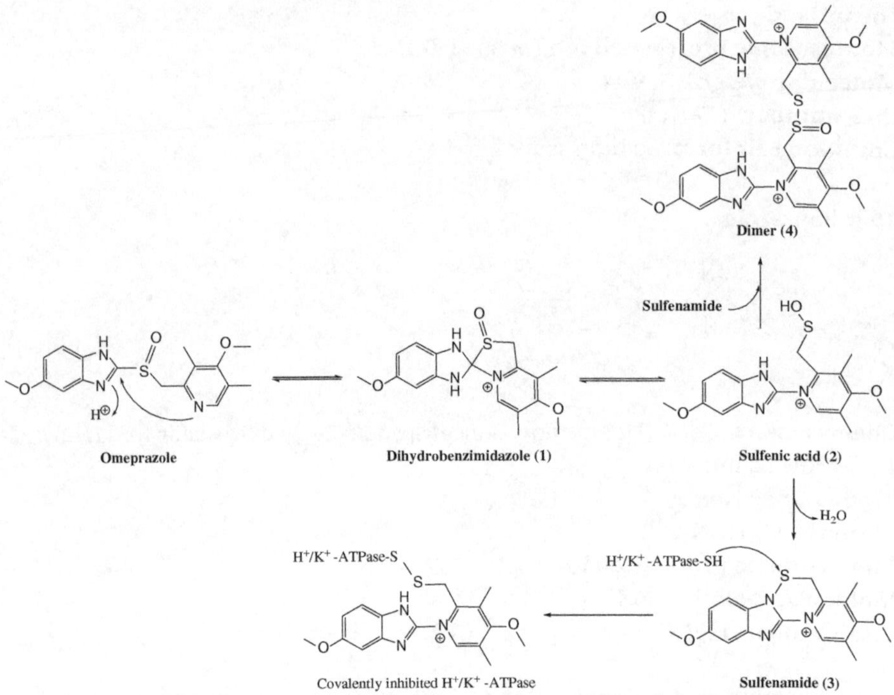

Scheme 252-257-1 Acidic degradation pathway of omeprazole per the studies of Brandstrom et al. [5] and Qaisi et al. [**6**] and its mode of action [4, 5]. *Source:* Brandstrom et al. [5] and Qaisi et al. [6].

omeprazole by Shankar et al. [8]. All other identified minor degradants were dimeric degradants, which were linked together by either mono-, di- (such as **8** and **8′**), or tri-sulfide bridges. Brandstrom et al. also reported many disulfide-linked dimeric degradants [5].

In the forced degradation study by Shankar et al. [8], omeprazole was also found to be very susceptible to acidic degradation; in 10 mM HCl at room temperature for 30 minutes, the majority of omeprazole decomposed to form three major degradants (**7′**,**9–10**, Scheme 252–257-3). The structural elucidation of these degradants was based on LC–MS and LC–MS/MS analysis. In their study, Shankar et al. proposed the structure of degradant **10**; nevertheless, due to the pseudo-symmetry of the molecule, the structure of **10′** would also be possible, as analogous degradants were present in sets of two isomers [5]. It would also be possible that both **10** and **10′** might be present, but could not be separated by the LC–MS method used in their study [8]. In studying degradation behaviors of omeprazole and lansoprazole in aquatic environments, the sulfide degradant (**9**) was observed as the major degradant for both drugs in aqueous solutions under either dark or simulated sunlight [9]; its formation mechanism remained unclear despite a few proposed ones.

The hydrolytic stress in neutral and alkaline solutions was carried out under quite harsh conditions, during which numerous degradants were generated [8].

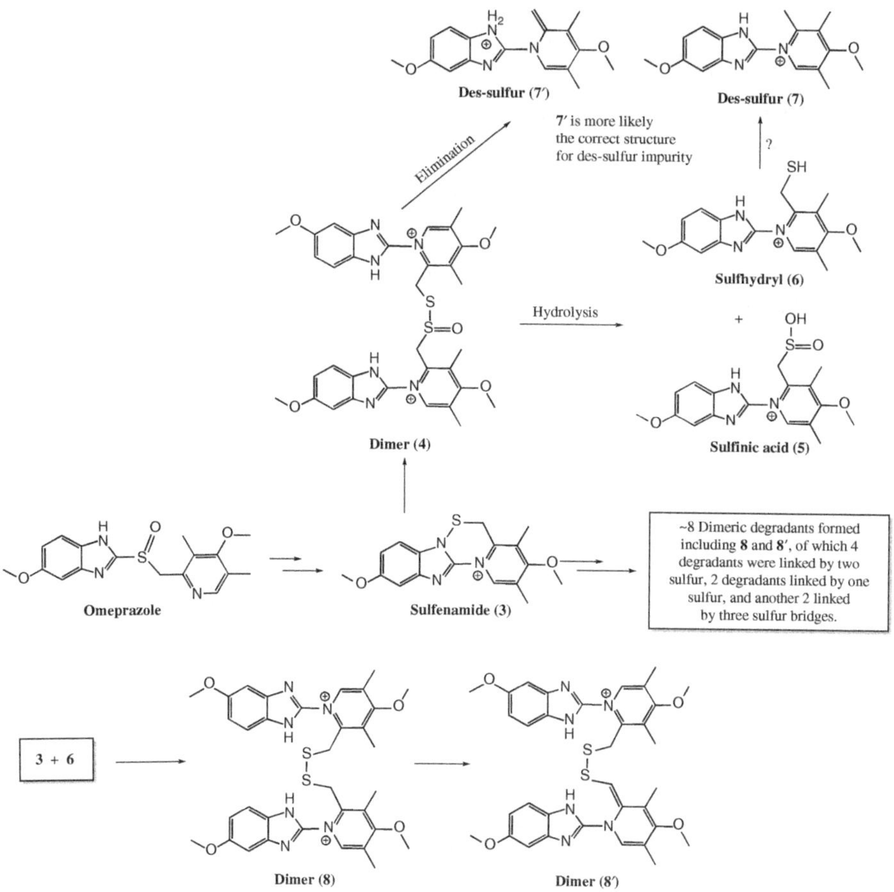

Scheme 252–257-2 Acidic degradation of omeprazole in 18 mM HCl solution at 10°C for up to 48 hours [7]. The formation mechanism of **6** is compiled by the present author, which is also partly based on the original study of Brandstrom et al. *Source:* Brandstrom et al. [5] and Roberts et al. [7].

Nevertheless, these degradation profiles would not appear to be relevant to the real degradation scenario of omeprazole. In oxidative stress with 3% hydrogen peroxide at room temperature for 24 hours, the predominant degradant was omeprazole sulfone (**11**).

In solid state formulations, the degradation pathways of omeprazole appear to be different from those in the solution state as described above. For example, in an omeprazole sample subjected to accelerated stability condition, two sets of isomers (H238/85 and degradants B/C) were observed as the major degradants and identified by the original developer of the drug (Scheme 252–257-4) [10]. The isomeric degradants H238/85 were also seen as the major degradants in enteric-coated pellets that were stored under the accelerated stability condition of 40°C/75%RH for 120 days [4]. Yet in another stability study of a commercially available omeprazole formulated product, Rajab et al. reported the observation of a number of

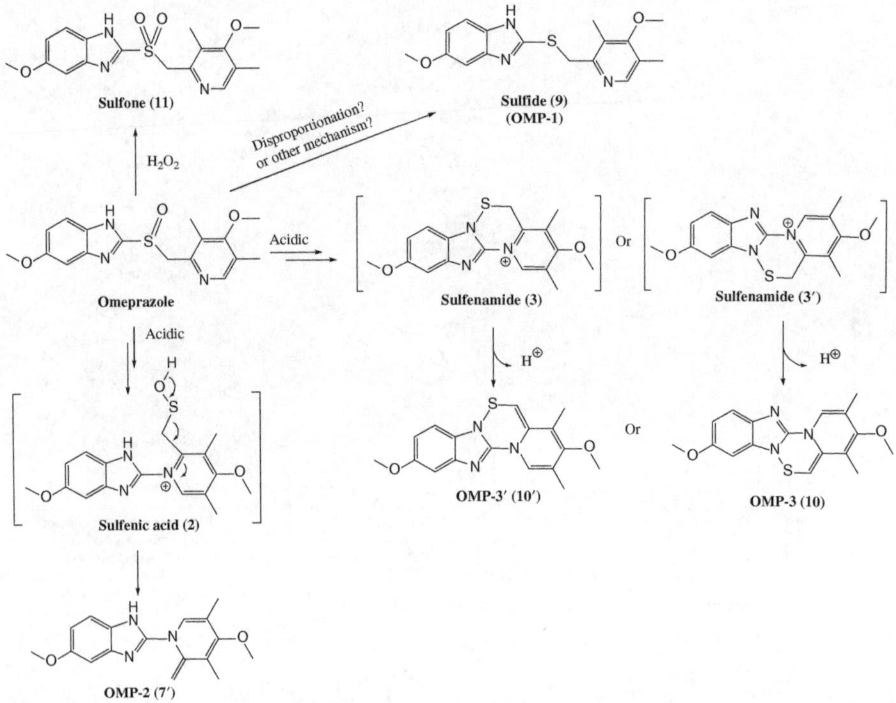

Scheme 252–257-3 Forced degradation of omeprazole performed by Shankar et al. under acidic (10 mM HCl at room temperature for 30 minutes) and oxidative conditions (3% hydrogen peroxide at room temperature for 24 hours), respectively [8]. The nomenclature of the degradants, OMP-1, OMP-2, and OMP-3, follows that of Shankar et al. The mechanisms of the formation of these three major acidic degradants (OMP-1, OMP-2, and OMP-3) are different from those proposed by Shankar et al. The intermediates **2** and **3** were previously reported by Qaisi et al. [6]; by analogy, sulfenamide 3′ might also be formed, and its deprotonation should give OMP-3. *Source:* Qaisi et al. [6] / with permission of Elsevier.

degradants, of which at least two major degradants had not been reported before [11]. These two major degradants, HDMP and MBT (Scheme 252–257-4), resulted from the cleavage of the omeprazole molecule, which was proposed to proceed through a thiooxirane and a pseudo-carbenoid species (not shown). According to the present author, the formation of the thiooxirane intermediate might be better explained by a 2,3-sigmatropic rearrangement mechanism, which is well-known for the transformation of allylic sulfoxides to allylic sulfenates [12]. From the thiooxirane intermediate, the oxygen-containing analogous degradant (MBO) could also be formed; as a matter of fact, MBO was observed from the degradation of lansoprazole in the solid state [10] as well as in an aquatic environment [9].

Overall, it appears that the degradation pathways of omeprazole are very complex. Even under similar acidic conditions [5–8], different temperatures and durations of the degradation could produce vastly different degradation profiles. Furthermore, the degradation profile in the solid state differed from that in the solution state.

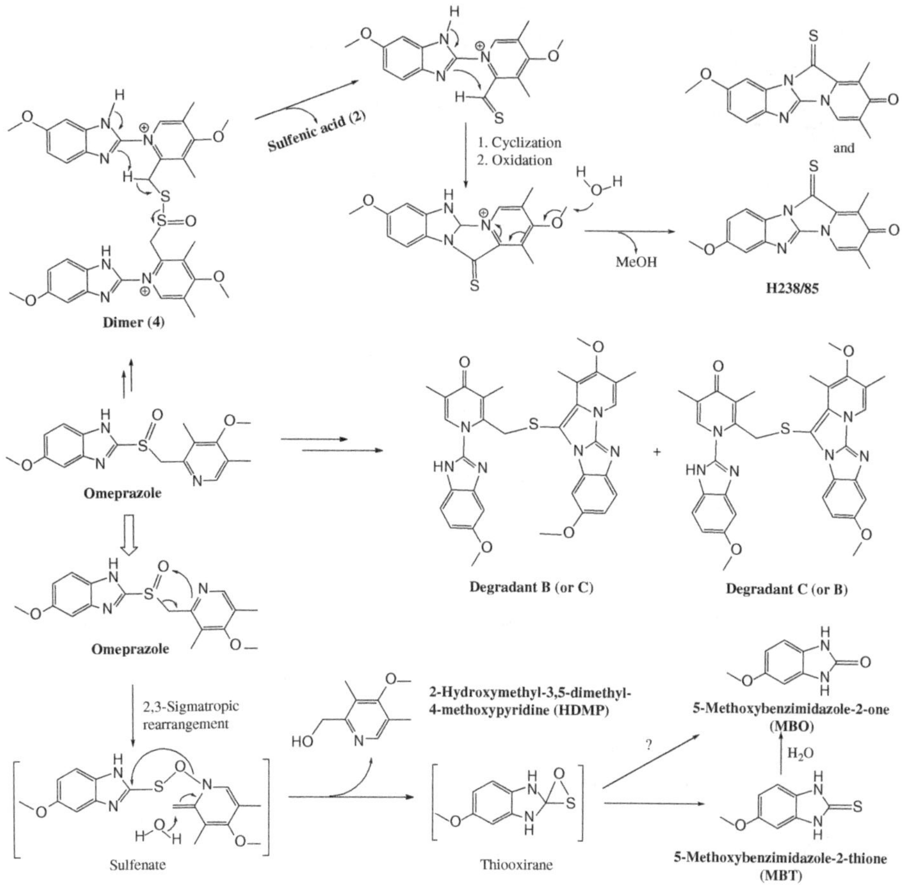

Scheme 252–257-4 Degradation pathways of omeprazole in solid state [4,10,11]. The early stage of the mechanism for the formation of HDMP and MBT is different from that of Rajab et al. [11], who proposed the involvement of a thiooxirane and a pseudo-carbenoid species (the latter is not shown).

Esomeprazole, approved for medical use in 2000, is the *S*-isomer of omeprazole, which is the chiral switch of the original racemic drug, omeprazole, to improve its pharmacokinetic properties [13]. Its degradation chemistry would be expected to be identical to that of omeprazole.

Lansoprazole is another PPI, which was developed by Takeda and approved in 1991 [14]. Its degradation chemistry is highly similar to that of omeprazole (Scheme 252–257-5) [9, 11, 15–17]. One acidic degradant reported for lansoprazole was designated as DP-5, and its structure was proposed by Shankar et al. [17]. It appears that the proposed structure of **DP-5** would be difficult to rationalize from a mechanistic perspective. On the other hand, the formation of **DP-5′**, an isomer of **DP-5**, could be easily explained through the hydrolysis of the trifluoroethoxy side chain of lansoprazole sulfide degradant (**DP-2**). For the family of the PPI drug molecules, hydrolysis of this alkoxy side chain (either a methoxy or trifluoroethoxy) is a

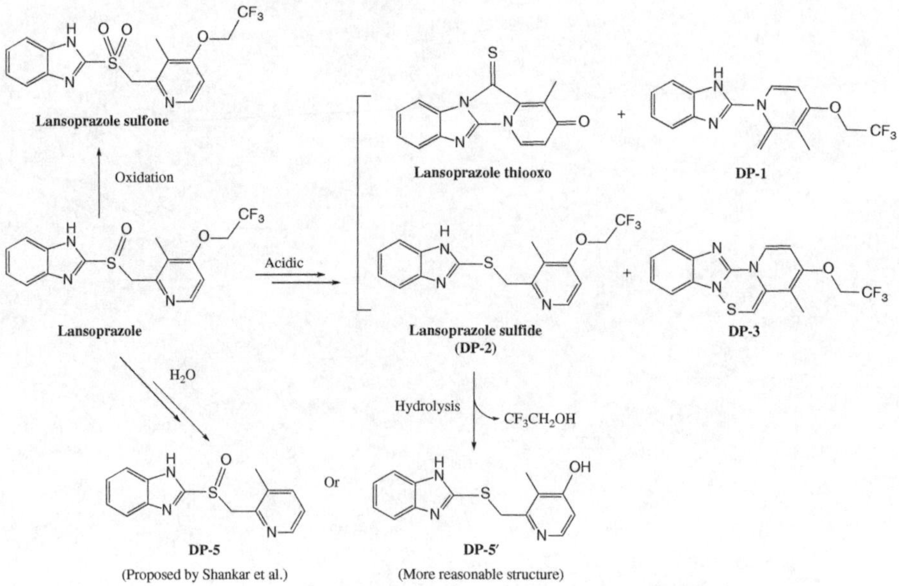

Scheme 252–257-5 Degradation pathways of lansoprazole under acidic, neutral, and oxidative conditions. The nomenclature of DP-1, DP-2, DP-3, and DP-5 follows that of Shankar et al. [17]. *Source:* Shankar et al. [17] / John Wiley & Sons.

well-known degradation pathway [4, 5, 10, 17]. Hence, the real degradant could likely be **DP-5′**, rather than **DP-5**.

The other members of the PPI family include pantoprazole, rabeprazole, and ilaprazole, which were discovered and/or developed respectively by Byk Gulden Lomberg Chemische Fabrik/Smith Kline & French [18], Eisai [19], and Il-Yang Pharmaceutical (a Korean pharmaceutical company) [20]. The degradation chemistry of these PPIs would be expected to be similar to that of omeprazole and lansoprazole, particularly under acidic conditions [5–9, 11, 15–17]. For example, Bhandi et al. reported that rabeprazole decomposed to yield hydrolytic and oxidative degradants in neutral, acidic, and hydrogen peroxide solutions, respectively (Scheme 252–257-6) [21]. In an accelerated study of a formulated rabeprazole product, Dev et al. identified three novel degradants using LC–MS and LC-NMR (Scheme 252–257-7) [22]. Not only were these degradants not reported for rabeprazole and its formulated products before, but also the types of these degradants were not reported for other PPI drug products. Although no specific formation mechanisms were provided by the original researchers, the formation of Impurity II and III could be rationalized through Baeyer-Villiger oxidation of the sulfone degradant, which seems to be present in the accelerated stability study. Nevertheless, a drawback for this mechanism is that no Baeyer-Villiger oxidation of a sulfone has been reported. On the other hand, Baeyer-Villiger oxidation of a particular sulfoxide has been reported [23].

In a photo-stress degradation of rabeprazole in methanol solution, it was found to degrade into the hydroxymethylpyridine degradant, benzimidazolone, and benzimidazole degradants. The formation of these degradants might go through a

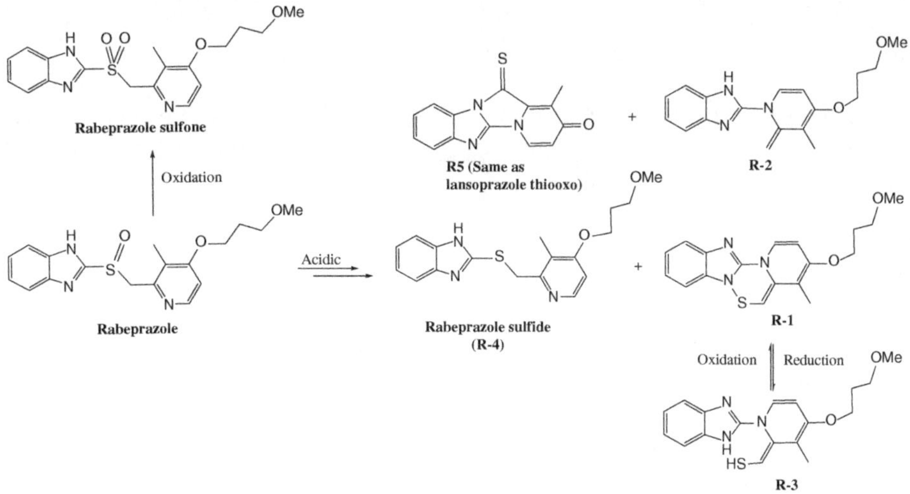

Scheme 252–257-6 Degradation pathways of rabeprazole under acidic, neutral, and oxidative conditions. The nomenclature of R-1, R-2, R-3, R-4, and R-5 follows that of Bhandi et al. [21]. *Source:* Bhandi et al. [21] / Royal Society of Chemistry.

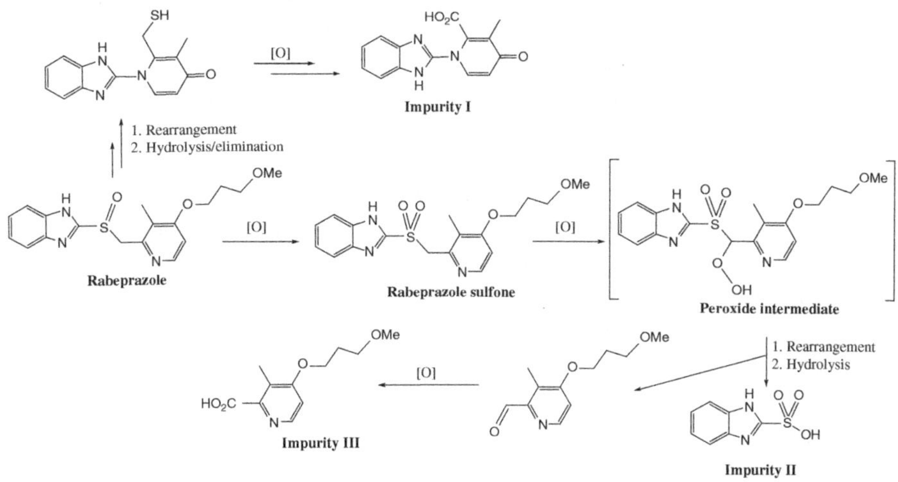

Scheme 252–257-7 Degradation pathway of rabeprazole in a formulated product.

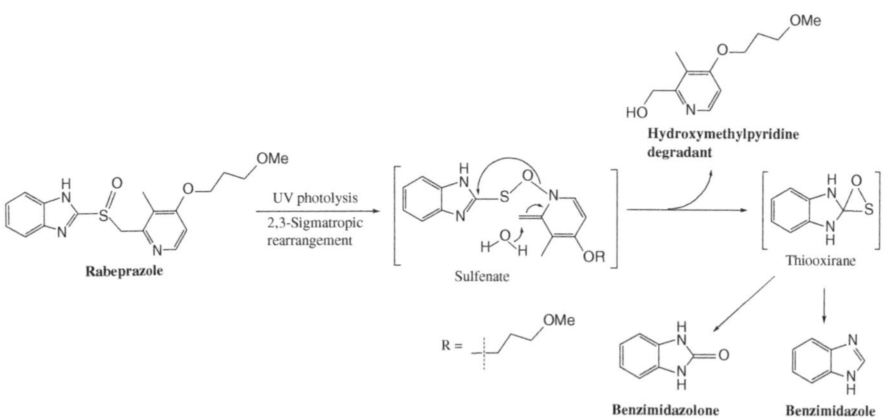

Scheme 252–257-8 Photochemical degradation of rabeprazole in methanol solution.

thiooxirane intermediate (Scheme 252–257-8), and the latter type of intermediate was proposed in the nonphotochemical degradation of omeprazole in the solid state [4, 10, 11].

References

1 Omeprazole. https://en.wikipedia.org/wiki/Omeprazole (accessed March 3 2024).
2 Strand, D.S., Kim, D., and Peura, D.A. (2017). 25 years of proton pump inhibitors: a comprehensive review. *Gut Liver* 11: 27–37.
3 Bruton, L.L. (1996). *Goodman and Gilmans: The Pharmacological Basis of Therapeutics*, 901. New York: McGraw-Hill.
4 Storpirtis, S. and Rodrigues, D. (1998). In vitro evaluation of dissolution properties and degradation products of omeprazole in enteric-coated pellets. *Drug Dev. Ind. Pharm.* 24: 1101–1107.
5 Brandstrom, A., Lindberg, P., Bergman, N.-A. et al. (1989). Chemical reactions of omeprazole and omeprazole analogues. I. A survey of the chemical transformations of omeprazole and its analogues. *Acta Chem. Scand. B* 43: 536–548.
6 Qaisi, A.M., Tutunji, M.F., and Tutunji, L.F. (2006). Acid decomposition of omeprazole in the absence of thiol: a differential pulse polarographic study at the static mercury drop electrode (SMDE). *J. Pharm. Sci.* 95: 384–391.
7 Roberts, J., McNaughtan, M.L., MacLachlan, J. et al. (2018). Identification of the acid-induced degradation products of omeprazole and 5-hydroxyomeprazole by high-resolution mass spectrometry. *Rapid Commun. Mass Spectrom.* 32: 929–941.
8 Shankar, G., Borkar, R.M., Udutha, S. et al. (2019). Identification and structural characterization of stress degradation products of omeprazole using Q-TOF-LC-ESI-MS/MS and NMR experiments: evaluation of toxicity of degradation products. *New J. Chem.* 43: 7294–7306.
9 DellaGreca, M., Iesce, M.R., Previtera, L. et al. (2006). Degradation of lansoprazole and omeprazole in the aquatic environment. *Chemosphere* 63: 1087–1093.
10 Jansson, S.-O. and Vessman, J. (1991). Trends in product analysis for research in the pharmaceutical industry. *Mikrochim. Acta* II: 103–112.
11 Rajab, A., Touma, M., Rudler, H. et al. (2013). Slow, spontaneous degradation of lansoprazole, omeprazole and pantoprazole tablets: isolation and structural characterization of the toxic antioxidants 3H-benzimidazole-2-thiones. *Pharmazie* 68: 749–754.
12 Colomer, I., Velado, M., Fernández de la Pradilla, R., and Viso, A. (2017). From allylic sulfoxides to allylic sulfenates: fifty years of a never-ending [2,3]-sigmatropic rearrangement. *Chem. Rev.* 117: 14201–14243.
13 Kalaitzakis, E. and Björnsson, E. (2007). A review of esomeprazole in the treatment of gastroesophageal reflux disease (GERD). *Ther. Clin. Risk Manag.* 3: 653–663.
14 Lansoprazole. https://en.wikipedia.org/wiki/Lansoprazole (accessed March 9 2024).
15 Ramulu, K., Rao, B.M., and Rao, N.S. (2013). Identification, isolation and characterization of potential degradation product in lansoprazole drug substance. *Raysayan J. Chem.* 6: 274–283.

16 Battu, S. and Pottabathini, V. (2015). Hydrolytic degradation study of lansoprazole, identification, isolation and characterisation of base degradation product. *Am. J. Anal. Chem.* 6: 145–155.

17 Shankar, G., Borkar, R.M., Udutha, S. et al. (2017). Forced degradation studies of lansoprazole using LC-ESI HRMS and ^{1}HNMR experiments: in vitro toxicity evaluation of major degradation products. *J. Mass Spectrom.* 52: 459–471.

18 Pantoprazole. https://en.wikipedia.org/wiki/Pantoprazole (accessed March 9 2024).

19 Rabeprazole. https://en.wikipedia.org/wiki/Rabeprazole (accessed March 9 2024).

20 Ilaprazole. https://en.wikipedia.org/wiki/Ilaprazole (accessed March 9 2024).

21 Bhandi, M.M., Borkar, R.M., Shankar, G. et al. (2016). Identification and characterization of stressed degradation products of rabeprazole using LC-ESI/MS/MS and H-NMR experiments: in vitro toxicity evaluation of major degradation products. *RSC Adv.* 6: 10719–10735.

22 Dev, R.V., Kiran, G.S.U., Subbaiah, B.V. et al. (2009). Identification of degradation products in stressed tablets of Rabeprazole sodium by HPLC-hyphenated techniques. *Magn. Reson. Chem.* 47: 443–448.

23 Ma, X., Liu, Y., Du, L. et al. (2020). Post-functionalization of dibenzothiophene to functionalized biphenyls via a photoinduced thia-Baeyer-Villiger oxidation. *Nat. Commun.* 11: 914. https://doi.org/10.1038/s41467-020-14522-7.

258: Ondansetron [WHO List of Essential Medicines]

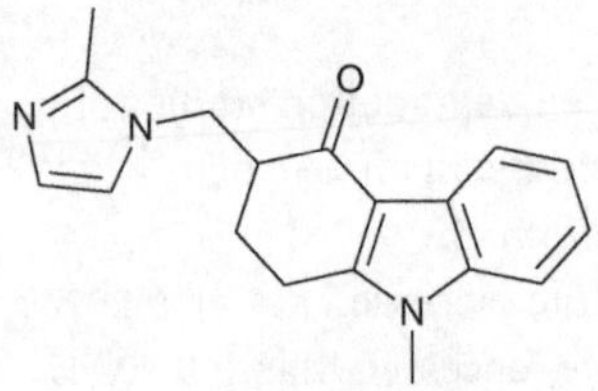

Chemical name: (*R,S*)-9-Methyl-3-[(2-methyl-1*H*-imidazol-1-yl)methyl]-2,3-dihydro-1*H*-carbazol-4(9*H*)-one
Trade name: Zofran
Formula: $C_{18}H_{19}N_3O$
Monoisotopic protonated ion (*m/z*): 294.1601
Molecular weight: 293.37
CAS number: 99614-02-5
Common salt form: HCl salt

Ondansetron, a highly selective serotonin 5-HT_3 receptor antagonist, is clinically used for the prevention of nausea and vomiting induced by chemotherapy, radiation, surgery, and other causes [1]. Chemically, it is a fused indole derivative with a conjugated keto moiety and an imidazole-*N*-methyl substituent alpha to the keto. The alpha-substituted keto part of the drug molecule would be susceptible to base-catalyzed retro-Michael addition, producing Impurity D (USP) and 2-methylimidazole (Scheme 258-1) [2]. The forced degradation was conducted by refluxing ondansetron in 1 M NaOH solution (water/methanol, 1/1, v/v) for 30 hours. Under this rather harsh condition, the majority of the drug substance was degraded and Impurity D was formed as the largest degradant (33.9% yield), in addition to the other four degradants which were not identified in the study. On the other hand, Talluri et al. also performed a forced degradation of ondansetron in 1 M NaOH solution, in which they identified another degradant (DP-1) [3], which apparently resulted from the Michael addition of Impurity D by hydroxide ion.

The same researchers also conducted photochemical stress of ondansetron in solution and identified three photodegradants (Scheme 258-2), DP-3, DP-4, and DP-5. DP-3 and DP-5 were proposed as degradants with oxidation occurring on the imidazole ring by singlet oxygen, while DP-4 was proposed as the enol form of ondansetron due to tautomerization.

In a stability study of ondansetron tablets, a novel degradant was observed, and the subsequent investigation with API-excipient compatibility study revealed that this impurity was caused by the reaction of the API with chloroacetic acid, an impurity in excipient sodium starch glycolate (Scheme 258-3) [4].

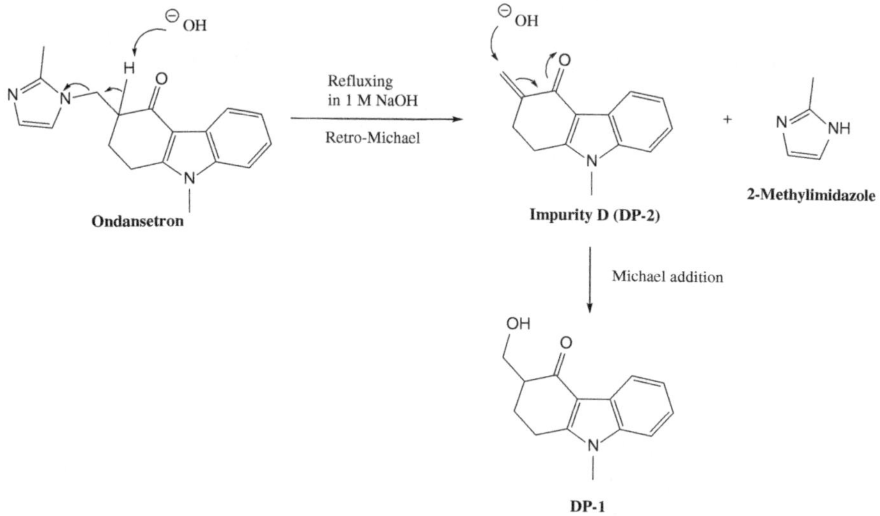

Scheme 258-1 Base-catalyzed degradation of ondansetron via retro-Michael addition.

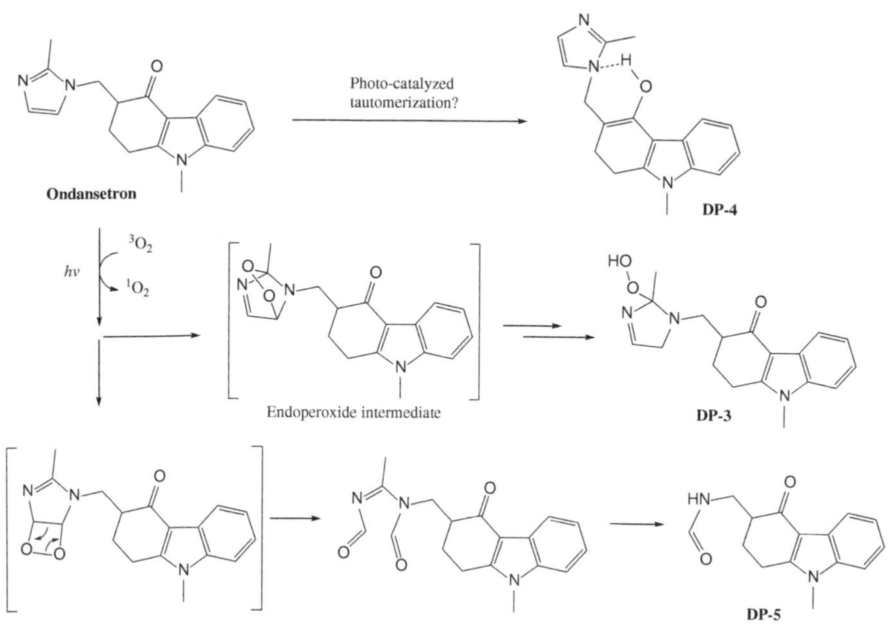

Scheme 258-2 Photochemical degradation of ondansetron in solution. The nomenclature of the degradants follows that of Talluri et al. [3]. *Source:* Talluri et al. [3] /John Wiley & Sons.

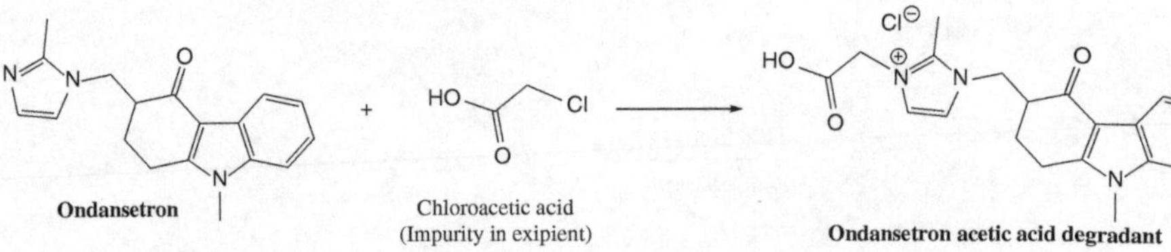

Scheme 258-3 Degradation of ondansetron tablets that contained sodium starch glycolate as one of the excipients.

References

1 Ondansetron. https://en.wikipedia.org/wiki/Ondansetron (accessed March 9 2024).

2 Kristiana, E., Saefumillah, A., and Budianto, E. (2020). Degradation of ondansetron: isolation and characterization impurity D ondansetron as A candidate reference standard impurity in drug. *AIP Conf. Proc.* 2242: 040054. https://doi.org/10.1063/5.0010657.

3 Talluri, M.V.N.K., Keshari, K.K., Kalariya, P.D., and Srinivas, R. (2015). Selective separation and characterization of the stress degradation products of ondansetron hydrochloride by liquid chromatography with quadrupole time-of-flight mass spectrometry. *J. Sep. Sci.* 38: 1625–1632.

4 Gandhi, H.M., Gollapalli, N.R., and Lilakar, J.K.D. (2018). Identification, isolation and characterization of unknown impurity in ondansetron drug product. *Int. J. Adv. Res. Ideas Innov. Tech.* 4: 1264–1271.

259–265: Opioids

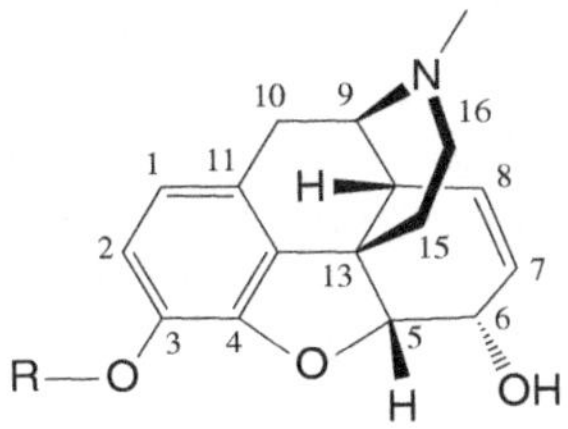

Morphine, R = H;
Codeine, R = Me.

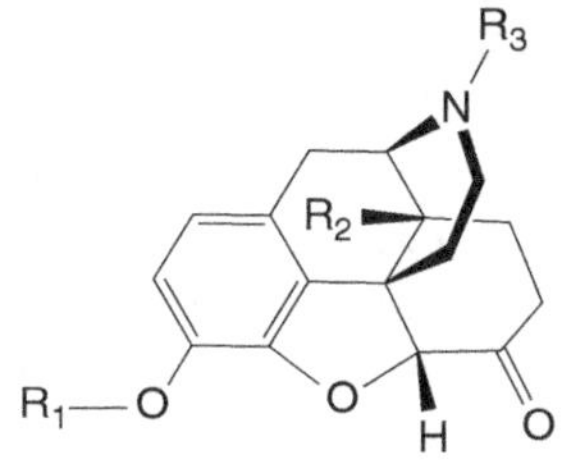

Oxymorphone, R_1 = H, R_2 = OH, R_3 = Me;
Oxycodone, R_1 = Me, R_2 = OH, R_3 = Me;
Hydromorphone, R_1 = H, R_2 = H, R_3 = Me;
Hydrocodone, R_1 = Me, R_2 = H, R_3 = Me;
Naloxone, R_1 = H, R_2 = OH, R_3 = Propen-3-yl.

Opium has been used as a pain reliever for millennia, long before morphine, its main ingredient, was isolated in the early 1800s. Today, morphine is utilized not only as an active pharmaceutical ingredient but also as a starting material for a number of its analogous drugs. Among the opioids that are covered in this monograph, they may be divided into different structural categories, e.g., those with 3-hydroxyl group versus those with 3-methoxyl group. The ones with 3-hydroxyl group would be susceptible to oxidative coupling to yield the corresponding dimeric degradants, e.g., the so-called pseudomorphine, while the ones with 3-methoxyl group would not undergo such oxidative degradation due to the blockage of the phenolic hydroxyl group, which prevents the formation of the phenolic hydroxyl free radical that eventually leads to the dimerization. On the other hand, those opioids with a 6-keto moiety could undergo a different type of dimerization process – aldol condensation, as in the case of oxycodone. Aside from their difference, all the opioids contain a tertiary amine functional group, which could be oxidized to form the corresponding *N*-oxide degradants, particularly under neutral to alkaline conditions.

259: Morphine [WHO List of Essential Medicines]

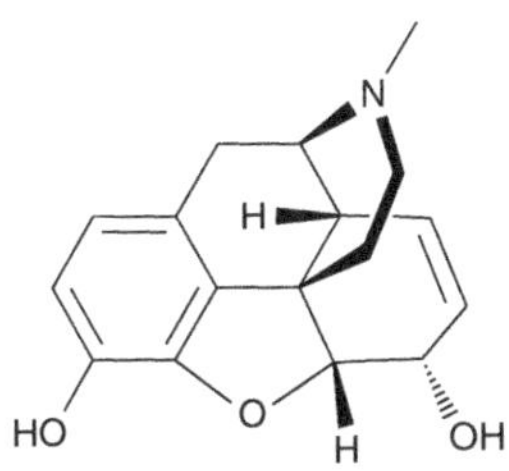

Chemical name: (*4R,4aR,7S,7aR,12bS*)-3-Methyl-2,3,4,4a,7,7a-hexahydro-1*H*-4,12-methano[1]benzofuro[3,2-*e*]isoquinoline-7,9-diol
Formula: $C_{17}H_{19}NO_3$
Monoisotopic protonated ion (m/z): 286.1438

Molecular weight: 285.34
CAS number: 57-27-2
Common Salt form: Sulfate, HCl salt

Morphine was isolated in 1804 from the opium poppy plant by Friedrich Sertürner, a German pharmacist who named the substance *morphium* after the Greek god of dreams, Morpheus, due to its effect of causing sleepiness [1]. On the other hand, opium has been known for approximately four millennia as a powerful pain reliever [2]. Today, morphine remains one of the most utilized analgesics for treating acute and chronic severe pain [3]. Several functional groups of morphine, such as the phenolic hydroxyl, the hydroxyl alpha to the vinyl moiety, and the tertiary amino group, are susceptible to oxidative degradation. More specifically, the oxidation of the phenolic functionality leads to the formation of pseudomorphine, a dimeric degradant via oxidative coupling, as the major degradant, and morphine-*N*-oxide (also known as genomorphine) as the minor degradant (Scheme 259–265-1) [4, 5]. The study by Vermeire and Remon also showed that the presence of oxygen and higher pH accelerated the degradation of morphine in solution [5].

Miolo et al. studied the photodegradation of morphine in solution and solid state [6], and found that while it was stable toward UVA irradiation, morphine decomposed significantly under UVB irradiation in solid state and solutions. In aqueous solution, degradation due to hydrolysis and photooxidation appeared to be the major photodegradation pathways, in addition to photo-reductive products. In methanol solution, three major degradants were observed, with each of them incorporating two oxygen atoms. Tentative structures for the degradants formed in both solutions were proposed solely based on accurate mass spectrometric analysis.

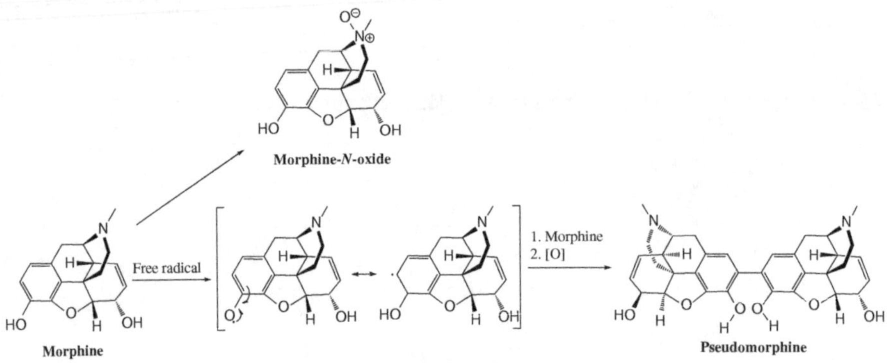

Scheme 259–265-1 Oxidative degradation of morphine.

260: Codeine [WHO List of Essential Medicines]

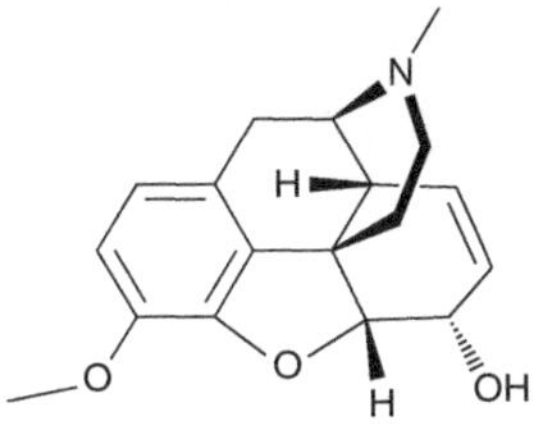

Chemical name: (5*α*,6*α*)-7,8-Didehydro-4,5-epoxy-3-methoxy-17-methylmorphinan-6-ol
Formula: $C_{18}H_{21}NO_3$
Monoisotopic protonated ion (*m/z*): 300.1594
Molecular weight: 299.37
CAS number: 76-57-3
Common Salt form: Sulfate, HBr, HCl salt

Codeine is a naturally occurring methylated morphine, which was discovered by Pierre Jean Robiquet, a French pharmacist and chemist who also had many other pioneering findings including the isolation and characterization of asparagine, the first amino acid to be identified [7, 8]. Codeine accounts for ~2% of opium and is the most commonly used opiate. Pharmacologically, it is a prodrug of morphine and it converts to morphine via in vivo demethylation. Due to the blocking of the 3-hydroxyl group by methylation, codeine would not be expected to undergo free-radical-mediated oxidation to form the corresponding dimeric degradant, dimethylpseudomorphine. Hence, its main oxidative degradant would be codeine-*N*-oxide (Scheme 259–265-2) [9].

Powell studied the degradation kinetics and pH-stability profile of codeine and found that it had a wide pH range of maximum stability between pH ~1 and 9 [10]. The study also confirmed the previous observations that phosphate buffer destabilized codeine: it decomposed ~20 times faster in a 50 mM, pH 7 phosphate buffer than in an unbuffered solution.

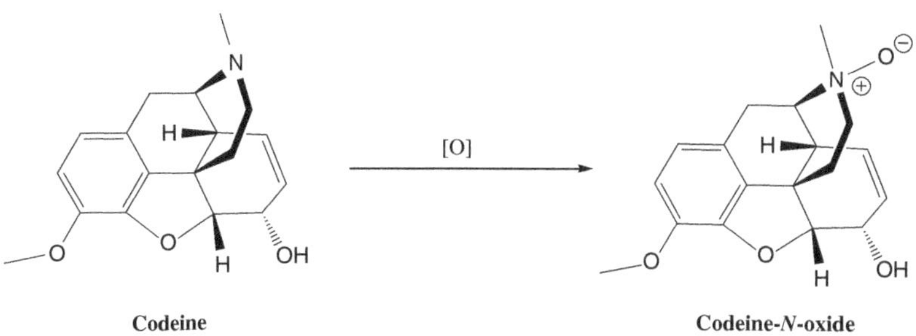

Scheme 259–265-2 Oxidative degradation of codeine.

261: Hydromorphone [WHO List of Essential Medicines]

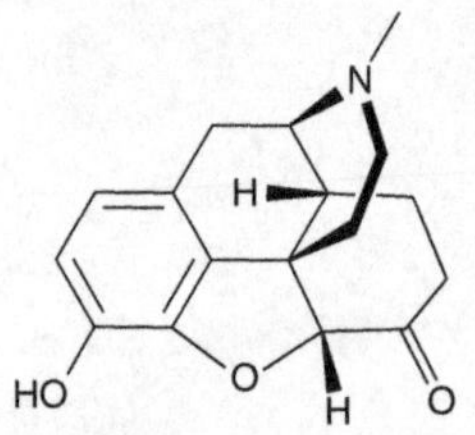

Chemical name: 4,5-α-Epoxy-3-hydroxy-17-methylmorphinan-6-one
Trade name: Dilaudid, Palladone, others
Formula: $C_{17}H_{19}NO_3$
Monoisotopic protonated ion (*m/z*): 286.1438
Molecular weight: 285.34
CAS number: 466-99-9
Common Salt form: HCl salt

Hydromorphone is an isomer of morphine, with the 6-hydroxy group oxidized to 6-keto and the 7,8-double bond reduced to a 1,2-substituted ethane moiety [11]. It can be made from morphine either by an isomerization or by a reductive/oxidative process. Because its two oxidizable functional groups, i.e., the 3-hydroxy and tertiary amino groups, are identical to those of morphine, its degradation chemistry would be expected to be the same or very similar to that of morphine (refer to Scheme 259–265-1) [12].

262: Oxymorphone

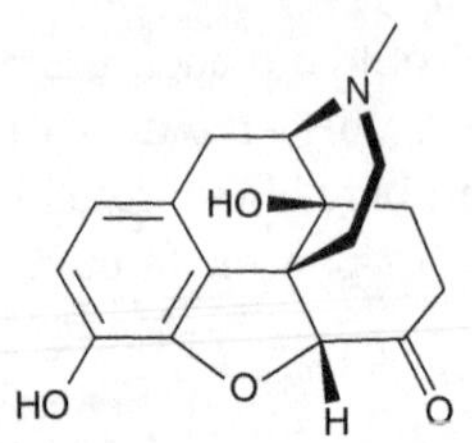

Chemical name: 4,5α-Epoxy-3,14-dihydroxy-17-methylmorphinan-6-one
Trade name: Numorphan, Numorphone, Opana, others
Formula: $C_{17}H_{19}NO_4$
Monoisotopic protonated ion (m/z): 302.1387
Molecular weight: 301.34
CAS number: 76-41-5
Common Salt form: HCl salt

Oxymorphone was approved for medical use in 1959, and some of its extended-release forms were removed from the market at the request of the FDA due to illicit use of the dosage forms [13]. Chemically, it is a hydroxylated derivative of hydromorphone, and thus its degradation chemistry would be expected to be the same or very similar to that of hydromorphone and morphine (refer to Scheme 259–265-1).

263: Hydrocodone

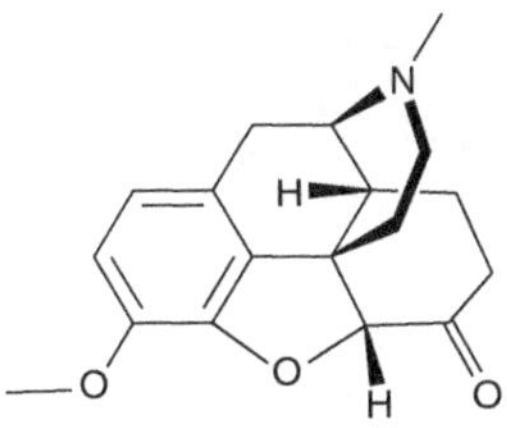

Chemical name: 4,5α-Epoxy-3-methoxy-17-methylmorphinan-6-one
Formula: $C_{18}H_{21}NO_3$
Monoisotopic protonated ion (*m/z*): 300.1594
Molecular weight: 299.37
CAS number: 125-29-1
Common Salt form: Bitartrate, HCl salt

Hydrocodone, also known as dihydrocodone, is a semi-synthetic opioid drug usually made from codeine. In addition to treating pain, it is also used as a cough suppressant, typically in combination with other medicines for similar indications [14]. Due to its structural similarity to codeine, its degradation chemistry would be expected to be similar to that of codeine (refer to Scheme 259–265-2).

264: Oxycodone [WHO List of Essential Medicines]

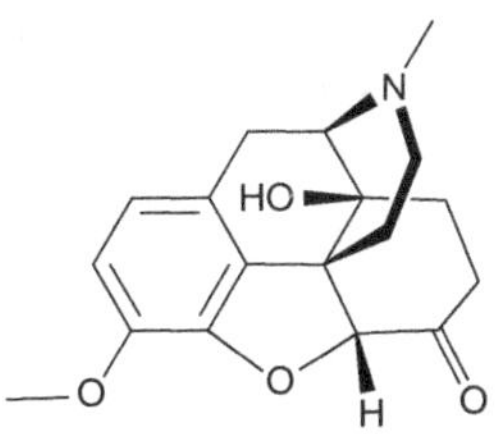

Chemical name: (*5R,9R,13S,14S*)-4,5α-Epoxy-14-hydroxy-3-methoxy-17-methylmorphinan-6-one
Trade name: Roxicodone, OxyContin
Formula: $C_{18}H_{21}NO_4$
Monoisotopic protonated ion (*m/z*): 316.1543
Molecular weight: 315.37
CAS number: 76-42-6
Common Salt form: HCl salt

Oxycodone is a semi-synthetic opioid drug, which is a methylated derivative of hydromorphone and a hydroxylated derivative of hydrocodone. Due to the blockage of the phenolic hydroxyl at the 3-position by the methyl group, oxidative dimerization would not be expected for oxycodone, which is the same scenario for hydrocodone. On the other hand, it can still undergo oxidation on its tertiary amino group

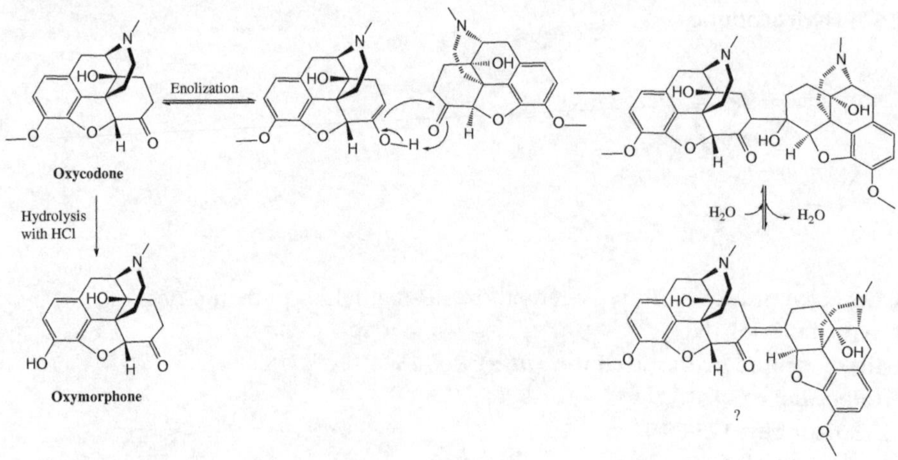

Scheme 259–265-3 Degradation pathways of oxycodone.

to produce the corresponding *N*-oxide [15]. Furthermore, Sitasuwan et al. reported that the 3-methoxyl moieties in opioids, such as oxycodone, hydrocodone, and codeine, could be hydrolyzed to produce the corresponding 3-hydroxyl derivatives by hydrochloric acid (Pathway a, Scheme 259–265-3) [16]. Nevertheless, such 3-O-demethylation would not proceed easily under nonenzymatic conditions, and thus, it would not be expected to be a meaningful degradation pathway under real-life scenarios. Even though 3-*O*-methylated opioids such as oxycodone would not undergo oxidative dimerization, a dimeric degradant of oxycodone was observed in a long-term stability study of an oxycodone hydrochloride injectable product [17]. The formation of this dimeric degradant was attributed to an aldol condensation reaction between two drug molecules under weakly acidic conditions (Pathway b, Scheme 259–265-3).

265: Naloxone [WHO List of Essential Medicines]

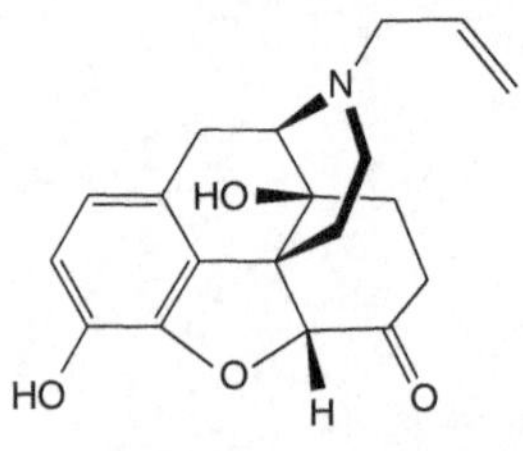

Chemicalname:(*4R,4aS,7aR,12bS*)-4a,9-Dihydroxy-3-(prop-2-en-1-yl)-2,3,4,4a,5,6-hexahydro-1*H*-4,12-methano[1]benzofurano[3,2-e]isoquinolin-7(7a*H*)-one
Trade name: Narcan, Evzio, Nyxoid, others
Formula: $C_{19}H_{21}NO_4$
Monoisotopic protonated ion (*m/z*): 328.1543
Molecular weight: 327.38
CAS number: 465-65-6
Common Salt form: HCl salt

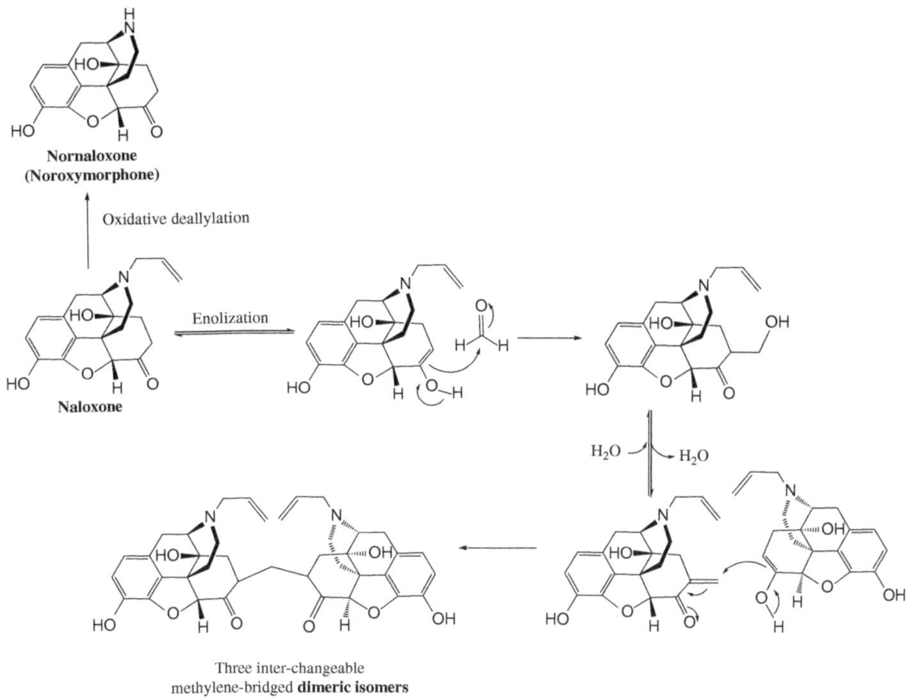

Scheme 259–265-4 Dealkylation and formaldehyde-triggered dimerization of naloxone.

In contrast to the opioids discussed in this monograph, naloxone is an opioid antagonist and as such, it is used clinically to reverse or reduce the pharmacological effects of opioids [18]. It was developed in the 1960s and approved for treating opioid overdose in the United States in 1971. Its structure is identical to oxymorphone, except that the *N*-methyl group of the latter is replaced by an allyl group. Thus, its degradation chemistry would be expected to be similar to that of oxymorphone, as well as hydromorphone and morphine with regard to the phenolic hydroxyl and tertiary amino moiety (refer to Scheme 259–265-1). Furthermore, this replacement seems to create another degradation pathway by which nornaloxone (noroxymorphone) is formed, resulting from the dealkylation of the tertiary amino moiety. Nornaloxone appeared to be the only degradant in certain expired naloxone products from first-responders' supplies [19]. In another study, a methylene-bridged dimeric degradant of naloxone was observed, the formation of which was attributed to formaldehyde present in an experimental oral film formulation of naloxone with buprenorphine (Scheme 259–265-4) [20].

References

1 Morphine. https://en.wikipedia.org/wiki/Morphine (accessed March 3 2024).

2 Norn, S., Kruse, P.R., and Kruse, E. (2005). History of opium poppy and morphine [in Danish]. *Dan Medicinhist Arbog.* 33: 171–184.

3 Raspopin, Y.S., Shifman, E.M., and Feoktistova, D.A. (2024). History and prospects for the use of morphine in clinical practice: literature review [in Russian]. *Reg. Manag. Anesth. Acute Pain Manag.* 18: 33–40. https://doi.org/10.17816/RA624003.

4 Vermeire, A. and Remon, J.P. (1997). The solubility of morphine and the stability of concentrated morphine solutions in glass, polypropylene syringes and PVC containers. *Int. J. Pharm.* 146: 213–223.

5 Vermeire, A. and Remon, J.P. (1999). Stability and compatibility of morphine. *Int. J. Pharm.* 187: 17–51.

6 Miolo, G., Tucci, M., Mazzoli, A. et al. (2016). Photostability of 6-MAM and morphine exposed to controlled UV irradiation in water and methanol solution: HRMS for the characterization of transformation products and comparison with the dry state. *J. Pharm. Biomed. Anal.* 126: 48–59.

7 Codeine. https://en.wikipedia.org/wiki/Codeine (accessed May 4 2024).

8 Pierre Jean Robiquet. https://en.wikipedia.org/wiki/Pierre_Jean_Robiquet (accessed May 4 2024).

9 Stubberud, K.P. and Åström, O. (1998). Separation of ibuprofen, codeine phosphate, their degradation products and impurities by capillary electrophoresis: II. Validation. *J. Chrom. A.* 826: 95–102.

10 Powell, M.F. (1986). Enhanced stability of codeine sulfate: effect of pH, buffer, and temperature on the degradation of codeine in aqueous solution. *J. Pharm. Sci.* 75: 901–903.

11 Hydromorphine. https://en.wikipedia.org/wiki/Hydromorphine (accessed May 5 2024).

12 Hildebrand, K.R., Elsberry, D.E., and Anderson, V.C. (2001). Stability and compatibility of hydromorphone hydrochloride in an implantable infusion system. *J. Pain Symp. Manag.* 22: 1042–1047.

13 Oxymorphone. https://en.wikipedia.org/wiki/Oxymorphone (accessed May 5 2024).

14 Hydrocodone. https://en.wikipedia.org/wiki/Hydrocodone (accessed May 5 2024).

15 Sonar, V.N., Parkin, S., and Crooks, P.A. (2012). Oxycodone *N*-oxide. *Acta Cryst.* C68: o436–o438.

16 Sitasuwan, P., Melendez, C., Marinova, M. et al. (2016). Degradation of opioids and opiates during acid hydrolysis leads to reduced recovery compared to enzymatic hydrolysis. *J. Anal. Toxicol.* 40: 601–607.

17 Zhuang, T., Jin, J., Ma, Y. et al. (2021). Structural elucidation and synthesis of a dimeric degradation impurity during long-term stability studies of oxycodone hydrochloride injection. *New J. Chem.* 45: 18651–18658.

18 Naloxone. https://en.wikipedia.org/wiki/Naloxone (accessed May 5 2024).

19 Pruyn, S., Frey, J., Baker, B. et al. (2019). Quality assessment of expired naloxone products from first-responders' supplies. *Prehosp. Emerg. Care* 23: 647–653.

20 Schmidt, P., Kolb, C., Reiser, A. et al. (2022). Isolation, identification and structural verification of a methylene-bridged naloxone "dimer" formed by formaldehyde. *J. Pharm. Sci.* 111: 1682–1689.

266: Oseltamivir

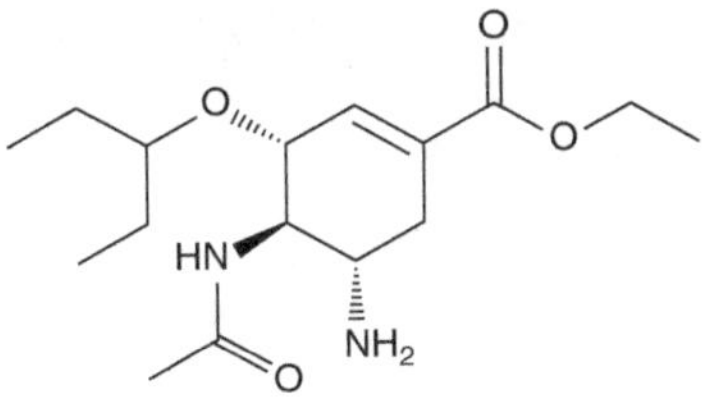

Chemical name: Ethyl (*3R,4R,5S*)-5-amino-4-acetamido-3-(pentan-3-yloxy)-cyclohex-1-ene-1-carboxylate
Trade name: Tamiflu
Formula: $C_{16}H_{28}N_2O_4$
Monoisotopic protonated ion (*m/z*): 313.2122
Molecular weight: 312.41
CAS number: 196618-13-0
Salt form: Phosphate

Oseltamivir was discovered by Gilead Sciences and then licensed to Roche in 1996 [1]. It is an inhibitor of neuraminidase, a viral surface enzyme that enables the virus to bud from infected cells and thus spread in the body [2]; it is used to treat and prevent influenza A and B, the viruses that cause influenza. Oseltamivir contains several potential hot spots susceptible to various types of degradation: the ethyl ester (hydrolysis), Michael acceptor (potentially Michael addition, but steric hindrance should decrease or suppress the reactivity), primary amine group (may subject to the Maillard reaction if excipients contain reducing sugars), and acetamide (hydrolysis and intra-molecular migration). In a forced degradation and kinetics study of oseltamivir in aqueous solution at 70°C, Oliyai et al. observed three degradants, oseltamivir carboxylic acid, isomer I and II, resulting respectively from hydrolysis of the ethyl ester, acyl migration, and both hydrolysis and migration (Scheme 266-1) [3]. In another study examining the stability of extemporaneously prepared oral solutions, Albert and Bockshorn only found oseltamivir carboxylic acid and isomer I as the degradants [4]. In another stress study, two additional degradants were formed [5], which are secondary hydrolytic degradants and hence would be unlikely to form under regular stability storage conditions.

In a photochemical degradation study under UV and simulated sunlight, oseltamivir was found to undergo photochemical Michael addition as the major photodegradation pathway, in addition to other pathways (Scheme 266-2) [6]. The formation of the photodegradant, TP312, seems to be rather unusual; the structure was proposed based on high-resolution LC–MS/MS data.

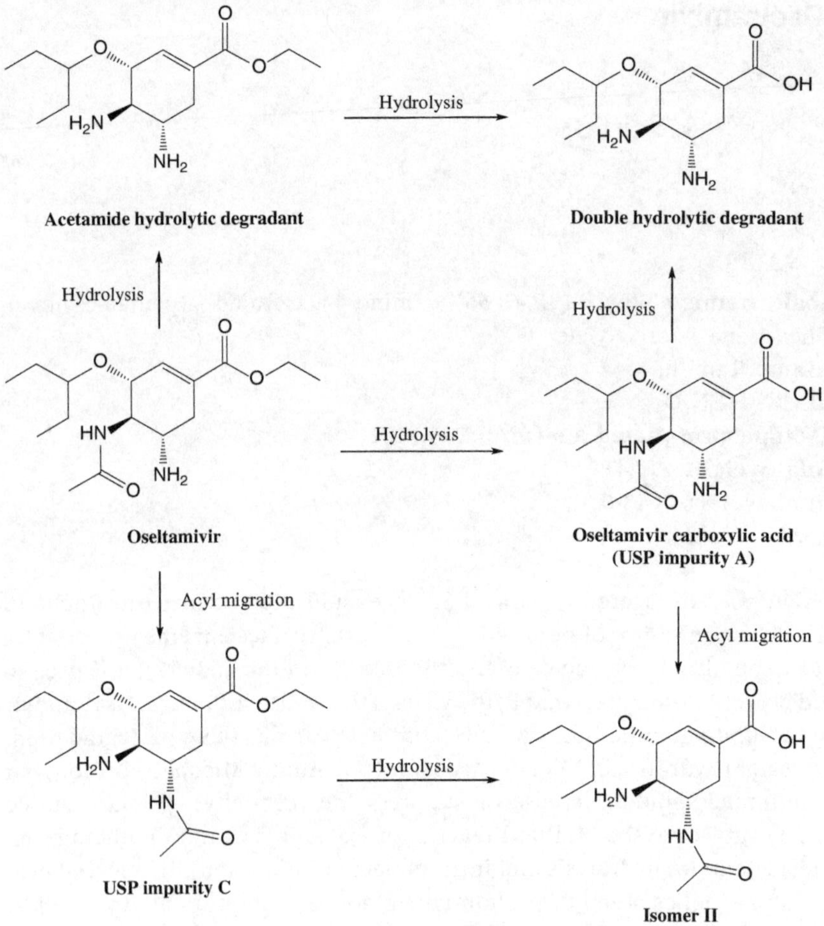

Scheme 266-1 Degradation pathways of oseltamivir.

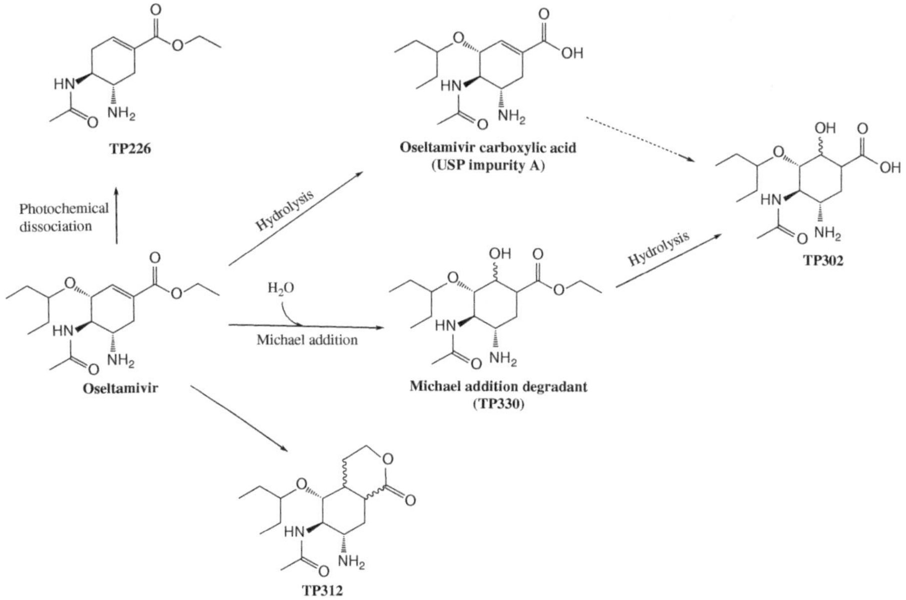

Scheme 266-2 Photodegradation pathways of oseltamivir.

References

1 Oseltamivir. https://en.wikipedia.org/wiki/Oseltamivir (accessed May 8 2024).

2 Air, G.M. and Laver, W.G. (1989). The neuraminidase of influenza virus. *Proteins* 6: 341–356.

3 Oliyai, R., Yuan, L.C., Dahl, T.C. et al. (1998). Biexponential decomposition of a neuraminidase inhibitor prodrug (GS-4104) in aqueous solution. *Pharm. Res.* 15: 1300–1304.

4 Albert, K. and Bockshorn, J. (2007). Chemical stability of oseltamivir in oral solutions. *Pharmazie* 62: 678–682.

5 Junwal, M., Sahu, A., Handa, T. et al. (2012). ICH guidance in practice: degradation behaviour of oseltamivir phosphate under stress conditions. *J. Pharm. Biomed. Anal.* 62: 48–60.

6 Goncalves, C., Perez, S., Osorio, V. et al. (2011). Photofate of oseltamivir (Tamiflu) and oseltamivir carboxylate under natural and simulated solar irradiation: kinetics, identification of the transformation products, and environmental occurrence. *Environ. Sci. Technol.* 45: 4307–4314.

267: Oxybutynin

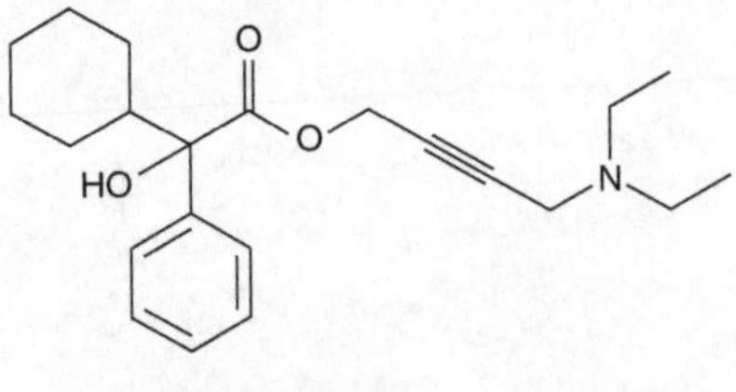

Chemical name: 4-Diethylaminobut-2-ynyl 2-cyclohexyl-2-hydroxy-2-phenylethanoate
Trade name: Ditropan, Gelnique, others
Formula: $C_{22}H_{31}NO_3$
Monoisotopic protonated ion (*m/z*): 358.2377
Molecular weight: 357.49
CAS number: 5633-20-5
Salt form: HCl salt

Oxybutynin is an anticholinergic agent, which is used primarily as the first-line treatment for urinary incontinence due to an overactive bladder [1]. It contains an ester and tertiary amine moieties, thus rendering it susceptible to hydrolytic and oxidative degradation, respectively (Scheme 267) [2–4]. In acidic to neutral pH at 37°C, oxybutynin was quite stable, but became increasingly unstable as pH increased

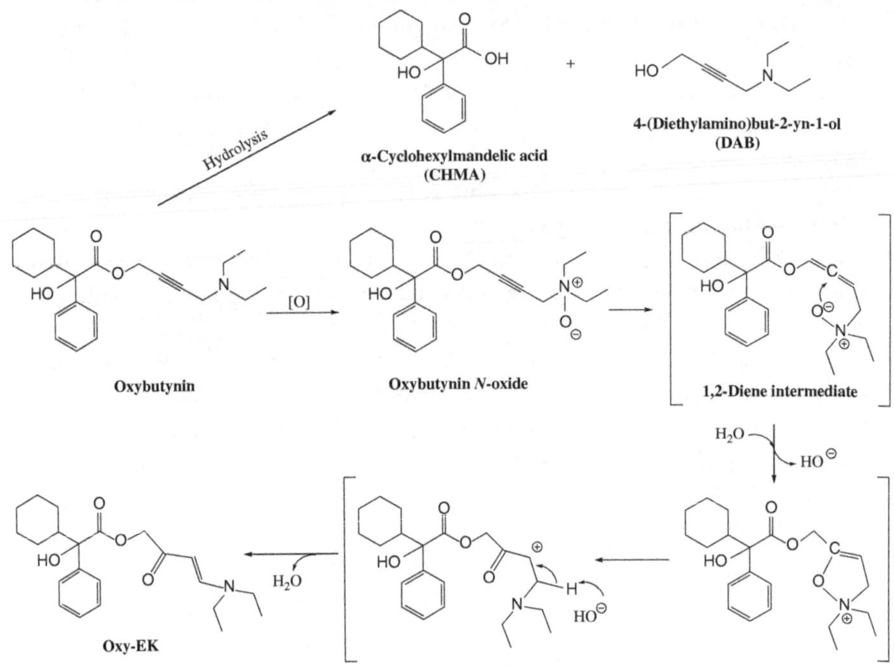

Scheme 267 Degradation pathways of oxybutynin.

at the alkaline pH region [2]. Furthermore, due to the presence of the alkyne functionality alpha to the tertiary amine (i.e., a tertiary propargylamine), Canavesi et al. found that oxybutynin *N*-oxide, the initially formed oxidative degradant in the presence of hydrogen peroxide, could undergo a rearrangement to produce the degradant (Oxy-EK) that contains an enamino ketone functionality [4]. Interestingly, the Oxy-EK degradant was also formed in the presence of a free-radical initiator, but the *N*-oxide degradant was absent in the reaction solution. As no specific formation mechanism was proposed by the original authors, a plausible mechanism is presented in Scheme 267 by the present author. The key for this mechanism would be the 1,2-diene intermediate, which could be formed via deprotonation of the methylene group next to the ester functionality.

References

1 Oxybutynin. https://en.wikipedia.org/wiki/Oxybutynin (accessed May 9 2024).
2 Miyamoto, E., Kawashima, S., Murata, Y. et al. (1994). Physico-chemical properties of oxybutynin. *Analyst* 119: 1489–1492.
3 El-Gindy, A. (2005). High performance liquid chromatographic determination of oxeladin citrate and oxybutynin hydrochloride and their degradation products. *Il Farmaco* 60: 689–699.
4 Canavesi, R., Aprile, S., Giovenzana, G.B. et al. (2016). New insights in oxybutynin chemical stability: identification in transdermal patches of a new impurity arising from oxybutynin *N*-oxide rearrangement. *Eur. J. Pharm. Sci.* 84: 123–131.

268–269: Oxymetazoline, Xylometazoline

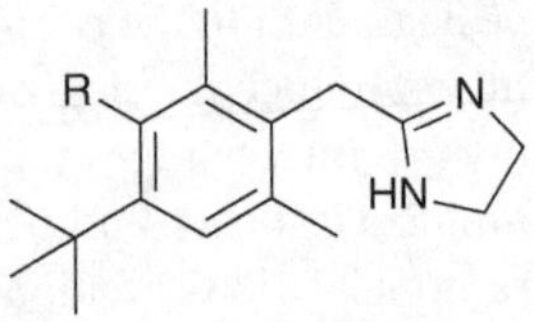

Oxymetazoline, R = –OH;
Xylometazoline, R = –H.

268: Oxymetazoline

Chemical name: 3-(4,5-Dihydro-1*H*-imidazol-2-ylmethyl)-2,4-dimethyl-6-tert-butylphenol
Trade name: Afrin, others
Formula: $C_{16}H_{24}N_2O$
Monoisotopic protonated ion (*m/z*): 261.1961
Molecular weight: 260.38
CAS number: 1491-59-4
Salt form: HCl salt

269: Xylometazoline [WHO List of Essential Medicines]

Chemical name: 2-[(4-tert-Butyl-2,6-dimethylphenyl)methyl]-4,5-dihydro-1*H*-imidazole
Trade name: Otrivin, Otrivine, others
Formula: $C_{16}H_{24}N_2$
Monoisotopic protonated ion (*m/z*): 245.2012
Molecular weight: 244.38
CAS number: 526-36-3
Salt form: HCl salt

Oxymetazoline and xylometazoline are sympathomimetic drugs and clinically used as topical decongestants and vasoconstrictors through stimulation of the α-receptors on blood vessels [1, 2]. Both drugs are usually available as a nasal spray or drops for treating nasal congestion, as well as eye drops for the treatment of drooping eyelids or allergic conjunctivitis. Both oxymetazoline and xylometazoline are imidazoline drugs containing a 4,5-dihydroimidazole ring, which are susceptible to typical hydrolytic degradation of amidines, particularly in the alkaline pH region (Scheme 268–269) [3–6]. Under severe hydrolytic conditions, further hydrolysis of the resulting amide linkage could occur [3]. A kinetics study by Stanisz showed that xylometazoline was most stable in pH 2.0–4.0 solutions [5]. On the other hand, a forced degradation study of oxymetazoline using 3-chloroperoxybenzoic acid (MCPBA) generated two isomeric oxidative degradants: the N^3-hydroxylated

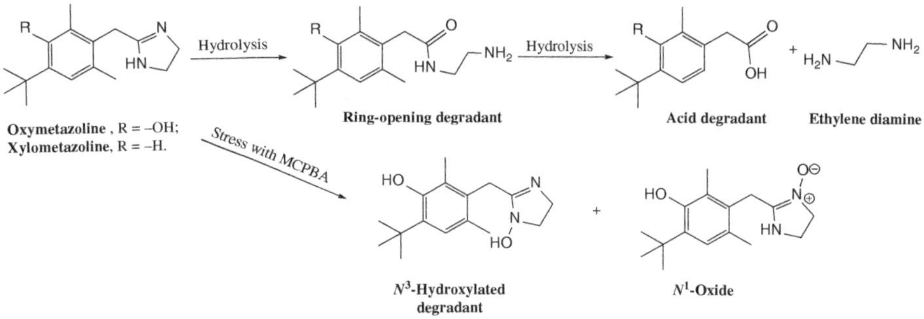

Scheme 268–269 Hydrolytic and oxidative degradation pathways of oxymetazoline and xylometazoline.

degradant and N^1-oxide, with the former as the major one, which was observed in a severely stressed commercial sample [6].

Challier et al. studied the photodegradation behaviors of oxymetazoline and xylometazoline in the presence of riboflavin (vitamin B_2), which is a photosensitizer and also a natural constituent of the eye [7]. Oxymetazoline was found to degrade photochemically four orders of magnitude faster than xylometazoline, apparently due to the presence of the hydroxyl group that renders oxymetazoline much more labile toward the reactive oxygen species (ROS) generated in the photosensitization by riboflavin.

References

1 Oxymetazoline. https://en.wikipedia.org/wiki/Oxymetazoline (accessed April 18 2024).
2 Xylometazoline. https://en.wikipedia.org/wiki/Xylometazoline (accessed April 18 2024).
3 Golander, Y. and DeWitte, W.J. (1985). Xylometazoline hydrochloride. *Anal. Prof. Drug Sub.* 14: 135–156.
4 Prajapati, N., Dalal, M., and Raj, H.A. (2016). Stability indicating liquid chromatographic method for estimation of xylometazoline hydrochloride in pharmaceutical dosage form. *Int. J. Pharm. Chem. Anal.* 3: 124–132.
5 Stanisz, B. (2003). Kinetics of hydrolysis of xylometazoline (XYL) in aqueous solution. *React. Kinet. Catal. Lett.* 80: 67–73.
6 Leader, I.P., Halket, J., and Finch, P. (2004). Synthesis and identification of two potential oxidation degradants of oxymetazoline. *Rapid Commun. Mass Spectrom.* 18: 1645–1654.
7 Challier, C., Mártire, D.O., García, N.A., and Criado, S. (2017). Visible light-mediated photodegradation of imidazoline drugs in the presence of riboflavin: possible undesired effects on imidazoline-based eye drops. *J. Photochem. Photobiol. A: Chem.* 332 (1): 399–405.

270–271: Paclitaxel, Docetaxel

270: Paclitaxel [WHO List of Essential Medicines]

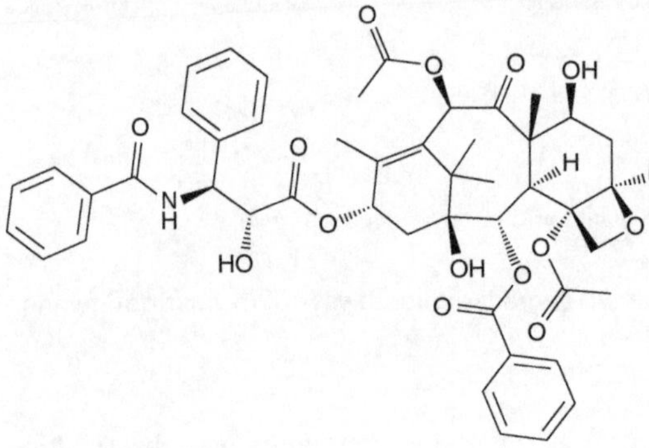

Chemicalname: (*1S,2S,3R,4S,7R,9S,10S,12R,15S*)-4,12-Bis(acetyloxy)-1,9-dihydroxy-15-({(*2R,3S*)-2-hydroxy-3-phenyl-3-[(phenylcarbonyl)amino]propanoyl}oxy)-10,14,17,17-tetramethyl-11-oxo-6-oxatetracyclo[11.3.1.$0^{3,10}$.$0^{4,7}$]eptadic-13-en-2-yl benzoate
Trade name: Taxol
Formula: $C_{47}H_{51}NO_{14}$
Monoisotopic protonated ion (*m/z*): 854.3382
Molecular weight: 853.92
CAS number: 33069-62-4

271: Docetaxel [WHO List of Essential Medicines]

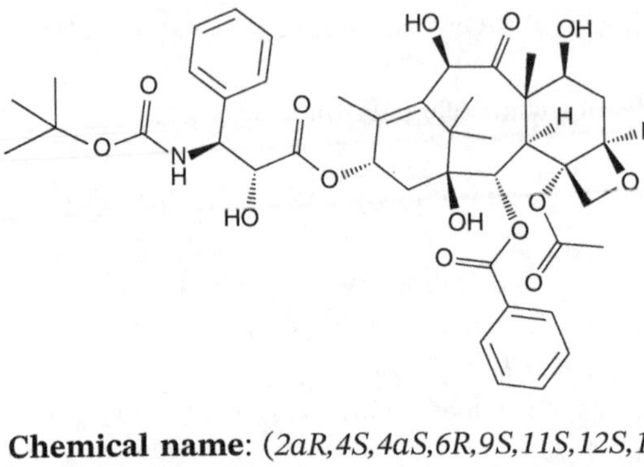

Chemical name: (*2aR,4S,4aS,6R,9S,11S,12S,12aR,12bS*)-12b-Acetoxy-9-(((*2R,3S*)-3-((tert-butoxycarbonyl)amino)-2-hydroxy-3-phenylpropanoyl)oxy)-4,6,11-trihydroxy-4a,8,13,13-tetramethyl-5-oxo-2a,3,4,4a,5,6,9,10,11,12,12a,12b-dodecahydro-1*H*-7,11-methanocyclodeca[3,4]benzo[1,2-b]oxet-12-yl benzoate
Trade name: Taxotere, Docecad, Docefrez, others
Formula: $C_{43}H_{53}NO_{14}$
Monoisotopic protonated ion (*m/z*): 808.3539
Molecular weight: 807.89
CAS number: 114977-28-5

Paclitaxel, originally known as taxol, is a chemotherapeutic drug, which is widely used to treat cancers of various types, including refractory ovarian cancer, the first indication approved by the US FDA in 1992 [1], as well as numerous other cancers such as esophageal cancer, breast cancer, lung cancer, cervical cancer, and pancreatic cancer [2]. The efforts leading to the discovery of taxol started in August 1962 when the USDA scientist Arthur Barclay collected samples from the bark of a Pacific yew tree, *Taxus brevifolia*, as part of a National Cancer Institute (NCI)-funded screening program headed by Jonathan Hartwell [1, 3]. After the extract of the bark showed anti-cancer activity in a KB cell assay, Barclay went back to the original collection site to collect 30 pounds of the bark 2 years later and sent it to Research Triangle Institute (RTI) at North Carolina where Monroe Wall and Mansukh Wani used a process of "bioactivity-directed fractionation" to identify bioactive compounds [1]. A crystalline compound was isolated from the bark extract in 1966, which was named taxol by Wall. In 1971, Wall, Wani, and their colleagues published the complete structure of taxol. In 1979, Susan Horwitz's group at Albert Einstein College of Medicine determined the anti-cancer mechanism of taxol, showing that taxol induces cell death by promoting the assembly of cell microtubules. Under a NCI's Cooperative Research and Development Agreement, Bristol Myers-Squibb took over the clinical development of taxol in 1989 and finally brought the drug into market in 1992 [1–4], which became a game changer for millions of cancer patients [5].

Paclitaxel has a complex tricyclic core structure with four relatively simple side chains linked via ester linkage. It is primarily converted to 7-*epi*-taxol, the thermodynamically more stable isomer due to epimerization, in both solid and solution states upon heating [6]. The epimerization at the 7-position is reversible through a retro-aldol/aldol process (Scheme 270–271-1) [7], which can be catalyzed by base but not acid [8]. Removal of the 10-acetyl group accelerated the epimerization process. On the other hand, all the four ester bonds are susceptible to hydrolytic degradation, with the C-13 ester bond being the most labile one toward hydrolysis in alkaline pH, yielding baccatin III, *N*-benzoyl-3-phenylisoserine, and 7-*epi*-baccatin III as the major hydrolytic degradants, in addition to minor hydrolytic degradants resulting from slower parallel and sequential hydrolytic degradation pathways (Scheme 270–271-1) [9]. Volk et al. performed a forced degradation study of paclitaxel under alkaline, oxidative (hydrogen peroxide), and photolytic conditions [10]. While the alkaline stress generated degradants that were consistent with those reported in the literature, 10-deacetylpaclitaxel was observed as the only degradant in hydrogen peroxide solution (Scheme 270–271-1), and an isomeric degradant was formed as the predominant photodegradant (Scheme 270–271-2), respectively [11]. It was found that amorphous paclitaxel, formulated in polymeric micelles based on hydrophobized hyaluronic acid, was more susceptible to the photoisomerization process [12].

Unlike paclitaxel, a natural product, docetaxel is a semi-synthetic compound made from 10-deacetylbaccatin III, the noncytotoxic precursor extracted from the needles of the European yew, *Taxus baccata*, which is more abundant than the Pacific yew [13]. Approved by the US FDA in 1994, it has the same mode of action as paclitaxel and thus, very similar indications for various cancers [14]. Chemically,

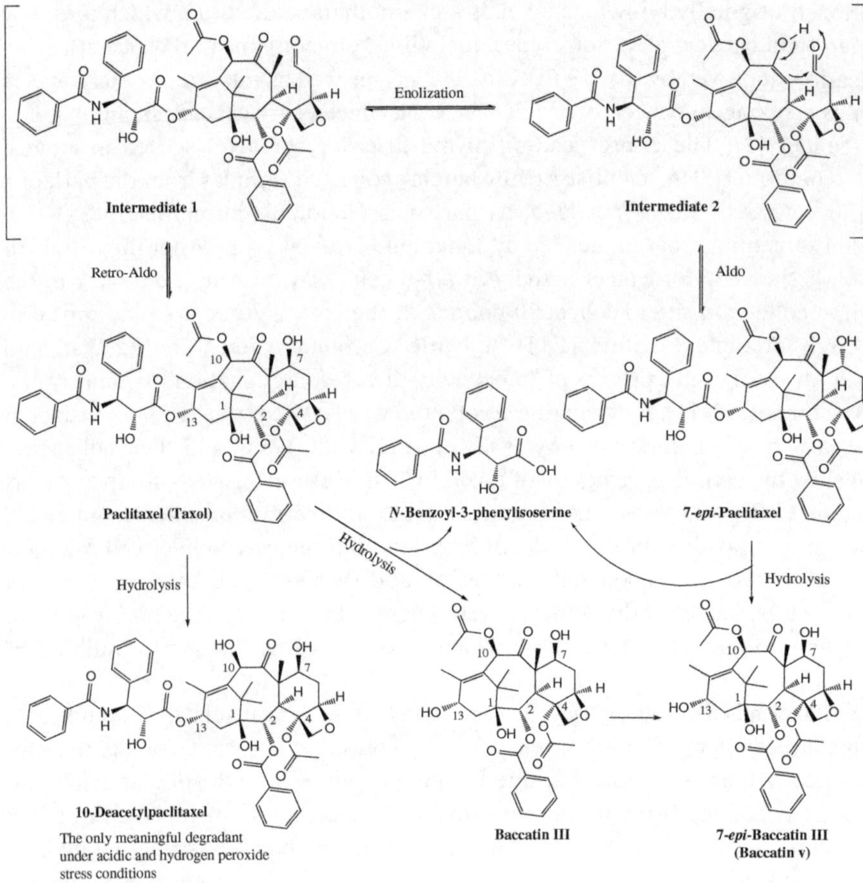

Scheme 270–271-1 Degradation pathways of paclitaxel under alkaline (50 mM NaCO3 solution), acidic (240 minutes in 2.4 M HCl solution), and oxidative (15 minutes in 7.5% hydrogen peroxide solution) stress conditions.

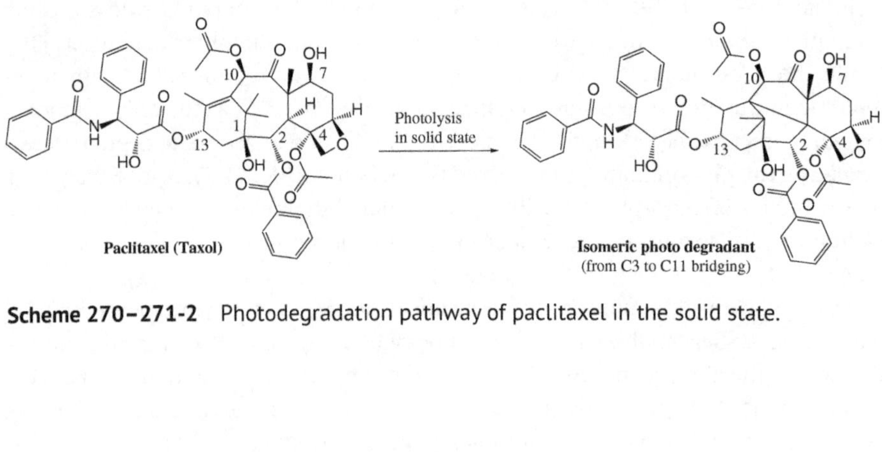

Scheme 270–271-2 Photodegradation pathway of paclitaxel in the solid state.

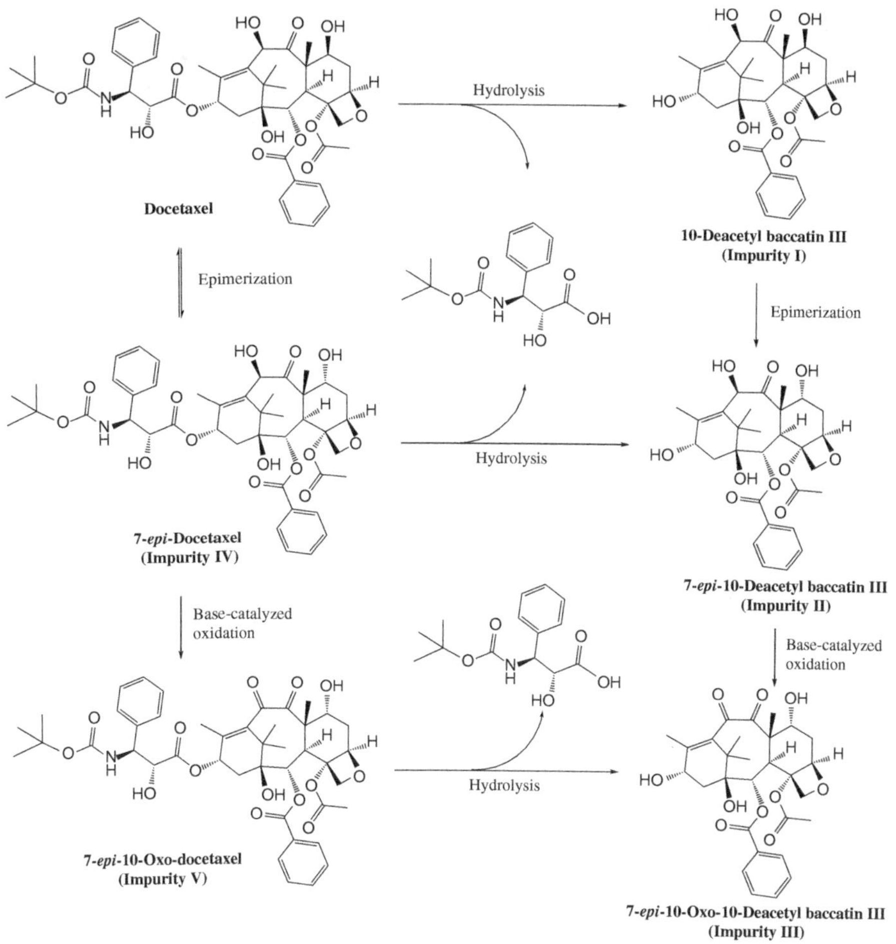

Scheme 270–271-3 Degradation pathways of docetaxel under forced degradation conditions; the pathways are proposed by the present author based on the degradation products identified by Kumar et al., and the numbering of the degradants follows that of Kumar et al. [15]. *Source:* Kumar et al. [15] / with permission of Elsevier.

docetaxel has the 10-deacetylbaccatin III core structure with an *N*-tert-BOC-3-phenylisoserine side chain on the 13-hydroxyl group. Due to the similarity between the two drug molecules, their degradation chemistry would be expected to be quite similar, except for the now exposed 10-hydroxyl group on the core of docetaxel. Indeed, in a forced degradation study, major degradation products were observed under the alkaline stress condition in which similar hydrolytic degradation and epimerization were observed (Scheme 270–271-3) [15]. Nevertheless, the exposed 10-hydroxyl group of docetaxel was found to undergo base-catalyzed oxidation to produce the corresponding 10-oxo degradants.

References

1 National Historic Chemical Landmark. Discovery of camptothecin ad taxol. https://www.acs.org/education/whatischemistry/landmarks/camptothecintaxol.html (accessed June 8 2024).

2 Paclitaxel. https://en.wikipedia.org/wiki/Paclitaxel (accessed May 18 2024).

3 Wani, M.C. and Horwitz, S.B. (2014). Nature as a remarkable chemist: a personal story of the discovery and development of Taxol®. *Anticancer Drugs* 25: 482–487.

4 Gallego-Jara, J., Lozano-Terol, G., Sola-Martínez, R.A. et al. (2020). A compressive review about Taxol®: history and future challenges. *Molecules* 25: 5986.

5 Arnst, J. (2020). When Taxol met tubulin. *J. Biol. Chem.* 295: 13994–13995.

6 MacEachern-Keith, G.J., Wagner Butterfield, L.J., and Incorvia Mattina, M.J. (1997). Paclitaxel stability in solution. *Anal. Chem.* 69: 72–77.

7 McLaughlin, J.L., Miller, R.W., Powell, R.G., and Smith, C.R. Jr. (1981). 19-Hydroxybaccatin III, 10-deacetylcephalomannine, and 10-deacetyltaxol: new antitumor taxanes from *Taxus wallichiana*. *J. Nat. Prod.* 44: 312–319.

8 Tian, J. and Stella, V.J. (2008). Degradation of paclitaxel and related compounds in aqueous solutions I: epimerization. *J. Pharm. Sci.* 97: 1224–1235.

9 Tian, J. and Stella, V.J. (2008). Degradation of paclitaxel and related compounds in aqueous solutions II: nonepimerization degradation under neutral to basic pH conditions. *J. Pharm. Sci.* 97: 3100–3108.

10 Volk, K.J., Hill, S.E., Kerns, E.H., and Lee, M.S. (1997). Profiling degradants of paclitaxel using liquid chromatography–mass spectrometry and liquid chromatography–tandem mass spectrometry substructural techniques. *J. Chrom. B* 696: 99–115.

11 Chen, S.H., Combs, C.M., Hill, S.E. et al. (1992). The photochemistry of taxol: synthesis of a novel pentacyclic taxol isomer. *Tetrahedron Lett.* 33: 7679–7680.

12 Šmejkalová, D., Nešporová, K., Hermannová, M. et al. (2014). Paclitaxel isomerisation in polymeric micelles based on hydrophobized hyaluronic acid. *Int. J. Pharm.* 466: 147–155.

13 Dev, R.V., Babu, J.M., Vyas, K. et al. (2006). Isolation and characterization of impurities in docetaxel. *J. Pharm. Biomed. Anal.* 40: 614–622.

14 Docetaxel. https://en.wikipedia.org/wiki/Docetaxel (accessed May 18 2024).

15 Kumar, D., Tomar, R.S., Deolia, S.K. et al. (2007). Isolation and characterization of degradation impurities in docetaxel drug substance and its formulation. *J. Pharm. Biomed. Anal.* 43: 1228–1235.

272–277: Penicillins

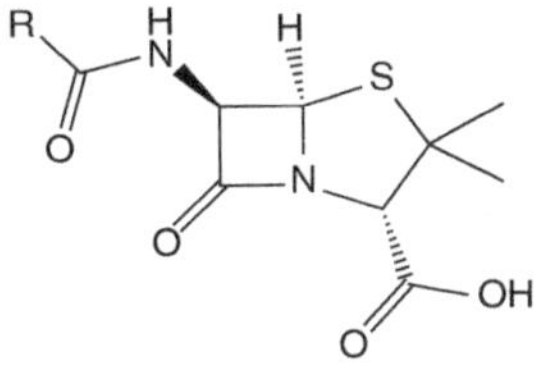

General structure of penicillins

The penicillin family of antibiotics was initially discovered “accidentally” by Alexander Fleming during August–September of 1928 when he noticed a blue-green mold inhibiting the growth of the bacteria *S. aureus* [1]. It took another 14 years, during which time Howard Florey, Ernst Chain, and their colleagues scaled up the penicillin preparation in the laboratory, which enabled them to perform various in vitro and in vivo experiments to demonstrate the efficacy of penicillin extract. The scale-up preparation also enabled Fleming to perform the most important clinical test in 1942, in which a patient with a fatal infection in the nervous system was cured [1]. During the period of 1943–1945, collaborative research aiming to elucidate the structure of penicillin was carried out by research groups in the United Kingdom and the United States, resulting in the finding that several different forms of penicillin exist, dependent upon the strain of the mold used and/or the cultural conditions. The original penicillin prepared by Florey and Chain at Oxford University was named penicillin F (or penicillin I) to commemorate the discovery of penicillin by Fleming. Another penicillin isolated from a different *Penicillium* mold was penicillin G, which turned out to be much easier to isolate and more stable than penicillin F [1, 2]. Penicillin G is still clinically used today, while all of the other natural penicillins are no longer used [3]. Since the elucidation of 6-aminopenicillanic acid (6-APA) as the core of the penicillin family [4], various semi-synthetic penicillins have been developed for clinical use [5, 6], including those covered in this monograph. The discovery and eventual mass production of penicillins is a landmark for modern medicines, for which Fleming, Florey, and Chain were awarded the Nobel Prize in Physiology or Medicine in 1945.

The structures of penicillins feature a fused bicyclic core (a highly constrained thiazolidine) that contains a constrained four-membered β-lactam ring. The β-lactam is unstable and its reactivity is the basis for the mode of action of penicillins. On the other hand, this reactivity is also the main reason for the degradation of penicillins, which starts with the breakage of the β-lactam. Deshpande et al. performed a survey of the degradation of β-lactam antibiotics [7]. Under acidic or neutral pH, the main degradation pathway is isomerization to produce pseudopenicillins (Scheme 272–277-1). Under alkaline pH, the β-lactam is hydrolyzed directly, which is followed by decarboxylation. Various further degradants could form depending on specific pH, temperature, and duration of the degradation.

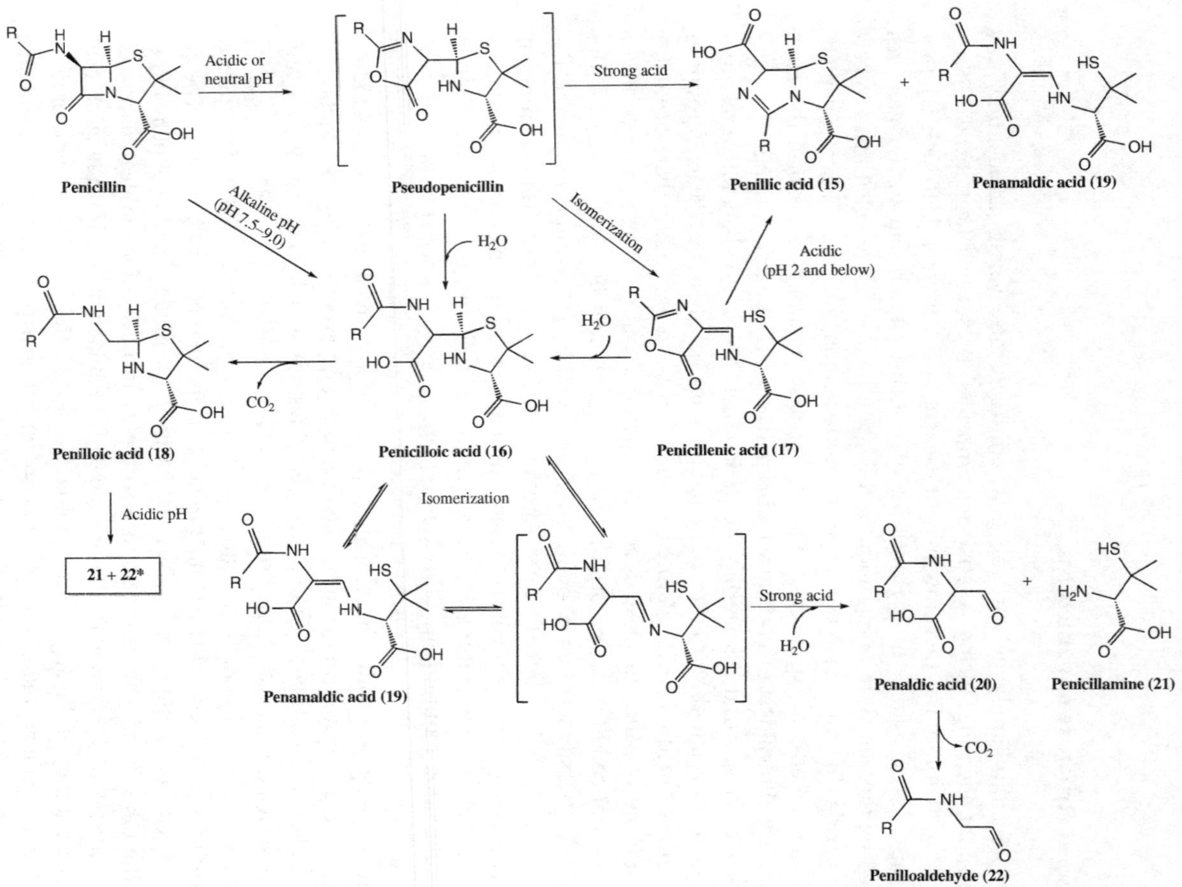

Scheme 272–277-1 Degradation pathways of penicillin in acidic and alkaline pH [7]. *Source:* Adapted from Deshpande et al. [7].

272: Benzylpenicillin [WHO List of Essential Medicines]

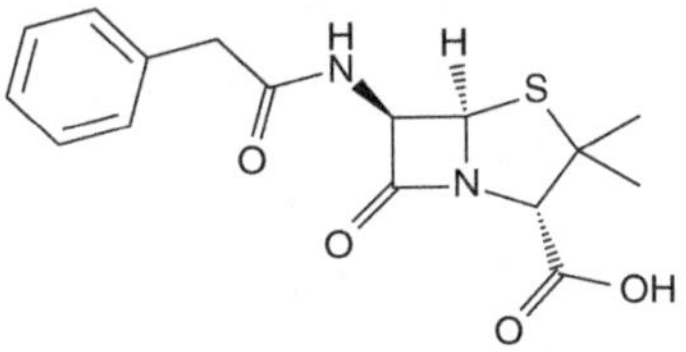

Chemical name: (2*S*,5*R*,6*R*)-3,3-Dimethyl-7-oxo-6-(2-phenylacetamido)-4-thia-1-azabicyclo[3.2.0]heptane-2-carboxylic acid
Formula: $C_{16}H_{18}N_2O_4S$
Monoisotopic protonated ion (*m/z*): 335.1060
Molecular weight: 334.39
CAS number: 61-33-6
Salt form: Potassium salt, sodium salt

Benzylpenicillin, also called penicillin G, penicillin II, or benzylpenicillinic acid, is the first natural penicillin antibiotic that is mass produced and distributed for antibacterial indications, particularly for Gram-positive infections [1, 8]. The main degradation pathway of benzylpenicillin is the hydrolytic cleavage of the β-lactam to produce benzylpenicilloic acid. At pH 2.5, benzylpenicilloic acid could undergo intramolecular condensation to give benzylpenillic acid (Scheme 272–277-2) [9]. Decomposition of benzylpenicillin at pH 2.5 could also proceed via rearrangement

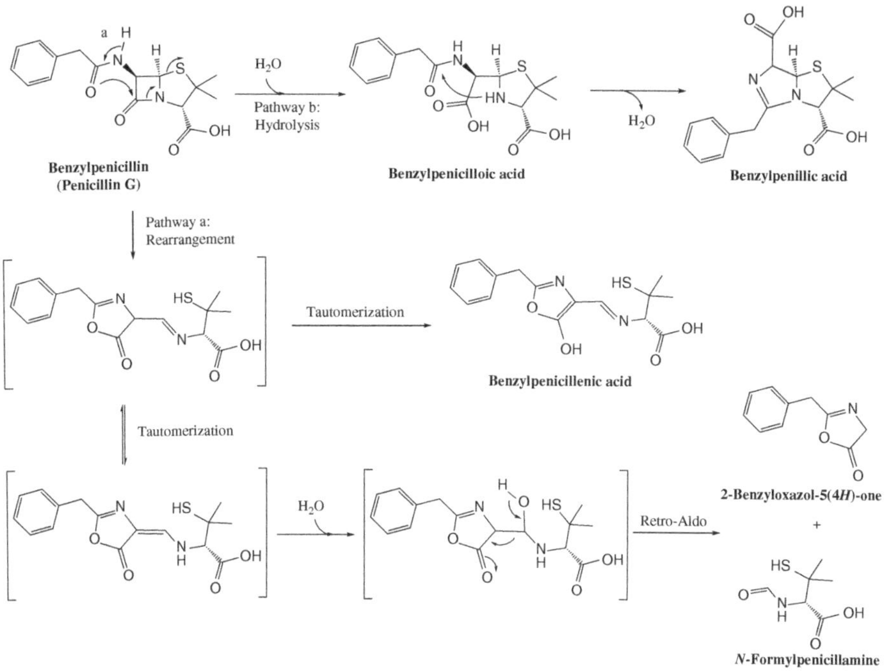

Scheme 272–277-2 Degradation pathways of benzylpenicillin in acidic pH solutions [7, 8].

and subsequent tautomerization, hydration, and retro-aldol degradation to yield benzylpenicillenic acid, *N*-formylpenicillamine, and presumably 2-benzyloxazol-5 (4*H*)-one. Benzylpenicillenic acid and its analogous degradants are believed to be the main determinants for causing the allergic reactions of benzylpenicillin and its analogs, through bio-conjugation with the lysine ε-amino groups of extracellular proteins [8]. Bird et al. reported that several penicillins decomposed to give *N*-formylpenicillamine, and at pH 5 benzylpenicillin would degrade to yield 30% *N*-formylpenicillamine [10].

273: Ampicillin [WHO List of Essential Medicines]

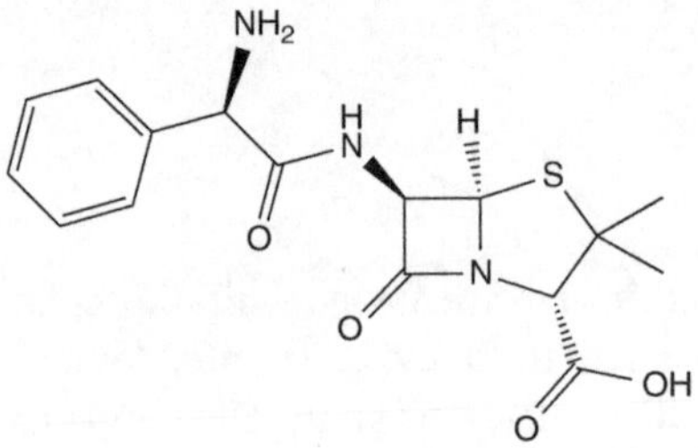

Chemicalname:(2*S*,5*R*,6*R*)-6-([(2*R*)-2-Amino-2-phenylacetyl]amino)-3,3-dimethyl-7-oxo-4-thia-1-azabicyclo[3.2.0]heptane-2-carboxylic acid
Trade name: Principen, others
Formula: $C_{16}H_{19}N_3O_4S$
Monoisotopic protonated ion (*m/z*): 350.1169
Molecular weight: 349.41
CAS number: 69-53-4
Salt form: Sodium salt

Ampicillin, introduced in 1961, is an amino derivative of benzylpenicillin, and because the amino group helps the drug to transport through the membrane of Gram-negative bacteria, it is also effective for treating Gram-negative bacterial infections [11]. Furthermore, the presence of the amino group also significantly enhances the bioavailability of the drug, rendering it available in oral dosage forms. On the other hand, the amino group causes additional degradation pathways (Scheme 272–277-3); other than that, the degradation pathways of ampicillin are similar to those outlined in Scheme 272–277-1 and Scheme 272–277-2 [12].

274: Amoxicillin [WHO List of Essential Medicines]

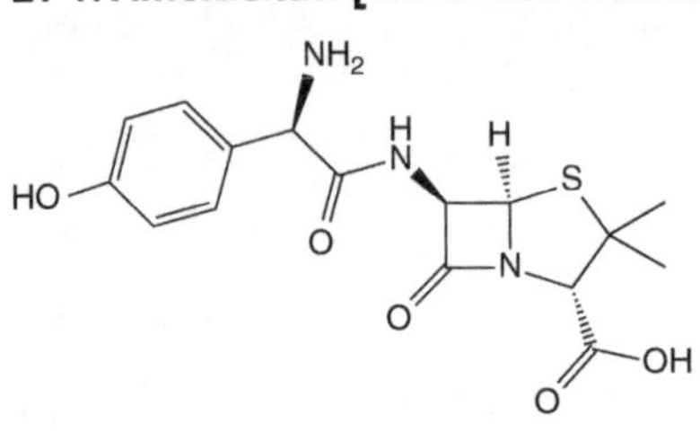

Chemical name: (*2S,5R,6R*)-6-{[(*2R*)-2-Amino-2-(4-hydroxyphenyl)acetyl]amino}-3,3-dimethyl-7-oxo-4-thia-1-azabicyclo[3.2.0]heptane-2-carboxylic acid
Trade name: Amoxil, Trimox, others
Formula: $C_{16}H_{19}N_3O_5S$
Monoisotopic protonated ion (*m/z*): 366.1118
Molecular weight: 365.40
CAS number: 26787-78-0
Salt form: Sodium salt

Amoxicillin, developed by Beecham in the United Kingdom and approved for medical use in 1972, is a widely used oral antibiotic [13] and also one of the most commonly prescribed antibiotics for children [14]. It is a hydroxyl derivative of ampicillin, with the hydroxyl group on the phenyl ring. In its formulated drug products, it is present in both the acid form (in the form of trihydrate) and the sodium salt form. Nowadays, it is often used in combination with clavulanic acid, a β-lactamase inhibitor, to overcome drug resistance toward penicillins [15]. Due to the structural similarity to ampicillin, the degradation chemistry of

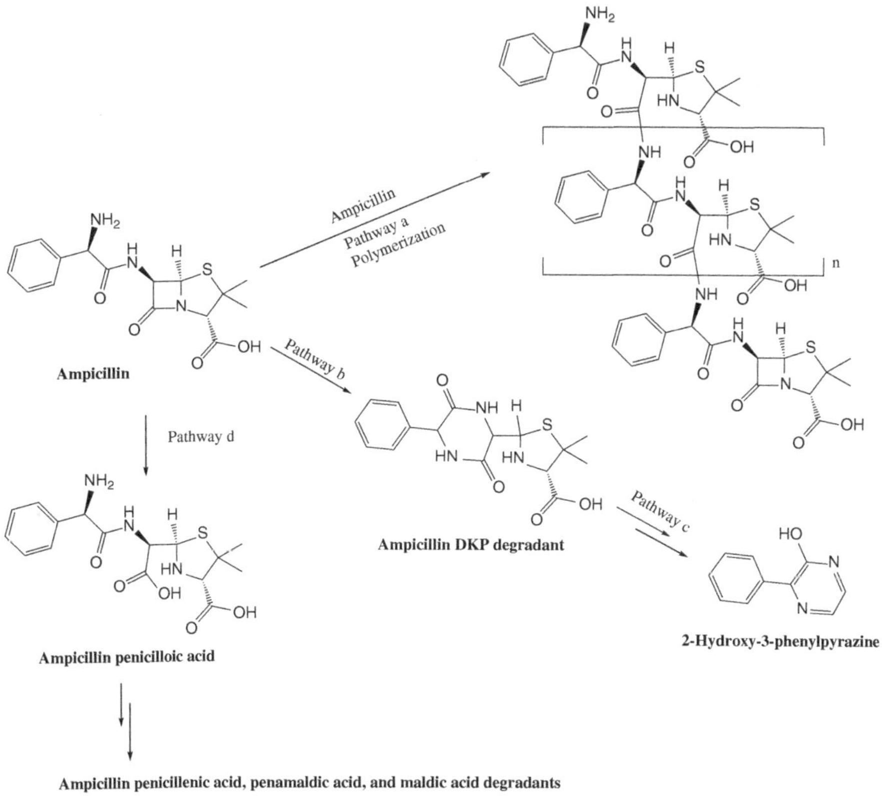

Scheme 272–277-3 Degradation pathways of ampicillin, based on those of Robinson-Fuentes et al. [12] with modification. *Source:* Adapted from Robinson - Fuentes et al. [12].

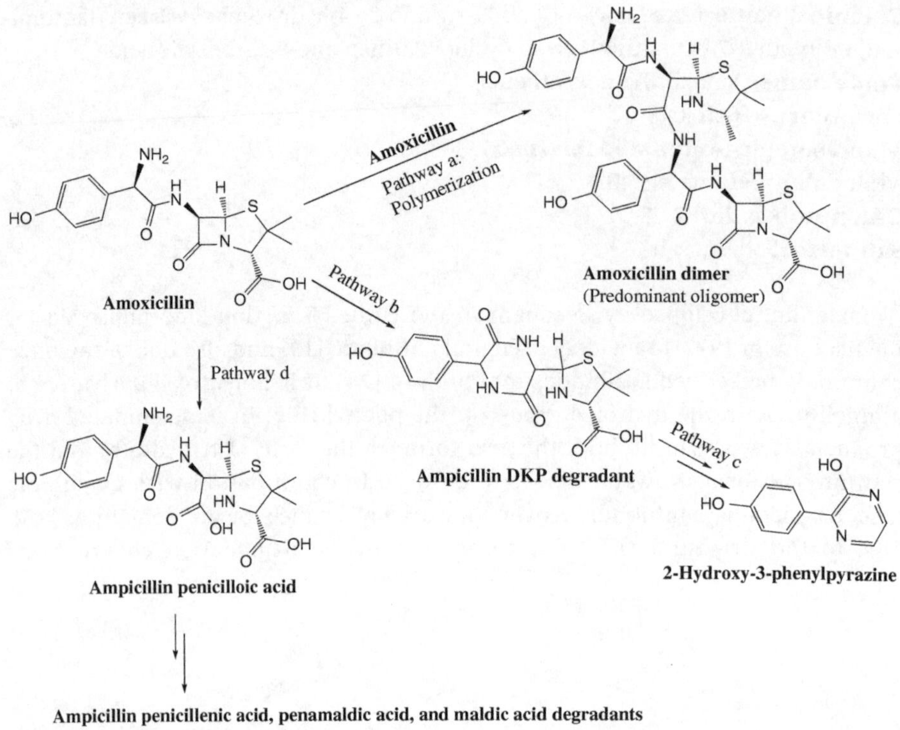

Scheme 272–277-4 Degradation pathways of ampicillin [18–20]. For the core structures of penicillenic acid, penamaldic acid, and maldic acid degradants, refer to Scheme 272–277-1.

amoxicillin would be expected to be similar to that of ampicillin, particularly under stress conditions in which analogous penicilloic acid, penilloic acid, DKP degradants, as well as their diastereomers were observed [16, 17]. In API samples of amoxicillin trihydrate, amoxicilloic acid (penicilloic acid) was found to be the main degradant, while in API samples of amoxicillin sodium, DKP degradant (piperazine-2,5-dione) and a dimeric degradant were also observed (Scheme 272–277-4) [18, 19]. In a photo stress degradation study in aqueous solutions, the majority of the degradants described above were observed during different stages of the photolysis [20].

275: Phenoxymethylpenicillin [WHO List of Essential Medicines]

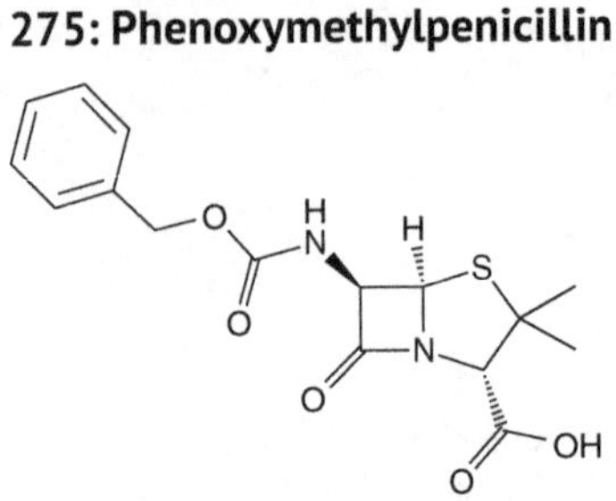

Chemical name: 3,3-Dimethyl-7-oxo-6-(2-phenoxyacetamido)-4-thia-1-azabicyclo [3.2.0]heptane-2-carboxylic acid
Trade name: Veetids, Apocillin, others
Formula: $C_{16}H_{18}N_2O_5S$
Monoisotopic protonated ion (*m/z*): 351.1009
Molecular weight: 350.39
CAS number: 87-08-1
Salt form: Potassium salt

Phenoxymethylpenicillin, also known as penicillin V and penicillin VK, was developed by Eli Lilly in 1948. It is a penicillin antibiotic indicated for several bacterial infections, particularly strep throat, otitis media, and cellulitis [21]. Due to its structural similarity to benzylpenicillin, their degradation chemistry would be expected to be very similar. In a study performed by Du et al. [22], penicilloic acid (3 diastereomers), penilloic acid (2 diastereomers), penicillenic acid, and two dimeric degradants were detected in commercial drug substance samples (Penicillin V for suspension) (Scheme 272–277-5).

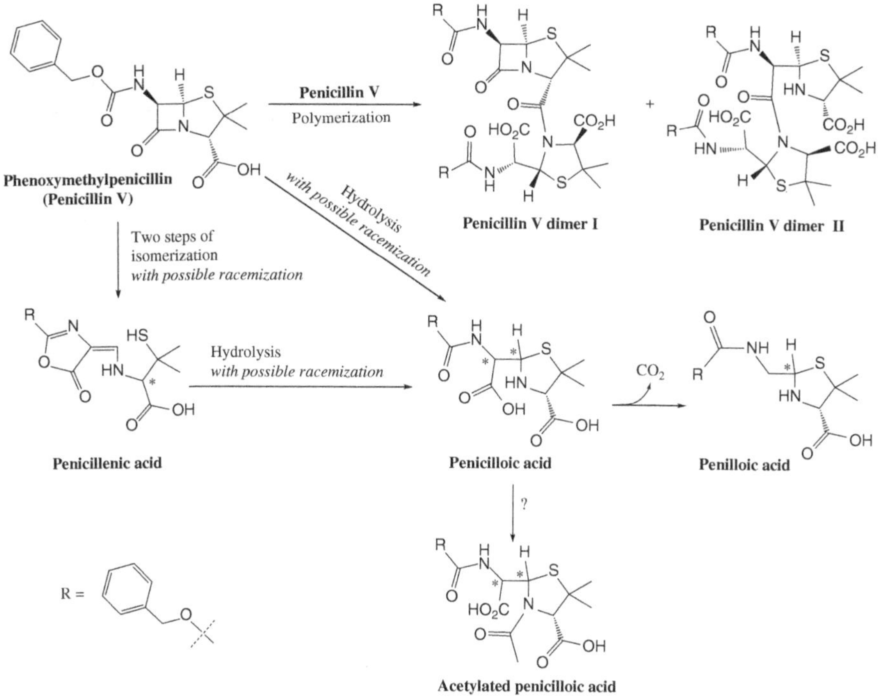

Scheme 272–277-5 Degradation products detected in one batch of penicillin raw material and three batches of Penicillin V for suspension per the study by Du et al. [22]. The source of acetylation was not clear. *Source:* Du et al. [22] / with permission of Elsevier.

276: Oxacillin

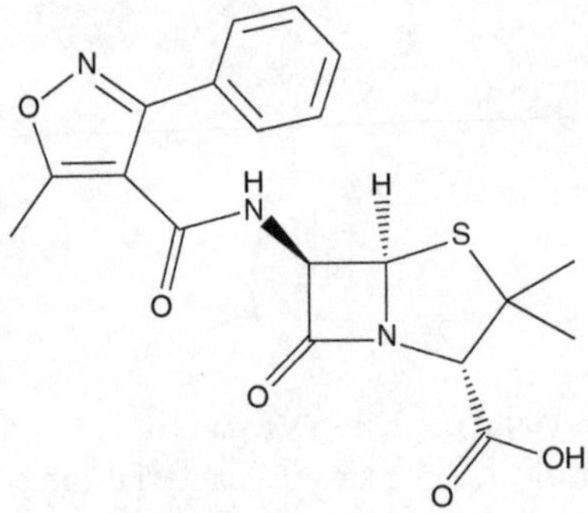

Chemical name: 3,3-Dimethyl-7-oxo-6-(2-phenoxyacetamido)-4-thia-1-azabicyclo [3.2.0]heptane-2-carboxylic acid
Trade name: Veetids, Apocillin, others
Formula: $C_{19}H_{19}N_3O_5S$
Monoisotopic protonated ion (*m/z*): 402.1118
Molecular weight: 401.44
CAS number: 87-08-1
Salt form: Sodium salt

Oxacillin is a penicillinase-resistant β-lactam antibiotic, which was developed by Beecham and approved for medical use in 1962 [23]. It is usually used in its sodium salt form in its formulated products. Due to its structural similarity with the penicillins that have already been described in this monograph, the degradation chemistry of oxacillin would be expected to be similar to the penicillins [24, 25]. For example, Ashline et al. reported the isolation and structural elucidation of novel thietan-2-one degradants of sodium nafcillin and sodium oxacillin, respectively, which turned out to be the corresponding penicillenic acid degradants (refer to Scheme 272–277-1).

277: Piperacillin [WHO List of Essential Medicines]

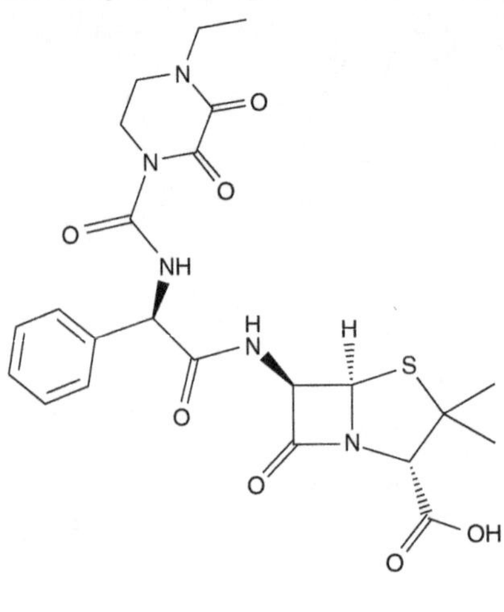

Chemical name: (*2S,5R,6R*)-6-{[(*2R*)-2-[(4-Ethyl-2,3-dioxo-piperazine-1-carbonyl) amino]-2-phenyl-acetyl]amino}-3,3-dimethyl-7-oxo-4-thia-1-azabicyclo[3.2.0] heptane-2-carboxylicacid
Trade name: Pipracil
Formula: $C_{23}H_{27}N_5O_7S$
Monoisotopic protonated ion (*m/z*): 518.1704
Molecular weight: 517.56
CAS number: 61477-96-1
Salt form: Sodium salt

Piperacillin is a β-lactam antibiotic of the ureidopenicillin class and it is typically used in combination with tazobactam, a β-lactamase inhibitor, particularly for treating serious, healthcare-associated infections [26]. It is a derivative of ampicillin and the incorporation of the ureido substituent on the side chain amino group enhances its activity against Gram-negative bacterial infections. Piperacillin was synthesized by Toyama Chemical of Japan in 1976, which was then introduced to the world market by Lederle [27]. Although the degradation chemistry of piperacillin resembles that of the structurally similar beta-lactam antibiotics, such as oxacillin, the ureido substituent could introduce new degradation pathways. In a stability study of a piperacillin/tazobactam combination by Viola et al., the ureido side chain was found to be hydrolyzed, while the rest of the drug molecule remained intact [28]. Furthermore, a degradant resulting from the ring expansion of the penicillin core was also detected. Among the more common degradants, a dimeric degradant, penicilloic acid, penicillenic acid, and penilloic acid degradants of piperacillin were also observed (Scheme 272–277-6).

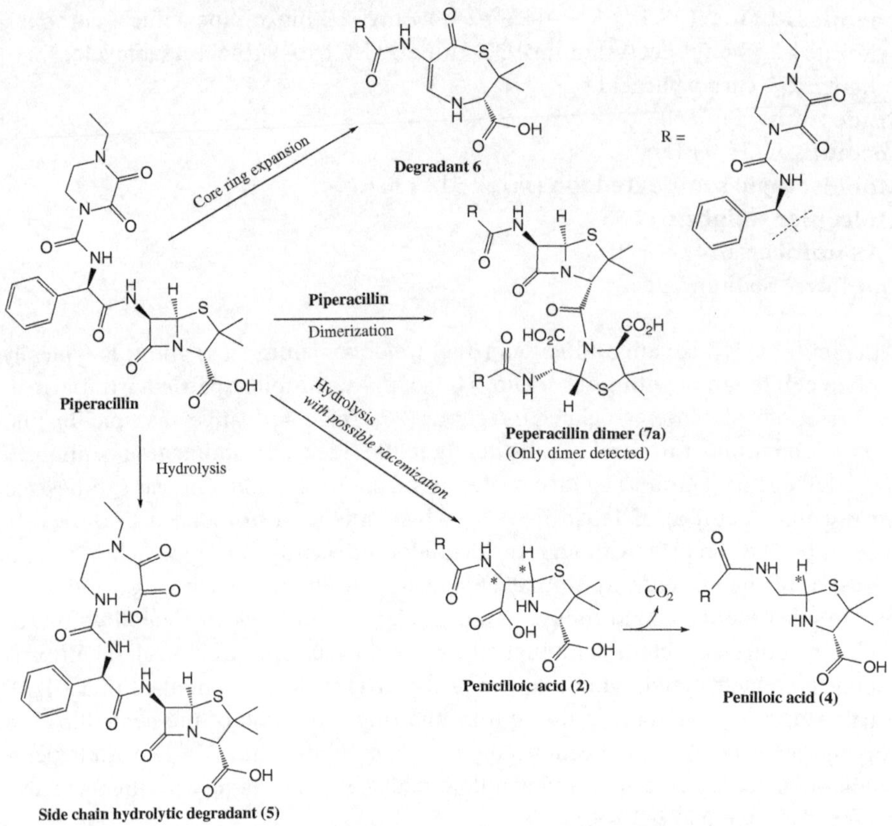

Scheme 272-277-6 Degradation pathways of piperacillin per study of Viola et al. [28]. The numbering of degradants follows that of Viola et al. *Source:* Viola et al. [28] / with permission of Elsevier.

References

1 Lalchhandama, K. (2021). History of penicillin. *Wiki J. Med.* 8: 3. https://doi.org/10.15347/wjm/2021.003.

2 Robinson, F.A. (1947). Chemistry of penicillin. *Analyst* 72 (856): 274.

3 Penicillin. https://en.wikipedia.org/wiki/Penicillin (accessed June 15 2024).

4 Hodgkin, D.C. (1949). The X-ray analysis of the structure of penicillin. *Adv. Sci.* 6: 85–89.

5 Sheehan, J.C. and Logan, K.R.H. (1959). A general synthesis of the penicillins. *J. Am. Chem. Soc.* 81: 5838–5839.

6 Batchelor, F.R., Doyle, F.P., Nayler, J.H., and Rolinson, G.N. (1959). Synthesis of penicillin: 6-aminopenicillanic acid in penicillin fermentations. *Nature* 183: 257–258.

7 Deshpande, A.D., Baheti, K.G., and Chatterjee, N.R. (2004). Degradation of β-lactam antibiotics. *Cur. Sci.* 87: 1684–1695.

8 Canzani, D. and Aldeek, F. (2017). Penicillin G's function, metabolites, allergy, and resistance. *J. Nutr. Hum. Health* 1: 28–40.

9 Kessler, D.P., Cushman, M., Ghebre-Sellassie, I. et al. (1983). Investigation of a proposed penicillin G acidic degradation scheme using high-pressure liquid chromatography and optimization techniques and mechanistic considerations. *J. Chem. Soc. Perkin Trans.* 2: 1699–1703.

10 Bird, A.E., Jennings, K.R., and Marshall, A.C. (1986). *N*-Formylpenicillamine and penicillamine as degradation products of penicillins in solution. *J. Pharmacy Pharmacol.* 38: 913–917.

11 Ampicillin. https://en.wikipedia.org/wiki/Ampicillin (accessed June 14 2024).

12 Robinson-Fuentes, V.A., Jefferies, T.M., and Branch, S.K. (1997). Degradation pathways of ampicillin in alkaline solutions. *J. Pharm. Pharmacol.* 49: 843–851.

13 de Haen, P. (1975). New products parade 1973–1974, USA. *Drug Intelli. Clin. Pharm.* 9: 504–512.

14 Amoxicillin. https://en.wikipedia.org/wiki/Amoxicillin (accessed June 14 2024).

15 Geddes, A.M., Klugman, K.P., and Rolinson, G.N. (2007). Introduction: historical perspective and development of amoxicillin/clavulanate. *Int. J. Antimicrob. Agents Suppl.* 2: S109–S112.

16 Nägele, E. and Moritz, R. (2005). Structure elucidation of degradation products of the antibiotic amoxicillin with ion trap MS^n and accurate mass determination by ESI TOF. *J. Am. Soc. Mass Spectrom.* 16: 1670–1676.

17 Perez-Parada, A., Aguera, A., del Mar Gomez-Ramos, M. et al. (2011). Behavior of amoxicillin in wastewater and river water: identification of its main transformation products by liquid chromatography/electrospray quadrupole time-of-flight mass spectrometry. *Rapid Commun. Mass Spectrom.* 25: 731–742.

18 De Pourcq, P., Hoebus, J., Roets, E. et al. (1985). Quantitative determination of amoxicillin and its decomposition products by high-performance liquid chromatography. *J. Chrom. A.* 321: 441–449.

19 Roets, E. and De Pourcq, P. (1984). Isolation and structure elucidation of ampicillin and amoxicillin oligomers. *J. Chrom. A.* 303: 117–129.

20 Arsand, J.B., Hoff, R.B., Jank, L. et al. (2018). Transformation products of amoxicillin and ampicillin after photolysis in aqueous matrices: identification and kinetics. *Sci. Total Environ.* 642: 954–967.

21 Phenoxymethylpenicillin. https://en.wikipedia.org/wiki/Phenoxymethylpenicillin (accessed June 14 2024).

22 Du, J., Chang, Y., Zhang, X., and Hu, C. (2020). Development of a method of analysis for profiling of the impurities in phenoxymethylpenicillin potassium based on the analytical quality by design concept combined with the degradation mechanism of penicillins. *J. Pharm. Biomed. Anal.* 186: 113309.

23 Oxacillin. https://en.wikipedia.org/wiki/Oxacillin (accessed June 14 2024).

24 Ashline, K.A., Attrill, R.P., Chess, E.K. et al. (1990). Isolation, structure elucidation, and synthesis of novel penicillin degradation products: thietan-2-ones. *J. Chem. Soc. Perkin Trans.* 2: 1559–1566.

25 Wu, Q.-Q., Liu, Y.-M., and Ni, M.-X. (2011). Analyzing impurities degraded by alkali in oxacillin sodium. *Chin. J. Pharm. Anal.* 31: 1102–1105.

26 Piperacillin. https://en.wikipedia.org/wiki/Piperacillin (accessed June 14 2024).
27 Kumazawa, J. and Yagisawa, M. (2002). The history of antibiotics: the Japanese story. *J. Infect. Chemother.* 8: 125–133.
28 Viola, A., Ferrazzano, L., Martelli, G. et al. (2019). Novel insights into the chemistry of an old medicine: a general degradative pathway for penicillins from a piperacillin/tazobactam stability study. *Eur. J. Pharm. Sci.* 136: 104957.

278–280: Phenylephrine, Pseudoephedrine, Ephedrine

278: Phenylephrine

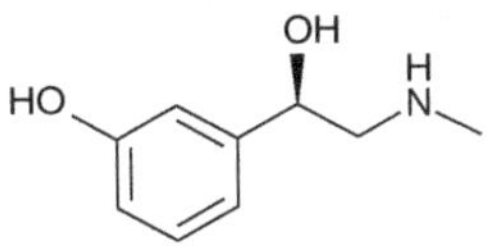

Chemical name: (*R*)-3-[-1-Hydroxy-2-(methylamino)ethyl]phenol
Trade name: Neo-synephrine, others
Formula: $C_9H_{13}NO_2$
Monoisotopic protonated ion (*m/z*): 168.1019
Molecular weight: 167.21
CAS number: 59-42-7
Salt form: HCl salt, Bitartrate

279: Pseudoephedrine

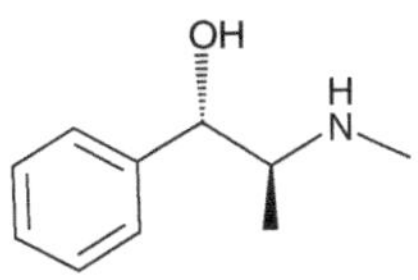

Chemical name: (*1S,2S*)-2-Methylamino-1-phenylpropan-1-ol
Formula: $C_{10}H_{15}NO$
Monoisotopic protonated ion (*m/z*): 166.1226
Molecular weight: 165.24
CAS number: 90-82-4
Salt form: HCl salt, Sulfate

280: Ephedrine [WHO List of Essential Medicines]

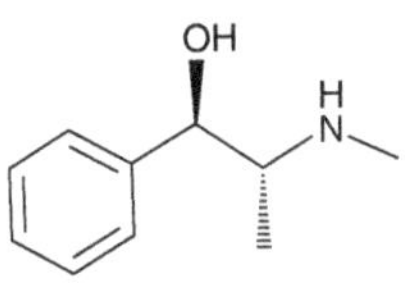

Chemical name: (*1R,2S*)-2-(Methylamino)-1-phenylpropan-1-ol
Formula: $C_{10}H_{15}NO$
Monoisotopic protonated ion (*m/z*): 166.1226
Molecular weight: 165.24
CAS number: 299-42-3
Salt form: HCl salt, Sulfate

Phenylephrine is a sympathomimetic drug primarily used as a nasal decongestant. Although it can be taken orally or as a nasal spray for that indication, a recent study by an independent advisory committee to the US FDA concluded that there is insufficient evidence to show that the oral administration of phenylephrine is effective in relieving nasal congestion [1]. Phenylephrine came into medical use in the 1930s, and more recently, it has become a substitute for pseudoephedrine, a nasal decongestant that has the potential for abuse. Despite its long history of medical use, there has been limited literature regarding the degradation chemistry of phenylephrine. The phenol ring of the drug molecule is an electron-rich moiety, and as such, it would be labile toward further oxidation. In a photodegradation study, phenylephrine was oxidized to form adrenaline (Scheme 278–280-1) [2]. In a stability study by Millard et al., phenylephrine was found to yield two isoquinolinone degradants [3, 4], which were most likely preceded by the two tetrahydroisoquinolinediols. The latter two intermediary degradants were also observed in the formulated products containing phenylephrine and drug molecules with similar structural features via phenolic cyclization [5]. The phenolic cyclization process consists of two steps: cyclization, in which the iminium intermediates are formed between the drug molecules and formaldehyde, a common impurity and contaminant [6], and the subsequent spontaneous oxidation.

The secondary amine functionality of phenylephrine would also chemically interact with other excipients (or their impurities) or counter ions of a drug molecule that is co-formulated with phenylephrine. For example, in tablet formulations of phenylephrine with aspirin, three acetylated phenylephrine degradants

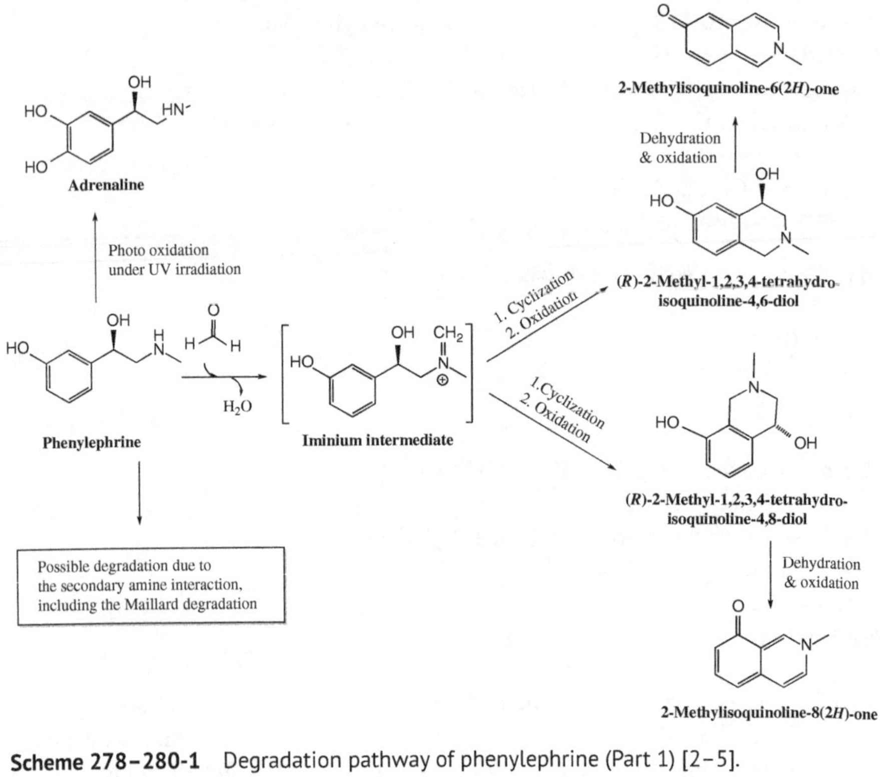

Scheme 278–280-1 Degradation pathway of phenylephrine (Part 1) [2–5].

were detected. At room temperature, it was primarily *N*-acetylphenylephrine, while at elevated temperatures, acetylation also occurred onto the phenylephrine's phenolic and hydroxyl groups, resulting in the formation of di- and tri-acetylated degradants, respectively (Scheme 278–280-2) [7]. In a combination product containing chlorpheniramine maleate, phenylephrine was found to react with the counter ion maleate [8], and the correct structure of this degradant was reported by Wong et al. as resulting from the Michael addition [9]. In another combination tablet formation containing acetaminophen (paracetamol), Dousa et al. reported the identification of two novel degradants formed between the drug molecule and 5-hydroxymethyl-2-furaldehyde (5-HMF), an aldehyde impurity commonly present in polysaccharides such as starch [10].

Pseudoephedrine is a natural product that is present in *Ephedra sinica*, also known as Ma Huang, along with ephedrine, its diastereomer. The medical use of Ma Huang dates back to over 5,000 years in China and over 2,000 years in the Middle East for treating fever, cough, and common colds among others [11]. The use of pure pseudoephedrine in modern times for treating nasal and sinus congestion started in the 1920s [12]. Due to the risk of potential abuse and the ease of converting to methcathinone, a controlled substance [13], its over-the-counter (OTC) status has been changed to behind-the-counter in the United States, and it is being partially replaced by phenylephrine, its structural analog.

On the other hand, ephedrine, which is a diastereomer of pseudoephedrine and available as a racemic mixture, was first isolated in 1885 and came into commercial use in 1926 [14]. Today, its sulfate salt is often used as a ready-to-use injectable for the prevention of low blood pressure during anesthesia. Its oral dosage forms are used for the temporary relief of shortness of breath, chest tightness, and wheezing due to asthma. Due to their structural similarity to phenylephrine, the degradation chemistry of pseudoephedrine and ephedrine would be expected to be similar to that of phenylephrine, particularly with regard to the reactivity of the secondary amine moieties (refer to Scheme 278–280-1 and -2). On the other hand, it has long been known that ephedrine and pseudoephedrine are susceptible to oxidative degradation [15–17], resulting in the formation of *R,S*- and *S*-methcathinone, respectively, as the primary degradants (Scheme 278–280-3). In a forced degradation study of a pseudoephedrine tablet formulation by Wu et al. [17], two additional degradants were reported, which would result from rather unusual oxidation of each methyl group, respectively. Typically, an *N*-carboxylic acid such as the degradant CAPP would not be very stable, particularly in isolated form. In the original study of ephedrine stability in solution performed in 1931, a 2:1 adduct between ephedrine and carbon dioxide (referred to as ephedrine carbonate by the authors) was isolated, and it would decompose on standing at room temperature to give back ephedrine [15]. Although no specific structure was provided in the 1931 study, a plausible structure is presented in Scheme 278–280-3, where the first ephedrine molecule covalently interacts with carbon dioxide from the air to yield the *N*-carboxylic acid, and then the latter forms a salt with the second ephedrine molecule.

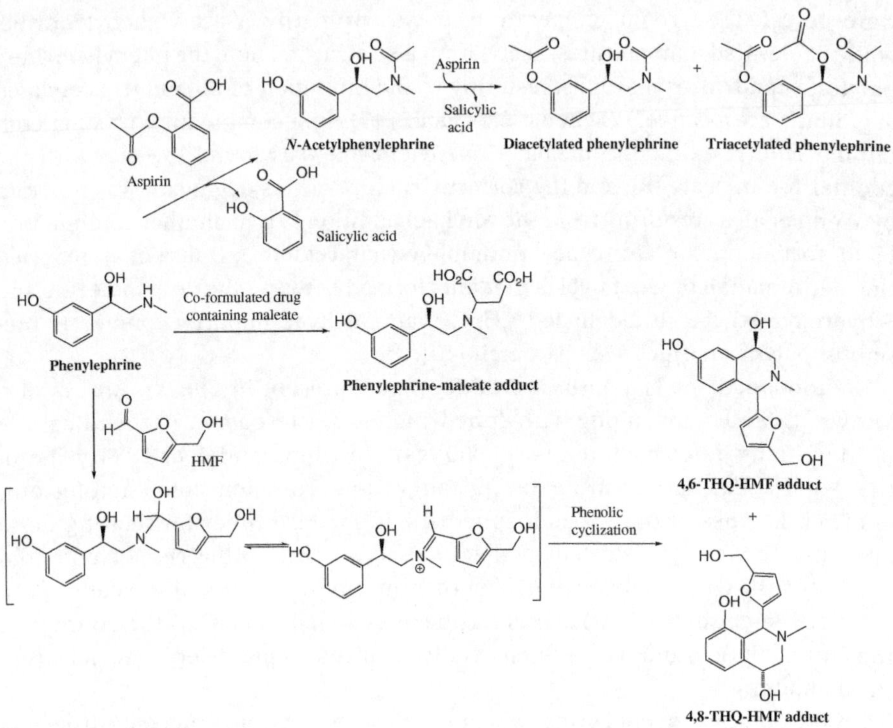

Scheme 278–280-2 Degradation pathway of phenylephrine (Part 2) [7–10].

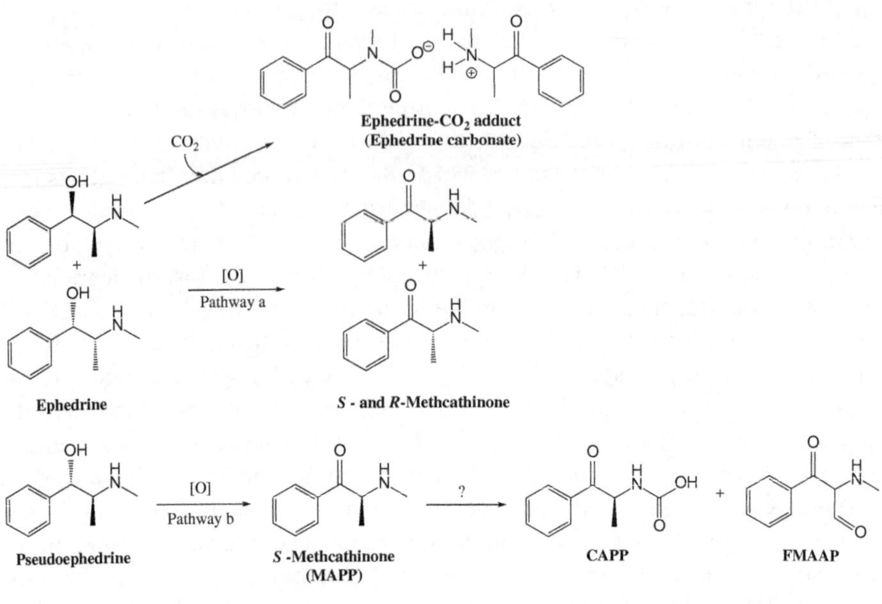

Scheme 278–280-3 Degradation pathways of ephedrine and pseudoephedrine. The nomenclature for the degradants MAPP, CAPP, and FMAAP follows that of Wu et al. [17]. *Source:* Wu et al. [17] / with permission of Elsevier.

References

1 Phenylephrine. https://en.wikipedia.org/wiki/Phenylephrine (accessed May 18 2024).
2 Luduena, F.P., Snyder, A.L., and Lands, A.M. (1963). Effect of ultra-violet irradiation of phenylephrine solutions. *J. Pharm. Pharmacol.* 15: 538–543.
3 Millard, B.J., Priaulx, D.J., and Shotton, E. (1973). The stability of aqueous solutions of phenylephrine at elevated temperatures: identification of the decomposition products. *J. Pharm. Pharmacol.* 25 (Suppl): 24P–31P.
4 Luduena, F.P., Snyder, A.L., and Lands, A.M. (1963). Effect of ultra-violet irradiation of phenylephrine solutions. *J. Pharm. Pharmacol.* 15: 538–543.
5 Chafetz, L., Turdiu, R., and R. (1987). Phenolic cyclization of epinephrine, metaproterenol, metaraminol, phenylephrine, and terbutaline with formaldehyde. *Pharm. Res.* 4: 158–161.
6 Li, M. (2012). Section 5.2. Degradation caused by impurity of excipients. In: *Organic Chemistry of Drug Degradation*. Cambridge: RSC Publishing.
7 Troup, A.E. and Mitchner, H. (1964). Degradation of phenylephrine hydrochloride in tablet formulations containing aspirin. *J. Pharm. Sci.* 53: 375–379.
8 Marin, A., Espada, A., Vidal, P., and Barbas, C. (2005). Major degradation product identified in several pharmaceutical formulations against the common cold. *Anal. Chem.* 77: 471–477.
9 Wong, J., Wiseman, L., Al-Mamoon, S. et al. (2006). Major degradation product identified in several pharmaceutical formulations against the common cold. *Anal. Chem.* 78: 7891–7895.
10 Dousa, M., Gibala, P., Havlícek, J. et al. (2011). Drug-excipient compatibility testing – identification and characterization of degradation products of phenylephrine in several pharmaceutical formulations against the common cold. *J. Pharm. Biomed. Anal.* 55: 949–956.
11 Głowacka, K. and Wiela-Hojenska, A. (2021). Pseudoephedrine – benefits and risks. *Int. J. Mol. Sci.* 22: 5146.
12 Kenny, K. (2017). Pseudoephedrine: uses, indications, and adverse effects. *Pharmacy Times* 83 (11): https://www.pharmacytimes.com/view/pseudoephedrine-uses-indications-and-adverse-effects (accessed July 10 2024.
13 American Chemical Society. Pseudoephedrine, molecule of the week archive. https://www.acs.org/molecule-of-the-week/archive/p/pseudoephedrine.html (accessed July 10 2024).
14 Ephedrine. https://en.wikipedia.org/wiki/Ephedrine (accessed July 14 2024).
15 Moore, E.E. and Moore, M.B. (1931). Effect of sunlight on ephedrine solutions. *Ind. Eng. Chem.* 23: 21–23.
16 Beckett, A.H., Jones, G.R., and Hollingsbee, D.A. (1978). Degradation of (-)-ephedrine in solution and during extraction with diethyl ether. *J. Pharm. Pharmac.* 30: 15–19.
17 Wu, N., Feng, W., Lin, E. et al. (2002). Quantitative and structural determination of pseudoephedrine sulfate and its related compounds in pharmaceutical preparations using high-performance liquid chromatography. *J. Pharm. Biomed. Anal.* 30: 1143–1155.

281: Pilocarpine [WHO List of Essential Medicines]

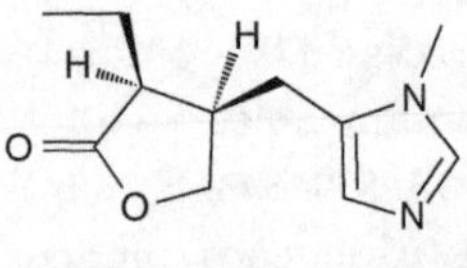

Chemical name: (3*S*,4*R*)-3-Ethyl-4-((1-methyl-1*H*-imidazol-5-yl)methyl) dihydrofuran-2(3*H*)-one
Trade name: Isopto Carpine, Salagen, Vuity, others
Formula: $C_{11}H_{16}N_2O_2$
Monoisotopic protonated ion (*m/z*): 209.1285
Molecular weight: 208.26
CAS number: 92-13-7
Salt form: HCl salt, Nitrate

Pilocarpine is a histidine alkaloid that was originally isolated from *Pilocarpus jaborandi*, a South American plant. It is an agonist of the cholinergic M-receptor and clinically used for treating glaucoma in the form of eye drops [1], and for treating dry mouth in the form of oral dosages [2]. The main degradation pathways (Scheme 281) of pilocarpine are epimerization (at the methylene position α to the carbonyl group) and hydrolysis (of the lactone ring) [3–8]. At pH 10, Bundgaard and Hansen found that the predominant degradant was pilocarpic acid, and while the epimerization was found to be reversible, the equilibrium was strongly in favor of isopilocarpine [4]. Although it might not be clearly described in the literature, the epimerization of pilocarpine would most likely proceed through enolization of the lactone ring. On the other hand, pilocarpic acid would not epimerize to isopilocarpic acid [4].

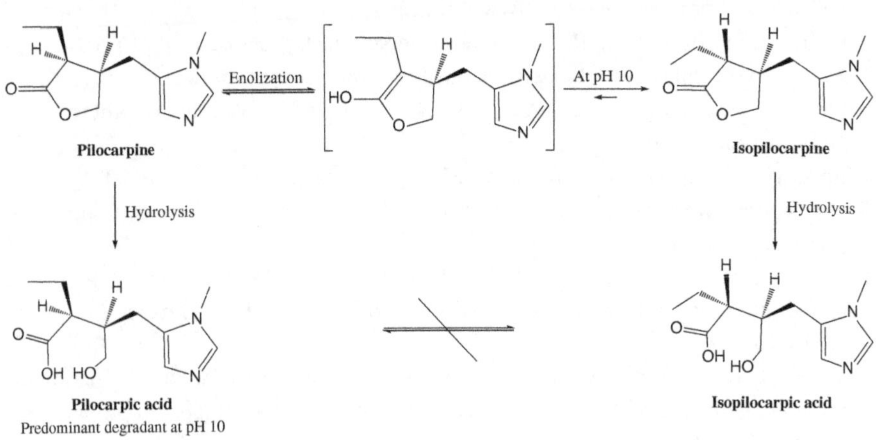

Scheme 281 Degradation pathways of pilocarpine.

References

1 Lin, Y.-H., Fang, L.-H., and Du, G.-H. (2018). Pilocarpine. In: *Natural Small Molecule Drugs from Plants*. Singapore: Springer https://doi.org/10.1007/978-981-10-8022-7_51.

2 Pilocarpine. https://en.wikipedia.org/wiki/Pilocarpine (accessed May 18 2024).

3 O'Donnell, J.J., Sandman, R., and Drake, M.V. (1980). Measurement of pilocarpine and its degradation products by high-performance liquid chromatography. *J. Pharm. Sci.* 69: 1096–1097.

4 Bundgaard, H. and Hansen, S.H. (1982). Hydrolysis and epimerization kinetics of pilocarpine in basic aqueous solution as determined by HPLC. *Int. J. Pharm.* 10: 281–289.

5 Yoshioka, S., Aso, Y., Shibazaki, T., and Uchiyama, M. (1986). Stability of pilocarpine ophthalmic formulations. *Chem. Pharm. Bull.* 34: 4280–4286.

6 Sternitzke, K.D., Fan, T.Y., and Dunn, D.L. (1992). High-performance liquid chromatographic determination of pilocarpine hydrochloride and its degradation products using a β-cyclodextrin column. *J. Chrom. A* 589: 159–164.

7 Fawcett, J.P., Tucker, I.G., Davies, N.M., and Ferguson, M.M. (1994). Formulation and stability of pilocarpine oral solution. *Int. J. Pharm. Pract.* 3: 14–18.

8 Fan, T.Y., Wall, G.M., Sternitzke, K. et al. (1996). Improved high-performance liquid chromatographic determination of pilocarpine and its degradation products in ophthalmic solutions. Importance of octadecylsilane column choice. *J. Chromatogr.* 740: 289–295.

282: Posaconazole

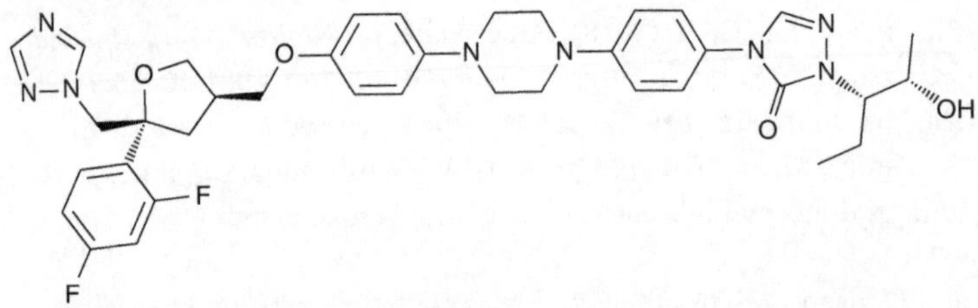

Chemicalname: 4-[4-[4-[4-[[(*3R,5R*)-5-(2,4-Difluorophenyl)tetrahydro-5-(1*H*-1,2,4-triazol-1-ylmethyl)-3-furanyl]methoxy]phenyl]-1-piperazinyl]phenyl]-2-[(*1S,2S*)-1-ethyl-2-hydroxypropyl]-2,4-dihydro-3*H*-1,2,4-triazol-3-one
Trade name: Noxafil, Posanol
Formula: $C_{37}H_{42}F_2N_8O_4$
Monoisotopic protonated ion (*m/z*): 701.3370
Molecular weight: 700.79
CAS number: 171228-49-2

Posaconazole is a triazole antifungal medication developed in the mid-2000s by Schering-Plough, which was merged into Merck & Co. in 2009. Its mode of action is to inhibit the fungal enzyme lanosterol 14-α-demethylase, which is vital for the formation of fungal cell walls [1]. Posaconazole is a long, somewhat linear molecule with a rigid central moiety consisting of an *N, N*-diphenyl-substituted piperazine building block. The substituted piperazine ring is susceptible to oxidative degradation. In 2000, a Schering-Plough research group published a forced degradation study of posaconazole in solid state, in which all four major degradants stemmed from the oxidative destruction of the piperazine ring (Scheme 282-1) [2, 3]. Later on, two more degradants were reported: one is a ring-broken degradant and the other ring-contracted degradant. The structure of the ring-contracted degradant (degradant **7**) was originally proposed by Zhong et al. based on high-resolution mass spectrometric data [4], and the corrected structure was shortly published by the same research group with comprehensive NMR analysis [5]. Due to the fact that no mechanism for the formation of degradant **7** was proposed by the original authors [5], a plausible mechanism is hereby proposed, which might proceed via an intramolecular Cannizzaro rearrangement (Scheme 282-1) [6].

In a forced degradation study of posaconazole in solution, both nitrogens of the substituted piperazine ring were found to be oxidized, resulting in the formation of the respective *N*-oxide degradants, as well as the *N,N*-dioxide degradant (Scheme 282-2) [7, 8]. Posaconazole was found to be stable under other conditions of forced degradation.

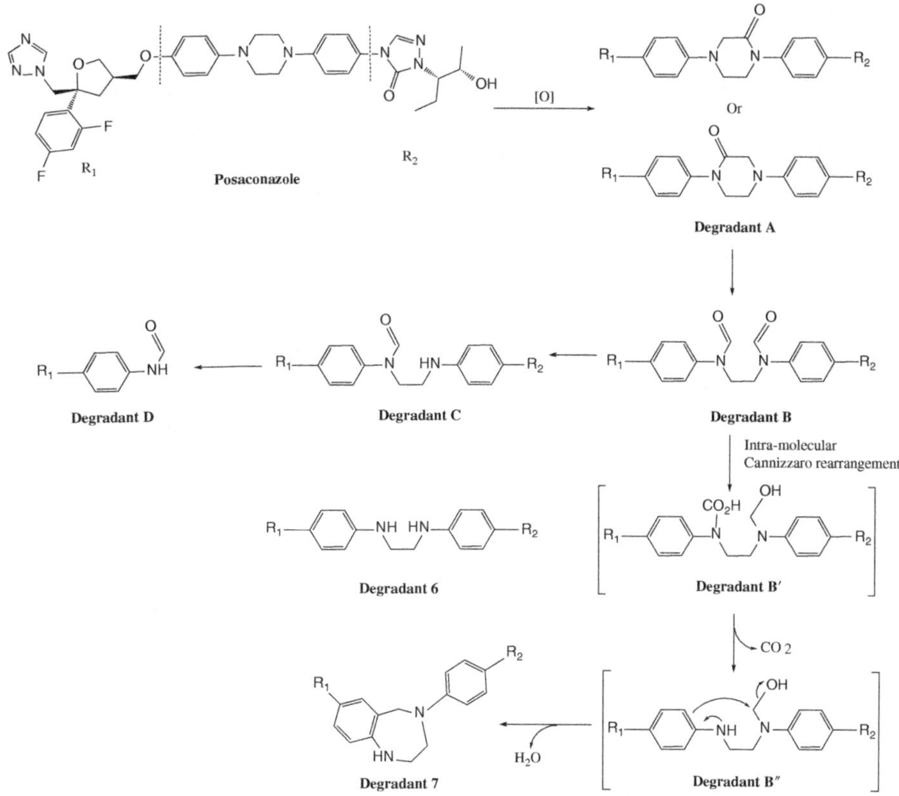

Scheme 282-1 Degradation of posaconazole in solid state under either a forced degradation condition (150°C, 12 hours) [2], which gave degradants A–D, or in a formulation under stability condition [4], which yielded degradants 6 and 7. The nomenclature of the degradants follows that of respective studies [2, 4]. The plausible formation mechanism of degradant 7 is proposed by the current author.

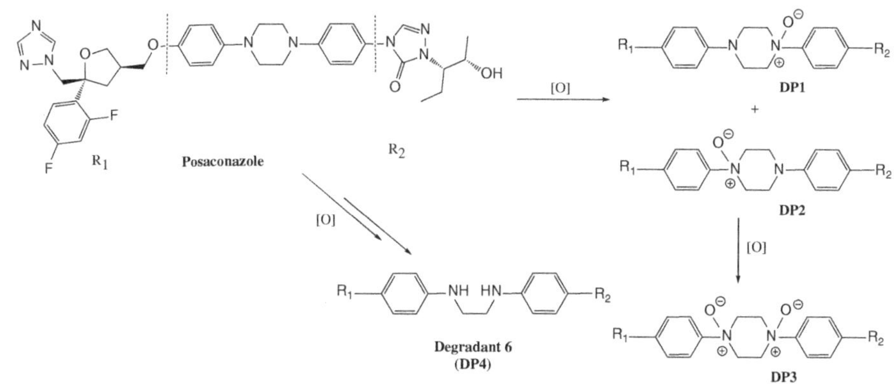

Scheme 282-2 Forced oxidative degradation of posaconazole in hydrogen peroxide solution [5, 7]. The nomenclature of the degradants follows that of Zhong et al. (Degradant 6) [5] and Yang et al. (DP1–DP4) [7], respectively. *Source:* Zhong et al. [5] and Yang et al. [7].

References

1 Greer, N.D. (2007). Posaconazole (Noxafil): a new triazole antifungal agent. *Proc. (Bayl Univ Med Cent).* 20: 188–196. https://doi.org/10.1080/08998280.2007.11928283.

2 Feng, W., Liu, H., Chen, G. et al. (2001). Structural characterization of the oxidative degradation products of an antifungal agent Scheme 56592 by LC-NMR and LC–MS. *J. Pharm. Biomed. Anal.* 25: 545–557.

3 Warrack, B.M., Goodenough, A.K., and Chen, G. (2010). Analysis of impurities and degradants in pharmaceuticals by high resolution tandem mass spectrometry and on-line H/D exchange LC/MS. *Am. Pharm. Rev.* 13: 20–27. https://www.americanpharmaceuticalreview.com/Featured-Articles/116627-Analysis-of-Impurities-and-Degradants-in-Pharmaceuticals-by-High-Resolution-Tandem-Mass-Spectrometry-and-On-line-H-D-Exchange-LC-MS/ (accessed July 13 2024).

4 Zhong, W., Yang, J., and Yang, X. (2011). Structural determination of an unknown drug product using CASI technology in a Fourier transform ion cyclotron resonance mass spectrometer. *Rapid Commun. Mass Spectrom.* 25: 3651–3655.

5 Zhong, W., Yang, X., Tong, W., and Martin, G.E. (2012). Structural characterization of a novel degradant of the antifungal agent posaconazole. *J. Pharm. Biomed. Anal.* 66: 40–49.

6 Li, M., Chen, B., Monteiro, S., and Rustum, A.M. (2009). Mechanism of base-catalyzed autooxidation of corticosteroids containing 20-keto-21-hydroxyl side chain. *Tetrahedron Lett.* 50: 4575–4581.

7 Yang, Y., Zhu, X., Zhang, F. et al. (2016). Stability-indicating HPLC method development and structural elucidation of novel degradation products in posaconazole injection by LC–TOF/MS, LC–MS/MS and NMR. *J. Pharm. Biomed. Anal.* 125: 165–177.

8 Ribeiro de Araújo, J.I., de Moura França, L., Soares, M.F.L.R. et al. (2019). Stability study and oxidative degradation kinetics of posaconazole. *Microchem. J.* 151: 104181.

283: Pralidoxime

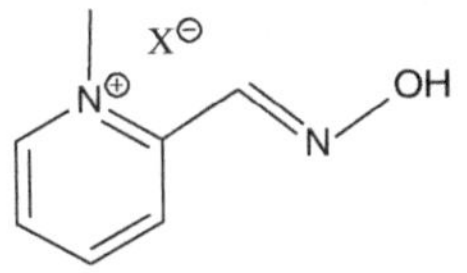

$X^- = Cl^-$, I^-, or $MeSO_3^-$

Chemical name: 2-[(Hydroxyimino)methyl]-1-methylpyridin-1-ium chloride (for the chloride salt)
Trade name: Atnaa, Duodote, Protopam
Formula: $C_7H_9N_2O^+$
Monoisotopic pyridinium ion (*m/z*): 137.070939
Molecular weight: 172.61 (for Pralidoxime chloride)
CAS number: 6735-59-7
Salt form: Chloride, Iodide, Mesylate

Pralidoxime is an antidote for the treatment of poisoning by organophosphates, typically in combination with atropine and diazepam or midazolam [1]. Organophosphates are covalent inhibitors of acetylcholinesterase, and pralidoxime reactivates the enzyme by cleaving the covalent linkage formed between the inhibitors and acetylcholinesterase. Pralidoxime is susceptible to degradation in aqueous solutions of both acidic and alkaline pH (Scheme 283) [2–5]. In a study of pralidoxime injection solution buffered at pH 2.5, six degradants were observed after storage at 80°C for 7–26 days, of which five were identified based on comparison to reference compounds, i.e., the major degradants **3**, **4**, **4′** and minor ones **2** and **6** [2]. Hydroxylamine and cyanide were detected in pralidoxime formulation [5]. Although a straightforward hydrolysis of pralidoxime should produce degradant **7** and hydroxylamine, degradant **7** was not observed in the studies performed by Prue et al. and Fyhr et al., respectively. It would be possible that degradant **7**, an aldehyde, would undergo disproportionation to yield simultaneously degradants **4** and **4′**, which appears to be consistent with the observation that both degradants **4** and **4′** are major degradation products [2].

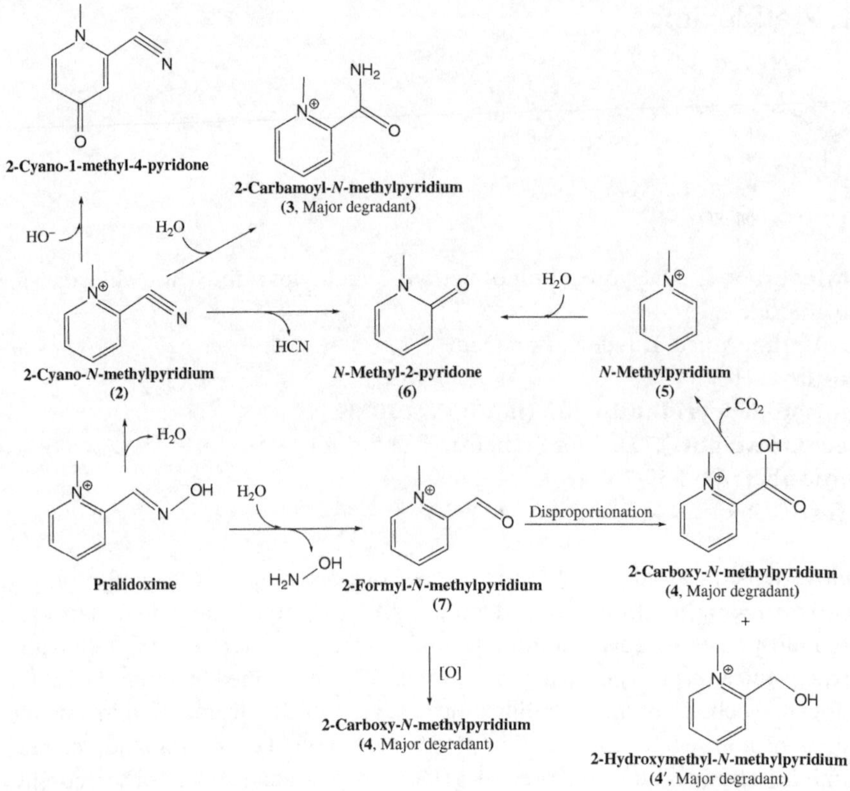

Scheme 283 Degradation chemistry of pralidoxime in aqueous solutions of different pH. The numbering of the degradants follows that of Prue et al. [2]. The summary of the degradation pathways is based on the literature cited in this monograph, and the disproportionation mechanism is proposed by the present author. *Source:* Prue et al. [2] / John Wiley & Sons.

References

1 Pralidoxime. https://en.wikipedia.org/wiki/Pralidoxime (accessed March 3 2024).

2 Prue, D.G., Johnson, R.N., and Kho, B.T. (1983). High-performance liquid chromatographic determination of pralidoxime chloride and its major decomposition products in injectable solutions. *J. Pharm. Sci.* 72: 751–756.

3 Utley, D. (1983). Determination of 2-hydroxyiminomethyl-1-methyl-pyridinium methanesulphonate (pralidoxime mesylate, P2S) and its degradation products in solution by liquid chromatography. *J. Chrom. A.* 265: 311–322.

4 Fyhr, P., Brodin, A., Ernerot, L., and Lindquist, J. (1986). Degradation pathway of pralidoxime chloride in concentrated acidic solution. *J. Pharm. Sci.* 75: 608–611.

5 Utley, D. (1990). Simultaneous determination of hydroxylamine and formulations containing pralidoxime salt by flow injection. *Analyst* 115: 1239–1242.

284: Praziquantel [WHO List of Essential Medicines]

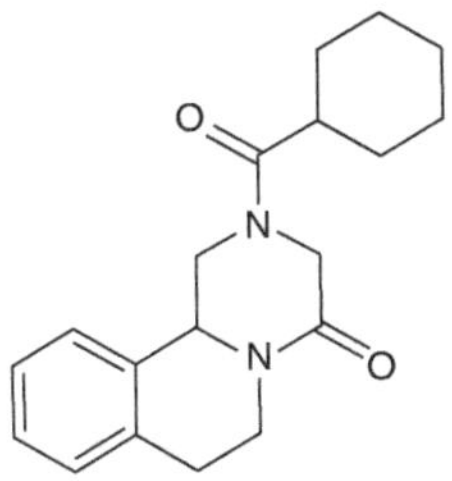

Chemical name: (*R,S*)-2-(Cyclohexylcarbonyl)-1,2,3,6,7,11b-hexahydro-4*H*-pyrazino[2,1-*a*]isoquinolin-4-one
Trade name: Biltricide
Formula: $C_{19}H_{24}N_2O_2$
Monoisotopic protonated ion (*m/z*): 313.1911
Molecular weight: 312.41
CAS number: 55268-74-1

Praziquantel is an anthelmintic medication for the treatment of the tropical diseases caused by parasitic worm infections [1–3]. It can be viewed as a piperazine derivative in which both nitrogens are acylated. While the two amide bonds are susceptible to hydrolytic degradation, particularly under forced degradation conditions [3], the main degradation pathways of praziquantel are of oxidative nature [2]. The structure of the drug molecule lacks polar functional groups, rendering it a biopharmaceutical classification system (BCS) class II drug, i.e., a drug with high permeability but low solubility [2]. In order to increase the solubility of the drug, various approaches have been tried, including using different cyclodextrins to form inclusion complexes [2] and mechanochemical activation [3]. During the process of evaluating the stability of praziquantel-cyclodextrin inclusion complexes by Spehar et al., a total of ten degradants were observed, among which seven had not been reported before (Scheme 284-1) [2]. The structure assignment of these degradants was primarily based on high-resolution mass spectrometric data, and as such, some of the proposed structures may be ambiguous. For example, degradant **X** may likely have the structure of **X'** instead, where the double bond is located at a different position, as proposed by Kumar and Chalannavar in their forced degradation study, in which it was called **DP4** [3]. Furthermore, Kumar and Chalannavar only observed four major degradants, i.e., **DP1–3** and **DP4**, among which **DP2** was not observed in the study by Spehar et al. [2]. Nevertheless, **DP2** could be an intermediate for degradant **III**. The fact that the forced degradation study projected only a very limited number of degradants may partially be attributed to the forced oxidative condition employed by Kumar and Chalannavar, where only hydrogen peroxide was used [3]. As hydrogen peroxide is usually a nucleophilic oxidative reagent, it apparently failed to produce the vast majority of the oxidative degradants that were likely formed via free-radical-mediated oxidation [4].

In another effort to improve the solubility of praziquantel, the drug substance was co-ground with either crospovidone or povidone to produce amorphous

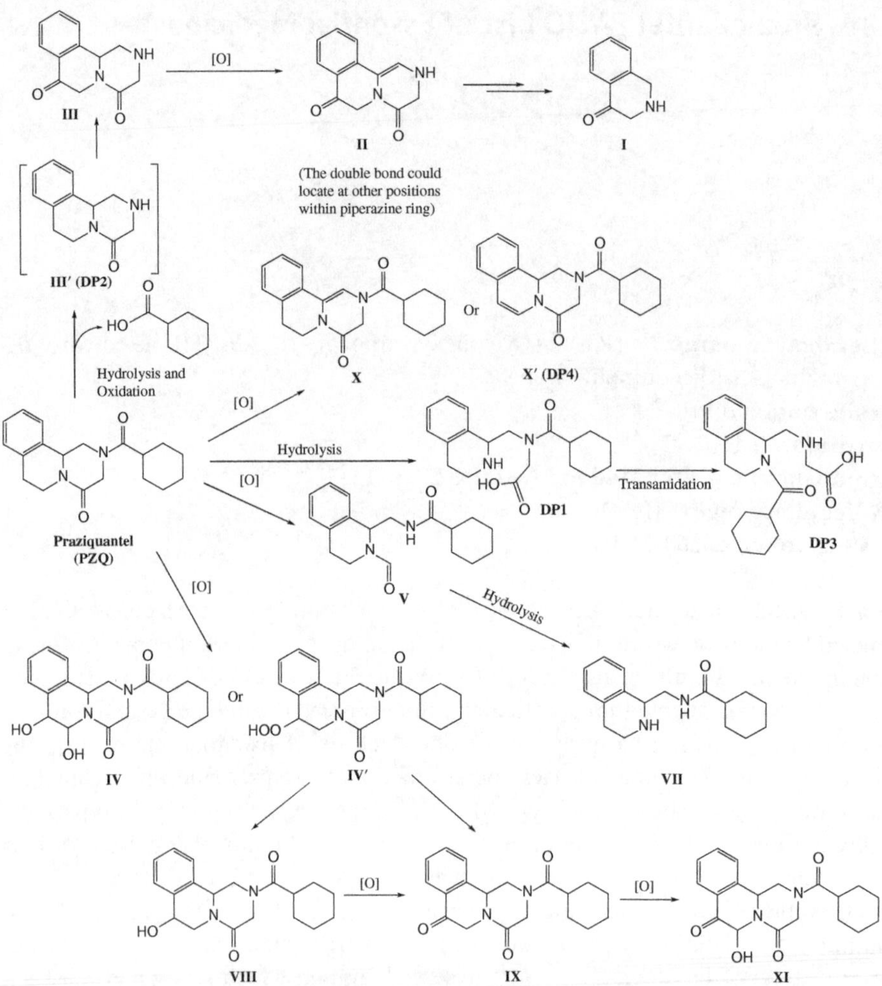

Scheme 284-1 Degradation products (I-XI) of praziquantel-cyclodextrin complexes under accelerated stability conditions, observed by Spehar et al. [2], and degradation products (DP1-DP4) of praziquantel in solutions under forced degradation conditions, observed by Kumar and Chalannavar [3]. The degradation pathways illustrated for degradants I-XI are proposed by the present author based on the results by Spehar et al. [2]. *Source:* Spehar et al. [2] and Kumar and Chalannavar [3].

dispersions [5]. Different degradation products were formed in the two processes, and two of the most prominent degradants were analyzed by GC–MS, LC–MS, and 1H NMR, and their structures proposed (Scheme 284-2).

In a photodegradation study of praziquantel with UV-C irradiation, a number of photodegradants were detected [6]; those that were formed in relatively early stages are presented in Scheme 284-3.

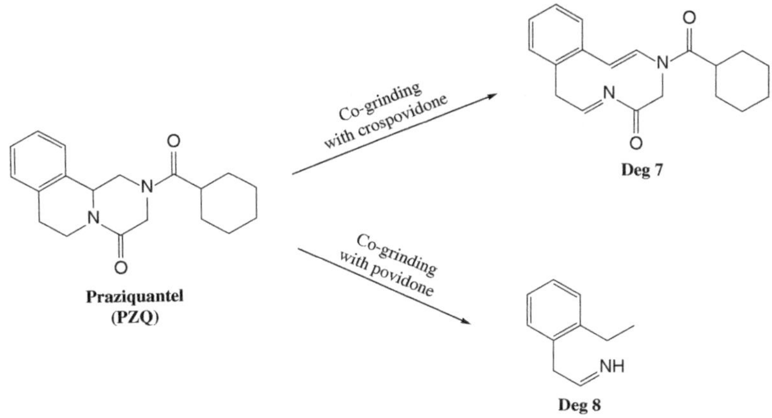

Scheme 284-2 Oxidative degradation of praziquantel during co-grinding with crospovidone or povidone. Only the structures of two major degradants were proposed based on the data from GC–MS, LC–MS, and 1H NMR analysis [5]. *Source:* Sagud et al. [5] / with permission of Elsevier.

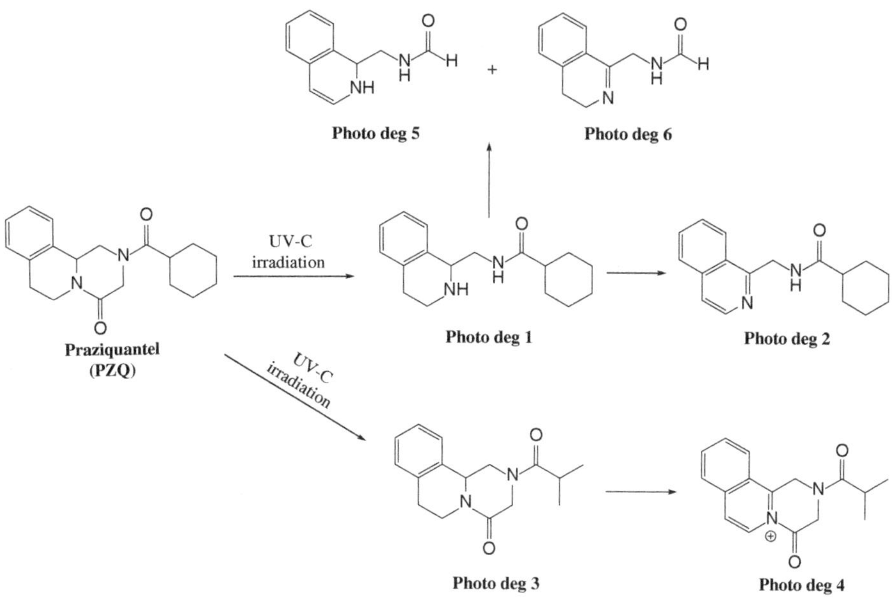

Scheme 284-3 Photodegradation of praziquantel in air-purged aqueous solution with UV-C irradiation.

References

1 Praziquantel. https://en.wikipedia.org/wiki/Praziquantel (accessed March 10 2024).

2 Spehar, T.K., Pocrnic, M., Klaric, D. et al. (2021). Investigation of praziquantel/cyclodextrin inclusion complexation by NMR and LC-HRMS/MS: mechanism, solubility, chemical stability, and degradation products. *Mol. Pharmaceutics* 18: 4210–4223.

3 Kumar, A. and Chalannavar, R.K. (2022). Characterization of the major degradation products of the praziquantel API by mass spectrometry: development and validation of a stability-indicating reversed phase UPLC method. *J. Liq. Chrom. Related Tech.* 45: 130–142.

4 Li, M. (2012). Section 3.2. Free radical-mediated autooxidation. In: *Organic Chemistry of Drug Degradation*. Cambridge: RSC Publishing.

5 Sagud, I., Zanolla, D., Perissutti, B. et al. (2018). Identification of degradation products of praziquantel during the mechanochemical activation. *J. Pharm. Biomed. Anal.* 159: 291–295.

6 Cizmic, M., Ljubas, D., Curkovic, L. et al. (2017). Kinetics and degradation pathways of photolytic and photocatalytic oxidation of the anthelmintic drug praziquantel. *J. Hazardous Mat.* 323 (Part A): 500–512.

285: Primaquine [WHO List of Essential Medicines]

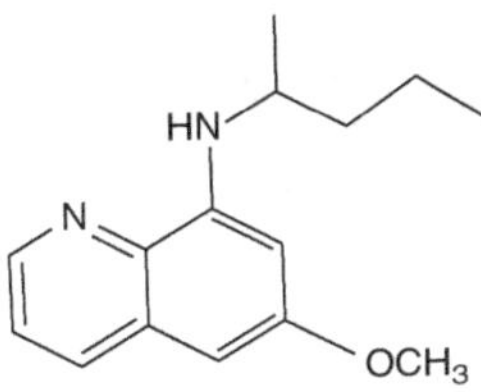

Chemical name: (*RS*)-*N*-(6-Methoxyquinolin-8-yl)pentane-1,4-diamine
Trade name: Neo-Quipenyl, Primachin
Formula: $C_{15}H_{21}N_3O$
Monoisotopic protonated ion (*m/z*): 260.1757
Molecular weight: 259.35
CAS number: 90-34-6
Salt form: Diphosphate

Primaquine is an anti-malaria drug that was first made by Robert Elderfield of Columbia University in the 1940s [1]. It is effective in treating malaria caused by the latent liver forms of *Plasmodium vivax* and *Plasmodium ovale*, which may survive after the treatment with chloroquine. Unlike the other actively used quinoline-based anti-malarial drugs, it is the only 8-amino-substituted derivative [2]. Furthermore, its side chain amino is a primary amine and thus, it would not undergo *N*-oxidation, as opposed to the tertiary amino groups in both chloroquine and hydroxychloroquine that would be oxidized to form the corresponding *N*-oxide degradants (refer to Monograph 221–222). In addition, the photochemical degradation of primaquine in the presence of oxygen was reported to occur exclusively at the side chain, in which the alkyl component of the side chain suffered various types of oxidative degradation along with deamination [2], resulting in the formation of two major degradants and six minor ones (Scheme 285). This photodegradation behavior is quite different from that of chloroquine and hydroxychloroquine, except for a similar type of dimeric degradants. In another study where the solution was purged with helium prior to UV irradiation, only two photodegradants were observed [3].

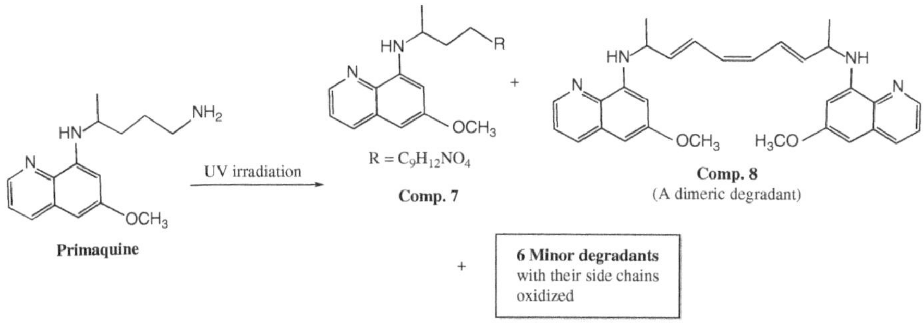

Scheme 285 Photooxidative degradation of primaquine.

References

1 Primaquine. https://en.wikipedia.org/wiki/Primaquine (accessed March 10 2024).
2 Kristensen, S., Grislingaas, A.-L., Greenhill, J.V. et al. (1993). Photochemical stability of biologically active compounds: V. Photochemical degradation of primaquine in an aqueous medium. *Int. J. Pharm.* 100: 15–23.
3 Kristensen, S., Nord, K., Orsteen, A.-L., and Tønnesen, H.H. (1998). Photoreactivity of biologically active compounds, XIV: influence of oxygen on light induced reactions of primaquine. *Pharmazie* 53: 98–103.

9

Reserpine to Zolpidem

286: Reserpine

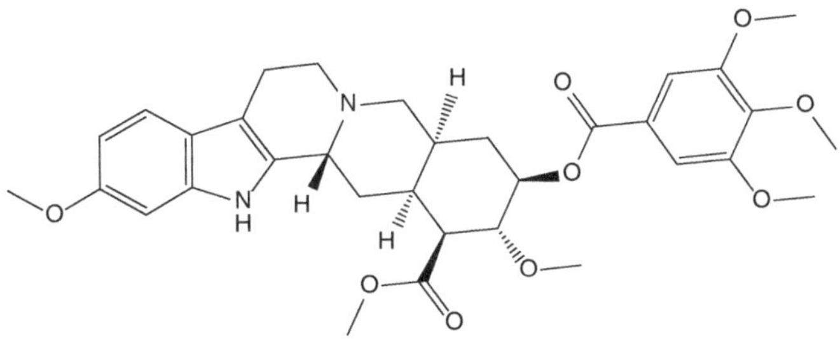

Chemical name: Methyl (3β,16β,17α,18β,20α)-11,17-dimethoxy-18-[(3,4,5-trimethoxybenzoyl)oxy]yohimban-16-carboxylate
Trade name: Serpasil, others
Formula: $C_{33}H_{40}N_2O_9$
Monoisotopic protonated ion (*m/z*): 609.2807
Molecular weight: 608.69
CAS number: 50-55-5

Reserpine is an indole alkaloid that was first isolated in 1952 by Muller et al. from the dried root of the medicinal plant *Rauwolfia serpentina*, commonly known as Indian snakeroot, according to the description by Slotkin [1]. One of its common ancient names means "madman's medicine," as the plant had been used for the treatment of insanity in India for centuries [2]. After its isolation in modern times, it was used as a first-generation antidepressant and played an important role in the treatment of schizophrenia [3]. Nowadays, it is mostly used for treating hypertension, particularly in combination with a thiazide diuretic, such as hydrochlorothiazide [2]. As a substituted indole derivative, both the 2-methine position and the double bond of the pyrrole moiety are quite susceptible to autooxidation. In aged solid dosage form of reserpine, 3,4-dehydroreserpine was found to be the major degradant (**2**, Scheme 286-1) [4]. Further oxidation of degradant **2** led to the formation of 3,4,5,6-tetradehydroreserpine, also known as lumireserpine (**3**). Degradants **2**

Compendium of Drug Degradation Pathways, First Edition. Min Li.

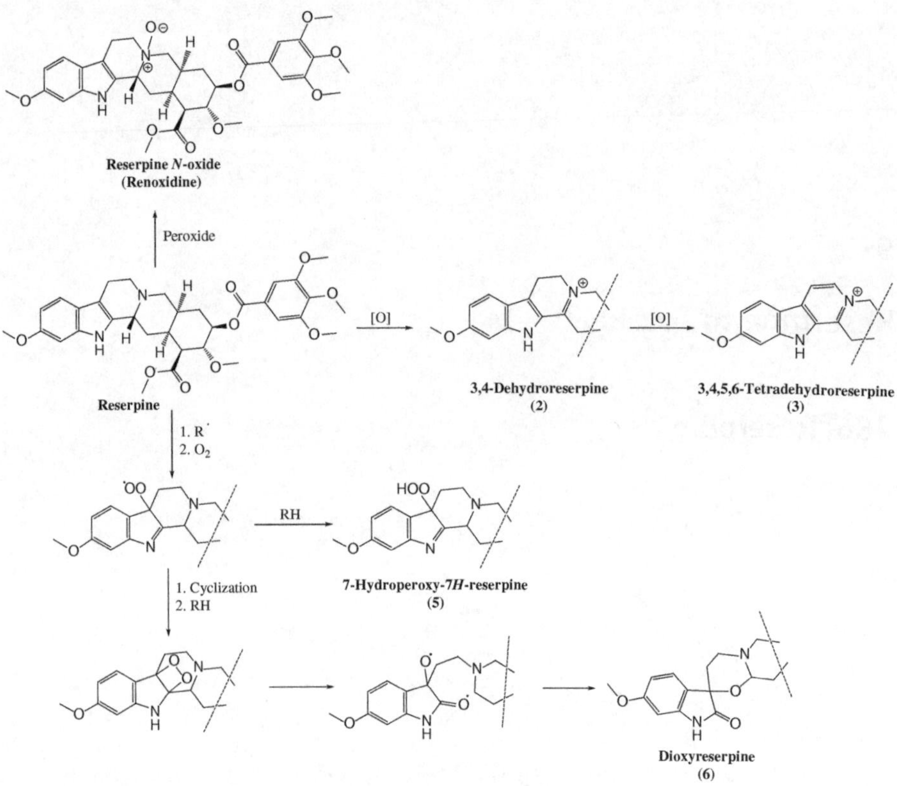

Scheme 286-1 Oxidative degradation of reserpine.

and **3** were also reported to be the main degradation products under photochemical degradation conditions [4, 5]. In another study by Awang et al., in which a solution of reserpine in chloroform was standing at room temperature and exposed to air, reserpine was found to form another two oxidative degradants (**5** and **6**) [6]. While the major degradant (**6**) has a rather unusual spiro structure, the minor one (**5**) was determined to be 7-hydroperoxy-7*H*-reserpine, a degradant resulting from the initial hydroperoxidation of reserpine. It was proposed by Awang et al. that both degradants **5** and **6** could be formed from a common hydroperoxidyl radical intermediate. Although the mechanism proposed by Awang et al. is mediated by free radicals, the same degradation pathway could also proceed via a non-free-radical-mediated mechanism. It is interesting to note that the *N*-oxide degradant of reserpine, also known as renoxidine, was not observed in the aforementioned autooxidation studies under various photochemical conditions. On the other hand, treatment of reserpine with peroxides produced reserpine *N*-oxide [7, 8].

Reserpine contains two ester functional groups, both of which are susceptible to hydrolytic degradation, with the hydrolysis of the trirmethoxybenzoate ester moiety being much more rapid than the methyl ester (Scheme 286-2) [9]. The final hydrolytic degradation product is reserpic acid. Reserpine was also reported to undergo epimerization at the 3-position to give rise to 3-isoreserpine in the presence of strong acid, heat, or light [10].

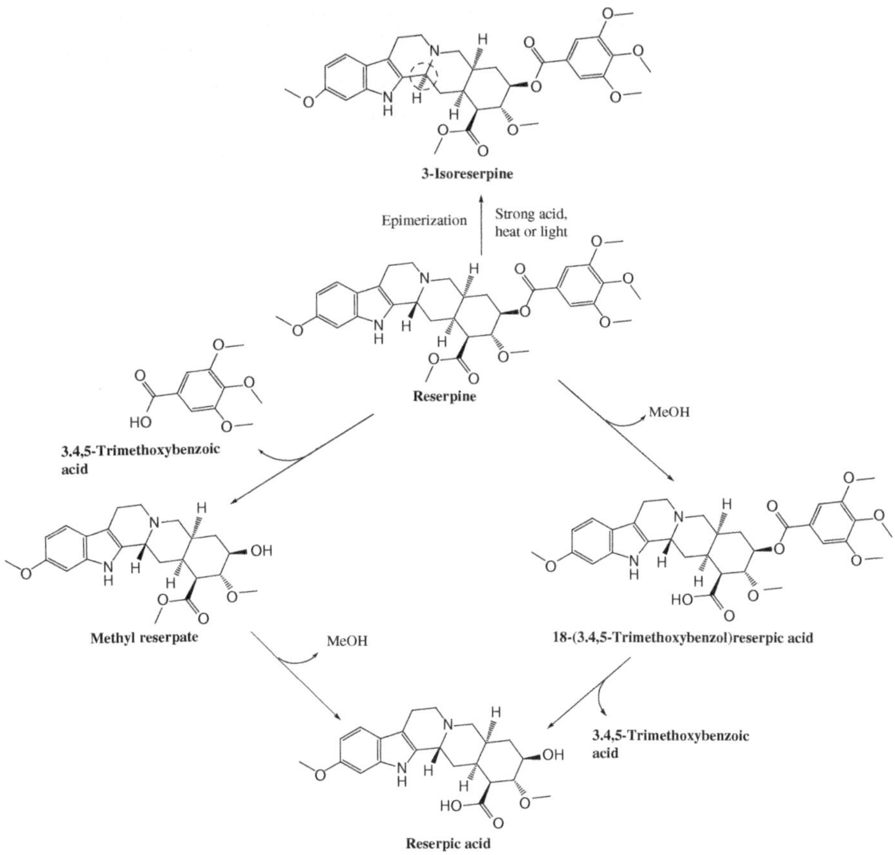

Scheme 286-2 Degradation of reserpine via hydrolysis and epimerization.

References

1 Slotkin, T.A. (1974). Reserpine. In: *Poisons of Plant Origin* (ed. L.L. Simpson and D.R. Curtis). Boston, MA: Springer https://doi.org/10.1007/978-1-4684-2943-5_1.

2 Reserpine. https://en.wikipedia.org/wiki/Reserpine (accessed March 10 2024).

3 López-Muñoz, F., Bhatara, V.S., Alamo, C., and Cuenca, E. (2004). Historical approach to reserpine discovery and its introduction in psychiatry. [Article in Spanish]. *Actas Esp. Psiquiatr.* 32: 387–395.

4 Vincent, A. and Awang, D.V.C. (1981). Determination of reserpine in pharmaceutical formulations by high performance liquid chromatography. *J. Liq. Chromotogr.* 4: 1651–1661.

5 Ljungberg, S. (1959). Effect of ultraviolet irradiation on reserpine. *J. Pharm. Belg.* 14: 115–125.

6 Awang, D.V.C., Dawson, B.A., Girard, M., and Vincent, A. (1990). The product of reserpine autoxidation. *J. Org. Chem.* 55: 4443–4448.

7 Court, W.E. and Timmins, P. (1975). The thin layer chromatographic-behaviour of some rauwolfia alkaloids. *Planta. Med.* 27: 319–329.

8 Dargel, E. and Mielck, J.B. (1990). HPLC-methods for separation and quantitation of reserpine and its main degradation products. *J. Liq. Chromotogr.* 13: 3973–3984.

9 Gaskell, A.J. and Joule, J.A. (1967). The acid catalyzed C3 epimerization of reserpine and deserpidine. *Tetrahedron* 23: 4053–4063.

10 Schirmer, R.E. (1975). *Reserpine. Anal. Prof. Drug Sub.* 4: 384–430.

287: Rifampicin [WHO List of Essential Medicines]

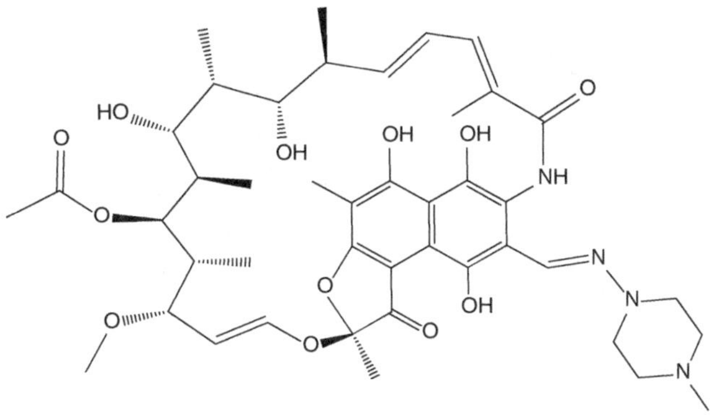

Chemical name: (7*S*,9*E*,11*S*,12*R*,13*S*,14*R*,15*R*,16*R*,17*S*,18*S*,19*E*,21*Z*)-2,15,17,27,29-Pentahydroxy-11-methoxy-3,7,12,14,16,18,22-heptamethyl-26-{(*E*)-[(4-methylpiperazin-1-yl)imino]methyl}-6,23-dioxo-8,30-dioxa-24-azatetracyclo[23.3.1.1^{4,7}.0^{5,28}]triaconta-1(28),2,4,9,19,21,25(29),26-octaen-13-yl acetate
Trade name: Rifadin, others
Formula: $C_{43}H_{58}N_4O_{12}$
Monoisotopic protonated ion (*m/z*): 823.4124
Molecular weight: 822.95
CAS number: 13292-46-1
Common salt form: Sodium salt

Rifampicin, also known as rifampin, is a semi-synthetic antibiotic that was developed at the Italian pharmaceutical company Dow-Lepetit Research Laboratories in the 1950s [1]. It is used as a first-line antituberculosis drug and most often in combination with other antituberculosis drugs such as isoniazid and pyrazinamide. It has two moieties that are susceptible to hydrolytic degradation. The hydrolysis of its hydrazone linkage yields 3-formylrifamycin (Scheme 287), which could react with isoniazid, another antituberculosis drug that is often co-formulated with rifampicin [2]. The interaction between 3-formylrifamycin and isoniazid has been described in the monograph for isoniazid in this book. On the other hand, the hydrolysis of the acetyl group of rifampicin gives rise to 25-desacetyl rifampicin [3]. Rifampicin also contains a highly substituted naphthalene ring and the vast majority of the substituents are electron-donating functional groups. Hence, the naphthalene moiety is electron-rich and thus susceptible to oxidative degradation to produce rifampicin quinone [3]; this degradation process can be inhibited by ascorbic acid [4]. In addition, the piperazine ring can also undergo oxidative degradation to give rifampicin *N*-oxide [5].

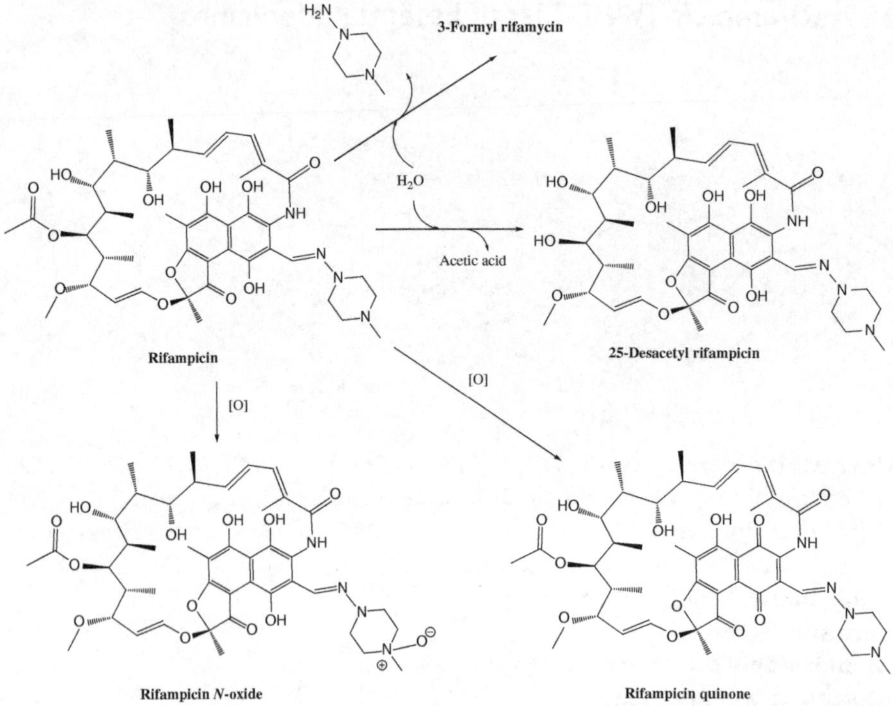

Scheme 287 Hydrolytic and oxidative degradation pathways of rifampicin. For the chemical interaction between 3-formyl rifamycin and isoniazid, refer to the monograph for isoniazid.

References

1 Sensi, P. (1983). History of the development of rifampin. *Rev. Infect. Dis.* 5 (Suppl 3): S402–S406.

2 Singh, S., Mariappan, T.T., Sharda, N. et al. (2000). The reason for an increase in decomposition of rifampicin in the presence of isoniazid under acid conditions. *Pharm. Pharmacol. Commun.* 6: 405–410.

3 Mohan, B., Sharda, N., and Singh, S. (2003). Evaluation of the recently reported USP gradient HPLC method for analysis of anti-tuberculosis drugs for its ability to resolve degradation products of rifampicin. *J. Pharm. Biomed. Anal.* 31: 607–612.

4 Maggi, N., Pasqualucci, C.R., Ballotta, R., and Sensi, P. (1966). Rifampicin: a new orally active rifamycin. *Chemother.* 11: 285–292.

5 Ripamonti, A., Ferrari, P., and Gallo, G.G. (1984). Structure identification of rifampicin *N*-oxide. Il Farmaco. *Ed. Sci.* 39: 806–809.

288–290: Risperidone, Paliperidone, Paliperidone Palmitate

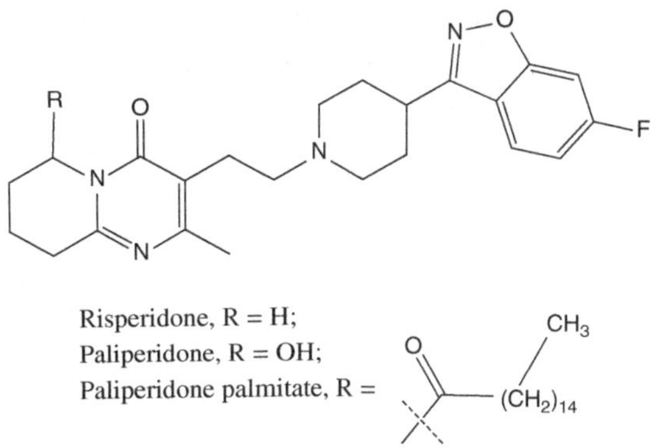

288: Risperidone [WHO List of Essential Medicines]

Chemical name: 3-[2-[4-(6-Fluoro-1,2-benzoxazol-3-yl)piperidin-1-yl]ethyl]-2-methyl-6,7,8,9-tetrahydropyrido[1,2-a]pyrimidin-4-one
Trade name: Risperdal, Okedi, others
Formula: $C_{23}H_{27}FN_4O_2$
Monoisotopic protonated ion (*m/z*): 411.2191
Molecular weight: 410.49
CAS number: 106266-06-2
Common salt form: Tartrate

289: Paliperidone

Chemical name: 3-{2-[4-(6-Fluoro-1,2-benzoxazol-3-yl)piperidin-1-yl]ethyl}-9-hydroxy-2-methyl-6,7,8,9-tetrahydropyrido[1,2-a]pyrimidin-4-one
Trade name: Invega
Formula: $C_{23}H_{27}FN_4O_3$
Monoisotopic protonated ion (*m/z*): 427.2140
Molecular weight: 426.49
CAS number: 144598-75-4

290: Paliperidone Palmitate

Chemical name: 3-{2-[4-(6-Fluoro-1,2-benzoxazol-3-yl)-1-piperidinyl]ethyl}-2-methyl-4-oxo-6,7,8,9-tetrahydro-4*H*-pyrido[1,2-a]pyrimidin-9-yl palmitate
Trade name: Invega Sustenna, Invega Trinza, Invega Hafyera
Formula: $C_{39}H_{57}FN_4O_4$
Monoisotopic protonated ion (*m/z*): 665.4437
Molecular weight: 664.91
CAS number: 199739-10-1

Risperidone is an atypical antipsychotic medication used primarily to treat schizophrenia and bipolar disorder. It was developed in Janssen Pharmaceutical Research in the late 1980s and was approved by the US FDA in 1993 [1, 2]. Risperidone is susceptible to oxidative degradation, particularly at its 9-position. In a tablet dosage form, 9-hydroxyrisperidone was found to be the major degradant after the tablets were stressed at 60°C for 1 month (Pathway a, Scheme 288–290-1) [3]. 9-Hydroxyrisperidone is also an active metabolite of risperidone and is designated as Impurity C in the European Pharmacopeia. It was later developed by Janssen into another antipsychotic drug, which is known as paliperidone [4]. Also present in the stressed tablets was the minor degradant, risperidone *N*-oxide [3], which has both the *cis*- and *trans*-forms [5] and the *cis*-form is specified in the US Pharmacopeia as Impurity Q. In another paper [6], Impurity P of risperidone, also known as bicyclorisperidone [7], was mentioned as a potential degradant as well, although it was not specified how this impurity was formed. Based on its structure, it appears that Impurity P could have been formed through an intramolecular rearrangement process (Pathway b, Scheme 288–290-1). Another degradation pathway via an intramolecular rearrangement was reported by Bharathi et al. [6], in which the benzisoxazole ring of risperidone was isomerized into benzoxazole (Pathway c, Scheme 288–290-1). This degradant was observed in a batch of risperidone tablets that were stored under the accelerated stability condition of 40°C/75%RH for 6 months. According to the authors, this degradant could also be formed under thermal stress conditions. This intramolecular isomerization is very similar to the photoisomerization of benzisothiazole–benzothiazole that has been reported for ziprasidone and lurasidone, respectively [8, 9].

Paliperidone is 9-hydroxyrisperidone, which was also developed by Janssen and approved by the US FDA for the treatment of schizophrenia in 2006 and later for schizoaffective disorder [10]. Due to its high structural similarity to risperidone, its

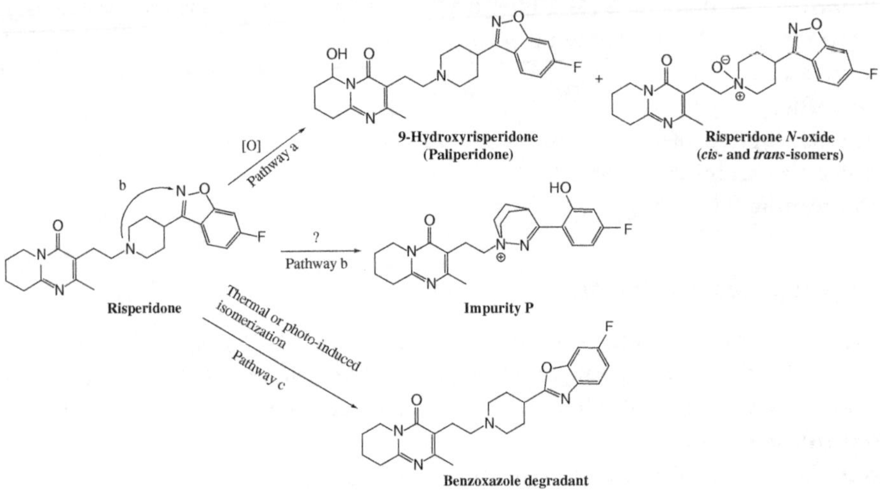

Scheme 288–290-1 Oxidative and isomerization degradation pathways of risperidone.

degradation chemistry would be expected to be largely similar. On the other hand, among the few reported forced degradation studies of paliperidone, two seem to be of reasonable quality. In a solution of concentrated HCl, which was subjected to 4 hours of refluxing, paliperidone was found to give the keto degradant (Impurity C), a specified impurity in the USP, and an unknown degradant as the major degradants (Scheme 288–290-2). It was surprising that an oxidative degradant (Impurity C) was formed under this acidic stress condition. Impurity A was also observed in this acidic stress, which is nonetheless an artificial degradant due to the use of the HCl solution [11]. While in a solution of 5 M NaOH, which was refluxed for 48 hours, an unknown impurity was found to be the major degradant. In the oxidative stress with 6% hydrogen peroxide at room temperature, the *N*-oxide degradant (Impurity D) was detected as the predominant degradant. In the apparent solid state forced degradation under thermal and ICH photostability conditions, no obvious degradation was observed.

In a photo stress study in solution with UVA and UVC irradiation, paliperidone was found to be converted to the dealkylated piperidine degradant (Impurity B) and its benzoxazole isomer [12]. The latter degradant was further degraded to form three additional degradants (Scheme 288–290-3).

Paliperidone palmitate is a prodrug of paliperidone that enables a sustained release of the active ingredient over a period of up to several months [13]. As usual, in the case of prodrugs, the active drug paliperidone is a degradant, resulting from the hydrolytic degradation of paliperidone palmitate. Under the stress of hydrogen peroxide, paliperidone palmitate *N*-oxide was formed as another major degradant (Scheme 288–290-4) [14, 15]. The photochemical degradation of paliperidone palmitate would be expected to be quite similar to that of paliperidone and risperidone, as they contain nearly identical chromophores.

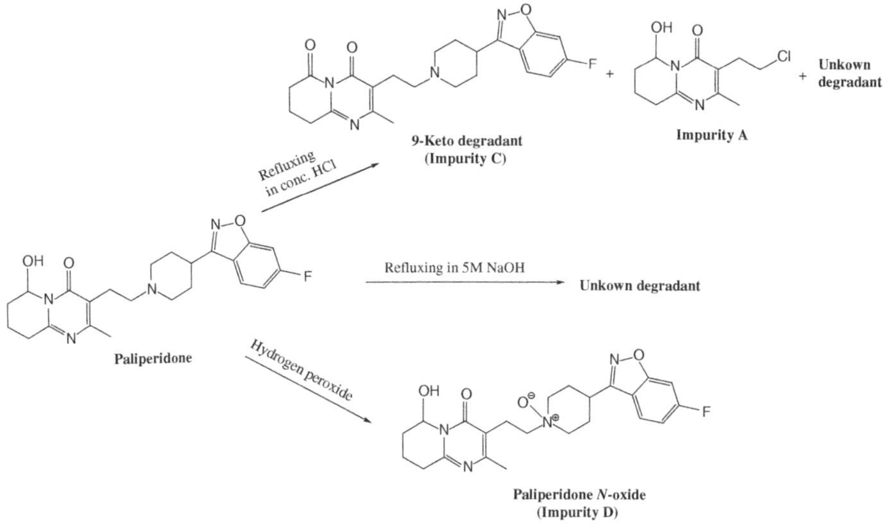

Scheme 288–290-2 Degradation of paliperidone in solutions under forced degradation conditions.

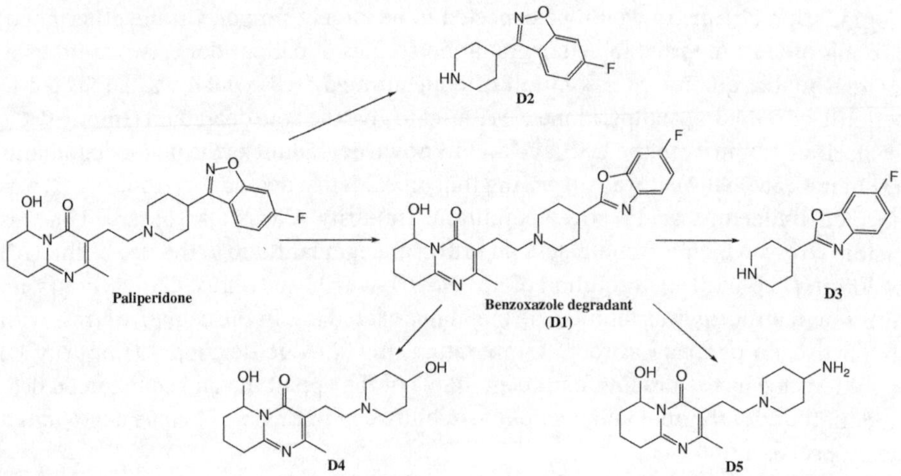

Scheme 288–290-3 Photochemical degradation of paliperidone under UVA and UVC irradiation. The nomenclature of the degradants follows that of Skibinski et al. [12]. *Source:* Skibinski et al. [12] /John Wiley & Sons.

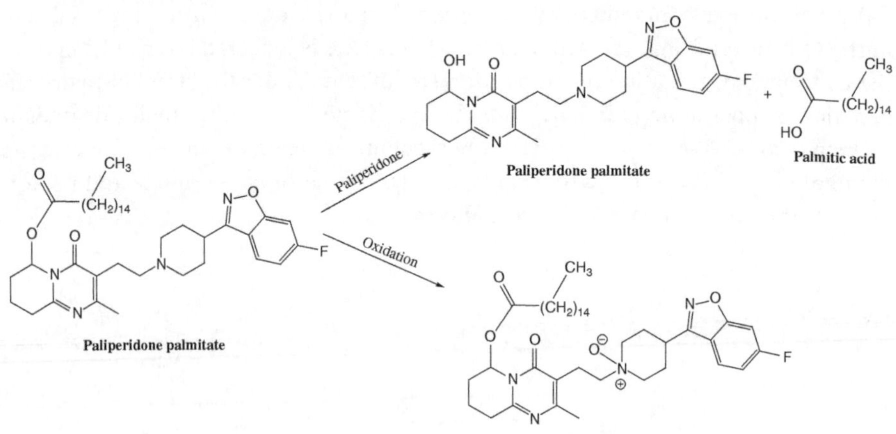

Scheme 288–290-4 Hydrolytic and oxidative degradation of paliperidone palmitate.

References

1 Colpaert, F.C. (2003). Discovering risperidone: the LSD model of psychopathology. *Nat. Rev. Drug Discov.* 2: 315–320.

2 Risperidone. https://en.wikipedia.org/wiki/Risperidone (accessed March 10 2024).

3 Tomar, R.S., Joseph, T.J., Murthy, A.S.R. et al. (2004). Identification and characterization of major degradation products of risperidone in bulk drug and pharmaceutical dosage forms. *J. Pharm. Biomed. Anal.* 36: 231–235.

4 Wu, Y.-J. (2012). Heterocycles and medicine: A survey of the heterocyclic drugs approved by the U.S. FDA from 2000 to present. In: *Progress in Heterocyclic Chemistry*, vol. 24 (ed. G.W. Gribble and J.A. Joule). Elsevier, Oxford.
5 El-Sherif, Z.A., El-Zeany, B., and El-Houssini, O.M. (2005). High performance liquid chromatographic and thin layer densitometric methods for the determination of risperidone in the presence of its degradation products in bulk powder and in tablets. *J. Pharm. Biomed. Anal.* 36: 975–981.
6 Bharathi, C., Chary, D.K., Kumar, M.S. et al. (2008). Identification, isolation and characterization of potential degradation product in risperidone tablets. *J. Pharm. Biomed. Anal.* 46: 165–169.
7 Germann, D., Kurylo, N., and Han, F. (2012). Risperidone. *Prof. Drug Sub. Excip. Relat. Methodol.* 37: 313–361.
8 Sharp, T.R., Leeman, K.R., Bryant, D.E., and Horan, G.J. (2003). On the photoisomerization of the benzisothiazole portion of ziprasidone. *Tetrahedron Lett.* 44: 1559–1561.
9 Wang, K.C., Guo, Q., Kuang, Z. et al. (2022). Structural elucidation of two novel degradants of lurasidone and their formation mechanisms under free radical-mediated oxidative and photolytic conditions via liquid chromatography photodiode array/ultraviolet-tandem mass spectrometry and one-dimensional/two-dimensional nuclear magnetic resonance spectroscopy. *J. Mass Spectrom.* 57 (7): e4871.
10 Minwalla, H.D., Wrzesinski, P., Desforges, A. et al. (2021). Paliperidone to treat psychotic disorders. *Neurol. Int.* 13: 343–358.
11 Jadhav, S.A., Landge, S.B., Choudhari, P.M. et al. (2011). Stress degradation behavior of paliperidone, an antipsychotic drug, and development of suitable stability-indicating RP-LC method. *Chrom. Res. Int.* 2011: 256812. https://doi.org/10.4061/2011/256812.
12 Skibinski, R., Komsta, L., and Inglot, T. (2016). Characterization of paliperidone photodegradation products by LC-Q-TOF multistage mass spectrometry. *Biomed. Chromatogr.* 30: 894–901.
13 Morris, M.T. and Tarpada, S.P. (2017). Long-acting injectable paliperidone palmitate: a review of efficacy and safety. *Psychopharmacol. Bull.* 47: 42–52.
14 Bindu, K.H., Dhekale, N.H., Suryanarayana, M.V., and Anjaneyulu, Y. (2012). A validated stability indicating UPLC method for simultaneous determination of assay, related substances, and degradation products of paliperidone palmitate active pharmaceutical ingredient and its pharmaceutical injection forms. *J. Liq. Chromat. Rel. Technol.* 35: 533–546.
15 Trivedi, R.K., Jain, P., Patel, M.C. et al. (2013). A rapid, stability indicating RP-UPLC method for determination of paliperidone palmitate in a depot injectable formulation. *J. Appl. Pharm. Sci.* 3: 087–092.

291: Ritonavir [WHO List of Essential Medicines]

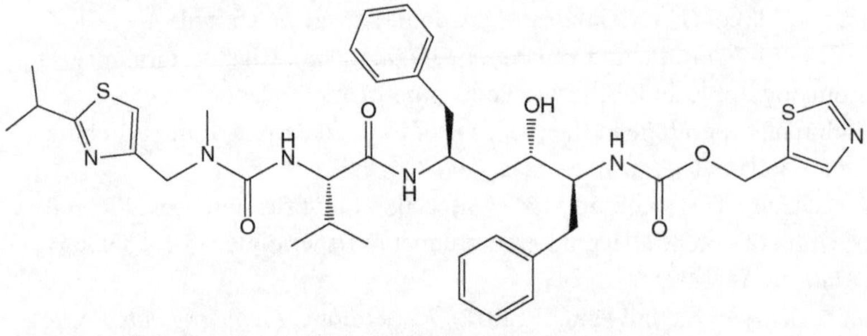

Chemical name: 1,3-Thiazol-5-ylmethyl *N*-[(2*S*,3*S*,5*S*)-3-hydroxy-5-[(2*S*)-3-methyl-2-{[methyl({[2-(propan-2-yl)-1,3-thiazol-4-yl]methyl})carbamoyl]amino}butanamido]-1,6-diphenylhexan-2-yl]carbamate
Trade name: Norvir
Formula: $C_{37}H_{48}N_6O_5S_2$
Monoisotopic protonated ion (*m/z*): 721.3200
Molecular weight: 720.95
CAS number: 155213-67-5

Ritonavir is an inhibitor of the protease of HIV, the virus that causes AIDS. This anti-AIDS drug was developed by Abbott Laboratories, the predecessor of AbbVie, and approved by the US FDA in 1996 [1]. It is used in combination with other antiviral agents, and since it is a potent inhibitor of CYP3A4, quantitatively the most important metabolizing enzyme [2], nowadays ritonavir is more commonly utilized as a pharmacokinetic booster for its companion drugs [3]. The most recent example is the anti-COVID-19 drug, Paxlovid, which contains nirmatrelvir, an inhibitor of the main protease of SARS-CoV-2, and ritonavir [4]. Ritonavir is a peptidomimetic that contains an amide, carbamate, and urea moieties, all of which could be susceptible to hydrolytic degradation. A forced degradation study reported five major degradation products under acidic, alkaline, and oxidative conditions (Scheme 291) [5]. The structural elucidation of the degradants was only based on LC–MS analysis and thus, considered tentative. It is interesting to note that no degradant would be produced from the direct hydrolysis of three hydrolysable moieties. Two out of the five major degradants were formed from intramolecular cyclization of the carbamate moiety via the attack by the nearby hydroxyl group. Part of the problems associated with these forced degradation studies are either too much degradation (and thus the degradants may not represent the real ones observed in normal stability studies) [5] or the proposed structures are gas phase fragments (such as carbocation), rather than usual compounds [6].

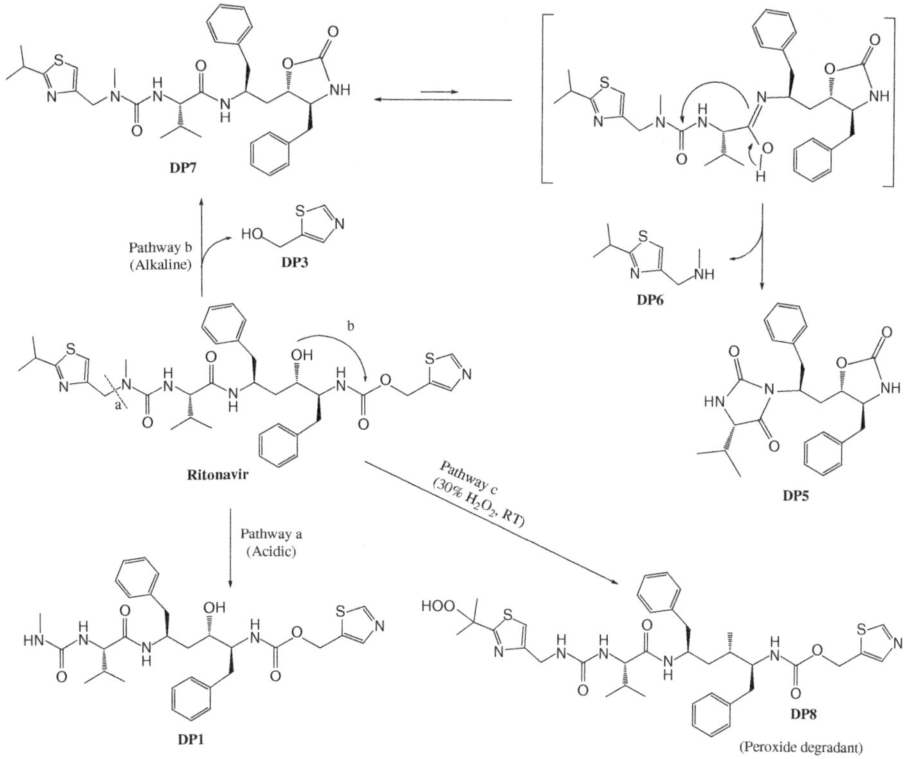

Scheme 291 Degradation of ritonavir under acidic, alkaline, and oxidative (30% hydrogen peroxide) conditions. The mechanism illustrated is based on that of Rao et al. [5] with modification. *Source:* Rao et al. [5] / with permission of Elsevier.

References

1 Ritonavir. https://en.wikipedia.org/wiki/Ritonavir (accessed August 6 2024).

2 Werk, A.N. and Cascorbi, I. (2014). Functional gene variants of CYP3A4. *Clin. Pharmacol. Thera.* 96: 340–348.

3 Jackson, A., Hill, A., Puls, R. et al. (2011). Pharmacokinetics of plasma lopinavir/ritonavir following the administration of 400/100 mg, 200/150 mg and 200/50 mg twice daily in HIV-negative volunteers. *J. Antimicrob. Chemothe.* 66: 635–640.

4 Lamb, Y.N. (2022). Nirmatrelvir plus ritonavir: first approval. *Drugs* 82: 585–591.

5 Rao, R.N., Ramachandra, B., Vali, R.M., and Raju, S.S. (2010). LC–MS/MS studies of ritonavir and its forced degradation products. *J. Pharm. Biomed. Anal.* 53: 833–842.

6 Tiwari, R.N. and Bonde, C.G. (2011). LC, LC-MS/TOF and MS^n studies for the separation, identification and characterization of degradation products of ritonavir. *Anal. Methods* 3: 1674–1681.

292: Rivaroxaban [WHO List of Essential Medicines]

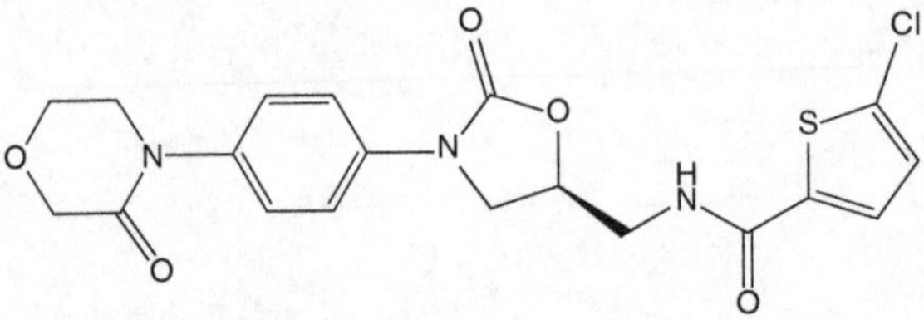

Chemical name: (*S*)-5-Chloro-*N*-{[2-oxo-3-[4-(3-oxomorpholin-4-yl)phenyl] oxazolidin-5-yl]methyl}thiophene-2-carboxamide
Trade name: Xarelto
Formula: $C_{19}H_{18}ClN_3O_5S$
Monoisotopic protonated ion (*m/z*): 436.0728
Molecular weight: 435.88
CAS number: 366789-02-8

Rivaroxaban is an inhibitor of Factor Xa, a serine protease in the blood coagulation cascade, which activates prothrombin into thrombin, an event that eventually leads to clot formation and wound closure [1]. It was originally developed by Bayer and became the first approved direct and orally available Factor Xa inhibitor by regulatory agencies, first by Health Canada and the European Medicines Agency in 2008, and then by the US FDA in 2011 [2]. The degradation chemistry of rivaroxaban is primarily of hydrolytic nature, as demonstrated by a few forced degradation studies [3-6]. Among these studies, the 3-oxomorpholine and 2-oxo-oxazolidine rings were hydrolytically degraded under acidic and alkaline stress conditions to yield

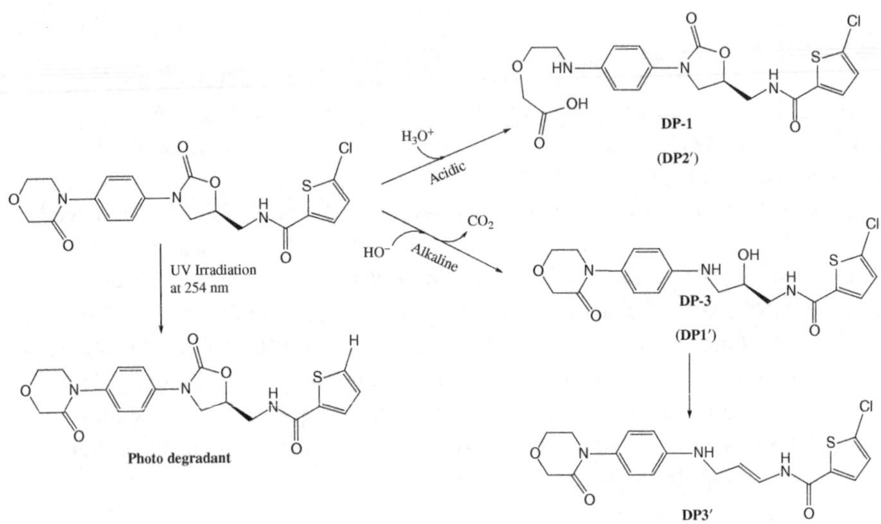

Scheme 292 Degradation of rivaroxaban under forced acidic, alkaline, and photolytic degradation conditions.

DP-1 and DP-3 (subsequently to DP3'), respectively (Scheme 292) [3]. The structures of the degradants reported from different studies were somewhat different, among which the two structures proposed by Ramisetti and Kuntamukkala (**DP-1** and **DP-3**) [3] and the one by Arous et al. (**DP3'**) [4] based on MS and NMR data appeared more reliable than others. UV irradiation of a rivaroxaban solution in acetonitrile at 254 nm produced the des-chloro degradant [5].

References

1 Núñez-Navarro, N.E., Santana, F.M., Parra, L.P., and Zacconi, F.C. (2019). Surfing the blood coagulation cascade: insight into the vital Factor Xa. *Curr. Med. Chem.* 26: 3175–3200.

2 Rivaroxaban. https://en.wikipedia.org/wiki/Rivaroxaban (accessed March 10 2024).

3 Ramisetti, N.R. and Kuntamukkala, R. (2014). Development and validation of a stability indicating LC-PDA-MS/MS method for separation, identification and characterization of process related and stress degradation products of rivaroxaban. *RSC Adv.* 4: 23155–23167.

4 Arous, B., Al-Mardini, M.A., Ghazal, H., and Al-Lahham, F. (2018). Stability-indicating method, for the determination of rivaroxaban and its degradation products, using LC-MS and TLC. *Res. J. Pharm. Tech.* 11: 1–9.

5 Wingert, N.R., dos Santos, N.O., Nunes, M.A.G. et al. (2016). Characterization of three main degradation products from novel oral anticoagulant rivaroxaban under stress conditions by UPLC-Q-TOF-MS/MS. *J. Pharm. Biomed. Anal.* 123: 10–15.

6 Nimje, H., Chavani, R., Pawari, S., and Deodhar, M. (2022). Development and validation of stability-indicating RP-HPLC method for rivaroxaban in tablet dosage form. *J. Res. Pharm.* 26: 1703–1712.

293: Rocuronium Bromide

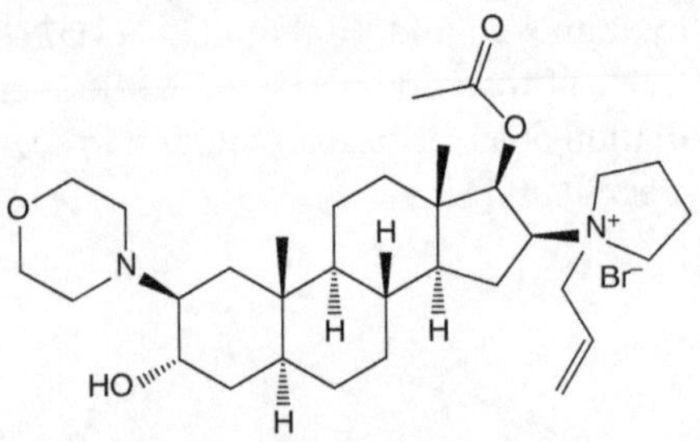

Chemical name: 1-((*2S,3S,5S,8R,9S,10S,13S,14S,16S,17R*)-17-Acetoxy-3-hydroxy-10,13-dimethyl-2-morpholinohexadecahydro-1*H*-cyclopenta[a]phenanthren-16-yl)-1-allylpyrrolidinium Bromide
Trade name: Zemuron, Esmeron
Formula: $C_{32}H_{53}BrN_2O_4$
Monoisotopic quaternary ammonium ion (*m/z*): 529.4000
Molecular weight: 609.69
CAS number: 119302-91-9

Rocuronium bromide is a nondepolarizing neuromuscular blocker, which is clinically used to provide skeletal muscle relaxation in general anesthesia or to maintain mechanical ventilation [1].

Designed to have a rapid onset and intermediate duration of action by Organon [2], it was approved by the US FDA in 1994. Chemically, it is an aminosteroid with four substituents at the 2-, 3-, 16-, and 17-positions, respectively. The 2-morpholine substituent was found to be susceptible to oxidative degradation, either under forced degradation conditions or normal storage conditions, leading to the formation of the degradants with partially or totally destructed morpholine ring (Scheme 293-1) [3–5].

El Houssini et al. performed a systematic forced degradation study on rocuronium bromide [4]. However, the hydrolytic and oxidative conditions used were quite harsh; hence, the majority of the degradants observed are likely artificial degradants, except perhaps for Degradant A (the deacetylated degradant). The latter was formed under both acidic and alkaline stress conditions, which was reported to be a potential metabolite of rocuronium, with 20 times less potency [6]. The majority of the degradants formed under harsh hydrolytic conditions are summarized in Scheme 293-2, in which the structure of Degradant E is different from that proposed by El Houssini et al.

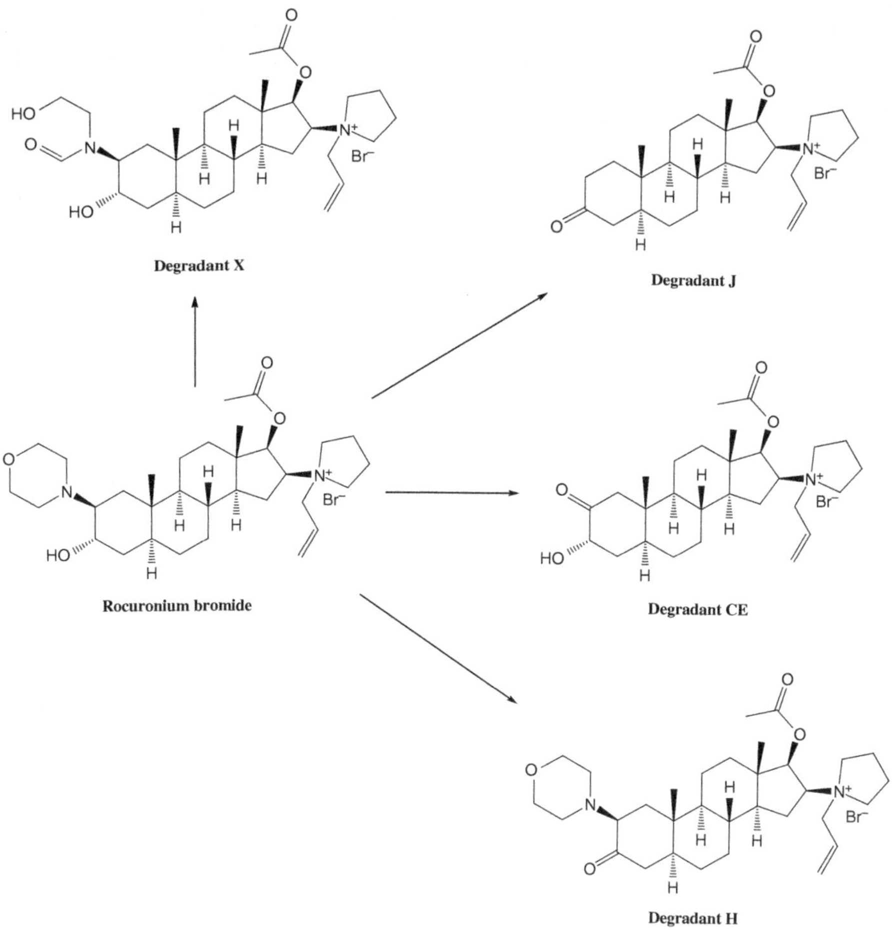

Scheme 293-1 Oxidative degradation pathways of rocuronium bromide. Degradant X was elucidated by Wegener et al. [3], Degradants H and J reported by El Houssini et al. [4], and Degradant CE elucidated by Guo et al. [5]. *Source:* El Houssini et al. [4] and Guo et al. [5].

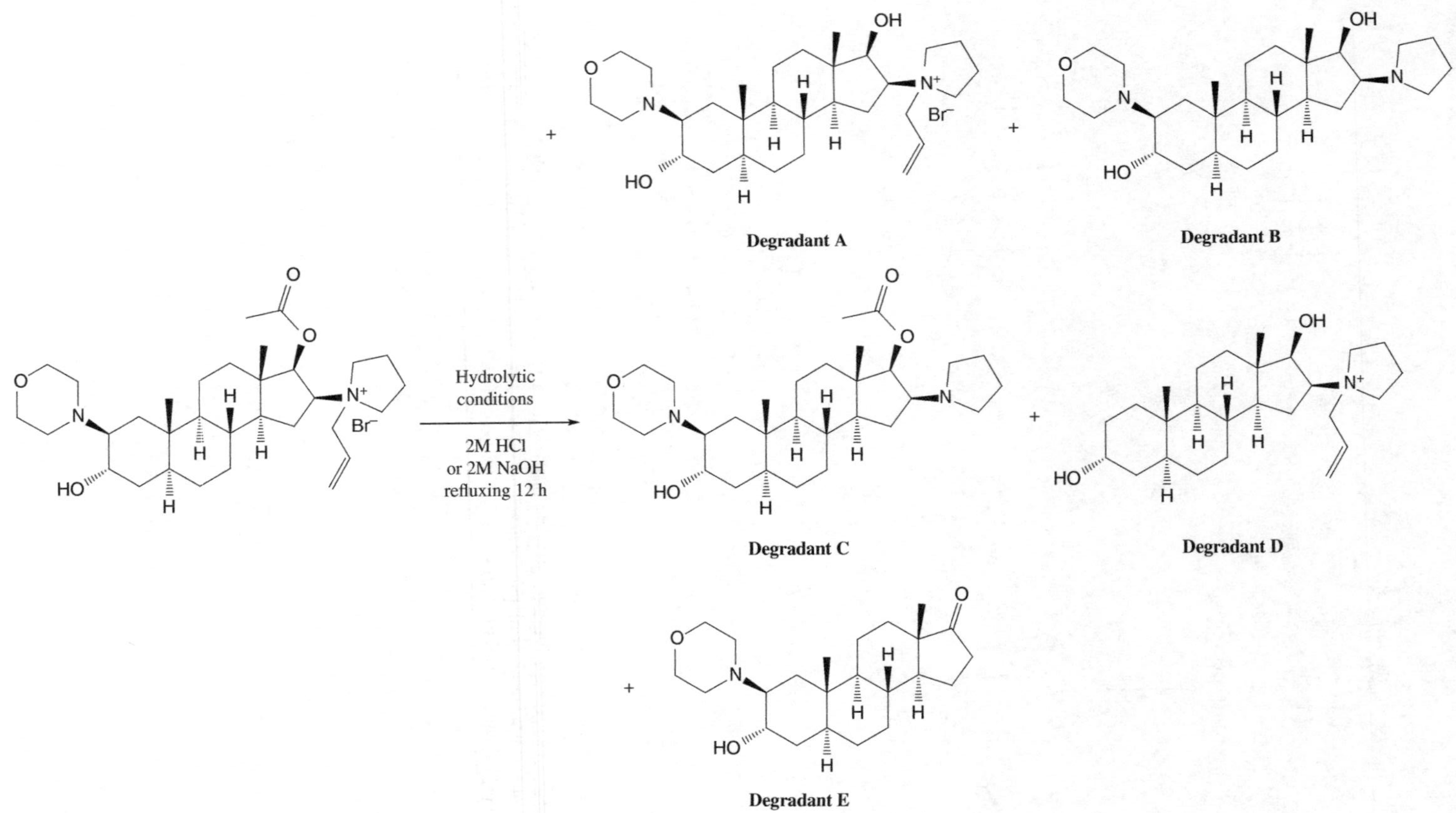

Scheme 293-2 Degradation of rocuronium bromide under hydrolytic stress conditions. The stress conditions were very harsh, and it seems that only Degradant A might be a real degradant that could be formed during a normal stability study or in storage.

References

1 Rocuronium bromide. https://en.wikipedia.org/wiki/Rocuronium_bromide (accessed August 27 2024).

2 Hunter, J.M. (1996). Rocuronium: the newest aminosteroid neuromuscular blocking drug. *Brit. J. Anaesth.* 76: 481–483.

3 Wegener, O., Harms, G., Volmer, D.A., and Hayen, H. (2015). Structural characterization of a degradation product of rocuronium using nanoelectrospray-high resolution mass spectrometry. *Drug Test. Anal.* 7: 773–779.

4 El Houssini, O.M., El-Rahman, M.K.A., Fahem, D.K., and Zaazaa, H.E. (2022). Application of ICH guidelines for studying the degradation behavior of rocuronium bromide coupled with stability-indicating RP-LC method. *J. Chromat. Sci.* 60: 217–223.

5 Guo, Q., Kuang Z., Lin, J., Li, M. Unpublished results at the Center of Excellence for Modern Analytical Technologies (CEMAT) of Huahai Pharmaceutical Co., Ltd. Degradant CE occurred at a relative retention time (RRT) of 1.09 in a HILIC HPLC method and its structure was elucidated by LC-multi stage high resolution mass spectrometric analysis.

6 Muir, A.W., Houston, J., Green, K.L. et al. (1989). Effects of a new neuromuscular blocking agent (Org 9426) in anaesthetized cats and pigs and in isolated nerve-muscle preparations. *Brit. J. Anaesthesia* 63: 400–410.

294: Rosuvastatin

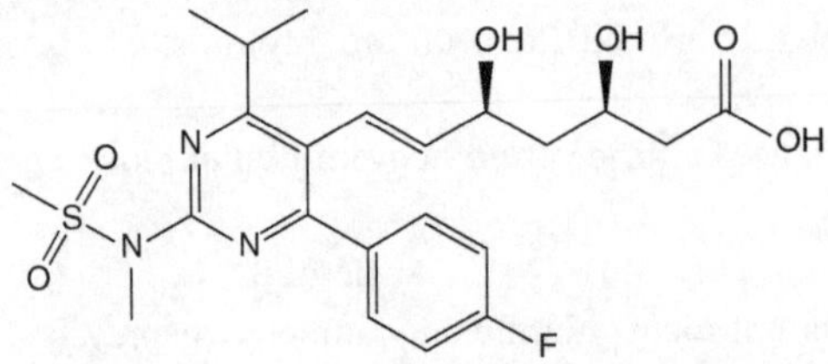

Chemical name: (*3R,5S,6E*)-7-[4-(4-Fluorophenyl)-2-(*N*-methylmethanesulfonamido)-6-(propan-2-yl)pyrimidin-5-yl]-3,5-dihydroxyhept-6-enoic Acid
Trade name: Crestor
Formula: $C_{22}H_{28}FN_3O_6S$
Monoisotopic protonated ion (*m/z*): 482.1756
Molecular weight: 481.54
CAS number: 287714-41-4
Common salt form: Calcium salt

Rosuvastatin is an inhibitor of 3-hydroxy-3-methylglutaryl-CoA reductase (often abbreviated as HMG-CoA reductase), the rate-limiting enzyme in cholesterol biosynthesis. It is clinically used to treat high cholesterol (hyperlipidemia) and thus lowers the risk of heart attacks and strokes in patients at high risk. Rosuvastatin was developed by AstraZeneca and approved by the US FDA in 2003 [1]. All statin drugs have a chiral 3,5-dihydroxy-6-heptanoic acid side chain or its cyclized lactone form, which is essential for their activities. For statin drugs that adopt the open acid form, such as rosuvastatin and atorvastatin, the heptanoic acid side chain tends to cyclize to produce the corresponding 6-membered lactone ring (Scheme 294-1) [2, 3]. Further stress under harsh acidic condition led to the formation of the dehydrated cyclized degradant [4]. Rosuvastatin was reported to be relatively stable under alkaline and thermal forced degradation conditions [2].

A number of studies have demonstrated that rosuvastatin is quite susceptible to photochemical degradation, and the major degradants are a pair of cyclized epimers (Scheme 294-2) [2, 3, 5, 6]. In a formulation containing meglumine, a secondary amine excipient, rosuvastatin was found to react with meglumine to form an amide adduct [3].

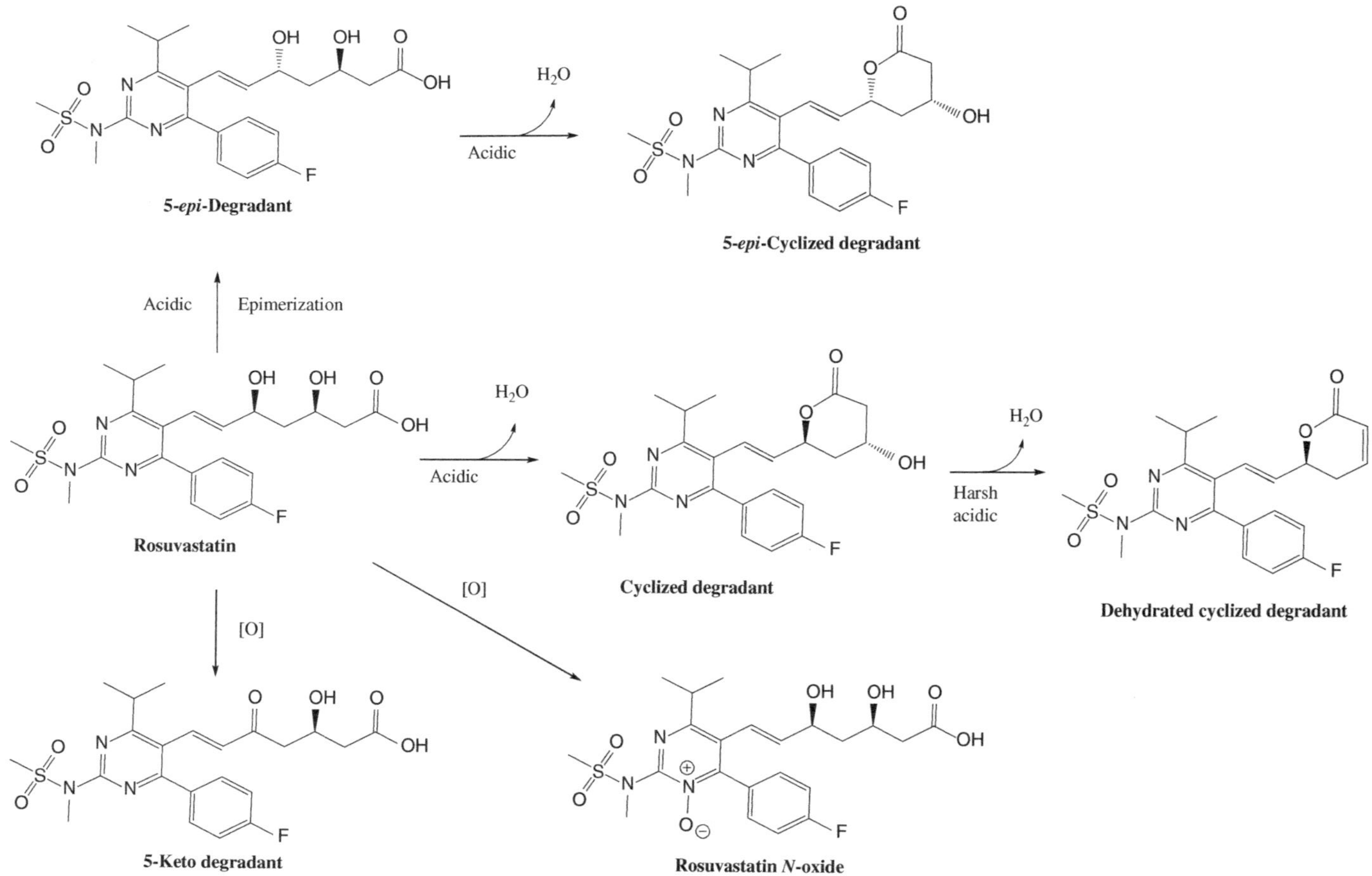

Scheme 294-1 Hydrolytic and oxidative degradation pathways of rosuvastatin.

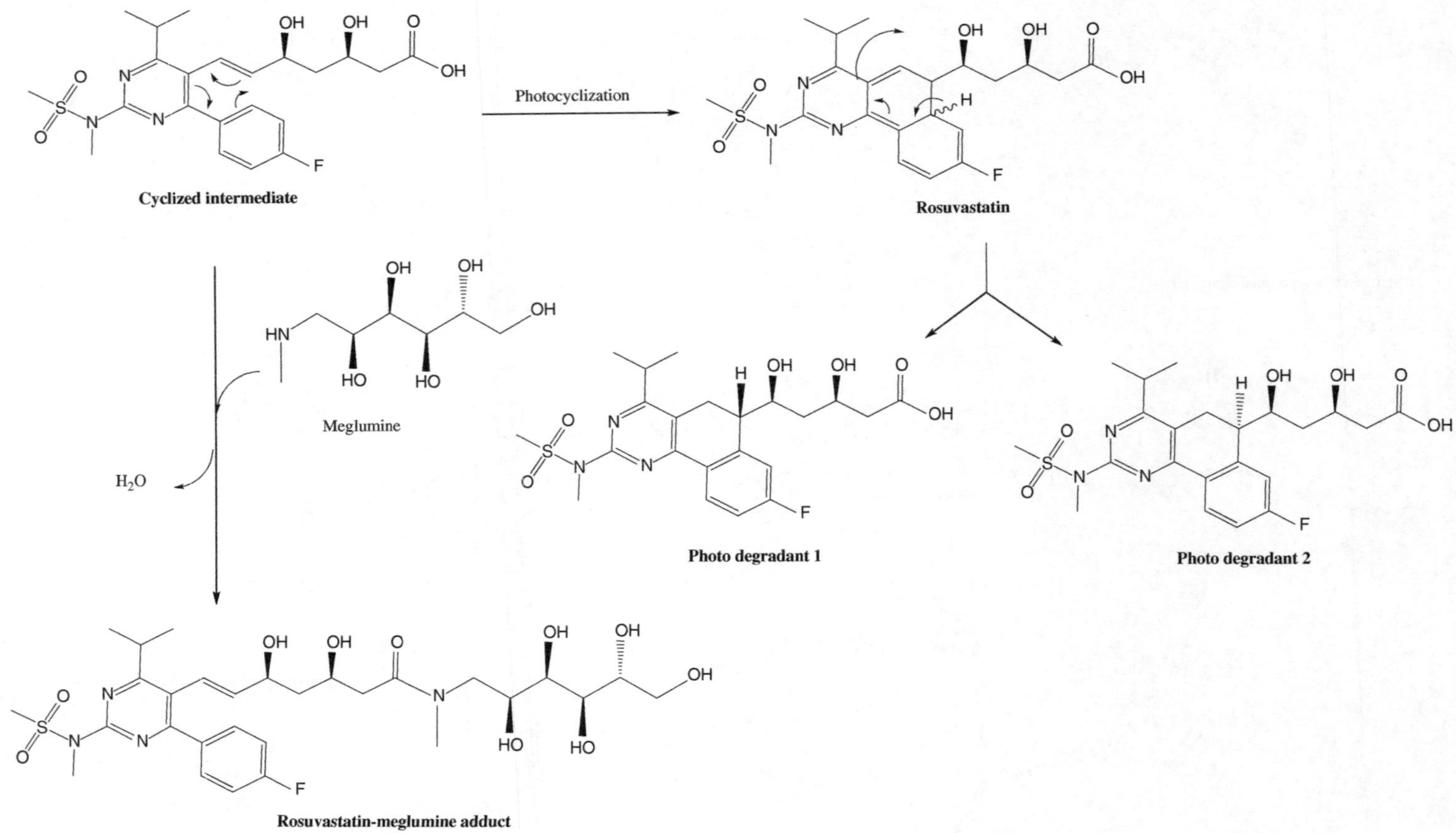

Scheme 294-2 Photochemical degradation of rosuvastatin and the adduct formation with meglumine, a secondary amine excipient.

References

1 Rosuvastatin. https://en.wikipedia.org/wiki/Rosuvastatin (accessed March 3 2024).

2 Khedr, A., Belal, F., Ibrahim, F., and Elawady, T. (2013). Analysis of rosuvastatin stress degradation behavior using liquid chromatography coupled to ultraviolet detection and electrospray ionization mass spectrometry. *Anal. Meth.* 5: 6494–6502.

3 Kishore, C.R.P. and Mohan, G.V.K. (2017). Structural identification and estimation of rosuvastatin calcium related impurities in rosuvastatin calcium tablet dosage form. *Anal. Chem. Res.* 12: 17–27.

4 Mostafa, N.M., Badawey, A.M., Lamie, N.T. et al. (2014). Selective chromatographic methods for the determination of rosuvastatin calcium in the presence of its acid degradation products. *J. Liq. Chromat. Rel. Technol.* 37: 2182–2196.

5 Litvic, M., Smic, K., Vinkovic, V., and Filipan-Litvic, M. (2013). A study of photodegradation of drug rosuvastatin calcium in solid state and solution under UV and visible light irradiation: the influence of certain dyes as efficient stabilizers. *J. Photochem. Photobiol. A: Chem.* 252: 84–92.

6 Borioni, A., Mammone, F.R., Risoluti, R. et al. (2023). Insight into the photolytic degradation products of Rosuvastatin: full chiral and structural elucidation and conversion kinetics by a combined chromatographic, spectroscopic and theoretical approach. *J. Pharm. Biomed. Anal.* 236: 115636.

295: Salbutamol [WHO List of Essential Medicines]

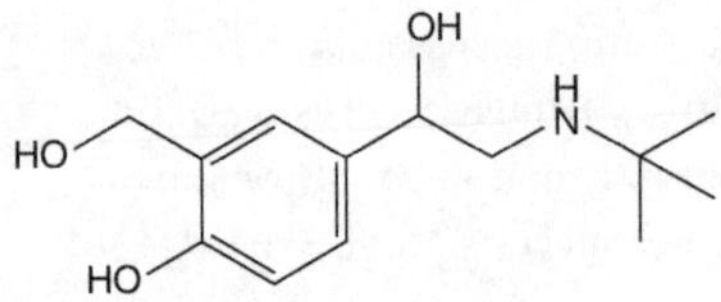

Chemical name: (*R*,*S*)-4-[2-(tert-Butylamino)-1-hydroxyethyl]-2-(hydroxymethyl) phenol
Trade name: Ventolin, Proventil, Others
Formula: $C_{13}H_{21}NO_3$
Monoisotopic protonated ion (*m/z*): 240.1594
Molecular weight: 239.32
CAS number: 18559-94-9
Common salt form: Sulfate

Salbutamol, also known as albuterol, is a short-acting β2-selective adrenergic receptor agonist, which causes the relaxation of airway smooth muscle [1]. It is clinically used for the treatment of asthma and chronic obstructive pulmonary disease (COPD). Salbutamol went onto the market in the United Kingdom in 1969 and was approved by the US FDA in 1982. It is a derivative of salicyl alcohol that contains a phenol core structure. Hence, the stability of salbutamol is pH dependent, and it was reported to be most stable at pH 3.5 [2, 3]. In a forced degradation study of salbutamol in a few acidic media, which were heated at 65°C or 85°C for various periods of time, approximately seven degradants were detected, the majority of which are oxidative degradants (Scheme 295). Although no specific mechanism was provided to rationalize the formation of these degradants in the original paper [2], it would be reasonable to presume that the formation of Degradants I, II, III, IV, and IV' was mediated by free-radical-mediated processes. For example, free-radical-mediated oxidation of salbutamol would lead to the Criegee intermediate [4], and its subsequent decomposition should give Degradant II, formaldehyde, and t-butylamine. Under elevated temperature, Degradant II could undergo decarboxylation to yield HMP. The latter could then couple with salbutamol to give Degradant VI, probably via the benzylic carbocation generated from salbutamol in acidic solutions. Degradant IV' was not mentioned in the original paper; instead Degradant IV was presented, which is an isomer of IV'. Mechanistically, Degradant IV' should be a reasonable intermediate leading to the formation of Degradants I, II, III, and t-butylaminoacetic acid. On the other hand, the dimeric degradant, although its structure was not proposed in the original paper, could be formed via the condensation of two molecules of salbutamol in acidic solutions. In an attempt to control the oxidative degradation of salbutamol, it seemed that the sulfonyl ester of salbutamol was formed as one of the degradants when sodium bisulfite or sodium metabisulfite was present in the solutions [3].

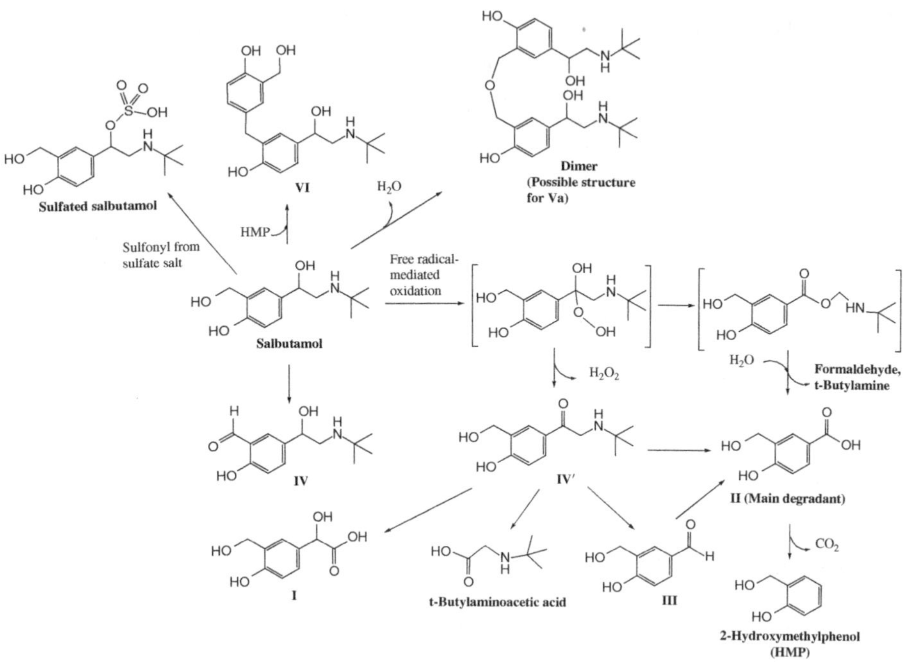

Scheme 295 Degradation pathways of salbutamol.

References

1 Salbutamol. https://en.wikipedia.org/wiki/Salbutamol (accessed October 25 2023).

2 Mälkki-Laine, L. and Bruins, A.P. (1994). Structural characterization of the decomposition products of salbutamol by liquid chromatography ionspray mass spectrometry. *J. Pharm. Biomed. Anal.* 12: 543–550.

3 Mälkki-Laine, L., Purra, K., Kähkönen, K., and Tammilehto, S. (1995). Decomposition of salbutamol in aqueous solutions. II. The effect of buffer species, pH, buffer concentration and antioxidants. *Int. J. Pharm.* 117: 189–195.

4 Li, M. (2012). Section 3.5.8. Oxidation of drugs containing alcohol, aldehyde, and ketone functionalities. In: *Organic Chemistry of Drug Degradation*. Cambridge: RSC Publishing.

296: Sildenafil

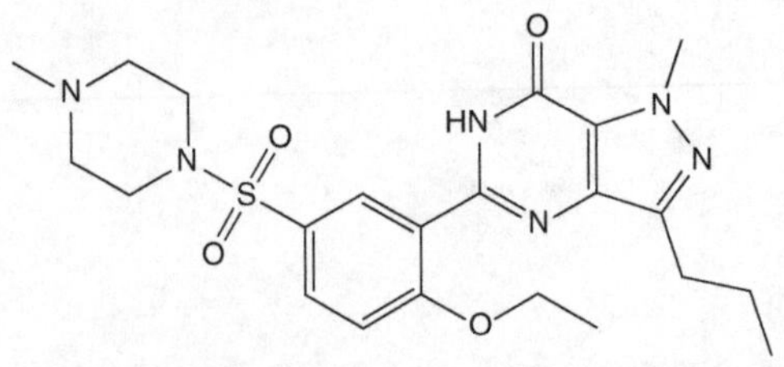

Chemical name: 5-{2-Ethoxy-5-[(4-methylpiperazin-1-yl)sulfonyl]phenyl}-1-methyl-3-propyl-1,6-dihydro-6*H*-pyrazolo[4,3-d]pyrimidin-7-one
Trade name: Viagra
Formula: $C_{22}H_{30}N_6O_4S$
Monoisotopic protonated ion (*m/z*): 475.2122
Molecular weight: 474.58
CAS number: 139755-83-2
Common salt form: Citrate

Sildenafil is a selective PDE5 inhibitor originally developed for treating angina by Pfizer. During clinical trials, it was found that sildenafil had limited efficacy for angina but a significant effect on erectile dysfunction. Hence, the drug was repurposed for treating erectile dysfunction, and this indication was approved by the US FDA in 1998. Since 2015, the indication of sildenafil has been expanded to include pulmonary arterial hypertension [1].

Although a few forced degradation studies have been reported, none seem to have performed in-depth structure elucidation of the degradants, except for the *N*-oxide degradant, which was fully characterized by LC–MS and NMR [1]. Two photochemical degradation studies were reported, in which the photodegradation occurred almost exclusively on the piperazine moiety, including its complete removal (Scheme 296) [2, 3]. On the other hand, it appeared that the pyrazolopyrimidine core of sildenafil is susceptible to hydrolysis under forced degradation conditions [4, 5]; nevertheless, the structures of the hydrolytic degradants were not elucidated.

The structure of vardenafil, another PDE5 inhibitor, is highly similar to that of sildenafil. Hence, its degradation chemistry would be expected to be very similar to that of sildenafil, particularly the photochemical and oxidative degradation pathways [3].

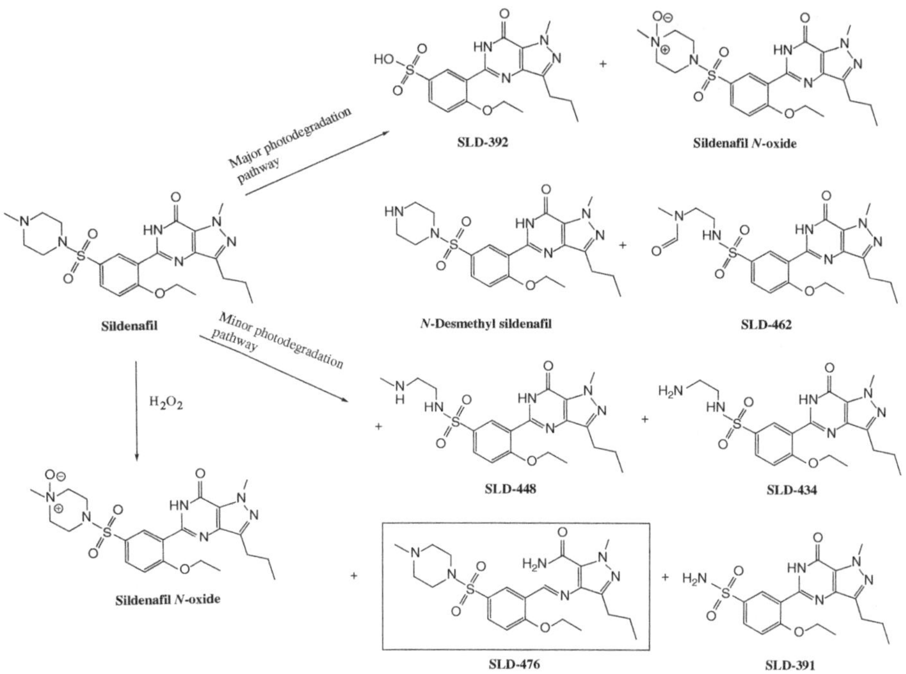

Scheme 296 Photochemical and oxidative degradation pathways of sildenafil [3]. The photochemical degradation occurred on the piperazine ring, except for SLD-476, where the pyrazolopyrimidine core was destroyed. The nomenclature for the SLD-series follows that of Herbert et al. in which the numeric numbers denote the nominal molecular weights of the corresponding degradants. *Source:* Herbert et al. [3] / with permission of Elsevier.

References

1. Abdel-Kader, M.S., Alam, P., Soliman, G.A. et al. (2021). Eco-friendly stability-indicating RP-HPTLC method for sildenafil analysis, characterization and biological evaluation of its oxidized stress degradation product. *Nature Sci. Rep.* 11: 15358.
2. Eichhorn, P., Pérez, S., Aceña, J. et al. (2012). Identification of phototransformation products of sildenafil (Viagra) and its N-demethylated human metabolite under simulated sunlight. *J. Mass Spectrom.* 47 (6): 701–711.
3. Herbert, L.P., Becker-Krail, D.B., and Cory, W.C. (2015). Persistent phototransformation products of vardenafil (Levitra®) and sildenafil (Viagra®). *Chemosphere* 134: 557–562.
4. Segall, A.I., Vitale, M.F., Perez, V.L. et al. (2000). Revered-phase HPLC determination of sildenafil citrate in the presence of its oxidative-induced degradation products. *J. Liq. Chromat. Rel. Technol.* 23: 1377–1386.
5. Atipairin, A., Woradechakul, C., Chee, K.S. et al. (2014). Method validation for determination of sildenafil citrate in extemporaneous oral suspension. *Int. J. Pharm. Pharm. Sci.* 6: 131–136.

297–298: Simvastatin, Lovastatin

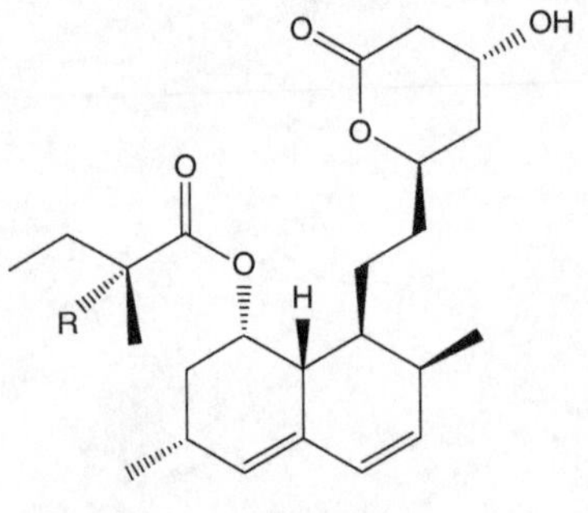

Simvastatin, R = Me;
Lovastatin, R = H.

297: Simvastatin

Chemical name: (*1S,3R,7S,8S,8aR*)-8-{2-[(*2R,4R*)-4-Hydroxy-6-oxotetrahydro-2*H*-pyran-2-yl]ethyl}-3,7-dimethyl-1,2,3,7,8,8a-hexahydronaphthalen-1-yl 2,2-dimethylbutanoate
Trade name: Zocor
Formula: $C_{25}H_{38}O_5$
Monoisotopic protonated ion (*m/z*): 419.2792
Molecular weight: 418.57
CAS number: 79902-63-9

298: Lovastatin [Mevinolin] [WHO List of Essential Medicines]

Chemical name: (*1S,3R,7S,8S,8aR*)-8-{2-[(*2R,4R*)-4-Hydroxy-6-oxotetrahydro-2*H*-pyran-2-yl]ethyl}-3,7-dimethyl-1,2,3,7,8,8a-hexahydronaphthalen-1-yl (*2S*)-2-methylbutanoate
Trade name: Mevacor
Formula: $C_{24}H_{36}O_5$
Monoisotopic protonated ion (*m/z*): 405.2636
Molecular weight: 404.55
CAS number: 75330-75-5

Lovastatin, a natural product originally isolated from the fungus *Aspergillus terreus* by Merck Sharp & Dohme Research Laboratories [1], is the first specific HMG CoA reductase inhibitor approved for treating hypercholesterolemia in 1987 [2]. It was developed based on the pioneering finding of the lead compound, mevastatin (also known as compactin), another natural product isolated from *Penicillium citrium* by Akira Endo of Sankyo Co., Ltd. in Japan in 1973, which went into clinical study, but the program was terminated later [3].

Lovastatin is the 3-methylated derivative of mevastatin, and it appeared that the introduction of this extra methyl group enabled lovastatin to become an approvable drug. On the other hand, simvastatin is a methylated derivative of lovastatin with the methyl group on the pentanoyl side-chain. Due to the steric hindrance imposed by this extra methyl group, the ester functional group of simvastatin becomes more resistant to hydrolysis either in vitro or in vivo [4]. For the critical 3,5-dihydroxy-6-heptanoic acid side chain, both simvastatin and lovastatin take the form of its cyclized lactones, and hence the lactones are pro-drugs. Thus, the open ring acid forms, which are active drugs, are considered degradation impurities of the respective drug molecules (Scheme 297–298-1). According to Ellison et al. of Merck Sharp & Dohme Research Laboratories [5], in the acidic pH range, the

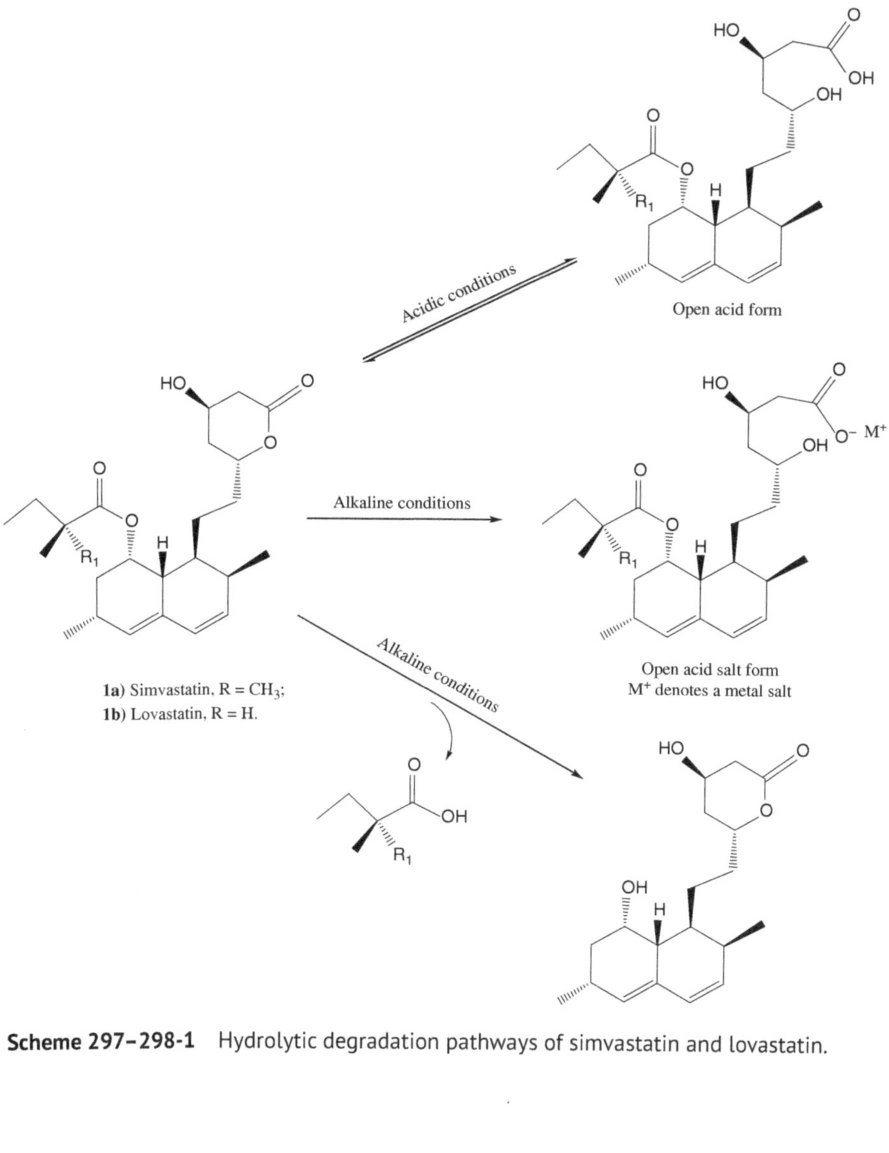

Scheme 297–298-1 Hydrolytic degradation pathways of simvastatin and lovastatin.

lactone and the open acid forms are in equilibrium, while in alkaline pH, the hydrolytic ring opening of the lactone forms is quick and irreversible.

In addition to the two moieties that are susceptible to hydrolytic degradation as mentioned above, the major degradation of simvastatin and lovastatin is of oxidative nature, which initially occurs on the α-positions of the embedded butadiene functionality via a free-radical-mediated process (Scheme 297–298-2). The carbon-based free radicals formed should readily react with molecular oxygen to produce hydroperoxide intermediates, which are not stable and subsequently decompose to yield a large number of low-level degradants [6]. It was reported that the oxidative degradation could be prevented by storage in an inert atmosphere [5].

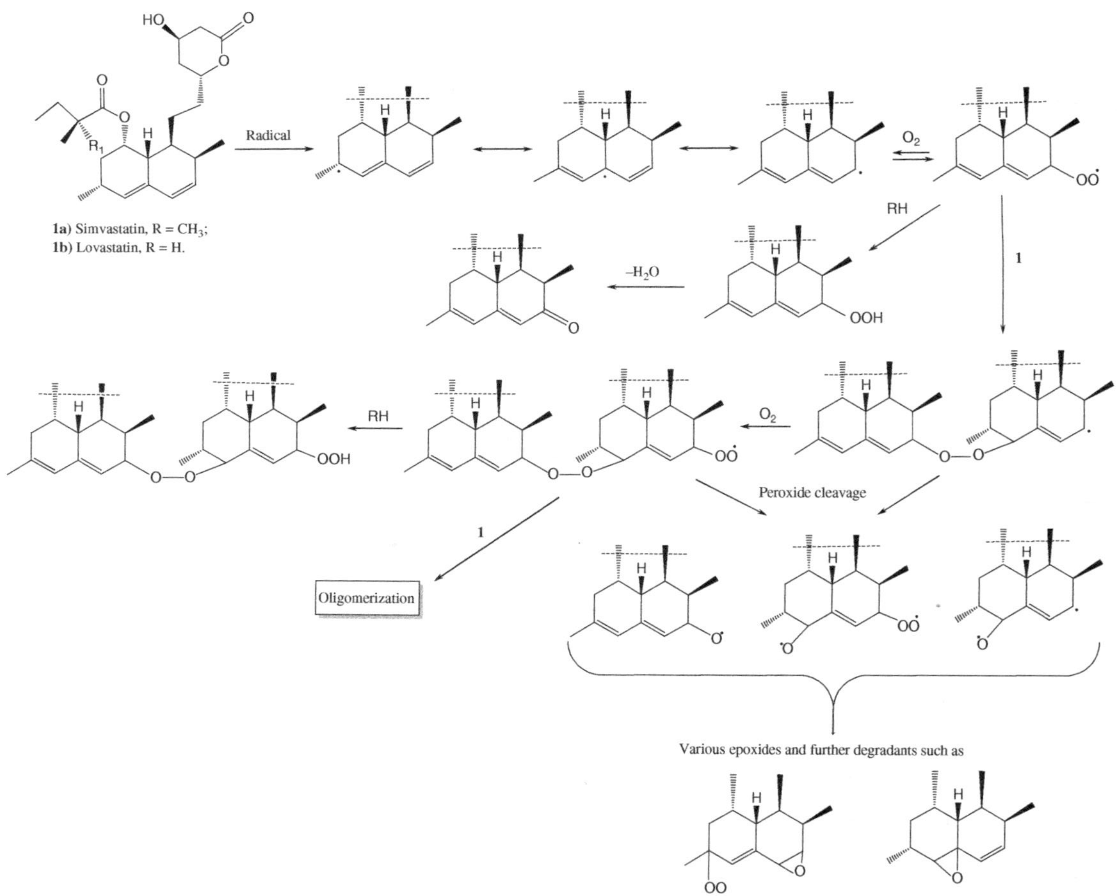

Scheme 297–298-2 Oxidative degradation pathways of simvastatin and lovastatin.

References

1 Alberts, A.W., Chen, J., Kuron, G. et al. (1980). Mevinolin: a highly potent competitive inhibitor of hydroxymethylglutaryl-coenzyme A reductase and a cholesterol-lowering agent. *Proc. Natl. Acad. Sci. USA* 77: 3957–3961.

2 Lovastatin. https://en.wikipedia.org/wiki/Lovastatin (accessed March 3 2024).

3 Endo, A., Kuroda, M., and Tsujita, Y. (1976). ML-236A, ML-236B, and ML-236C, new inhibitors of cholesterogenesis produced by *Penicillium citrinium*. *J. Antibio.* 29: 1346–1348.

4 Vree, T.B., Dammers, E., Ulc, I. et al. (2003). Differences between lovastatin and simvastatin hydrolysis in healthy male and female volunteers. *Sci. World J.* 3: 1332–1343.

5 Ellison, D.K., Moore, W.D., and Petts, C.R. (1993). Simvastatin. *Anal. Prof. Drug Sub. Excip.* 22: 359–388.

6 Smith, G.B., DiMichele, L., Colwell, L.F. Jr. et al. (1993). Autooxidation of simvastatin. *Tetrahedron* 49: 4447–4462.

299: Sitagliptin

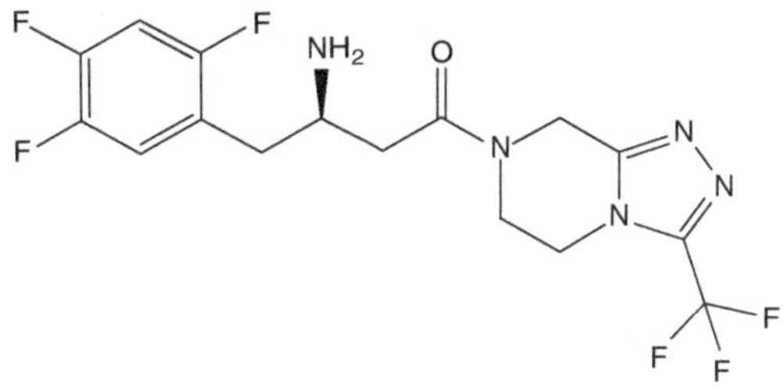

Chemical name: (*R*)-4-Oxo-4-[3-(trifluoromethyl)-5,6-dihydro[1,2,4]triazolo[4,3-a]pyrazin-7(8*H*)-yl]-1-(2,4,5-trifluorophenyl)butan-2-amine
Trade name: Januvia
Formula: $C_{16}H_{15}F_6N_5O$
Monoisotopic protonated ion (*m/z*): 408.1254
Molecular weight: 407.32
CAS number: 486460-32-6
Common salt form: Phosphate

Sitagliptin, developed by Merck & Co., Inc., was the first dipeptidyl peptidase IV (DPP4) inhibitor approved by the US FDA in 2006 to treat type II diabetes [1, 2]. It contains an amide linkage in the middle of the drug molecule, which is susceptible to hydrolytic degradation. The cleavage of the amide bond under both acidic and alkaline conditions was observed in several forced degradation studies, most of which reported both hydrolytic degradants, i.e., (*R*)-3-amino-4-(2,4,5-trifluorophenyl) butanoic acid (DP-1) and 3-(trifluoromethyl)-5,6,7,8-tetrahydro-[1,2,4]triazolo[4,3-*a*]pyrazine (DP-2) (Scheme 299) [3–8]. In a thermal stress study of sitagliptin in solid state, two deaminated degradants (DP-3 and DP-4) were detected [5], which were also observed in a compatibility study between sitagliptin and magnesium stearate, as well as in an alkaline stress solution and an oxidative stress solution of sitagliptin, respectively [8]. It is worth noting that most of the reported oxidative

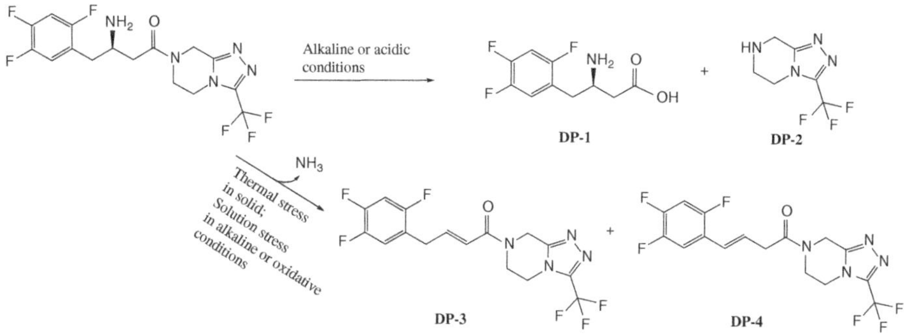

Scheme 299 Degradation of sitagliptin via hydrolytic and deaminative pathways under stress conditions.

stress studies utilized hydrogen peroxide solutions, and the resulting solutions were heated at quite high temperatures, such as 70°C or 80°C, for extended periods [6]. Such conditions seem to be too harsh, which tend to produce artificial degradants.

Sitagliptin also possesses a primary amino group, which would be expected to interact chemically with excipients containing functional groups such as Michael acceptors and reducing sugars via Michael addition and Maillard degradation, respectively [9]. In the API-excipient compatibility study performed by Gumieniczek et al. [8], sitagliptin was found to show potent interaction with fumaric acid (a Michael acceptor) and lactose (a reducing sugar), based on infrared spectroscopic evidence. Sitagliptin is often used in fixed combination with metformin, e.g., the drug product Janumet®. Peraman et al. performed an open-dish stress on Janumet tablets and found that more degradants were observed than those combined from the individually stressed sitagliptin and metformin samples, of which only DP-1 and DP-2 were identified for sitagliptin per comparison with reference compounds [4].

References

1 Florentin, M., Kostapanos, M.S., and Papazafiropoulou, A.K. (2022). Role of dipeptidyl peptidase 4 inhibitors in the new era of antidiabetic treatment. *World J. Diabetes* 13: 85–96.

2 Sitagliptin. https://en.wikipedia.org/wiki/Sitagliptin (accessed March 12 2024).

3 El-Bagary, R.I., Elkady, E.F., and Ayoub, B.M. (2011). Liquid chromatographic determination of sitagliptin either alone or in ternary mixture with metformin and sitagliptin degradation product. *Talanta* 85: 673–680.

4 Peraman, R., Gowra, C.S., Reddy, Y.P., and Peruru, K.K. (2013). Stability-indicating RP-HPLC method for simultaneous determination of metformin hydrochloride and sitagliptin phosphate in dosage forms. *Chromatogr.* https://doi.org/10.1007/s10337-013-2525-4.

5 Sonune, D.P. and Mone, M.K. (2013). Isolation, characterization of degradation products of sitagliptin and development of validated stability-indicating HPLC assay method for sitagliptin API and tablets. *Int. J. Pharm. Sci. Res.* 4: 3494–3503.

6 Vishnuvardhan, C., Radhakrishnanand, P., Navalgund, S.G., and Satheeshkumar, N. (2014). Liquid chromatography/electrospray ionisation tandem mass spectrometric study of sitagliptin and its stressed degradation products. *Drug Res.* 64: 668–674.

7 Lange, A.D.C., Miron, D., Sfair, L.L., and Schapoval, E.E.S. (2023). Sitagliptin phosphate: isolation and identification of two degradation products formed under acid conditions and determination of in vitro cytotoxicity. *Drug Anal. Res. Porto. Alegre.* 7: 41–45.

8 Gumieniczek, A., Berecka, A., Mroczek, T. et al. (2019). Determination of chemical stability of sitagliptin by LC-UV, LC-MS and FT-IR methods. *J. Pharm. Biomed. Anal.* 164: 789–807.

9 Li, M. (2012). Drug–excipient interactions and adduct formation. In: *Organic Chemistry of Drug Degradation*, 150–164. Cambridge: RSC Publishing.

300: Sofosbuvir [WHO List of Essential Medicines]

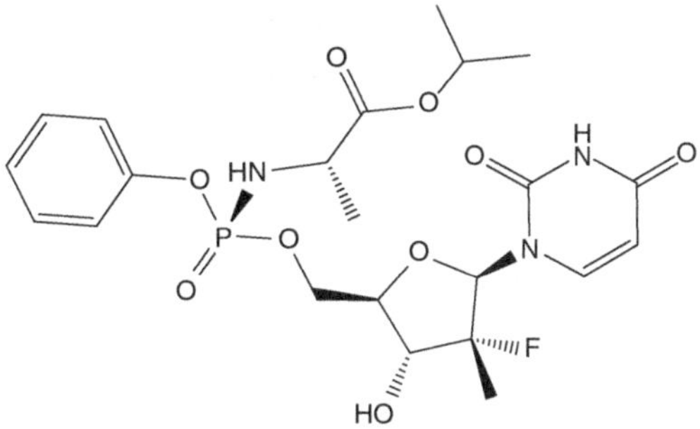

Chemical name: Isopropyl (2*S*)-2-{[(*S*)-{[(2*R*,3*R*,4*R*,5*R*)-5-(2,4-dioxo-3,4-dihydro-1(2*H*)-pyrimidinyl)-4-fluoro-3-hydroxy-4-methyltetrahydro-2-furanyl]methoxy}(phenoxy) phosphoryl]amino}propanoate
Trade name: Sovaldi
Formula: $C_{22}H_{29}FN_3O_9P$
Monoisotopic protonated ion (*m/z*): 530.1698
Molecular weight: 529.46
CAS number: 1190307-88-0

Sofosbuvir, also known as PSI-7977 or GS-7977, is a breakthrough medication for the treatment of hepatitis C virus (HCV) infection, with significantly higher curing rates than previous therapies but with much less side effects [1]. It is a prodrug and its active metabolite, GS-461203 (2′-deoxy-2′-α-fluoro-β-C-methyluridine-5′-triphosphate), is a defective substrate for HCV's NS5B RNA-dependent RNA polymerase (RdRp), thus inhibiting the viral RNA synthesis and viral replication [2, 3]. It was originally invented by Michael J. Sofia and his research team at Pharmasset, Inc. [3], a New Jersey biotech company that was acquired by Gilead Sciences for the antiviral asset in 2011. Sofosbuvir combination with ribavirin was first approved for oral dual therapy of HCV genotypes 2 and 3 [2]. Its combinations with other antiviral agents, e.g., ledipasvir and velpatasvir (both are NS5A inhibitors), were subsequently approved to cover other HCV genotypes [4].

The major degradation of sofosbuvir is of hydrolytic nature: the isopropyl ester, phenyl phosphate, and phosphoryl amide functionalities can be hydrolyzed respectively or successively to yield a number of hydrolytic degradants, depending on the pH of the media and duration of the hydrolysis (Scheme 300-1) [5–8]. The activation energies for hydrolytic degradation of sofosbuvir in solutions were found to be 77.9–79.5 kJ/mol [9]. On the other hand, the activation energy for the solid state decomposition was determined to be 115.3 kJ/mol, indicating that sofosbuvir has good thermal stability in the solid state under routine temperature [10].

In an oxidative stress of sofosbuvir in 6% hydrogen peroxide at room temperature for 19 days, Nebsen and Elzanfaly reported an oxidative degradant resulting from the oxidation of the sugar ring (Scheme 300-2) [6]. They also performed a photochemical forced degradation study and found that the photo reactions all occurred on the isopropyl ester side chain, which is somewhat surprising as the ester group is distant away from the chromophore (the phenol group).

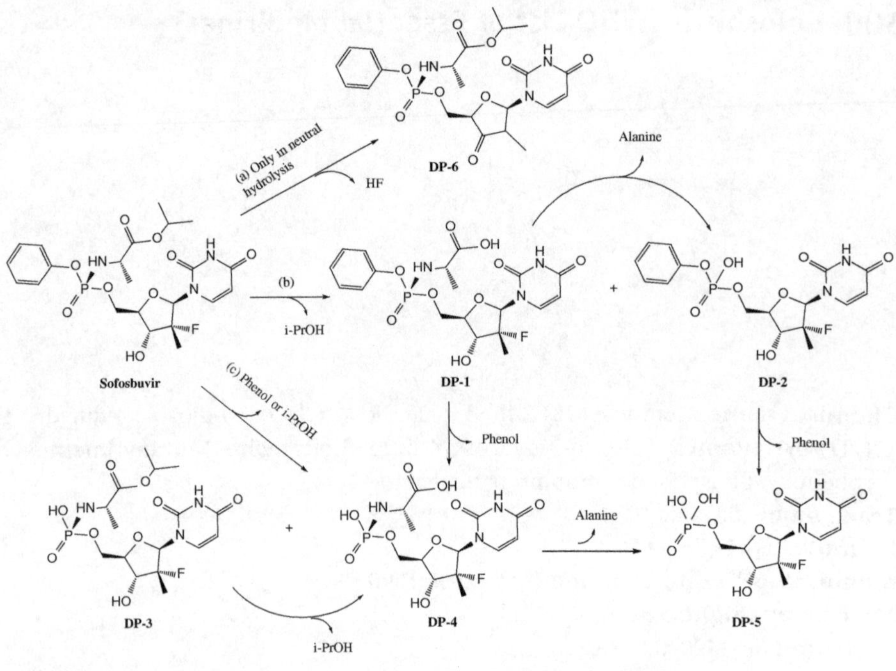

Scheme 300-1 Degradation of sofosbuvir to produce the hydrolytic degradants (DP-1–DP-5) in either acidic or alkaline solutions [4–6, 8] and to produce the elimination degradant (DP-6) in neutral solution [7]. The occurrence of a specific pathway or the distribution of the degradants can vary depending on the pH of the media and or duration of the forced degradation.

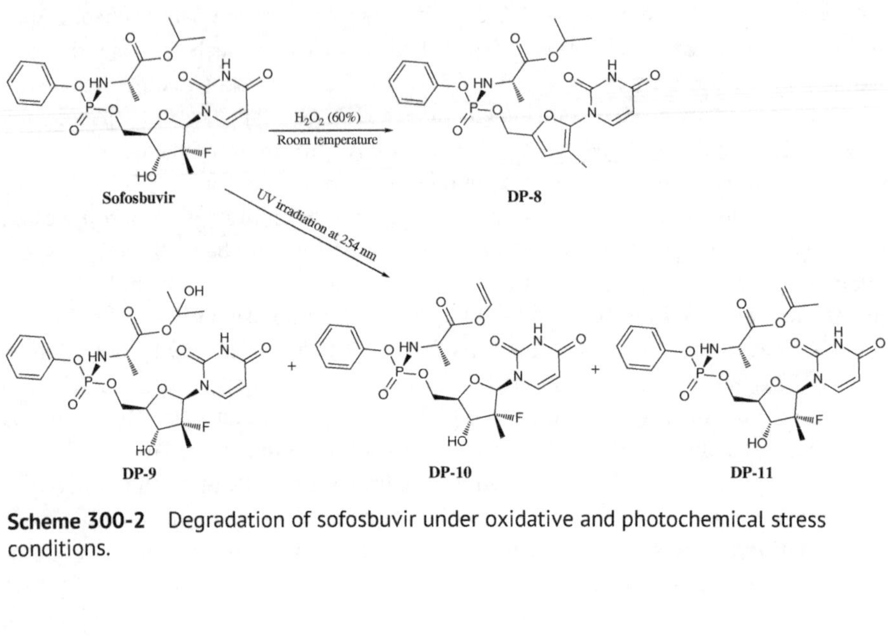

Scheme 300-2 Degradation of sofosbuvir under oxidative and photochemical stress conditions.

References

1 McQuaid, T., Savini, C., and Seyedkazemi, S. (2015). Sofosbuvir, a significant paradigm change in HCV treatment. *J. Clin. Transl. Hepatol.* 3: 27–35.

2 Sofosbuvir. https://en.wikipedia.org/wiki/Sofosbuvir (accessed March 12 2024).

3 Sofia, M.J., Bao, D., Chang, W. et al. (2010). Discovery of a β-D-2′-Deoxy-2′-α-fluoro-2′-β-C-methyluridine nucleotide prodrug (PSI-7977) for the treatment of hepatitis C virus. *J. Med. Chem.* 53: 7202–7218.

4 Greig, S.L. (2016). Sofosbuvir/Velpatasvir: a review in chronic hepatitis C. *Drugs* 76: 1567–1578.

5 Pottabathini, V., Gugulothu, V., Kaliyaperumal, M., and Battu, S. (2016). Identification, isolation and structure confirmation of forced degradation products of sofosbuvir. *Am. J. Anal. Chem.* 7: 797–815.

6 Nebsen, M. and Elzanfaly, E.S. (2016). Stability-indicating method and LC–MS-MS characterization of forced degradation products of sofosbuvir. *J. Chromatogr. Sci.* 54: 1631–1640.

7 Swain, D., Samanthula, G., Shweta Bhagat, S. et al. (2016). Characterization of forced degradation products and *in silico* toxicity prediction of sofosbuvir: a novel HCV NS5B polymerase inhibitor. *J. Pharm. Biomed. Anal.* 120: 352–363.

8 Zaman, B. and Hassan, W. (2020). Development of stability indicating HPLC-UV method for determination of process impurities and degradation products in sofosbuvir and velpatasvir tablets. *Pharm. Chem. J.* 54: 1295–1305.

9 Petřík, J., Heřt, J., Řezanka, P. et al. (2020). Development of HPLC method for the purity test by design of experiments and determination of activation energy of hydrolytic degradation reactions of sofosbuvir. *Curr. Pharm. Anal.* 16: 976–987.

10 Wang, X.-J. and You, J.-Z. (2017). Study on the thermal decomposition of sofosbuvir. *J. Anal. Appl. Pyrolysis* 123: 376–384.

301–302: Sumatriptan, Rizatriptan

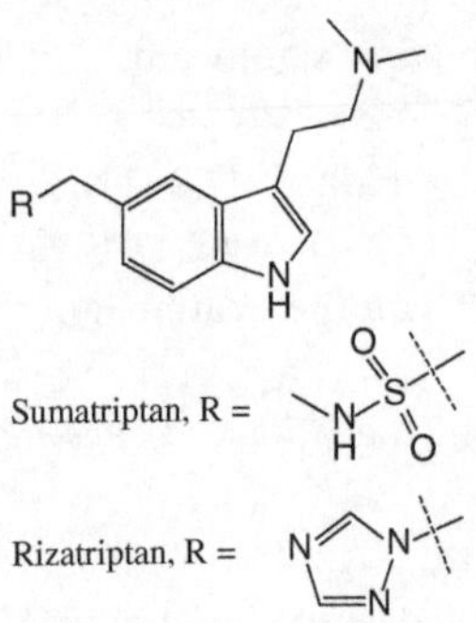

301: Sumatriptan [WHO List of Essential Medicines]

Chemical name: 1-[3-(2-Dimethylaminoethyl)-1*H*-indol-5-yl]-*N*-methyl-methane sulfonamide
Trade name: Imitrex, Onzetra, Tosymra
Formula: $C_{14}H_{21}N_3O_2S$
Monoisotopic protonated ion (*m/z*): 296.1427
Molecular weight: 295.40
CAS number: 103628-46-2
Common salt form: Hemisulfate, Succinate

302: Rizatriptan

Chemical name: *N,N*-Dimethyl-2-[5-(1*H*-1,2,4-triazol-1-ylmethyl)-1*H*-indol-3-yl] ethanamine
Trade name: Maxalt
Formula: $C_{15}H_{19}N_5$
Monoisotopic protonated ion (*m/z*): 270.1713
Molecular weight: 269.35
CAS number: 145202-66-0
Common salt form: Benzoate

Sumatriptan is a second-generation serotonin 5-$HT_{1B/1D}$ receptor agonist. Developed by Glaxo, it is the first triptan drug approved for clinical treatment of migraine headaches in 1991 [1]. It is now available in several dosage forms, including oral, nasal spray, and subcutaneous injectable forms. Sumatriptan is an indole derivative with a side chain tertiary amine and methylsulfamoyl groups. The indole ring and the tertiary amino group would be susceptible to oxidative degradation. In the forced degradation study by Xu et al. in 2001 [2], six main oxidative degradants were reported; nevertheless, based on the LC–MS/MS results and degradation chemistry principles, it was only possible for the authors to propose a definite structure (the *N*-oxide degradant) for the most abundant degradant (Scheme 301–302-1). The

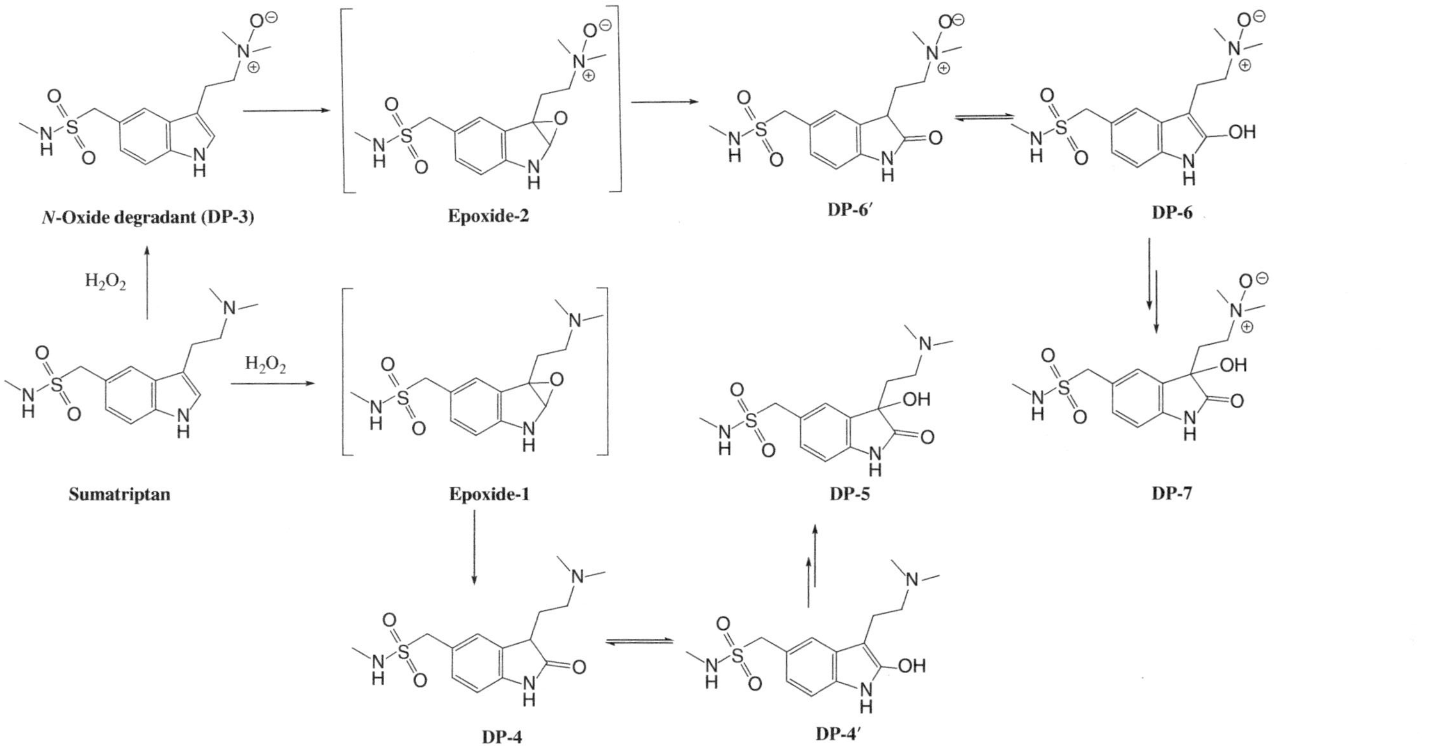

Scheme 301–302-1 Oxidative degradation pathways of sumatriptan. The nomenclature of DP-3 through DP-7 follows that of Udutha et al. [3]. The mechanism proposed here is somewhat different from Udutha et al. in that the epoxide intermediates were not mentioned in the original paper.

other oxidative degradants were found to contain one, two, or three extra oxygens, as compared to sumatriptan, and it appeared that all the extra oxygens would be on the indole ring according to their MS^2 fragmentation pathways. It needs to be noted that the oxidative stress condition used in this study was somewhat harsh (3%, 90°C, 30 minutes). In another oxidative stress study in 3% hydrogen peroxide at room temperature for 10 days, five oxidative degradants were reported with proposed structures [3]. Among the oxidative degradants, the most abundant one (DP-7) had three extra oxygens incorporated at both the tertiary amine and indole moieties. The structures of DP-7 and DP-3 (the *N*-oxide degradant) were further confirmed by NMR analysis. Although it was not mentioned by the original authors, it would be reasonable to assume that the oxidation of the indole ring would proceed via the two epoxide intermediates, which tend to be unstable at room temperature and should spontaneously rearrange to the corresponding 2-oxo degradants [4] (Scheme 301–302-1). The latter 2-oxo degradants are the tautomers of the corresponding 2-hydroxyindole degradants.

Other than the oxidative degradation, sumatriptan was found to be susceptible to degradation under somewhat harsh acidic or alkaline conditions. Most of the degradation occurred at the methylsulfamoyl side chain, yielding DP-1 and DP-2 in refluxing 1 N NaOH solution for up to 72 hours, [3] in 0.1 N NaOH at 90°C for 10 hours, or in 0.1 N HCl at 90°C for 10 hours (Scheme 301–302-2) [2]. Under the latter acidic condition, the 3-vinyl degradant was observed as the major degradant, resulting from the elimination of the side chain amino group [2]. In the stress solution of 0.1 N NaOH at 90°C for 10 hours, desmethyl degradant was detected as a minor degradant [2]. In a photolytic stress study of sumatriptan in solid state, a dimeric degradant (DP-8) was formed after exposure to 200 Whm^{-2} UV lighting in a photo stability chamber for 1 week [3].

Rizatriptan is another second-generation serotonin 5-$HT_{1B/1D}$ receptor agonist, which was developed by Merck & Co., Inc. and approved in 1998. It is clinically used for the treatment of migraine through inhibiting inflammation and dural vasodilation [5, 6]. It differs from sumatriptan only in the 5-position of the indole ring: the triazole replaces the methylsulfamoyl substituent. Hence, it would be expected that the degradation chemistry of rizatriptan should be similar to that of sumatriptan, particularly with regard to the tertiary amino side chain. For example, Rao et al. reported the *N*-oxide degradant of rizatriptan after its treatment with 3% hydrogen peroxide at 60°C for 2 hours (Scheme 301–302-3) [7]. In a formulation containing aspartame, which is a methyl ester, the quaternary ammonium degradant was formed, with aspartame as the methyl donor onto the tertiary amine moiety [8].

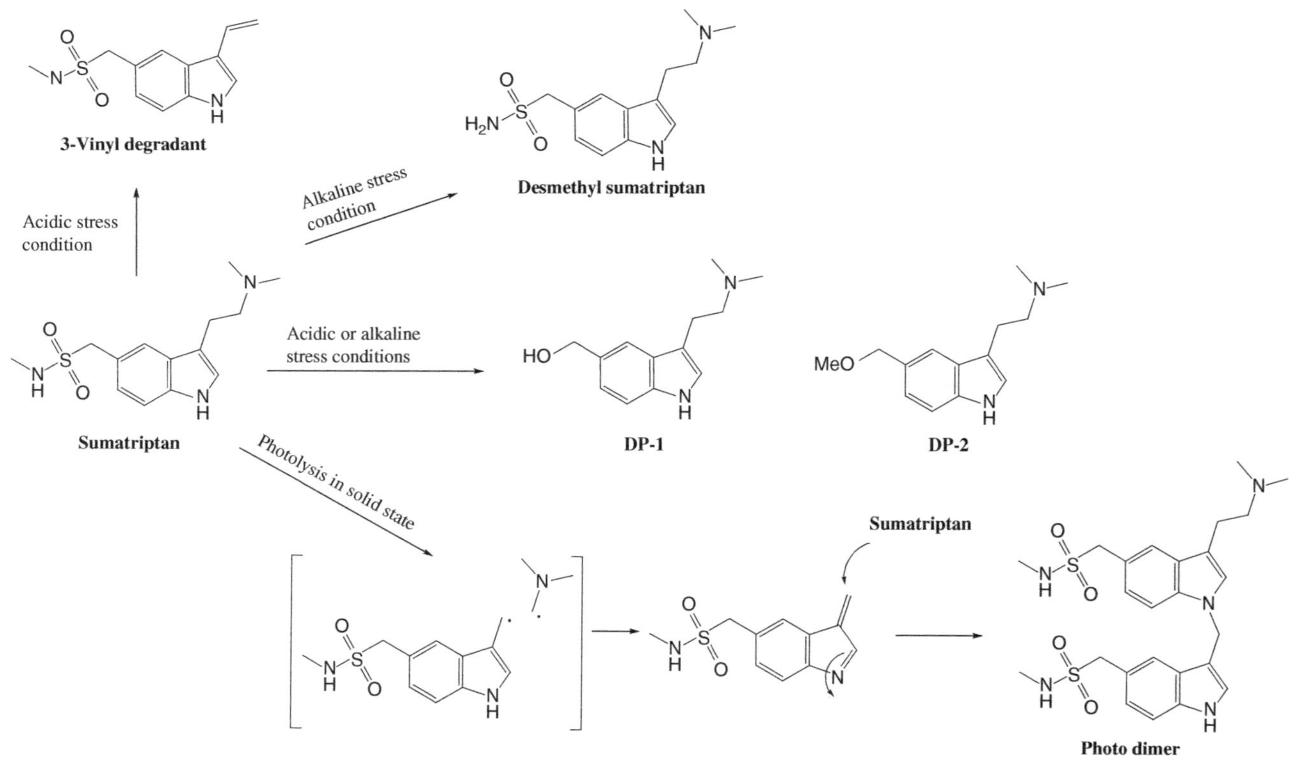

Scheme 301–302-2 Degradation pathways of sumatriptan under acidic, alkaline, and photolytic conditions.

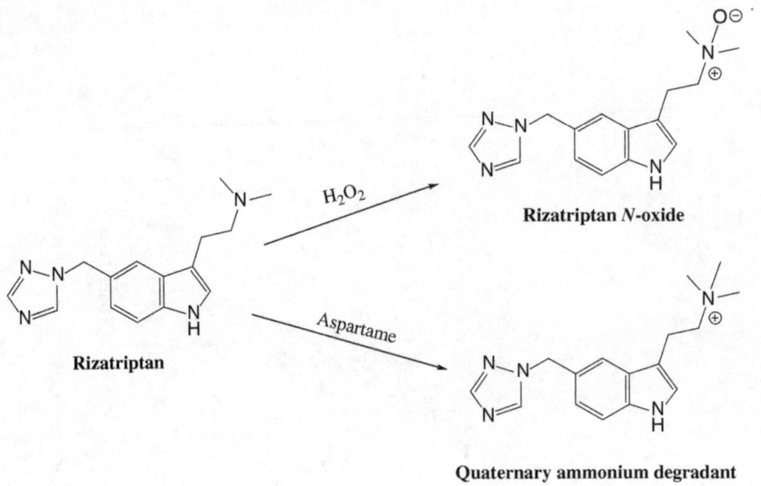

Scheme 301-302-3 Degradation of rizatriptan due to the tertiary amine moiety.

References

1 Sumatriptan. https://en.wikipedia.org/wiki/Sumatriptan (accessed August 18 2024).

2 Xu, X., Bartlett, M.G., J.T., and Stewart, J.T. (2001). Determination of degradation products of sumatriptan succinate using LC-MS and LC-MS-MS. *J. Pharm. Biomed. Anal.* 26: 367–377.

3 Udutha, S., Shankar, G., Borkar, R.M. et al. (2018). Identification and Characterization of stress degradation products of sumatriptan succinate by using LC/Q-TOF-ESI-MS/MS and NMR: toxicity evaluation of degradation products. *J. Mass Spectrom.* 53: 963–975.

4 Li, M. (2012). Section 3.5.10. Oxidation of heterocyclic aromatic rings. In: *Organic Chemistry of Drug Degradation*. Cambridge: RSC Publishing.

5 Hargreaves, R.J., Lines, C.R., Rapoport, A.M. et al. (2009). Ten years of rizatriptan: from development to clinical science and future directions. *Headache* 49 (Suppl 1): S3–S20.

6 Williamson, D.J., Hill, R.G., Shepheard, S.L., and Hargreaves, R.J. (2001). The anti-migraine 5-$HT_{1B/1D}$ agonist rizatriptan inhibits neurogenic dural vasodilation in anaesthetized guinea-pigs. *Br. J. Pharmacol.* 133: 1029–1034.

7 Rao, B.M., Sangaraju, S., Srinivasu, M.K. et al. (2006). Development and validation of a specific stability indicating high performance liquid chromatographic method for rizatriptan benzoate. *J. Pharm. Biomed. Anal.* 41: 1146–1151.

8 Zecevic, M., Jocic, B., Zivanovic, L., and Protic, A. (2008). Application of multicriteria methodology in the development of improved RP-LC-DAD for determination of rizatriptan and its degradation products. *Chromatographia* 68: 911–918.

303: Tacrolimus [WHO List of Essential Medicines]

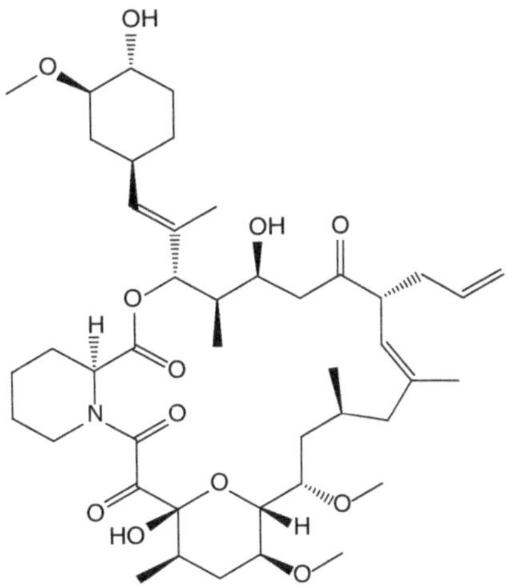

Chemical name: (*1R,9S,12S,13R,14S,17R,18E,21S,23S,24R,25S,27R*)-1,14-Dihydroxy-12-[(*E*)-1-[(*1R,3R,4R*)-4-hydroxy-3-methoxycyclohexyl]prop-1-en-2-yl]-23,25-dimethoxy-13,19,21,27-tetramethyl-17-prop-2-enyl-11,28-dioxa-4-azatricyclo[22.3.1.04,9]octacos-18-ene-2,3,10,16-tetrone
Trade name: Prograf, Advagraf, Protopic
Formula: $C_{44}H_{69}NO_{12}$
Monoisotopic protonated ion (*m/z*): 804.4893
Molecular weight: 804.03
CAS number: 104987-11-3

Tacrolimus, also known as FK-506 or Fujimycin, is an immunosuppressive drug, which is clinically used to prevent organ rejection for patients undergoing allogenic organ transplant [1]. It is also used as a topic formulation to treat eczema and psoriasis. Tacrolimus is a natural product isolated by Tanaka et al. from the bacterium *Streptomyces tsukubensis* in 1987 [2]. Chemically, it is a macrolide lactone, and the lactone ester linkage can be susceptible to hydrolytic degradation. Nevertheless, it seems that the hydrolysis only occurred meaningfully under strong acidic or alkaline stress conditions [3]. Under the latter stress conditions, dehydration could also take place either as a concurrent degradation pathway or a subsequent one to yield the diene or hydrolytic diene degradants (Scheme 303-1). Such hydrolytic degradation was also reported for similar macrolide lactones, e.g., everolimus and sirolimus [4]. Upon heating, tacrolimus can undergo a sigmatropic shift to produce the regio-isomer [5], which can also dehydrate to yield the corresponding regio-isomer diene. Both the 4-methyl and 8-(3-propenyl) groups can epimerize to afford the corresponding epimeric degradants, 4-epimer diene and 8-epimer [6, 7], which are most likely mediated via the respective keto enolization processes.

Tacrolimus also contains an amide bond with rotational barrier, and hence, it can exist as both the *cis*- and *trans*-forms. In solid state, tacrolimus exists as the

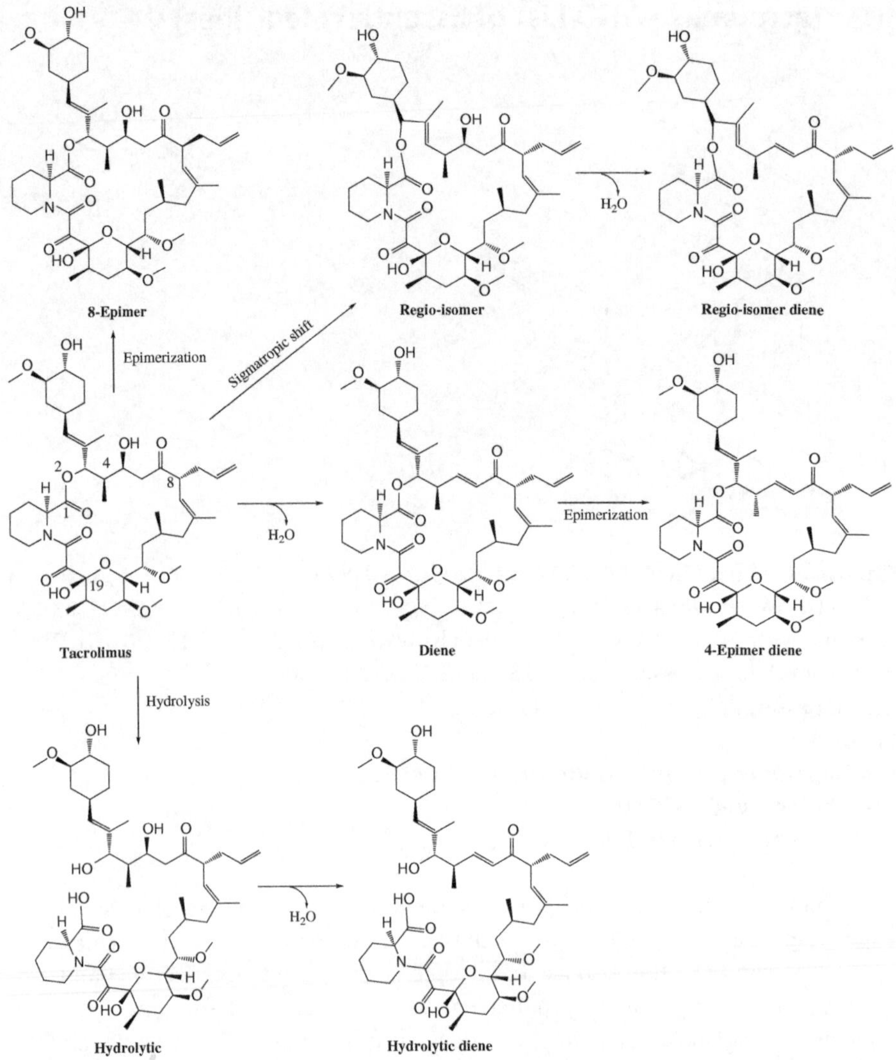

Scheme 303-1 Degradation pathways of tacrolimus. The numbering of the tacrolimus core follows that of Skytte et al. [5]. *Source:* Skytte et al. [5] / with permission of Elsevier.

trans-form and *in vivo* the vast majority of it would quickly convert to the pharmacologically active *cis*-form, based on a study by Namiki et al., indicating that ~92% of tacrolimus was present in its *cis*-form, with its 20-hydrate and 19-epimer at 1% and 7%, respectively (Scheme 303-2) [8]. It is worth noting that both the 20-hydrate and 19-epimer were referred to as the "tautomeric" forms of tacrolimus in the paper by Namiki et al., which does not seem to be accurate.

In a tacrolimus-excipient compatibility study by Peterka et al. [9], a novel degradant was identified as tacrolimus alpha-hydroxy acid, which was attributed to the

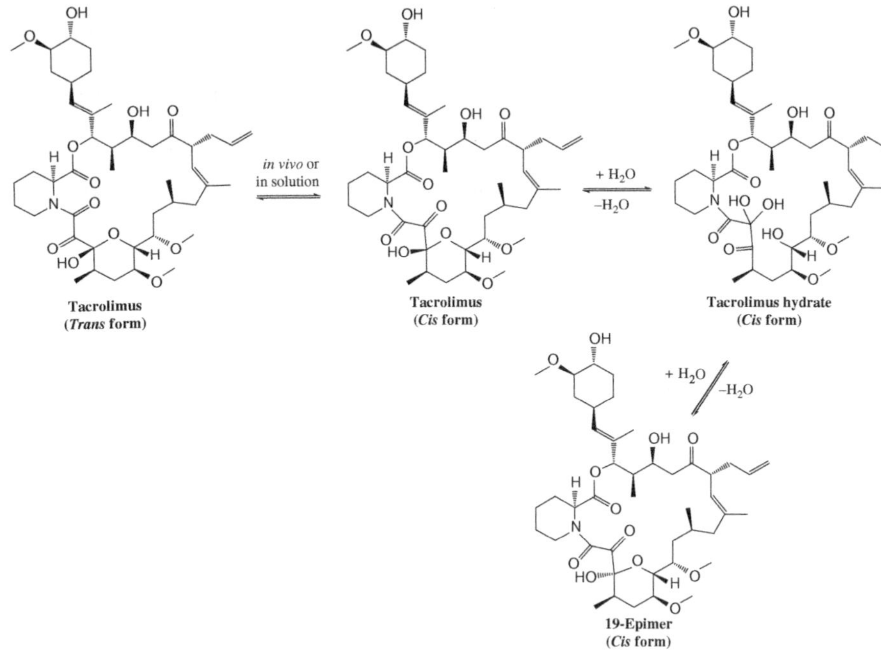

Scheme 303-2 *Trans/cis* conversion, formation of tacrolimus hydrate and 19-epimer *in vivo* or in solution.

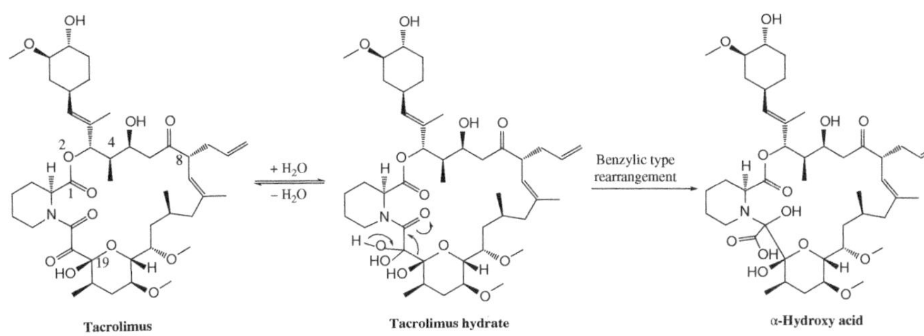

Scheme 303-3 Formation of tacrolimus α-hydroxy acid via benzylic type rearrangement.

presence of magnesium stearate (Scheme 303-3). A later study revealed that this degradant could also be generated in the forced degradation of tacrolimus with a mixture of 5 mM NaOH and acetonitrile (1:3.5, v/v)[1].

1 Wang, J., Tian, Y., Lin, J., Li, M. Unpublished results at the present author's laboratory.

References

1 Tacrolimus. https://en.wikipedia.org/wiki/Tacrolimus (accessed March 13 2024).
2 Tanaka, H., Kuroda, A., Marusawa, H. et al. (1987). Structure of FK506, a novel immunosuppressant isolated from *Streptomyces*. *J. Am. Chem. Soc.* 109: 5031–5033.
3 Prajapati, M., Eiriksson, F.F., and Loftsson, T. (2020). Stability characterization, kinetics and mechanism of tacrolimus degradation in cyclodextrin solutions. *Int. J. Pharm.* 586: 119579.
4 Ilichev, Y.V., Alquier, L., and Maryanoff, C.A. (2007). Degradation of rapamycin and its ring opened isomer: role of base catalysis. *Arkivoc.* 12: 110–131.
5 Skytte, D.M., Jaroszewski, J.W., Johansen, K.T. et al. (2013). Some transformations of tacrolimus, an immunosuppressive drug. *Eur. J. Pharm. Sci.* 48: 514–522.
6 Skak, N., Holm, P. (2017). Stabilized tacrolimus composition. US Patent 9,549,918 B2. Filed 17 February 2011. Granted 24 January 2017.
7 Skytte, D.M., Frydenvang, K., Hansen, L. et al. (2010). Synthesis and characterization of an epimer of tacrolimus, an immunosuppressive drug. *J. Nat. Prod.* 73: 776–779.
8 Namiki, Y., Fujiwara, A., Kihara, N. et al. (1995). Determination of the immunosuppressive drug tacrolimus in its dosage forms by high-performance liquid chromatography. *Chromatographia* 40: 253–258.
9 Peterka, T.R., Grahek, R., Hren, J. et al. (2015). Solid state compatibility study and characterization of a novel degradation product of tacrolimus in formulation. *J. Pharm. Biomed. Anal.* 110: 67–75.

304: Tadalafil

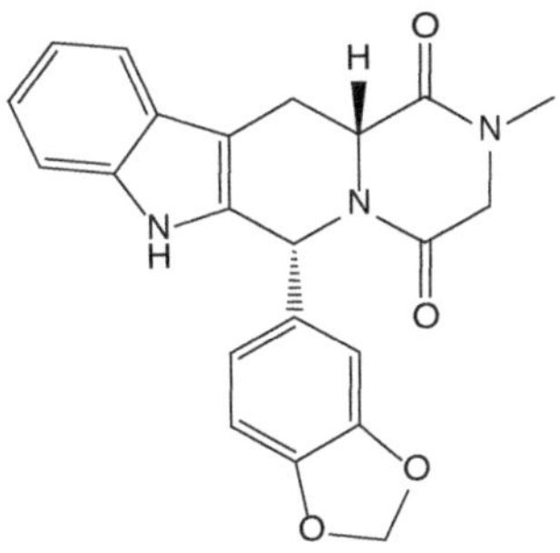

Chemical name: (*2R,8R*)-2-(1,3-Benzodioxol-5-yl)-6-methyl-3,6,17-triazatetracyclo [8.7.0.03,8.011,16]heptadeca-1(10),11,13,15-tetraene-4,7-dione
Trade name: Cialis
Formula: $C_{22}H_{19}N_3O_4$
Monoisotopic protonated ion (*m/z*): 390.1448
Molecular weight: 389.41
CAS number: 171596-29-5

Tadalafil is a highly selective phosphodiesterase 5 (PDE5) inhibitor clinically used to treat erectile dysfunction, benign prostatic hyperplasia, and pulmonary arterial hypertension [1, 2]. It was developed by Eli Lilly and approved by the US FDA in 2003.

Chemically, tadalafil is an indole derivative with an extended fused ring system that contains a diketopiperazine (DKP) moiety. The 6-membered DKP ring is usually hydrolytically stable but can be hydrolyzed under harsh acidic and alkaline stress conditions (Scheme 304) [3], which typically does not resemble a real degradation scenario. On the other hand, Temussi et al. reported that tadalafil underwent epimerization on the chiral methylene of the DKP ring at pH 12 to produce C12a-epimer [4]. The same authors also studied the photochemical degradation behavior of tadalafil in solution under UV irradiation and observed C6-epimer and two diastereo photo-hydrolytic degradants as the main degradants. In an oxidative forced degradation study of tadalafil in ~4% hydrogen peroxide in methanol at room temperature for 24 hours, it was found to decompose at ~2% [5]. Another stress study also showed that tadalafil had a certain degree of degradation under oxidative condition [6]. Nevertheless, no structural elucidation of the oxidative degradant was performed in either study. It seems likely that the oxidative degradation would occur on the indole ring, as it is known to be susceptible to oxidation by hydrogen peroxide [7].

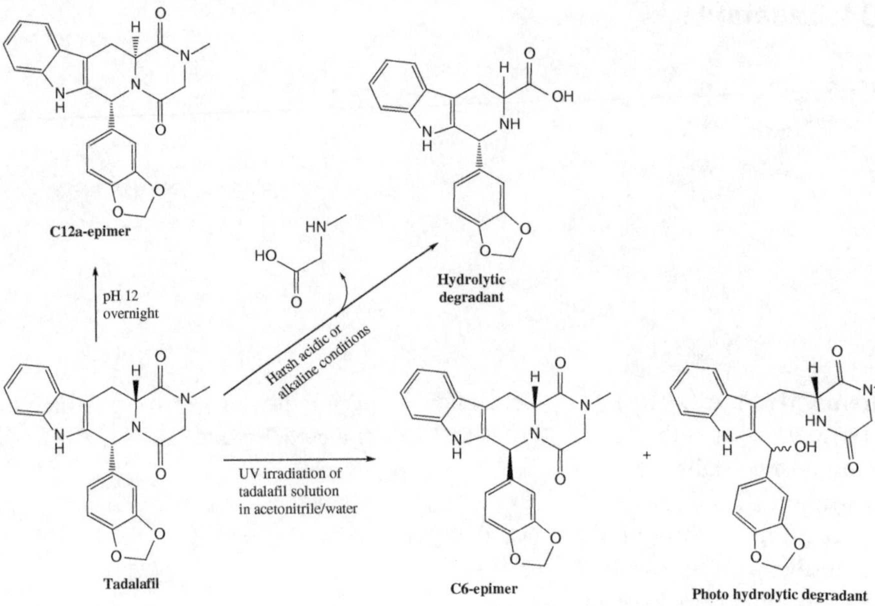

Scheme 304 Hydrolytic and photochemical degradation pathways of tadalafil.

References

1 Ahmed, N.S. (2019). Tadalafil: 15 years' journey in male erectile dysfunction and beyond. *Drug Dev. Res.* 80: 683–701.
2 Tadalafil. https://en.wikipedia.org/wiki/Tadalafil (accessed March 13 2024).
3 Rezk, M.R., Abdel-Moety, E.M., Wadie, M., and Tantawy, M.A. (2020). Stability assessment of Tamsulosin and Tadalafil co-formulated in capsules by two validated chromatographic methods. *J. Sep. Sci.* 44 (2): 530–538.
4 Temussi, F., Bassolino, G., Cermola, F. et al. (2010). Investigation on the phototransformation of tadalafil in aqueous media. 6-Epimerization vs. solvent trapping reaction. *Photochem. Photobiol. Sci.* 9: 1139–1144.
5 Pinto, M.A., Nicorena, K.F., Machado, M.M. et al. (2022). Tadalafil and sildenafil illicit association: stability indicating HPLC method, photodegradation kinetic and toxicological studies. *Braz. J. Pharm. Sci.* 58: e19491.
6 Rao, D.V.S., Radhakrishnan, P., and Himabindu, V. (2008). Stress degradation studies on tadalafil and development of a validated stability-indicating LC assay for bulk drug and pharmaceutical dosage form. *Chromatographia* 67: 183–188.
7 Li, M. (2012). Section 3.5.2. Double bonds susceptible to addition by hydroperoxides. In: *Organic Chemistry of Drug Degradation*. Cambridge: RSC Publishing.

305: Tamoxifen [WHO List of Essential Medicines]

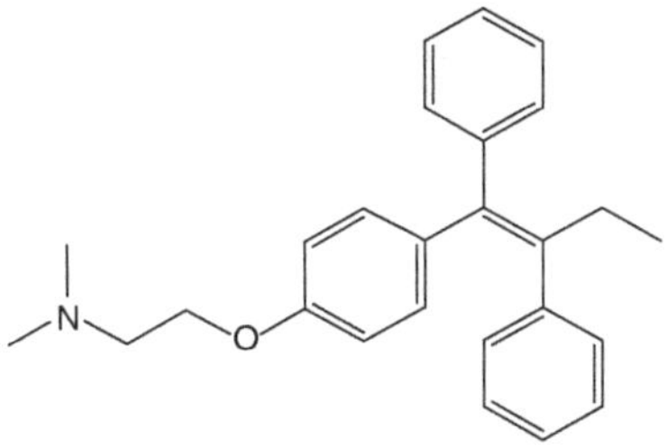

Chemical name: 2-[4-[(*Z*)-1,2-Diphenylbut-1-enyl]phenoxy]-*N*, *N*-dimethylethanamine
Trade name: Nolvadex, Others
Formula: $C_{26}H_{29}NO$
Monoisotopic protonated ion (*m/z*): 372.2322
Molecular weight: 371.52
CAS number: 10540-29-1
Common salt form: Citrate

Tamoxifen is a selective estrogen receptor modulator clinically used to treat and prevent breast cancer [1, 2]. It was originally synthesized by Dora Richardson in 1962 (code-named ICI-46,474), as part of a research team headed by Arthur L. Walpole at ICI Pharmaceuticals. The team's original goal was to develop a morning-after contraceptive pill; nevertheless, it turned out to be ineffective as a contraceptive pill. Following on the advice by Walpole, V. Craig Jordan, who studied tamoxifen for his doctoral research at the University of Leeds in the early 1970s, went to Worcester Foundation for Experimental Biology to continue studying tamoxifen and eventually determined that tamoxifen is effective for treating breast cancer. Jordan's work triggered the process that led to the US FDA's approval of tamoxifen as an anti-breast cancer drug in 1977 [3, 4].

Chemically, tamoxifen can be considered a derivative of stilbene, and its central carbon–carbon double bond is susceptible to photochemical *cis–trans* isomerization. Due to conjugation to the three phenyl substituents, tamoxifen can undergo photochemical *cis-trans* isomerization under the irradiation by UV and sunlight (Scheme 305-1) as its UV absorption extends to 320 nm, to produce the *E*-isomer [5, 6]. While tamoxifen (in *Z*-configuration) is an anti-estrogen, its regio *E*-isomer is a full estrogen agonist [7]. In the presence of air, several photo-oxidative degradants were also observed, among which the two phenanthrene degradants were formed via intra-molecular crosslinking, while the other two degradants apparently resulted from the oxidative destruction of the central carbon–carbon double bond [5].

Tamoxifen also contains a tertiary amino moiety that should be susceptible to oxidative degradation to yield the *N*-oxide degradant (Scheme 305-2) [8]. Although it appears that the *N*-oxide degradant has not been specifically reported [9], the analogous α-acetoxytamoxifen *N*-oxide was readily prepared by treating α-acetoxytamoxifen with hydrogen peroxide [10].

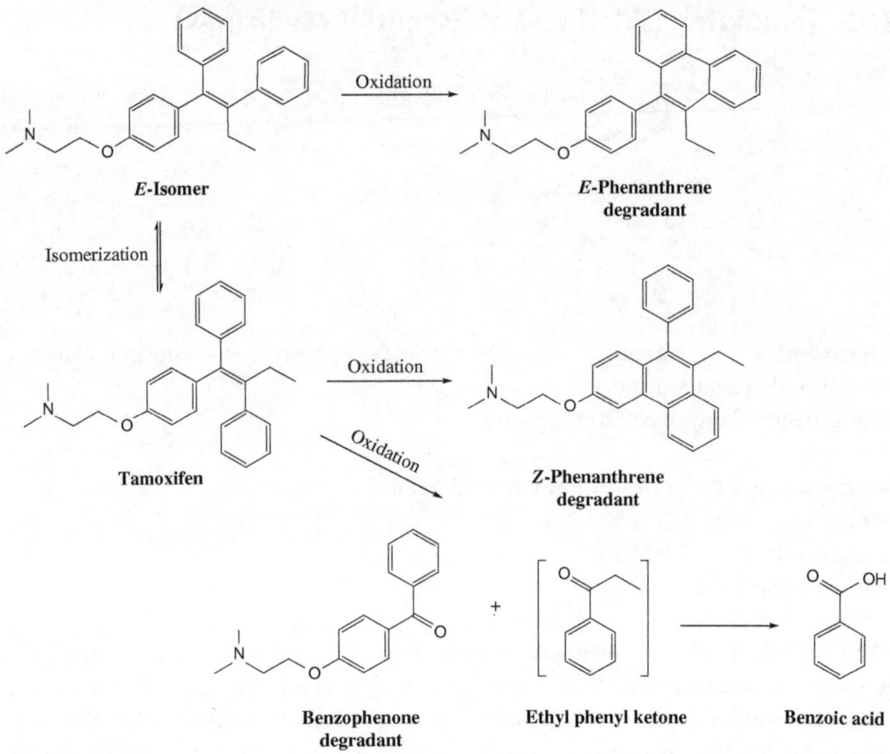

Scheme 305-1 Photochemical degradation pathways of tamoxifen in aqueous methanol solution exposed to air. When the photolysis was carried out in argon atmosphere, only the *E*-isomer was formed as the predominant degradant. *Source:* DellaGreca et al. [5] / with permission of Elsevier.

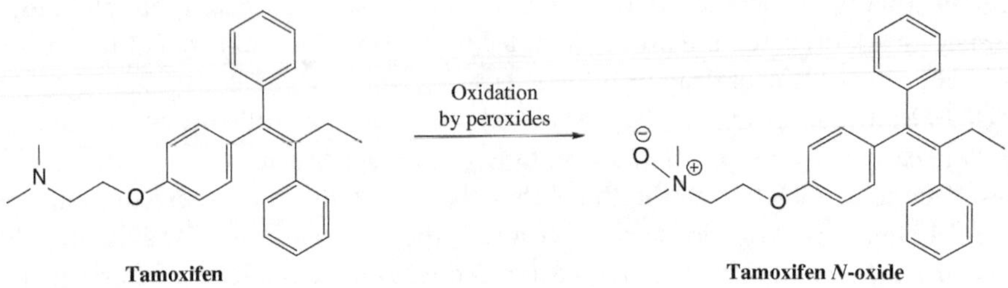

Scheme 305-2 Probable oxidative degradation of tamoxifen by peroxides.

References

1 Tamoxifen. https://en.wikipedia.org/wiki/Tamoxifen (accessed March 13 2024).

2 Maximov, P.Y., McDaniel, R.E., and Jordan, V.C. (2013). *Tamoxifen: Pioneering Medicine in Breast Cancer*. New York: Springer, Basel.

3 MD Anderson News Release June 12, 2024 (2024). MD Anderson celebrates the life and legacy of V. Craig Jordan, Ph.D. https://www.mdanderson.org/newsroom/md-anderson-celebrates-the-life-and-legacy-of-v-craig-jordan-PhD.h00-159698334.html#:~:text=Pharmacologist%20V.%20Craig%20Jordan,%20Ph.D.,%20a (accessed October 2 2024).

4 Abderrahman, B. (2024). V. Craig Jordan obituary: pharmacologist who revolutionized breast cancer treatments. https://www.nature.com/articles/d41586-024-02448-9#:~:text=The UK–US pharmacologist V. Craig Jordan (accessed October 2 2024).

5 DellaGreca, M., Iesce, M.R., Isidori, M. et al. (2007). Phototransformation products of tamoxifen by sunlight in water. Toxicity of the drug and its derivatives on aquatic organisms. *Chemosphere* 67: 1933–1939.

6 Jalonen, H.G. (1988). Simultaneous determination of tamoxifen citrate and its E isomer impurity in bulk drug and tablets by high-performance liquid chromatography. *J. Pharm. Sci.* 77: 810–813.

7 Jordan, V.C. and Murphy, C.S. (1990). Endocrine pharmacology of antiestrogens as antitumor agents. *Endocr. Rev.* 11: 578–610.

8 Li, M. (2012). Section 3.5.3.1. Formation of *N*-oxides via nucleophilic attack. In: *Organic Chemistry of Drug Degradation*. Cambridge: RSC Publishing.

9 Shete, H. and Patravale, V. (2013). Long chain lipid based tamoxifen NLC. Part I: preformulation studies, formulation development and physicochemical characterization. *Int. J. Pharm.* 454: 573–583.

10 Umemoto, A., Monden, Y., Komaki, K. et al. (1999). Tamoxifen-DNA adducts formed by α-acetoxytamoxifen *N*-oxide. *Chem. Res. Toxicol.* 12: 1083–1089.

306: Temozolomide

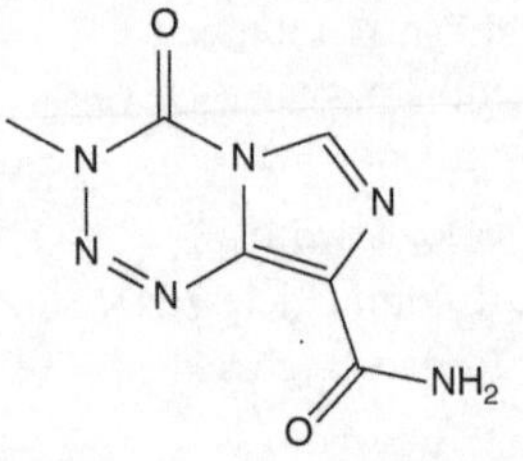

Chemical name: 3,4-Dihydro-3-methyl-4-oxoimidazo[5,1-d]-1,2,3,5-tetrazine-8-carboxamide
Trade name: Temodar, Temodal
Formula: $C_6H_6N_6O_2$
Monoisotopic protonated ion (*m/z*): 195.0625
Molecular weight: 194.15
CAS number: 85622-93-1

Temozolomide is an antitumor medication, which is primarily used for treating brain tumors; it is a first-line treatment for glioblastoma and a second-line treatment for astrocytoma [1]. It was designed as a prodrug of the putative active metabolite of dacarbazine, a prior approved anti-cancer drug, by Malcolm Stevens' group at Aston University in Birmingham, England, in 1987 [2]. The drug candidate, originally code-named CCRG 81045, was further developed by Schering-Plough, with the code name of SCH 52365, and received market approval in 1999 [1, 3].

Upon hydrolytic cleavage, temozolomide releases the active drug, MTIC (a triazene derivative), which further decomposes to AIC and methyldiazonium ion. The diazonium ion, itself a strong methylation agent, along with methyl cation, its ultimate decomposition product, methylates tumor DNA to suppress tumor growth. This activation pathway is also the main degradation pathway of temozolomide (Scheme 306), in which the detectable degradant is AIC [4–6].

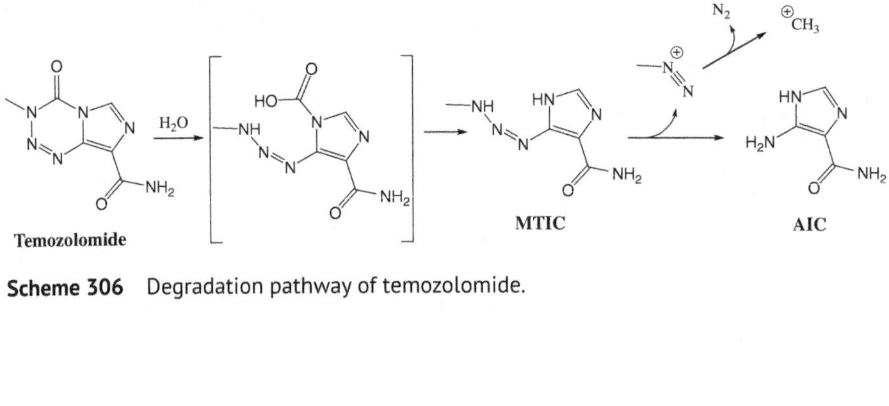

Scheme 306 Degradation pathway of temozolomide.

References

1 Temozolomide. https://en.wikipedia.org/wiki/Temozolomide (accessed October 3 2024).

2 Stevens, M.F.G., Hickman, J.A., Langdon, S.P. et al. (1987). Antitumor activity and pharmacokinetics in mice of 8-carbamoyl-3-methylimidazo[5,l-*d*]-l,2,3,5-tetrazin-4(3*H*)-one (CCRG 81045; M & B 39831), a novel drug with potential as an alternative to dacarbazine. *Cancer Res.* 47: 5846–5852.

3 Liang, L., Nale, L., James, L. et al. (2007). Dose intense schedules of temozolomide (SCH 52365) achieve better efficacy than continuous low dose schedules in human brain cancer and melanoma models. *Cancer Res.* https://aacrjournals.org/cancerres/article/67/9_Supplement/3195/538778 (accessed October 3 2024.

4 Saravanan, G., Ravikumar, M., Jadhav, M.J. et al. (2007). A stability-indicating LC assay and degradation behavior of temozolomide drug substances. *Chromatographia* 66: 291–294.

5 Lopes, I.C., de Oliveira, S.C.B., and Oliveira-Brett, A.M. (2013). Temozolomide chemical degradation to 5-aminoimidazole-4-carboxamide – electrochemical study. *J. Electroanal. Chem.* 704: 183–189.

6 Jedynak, L., Puchalska, M., Zezula, M. et al. (2013). Stability of sample solution as a crucial point during HPLC determination of chemical purity of temozolomide drug substance. *J. Pharm. Biomed. Anal.* 83: 19–27.

307: Tenofovir Disoproxil [WHO List of Essential Medicines]

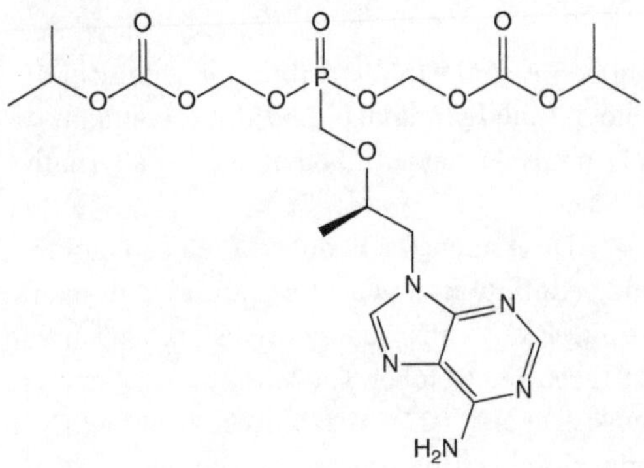

Chemical name: Bis{[(isopropoxycarbonyl)oxy]methyl} ({[(2*R*)-1-(6-amino-9*H*-purin-9-yl)-2-propanyl]oxy}methyl)phosphonate
Trade name: Viread
Formula: $C_{19}H_{30}N_5O_{10}P$
Monoisotopic protonated ion (*m/z*): 520.1803
Molecular weight: 519.45
CAS number: 201341-05-1
Common salt form: Fumarate

Tenofovir disoproxil, developed by Gilead Sciences, is an antiviral medication used to treat chronic hepatitis B and to prevent and treat HIV/AIDS [1]. It is an ester prodrug of tenofovir, which is a nucleotide analog that inhibits the reverse transcriptase of retroviruses. The active metabolite, tenofovir, is released upon metabolic activation via hydrolytic enzymes.

The carbonate moieties of tenofovir disoproxil are susceptible to hydrolytic degradation, particularly under alkaline conditions, leading to the formation of tenofovir monoester and tenofovir, respectively (Scheme 307-1) [2].

On the other hand, the amino group of tenofovir disoproxil would be expected to undergo the Maillard reaction with excipients containing reducing sugars such as lactose (Scheme 307-2) [3, 4].

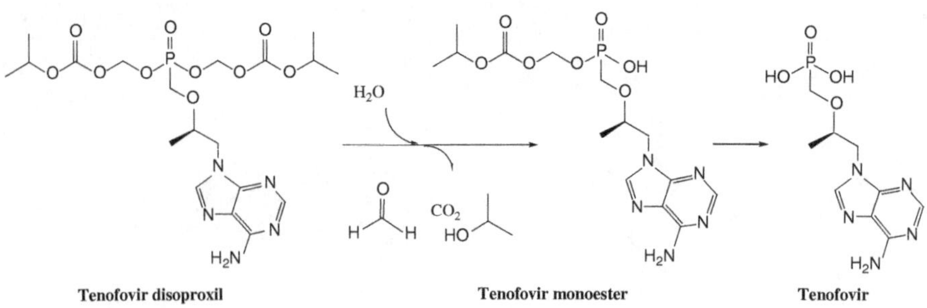

Scheme 307-1 Hydrolytic degradation pathway of tenofovir disoproxil.

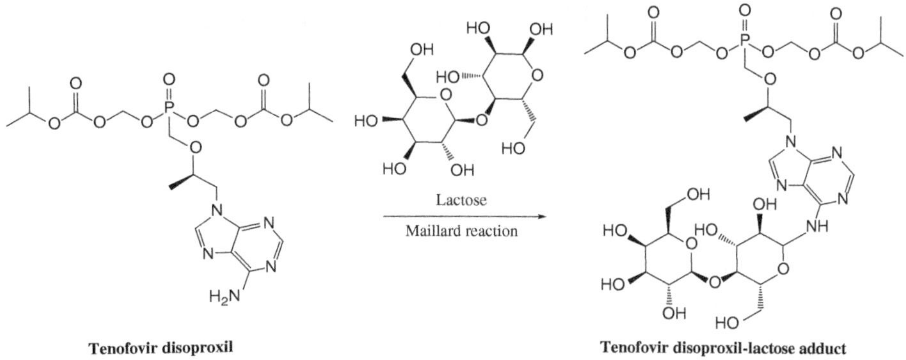

Scheme 307-2 Maillard degradation of tenofovir disoproxil in the presence of lactose, a commonly used excipient containing reducing sugar.

References

1 Tenofovir disoproxil. https://en.wikipedia.org/wiki/Tenofovir_disoproxil (accessed October 3 2024).

2 Ashenafi, D., Varalaxmi Chintam, V., van Veghel, D. et al. (2010). Development of a validated liquid chromatographic method for the determination of related substances and assay of tenofovir disoproxil fumarate. *J. Sep. Sci.* 33: 1708–1716.

3 Lai, X., Chen, X., Li, M. et al. (2022). Purification and mass spectrometry study of Maillard reaction impurities in five acyclic nucleoside antiviral drugs. *J. Pharm. Biomed. Anal.* 212: 114637.

4 Kurmi, M., Sahu, A., Ladumor, M.K. et al. (2019). Stability behaviour of antiretroviral drugs and their combinations. 9: identification of incompatible excipients. *J. Pharm. Biomed. Anal.* 166: 174–182.

308: Terbinafine [WHO List of Essential Medicines]

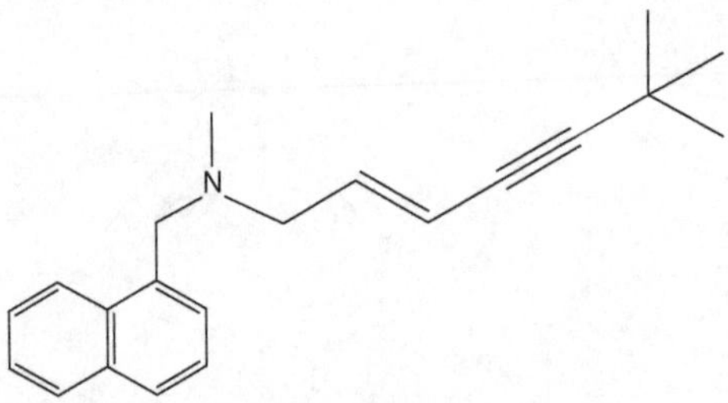

Chemical name: (*2E*)-*N*,6,6-Trimethyl-*N*-(1-naphthylmethyl)-2-hepten-4-yn-1-amine
Trade name: Lamisil, Terbin, others
Formula: $C_{21}H_{25}N$
Monoisotopic protonated ion (*m/z*): 292.2060
Molecular weight: 291.44
CAS number: 91161-71-6
Common salt form: HCl salt

Terbinafine is an antifungal medication in the family of allylamines that inhibit the fungal enzyme squalene epoxidase [1, 2]. It was developed in 1979 based on naftifine, the first allylamine antifungal drug [3] and approved by the US FDA in 1996.

Despite its relatively long history, there seems to be no in-depth study with regard to its degradation chemistry. A photodegradation study of terbinafine performed by Kryczyk-Poprawa et al. appeared to have examined the majority of the photodegradants, and yet, the structural elucidation of the degradants relied only on LC–MS analysis. Upon irradiation by UVA light, the major photodegradation pathways appeared to be the two oxidative cleavage routes (Scheme 308-1). Another significant photooxidative degradation led to the formation of hydroxyterbinafine. Photoisomerization leading to the formation of *Z*-terbinafine was also observed. For the irradiation on the cream formulated terbinafine, two additional photoisomers were also detected [4]. In an accelerated stability study of formulated terbinafine by Matysova et al., EP Impurity A was detected as the only degradant [5]. The formation of EP Impurity A should be accompanied by the simultaneous occurrence of the allylic aldehyde degradant (TBF-A) or a structurally closely related degradant. It is interesting to note that the allylic aldehyde degradant was identified as a reactive metabolite of terbinafine, which is implicated in the probable idiosyncratic liver toxicity of terbinafine [6].

Terbinafine contains a tertiary amino moiety, and as such, terbinafine *N*-oxide would be expected to be a likely degradant (Scheme 308-2) [7]. Nevertheless, it seems that the *N*-oxide degradant has not been reported so far, despite the fact that it is commercially available as a reference material.

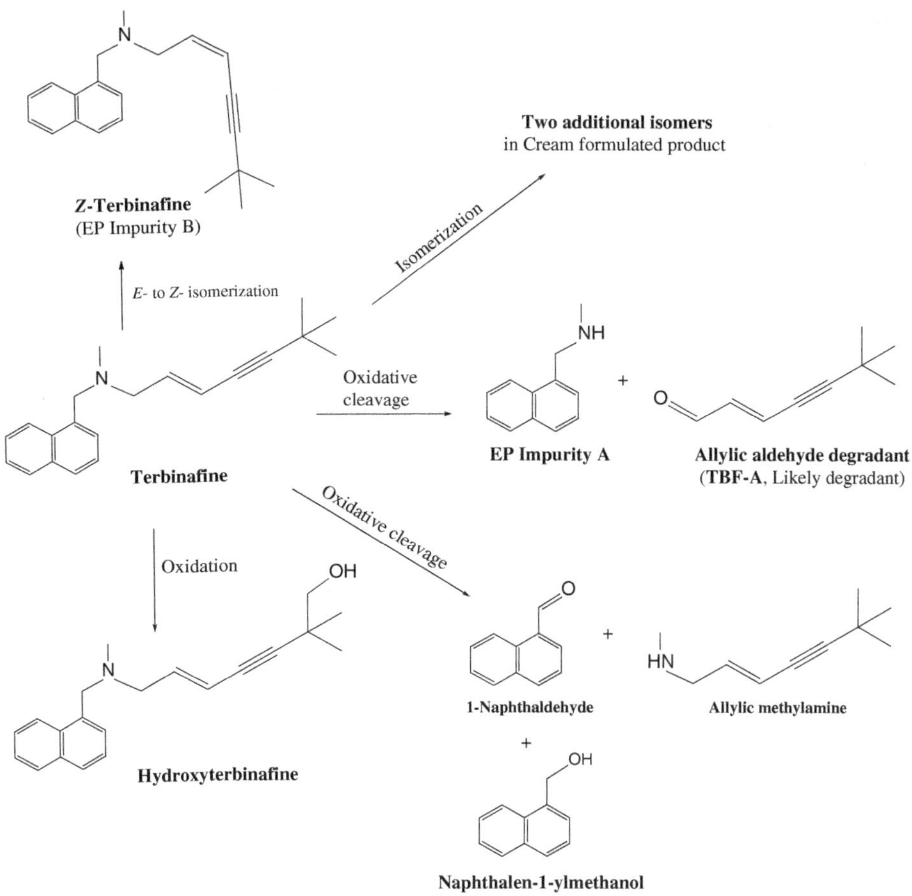

Scheme 308-1 Photodegradation pathway of terbinafine based on the results of Kryczyk-Poprawa et al. [4] with modification. *Source:* Kryczyk-Poprawa et al. [4] /John Wiley & Sons.

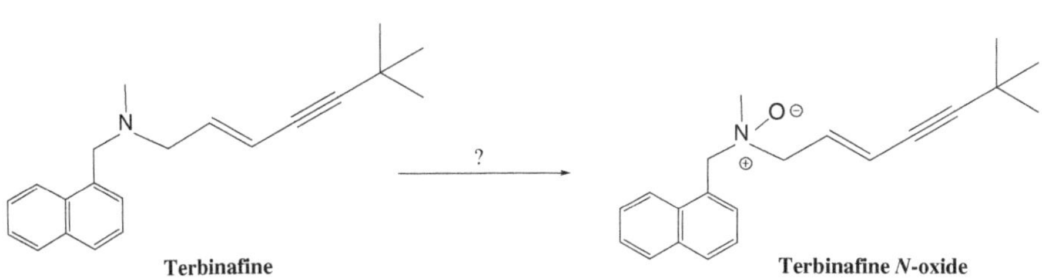

Scheme 308-2 Likely *N*-oxidation of terbinafine.

References

1 Terbinafine. https://en.wikipedia.org/wiki/Terbinafine (accessed August 27 2024).

2 Ryder, N.S. (1992). Terbinafine: mode of action and properties of the squalene epoxidase inhibition. *Br. J. Dermatol.* 126 (Suppl 39): 2–7.

3 Zhang, A.Y., Camp, W.L., and Elewski, B.E. (2007). Advances in topical and systemic antifungals. *Dermatol. Clin.* 25: 165–183.

4 Kryczyk-Poprawa, A., Zmudzki, P., Koczurkiewicz, P. et al. (2019). *Photochem. Photobiol.* 95: 911–923.

5 Matysova, L., Solich, P., Mareka, P. et al. (2006). Separation and determination of terbinafine and its four impurities of similar structure using simple RP-HPLC method. *Talanta* 68: 713–720.

6 Iverson, S.L. and Uetrecht, J.P. (2001). Identification of a reactive metabolite of terbinafine: insights into terbinafine-induced hepatotoxicity. *Chem. Res. Toxicol.* 14: 175–181.

7 Li, M. (2012). Section 3.5.3.1. Formation of *N*-oxides via nucleophilic attack. In: *Organic Chemistry of Drug Degradation*. Cambridge: RSC Publishing.

309–310: Vinblastine, Vincristine

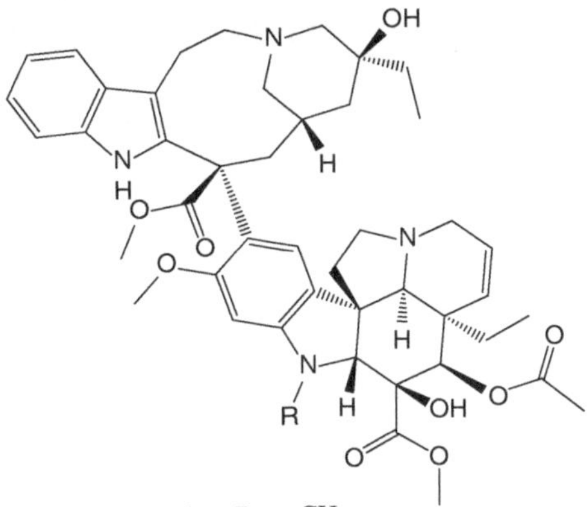

Vinblastine, R = $-CH_3$;
Vincristine, R = Formyl.

309: Vinblastine [WHO List of Essential Medicines]

Chemical name: Dimethyl (2β,3β,4β,5α,12β,19α)-15-[(*5S,9S*)-5-ethyl-5-hydroxy-9-(methoxycarbonyl)-1,4,5,6,7,8,9,10-octahydro-2*H*-3,7-methanoazacycloundecino [5,4-b]indol-9-yl]-3-hydroxy-16-methoxy-1-methyl-6,7-didehydroaspidospermidine-3,4-dicarboxylate
Trade name: Velban
Formula: $C_{46}H_{58}N_4O_9$
Monoisotopic protonated ion (*m/z*): 811.4277
Molecular weight: 810.99
CAS number: 865-21-4
Common salt form: Sulfate

310: Vincristine [WHO List of Essential Medicines]

Chemical name: (*3aR,3a1R,4R,5S,5aR,10bR*)-Methyl 4-acetoxy-3a-ethyl-9-((*5S,7S,9S*)-5-ethyl-5-hydroxy-9-(methoxycarbonyl)-2,4,5,6,7,8,9,10-octahydro-1*H*-3,7-methano[1]azacycloundecino[5,4-b]indol-9-yl)-6-formyl-5-hydroxy-8-methoxy-3a,3a1,4,5,5a,6,11,12-octahydro-1*H*-indolizino[8,1-cd]carbazole-5-carboxylate
Trade name: Oncovin
Formula: $C_{46}H_{56}N_4O_{10}$
Monoisotopic protonated ion (*m/z*): 825.4069
Molecular weight: 824.97
CAS number: 57-22-7
Common salt form: Sulfate

Vinblastine and vincristine are structurally highly related *Vinca* alkaloids, which are isolated from the Madagascar periwinkle plant *Vinca rosea (Catharanthus roseus)*

[1, 2]. The plant has been used as a folk remedy for centuries. Both drugs have the same pharmacological mode of action by interfering with the microtubule assembly, thus inhibiting cell division. Their indications include the treatment of cancers such as Hodgkin's lymphoma, lung cancer, bladder cancer, and leukemia. Vinblastine was first isolated by Robert Noble and Charles Thomas Beer at the University of Western Ontario in 1958 [1] and approved by the US FDA in 1965. On

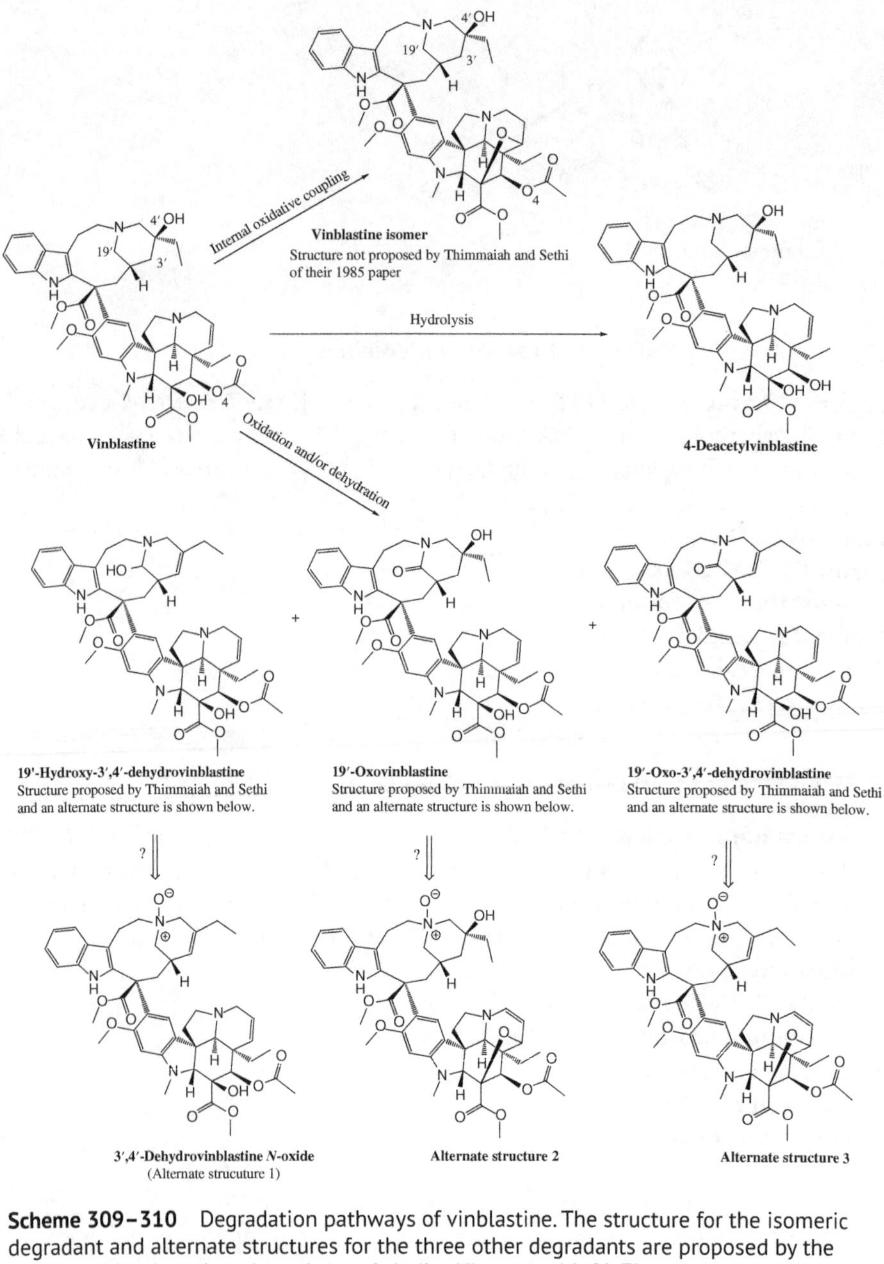

Scheme 309–310 Degradation pathways of vinblastine. The structure for the isomeric degradant and alternate structures for the three other degradants are proposed by the present author based on degradants of similar *Vinca* steroids [6, 7].

the other hand, vincristine, also known as leurocristine, was isolated in 1961 and approved by the US FDA in 1963 [3].

Vinblastine and vincristine each contain three ester functional groups, and thus are expected to be susceptible to hydrolytic degradation, in addition to other types of degradation. In a pH 8.8 buffer solution that was heated at 37°C for 72 hours, five degradants were observed [4], among which 4-deacetylvinblastine is a hydrolytic degradant and its structure was further confirmed via the synthesis of the impurity according to a previously published procedure [5]. For the remaining four degradants, three resulted from oxidation and or dehydration, and their structures were proposed solely based on mass spectrometric analysis by the original authors (Scheme 309–310), while the structure of the isomeric degradant was not proposed [4]. Based on the degradation patterns of similar *Vinca* steroids [6, 7], the structure for the isomeric degradant and alternate structures for the three other degradants are proposed by the present author.

Vincristine, also known as leurocristine, is structurally highly similar to vinblastine, and the only difference lies in the substituent on the nitrogen of the lower tryptophan moiety. Hence, its degradation chemistry would be expected to be highly similar to that of vinblastine. In a study by Sethi and Thimmaiah [8], degradants due to hydrolysis at the 4-position were observed; on the other hand, the structures of certain degradants proposed by them appear to be unlikely to occur under in vitro degradation conditions used in the study. Vendrig et al. studied the degradation kinetics of vincristine in aqueous solutions and found that it was most stable at pH 4.8, under which condition the half-life of vincristine was found to be 136 hours [9].

References

1 Vinblastine. https://en.wikipedia.org/wiki/Vinblastine (accessed March 15 2024).
2 Vincristine. https://en.wikipedia.org/wiki/Vincristine (accessed March 15 2024).
3 Neuss, N., Gorman, M., Boaz, H.E., and Cone, N.J. (1962). Vinca alkaloids. XI. Structures of leurocristine (LCR) and vincaleukoblastine (VLB). *J. Am. Chem. Soc.* 84: 1509–1510.
4 Thimmaiah, K.N. and Sethi, V.S. (1985). Chemical characterization of the degradation products of vinblastine dihydrogen sulfate. *Cancer Res.* 45: 5382–5385.
5 Hargrove, W.W. Novel acyl derivatives of desacetylvinkaleukoblastine and process for their preparation. U.S. Patent 3,392,173. Filed 9 March 1964. Granted 9 July 1968.
6 Sima Sariaslani, F., Duffel, M.W., and Rosazza, J.P. (1985). One-electron oxidation of vindoline and 16-*O*-acetylvindoline catalyzed by peroxidase. *J. Med. Chem.* 28: 629–633.
7 Thimmaiah, K.N., Lloyd, W.D., and Sethi, V.S. (1990). Chemical characterization of the in vitro degradation products of vindesine sulfate. *Microchem. J.* 42: 115–120.
8 Sethi, V.S. and Thimmaiah, K.N. (1985). Structural studies on the degradation products of vincristine dihydrogen sulfate. *Cancer Res.* 45: 5386–5389.
9 Vendrig, D.E.M.M., Beijnen, J.H., van der Houwen, O.A.G.J., and Holthuis, J.J.M. (1989). Degradation kinetics of vincristine sulphate and vindesine sulphate in aqueous solutions. *Int. J. Pharm.* 50: 189–196.

311: Zidovudine [WHO List of Essential Medicines]

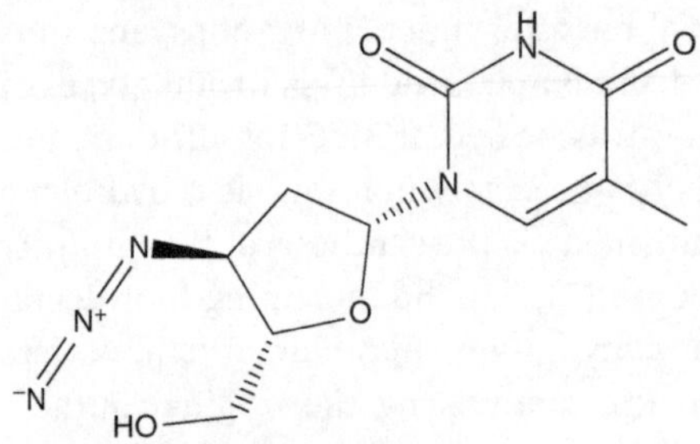

Chemicalname:1-[(*2R,4S,5S*)-4-Azido-5-(hydroxymethyl)oxolan-2-yl]-5-methylpyrimidine-2,4-dione
Trade name: Retrovir, others
Formula: $C_{10}H_{13}N_5O_4$
Monoisotopic protonated ion (*m/z*): 268.1040
Molecular weight: 267.25
CAS number: 30516-87-1

Zidovudine, also known as azidothymidine or AZT, is a nucleoside analog inhibitor of the HIV reverse-transcriptase, which is the first antiviral drug approved to treat HIV/AIDS [1]. It was originally synthesized by Jerome Horwitz's group at Wayne State University in 1964 in an effort to develop a treatment for cancers caused by retroviruses [2]. In 1984, just a few years after AIDS was identified in the United States, Marty St. Clare, a virologist at then Burroughs-Wellcome (eventually became part of GSK), started to screen compounds for anti-AIDS treatment and found zidovudine was effective in inhibiting HIV approximately 5 months into the screening program [3].

Zidovudine is susceptible to hydrolytic and photochemical degradation, due to the presence of the glycosidic linkage and the azido group, respectively. In a forced degradation study in 2 M HCl solution at 80°C for 72 hours, two degradants (thymine and desazido degradant) were detected, both of which were rationalized as hydrolytic degradants (Scheme 311-1) [4]. On the other hand, Dunge et al. performed forced degradation under conditions of hydrolysis, oxidation, and photolysis; thymine was found to be the major degradant under all these conditions [5]. In a later study, the same group also reported two minor acidic degradants (Z7 and Z9) being isomeric to zidovudine [6]. Furthermore, many photochemical degradants were observed in both solution and solid-state photolysis, among which Z1, Z2, Z3, and Z8 were presumably mediated via the nitrene intermediate, and Z4–Z6 resulting from the photo-dimerization pathway.

As zidovudine is also commonly used in combination with other antiviral drugs, degradation due to drug–drug interaction is also possible. Kurmi et al. reported two degradants resulting from a 3+2 cycloaddition reaction between zidovudine and efavirenz (Scheme 311-2) [7]. Another degradant due to a similar 3+2 cycloaddition mechanism was found between zidovudine and acesulfame potassium, an artificial sweetener (Scheme 311-3) [8].

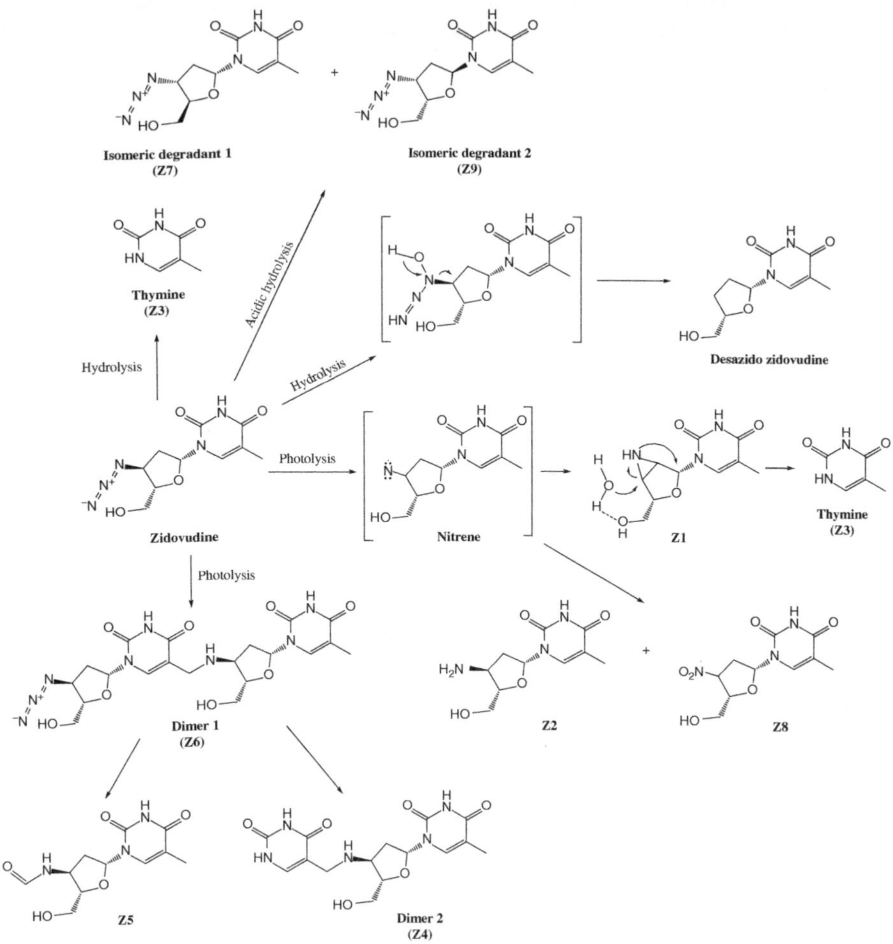

Scheme 311-1 Hydrolytic and photochemical degradation pathways of zidovudine.

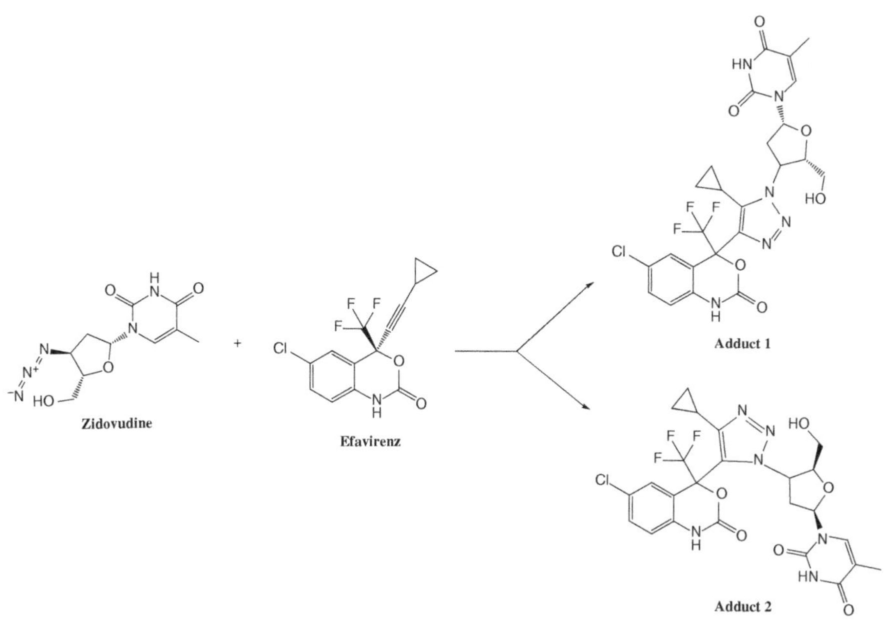

Scheme 311-2 Degradation of zidovudine due to chemical interaction with efavirenz via a 3 + 2 cycloaddition process. *Source:* Kurmi et al. [7] / with permission of Elsevier.

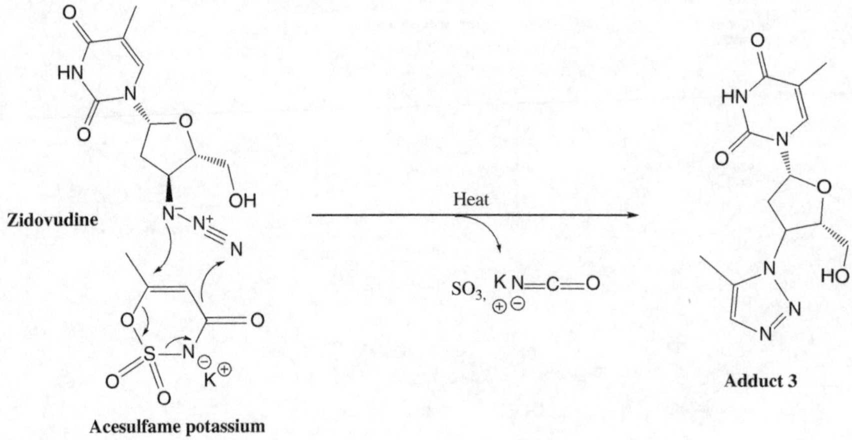

Scheme 311-3 Degradation of zidovudine due to chemical interaction with acesulfame potassium via a 3 + 2 cycloaddition process. *Source:* Adapted from Aparna et al. [8].

References

1 Zidovudine. https://en.wikipedia.org/wiki/Zidovudine (accessed March 3 2024).

2 Florez, C.V. *Zidovudine or Azidothymidine (AZT)*. Embryo Project Encyclopedia (2020-06-30) ISSN: 1940-5030. https://hdl.handle.net/10776/13154.

3 GSK. The 40 year fight against HIV. https://www.gsk.com/en-gb/behind-the-science-magazine/the-40-year-fight-against-hiv/. (accessed 4 July 2025).

4 Devrukhakar, P.S., Shankar, M.S., Shankar, G., and Srinivas, R. (2017). A stability-indicating LC–MS/MS method for zidovudine: identification, characterization and toxicity prediction of two major acid degradation products. *J. Pharm. Anal.* 7: 231–236.

5 Dunge, A., Chakraborti, A.K., and Singh, S. (2004). Mechanistic explanation to the variable degradation behaviour of stavudine and zidovudine under hydrolytic, oxidative and photolytic conditions. *J. Pharm. Biomed. Anal.* 35: 965–970.

6 Kurmi, M., Sahu, A., Tiwari, S.K., and Singh, S. (2017). Stability behaviour of antiretroviral drugs and their combinations. 6: evidence of formation of potentially toxic degradation products of zidovudine under hydrolytic and photolytic conditions. *RSC Adv.* 7: 18803.

7 Kurmi, M., Sahu, A., Balhara, A. et al. (2020). Stability behaviour of antiretroviral drugs and their combinations. 11: characterization of interaction products of zidovudine and efavirenz, and evaluation of their anti HIV-1 activity, and physiochemical and ADMET properties. *J. Pharm. Biomed. Anal.* 178: 112911.

8 Aparna, P., Rao, S.V., Thomas, K.M. et al. (2010). Identification, isolation, and characterization of potential degradation products in a triple combination lamivudine, zidovudine, and nevirapine tablet for oral suspension. *Pharmazie* 65: 331–335.

312: Zolpidem

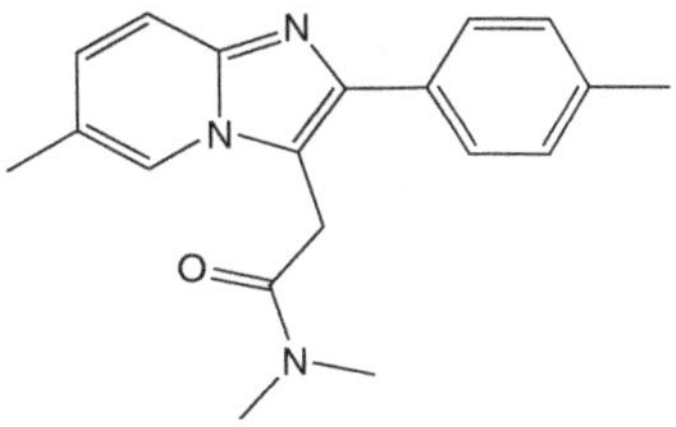

Chemical name: *N,N*-Dimethyl-2-[6-methyl-2-(4-methylphenyl)imidazo[1,2-a] pyridin-3-yl]acetamide
Trade name: Ambien
Formula: $C_{19}H_{21}N_3O$
Monoisotopic protonated ion (*m/z*): 308.1757
Molecular weight: 307.40
CAS number: 82626-48-0
Common salt form: Tartrate

Zolpidem is a $GABA_A$ receptor agonist, which is clinically used for the short-term treatment of insomnia. In the early 1970s, Jean-Pierre Kaplan, a scientist at the French pharmaceutical company Synthélabo (which eventually became part of Sanofi), started to conduct research on GABA, or gamma-aminobutyric acid, along with fellow scientist George Pascal, which led to the discovery of zolpidem [1]. It was first approved in Europe in 1988 and then in the United States in 1992 [2].

Chemically, zolpidem is a dimethylamide derivative and hence, it can be hydrolyzed to form zolpacid and dimethylamine (Scheme 312). It was also reported to be susceptible to photochemical degradation to yield oxozolpidem, zolpaldehyde, and zolpyridine as the photochemical degradants [3, 4].

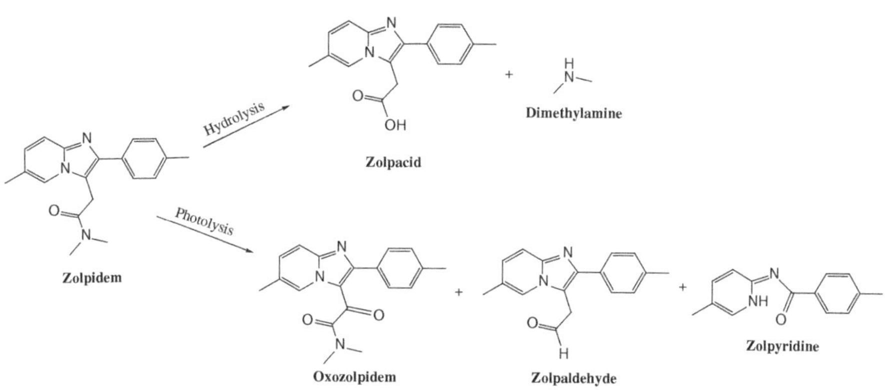

Scheme 312 Hydrolytic and photochemical degradation pathways of zolpidem.

References

1 The museum of lost things. https://www.museumoflost.com/ambien/#:~:text=Jean-Pierre%20Kaplan%20was%20a%20Paris%20born%20biochemist%2C%20who, developing%20a%20revolutionary%20new%20type%20of%20sleeping%20tablet (accessed October 12 2024).
2 Zolpidem. https://en.wikipedia.org/wiki/Zolpidem (accessed March 15 2024).
3 Malesevic, M., Zivanovic, L., Protic, A., and Jovic, Z. (2011). Multiobjective optimization approach in evaluation of chromatographic behaviour of zolpidem tartrate and its degradation products. *Chromatographia* 74: 197–208.
4 Malesevic, M., Zivanovic, L., Protic, A. et al. (2014). Stress degradation studies on zolpidem tartrate using LC–DAD and LC–MS methods. *Acta. Chromatographica* 26: 81–96.

Index

Drugs in **bold** are those covered in monographs (with their first page numbers followed immediately), the ones in *italic* are their trade names, while the rest are drugs mentioned in the book.

Compendium of Drug Degradation Pathways, First Edition. Min Li.

p

q

r